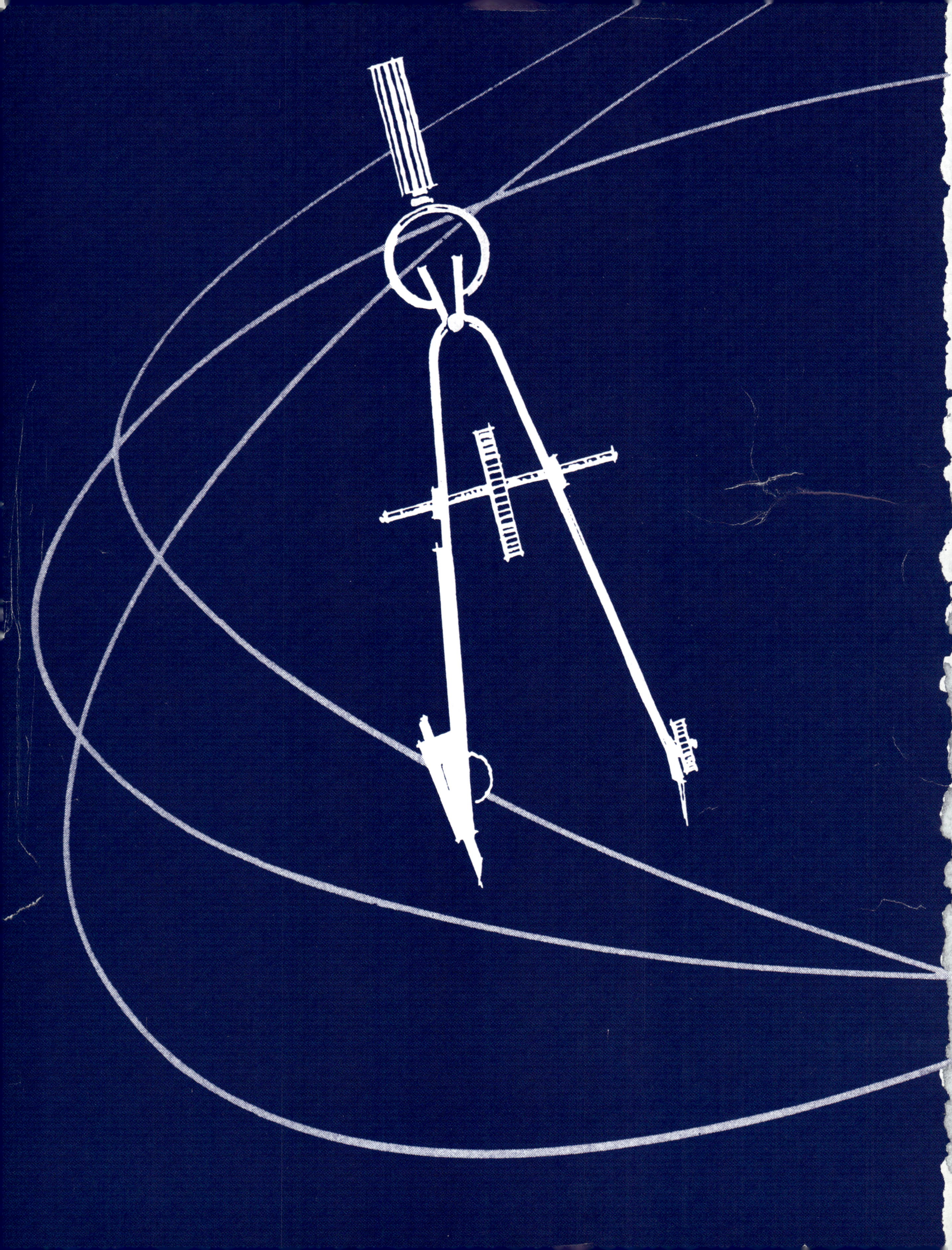

Automobile Design Graphics

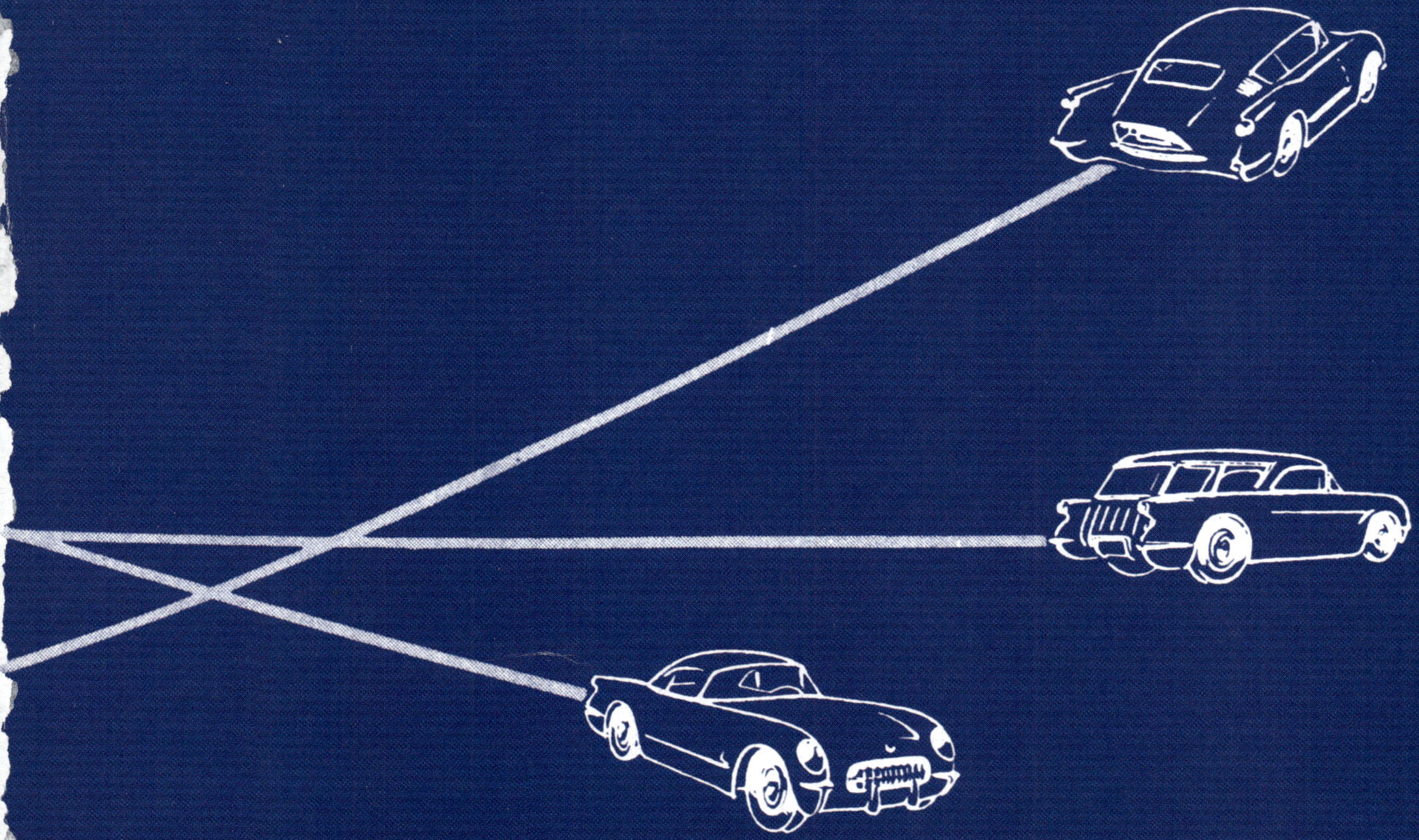

A Visual History from the Golden Age to the Gas Crisis 1900–1973

Automobile Design

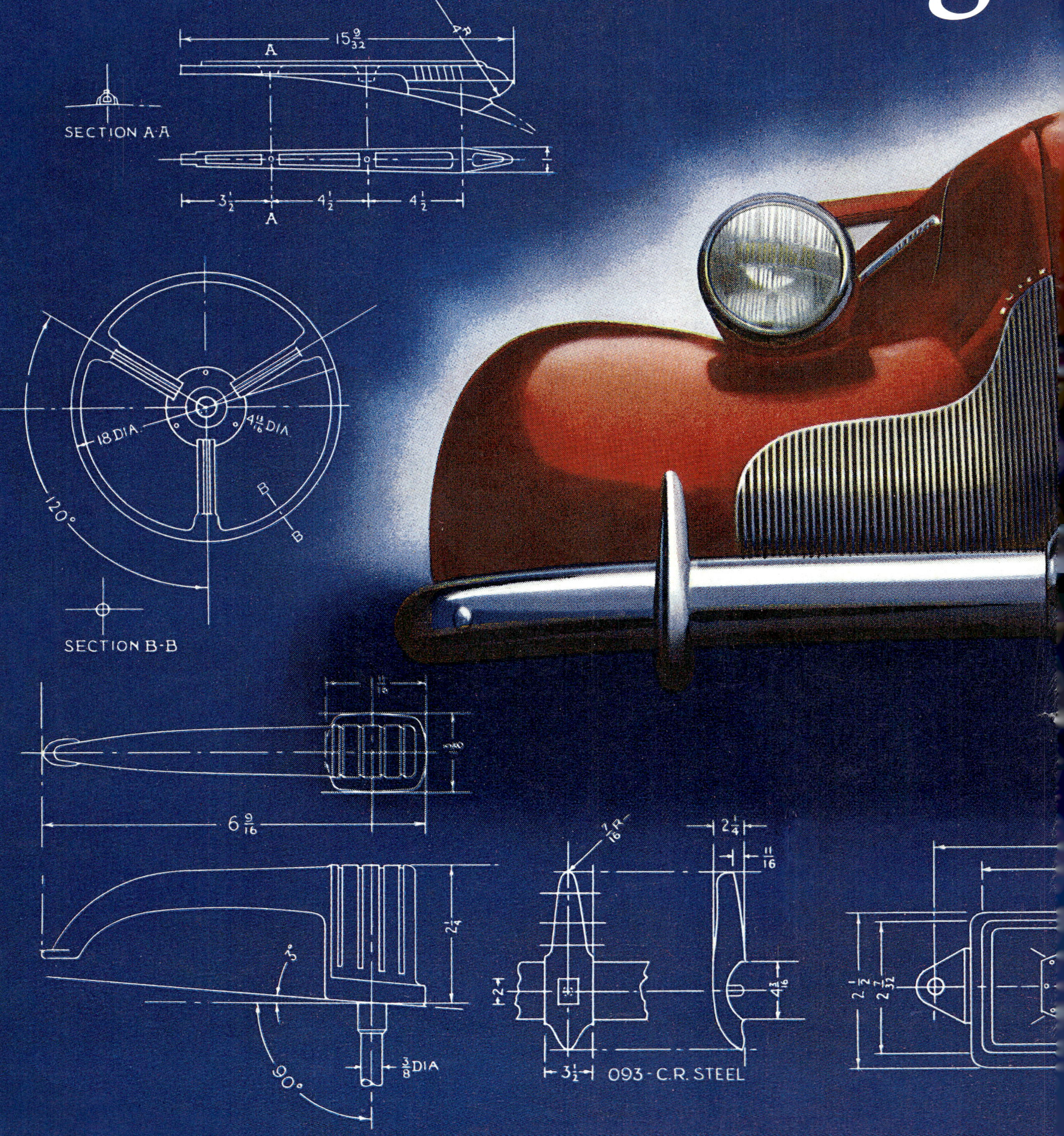

Graphics

Jim Heimann (Ed.)
Steven Heller
Jim Donnelly

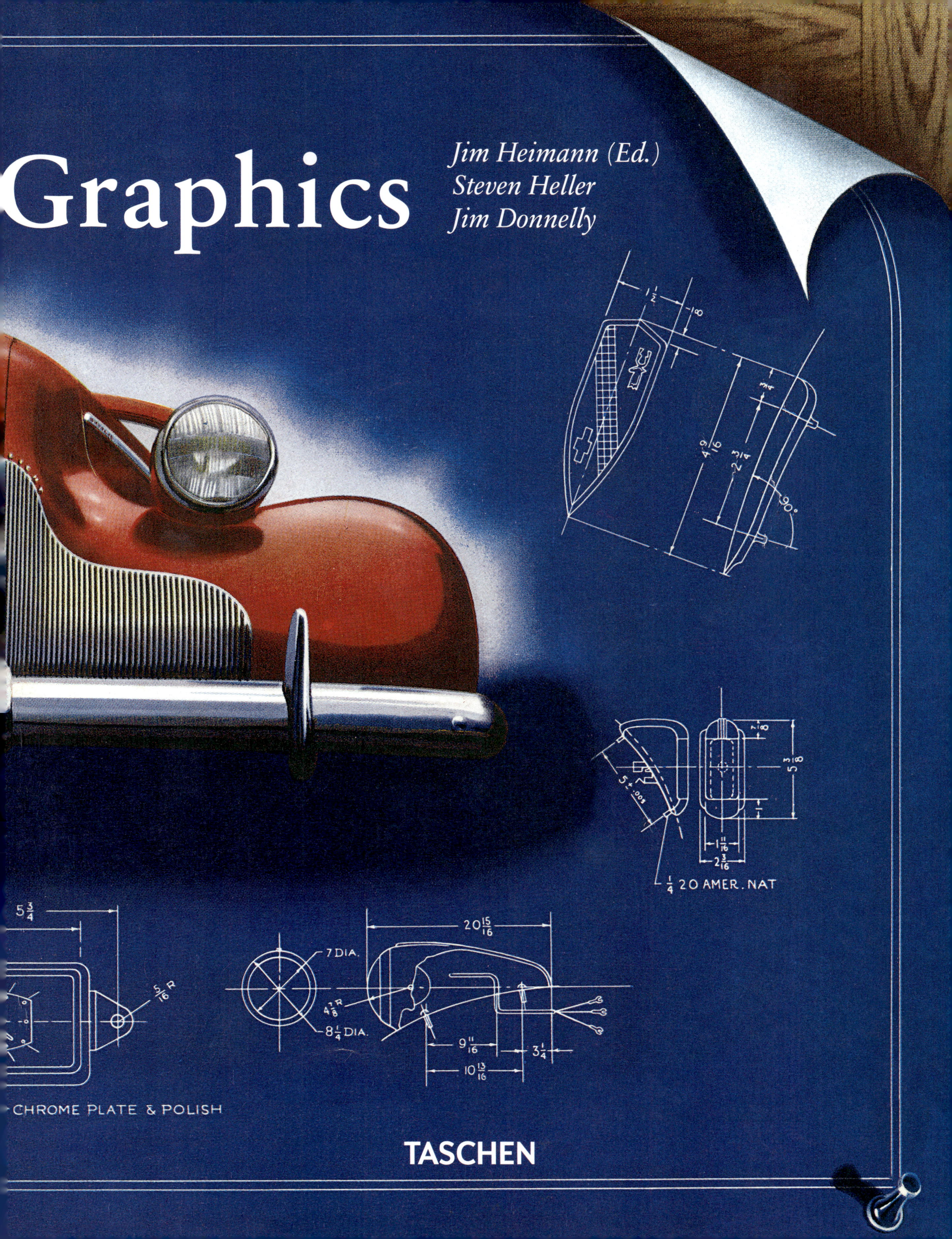

TASCHEN

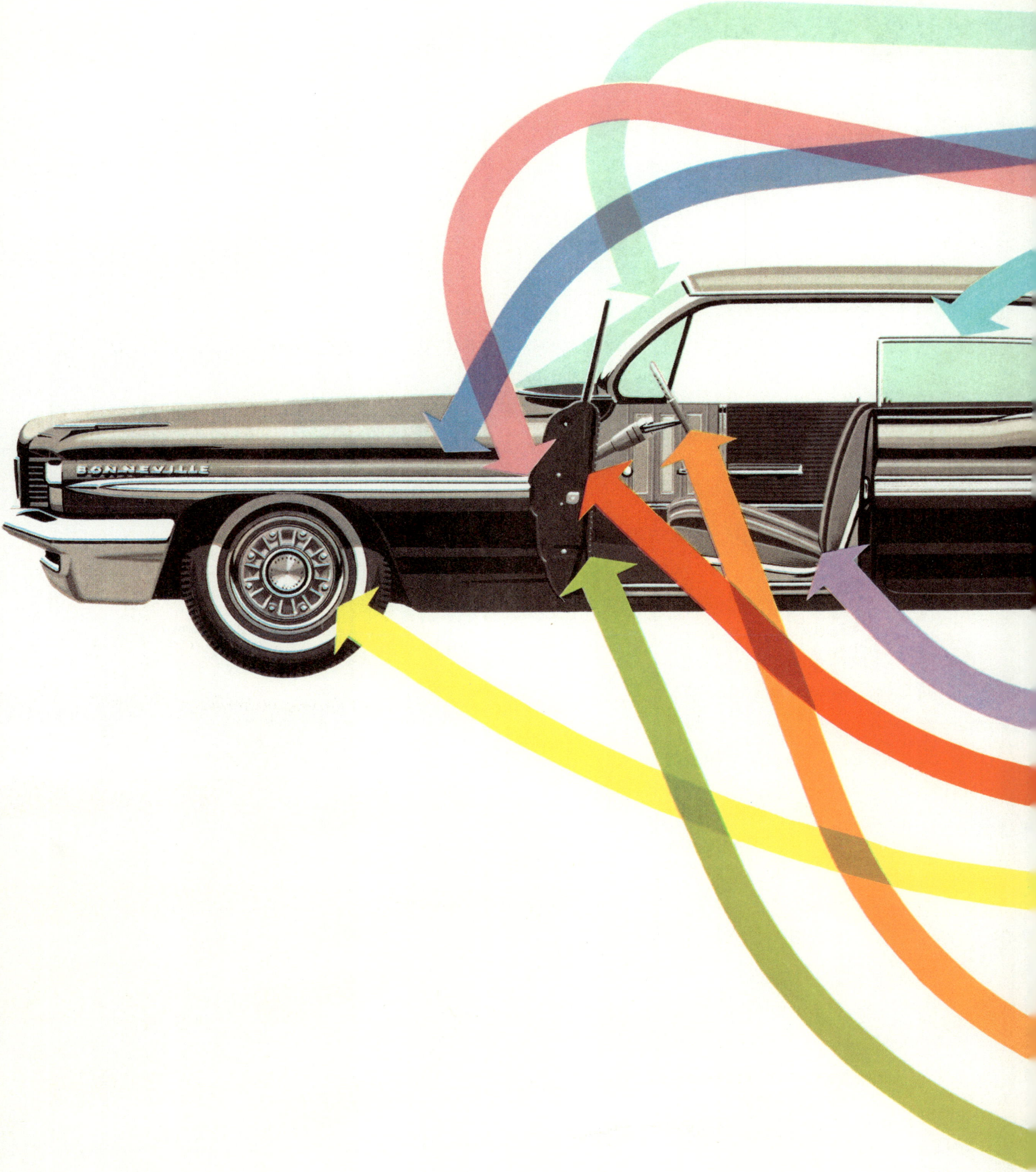
BONNEVILLE

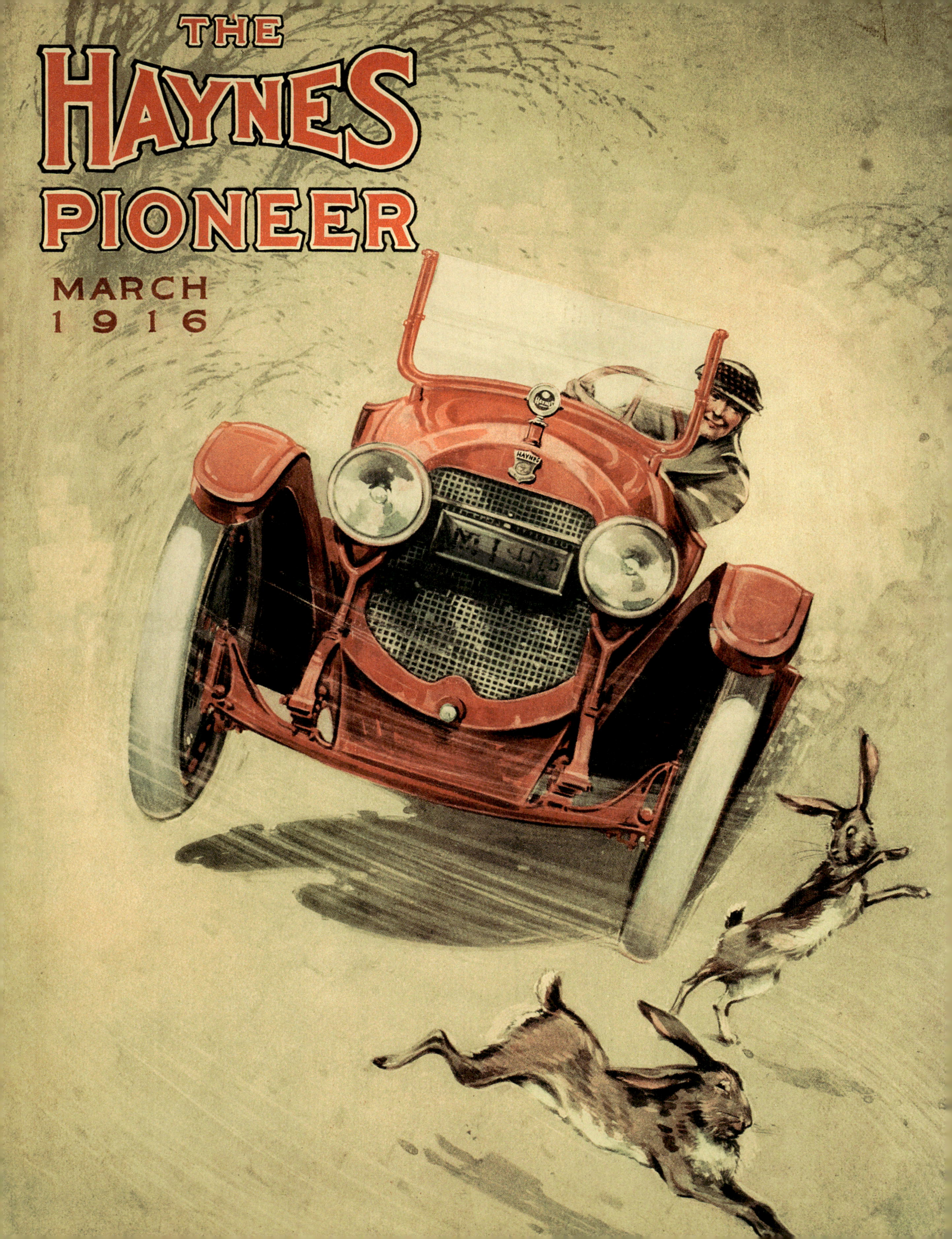
THE HAYNES PIONEER
MARCH
1916
HAYNES

JIM DONNELLY

Creating Desire, One Page at a Time

An automaker could convince you that you were smart, a good spouse, and a hero to your kids if you picked its car. The brochures were powerful tales about the automobiles, and ourselves.

Physically, an automobile is pieced together from precisely formed steel, glass, and rubber. The building process can be distilled to visceral sensations—repetitive, deafening, the hammers of the gods of fire and war: Foundries glowing orange, stamping mills crashing, power tools screeching, foremen yelling to hurry all these tasks.

Those images are the polar opposite of the gentle, very personal art of suggestion that encourages a person to buy the car after it's been forged by such primal, elemental forces. This form of persuasion is like walking through a narrow hallway lined with mirrors. Stripped bare, the function of automobile advertising is to hold up a looking glass into which the fascinated customer peers, seeing a car reflected back at him—along with a satisfied version of himself. The message, even if expressed through something as humble as a dealer brochure, was that buying the product could change your life for the better.

An accepted business guideline dictates that the more expensive the purchase, the harder the seller must work to close the deal. In the early era of the automobile, most mortals made only three major investments in their lifetime: their home, their own funeral, and a new car. The car and home were optional, so automakers had to make the automobile purchase seem essential. It was economic Darwinism of the most unforgiving sort: According to available records, something on the order of 5,000 companies produced cars between 1900 and 1930 in the United States alone. Those that were successful at selling their cars, and outlasted the Great Depression, went on to build cars by the millions. Those that failed to sell joined thousands of other firms that built only a handful of them.

For all the automobile's increasing complexity and performance, the giant factories in Detroit and elsewhere relied on simple marketing: the illustrated sales brochure that awaited buyers who strolled into a showroom. The strategy: Reach buyers and keep them thinking—constantly—about how the car could change them. Rarely have human ambitions—needs, so deep and personal—been massaged so effectively, for so long, through such a rudimentary medium.

At its simplest, a new-car brochure was an inexpensive catalog. It was an indispensible tool for the dealer. True, the foldout giveaways might end up in a wastebasket, but they also might linger on the kitchen table or beside the easy chair. And stay there, thumbed through again and again. At leisure, readers could visualize how a car might change them. They could imagine themselves in the brochure's inviting artwork, driving lazily through the countryside with the family, just because they felt like going. They could become that attractive, smiling person in the brochure. Who wouldn't want to take such a pleasurable dream for a spin?

Yes, a dream. At first, cars were only for the very wealthy as they emerged from the Gilded Age, expensive toys able to send them bumping along on a smoky, clattering joyride. The motorist's quest was to get back to the estate without ramming anything or sustaining an embarrassing breakdown. As the 20th century dawned, men led by Ransom Eli Olds and Henry Ford already knew that the automobile could never achieve its transformative social and business potential unless millions of ordinary people could afford one. The millions agreed. Everyday workers saw the automobile as nothing less than a tool of personal liberation. They hungered for freedom to move on their own schedule, escape from barnyard drudgery, and learn in person what lay beyond the end of the jam-packed trolley's line.

The early Winton, a motoring and racing standout assembled in Cleveland that dated to the turn of the 20th century, made these

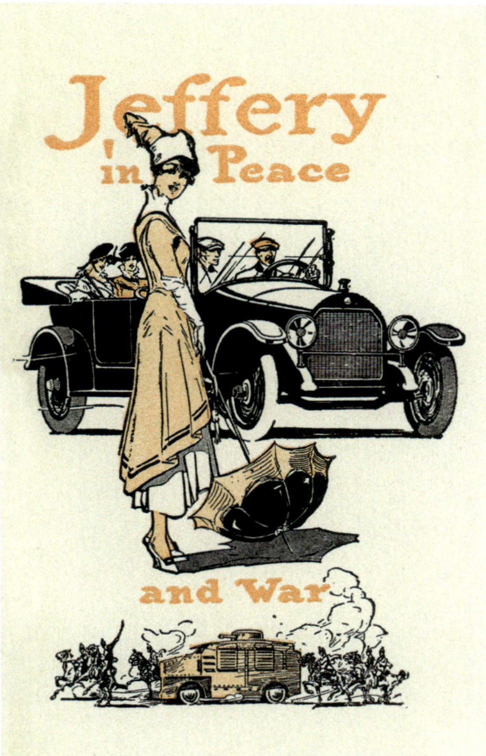

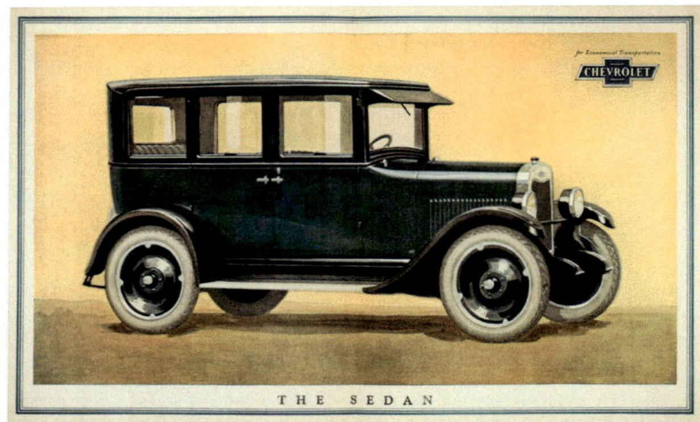

Chevrolet 1920 *(left)*

Ford 1915 *(opposite)*

Haynes Light Twelve 1916 *(page 6)*

Jeffery 1915 *(left)*

Automobile salon 1928 *(below)*

enticing claims: "DISPENSE WITH A HORSE. No odor. Costs less than 1/2 cent per mile to operate." You could find the smudgy print advertisements in the back pages of the local daily broadsheet, or in magazines such as the *Police Gazette* or *The Sporting News*. Substantially, ads like this, in tandem with the newly emerging sales brochure, focused on themes that prospective "automobilists" could grasp. Drive and you were at the forefront of land-based adventure, despite any pretenses toward practicality. The 1899 Locomobile bragged about "Speed! Reliability! Safety!" The Bates car, of Lansing, Michigan, lyrically pledged, "Use a Bates and keep your dates." As early as 1905, the new saying was "Don't experiment, just buy a Ford," which demonstrates that even that early, Ford embraced value and sturdy reliability. Still, pre–World War I car buyers were the space tourists of their day.

Early brochures roughly paralleled the arrival of the local seller. An early example of an enduring treatment can be found in a

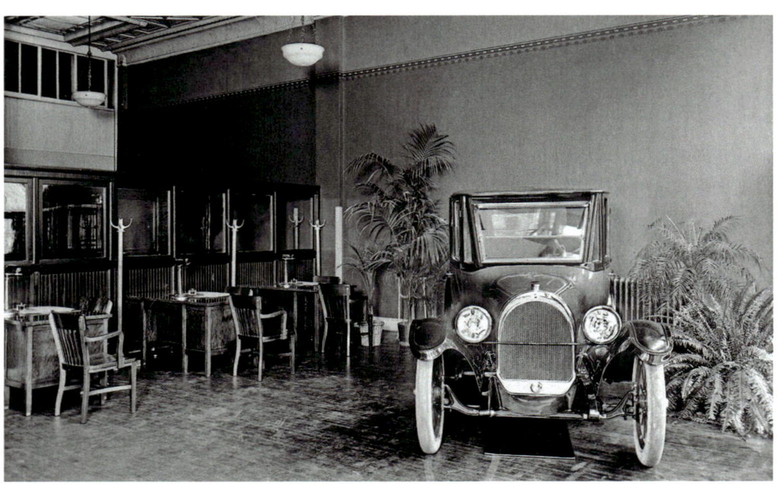

26-page booklet produced in 1916 by the Thomas B. Jeffery Company of Kenosha, Wisconsin, which later became Nash and, still later, American Motors Corporation. The inks are black and light orange. Ladies with parasols are tall in the rear seat of a Jeffery Four, guarding their millinery against the breeze as it outsprints a greyhound. Other drawings show the interior fittings, as a painting depicts a Jeffery at a dressage competition. The imagery was obvious.

When the smoke and mustard gas clouds of World War I finally cleared over Europe, marketing sophistication was already evolving at a rapid pace. Boasting in vague terms about reliability and performance was no longer enough. Mass production and easy credit put tens of millions into the car market. The get-big-or-die reality was already squeezing automakers out of the business by the hundreds every year, making extinct all but those that could claim a measure of high material quality, styling, power, or other prestige beyond that of the mass-scale builders. For the giants that remained at the top of the industry, led by Ford and General Motors, the stakes were staggering: competition for millions of dollars. They had to find new messages, and media, to move their vast amounts of metal. The brochure got their message into a lot of new hands.

History demonstrates clearly that General Motors, having twice emerged from fiscal collapse by 1920, won the early battle for buyers' minds. The campaign had little to do directly with GM's cars or production methods. During the 1920s, GM's president was a financial and organizational genius named Alfred P. Sloan. He innately understood that GM was in the business of increasing demand (as opposed to simply building cars) and making profits (again, not cars). To Sloan, cars were merely the means to an end. His eternal brilliance was his ability to play to human needs and desires.

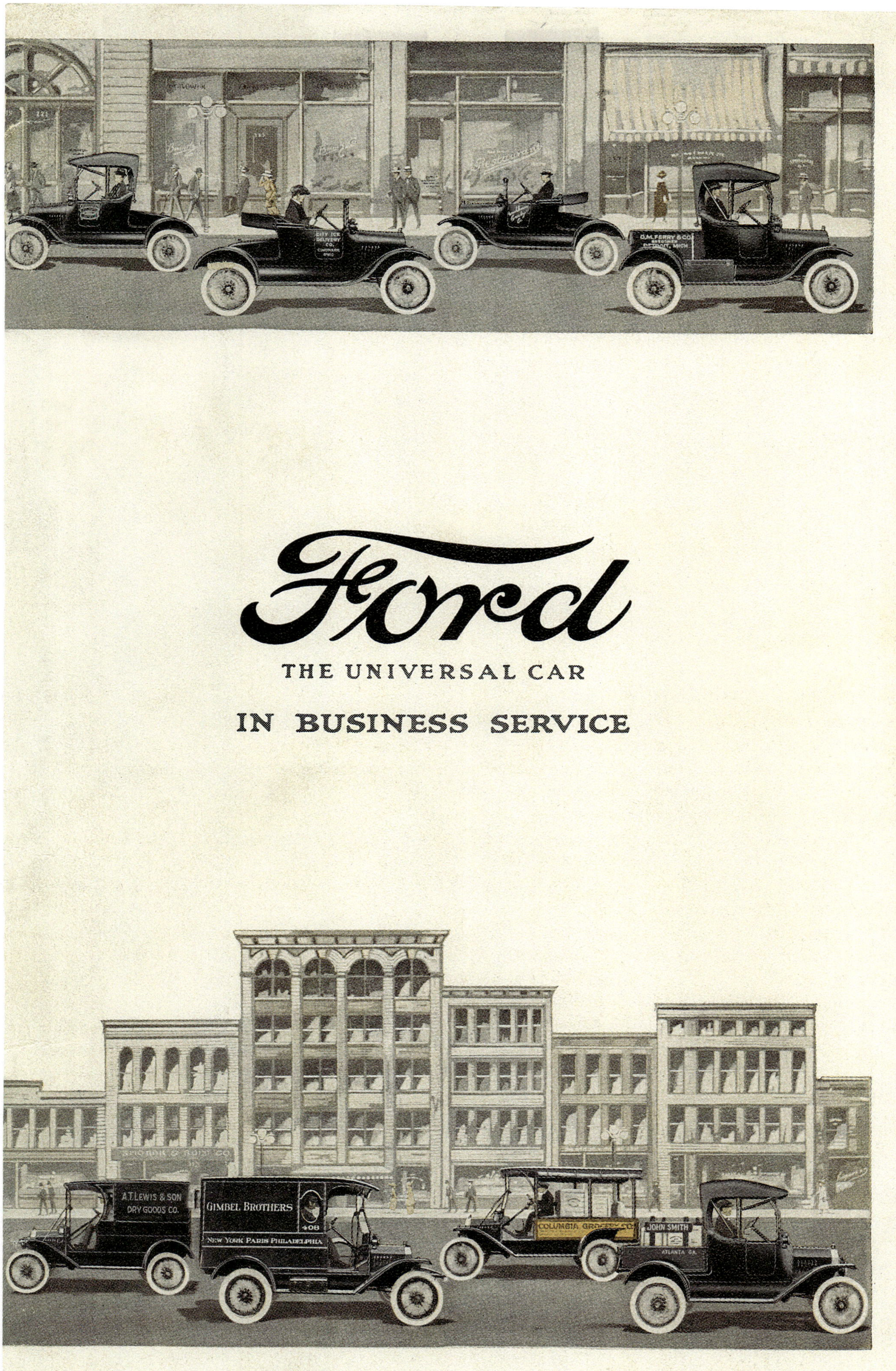
Ford
THE UNIVERSAL CAR
IN BUSINESS SERVICE
A.T. LEWIS & SON
DRY GOODS CO.
GIMBEL BROTHERS
NEW YORK PARIS PHILADELPHIA
JOHN SMITH
ATLANTA GA.

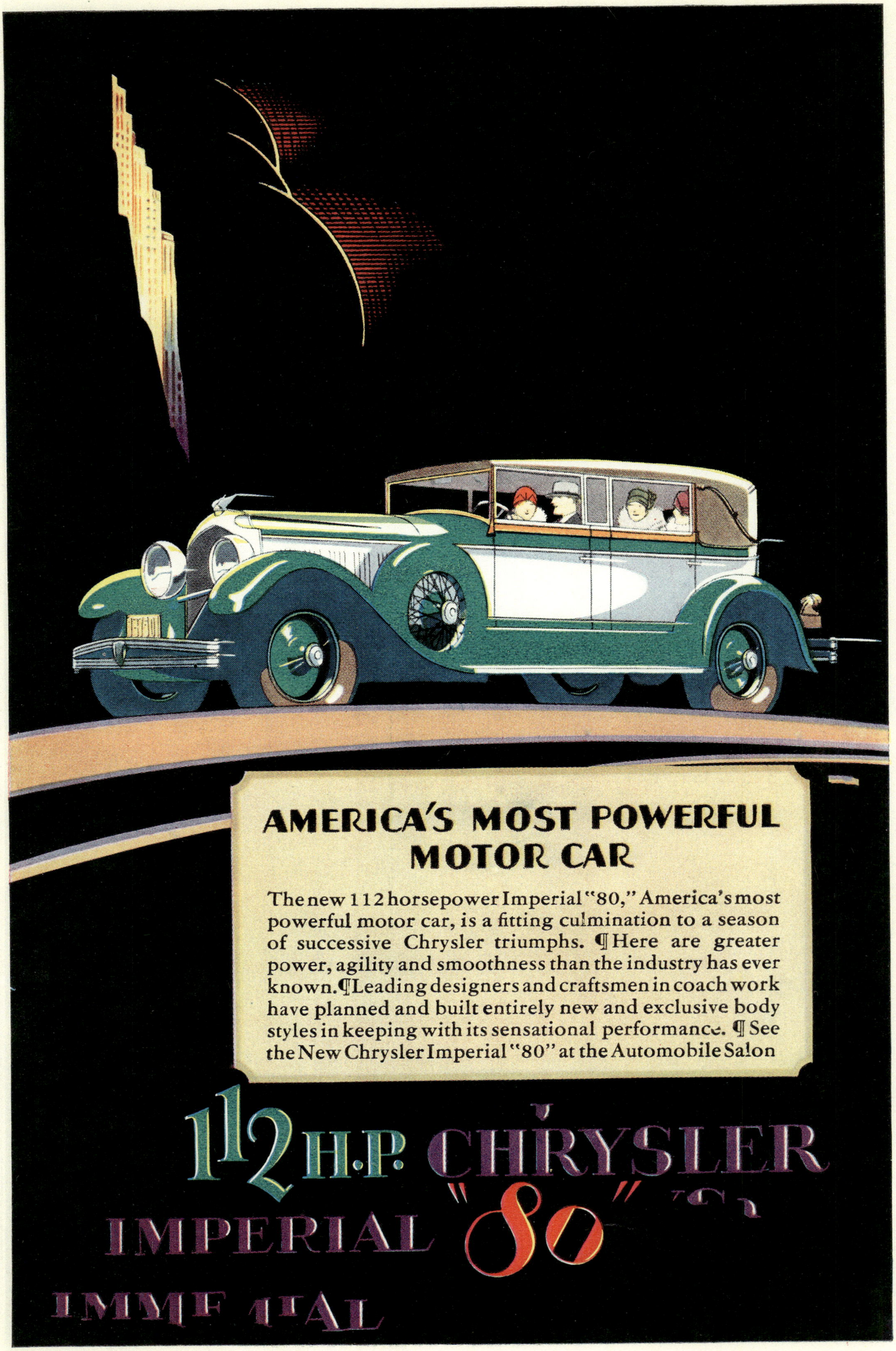

AMERICA'S MOST POWERFUL MOTOR CAR
The new 112 horsepower Imperial "80," America's most powerful motor car, is a fitting culmination to a season of successive Chrysler triumphs. ¶ Here are greater power, agility and smoothness than the industry has ever known. ¶ Leading designers and craftsmen in coach work have planned and built entirely new and exclusive body styles in keeping with its sensational performance. ¶ See the New Chrysler Imperial "80" at the Automobile Salon
112 H.P. CHRYSLER
IMPERIAL "80"

Chrysler Imperial 80 1928 *(opposite)*

Lexington & Concord Sight-Seeing Co. 1907 *(right)*

Studebaker 1929 *(far right)*

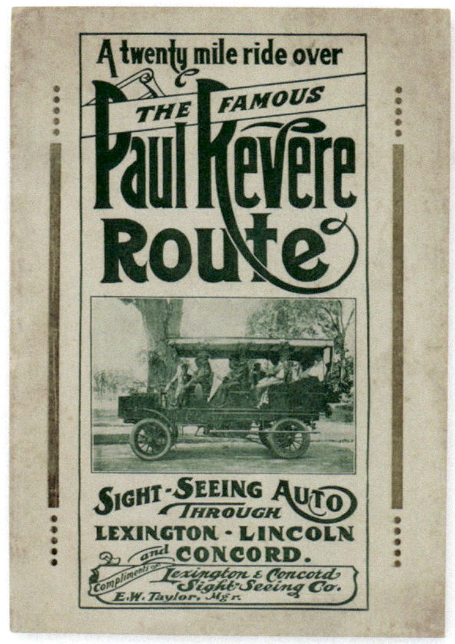

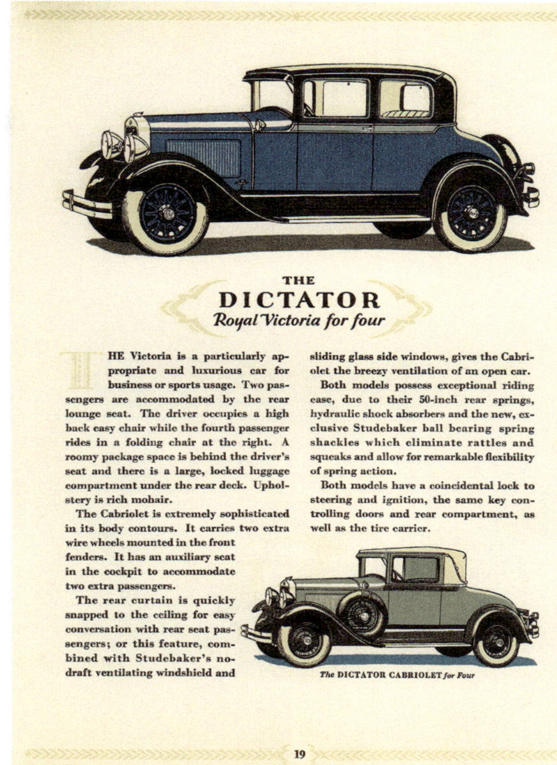

Sloan dictated that a Chevrolet, for instance, should undergo minor appearance changes every year, however superficial. The advertising (by this time, it had largely gravitated to hugely circulated magazines such as *Harper's* and *The Saturday Evening Post*) came to breathlessly alert readers to the new Chevrolets, which could be viewed at their nearest retailer, traditionally at the beginning of fall. The practice escalated to the point where the lightly updated "new" cars were transported to dealers under wraps, and kept in the showroom covered, until the appointed date. When the covers came off the cars, racks filled with freshly printed brochures were right next to them, the catalogs free and full of extravagant copy about what was new.

Sloanism, as it came to be reverently termed, was founded in part on an unsubtle play to everyman's desire to one-up the guy next door by having a newer, more modern GM car. To that end, beginning in 1924, Sloan also decreed a monumental reorganization of GM into a price-based hierarchy of makes, ranging from Chevy at the bottom to Cadillac at the top. Not only that, he created a new internal Art & Colour staff (the British spelling of "color" was an affectation that GM adopted for its prestige) led by a onetime carriage maker turned custom-body builder named Harley Earl, who created tapestries of paint and fabric, richer in proportion to price, for GM's new model range. The idea was that a Chevrolet buyer could view himself as a climber and thus aspire to a Pontiac, Buick, or Cadillac rather than jumping to Chrysler, and thus keep the profits on GM's balance sheet.

Linking the product to visceral superiority—money! success! envy! sexiness! the latest of everything!—became the overarching goal of car advertising between the world wars. It seemed as if brochure copy extolling the car itself diminished in inverse proportion to the number of cars being sold and their price. That's not to suggest that advertising ignored technical advances, because in the years that spanned the wars, there were plenty; they were legitimately important and lushly illustrated. Innovations included Ford's flathead V-8 in cheap cars, the first workable automatic transmissions, independent front suspension, plus the oft-overlooked creation (by GM) of the full-steel turret top for closed cars. The Depression, while chilling output, had the unintended consequence of making the car brochure into something of a short encyclopedia on the product. The brochures praised technical innovations plus longevity, ease of maintenance, and low repair costs, instilling customer confidence that purchasing a new car was a thrifty and worthwhile investment.

Very often, they focused more on people than cars. Noticeably, women continued making their way into automotive ads, and without a male to guide them. Cars had already been marketed directly to women, such as the Detroit and Baker electrics, and then moving on to the first self-starter, offered by Cadillac in 1912, to make cars easier and more comfortable to drive.

It's key to understand how most people lived in the 1920s and 1930s. There was no television. Newspapers were published in the afternoon. After work and dinner, families might pass the brochures around and talk about how rapidly cars changed, gained appointments, were growing faster and more reliable. Wives scanned the brochures. They excited children. Sometimes, kids cut out the cars to make little paper toys. The car bored into the family's consciousness and stayed put.

World War II saw the mega-builders of Detroit end car production in early 1942 and shift immediately to government-controlled output of military matériel on a colossal scale. The car companies were tasked with maintaining morale among both workers and their customers.

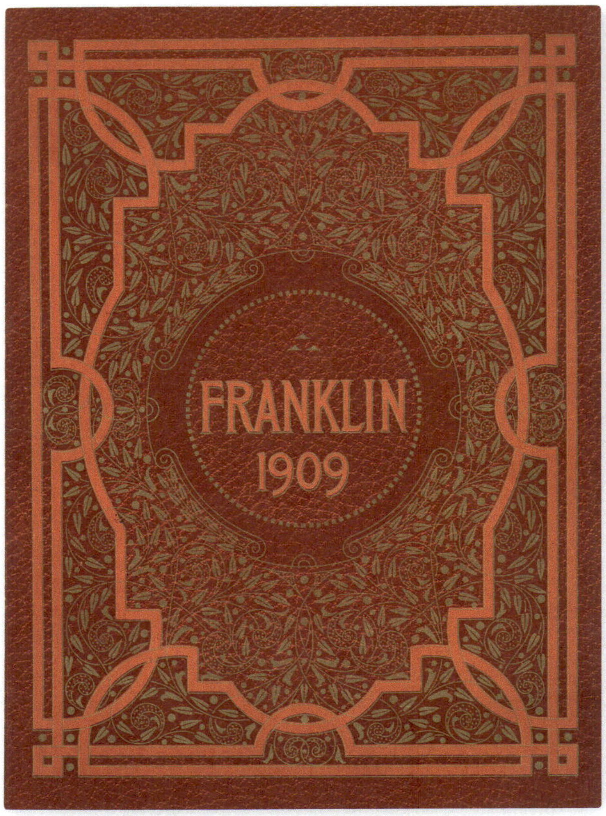

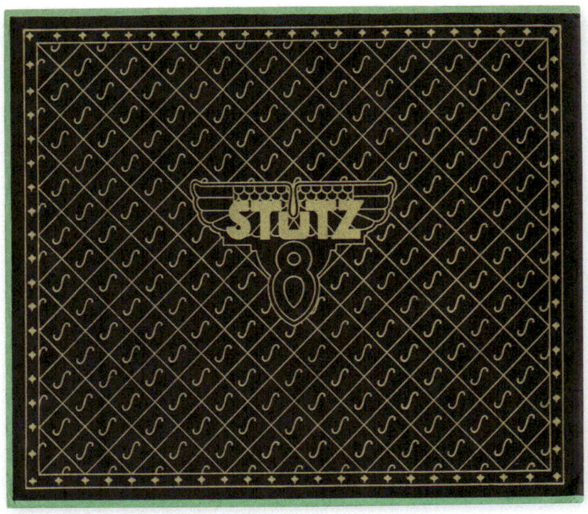

Franklin 1909 *(far left)*

Stutz 1927 *(left)*

Automobile salon 1928 *(opposite)*

Detroit's messages to America were evocative, patriotic, and, since there were no cars to be sold, became the industry themselves. With legitimate justification, Detroit proclaimed its war contributions and promised great cars once the shooting ended. More than one car builder's ads showed a G.I. bivouacked in some bombed-out wasteland, thinking dreamily of coming home to a new car. "There's a Ford in your future," said the best-remembered promise. Back at home, most new-car dealerships shifted to selling suddenly precious used cars out of their showrooms, plus repairs and tires. Rationing limited the number of miles people could drive anyway. Wartime steel drives condemned countless classic cars with poor fuel economy to junkyards.

The promise that a Ford, or something else, was in the future came true. The postwar world of cars went to new realms of excess. Returning servicemen, flush with combat bonuses, wanted cars right now. As a society, America had had its fill of thrift, austerity, and seriousness. The cars? Detroit was obsessed with ever-more-baroque styling, combinations of fins and tunnels and bullet front ends, which ladled on poundage and mandated ever-thirstier engines with enough horsepower to get everything moving. In turn, the advertising focused on smaller novelties, sometimes with no real significance, starting with selections of paints and fabrics in dizzying numbers of combinations.

Privation, fighting, and misery were banished. So might be taste and forthright presentation. PowerPilot, Whirlaway, TorsionAire, and Airflyte joined the language of auto features. The pitches largely came from agencies. An early entrant was Young & Rubicam, which wrote Packard copy. J. Walter Thompson, which could trace its history to the 1800s, had a long relationship with Ford. Campbell-Ewald held the most prized account in the industry, Chevrolet, for 91 years.

These firms, and others, were tasked with pulling car buyers back to reason when the great wallow in extravagance that reached its zenith in the late 1950s became unseemly. From 1960 on, Americans were pulled to less excessive cars — from Europe, from Japan and, eventually, again from the United States. The ad people now had to convince buyers that less was more. Their success was amazing. The wry campaign that the Doyle Dane Bernbach agency ran for Volkswagen in these years was most notably and memorably brilliant, as were the Volkswagen brochures, even more so once the first OPEC oil crunch hit in the 1970s.

An automaker could convince you that you were smart, a good spouse, and a hero to your kids if you picked its car. The brochures were powerful tales about the automobiles, and ourselves. A good ad can grip our souls, whether it's for a car, shampoo, or a smartphone. Unlike a billboard or Internet pop-up, a brochure could keep the message right in front of you, whenever you wanted to see it. They took you, in art and words, dancing across a psychic threshold. The story of the automobile's evolution in terms of engineering and styling has been thoroughly told. The story of the auto sales brochure is equally important. Engineers made cars better, but advertising, via these pamphlets, is what made buyers crave them.

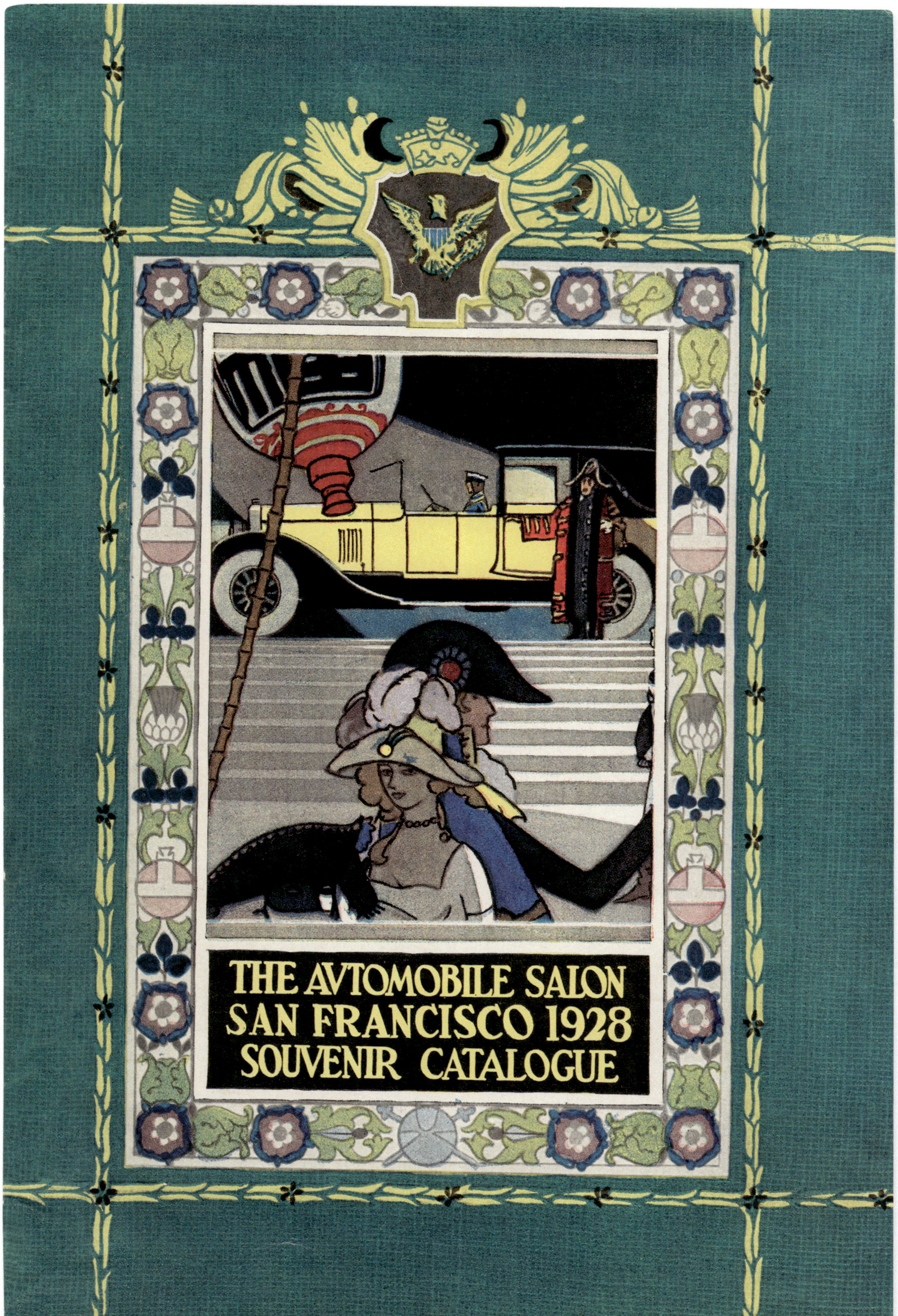
THE AVTOMOBILE SALON
SAN FRANCISCO 1928
SOUVENIR CATALOGUE

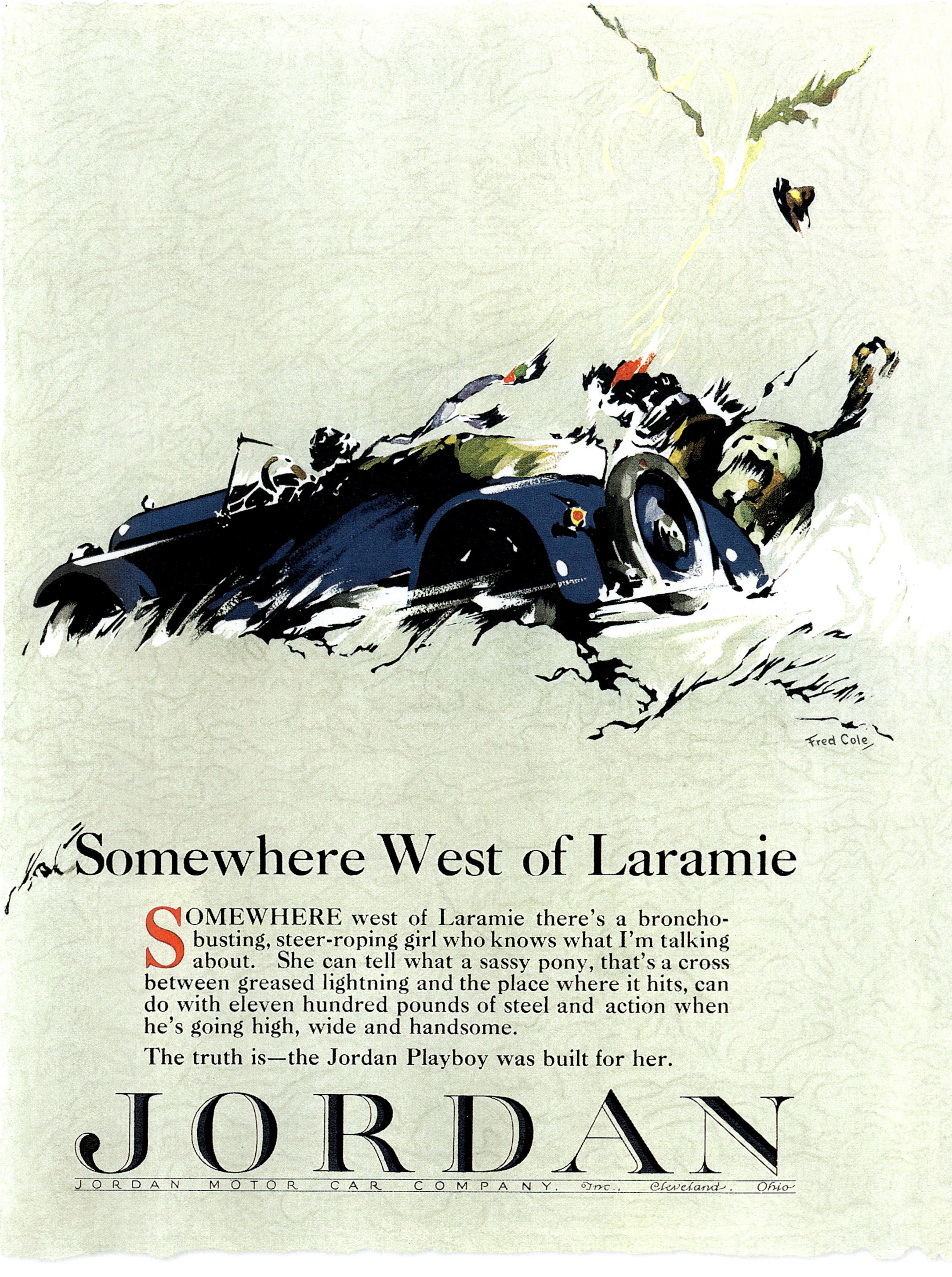
Fred Cole
Somewhere West of Laramie
SOMEWHERE west of Laramie there's a broncho-busting, steer-roping girl who knows what I'm talking about. She can tell what a sassy pony, that's a cross between greased lightning and the place where it hits, can do with eleven hundred pounds of steel and action when he's going high, wide and handsome.
The truth is—the Jordan Playboy was built for her.
JORDAN
JORDAN MOTOR CAR COMPANY, Inc., Cleveland, Ohio

JIM DONNELLY

Verlangen erwecken – Seite für Seite

Ein Automobilhersteller konnte Sie davon überzeugen, dass Sie klug waren, ein guter Ehemann oder ein Held in den Augen Ihrer Kinder, wenn Sie sein Auto kauften. Die Broschüren erzählten hochinteressante Geschichten über das Auto und über uns selbst.

Rein physisch betrachtet, wird ein Automobil aus präzise geformtem Stahl, aus Glas und Gummi zusammengesetzt. Der Produktionsprozess selbst lässt sich auf urtümlich anmutende Eindrücke reduzieren, die sich ständig wiederholen und ohrenbetäubend laut sind wie Hammerschläge der Feuer- und Kriegsgötter: rot glühende Schmieden, krachende Stampfwerke, kreischende Werkzeuge, brüllende Vorarbeiter, die zur Eile antreiben.

Solche Bilder stehen der sanften, sehr persönlichen Überzeugungskunst diametral entgegen, mit der Menschen ermutigt werden sollen, das Auto zu kaufen, nachdem es mit derart elementaren Urkräften produziert wurde. Diese Art der Überzeugung gleicht dem Gang durch einen sehr schmalen verspiegelten Korridor. Eigentlich besteht die Aufgabe der Automobilwerbung darin, dem Kunden einen Spiegel vorzuhalten, in dem er ein Auto sieht – und ein sehr zufriedenes Abbild seiner selbst. Die Botschaft, auch wenn sie über ein so bescheidenes Medium wie eine Händlerbroschüre vermittelt wurde, bestand darin, dass der Kauf des Produkts das Leben des Käufers verbessern könne.

In der Geschäftswelt geht am allgemein davon aus, dass ein Verkäufer umso härter arbeiten muss, je höher der Kaufpreis des Objekts ist. In der Frühzeit des Automobils tätigten Normalsterbliche im Lauf ihres Lebens nur drei große Investitionen: ein Eigenheim, die eigene Beerdigung und ein neues Auto. Auto und Eigenheim musste man nicht unbedingt haben, also mussten die Automobilhersteller den Verbraucher davon überzeugen, dass ihr Produkt doch unverzichtbar sei. Damals herrschte ein gnadenloser Wirtschaftsdarwinismus: Den verfügbaren Aufzeichnungen zufolge stellten allein in den USA zwischen 1900 und 1930 rund 5000 Firmen Automobile her. Diejenigen, die Erfolg hatten – und die Weltwirtschaftskrise überdauerten –, bauten Millionen davon, und die Tausende, die nur eine Handvoll produzierten, blieben erfolglos.

Trotz wachsender Komplexität und Leistung ihrer Erzeugnisse vertrauten die großen Hersteller in Detroit und anderswo bei der Vermarktung stets auf Einfaches: bebilderte Verkaufsbroschüren, die den Käufer in den Ausstellungsräumen erwarteten. Die Strategie war, den Käufer zu erreichen und ihn ohne Unterlass darüber nachdenken zu lassen, wie ihn das Auto verändern könnte. Selten wurden menschliche Ambitionen – so tief sitzende, persönliche Bedürfnisse – so wirksam und dauerhaft durch ein derart schlichtes Medium geprägt.

In ihrer einfachsten Form war eine Broschüre für neue Autos ein preiswerter Katalog. Sie war ein unverzichtbares Verkaufsinstrument der Händler. Zwar konnten die kostenlosen Broschüren einfach im Papierkorb landen, vielleicht blieben sie aber auch auf dem Küchentisch oder neben dem Wohnzimmersessel liegen, wo sie dann wieder und wieder durchgeblättert wurden. Leser konnten sich in aller Ruhe vorstellen, wie das Auto ihr Leben verändern würde. Sie konnten sich selbst in die einladend gestalteten Bilder der Broschüre hineinversetzen, wie sie die Familie nur zum Spaß über Land kutschierten. Sie selbst konnten die gut aussehende, lächelnde Person in der Werbung sein. Wer hätte nicht Lust, einen solch angenehmen Traum einmal probezufahren?

Ja, es war ein Traum. Im „Gilded Age", der großen Wirtschaftsblüte in den USA um die vorletzte Jahrhundertwende, kamen Autos nämlich nur für die Wohlsituierten infrage: teures Spielzeug, mit dem sie qualmend und scheppernd ihre Spritztouren auf holprigen Pisten unternehmen konnten. Ziel des Kraftfahrers war es damals, das heimische Anwesen wieder zu erreichen, ohne mit irgendetwas kollidiert zu sein oder eine peinliche Panne erlitten zu haben. Im frühen 20. Jahrhundert war Visionären wie Ransom Eli Olds und Henry Ford bereits klar, dass das Automobil sein die Gesellschaft und die Wirtschaft umwälzendes Potenzial niemals würde verwirklichen können, solange sich nicht Millionen von Normalbürgern eines leisten konnten. Und die Millionen stimmten ihnen zu. Gewöhnliche Arbeiter erkannten im Automobil nichts Geringeres als das Mittel zu ihrer persönlichen

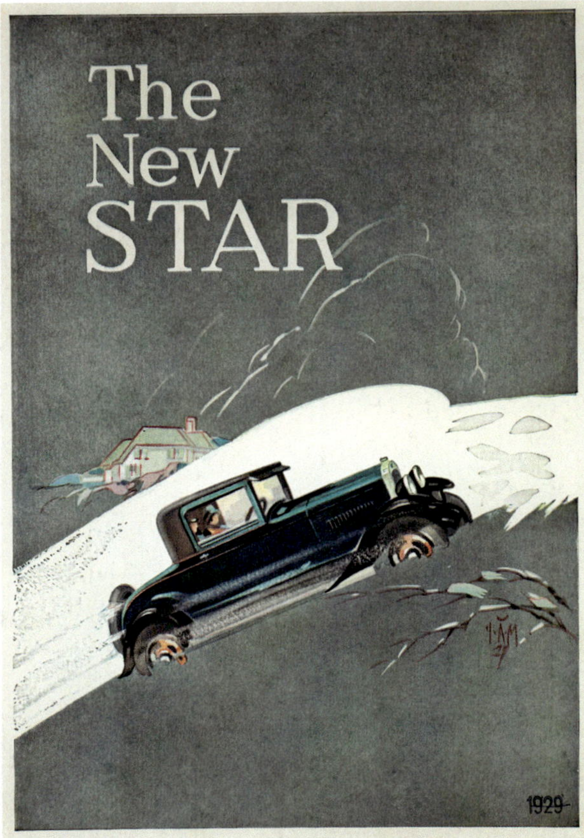

Jordan 1926 *(page 14)*

Star 1929 *(left)*

Paige Jewett 1927 *(opposite)*

Befreiung. Sie lechzten förmlich nach der Freiheit, sich nach ihrem eigenen Fahrplan fortzubewegen, der Plackerei auf der Farm zu entfliehen und herauszufinden, was jenseits der Endstation der überfüllten Straßenbahn lag.

Für den frühen Winton, ein Freizeit- und Rennsportgefährt, das um die Wende zum 20. Jahrhundert in Cleveland gebaut wurde, wurde mit diesem verlockenden Versprechen geworben: „VERZICHTEN SIE AUF DAS PFERD. Geruchlos. Betriebskosten unter einem ½ Cent pro Meile." Solche Anzeigen fand man auf den fleckigen letzten Seiten der Lokalzeitungen oder in Zeitschriften wie *Police Gazette* oder *The Sporting News*. Im Grunde konzentrierten sich diese Annoncen genau wie die gerade aufkommenden Werbebroschüren auf Themen, die potenzielle Automobilisten nachvollziehen konnten. Als Autofahrer gehörte man zu den Pionieren des großen Abenteuers zu Lande – praktischer Nutzen hin oder her. Der Locomobile von 1899 brüstete sich mit „Geschwindigkeit! Zuverlässigkeit! Sicherheit!". Für den Bates aus Lansing (Michigan) reimten die Werbetexter: „Use a Bates and keep your dates." („Fahren Sie einen Bates, und aus Ihren Verabredungen wird was.") Schon 1905 hieß es: „Experimentieren Sie nicht herum, kaufen Sie einfach einen Ford", woran sich zeigt, dass Ford bereits damals für dauerhaften Wert und robuste Zuverlässigkeit stand. Dennoch: Autokäufer vor dem Ersten Weltkrieg waren die Weltraumtouristen ihrer Zeit.

Etwa zur gleichen Zeit wie die ersten Broschüren kamen auch die Händler vor Ort auf. Ein frühes Beispiel für ein erfolgreiches Werbemuster findet man in einem 26-Seiten-Heft, das 1916 die Thomas B. Jeffery Company aus Kenosha (Wisconsin) herausgab, aus der später Nash und noch später die American Motors Corporation wurde. Die Druckfarben sind Schwarz und Hellorange. Auf den Rücksitzen eines Jeffery Four sitzen Damen mit Sonnenschirmen, die ihre Hüte im Fahrtwind festhalten, während das Fahrzeug einen Windhund überholt. Andere Zeichnungen zeigen die Innenausstattung, während ein Gemälde den Jeffery bei einem Pferdedressurwettbewerb darstellt. Die Bildsprache war eindeutig.

Als sich über Europa der Rauch und die Senfgaswolken des Ersten Weltkriegs langsam legten, wurden die Vermarktungstechniken rasch raffinierter. Schwammige Versprechungen über Leistung und Zuverlässigkeit reichten nicht mehr aus. Die Massenproduktion und die leichte Verfügbarkeit von Krediten machten Autos nun für zig Millionen Menschen erschwinglich. „Wachs oder stirb" war das Motto, das alljährlich Hunderte Automobilhersteller vom Markt verdrängte und all jene verschwinden ließ, die ihre Produkte nicht durch hohe Materialgüte, außergewöhnliches Design, starke Motoren oder andere prestigeträchtige Merkmale aus der Masse der Produkte hervorheben konnten. Für die verbleibenden Branchengiganten, allen voran Ford und General Motors, stand viel auf dem Spiel: Sie konkurrierten um Millionenbeträge. Sie mussten neue Werbeaussagen und neue Medien finden, um ihre Blechlawinen ins Rollen zu bringen. Mit der Broschüre gelangten ihre Aussagen in viele neue Hände.

Es ist historisch belegt, dass General Motors – ein Unternehmen, das bis 1920 bereits zweimal kurz vor der Pleite gestanden hatte – im Kampf um die Sympathien der Käufer anfänglich in Führung lag. Der Werbefeldzug hatte wenig mit den Autos von GM oder den Herstellungsmethoden zu tun. In den 1920er-Jahren wurde das Unternehmen von einem Finanz- und Organisationsgenie namens Alfred P. Sloan geführt. Er verstand intuitiv, dass GM die Nachfrage ankurbeln (statt einfach nur Autos zu bauen) und Profit (nicht Autos) machen musste. Autos waren für Sloan nur Mittel zum Zweck. Seine wahre Genialität lag in der Fähigkeit, menschliche Wünsche und Bedürfnisse auszunutzen.

Sloan verfügte beispielsweise, dass die Chevrolet-Modelle jedes Jahr anders aussehen sollten, und sei es auch nur noch so oberflächlich. Atemlos ließ die Werbung (die sich in dieser Zeit meist in auflagenstar-

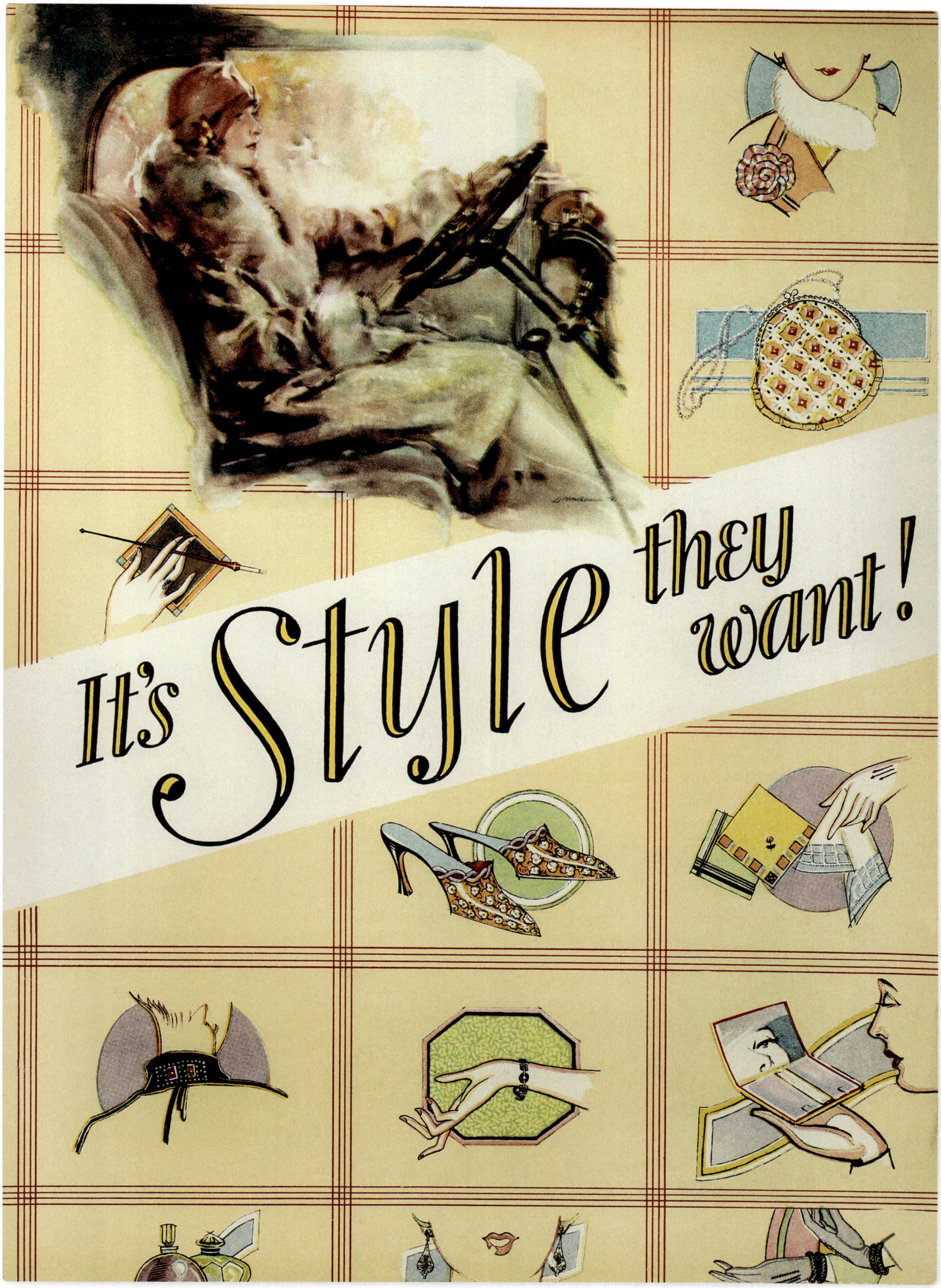
It's Style they want!

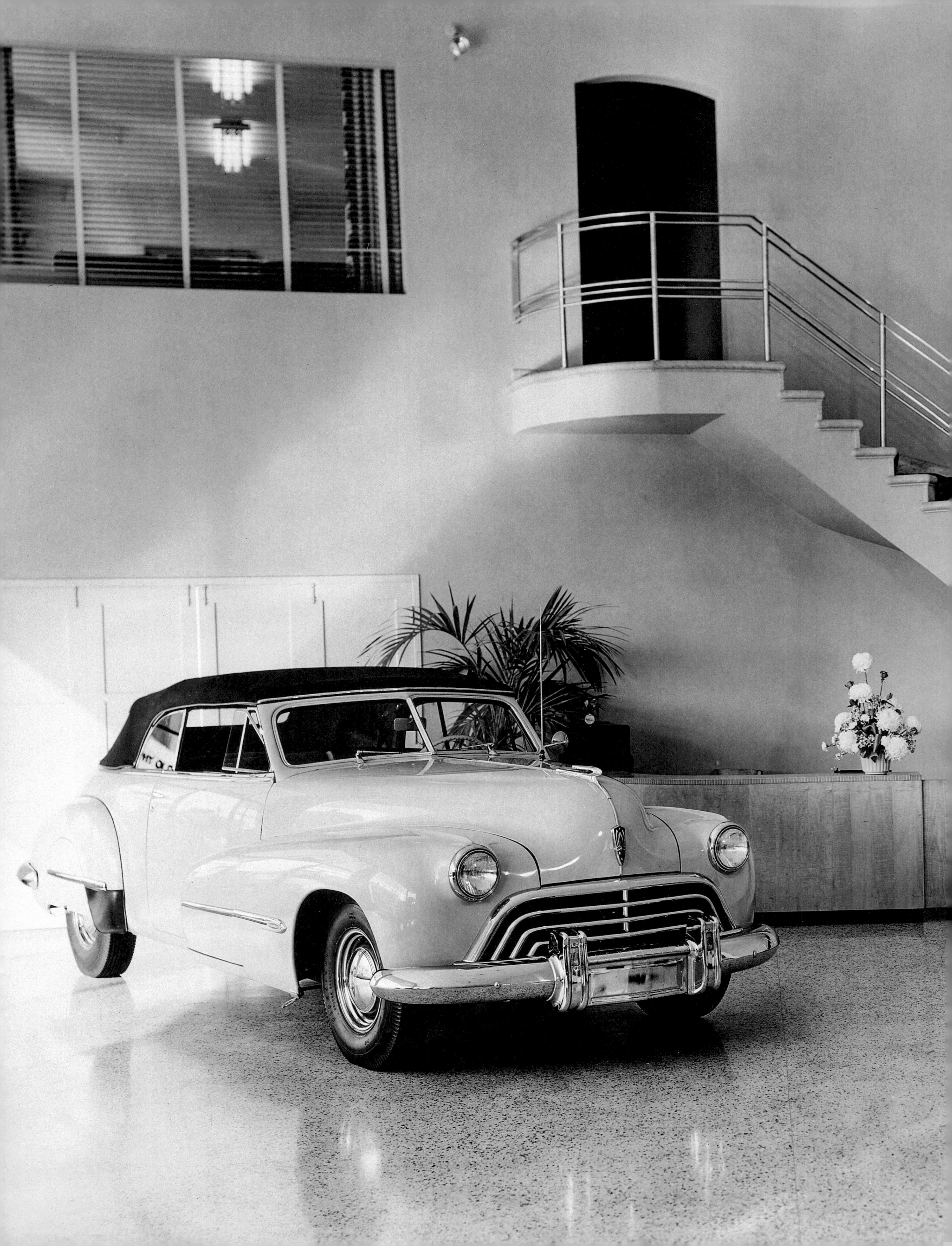

Oldsmobile showroom 1947 *(opposite)*

Lincoln 1933 *(right)*

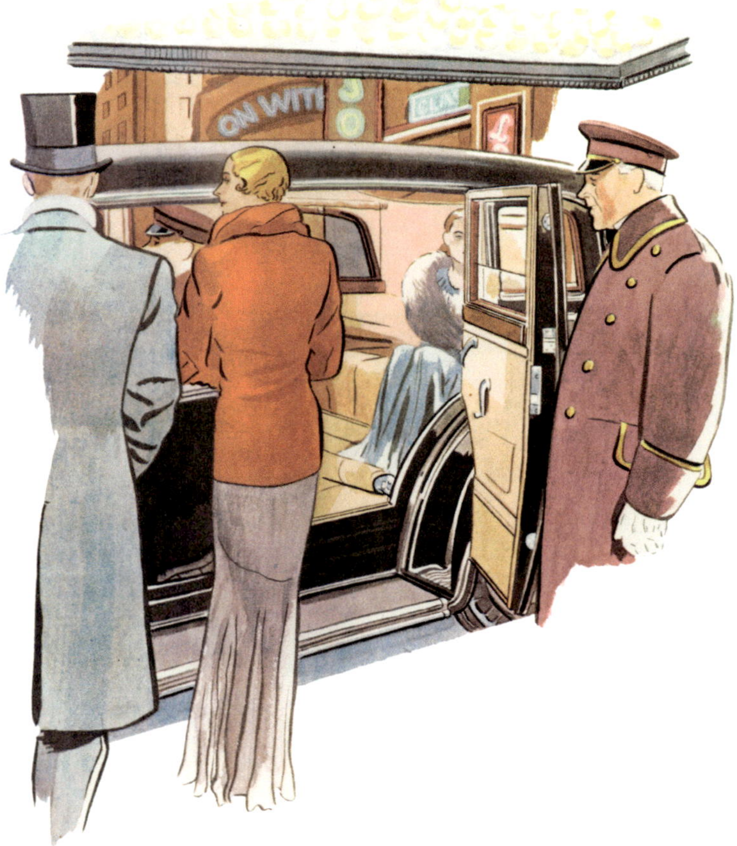

ken Zeitschriften wie *Harper's* und *The Saturday Evening Post* fand) den Leser wissen, dass er die „neuen" Chevys (traditionell am Herbstanfang) beim Händler um die Ecke besichtigen konnte. Das ging so weit, dass man die leicht modifizierten „neuen" Modelle unter Planen zum Händler transportierte und bis zum offiziellen Termin auch im Ausstellungsraum verdeckt hielt. Wurden die Wagen dann enthüllt, fand man gleich neben den Autos Ständer mit druckfrischen Gratisbroschüren, in denen mit blumigen Worten alle Neuheiten beschrieben wurden.

Der „Sloanismus", wie man diese Art der Vermarktung ehrerbietig nannte, beruhte zum Teil auf dem menschlichen Bedürfnis, den Nachbarn zu übertreffen, indem man einen neueren, moderneren GM fuhr. Zu diesem Zweck ordnete Sloan 1924 auch eine grundlegende Umstrukturierung des Unternehmens an, indem jeder GM-Marke ein bestimmtes Preissegment zugewiesen wurde. Die Hierarchie begann mit Chevrolet auf der untersten Stufe und endete bei Cadillac an der Spitze. Außerdem rief Sloan eine neue Abteilung ins Leben, die für Farbe und Gestaltung zuständig war. Geleitet wurde sie von Harley Earl, einem ehemaligen Wagenbauer, der mittlerweile Karosserien nach Maß herstellte. Er schuf nun für die neuen GM-Modelle Stoff- und Farbpaletten, die mit steigenden Preisen immer edler wurden. Dahinter stand die Idee, dass sich der Chevrolet-Käufer seinen sozialen Aufstieg verbildlichen konnte, indem er danach strebte, zu einem Pontiac, Buick oder Cadillac zu wechseln – statt etwa zum Konkurrenten Chrysler. So blieb der Gewinn im Haus.

Die Verknüpfung der Produkte mit einer gefühlten Überlegenheit – Geld! Erfolg! Neid! Erotik! Von allem das Neueste! – wurde zum übergreifenden Ziel der Autowerbung zwischen den Weltkriegen. Es schien, als nähme die eigentliche Beschreibung der Fahrzeuge mit wachsenden Verkaufszahlen und steigenden Preisen immer weniger Platz ein. Das soll nicht heißen, dass die Werbung technische Fortschritte ignorierte, denn davon gab es viele in den Zwischenkriegsjahren – sie wurden zu Recht wichtig genommen und opulent illustriert. Zu den Neuerungen gehörten der V8-Seitenventilmotor von Ford in preiswerten Fahrzeugen, das erste voll funktionsfähige Automatikgetriebe, die Einzelradaufhängung an der Vorderachse sowie – häufig übergangen – die erste Autodachkonstruktion aus Vollstahl für geschlossene Fahrzeuge von GM. Während die Wirtschaftskrise den Absatz stagnieren ließ, machte sie aus der Broschüre unabsichtlich ein kleines Nachschlagewerk über das Auto. Die Broschüren priesen neben den technischen Neuerungen auch an, wie langlebig, wartungsarm und preiswert in der Reparatur die Autos waren, um den Kunden davon zu überzeugen, dass der Kauf eines neuen Fahrzeugs eine lohnenswerte und sinnvolle Investition war.

Oft gingen die Texte mehr auf die Menschen als auf die Autos ein. Vor allem Frauen nahmen immer mehr Raum in den Automobilanzeigen ein, ohne dass ihnen ein männlicher Begleiter helfend zur Seite stand. Bereits früh waren Autos auch direkt für ein weibliches Publikum vermarktet worden, wie zum Beispiel die Elektroautos von Detroit und Baker und später der erste Selbstanlasser, den Cadillac 1912 auf den Markt brachte, damit Autos leichter und bequemer zu fahren waren.

Es ist wichtig zu verstehen, wie die meisten Menschen in den 1920er- und 30er- Jahren lebten. Es gab kein Fernsehen. Zeitungen erschienen nachmittags. Nach der Arbeit und dem Abendessen reichte man die Broschüren vielleicht in der Familie herum, sprach darüber, wie schnell sich die Autos veränderten, die Ausstattung verbesserte, die Fahrzeuge schneller und zuverlässiger wurden. Ehefrauen überflogen die Broschüren. Kinder fanden sie aufregend und schnitten manchmal die Autobilder aus, um sich kleine Papierspielzeuge daraus zu basteln. Das Automobil drang in das Bewusstsein der Familie ein und setzte sich dort fest.

Im Zweiten Weltkrieg, Anfang 1942, stellten die Megafabrikanten in Detroit die Autoherstellung ein und stiegen unter Aufsicht der Regierung sofort auf die Produktion von Kriegsmaterial in gigan-

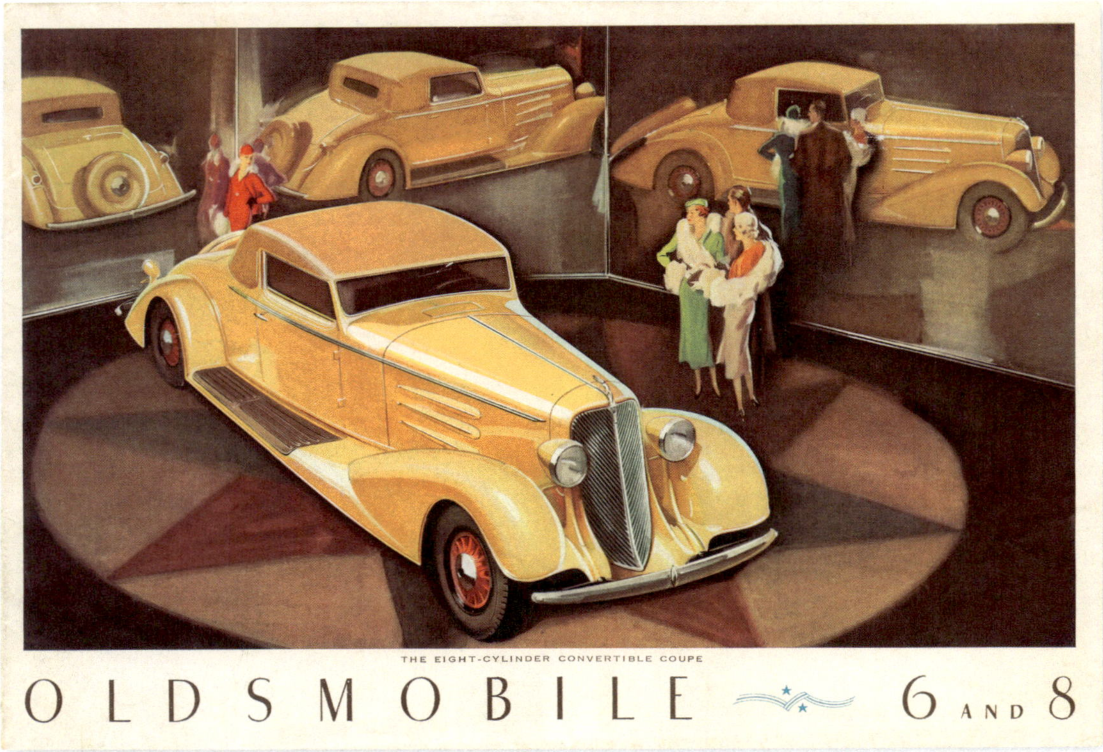

Oldsmobile 1933 *(left)*

Hupmobile 1934 *(opposite)*

tischem Umfang um. Es oblag den Automobilherstellern, die Moral sowohl der Belegschaft als auch der Kundschaft aufrechtzuerhalten. Da es keine Autos zu verkaufen gab, bewarb sich die Industrie in aufrüttelnden, patriotischen Botschaften selbst, indem sie zu Recht ihren Beitrag zum Kriegseinsatz hervorhob und den Bürgern großartige Autos versprach, sobald der Krieg beendet sei. Ein beliebtes Motiv war etwa der G.I., der in einer ausgebombten Einöde biwakierte und von einem neuen Auto träumte, das ihn bei seiner Rückkehr in die Heimat erwartete. Eine Werbung versprach: „Es gibt einen Ford in deiner Zukunft." An der Heimatfront stellten sich die meisten Neuwagenhändler auf die plötzlich kostbar gewordenen Gebrauchtfahrzeuge sowie Reparaturen und Reifenverkauf um. Die Kraftstoffrationierung setzte der Beweglichkeit der Menschen ohnehin Grenzen. Und zur Rückgewinnung des kriegswichtigen Stahls landeten zahllose Altautos mit allzu hohem Verbrauch auf dem Schrottplatz.

Die Zusage, dass es einen Ford – oder ein anderes Auto – in der Zukunft gebe, sollte sich bewahrheiten. In der Nachkriegszeit erlebte die Welt des Automobils in den USA ganz neue Auswüchse. Die heimkehrenden Soldaten, die ordentliche Kampfzulagen erhalten hatten, wollten sofort neue Autos. Die amerikanische Gesellschaft hatte genug von Sparsamkeit, Zurückhaltung und Seriosität. Und das sah man den Autos an: Detroit überschlug sich mit immer barockerem Design mit Flossen, Tunneln und „Raketenköpfen", die das Gewicht in die Höhe trieben und nach immer durstigeren Motoren verlangten, die genügend Leistung brachten, um alles in Fahrt zu bringen. Die Werbung hingegen konzentrierte sich auf unscheinbare Neuerungen von zum Teil zweifelhaftem praktischen Wert, wie etwa ein breites Angebot an Farben und Stoffen in einer schwindelerregenden Zahl von Kombinationen.

Entbehrung, Kampf und Elend wurden verbannt – und mit ihnen wohl auch der gute Geschmack und die offene, ehrliche Darstellung. Unter den Extras fand man nun merkwürdige Begriffe wie „Power-Pilot", „Whirlaway", „TorsionAire" und „Airflyte". Die Texte stammten weitgehend aus Werbeagenturen. Zu den Ersten, die hier mitmischten, gehörten Young & Rubicam, die für die Packard-Werbung verantwortlich zeichneten. J. Walter Thompson, eine Agentur, deren Wurzeln bis ins 19. Jahrhundert zurückreichten, arbeitete lange für Ford. Und Campbell-Ewald konnte den begehrtesten Kunden der Branche, Chevrolet, ganze 91 Jahre lang halten.

Diese und andere Firmen mussten denn auch die Autokäufer wieder zu Verstand bringen, als das große Schwelgen in extravagantem Schnickschnack, das Ende der 1950er-Jahre seinen Höhepunkt erreicht hatte, nicht mehr ziemlich erschien. Von den 1960er-Jahren an zog es die Amerikaner zu weniger exzessiven Autos, zunächst aus Europa und Japan und später auch wieder aus den USA. Die Werbefachleute mussten die Käufer nun davon überzeugen, dass weniger mehr war. Und sie waren erstaunlich erfolgreich. Die verschmitzte Kampagne der Werbeagentur Doyle Dane Bernbach für Volkswagen war ebenso genial und erinnerungswürdig wie die VW-Broschüren, insbesondere nach der ersten großen OPEC-Ölkrise in den 1970er-Jahren.

Ein Automobilhersteller konnte Sie davon überzeugen, dass Sie klug waren, ein guter Ehemann oder ein Held in den Augen Ihrer Kinder, wenn Sie sein Auto kauften. Die Broschüren erzählten hochinteressante Geschichten über das Auto und über uns selbst. Eine gute Anzeige spricht uns tief in unserem Innern an – gleichgültig, ob sie für ein Auto, ein Shampoo oder ein Smartphone wirbt. Anders als eine Plakatwand oder ein Pop-up im Internet war die Werbebotschaft einer Broschüre immer direkt vor unserer Nase und konnte jederzeit angesehen werden. Broschüren hoben uns mit Wort und Bild über eine psychische Schwelle hinweg. Die Evolutionsgeschichte des Automobils aus der Sicht von Technik und Design wurde schon erschöpfend dargelegt, doch die Geschichte der Automobilbroschüre ist gleichermaßen wichtig. Ingenieure schufen und verfeinerten das Auto, doch erst die Werbung schuf durch diese Hefte das Verlangen, es zu besitzen.

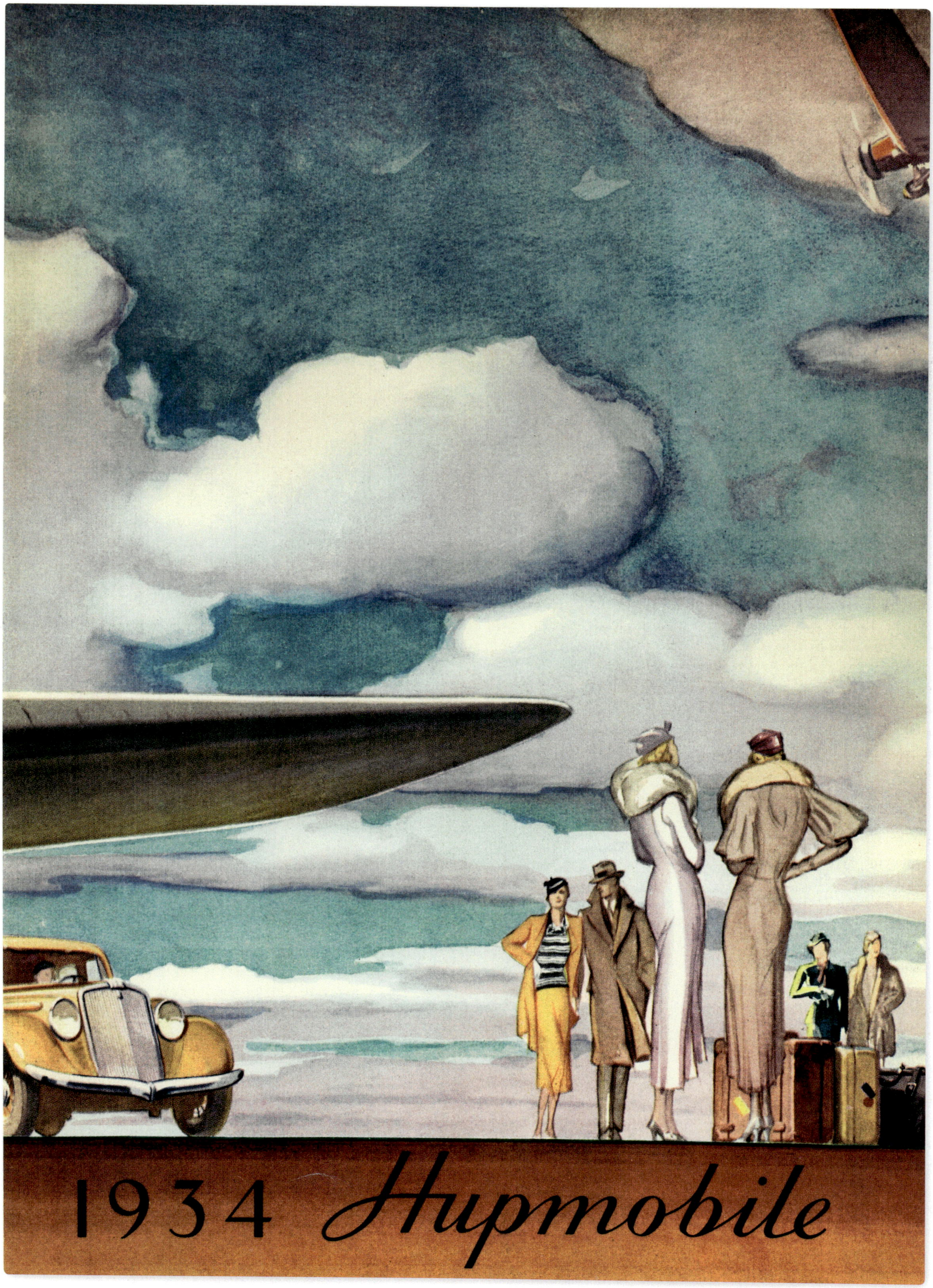
1934 Hupmobile

OUT OF THE *Needs of the Nation*

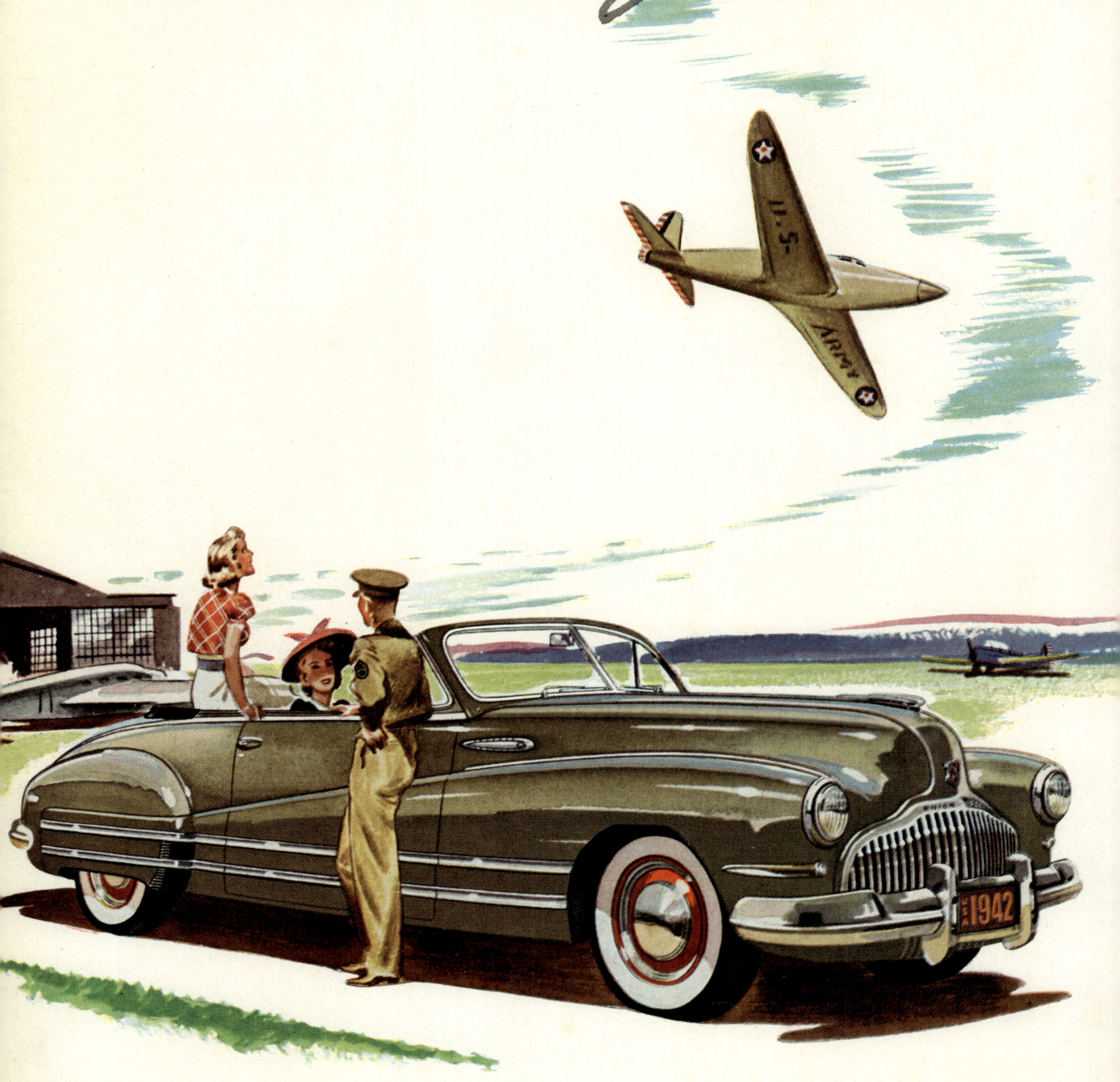

JIM DONNELLY

Du désir à chaque page

Il fut un temps où un constructeur automobile avait le pouvoir de vous convaincre que vous étiez quelqu'un d'intelligent, un bon mari et un héros auprès de vos enfants, dès lors que vous achetiez telle ou telle voiture. Les catalogues avaient le charme de récits merveilleux qui nous parlaient de l'automobile et de nous-mêmes.

D'un point de vue purement organique, l'automobile est constituée d'acier, de verre et de caoutchouc soigneusement mêlés au fil de gestes élémentaires, aussi répétitifs qu'assourdissants : le martèlement du fer par les dieux de la Guerre et du Feu. Rougeoiement des forges, fracas de l'emboutissage, stridence des machines-outils, clameurs des contremaîtres pour hâter les cadences… Il est loin l'art de la suggestion, subtil et personnalisé, qui doit pourtant servir à promouvoir l'achat du véhicule pétri par tant de forces primitives.

Imaginez un labyrinthe de glaces dans lequel, pour mieux le convaincre, on entraînerait le consommateur ; réduite à sa fonction première, la publicité automobile revient en effet à lui tendre un miroir où il peut se contempler, fasciné par le double reflet qui lui est renvoyé. Celui de la voiture, bien sûr, mais aussi son propre portrait. Car le message est clair, et peu importe si la brochure qui le propage est des plus ordinaires : « Acheter cette voiture va vous changer la vie. »

À en croire l'une des règles d'or du marketing, plus un achat est coûteux, plus le vendeur doit batailler pour conclure le marché. Il est vrai qu'aux tout débuts de l'automobile, les occasions d'investissement important, dans la vie d'un homme, se comptaient sur les doigts d'une main : l'acquisition du logement, le financement des obsèques… et l'achat de la voiture. Le logement et la voiture étaient même assez facultatifs, aussi les constructeurs se sont-ils attachés à faire de l'achat automobile un acte essentiel. L'heure était au darwinisme économique, c'est-à-dire à la sélection naturelle – qui fut impitoyable : d'après les archives disponibles, rien qu'aux États-Unis quelque 5 000 entreprises ont fabriqué des automobiles entre 1900 et 1930. Celles qui réussirent à en vendre et qui survécurent à la Grande Dépression continuèrent sur leur lancée, jusqu'à en produire des millions ; les autres rejoignirent la cohorte des marques éphémères qui n'ont créé que quelques modèles.

Alors que progressaient la sophistication et les performances de l'automobile, le marketing des usines géantes de Detroit, de son côté, s'évertuait à cultiver la simplicité : celle des brochures illustrées qui attendaient les visiteurs des showrooms. Leur stratégie : dans un premier temps, toucher l'acheteur, puis le convaincre qu'avec une telle voiture, sa vie ne pouvait que changer. Rarement les ambitions humaines – les besoins les plus profonds et les plus intimes – auront été manipulées avec tant d'efficacité, et pendant si longtemps, à travers un médium aussi rudimentaire. Dans sa version basique, la brochure consacrée à un nouveau modèle se présentait sous la forme d'un catalogue bon marché. Pour le vendeur, c'était un outil indispensable. Certes, comme tout dépliant gratuit, il courait le risque de finir ses jours au fond d'une corbeille à papier, mais ses chances n'étaient pas moindres de séjourner un bon moment sur la table de la cuisine ou l'accoudoir du fauteuil et d'y être feuilleté, parcouru, consulté, examiné sous toutes les coutures. Le lecteur pouvait ainsi tout à sa guise se représenter la façon dont sa future voiture allait vraiment changer sa vie ; l'occasion, pour lui, de se projeter dans la mise en page – évidemment alléchante – et de s'imaginer en famille en train de sillonner la campagne au gré de ses envies ; rien ne lui interdisait non plus de s'identifier au personnage séduisant qui s'affichait tout sourire sur les images du catalogue. Comment résister à la tentation d'embarquer pour un voyage de rêve dans un univers aussi glamour ?

Car il est bien question de rêve. Jouets de luxe à peine sortis de l'Âge d'or, les voitures, à l'origine, n'étaient destinées qu'aux très riches qu'elles promenaient cahin-caha dans d'improbables virées enfumées et cliquetantes ; le Graal de l'automobiliste, alors, était de réussir à regagner son domaine sans avoir froissé de tôle ni subi quelque panne humiliante. Puis arriva le 20e siècle, et avec lui le temps des pionniers. Très tôt des hommes comme Ransom Eli Olds et Henry Ford prirent conscience que l'automobile ne réaliserait son potentiel de transformation sociale et économique qu'à condition de se rendre accessible aux millions de citoyens ordinaires. Et les millions de citoyens ordinaires étaient du même avis. Ouvriers et employés accueillirent l'automobile

Buick 1942 *(page 22)*

Ford 1940 *(right)*

Ford assembly line, Dearborn, Michigan, 1936 *(opposite)*

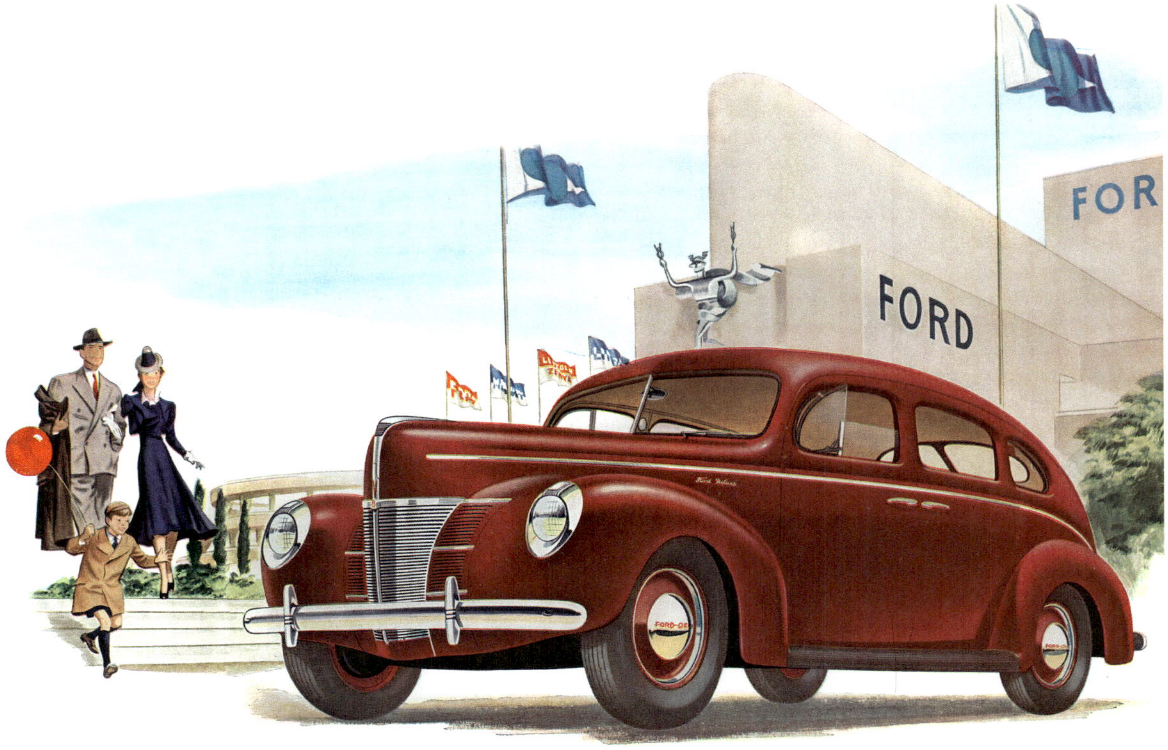

comme l'instrument de leur libération individuelle. Ils avaient soif de liberté de mouvement – enfin seuls maîtres à bord –, soif d'échapper à leur labeur de bêtes de somme, soif de découvrir *de visu* ce qui se cachait là-bas, au-delà du terminus du tramway surpeuplé.

Ainsi la première Winton, un prototype de course assemblé à Cleveland au tournant du 20e siècle, adressait-elle le message suivant : « PASSEZ-VOUS DE CHEVAL. Aucune odeur. Moins d'un demi-cent le mile. » Dans les dernières pages des quotidiens locaux ou dans des périodiques tels la *Police Gazette* ou *The Sporting News*, on pouvait tomber sur des annonces mal imprimées qui développaient, comme les premières brochures commerciales, une argumentation intelligible à tous les futurs *automobilistes* : prendre le volant, c'était rejoindre l'avant-garde d'une aventure routière dont la commodité n'était en rien exclue. Illustrations : la Locomobile de 1899, qui vantait la trinité de ses vertus (« Vitesse ! Fiabilité ! Sécurité ! »), ou encore la Bates conçue à Lansing (Michigan), carrément lyrique : « Roulez en Bates et arrivez à l'heure ». Dès 1905, l'adage en vogue invitait à la sagesse : « Évitez les expérimentations, achetez simplement une Ford » – témoignage du précoce attachement de la marque à la valeur « robustesse ». Reste qu'avant la Première Guerre mondiale, acheter une automobile revenait tout de même un peu à jouer les astronautes de l'époque.

Les premiers catalogues apparaîtront peu ou prou au même moment que les concessionnaires locaux. Pour avoir une idée de la présentation alors la plus coutumière, il suffit de consulter le livret de 26 pages réalisé en 1916 par l'entreprise Thomas B. Jeffery, basée à Kenosha (Wisconsin), qui deviendra Nash puis, plus tard, American Motors Corporation. Le trait noir et orangé y croque des élégantes au parasol installées à l'arrière d'une Jeffery Four, où elles protègent leur coiffure de la brise tandis que le bolide dépasse un lévrier. D'autres dessins décrivent les aménagements intérieurs et une illustration peinte met en scène ladite Jeffery lors d'un concours de dressage. Le symbolisme ainsi recherché est évident.

Quand les vapeurs de gaz moutarde du premier conflit mondial finissent par se dissiper, le marketing peut passer à la vitesse supérieure. Célébrer en termes vagues la fiabilité et la performance d'un véhicule ne suffit plus. La production de masse et les facilités de crédit drainent vers le marché automobile des dizaines de millions de nouveaux adeptes. Et chaque année, des centaines de constructeurs soumis à la pression du « get-big-or-die » (grossir ou périr) sortent du jeu.

Les seuls à échapper à l'extinction sont ceux qui peuvent s'appuyer sur la reconnaissance d'un haut niveau de qualité – en termes de technologie, de style, de performances – ou sur tout autre indice valorisant capable de les démarquer de la production de masse. Pour les géants du secteur, emmenés par Ford et General Motors, les enjeux, vertigineux, s'évaluent en millions de dollars ; s'ils veulent écouler toute la masse de métal qui sort chaque jour de leurs chaînes de montage, il leur faut impérativement inventer de nouveaux messages et de nouveaux médias. Les catalogues vont désormais circuler dans de nouvelles mains, de plus en plus nombreuses.

General Motors, qui avait échappé par deux fois à la faillite avant 1920, n'en va pas moins remporter la première manche de cette conquête psychologique du client. La campagne victorieuse aura pourtant peu à voir avec les méthodes de production et les modèles maison. À l'époque, l'entreprise est présidée par un génie de la finance et de l'organisation nommé Alfred P. Sloan. D'instinct, Sloan a compris que General Motors ne devait pas se limiter à produire des véhicules, mais aussi et surtout à augmenter la demande et à réaliser des profits. À ses yeux, les voitures sont des moyens servant à atteindre un objectif. Avec un brio qui ne se démentira jamais, Sloan saura jouer comme personne des besoins et des désirs du consommateur.

Ainsi Sloan décidera-t-il de faire subir chaque année aux Chevrolet quelques légères modifications d'apparence. La publicité, alors, a déjà pris ses aises dans les pages des magazines de très grande diffusion comme *Harper's* et le *Saturday Evening Post*. Elle va donc se faire un

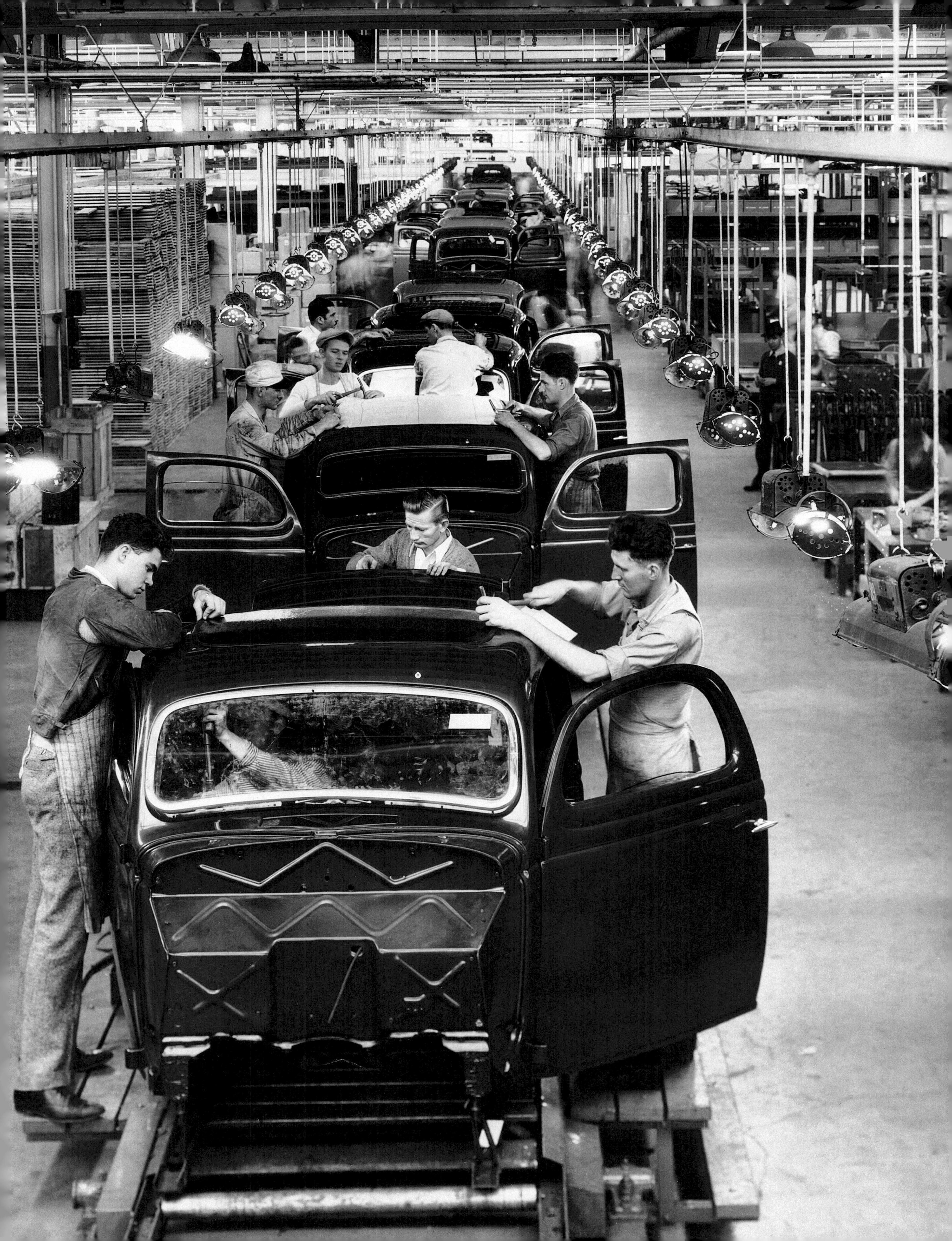

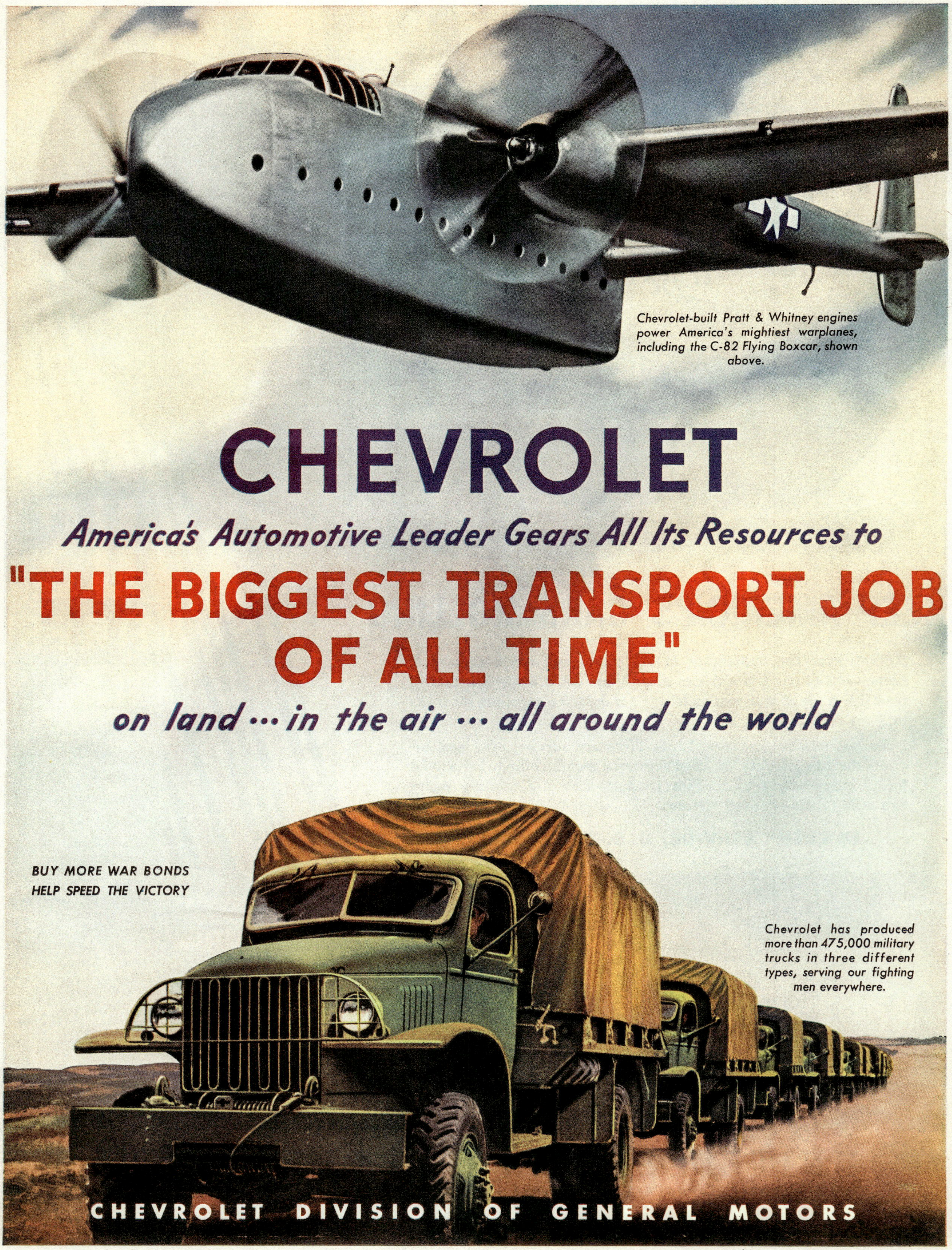
Chevrolet-built Pratt & Whitney engines power America's mightiest warplanes, including the C-82 Flying Boxcar, shown above.
CHEVROLET
America's Automotive Leader Gears All Its Resources to
"THE BIGGEST TRANSPORT JOB OF ALL TIME"
on land ··· in the air ··· all around the world
BUY MORE WAR BONDS
HELP SPEED THE VICTORY
Chevrolet has produced more than 475,000 military trucks in three different types, serving our fighting men everywhere.
CHEVROLET DIVISION OF GENERAL MOTORS

Chevrolet 1945 *(opposite)*

Willys Jeep 1944 *(right)*

Cadillac 1944 *(far right)*

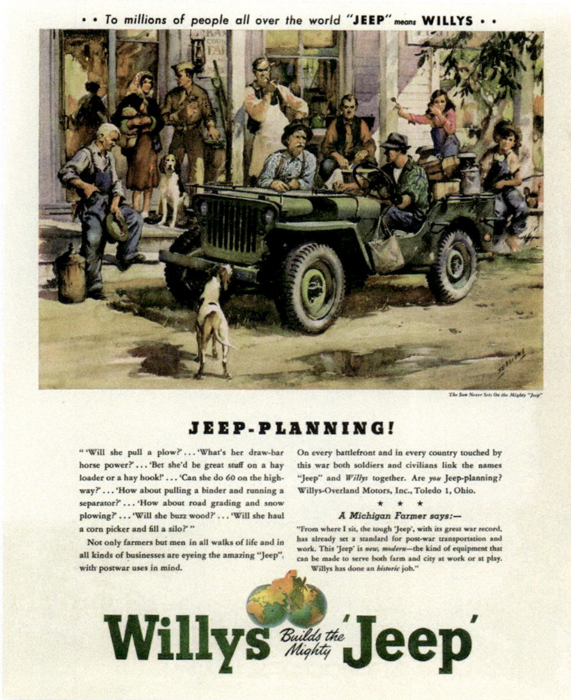

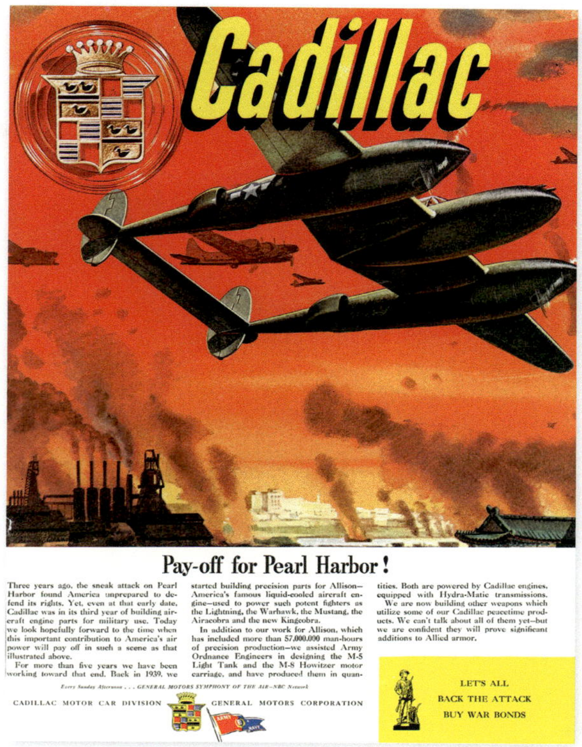

plaisir, généralement en début d'automne, d'alerter leurs lecteurs de l'arrivée imminente des nouvelles Chevy chez le concessionnaire le plus proche. L'exercice ira jusqu'à dissimuler aux regards, lors de leur transport, ces « nouveaux » modèles superficiellement retouchés et à les recouvrir d'une bâche dans les showrooms jusqu'à la date prévue pour leur présentation. Lorsque le voile se lève, le public peut découvrir des porte-revues garnis de catalogues fraîchement imprimés et gratuits. Disposés juste à côté des voitures, ils sont truffés d'articles dithyrambiques qui en détaillent toutes les nouveautés.

Le sloanisme – puisque c'est ainsi qu'on baptisa cette efficace stratégie – reposait en partie sur la manipulation assumée de l'émulation. Comment? En incitant le client à acheter une voiture encore plus récente et plus moderne que celle de son voisin. Dans le même esprit, Sloan engagera à partir de 1924 une profonde réorganisation de GM autour d'une hiérarchie des marques désormais édifiée sur les prix (Chevrolet en bas, Cadillac tout en haut). Il dotera en outre l'entreprise d'un nouveau département Art & Colour (préférant l'anglicisme « colour » à l'américain « color », prestige oblige). À la tête de ce studio de style, un ancien carrossier devenu fabricant de châssis sur mesure : Harley Earl. C'est lui qui imaginera les palettes de couleurs et de tissus destinées à enrichir la nouvelle gamme GM sans ruiner le consommateur. Il s'agit de permettre au client Chevrolet d'envisager une montée en gamme en restant fidèle au groupe : mieux vaut le guider vers Pontiac, Buick ou Cadillac plutôt que vers la concurrence, par exemple Chrysler. Ainsi le bilan de la GM peut-il rester bénéficiaire.

Pendant l'entre-deux-guerres, la publicité automobile va s'appliquer à associer le produit à tout ce qui excite le désir de supériorité : l'argent, le succès, l'envie, l'attraction sexuelle, la dernière nouveauté… L'espace consenti au descriptif détaillé des modèles donnera même l'impression de diminuer en proportion inverse du nombre de voitures vendues et de leur prix. Mais la publicité ne passera pas pour autant à côté des innovations technologiques de la période, aussi nombreuses que déterminantes; elle s'efforcera même de les illustrer abondamment : c'est d'alors que date l'apparition sur des véhicules bon marché du moteur Ford V8 à soupapes latérales, des premières transmissions automatiques efficaces et de la suspension avant indépendante, sans oublier la création – trop souvent oubliée – du premier toit en acier d'une seule pièce pour voitures fermées (une trouvaille de la GM).

La Grande Dépression, si elle affecte sérieusement la production, va avoir une conséquence inattendue : la métamorphose des catalogues en petites encyclopédies dédiées. Avancées technologiques, longévité, facilité d'entretien et réduction des coûts de réparation sont mis en avant. Le client doit être convaincu que l'achat de sa nouvelle voiture sera un investissement aussi utile qu'économique. Le plus souvent, d'ailleurs, les brochures privilégient moins le véhicule lui-même que ses utilisateurs. Évolution significative, les femmes continuent de se frayer un chemin dans les annonces automobiles, sans mâle pour les guider : depuis quelque temps déjà, certains modèles sont clairement conçus à leur intention, comme les Detroit et Baker électriques. C'est aussi le cas des premiers starters automatiques proposés par Cadillac en 1912, censés rendre la conduite plus facile et plus confortable.

Il faut se représenter le quotidien dans les années 1920 et 1930. La télévision n'existe pas. Les journaux paraissent l'après-midi. Après le travail et le dîner, les familles ont tout le temps de s'échanger les catalogues automobiles et de commenter les incessantes nouveautés, les progrès de la vitesse et de la fiabilité. Les mères dévorent les brochures, les enfants s'enthousiasment à leur lecture. Parfois même ils en découpent les pages pour fabriquer des autos en papier. La voiture s'incruste dans l'imaginaire familial pour ne plus en décoller.

À partir de 1942, la Seconde Guerre mondiale contraindra les géants de Detroit à suspendre leur production automobile civile. Immédiatement, ils se tournent vers la fabrication de matériel militaire, à une échelle colossale et sous le contrôle du gouvernement.

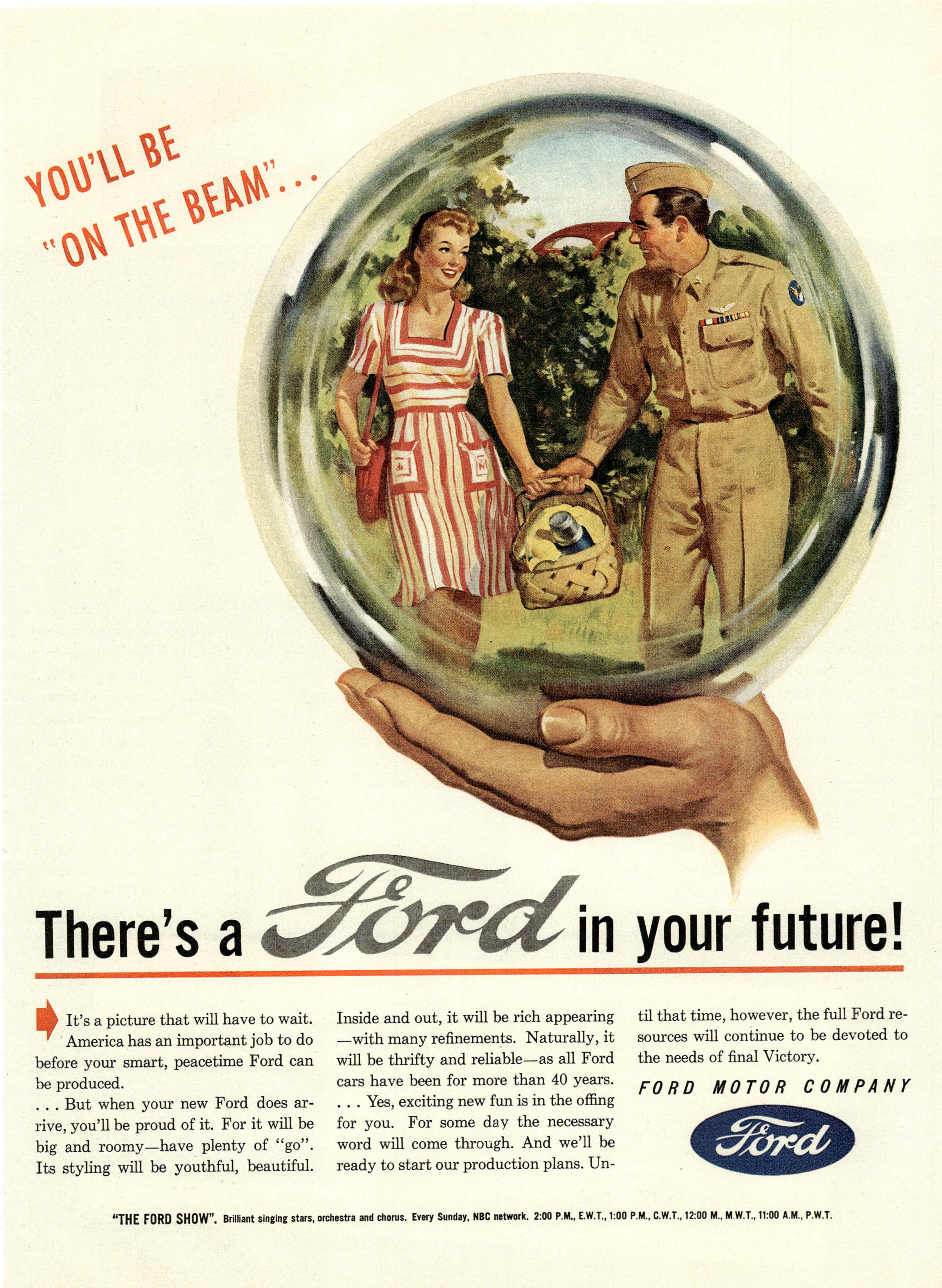
YOU'LL BE "ON THE BEAM"...
There's a Ford in your future!
It's a picture that will have to wait. America has an important job to do before your smart, peacetime Ford can be produced.
... But when your new Ford does arrive, you'll be proud of it. For it will be big and roomy—have plenty of "go". Its styling will be youthful, beautiful. Inside and out, it will be rich appearing—with many refinements. Naturally, it will be thrifty and reliable—as all Ford cars have been for more than 40 years.
... Yes, exciting new fun is in the offing for you. For some day the necessary word will come through. And we'll be ready to start our production plans. Until that time, however, the full Ford resources will continue to be devoted to the needs of final Victory.
FORD MOTOR COMPANY
Ford
"THE FORD SHOW". Brilliant singing stars, orchestra and chorus. Every Sunday, NBC network. 2:00 P.M., E.W.T., 1:00 P.M., C.W.T., 12:00 M., M.W.T., 11:00 A.M., P.W.T.

Ford 1944 *(opposite)*

Ford 1946 *(right)*

General Motors 1954 *(following spread)*

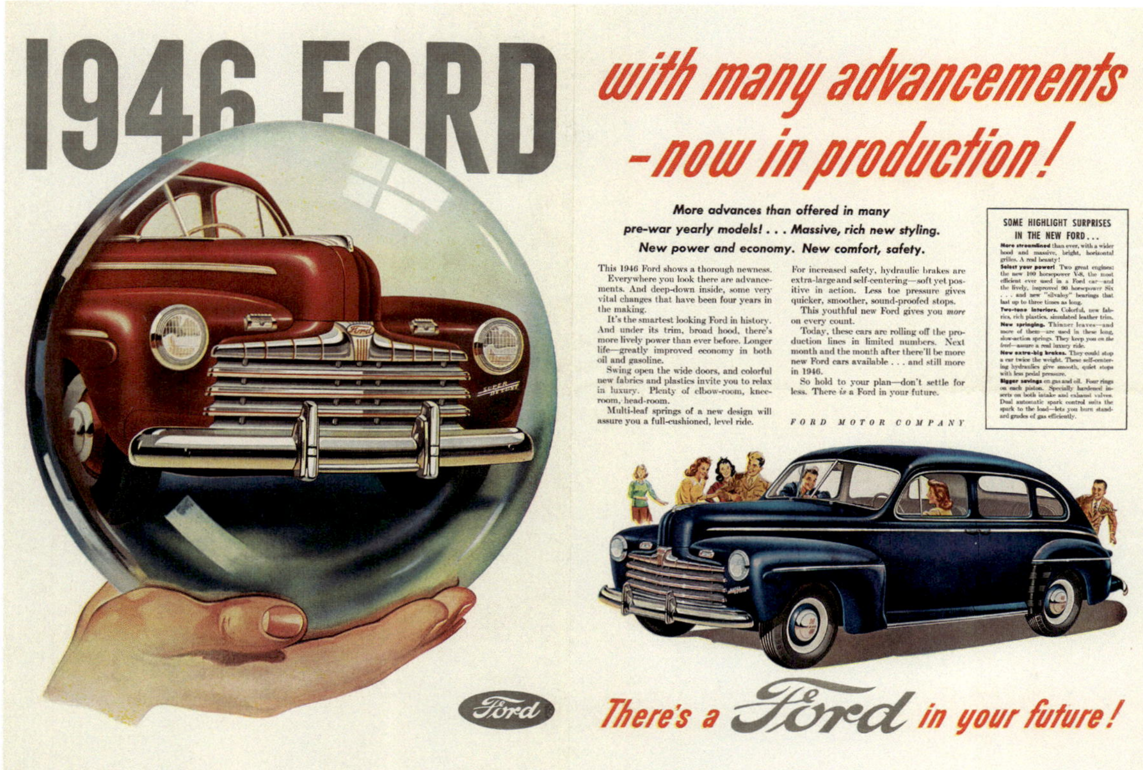

Les constructeurs ont alors le devoir de préserver le moral de leurs employés et de leurs clients. La communication automobile en direction du peuple américain va se faire évocatrice, patriotique et, dans la mesure où il n'y a plus de voitures à vendre, devenir l'essentiel de l'activité du secteur. À bon droit, Detroit exalte sa contribution à l'effort de guerre et promet, une fois qu'elle sera terminée, l'arrivée de voitures formidables. Des publicités de l'époque montrent un G.I. bivouaquant sous les bombes dans un paysage dévasté, rêvant au retour dans ses foyers… et à sa nouvelle voiture. « Il y a une Ford dans votre avenir », assure la promesse la mieux entendue du moment. Au pays, la plupart des garages qui commercialisaient des voitures neuves se mettent à écouler de l'occasion, plus précieuse que jamais; ils se lancent aussi dans la réparation et le commerce de pneumatiques. De toute façon, le rationnement limite le kilométrage autorisé. La course à l'acier de ces temps de guerre envoie également à la casse d'innombrables véhicules déjà trop vieux, coupables de consommer au-delà du raisonnable.

Et la promesse va se réaliser : c'est bien une Ford qui point à l'horizon, et elle n'est pas seule. Le monde automobile de l'après-guerre va gravir en peu de temps de nouveaux sommets d'exubérance. Les mobilisés sont de retour, les poches gonflées par les primes gagnées au front; ils veulent des voitures, et tout de suite. L'Amérique elle-même a eu son lot de parcimonie, d'austérité et de sérieux. À elle aussi, maintenant, il faut des voitures. Mais lesquelles? Detroit se laisse griser par des stylismes de plus en plus extravagants où cohabitent ailerons, tunnels et capots en forme d'obus… On voit surgir des machines qui pèsent des tonnes et exigent des moteurs toujours plus gourmands, assez puissants pour les déplacer. De son côté la publicité se focalise sur des expédients plus modestes, parfois même tout à fait insignifiants; ainsi ces combinaisons de couleurs et de tissus assez nombreuses pour vous donner le tournis. Les privations, le combat et la misère ne sont plus de saison; le bon goût et la sobriété non plus. C'est une époque où le vocabulaire automobile s'enrichit de nouvelles appellations ronflantes, du genre PowerPilot, Whirlaway, TorsionAire ou Airflyte (la fierté de la marque Nash). Pour l'essentiel, les argumentaires proviennent des agences. Young & Rubicam, qui signe celui de la Packard, a été l'une des premières à opérer sur ce marché. J. Walter Thompson, dont les origines remontent au début du 21e siècle, entretient une longue relation avec Ford. Campbell-Ewald détient de son côté le budget le plus convoité du secteur : celui de Chevrolet, depuis 1919.

L'excentricité débridée atteindra son apogée à la fin des années 1950. Quand elle tournera à l'incongruité, ces agences, parmi d'autres, auront à cœur de ramener les automobilistes à la raison. À partir de 1960, la tendance est aux modèles moins excessifs, venus d'Europe, du Japon puis de nouveau des États-Unis. Les publicitaires doivent désormais persuader le consommateur que moins, c'est plus. Et ils y parviennent avec éclat : la campagne ironique et brillante réalisée pour Volkswagen par l'agence Doyle Dane Bernbach laissera une trace mémorable, de même que les catalogues VW. Leur impact sera même encore plus remarquable après le premier choc pétrolier, au début des années 1970.

Il fut un temps où un constructeur automobile avait le pouvoir de vous convaincre que vous étiez quelqu'un d'intelligent, un bon mari et un héros auprès de vos enfants, dès lors que vous achetiez telle ou telle voiture. Les catalogues avaient le charme de récits merveilleux qui nous parlaient de l'automobile et de nous-mêmes. On sait bien qu'une publicité réussie – que ce soit pour une voiture, un shampooing ou un smartphone – est capable de nous tourner la tête. Mais à la différence d'une affiche ou d'un pop-up Internet, le catalogue rapproche le message et nous le rend disponible à volonté. Par l'image et le texte, il nous emporte et nous invite à franchir un seuil psychologique. La grande histoire de l'évolution automobile, pour ce qui touche au style et à la technologie, a déjà été racontée par le menu. L'histoire des brochures et catalogues commerciaux n'est pas moins importante. Améliorer les voitures, bien sûr, c'était le rôle des ingénieurs. Mais la publicité, grâce à ces documents, savait seule déclencher l'envie de les acheter.

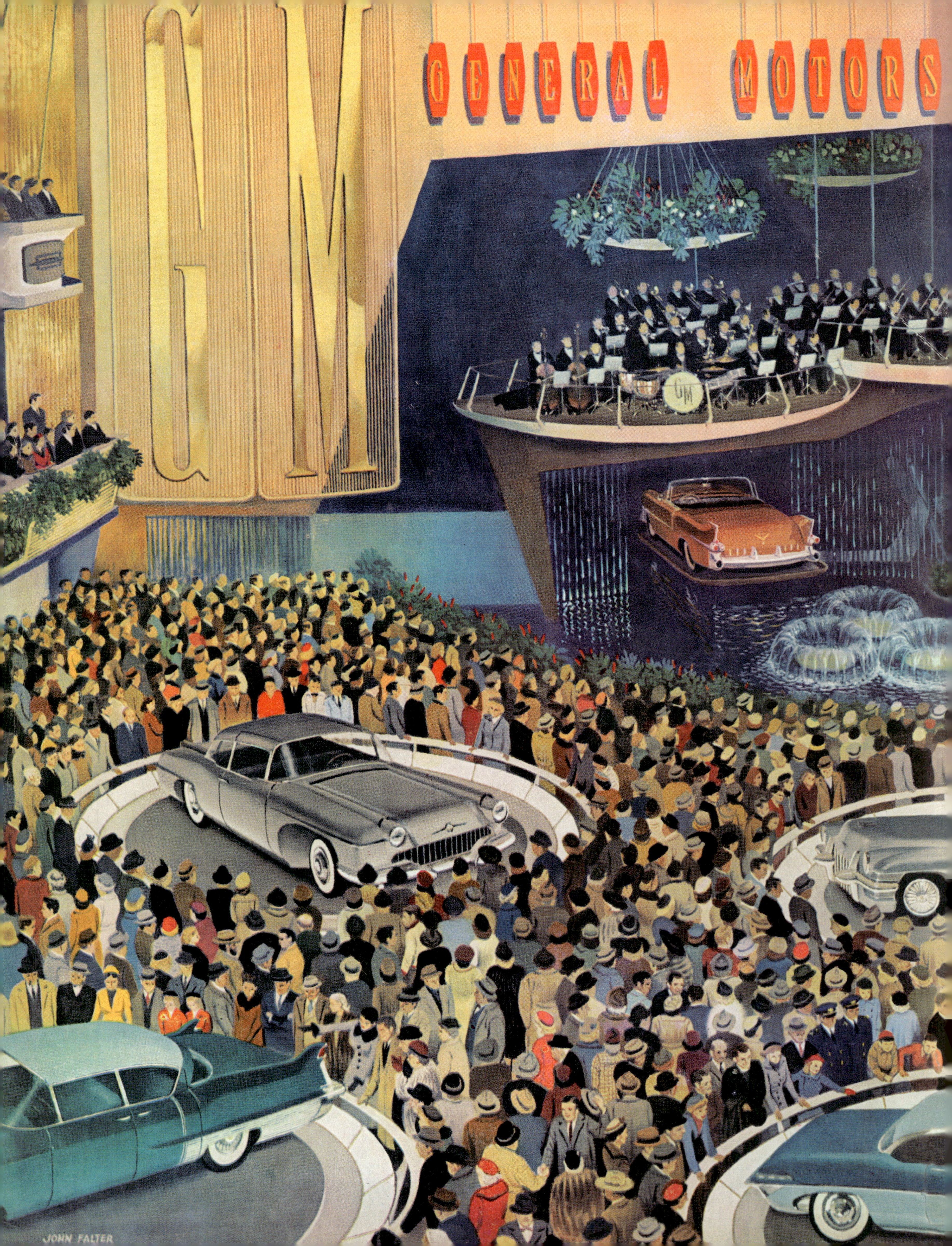
GM
GENERAL MOTORS
JOHN FALTER

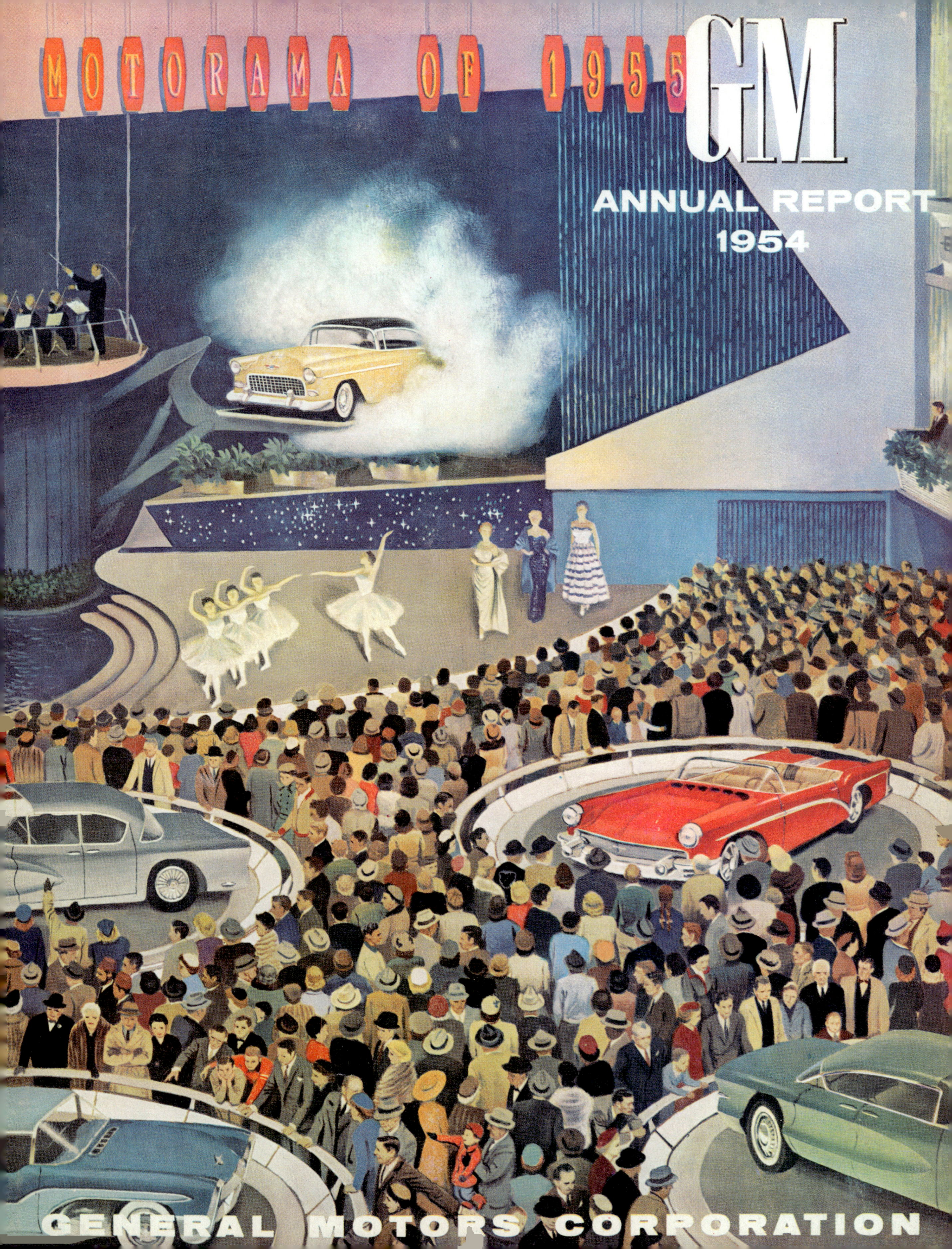

GM

ANNUAL REPORT 1954

GENERAL MOTORS CORPORATION

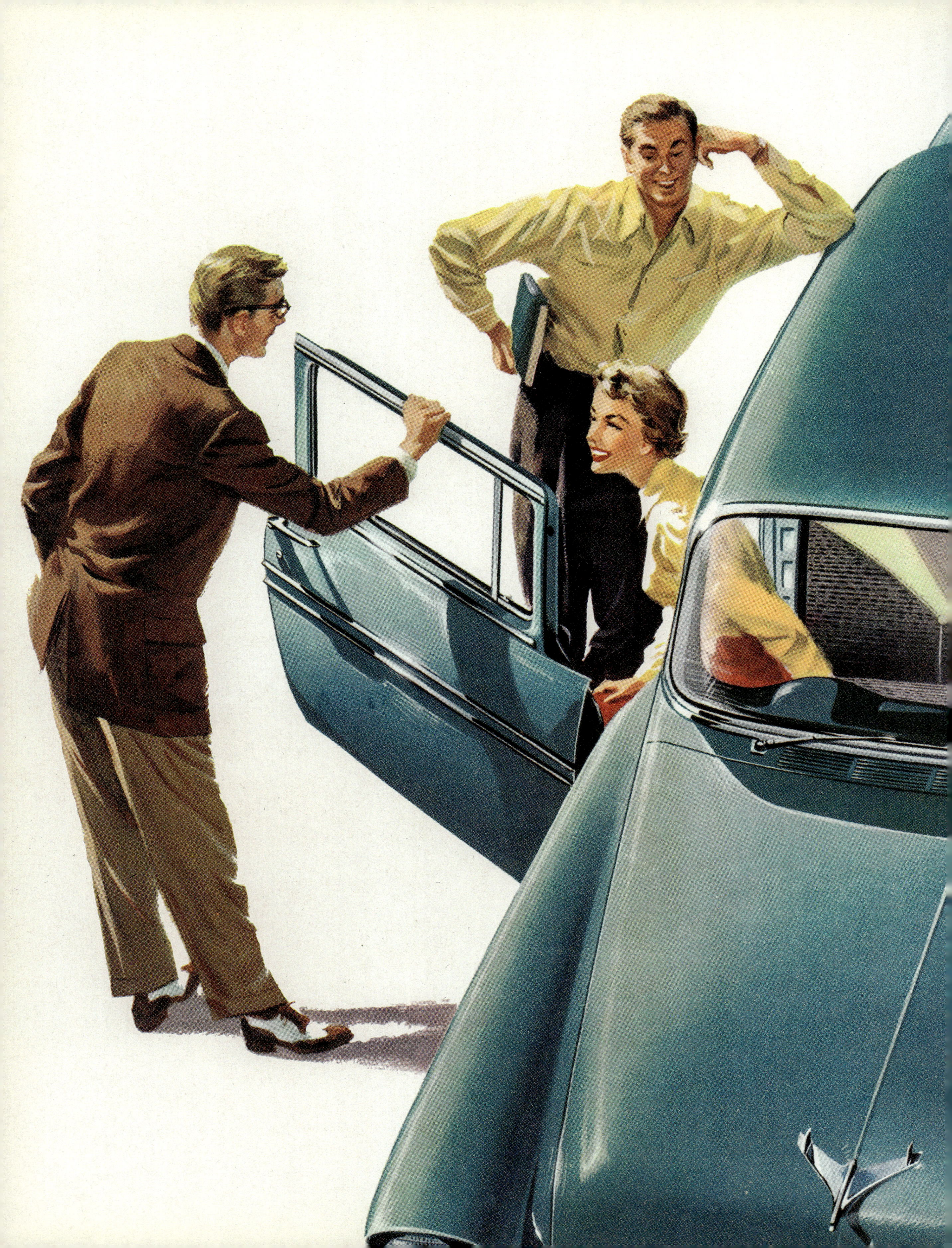

STEVEN HELLER

Auto Brochures and Sublimations of Desire

Americans were seduced into accepting the automobile as the embodiment of modern dreams, urges, and passions. Brochures promised a better life through automotive pleasure.

Lavishly produced, luxuriously designed with stunning full-color gatefolds, tipped-in colored paint chips, and plush upholstery swatches, automotive sales brochures were the V-8s of printing. Designed to spark primal urges, they combined movie-star magazine fantasy with business prospectus functionality. They sold unbridled ingenuity, innovative engineering, and the American dream—all between the covers.

Roadside billboards and full-page magazine advertisements captured attention and directed consumers into glass-walled showrooms across the nation. Car brochures turned on ignitions of desire and floored the senses. In addition to their visual allure, they bathed cars in a golden glow, emphasizing style and fashion yet balancing the technological with the aesthetic. Like a Times Square peep show (and often a Broadway extravaganza), these booklets pulled the curtains back to reveal all the stunning extras as though they were standard equipment. After seeing the options laid out on pristine pages—sometimes separated by translucent faux parchment slip-sheets and other special printing effects—who could resist leather seats, air conditioning, and other amenities that inflated the bottom line? A test drive was meant to trigger uncontrollable yearning; the brochure reminded consumers that pleasure was within their grasp.

Early brochure designers—most of them anonymous commercial artists and typographers—initially saw their role as informing the consumer through straightforward manuals. Soon, however, practicality was not enough. Mastery of the craft *and* art of titillation was essential. The 1927 Stutz 8 brochure, with its Egyptian-influenced logo and Celtic-inspired decorative cover design—a reference to Art Nouveau—wed old to new. This design seduced the customer into thinking of the car as a palace on wheels.

The language of futurism had yet to be employed, but the aura of fantasy triggered yearning. Playing on wish fulfillment was inevitable and the idea that buying *this* car would make one powerful was an effective ploy: The brochure for the Studebaker and Erskine Motor Car's the President, the Commander, and the Dictator—yes, the Dictator—was a masterpiece of marketing hubris: you can rule the world in your Studebaker. It was also commonplace for auto makers to inject as much luster as possible into even the most quotidian equipment: The 1932 Oldsmobile promotion implied that the Art Deco or Moderne–styled features and appointments, including lights, ashtrays, and door handles, were veritable crown jewels.

These feasts for the eye were enticing yet ephemeral—paper Cinderellas. When the clock struck midnight, as it were, the new auto models made the previous year's brochure obsolete and disposable. Regardless, auto manufacturers spent extravagant amounts annually. High-priced cars and wealthy customers demanded lavish design and production values, while more modest—yet still alluring and imaginative—brochures targeted the masses. Brochures were so integral to marketing strategy, and what is now called brand awareness, it is no surprise that their design was given a high priority—although most were produced by journeyman designers, not the biggest design stars of their day.

Advertising was still the foremost game. Firms like N. W. Ayer and J. Walter Thompson, and art directors like Wallace Elton (Ford) and Paul Rand (Kaiser Fraser) focused almost exclusively on ad campaigns. (Rand, the leading Modernist, created a classic 1951 brochure cover for the company, but that was an anomaly.)

Whoever the designers were, little effort was spared in brochures when it came to resplendent layouts, stunning illustrations, dramatic photographs, and grand typography. And why not? These were consumer manifestos promoting auto-volution (yearly changes in style and form). Detroit was not just selling machines; it was espousing the American ethos of economic superiority, rarely found in car brochures outside the United States.

Chevrolet 1956 *(page 32)*

Lincoln 1954 *(right)*

Cadillac 1954 *(opposite)*

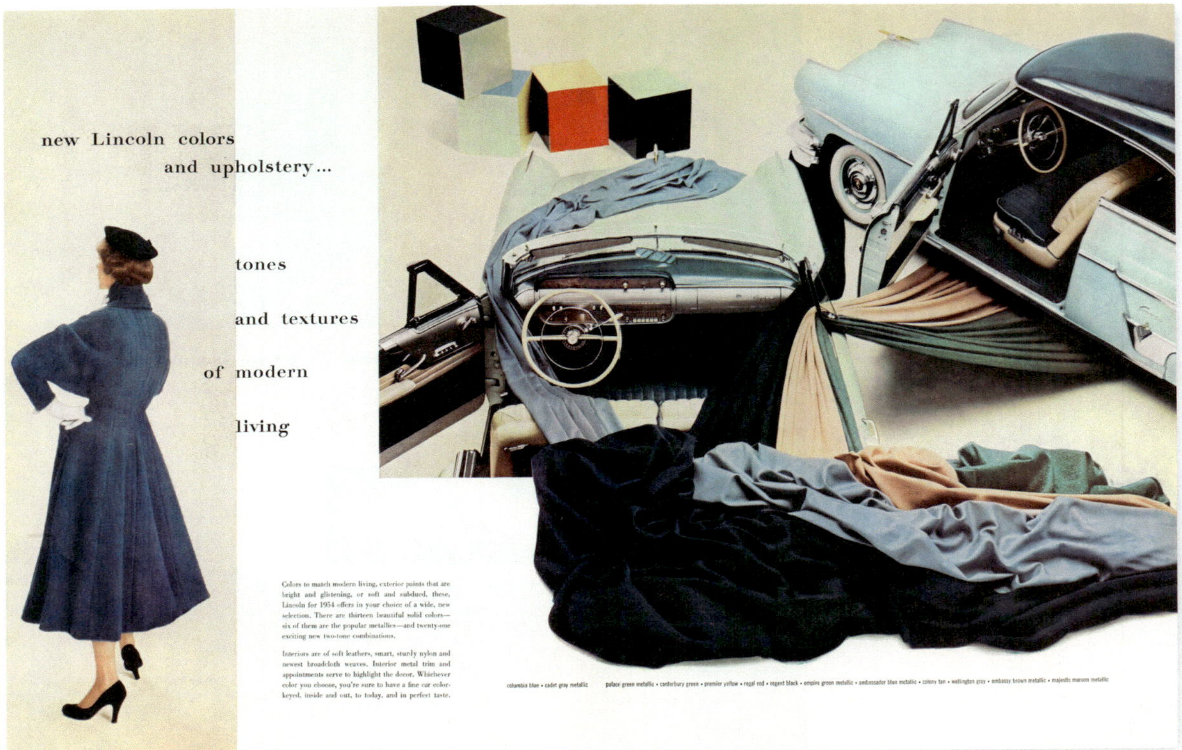

Still, a car was a machine. Not just some patent-medicine elixir, these machines needed to be framed as efficient and reliable—paragons of engineering perfection. American drivers had to be convinced that behind the wheel they would experience supreme confidence, whether in the thick of traffic, rolling along country roads, or motoring in America's great mountain country. And how did the automakers accomplish this alchemy? Through successful pitches like this: "The brilliant engineering that brought you Torsion-Level Ride now brings you THE GREATEST PACKARD OF THEM ALL," read the brochure for the 1956 models, which included the Patrician, the Caribbean, and the 400. Tony words replaced routine mechanical descriptions.

Superlatives have long been the standard equipment of automobile promotion. The intent was to pile pleasure upon pleasure—illusion upon illusion. Copywriters were masters of hyperbole: "In styling, engineering and performance...in pride of ownership and progress... in every way you choose to look at it...this is the greatest Packard of them all." Autos large and small, whatever the model and whichever the style, possessed irresistible appeal because they were spiced with elegance and flair. Every car was doubtless the most beautiful interpretation of its body type on the highways.

Efficiency as an invaluable automotive attribute was critical to the success of car sales. But these brochures were designed to transcend the commonplace notions of function in favor of form. Some offered exaltations that imbued certain automobiles with manufactured heritages. "In America's mad industrial drive for greater production volume," chided the copywriter for the 1937 Pierce-Arrow brochure, "motor car individuality has been almost completely, if not wantonly, sacrificed. The heritage of character by which nearly every car in *motordom* could be identified a few years ago has been forgotten or disregarded in the stampede to mass production. Today, most any automobile is just another unit of transportation moving with the mob. Identity has been swept aside in the surge to 'uniformity.'"

Heritage meant individuality. Uniformity was a sin that was hardly absolved by the auto industry's assembly-line production innovations. Nonetheless, in its 1937 brochure (and others that followed) Pierce-Arrow strove through hubris to be distinctive. And like others before and after, never compromised with the mediocre.

Comfort was as important as efficiency, and arguably the most effective promotional hook on which to hang a sales pitch, especially for the secondary car consumer: women. From the early days of automotive consumption well into the late 20th century, men paid the bills. But middle-class, white women, who regularly were featured in stylized drawings and photographs in ads and brochures, more or less determined what their husbands would purchase. For women, comfort was more significant than engineering per se, so in copy and image a balance had to be struck. In the headline "Big Car Comfort... Plus Economy" the former phrase is aimed at the ladies and the latter toward the men. Read this and think of its implications: "It isn't the size of the 1940 Mercury alone that makes it so comfortable. A wide body makes roominess, of course, but it is the way the car is designed and constructed that gives such a buoyant restful ride. Comfort is built in..." In these sentences men get "the size," while women get "so comfortable." Men get the "designed and constructed"; women get the "buoyant restful ride."

As varied as the brochures looked, standards existed regarding the subsets of material provided by the salesman. From the "sheet," a one-page

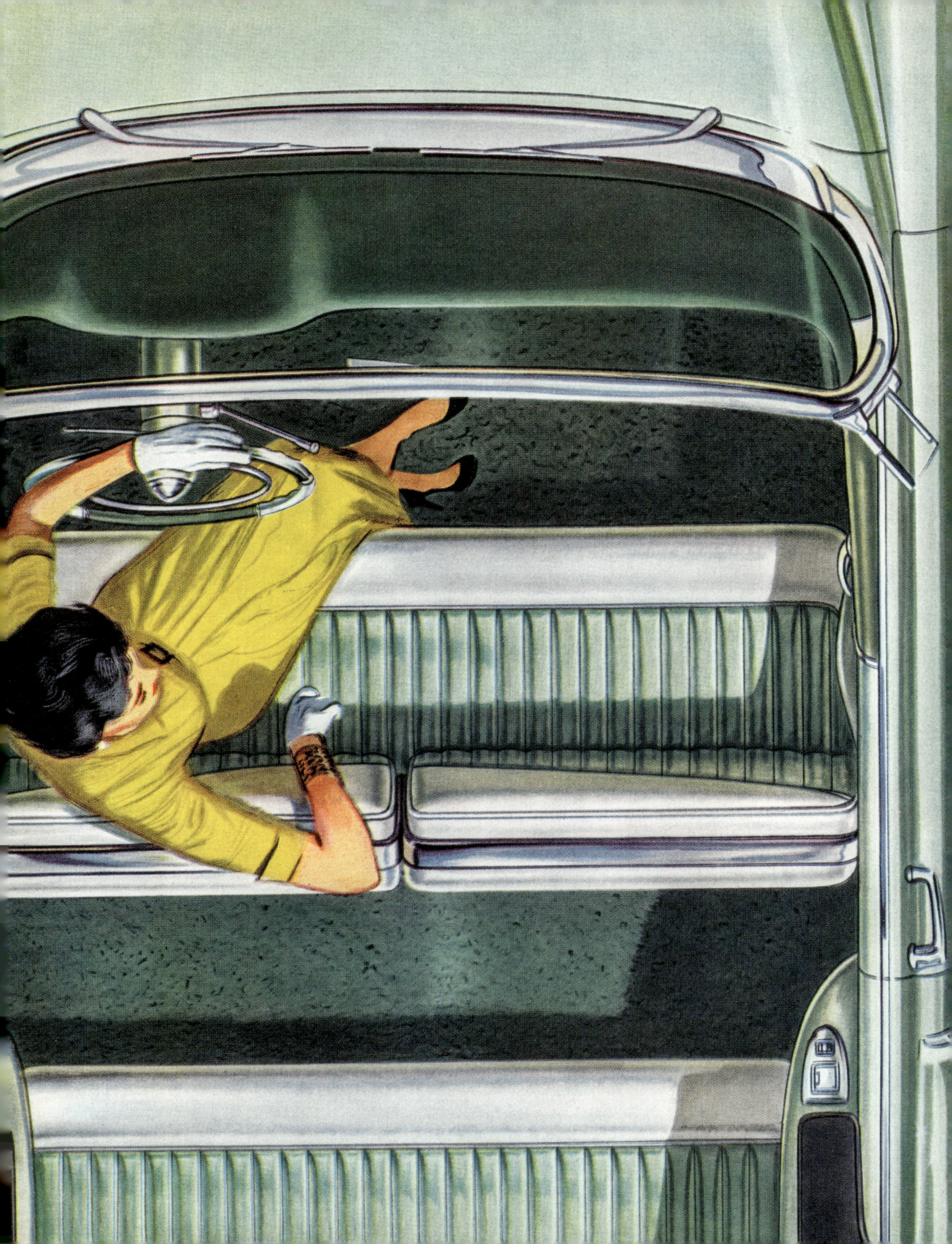

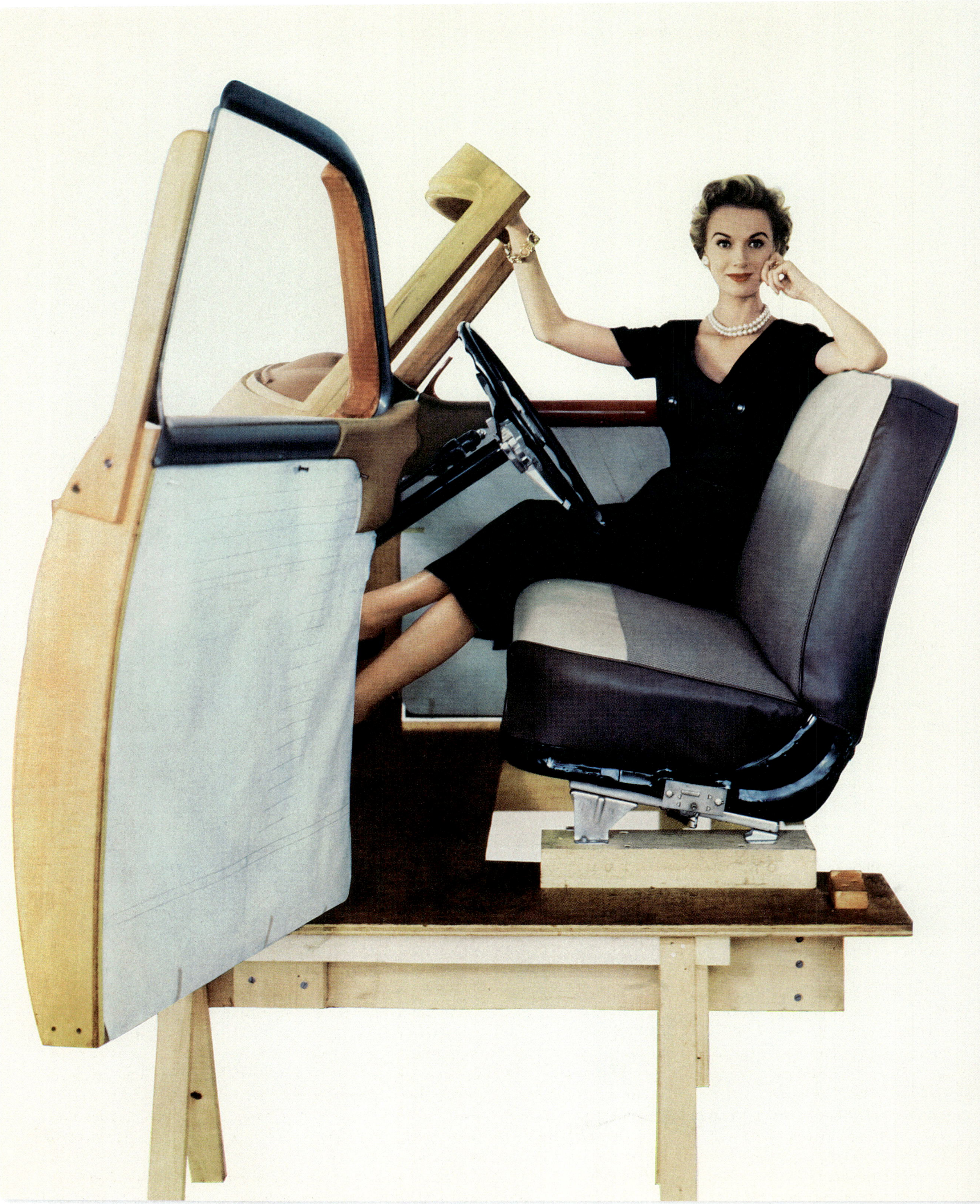

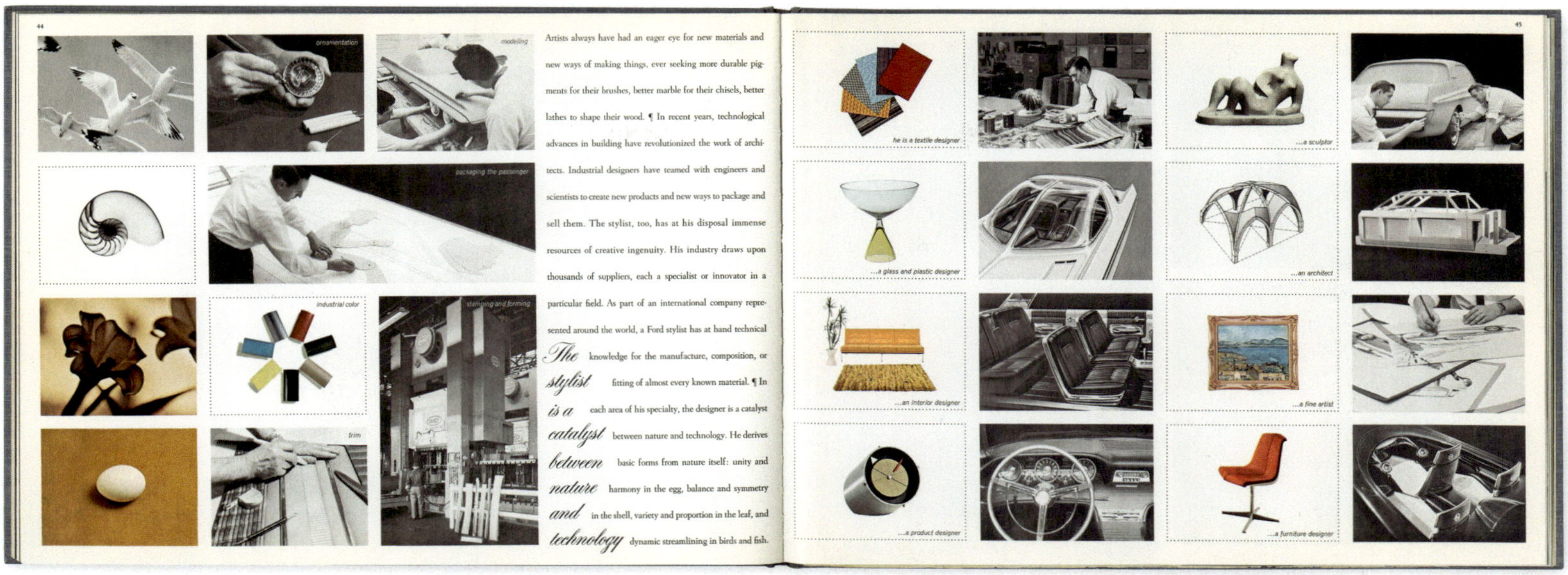

Chrysler 1953 *(opposite)*

Ford 1963 *(above)*

description; to the "folder," an accordion-folded sheet that could be as large as a poster; to the primary "catalog," which showcases all the lustrous auto attributes; to the "prestige," the largest and most exuberant booklet of the year, and the most coveted too.

Art directors from agencies routinely gave their concepts to the Detroit art studios—the workhorses of design. Work would pass back and forth until deadline time. The studios and the agencies were separate business—the studios competing for agency work, and the agencies competing for car accounts from the automobile companies. In the end, the engineers at the factory always had the last word on how the cars were presented, as though design was an afterthought.

Legions of agency copywriters in Detroit, Philadelphia, and New York kept the automobile myth rolling and armies of bullpen artists, expert in the art of rendering precision highlights, embedded the mythic image of the car in the public's dreams. Drawing and painting was so much more heroic and romantic than even the most impressive photography. Illustration was expressive, representative, and surreal. The 1920s-era REO Flying Cloud booklet uses a bold woodcut of a speeding car on its cover, so artful that an Italian Futurist or English Vorticist would be proud to take credit for doing it. The precisionist interior drawings, by contrast, are so real they could be mistaken for photos. The brochure for the Pontiac Silver Streak Sixes and Eights for 1936 was a veritable gallery of pastel and gouache virtuosity. The Cadillac for 1954 brochure was awash with airbrushing so shiny that it veritably blinded the reader with chrome reflections. The 1957 Custom and Custom 300 were so radically elongated, each looked more like a Chris Craft boat than a car. Even the 1957 Ford Edsel, the ill-fated model for which "the emphasis is on performance; the accent is on elegance," was rendered with such exaggerated virtue (length was indeed everything) it is now incredible to realize it was such a marketing failure that the noun "edsel" became a synonym for product disaster.

The automotive renderers from the '30s and beyond, among them Paul Gerding for DeSoto, John Reinhart for Ford, Stephen Pojanski for Buick, Dom Pacitti for Plymouth, Robert Cadaret for Chevrolet and GM, and the team of Art Fitzpatrick and Van Kaufman for Pontiac, were pinup artists—except with cars. Their respective talents for exaggerating the American dream machine captured a sense of majesty and wonder.

The drawn, painted, and airbrushed cars that adorned such brochures as the 1935 Nash Lafayette, 1937 Oldsmobile, and 1938 Nash were not always sexy; sometimes they evoked the "feel" of children's-book illustrations. This may not have been conscious on the artists' part, but in subtle ways this helped to give American autos anthropomorphic personalities. The cover of the Oldsmobile Six and Eight brochure features two gleaming futuristic grilles on red and green cars, composed like the prows of great ocean liners, suggesting children's tales. Then on the inside cover are watercolor vignettes in which the cars appear to be members of the families, like a friend or pet. The cars appear throughout the booklet in scenes designed to imply an inextricable relationship between the characters. Car, driver, and passenger are a nuclear family. Engineers and industrial designers are kings of motordom, but art and artists have long contributed significantly to car culture. In the "1938 Great News From Nash," an unnamed "famous artist" provides a testimonial: "Here is convincing proof that individuality in automobile styling can be achieved without resort to freakish design. Here is streamlining without an ugly or awkward detail...Like a perfect piece of sculpture, the new Nash is beautiful from every angle." Presumably this famous artist was seduced by the brilliant airbrush and gouache renderings of the mammoth Ambassador Eight Coupe, Lafayette and

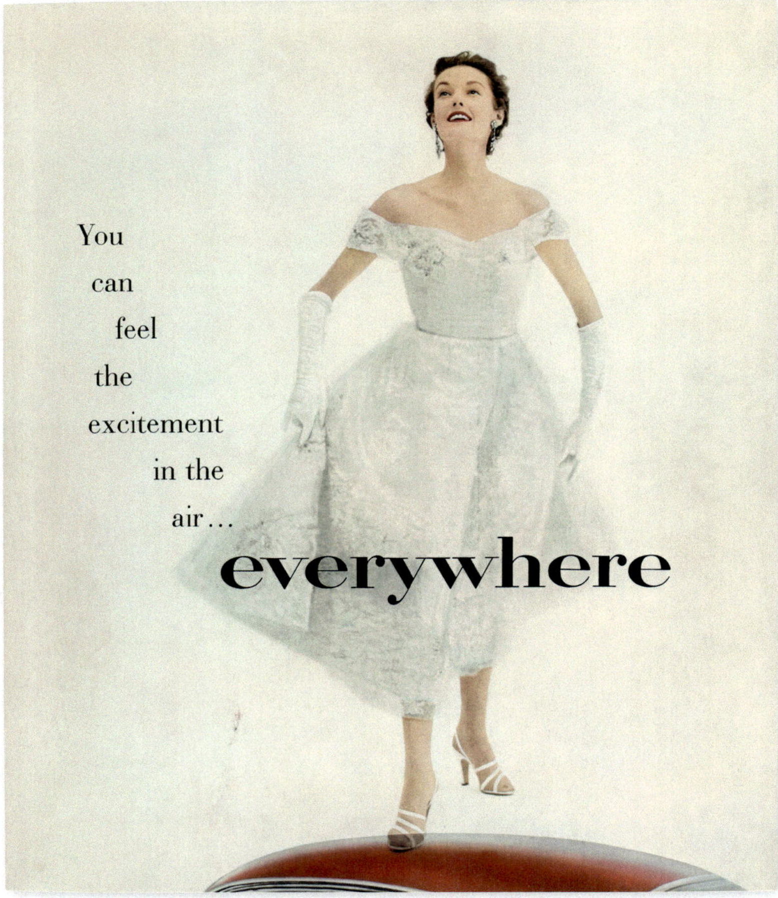

Chrysler 1953 *(left)*

Raymond Loewy, car designer *(opposite)*

Victoria Sedan in mouthwatering reds, blues, greens, and yellows. So captivating were these shades, like lipstick these mechanical sirens beckoned the susceptible.

During World War II American automotive production was geared toward defeating the Axis with jeeps, trucks, and tanks, so few lavish brochures were produced for consumers. After victory, the rosy postwar future was the alluring force for enhancing automobile consumption. That is when Futurism kicked in with a vengeance. The owner's manual for 1949 Oldsmobile was covered with futuristic-looking, airbrushed globes around which cars orbited. The hood nameplate was also a ringed globe that combined Saturn and a gyroscope. Beautiful women in fashion-model poses routinely filled brochures, but so did visions of things to come. Under the title "Oh! Those '56 Oldsmobiles!" the leitmotif is rocketry, modernity, and luxury. The logo and headpiece for the Super 88 was a soaring rocket, which was echoed by a "smartly fashioned airfoil grille" that resembled a jet engine. The copy was equally space oriented: "The Oldsmobile for '56...flashes ahead—with the smoothly powerful new Rocket T-350 Engine and powerfully smooth new *Jetaway* Hydra-Matic Drive." With space as theme, copywriters had endless opportunities for metaphor and analogy.

At the start of the '60s, following the flurry of fin-mania in the late '50s, engineers switched gears. They replaced massiveness with sleekness. Auto illustrators likewise began developing a subtle new language rooted in exaggerated perspective. While accurately capturing the details, cars had to suggest a certain kind of monumentalism. They had to fit neatly into the urban and rural landscape (preferably a Mies van der Rohe modern landscape) to suggest unity with the sophisticated, contemporary architectural environment. The portfolio-styled "prints" in the 1961 Oldsmobile Classic brochure shows these new Olds with the backdrops of glass-walled modern buildings; the auto was an essential accoutrement of modernity.

Americans were seduced into accepting the automobile as the embodiment of modern dreams, urges, and passions. Brochures promised a better life through automotive pleasure. They were fairy-tale books that sublimated the daily grind. Yet cars were not free from frustrations. Once on the road, maintenance headaches brought the dreamer down to earth. An automobile was an investment (albeit one that would greatly depreciate), so when the visual pyrotechnics faded from memory the reality of automotive ownership was subject to the vicissitudes of daily life. Today, these brochures are reminders of how exacting and convincing the illusion was and how everyone was more than happy to buy into it.

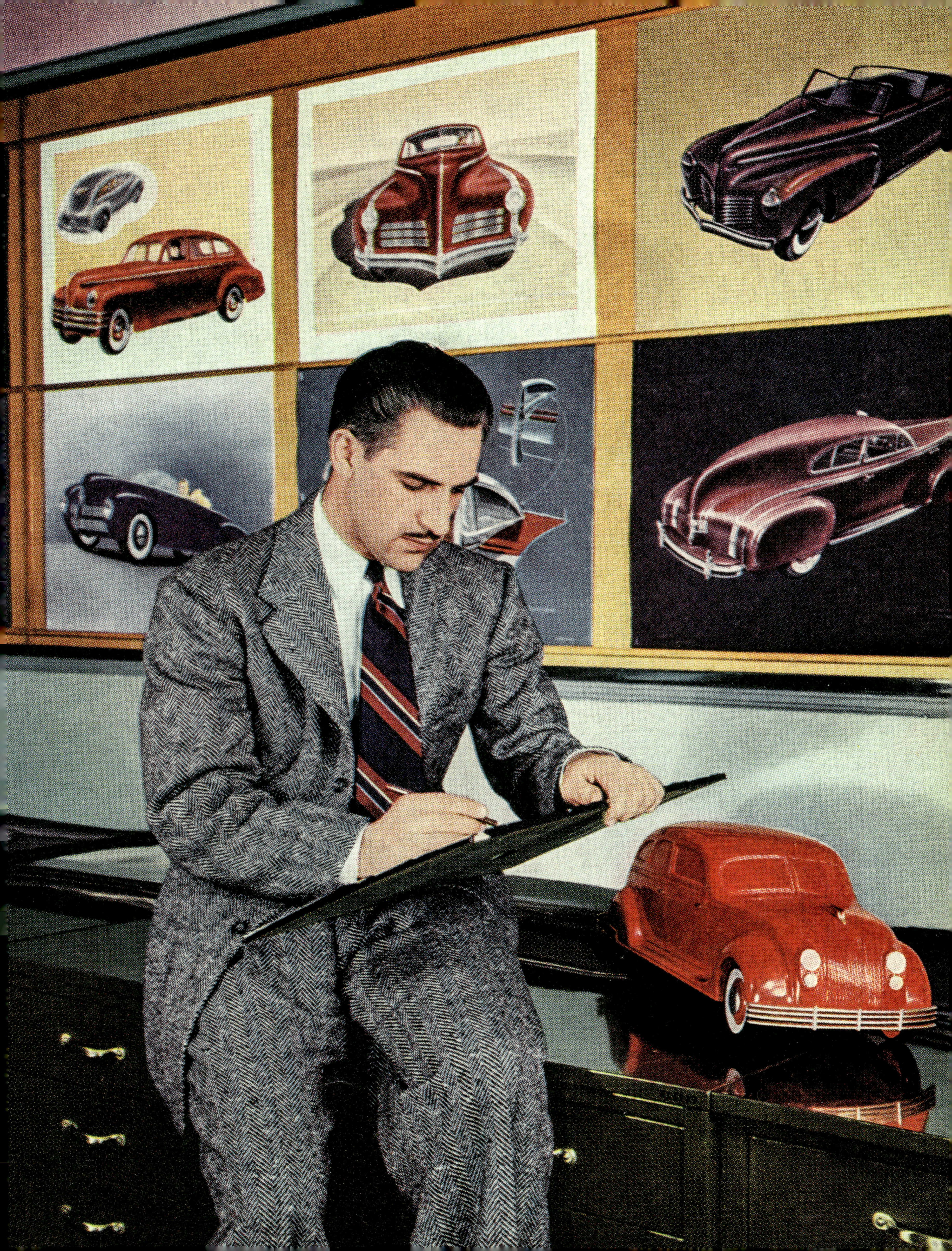

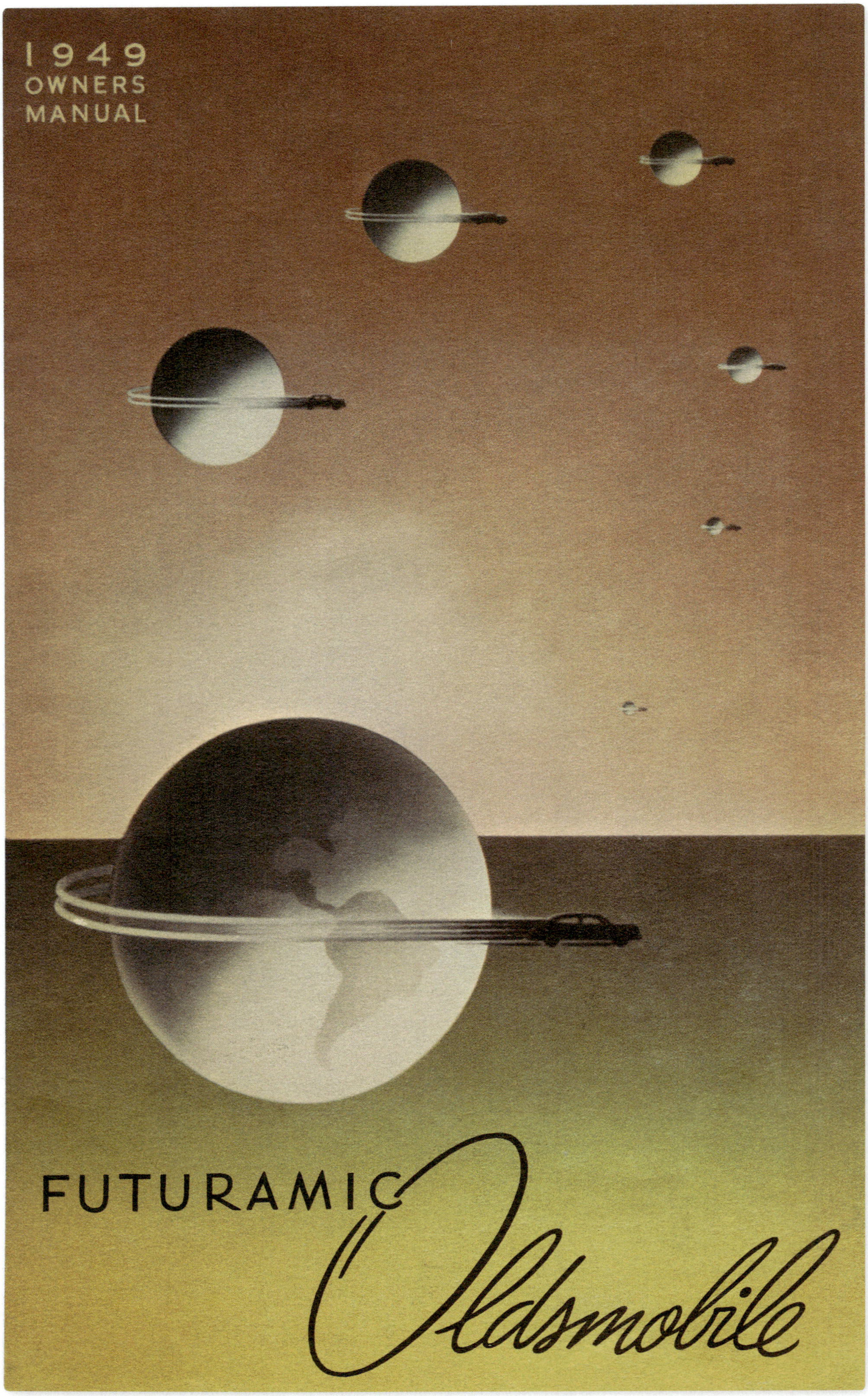
1949
OWNERS
MANUAL
FUTURAMIC
Oldsmobile

STEVEN HELLER

Automobilprospekte und die Sublimierung des Verlangens

Die Amerikaner wurden dazu verführt, das Automobil als Verkörperung moderner Träume, Bedürfnisse und Leidenschaften zu betrachten. Prospekte versprachen ein besseres Leben durch automobilen Genuss.

Aufwendig produziert, üppig bebildert, mit atemberaubenden, vollfarbigen Ausklappseiten, eingeklebten Farb- und Stoffmustern – Autobroschüren waren der V8 der Druckindustrie. Sie sollten Urtriebe ansprechen und verknüpften die Fantasien der Filmstarmagazine mit der Funktionalität von Verkaufsprospekten. Sie verkauften ungebremsten Einfallsreichtum, innovative Technik und den amerikanischen Traum.

Plakatwände und ganzseitige Anzeigen in Zeitschriften erregten die Aufmerksamkeit der Verbraucher und lenkten deren Schritte in die verglasten Ausstellungsräume im ganzen Land. Die Werbeschriften für Automobile dienten dem Verlangen als Anlasser und brachten die Sinne auf Touren. Neben ihren optischen Reizen verpassten sie dem Auto einen Glorienschein, hoben Mode und Design hervor und fanden gleichzeitig ein ausgewogenes Verhältnis von Technik und Ästhetik. Wie eine Peepshow am Times Square (oder häufig auch eine Revue am Broadway) lüfteten diese Heftchen den Vorhang und enthüllten atemberaubende Extras, als gehörten sie zur Grundausstattung. Wer konnte Ledersitzen, Klimaanlage und anderen Nettigkeiten, die den Preis ordentlich in die Höhe trieben, schon widerstehen, nachdem sie ihm erst einmal auf Hochglanzseiten präsentiert worden waren, manchmal noch durch Zwischenblätter aus Pergamentimitat und ähnliche drucktechnische Spezialeffekte hervorgehoben? Eine Probefahrt sollte unstillbare Sehnsucht schaffen und der Prospekt den Verbraucher daran erinnern, dass solche Freuden zum Greifen nah waren.

Die frühen Werbeschriften wurden oft von namenlosen Gebrauchsgrafikern oder Schriftsetzern gestaltet, die ihre Aufgabe anfangs darin sahen, die Verbraucher durch überschaubare Handbücher über das Produkt zu informieren. Bald schon reichte der praktische Nutzen jedoch nicht mehr aus. Neben die Handwerkskunst trat die Kunst der Verführung. Mit ihrem ägyptisch beeinflussten Markenemblem und den keltischen Anklängen auf dem jugendstilhaft gestalteten Umschlag verband die Broschüre für den Stutz 8 von 1927 das Alte mit dem Neuen: Dem Kunden wurde der Eindruck vermittelt, bei dem Fahrzeug handele es sich um einen Palast auf Rädern.

Die Sprache des Futurismus kam noch nicht zum Ausdruck, doch die Aura der Fantasie weckte bereits Sehnsüchte. Das Spiel mit den Wunschträumen war unvermeidlich, und die Vorstellung, der Kauf gerade *dieses* Autos verleihe Macht, war ein wirksames Konzept: Der Prospekt für den President, den Commander und den Dictator von Studebaker and Erskine war ein Meisterwerk anmaßenden Marketings. Es war auch durchaus üblich, dass Autohersteller selbst der profansten Ausstattung so viel Glanz wie möglich verpassten: Die Werbung für den Oldsmobile von 1932 tat geradezu so, also ob die Scheinwerfer, Aschenbecher und Türgriffe im Stil der Moderne oder des Art déco die reinsten Kronjuwelen wären.

Ein solcher Augenschmaus war zwar verlockend, aber auch kurzlebig. Wenn die neuen Modelle auf den Markt kamen, waren die Prospekte des Vorjahres Schnee von gestern. Und dennoch gaben die Automobilhersteller jedes Jahr ein Vermögen dafür aus. Teure Autos und wohlhabende Kunden verlangten nach opulenter Aufmachung und hochwertigem Druck, während die Masse mit bescheideneren – aber immer noch reiz- und fantasievollen – Broschüren bedacht wurde. Die Werbeschriften waren ein so fester Bestandteil der Vermarktungsstrategie und dessen, was man heutzutage als Markenbewusstsein bezeichnet, dass der hohe Stellenwert, den man ihnen beimaß, nicht überrascht – wenngleich die meisten von einfachen Atelierangestellten gestaltet wurden und nicht von den Stardesignern der damaligen Zeit.

Im Vordergrund stand noch immer die Werbung. Agenturen wie N.W. Ayer und J. Walter Thompson und Artdirektoren wie Wallace Elton (Ford) und Paul Rand (Kaiser Fraser) befassten sich fast ausschließlich mit Anzeigenkampagnen. (Rand, der führende Modernist, schuf 1951 einen mittlerweile klassischen Prospekttitel für das Unternehmen, aber das war eine Ausnahme.)

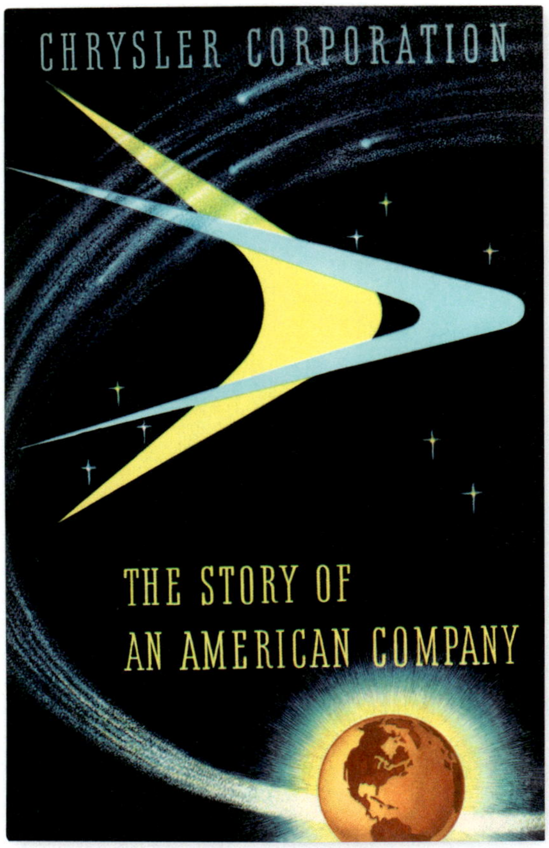

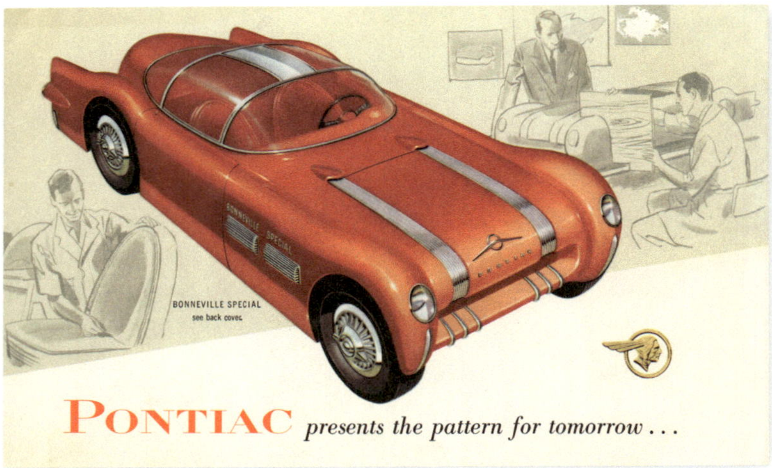

Unabhängig vom Designer wurden keine Kosten gescheut, wenn es um prächtiges Layout, atemberaubende Bilder, packende Fotos und eindrucksvolle Typografie ging. Und warum auch nicht? Dies waren schließlich Verbrauchermanifeste, die die Evolution des Automobils bewarben. Detroit verkaufte nicht einfach Maschinen, sondern vertrat das amerikanische Ethos der wirtschaftlichen Überlegenheit, das man selten in Autoprospekten außerhalb der USA wiederfand.

Dennoch war das Auto eine Maschine, kein Zaubertrank, keine Wundermedizin. Man musste es als wirtschaftlich und zuverlässig darstellen. Die Autofahrer mussten davon überzeugt werden, dass sie ihrem Fahrzeug vertrauen konnten, ob im dichten Stadtverkehr, auf der Landstraße oder in den großartigen Berglandschaften der USA – zum Beispiel mit Werbesprüchen wie diesem: „Die hervorragende Ingenieurskunst, die Ihnen die Drehstabfederung bescherte, beschert Ihnen nun DEN GROSSARTIGSTEN PACKARD VON ALLEN." So war es im Prospekt für die Modelle des Baujahrs 1956 zu lesen, zu denen der Patrician, der Caribbean und der 400 gehörten. Blumige Worte traten nun an die Stelle banaler Beschreibungen der Mechanik.

Superlative waren schon immer das Werkzeug der Automobilwerbung. Das Ziel war es stets, ein Lustgefühl durch das nächste zu übertrumpfen, eine Illusion durch die nächste. Die Werbetexter waren Meister der Übertreibung: „Im Design, in der Technik und in der Leistung … im Besitzerstolz und in der Fortschrittlichkeit … aus welchem Winkel Sie ihn auch betrachten … das ist der großartigste Packard von allen." Große wie kleine Autos besaßen einen unwiderstehlichen Reiz, weil sie mit Eleganz und Flair gewürzt waren. Jedes Fahrzeug war ohne Zweifel das schönste seiner jeweiligen Karosserieform auf der Straße.

Die Wirtschaftlichkeit als wertvolles Attribut eines Autos war für dessen Absatz von entscheidender Bedeutung. Doch die Prospekte waren so angelegt, dass die Form stets die schnöde Funktion verdrängte. Bisweilen las man Lobpreisungen, die bestimmte Autos mit erfundenen Traditionen ausstatteten. „Im irrwitzigen Wettlauf der amerikanischen Industrie um größere Produktionsmengen", rügte der Texter der Werbebroschüre für den Pierce-Arrow von 1937, „hat man die Individualität des Kraftfahrzeugs leichtfertig, wenn nicht gar mutwillig nahezu vollkommen geopfert. Die Tradition des Charakters, an der man zuvor noch beinahe jedes Fahrzeug in der Welt der Automobile erkennen konnte, geriet in Vergessenheit oder blieb in der wilden Hektik der Massenfertigung auf der Strecke. Heutzutage ist fast jedes Automobil bloß noch ein Fortbewegungsmittel, das in der Menge untergeht. Die Identität wurde von der Woge der ‚Einheitlichkeit' hinweggefegt."

Tradition bedeutete Individualität. Uniformität war eine Sünde, und die Neuerung der Fließbandproduktion in der Autoindustrie schaffte hier kaum Abhilfe. Dennoch profilierte sich Pierce-Arrow in der Broschüre für 1937 durch Hybris und ließ sich, wie andere davor und danach, auf keinen Kompromiss mit dem Mittelmaß ein.

Komfort war ebenso wichtig wie Wirtschaftlichkeit und möglicherweise das wirksamste Verkaufsargument, insbesondere für die Nutzer eines Zweitwagens: Frauen. Von den frühesten Tagen des Autokaufs bis weit ins späte 20. Jahrhundert hinein bezahlten die Männer die Rechnungen. Doch die weißen Frauen aus der amerikanischen Mittelschicht, die regelmäßig in stilisierten Zeichnungen und auf Fotos in Anzeigen und Prospekten auftauchten, bestimmten oft, was ihre Ehemänner kauften. Für Frauen war Komfort wichtiger als Technik an sich, sodass man in Text und Bild eine gewisse Ausgewogenheit erzielen musste. In der Überschrift „Der Komfort eines großen Wagens … plus Wirtschaftlichkeit" richtet sich der erste Teil an die Dame und der zweite an den Herrn. Achten Sie auf die Nuancen: „Der Mercury des Baujahrs 1940 ist nicht allein dank seiner Größe so komfortabel. Eine breite Karosserie schafft selbstverständlich Geräumigkeit, doch erst das Design und die Bauweise des Fahrzeugs sorgen für ein solch be-

Oldsmobile 1949 *(page 40)*

Chrysler 1955 *(opposite left)*

Pontiac 1954 *(opposite right)*

General Motors 1951 *(right)*

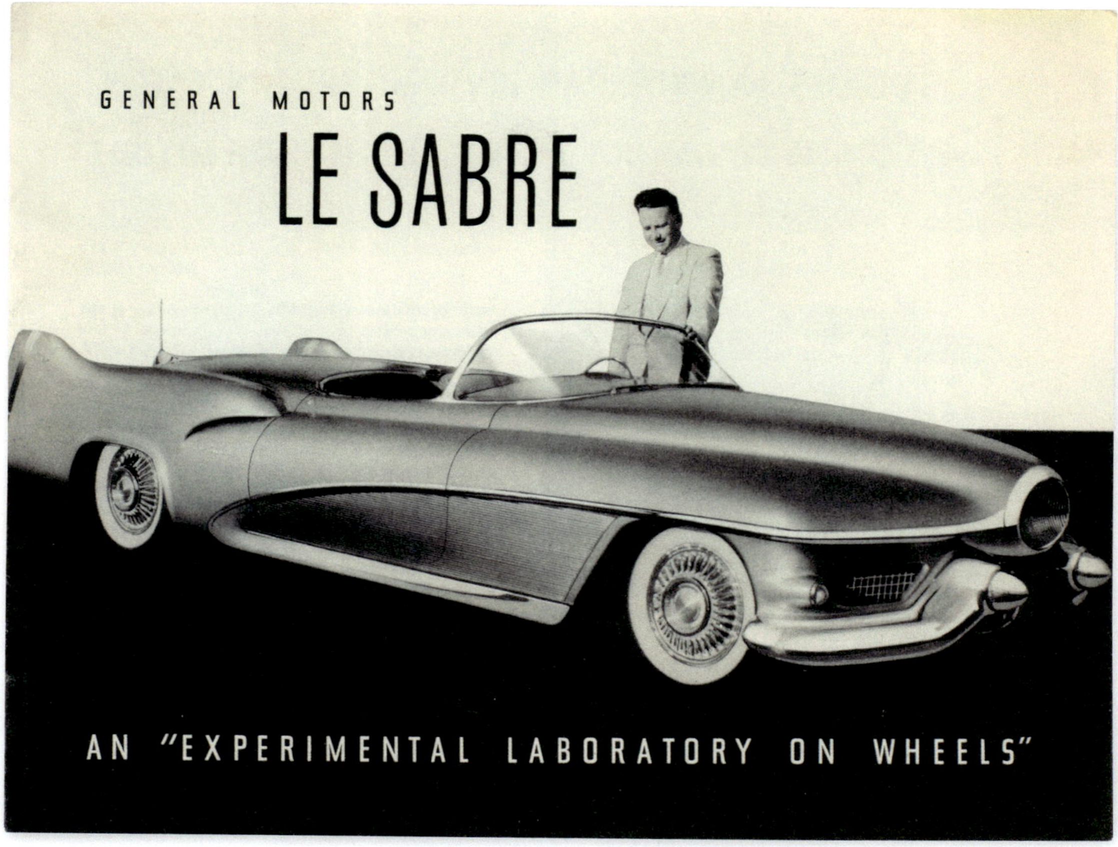

schwingtes, erholsames Fahrgefühl. Der Komfort ist eingebaut …" Hier lesen die Männer „Größe", die Frauen „komfortabel" – die Männer „Design und Bauweise", die Frauen „erholsames Fahrgefühl".

So unterschiedlich die Prospekte auch aussahen, gab es doch gewisse Regeln für die jeweiligen Materialien, die dem Autohändler zur Verfügung gestellt wurden. Diese reichten von einer einseitigen Beschreibung über ein ziehharmonikagefaltetes Blatt (mitunter auffaltbar bis zur Plakatgröße) und den Hauptkatalog, der all die wunderbaren Merkmale des Fahrzeugs aufführte, bis hin zum Hochglanzprospekt, der aufwendigsten und daher auch begehrtesten Broschüre des Jahres.

Die Artdirektoren in den Werbeagenturen gaben ihre Konzepte üblicherweise an die Detroiter Grafikateliers weiter, die Arbeitspferde des Designs. Die Arbeiten gingen dann bis zum Abgabetermin hin und her. Grafiker und Agenturen waren getrennte Unternehmen: Die Ateliers buhlten um die Aufträge der Agenturen, während die Agenturen im Wettbewerb um die Aufträge der Automobilindustrie standen. Am Ende hatten stets die Ingenieure des Herstellers das letzte Wort – als ob die Gestaltung nebensächlich wäre.

Legionen von Werbetextern in den Agenturen von Detroit, Philadelphia und New York hielten den Automobilmythos lebendig und Heerscharen von Gebrauchsgrafikern pflanzten das mythische Image vom Automobil in die Träume der Verbraucher. Gemalte und gezeichnete Bilder waren so viel heroischer als jedes noch so eindrucksvolle Foto. Auf dem Titel des Prospekts für den REO Flying Cloud aus den 1920er-Jahren findet man den kühnen Holzschnitt eines rasenden Autos, der so kunstvoll ist, dass ein italienischer Futurist oder ein englischer Vortizist sicherlich gerne seinen Namen dafür hergegeben hätte. Die detailgetreuen Zeichnungen der Innenausstattung hingegen sind so realistisch, dass man sie für Fotografien halten könnte. Die Broschüre für die Modelle Six und Eight des Pontiac Silver Streak aus dem Baujahr 1936 ist eine wahre Galerie der Pastell- und Gouachemalerei. Die Airbrush-Arbeiten im Prospekt für den Cadillac von 1954 glänzen so sehr, dass sie den Leser mit Chromspiegelungen förmlich blendeten. Der Custom und der Custom 300 von 1957 waren in der Darstellung so gestreckt, dass sie eher nach einem Boot von Chris-Craft als nach einem Auto aussahen. Und angesichts der übertriebenen Darstellung des unglückseligen Ford Edsel von 1957, bei dem angeblich „die Betonung auf Leistung, der Akzent auf Eleganz" lag, ist es rückblickend kaum vorstellbar, dass „edsel" im Englischen aufgrund des Schicksals dieses Modells zum Synomym für einen Produktflop wurde.

Die Automobilkünstler der 1930er-Jahre – dazu zählten Paul Gerding für DeSoto, John Reinhart für Ford, Stephen Pojanski für Buick, Dominic Pacitti für Plymouth, Robert Caderet für Buick und das Team Art Fitzpatrick und Van Kaufman für Pontiac – waren im Grunde Pin-up-Künstler, nur waren ihre Motive eben Autos. Ihre jeweiligen Talente, die amerikanische Traummaschine zu (über)zeichnen, bannten ein Gefühl der Erhabenheit und des Staunens aufs Papier.

Die Autos in Broschüren wie jenen für den Nash Lafayette von 1935, den Oldsmobile Baujahr 1937 und den 1938er-Nash waren nicht immer „sexy" – manchmal glichen sie eher den Bildern in einem Kinderbuch. Vielleicht war dies nicht immer eine bewusste Entscheidung seitens der Künstler, doch half es auf subtile Art, den Autos anthropomorphe Persönlichkeiten zu verleihen. Das Titelbild des Prospekts für die Oldsmobile-Modelle Six und Eight zeigt zwei glänzende futuristische Kühlergrille an roten und grünen Autos, die wie der Bug eines Ozeanriesen gestaltet sind und an Kindergeschichten erinnern. Auf der Umschlaginnenseite stößt man dann auf kleine Aquarellvignetten, die die Autos als Familienmitglieder darstellen, wie einen Freund oder ein Haustier. Im ganzen Heft erscheinen die Autos in Szenen, die eine untrennbare, innige Beziehung zwischen den Charakteren andeuten sollen: Auto, Fahrer und Passagiere bilden eine Kernfamilie.

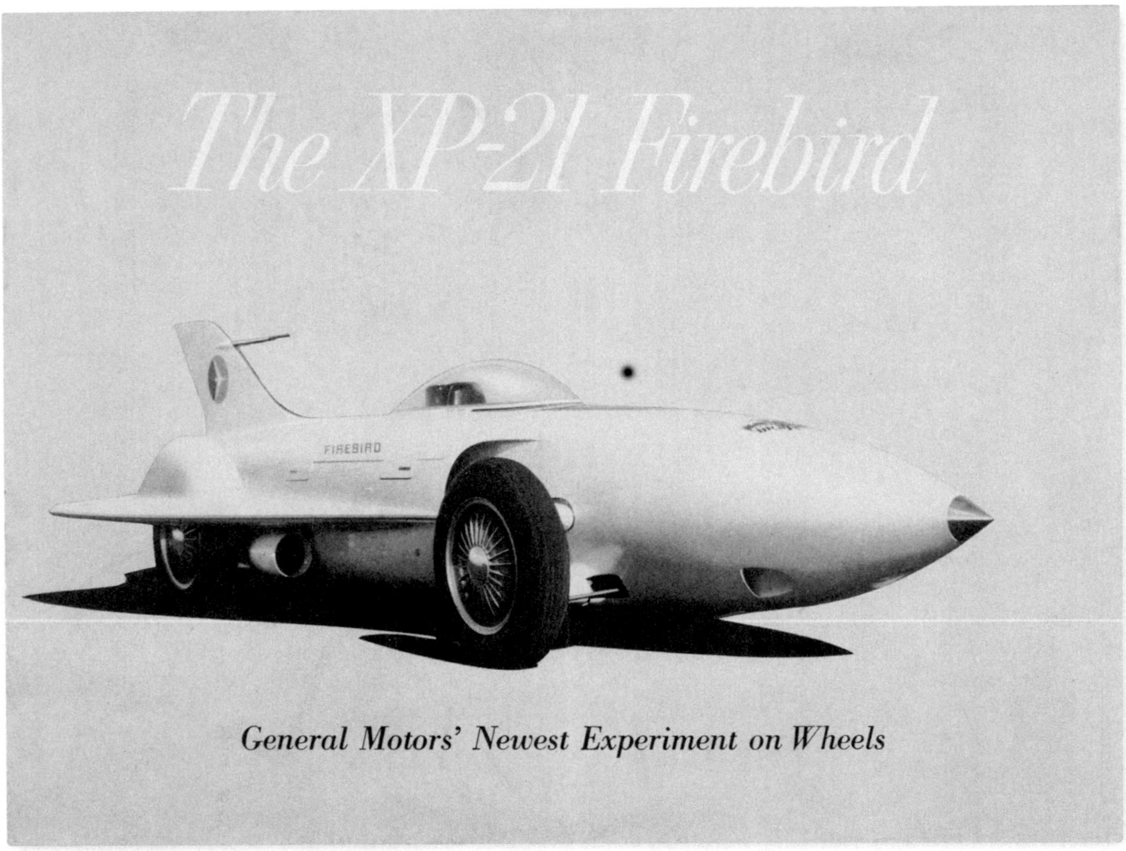

General Motors 1953 *(left)*

Volkswagen 1959 *(opposite)*

Ingenieure und Karosseriedesigner mögen zwar die Könige der Autowelt sein, doch haben Künstler stets einen beträchtlichen Beitrag zur Automobilkultur geliefert. In den „1938 Great News from Nash" wird ein namenloser „berühmter Künstler" mit folgenden Worten zitiert: „Hier haben wir den überzeugenden Beweis, dass Individualität im Automobildesign zu erreichen ist, ohne auf verrückte Ideen zurückzugreifen. Hier wurde stromlinienförmiges Design ohne hässliche Details geschaffen … Wie eine perfekte Skulptur ist der neue Nash aus jedem Betrachtungswinkel schön." Anscheinend wurde dieser berühmte Künstler von den in Gouachetechnik gemalten und retuschierten Darstellungen des monströsen Ambassador Eight Coupe, Lafayette und Victoria Sedan in anregenden Rot-, Blau-, Grün- und Gelbtönen verführt. In der Tat waren diese Farben derart einnehmend, dass sich anfällige Käufer von diesen mechanischen Sirenen anlocken ließen.

Während des Zweiten Weltkriegs stellte sich die amerikanische Automobilproduktion darauf um, die Achsenmächte mit Jeeps, Lkws und Panzern in die Knie zu zwingen, und so wurden nur wenige Hochglanzprospekte für die Verbraucher herausgegeben. Nach dem Sieg der Alliierten war die rosige Zukunft der Nachkriegsjahre die treibende Kraft hinter dem wachsenden Automobilkonsum. An dieser Stelle trat der Futurismus mit Macht auf den Plan. Die Betriebsanleitung für den Oldsmobile von 1949 war von futuristisch anmutenden, von Automobilen umschwirrten Globen übersät. Der Namenszug auf dem Kühler bestand ebenfalls aus einem umringten Globus, der den Planeten Saturn mit einem Kreiselgerät kombinierte. Hübsche Frauen in Fotomodellposen gehörten ebenso zum Standardinventar der Prospekte wie Zukunftsvisionen. Unter der Überschrift „Oh! Diese 56er-Oldsmobiles!" waren die Leitmotive Raketen, Modernität und Luxus. Emblem und Kühlerfigur des Super 88 war eine aufsteigende Rakete, deren Symbolik sich im „aerodynamischen Kühlergrill" wiederfand, der einem Düsentriebwerk glich. Auch der Text war ganz auf Raumfahrt zugeschnitten: „Der Oldsmobile für 1956 … rauscht nach vorn – mit einem geschmeidig kraftvollen neuen Rocket-T-350-Motor und dem kraftvoll geschmeidigen neuen Jetaway Hydra-Matic Drive." Rund um das Thema Raumfahrt ergaben sich für die Texter endlose Möglichkeiten für Metaphorik und Analogie.

Anfang der 1960er-Jahre, nach der Heckflossenmanie der späten 50er, schalteten die Ingenieure einen Gang herunter. Schnittigkeit trat nun an die Stelle von Masse. Auch die Illustratoren entwickelten eine subtile neue Bildsprache, die in einer übertriebenen perspektivischen Darstellung wurzelte. Während die Feinheiten detailliert ins Bild gerückt werden sollten, mussten die Fahrzeuge insgesamt monumental erscheinen. Sie mussten sich nahtlos in die städtische wie in die ländliche Umgebung einpassen (vorzugsweise eine moderne Landschaft à la Mies van der Rohe), um eine Einheit mit der eleganten zeitgenössischen Architektur zu suggerieren. Im Prospekt für den Oldsmobile Classic von 1961 zeigen die „Drucke" im Stil einer Künstlermappe die neuen Olds-Modelle vor den Glasfassaden moderner Gebäude: Das Auto war ein unverzichtbares Ausstattungsmerkmal der Moderne.

Die Amerikaner wurden dazu verführt, das Automobil als Verkörperung moderner Träume, Bedürfnisse und Leidenschaften zu betrachten. Prospekte versprachen ein besseres Leben durch automobilen Genuss. Sie waren Märchenbücher, die den Alltag verdrängten. Und dennoch hielten Autos auch Enttäuschungen bereit. Auf der Straße brachten Pannen und Reparaturen die Träumer wieder auf den Boden der Tatsachen zurück. Das Auto war eine Investition (die allerdings rasch an Wert verlor), und so war die Realität des Autobesitzes, wenn das anfängliche visuelle Feuerwerk verblasste, den Wechselfällen des täglichen Lebens ausgesetzt. Heute sind die Broschüren Erinnerungsstücke, die zeigen, wie berechnet und überzeugend die Illusion dereinst war und wie jeder Käufer nur allzu gern bereit war, ihr zu glauben.

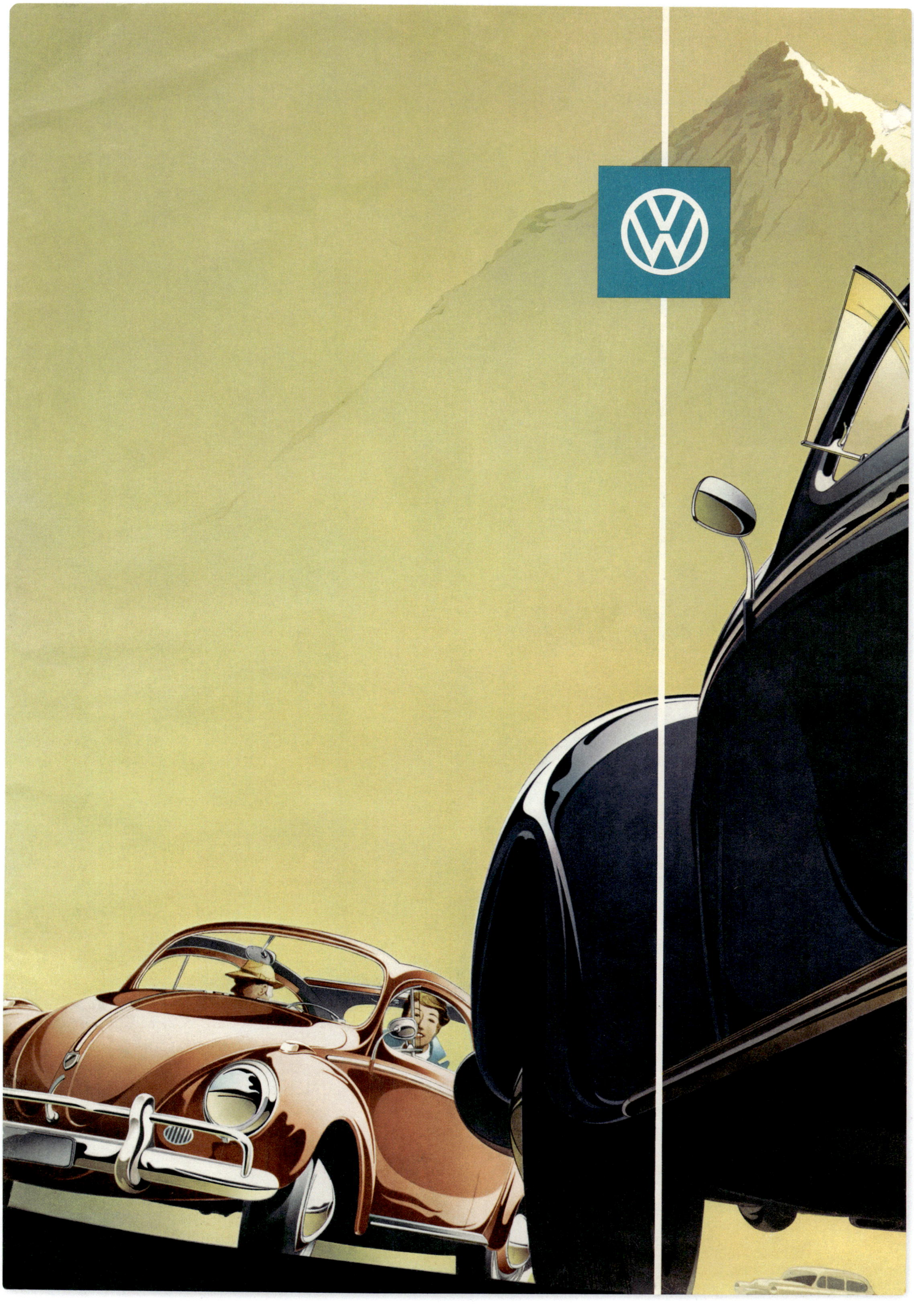

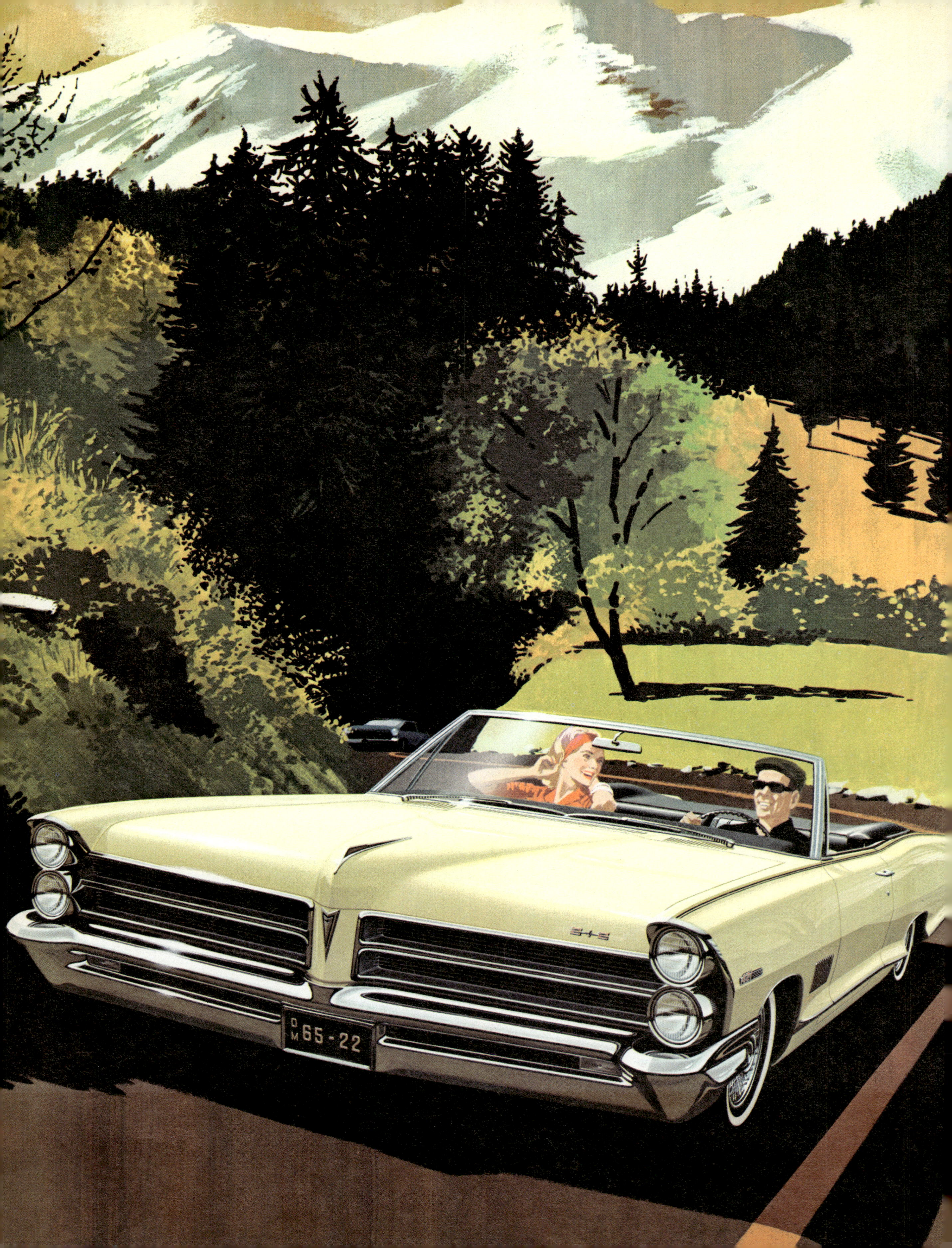
D
M
65-22

STEVEN HELLER

L'art de la sublimation

Les Américains ont été conduits à considérer l'automobile comme l'incarnation de tous les rêves et passions modernes. À travers le plaisir motorisé, les catalogues promettaient une vie meilleure.

Souvent des échantillons de peinture ou de tissu étaient glissés dans les catalogues automobiles métamorphosés en dépliants étourdissants. Luxe, magnificence et volupté : dans leur genre, ils faisaient figure de mécaniques de pointe. Auprès des lecteurs, ils étaient supposés déclencher toutes sortes d'élans primordiaux. Alliant le glamour des magazines de cinéma à un pragmatisme de prospectus commercial, ils vendaient de l'innocence (débridée), de l'ingénierie (avancée) et le rêve américain, encore et toujours.

À travers tout le pays, panneaux d'affichage et annonces presse aimantaient les regards et guidaient le futur client vers les néons des showrooms aux murs de verre, tandis que les catalogues étaient chargés d'enflammer le désir et d'aiguillonner les sens. Par-delà leur apparence soignée, ils s'évertuaient à nimber l'automobile d'une aura majestueuse, à en souligner le style, la classe, l'esthétique, en contrepoint de son excellence technologique ; comme à Broadway ou dans un peep-show de Times Square, ils levaient leur rideau sur quantités d'options sensationnelles, présentées aussi naturellement que l'eussent été des équipements standard. À tourner ces pages impeccables où venaient parfois s'intercaler, entre autres effets d'impression, des feuilles translucides en simili-parchemin, comment résister aux sièges en cuir, à la climatisation, à tous ces agréments qui, l'air de rien, relevaient la facture finale ? En général, un petit tour au volant permettait de générer chez le client potentiel une appétence irrépressible ; la brochure servait ensuite à lui rappeler que le plaisir était à sa portée.

À l'origine, les premiers auteurs de ces brochures – dont une majorité d'illustrateurs commerciaux et de typographes anonymes – se faisaient de leur rôle une idée fonctionnelle : informer le consommateur au moyen de documents directs et accessibles. Bientôt, cependant, la fonctionnalité n'a plus suffi ; maîtriser l'art de la stimulation était devenu essentiel. Ainsi, dès 1927, le catalogue de la Stutz 8, avec son logo d'ascendance égyptienne et l'influence celtique de sa couverture – allusion à l'Art nouveau–, mariait-t-il l'ancien et le nouveau ; le lecteur devait être amené à assimiler cette berline à un palais mobile. Le langage futuriste n'était pas encore de rigueur, mais la touche fantasmatique contribuait à susciter l'envie. Il était inévitable qu'on en vînt à instrumentaliser l'accomplissement du désir : conditionner la puissance personnelle à l'achat de *cette* voiture, et pas d'une autre, avait beau relever du subterfuge, ça marchait. En matière d'arrogance marketing, le catalogue Studebaker and Erskine qui présente les modèles President, Commander et Dictator – oui, vous avez bien lu, Dictator – tient du chef-d'œuvre : au volant de votre Studebaker, vous êtes le maître du monde. Les constructeurs automobiles prennent l'habitude de rehausser leurs équipements, si ordinaires soient-ils, de tout le lustre souhaitable. Le cru 1932 des Oldsmobile, par exemple, exhibe ses aménagements Art déco ou modern style (phares, cendriers, poignées de porte…) comme autant d'ornements inestimables.

Telles des Cendrillons de papier, ces agapes visuelles étaient aussi séduisantes qu'éphémères : il suffisait que sonnent les douze coups de minuit, c'est-à-dire que surgissent les nouveaux modèles, pour condamner à l'obsolescence et au rebut les catalogues précédents. Les constructeurs investissaient chaque année des sommes extravagantes : proposer des véhicules dispendieux à une riche clientèle requiert prodigalité et valeur ajoutée (les brochures plus modestes, quoique toujours attrayantes et imaginatives, ciblaient d'abord les masses). Les catalogues participaient donc intimement de la stratégie marketing et de ce qu'on appellerait aujourd'hui la notoriété de la marque ; il n'est guère étonnant que leur conception ait été envisagée avec le plus extrême sérieux, même si la plupart d'entre eux furent réalisés par des concepteurs qui n'étaient pas, alors, les stars de leur spécialité.

La publicité continuait à prévaloir. Des agences comme N.W. Ayer, J. Walter Thompson et des directeurs artistiques comme Wallace Elton (Ford) ou Paul Rand (Kaiser Fraser) se consacraient presque exclusivement aux campagnes publicitaires ; Rand, le premier graphiste à introduire l'art moderne européen dans la communication d'entreprise, créa

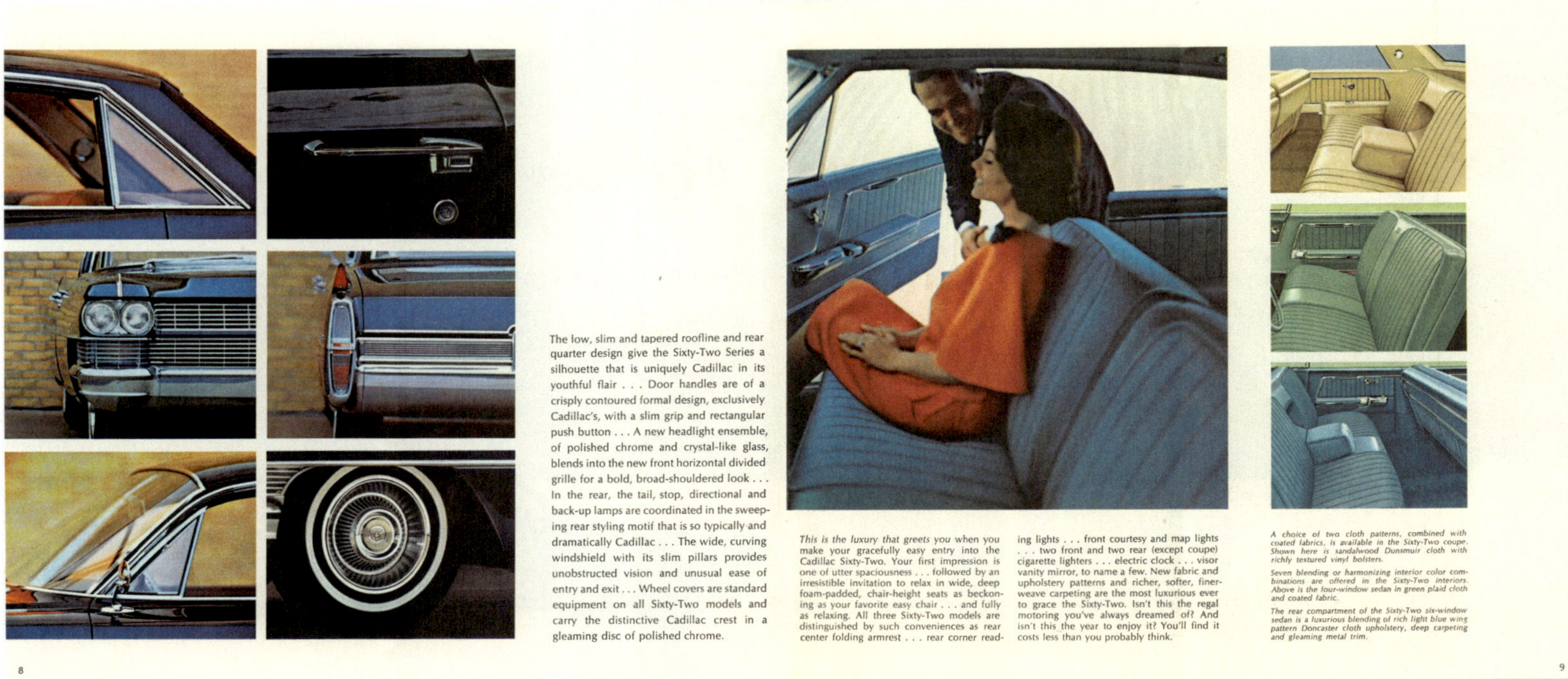

Pontiac 1965 *(page 46)*

Cadillac 1964 *(above)*

Ford 1969 *(opposite)*

bien pour Kaiser une mémorable couverture (en 1951), mais elle reste l'exception qui confirme la règle. Néanmoins, quelle que fût l'identité des designers, les catalogues ne lésinaient pas sur les mises en pages éblouissantes, les illustrations saisissantes, les photos spectaculaires et les typographies grandioses ; pourquoi en eût-il été autrement? Il s'agissait après tout de plaquettes destinées au client et célébrant l'« autovolution », autrement dit l'ensemble des modifications apportées chaque automne au style et à la configuration des modèles. Detroit ne se contentait pas de vendre des voitures, mais épousait le concept on ne peut plus américain de supériorité économique, beaucoup moins perceptible dans les brochures automobiles du reste du monde.

Mais une voiture reste une mécanique. Et au contraire d'une quelconque poudre de perlimpinpin, une mécanique doit donner des gages d'efficacité et de fiabilité, critères de perfection conceptuelle. Il fallait donc convaincre les automobilistes américains qu'une fois au volant, ils éprouveraient un sentiment de confiance totale en toutes circonstances : circulation dense, petite route de campagne, haute montagne. Les constructeurs ont donc usé de force arguments porteurs tels que : « L'ingénierie brillante qui vous a offert la suspension à barres de torsion et correcteur automatique d'assiette vous présente aujourd'hui LA MEILLEURE PACKARD JAMAIS CONÇUE. » Ce catalogue de 1956 dévoilait également la Patrician, la Caribbean et la 400. Le poids des mots s'y substituait aux habituelles descriptions mécaniques.

La communication automobile n'a jamais été avare de superlatifs. Empiler les couches de plaisir et d'illusion exige en effet de la part des rédacteurs, passés maîtres en la matière, un sens achevé de l'hyperbole : « Sous l'angle de l'aérodynamique, de la technologie ou de la performance… En termes d'investissement et de progrès… D'où que vous la regardiez… C'est la meilleure Packard jamais conçue. » Petites ou grandes, indépendamment du modèle ou du style, les automobiles exerçaient alors une attraction irrésistible car auréolées de panache et d'élégance. Sur les autoroutes, chaque voiture était à coup sûr la plus belle incarnation de sa catégorie. Bien sûr l'efficacité, qualité sans égale, on l'a dit, conditionnait le succès commercial, mais c'était justement la vocation de ces catalogues que de transcender en faveur de la forme les représentations habituellement associées à la fonctionnalité. Certains d'entre eux versaient dans un lyrisme adapté à l'héritage particulier du modèle évoqué. Ainsi la brochure consacrée à la Pierce-Arrow de 1937 : « La course folle de l'industrie américaine pour produire toujours plus a sacrifié presque entièrement, pour ne pas dire gratuitement, l'individualité automobile. Le tempérament hérité par lequel on pouvait, il y a quelques années encore, identifier quasiment toutes les voitures du *règne automobile* a été oublié ou ignoré dans cette ruée vers la production de masse. Aujourd'hui chaque voiture, ou presque, n'est qu'un moyen de transport parmi d'autres, noyé dans la foule. L'identité a été balayée par la course à l'"uniformité". » Le patrimoine, c'était la singularité – et l'uniformité un péché que toutes les innovations sorties des chaînes de production ne pouvaient suffire à absoudre. Dans son catalogue de 1937 (et dans d'autres qui suivront), Pierce-Arrow décida alors de faire de l'orgueil un signe distinctif. Et comme d'autres marques, avant et après, ne transigera jamais avec la médiocrité.

Le confort était aussi important que l'efficience, et sans doute le meilleur hameçon promotionnel auquel suspendre un argumentaire commercial, surtout dans le cas de certain consommateur « secondaire » : la femme. Des premières heures de l'âge automobile jusqu'à la fin du 20 siècle, les hommes ont payé les factures. Mais les femmes blanches de la classe moyenne, régulièrement représentées par l'iconographie des publicités et des catalogues, décidaient plus ou moins de ce que leurs maris allaient acheter. Et pour elles, le confort revêtait plus d'importance que la technologie ; aussi était-il impératif, par le texte et l'image, d'établir un équilibre. Le début du slogan « Le confort d'une grosse voiture… plus l'économie » est un clin d'œil aux femmes alors que le dernier mot s'adresse aux hommes. De même lira-t-on la phrase

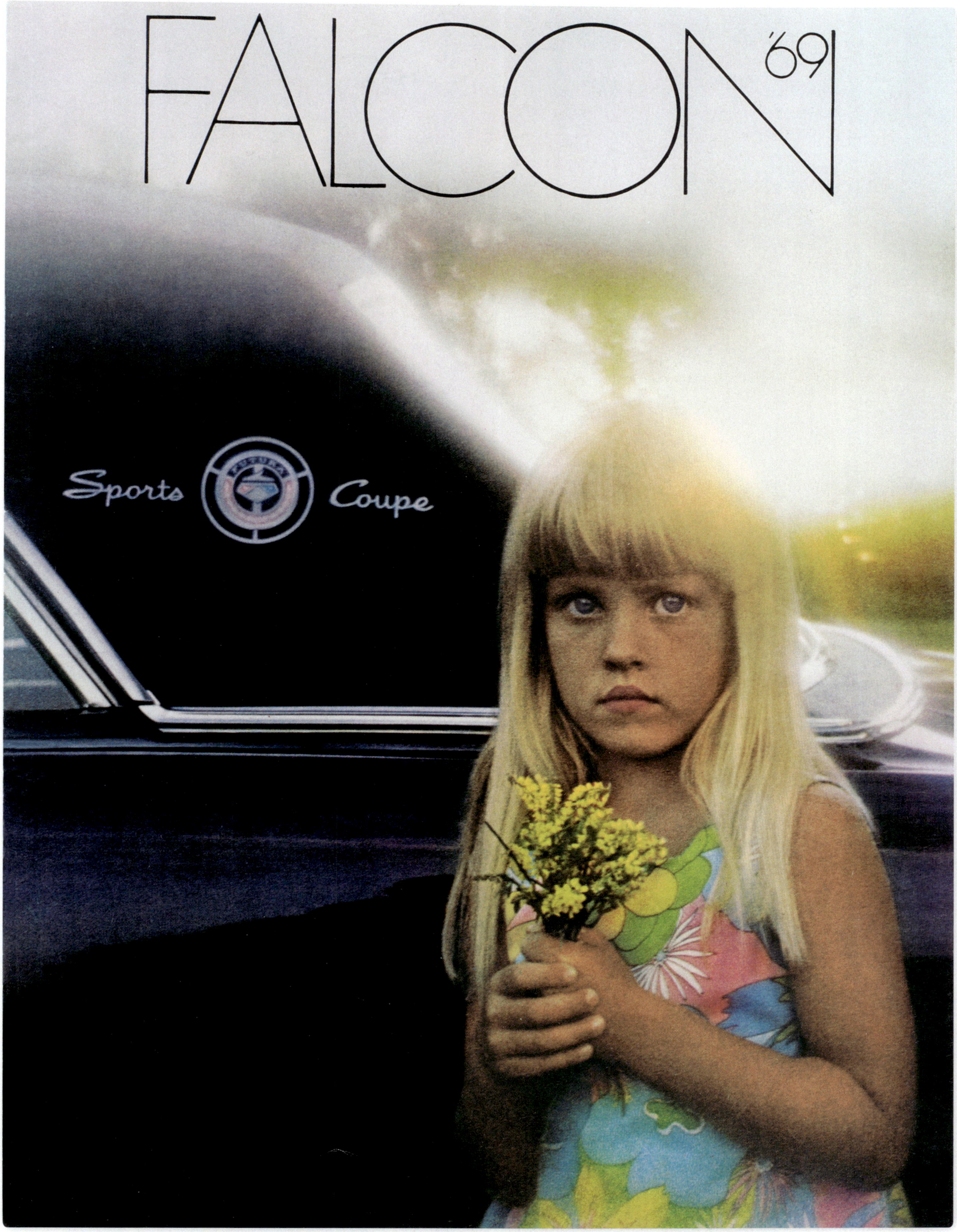
FALCON '69
Sports
FUTURA
Coupe

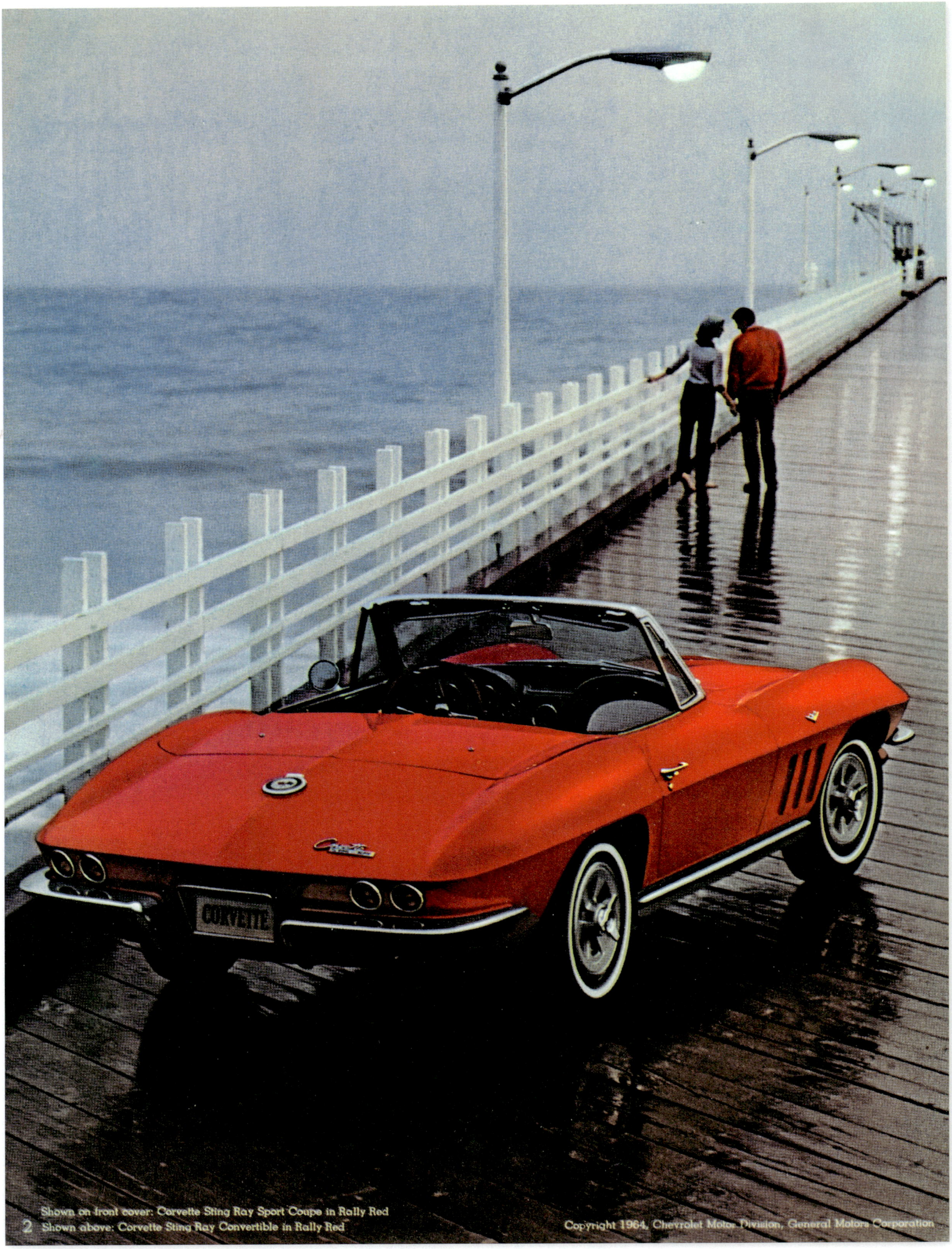
CORVETTE
Shown on front cover: Corvette Sting Ray Sport Coupe in Rally Red
2 Shown above: Corvette Sting Ray Convertible in Rally Red
Copyright 1964, Chevrolet Motor Division, General Motors Corporation

Chevrolet 1965 *(opposite)*

Chevrolet 1968 *(right)*

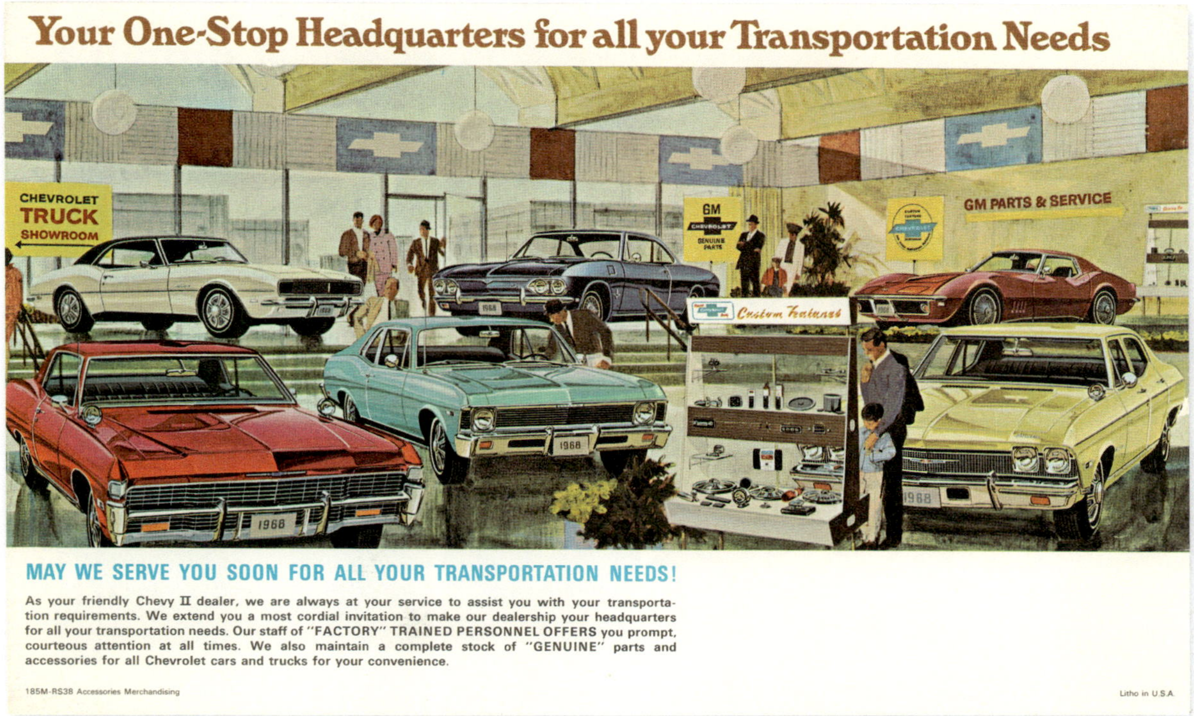

suivante à la lumière de ce qu'elle insinue : « Ce ne sont pas seulement les dimensions de la Mercury 1940 qui la rendent si confortable. Bien sûr un châssis large crée du volume, mais c'est la façon dont la voiture a été dessinée et construite qui lui vaut d'être si agréable à conduire et si reposante. Le confort est livré en série… » Dans ces phrases les hommes entendent « les dimensions » et les femmes « si confortable ». Les hommes s'arrêtent sur « dessinée et construite », les femmes sur « agréable à conduire et reposante ».

Si variés fussent les catalogues, leur contenu respectait tout de même certaines constantes formelles. L'éventail allait ainsi du feuillet descriptif d'une page (« sheet ») à la page pliée en accordéon (« folder »), qui pouvait atteindre le format d'une affiche. Les catalogues avaient leur version basique, florilège de tous les mérites disponibles, et leur déclinaison « prestige », en l'espèce la brochure la plus épaisse et la plus flamboyante de l'année, également la plus recherchée. En général, les directeurs artistiques des agences transmettaient leurs concepts aux studios de création de Detroit (les vrais soutiers du métier) puis les allers-retours se succédaient jusqu'à l'ultime *deadline*. Agences et studios, de fait, ne se marchaient pas sur les pieds : les studios bataillaient pour que les agences leur confient du travail, pendant que les agences s'arrachaient les budgets publicitaires des constructeurs automobiles. Les ingénieurs des marques, de toute façon, avaient toujours le dernier mot sur la façon de présenter les modèles, comme si la partie créative du processus était finalement très accessoire.

À Detroit, Philadelphie et New York, des légions de rédacteurs alimentèrent ainsi la mythologie automobile tandis que des escadrons d'artistes maison, experts dans l'art de restituer les détails les plus précis, ancraient l'imagerie au plus profond des rêves collectifs : dessin et peinture étaient tellement plus épiques, plus romantiques, que même la plus impressionnante des photographies ! L'illustration permettait de jouer à la fois sur l'expressivité, la fidélité et la dimension surréelle ; témoin, la couverture de la brochure de la REO Flying Cloud (fin des années 1920), où la voiture lancée à pleine vitesse et gravée en relief est d'une virtuosité qu'un futuriste italien ou un vorticiste anglais ne renieraient pas. *A contrario*, la précision des illustrations intérieures pourrait les faire passer pour des photos. Le catalogue des Pontiac Silver Streak 6 et 8 (1936) est un régal de dextérité au pastel et à la gouache. Le bain d'aérographe auquel a manifestement été soumise la brochure de la Cadillac de 1954 en fait briller les chromes jusqu'à aveugler le lecteur. Quant aux Custom et Custom 300 de 1957, leur allongement outrancier leur donne des airs de Chris Craft plus que d'automobile. L'infortunée Edsel lancée par Ford en cette année 1957 – « l'accent a été placé sur la performance, et l'élégance soulignée » – était représentée avec tant d'esbroufe (toujours cette fameuse longueur) qu'on peine aujourd'hui à croire qu'elle fut un échec commercial. Le seul nom de « Edsel » est même devenu synonyme de désastre industriel dans la culture populaire américaine.

Paul Gerding (pour DeSoto), John Reinhart (Ford), Stephen Pojanski (Buick), Dominic Pacitti (Plymouth), Robert Cadaret (Buick), le duo Art Fitzpatrick-Van Kaufman (Pontiac)… Les illustrateurs automobiles des années 1930 et au-delà auraient pu se montrer beaucoup plus racoleurs qu'ils ne le furent – ils disposaient pour cela du savoir-faire requis ; or, entre leurs mains, la machine à rêve américaine, sublimée, joignait l'enchantement à la solennité. Dessinées, peintes ou colorées à l'aérographe, les stars des catalogues comme la Nash Lafayette de 1935, l'Oldsmobile de 1937 et la Nash de 1938 n'étaient pas toujours des plus sexy ; elles rappelaient même plutôt l'atmosphère des livres pour enfants. Un anthropomorphisme subtil, peut-être inconscient de la part des artistes, contribua ainsi à forger la « personnalité » des voitures américaines. Les calandres rutilantes et futuristes des Oldsmobile 6 et 8, en couverture de la brochure de 1938, évoquent la proue de deux paquebots ; le conte pour enfants n'est pas loin. Sur les vignettes à l'aquarelle de la deuxième de couverture, les autos se

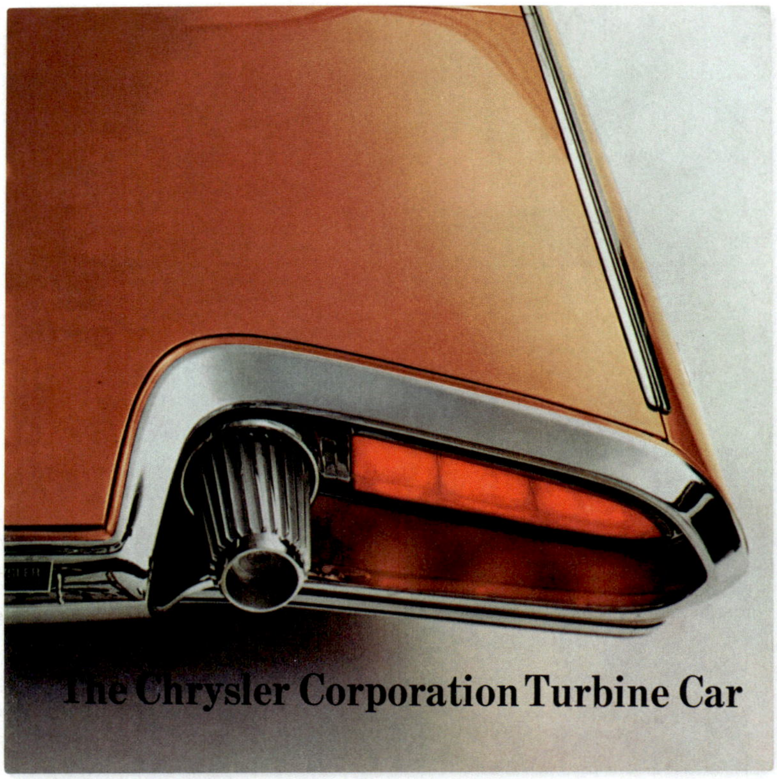

Chrysler 1966 *(left)*

De Tomaso Pantera 1972 *(opposite)*

présentent comme des membres du foyer, amies proches ou animaux de compagnie. Les scènes qui se succèdent au fil du document suggèrent à chaque fois les liens entrecroisés unissant les personnages. Voiture, conducteur et passager forment une famille nucléaire.

Les ingénieurs et les designers industriels règnent sur le monde motorisé et, cependant, la culture automobile doit aussi beaucoup au dessin et aux dessinateurs. Dans le catalogue intitulé « Great News From Nash » (1938), un « célèbre artiste » anonyme livre son témoignage : « Voici la preuve éloquente qu'il est possible de parvenir à un stylisme original sans recourir nécessairement à un design délirant. Cette aérodynamique-là ne s'embarrasse d'aucun détail disgracieux ou maladroit… Comme une sculpture parfaite, la nouvelle Nash est belle sous tous les angles. » Probablement ce célèbre artiste a-t-il été séduit par la brillante restitution, à la gouache et à l'aérographe, des éléphantesques Ambassador Eight, LaFayette et Victoria, aux livrées rutilantes, azur, verte et jaune, plus éclatantes l'une que l'autre. De ces jeux d'ombres et de couleurs, comme un rouge envoûtant à leurs lèvres, les sirènes mécaniques faisaient chavirer les cœurs sensibles.

Pendant la Seconde Guerre mondiale, la production automobile américaine participa, grâce aux jeeps, camions et chars d'assaut, à l'anéantissement des puissances de l'Axe. L'époque ne se prêtait pas vraiment aux brochures de luxe. En revanche, une fois la victoire acquise, les promesses de l'après-guerre avaient tout pour favoriser l'essor du marché automobile. Le futurisme pouvait revenir à la mode, comme le confirment le manuel d'utilisation de l'Oldsmobile 1949, ses planètes peintes à l'aérographe et les voitures qui gravitent autour d'elles. Le globe ceint d'un anneau, en guise d'insigne de capot, suggère à la fois Saturne et un gyroscope. Les jolies femmes ne sont plus seules à poser : désormais les brochures exposent également les représentations que l'on se fait de l'avenir. Conquête spatiale, luxe et modernité reviennent comme autant de leitmotivs. Le logo et l'en-tête du catalogue de la Super 88 montrent une fusée en plein envol, à laquelle fait écho la « grille d'aération élégamment dessinée » censée rappeler un moteur d'avion. Le texte est lui aussi d'inspiration spatiale : « L'Oldsmobile de 56 […] lance des éclairs – grâce à la puissance souple de son nouveau moteur Rocket T-350 et à la souplesse puissante de sa nouvelle transmission Jetaway Hydra-Matic. » L'espace offrait aux rédacteurs profusion de métaphores et d'analogies possibles.

Le début des années 1960 signa le déclin de la folie du tout-aileron, qui avait marqué la fin de la décennie précédente, et un changement de braquet : aux volumes massifs les ingénieurs substituèrent le raffinement. Les illustrateurs commencèrent alors à développer un nouveau langage tout en finesse qui plonge ses racines dans la perspective exagérée : si la représentation des détails reste minutieuse, les voitures se doivent d'exprimer une certaine majesté. En s'intégrant clairement dans le paysage urbain ou rural (de préférence un paysage moderne à la Mies Van der Rohe), symbolisant en cela l'unité de la machine avec son environnement architectural, contemporain et sophistiqué. Les nouveaux modèles qui jalonnent le catalogue Oldsmobile Classic de 1961 paradent devant les façades transparentes de constructions modernes ; l'automobile demeure un attribut essentiel de la modernité.

Les Américains ont été conduits à considérer l'automobile comme l'incarnation de tous les rêves et passions modernes. À travers le plaisir motorisé, les catalogues promettaient une vie meilleure, tenant lieu de contes de fées et voués à sublimer la routine ordinaire. Ce monde idéal, pourtant, n'était pas tout à fait exempt d'écueils. Les questions d'entretien, par exemple, avaient tôt fait de ramener le rêveur sur terre. Une voiture, c'était un investissement (de ceux qui se déprécient vite, et lourdement). Lorsque l'éblouissement visuel commençait à s'estomper, il n'y avait pas de raison que la possession d'une automobile échappât aux vicissitudes de la vie quotidienne. Ces catalogues, aujourd'hui, viennent nous rappeler combien l'illusion sut se faire convaincante. À quel point, aussi, chacun était heureux d'en être le jouet.

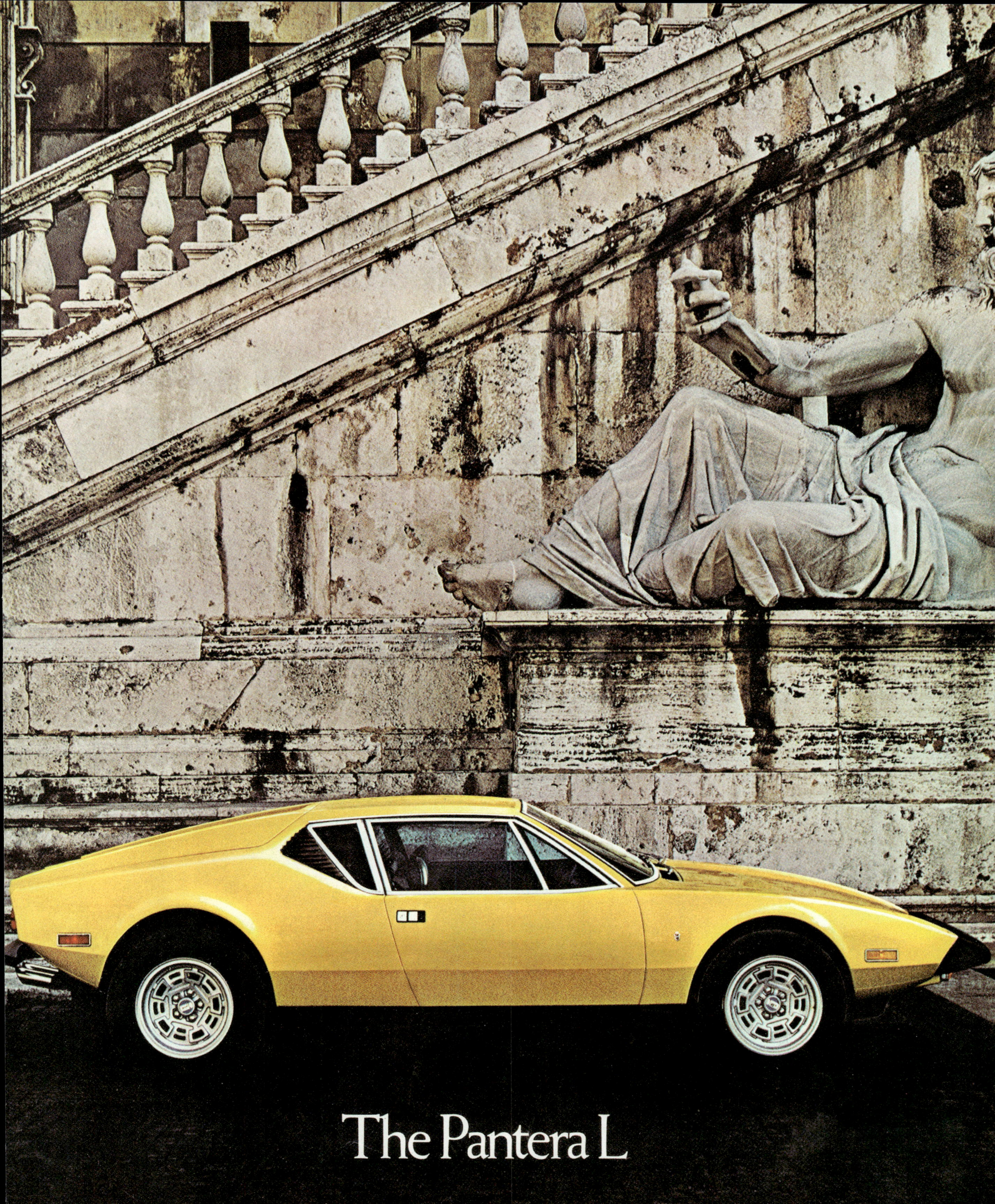
The Pantera L

Igniting the Desire
1900–1929

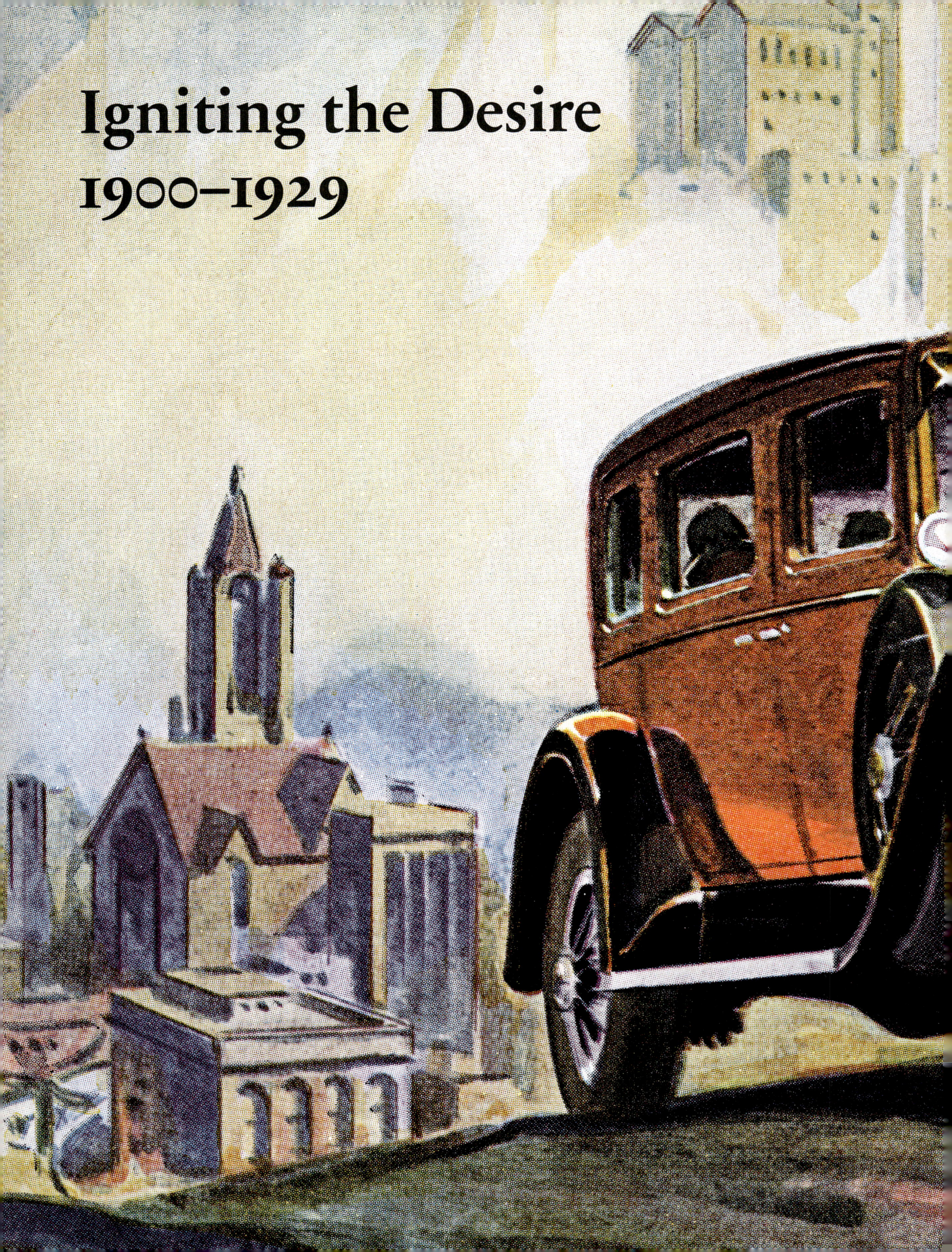

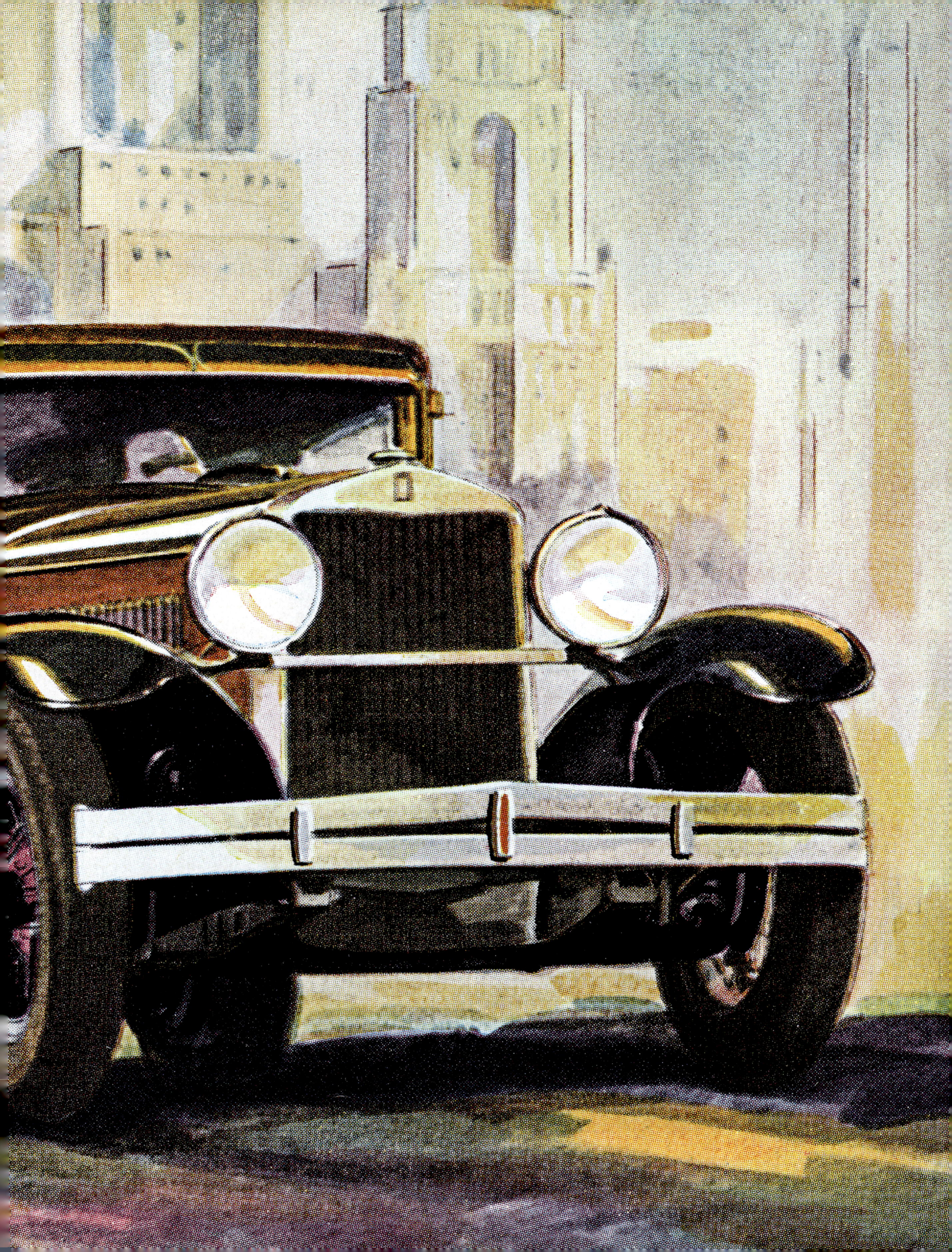

Alco

Alco Touring Car 1913

Alluding to the Alco's standing in society, this car was built by a producer of railroad locomotives. Buyers respected the brutal power of its massive engines but not enough of them bought the cars, and by 1914 Alco ceased production.

Dieses Automobil wurde von einem Lokomotivenhersteller gebaut, und diese Herkunft spiegelte sich auch im Ruf wider, den Alco hatte: Die Kunden waren beeindruckt von der enormen Kraft der Motoren, aber letzten Endes gab es zu wenig Käufer, und 1914 stellte Alco den Automobilbau ein.

Cette voiture, manifestement garante d'un certain statut social, avait été fabriquée par un constructeur de locomotives. Ses acheteurs apprécieront la puissance brute de ses motorisations impressionnantes, mais ne seront pas assez nombreux, et en 1914 Alco mettra un terme à sa production.

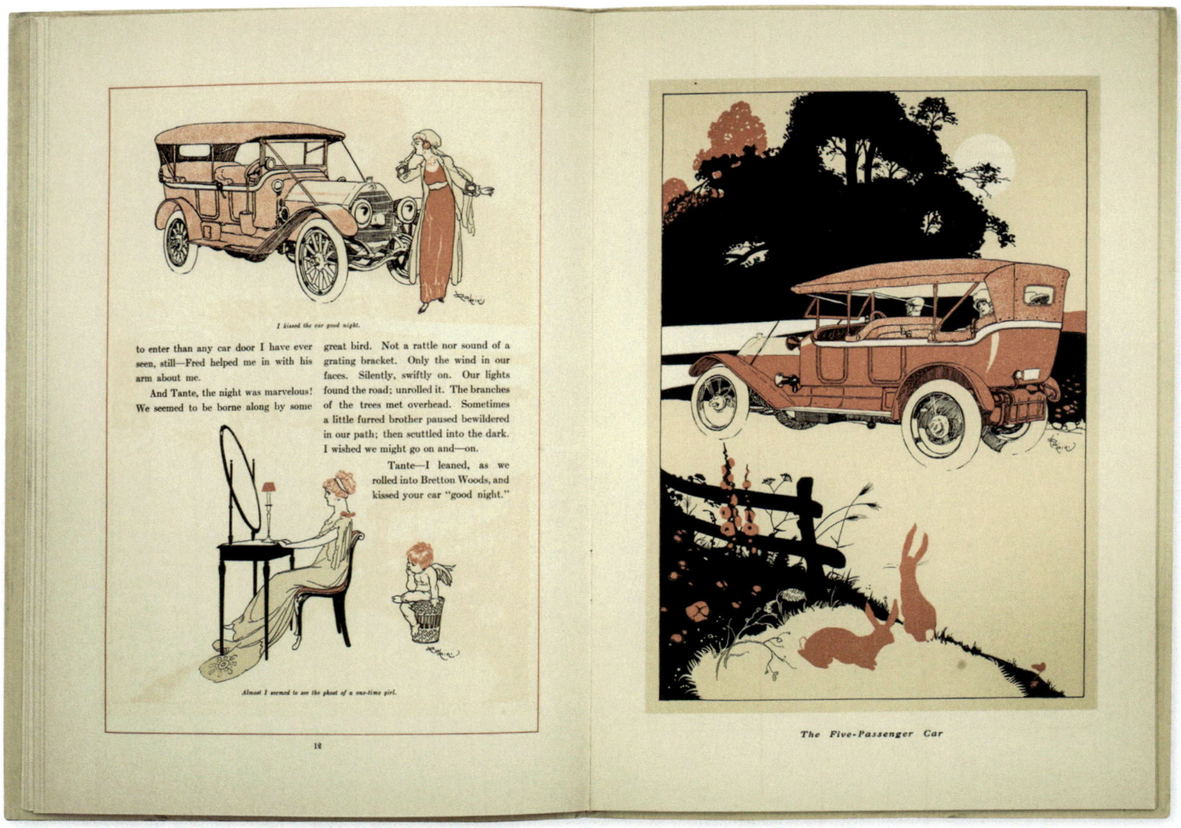

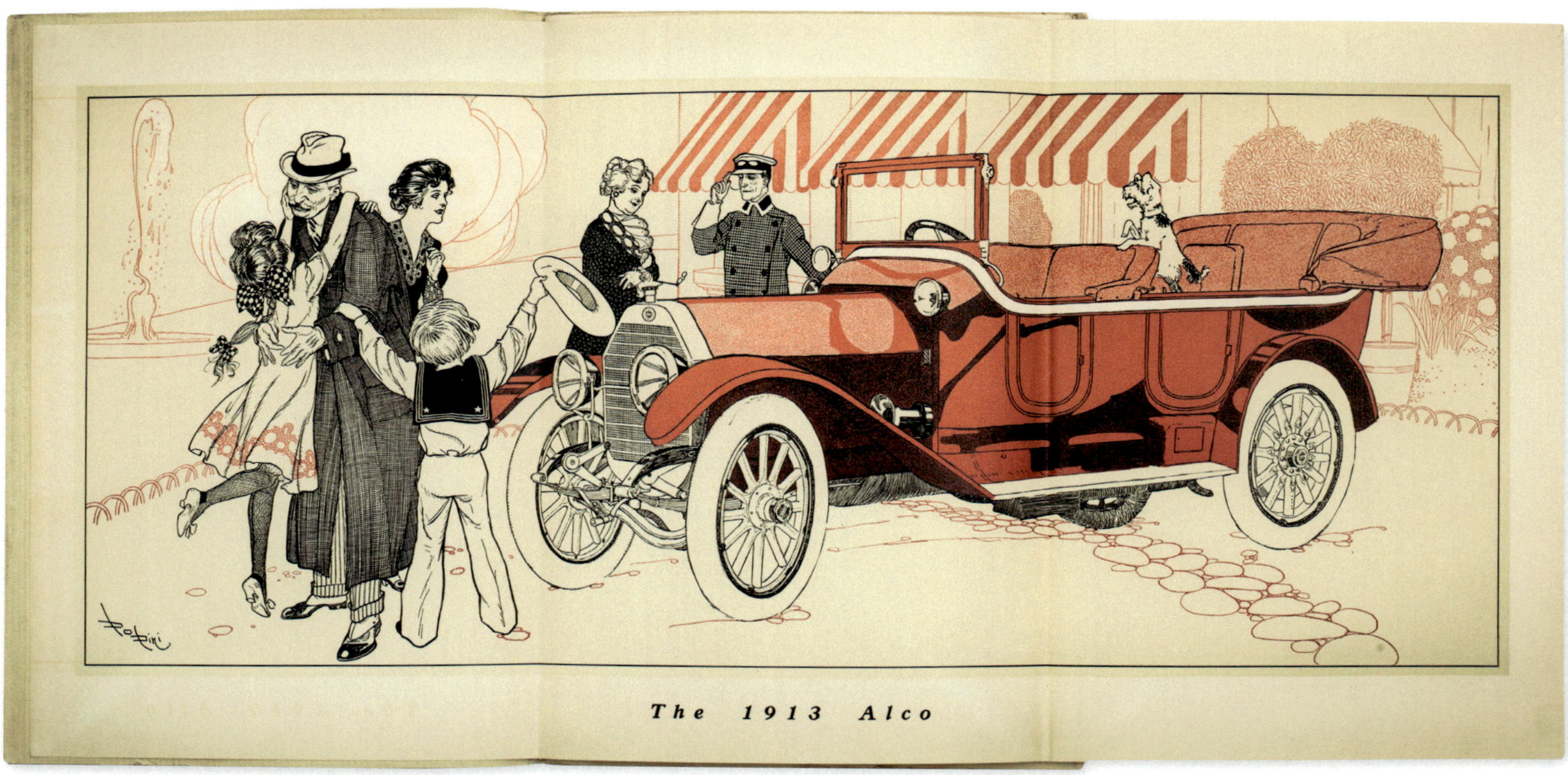

Baker Electric

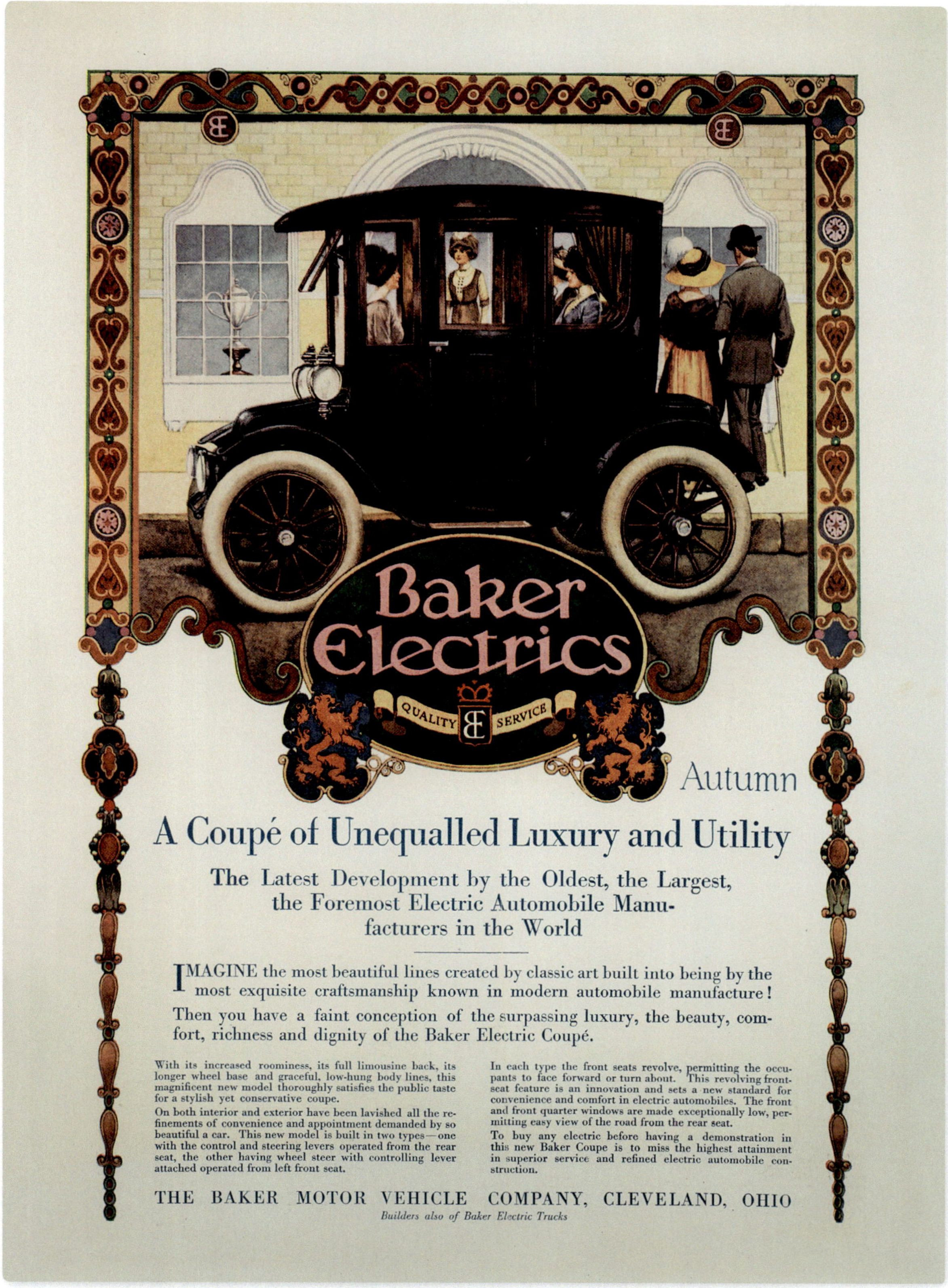

Baker Electric 1909

One of the first cars aimed at women drivers, the Baker Electric outsold all other electric cars in the United States. Baker eventually merged with Rauch & Land and went into auto bodywork. The primary appeal of the electric car, especially where women were concerned, was respite from hand-cranking the engine and adding smelly gasoline and oil. The introduction of the self-starter by Cadillac, and later others, coupled with their short range ended electric cars' viability, although better engineered versions are making something of a comeback today.

Als eines der ersten Autos, das speziell auf weibliche Autofahrer abzielte, verkaufte sich der Baker Electric besser als jedes andere Elektroauto in den Vereinigten Staaten. Baker fusionierte schließlich mit Rauch & Land und ging in den Karosseriebau. Der hauptsächliche Reiz am Elektroauto – insbesondere für Frauen – war, dass zum Anlassen keine Handkurbel bedient und zum Fahren kein übel riechendes Benzin oder Öl nachgefüllt werden musste. Die Einführung des elektrischen Anlassers durch Cadillac und später auch durch andere Hersteller machte den Elektroautos, die nur eine sehr begrenzte Reichweite hatten, vorerst den Garaus, aber heutzutage erlebt das E-Fahrzeug ja in technisch etwas ausgereifterer Form eine Art Comeback.

La Baker Electric, qui surclassait toutes ses concurrentes électriques aux États-Unis, a été l'une des premières voitures conçues en direction des femmes. Baker finira par fusionner avec Rauch & Land et se consacrera à la carrosserie auto. Pour les conductrices, un véhicule électrique avait d'abord l'avantage de faire oublier le démarrage à la manivelle et les odeurs d'essence ou d'huile. L'introduction du démarreur par Cadillac puis d'autres constructeurs et leur autonomie insuffisante porteront un coup fatal aux véhicules électriques, même si le principe, servi par des modèles mieux motorisés, connaît aujourd'hui un retour en grâce.

Cadillac

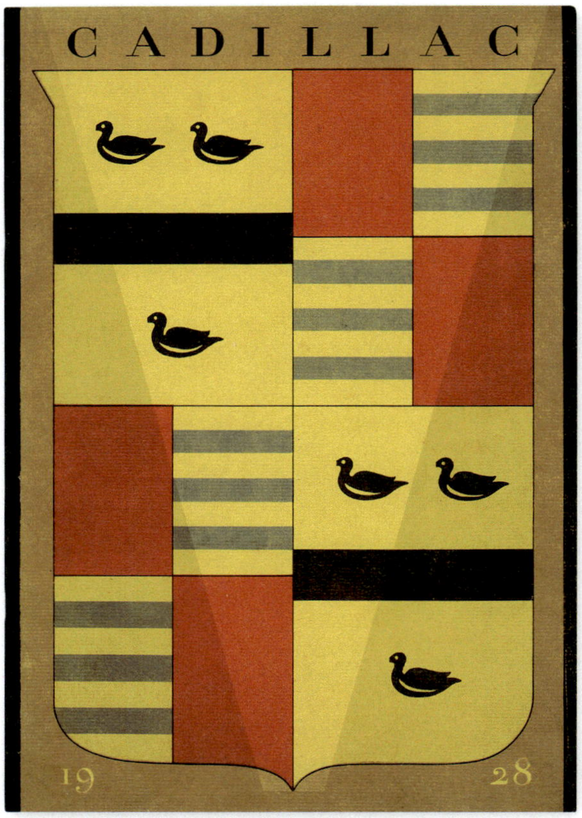

Cadillac 1928 *(this spread)*

A new Cadillac perfectly fits in southern Florida, which was being developed for luxury tourism in the 1920s. Originally conceived by Henry M. Leland (who also introduced its longtime competitor, Lincoln), Cadillac was one of America's first true luxury cars, pioneering the self-starter and engines that eventually encompassed 16 cylinders. It remains at the pinnacle of luxury automobiles to this day. The illustration depicts a very early version of the Cadillac crest, which was originally the coat of arms for Antoine de la Mothe Cadillac, the French explorer who founded Detroit around 1700.

Ein neuer Cadillac passte perfekt in den Süden Floridas, der in den 1920er-Jahren als Reiseziel für betuchte Touristen ausgebaut wurde. Der Cadillac, der ursprünglich von Henry M. Leland entworfen wurde (von dem auch der langjährige Hauptkonkurrent Lincoln stammte), war eines der ersten echten Luxusautomobile Amerikas, der Pionier des Selbstanlassers und schließlich mit 16 Zylindern motorisiert. Bis heute ist er der Inbegriff des Luxusautos geblieben. Die Abbildung zeigt eine sehr frühe Version des Cadillac-Emblems, das ursprünglich tatsächlich das Wappen von Antoine de la Mothe Cadillac war, dem französischen Abenteurer, der 1701 das Fort gegründet hatte, aus dem später die Autostadt Detroit hervorging.

Une nouvelle Cadillac : rien n'est plus représentatif du sud de la Floride, qui dans ces années 1920 s'offrait résolument au tourisme de luxe. Conçue par Henry M. Leland (à qui l'on doit également la Lincoln, qui sera longtemps sa concurrente), la Cadillac était l'une des premières vraies voitures de luxe américaines, qui ouvrit la voie au démarreur et à des moteurs bientôt bardés de seize cylindres… Sur ce segment du luxe, la marque a su rester au sommet jusqu'à nos jours. On voit ici l'une des toutes premières versions de l'écusson Cadillac, inspiré des armoiries de la famille d'Antoine de la Mothe Cadillac, l'explorateur français qui fonda la ville de Detroit vers 1700.

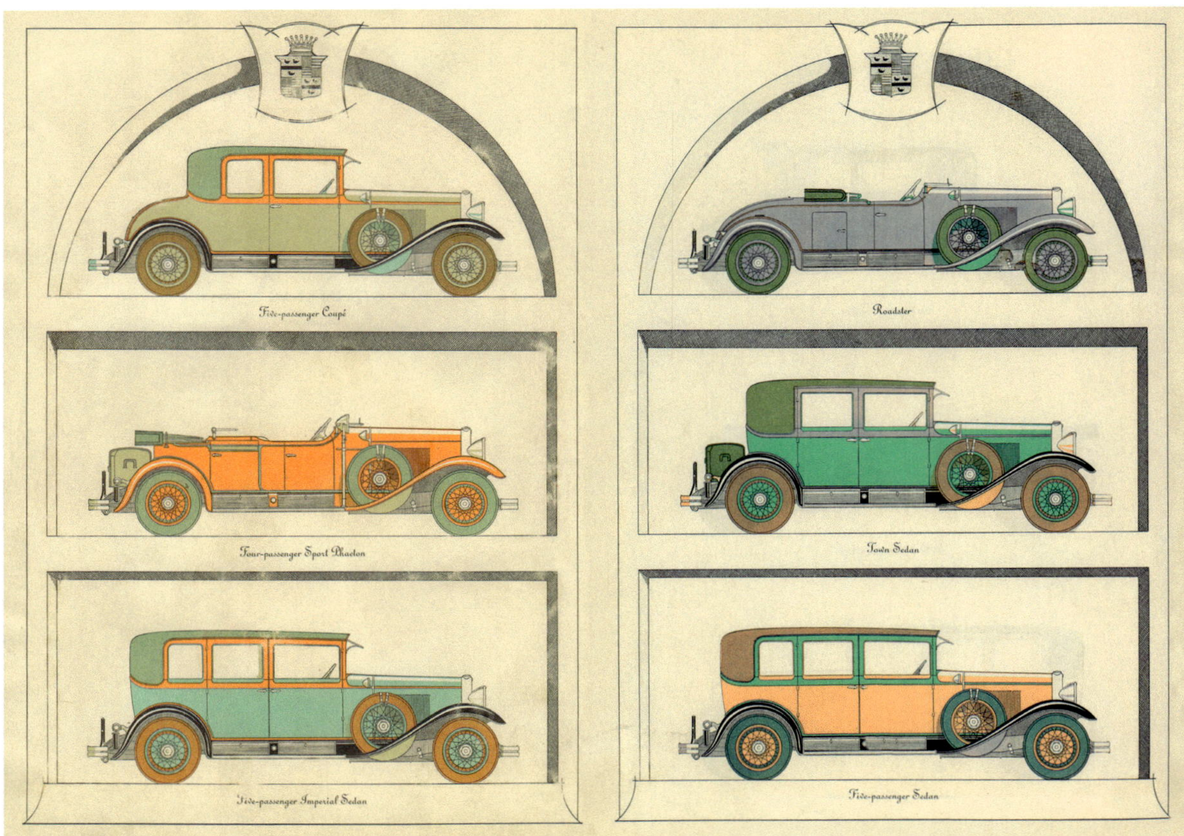

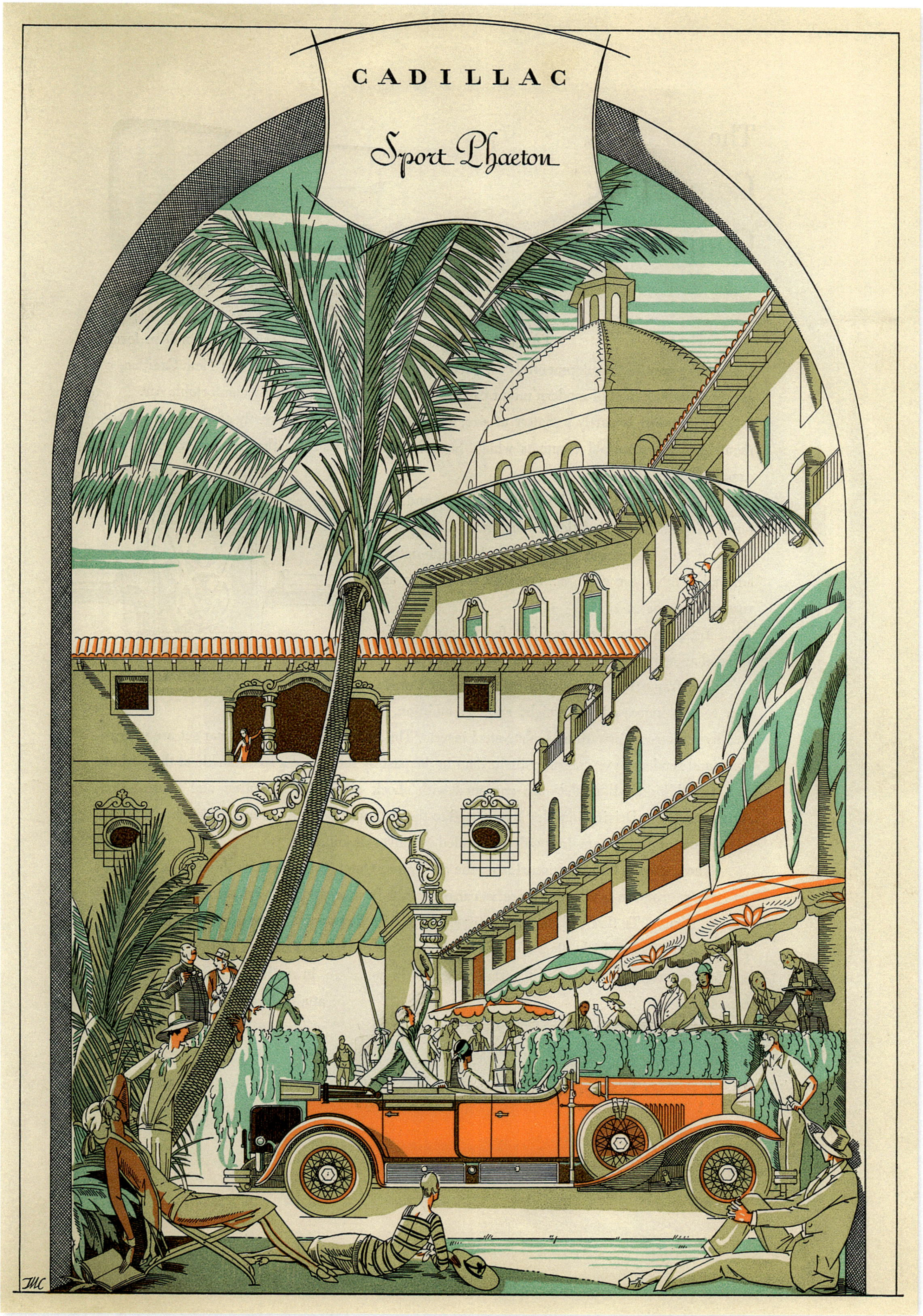
CADILLAC
Sport Phaeton

Cartercar

Cartercar 1915 *(right)*

Created by Byron Carter, the Cartercar was built with friction drive; it had no conventional transmission. William C. Durant absorbed it into his new General Motors after Carter died suddenly. When Durant was fired, Cartercar's factory built Oakland (later Pontiac) engines.

Das von Byron Carter erfundene Cartercar besaß kein herkömmliches Getriebe, sondern einen Friktionsantrieb. Nach Carters plötzlichem Tod übernahm William C. Durant das Auto in seine neue Firma General Motors, und nachdem Durant entlassen worden war, baute die Cartercar-Fabrik Motoren für Oakland (später Pontiac).

Créée par Byron Carter, la Cartercar, dépourvue de transmission conventionnelle, disposait d'un entraînement par friction Après la mort soudaine de Carter, William Durant absorba la marque au sein de la General Motors tout juste créée. Quand il fut évincé, l'usine des Cartercar se mit à fabriquer des moteurs pour Oakland (dont sera issue Pontiac).

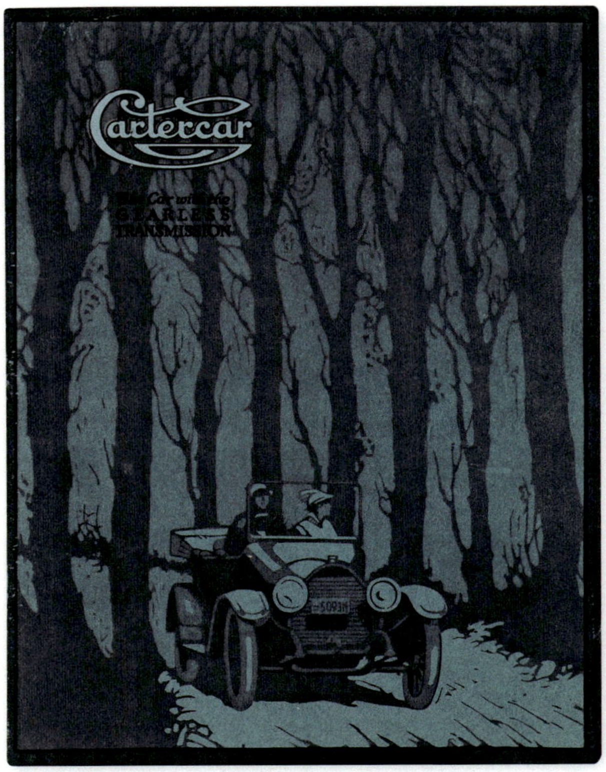

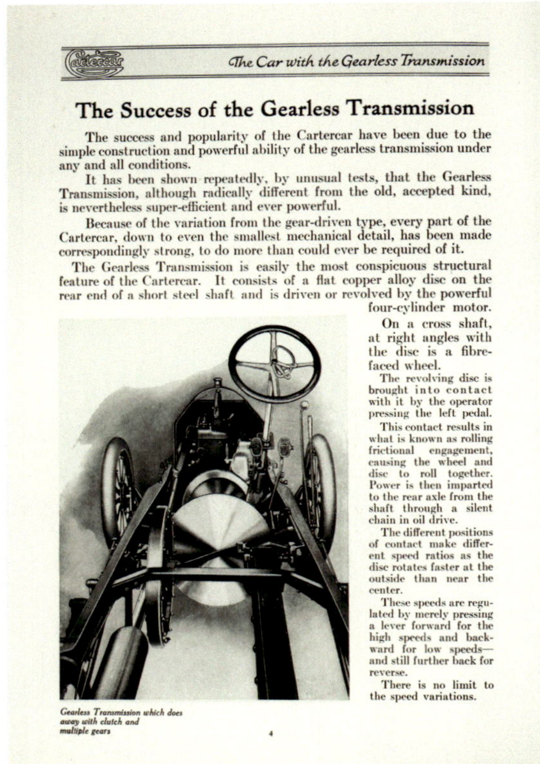

The Car with the Gearless Transmission

The Success of the Gearless Transmission

The success and popularity of the Cartercar have been due to the simple construction and powerful ability of the gearless transmission under any and all conditions.

It has been shown repeatedly, by unusual tests, that the Gearless Transmission, although radically different from the old, accepted kind, is nevertheless super-efficient and ever powerful.

Because of the variation from the gear-driven type, every part of the Cartercar, down to even the smallest mechanical detail, has been made correspondingly strong, to do more than could ever be required of it.

The Gearless Transmission is easily the most conspicuous structural feature of the Cartercar. It consists of a flat copper alloy disc on the rear end of a short steel shaft and is driven or revolved by the powerful four-cylinder motor.

On a cross shaft, at right angles with the disc is a fibre-faced wheel.

The revolving disc is brought into contact with it by the operator pressing the left pedal.

This contact results in what is known as rolling frictional engagement, causing the wheel and disc to roll together. Power is then imparted to the rear axle from the shaft through a silent chain in oil drive.

The different positions of contact make different speed ratios as the disc rotates faster at the outside than near the center.

These speeds are regulated by merely pressing a lever forward for the high speeds and backward for low speeds—and still further back for reverse.

There is no limit to the speed variations.

Gearless Transmission which does away with clutch and multiple gears

4

Chadwick

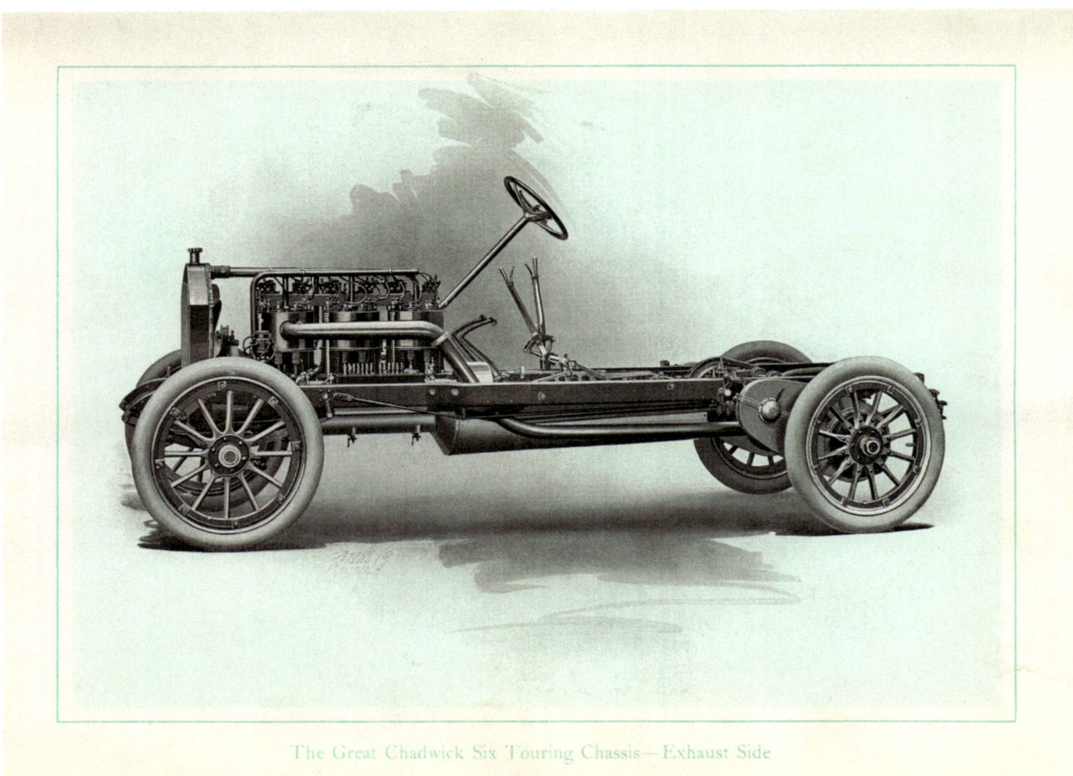

Chadwick 1910 *(left)*

This was America's first true ultra-high-performance car. Built in Pennsylvania, the premium Chadwick used a monstrous six-cylinder T-head engine with 706 cubic inches, and could easily reach 80 mph. Most were raced, and only two are known to still exist.

Dies war das erste echte Hochgeschwindigkeitsautomobil in Amerika. Der in Pennsylvania gebaute Chadwick besaß einen monströsen T-Kopf-Motor mit unglaublichen 11½ Litern Hubraum und erreichte damit locker eine Spitzengeschwindigkeit von 130 km/h. Die meisten dieser Fahrzeuge kamen in Rennen zum Einsatz, und soweit man weiß, existieren nur noch zwei Exemplare.

Voici la première vraie voiture à très hautes performances. Fabriquée en Pennsylvanie, la Chadwick haut de gamme recevait un monstrueux moteur six cylindres à soupapes latérales en T d'une cylindrée de 11 570 cm^3, qui atteignait facilement les 130 km/h. Il ne resterait que deux exemplaires de ces modèles dont la plupart ont été engagés en compétition.

Chalmers

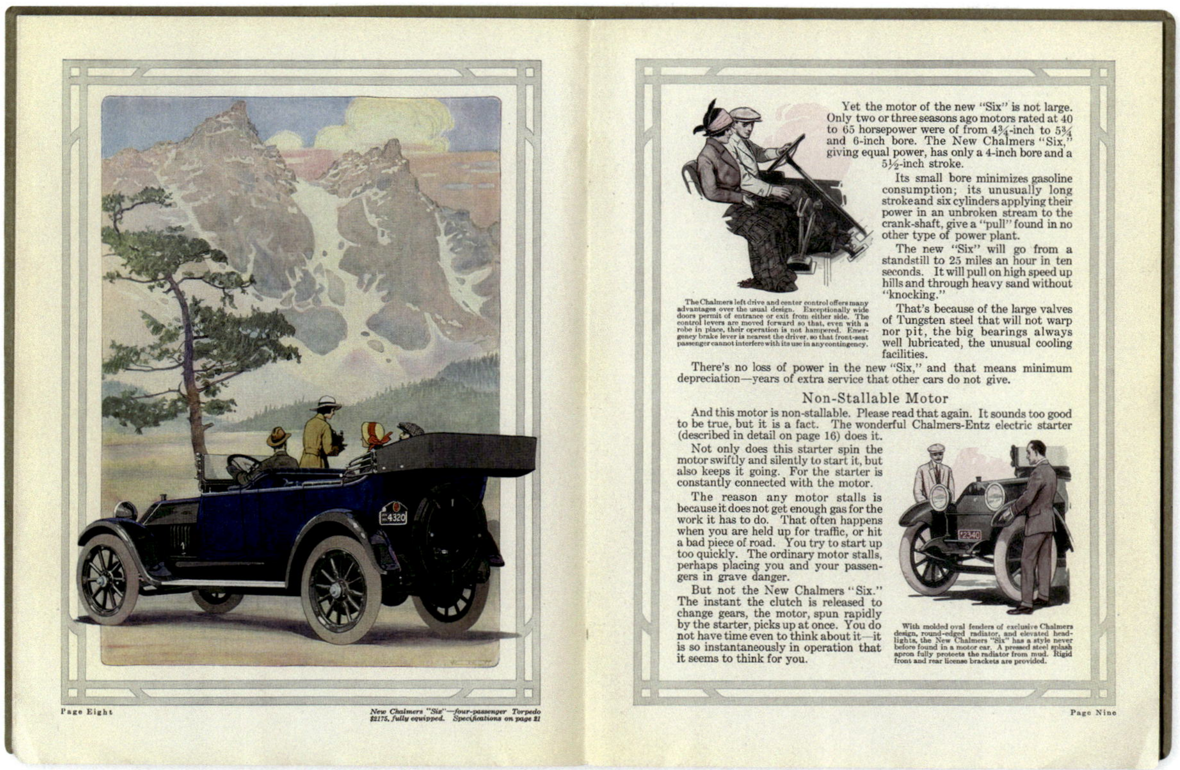

Chalmers 1914 *(left and below left)*, **1919** *(below right)*

The schematics illustrate a major Chalmers advantage: the curved, exhaust-heated intake manifold that fed fuel to its six-cylinder engine.

Die technische Zeichnung zeigt einen der großen Vorteile, die der Chalmers besaß: den abgasbeheizten Ansaugkrümmer, der dem Sechszylindermotor den Treibstoff zuführt.

Les schémas techniques ci-dessous décrivent l'un des atouts majeurs de la Chalmers : le collecteur d'admission incurvé et chauffé par l'échappement qui alimentait en essence son moteur six cylindres.

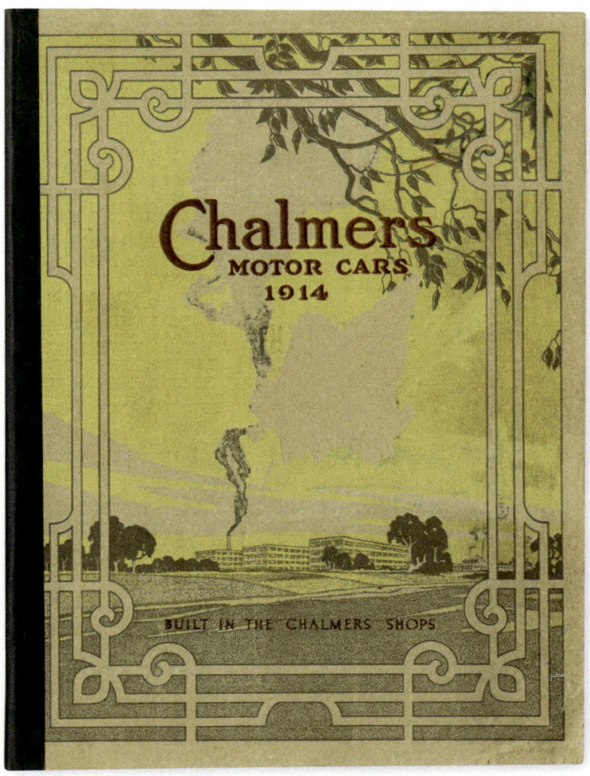

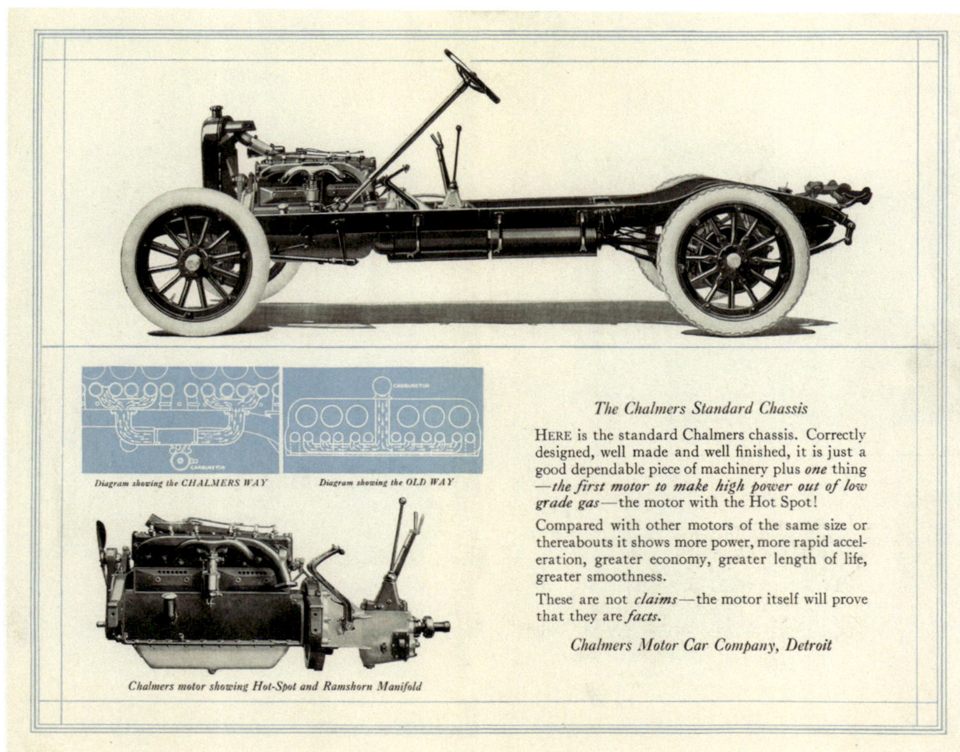

Chandler

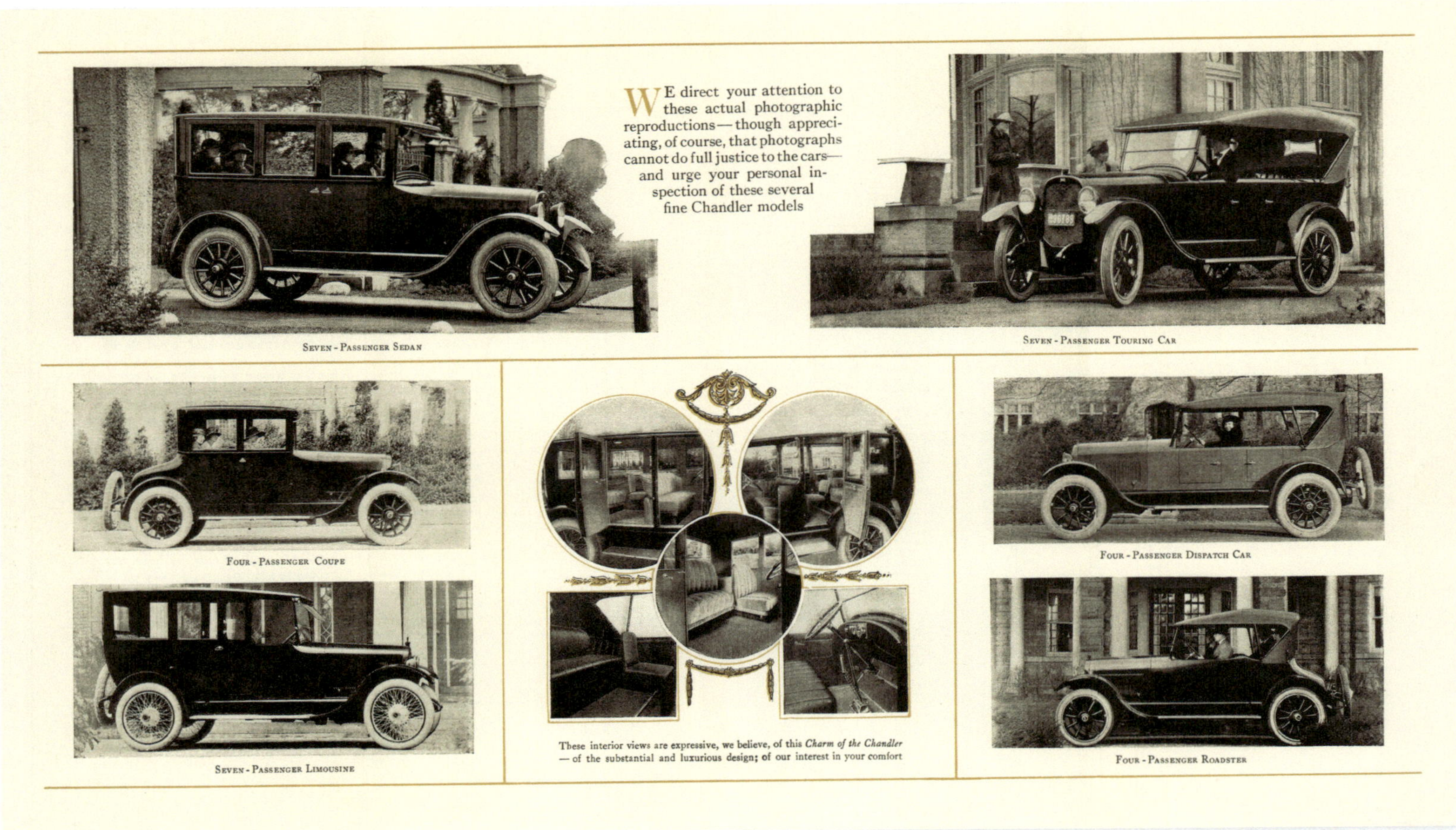

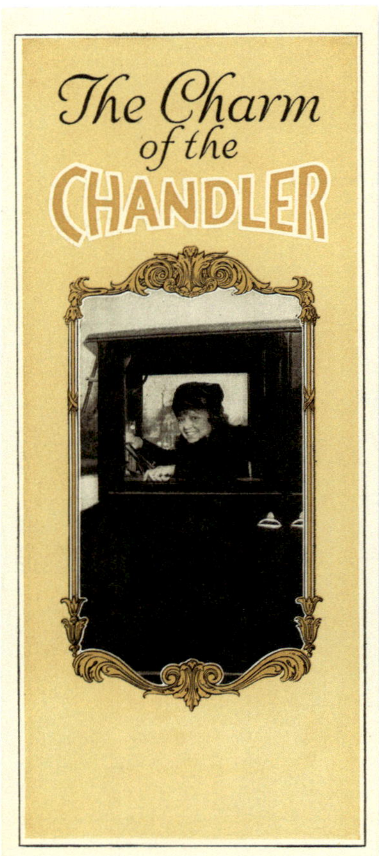

Chandler 1920

Playful accommodations are emphasized here. Though designed to mimic the Lozier, the Chandler was sophisticated in its own right. In the 1920s it had a fully synchronized transmission. The company ran into money trouble and was absorbed by Hupmobile in 1928.

Hier wird die verspielte Ausstattung betont. Wenngleich der Chandler nach dem Vorbild des Lozier entworfen wurde, besaß er doch einen ganz eigenen Charakter. In den 1920er-Jahren erhielt er ein voll synchronisiertes Getriebe. Als der Hersteller in finanzielle Nöte geriet, wurde er 1928 von Hupmobile übernommen.

Ici l'accent est mis sur les aménagements ludiques. Si elle a été conçue pour copier la Lozier, la Chandler n'en était pas moins dûment sophistiquée. Elle bénéficiait ainsi d'une transmission intégralement synchronisée. Mais l'entreprise rencontra des difficultés financières et se retrouva absorbée par Hupmobile en 1928.

Chevrolet

Chevrolet 1920 *(right)*

Chevrolet equipped more expensive cars and trucks with colors, capacity, and comfort. In 1927 Chevrolet was making a serious run at Ford's sale dominance, which it eventually outsold for decades on end. One of the great keys to Chevrolet's historic success was its adoption of a reliable, but economical, six-cylinder engine beginning in 1929, a real advance for a low-priced car.

Chevrolet stattete teurere Pkw und Lkw mit Farbe, Raum und Komfort aus. So wurde die Marke 1927 zu einem ernst zu nehmenden Herausforderer für Marktführer Ford, den Chevrolet schließlich überholte und jahrzehntelang in Schach hielt. Einer der Schlüssel zu diesem historischen Erfolg war der Einbau eines zuverlässigen und zugleich sparsamen Sechszylindermotors ab 1929 – ein wirklicher Fortschritt für ein preiswertes Auto.

Les camions et automobiles haut de gamme de Chevrolet ne cessaient de gagner en puissance, en confort et en couleurs. En 1927 la marque commençait même à battre sérieusement en brèche la position commerciale dominante de Ford et, *in fine*, dépassera son rival pour plusieurs décennies. L'une des clés de ce succès historique fut l'adoption à partir de 1929 d'un moteur six cylindres fiable et économique, une authentique avancée pour une voiture bon marché.

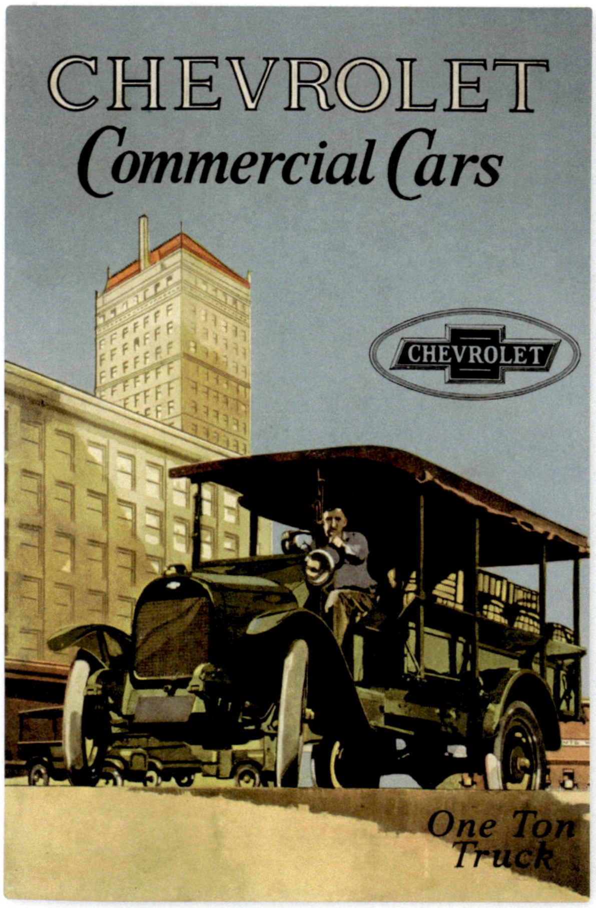

Chrysler

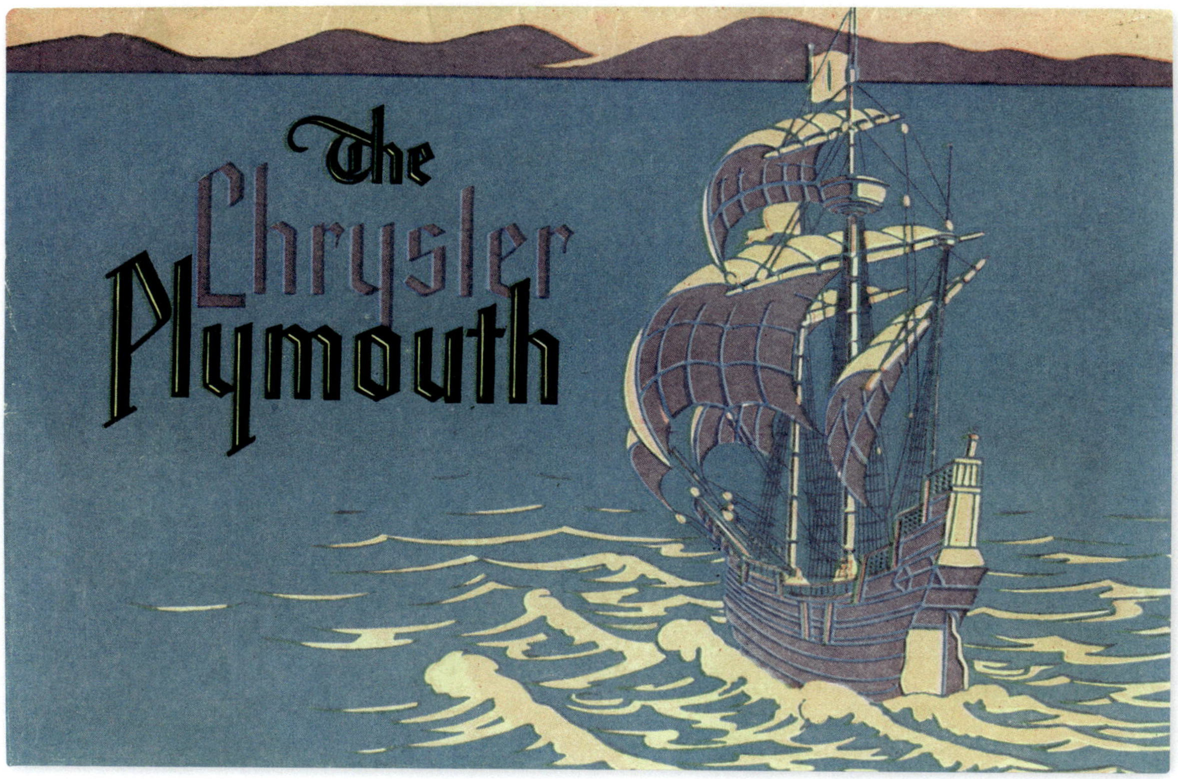

Chrysler Plymouth 1928 *(left)*

This brand-new entry into the ferociously competitive low-price field replaced the earlier four-cylinder Chrysler 52. The lack of a car illustration didn't hurt its sales. By 1929 the "Chrysler" appellation was dropped.

Dieser Neuling im hart umkämpften Niedrigpreissegment ersetzte das Vorgängermodell Chrysler 52 mit Vierzylindermotor. Dass er in der Werbung gar nicht abgebildet wurde, tat dem Verkaufserfolg keinen Abbruch. Von 1929 an verzichtete man auch auf den Zusatz „Chrysler".

Cette toute dernière venue sur le terrain férocement disputé de l'entrée de gamme remplaçait la Chrysler 52 quatre cylindres. L'absence d'automobile sur l'illustration ne contraria pas ses ventes. L'appellation « Chrysler » sera abandonnée en 1929.

Cleveland

Cleveland 1907

In a marketing ploy, the glowing distant image of the Eiffel Tower refers to a Cleveland car that was shipped to France for a 3,200 mile trouble-free trek across France and Spain. The Cleveland was founded in its namesake Ohio city in 1904 by E.J. Pennington, who had previously built the steam-powered Tractmobile. Its chassis was built by Garford, which later went on to successfully build trucks. The Cleveland's high price, $2,800, led to its demise in 1909.

Der strahlende Eiffelturm am Horizont verweist auf einen Cleveland, den man als Marketinggag nach Frankreich verschiffte und 5 000 Kilometer weit ohne Pannen durch Frankreich und Spanien fahren ließ. Die Automarke wurde 1904 von E.J. Pennington in jener Stadt in Ohio gegründet, deren Namen sie trug. Der Firmengründer hatte zuvor das dampfgetriebene Tractmobile gebaut. Das Fahrgestell stammte von der Firma Garford, die später erfolgreich Lkw baute. Der hohe Kaufpreis von 2 800 Dollar führte aber bereits 1909 zum frühzeitigen Aus für den Cleveland.

L'éclat rayonnant de la tour Eiffel renvoie à l'habile opération de marketing qui avait transporté la Cleveland outre-Atlantique, où elle avait effectué un périple de 5 000 kilomètres sans ennuis (entre la France et l'Espagne). La firme Cleveland avait été fondée en 1904 dans la ville de l'Ohio du même nom par E.J. Pennington, déjà créateur de la Tractmobile à vapeur. Son châssis était de fabrication Garford, qui se lancera plus tard avec succès dans la construction de poids lourds. Le prix élevé de la Cleveland, 2 800 dollars, causera sa perte en 1909.

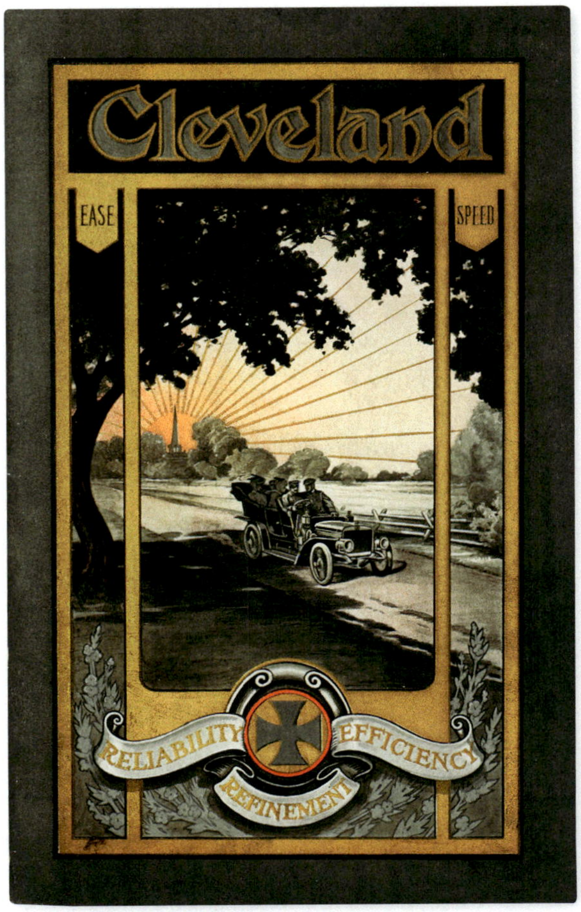

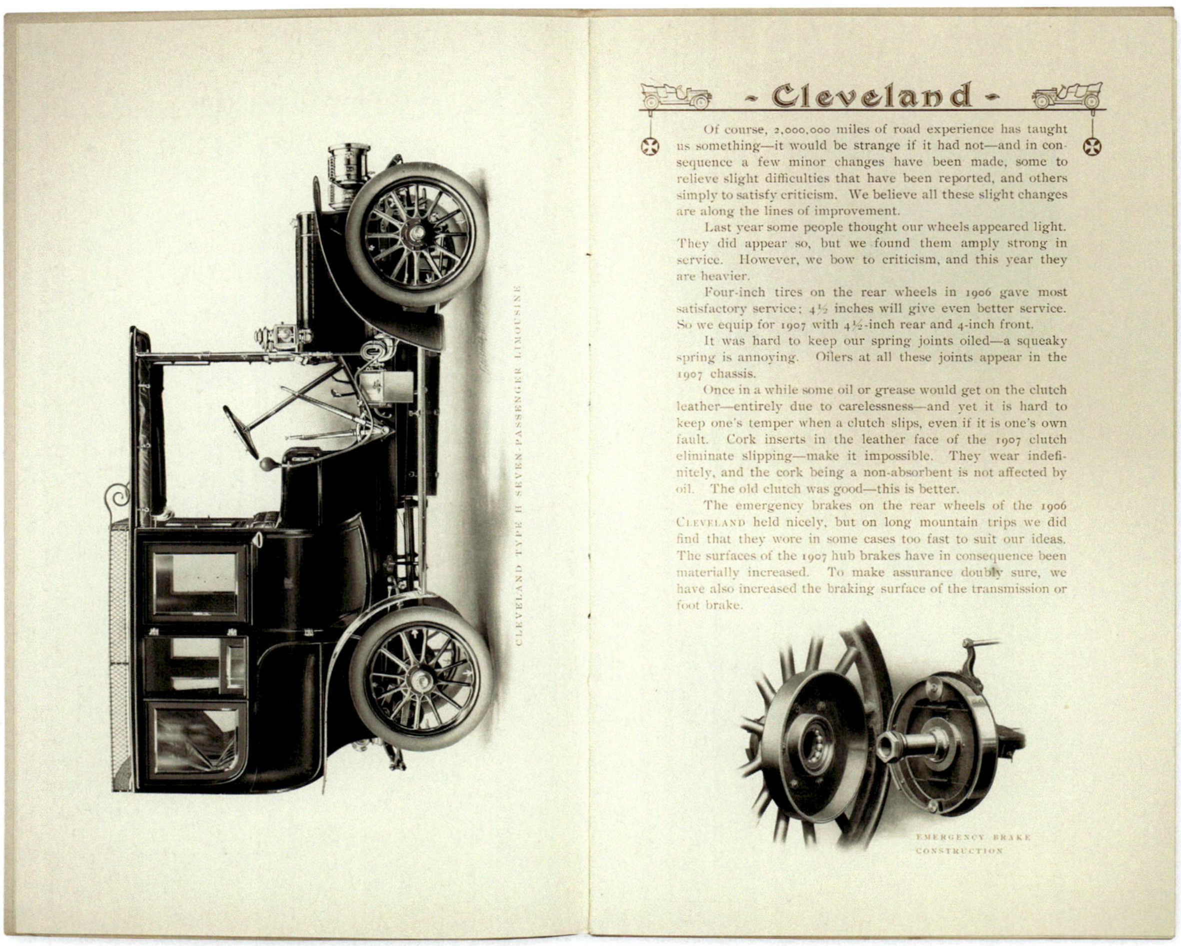

- Cleveland -

Of course, 2,000,000 miles of road experience has taught us something—it would be strange if it had not—and in consequence a few minor changes have been made, some to relieve slight difficulties that have been reported, and others simply to satisfy criticism. We believe all these slight changes are along the lines of improvement.

Last year some people thought our wheels appeared light. They did appear so, but we found them amply strong in service. However, we bow to criticism, and this year they are heavier.

Four-inch tires on the rear wheels in 1906 gave most satisfactory service; 4½ inches will give even better service. So we equip for 1907 with 4½-inch rear and 4-inch front.

It was hard to keep our spring joints oiled—a squeaky spring is annoying. Oilers at all these joints appear in the 1907 chassis.

Once in a while some oil or grease would get on the clutch leather—entirely due to carelessness—and yet it is hard to keep one's temper when a clutch slips, even if it is one's own fault. Cork inserts in the leather face of the 1907 clutch eliminate slipping—make it impossible. They wear indefinitely, and the cork being a non-absorbent is not affected by oil. The old clutch was good—this is better.

The emergency brakes on the rear wheels of the 1906 Cleveland held nicely, but on long mountain trips we did find that they wore in some cases too fast to suit our ideas. The surfaces of the 1907 hub brakes have in consequence been materially increased. To make assurance doubly sure, we have also increased the braking surface of the transmission or foot brake.

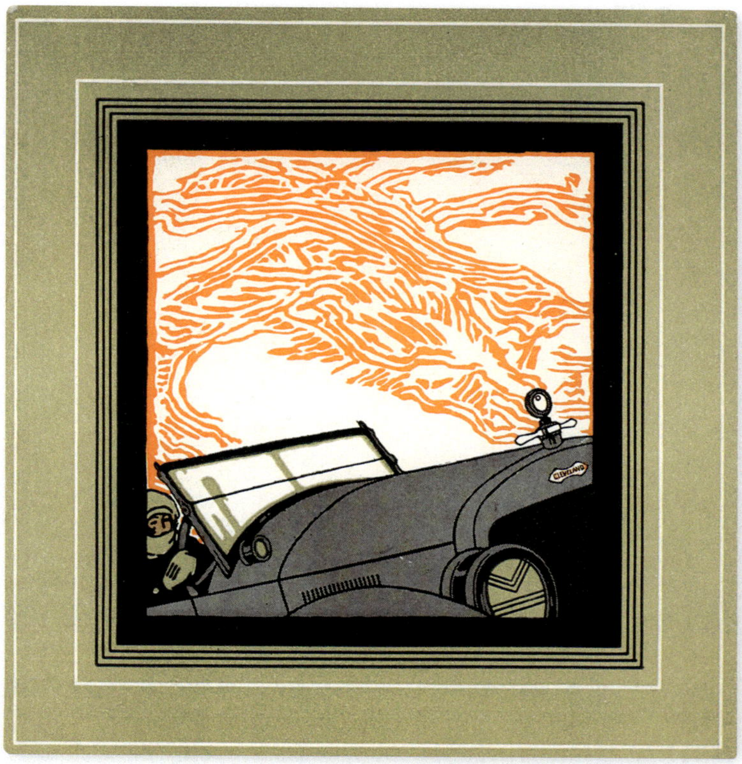

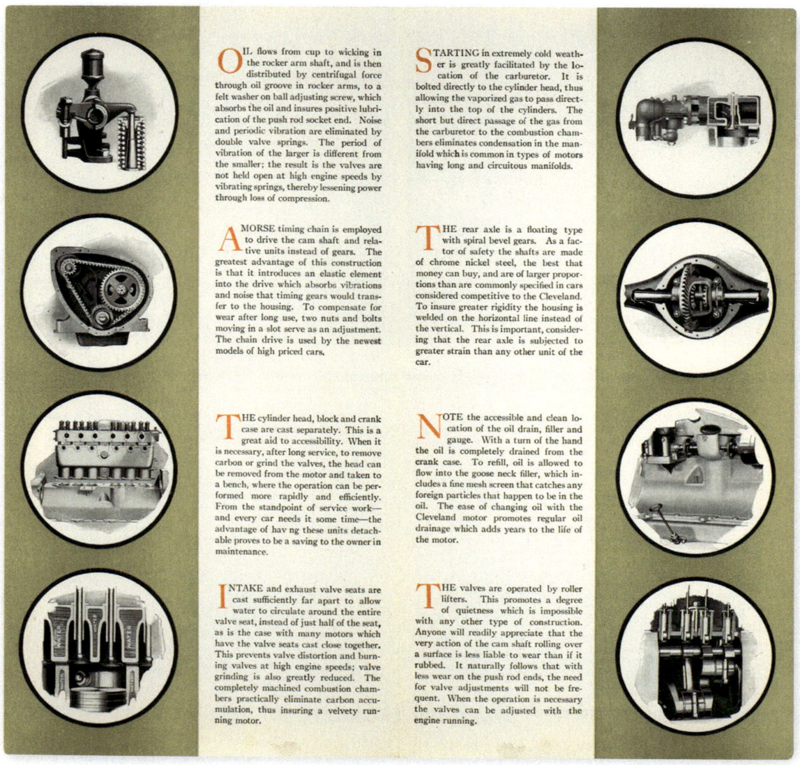

Cleveland Six 1924

Unrelated to the earlier Cleveland, the Cleveland Six was a smaller version of a Chandler. Both shared their straight-six engine. Before its demise in 1926, the Cleveland Six set many records with its reliable high-speed engine.

Der Cleveland Six, der nichts mit dem früheren Automobil gleichen Namens zu tun hatte, war eine kleinere Version des Chandler. Beide waren mit einem Sechszylinderreihenmotor ausgestattet. Vor seinem Ende im Jahre 1926 stellte der Cleveland Six mit seinem zuverlässigen Hochgeschwindigkeitsmotor zahlreiche Rekorde auf.

Dépourvue de liens de parenté avec la précédente Cleveland, la Cleveland Six était une version plus petite de la Chandler, dont elle partageait le moteur six cylindres en ligne. Avant sa disparition en 1926, la Cleveland Six a eu le temps de battre de nombreux records grâce à la vitesse et à la fiabilité de sa motorisation.

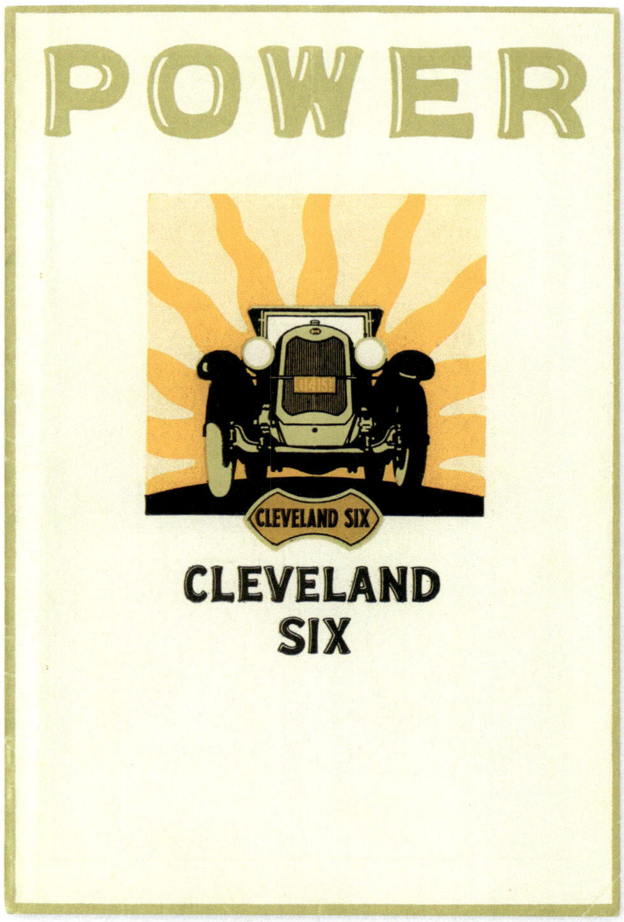

Cole

Cole Eight 1919 *(right)*

Based in Indianapolis, the Cole was an "assembled" car, meaning that the company purchased premium components from outside suppliers and assembled them into a car.

Der Cole aus Indianapolis (Indiana) war ein „Baukastenauto“: Der Hersteller kaufte hochwertige Bauteile von externen Zulieferern ein und baute sie zu einem Auto zusammen.

Basée à Indianapolis, la Cole était une automobile « assemblée », c'est-à-dire que l'entreprise avait acheté à des fournisseurs extérieurs ses principales composantes afin de les assembler et d'en faire une voiture.

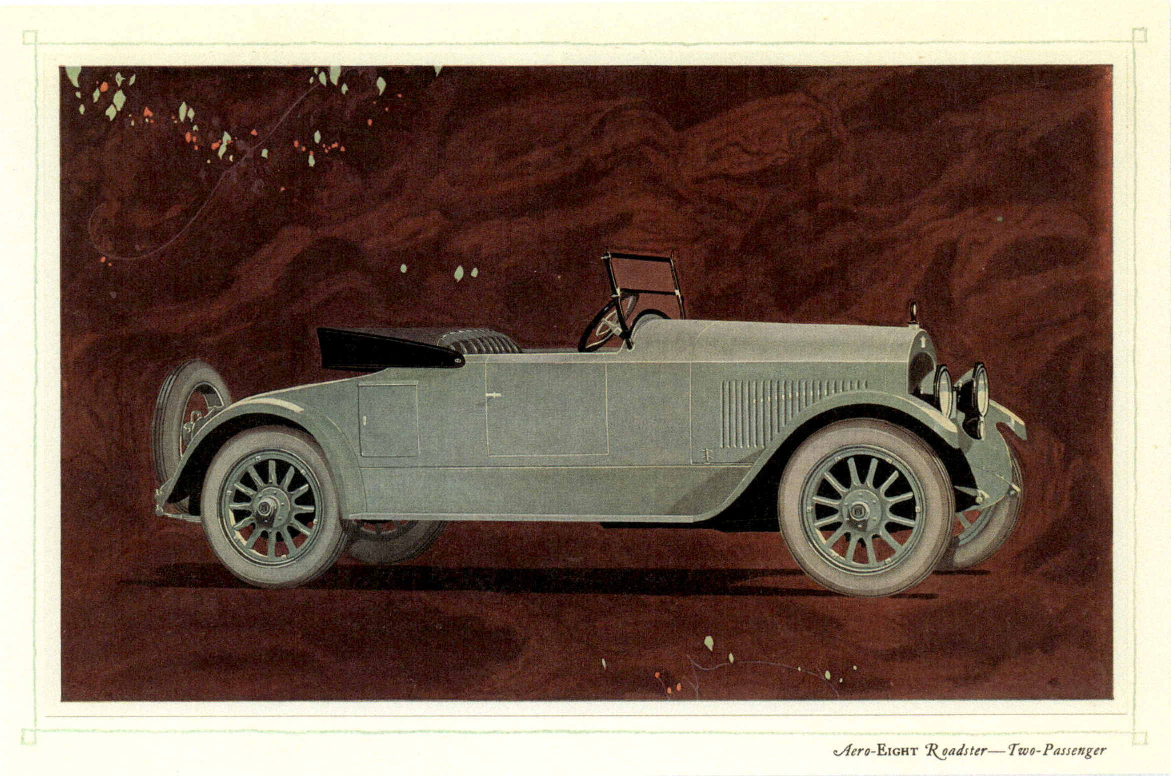

Columbia

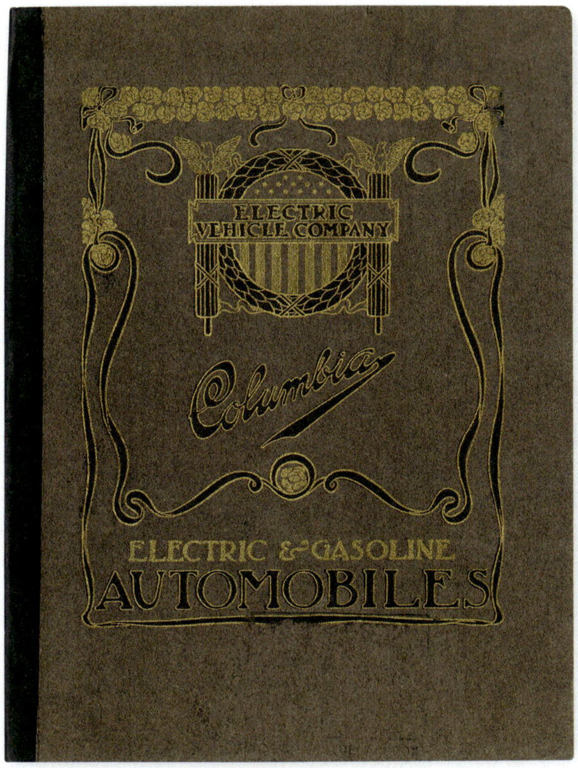

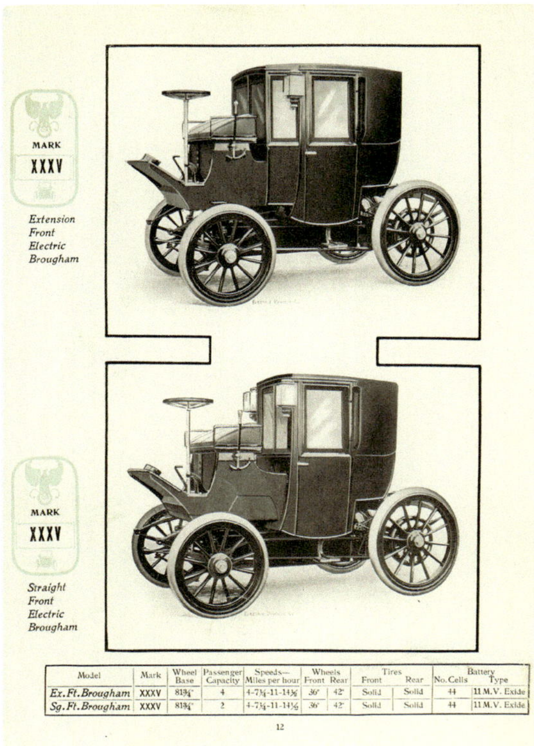

Model	Mark	Wheel Base	Passenger Capacity	Speeds—Miles per hour	Wheels Front	Wheels Rear	Tires Front	Tires Rear	Battery No. Cells	Battery Type
Ex.Ft.Brougham	XXXV	81¾"	4	4-7¼-11-14½	36"	42"	Solid	Solid	44	11 M.V. Exide
Sg.Ft.Brougham	XXXV	81¾"	2	4-7¼-11-14½	36"	42"	Solid	Solid	44	11 M.V. Exide

Columbia Electric 1904 *(left)*,
Columbia Six 1922 *(opposite)*

Colonel Albert Pope was the largest builder of bicycles in the United States when he decided to get into powered vehicles in 1897. The electric side of the firm collapsed two years later, likely a consequence of trying to market 37 different models.

Oberst Albert Pope war der größte Fahrradhersteller in den Vereinigten Staaten, als er sich 1897 entschloss, in das Kraftfahrzeuggeschäft einzusteigen. Der Teil des Unternehmens, der Elektroautos herstellte, ging bereits nach zwei Jahren wieder unter, was möglicherweise auch darauf zurückzuführen war, dass Pope versuchte, 37 verschiedene Modelle zu verkaufen.

Le colonel Albert Pope, alors le plus grand fabricant de bicyclettes des États-Unis, décida en 1897 de s'engager dans la construction de véhicules à moteur. La branche électrique de la firme s'écroula deux ans plus tard, probablement à la suite du projet consistant à lancer trente-sept modèles différents.

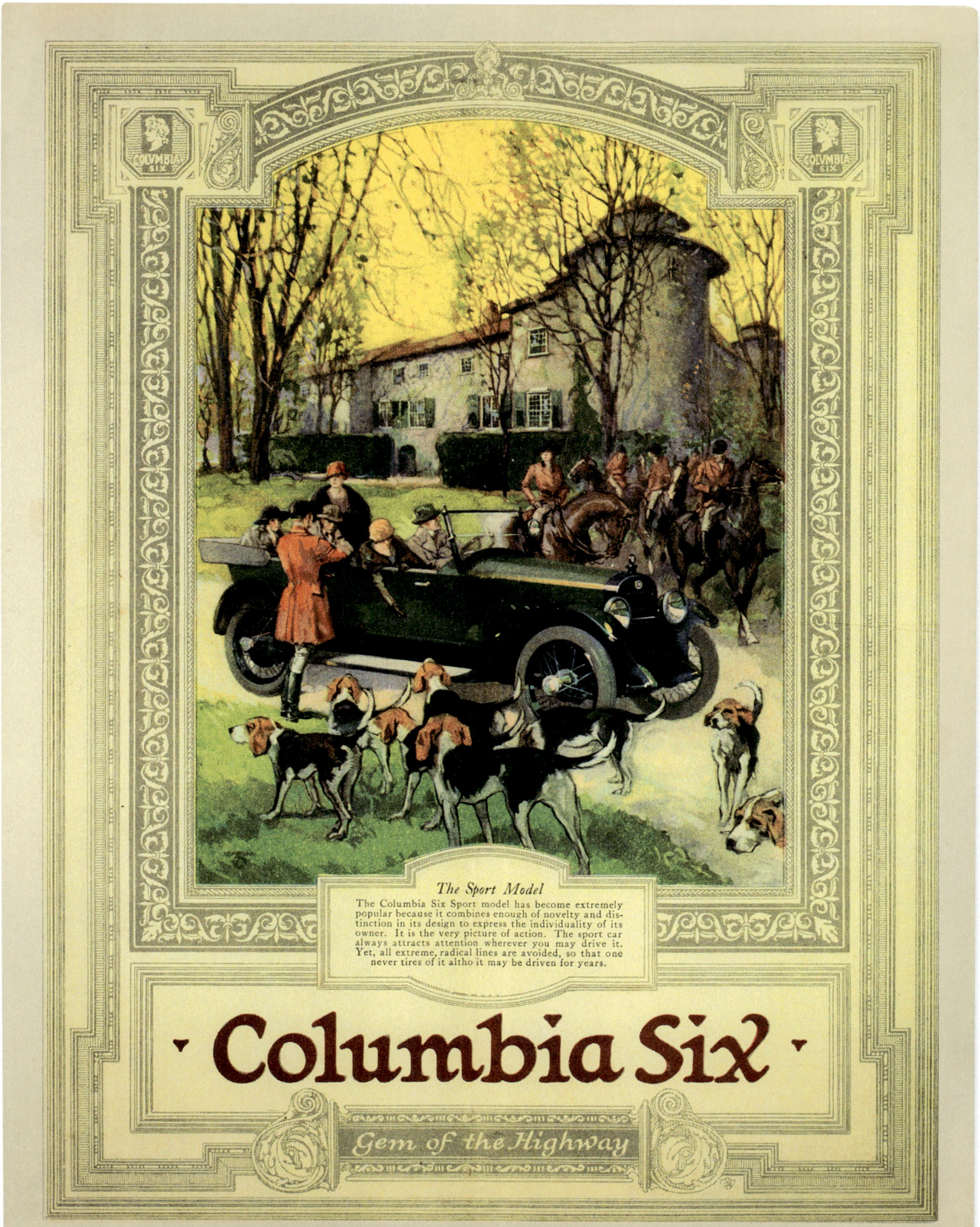
COLUMBIA SIX
COLUMBIA SIX
The Sport Model
The Columbia Six Sport model has become extremely popular because it combines enough of novelty and distinction in its design to express the individuality of its owner. It is the very picture of action. The sport car always attracts attention wherever you may drive it. Yet, all extreme, radical lines are avoided, so that one never tires of it altho it may be driven for years.
Columbia Six
Gem of the Highway

Detroiter

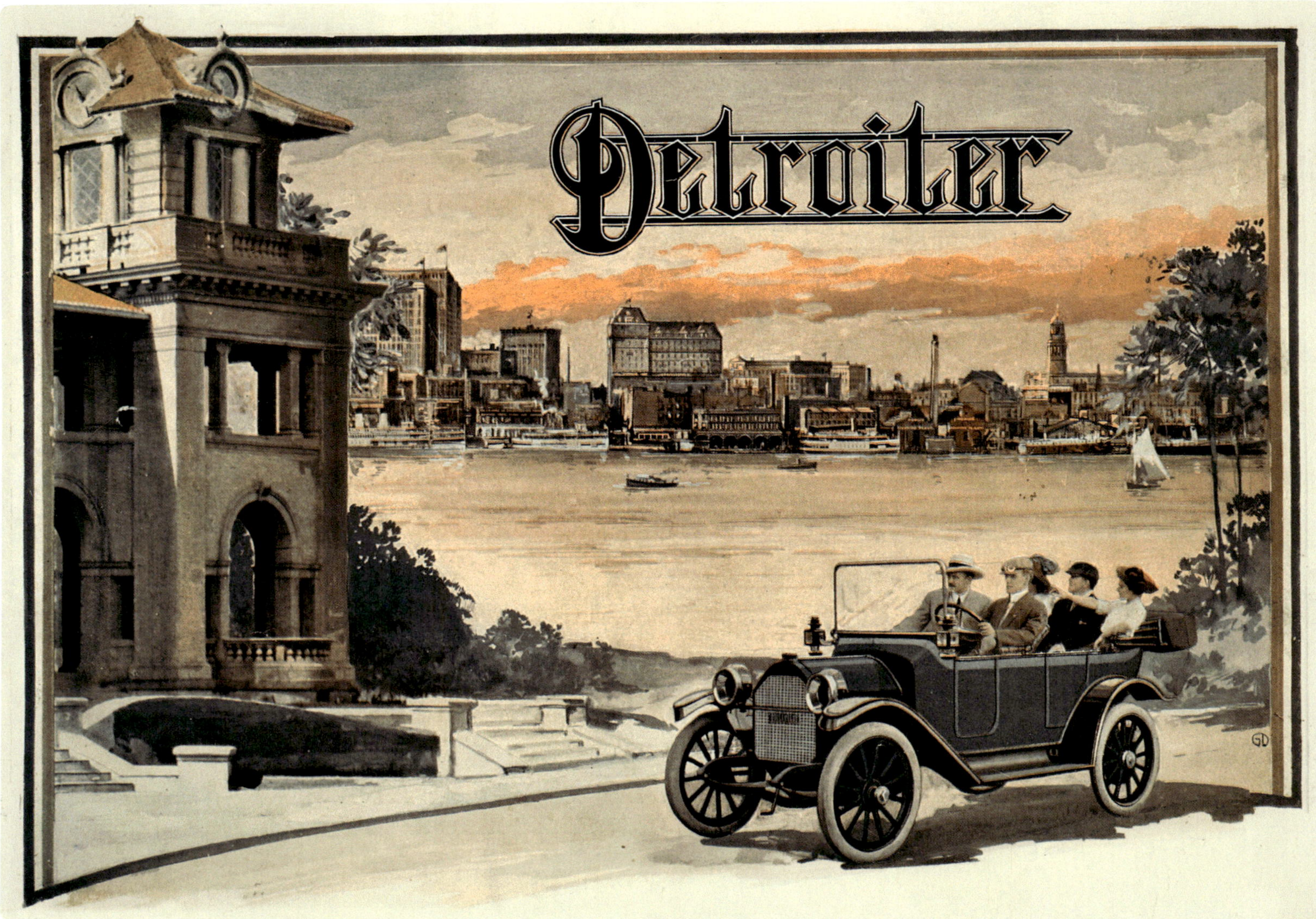

Detroiter A-3 1914 *(above)*, **1913** *(right)*

Many early cars incorporated "Detroit" into their names. Like the Cole and Columbia Six, the Detroiter was an "assembled" car, a cost-saving strategy that bought several years for many of the smaller makes to survive without incurring huge development costs.

Viele frühe amerikanische Automobile bauten den Namen „Detroit" in ihre Markenbezeichnung ein. Wie der Cole und der Columbia Six war auch der Detroiter ein „Baukastenauto" – eine kostensparende Strategie, die vielen kleineren Autobauern ermöglichte, mehrere Jahre ohne viel Entwicklungsaufwand zu überleben.

En ces temps pionniers, nombre d'automobiles incorporaient à leur nom le mot « Detroit ». Comme la Cole et la Columbia Six, la Detroiter était une voiture « assemblée », stratégie économique qui permit à beaucoup de petites marques de survivre plusieurs années sans avoir à assumer d'énormes coûts de développement.

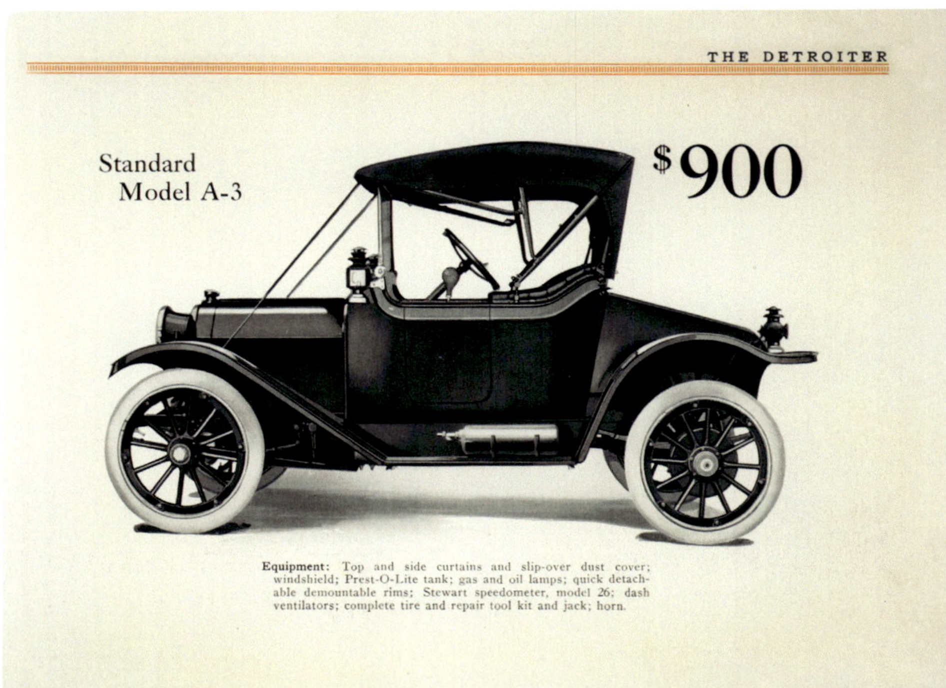

Dodge Brothers

Dodge Brothers 1915 *(right)*

Though not as common as Ford's Model T, the early Dodge Brothers cars were still profoundly simple, affordable, and enjoyed very strong buyer appeal.

Wenngleich sie nicht so verbreitet waren wie das Model T von Ford, so waren die frühen Fahrzeuge der Brüder Dodge doch auch ungemein schlicht, erschwinglich und bei den Kunden äußerst beliebt.

Même si on en croisait moins souvent que des Ford T, les premières Dodge Brothers n'en étaient pas moins profondément simples et abordables, et jouissaient d'un fort pouvoir de séduction commercial.

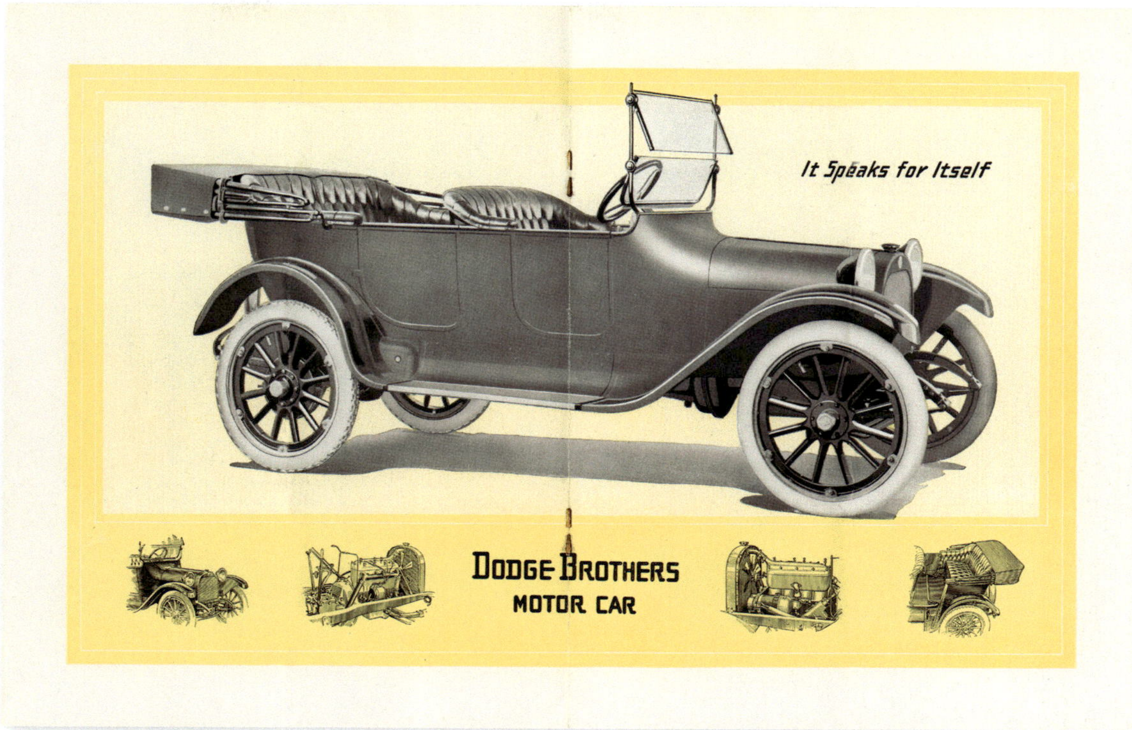

Erskine

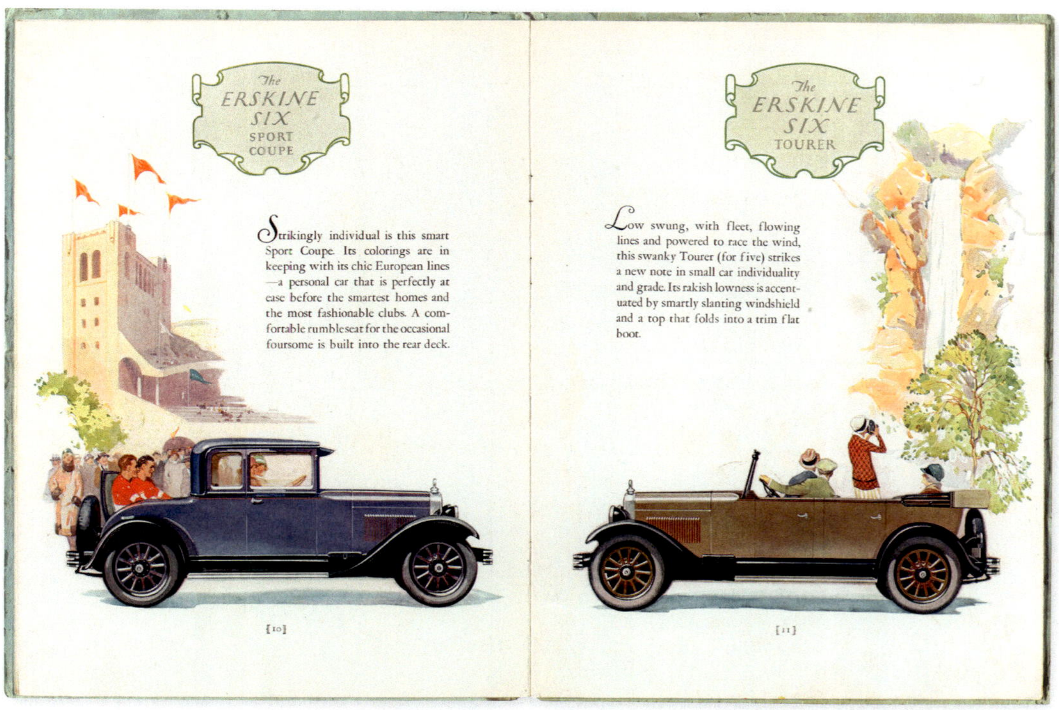

Erskine Six 1927 *(left)*

The Erskine was built by Studebaker between 1927 and 1930. They had a European flair as their small sixes were measured in liters.

Der Erskine wurde zwischen 1927 und 1930 von Studebaker gebaut. Dass man den Hubraum in Litern maß, sollte diesen Modellen ein europäisches Flair verleihen.

La Erskine a été fabriquée par Studebaker entre 1927 et 1930. La touche européenne était notamment soulignée par la mesure en centimètres cubes de la capacité de son petit moteur six cylindres.

Essex

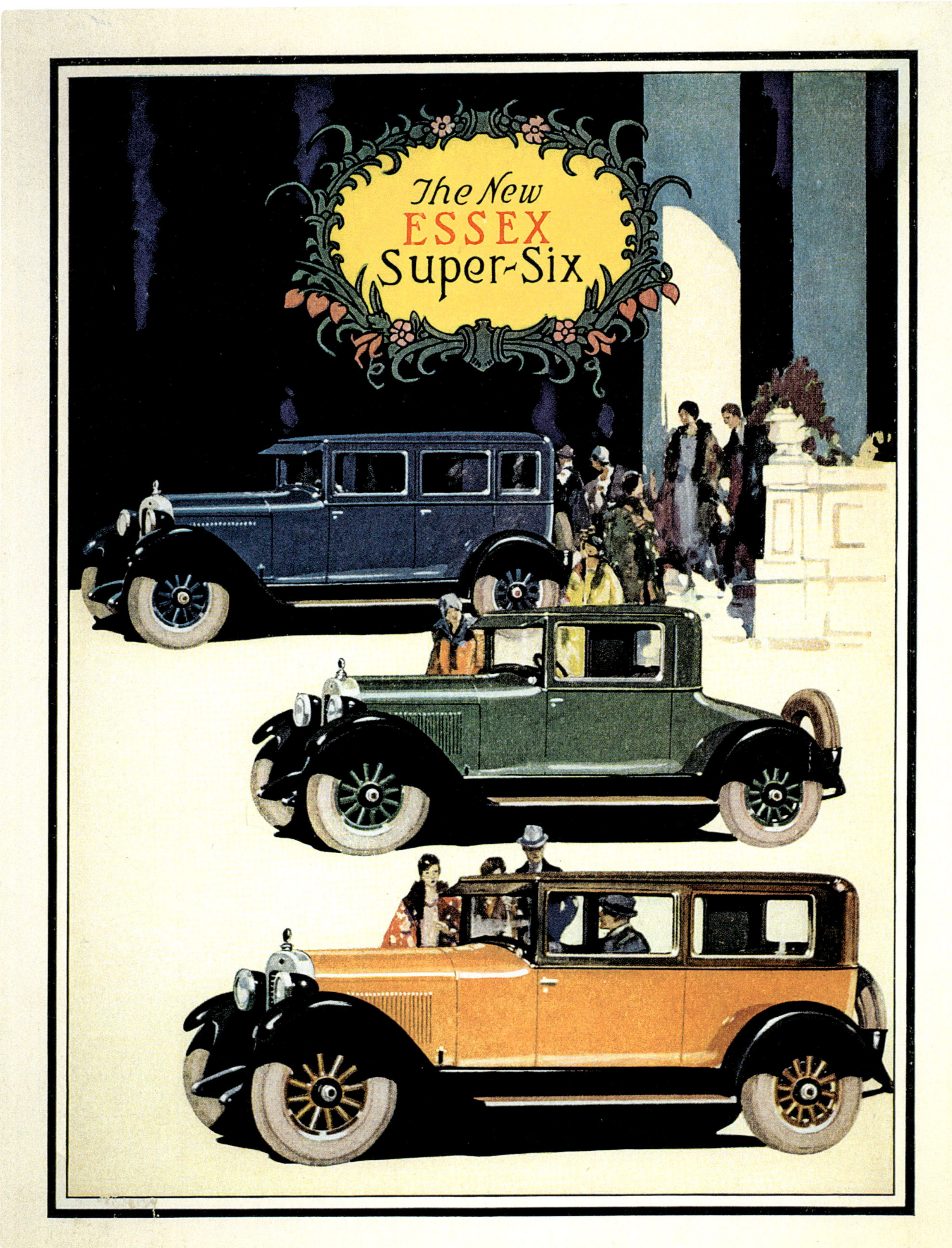

Essex Super Six 1927

The bright colors hint at the Essex's role as a more affordable version of the tiny Hudson. In 1934 the Hudson dropped the Essex name and called its lower-end car the Terraplane.

Die grellen Farben lassen erkennen, dass der Essex eine etwas preisgünstigere Version des winzigen Hudson sein sollte. Im Jahre 1934 gab Hudson die Bezeichnung „Essex" auf und nannte das Einstiegsmodell schlicht „Terraplane".

Ses couleurs vives présentent l'Essex comme une version plus abordable de la petite Hudson. En 1934 Hudson abandonna le nom « Essex » et baptisa son modèle d'entrée de gamme « Terraplane ».

Ford

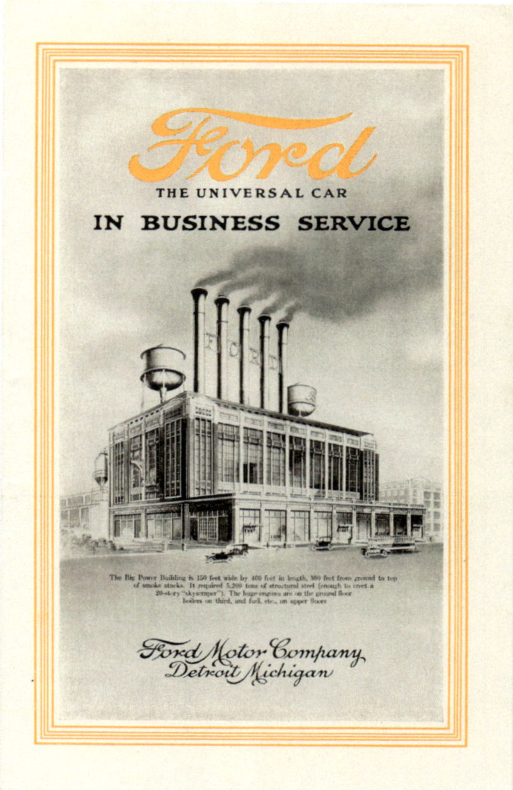

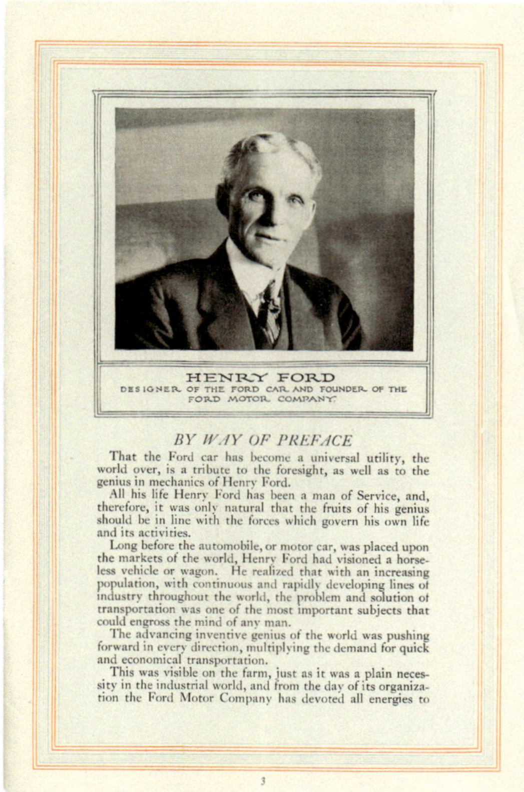

HENRY FORD
DESIGNER OF THE FORD CAR AND FOUNDER OF THE FORD MOTOR COMPANY

BY WAY OF PREFACE

That the Ford car has become a universal utility, the world over, is a tribute to the foresight, as well as to the genius in mechanics of Henry Ford.

All his life Henry Ford has been a man of Service, and, therefore, it was only natural that the fruits of his genius should be in line with the forces which govern his own life and its activities.

Long before the automobile, or motor car, was placed upon the markets of the world, Henry Ford had visioned a horseless vehicle or wagon. He realized that with an increasing population, with continuous and rapidly developing lines of industry throughout the world, the problem and solution of transportation was one of the most important subjects that could engross the mind of any man.

The advancing inventive genius of the world was pushing forward in every direction, multiplying the demand for quick and economical transportation.

This was visible on the farm, just as it was a plain necessity in the industrial world, and from the day of its organization the Ford Motor Company has devoted all energies to

3

Ford 1920 *(left)*, **1915** *(below)*

Near the end of its life, the Model T could be ordered with a self-starter to replace the wrist-snapping hand crank. Henry Ford did not invent the assembly line—for that, credit goes to Ransom Eli Olds—but Ford took the process to unheard-of heights at his Highland Park and River Rouge assembly plants. Indisputably, Ford was able to put the world on wheels due to his fast production and incessant cost-cutting.

Gegen Ende seines Produktionszeitraums konnte das Model T mit einem Anlasser bestellt werden, der die leidige Handkurbel überflüssig machte. Henry Ford war nicht, wie viele glauben, der Erfinder des Fließbands (diese Ehre gebührt Ransom Eli Olds), doch Ford führte dieses Verfahren in seinen Fertigungswerken in Highland Park und River Rouge großflächig ein. Es ist nicht abzustreiten, dass Ford die Welt durch seine rasche Produktion und seine unentwegten Sparmaßnahmen mobilisierte.

Vers la fin de sa vie, la Ford Model T pouvait être livrée avec un démarreur qui remplaçait la manivelle briseuse de poignet. Si ce n'est pas Henry Ford, mais Ransom Eli Olds, qui inventa la chaîne d'assemblage, il porta néanmoins le système à des niveaux jusqu'alors inédits dans ses usines de Highland Park et River Rouge. Grâce aux cadences de sa production et à d'incessantes compressions des coûts, Ford a mis le monde entier au volant.

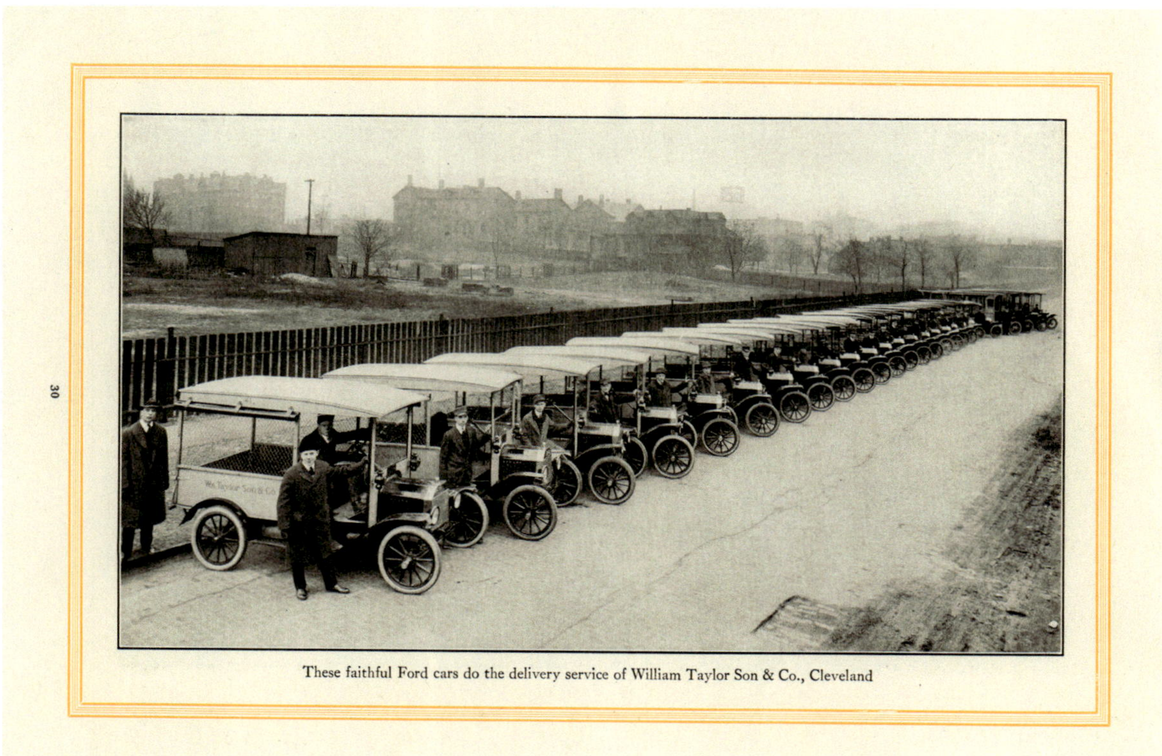

These faithful Ford cars do the delivery service of William Taylor Son & Co., Cleveland

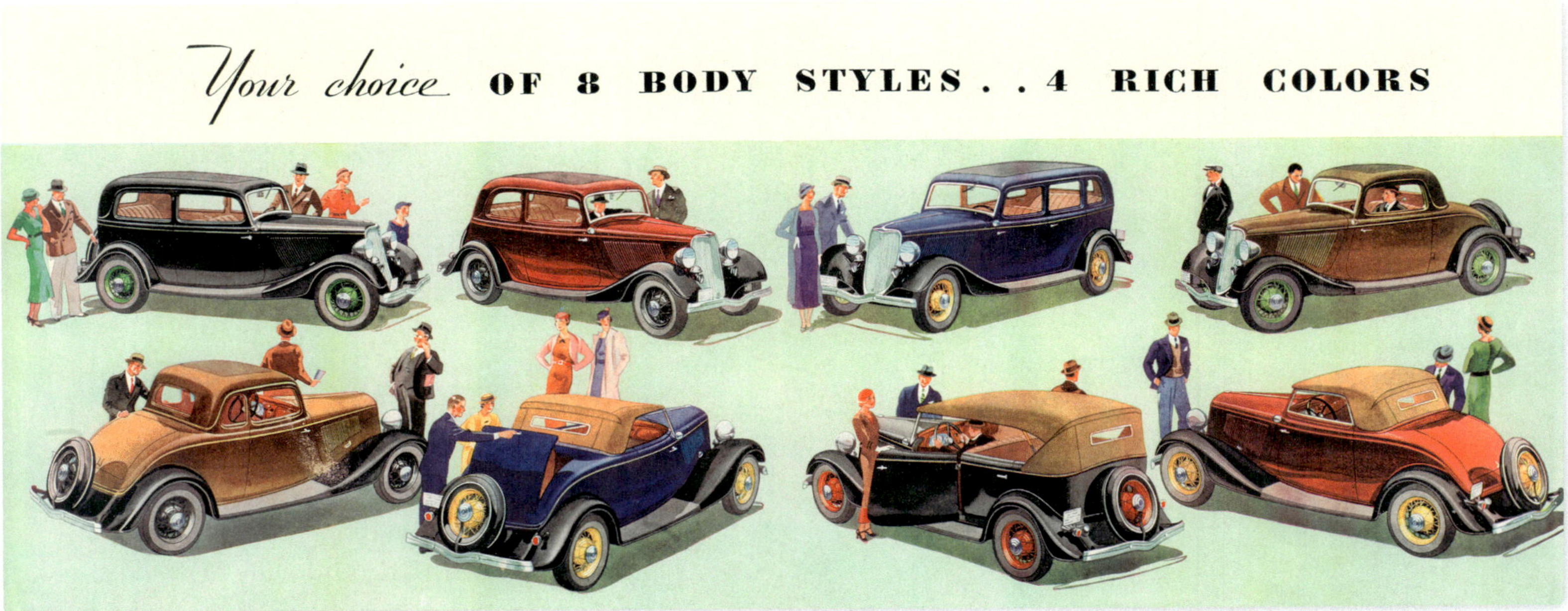

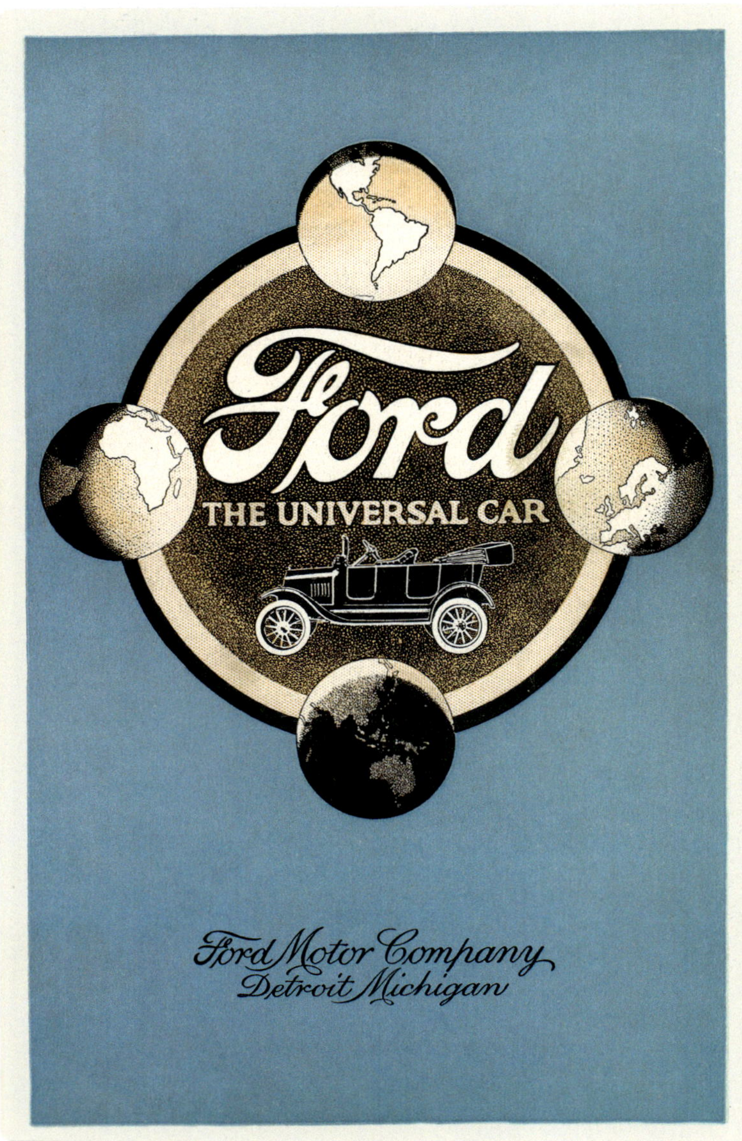

Ford 1934 *(above)*, **1920** *(left)*, **1927** *(opposite)*

By 1920 Henry Ford was already a preeminent industrialist in the world. His philosophy and worldview were in synch with the simplicity of his cars, which completely changed the way Americans lived and turned Ford into an immortal in the annals of global industry. American society was transformed by the low-priced automobile, thanks to Ford's visionary decision to make his cars affordable, simple, and reliable. After the final Model Ts were built in 1927, Ford shut down his plants to ready the Model A, a comparatively advanced car that likewise sold in the millions.

Schon 1920 galt Henry Ford weltweit als herausragender Unternehmer. Seine Philosophie und seine Weltanschauung passten zur Schlichtheit seiner Automobile, die die Lebensweise der Amerikaner radikal veränderten und Ford seinen Platz in den Annalen der Weltwirtschaft sicherten. Die amerikanische Gesellschaft wurde durch das erschwingliche Automobil völlig umgekrempelt – dank Fords weitsichtiger Entscheidung, seine Fahrzeuge preisgünstig, unkompliziert und zuverlässig zu halten. Nachdem 1927 die letzten Model Ts vom Band gelaufen waren, legte Ford seine Werke für ein paar Monate still, um die Produktion des Model A vorzubereiten, eines vergleichsweise fortschrittlichen Autos, das ebenfalls millionenfach verkauft wurde.

En 1920 Henry Ford était déjà un industriel d'importance mondiale. Sa philosophie et sa vision du monde correspondaient parfaitement à la simplicité de ses voitures, qui ont complètement modifié le mode de vie des Américains et assuré à Ford l'immortalité dans l'histoire industrielle mondiale : c'est en effet grâce à sa décision visionnaire de produire des modèles économiques, simples et fiables que la société américaine a été transformée par ces automobiles à petit prix. Quand les dernières Ford T auront été fabriqués en 1927, la marque fermera 23 usines pendant les sept mois de gestation de la Ford A, un véhicule comparativement plus évolué qui se vendra également en millions d'exemplaires.

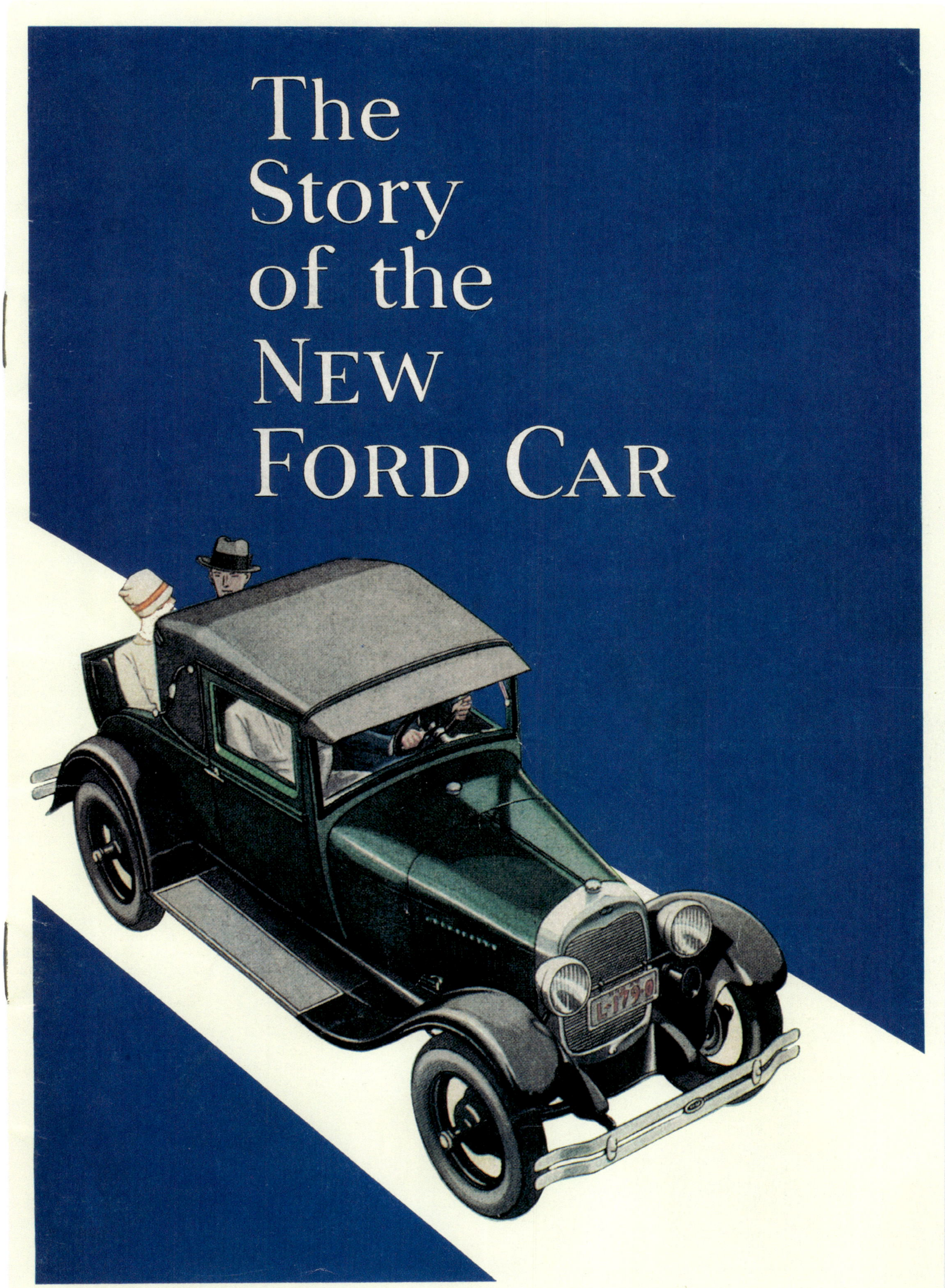
The
Story
of the
NEW
FORD CAR

Franklin

Franklin 1906 *(right)*, **1909** *(below)*

A technologically savvy person could treat this catalog like an engineering textbook. The arabesque pattern of the cover suggests wealth and exoticism. The detailed drawings depict Franklin's signature air-cooled engine, with four or six finned cylinders atop a separate crankcase. The Franklin wasn't the first American car with an air-cooled engine, which eliminated the need for a conventional radiator and fan. But the engine's designer, John Wilkinson, took it to Franklin after an earlier automotive partner in New York refused to compensate him for it. Franklin, based in Syracuse, New York, is still considered one of the most innovative and desirable of early American cars.

Für jemanden mit technischem Verständnis war dieser Katalog wie ein Lehrbuch des Automobilbaus. Die arabesken Muster auf dem Umschlag deuten Reichtum und Exotik an. Die detaillierten Zeichnungen zeigen den für den Franklin typischen luftgekühlten Motor mit vier oder sechs verrippten Zylindern über einem separaten Kurbelwellengehäuse. Der Franklin war nicht das erste amerikanische Auto mit einem luftgekühlten Motor, der den herkömmlichen Kühler und Kühlerventilator überflüssig machte, aber John Wilkinson, der Entwickler dieses Motors, ging damit zu Franklin, nachdem sich ein früherer Partner geweigert hatte, ihn für seine Erfindung zu entlohnen. Der Franklin, hergestellt in Syracuse im Bundesstaat New York, gilt als eines der innovativsten und begehrenswertesten frühen amerikanischen Automobile.

Les arabesques de la couverture de ce catalogue, qui a tout du manuel d'ingénierie, suggèrent toutefois l'opulence et l'exotisme. Ses dessins détaillés décrivent le moteur refroidi par air qui caractérise la Franklin, avec ses quatre ou six cylindres à ailettes au-dessus d'un carter séparé. Née à Syracuse, dans l'État de New York, la Franklin ne fut pas la première voiture américaine à moteur ainsi refroidi, qui permettait de se passer de radiateur et de ventilateur conventionnels. Mais si le concepteur du moteur, John Wilkinson, le proposa à Franklin, c'est que son partenaire initial, un constructeur new-yorkais, lui refusait toute indemnité. La Franklin reste considérée comme l'une des plus novatrices et des plus désirables des premières automobiles américaines.

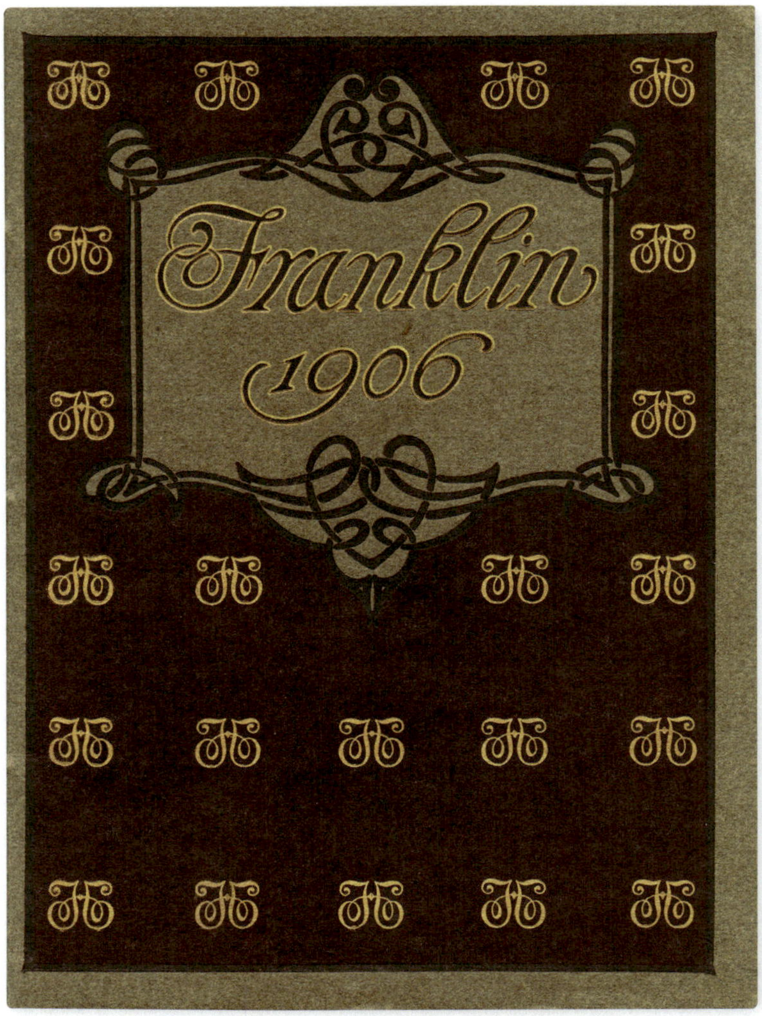

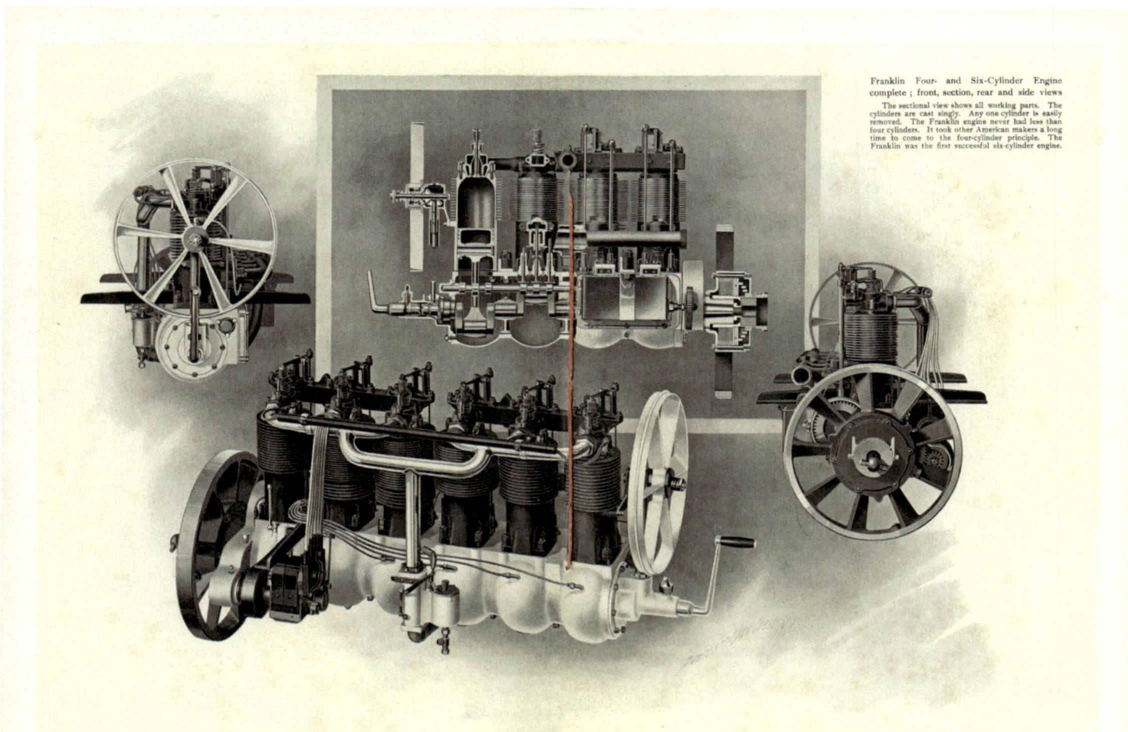

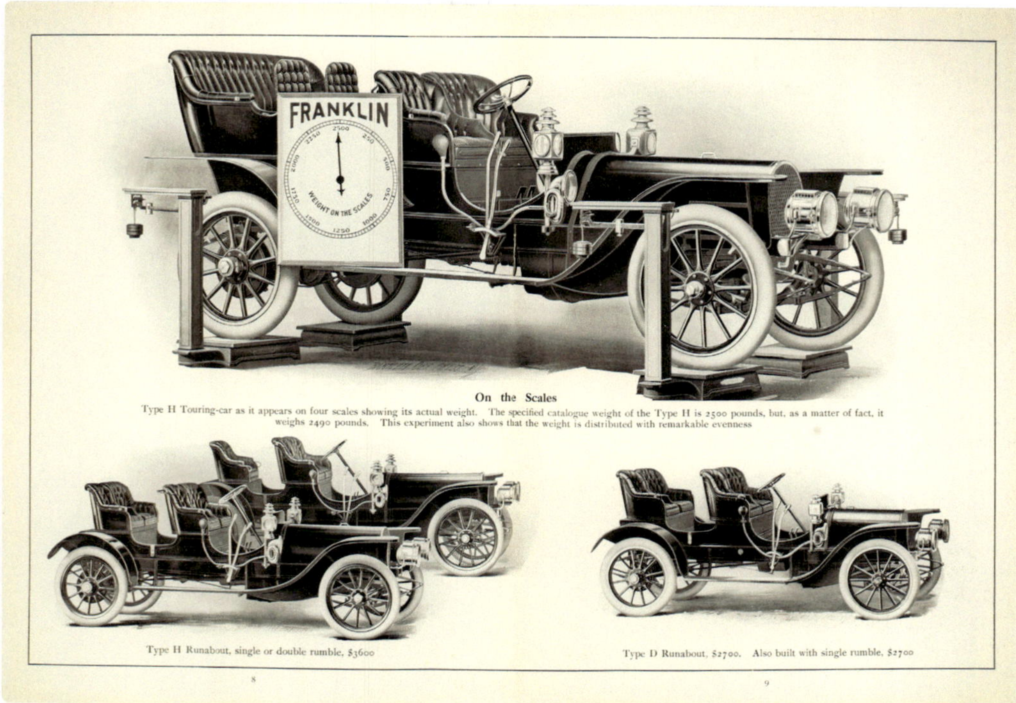

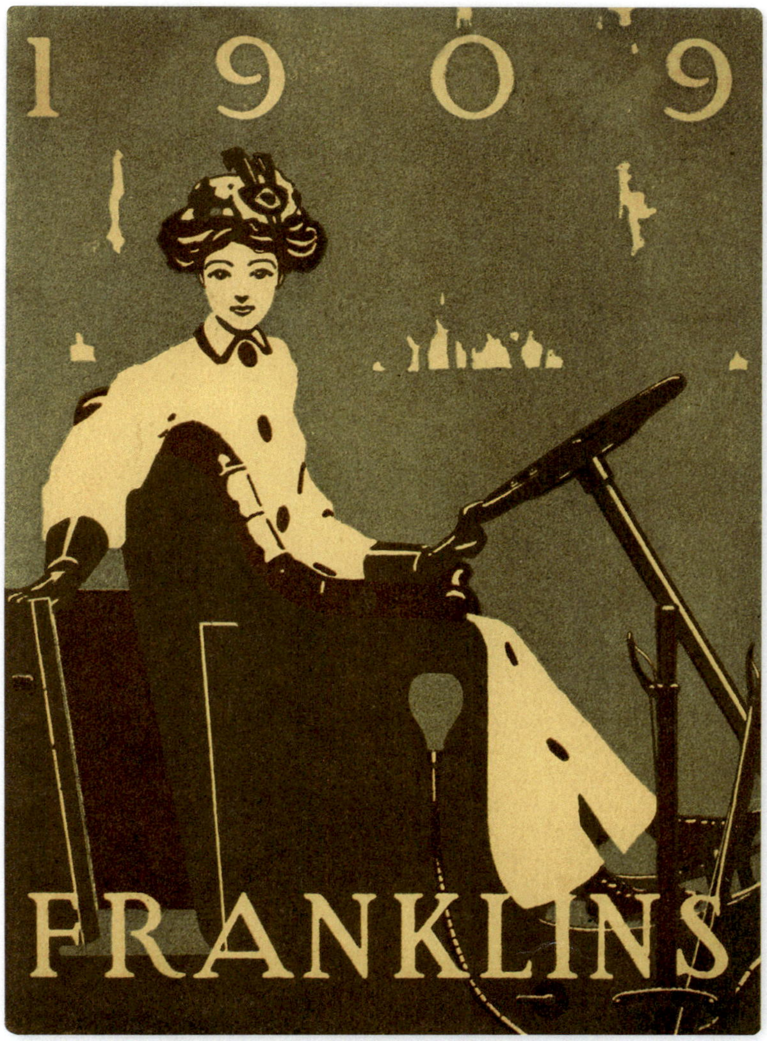

Franklin 1909

The Franklin was one of the great, advanced American classics. Under the 1909 cars' barrel hoods was an innovative flywheel-driven fan for engine cooling.

Der Franklin war einer der großen, fortschrittlichen amerikanischen Automobilklassiker. Unter der lang gestreckten fassförmigen Motorhaube des Modells aus dem Jahre 1909 saß ein innovativer schwungradgetriebener Ventilator zur Kühlung des Motors.

La Franklin fut non seulement l'un des grands classiques de l'automobile américaine, mais aussi l'une des voitures les plus novatrices. Sous le capot des modèles de l'année 1909 nichait par exemple un ventilateur à volant d'inertie destiné à refroidir le moteur.

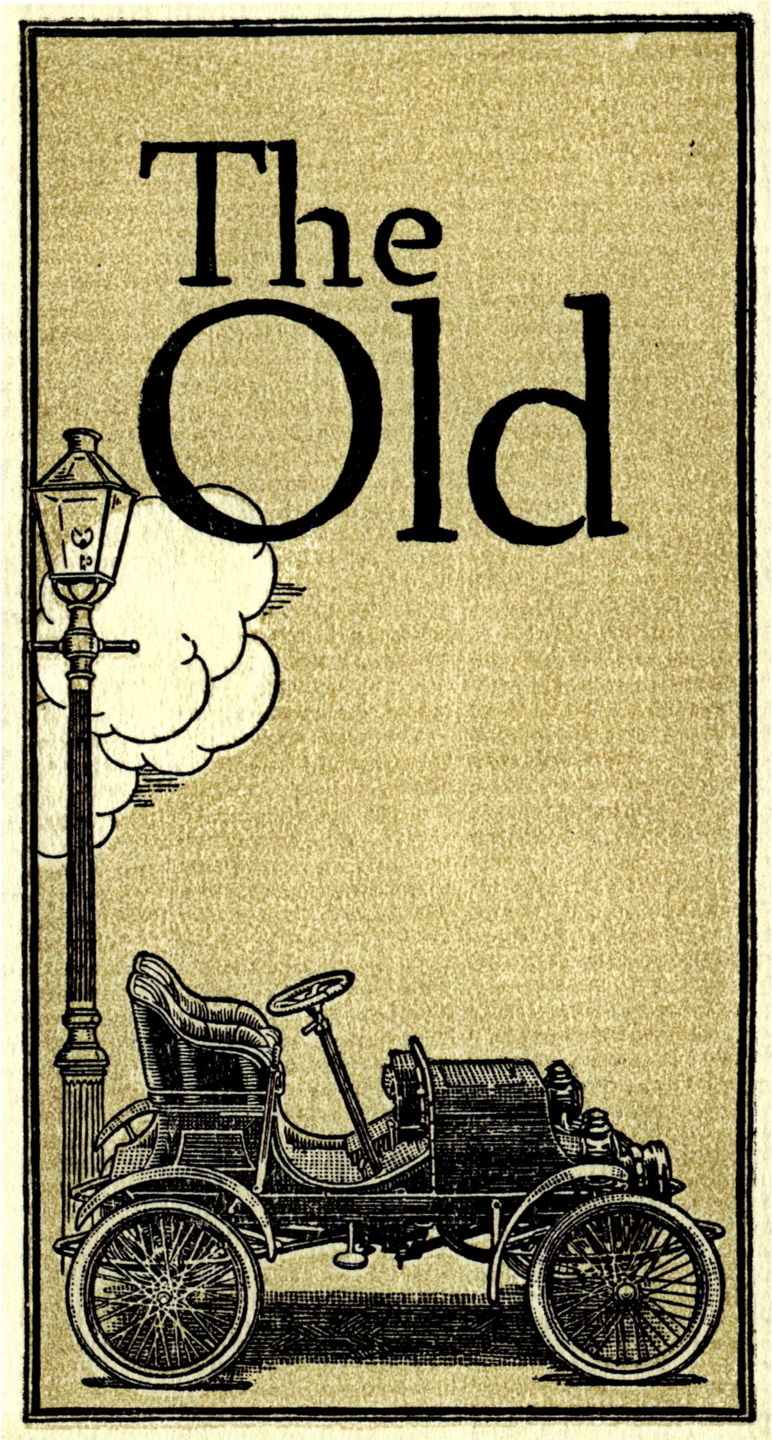

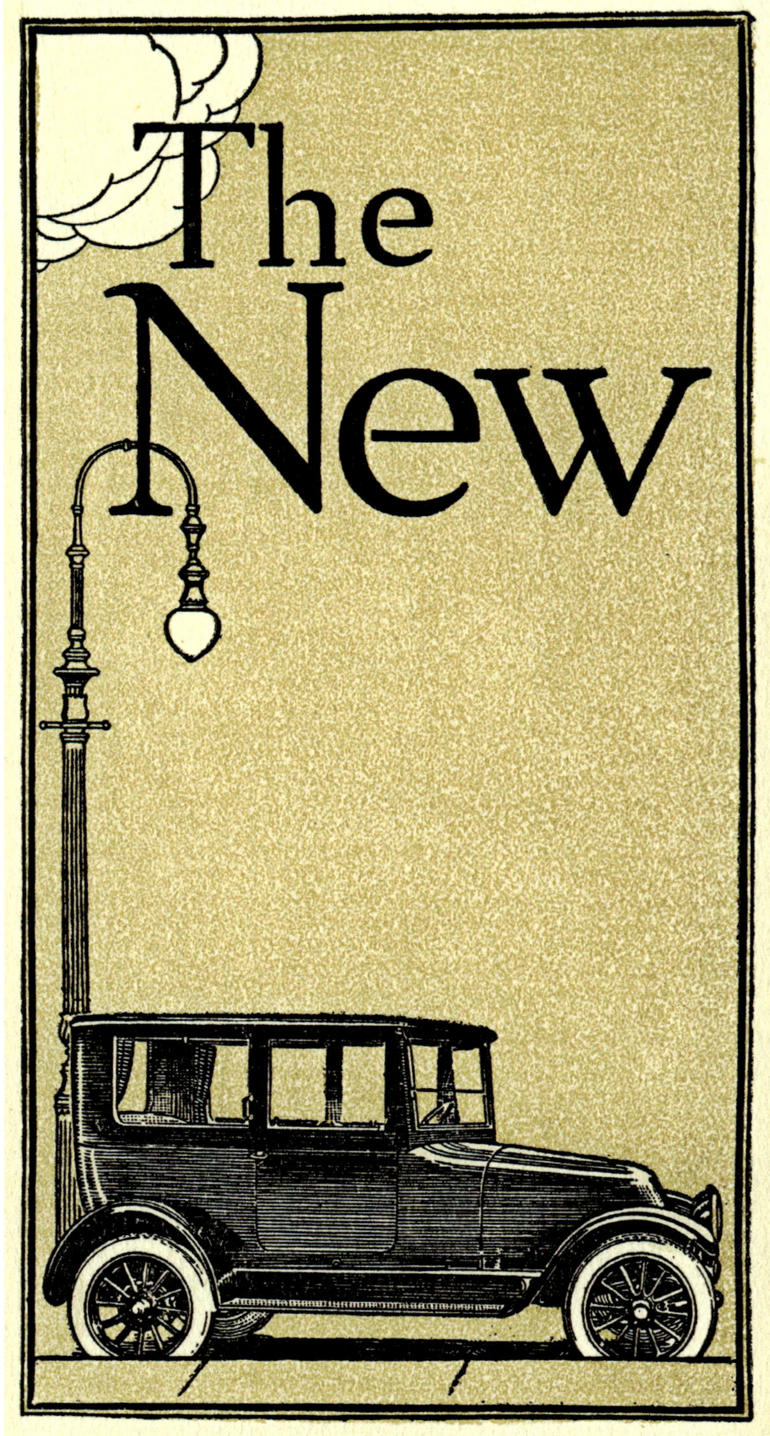

Franklin 1919

Franklin's rapid incorporation of the Renault-style hood is apparent here. More dramatic changes came later. Franklin's adoption of air cooling led to a number of "radiator shell" designs that emphasized the lack of a conventional cooling system. They would become more pronounced and incorporate a huge fan that propelled air over the engine and its built-in cooling fins.

Hier erkennt man sehr deutlich, dass Franklin rasch die Form der Motorhaube von Renault übernahm. Später folgten noch drastischere Veränderungen. Der Einsatz einer Luftkühlung führte zu einer Reihe von Kühlerverkleidungen, deren Design das Fehlen eines herkömmlichen Kühlsystems unterstrich. Sie wurden immer ausgeprägter und umfassten schließlich ein riesiges Kühlgebläse, das Luft über den Motor und dessen eingebaute Kühlrippen blies.

Avec ses capots plongeants, Franklin s'est très tôt approprié le style Renault, comme on le voit ici. Des changements plus spectaculaires apparaîtront plus tard. L'adoption du refroidissement par air généra plusieurs dessins de « calandres de radiateur » fort symboliques puisqu'elles ne cachaient de fait aucun radiateur. Elles finiront par devenir plus imposantes pour incorporer un énorme ventilateur propulsant l'air sur le moteur et ses ailettes de refroidissement intégrées.

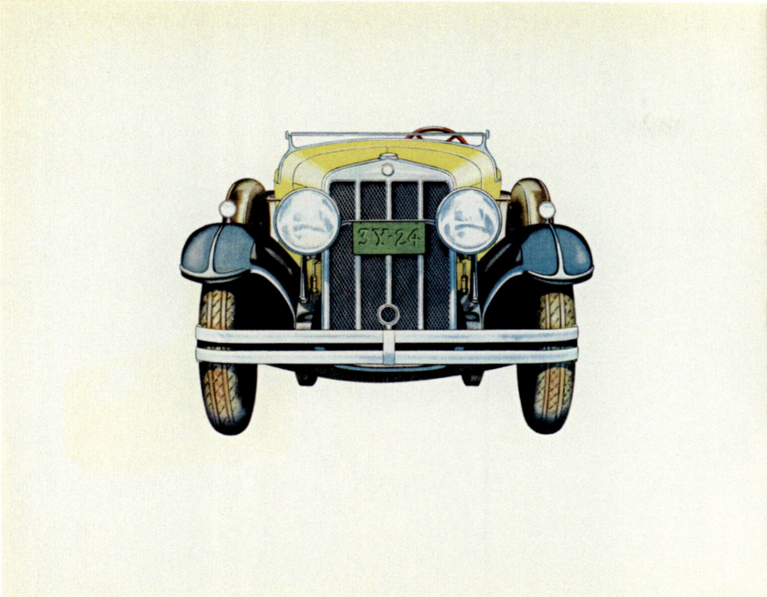

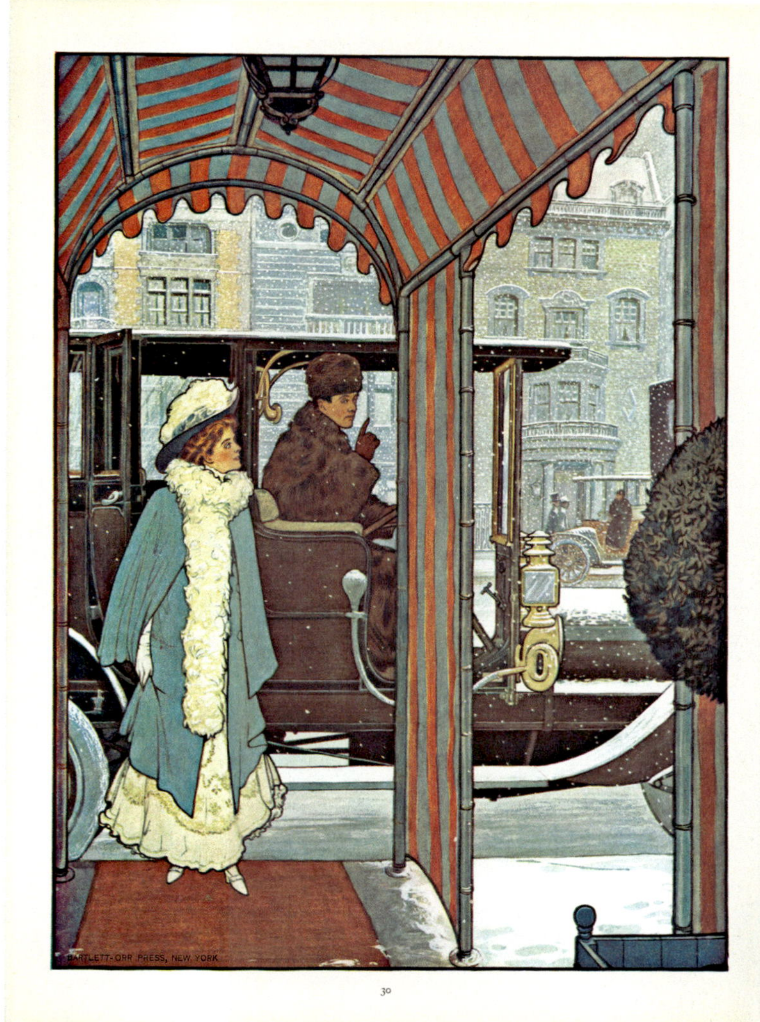

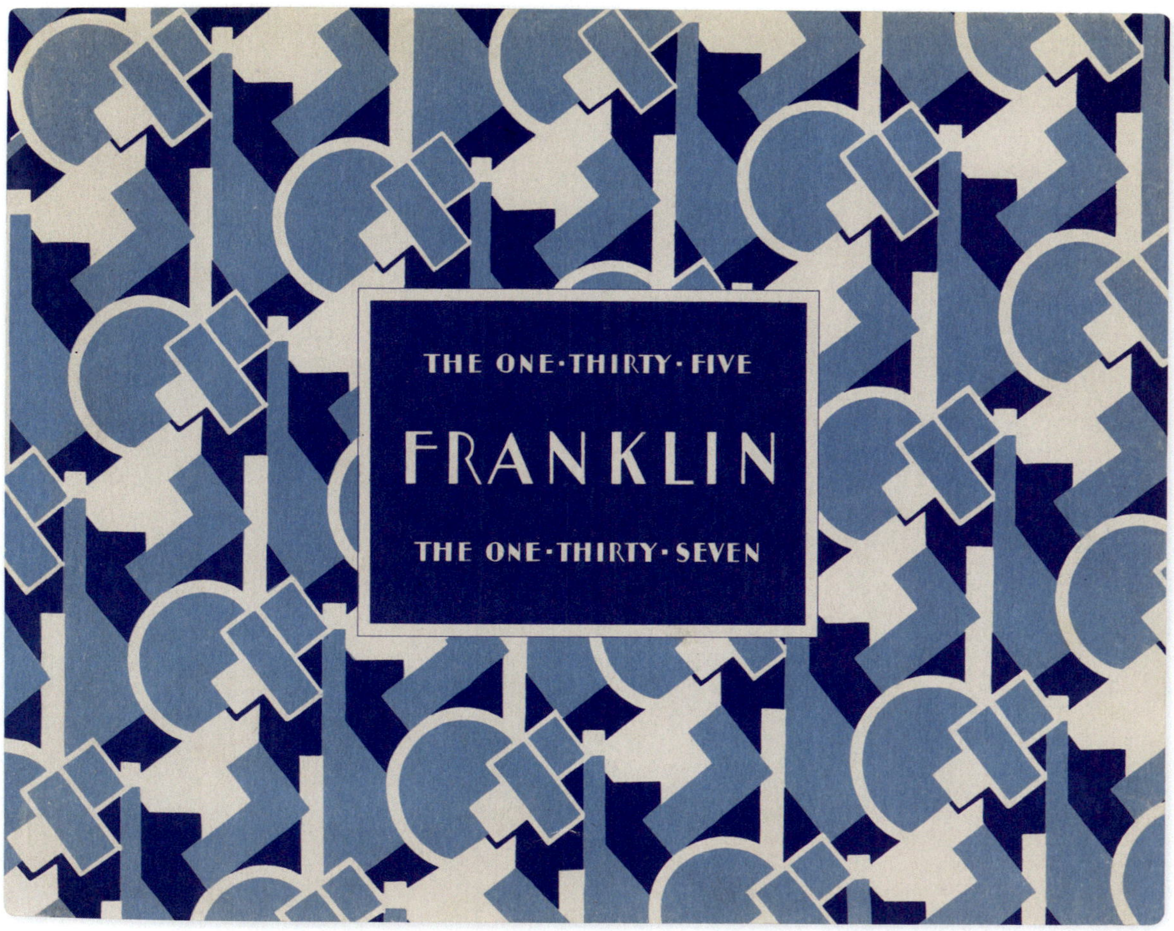

Franklin 1929 *(above left and left)*, **1909** *(above)*

Franklin introduced the new 135 and 137 models in 1929, the first creations of new designer Raymond Dietrich. Sales zoomed to over 14,000, a record. Unfortunately, the Franklin's steep pricing killed the company by 1934.

Franklin stellte die Modelle 135 und 137 im Jahre 1929 vor. Es waren die ersten Modellentwürfe, die auf den neuen Konstrukteur Raymond Dietrich zurückgingen. Der Absatz erreichte die Rekordzahl von 14 000 Fahrzeugen, doch der hohe Verkaufspreis brachte 1934 das Aus für Franklin.

Franklin présenta en 1929 ses nouveaux modèles 135 et 137, les premières créations du nouveau designer Raymond Dietrich. Les ventes s'envolèrent pour dépasser les 14 000 unités, un record. Hélas la politique tarifaire élevée pratiquée par Franklin aura raison de l'entreprise en 1934.

Garford

Garford 1912 *(right)*

In Elyria, Ohio, Garford began as a parts supplier before eventually building cars that were too expensive to succeed. The company that bought Garford, Willys-Overland, enjoyed later success building trucks.

Garford begann als Automobilzuliefe-rer in Elyria (Ohio), bevor die Firma schließlich selbst Autos baute, die aber zu teuer waren, um sich durchzusetzen. Das Unternehmen Willys-Overland, das Garford aufkaufte, war später als Lkw-Hersteller erfolgreich.

À Elyria, dans l'Ohio, Garford débuta comme fournisseur de pièces détachées avant de finir par construire des voitures qui coûteront trop cher pour rencontrer le succès. L'entreprise qui racheta Garford, Willys-Overland, aura un destin plus heureux avec ses camions et ses Jeep.

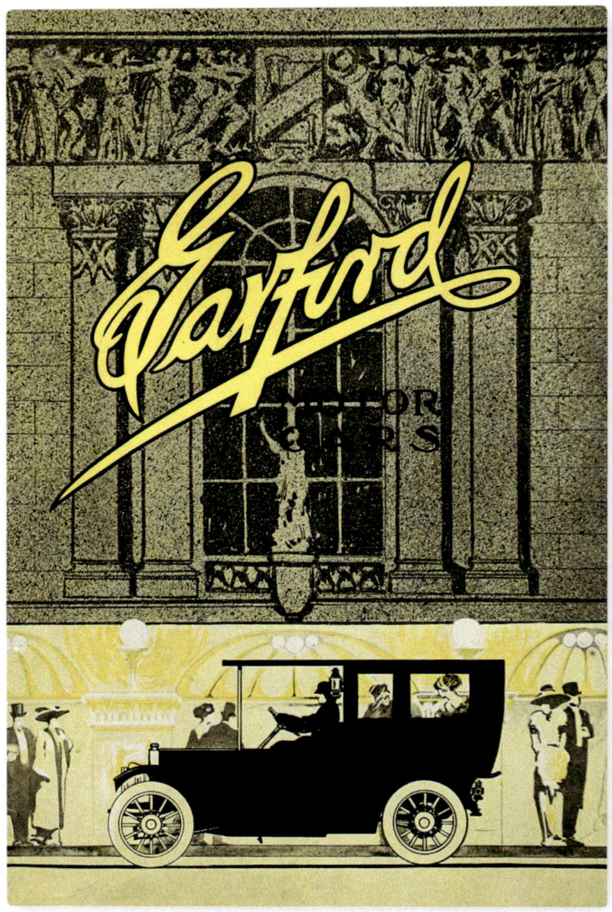

Gray

Gray 1925 *(right)*

The Gray attempted to compete with Ford's impossibly cheap Model T. In spite of flowery advertising prose, only around 1,700 were sold. Gray was founded in 1922 by Frank Beall, formerly vice president of Packard, who began building the car in Detroit.

Der Gray versuchte, mit dem sagenhaft bil-ligen Model T von Ford mitzuhalten, doch trotz blumiger Werbesprüche wurden nur rund 1700 Stück verkauft. Gray wurde 1922 von Frank Beall gegründet, ehemals Vice President bei Packard, der das Auto in Detroit zu bauen begann.

La Gray nourrissait l'ambition de rivaliser avec la Ford T et son prix incroyablement modique. En dépit de sa prose publici-taire fleurie, il ne s'en vendit que quelque 1 700 exemplaires. Gray avait été fondé en 1922 par Frank Beall, précédemment vice-président de Packard, la marque qui avait assuré les débuts de la fabrication de la voiture à Detroit.

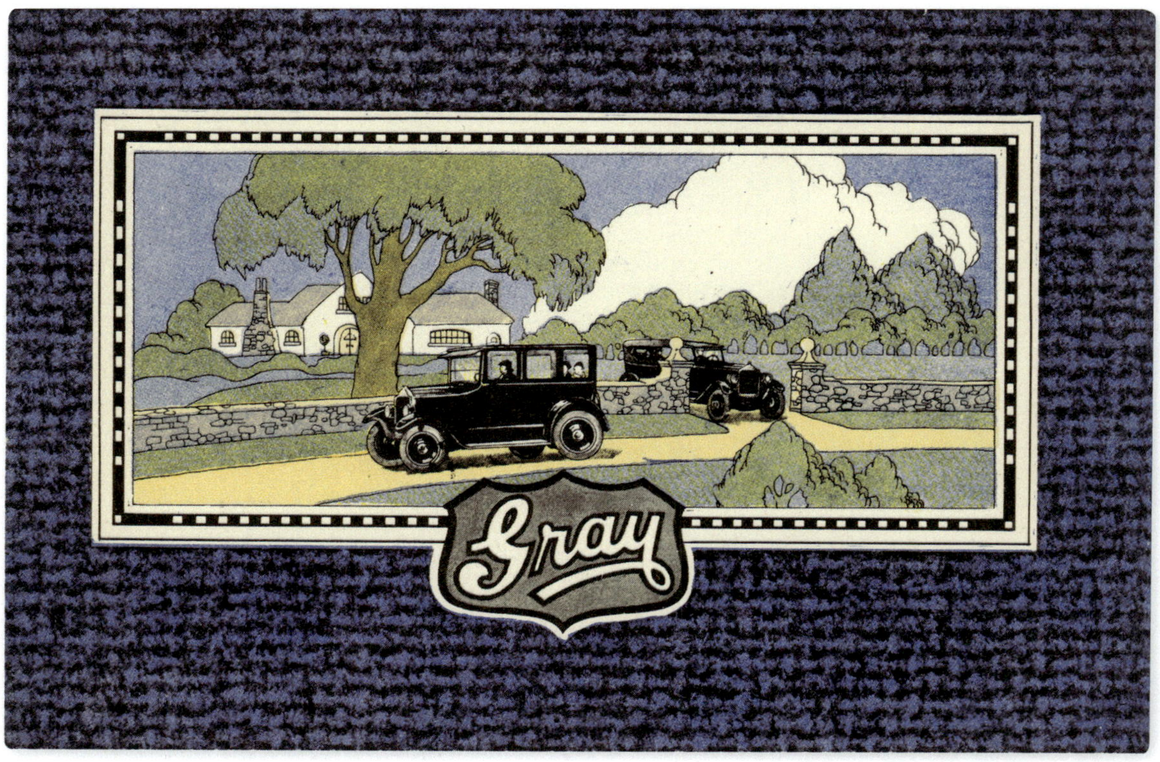

Hupmobile

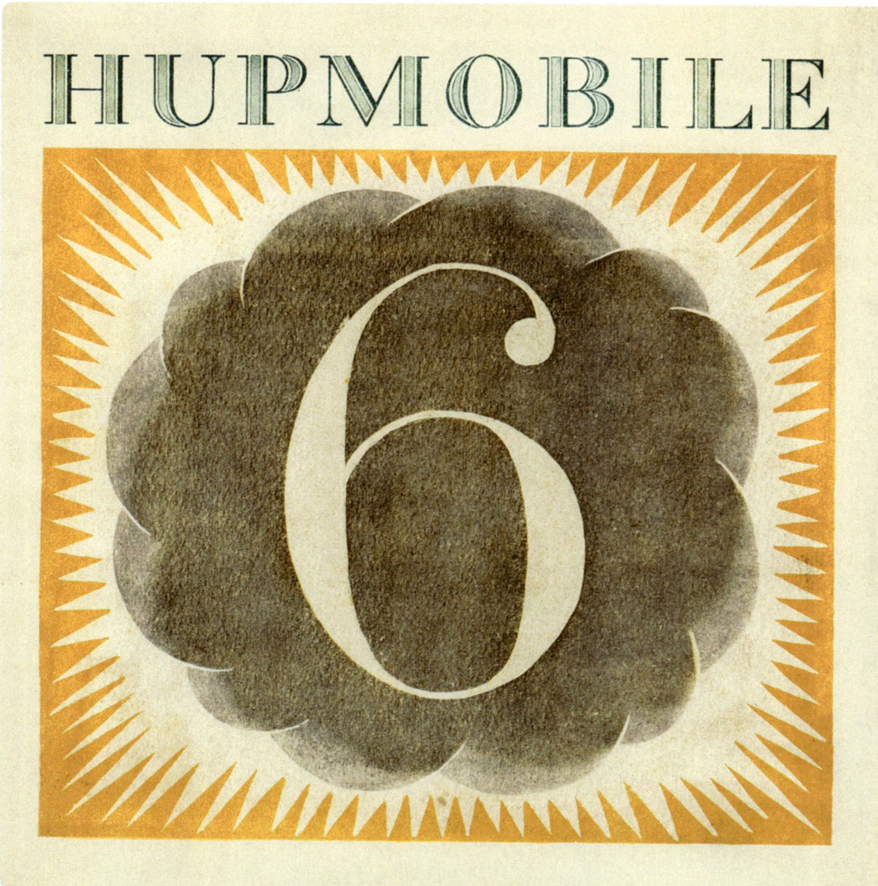

Hupmobile 1928 *(left)*,
1929 *(below and following)*

From his company's outset in 1909, Robert Craig Hupp demanded that his cars be both strong performers and attractive. Both the six- and eight-cylinder models had four-wheel brakes by the late 1920s, and styling by Amos Northup, who had crafted the lovely Wills Sainte Claire.

Von der Firmengründung im Jahre 1909 an verlangte Robert Craig Hupp, dass seine Autos sowohl leistungsstark als auch attraktiv sein sollten. Sowohl die Sechs- als auch die Achtzylindermodelle waren Ende der 1920er-Jahre mit Vierradbremsen ausgestattet, und die Karosserien wurden von Amos Northup gestaltet, der auch den anmutigen Wills Sainte Claire entworfen hatte.

Dès le démarrage de son entreprise en 1909, Robert Craig Hupp exigea que ses voitures soient aussi performantes que séduisantes. À la fin des années 1920 les modèles à six et huit cylindres disposaient de freins aux quatre roues et d'un design signé Amos Northup, créateur de la jolie Wills Sainte Claire.

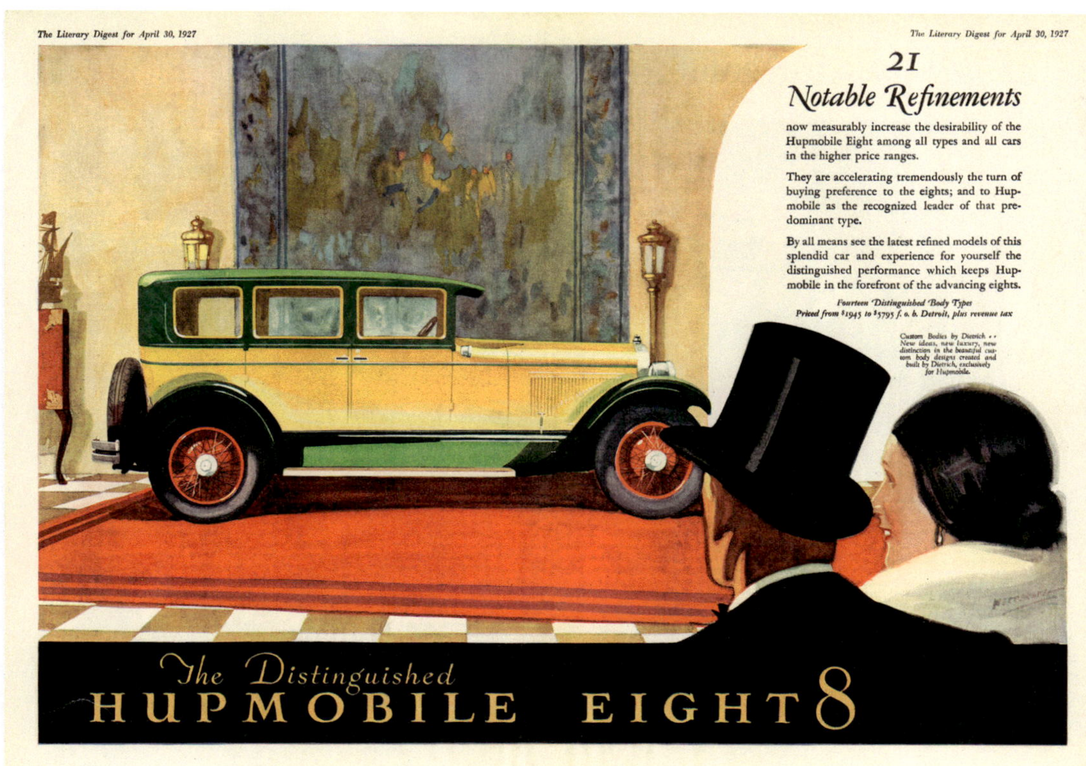

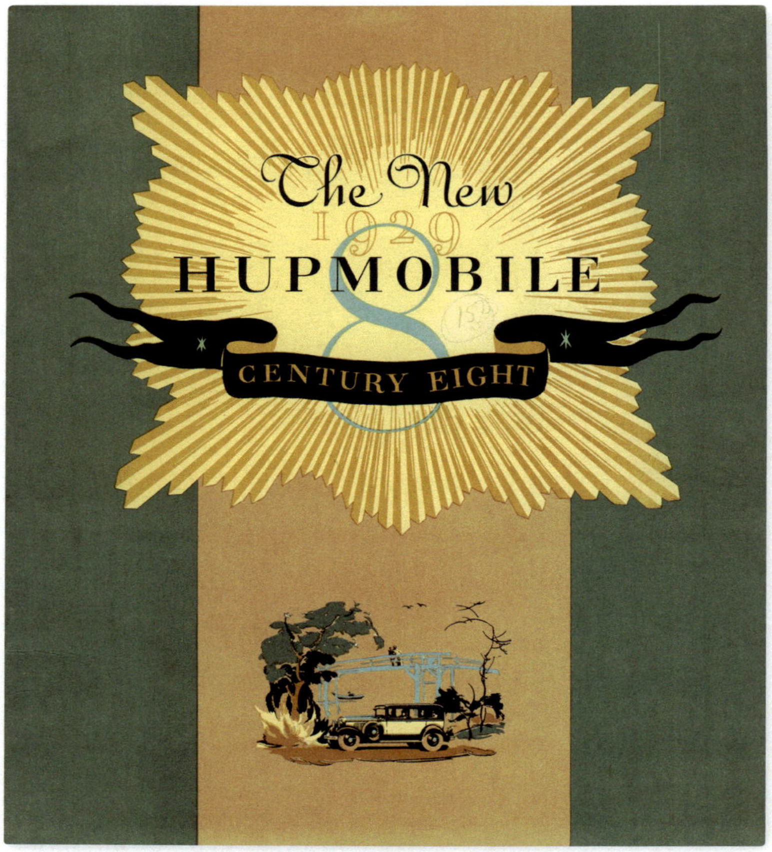

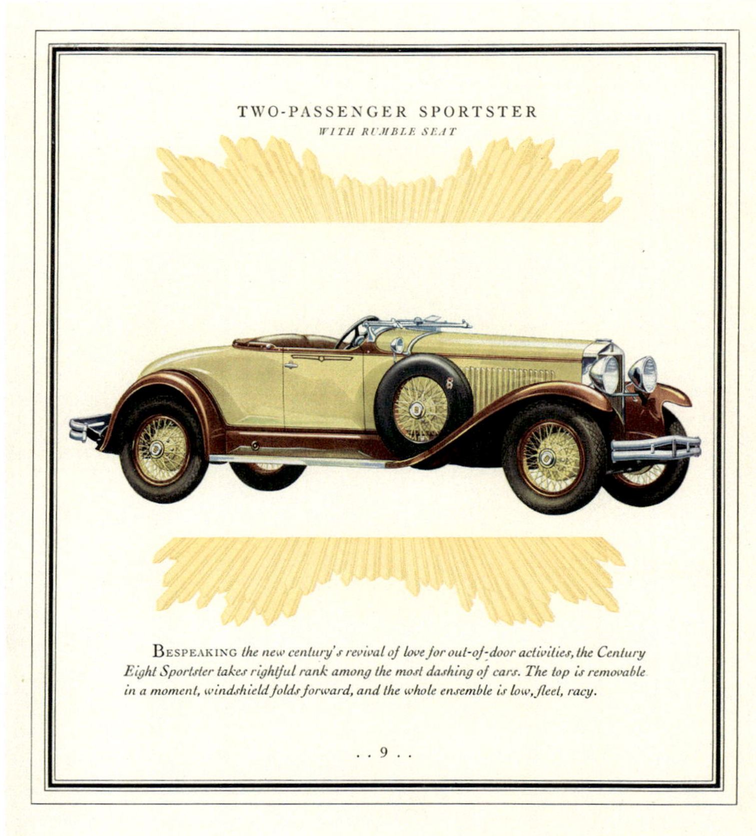

Jeffery

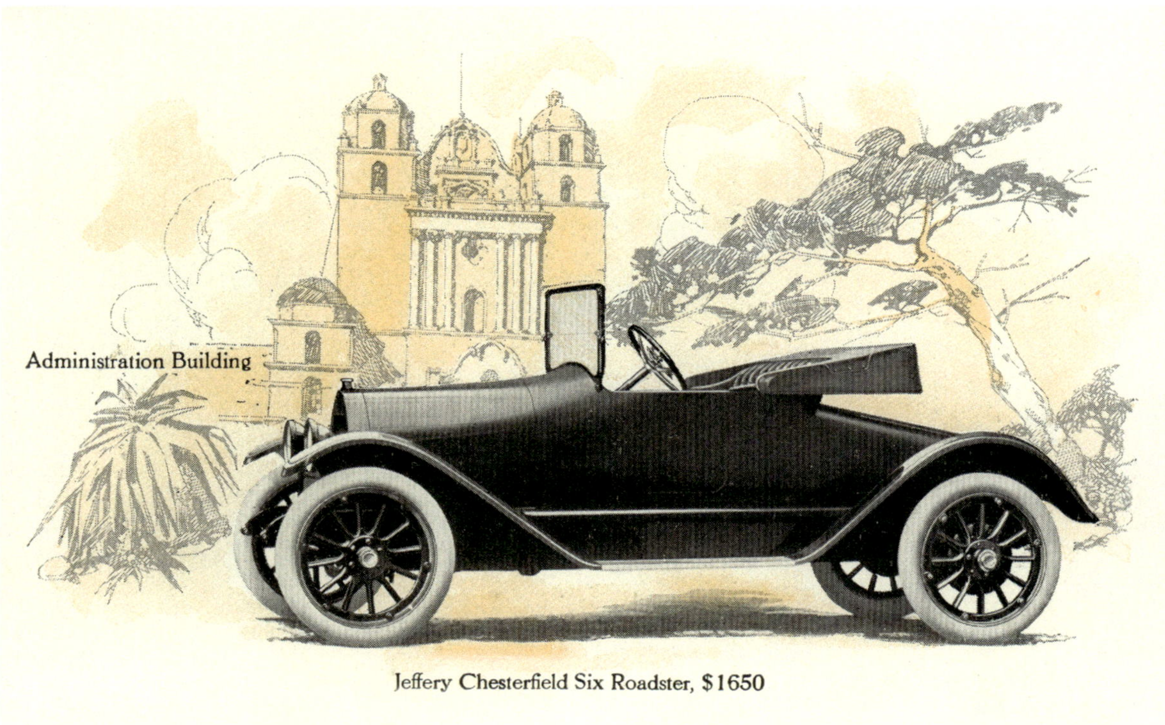

Jeffery Chesterfield Six 1919 *(left)*

Originally, the Jeffery had a soaring history. Charles Jeffery took over the company when his father, Thomas, who founded the firm, died in 1910. In 1916 Charles Jeffery decided to retire and sold the company to Charles Nash.

Ursprünglich ging es mit Jeffery steil bergauf. Nachdem sein Vater, Firmengründer Thomas Jeffery, 1910 verstorben war, übernahm Sohn Charles das Unternehmen, setzte sich aber bereits 1916 zur Ruhe und verkaufte es an Charles Nash.

La Jeffery n'aura pas vécu très longtemps. Charles Jeffery prit les rênes de l'entreprise en 1910 à la mort de son père Thomas, qui en était le fondateur. Six ans plus tard Charles décida de se retirer du jeu et vendit l'entreprise à Charles Nash.

King

King Eight 1916 *(right)*, **1917** *(below)*

The King Eight was among the first with left-hand steering and a cast engine block. The company was founded in Detroit in 1911 and survived its final year, 1923, in Buffalo, New York.

Der King Eight gehörte zu den ersten Automobilen mit Linkslenkung und einem Motorblock aus Gussmetall. Das Unternehmen wurde 1911 in Detroit (Michigan) gegründet und schloss 1923 in Buffalo (New York).

La King Eight figure parmi les premières automobiles à conduite à gauche avec bloc moteur en fonte. L'entreprise avait été fondée à Detroit en 1911 et vécut sa dernière année, 1923, à Buffalo, dans l'État de New York.

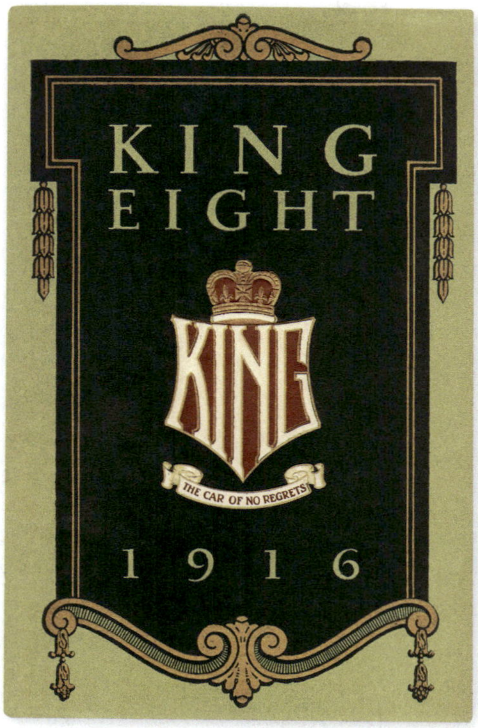

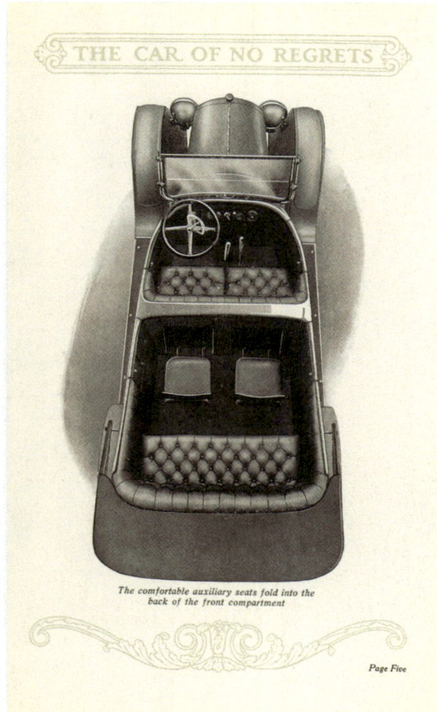

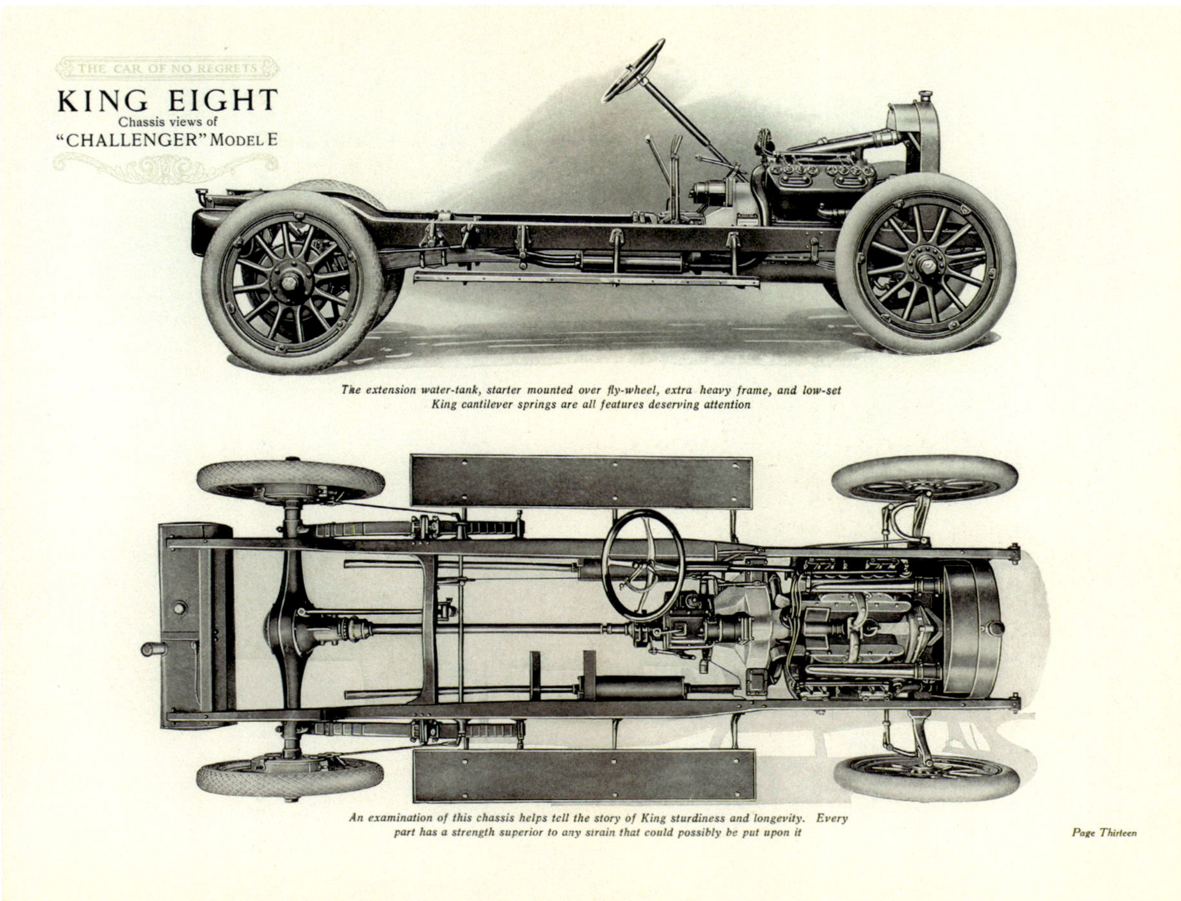

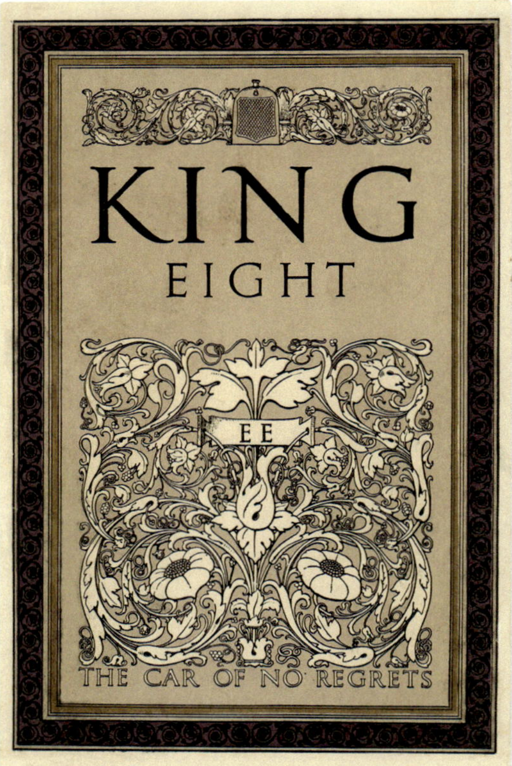

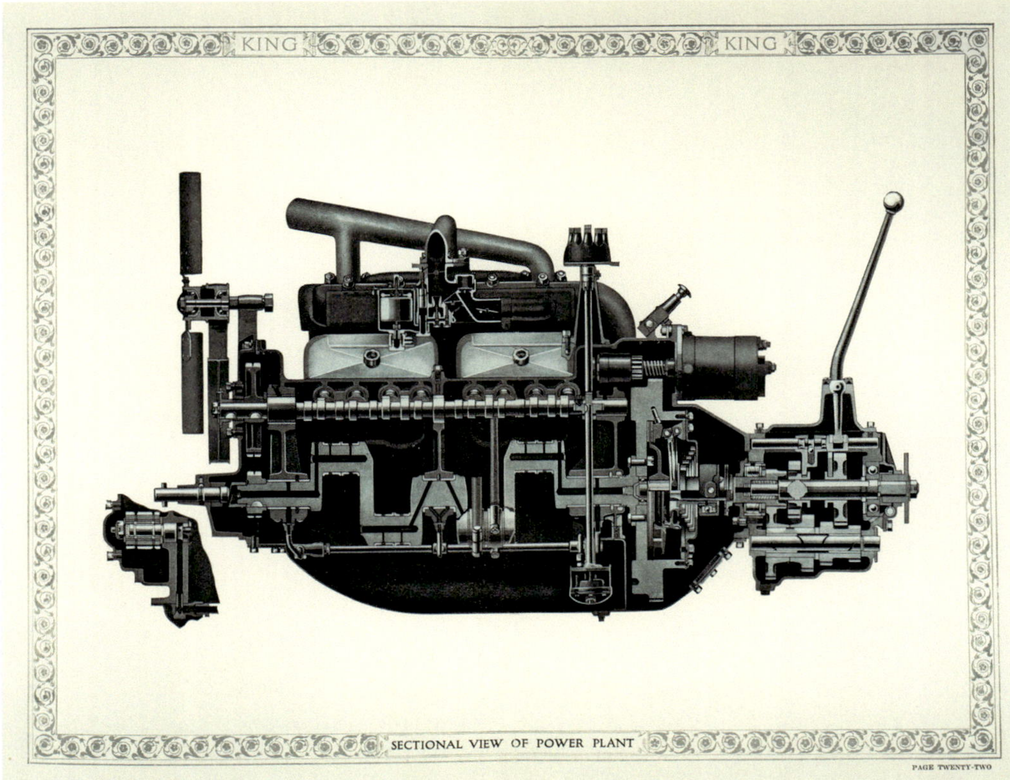

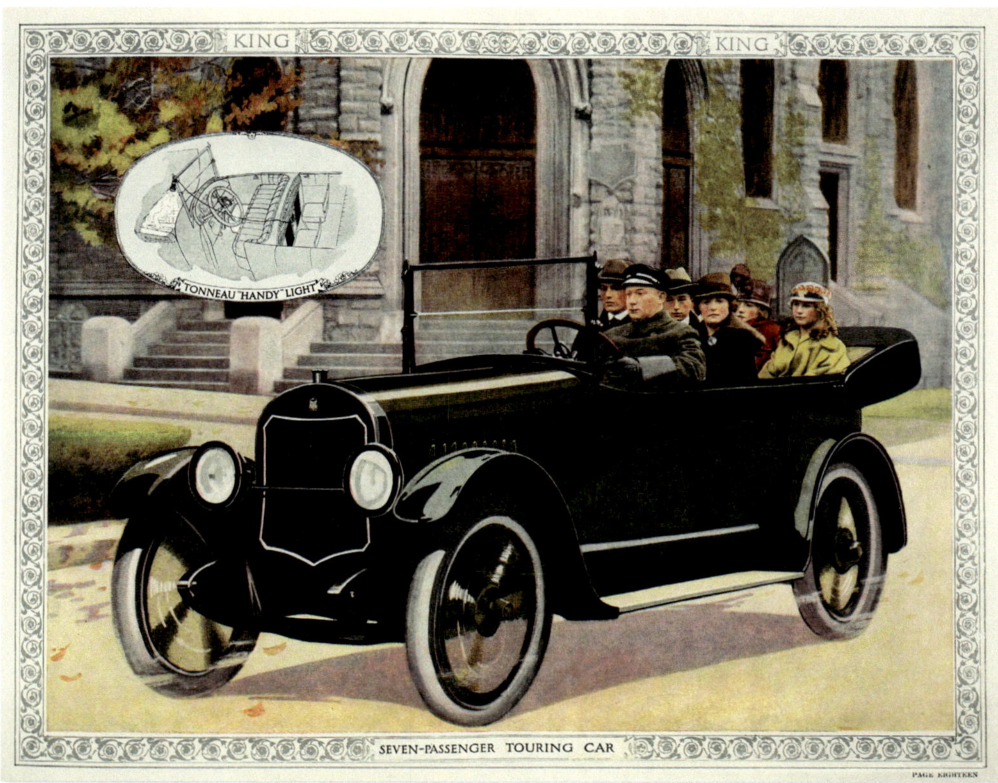

King Eight 1917 *(this spread)*

In the race to get an affordable V-8 engine on the market, King was beaten by Cadillac. That spelled doom for the company, which failed in 1923, although King still boasted that its eight-cylinder car, which arrived too late, could convincingly handle a steep mountain grade.

Im Wettlauf, einen erschwinglichen V8-Motor auf den Markt zu bringen, wurde King von Cadillac geschlagen. Damit war das Schicksal des Herstellers besiegelt, der 1923 in Konkurs ging – obwohl King noch damit prahlte, sein Achtzylinderautomobil, das zu spät herauskam, könne steile Bergstraßen mühelos bewältigen.

Battu par Cadillac dans la course au lancement d'un moteur V8 à prix raisonnable, King ne s'en relèvera pas et cessera ses activités en 1923. La marque n'en prétendait pas moins que son huit cylindres pouvait sans mal venir à bout d'une pente montagneuse escarpée. Hélas il arriva trop tard.

KING

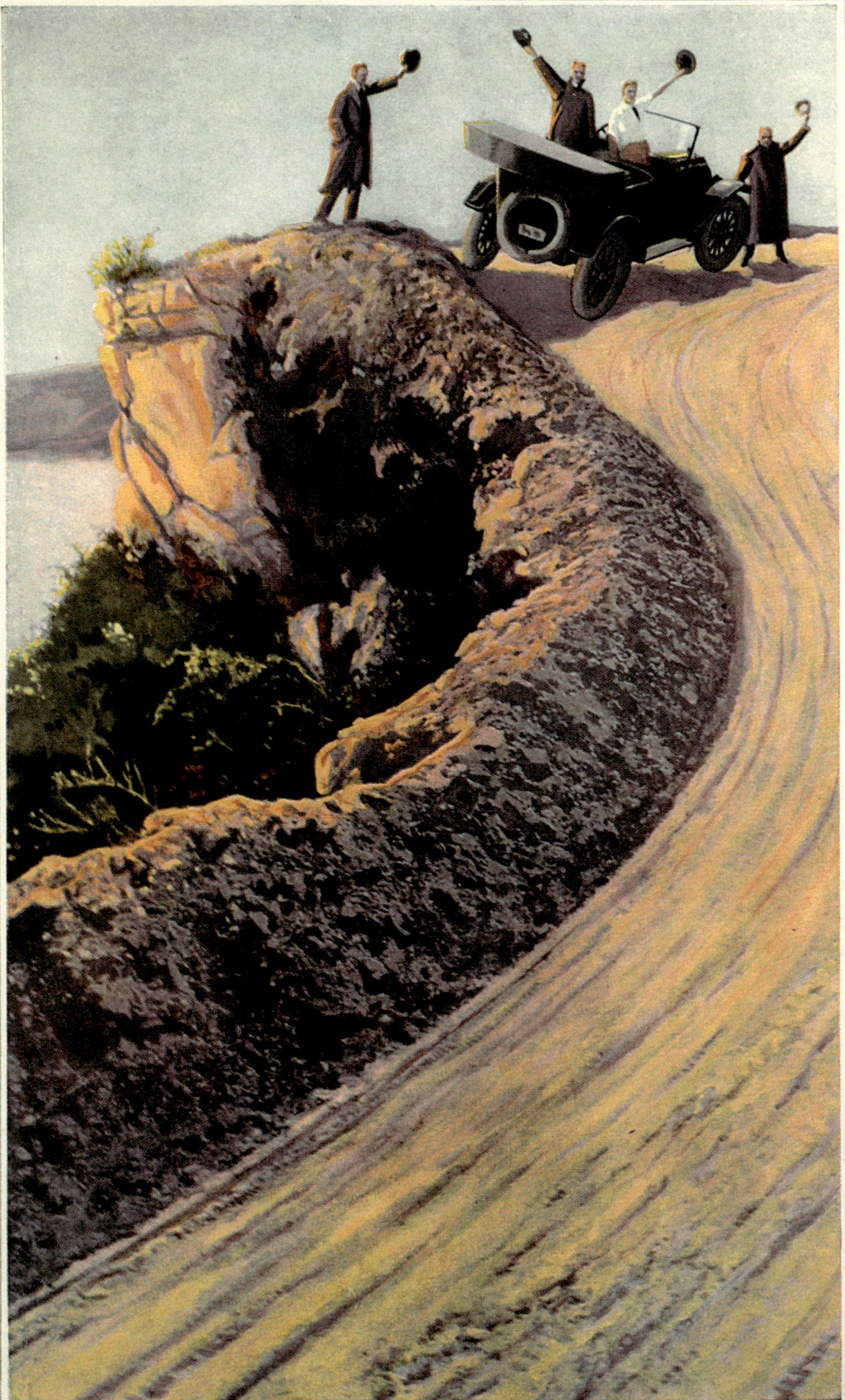

KING ASCENDING MT. WILSON SEALED IN HIGH GEAR—AT DEVIL'S ELBOW

PAGE TWO

Knox

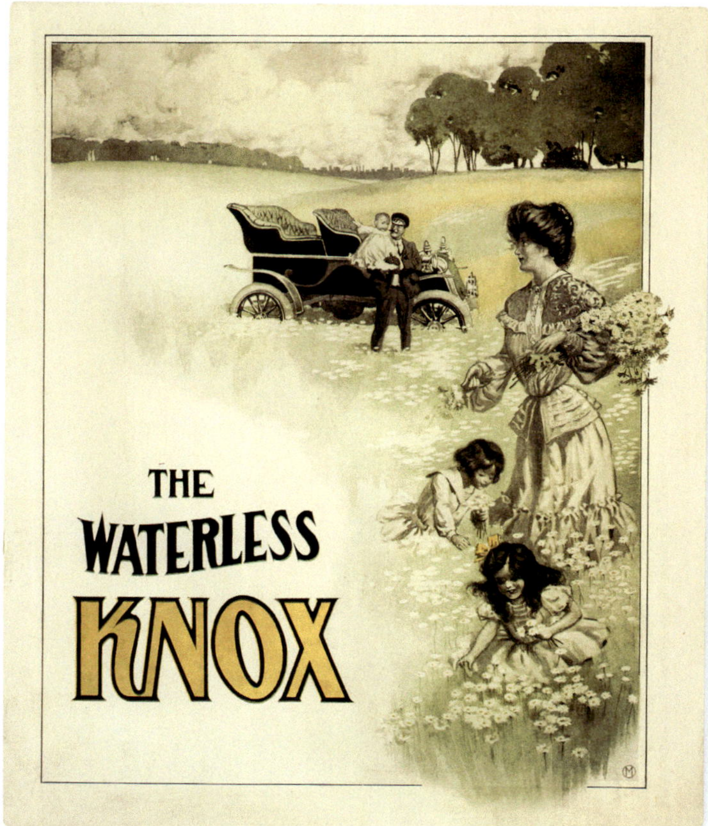

Knox Tudor Touring 1904

"Waterless" referred to the Knox's two-cylinder, air-cooled engine, nicknamed "Old Porcupine" because inventor Harry Knox used pins stuck in the cylinder barrels, not the more typical fins, to facilitate cooling. Air-cooled engines dated back to the earliest cars, including those from Daimler in Germany, but it was soon realized that water-cooled engines were more efficient since they didn't waste thermal heat. A notable exception is the original Volkswagen, whose Beetle was one of the world's most successful cars.

Das Wort *waterless* in der Anzeige bezieht sich auf den luftgekühlten Zweizylindermotor des Knox, den man scherzhaft als „altes Stachelschwein" bezeichnete, weil sein Erfinder Harry Knox zur Unterstützung der Luftkühlung statt der üblichen Kühlrippen Stifte verwendete, die in den Zylindermantel gesteckt waren. Luftgekühlte Motoren gab es bereits in den frühesten Automobilen – auch jenen, die Daimler in Deutschland baute –, aber man erkannte sehr bald, dass flüssigkeitsgekühlte Motoren wirtschaftlicher arbeiteten, da sie keine Wärme verschwendeten. Eine namhafte Ausnahme bildeten Fahrzeuge von Volkswagen, darunter der Original-„Käfer", der zu den erfolgreichsten Autos der Welt gehörte.

Le terme *waterless* renvoyait au moteur bicylindre refroidi par air de la Knox, surnommé « le vieux porc-épic » parce que pour faciliter le refroidissement, son inventeur Harry Knox utilisait des chevilles de 5 cm fixées aux manchons de cylindre, et non les habituelles ailettes de la culasse. Le principe du refroidissement par air remontait aux premières automobiles, dont les Daimler allemandes, mais il s'avéra vite que les moteurs à refroidissement liquide, exempts de pertes thermiques, étaient plus efficaces. À une exception notable, la première Volkswagen, cette Coccinelle qui sera l'une des voitures les plus vendues au monde.

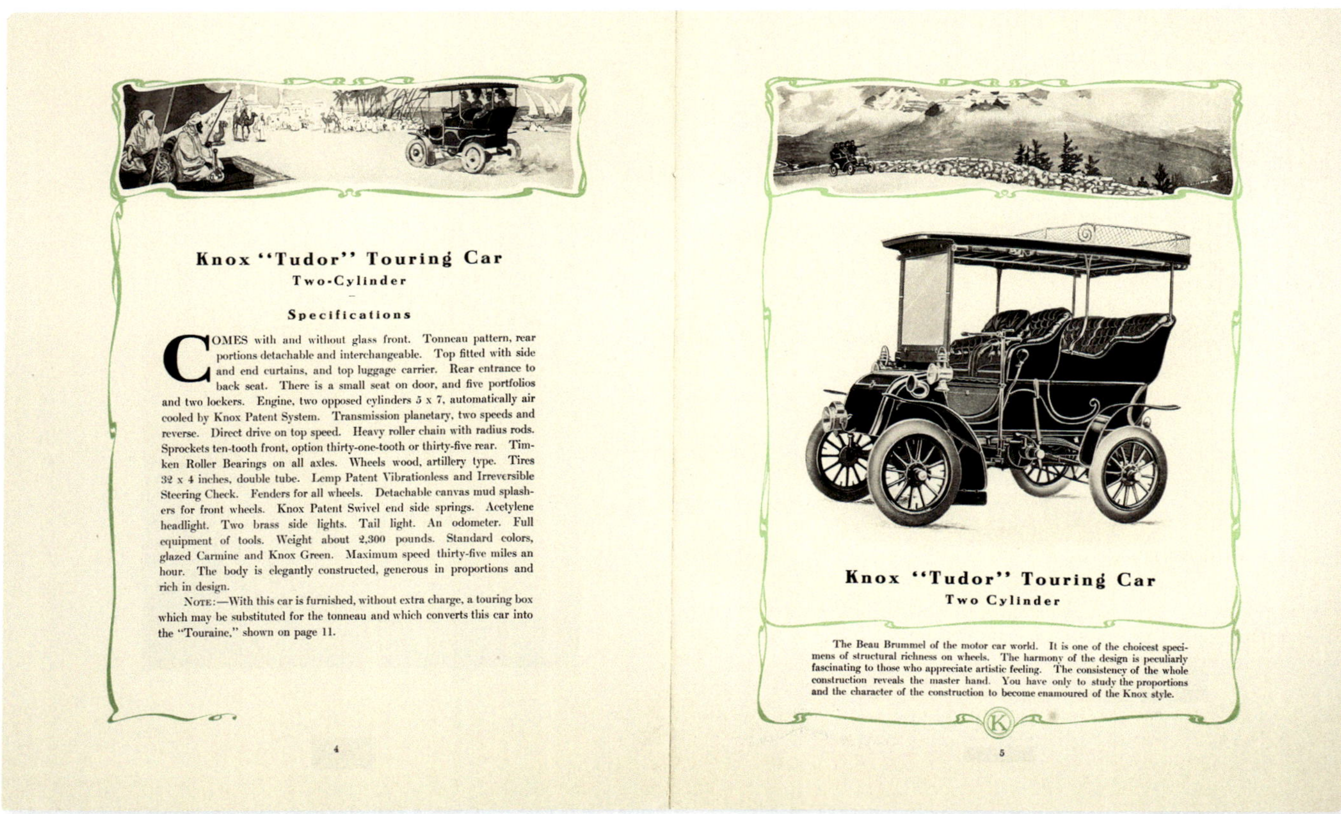

Knox "Tudor" Touring Car
Two-Cylinder

Specifications

COMES with and without glass front. Tonneau pattern, rear portions detachable and interchangeable. Top fitted with side and end curtains, and top luggage carrier. Rear entrance to back seat. There is a small seat on door, and five portfolios and two lockers. Engine, two opposed cylinders 5 x 7, automatically air cooled by Knox Patent System. Transmission planetary, two speeds and reverse. Direct drive on top speed. Heavy roller chain with radius rods. Sprockets ten-tooth front, option thirty-one-tooth or thirty-five rear. Timken Roller Bearings on all axles. Wheels wood, artillery type. Tires 32 x 4 inches, double tube. Lemp Patent Vibrationless and Irreversible Steering Check. Fenders for all wheels. Detachable canvas mud splashers for front wheels. Knox Patent Swivel end side springs. Acetylene headlight. Two brass side lights. Tail light. An odometer. Full equipment of tools. Weight about 2,300 pounds. Standard colors, glazed Carmine and Knox Green. Maximum speed thirty-five miles an hour. The body is elegantly constructed, generous in proportions and rich in design.

NOTE:—With this car is furnished, without extra charge, a touring box which may be substituted for the tonneau and which converts this car into the "Touraine," shown on page 11.

4

Knox "Tudor" Touring Car
Two Cylinder

The Beau Brummel of the motor car world. It is one of the choicest specimens of structural richness on wheels. The harmony of the design is peculiarly fascinating to those who appreciate artistic feeling. The consistency of the whole construction reveals the master hand. You have only to study the proportions and the character of the construction to become enamoured of the Knox style.

5

Lexington

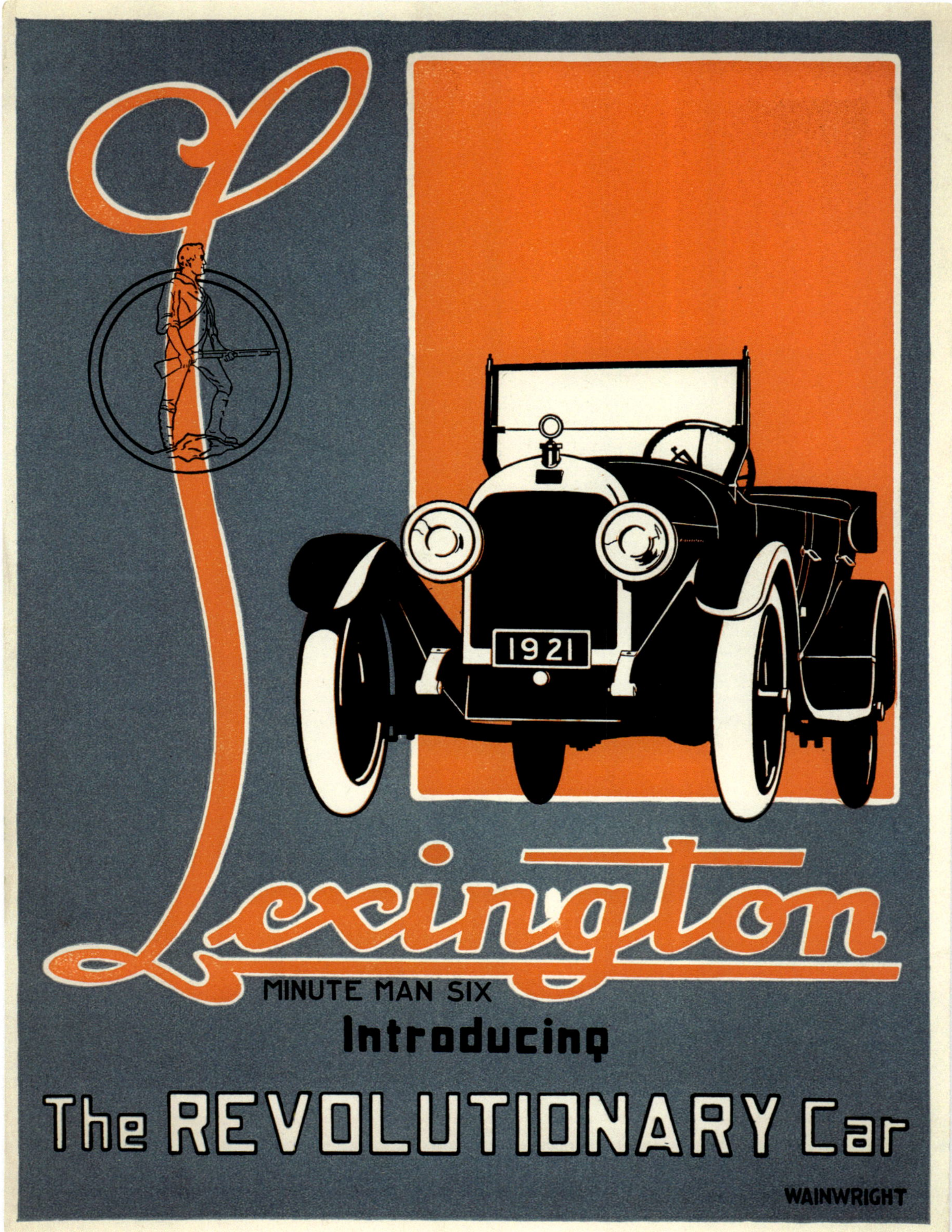

Lexington Minute Man Six 1921

How "revolutionary" was the Lexington? Not very. This was an "assembled" car, first using Ansted engines, then Continental.

Wie „revolutionär" war der Lexington wirklich? Nicht besonders. Es handelte sich um ein „Baukastenauto", das zuerst Motoren von Ansted und später von Continental verwendete.

La Lexington fut-elle vraiment « révolutionnaire » ? Pas tant que ça. C'était une voiture « assemblée », propulsée d'abord par des moteurs Ansted puis par des Continental.

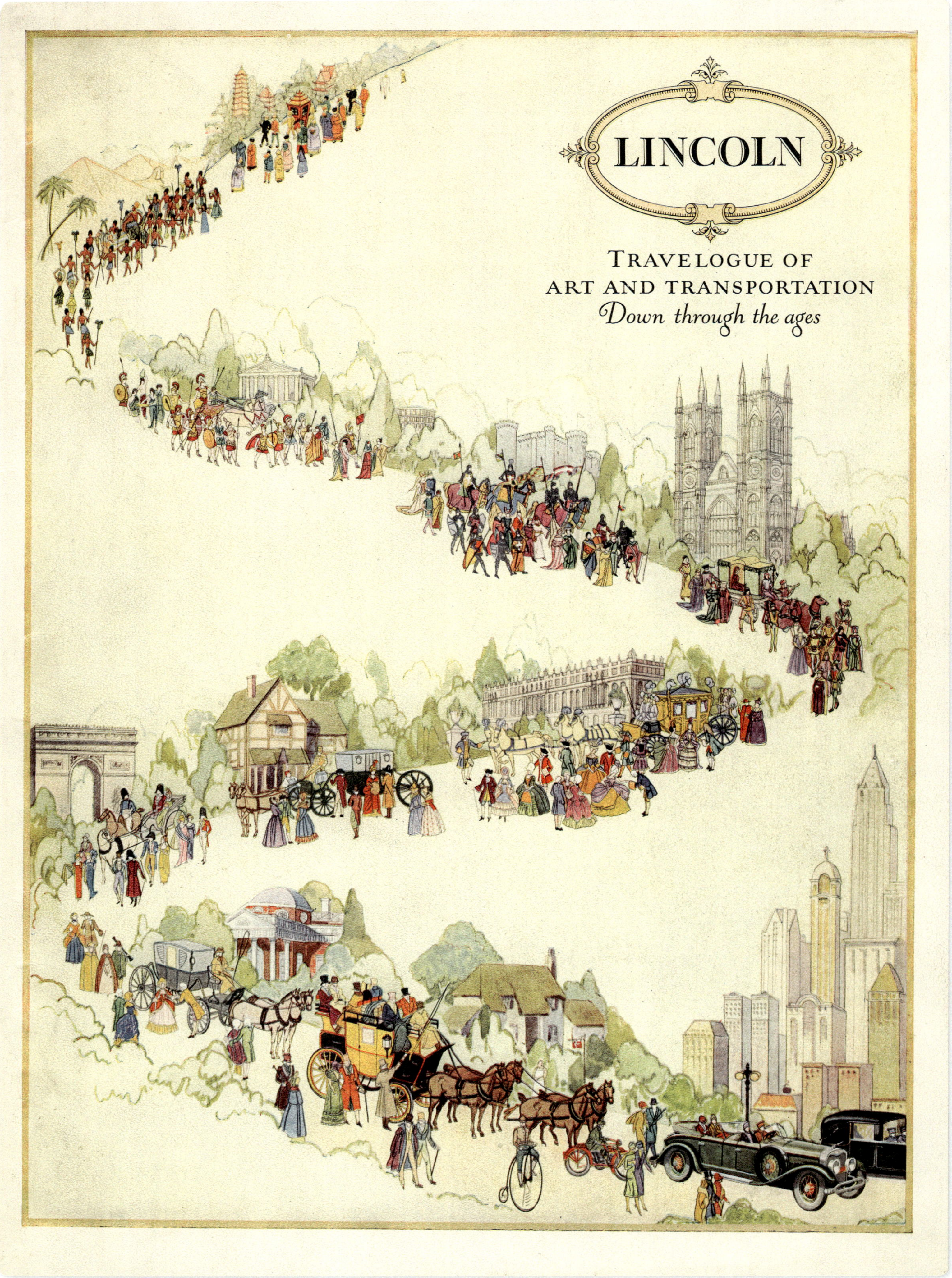
LINCOLN
TRAVELOGUE OF
ART AND TRANSPORTATION
Down through the ages

Lincoln

Lincoln 1927 *(this spread)*

For many years, Lincoln didn't embrace annual model changes. Still, dozens of body styles were offered, some through coachbuilders, including Brunn, Dietrich, and LeBaron. The brilliant machinist Henry M. Leland founded both Lincoln and its lifelong rival, Cadillac, founding Lincoln in 1922 before the Ford Motor Company acquired it two years later. Henry Ford wasn't one for luxury or styling, until the profits associated with a premium brand like Lincoln softened his opinion.

Viele Jahre lang widersetzte sich Lincoln dem Diktat des jährlichen Modellwechsels. Dennoch bot der Hersteller Dutzende von Varianten an, davon einige über Karosseriebauer wie Brunn, Dietrich und LeBaron. Der geniale Maschinenbauer Henry M. Leland gründete sowohl Lincoln als auch dessen ewigen Rivalen Cadillac. Lincoln erblickte 1922 das Licht der Welt und wurde zwei Jahre später von der Ford Motor Company übernommen. Eigentlich hielt Henry Ford nicht viel von Eleganz und Luxus, aber als er erkannte, wie hoch die Gewinne waren, die man mit einer Luxusmarke wie Lincoln erzielen konnte, wandelte sich seine Einstellung.

Pendant de nombreuses années, Lincoln résista à la vogue des changements de modèle annuels. Des dizaines de styles de robes différents n'en étaient pas moins proposés, certains par l'intermédiaire de carrossiers tels que Brunn, Dietrich et LeBaron. Brillant ingénieur motoriste, Henry M. Leland fonda à la fois Lincoln (1922) et son rival de toujours, Cadillac, avant de céder Lincoln à Ford en 1924. Henry Ford n'était guère porté sur le luxe et le style mais les bénéfices réalisés grâce à une marque haut de gamme comme Lincoln infléchirent quelque peu son jugement en la matière.

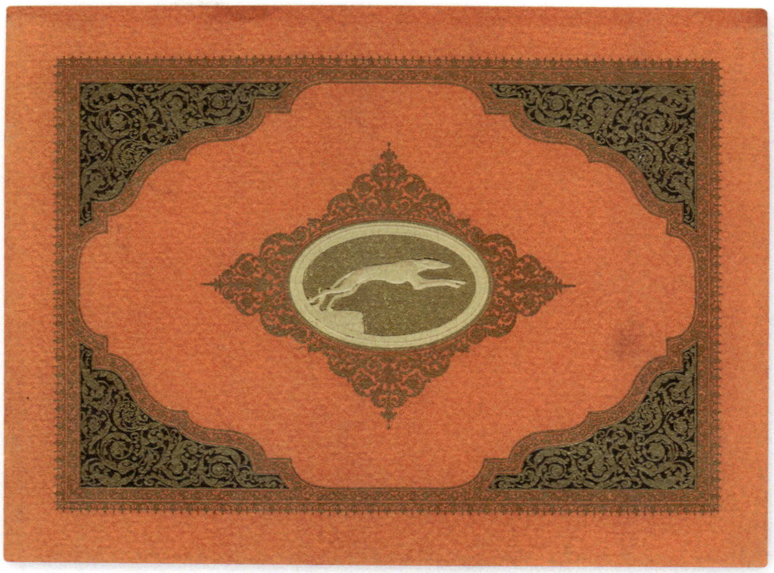

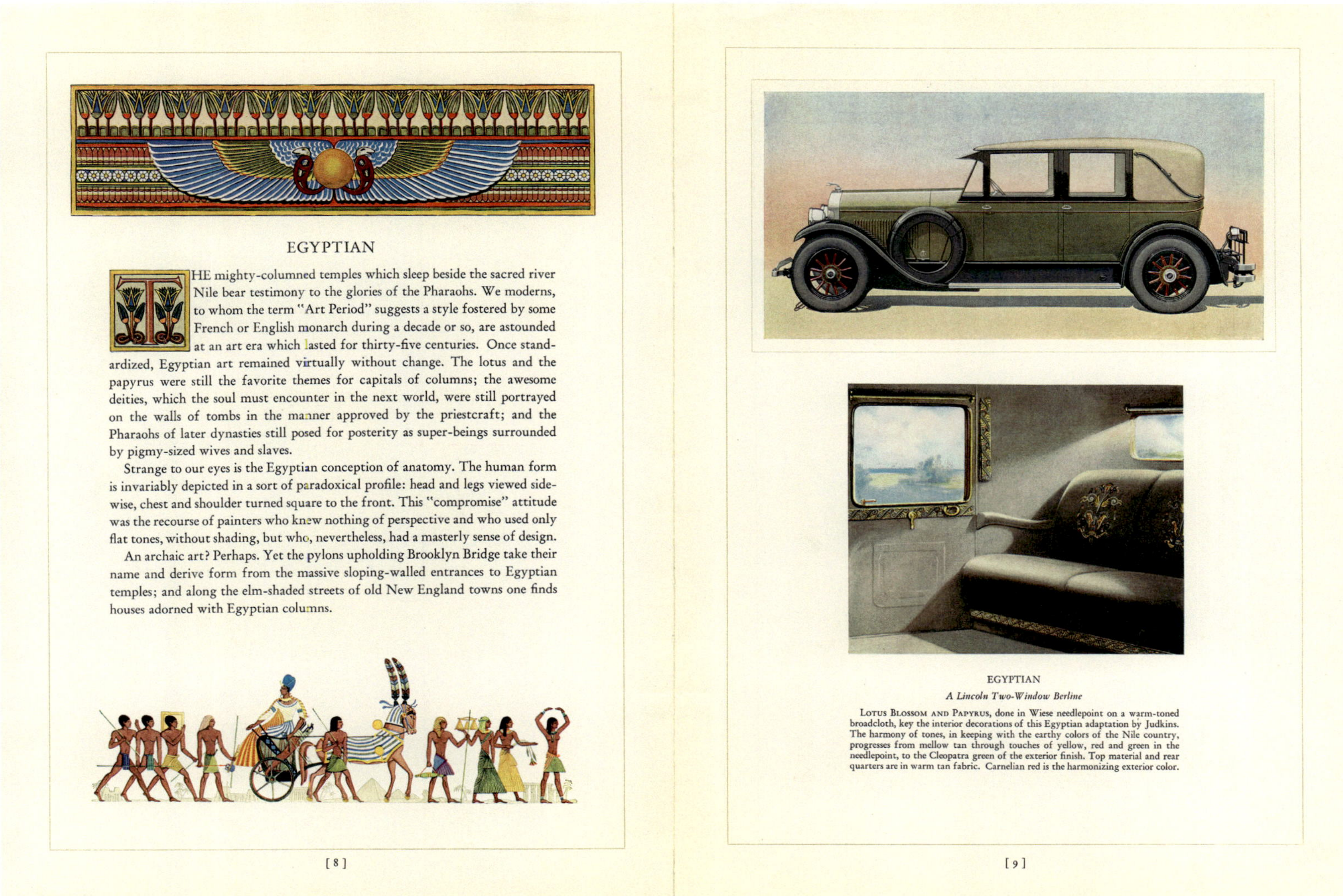

EGYPTIAN

THE mighty-columned temples which sleep beside the sacred river Nile bear testimony to the glories of the Pharaohs. We moderns, to whom the term "Art Period" suggests a style fostered by some French or English monarch during a decade or so, are astounded at an art era which lasted for thirty-five centuries. Once standardized, Egyptian art remained virtually without change. The lotus and the papyrus were still the favorite themes for capitals of columns; the awesome deities, which the soul must encounter in the next world, were still portrayed on the walls of tombs in the manner approved by the priestcraft; and the Pharaohs of later dynasties still posed for posterity as super-beings surrounded by pigmy-sized wives and slaves.

Strange to our eyes is the Egyptian conception of anatomy. The human form is invariably depicted in a sort of paradoxical profile: head and legs viewed sidewise, chest and shoulder turned square to the front. This "compromise" attitude was the recourse of painters who knew nothing of perspective and who used only flat tones, without shading, but who, nevertheless, had a masterly sense of design.

An archaic art? Perhaps. Yet the pylons upholding Brooklyn Bridge take their name and derive form from the massive sloping-walled entrances to Egyptian temples; and along the elm-shaded streets of old New England towns one finds houses adorned with Egyptian columns.

[8]

EGYPTIAN

A Lincoln Two-Window Berline

LOTUS BLOSSOM AND PAPYRUS, done in Wiese needlepoint on a warm-toned broadcloth, key the interior decorations of this Egyptian adaptation by Judkins. The harmony of tones, in keeping with the earthy colors of the Nile country, progresses from mellow tan through touches of yellow, red and green in the needlepoint, to the Cleopatra green of the exterior finish. Top material and rear quarters are in warm tan fabric. Carnelian red is the harmonizing exterior color.

[9]

Marmon

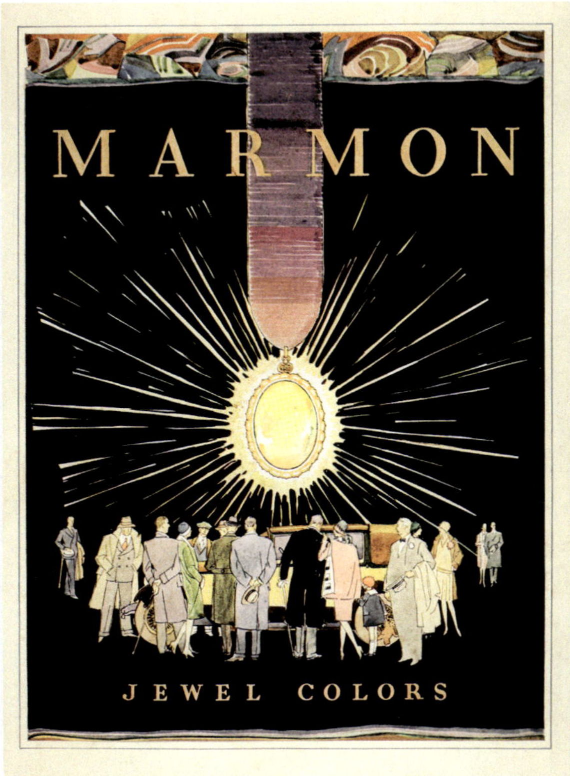

Marmon 68, 78 1927 *(this spread)*

The 78 was a step in Marmon's plan to offer straight eights exclusively, several years before the Depression upended the firm. From the outset, Marmon was a premium luxury and performance car, built in Indianapolis. A Marmon Wasp won the first Indianapolis 500 in 1911. These were some of the grandest cars in history, with a magnificent V-16 engine at the end, which came in 1933 during the depths of the Depression.

Der 78 war ein Schritt im Plan von Marmon, ausschließlich Fahrzeuge mit Achtzylinderreihenmotoren anzubieten – bis die Weltwirtschaftskrise das Unternehmen ein paar Jahre später in die Knie zwang. Von Anfang an war der Marmon, der in Indianapolis (Indiana) gebaut wurde, ein leistungsstarker Wagen der Oberklasse. Ein Marmon Wasp gewann 1911 das erste 500-Meilen-Rennen von Indianapolis. Diese Automobile gehörten zu den beeindruckendsten Fahrzeugen aller Zeiten – gegen Ende mit einem stattlichen V16-Motor ausgestattet, der ausgerechnet 1933 auf den Markt kam, am Tiefpunkt der Weltwirtschaftskrise.

Le modèle 78 constitua un jalon dans le projet de Marmon, qui consistait à proposer exclusivement des moteurs à huit cylindres en ligne, plusieurs années avant que la Grande Dépression ne mène la marque à sa perte. À l'origine Marmon était une automobile de luxe et très performante, fabriquée à Indianapolis. C'est d'ailleurs une Marmon Wasp qui remporta les premiers « Indy 500 » en 1911. Cette voiture reste comme l'une des plus prestigieuses de l'histoire automobile, avec au bout de sa course qui interviendra en 1933, au cœur de la crise, l'adoption d'un magnifique moteur V16.

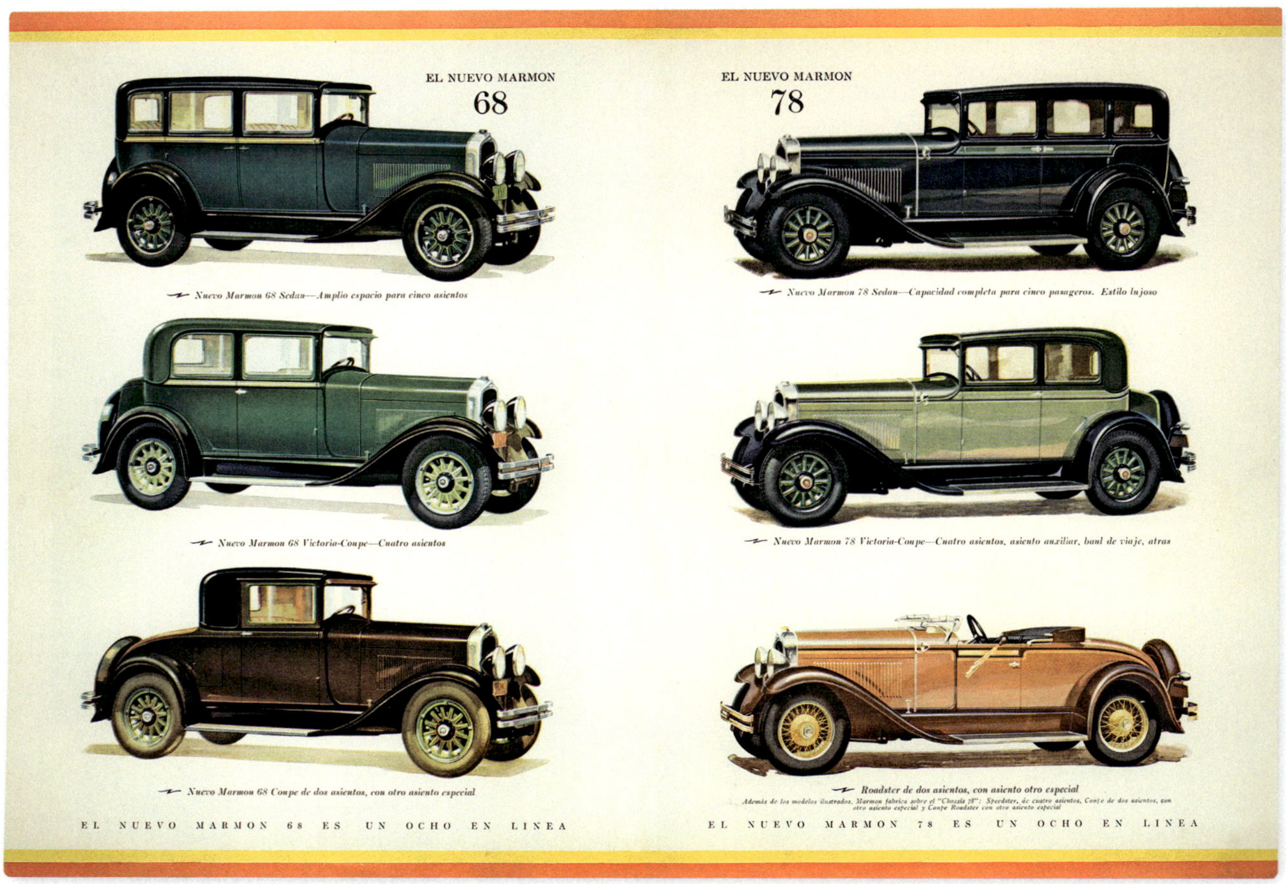

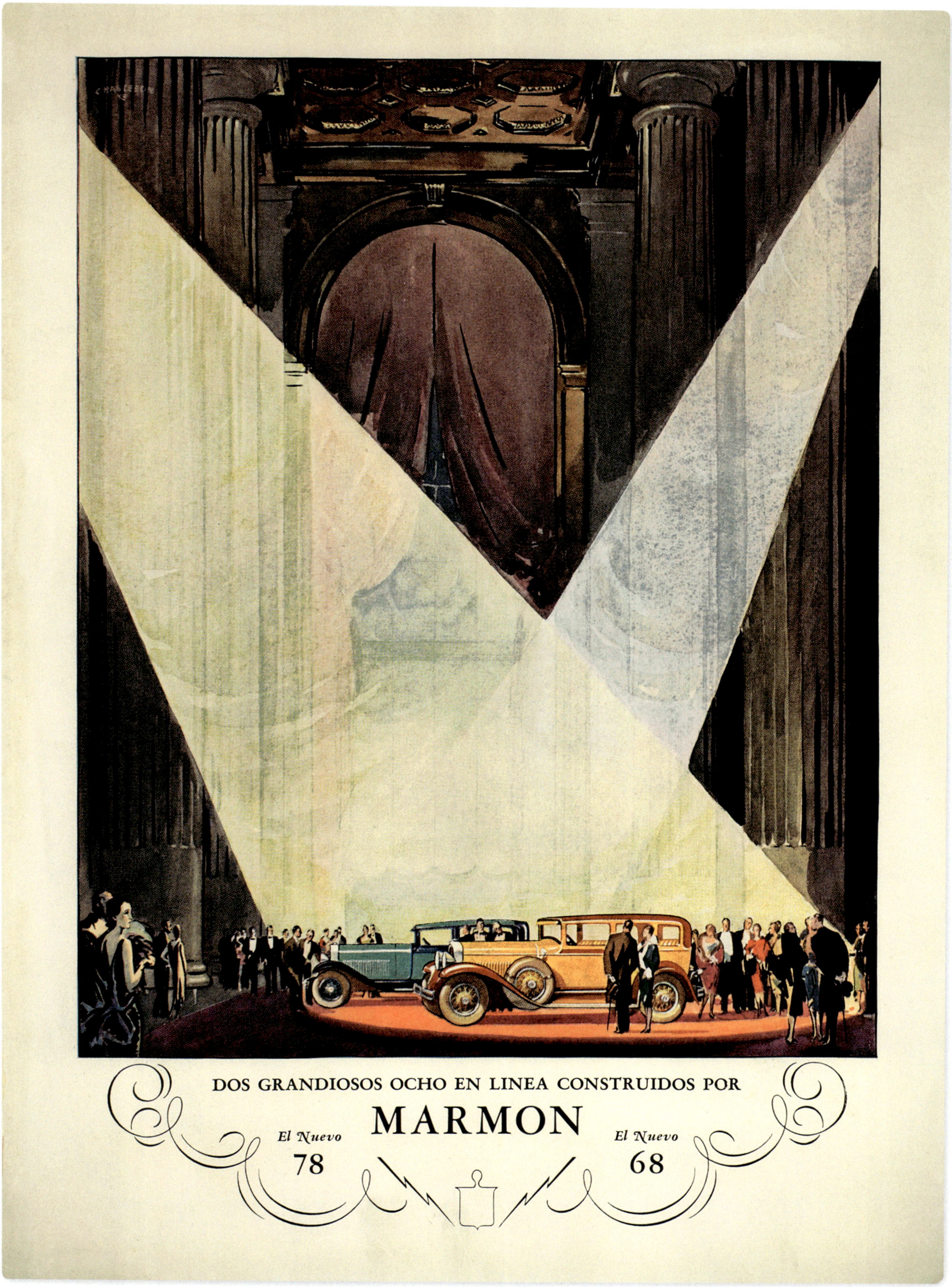
DOS GRANDIOSOS OCHO EN LINEA CONSTRUIDOS POR
MARMON
El Nuevo
78
El Nuevo
68

Maxwell

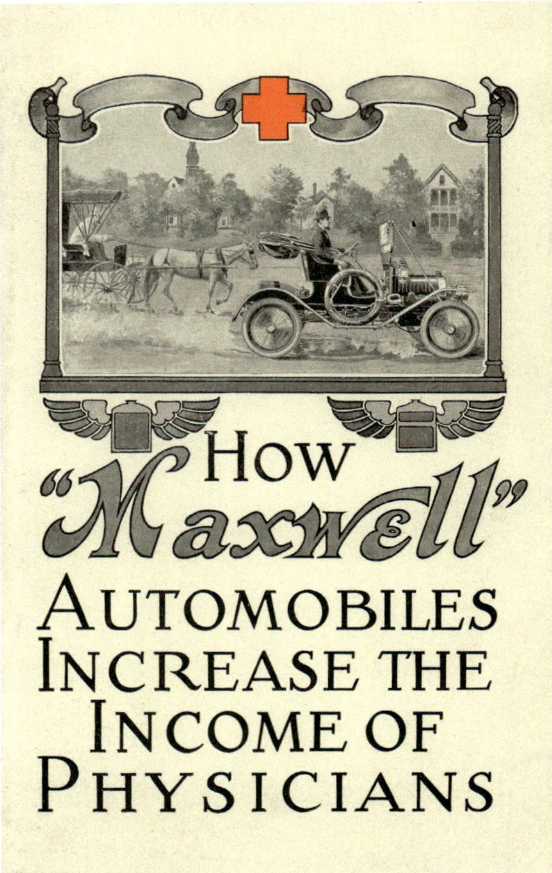

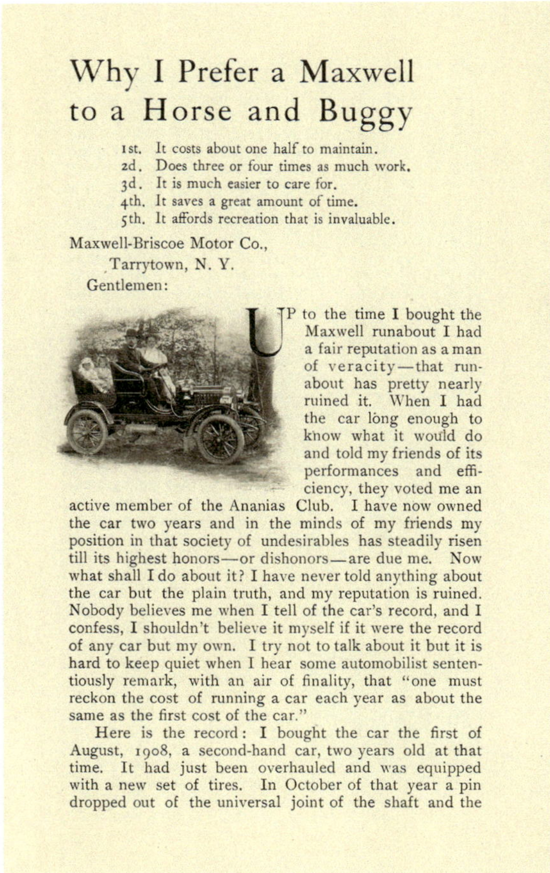

Why I Prefer a Maxwell to a Horse and Buggy

1st. It costs about one half to maintain.
2d. Does three or four times as much work.
3d. It is much easier to care for.
4th. It saves a great amount of time.
5th. It affords recreation that is invaluable.

Maxwell-Briscoe Motor Co.,
Tarrytown, N. Y.
Gentlemen:

Up to the time I bought the Maxwell runabout I had a fair reputation as a man of veracity—that runabout has pretty nearly ruined it. When I had the car long enough to know what it would do and told my friends of its performances and efficiency, they voted me an active member of the Ananias Club. I have now owned the car two years and in the minds of my friends my position in that society of undesirables has steadily risen till its highest honors—or dishonors—are due me. Now what shall I do about it? I have never told anything about the car but the plain truth, and my reputation is ruined. Nobody believes me when I tell of the car's record, and I confess, I shouldn't believe it myself if it were the record of any car but my own. I try not to talk about it but it is hard to keep quiet when I hear some automobilist sententiously remark, with an air of finality, that "one must reckon the cost of running a car each year as about the same as the first cost of the car."

Here is the record: I bought the car the first of August, 1908, a second-hand car, two years old at that time. It had just been overhauled and was equipped with a new set of tires. In October of that year a pin dropped out of the universal joint of the shaft and the

Maxwell 1911 *(above)*, **1916** *(right)*

A doctor's testimonial that a Maxwell was so reliable that he could visit more patients is the foundation of this brochure text. The Maxwell went on to be reinvented as the first four-cylinder Chrysler.

In dieser Broschüre erzählt ein Arzt, dass er den zuverlässigen Maxwell einem Pferdewagen vorziehe, weil er damit mehr Hausbesuche machen könne. Der Maxwell kam später in anderem Kleid noch einmal als erster Vierzylinder von Chrysler auf den Markt.

Le témoignage de ce médecin, sur lequel s'appuie sur le catalogue, est formel : sa Maxwell est si fiable qu'il peut rendre visite à davantage de patients qu'auparavant. La Maxwell sera réinventée sous les traits de la première Chrysler à quatre cylindres.

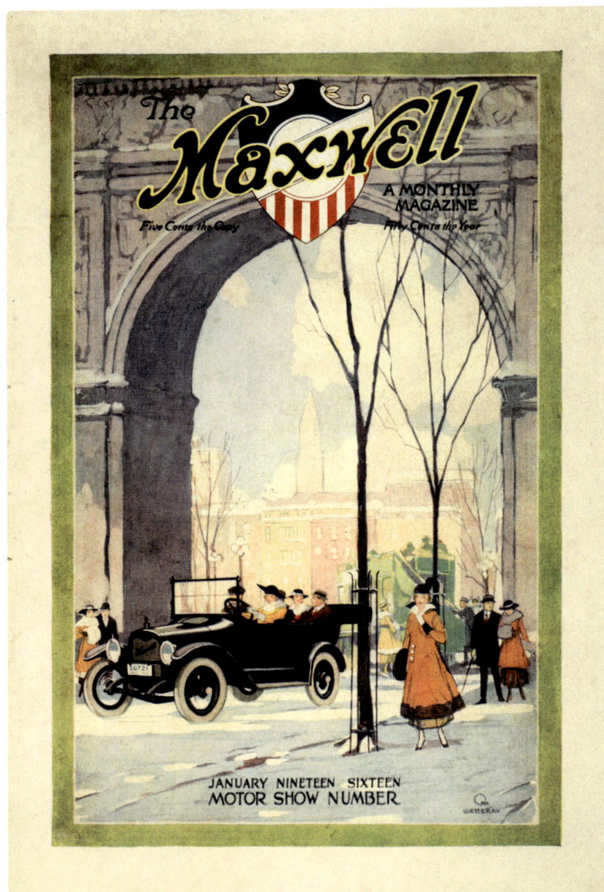

Michigan

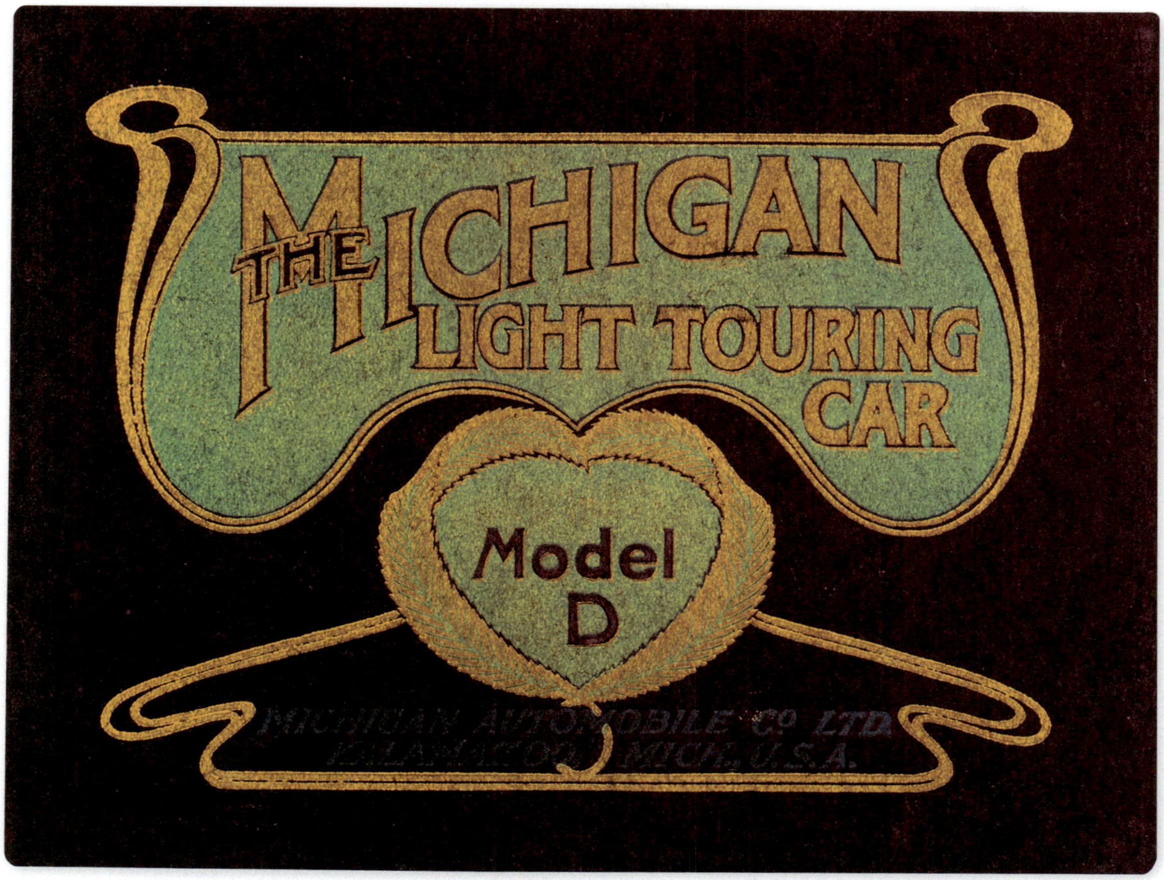

Michigan Model D 1905 *(left)*

In Kalamazoo, Michigan, a pair of brothers who made washboards and another pair who made bicycles decided to build cars. This two-cylinder Model D was built after the brothers split. The original Michigan from 1905 disappeared in 1907 when the firm's owners decided to concentrate on making parts.

In Kalamazoo (Michigan) gab es zwei Brüder, die Waschbretter herstellten, und zwei weitere, die Fahrräder produzierten. Dann entschlossen sie sich, gemeinsam Autos zu bauen. Dieses Model D mit zwei Zylindern wurde produziert, nachdem die Brüder schon wieder getrennte Wege gingen. Das ursprüngliche Unternehmen aus dem Jahr 1905 verschwand 1907 vom Markt, als die Firmeneigner entschieden, fortan nur noch Teile zu fertigen.

À Kalamazoo, dans le Michigan, deux frères qui fabriquaient des planches à laver et deux autres qui assemblaient des bicyclettes décidèrent un beau jour de construire des automobiles. Cette Model D bicylindre fut construite après la séparation des deux duos. La Michigan originale de 1905 disparut en 1907 quand les propriétaires de la firme choisirent de se concentrer sur la fabrication de pièces détachées.

Moon

Moon 6-40 Sport Sedan 1924 *(right)*

The Moon, which originated in St. Louis, used Continental power and, after 1924, a radiator shell inspired by Rolls-Royce. The company was named for its founder, Joseph W. Moon, a former buggy builder, and not for the extraterrestrial body.

Der Moon, der aus St. Louis (Missouri) kam, verwendete Continental-Motoren und ab 1924 eine von Rolls-Royce inspirierte Kühlerverkleidung. Das Unternehmen war nach seinem Gründer Joseph W. Moon benannt, einem ehemaligen Pferdewagenbauer, und nicht etwa nach dem Erdtrabanten.

La Moon, née à Saint-Louis, dans le Missouri, utilisait un moteur Continental et, après 1924, une calandre de radiateur inspirée par Rolls-Royce. L'entreprise ne portait pas le nom de la Lune, mais celui de son fondateur Joseph W. Moon, un ancien fabricant d'attelages pour chevaux.

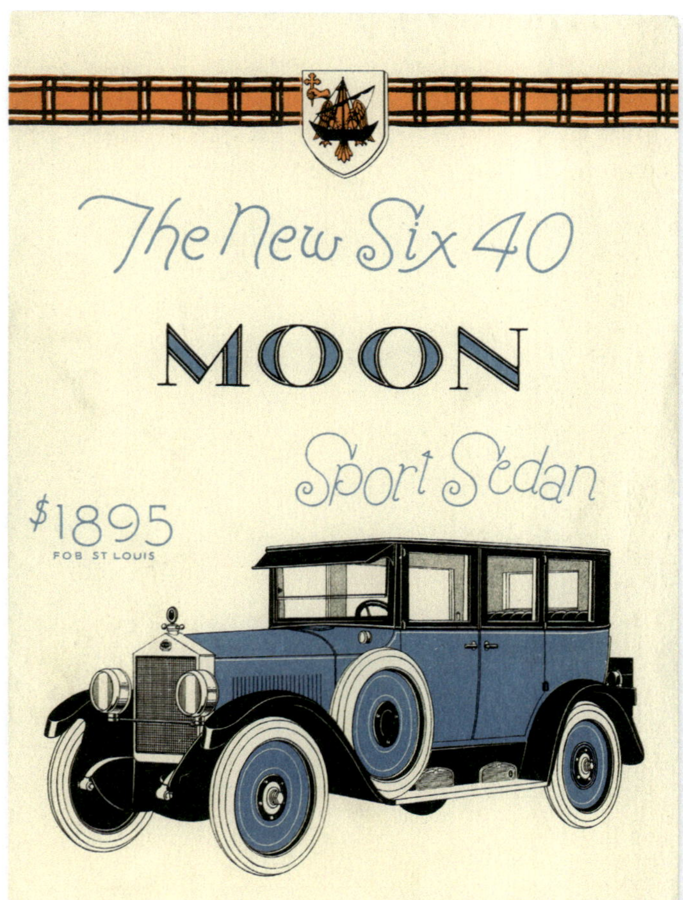

Oakland

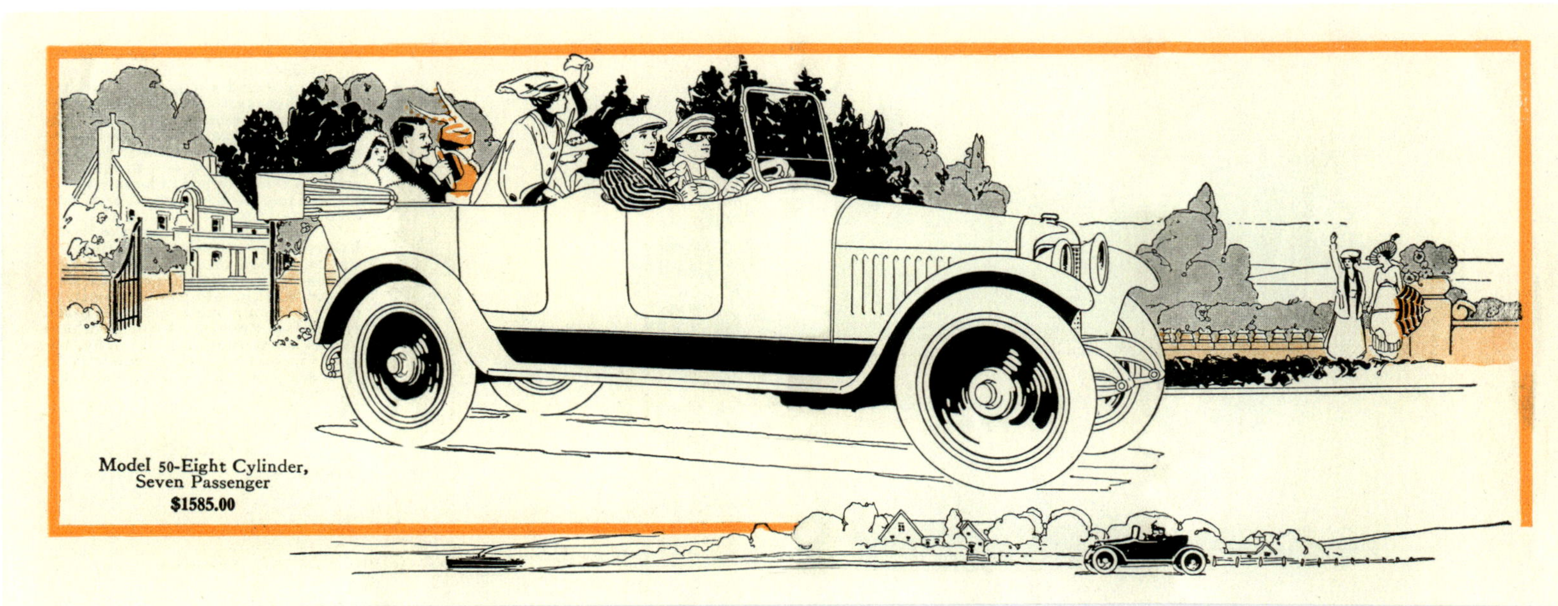

Oakland 1916 *(above)*, **1926** *(below and opposite)*

The Oakland offered a variety of engines before finally settling on an affordably priced V-8. A new brand, Pontiac, eclipsed the Oakland in 1931. The Oakland was conceived by Edward Murphy, builder of the Pontiac buggy, and Alanson Brush, an early industry pioneer who'd been heavily involved in the development of Henry M. Leland's original Cadillac.

Der Oakland bot viele Motorisierungen an, bevor man sich auf einen erschwinglichen V8-Motor festlegte. Im Jahre 1931 wurde der Oakland allerdings von der neuen Marke Pontiac überschattet. Entworfen hatten ihn Edward Murphy, der Konstrukteur des Pontiac-Pferdewagens, und Alanson Brush, ein früher Industriepionier, der auch stark an der Entwicklung von Henry M. Lelands ursprünglichem Cadillac beteiligt gewesen war.

L'Oakland proposa plusieurs types de motorisations avant d'opter pour un V8 très abordable, conçu par Edward Murphy, constructeur des voitures hippomobiles Pontiac, et Alanson Brush, pionnier de l'histoire automobile, qui avait été fortement impliqué dans le développement de la Cadillac originelle d'Henry M. Leland. Une nouvelle marque, Pontiac, éclipsa Oakland en 1931.

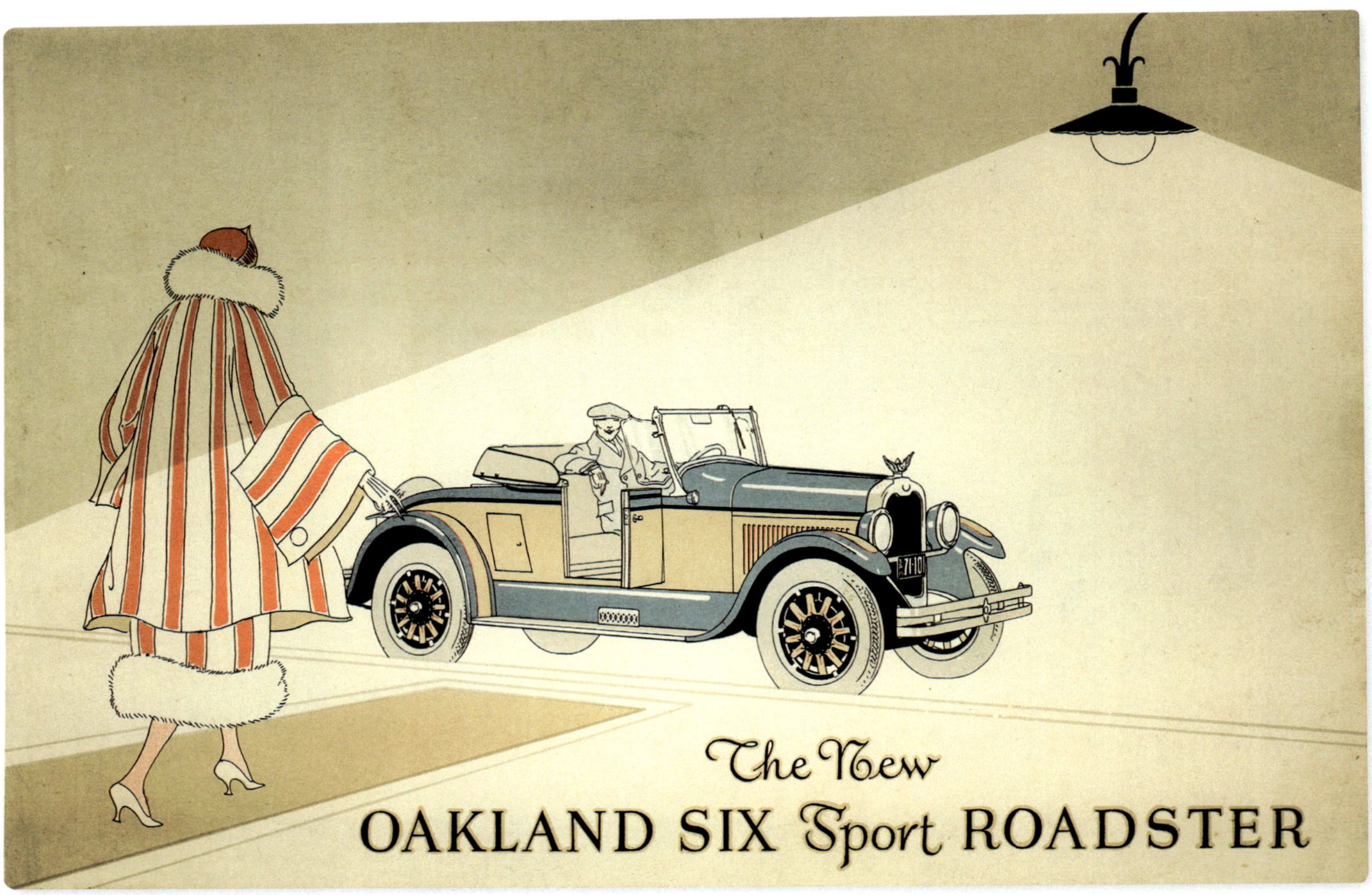

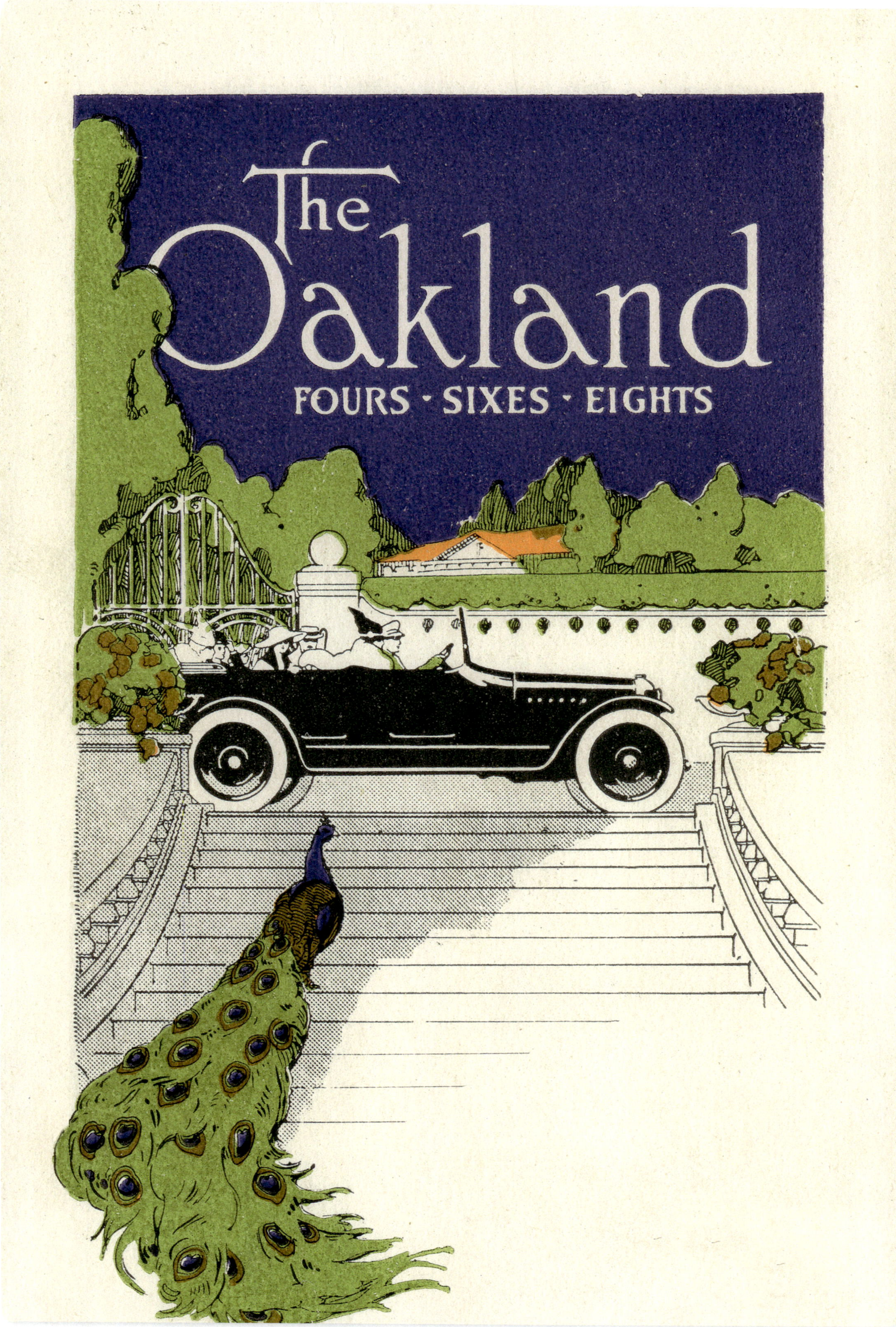
The Oakland
FOURS · SIXES · EIGHTS

Packard

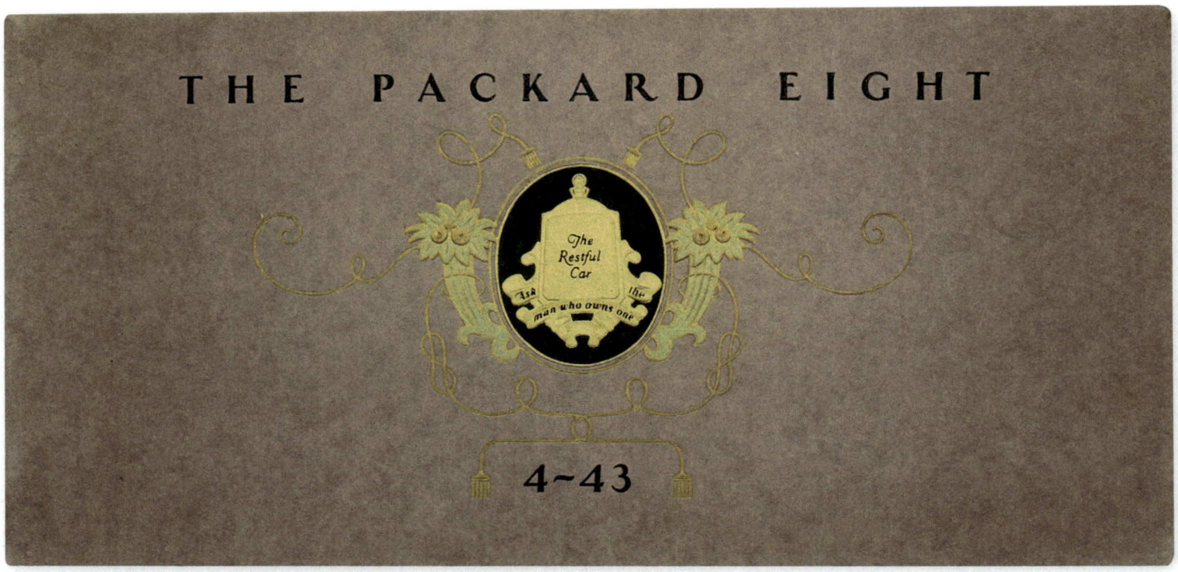

Packard 1928

Packard was founded in Warren, Ohio, in 1899 by James Ward Packard, who was so dissatisfied with a Winton he had bought that he decided to build his own car. That origin is now part of automotive folklore although it did actually occur. The firm moved to Detroit in 1903 and became fabled as one of the "Three Ps" of American luxury cars, the other two being Pierce-Arrow and Peerless. Packard generally gets credit for building the first automobile equipped with air conditioning, in 1940.

Weil James Ward Packard mit einem Winton, den er gekauft hatte, unzufrieden war, gründete er 1899 in Warren (Ohio) die Firma, die seinen Namen trug, und baute sein eigenes Auto. Diese Anekdote gehört inzwischen zur automobilen Folklore, aber sie hat sich wirklich so zugetragen. Die Firma zog 1903 nach Detroit um und machte sich einen Namen als eines der „drei Ps" der amerikanischen Luxusautomobile, neben Pierce-Arrow und Peerless. Packard schreibt man gemeinhin zu, 1940 das erste Auto mit Klimaanlage gebaut zu haben.

La marque Packard fut fondée en 1899 à Warren, dans l'Ohio, par James Ward Packard, qui était si mécontent de sa Winton personnelle qu'il décida alors de fabriquer sa propre voiture. Cette anecdote authentique appartient désormais au folklore de l'histoire automobile. La firme déménagea à Detroit en 1903 et accéda au rang de mythe parmi les « trois P » de la voiture de luxe américaine – aux côtés de Pierce-Arrow et Peerless. On crédite généralement Packard de la fabrication du premier modèle équipé de la climatisation, en 1940.

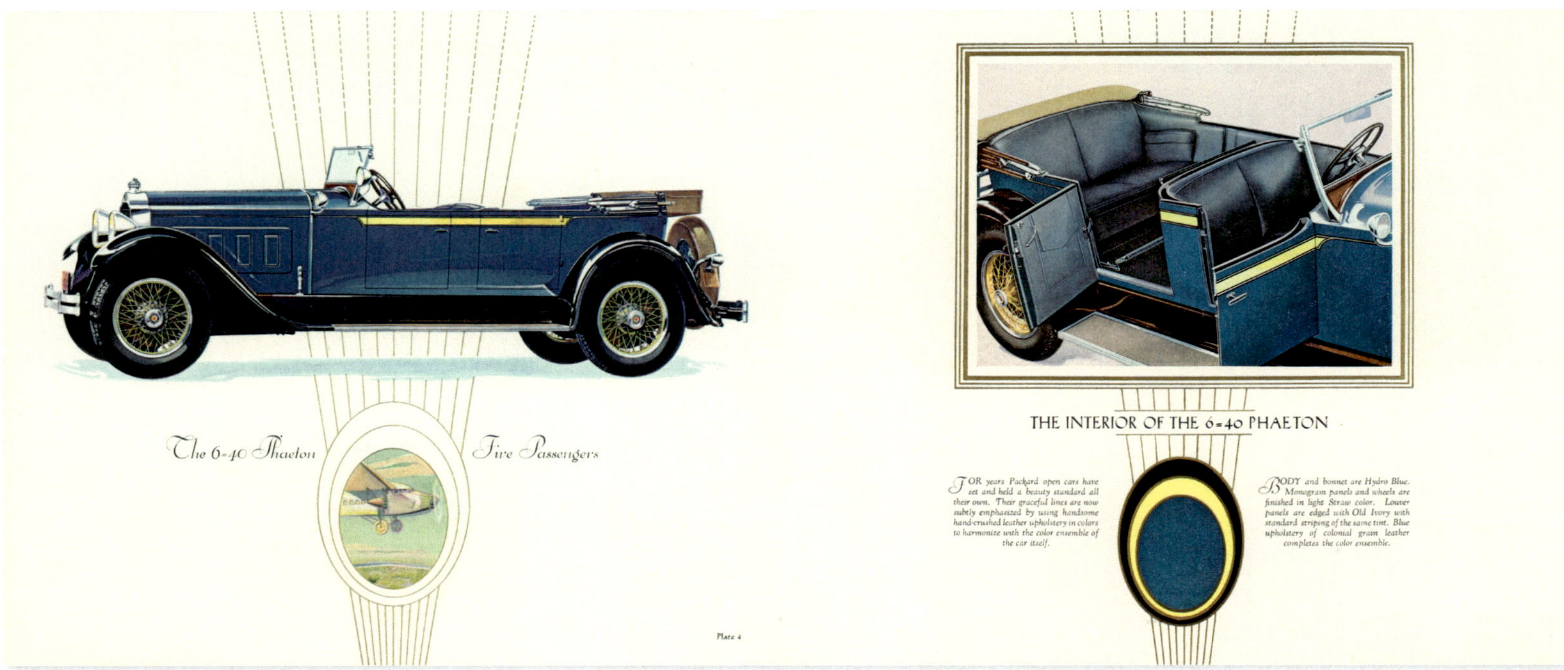

Paige

Paige 1927

Numerous body styles supported Paige's slogan, "The Most Beautiful Car in America." Additionally, it held a Daytona Beach speed record. The brand eventually became the Graham-Paige, but before that, when it was headed by Harry Jewett, the car was known as the Paige-Detroit. Jewett, who had also brought out a cheaper self-named car, sold out in 1928 to the Graham Brothers, who had a strong alliance with Dodge, prompting the name change. The Graham automobile lasted until 1940.

Zahlreiche Karosserievarianten stützten den Werbeslogan von Paige, „das schönste Auto in Amerika" zu bauen. Darüber hinaus hielt es den Geschwindigkeitsrekord auf der Rennstrecke von Daytona Beach. Letztendlich wurde die Marke in Graham-Paige umbenannt, doch bevor es so weit war, war sie – unter der Firmenleitung von Harry Jewett – als Paige-Detroit bekannt. Jewett, der unter eigenem Namen auch ein billigeres Auto auf den Markt gebracht hatte, verkaufte seine Firma 1928 an die Brüder Graham, die eng mit Dodge zusammenarbeiteten. Graham-Automobile wurden bis 1940 gebaut.

De nombreuses carrosseries différentes s'efforçaient d'appuyer le slogan choisi pour la Paige : « La plus belle voiture d'Amérique. » Elle détenait qui plus est un record de vitesse obtenu à Daytona. Avant d'être vendue en 1927 aux frères Graham, proches de Dodge, et de devenir Graham-Paige puis Graham, la marque s'est appelée Paige-Detroit quand elle était détenue par Harry Jewett, qui avait également lancé un modèle peu onéreux portant son nom. Les Graham seront construites jusqu'en 1940.

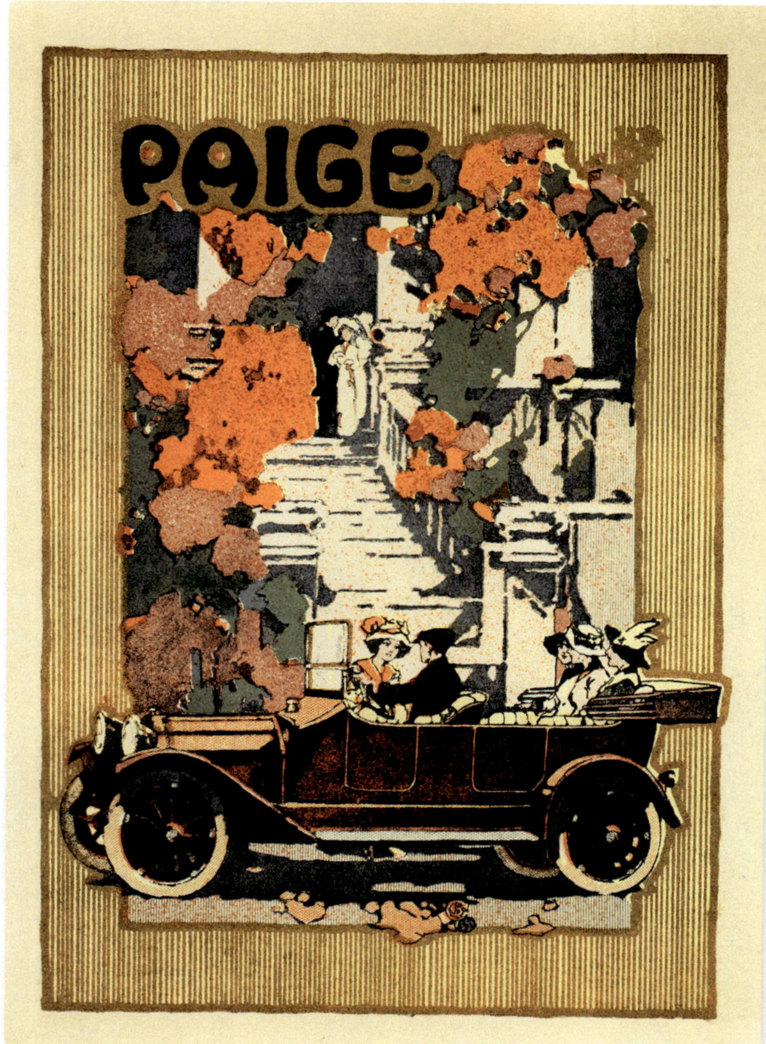

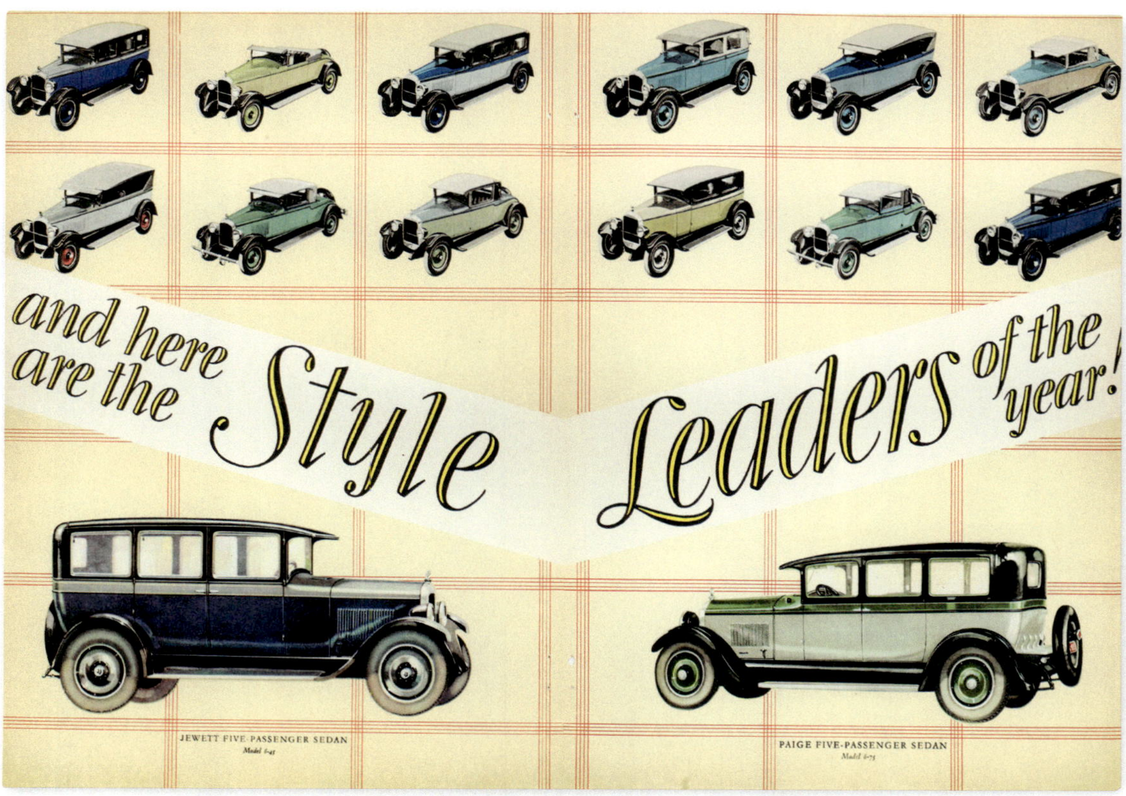

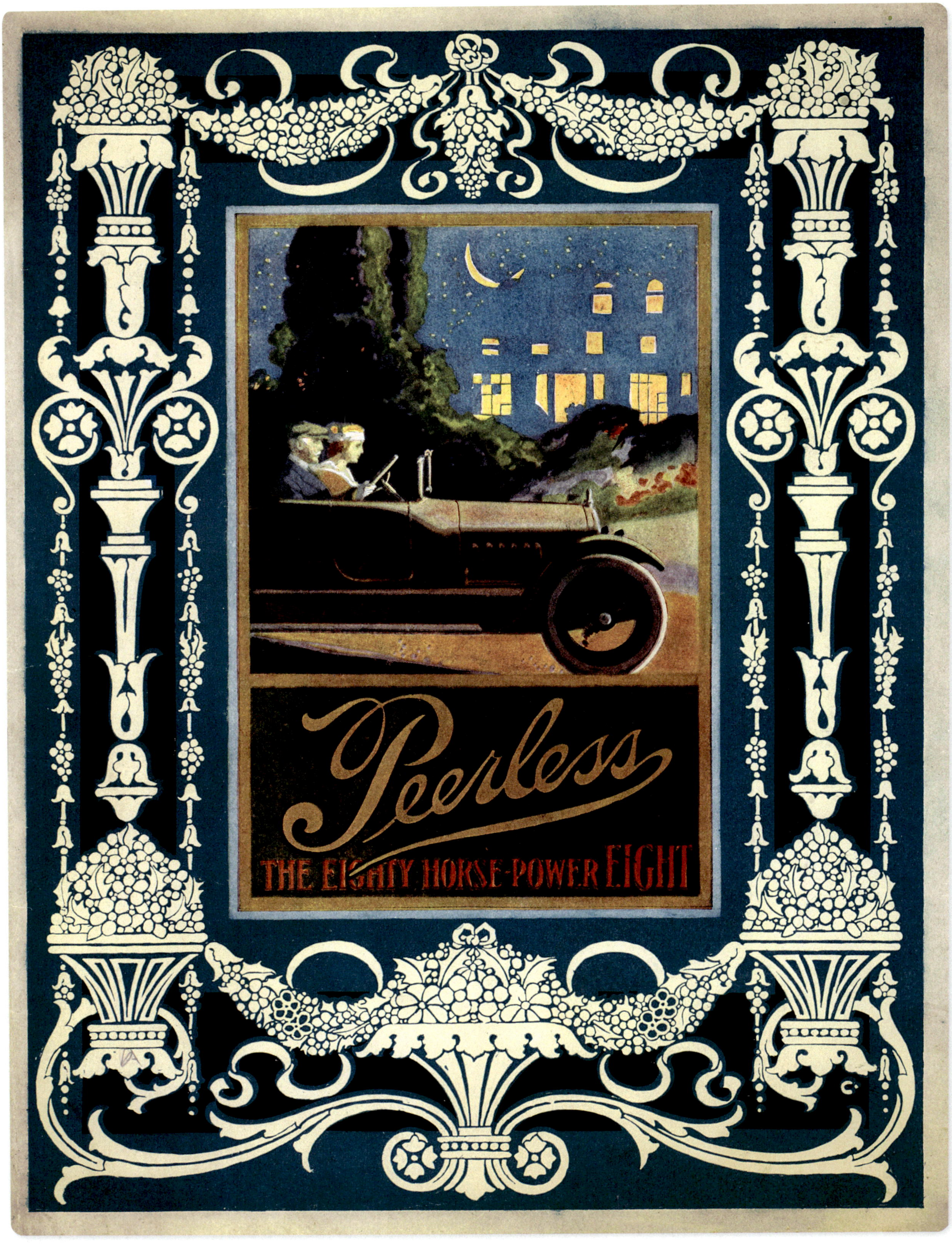
Peerless
THE EIGHTY HORSE-POWER EIGHT

Peerless

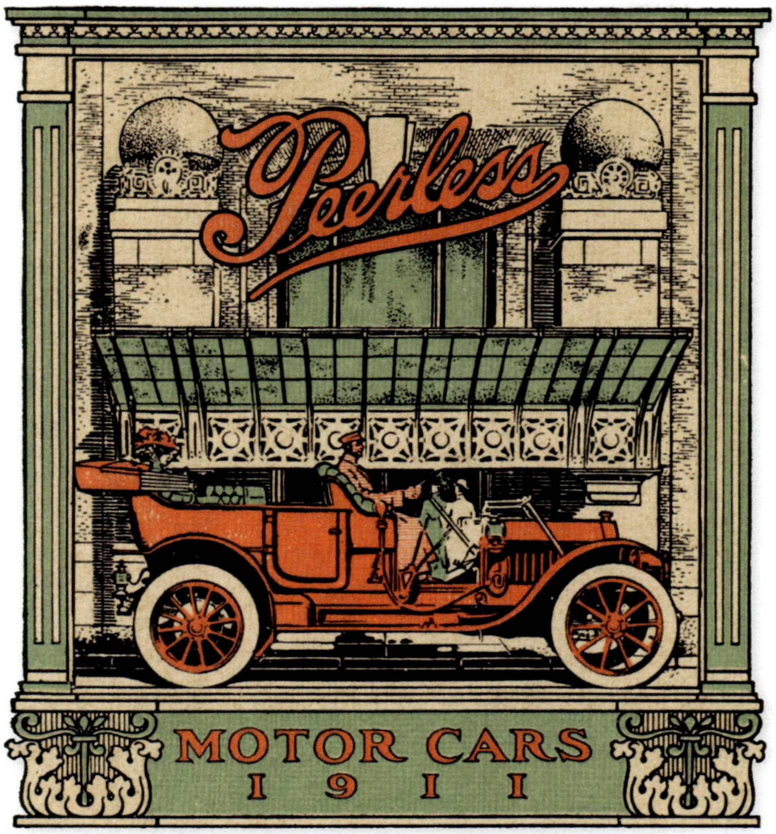

Peerless 1911 *(this spread)*

Largely forgotten today, the Peerless was a titan of an automobile, aiming for prestige from the start. It had an early big six and a V-8 engine by 1916, wowing consumers. Peerless actually began in 1900 from a company that had earlier made laundry wringers. Based in Cleveland, it expanded from bicycles to motorized buggies. The Peerless then blossomed into one of the great early American luxury cars until the Depression ended auto production in 1931 and sent the firm into a new business: brewing beer and ales when Prohibition ended.

Der heute weitgehend in Vergessenheit geratene Peerless war ein Titan unter den Automobilen, der von Anfang an auf Prestige ausgelegt war. Schon früh war er mit einem großen Sechszylinder- und ab 1916 mit einem V8-Motor ausgestattet, der die Autofahrer verzückte. Peerless ging 1900 aus einer Firma hervor, die zuvor Wäschemangeln hergestellt hatte. In Cleveland (Ohio) erweiterte sie ihre Produktpalette zunächst von Fahrrädern auf Motorwagen. Später blühte der Peerless zu einem der ersten großen amerikanischen Luxusautomobile auf, bis die Weltwirtschaftskrise die Autoproduktion 1931 jäh beendete und ein neues Zeitalter für das Unternehmen einläutete: Mit dem Ende der Prohibition begann die Firma, Bier zu brauen.

Très largement oubliée aujourd'hui, la Peerless était dans son genre une vraie géante, affamée de prestige dès sa naissance : tôt pourvue d'un gros six cylindres, elle héritera en 1916 d'un V8 qui en mettra plein les yeux aux acheteurs potentiels. Peerless est né en 1900 sur les bases d'une société qui avait fabriqué des essoreuses. Basée à Cleveland, la marque passa aux bicyclettes puis aux voitures motorisés. La Peerless deviendra par la suite l'une des premières grandes voitures de luxe américaines, jusqu'à ce que la crise économique ne mette un terme à sa production en 1931 et n'oriente l'entreprise vers un nouveau business : la bière, après la fin de la Prohibition.

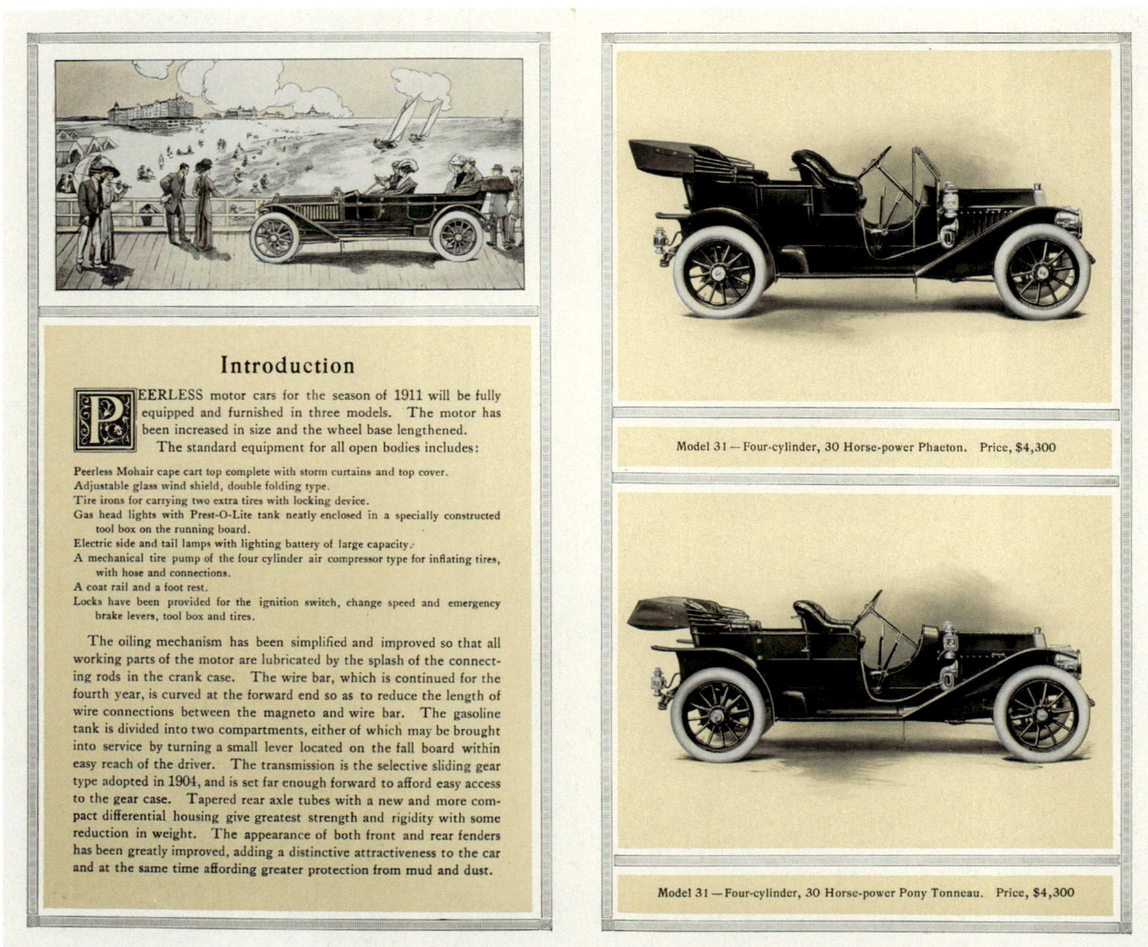

Introduction

PEERLESS motor cars for the season of 1911 will be fully equipped and furnished in three models. The motor has been increased in size and the wheel base lengthened.

The standard equipment for all open bodies includes:

Peerless Mohair cape cart top complete with storm curtains and top cover.
Adjustable glass wind shield, double folding type.
Tire irons for carrying two extra tires with locking device.
Gas head lights with Prest-O-Lite tank neatly enclosed in a specially constructed tool box on the running board.
Electric side and tail lamps with lighting battery of large capacity.
A mechanical tire pump of the four cylinder air compressor type for inflating tires, with hose and connections.
A coat rail and a foot rest.
Locks have been provided for the ignition switch, change speed and emergency brake levers, tool box and tires.

The oiling mechanism has been simplified and improved so that all working parts of the motor are lubricated by the splash of the connecting rods in the crank case. The wire bar, which is continued for the fourth year, is curved at the forward end so as to reduce the length of wire connections between the magneto and wire bar. The gasoline tank is divided into two compartments, either of which may be brought into service by turning a small lever located on the fall board within easy reach of the driver. The transmission is the selective sliding gear type adopted in 1904, and is set far enough forward to afford easy access to the gear case. Tapered rear axle tubes with a new and more compact differential housing give greatest strength and rigidity with some reduction in weight. The appearance of both front and rear fenders has been greatly improved, adding a distinctive attractiveness to the car and at the same time affording greater protection from mud and dust.

Model 31 — Four-cylinder, 30 Horse-power Phaeton. Price, $4,300

Model 31 — Four-cylinder, 30 Horse-power Pony Tonneau. Price, $4,300

Pierce-Arrow

Pierce Great Arrow 1907

This Buffalo-built car was already becoming a legend, and the brochure proudly displayed the stoutness of its chassis and its new straight-six with an impressive 65 hp. Two years later it would be renamed the Pierce-Arrow. As the automotive industry gained its footing, it began to offer specific lines of cars based on price. The Pierce-Arrow swiftly rose to the top tier of American automobiles based on luxury and exclusivity. Other such makes included Packard, Peerless, Cadillac, and, later, Ford's Lincoln and Chrysler's Imperial.

Dieses in Buffalo (New York) gebaute Auto war schon damals eine Legende, und die Broschüre verweist stolz auf das solide Fahrgestell und den neuen Sechszylinderreihenmotor, der eindrucksvolle 88 kW lieferte. Zwei Jahre später wurde die Marke in Pierce-Arrow umgetauft. Als die Automobilindustrie langsam Fuß fasste, begann man unterschiedliche Baureihen in verschiedenen Preissegmenten anzubieten. Der Pierce-Arrow stieg, gemessen an Luxus und Exklusivität, rasch in die oberste Schicht amerikanischer Automobile auf. Andere Marken in dieser Kategorie waren Packard, Peerless, Cadillac und später Lincoln von Ford und Imperial von Chrysler.

La légende, déjà, commençait à auréoler cette voiture construite à Buffalo, dont le catalogue vantait fièrement la robustesse de son châssis et de son nouveau moteur six cylindres en ligne – qui développait tout de même 65 chevaux-vapeur ! Deux ans plus tard, elle sera rebaptisée Pierce-Arrow. L'industrie automobile, à mesure qu'elle trouvait ses marques, commençait à proposer des gammes de modèles fondées sur le prix. La Pierce-Arrow accéda ainsi rapidement à la classe supérieure des voitures américaines revendiquant un positionnement de luxe et d'exclusivité. Elle y côtoiera d'autres marques – Packard, Peerless, Cadillac puis, plus tard, Lincoln (Ford) et Imperial (Chrysler).

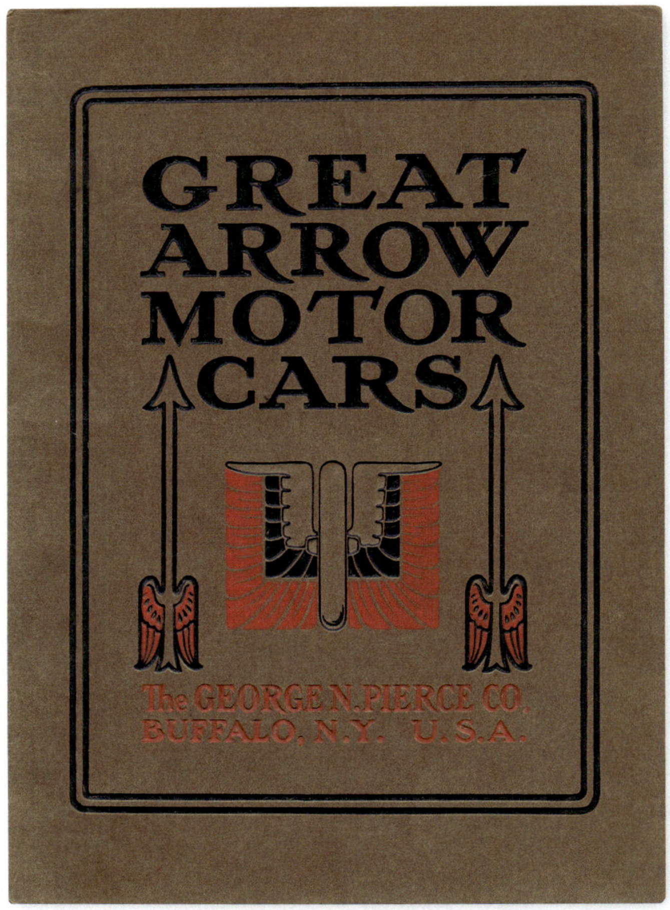

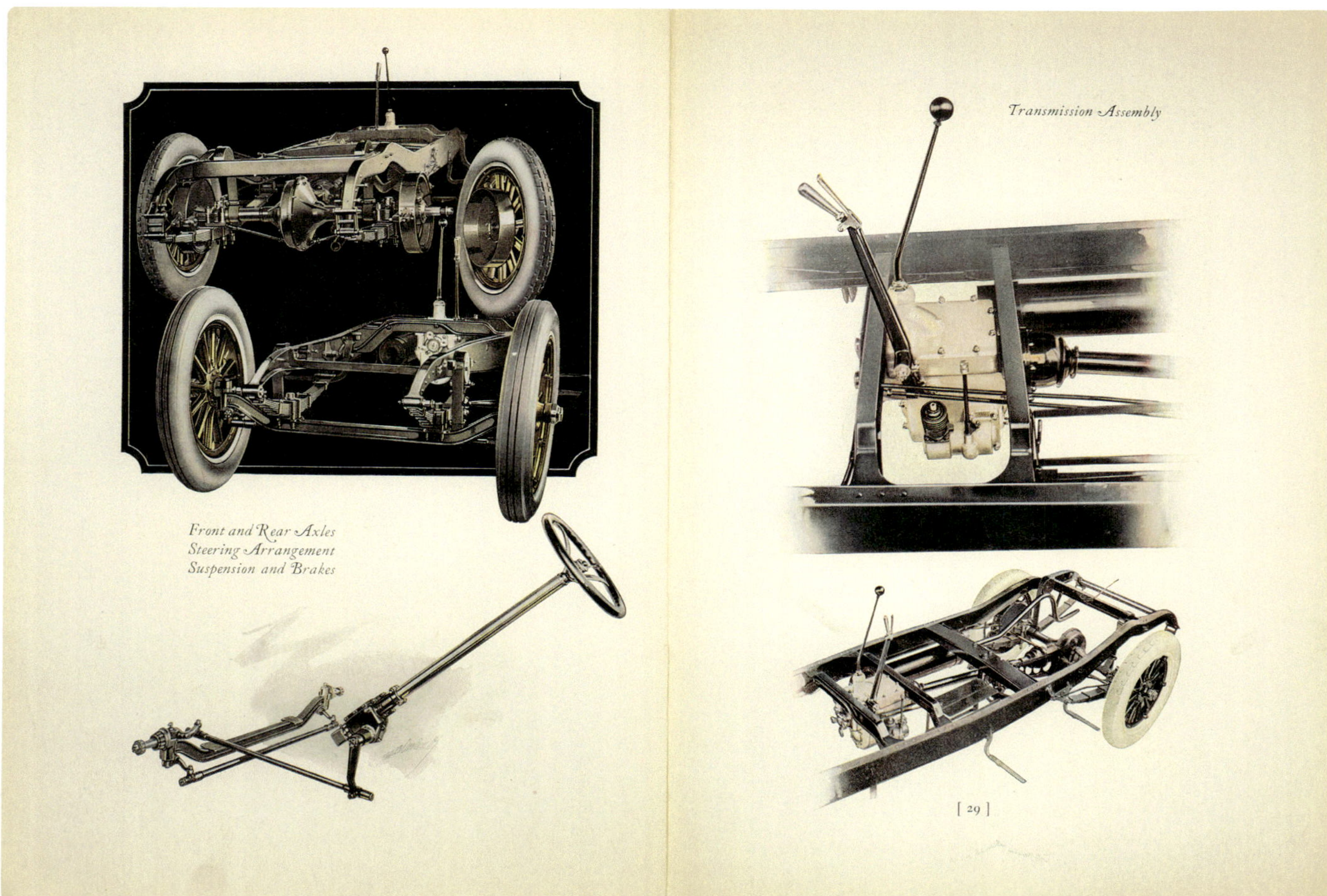

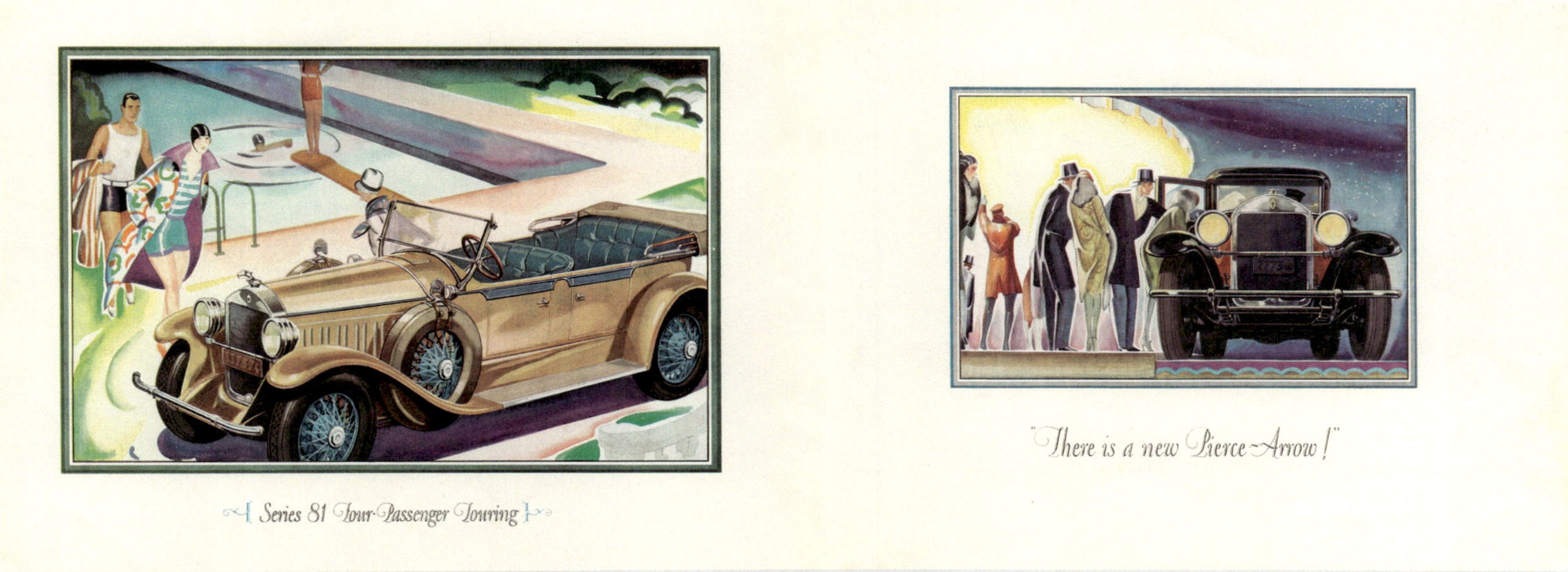

Pierce-Arrow 81 1928

For those who have never seen one up close, it's difficult to appreciate the size, grandeur, and quality of the early Pierce-Arrows. It becomes even more impressive when you consider that the company founder started out making birdcages in Buffalo, New York. Today Pierce-Arrow is revered as one of the most exclusive and best-engineered automobiles of the great American classic era. Unfortunately, the Depression claimed it as well, in 1938.

Wer sie noch nie aus der Nähe gesehen hat, kann die Größe, Pracht und Qualität der frühen Pierce-Arrows kaum einschätzen. Noch eindrucksvoller wird es, wenn man bedenkt, dass der Firmengründer als Hersteller von Vogelkäfigen in Buffalo (New York) angefangen hatte – vom Vogelbauer zum Autobauer. Heutzutage gilt der Pierce-Arrow als eines der exklusivsten und ausgereiftesten Automobile aus der Ära der amerikanischen Klassiker. Bedauerlicherweise fiel auch er 1938 der Weltwirtschaftskrise zum Opfer.

Qui n'a jamais eu l'occasion d'admirer les premières Pierce-Arrow de près peut difficilement prendre la mesure de leur majesté, de leurs dimensions et de leurs qualités. Surtout lorsqu'on sait que le fondateur de l'entreprise avait démarré dans la vie en fabriquant des cages à oiseau à Buffalo. Aujourd'hui, le nom de Pierce-Arrow est célébré comme celui de l'une des voitures les plus « exclusives » et les mieux conçues de cet âge d'or américain. Hélas, d'elle aussi la crise aura raison, et la marque disparaîtra en 1938.

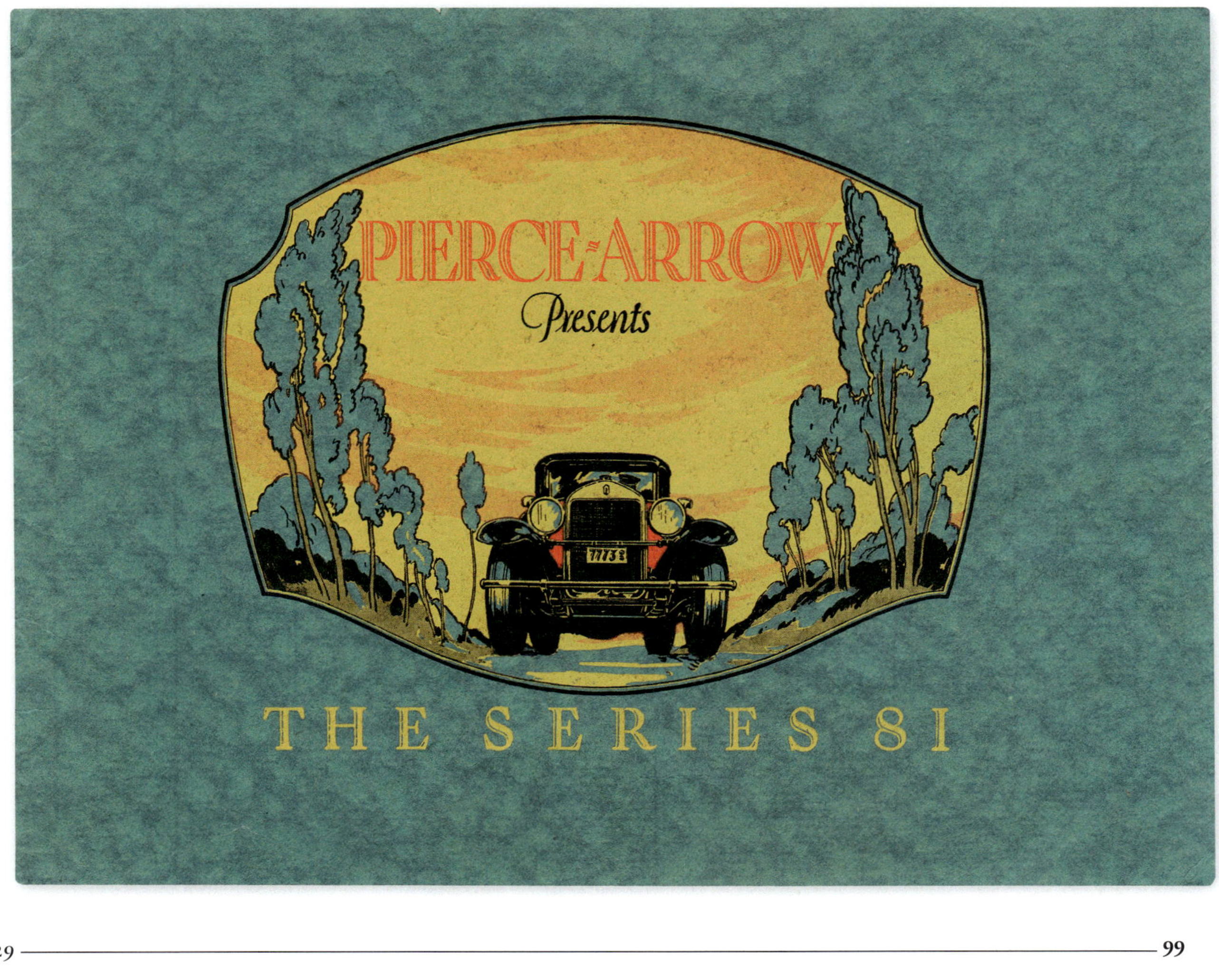

Pullman

Pullman 1917

It was no coincidence that the cover visually links the company to its luxury railroad cars. The first Pullman had six wheels, and though later editions were more conventional, they were too costly. The firm was gone by 1917.

Es war kein Zufall, dass das Titelblatt eine Verbindung zu den luxuriösen Eisenbahnwaggons des gleichen Unternehmens herstellte, die bis heute diesen Namen tragen. Der erste Pullman besaß noch sechs Räder, aber auch wenn spätere Auflagen eher dem gewohnten Aussehen eines Autos entsprachen, waren sie doch zu teuer, und die Firma war 1917 als Autobauer wieder vom Markt verschwunden.

Ce n'est pas un hasard si la couverture de cette brochure renvoie l'entreprise à ses luxueux wagons de chemin de fer. La première automobile Pullman avait six roues, et même si les versions suivantes seront plus conventionnelles, elles coûteront trop cher. La firme disparaîtra en 1917.

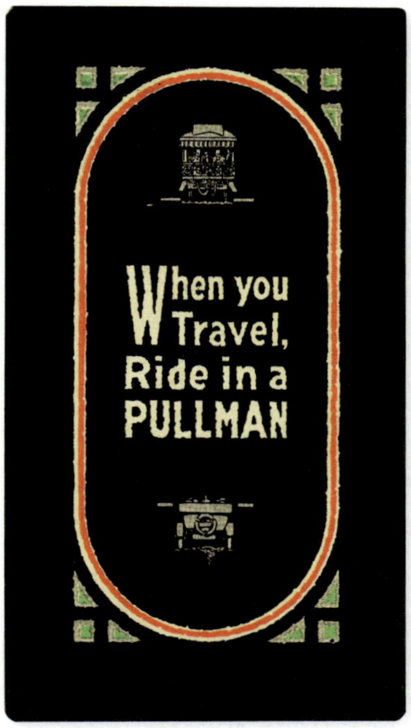

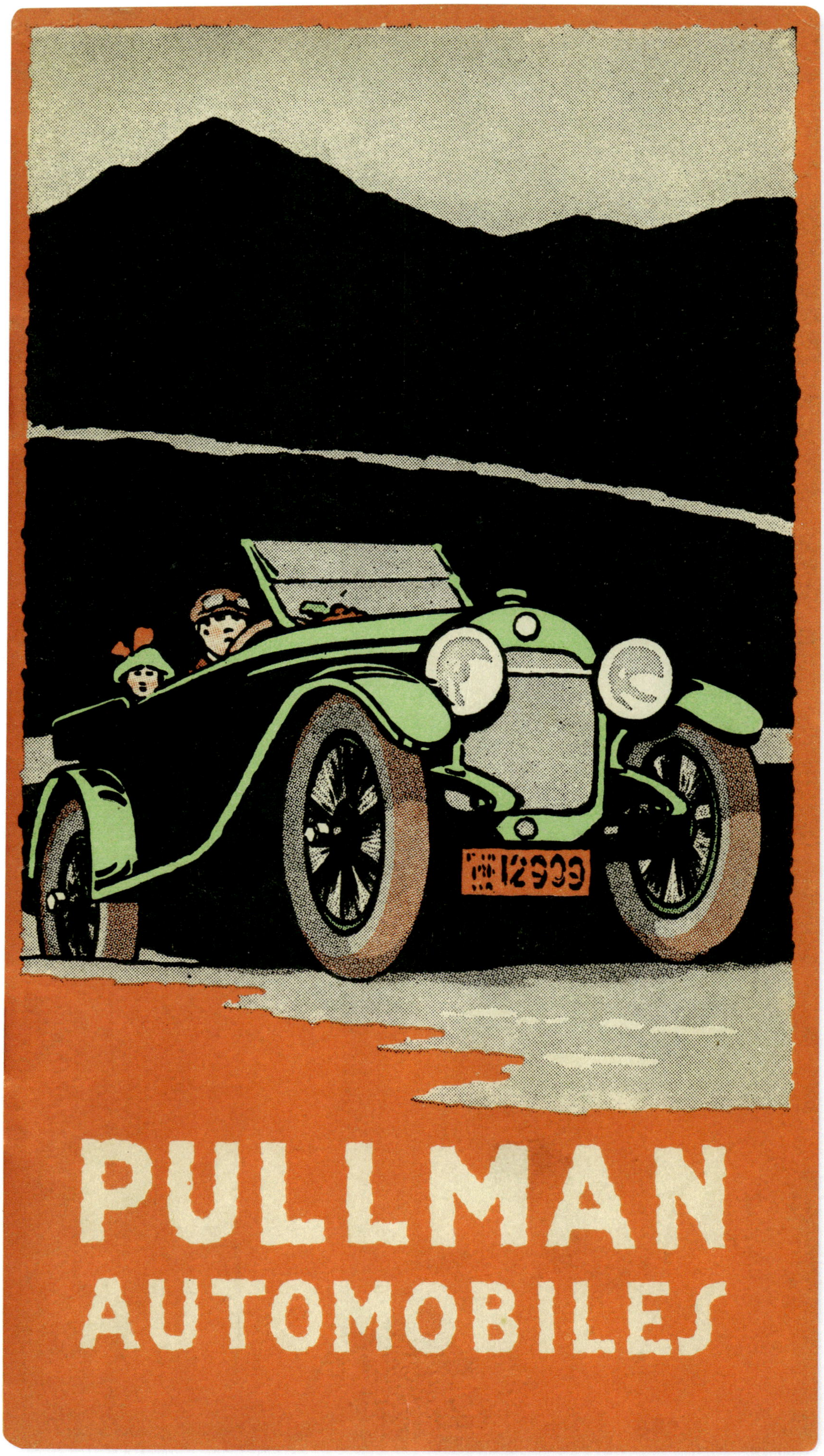

REO

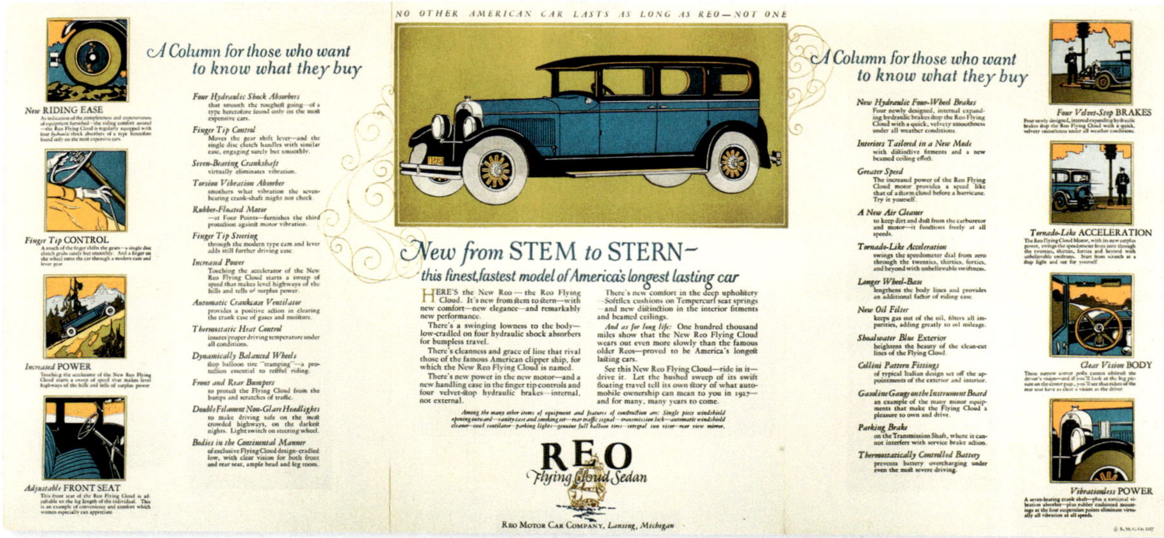

REO Flying Cloud 1929

Ransom Eli Olds had no legal claim to the name "Oldsmobile" after leaving the company in 1905, so he went on to found REO, named for his initials, that same year, and his reputation was strong enough to underpin it. Among other things, REO introduced the centrally located floor gearshift to the industry. The REO Flying Cloud is one of the great cars of the late 1920s, with a robust six-cylinder engine that boasted seven main bearings, and was the first to use Lockheed's new expanding brakes for automobiles.

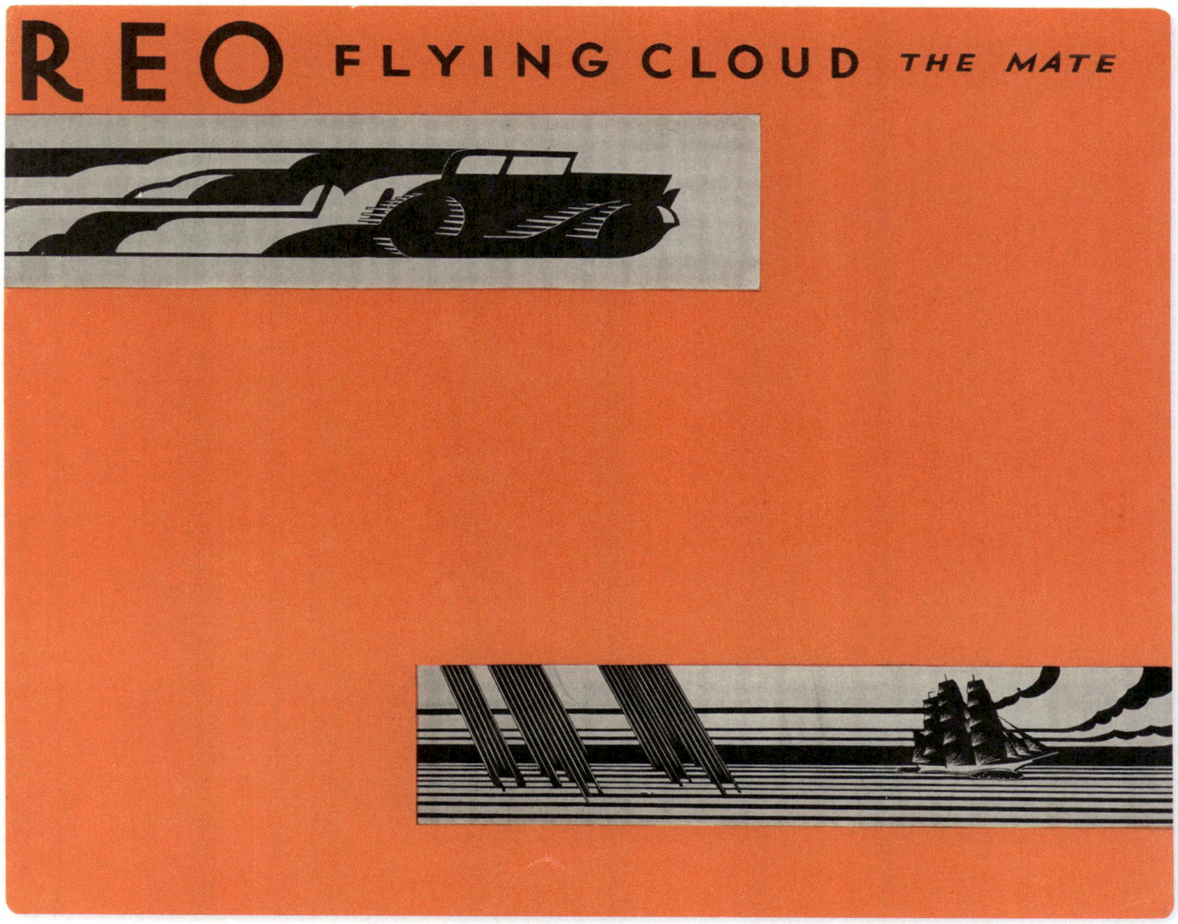

Als er die nach ihm benannte Firma Oldsmobile 1905 verließ, durfte Ransom Eli Olds den Markennamen nicht mitnehmen, und so gründete er noch im gleichen Jahr die Firma REO, deren Name sich aus seinen Initialen zusammensetzte, und schaffte es so, seinen guten Ruf in das neue Unternehmen hinüberzuretten. Unter anderem führte REO die mittig am Fahrzeugboden angebrachte Gangschaltung in den Automobilbau ein. Der REO Flying Cloud mit seinem robusten Sechszylindermotor, der mit sieben Hauptlagern ausgestattet war, gehörte zu den großartigsten Fahrzeugen der späten 1920er-Jahre und war das erste Auto, das die neuen Backenbremsen von Lockheed verwendete.

Après avoir quitté l'entreprise en 1905, Ransom Eli Olds ne détenait aucun droit légal lui permettant d'utiliser le nom « Oldsmobile », aussi décida-t-il cette même année de fonder REO, acronyme de ses initiales. Sa réputation était assez bien établie pour justifier ce choix. Entre autres innovations, REO introduira le positionnement au plancher du levier de vitesses. La REO Flying Cloud est l'une des voitures marquantes de la fin des années 1920, avec son robuste moteur six cylindres qui pouvait compter sur sept paliers de vilebrequin. Elle sera la première à utiliser les nouveaux freins hydrauliques Lockheed conçus spécialement pour les automobiles.

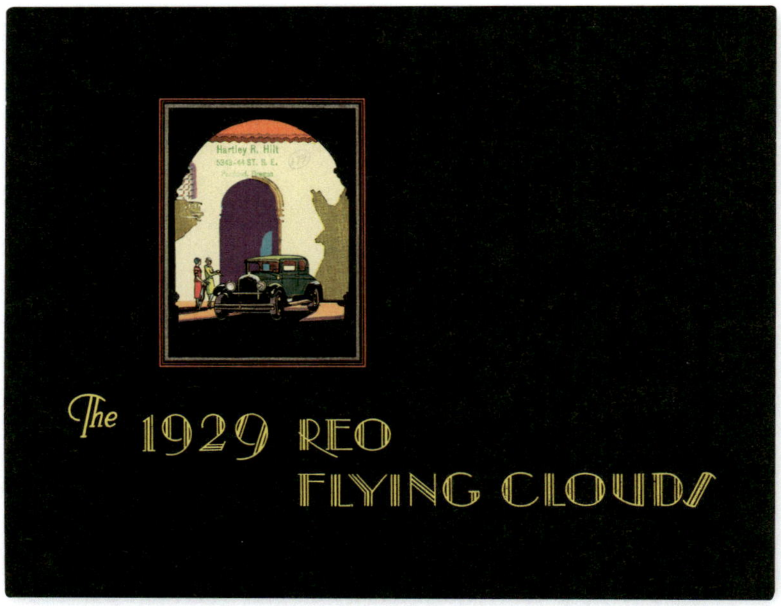

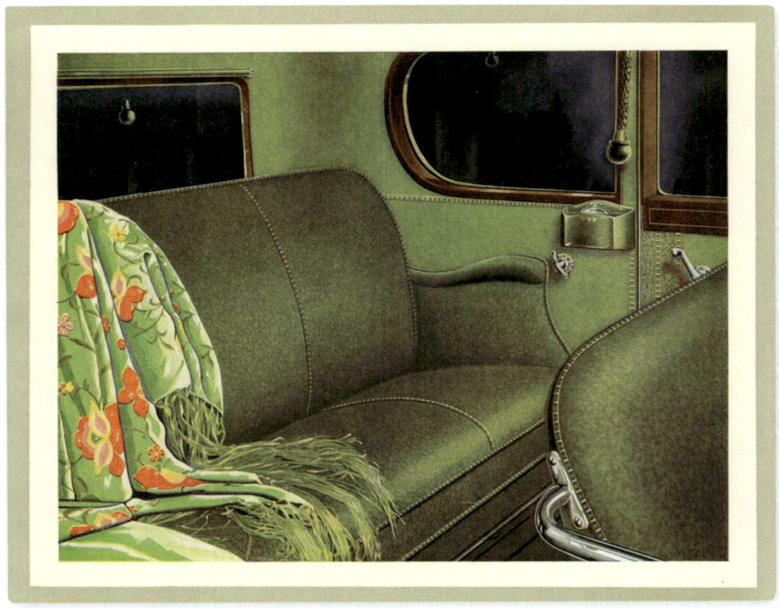

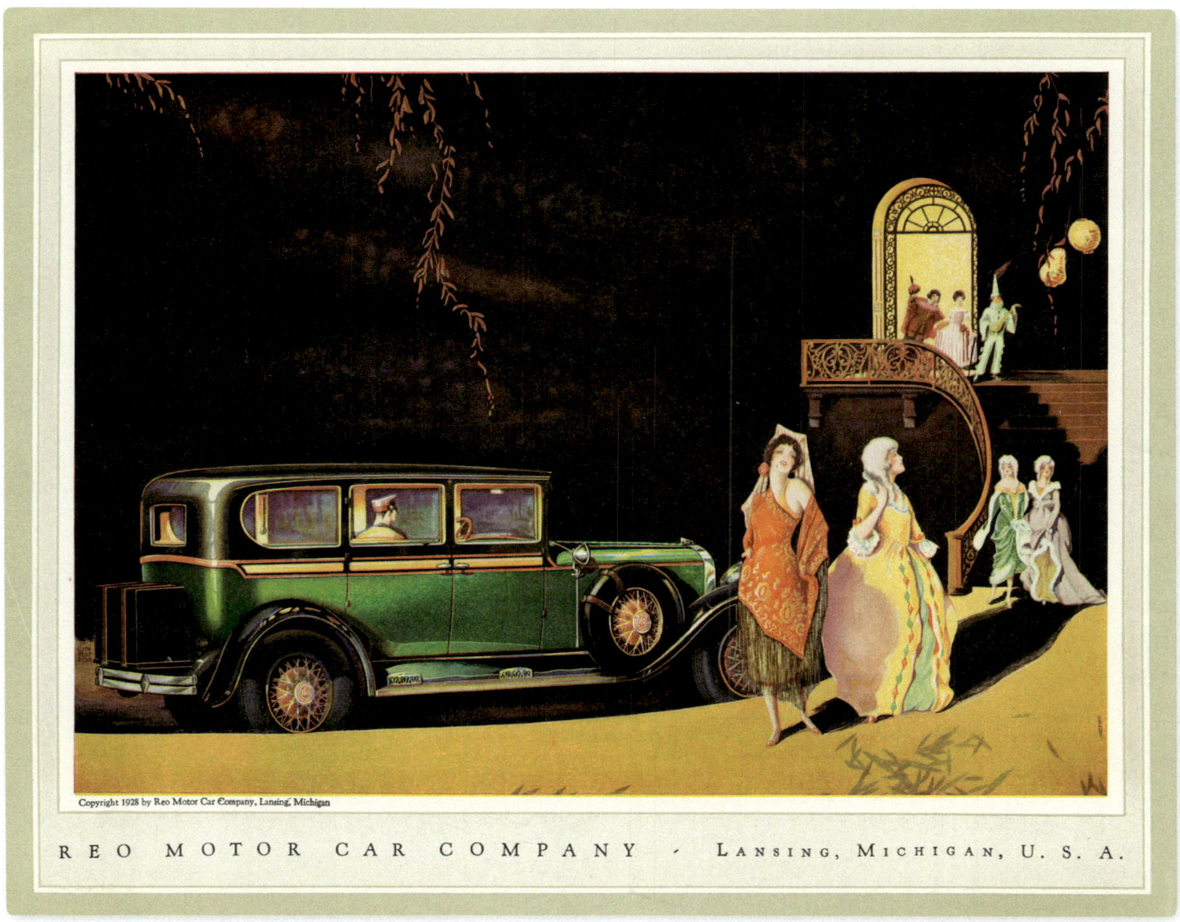

REO Flying Cloud 1929

REO lasted well into the Depression, until 1936, due in large part to the Flying Cloud. It was a smart-looking, durable car whose advertising matched its personal panache. The backseat area included a vase holder.

Dass REO bis 1936 existierte und den größten Teil der Wirtschaftskrise überdauerte, verdankt das Unternehmen hauptsächlich dem Flying Cloud. Es war ein schickes, langlebiges Fahrzeug, das der Extravaganz seiner Werbung durchaus gerecht wurde. Im Fond gab es sogar einen Blumenvasenhalter.

Si REO résista bien à la crise jusqu'en 1936, c'est en grande partie grâce à la Flying Cloud : une voiture aussi élégante que fiable dont la publicité cadrait très bien avec l'éclat qu'elle dégageait. Le compartiment arrière comprenait un support de vase.

Rollin

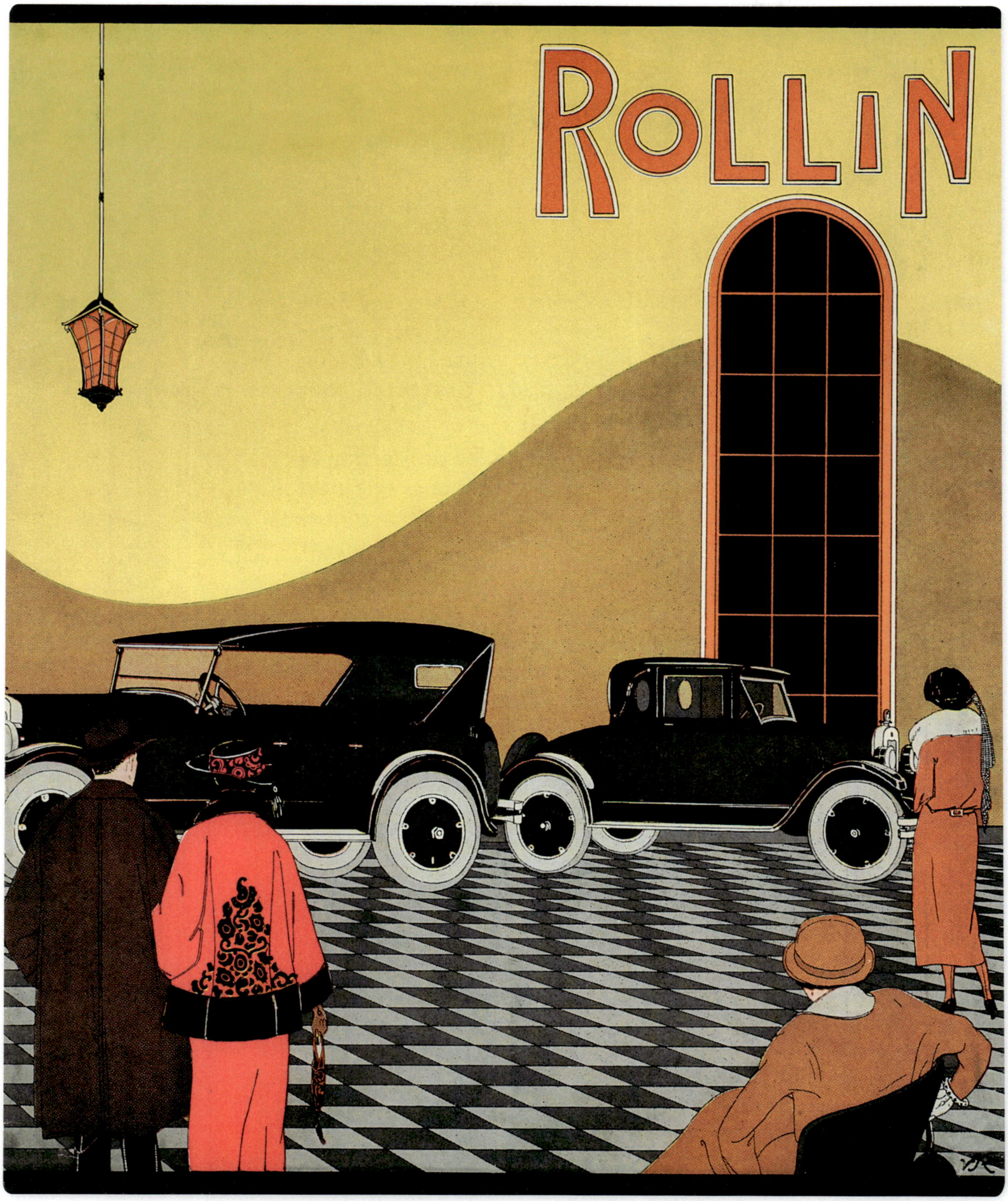

Rollin 1925

A talented team, led by engineer Fred M. Zeder, created the Rollin with a high-compression engine and aluminum connecting rods. Despite being a forerunner of the luxury compacts that would emerge 50 years later, the Rollin was too small for its time and the Cleveland-based company was gone after only one year.

Die talentierte Mannschaft des Ingenieurs Fred M. Zeder schuf den Rollin mit hochverdichtendem Motor und Pleuelstangen aus Aluminium. Obwohl man ihn als Vorläufer der Luxuskleinwagen betrachten könnte, die ein halbes Jahrhundert später auf den Markt kamen, war der Rollin für seine Zeit einfach zu klein, und das Unternehmen aus Cleveland, Ohio, war nach nur einem Jahr vom Markt weg.

Une équipe talentueuse emmenée par l'ingénieur Fred M. Zeder créa la Rollin, équipée d'un moteur à taux de compression élevé et de bielles en aluminium. Annonciatrice des modèles de luxe compacts qui apparaîtront cinquante ans plus tard, la Rollin était cependant trop petite pour son temps, et l'entreprise de Cleveland plia boutique après seulement un an d'existence.

Royal Tourist

Royal Tourist 1907 *(this spread)*

The Royal Tourist was built in Cleveland between 1904 and 1911, with a Robert Jardine design and uncommonly early shaft drive.

Der Royal Tourist wurde zwischen 1904 und 1911 in Cleveland (Ohio) gebaut, entworfen von Robert Jardine und ausgestattet mit einer für diese frühe Zeit noch ungewöhnlichen Antriebswelle.

La Royal Tourist fut construite à Cleveland entre 1904 et 1911, conçue par l'ingénieur français Robert Jardine et dotée d'une des rares et premières transmissions par arbre.

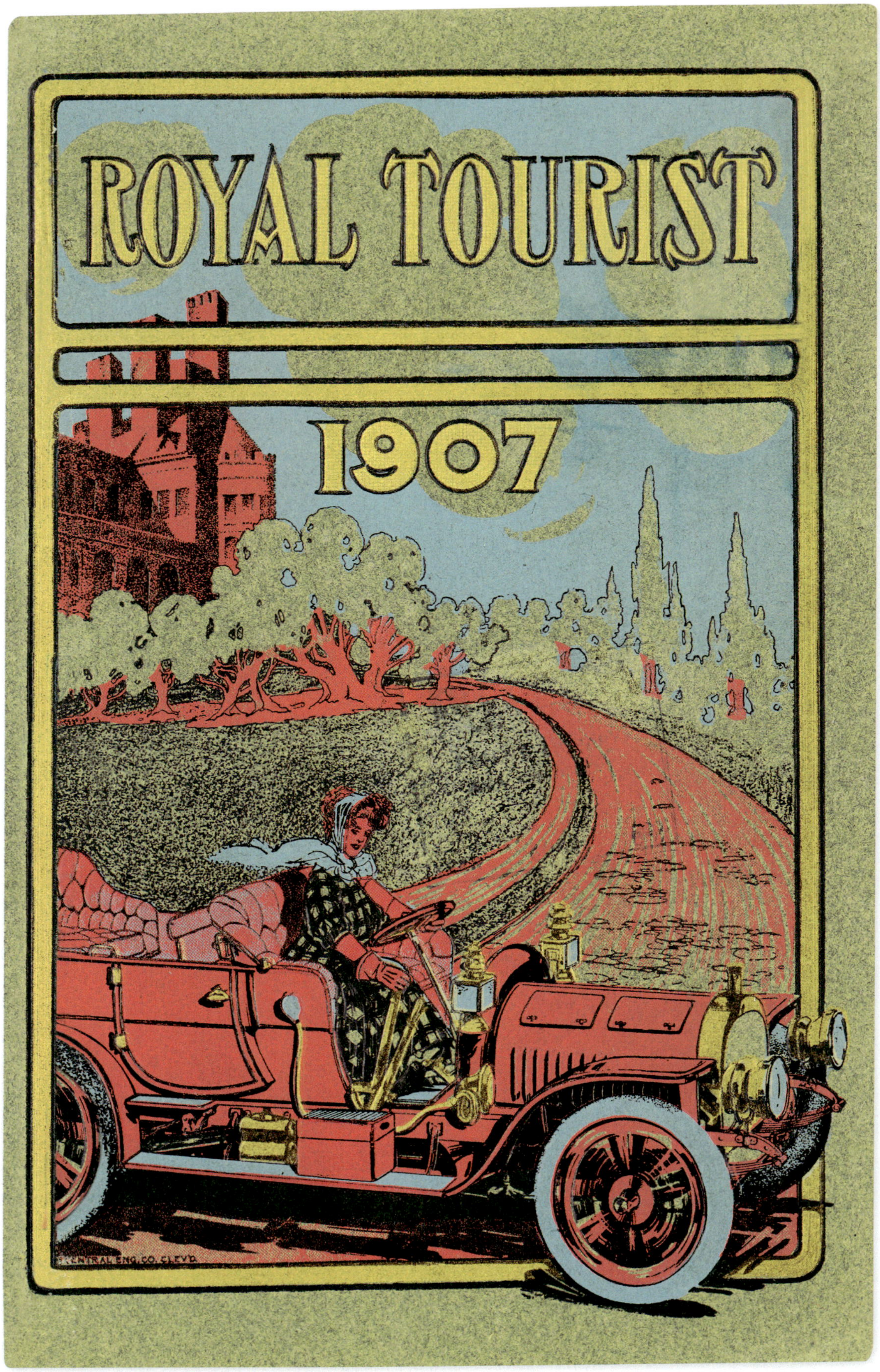

20

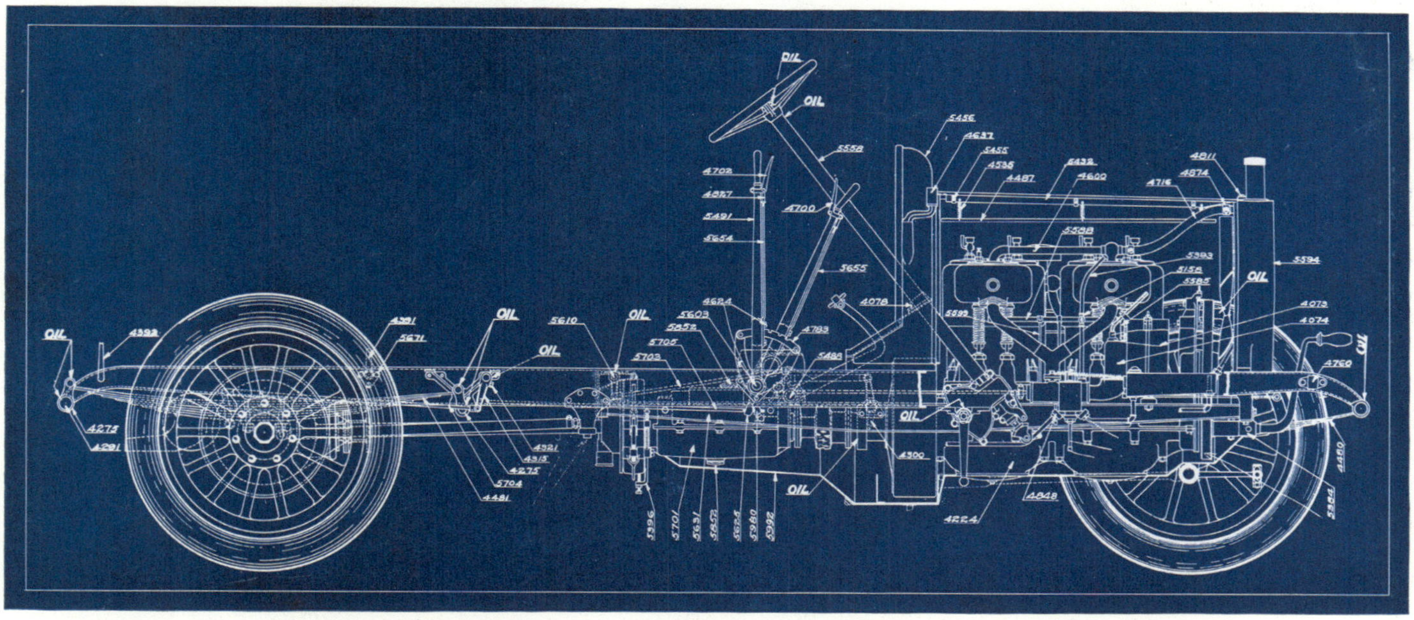

3711 Top radiator hose connection.
4033 Gasoline tank for limousine body.
4041 Yoke for steering con. rod.
4050 Washer for clutch pedal lever.
4078 Flange for steering column.
4082 Clip for gas pipe.
4205 Front spring shackle bolt.
4275 Shackle for rear spring.
4276 Shackle for front spring.
4290 Shackle bolt.
4291 Shackle bolt.
4300 Clutch fork.
4315 Rear brake lever.
4317 Pawl for sprag.
4318 Pawl bracket.
4321 Brake lever pin.
4323 Anchor pin for pawl.
4332 Washer for ⅜-in. bolt.
4373 Screw for pedal pads.
4374 Hood liner, R. H.
4375 Hood liner, L. H.
4382 Spacer on pedal shaft.

4583 Flange on exhaust pip.e
4606 Curved tube on water pipe.
4612 Cover for gasoline tank inlet.
4614 Bracket for coil box.
4624 Guide screw in hand lever.
4627 Shield for coil wires.
4636 Washer for side Pan.
4637 Oil lamp bracket.
4657 Grease cup.
4663 Oil pan stud holding spark gear cover.
4686 Washer for centre pan.
4700 Release lever for 5492.
4701 Release lever spring.
4702 Release lever for 4094.
4703 Release lever spring for 4702.
4716 Hose, for radiator return water.
4744 Union.
4746 Button head screw for lamp bracket.
4755 Socket wrench for spark plugs.
4756 Lifter for inlet valves.
4760 Bracket for lamp hanger.
4783 Quadrant for change gear.

5225 Clevis pin for truss rod.
5254 Auxiliary step.
5271 ⅛ inch oiler.
5278 ⅛ inch close nipple.
5280 ⅛ inch female union.
5281 ⅛ cap.
5282 ⅛ inch coupling.
5284 Glass for oiler sight feed.
5346 Muffler.
5347 End piece for muffler.
5348 Elbow for muffler.
5349 Rod for muffler.
5373 T for Presto Light connection.
5375 Frame, 114 inch wheel base.
5397 Shackle for front spring.
5398 Front spring spacer, L. H.
5399 Front spring spacer, R. H.
5400 Mud guard bracket.
5432 Hood rail.
5455 Support for top of hood.
5456 Dash.
5460 Brake pedal.

5610 Shaft for 5609.
5614 Bracket for trans. brake lever.
5625 Brake lever on hand brake shaft.
5627 Brake rod bracket, L. H.
5628 Brake rod bracket, R. H.
5633 Brake equalizer lever.
5634 Brake shaft (rear).
5635 Brake equalizer tube.
5636 Collar for equalizer tube.
5654 Latch for change gear lever.
5655 Latch for brake lever.
5659 Rear spring shackle shaft.
5665 Cross brake rod lever.
5669 Pin in rear of push rod.
5671 Front socket for rear fender iron.
5675 Hood door.
5676 Hood, R. H.
5677 Hood, L. H.
5703 Transmission brake rod.
5704 Hand brake rod.
5706 Pin in front of push rod.
5718 Gasoline tank.

4477 Female rod end.
4478 Spring for 4041.
4480 Front springs.
4481 Rear springs.
4482 McCord oiler.
4487 Tension rod for hood.
4527 Flange for hose con. tube.
4529 Flange for water inlet.
4535 Spring for tension rod.
4537 Hook for tension rod.
4538 Door handle for hood.
4539 Spring for door catch on hood.
4540 Base for door catch on hood.
4541 Door catch on hood.
4542 Handle for hood on hood.
4543 Nut for 4542 on hood.
4544 Catch for lock pin on hood.
4545 Lock pin on hood.
4553 Enameled plate for oiler.

4862 Yoke end, ⅜-in. rod.
4864 Filling flange for gasoline tank.
4865 Spring for locking pin on hood.
4874 Hose clamps.
4877 Grease cup.
4880 Oiler, 5-16 x 32 th'ds.
4903 Collar for clutch shaft.
5011 Lamp bracket (limousine body).
5127 Nut for brake rod.
5147 Plunger nut for carburetor flooder.
5148 Plunger knob for carburetor flooder.
5149 Plunger for carburetor flooder.
5150 Plunger case for carburetor flooder.
5151 Plunger case for carburetor flooder.
5152 Plunger for carburetor flooder.
5153 Plunger spring for carburetor flooder.
5154 Plunger spring for carburetor flooder.
5178 Door handle.
5223 Clevis pin for brake rod.

5478 Clutch cam.
5479 Clutch rod.
5480 Clutch cam roller.
5481 Throttle lever.
5486 Clutch operating lever.
5487 Clutch shaft bearing.
5488 Pedal shaft.
5491 Change gear lever.
5492 Hand brake lever.
5507 Adjustment sleeve.
5510 Strut rod.
5536 Throttle rod.
5594 Radiator.
5601 Brake shaft bracket.
5602 Brake lever sleeve.
5603 Hand brake shaft.
5605 Clutch shaft, L. H., side.
5606 Clutch shaft, R. H., side.
5609 Short trans. brake lever.

5770 Screw for pedal shaft bracket.
5772 Screw for trans. brake lever bracket.
5778 Screw for brake rod bracket.
5780 Quadrant bracket screw.
5791 Screw for torsion buffer and brake shaft bracket.
5794 Screw for clutch pedal.
5796 Screw for hand lever shaft sector.
5807 Screw for torsion buffer.
5808 Screw for brake rod lever.
5809 Screw for brake shaft bracket.
5812 Screw for clutch cam.
5817 Screw for exhaust pipe to muffler.
5819 Screw for torque rod socket.
5832 Stud for coil box bracket.
5835 Screw for running board—short.
5836 Screw for running board—long.

(Continued on next page)

21

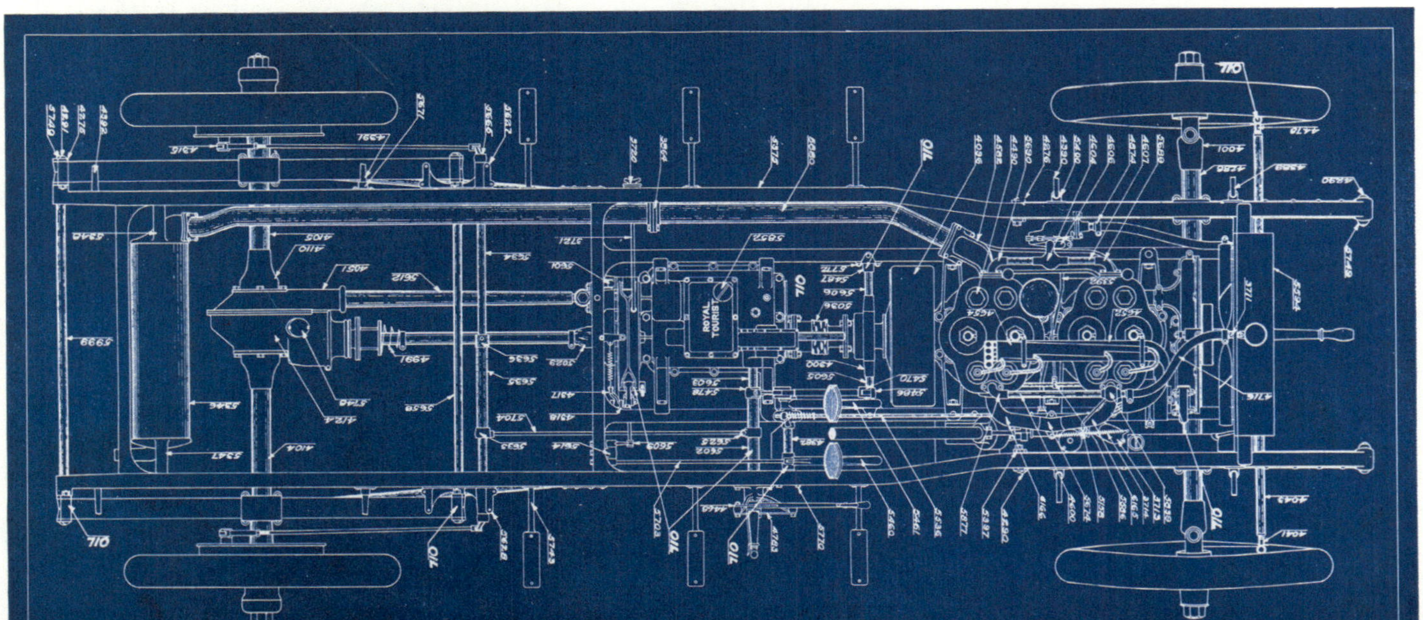

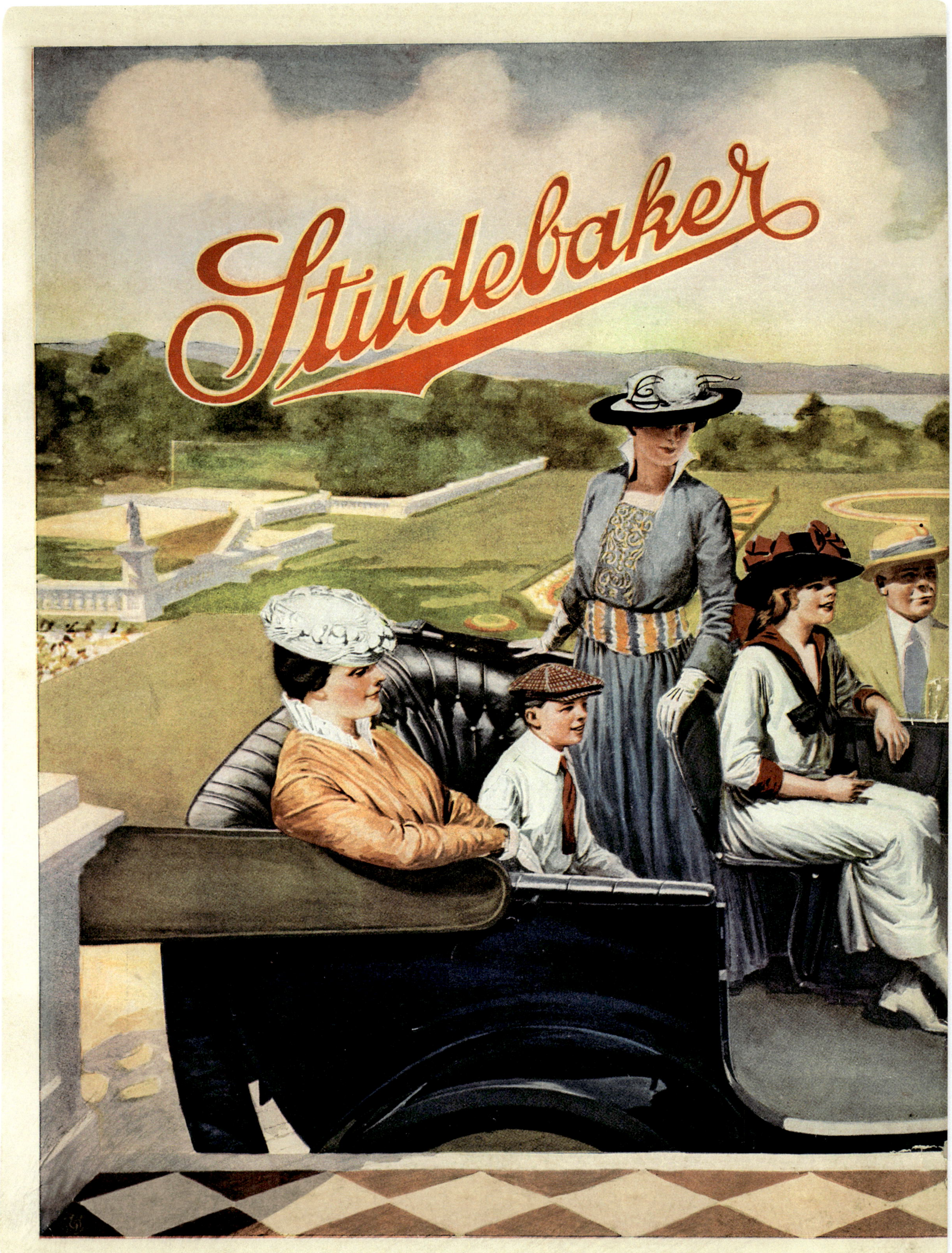
Studebaker

Studebaker

Studebaker 1916 *(this spread)*

Studebaker is one of the great names in American transportation, founded in 1852. Even their early cars were robustly engineered, here with a Hotchkiss-type drive, removable axles, and four-wheel brakes that were easily replaced. The Studebaker was never marketed solely on price, but it did famously produce some low-priced models such as the Scotsman and the Lark.

Studebaker, 1852 von einer deutschstämmigen Familie in South Bend (Indiana) gegründet, gehört zu den großen Namen im amerikanischen Fahrzeugbau. Selbst die frühen Automobile waren schon äußerst robust – hier mit Hotchkiss-Antrieb, ausbaubaren Radachsen und Vierradbremsen, die sich leicht austauschen ließen. Der Studebaker wurde zu keiner Zeit ausschließlich über den Preis vermarktet, aber dennoch produzierte die Firma einige berühmt gewordene Einstiegsmodelle wie den Scotsman und den Lark.

Fondé en 1852, Studebaker est l'un des grands noms de l'histoire américaine des transports. Ses premières voitures étaient déjà solides, comme on le voit ici avec cette transmission de type Hotchkiss, ces essieux amovibles et ces freins aux quatre roues faciles à remplacer. Le prix ne fut jamais l'élément prépondérant de la politique commerciale de Studebaker, mais la marque a produit des modèles économiques qui ont eu leur petit succès, comme la Scotsman et la Lark.

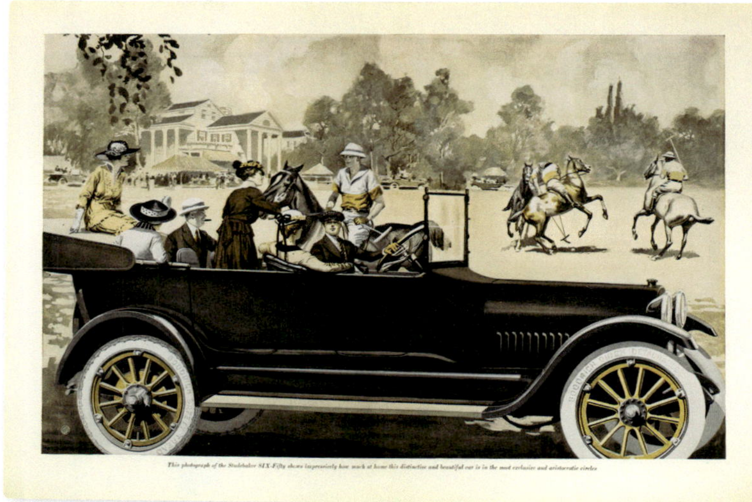

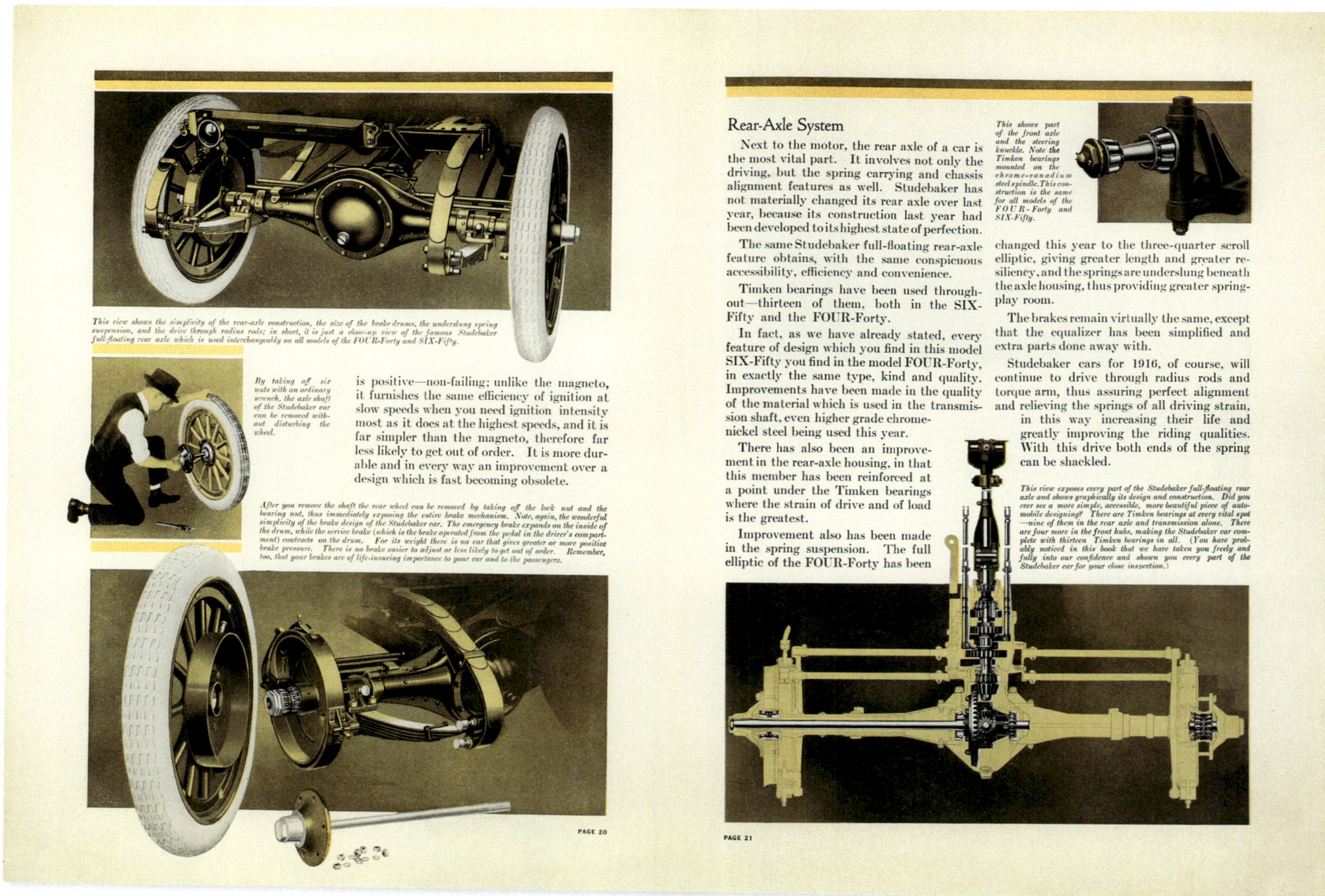

This view shows the simplicity of the rear-axle construction, the size of the brake drums, the underslung spring suspension, and the drive through radius rods; in short, it is just a close-up view of the famous Studebaker full-floating rear axle which is used interchangeably on all models of the FOUR-Forty and SIX-Fifty.

By taking off six nuts with an ordinary wrench, the axle shaft of the Studebaker car can be removed without disturbing the wheel.

is positive—non-failing; unlike the magneto, it furnishes the same efficiency of ignition at slow speeds when you need ignition intensity most as it does at the highest speeds, and it is far simpler than the magneto, therefore far less likely to get out of order. It is more durable and in every way an improvement over a design which is fast becoming obsolete.

After you remove the shaft the rear wheel can be removed by taking off the lock nut and the bearing nut, thus immediately exposing the entire brake mechanism. Note, again, the wonderful simplicity of the brake design of the Studebaker car. The emergency brake expands on the inside of the drum, while the service brake (which is the brake operated from the pedal in the driver's compartment) contracts on the drum. For its weight there is no car that gives greater or more positive brake pressure. There is no brake easier to adjust or less likely to get out of order. Remember, too, that your brakes are of life-insuring importance to your car and to the passengers.

PAGE 20

Rear-Axle System

Next to the motor, the rear axle of a car is the most vital part. It involves not only the driving, but the spring carrying and chassis alignment features as well. Studebaker has not materially changed its rear axle over last year, because its construction last year had been developed to its highest state of perfection.

The same Studebaker full-floating rear-axle feature obtains, with the same conspicuous accessibility, efficiency and convenience.

Timken bearings have been used throughout—thirteen of them, both in the SIX-Fifty and the FOUR-Forty.

In fact, as we have already stated, every feature of design which you find in this model SIX-Fifty you find in the model FOUR-Forty, in exactly the same type, kind and quality. Improvements have been made in the quality of the material which is used in the transmission shaft, even higher grade chrome-nickel steel being used this year.

There has also been an improvement in the rear-axle housing, in that this member has been reinforced at a point under the Timken bearings where the strain of drive and of load is the greatest.

Improvement also has been made in the spring suspension. The full elliptic of the FOUR-Forty has been changed this year to the three-quarter scroll elliptic, giving greater length and greater resiliency, and the springs are underslung beneath the axle housing, thus providing greater spring-play room.

The brakes remain virtually the same, except that the equalizer has been simplified and extra parts done away with.

Studebaker cars for 1916, of course, will continue to drive through radius rods and torque arm, thus assuring perfect alignment and relieving the springs of all driving strain, in this way increasing their life and greatly improving the riding qualities. With this drive both ends of the spring can be shackled.

This shows part of the front axle and the steering knuckle. Note the Timken bearings mounted on the chrome-vanadium steel spindle. This construction is the same for all models of the FOUR-Forty and SIX-Fifty.

This view exposes every part of the Studebaker full-floating rear axle and shows graphically its design and construction. Did you ever see a more simple, accessible, more beautiful piece of automobile designing? There are Timken bearings at every vital spot—nine of them in the rear axle and transmission alone. There are four more in the front hubs, making the Studebaker car complete with thirteen Timken bearings in all. (You have probably noticed in this book that we have taken you freely and fully into our confidence and shown you every part of the Studebaker car for your close inspection.)

PAGE 21

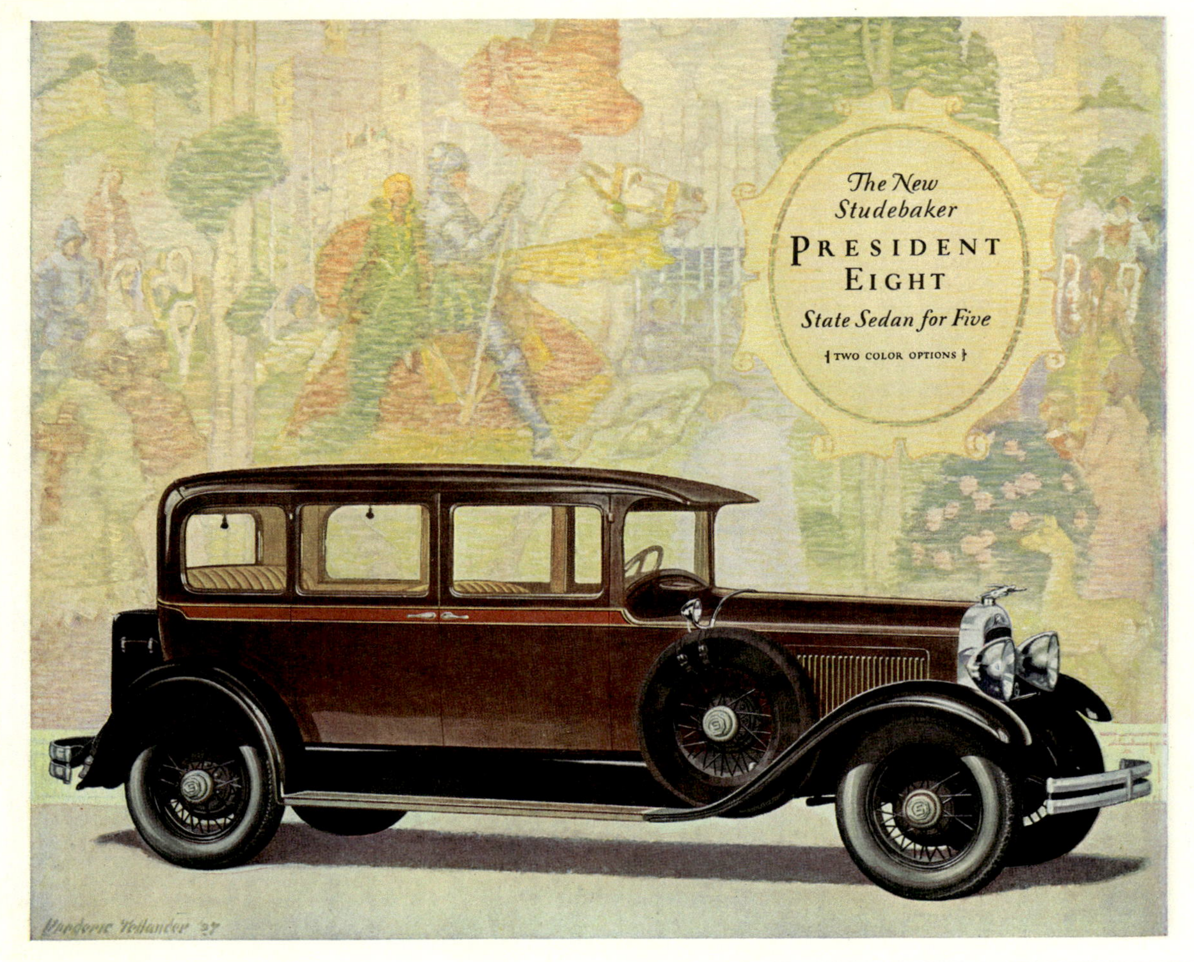

Studebaker President Eight 1928 *(above)*, **Studebaker-Erskine 1927** *(right)*

The President Eight's interior fittings included heavy broadcloths, etched-silver door trim, and a walnut dash, making it a great classic of the 1920s.

Mit der Innenausstattung des President Eight, zu der schwere Walkstoffe, Türleisten aus geätztem Silber und ein Armaturenbrett aus Nussbaum gehörten, wurde er zu einem prachtvollen Automobilklassiker der 1920er-Jahre.

L'habitacle de la President Eight réunissait une sellerie douillettement matelassée, des garnitures de portière en argent gravé et un tableau de bord en noyer qui en font l'une des grandes classiques des années 1920.

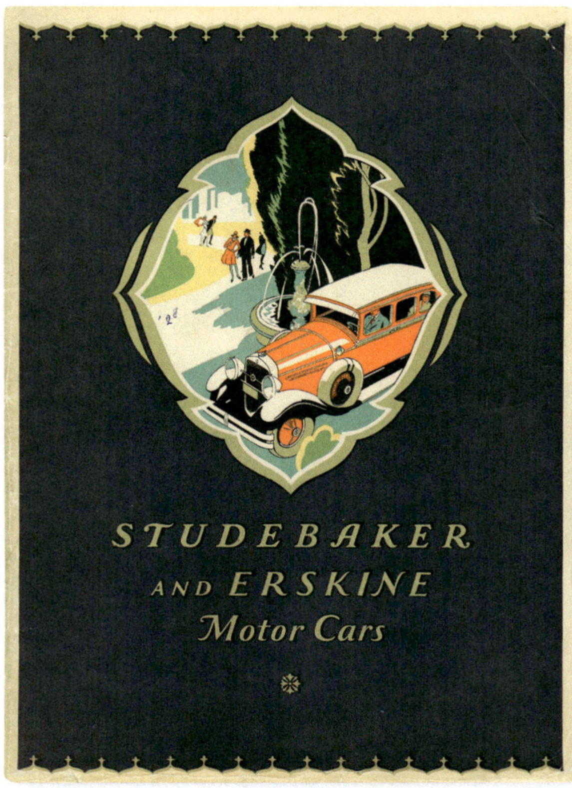

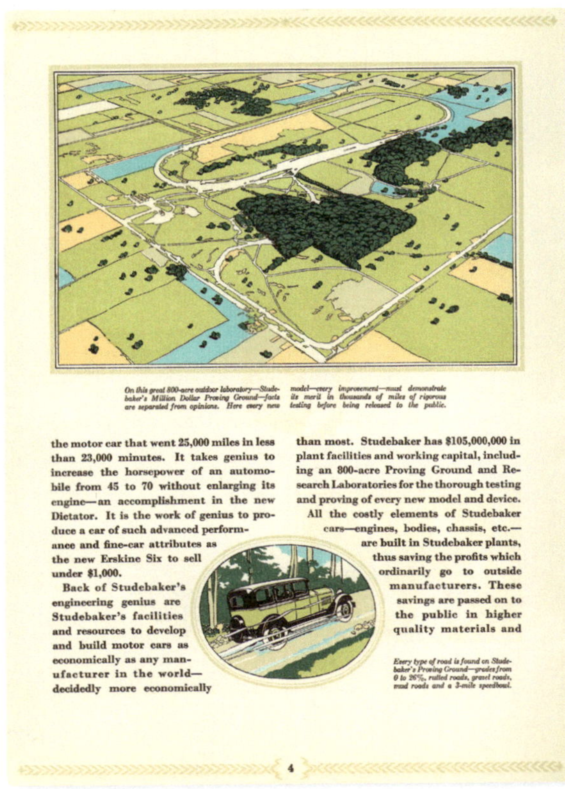

On this great 800-acre outdoor laboratory—Studebaker's Million Dollar Proving Ground—facts are separated from opinions. Here every new model—every improvement—must demonstrate its merit in thousands of miles of rigorous testing before being released to the public.

the motor car that went 25,000 miles in less than 23,000 minutes. It takes genius to increase the horsepower of an automobile from 45 to 70 without enlarging its engine—an accomplishment in the new Dictator. It is the work of genius to produce a car of such advanced performance and fine-car attributes as the new Erskine Six to sell under $1,000.

Back of Studebaker's engineering genius are Studebaker's facilities and resources to develop and build motor cars as economically as any manufacturer in the world—decidedly more economically than most. Studebaker has $105,000,000 in plant facilities and working capital, including an 800-acre Proving Ground and Research Laboratories for the thorough testing and proving of every new model and device.

All the costly elements of Studebaker cars—engines, bodies, chassis, etc.—are built in Studebaker plants, thus saving the profits which ordinarily go to outside manufacturers. These savings are passed on to the public in higher quality materials and

Every type of road is found on Studebaker's Proving Ground—grades from 0 to 26%, rutted roads, gravel roads, mud roads and a 3-mile speedbowl.

4

Stutz

Stutz 1928

Stutz, which produced the Black Hawk, was founded by Harry C. Stutz in 1911. The Indianapolis-based firm became legendary almost at once, having introduced the immortal Bearcat, one of the first American performance cars, in 1912. The Black Hawk was similarly certified as the fastest production car of its day. The company folded in 1938.

Die Firma Stutz, die den Black Hawk produzierte, war 1911 von Harry C. Stutz gegründet worden. Das Unternehmen mit Sitz in Indianapolis (Indiana) wurde praktisch auf der Stelle zur Legende, nachdem es 1912 den Bearcat auf den Markt gebracht hatte, einen der ersten amerikanischen Sportwagen. Dem Black Hawk wurde sogar bescheinigt, das schnellste Serienauto seiner Zeit gewesen zu sein. Die Stutz Motor Company wurde 1938 aufgelöst.

La marque Stutz, qui produisait la Black Hawk, a été fondée par Harry C. Stutz en 1911. Située à Indianapolis, l'entreprise est presque immédiatement entrée dans la légende avec le lancement en 1912 de l'immortelle Bearcat, l'une des premières voitures de sport américaines. La Black Hawk a également été reconnue comme la plus véloce des voitures de production de son époque. L'entreprise a disparu en 1938.

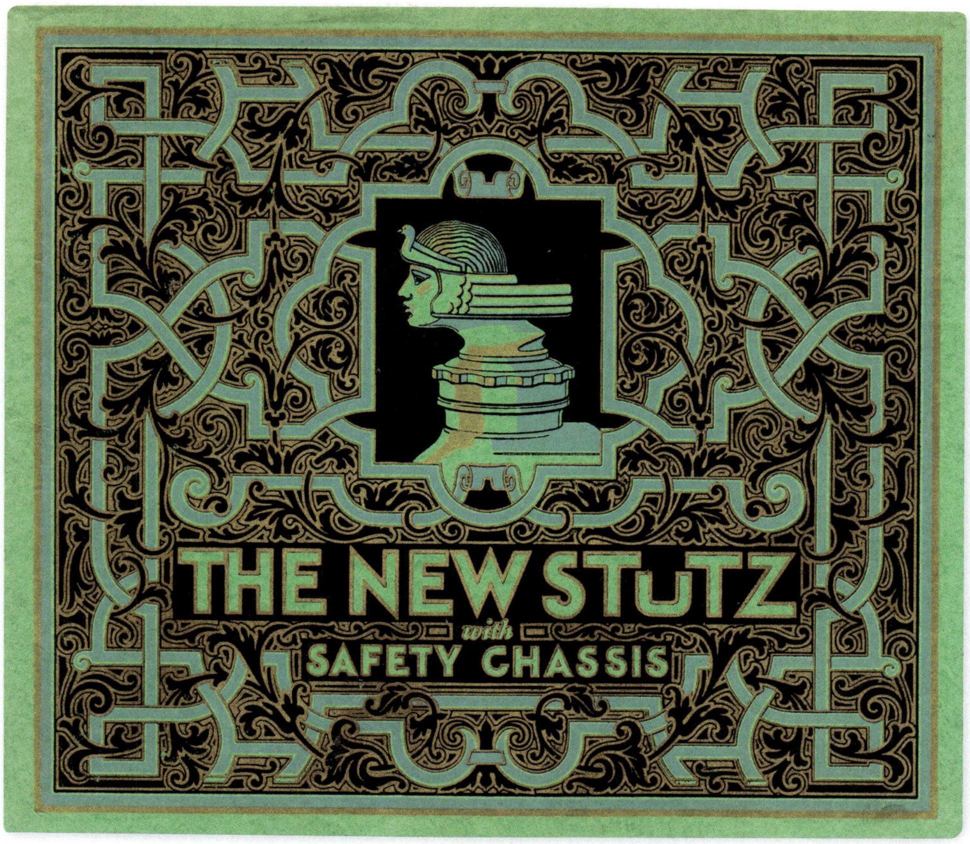

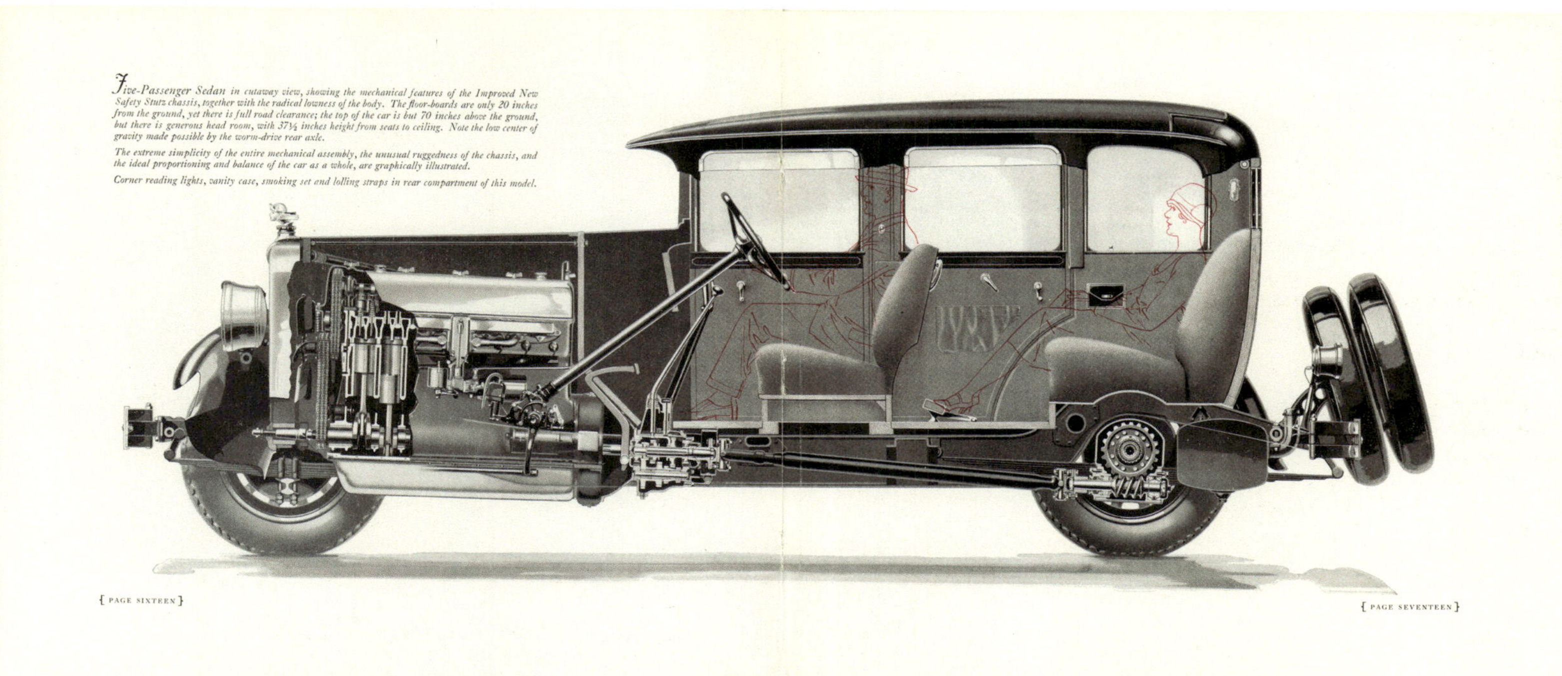

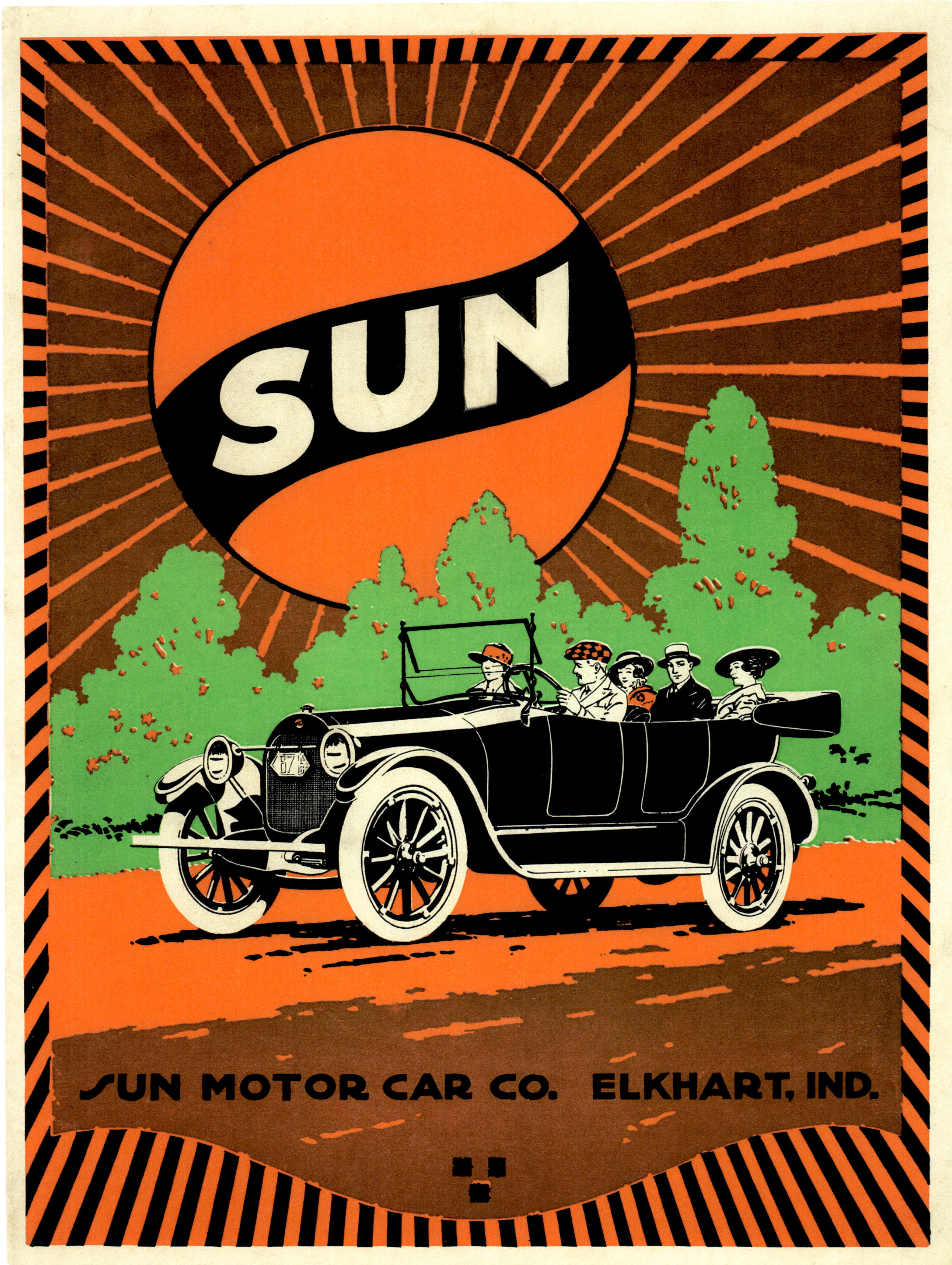
SUN
SUN MOTOR CAR CO. ELKHART, IND.

Sun

Sun 1916 *(this spread)*

Based in Elkhart, Indiana, the Sun's light went out after it built just 337 cars. It was a costly car, priced at around $1,100. Sun over-ordered engines and other materials and quickly went bankrupt.

Das Licht des Sun, der in Elkhart, Indiana, produziert wurde, verlosch schon nach dem Bau von nur 337 Fahrzeugen. Mit einem Listenpreis von fast 1 100 Dollar war er für seine Zeit ein teures Auto. Der Hersteller hatte zu viele Motoren und andere Teile auf Vorrat bestellt und wurde daraufhin rasch zahlungsunfähig.

Basées à Elkhart, dans l'Indiana, cette Sun et ses lumières se sont éteintes après la construction de 337 exemplaires seulement. C'était une voiture chère, vendue près de 1 100 dollars. Pour avoir commandé trop de moteurs et matériaux divers, Sun fit rapidement faillite.

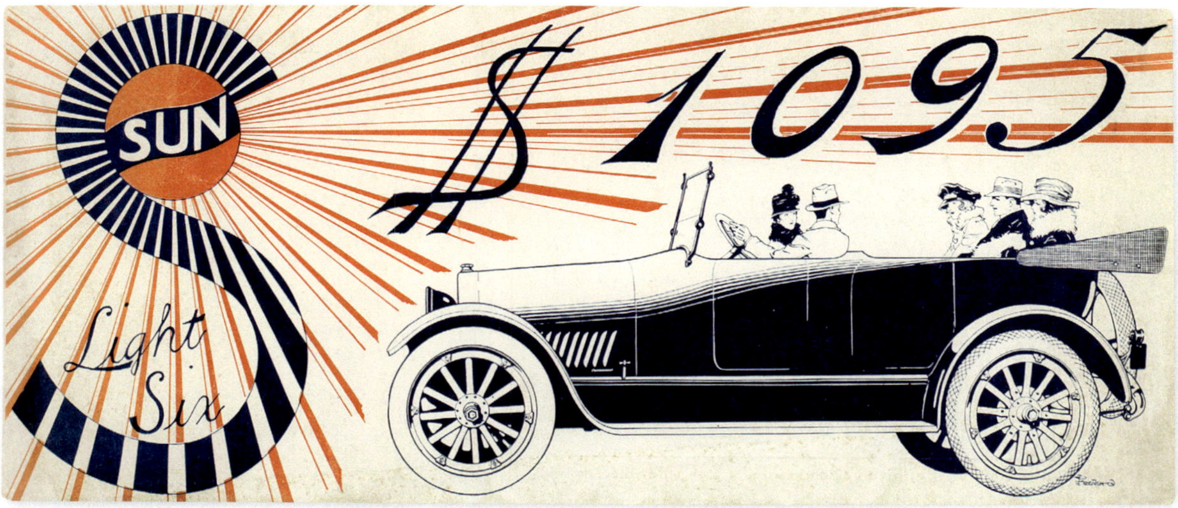

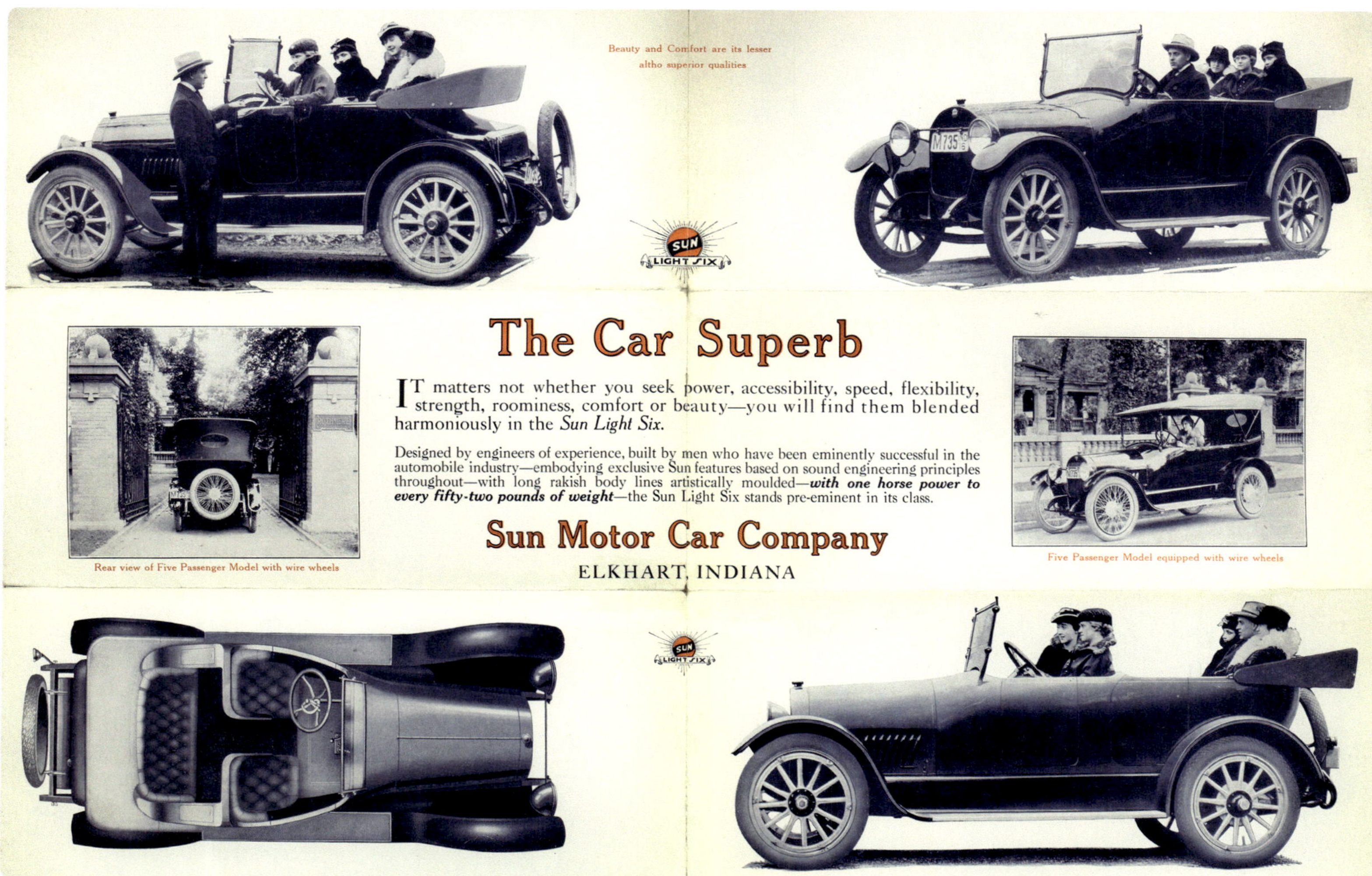

White

White 1908 *(right)*, **1915**
(below and opposite)

Most people associate Stanley with early steam cars, but at this time White was outselling Stanley by nearly two to one. Its powerful engine was conventionally positioned under a front hood. The White steamer was the first car driven by a president.

Die meisten Menschen bringen frühe Dampfautomobile mit der Marke Stanley in Verbindung, doch zu dieser Zeit verkaufte White fast doppelt so viele wie der berühmte Konkurrent. Der leistungsstarke Motor saß in der üblichen Position unter einer Haube im vorderen Teil. Das White-Dampfauto war das erste von einem US-Präsidenten gefahrene Automobil.

Aux véhicules à vapeur on aurait plutôt tendance à associer le nom de Stanley, mais à cette époque White vendait pratiquement deux fois plus de voitures. Son puissant moteur était, de façon très conventionnelle, placé sous le capot avant. La White à vapeur sera la première automobile conduite par un président des États-Unis.

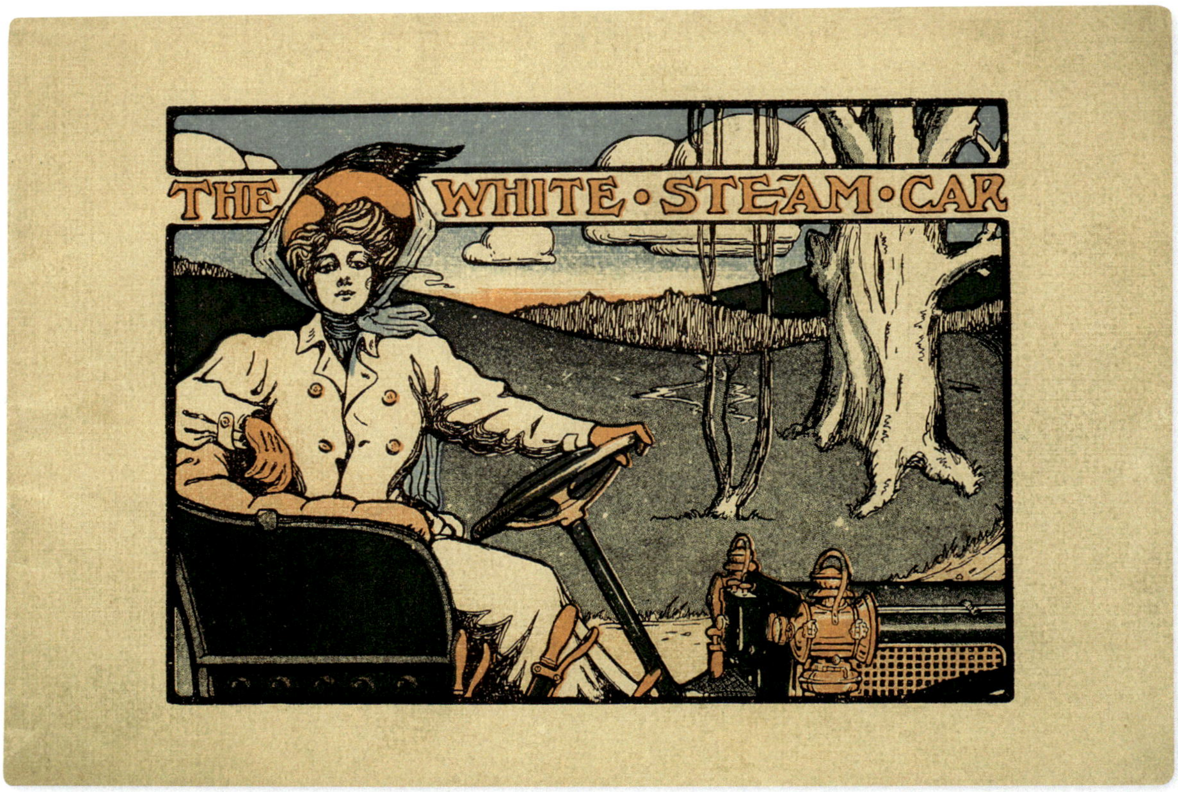

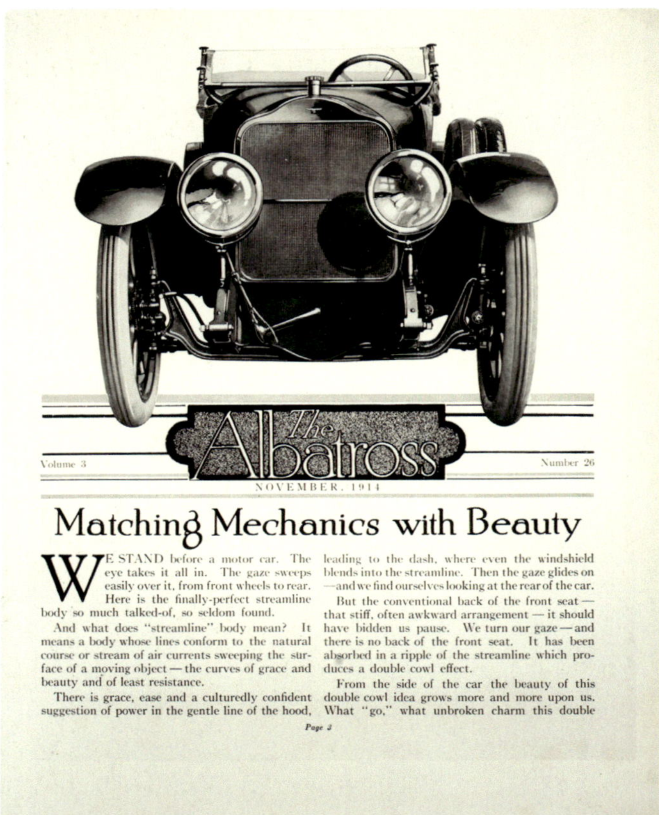

The Albatross

Volume 3 — Number 26

NOVEMBER, 1914

Matching Mechanics with Beauty

WE STAND before a motor car. The eye takes it all in. The gaze sweeps easily over it, from front wheels to rear. Here is the finally-perfect streamline body so much talked-of, so seldom found.

And what does "streamline" body mean? It means a body whose lines conform to the natural course or stream of air currents sweeping the surface of a moving object — the curves of grace and beauty and of least resistance.

There is grace, ease and a culturedly confident suggestion of power in the gentle line of the hood, leading to the dash, where even the windshield blends into the streamline. Then the gaze glides on —and we find ourselves looking at the rear of the car.

But the conventional back of the front seat — that stiff, often awkward arrangement — it should have bidden us pause. We turn our gaze — and there is no back of the front seat. It has been absorbed in a ripple of the streamline which produces a double cowl effect.

From the side of the car the beauty of this double cowl idea grows more and more upon us. What "go," what unbroken charm this double

Page 3

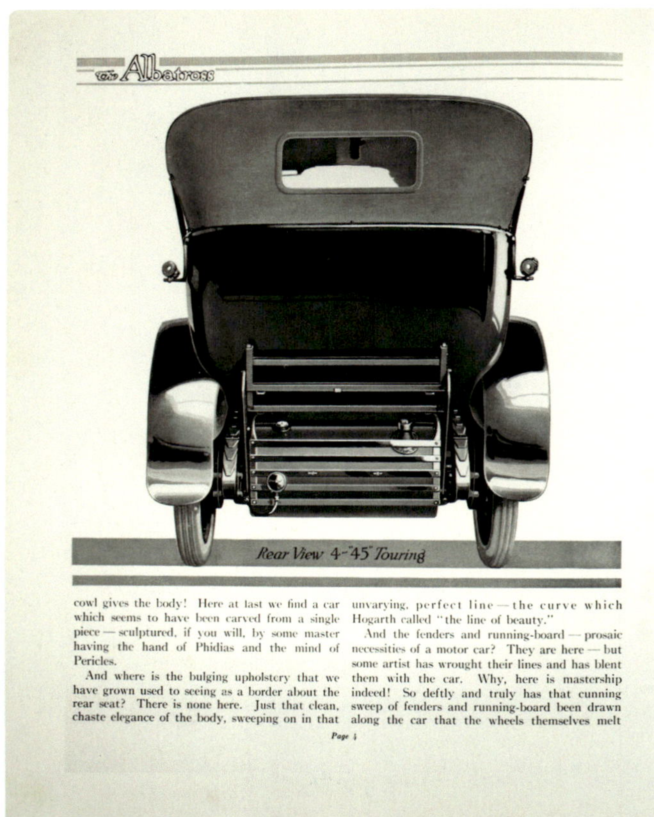

The Albatross

Rear View 4-"45" Touring

cowl gives the body! Here at last we find a car which seems to have been carved from a single piece — sculptured, if you will, by some master having the hand of Phidias and the mind of Pericles.

And where is the bulging upholstery that we have grown used to seeing as a border about the rear seat? There is none here. Just that clean, chaste elegance of the body, sweeping on in that unvarying, perfect line — the curve which Hogarth called "the line of beauty."

And the fenders and running-board — prosaic necessities of a motor car? They are here — but some artist has wrought their lines and has blent them with the car. Why, here is mastership indeed! So deftly and truly has that cunning sweep of fenders and running-board been drawn along the car that the wheels themselves melt

Page 4

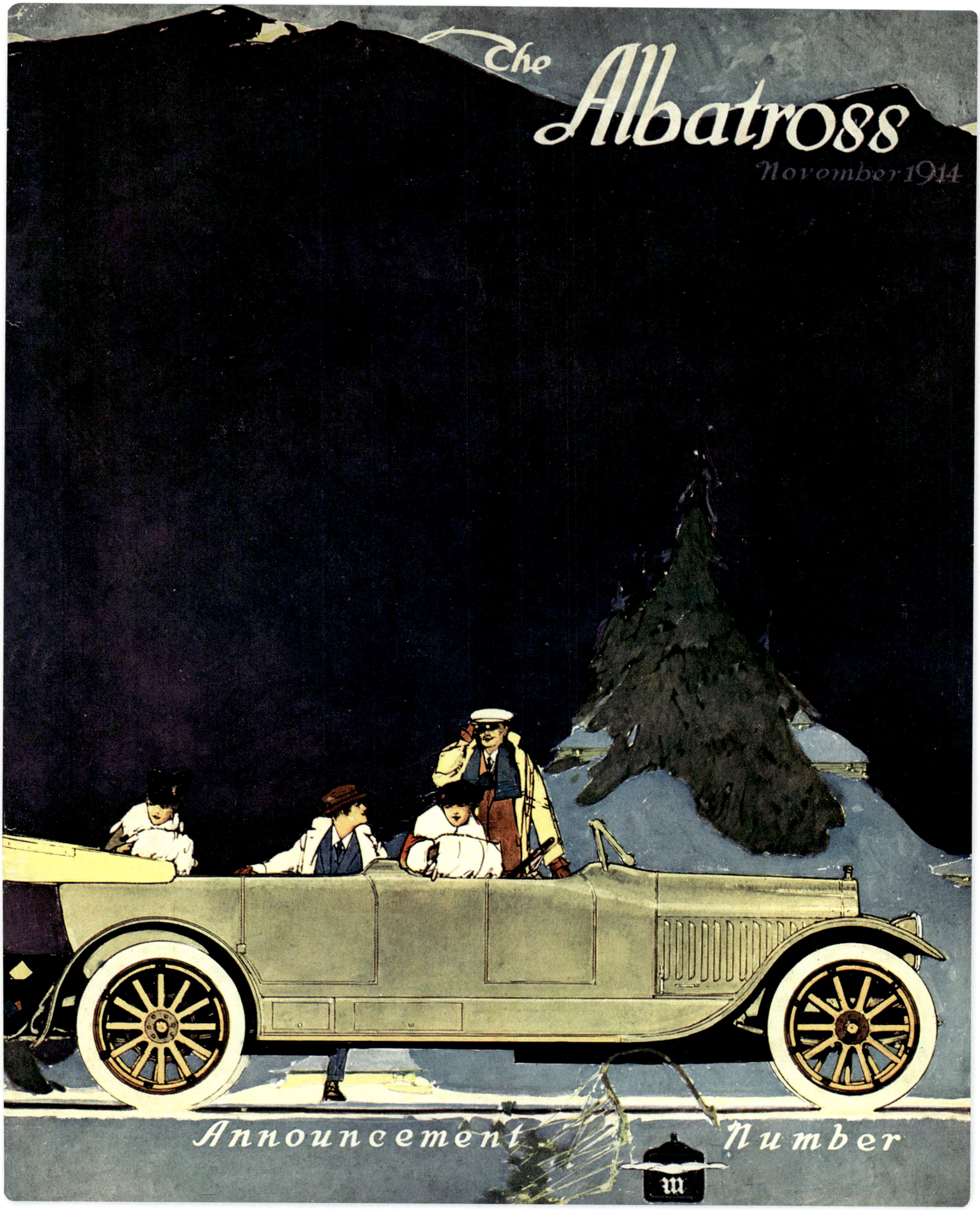
The Albatross
November 1914
Announcement Number

Willys-Overland

Willys-Overland 1913 *(this spread)*

Willys-Overland produced a huge array of four-cylinder cars such as this fishbowl-looking coupe. Some of the Willys cars from the 1930s to 1940 became prized by drag racers for their lightness. Willys-Overland is most famous, though, for producing the Jeep during and after World War II.

Willys-Overland produzierte eine breite Palette an Vierzylinderautomobilen, so wie dieses einem Fischglas ähnelnde Coupé. Einige der Fahrzeuge, die Willys in den 1930er-Jahren baute, waren wegen ihrer Leichtigkeit bei Beschleunigungsrennen sehr beliebt. Am bekanntesten aber wurde Willys-Overland aber als Hersteller des Jeep während des Zweiten Weltkriegs und danach.

Willys-Overland a produit une ribambelle de voitures à quatre cylindres, à l'instar de ce coupé aux faux airs de bocal à poissons. Grâce à leur légèreté, certains modèles des Willys des années 1930 à 1940 seront prisés des amateurs de *hot rods*. C'est toutefois pour sa Jeep, pendant et après la Seconde Guerre mondiale, que Willys-Overland restera célèbre.

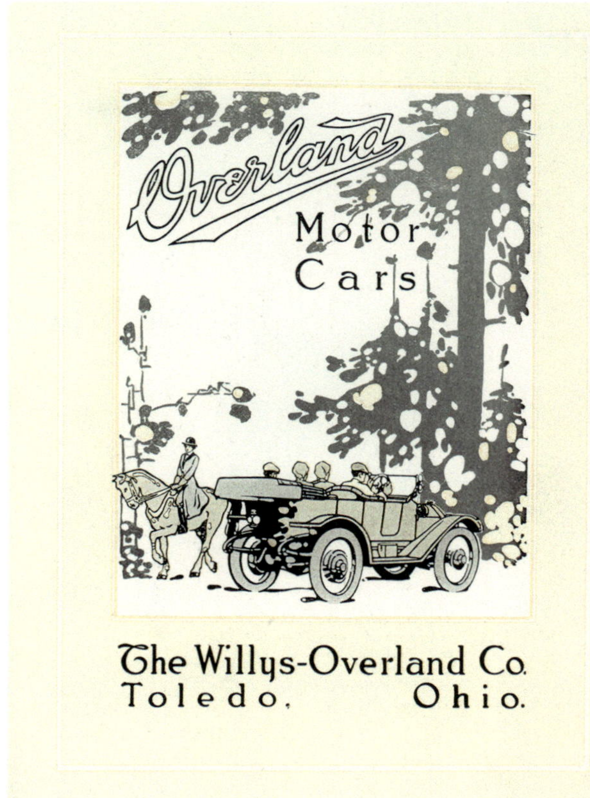

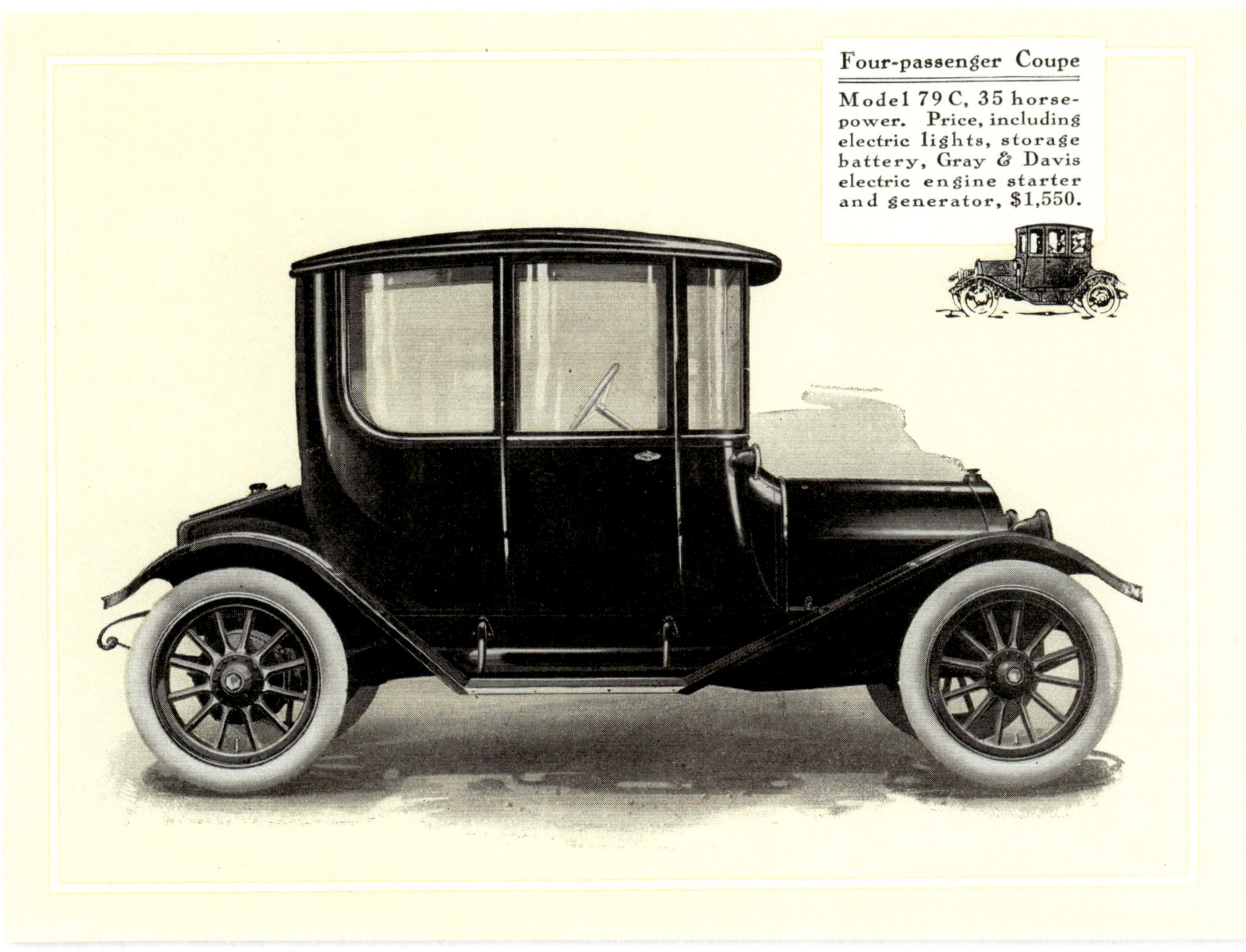

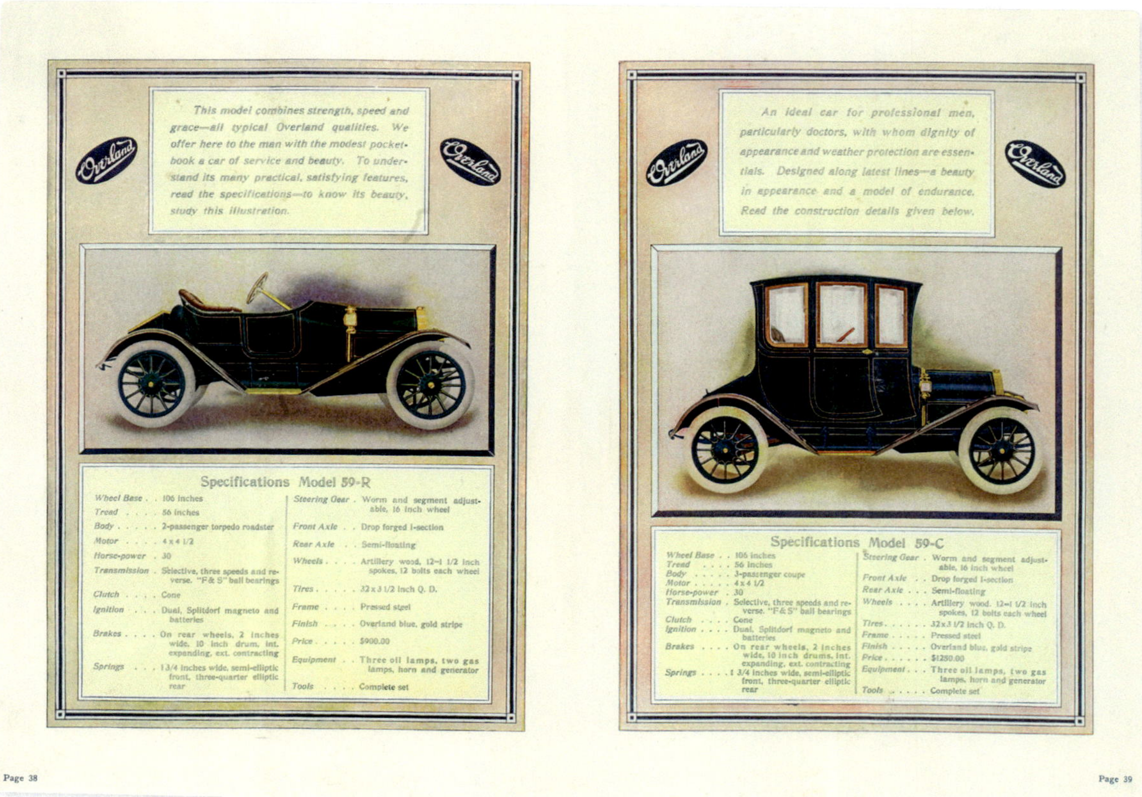

This model combines strength, speed and grace—all typical Overland qualities. We offer here to the man with the modest pocket-book a car of service and beauty. To understand its many practical, satisfying features, read the specifications—to know its beauty, study this illustration.

Specifications Model 59-R

Wheel Base	106 inches	*Steering Gear*	Worm and segment adjustable, 16 inch wheel
Tread	56 inches	*Front Axle*	Drop forged I-section
Body	2-passenger torpedo roadster	*Rear Axle*	Semi-floating
Motor	4 x 4 1/2	*Wheels*	Artillery wood, 12–1 1/2 inch spokes, 12 bolts each wheel
Horse-power	30	*Tires*	32 x 3 1/2 inch Q. D.
Transmission	Selective, three speeds and reverse. "F & S" ball bearings	*Frame*	Pressed steel
Clutch	Cone	*Finish*	Overland blue, gold stripe
Ignition	Dual, Splitdorf magneto and batteries	*Price*	$900.00
Brakes	On rear wheels, 2 inches wide, 10 inch drum, int. expanding, ext. contracting	*Equipment*	Three oil lamps, two gas lamps, horn and generator
Springs	1 3/4 inches wide, semi-elliptic front, three-quarter elliptic rear	*Tools*	Complete set

Page 38

An ideal car for professional men, particularly doctors, with whom dignity of appearance and weather protection are essentials. Designed along latest lines—a beauty in appearance and a model of endurance. Read the construction details given below.

Specifications Model 59-C

Wheel Base	106 inches	*Steering Gear*	Worm and segment adjustable, 16 inch wheel
Tread	56 inches	*Front Axle*	Drop forged I-section
Body	3-passenger coupe	*Rear Axle*	Semi-floating
Motor	4 x 4 1/2	*Wheels*	Artillery wood, 12–1 1/2 inch spokes, 12 bolts each wheel
Horse-power	30	*Tires*	32 x 3 1/2 inch Q. D.
Transmission	Selective, three speeds and reverse. "F & S" ball bearings	*Frame*	Pressed steel
Clutch	Cone	*Finish*	Overland blue, gold stripe
Ignition	Dual, Splitdorf magneto and batteries	*Price*	$1250.00
Brakes	On rear wheels, 2 inches wide, 10 inch drums, int. expanding, ext. contracting	*Equipment*	Three oil lamps, two gas lamps, horn and generator
Springs	1 3/4 inches wide, semi-elliptic front, three-quarter elliptic rear	*Tools*	Complete set

Page 39

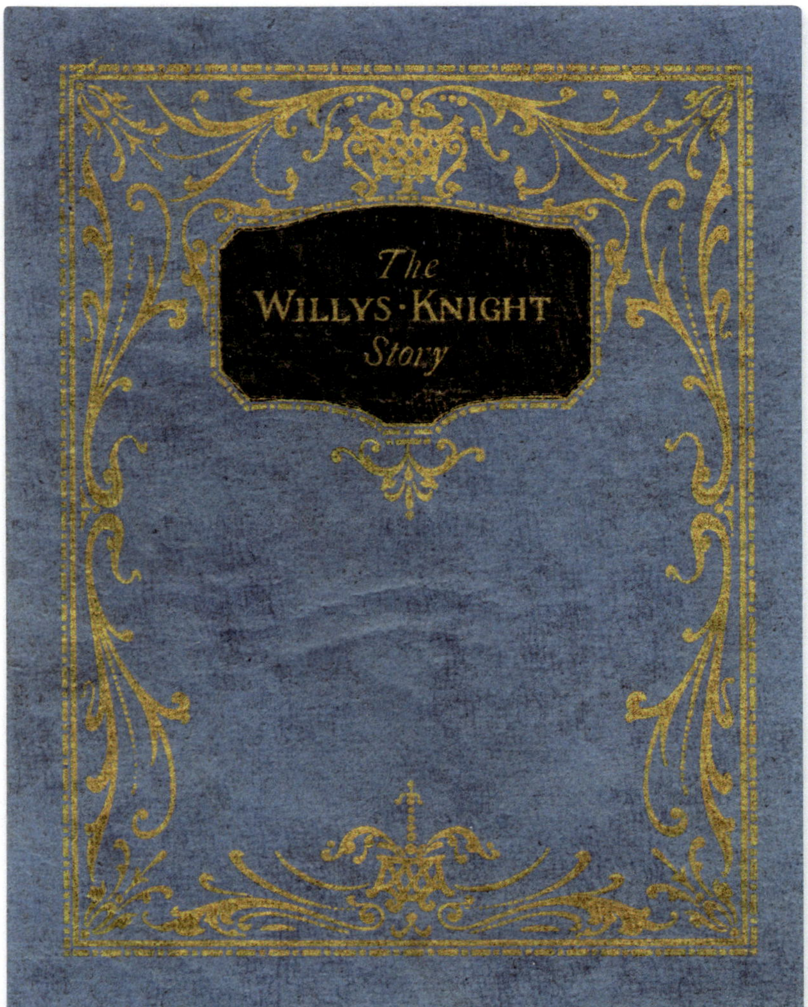

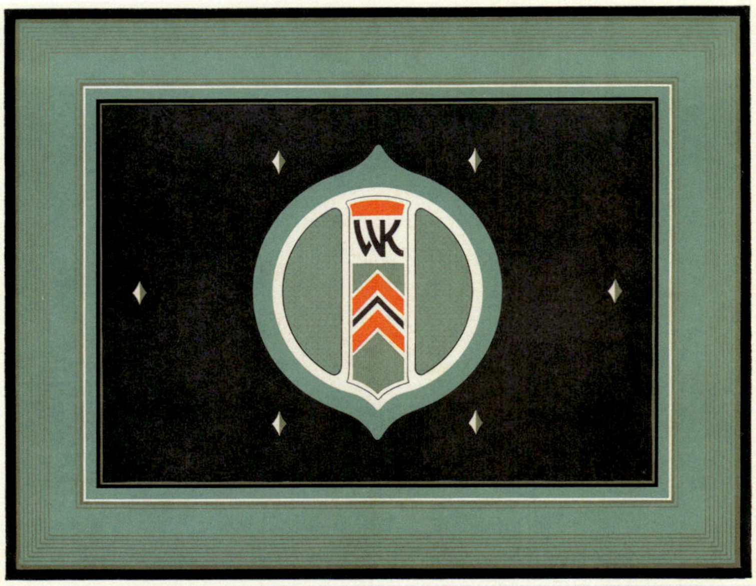

Willys-Knight Great Six 1929
(this spread)

All sorts of luxury trappings were applicable when it came to the sleeve-valve Willys-Knight. Many of these cars, especially the roadsters, are now prized by modern collectors. The "Knight" in its name refers to the sleeve-valve engine first developed by an engineer named Knight, which Willys bought under license. These were innovative, lavish cars.

Für den Willys-Knight mit Schiebermotor waren alle erdenklichen Luxusextras erhältlich. Viele dieser Fahrzeuge – vor allem die Roadster – sind heutzutage bei Sammlern äußerst begehrt. Der Namenszusatz „Knight" bezieht sich auf den Tüftler Charles Yale Knight, den Erfinder der Schiebermotors, für den Willys eine Lizenz erworben hatte. Diese Automobile waren innovative und noble Fahrzeuge.

Autour de la Willys-Knight à moteur sans soupapes flottait naturellement un parfum de luxe. Nombre de ces voitures, surtout les roadsters, sont aujourd'hui très prisées des collectionneurs. Le « Knight » renvoie à ce fameux moteur à chemises dites louvoyantes imaginé par un ingénieur ainsi nommé, dont Willys racheta la licence. Ces automobiles étaient aussi novatrices que somptueuses.

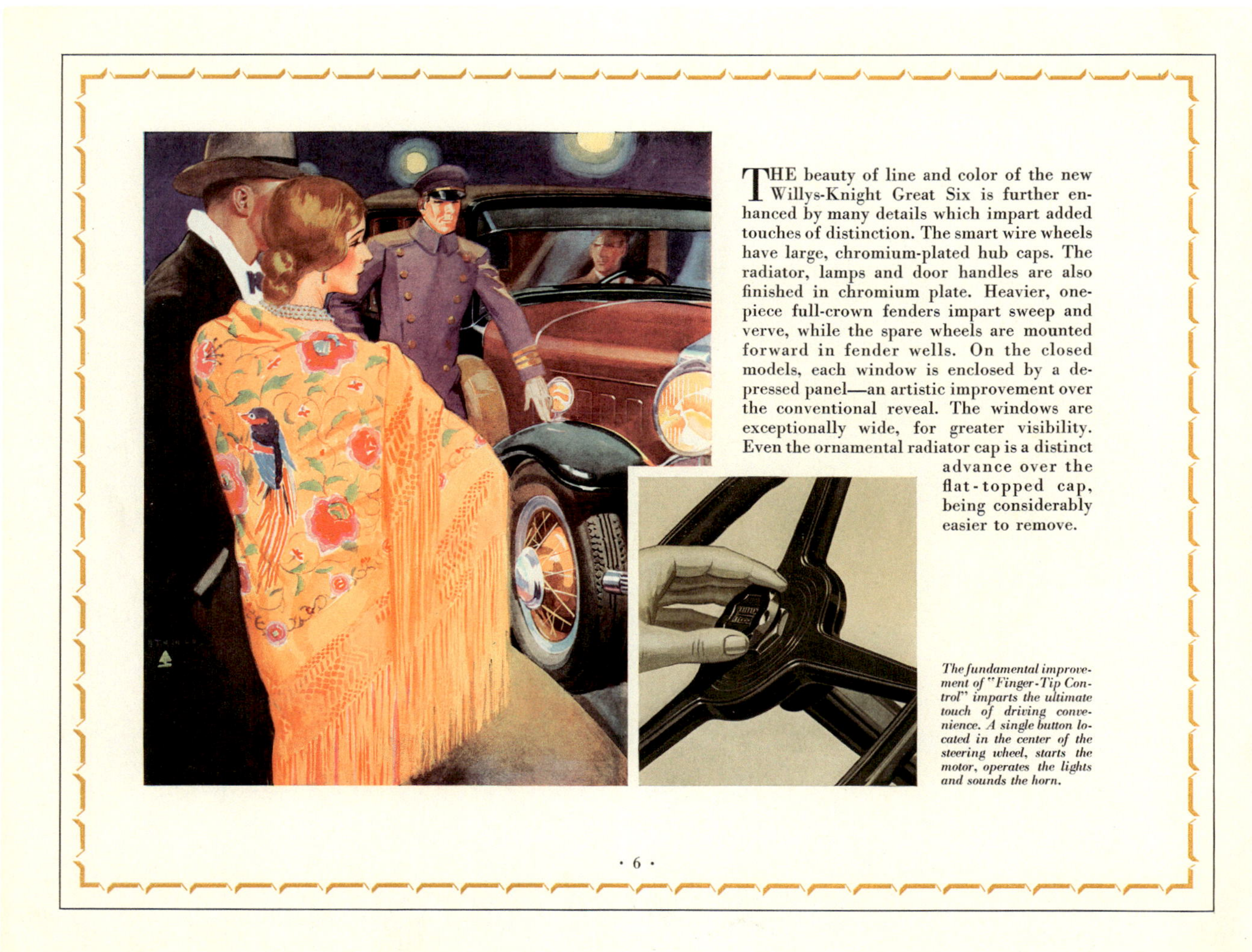

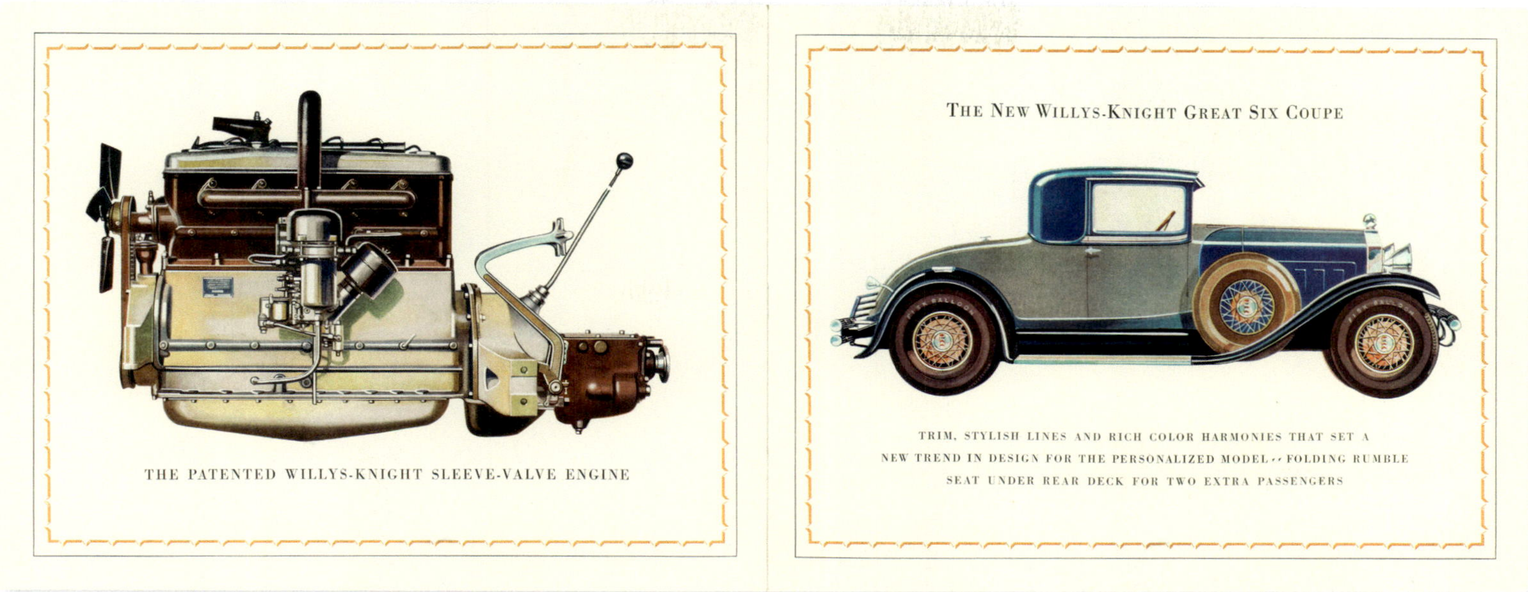
THE PATENTED WILLYS-KNIGHT SLEEVE-VALVE ENGINE
THE NEW WILLYS-KNIGHT GREAT SIX COUPE
TRIM, STYLISH LINES AND RICH COLOR HARMONIES THAT SET A NEW TREND IN DESIGN FOR THE PERSONALIZED MODEL -- FOLDING RUMBLE SEAT UNDER REAR DECK FOR TWO EXTRA PASSENGERS

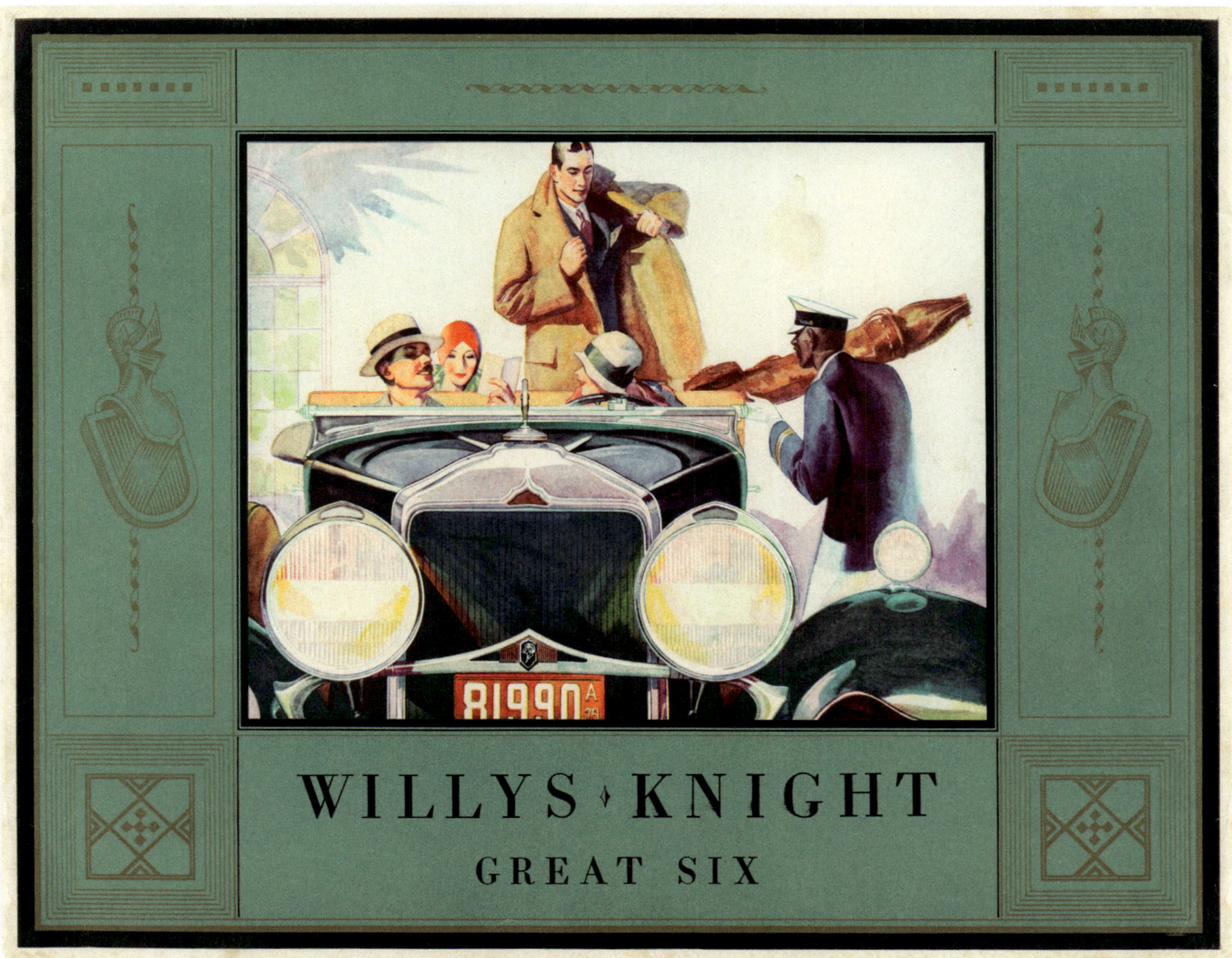
WILLYS KNIGHT
GREAT SIX

Dreams of Power and Joy
1930–1946

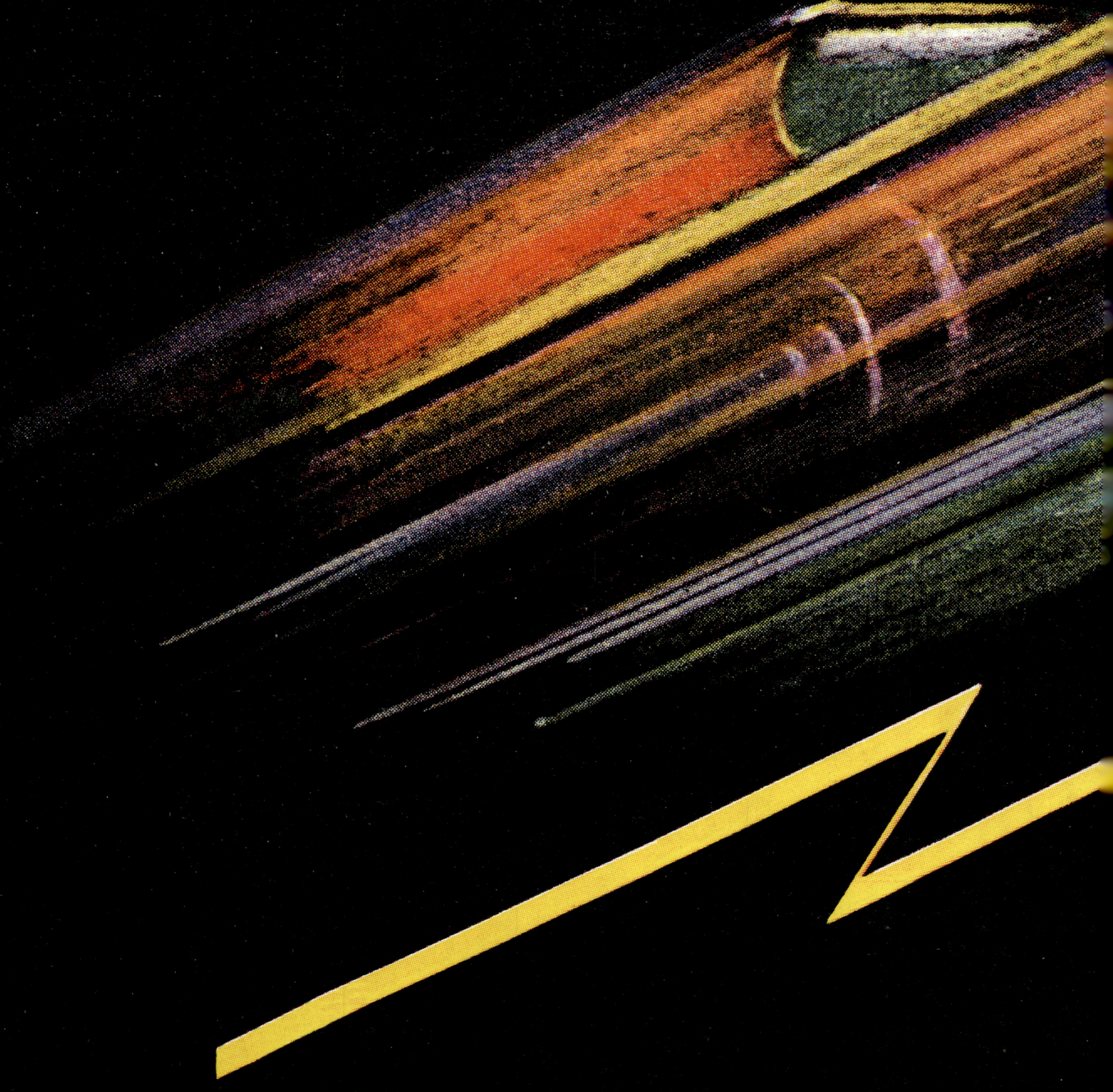

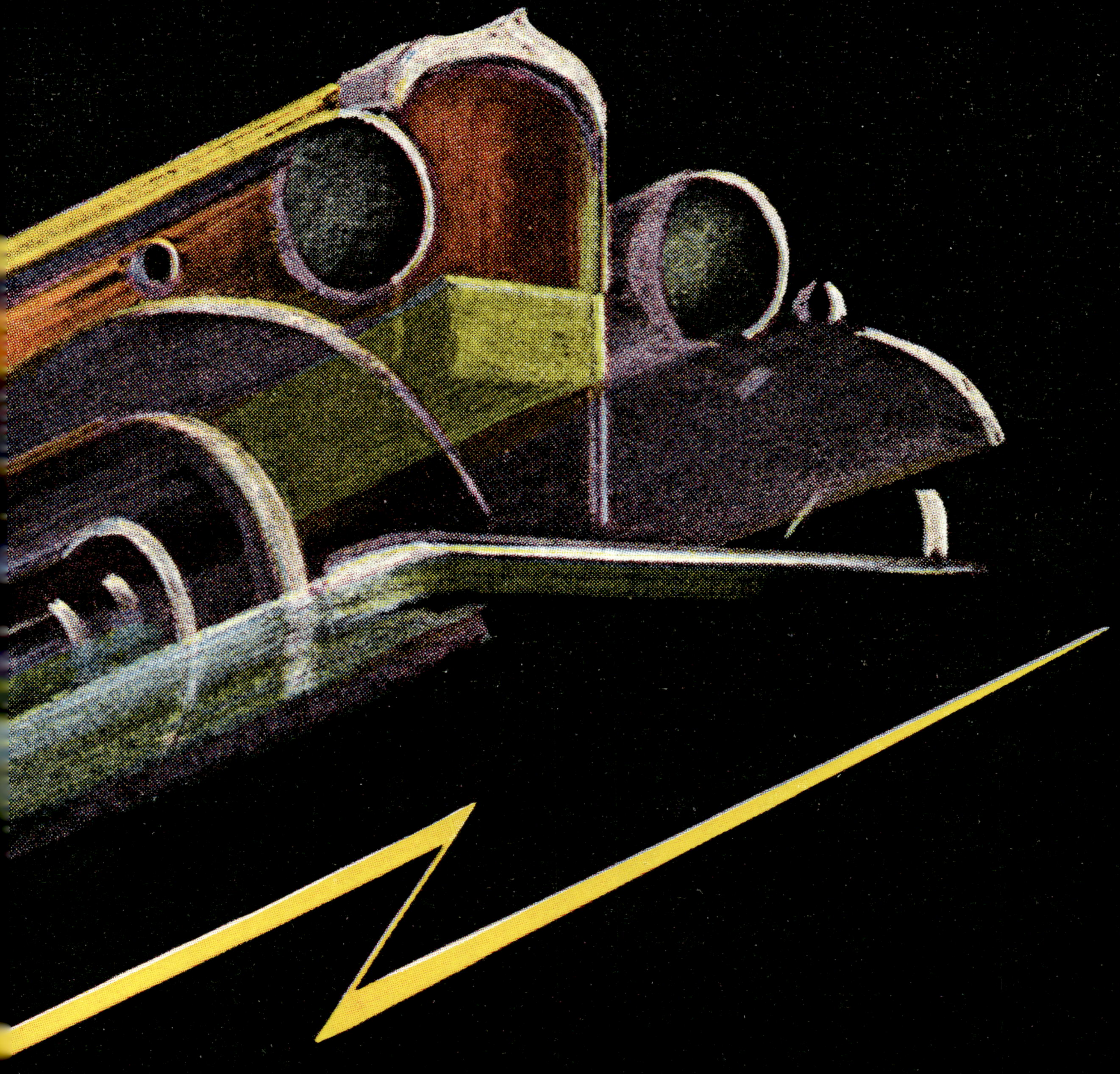

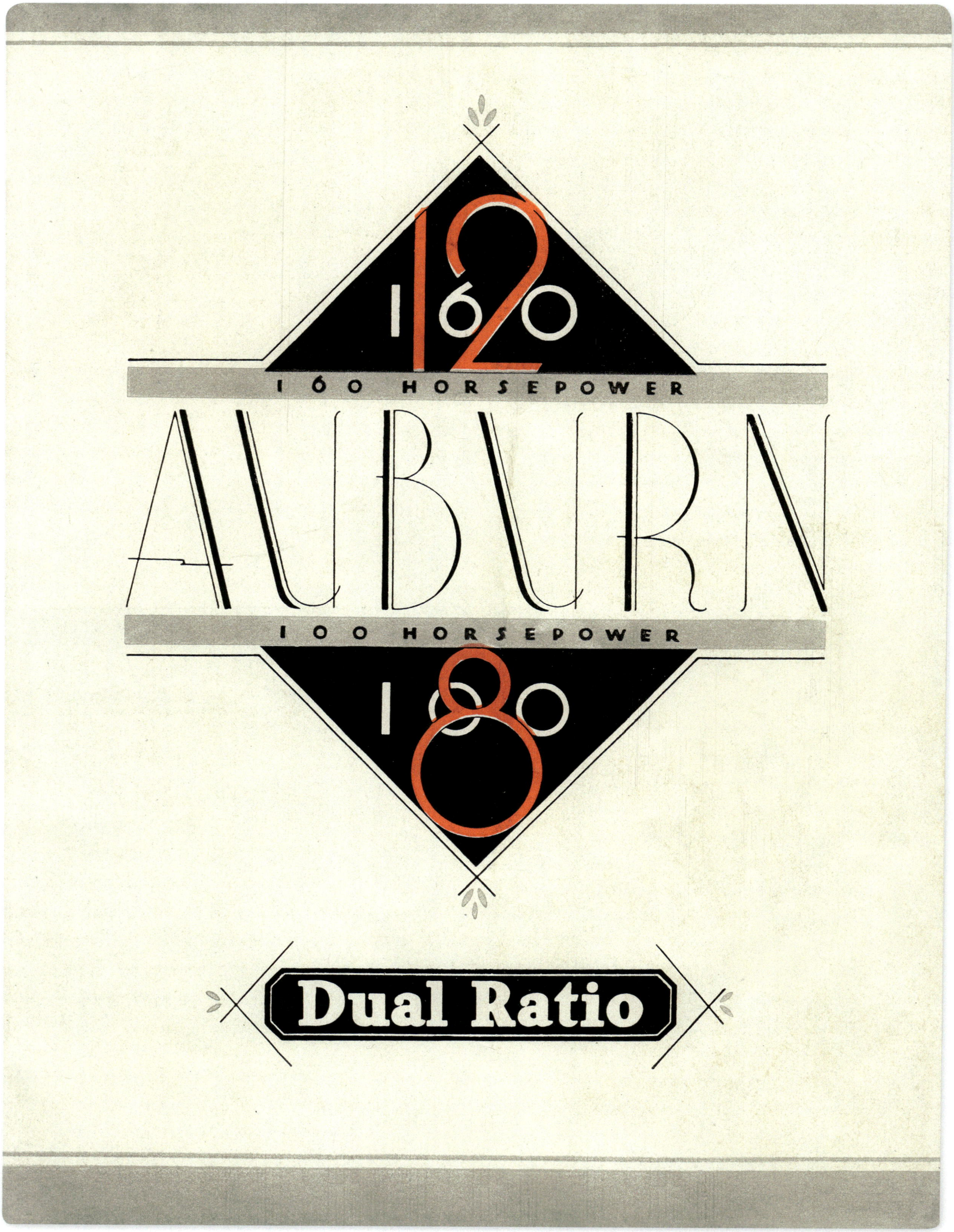
12
160
160 HORSEPOWER
AUBURN
100 HORSEPOWER
8
100
Dual Ratio

Auburn

Auburn 1932 *(opposite and right)*, **1935** *(below)*

In 1932 Auburn offered a Lycoming V-12, the least expensive 12-cylinder car on the market. It followed up with the spectacular Schwitzer-supercharged Speedster for 1935. Originally a spin-off from an Indiana carriage maker, Auburn, named for its hometown, was one of the most prestigious cars in the years that led up to the Great Depression. Ironically, that economic collapse doomed the manufacturer.

Im Jahre 1932 brachte Auburn mit dem Lycoming V12 das damals preisgünstigste V12-Automobil auf den Markt. Drei Jahre später folgte der aufsehenerregende Speedster mit Schwitzer-Turbolader. Auburn ging aus einem Wagenbauunternehmen hervor, das nach der gleichnamigen Stadt in Indiana benannt war, in der es seinen Sitz hatte. In den Jahren vor der Weltwirtschaftskrise war der Auburn eines der prestigeträchtigsten Automobile, doch der wirtschaftliche Zusammenbruch des Landes trieb auch seinen Hersteller in den Ruin.

En 1932, Auburn dévoila sa Lycoming V12, la moins chère du marché sur le segment des douze cylindres. Suivra en 1935 la spectaculaire Speedster à compresseur Schwitzer. Auburn, qui porte le nom de sa ville de naissance, doit son existence à un charron de l'Indiana, fabricant de véhicules hippomobiles. Elle deviendra l'une des marques les plus prestigieuses dans les années qui précèderont la Grande Dépression, avant d'être emportée à son tour par la crise économique.

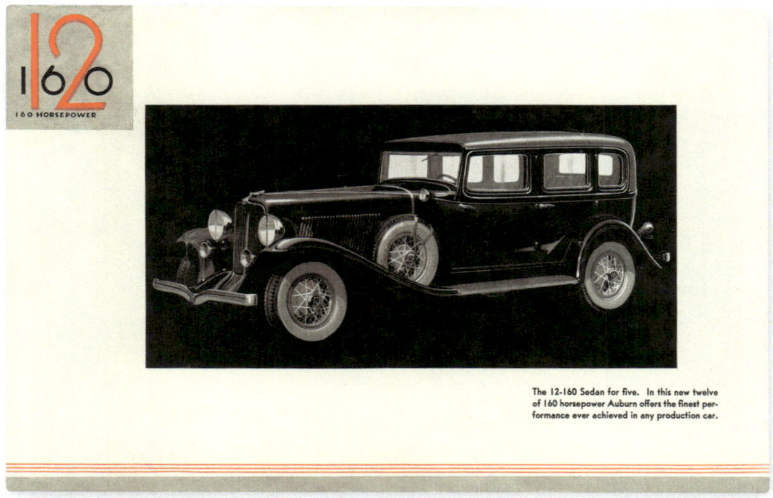

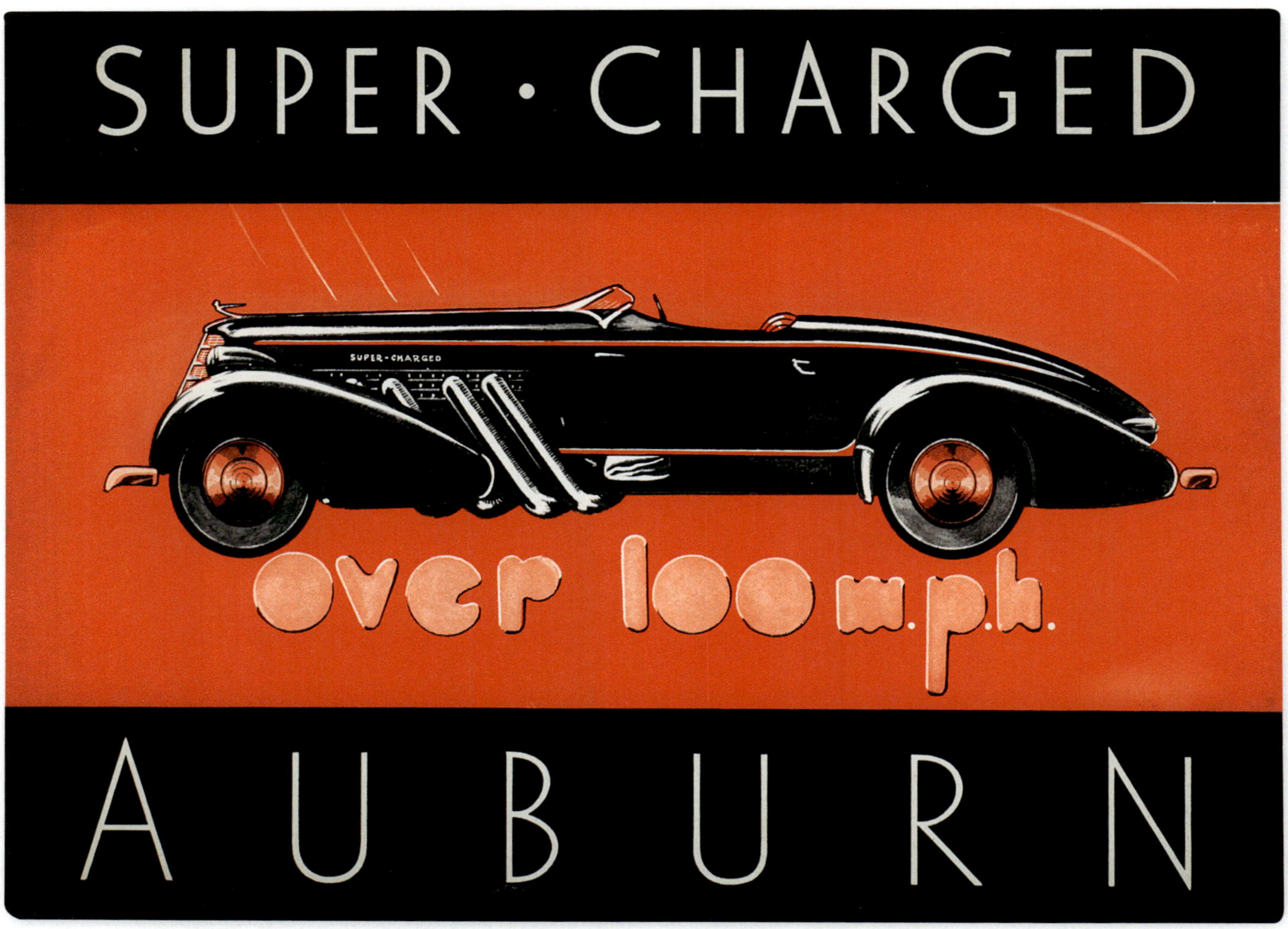

Buick

Buick 1930 *(this spread)*

Alfred P. Sloan was one the great geniuses of automotive assembly and marketing, running General Motors from the 1920s into the 1960s. He ordered the creation of GM's Art & Colour section, among the first dedicated styling staffs, to build individual identities for the corporation's myriad brands. His grand scheme placed Buick just below LaSalle and Cadillac on the prestige scale, and the Art & Colour section responded with paintings that made a Buick apropos at the loftiest social event.

Alfred P. Sloan, der General Motors von den 1920er- bis in die 1960-Jahre leitete, gehörte zu den großen Genies im Bau und bei der Vermarktung von Automobilen. Unter seiner Ägide wurde die Sektion für Art & Colour bei GM – eine der ersten Abteilungen, die ausschließlich für das Fahrzeugstyling zuständig war – damit beauftragt, jeder der unzähligen Marken des Konzerns eine unverwechselbare Identität zu verschaffen. Nach seinem Plan wurde Buick auf der Prestigeskala knapp unterhalb von LaSalle und Cadillac angesiedelt, und Art & Colour reagierte darauf mit Lackierungen, in denen man auch bei den vornehmsten gesellschaftlichen Ereignissen mit einem Buick vorfahren konnte.

Alfred P. Sloan, génie de l'industrie et du marketing automobiles, dirigea General Motors des années 1920 jusqu'aux Sixties et y présida à la création du département Art & Colour, l'un des premiers bureaux de style. Objectif : forger des identités individuelles pour chacune des nombreuses marques de la GM. Son ambitieux projet plaçant ainsi Buick juste au-dessous de LaSalle et Cadillac sur l'échelle du prestige, Art & Colour répondit par des illustrations qui associaient naturellement la Buick aux événements mondains les plus élégants.

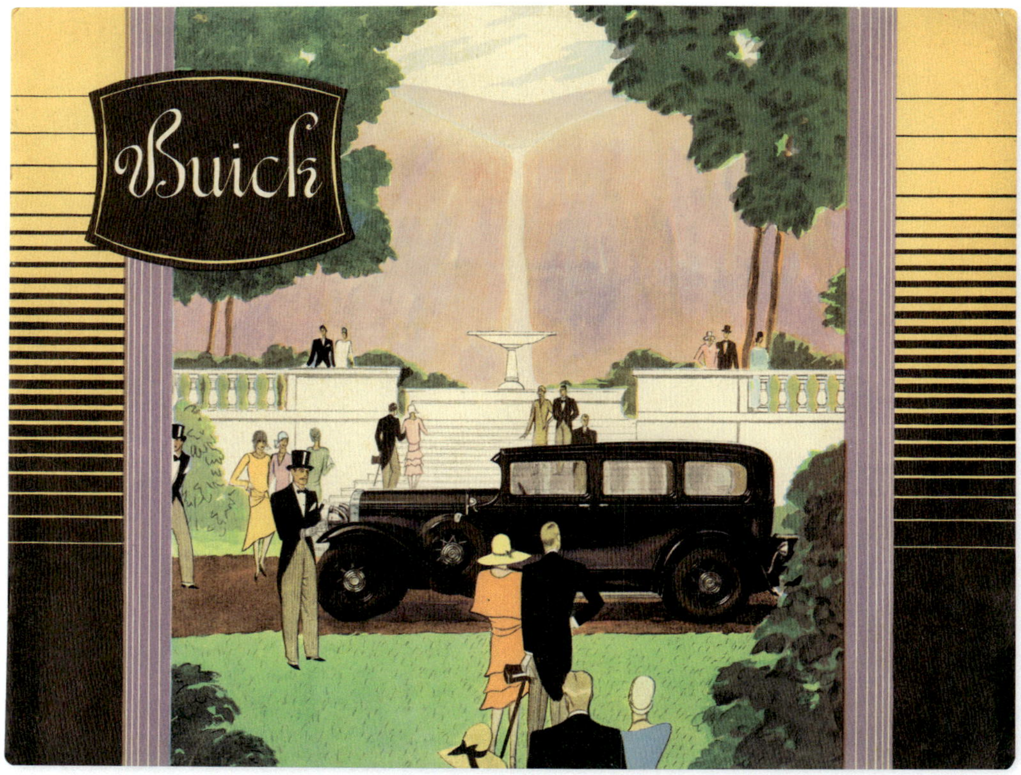

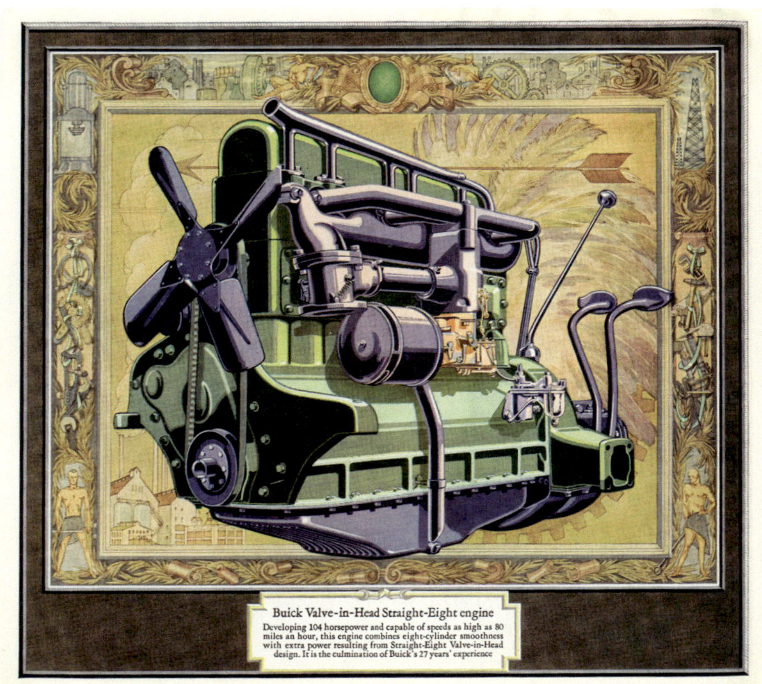

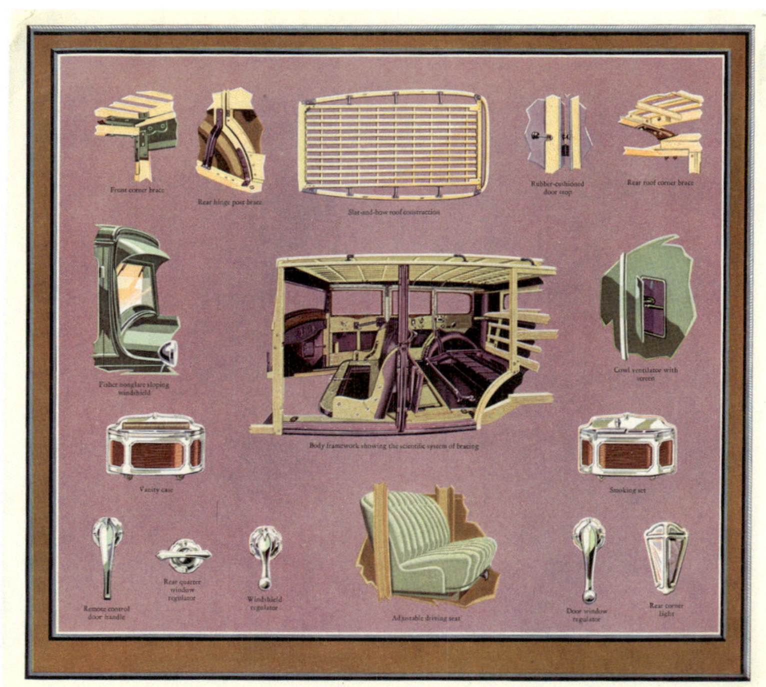

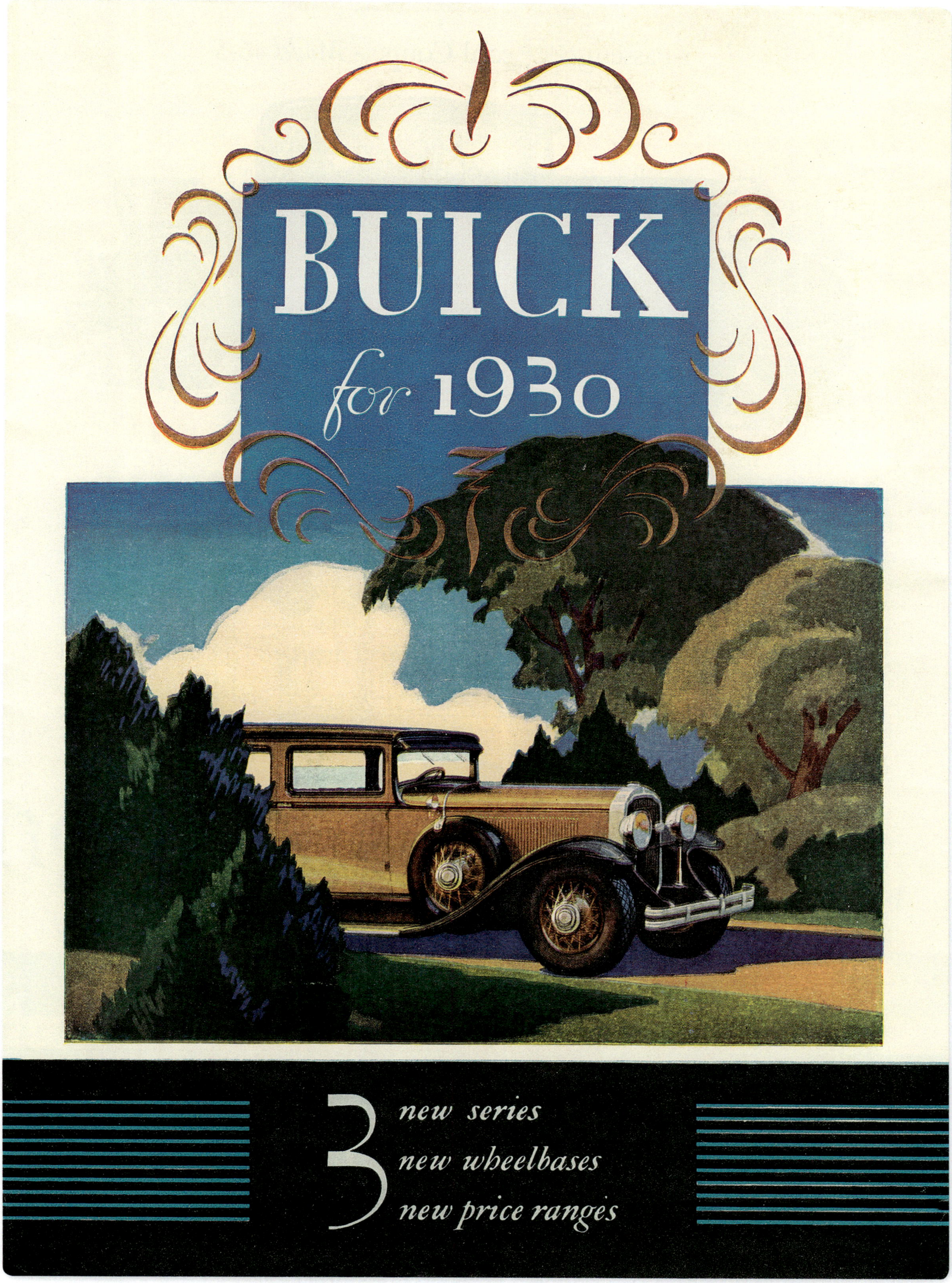
BUICK
for 1930
3 new series
new wheelbases
new price ranges

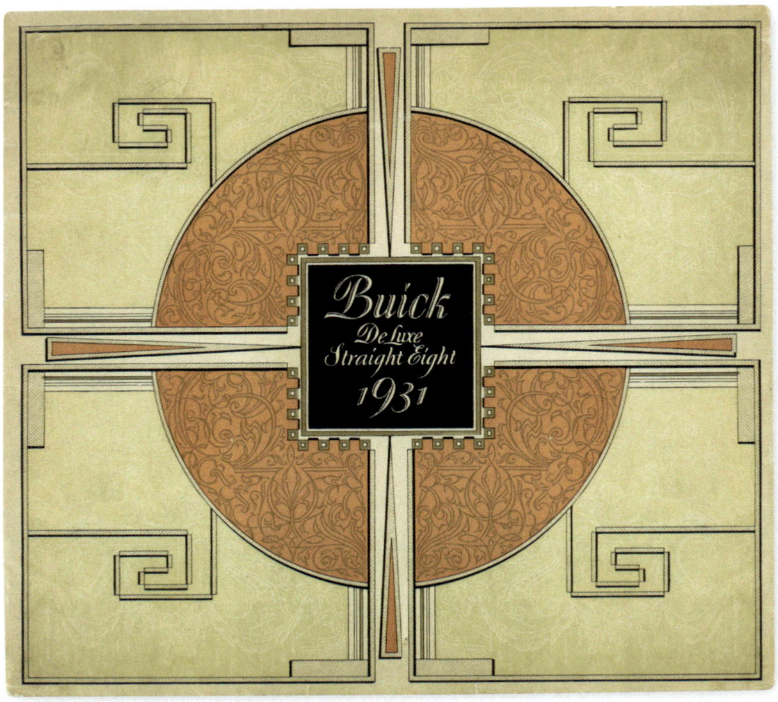

Buick Eight 1931 *(left)*, **1936** *(below)*

Harley Earl, who went from creating custom coaches in Los Angeles to running GM's first image factory, innately knew that color and styling meant drama when you were trying to convince the public that a carried-over car was actually very new. Earl directed the styling efforts of GM's Art & Colour section for decades, inventing the annual model change, which often meant an update in trim, that guaranteed the firm's awe-inspiring success.

Harley Earl begann seine Karriere im Stellmacherunternehmen seines Vaters in Los Angeles, wo er maßgefertigte Karosserien entwarf, und wurde schließlich Chefkonstrukteur in der ersten Imagefabrik von GM. Er verstand es, Farbe und Gestaltung zu nutzen, um der Öffentlichkeit weiszumachen, dass hinter einer schlichten Neuauflage ein neues Autos steckte. Earl leitete jahrzehntelang die Stylingarbeiten in der Sektion für Art & Colour und führte den jährlichen Modellwechsel ein, der häufig mit jenen kosmetischen Maßnahmen einherging, die dem Unternehmen seinen erstaunlichen Erfolg garantierten.

Harley Earl était passé de la création de carrosseries sur mesure, à Los Angeles, à la direction du premier département « design » de la General Motors. Il savait d'instinct que la couleur et le style jouent un rôle décisif quand il s'agit de convaincre le public du caractère innovant d'un ancien modèle. Earl dirigea donc pendant plusieurs décennies les efforts stylistiques du bureau Art & Colour de la GM où il inventa notamment le renouvellement annuel des modèles, qui instituait une actualisation systématique des habillages et assura à la firme son impressionnant succès.

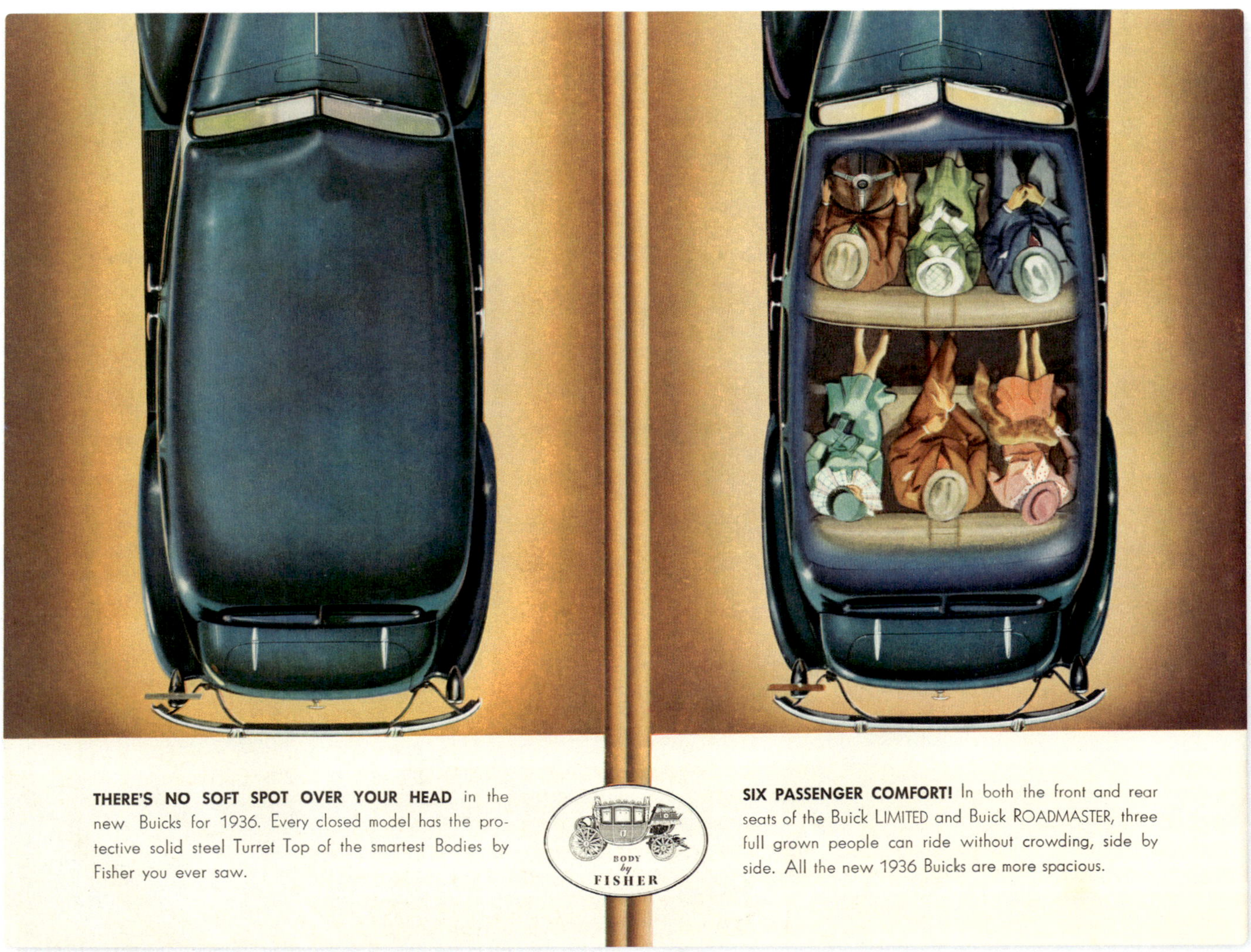

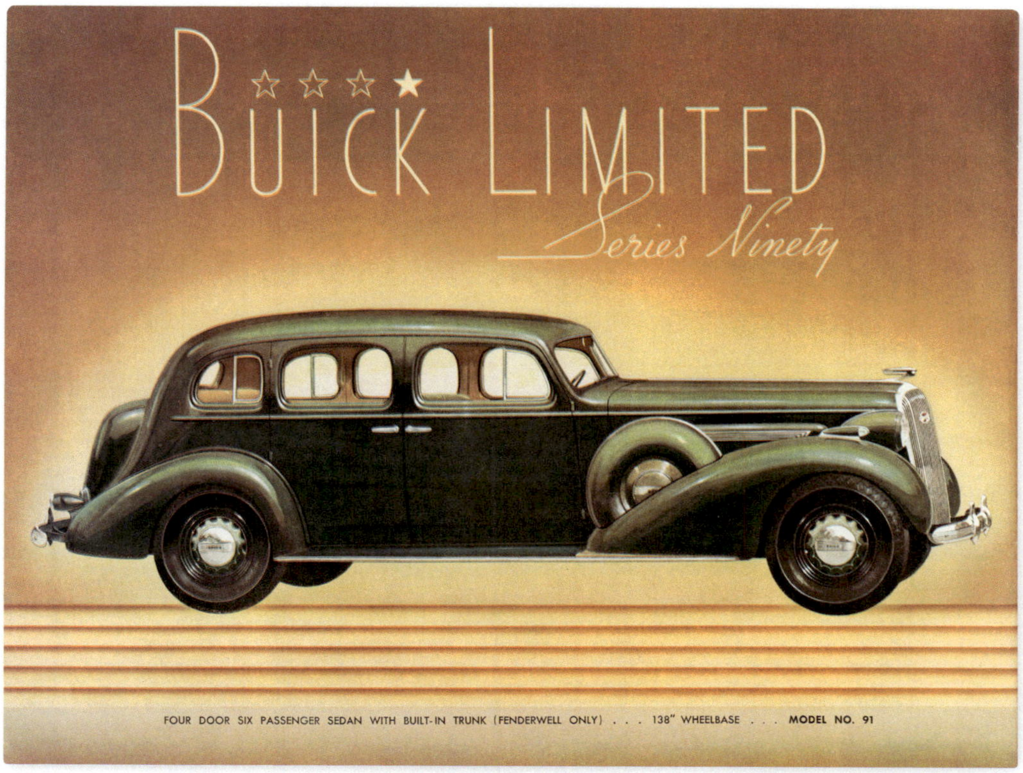

Buick Series 90 1936

A new division general manager, Harlow Curtice, ordered the Buick lineup to be fully revised for the 1936 model year. Its flagship Series 90 model, which became known as the Limited under Buick's new model nomenclature, got an all-new Fisher body with a steel turret top that did away with the former fabric insert.

Ein neuer Geschäftsführer bei Buick namens Harlow Curtice ordnete für das Modelljahr 1936 eine Generalüberholung der Modellpalette an. Das Flaggschiff, der Series 90, wurde unter der neuen Nomenklatur zum Limited und erhielt eine völlig neue Fisher-Karosserie mit Ganzstahldachkonstruktion anstelle des Stoffverdecks seines Vorgängers.

Nommé à la tête de la marque Buick, Harlow Curtice ordonna une révision complète de la gamme pour l'année modèle 1936. Son fleuron, la Series 90, qui se fera connaître sous le nom de Limited selon la nouvelle nomenclature adoptée, reçut ainsi une toute nouvelle carrosserie Fisher avec un toit en acier d'une seule pièce dit « turret top », qui remplaçait la toile et le bois précédemment utilisés.

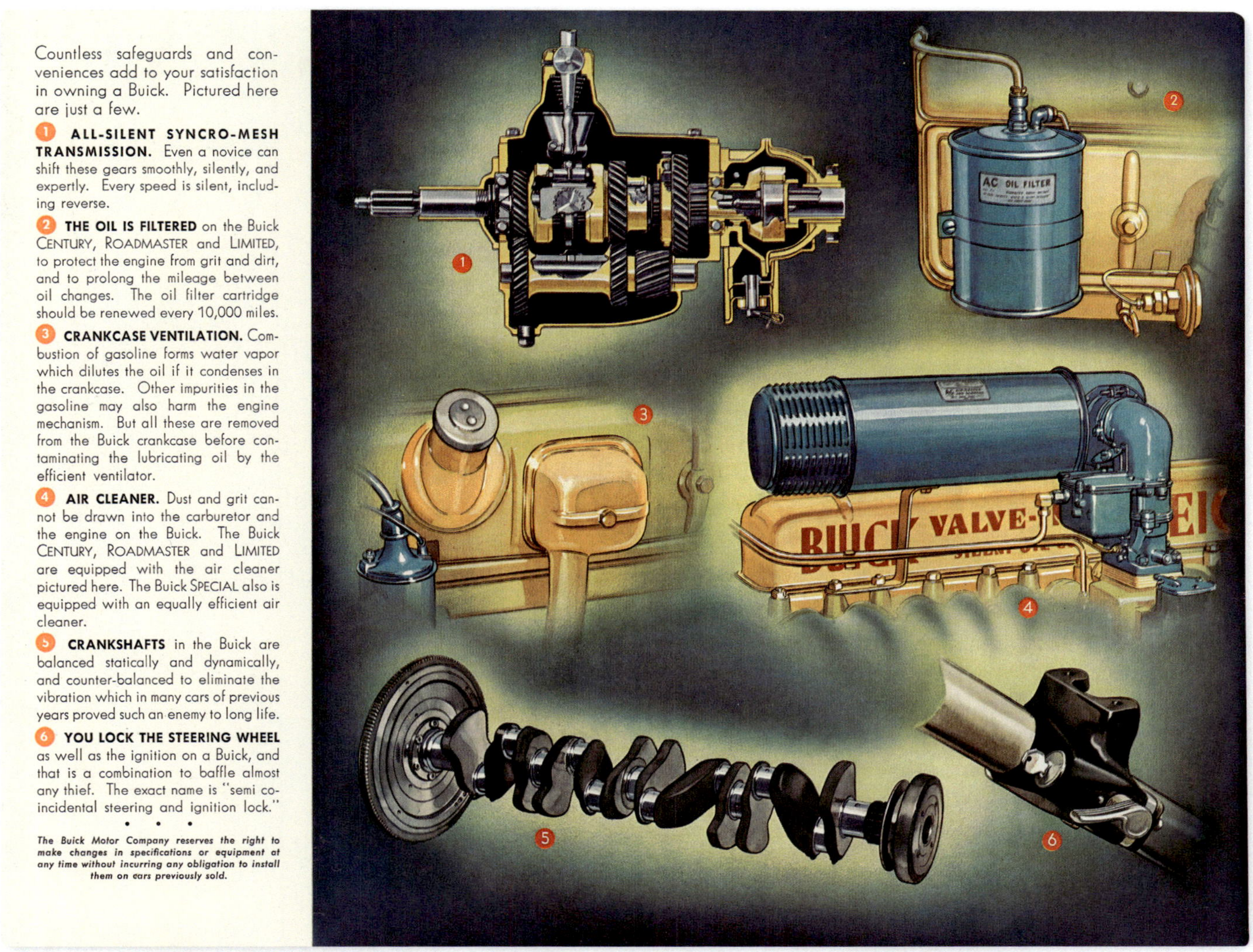

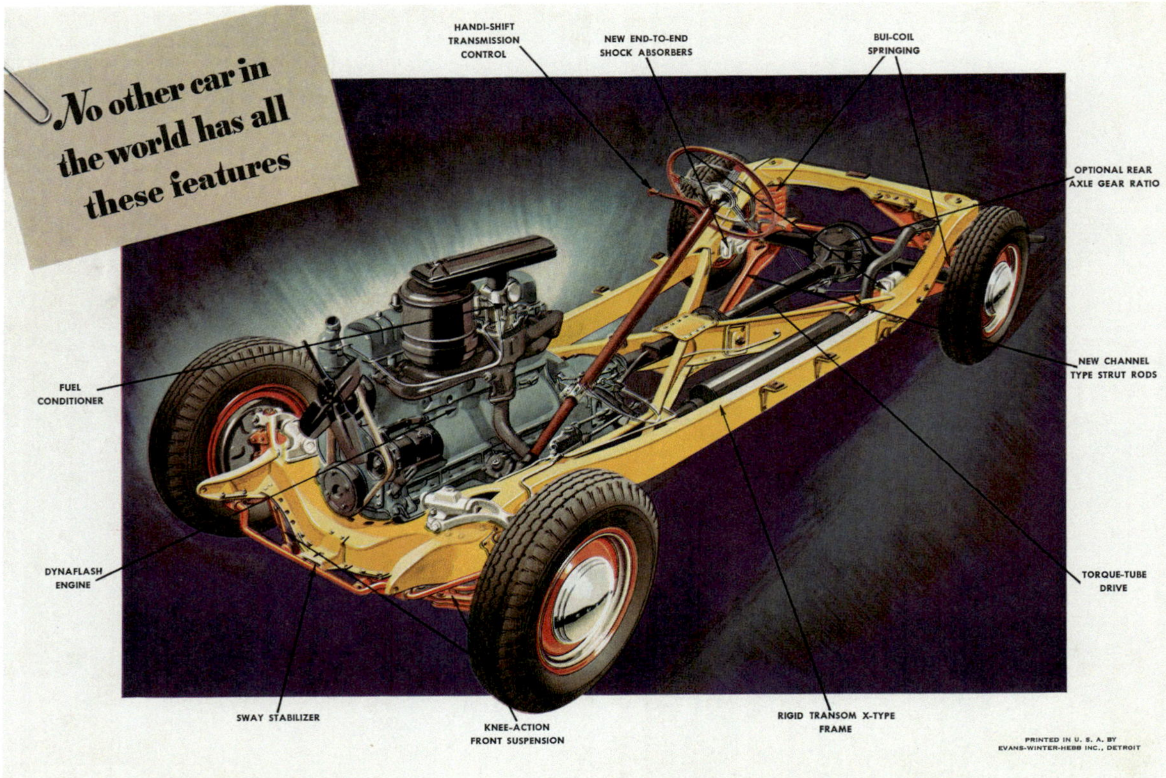

Buick 1939, 1940 *(opposite)*

Buick built its four millionth car in 1940, its best sales year ever. That record fell in 1941 when new bodies were introduced. Buick auto production ended early in 1942 due to World War II, and resumed production again in 1946.

Buick baute im Jahre 1940, dem absatzstärksten in der damaligen Firmengeschichte, sein viermillionstes Fahrzeug. Der Rekord wurde aber bereits im nächsten Jahr eingestellt, als neue Karosserieformen eingeführt wurden. Die Automobilproduktion bei Buick wurde 1942 aufgrund des Zweiten Weltkriegs vorübergehend eingestellt und erst 1946 wieder aufgenommen.

Buick a construit sa quatre millionième voiture en 1940, l'année de ses meilleures performances commerciales. Ce record tombera dès 1941 avec l'introduction de nouvelles carrosseries. La production automobile de Buick s'interrompit début 1942 à cause de la Seconde Guerre mondiale, pour redémarrer en 1946.

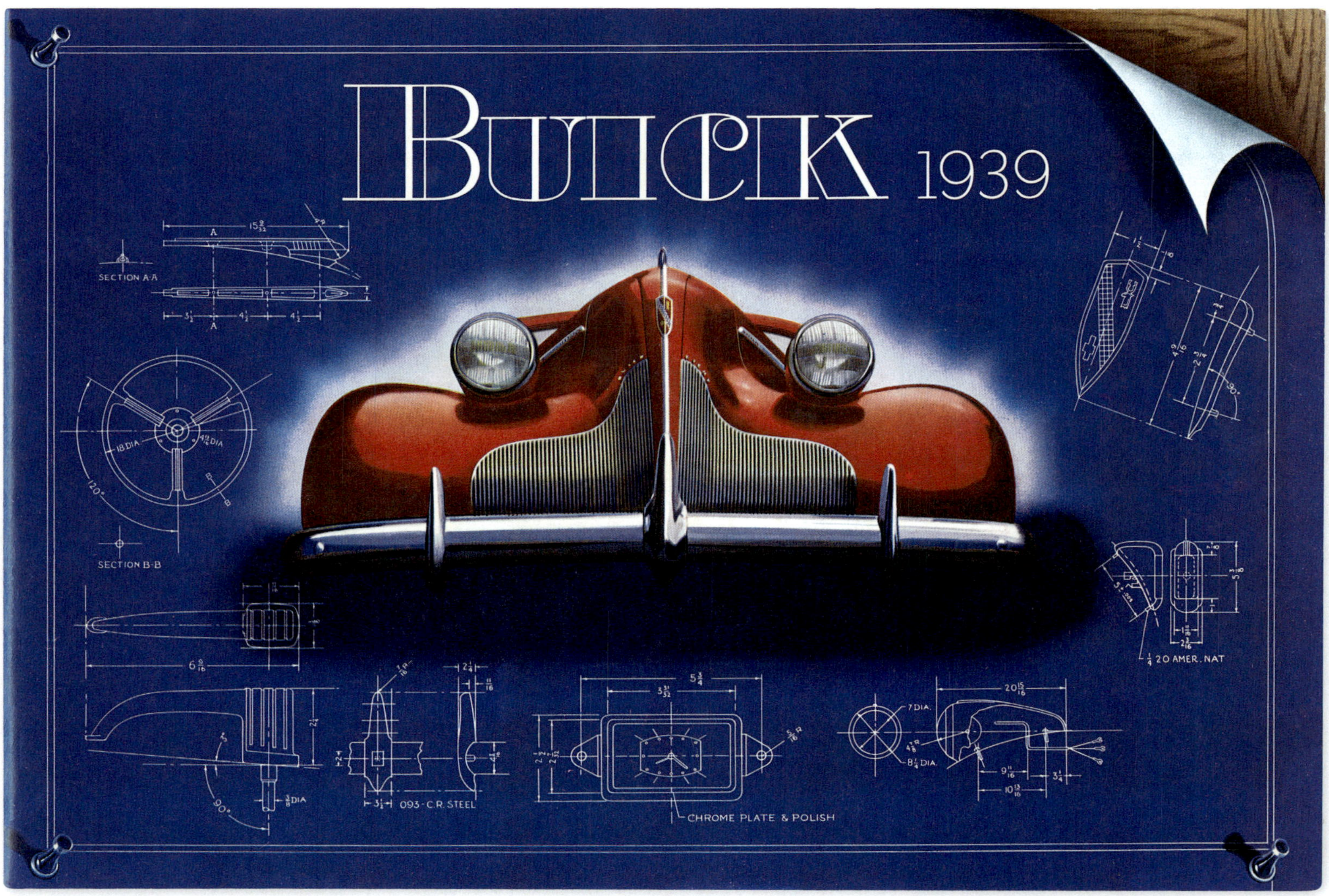

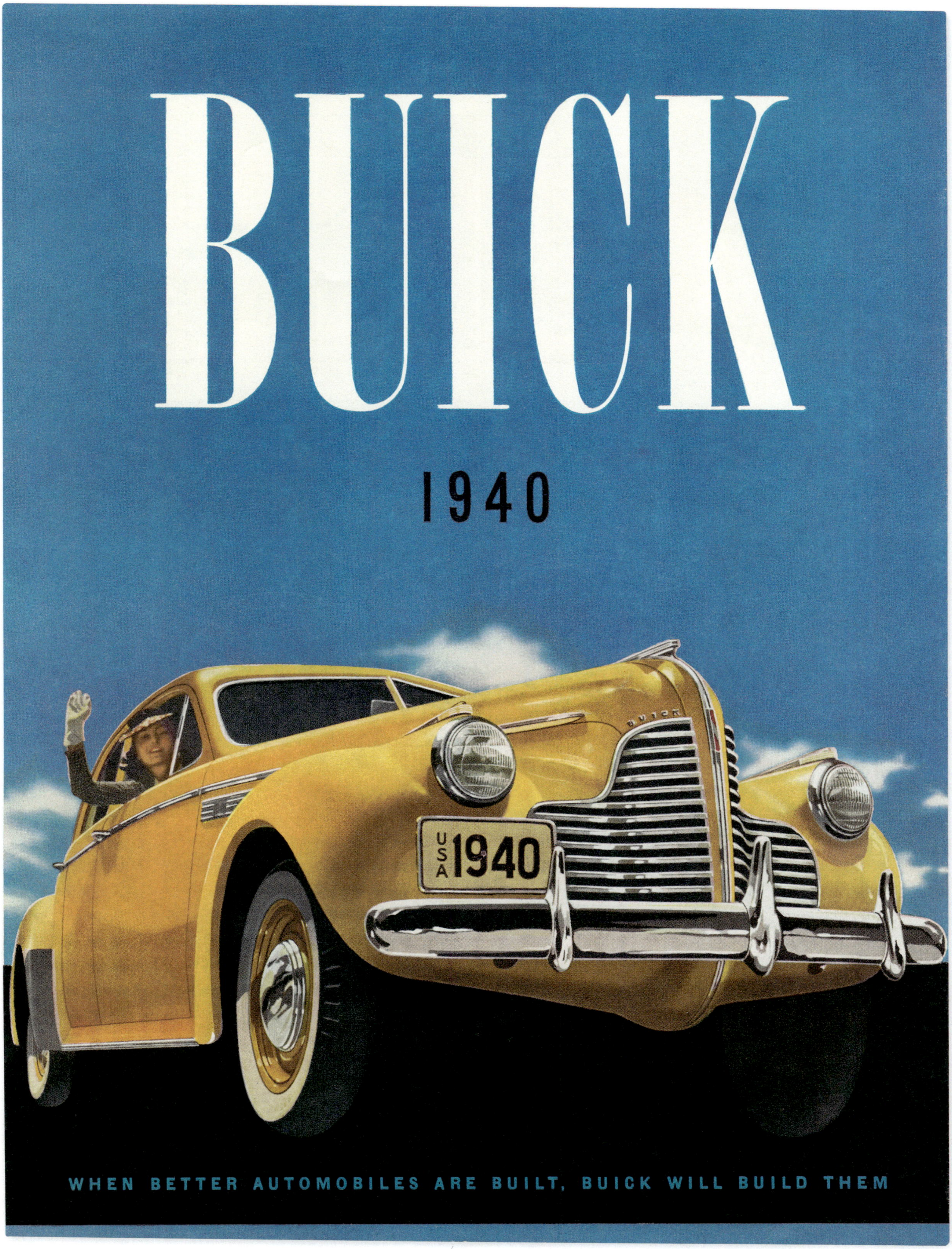
BUICK
1940
USA 1940
BUICK
WHEN BETTER AUTOMOBILES ARE BUILT, BUICK WILL BUILD THEM

Cadillac

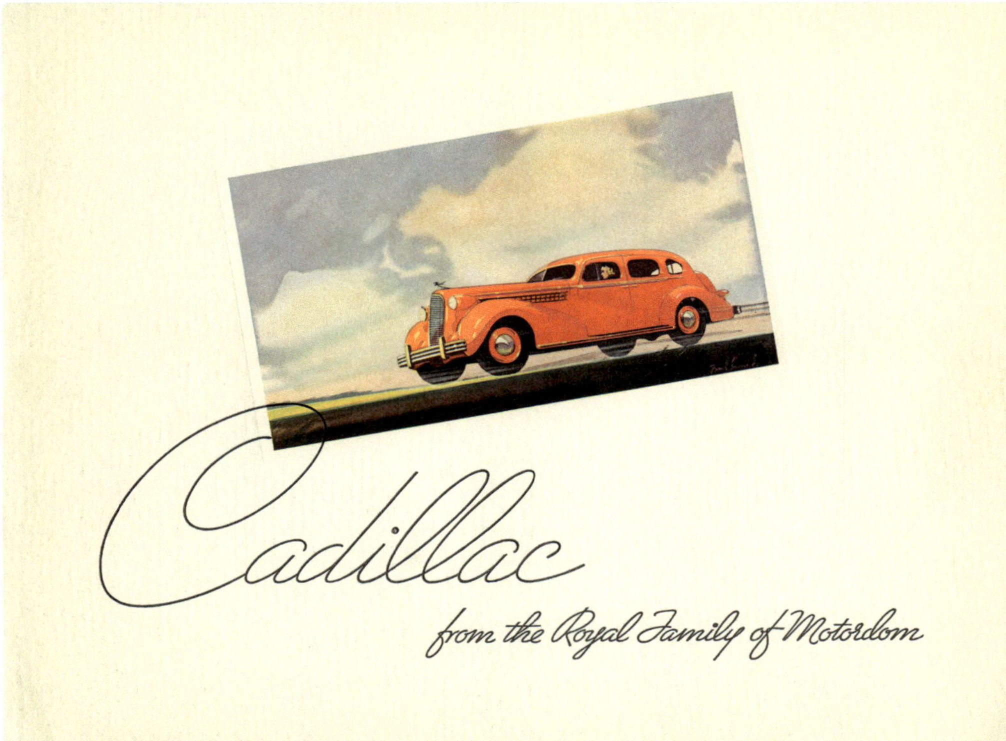

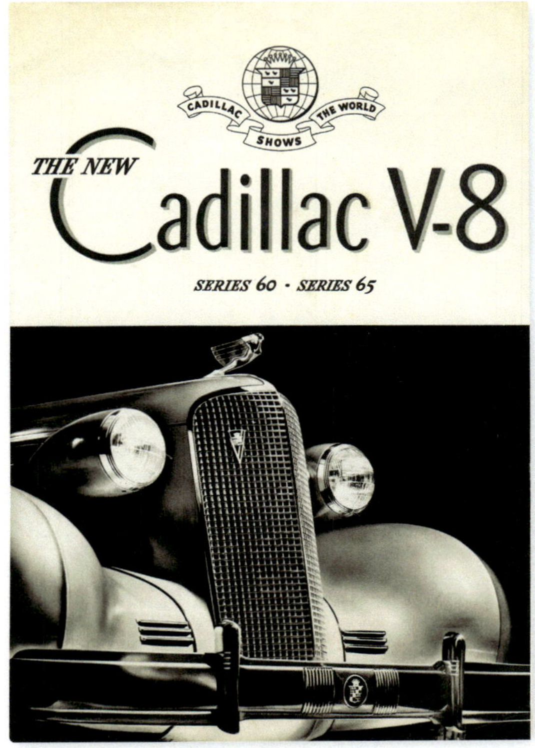

Cadillac 1939 *(this spread)*

The single louver on the hood sides identified the 1939 Cadillacs. These cars rode on a 141-inch wheelbase, but the coupe drawing somewhat exaggerates its length, a technique used by the manufacturers to dramatize scale.

Am einzelnen Lüftungsgitter an den Seiten der Motorhaube konnte man die Cadillacs von 1939 erkennen. Diese Autos hatten zwar einen Radstand von 3 581 Millimetern, doch in der Zeichnung des Coupés ist die Fahrzeuglänge übertrieben dargestellt – ein gern genutzter Kniff der Hersteller, um ihre Autos noch eindrucksvoller erscheinen zu lassen.

Les ouïes disposées de part et d'autre du capot permettent d'identifier les Cadillac de 1939. Ces modèles affichaient un empattement de 3,58 mètres mais l'illustration de la version coupé exagère quelque peu cette longueur, une technique employée par les constructeurs pour amplifier l'effet visuel du produit.

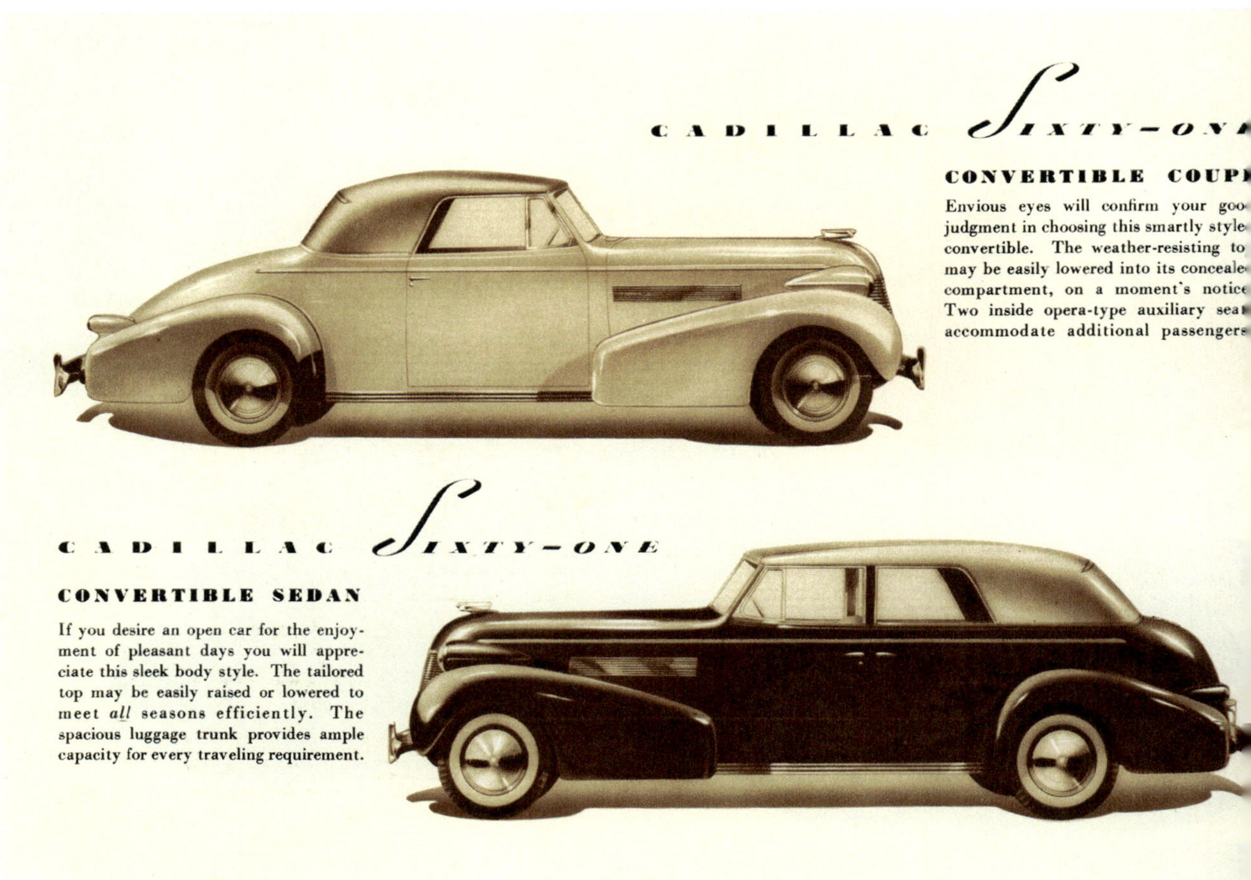

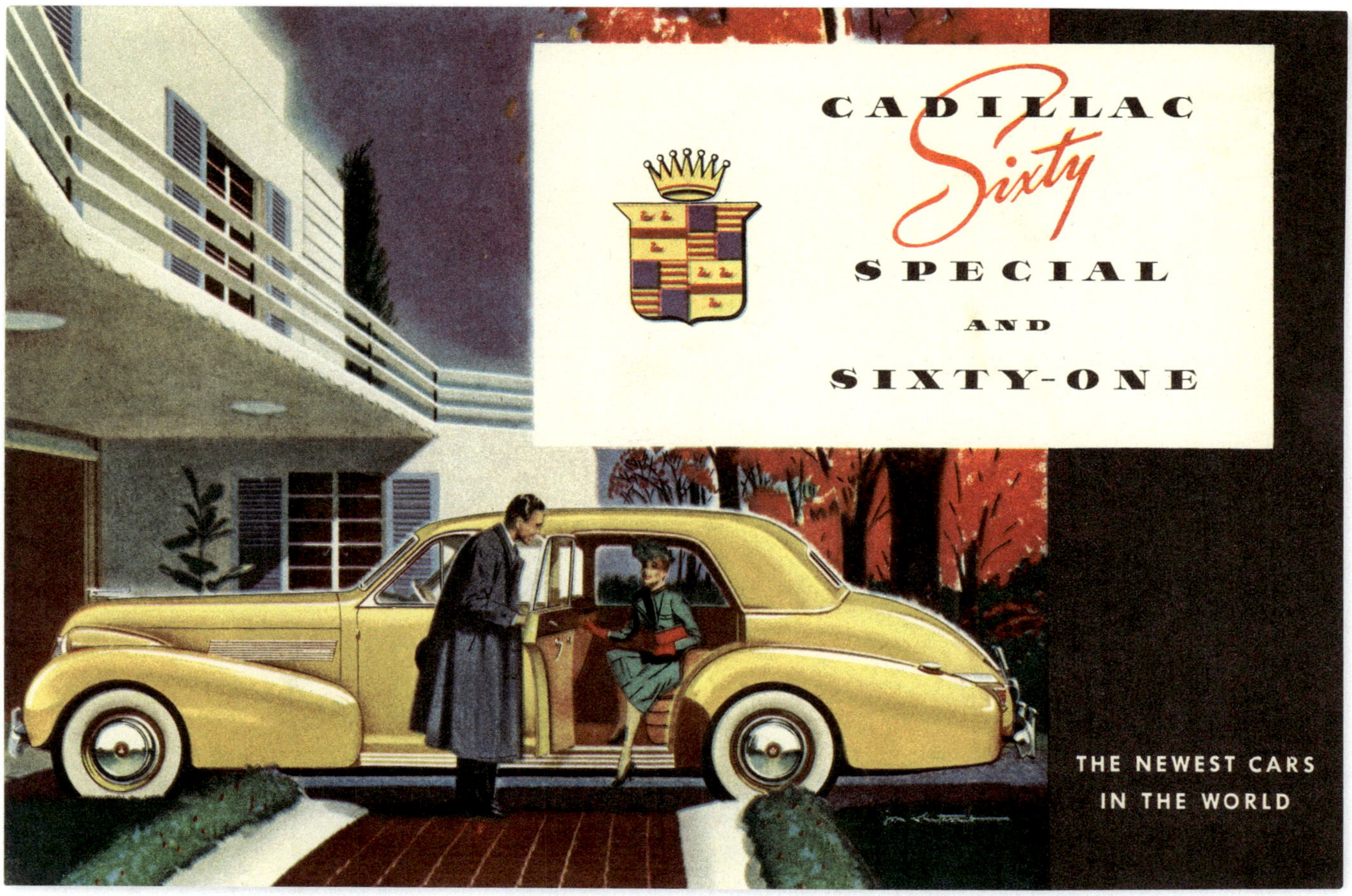

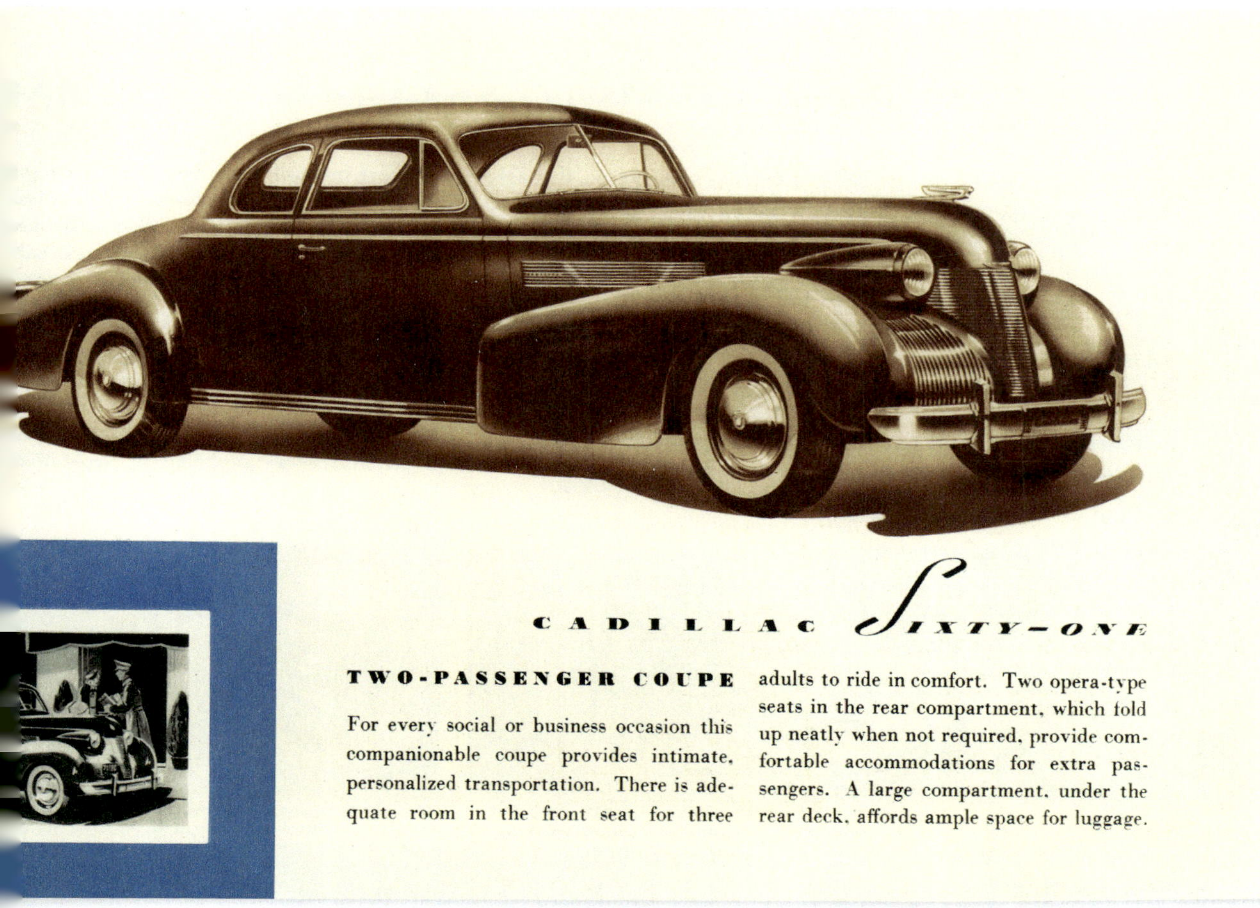

CADILLAC *Sixty-One*

TWO-PASSENGER COUPE

For every social or business occasion this companionable coupe provides intimate, personalized transportation. There is adequate room in the front seat for three adults to ride in comfort. Two opera-type seats in the rear compartment, which fold up neatly when not required, provide comfortable accommodations for extra passengers. A large compartment, under the rear deck, affords ample space for luggage.

Chevrolet

Chevrolet Master Six 1935 *(right)*,
De Luxe 1936 *(below)*, **1935** *(opposite)*

By the 1930s Chevrolet was ensconced in its identity as a builder of stout and practical cars.

In den 1930er-Jahren hatte sich Chevrolet seinen Ruf als Hersteller robuster und praktischer Automobile schon gesichert.

Tout au long des années 1930, Chevrolet est resté confortablement installé dans son image de constructeur de voitures robustes et fonctionnelles.

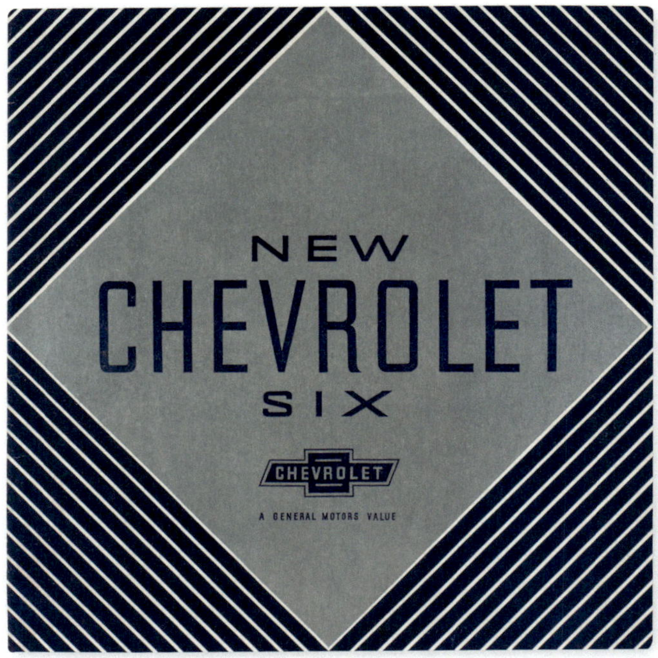

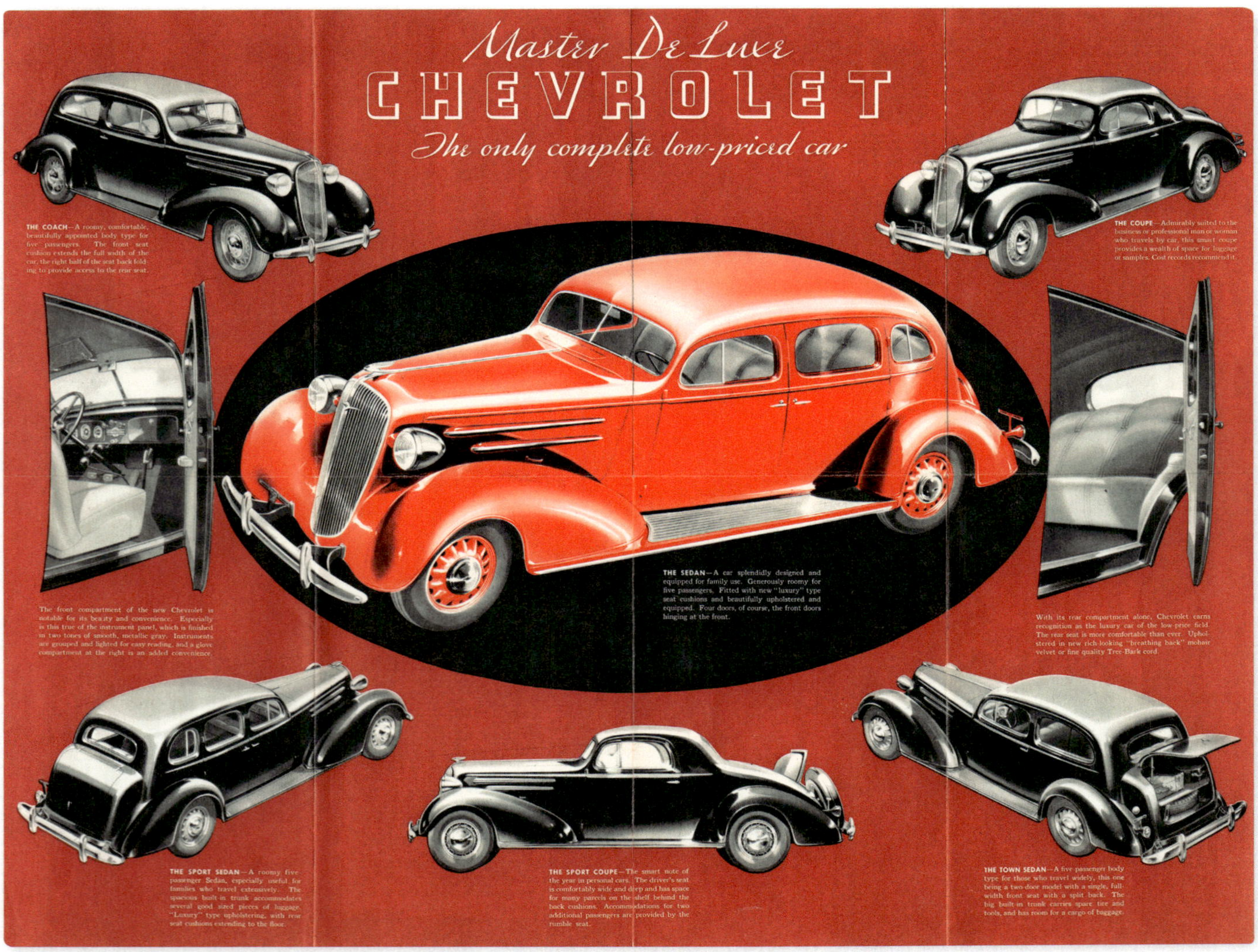

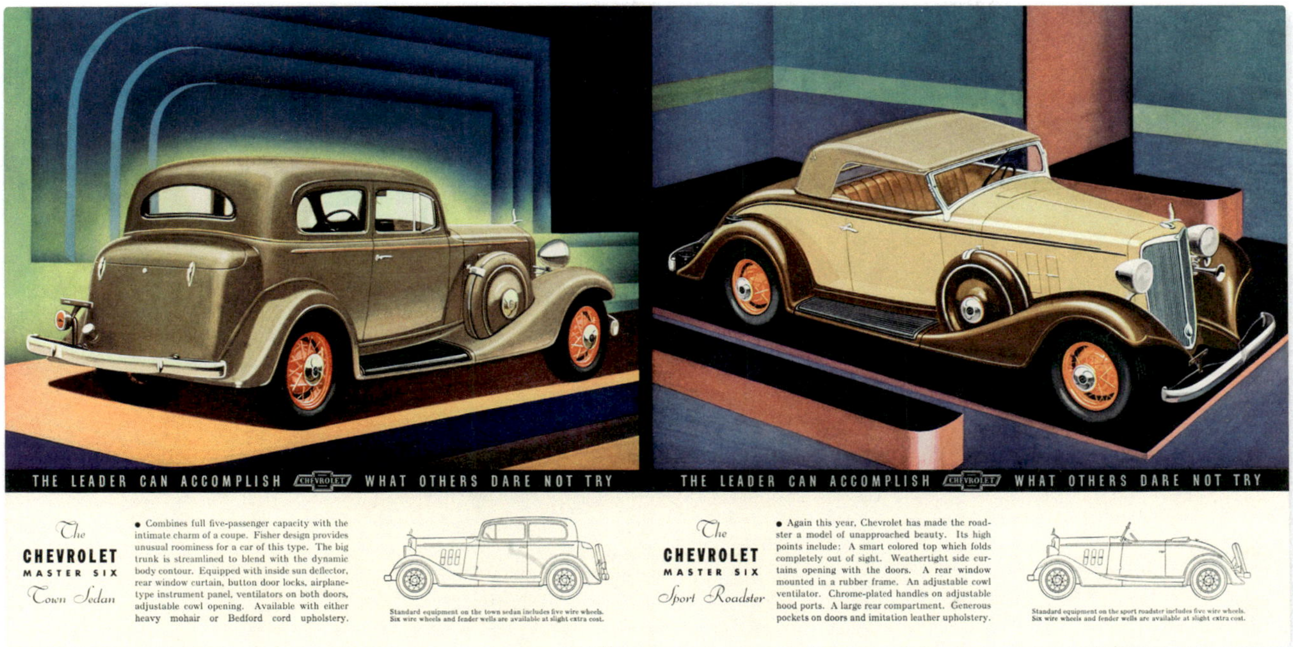
THE LEADER CAN ACCOMPLISH
CHEVROLET
WHAT OTHERS DARE NOT TRY
The CHEVROLET MASTER SIX Town Sedan
● Combines full five-passenger capacity with the intimate charm of a coupe. Fisher design provides unusual roominess for a car of this type. The big trunk is streamlined to blend with the dynamic body contour. Equipped with inside sun deflector, rear window curtain, button door locks, airplane-type instrument panel, ventilators on both doors, adjustable cowl opening. Available with either heavy mohair or Bedford cord upholstery.
Standard equipment on the town sedan includes five wire wheels. Six wire wheels and fender wells are available at slight extra cost.
THE LEADER CAN ACCOMPLISH
CHEVROLET
WHAT OTHERS DARE NOT TRY
The CHEVROLET MASTER SIX Sport Roadster
● Again this year, Chevrolet has made the roadster a model of unapproached beauty. Its high points include: A smart colored top which folds completely out of sight. Weathertight side curtains opening with the doors. A rear window mounted in a rubber frame. An adjustable cowl ventilator. Chrome-plated handles on adjustable hood ports. A large rear compartment. Generous pockets on doors and imitation leather upholstery.
Standard equipment on the sport roadster includes five wire wheels. Six wire wheels and fender wells are available at slight extra cost.

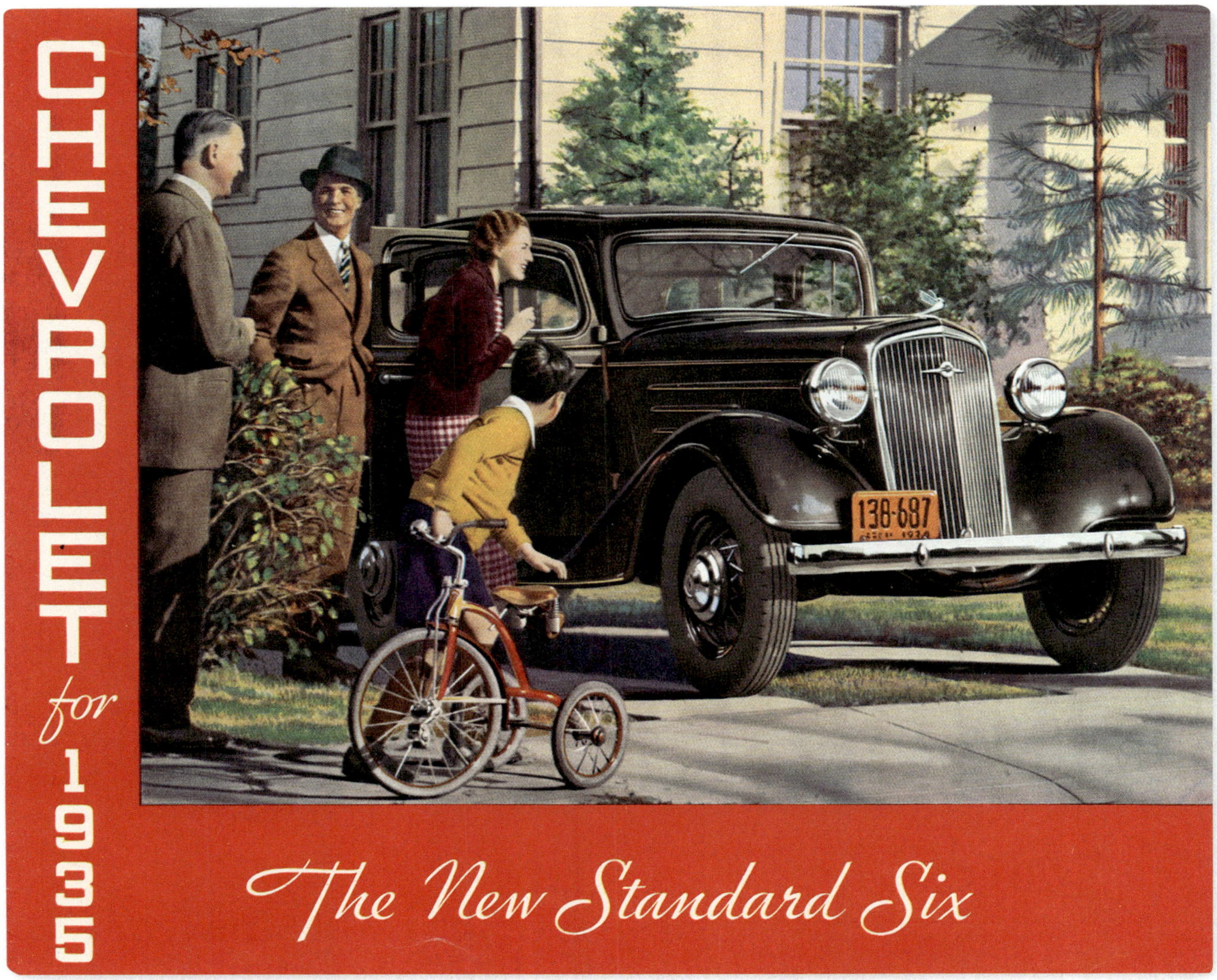
CHEVROLET for 1935
138-687
1934
The New Standard Six

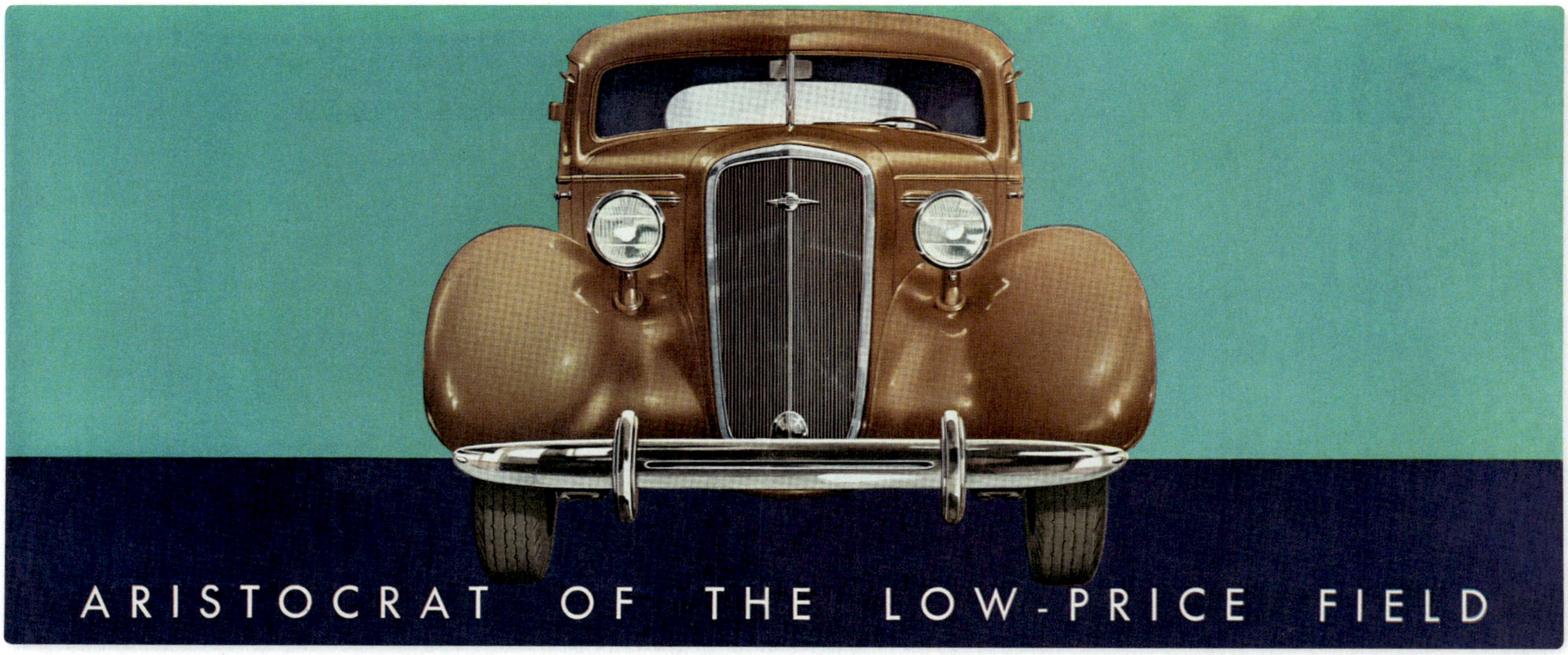

Chevrolet 1935

Chevrolet could credibly claim its aristocracy. It used inline six-cylinder engines exclusively from 1929 through 1954, and commanded the sales race for virtually that entire period. The horizontal hood louvers are a gradual move toward streamlining.

Chevrolet konnte sich zu Recht als „Adliger der unteren Preisklasse“ bezeichnen. Von 1929 bis 1954 verbaute man ausschließlich Sechszylinderreihenmotoren und führte damit praktisch über den gesamten Zeitraum die Absatzstatistiken an. Die waagerechten Lüftungsöffnungen in der Motorhaube deuten einen allmählichen Übergang zur Stromlinienform an.

Chevrolet pouvait à juste titre revendiquer son caractère aristocratique. De 1929 à 1954, la marque s'appuya en effet exclusivement sur des moteurs à six cylindres en ligne et domina le marché pendant la quasi-totalité de la période. Les ouïes de capot horizontales traduisent l'importance que prendra progressivement l'aérodynamique.

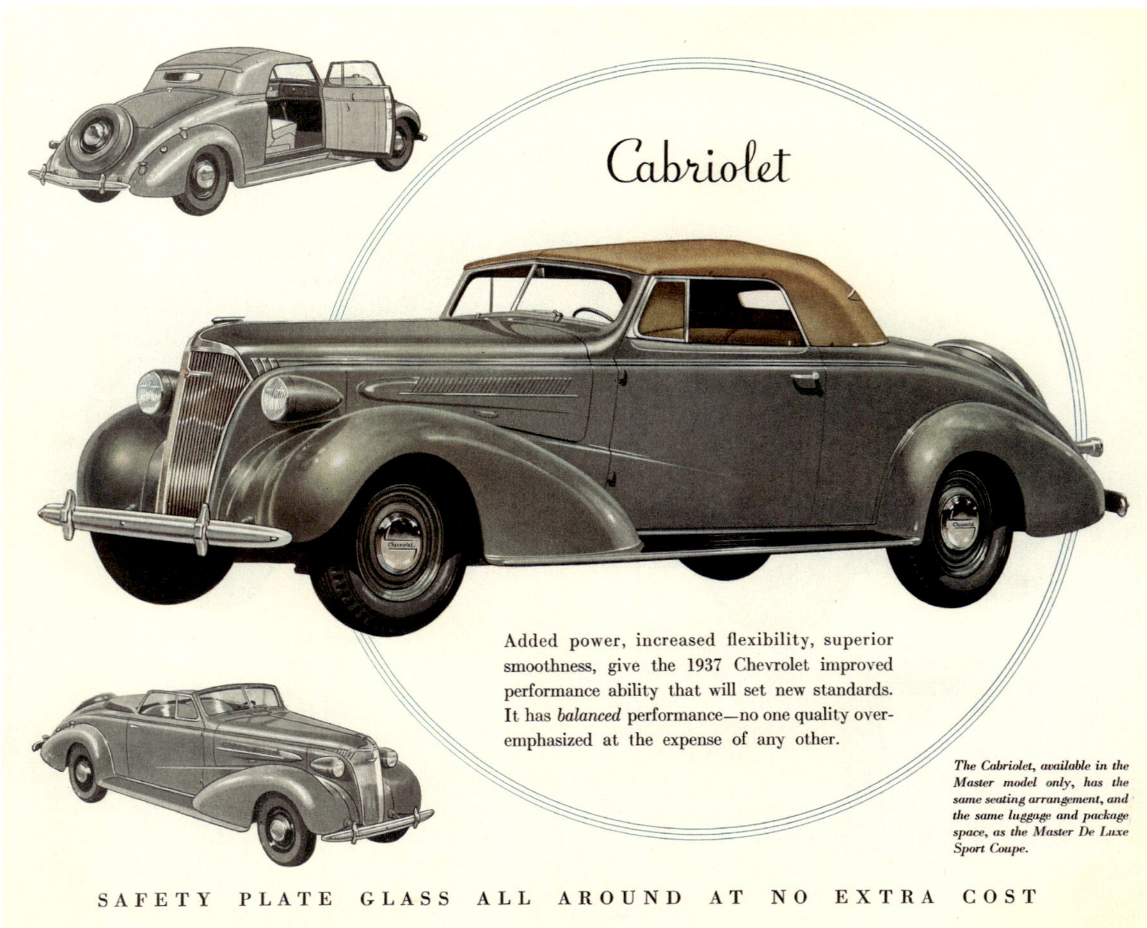

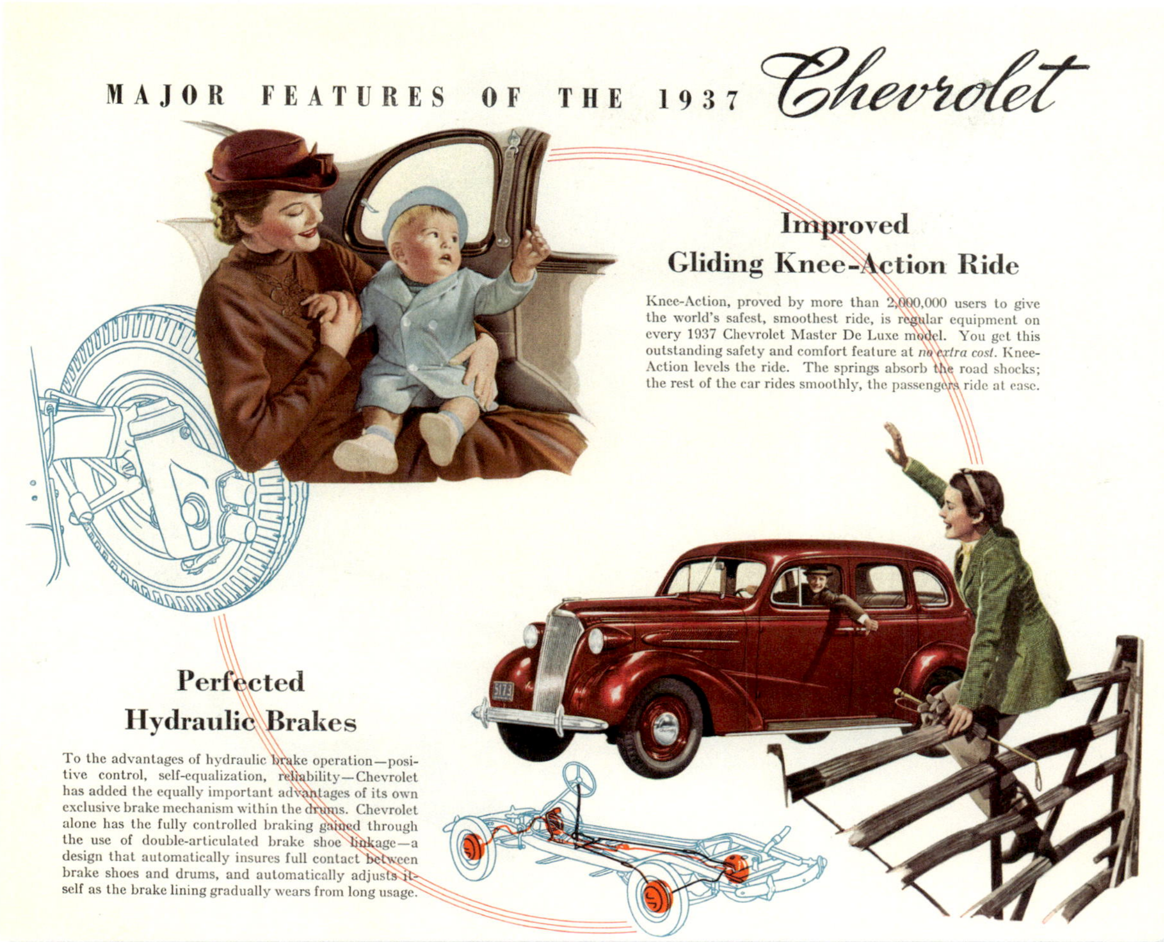

Chevrolet Master Deluxe 1937 *(left)*, **1935** *(below)*

The artwork suggests Chevrolet's adoption of the all-steel turret top and its attendant safety. While Chevrolet was far from the first to adopt hydraulic brakes, its sale in 1935 of 975,000 cars so equipped by Bendix made that newsworthy.

Die Zeichnungen verweisen auf den Umstieg zu einem – sicheren – Ganzstahldach. Und obwohl Chevrolet keinesfalls der erste Autohersteller war, der hydraulische Bremsen einsetzte, war diese Tatsache 1935 eine Meldung wert, als man 975 000 Fahrzeuge mit solchen Bremsen des Zulieferers Bendix verkaufte.

L'illustration rappelle ici l'adoption par Chevrolet du toit en acier d'une seule pièce et la sécurité qu'il procure. Chevrolet était loin d'être le premier constructeur à adopter le freinage hydraulique, mais la vente en 1935 de 975 000 voitures ainsi équipées par Bendix constitua tout de même un événement notable.

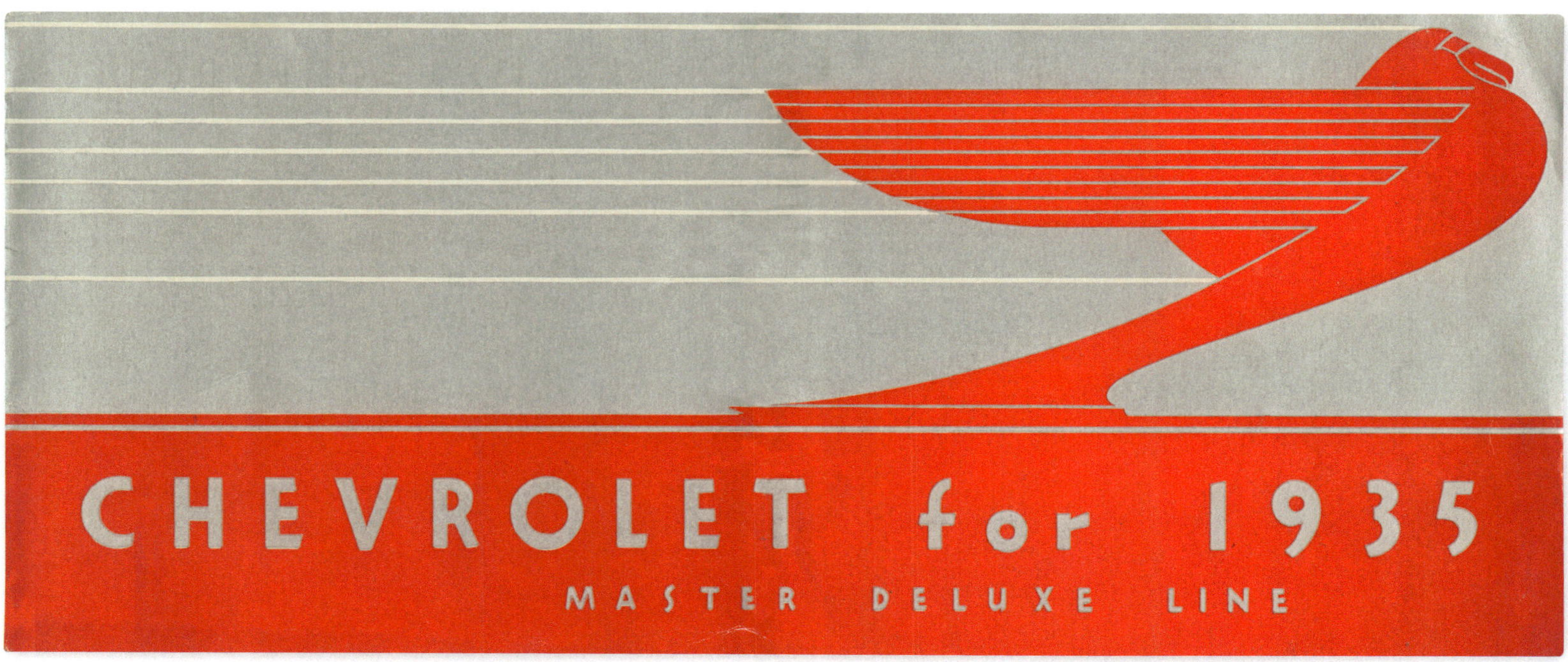

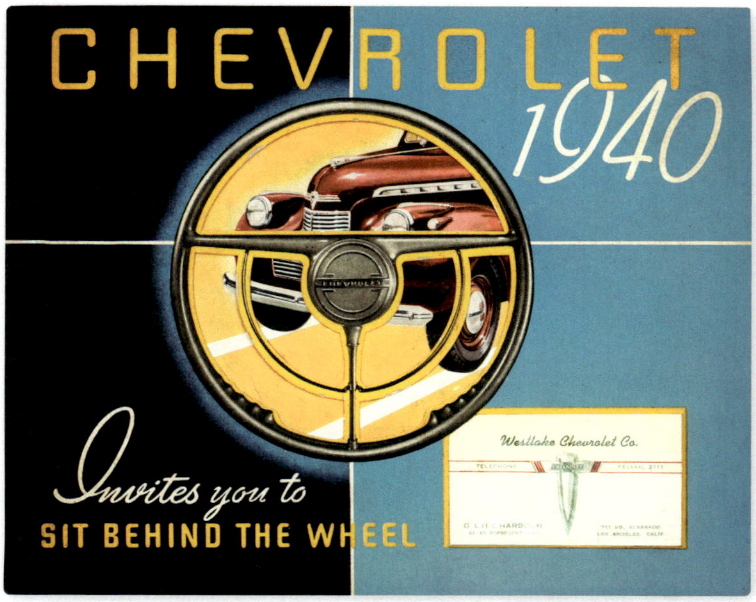

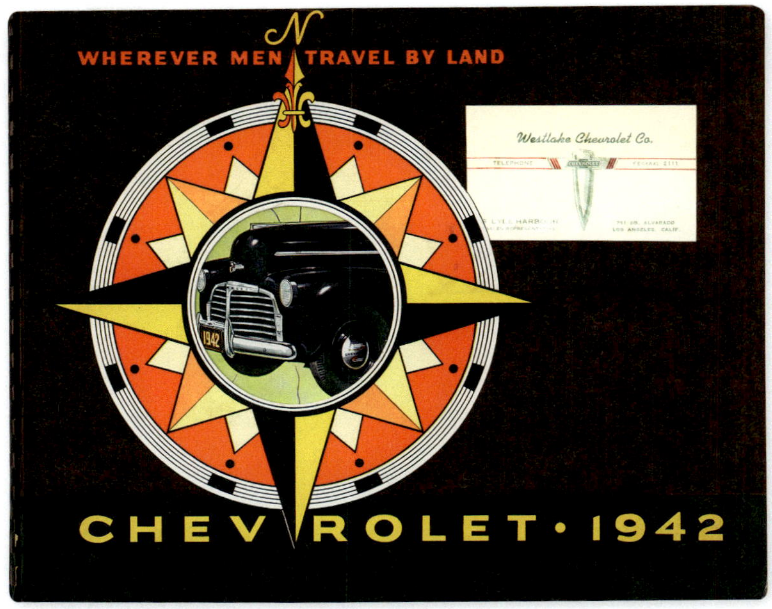

Chevrolet 1940 *(above and right)*, **1942** *(above right and opposite)*

With men off to war, more drivers than ever would be women. Ease of driving, indicated by the vacuum-operated gearshift, was a huge selling point. Mechanical reliability was even bigger. With civilian auto production halted, it was imperative that existing cars be kept operating during the war. This explains why these dealer manuals, which featured elaborate technical drawings and moving parts, communicated to motorists that their cars would indeed keep running for the duration.

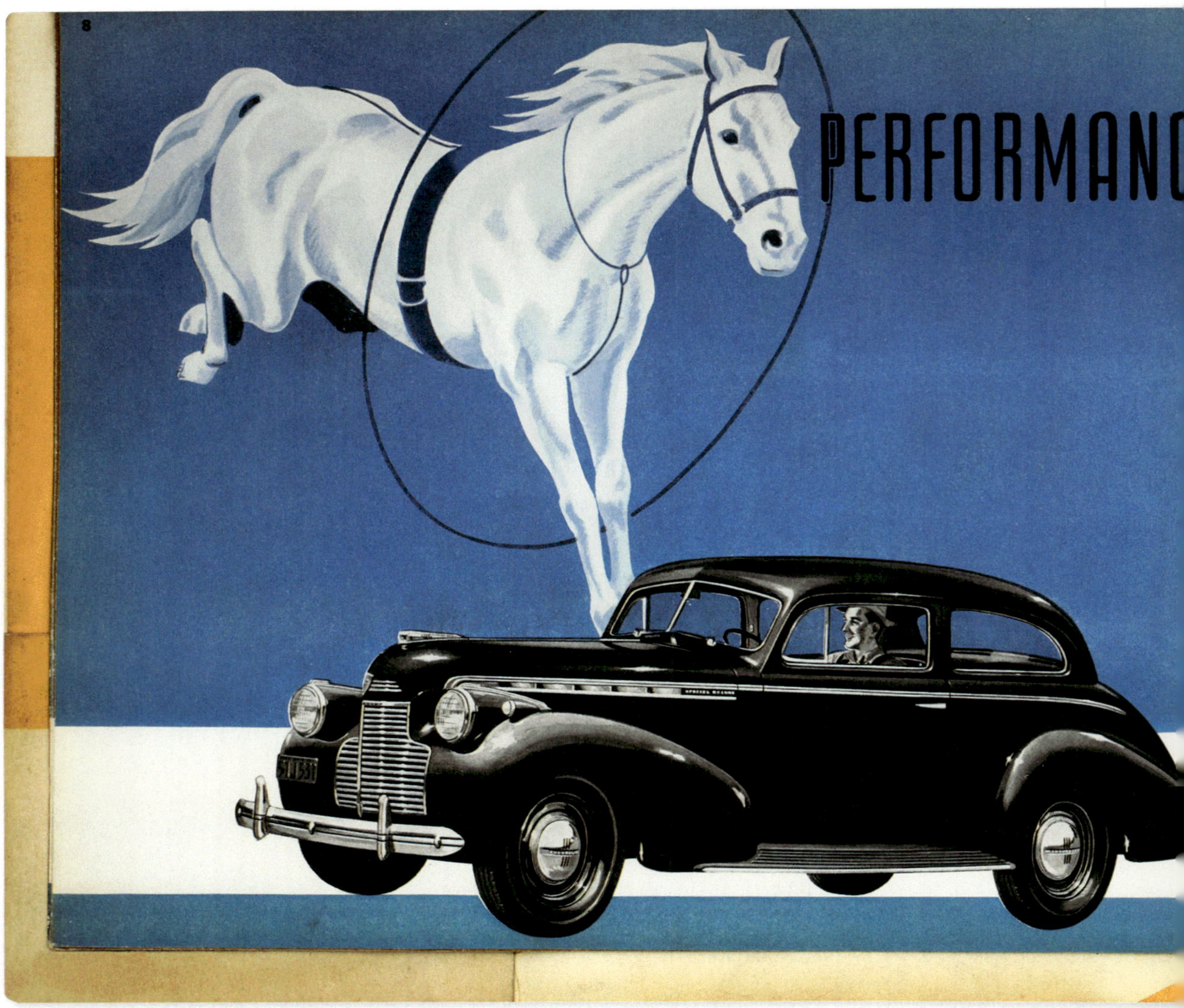

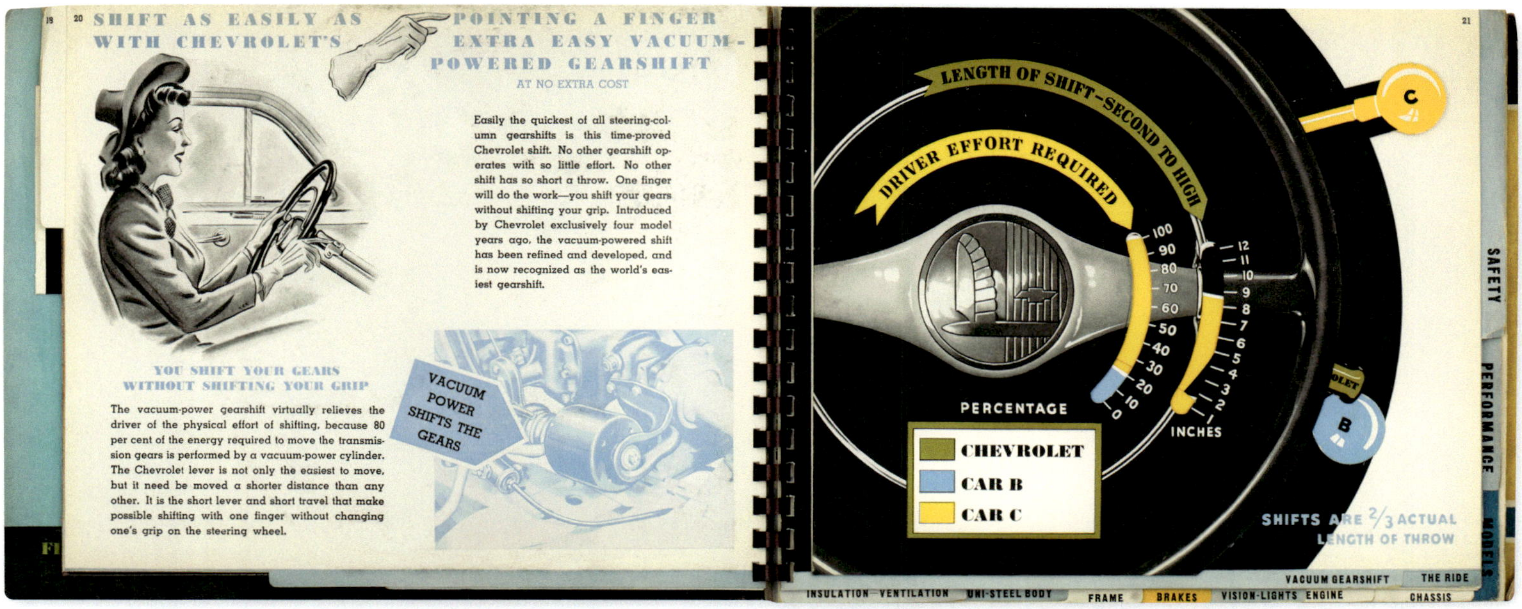

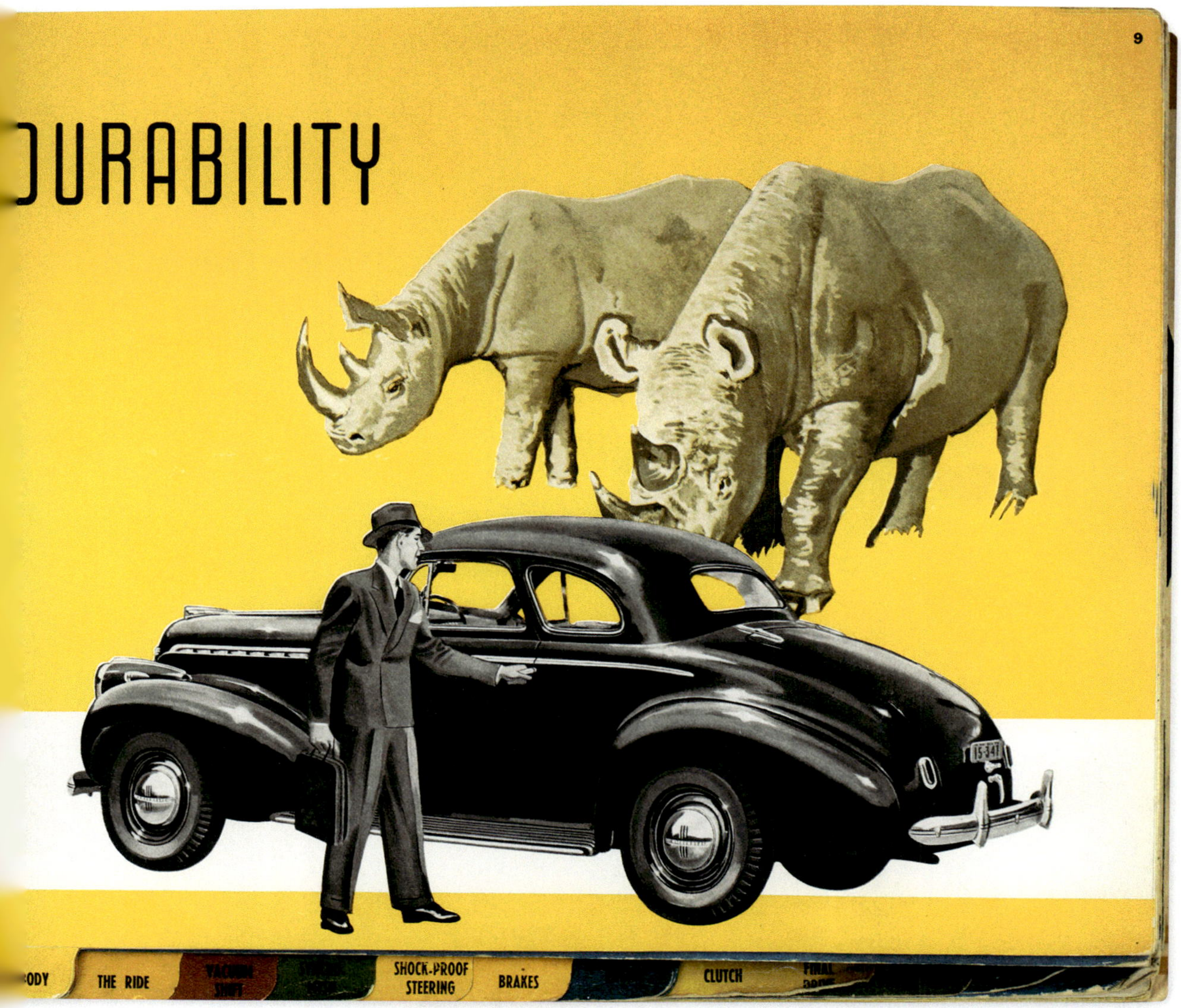

Als die Männer in den Krieg zogen, saßen in der Heimat mehr Frauen als je zuvor hinter dem Steuer. Erleichterungen beim Fahren, wie etwa eine Servoschaltung, waren nun wichtige Verkaufsargumente, aber ein noch größeres war die mechanische Zuverlässigkeit. Da die Produktion von Zivilfahrzeugen auf Eis lag, war es unbedingt notwendig, dass die vorhandenen Autos während der Kriegsjahre fahrtüchtig blieben. Das erklärt, weshalb diese Händlerhandbücher mit ihren detaillierten technischen Zeichnungen und sogar beweglichen Teilen dem Kraftfahrer das Gefühl geben sollten, dass ihre Autos den Krieg tatsächlich überstehen würden.

Une fois les hommes partis à la guerre, les conductrices se retrouvaient plus nombreuses que jamais. La facilité de conduite, ici évoquée par le levier de vitesses assisté par pompe à vide, représentait un argument commercial massue. Et la fiabilité mécanique, davantage encore. Puisque la production de véhicules civils était à l'arrêt, il fallait impérativement que les modèles existants puissent continuer à être utilisés pendant la guerre. C'est pourquoi ces manuels proposés par les concessionnaires, avec force schémas techniques et éléments mobiles, assuraient aux automobilistes que leur voiture tiendrait sur la durée.

Chrysler

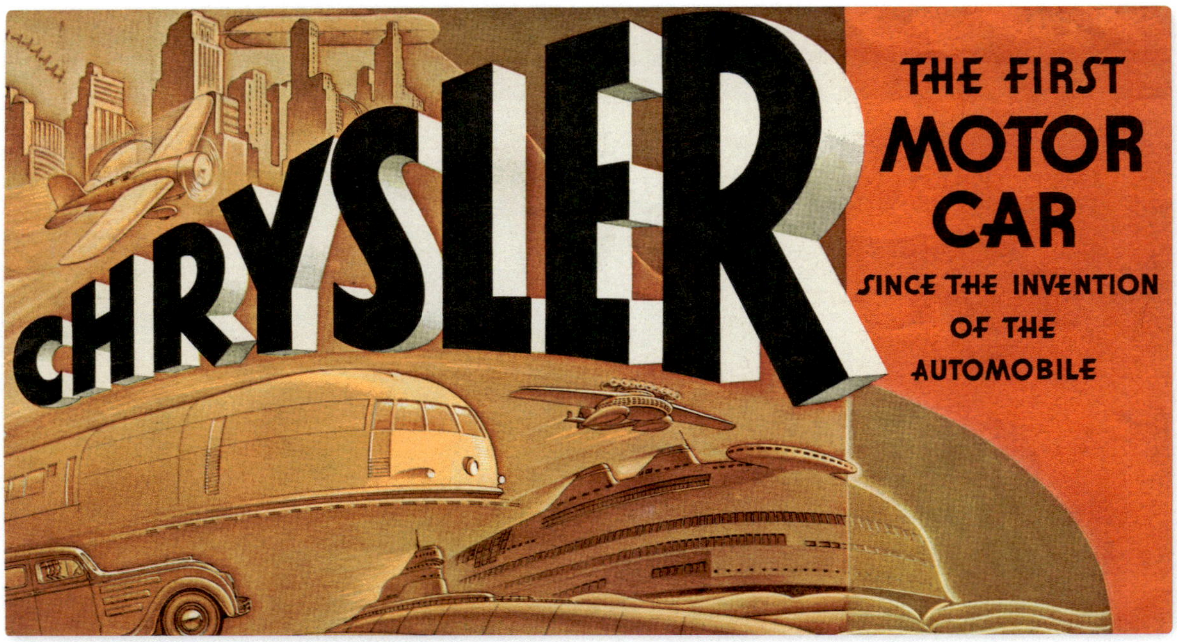

Chrysler Airflow 1934 *(this spread)*

While not massively successful, the Airflow was certainly innovative and strong. It likened itself to massive zeppelins and the Union Pacific M-10000 Streamliner train, which, ironically, used an engine from a General Motors division, Winton.

Auch wenn er nicht sonderlich erfolgreich war, so war der Airflow doch gewiss innovativ und leistungsstark. Er suggerierte einen Vergleich mit den gewaltigen Luftschiffen der damaligen Zeit und mit dem Streamliner-Zug M-10 000 der Union-Pacific-Eisenbahn, der ironischerweise mit einem Motor von Winton fuhr, einem Tochterunternehmen des Konkurrenten General Motors.

Si elle n'a pas rencontré un succès colossal, l'Airflow était assurément novatrice et costaude. Elle se comparait elle-même aux massifs zeppelins et au train aérodynamique M-10 000 « Streamliner » lancé par l'Union Pacific début 1934, lequel utilisait un moteur Winton, filiale de General Motors.

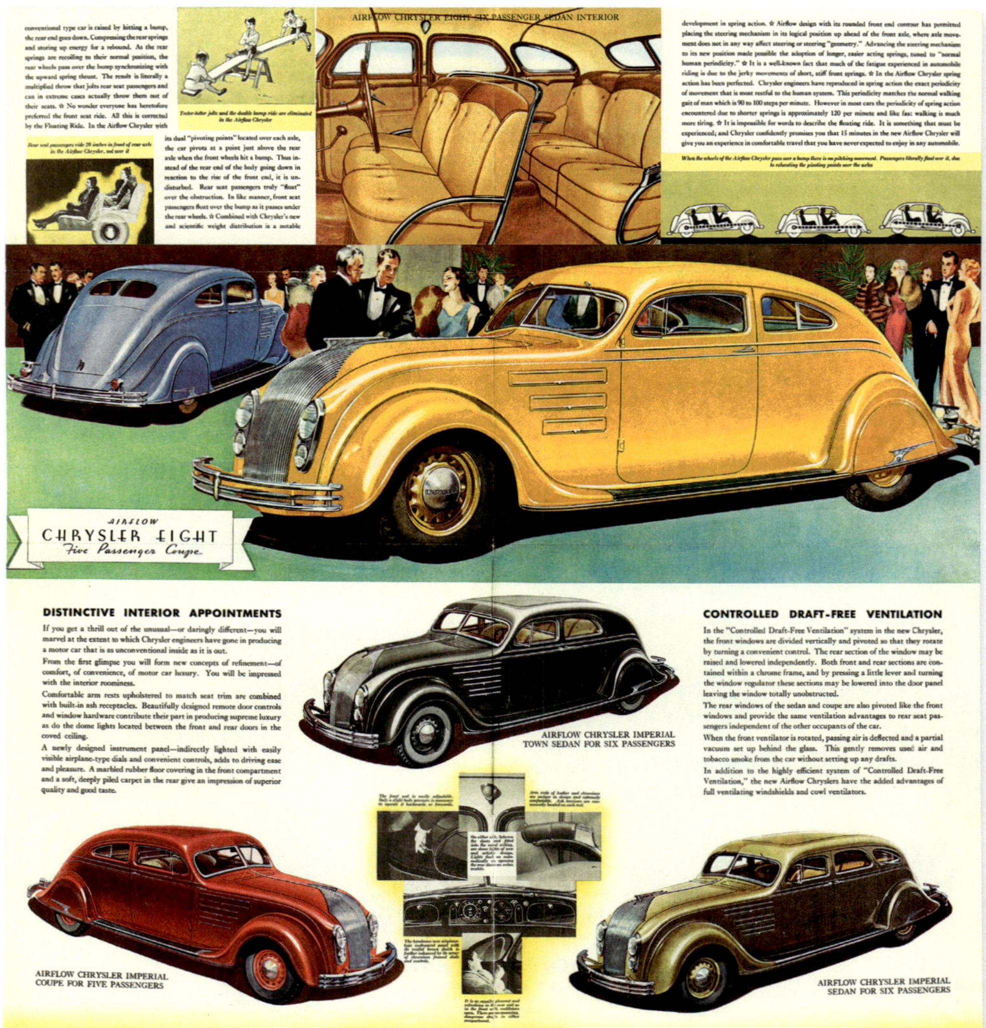

A New Era of Transportation
BRINGS FUNCTIONAL DESIGN

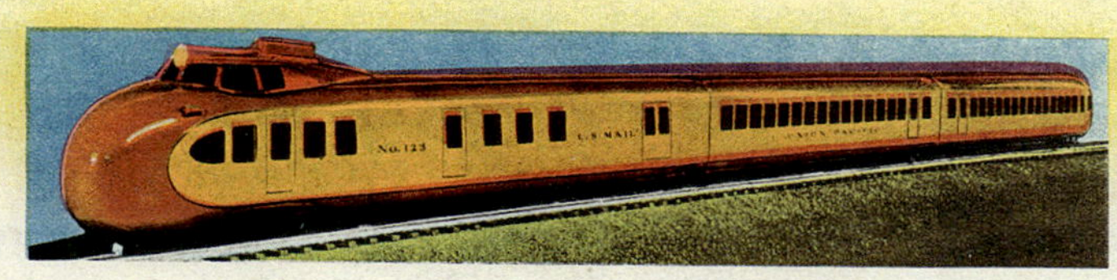

New era in railroad transportation—Streamlined passenger train

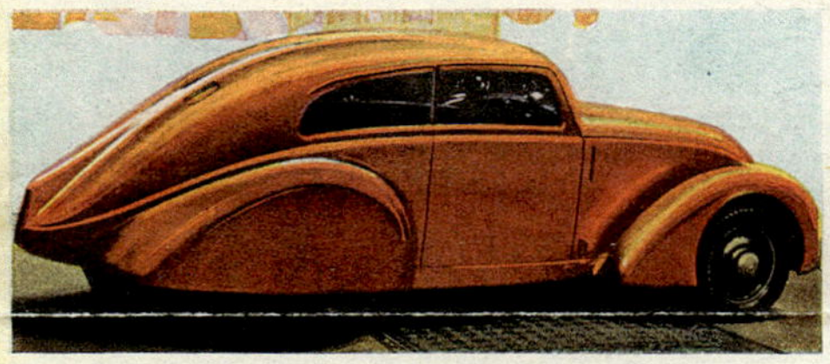

Automobile streamlining on the continent as featured at recent Berlin Auto Show

A new era of transportation has arrived—an era whose influence is revolutionizing transportation in every field, changing all precedent and obsoleting all the hide-bound traditions inherited from a horse-and-buggy age. It is an era that is fashioning all modes of transportation to a better performance of their functions. ☆ Exhaustive research has demonstrated the tremendous inefficiency and extravagant waste of power of conventionally shaped vehicles in battling wind resistance. Just put your hand out of a car window going at 30 miles per hour—then do the same at 60. Notice the tremendous difference in air pressure on your hand. From this simple test some idea is gained of the resistance encountered in propelling the conventional car over the highways. At 50 miles per hour the conventional car of the old horse-and-buggy design will waste approximately half its fuel overcoming wind resistance. ☆ When nature forms her creatures for speed, she shapes them in smooth lines from a rounded head to a tapering tail in order to overcome the resistance of their environment, whether it be air or water. ☆ Chrysler engineers have departed from the conventional and have designed the new Airflow Chrysler in accordance with this basic natural law. In these daring new models the lines of the gracefully rounded front end sweep smoothly over the flowing curves of the body to the sloping, tapering rear. ☆ All of the air disturbing appendages, fenders, head lamps, horns—have been covered and the new Airflow Chrysler slips through the air so perfectly that wind roar, present in conventional cars at high speed, is eliminated.

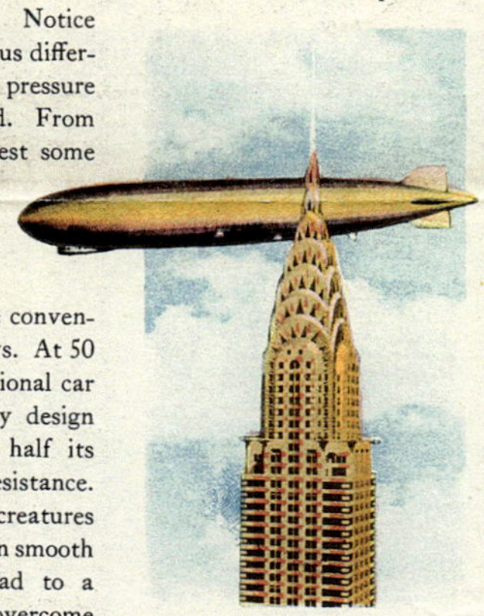

Ultra modern transportation and architecture—Graf Zeppelin over Chrysler Building

-and CHRYSLER brings

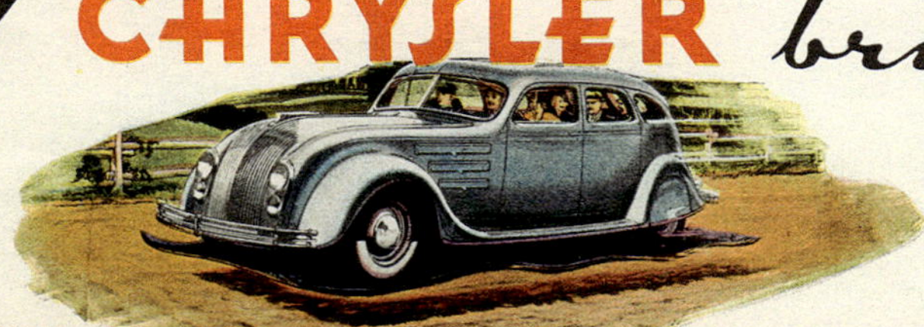

FLOATING RIDE

WITH the advent of the Airflow Chrysler comes a new and ideal ride—the Floating Ride, which makes the Airflow Chrysler truly a new form of transportation. ☆ In this new Airflow Chrysler all roads assume the smoothness of glass. You actually travel over dirt roads, gravel, macadam, concrete or asphalt at speeds up to 80 and 90 miles an hour, yet so smooth and restful is the character of the ride that you are completely relaxed . . . untired even by hundreds of miles of continuous travel. ☆ For the first time it is possible to write a letter, read a magazine or study a road map without having the type dance before your eyes. Or you can lay your head back on the seat and take a nap in complete comfort. ☆ This phenomenal Floating Ride is the result of the relocation of weight masses—longer, easier-acting springs, and the rigid, bridge-type, girder-trussed frame construction. ☆ The Airflow contour, aside from giving the new Chrysler a projectile-like passage through the air, makes possible for the first time a frame that is an integral part of the body, offering full protection to the occupants. This new girder-trussed frame is designed like a bridge with longitudinal, vertical and diagonal girders extending from the front to the back of the body, up and over the top of the body. In the Airflow Chrysler you actually ride with part of the frame above your head. The result of this new construction is the safest and most rigid car built. ☆ Passengers in the new Airflow Chryslers are now seated inside and under the frame, cradled between the wheels. There's no choice of seats any more. The back seat is located a full 20 inches ahead of the rear axle away from its jolt-producing influence. ☆ In the new Airflow Chrysler, the motor weight mass is moved forward over the front axle, locating new "centers of percussion" or pivoting points when the wheels encounter road inequalities. In the Airflow Chrysler, due to its new scientific weight distribution, these pivotal points of road shock are directly over the front and rear axles. This revolutionary design makes front spring action entirely independent of rear spring action, and literally cradles passengers over the worst roads without tremor or jolt. In the conventional car, due to improper weight distribution the "pivoting points" are located much closer together concentrating the weight towards the center of the car, thus giving a riding effect that is almost identical with sitting on the ends of the familiar "teeter-totter." ☆ Here are the facts of how you ride in your present car. When the front end of the

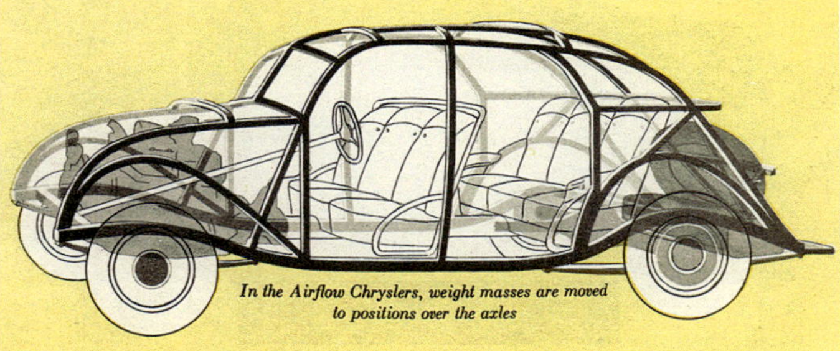

In the Airflow Chryslers, weight masses are moved to positions over the axles

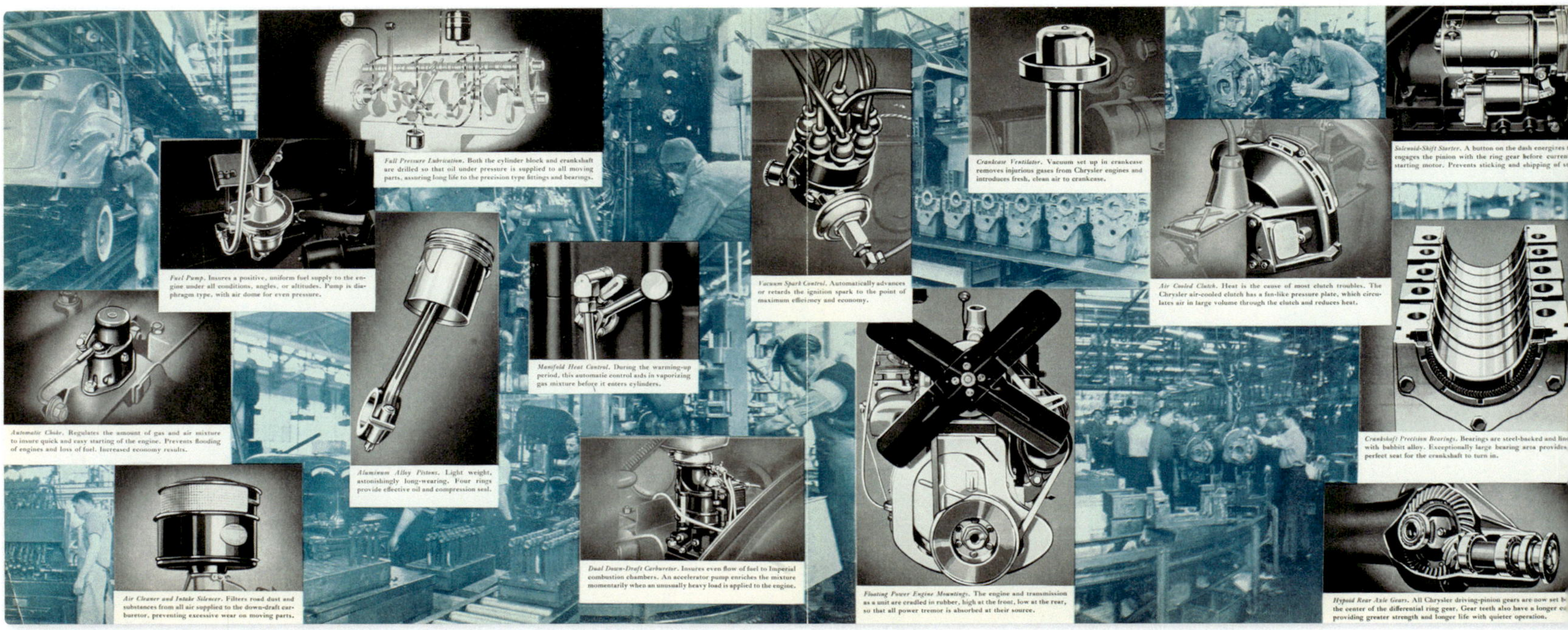

Chrysler 1938 *(this spread)*

Did buyers really understand how Fluid Drive's interacting hydraulic vanes worked? Probably not, but there were plenty of dealer-installed extras, as delineated in this heavily illustrated, highly explanatory dealer manual from the same year.

Ob die Käufer wirklich verstanden, wie die Hydraulikkupplung des sogenannten Fluid Drive funktionierte? Wahrscheinlich nicht, aber es gab zahlreiche Extras, die der Händler verkaufen konnte, wie dieses reich bebilderte und sehr erklärungsfreudige Händlerhandbuch aus dem gleichen Jahr zeigt.

Les propriétaires de cette Chrysler comprenaient-ils vraiment le fonctionnement de son système d'embrayage hydraulique Fluid Drive ? C'est peu probable, mais les concessionnaires avaient la fibre pédagogique, comme le montre ce manuel richement illustré et hautement explicatif datant de cette même année 1938.

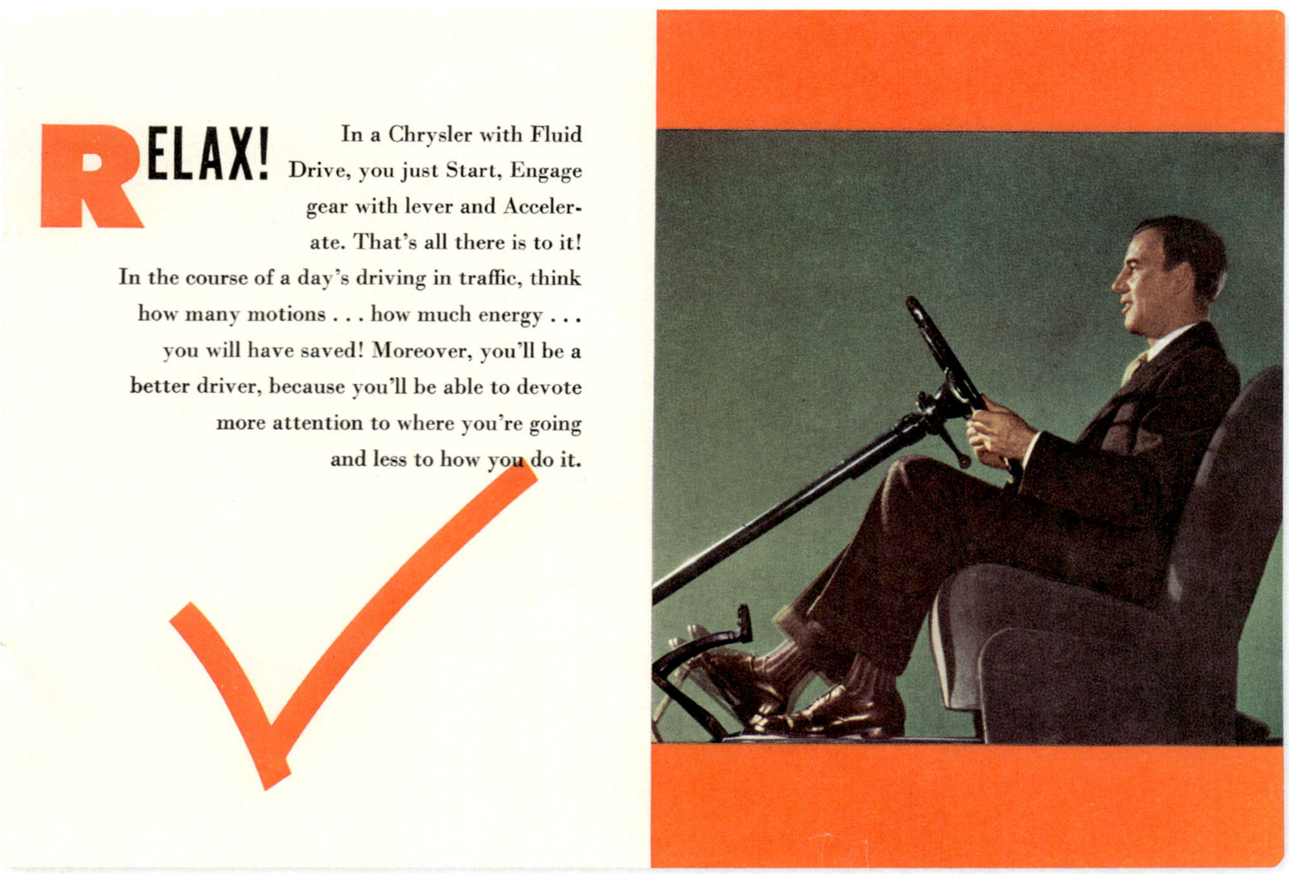

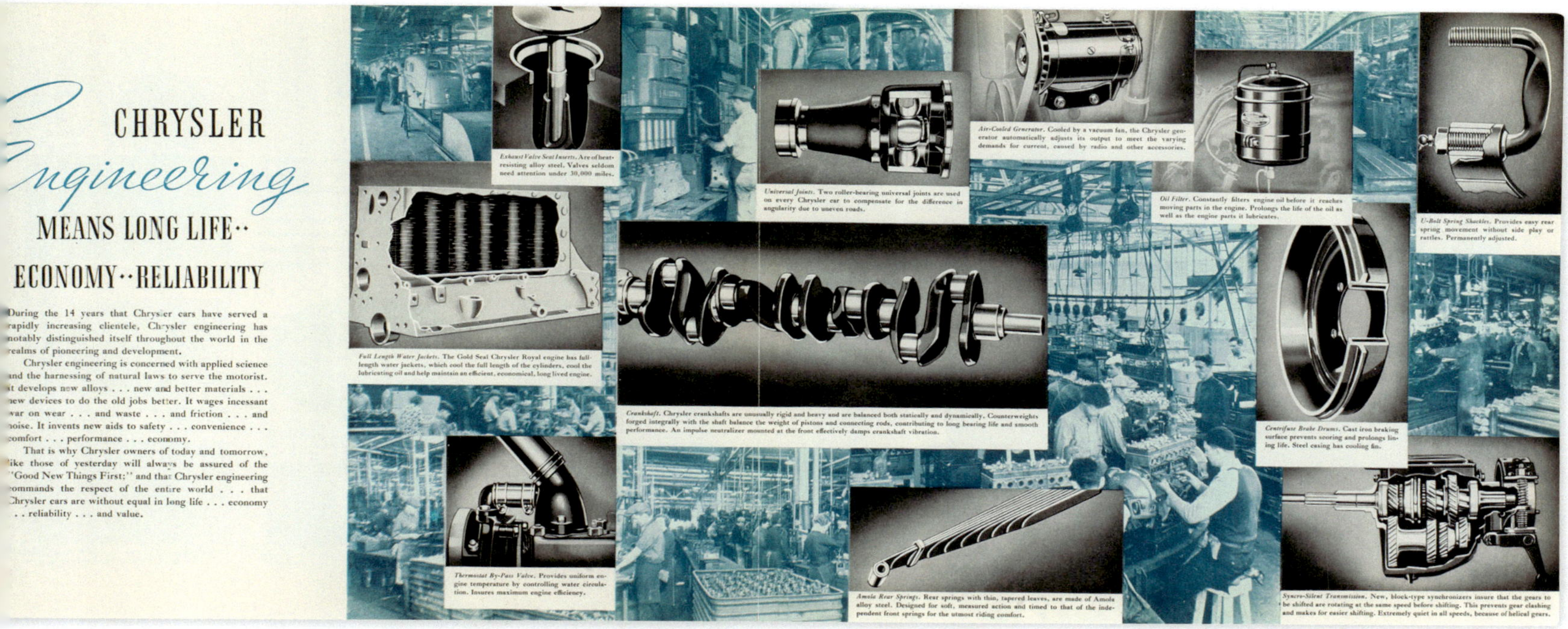
CHRYSLER
Engineering
MEANS LONG LIFE··
ECONOMY··RELIABILITY
During the 14 years that Chrysler cars have served a rapidly increasing clientele, Chrysler engineering has notably distinguished itself throughout the world in the realms of pioneering and development.
Chrysler engineering is concerned with applied science and the harnessing of natural laws to serve the motorist. It develops new alloys . . . new and better materials . . . new devices to do the old jobs better. It wages incessant war on wear . . . and waste . . . and friction . . . and noise. It invents new aids to safety . . . convenience . . . comfort . . . performance . . . economy.
That is why Chrysler owners of today and tomorrow, like those of yesterday will always be assured of the "Good New Things First;" and that Chrysler engineering commands the respect of the entire world . . . that Chrysler cars are without equal in long life . . . economy . . . reliability . . . and value.
Exhaust Valve Seat Inserts. Are of heat-resisting alloy steel. Valves seldom need attention under 30,000 miles.
Universal Joints. Two roller-bearing universal joints are used on every Chrysler car to compensate for the difference in angularity due to uneven roads.
Air-Cooled Generator. Cooled by a vacuum fan, the Chrysler generator automatically adjusts its output to meet the varying demands for current, caused by radio and other accessories.
Oil Filter. Constantly filters engine oil before it reaches moving parts in the engine. Prolongs the life of the oil as well as the engine parts it lubricates.
U-Bolt Spring Shackles. Provides easy rear spring movement without side play or rattles. Permanently adjusted.
Full Length Water Jackets. The Gold Seal Chrysler Royal engine has full-length water jackets, which cool the full length of the cylinders, cool the lubricating oil and help maintain an efficient, economical, long lived engine.
Crankshaft. Chrysler crankshafts are unusually rigid and heavy and are balanced both statically and dynamically. Counterweights forged integrally with the shaft balance the weight of pistons and connecting rods, contributing to long bearing life and smooth performance. An impulse neutralizer mounted at the front effectively damps crankshaft vibration.
Centrifuse Brake Drums. Cast iron braking surface prevents scoring and prolongs lining life. Steel casing has cooling fin.
Thermostat By-Pass Valve. Provides uniform engine temperature by controlling water circulation. Insures maximum engine efficiency.
Amola Rear Springs. Rear springs with thin, tapered leaves, are made of Amola alloy steel. Designed for soft, measured action and timed to that of the independent front springs for the utmost riding comfort.
Syncro-Silent Transmission. New, block-type synchronizers insure that the gears to be shifted are rotating at the same speed before shifting. This prevents gear clashing and makes for easier shifting. Extremely quiet in all speeds, because of helical gears.

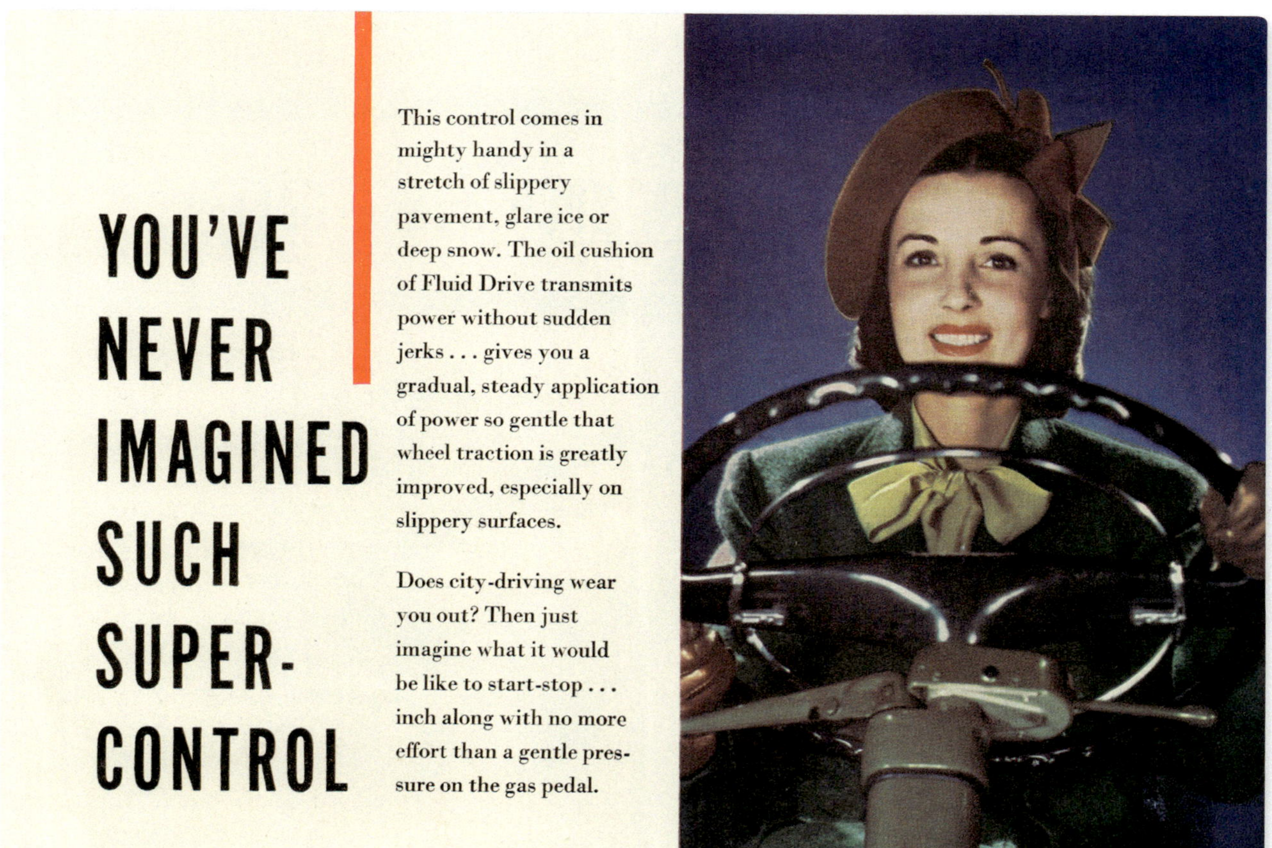
YOU'VE NEVER IMAGINED SUCH SUPER-CONTROL
This control comes in mighty handy in a stretch of slippery pavement, glare ice or deep snow. The oil cushion of Fluid Drive transmits power without sudden jerks . . . gives you a gradual, steady application of power so gentle that wheel traction is greatly improved, especially on slippery surfaces.
Does city-driving wear you out? Then just imagine what it would be like to start-stop . . . inch along with no more effort than a gentle pressure on the gas pedal.

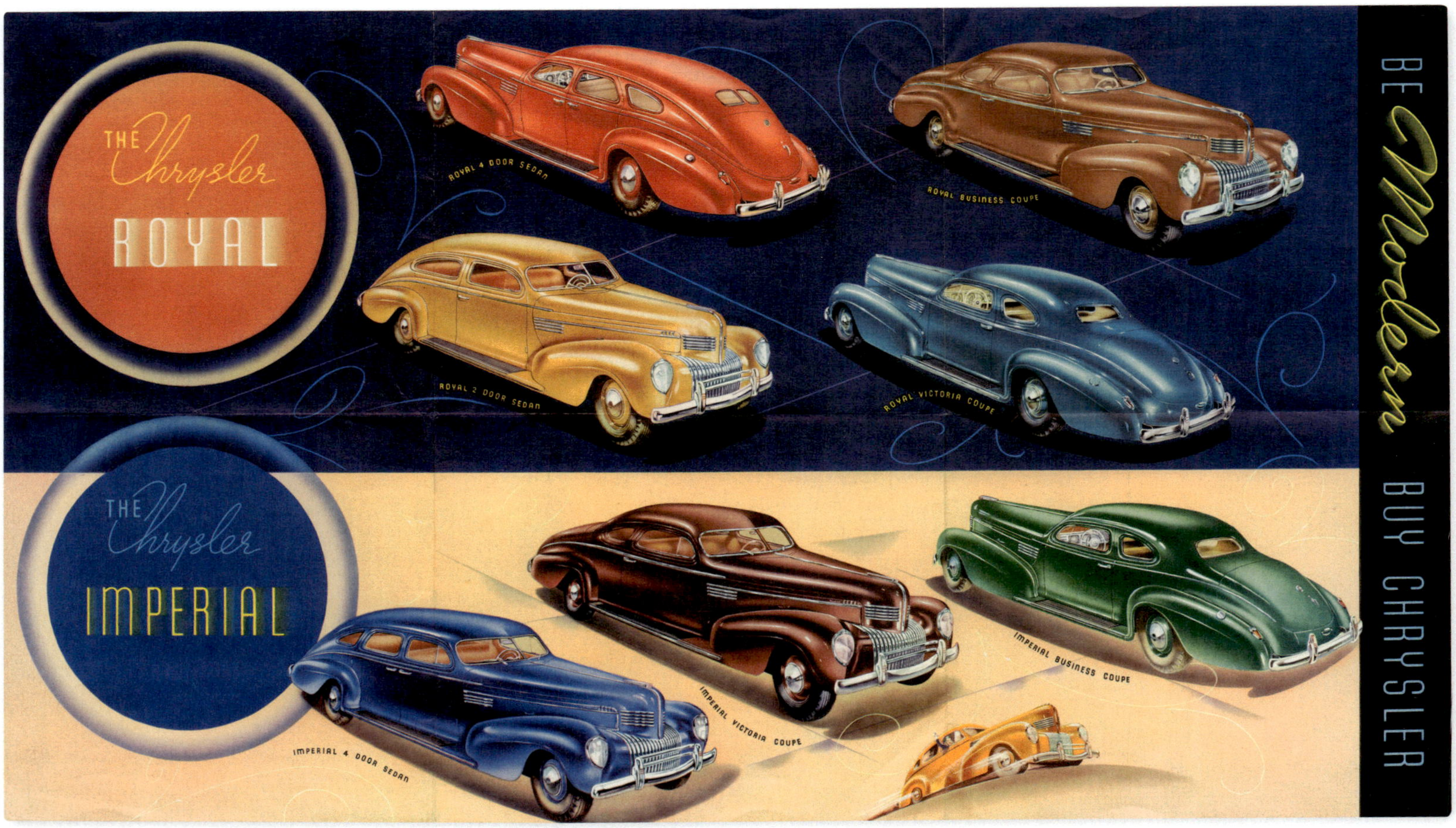
THE Chrysler
ROYAL
ROYAL 4 DOOR SEDAN
ROYAL BUSINESS COUPE
ROYAL 2 DOOR SEDAN
ROYAL VICTORIA COUPE
THE Chrysler
IMPERIAL
IMPERIAL 4 DOOR SEDAN
IMPERIAL VICTORIA COUPE
IMPERIAL BUSINESS COUPE
BE Modern
BUY CHRYSLER

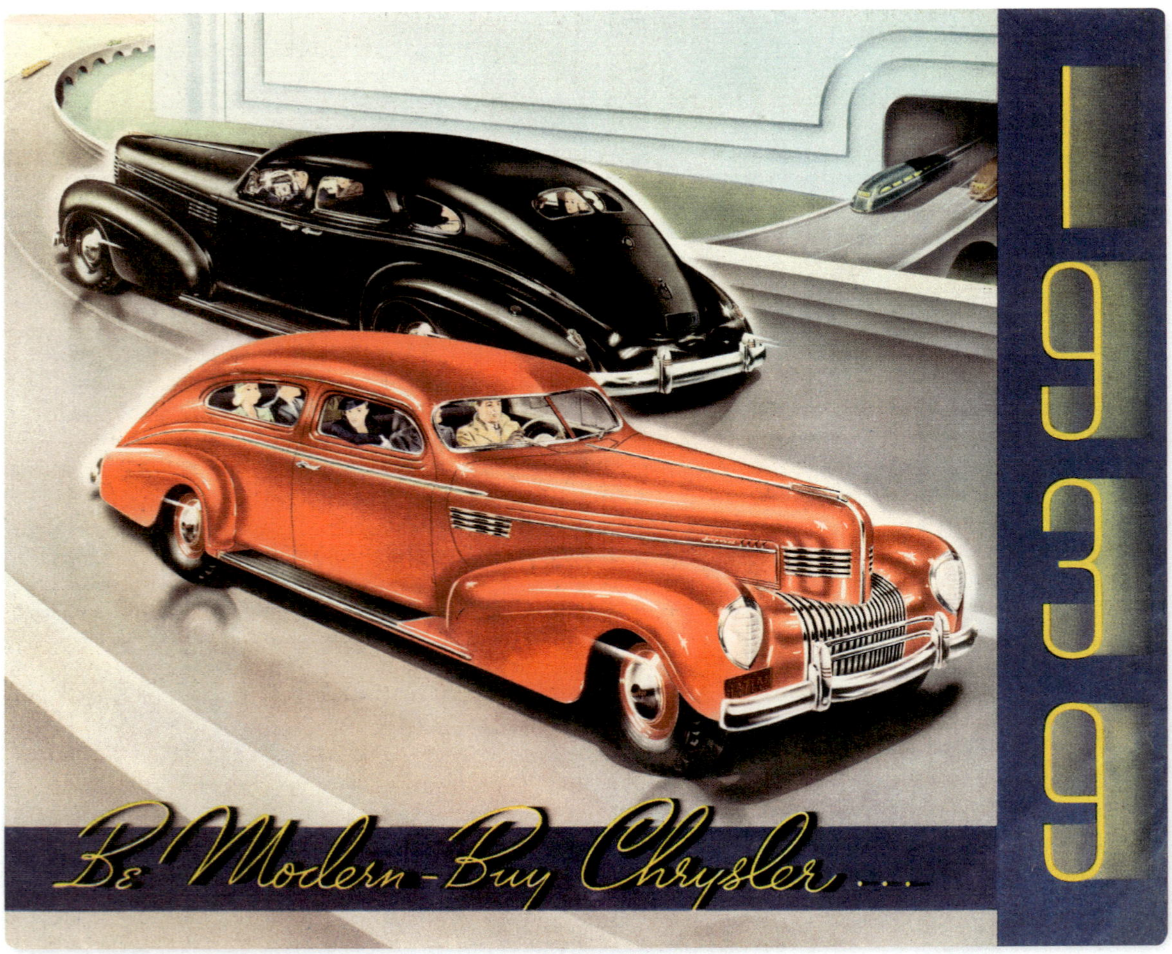
1939
Be Modern-Buy Chrysler...

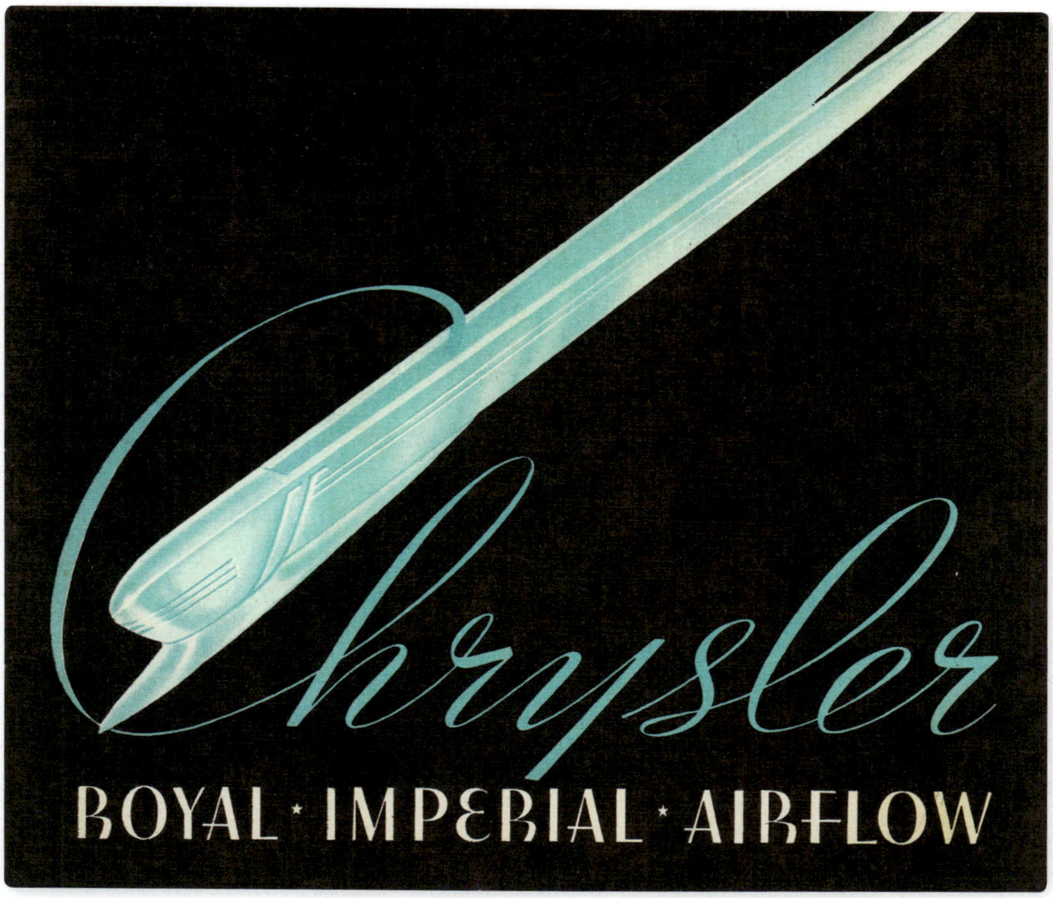

Chrysler Imperial 1939 *(opposite)*, **1937** *(above)*, **1940** *(below)*

Chrysler gave up on the Airflow design after 1937, shifting to comparatively conventional luxury cars. The standard Imperial engine was a 110 hp straight-eight and likely had Chrysler's new Fluid Drive semiautomatic transmission.

Chrysler gab das Airflow-Design nach 1937 auf und verlagerte sich fortan auf vergleichsweise gewöhnliche Luxusautos. Der Standardmotor des Imperial war ein Achtzylinderreihenmotor mit 149 Kilowatt Leistung, und er wurde wohl in den meisten Fällen mit der neuen halbautomatischen „Fluid Drive"-Schaltung von Chrysler bestellt.

Chrysler renonça au style Airflow après 1937 pour passer à des modèles de luxe plus conventionnels. L'Imperial fut dotée en série d'un moteur à huit cylindres en ligne développant 110 chevaux ainsi que de la nouvelle transmission semi-automatique Fluid Drive.

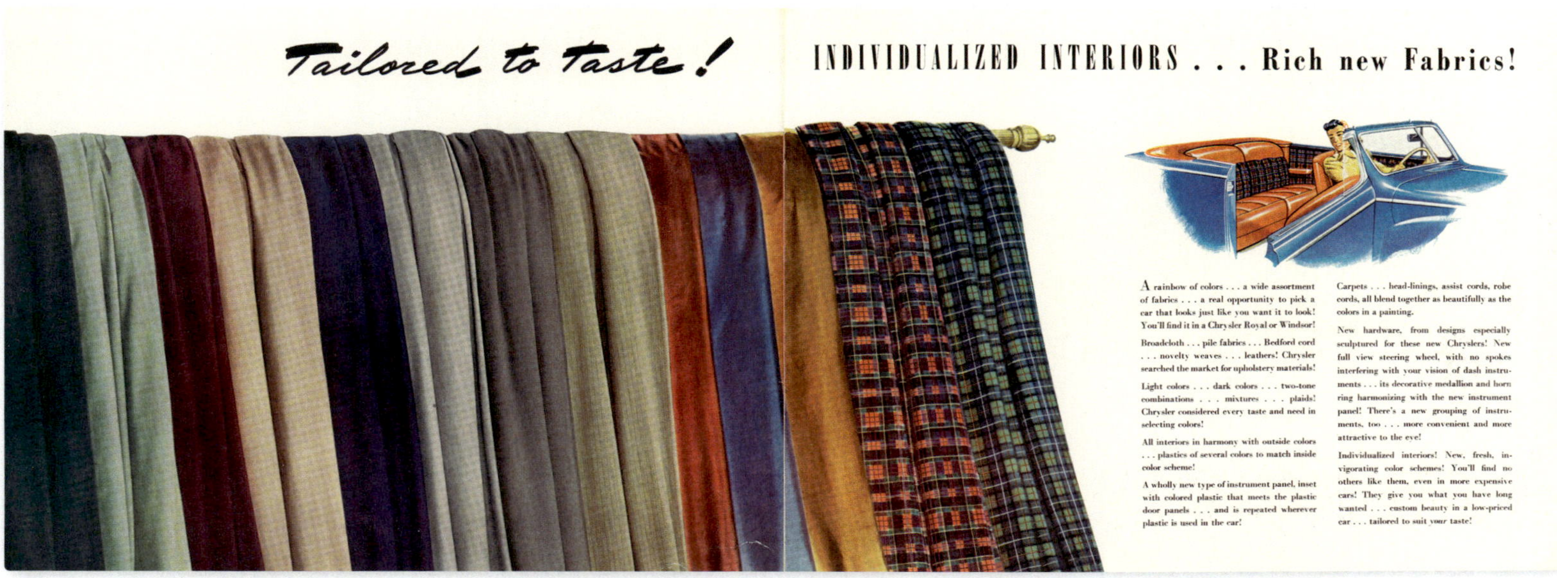

Continental

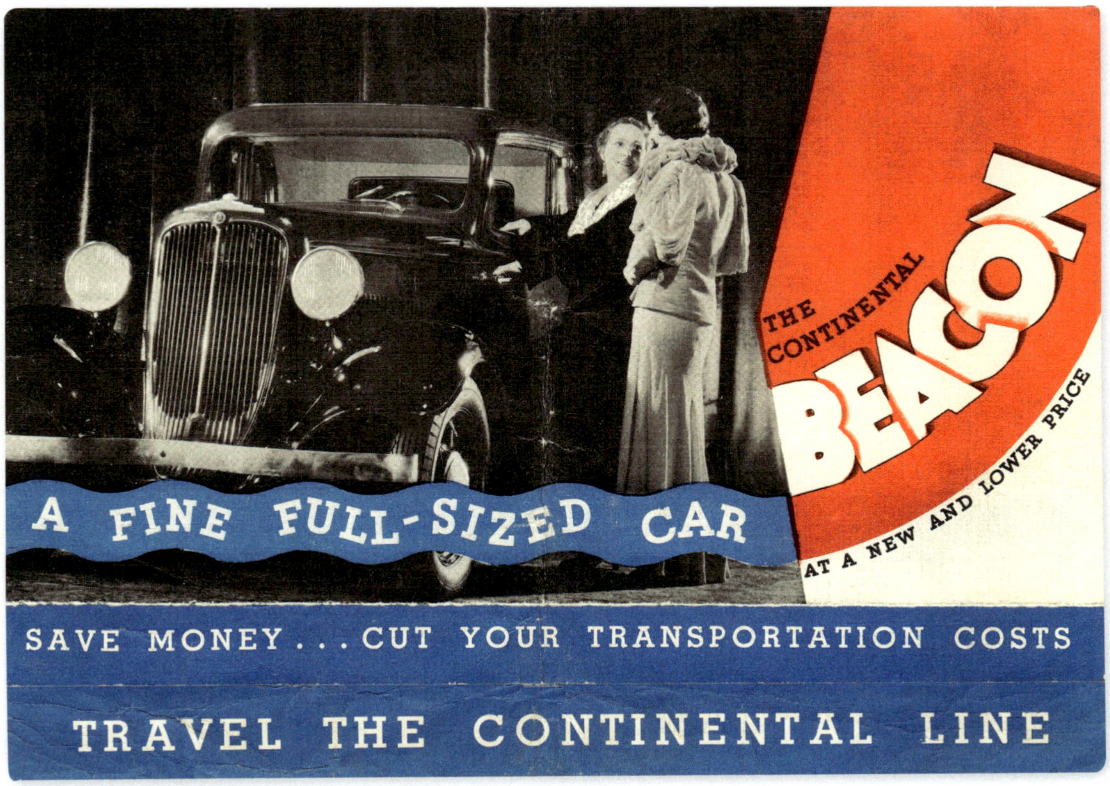

Continental Beacon 1934

Continental did famously well supplying engines for many manufacturers but flopped horribly building cars on its own. The Beacon was supposedly America's cheapest car, costing $380. Only 953 were produced.

Continental war als Zulieferer, der Autobauer mit Motoren versorgte, äußerst erfolgreich, versagte aber kläglich bei der Herstellung eigener Fahrzeuge. Der Beacon war mit einem Verkaufspreis von 380 Dollar angeblich das billigste Auto in Amerika. Trotzdem wurden nur 953 Stück produziert.

La firme Continental a connu un grand succès en fournissant ses moteurs à de nombreux constructeurs, mais elle a connu un revers retentissant lorsqu'elle a voulu fabriquer ses propres voitures. La Beacon se voulait la voiture la moins chère des États-Unis : elle ne coûtait de fait que 380 dollars. Seuls 953 exemplaires sortirent des chaînes de montage.

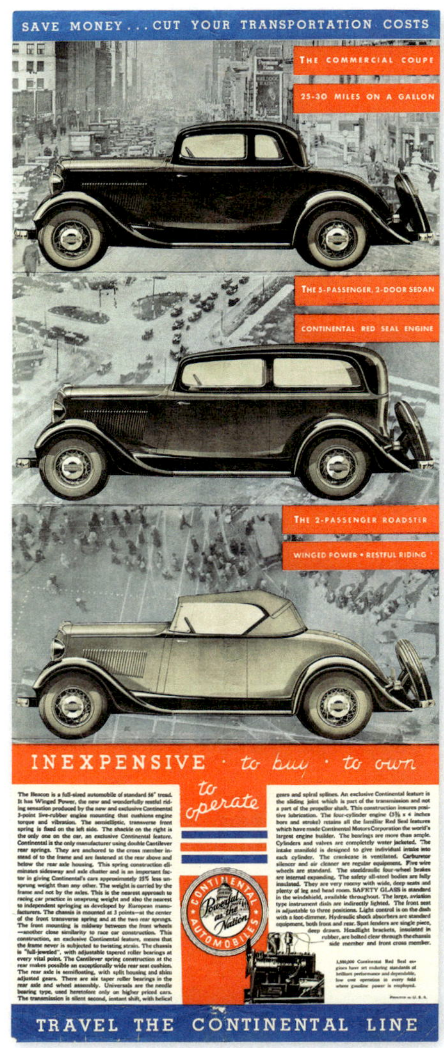

De Soto

De Soto Eight 1931

In late 1928 De Soto introduced its 1929 model as a mid-priced Chrysler product with a six-cylinder engine, but quickly blossomed. It gained a 70 hp straight-eight engine and a delicately shallow radiator shell that augmented its hood length. De Soto gained, and kept, its popularity. Named for the Spanish conquistador, more than two million De Sotos were built before the nameplate was dropped early in 1961.

Ende 1928 stellte De Soto sein Modell für das Jahr 1929 als Chrysler mittlerer Preislage mit Sechszylindermotor vor, doch daraus wurde bald mehr. Es erhielt einen Achtzylinderreihenmotor mit 95 kW und eine grazil flache Kühlerabdeckung, die die Motorhaube optisch länger wirken ließ. De Soto, benannt nach dem spanischen Eroberer, eroberte immer mehr Käuferherzen und verkaufte über zwei Millionen Fahrzeuge, bis die Marke Anfang 1961 eingestellt wurde.

Fin 1928, De Soto présenta son modèle 1929 comme un produit Chrysler de milieu de gamme à six cylindres qui ne tarda pas à se voir attribuer diverses prérogatives supplémentaires. Par exemple un moteur à huit cylindres en ligne et une calandre de radiateur tout en finesse qui accroissait la longueur du capot. De Soto conquit ainsi sa popularité pour la conserver longtemps. Portant le nom d'un conquistador espagnol, plus de deux millions de De Soto seront construites avant l'arrêt de la marque début 1961.

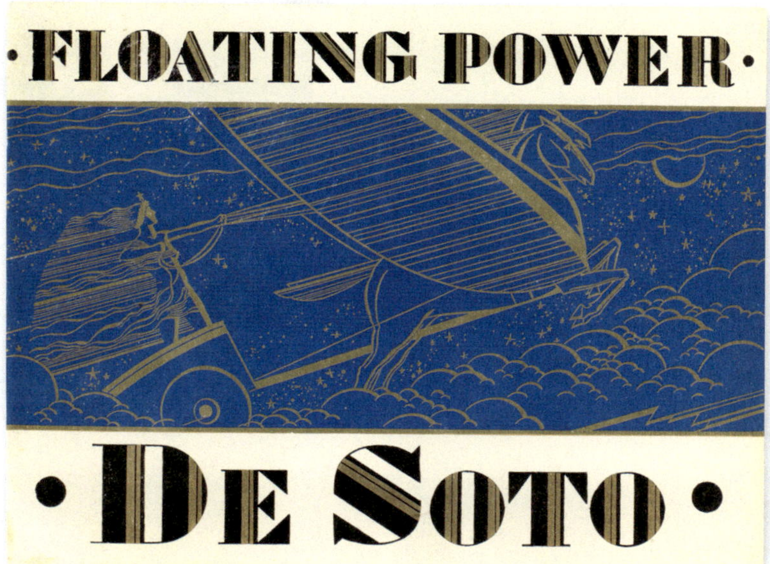

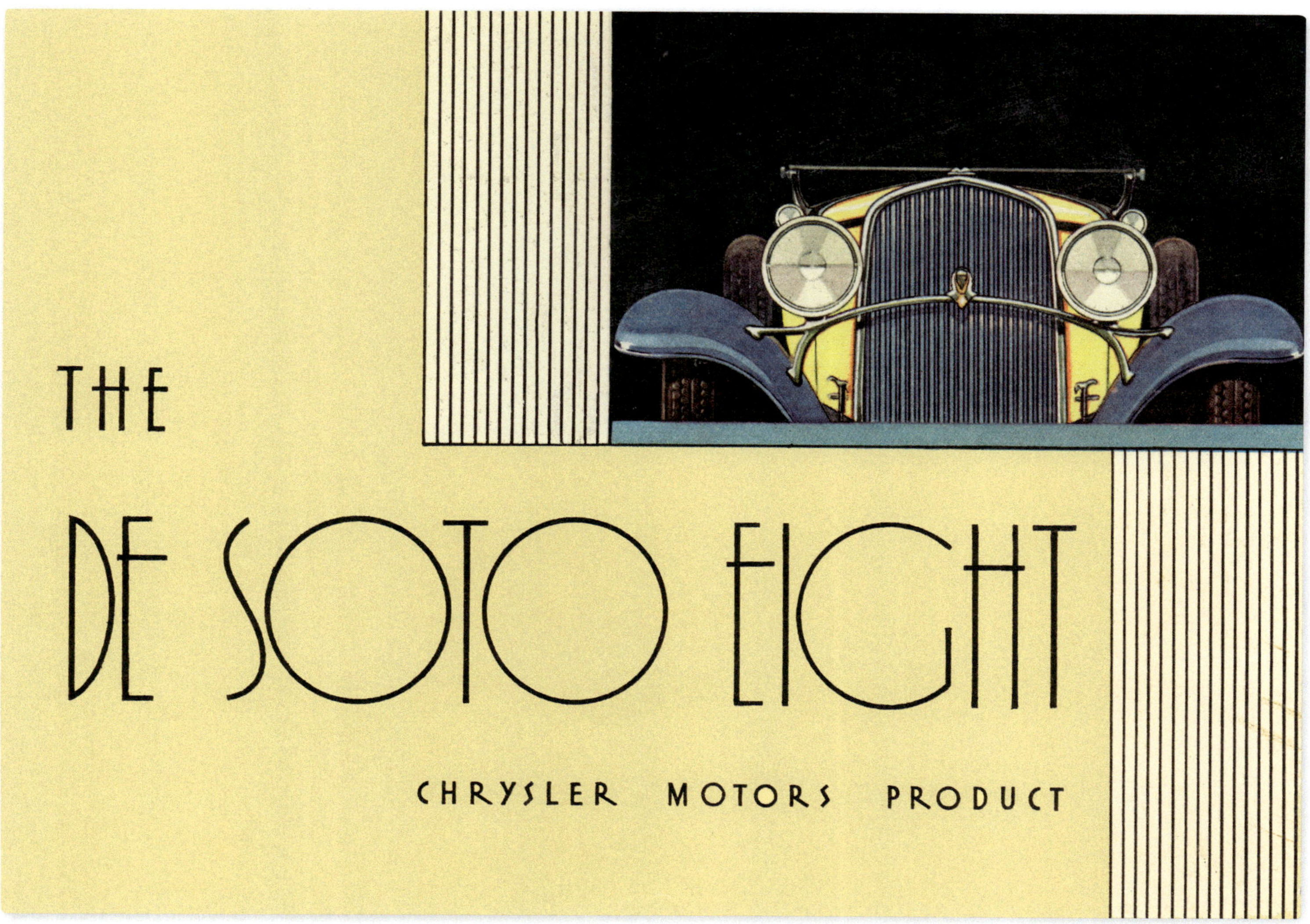

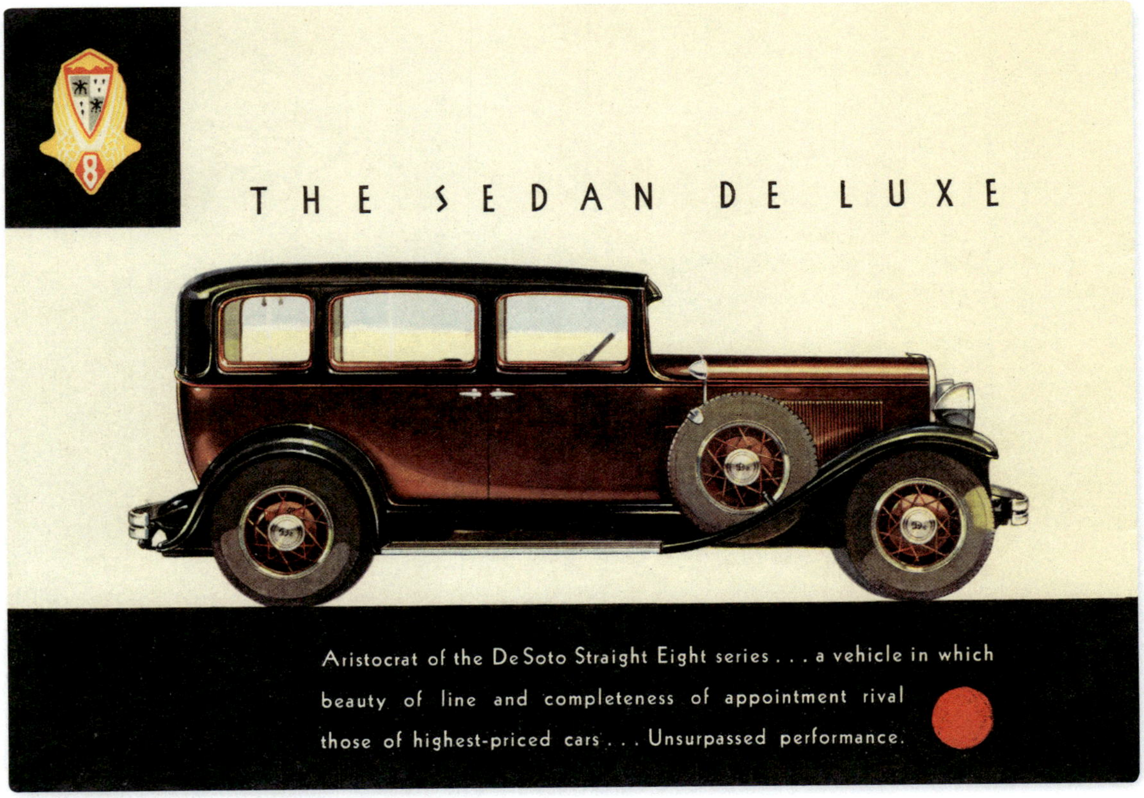
THE SEDAN DE LUXE
Aristocrat of the De Soto Straight Eight series . . . a vehicle in which beauty of line and completeness of appointment rival those of highest-priced cars . . . Unsurpassed performance.

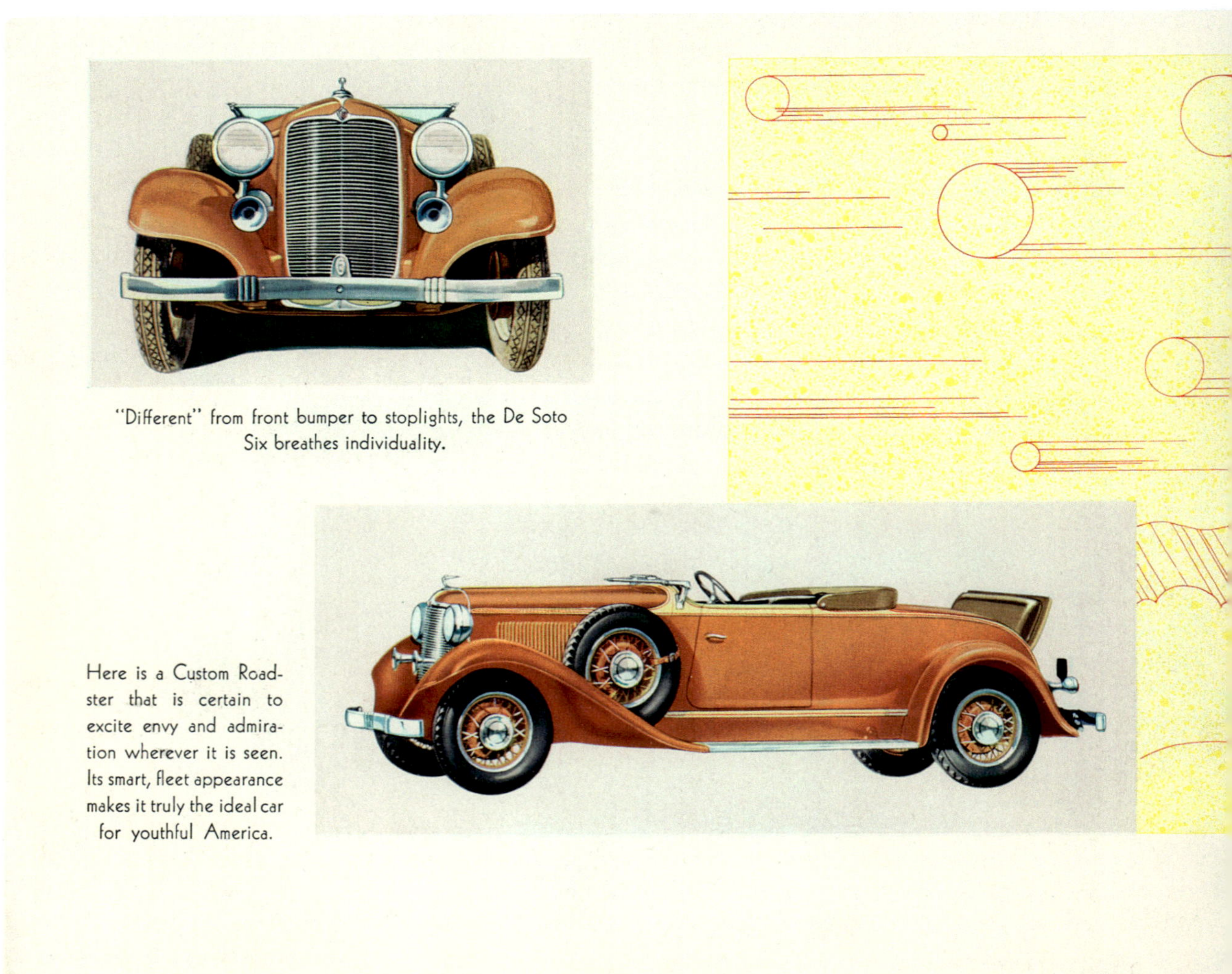
"Different" from front bumper to stoplights, the De Soto Six breathes individuality.
Here is a Custom Roadster that is certain to excite envy and admiration wherever it is seen. Its smart, fleet appearance makes it truly the ideal car for youthful America.

De Soto 1942 *(this spread)*

More than most American cars, immediate prewar De Sotos sported extravagant chromed grilles. Vertical grille pieces augmented De Soto's novel hidden headlamps.

Im Unterschied zu den meisten amerikanischen Autos dieser Zeit besaßen die De Sotos unmittelbar vor dem Zweiten Weltkrieg einen extravaganten verchromten Kühlergrill. Das vertikale Gitter hob die neuartigen versteckten Scheinwerfer des De Soto noch mehr hervor.

Plus que la plupart des voitures américaines, les De Soto lancées juste avant l'entrée en guerre des États-Unis arboraient de spectaculaires calandres chromées, comme cette grille en cascade qui renforçait l'effet de ses phares escamotables électriquement, une grande première sur le Nouveau Continent.

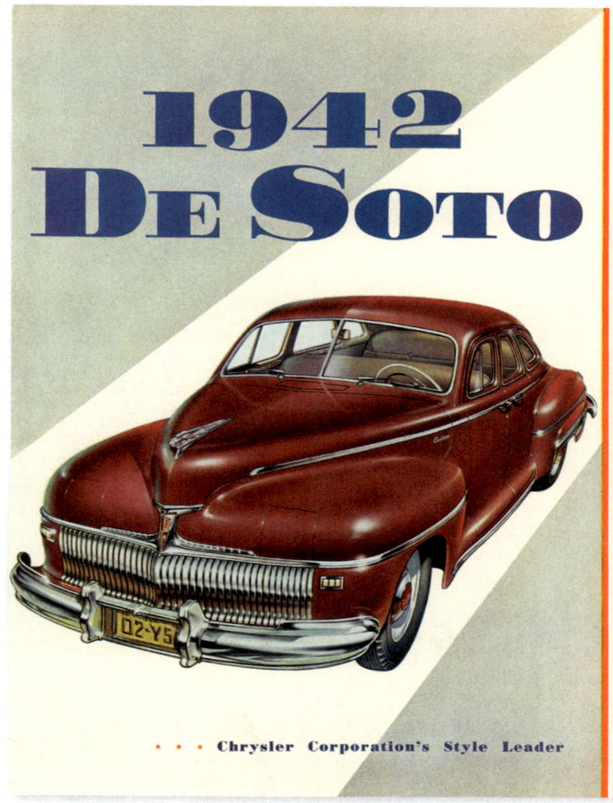

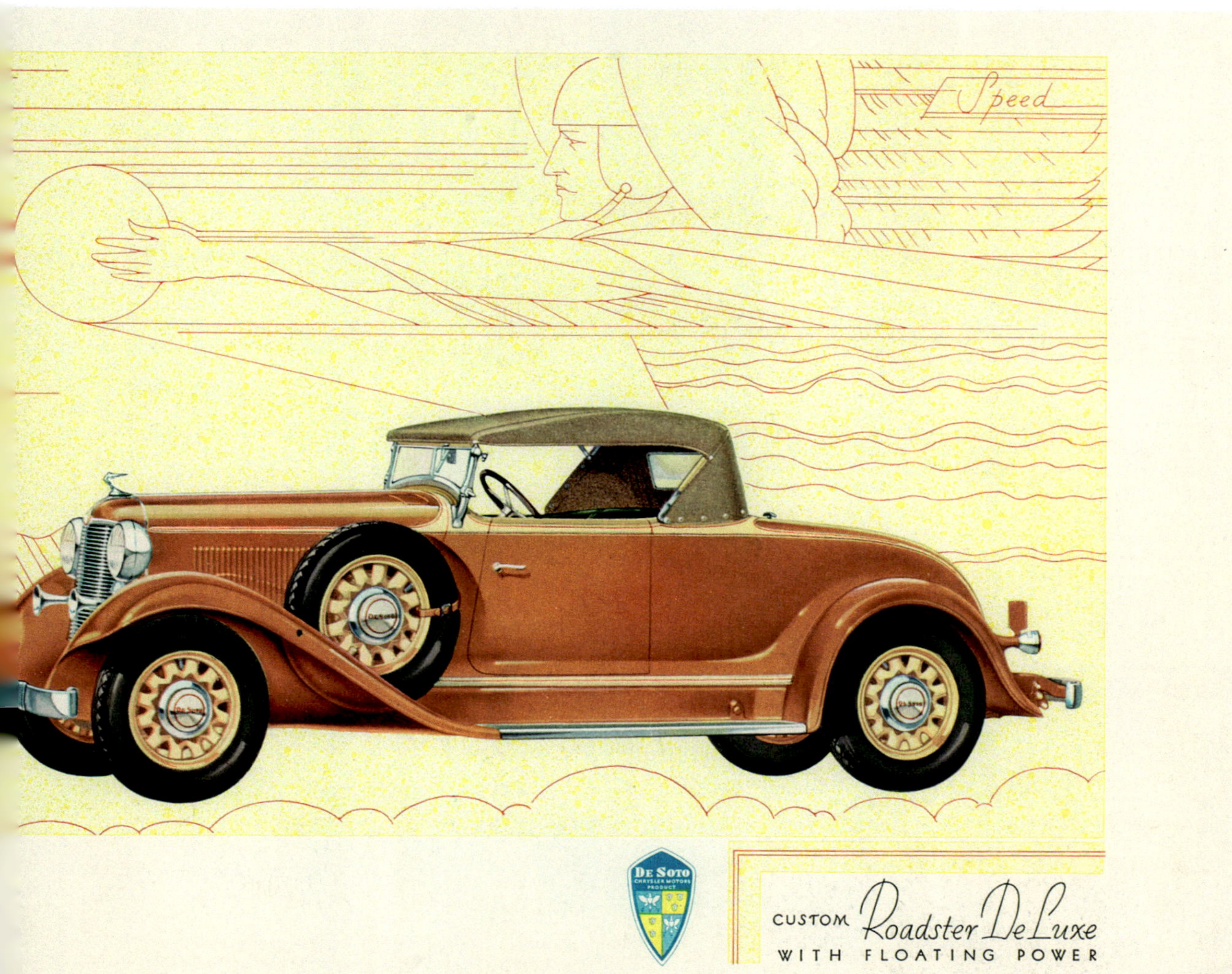

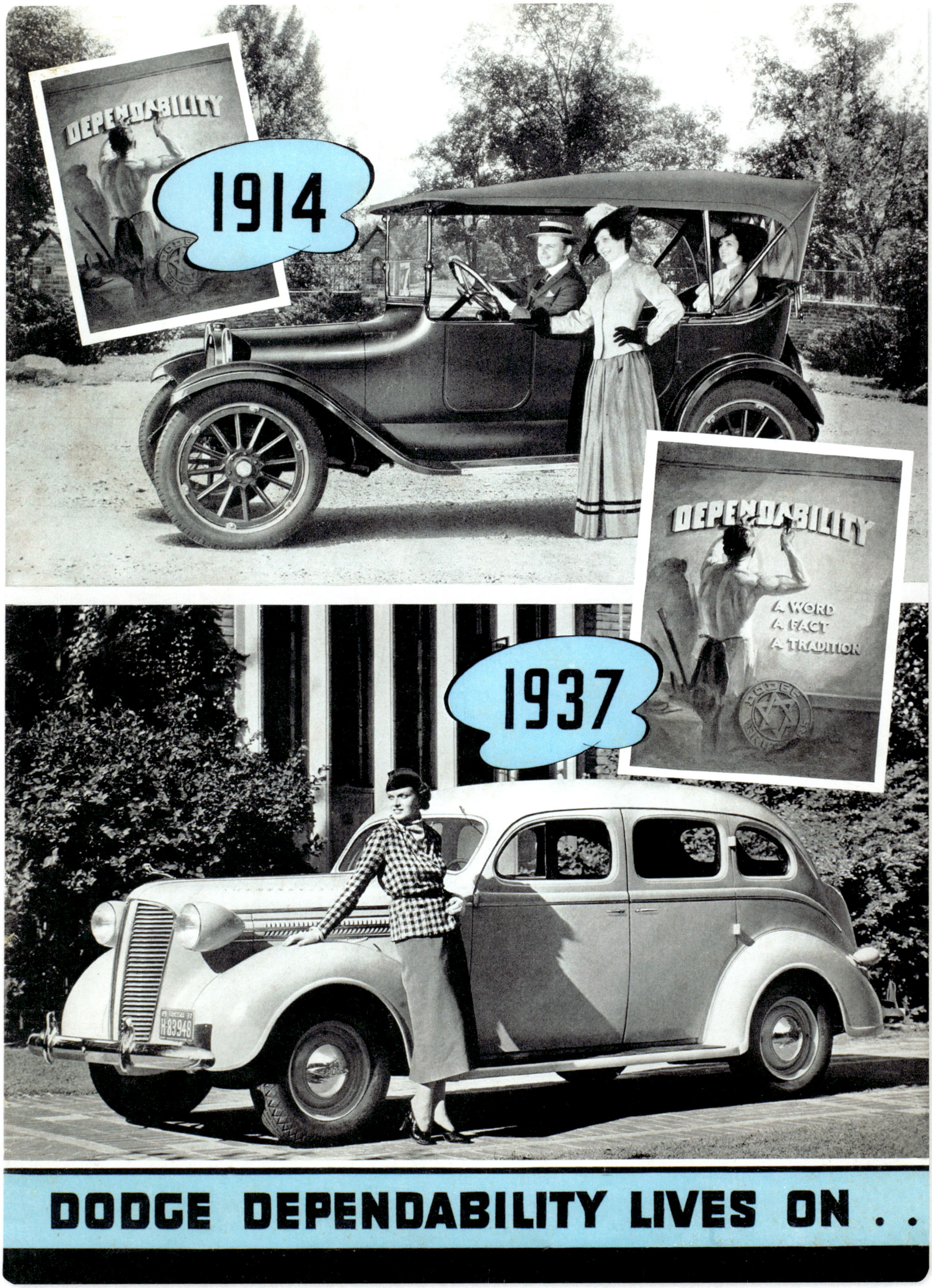
1914
DEPENDABILITY
DEPENDABILITY
A WORD
A FACT
A TRADITION
1937
H-83948
DODGE DEPENDABILITY LIVES ON . .

Dodge

Dodge 1937 *(opposite)*, **Dodge Six 1938** *(right)*, **Luxury Liner 1939** *(below)*

"Hollywood" aspirations notwithstanding, Dodges were fairly straightforward cars whose updates included dual tail lamps, dual sun visors, rubber insulation around the windshield, and a horn ring. About 295,000 were produced for 1937, a total that skidded to about 79,000 annual units in 1939 as war loomed.

Obwohl man sich hier mit Hollywood-glamour zu schmücken versucht, war der Dodge doch ein Auto ohne sonderlich viel Schnickschnack. Zu den Neuerungen gehörten doppelte Heckleuchten, beidseitige Sonnenblenden, Gummiabdichtungen um die Windschutzscheibe und ein Hupring. Für 1937 wurden noch rund 295 000 Fahrzeuge produziert, aber zwei Jahre später, als der Krieg drohte, waren es nur noch 79 000.

Nonobstant leurs rêves de gloire hollywoodienne, les Dodge étaient des voitures globalement sans histoires, dont les actualisations comprenaient par exemple des doubles feux arrière, des doubles pare-soleil, de la mousse isolante autour du parebrise et un cercle de klaxon au volant. Près de 295 000 voitures ont été produites en 1937, un total annuel qui chuta à 79 000 environ en 1939, alors s'amoncelaient les nuages de la guerre.

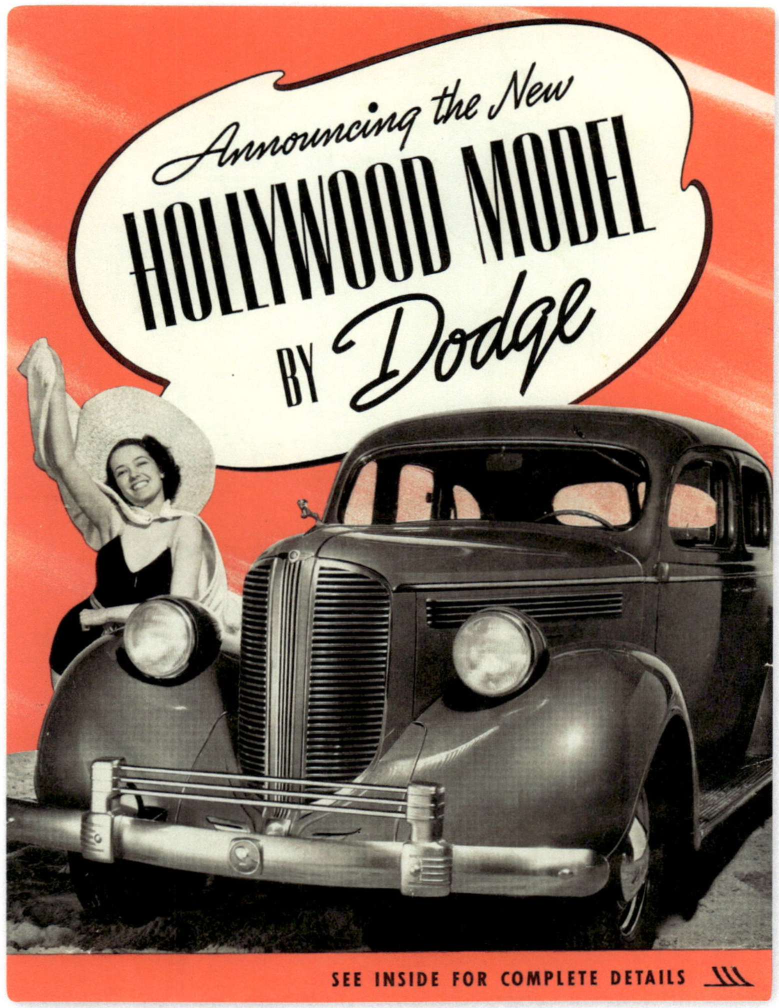

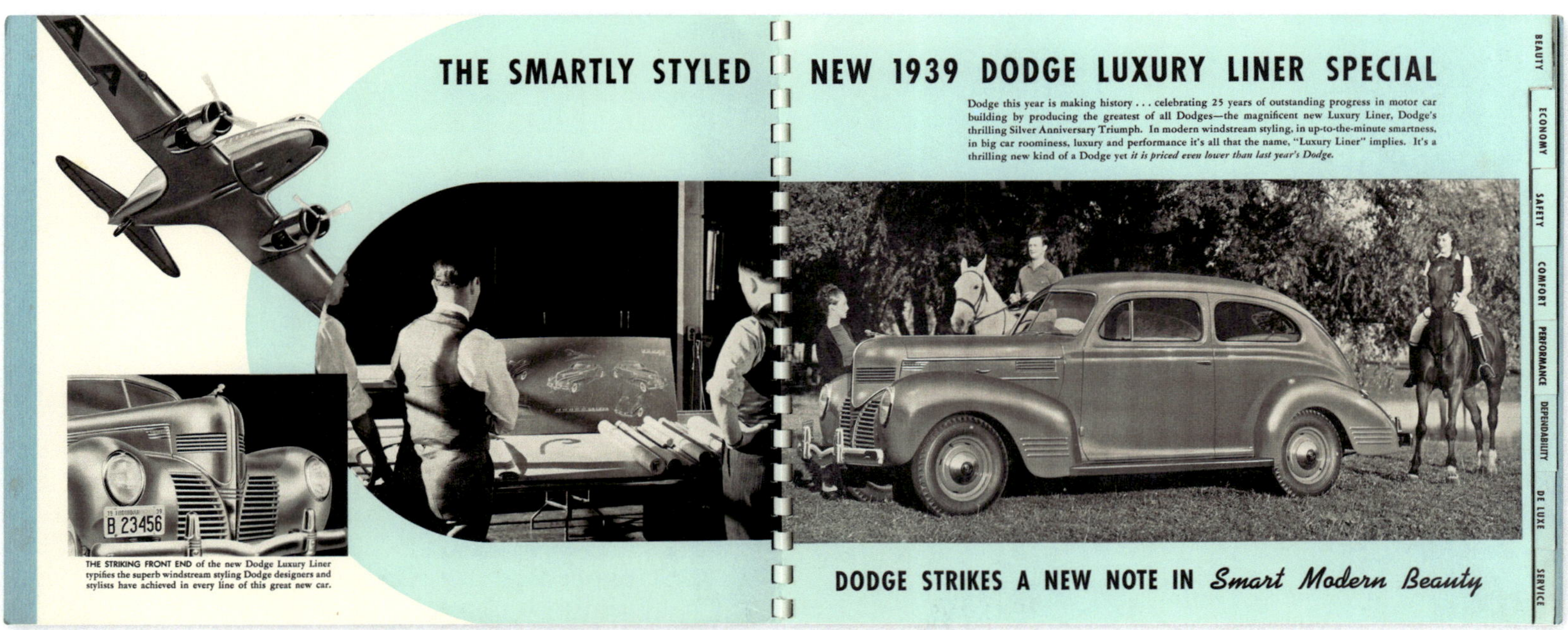

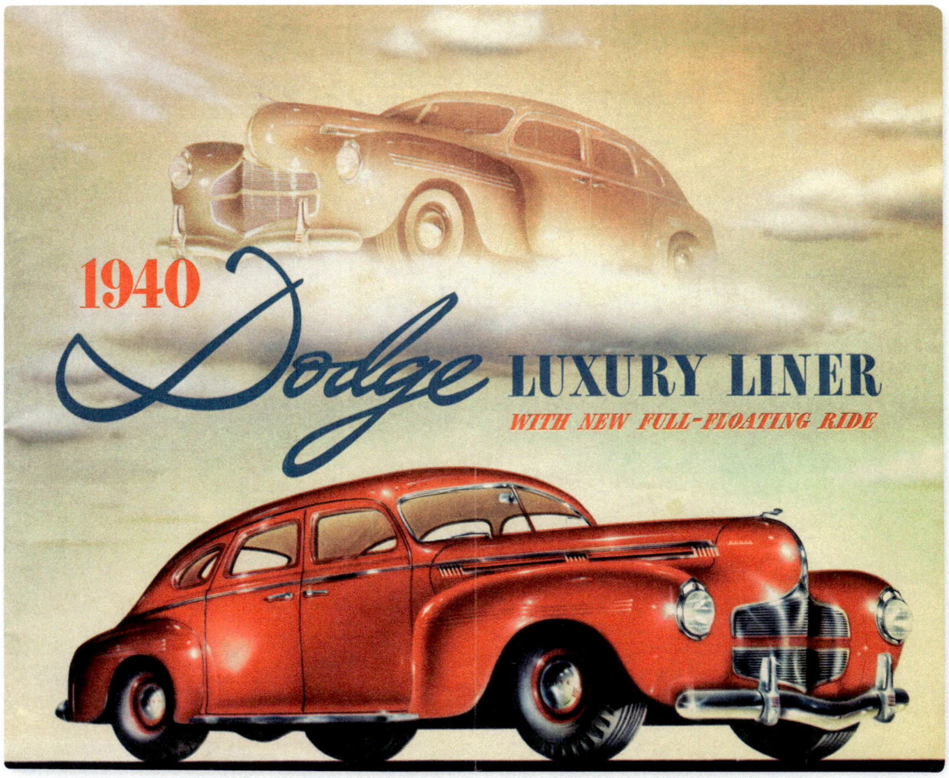

Dodge Luxury Liner 1940 *(left and below)*, **1941** *(opposite)*

It was big news when Chrysler's Fluid Drive went into the moderately priced Dodge. The simplified Fluid Drive diagram and operating instructions were actually quite informative.

Es war eine große Neuigkeit, als das „Fluid Drive"-Schaltgetriebe von Chrysler auch in dem preisgünstigeren Dodge angeboten wurde. Die schematische Darstellung des Fluid Drive und die Betriebsanleitung waren tatsächlich recht informativ.

La nouvelle fit du bruit lorsque cette Dodge plutôt bon marché se vit équiper de l'embrayage hydraulique « Fluid Drive » de Chrysler. Le schéma simplifié et le mode d'emploi du dispositif Fluid Drive étaient de fait très précis.

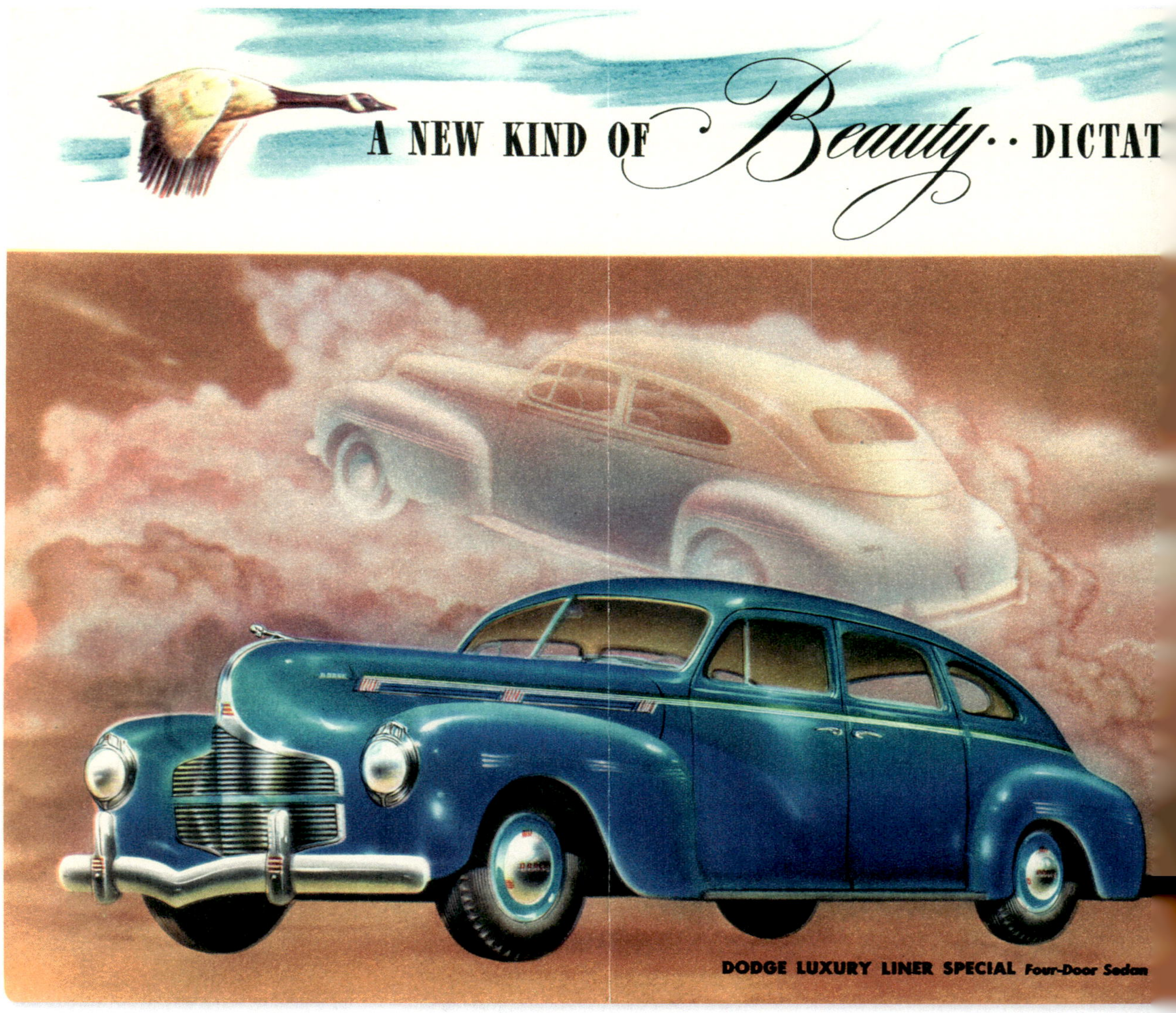

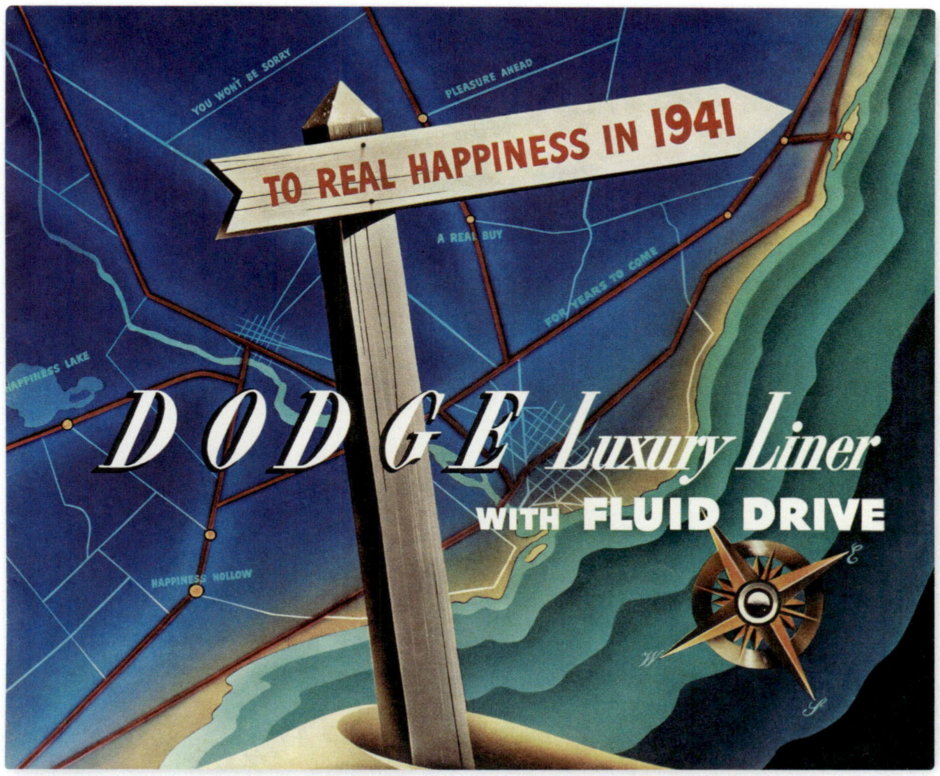
TO REAL HAPPINESS IN 1941
YOU WON'T BE SORRY
PLEASURE AHEAD
A REAL BUY
FOR YEARS TO COME
HAPPINESS HOLLOW
DODGE Luxury Liner
WITH FLUID DRIVE

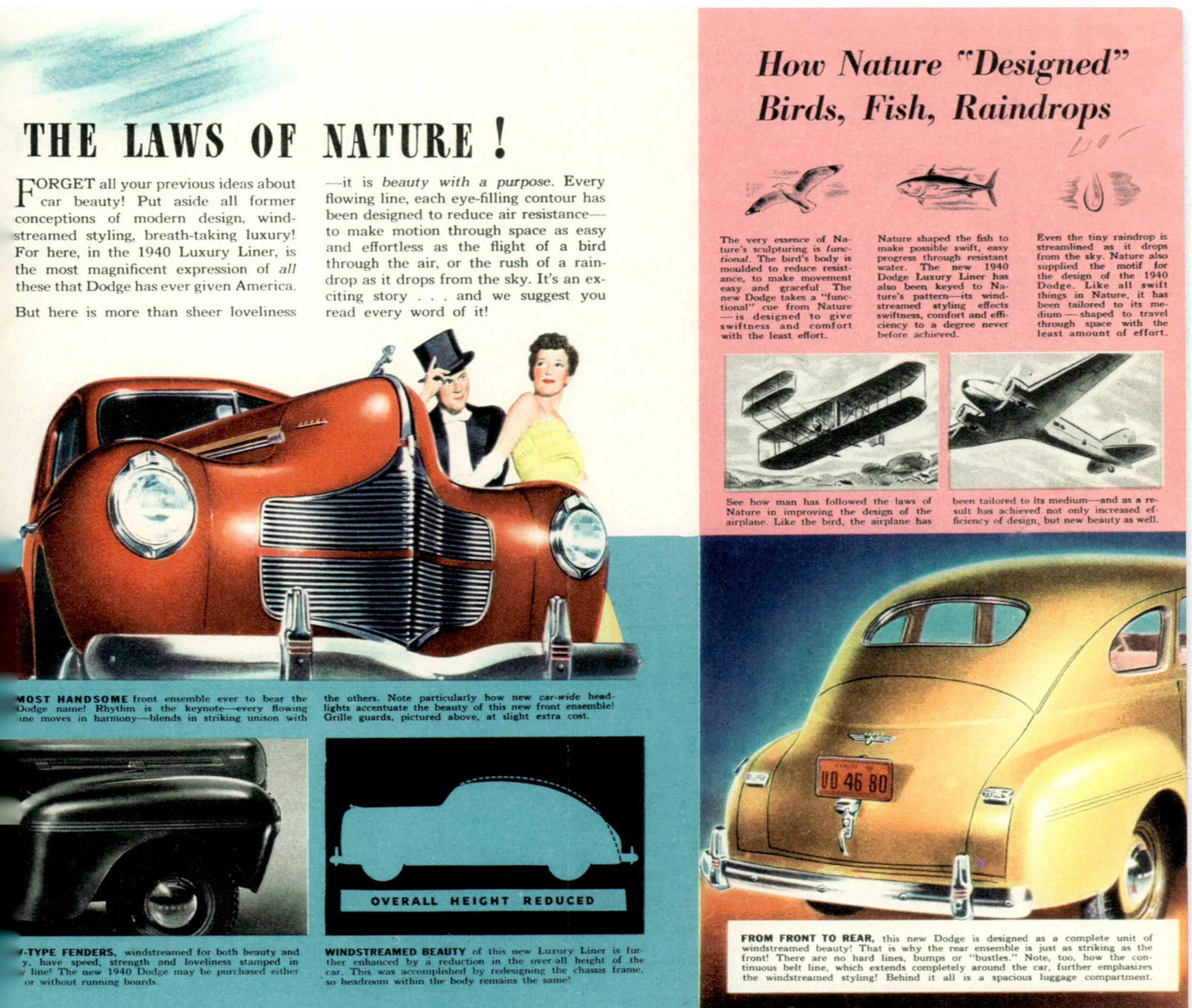
THE LAWS OF NATURE!
FORGET all your previous ideas about car beauty! Put aside all former conceptions of modern design, windstreamed styling, breath-taking luxury! For here, in the 1940 Luxury Liner, is the most magnificent expression of *all* these that Dodge has ever given America.
But here is more than sheer loveliness—it is *beauty with a purpose.* Every flowing line, each eye-filling contour has been designed to reduce air resistance—to make motion through space as easy and effortless as the flight of a bird through the air, or the rush of a raindrop as it drops from the sky. It's an exciting story . . . and we suggest you read every word of it!
MOST HANDSOME front ensemble ever to bear the Dodge name! Rhythm is the keynote—every flowing ine moves in harmony—blends in striking unison with the others. Note particularly how new *car-wide* headlights accentuate the beauty of this new front ensemble! Grille guards, pictured above, at slight extra cost.
OVERALL HEIGHT REDUCED
-TYPE FENDERS, windstreamed for both beauty and y, have speed, strength and loveliness stamped in line! The new 1940 Dodge may be purchased either or without running boards.
WINDSTREAMED BEAUTY of this new Luxury Liner is further enhanced by a reduction in the over-all height of the car. This was accomplished by redesigning the chassis frame, so headroom within the body remains the same!
How Nature "Designed" Birds, Fish, Raindrops
The very essence of Nature's sculpturing is *functional.* The bird's body is moulded to reduce resistance, to make movement easy and graceful. The new Dodge takes a "functional" cue from Nature—is designed to give swiftness and comfort with the least effort.
Nature shaped the fish to make possible swift, easy progress through resistant water. The new 1940 Dodge Luxury Liner has also been keyed to Nature's pattern—its windstreamed styling effects swiftness, comfort and efficiency to a degree never before achieved.
Even the tiny raindrop is streamlined as it drops from the sky. Nature also supplied the motif for the design of the 1940 Dodge. Like all swift things in Nature, it has been tailored to its medium—shaped to travel through space with the least amount of effort.
See how man has followed the laws of Nature in improving the design of the airplane. Like the bird, the airplane has been tailored to its medium—and as a result has achieved not only increased efficiency of design, but new beauty as well.
FROM FRONT TO REAR, this new Dodge is designed as a complete unit of windstreamed beauty! That is why the rear ensemble is just as striking as the front! There are no hard lines, bumps or "bustles." Note, too, how the continuous belt line, which extends completely around the car, further emphasizes the windstreamed styling! Behind it all is a spacious luggage compartment.

Essex

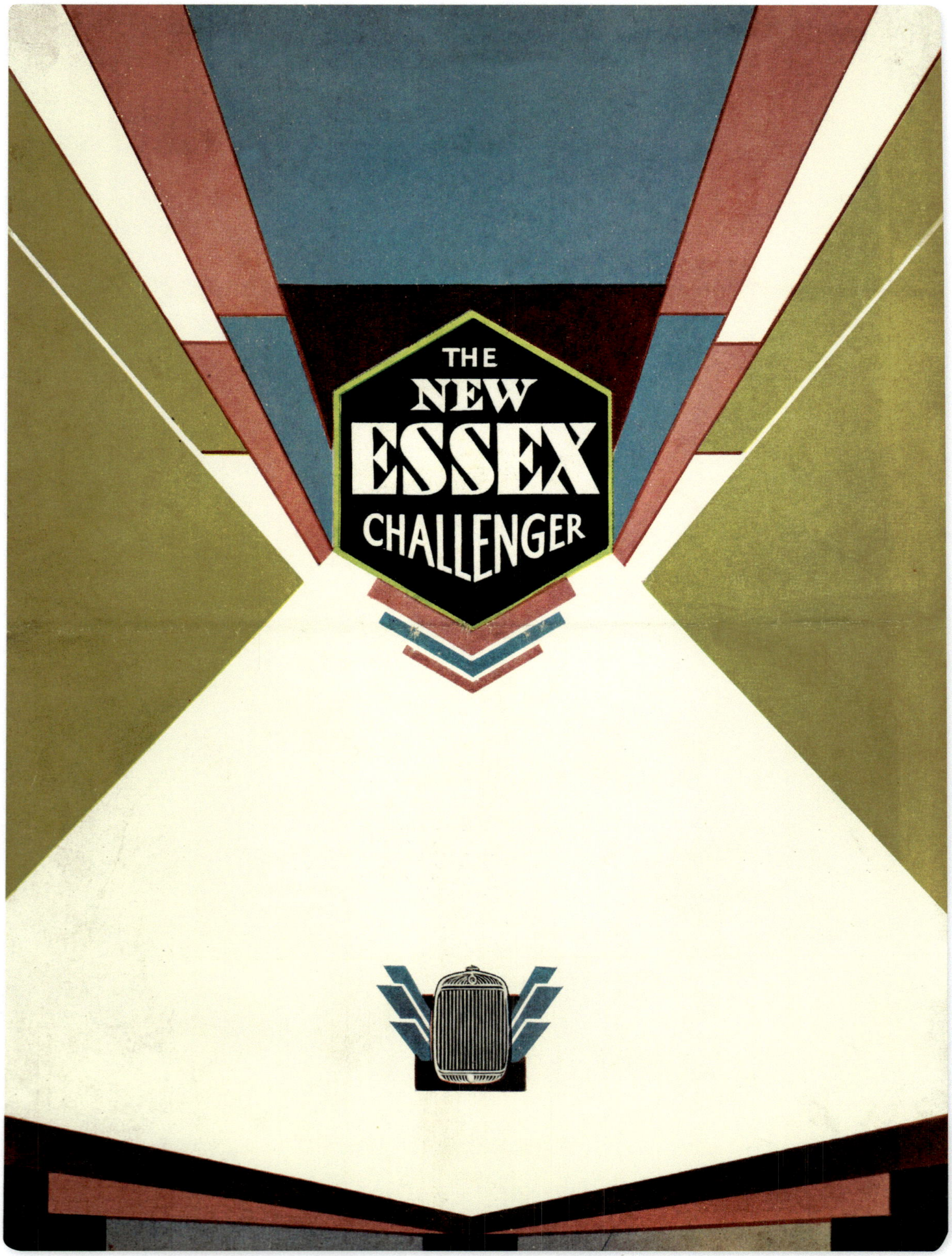

Essex 1930 *(this spread)*

Named for a region of England, the Essex was a less expensive sub-range of the Hudson, introduced in 1919. It was a good car, with bright appearance options, four-cylinder economy, and booming sales above 300,000 units. In 1932 Hudson renamed it the Terraplane.

Bei dem nach der englischen Grafschaft benannten Essex, der 1919 auf den Markt kam, handelte es sich um einen preiswerteren Ableger des Hudson. Es waren solide Autos in leuchtenden Farben und mit sparsamen Vierzylindermotoren, von denen über 300 000 Stück abgesetzt wurden. Hudson benannte das Fahrzeug 1932 in Terraplane um.

Lancée en 1919, Essex – du nom d'un comté du Royaume-Uni – était une sous-marque à prix modéré de Hudson. Il s'agissait d'une bonne voiture qui proposait d'intelligentes options esthétiques et le caractère économique d'un quatre cylindres. Il s'en écoula plus de 300 000 exemplaires. En 1932 Hudson la rebaptisa Terraplane.

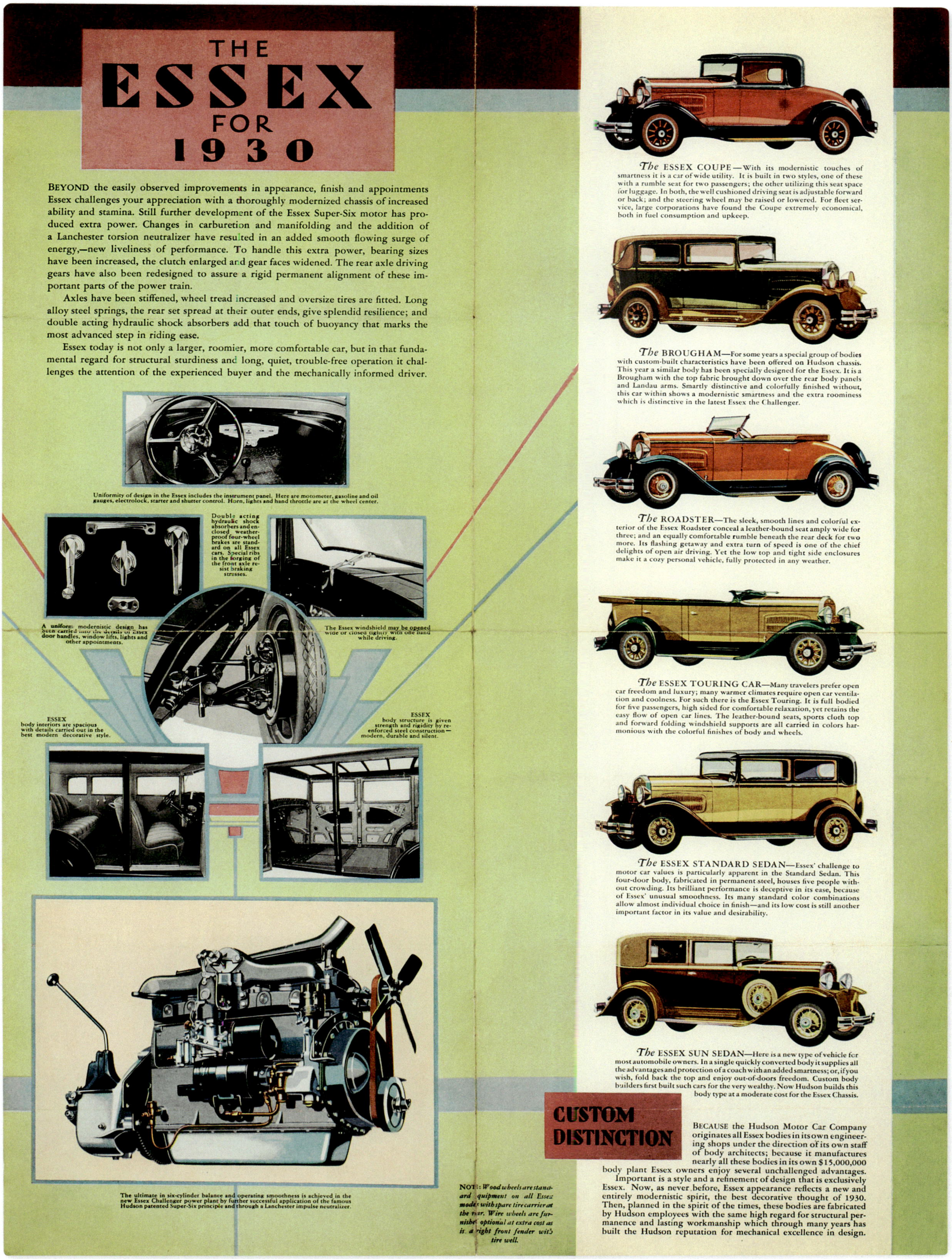

THE ESSEX FOR 1930

BEYOND the easily observed improvements in appearance, finish and appointments Essex challenges your appreciation with a thoroughly modernized chassis of increased ability and stamina. Still further development of the Essex Super-Six motor has produced extra power. Changes in carburetion and manifolding and the addition of a Lanchester torsion neutralizer have resulted in an added smooth flowing surge of energy,—new liveliness of performance. To handle this extra power, bearing sizes have been increased, the clutch enlarged and gear faces widened. The rear axle driving gears have also been redesigned to assure a rigid permanent alignment of these important parts of the power train.

Axles have been stiffened, wheel tread increased and oversize tires are fitted. Long alloy steel springs, the rear set spread at their outer ends, give splendid resilience; and double acting hydraulic shock absorbers add that touch of buoyancy that marks the most advanced step in riding ease.

Essex today is not only a larger, roomier, more comfortable car, but in that fundamental regard for structural sturdiness and long, quiet, trouble-free operation it challenges the attention of the experienced buyer and the mechanically informed driver.

Uniformity of design in the Essex includes the instrument panel. Here are motometer, gasoline and oil gauges, electrolock, starter and shutter control. Horn, lights and hand throttle are at the wheel center.

Double acting hydraulic shock absorbers and enclosed weatherproof four-wheel brakes are standard on all Essex cars. Special ribs in the forging of the front axle resist braking stresses.

A uniform modernistic design has been carried into the details of Essex door handles, window lifts, lights and other appointments.

The Essex windshield may be opened wide or closed tightly with one hand while driving.

ESSEX body interiors are spacious with details carried out in the best modern decorative style.

ESSEX body structure is given strength and rigidity by reenforced steel construction—modern, durable and silent.

The ultimate in six-cylinder balance and operating smoothness is achieved in the new Essex Challenger power plant by further successful application of the famous Hudson patented Super-Six principle and through a Lanchester impulse neutralizer.

NOTE: *Wood wheels are standard equipment on all Essex models with spare tire carrier at the rear. Wire wheels are furnished optional at extra cost as is a right front fender with tire well.*

The ESSEX COUPE—With its modernistic touches of smartness it is a car of wide utility. It is built in two styles, one of these with a rumble seat for two passengers; the other utilizing this seat space for luggage. In both, the well cushioned driving seat is adjustable forward or back; and the steering wheel may be raised or lowered. For fleet service, large corporations have found the Coupe extremely economical, both in fuel consumption and upkeep.

The BROUGHAM—For some years a special group of bodies with custom-built characteristics have been offered on Hudson chassis. This year a similar body has been specially designed for the Essex. It is a Brougham with the top fabric brought down over the rear body panels and Landau arms. Smartly distinctive and colorfully finished without, this car within shows a modernistic smartness and the extra roominess which is distinctive in the latest Essex the Challenger.

The ROADSTER—The sleek, smooth lines and colorful exterior of the Essex Roadster conceal a leather-bound seat amply wide for three; and an equally comfortable rumble beneath the rear deck for two more. Its flashing getaway and extra turn of speed is one of the chief delights of open air driving. Yet the low top and tight side enclosures make it a cozy personal vehicle, fully protected in any weather.

The ESSEX TOURING CAR—Many travelers prefer open car freedom and luxury; many warmer climates require open car ventilation and coolness. For such there is the Essex Touring. It is full bodied for five passengers, high sided for comfortable relaxation, yet retains the easy flow of open car lines. The leather-bound seats, sports cloth top and forward folding windshield supports are all carried in colors harmonious with the colorful finishes of body and wheels.

The ESSEX STANDARD SEDAN—Essex' challenge to motor car values is particularly apparent in the Standard Sedan. This four-door body, fabricated in permanent steel, houses five people without crowding. Its brilliant performance is deceptive in its ease, because of Essex' unusual smoothness. Its many standard color combinations allow almost individual choice in finish—and its low cost is still another important factor in its value and desirability.

The ESSEX SUN SEDAN—Here is a new type of vehicle for most automobile owners. In a single quickly converted body it supplies all the advantages and protection of a coach with an added smartness; or, if you wish, fold back the top and enjoy out-of-doors freedom. Custom body builders first built such cars for the very wealthy. Now Hudson builds this body type at a moderate cost for the Essex Chassis.

CUSTOM DISTINCTION

BECAUSE the Hudson Motor Car Company originates all Essex bodies in its own engineering shops under the direction of its own staff of body architects; because it manufactures nearly all these bodies in its own $15,000,000 body plant Essex owners enjoy several unchallenged advantages.

Important is a style and a refinement of design that is exclusively Essex. Now, as never before, Essex appearance reflects a new and entirely modernistic spirit, the best decorative thought of 1930. Then, planned in the spirit of the times, these bodies are fabricated by Hudson employees with the same high regard for structural permanence and lasting workmanship which through many years has built the Hudson reputation for mechanical excellence in design.

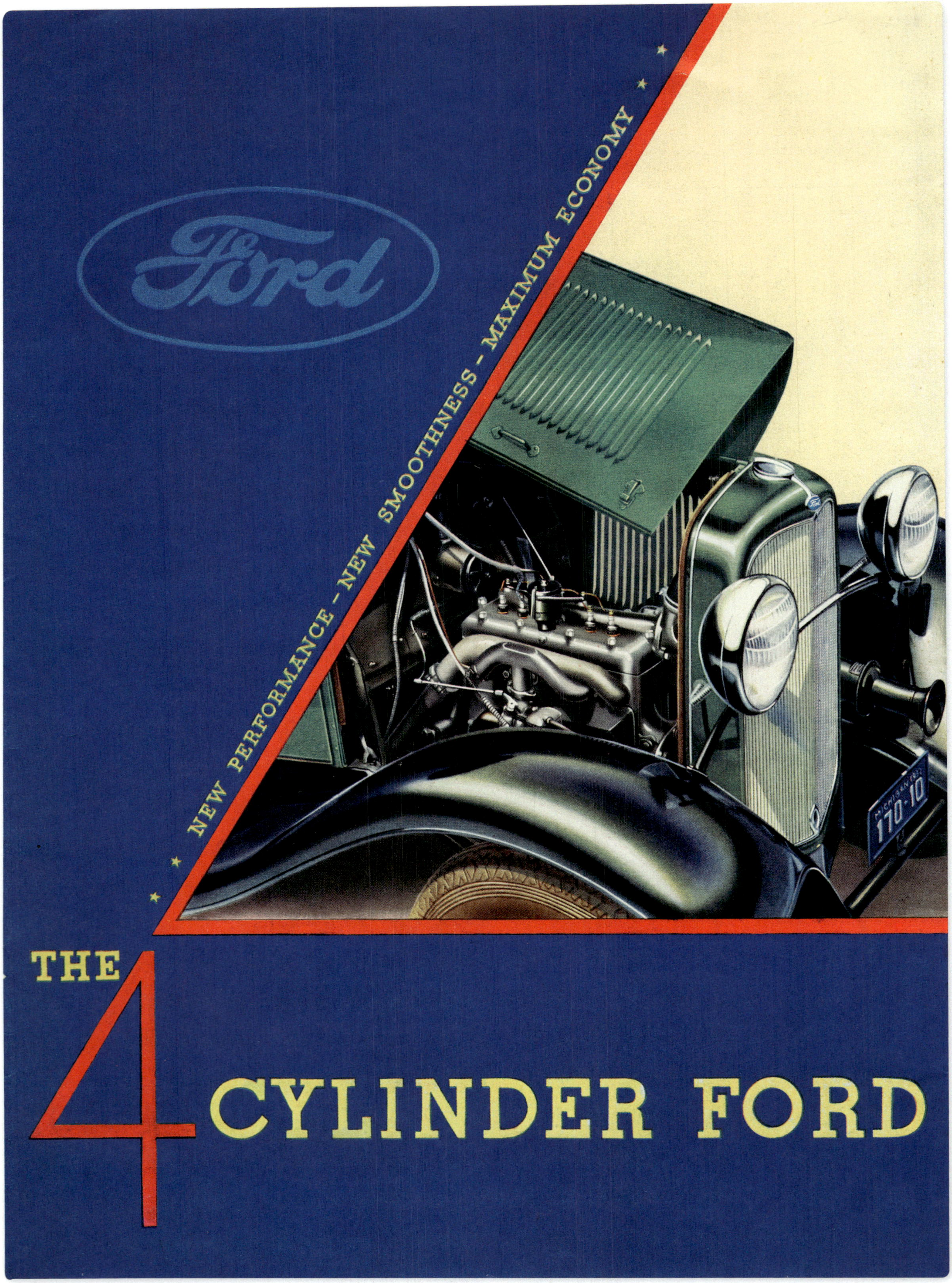
Ford
NEW PERFORMANCE - NEW SMOOTHNESS - MAXIMUM ECONOMY
THE 4 CYLINDER FORD

Ford

Ford V-8 1932 *(opposite)*, **1935** *(right)*, **1936** *(below)*

Ford's development of an affordable, reliable V-8 in 1932 was one of the prewar industry's most important advances. By 1935 two million of the 85 hp cars had been sold. Some still preferred the four-cylinder Model B, but it vanished after 1934.

Die Entwicklung eines erschwinglichen und zuverlässigen V8-Motors bei Ford im Jahre 1932 war einer der größten Fortschritte in der Automobilindustrie vor dem Zweiten Weltkrieg. Bis 1935 wurden zwei Millionen der 115 kW starken Autos abgesetzt. Einige Käufer bevorzugten immer noch das Model B mit seinen vier Zylindern, doch dessen Produktion wurde 1934 eingestellt.

En 1932, le développement par Ford d'un moteur V8 aussi fiable qu'abordable représenta l'une des avancées les plus importantes de l'industrie automobile de la période de l'avant-guerre. Trois ans plus tard, deux millions d'exemplaires de ce modèle développant 85 chevaux auront été vendues. Certains conducteurs continuaient de leur préférer la Ford B à quatre cylindres, mais celle-ci disparut après 1934.

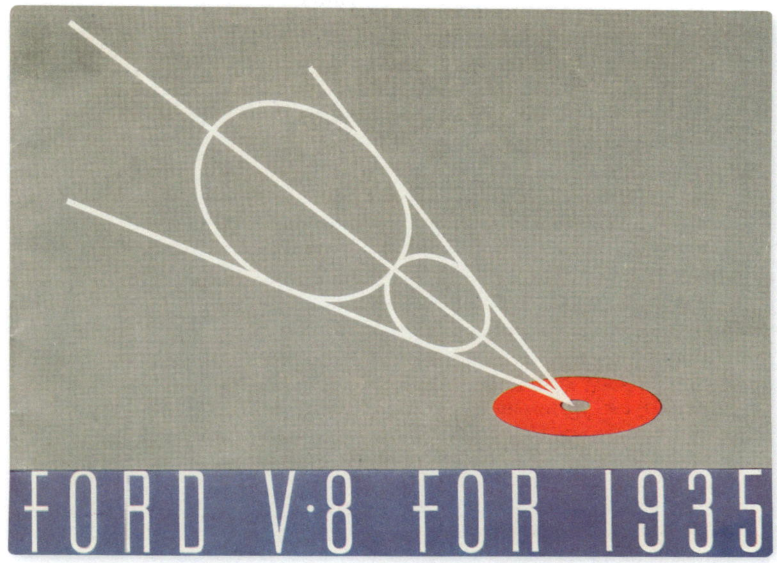

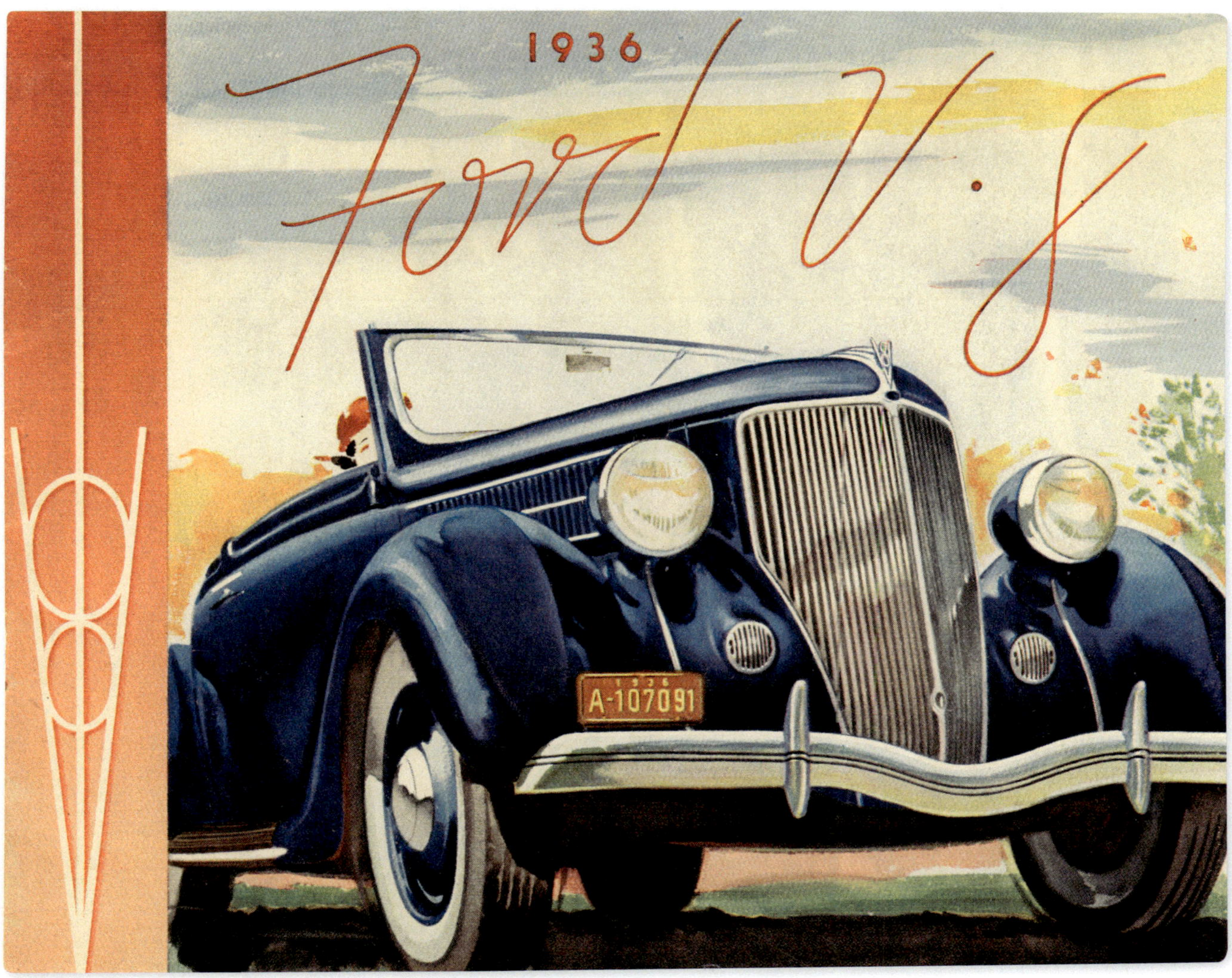

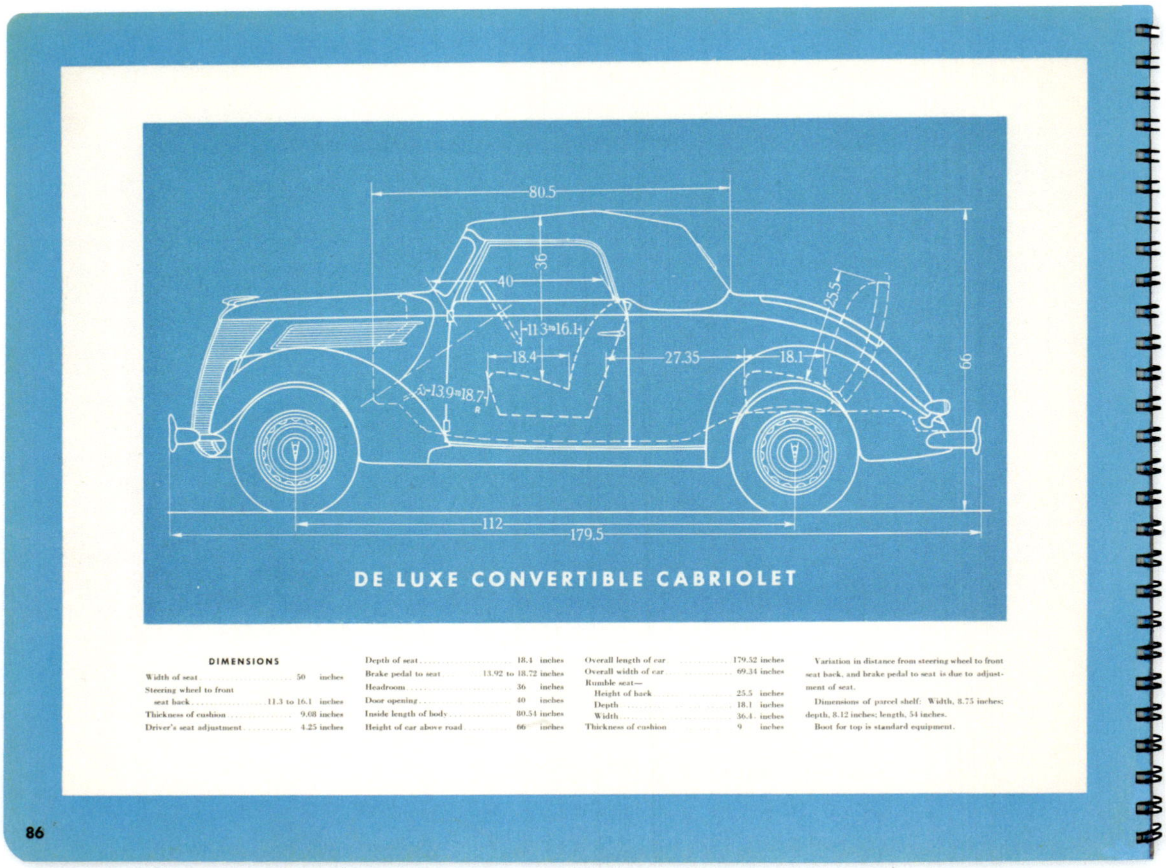

Ford 1937 *(this spread)*

Part of the reason for Ford's strong appeal was the action that Edsel Ford, Henry's only son, took in establishing a styling department, largely without his father's support. Artists like E.T. Gregorie created cars that could compete with GM's raciness.

Dass die Fords bei den Käufern so gut ankamen, lag zum Teil auch daran, dass Edsel Ford, Henrys einziger Sohn, ganz ohne väterliche Unterstützung eine Stylingabteilung auf die Beine gestellt hatte. Dort entwarfen Künstler wie E.T. Gregorie Automobile, die mit den schnittigen Fahrzeugen des Konkurrenten GM mithalten konnten.

Si les Ford plaisaient autant, c'était notamment grâce à l'initiative du fils unique d'Henry Ford, Edsel, qui avait créé un bureau de style en se passant largement du soutien paternel. Ainsi, des artistes comme E.T. Gregorie ont-ils pu imaginer des voitures capables de rivaliser avec l'audace créative de General Motors.

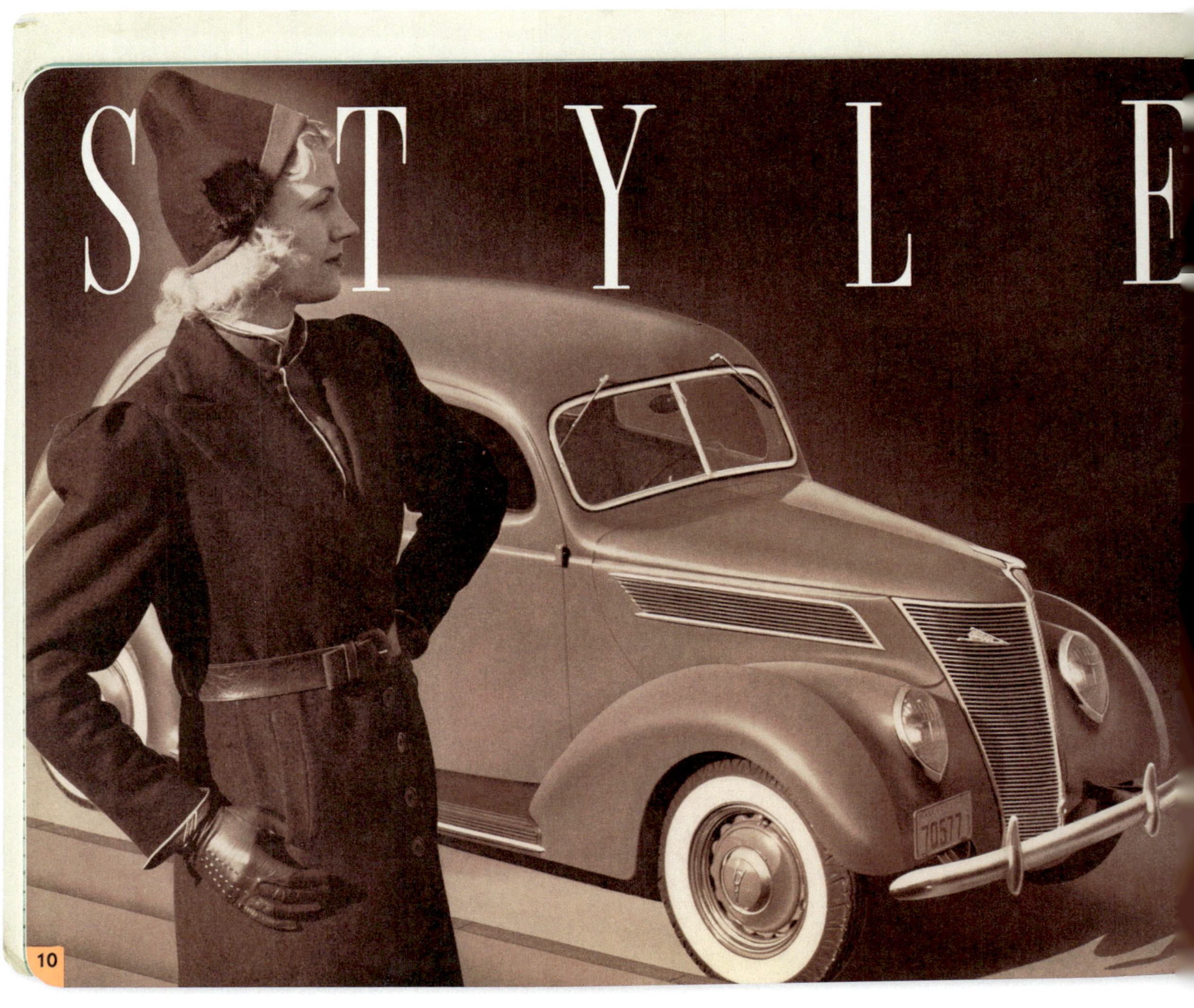

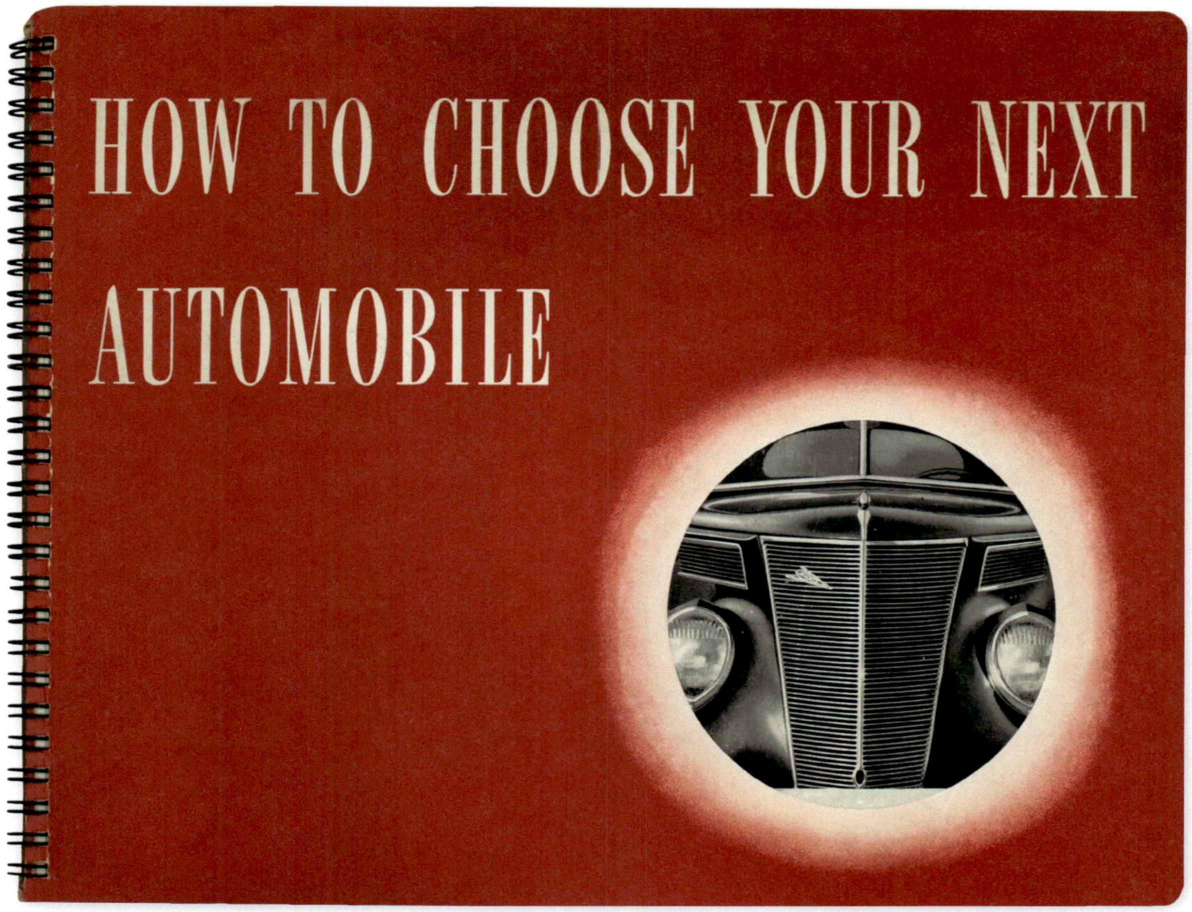
HOW TO CHOOSE YOUR NEXT AUTOMOBILE

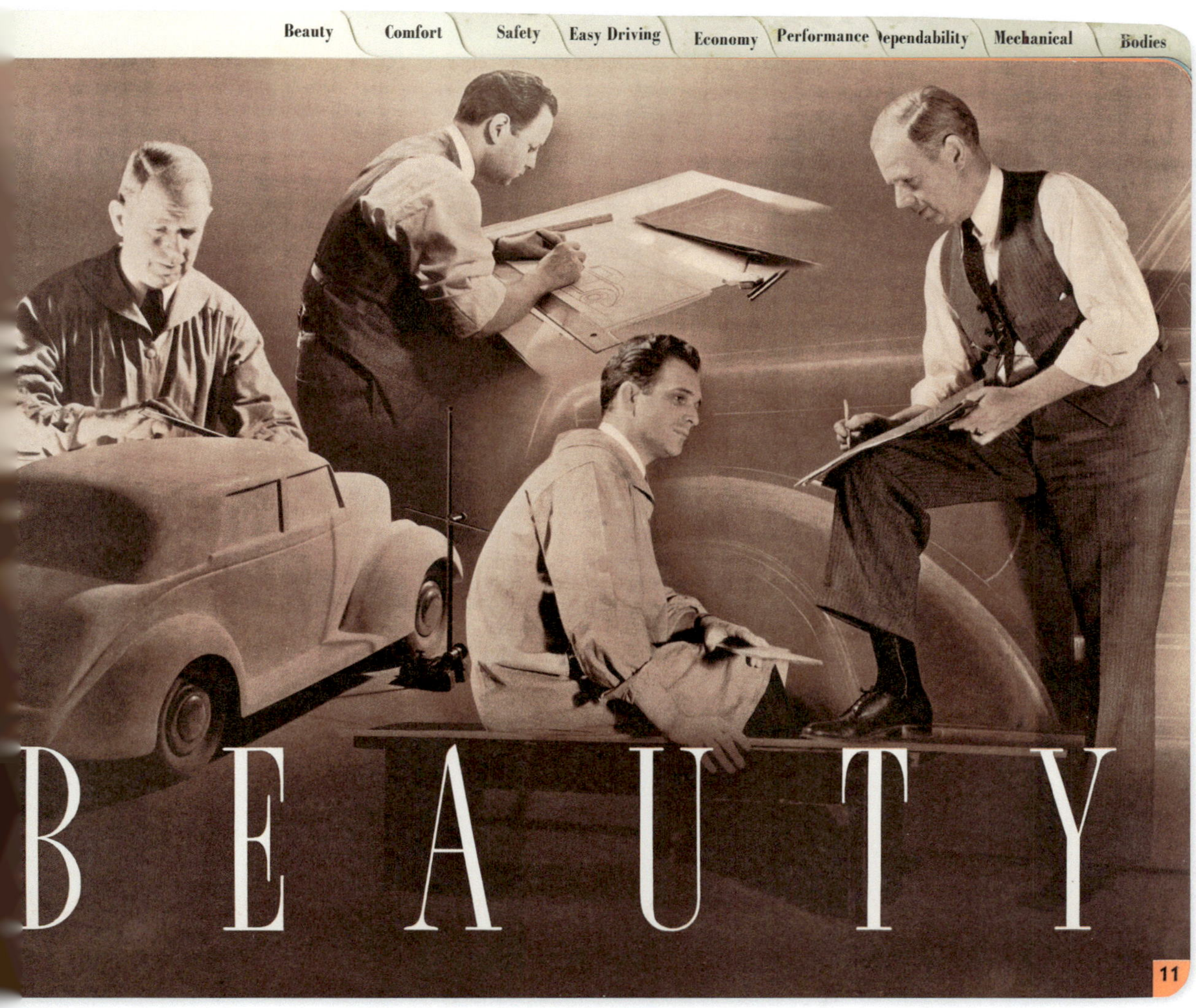
Beauty
Comfort
Safety
Easy Driving
Economy
Performance
ependability
Mechanical
Bodies
BEAUTY
11

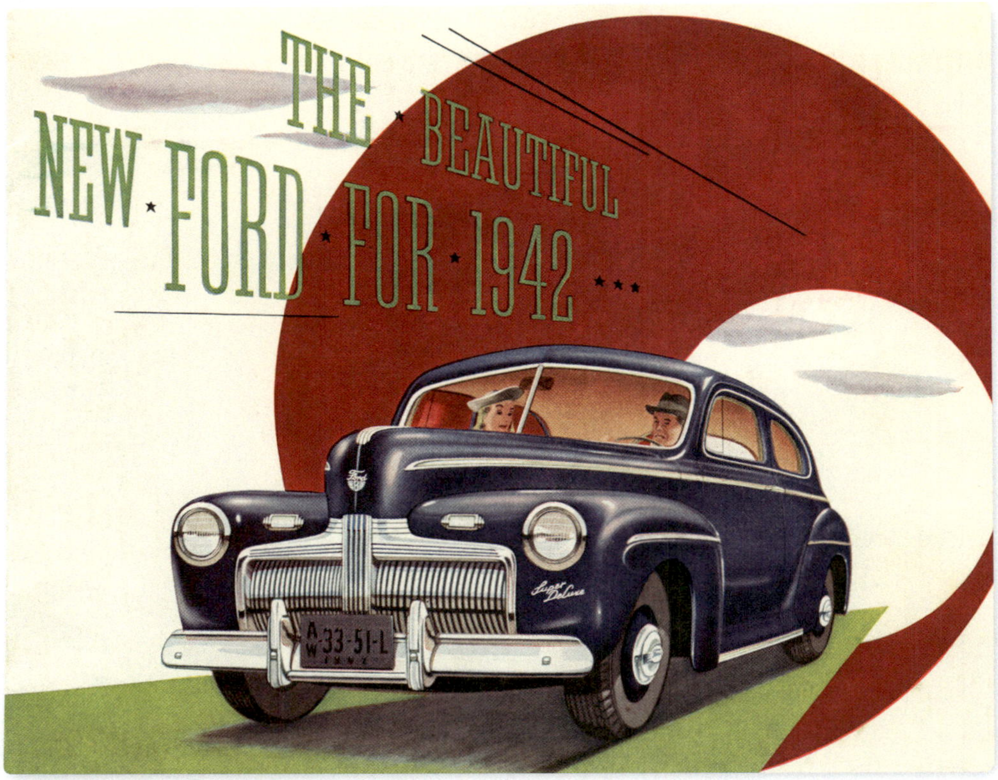

Ford 1942 *(above)*, **1940** *(right)*

Dealer catalogs spelled out the 1940 Ford in terms of its power, interior space, utility, safety, and, perhaps unexpectedly, flashes of color from a growing upholstery selection. The 1940 models are among the most iconic Fords ever built.

Händlerkataloge stellten bei den Fords des Modelljahres 1940 die Leistung, den Innenraum, die Nützlichkeit, die Sicherheit und – wahrscheinlich ganz unerwartet – die Farbtupfer in einer wachsenden Auswahl an Sitzbezügen heraus. Mehr als alle anderen Fords wurden die Modelle dieses Jahrgangs zum Kult.

Les catalogues des concessionnaires inventoriaient les qualités de la Ford de 1940 à l'aune de sa puissance, de son habitabilité, de sa fonctionnalité, de sa sécurité et, peut-être de manière plus inattendue, d'une débauche de couleurs rendue possible par un choix de selleries et de garnitures toujours plus vaste. Les modèles de 1940 figurent parmi les plus emblématiques que Ford ait jamais produits.

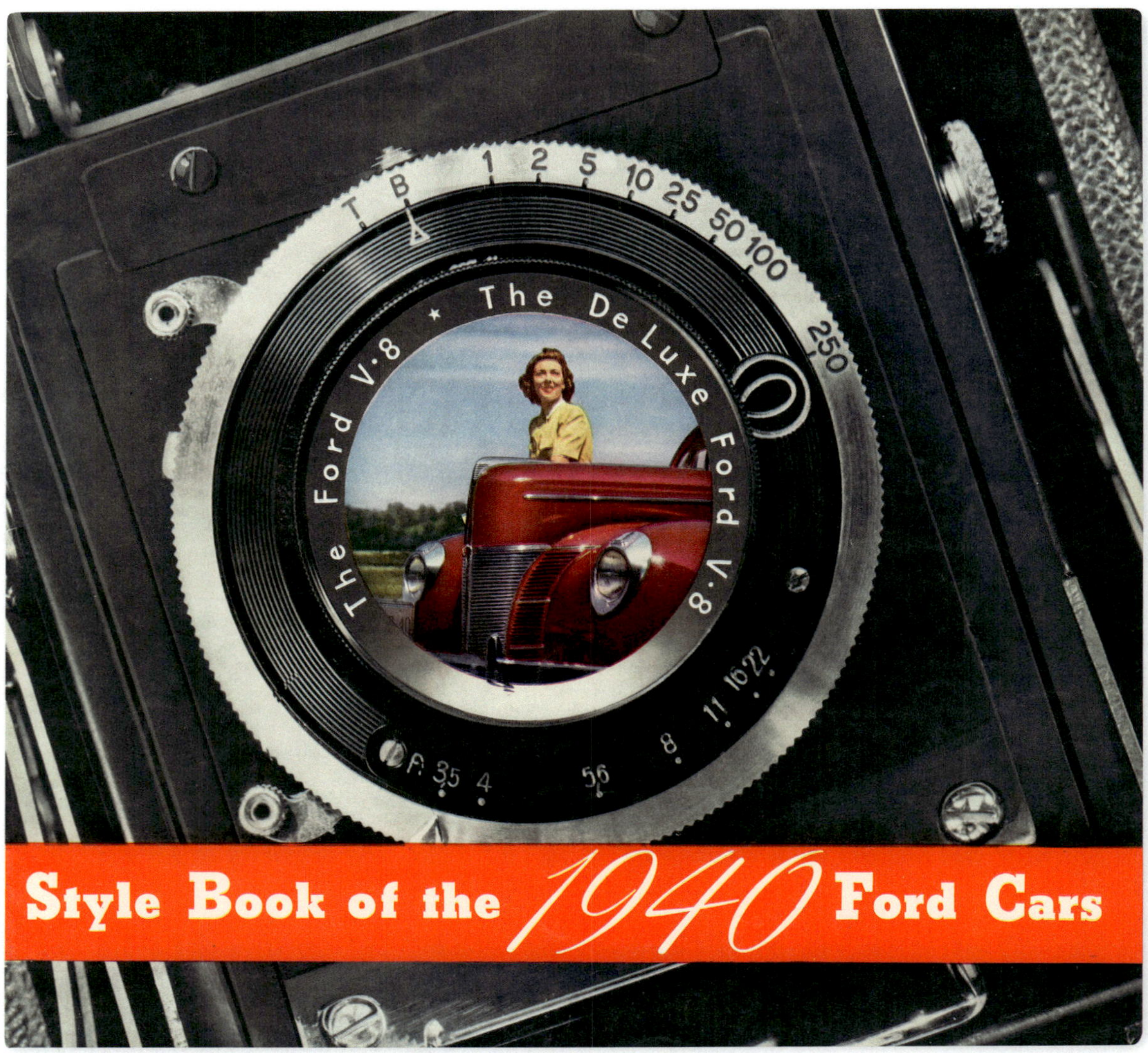

Ford 1939

These commissioned brochures, and accompanying billboards, were a standard part of Ford advertising for a generation: an almost Norman Rockwell–like illustration and a quick slogan. "Watch the Fords Go By" was by far the most famous of those sayings.

Diese im Auftrag von Ford produzierten Broschüren und die dazugehörigen Plakatwände gehörten eine Generation lang zum Standardrepertoire der Ford-Werbung: eine Abbildung im Stil von Norman Rockwell und ein einprägsamer Werbespruch. Der bei Weitem berühmteste unter ihnen war sicherlich: „Watch the Fords go by“ – „Schau, wie die Fords vorbeifahren!“

Pendant toute une génération, ces catalogues Ford et les panneaux qui les accompagnaient ont scrupuleusement respecté les codes de la stratégie de communication maison : une illustration rappelant Norman Rockwell et un slogan concis. « Watch the Fords Go By » (« Regardez passer les Ford ») était de loin la plus fameuse de ces exhortations.

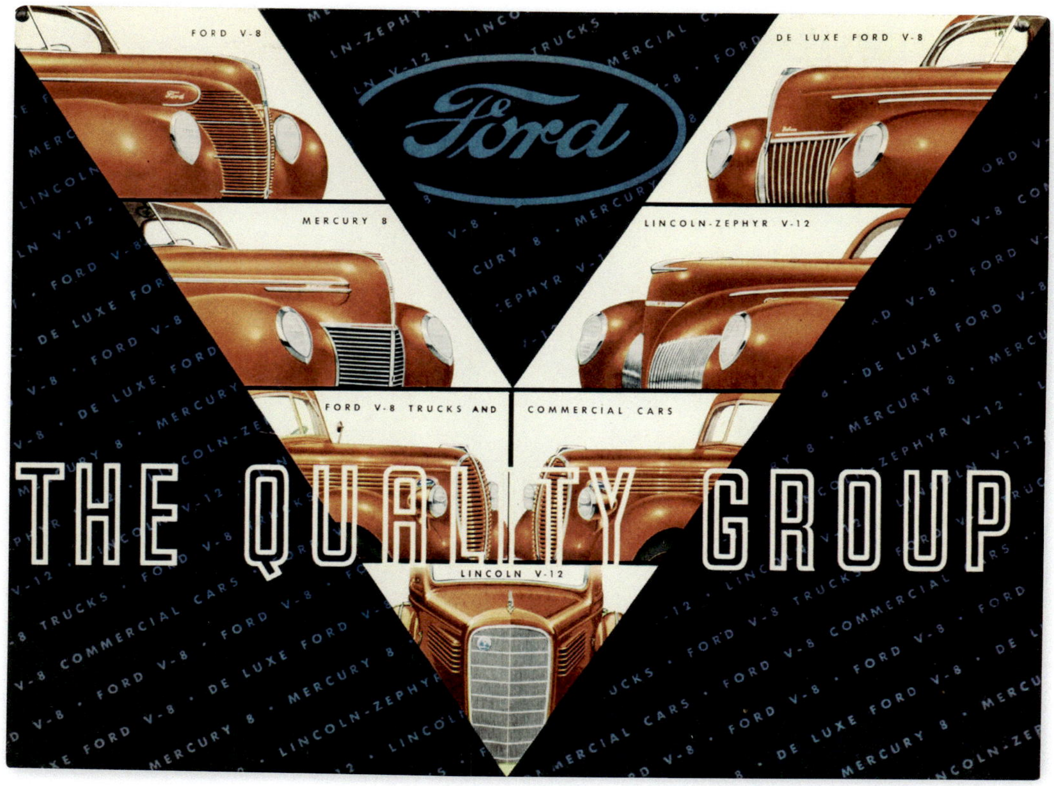

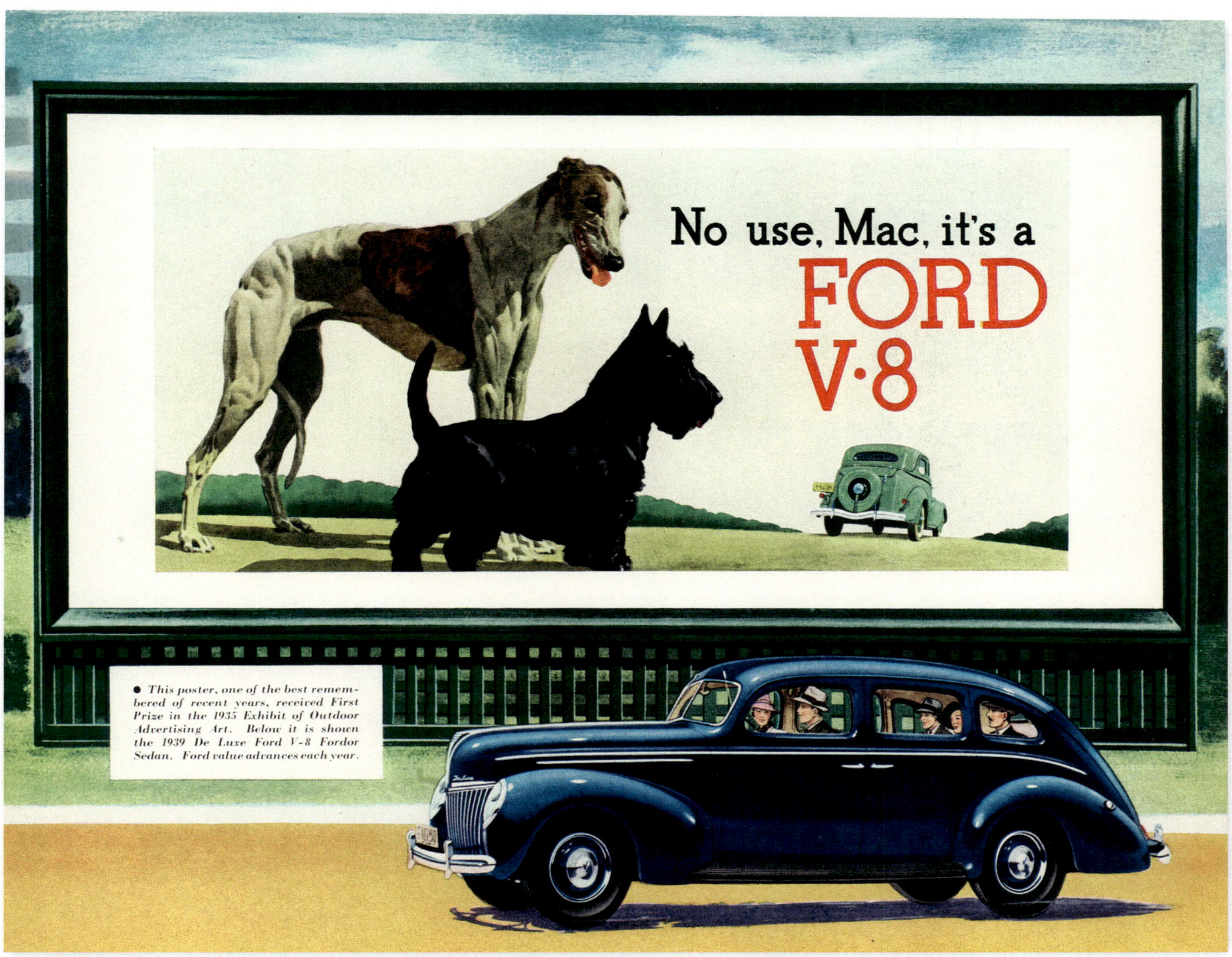

Graham

Graham 1930 *(right)*, **1933** *(below)*, **1939** *(opposite)*

The three Graham brothers (represented on the badge) built trucks before selling out to Dodge, then took over Paige. In 1930 they dropped the Paige name and continued with costly cars, some supercharged and unevenly styled despite the high-toned look of its brochure.

Die drei Graham-Brüder (abgebildet auf dem Firmenemblem) bauten Lkw, bevor sie ihr Unternehmen an Dodge verkauften und anschließend Paige übernahmen. Im Jahre 1930 ließen sie den Namen „Paige" fallen und machten mit teuren Automobilen weiter, von denen einige mit Turbomotoren ausgestattet und – trotz der hochwertig aufgemachten Broschüre – recht uneinheitlich im Styling waren.

Les trois frères Graham (représentés sur le badge) fabriquaient des camions avant de vendre leur entreprise à Dodge puis de racheter la firme Paige. En 1930, ils se délestèrent du nom de Paige pour continuer leur aventure avec des véhicules haut de gamme – dotés dans certains cas d'un compresseur centrifuge – et d'une inégale élégance, en dépit de la tonalité un rien pompeuse du catalogue.

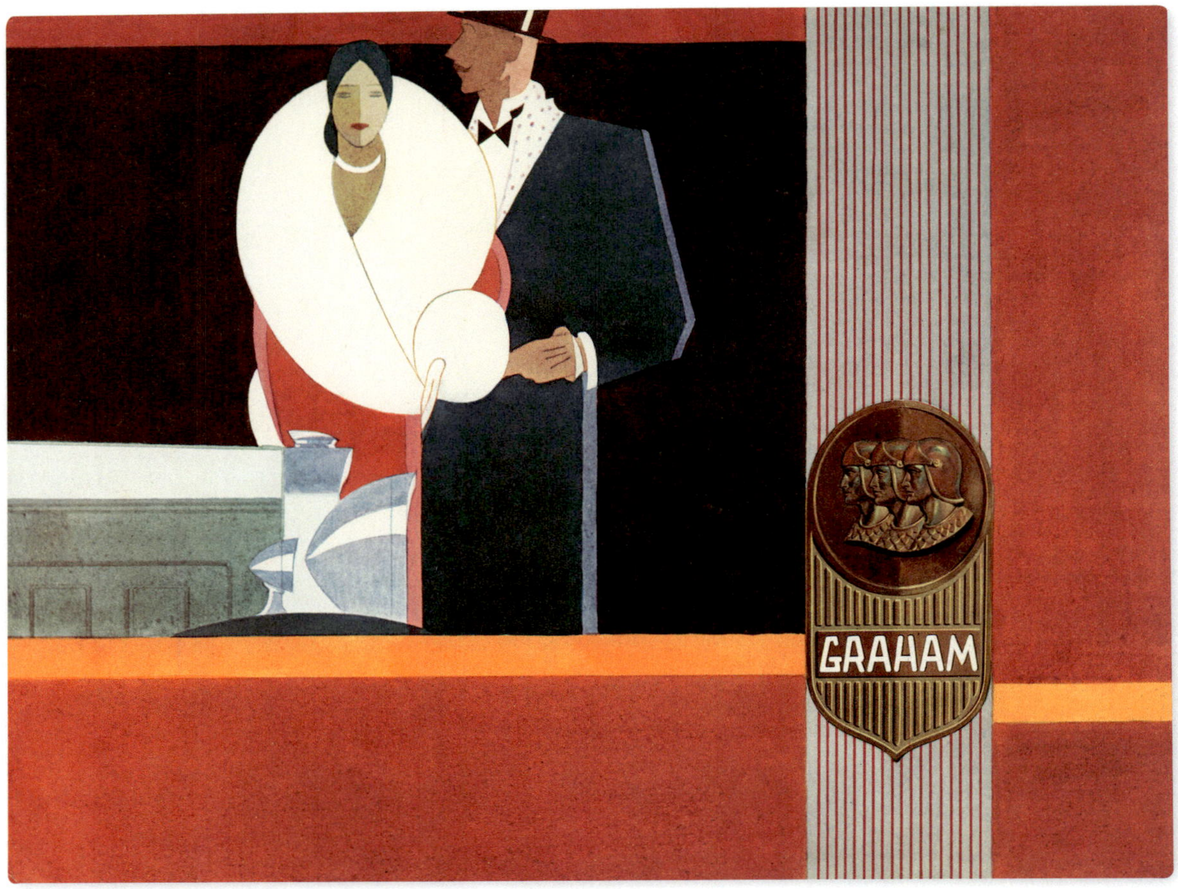

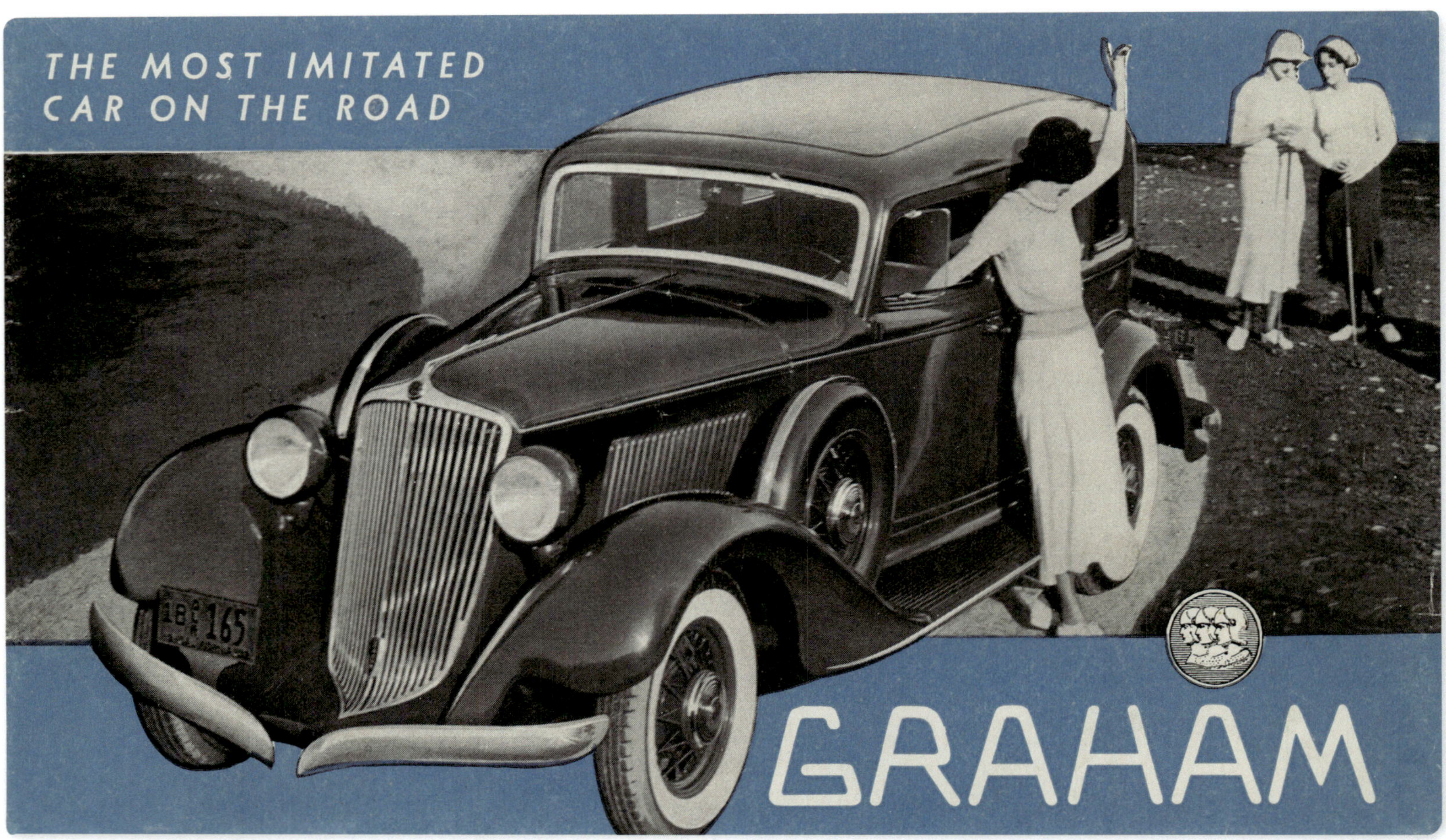

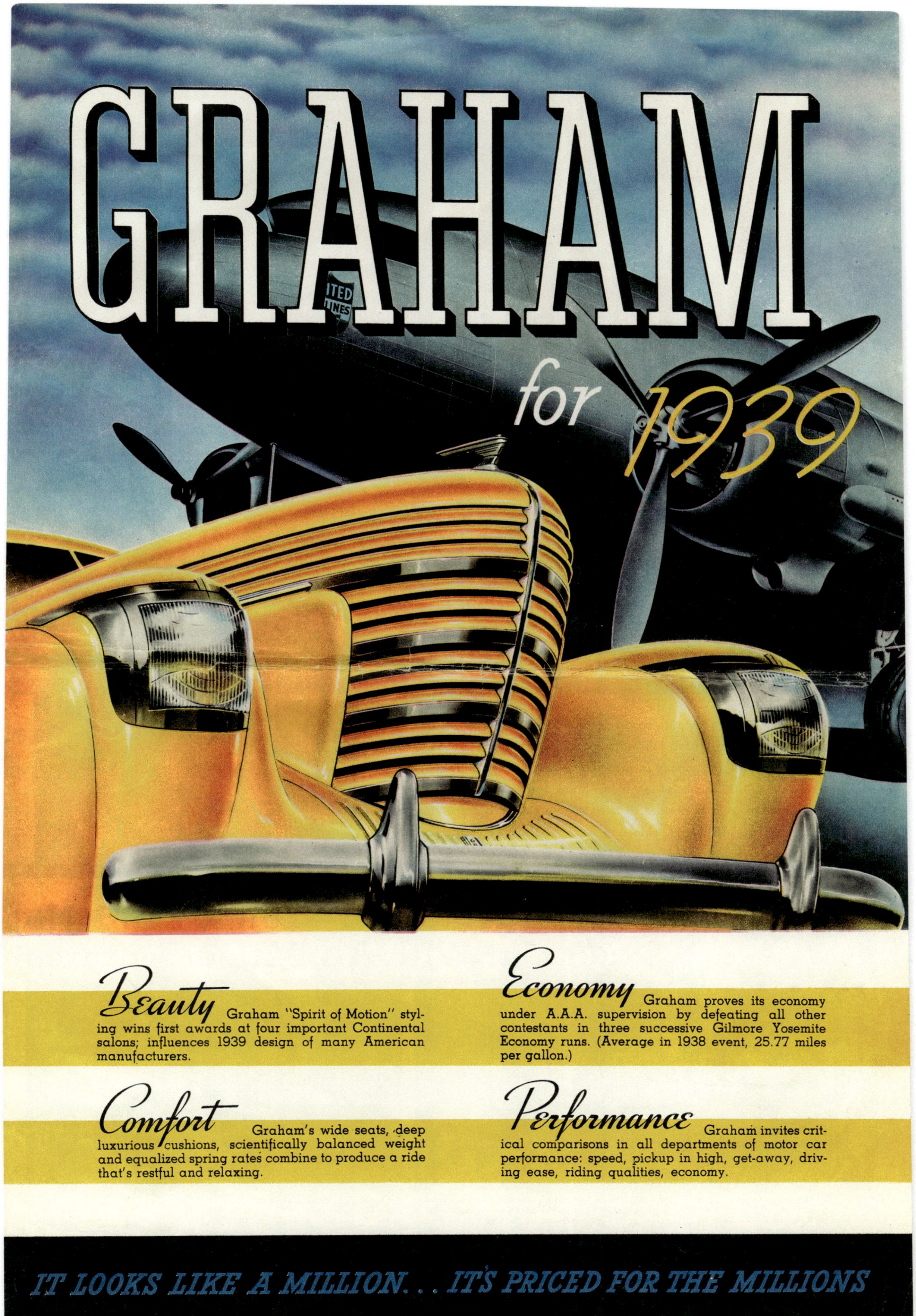
GRAHAM
for 1939
Beauty Graham "Spirit of Motion" styling wins first awards at four important Continental salons; influences 1939 design of many American manufacturers.
Economy Graham proves its economy under A.A.A. supervision by defeating all other contestants in three successive Gilmore Yosemite Economy runs. (Average in 1938 event, 25.77 miles per gallon.)
Comfort Graham's wide seats, deep luxurious cushions, scientifically balanced weight and equalized spring rates combine to produce a ride that's restful and relaxing.
Performance Graham invites critical comparisons in all departments of motor car performance: speed, pickup in high, get-away, driving ease, riding qualities, economy.
IT LOOKS LIKE A MILLION... IT'S PRICED FOR THE MILLIONS

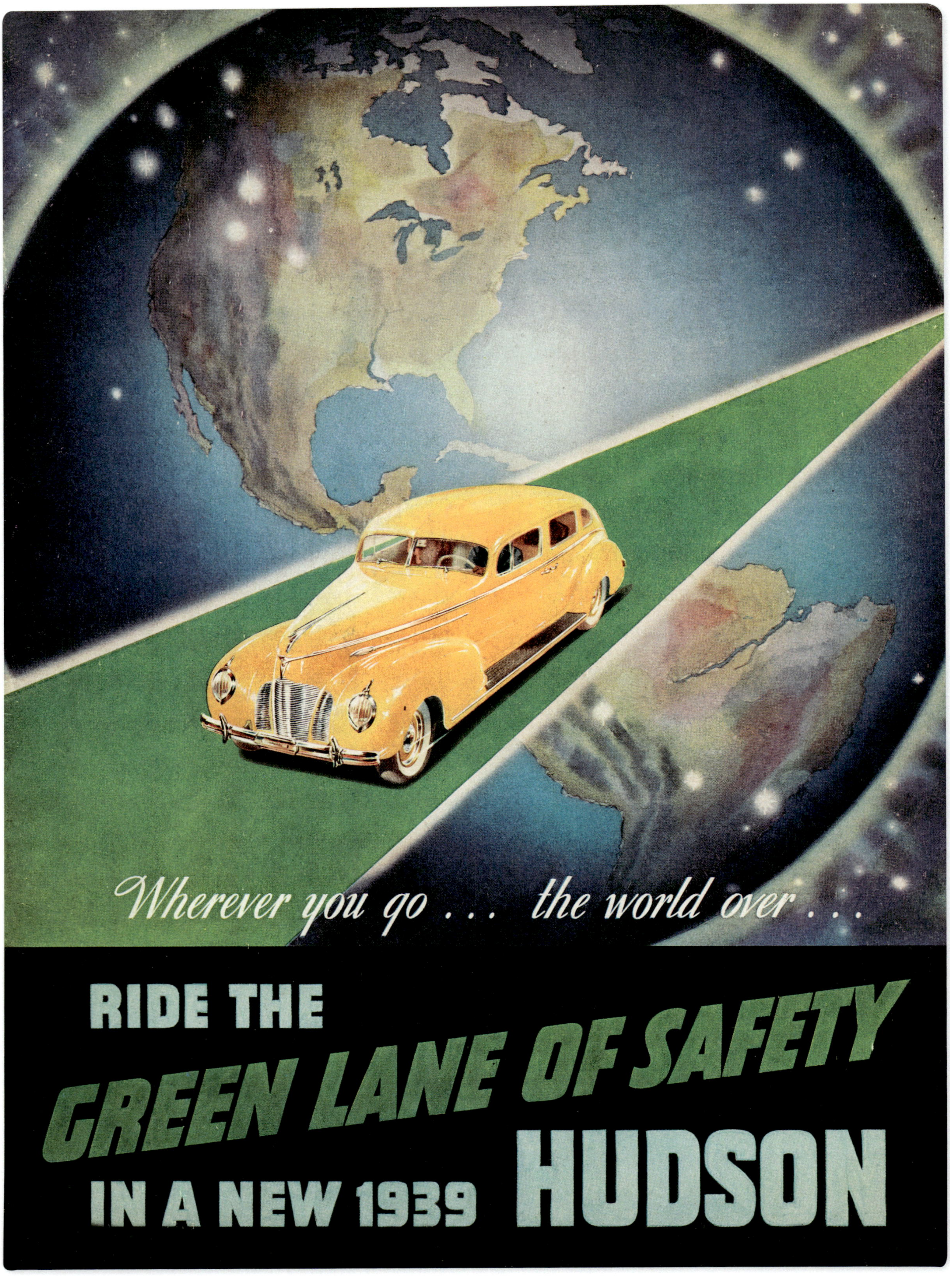

Wherever you go . . . the world over . . .
RIDE THE
GREEN LANE OF SAFETY
IN A NEW 1939 HUDSON

Hudson

Hudson 1939 *(opposite and below)*,
1938 *(right)*

The "read all about it" newspaper treatment of this ad, like its inspiration, catches the eye. The Selective Automatic Shift, earlier called the Electric Hand, is more a preselected shift than a true automatic transmission.

Die Schlagzeilenaufmachung dieser Anzeige war natürlich – ebenso wie das Auto, das sie anpries – ein Blickfang. „Selective Automatic Shift" („selektive automatische Schaltung"), früher „Electric Hand" („elektrische Hand") genannt, war eher eine Gangschaltung mit Vorwahl als ein echtes Automatikgetriebe.

Le traitement d'inspiration journalistique de cette publicité attire l'œil. Le dispositif baptisé « Selective Automatic Shift », qui s'est d'abord appelé « Electric Hand » (« la main électrique »), correspond davantage à un système de présélection des vitesses qu'à une véritable transmission automatique.

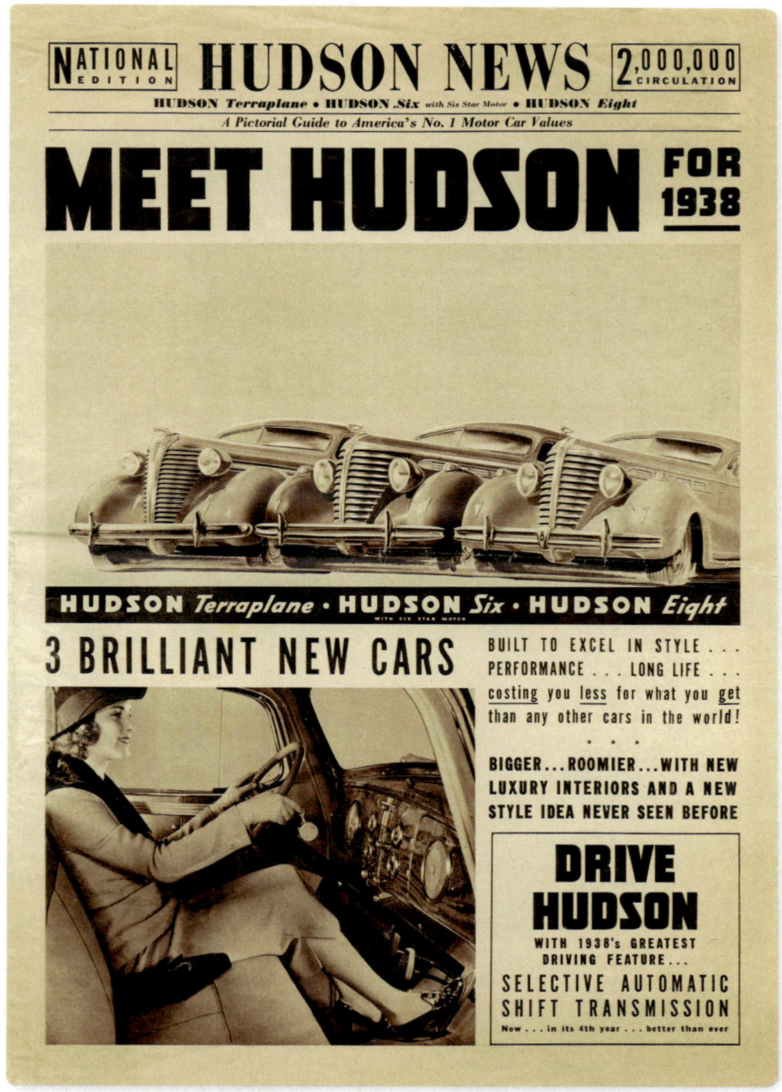

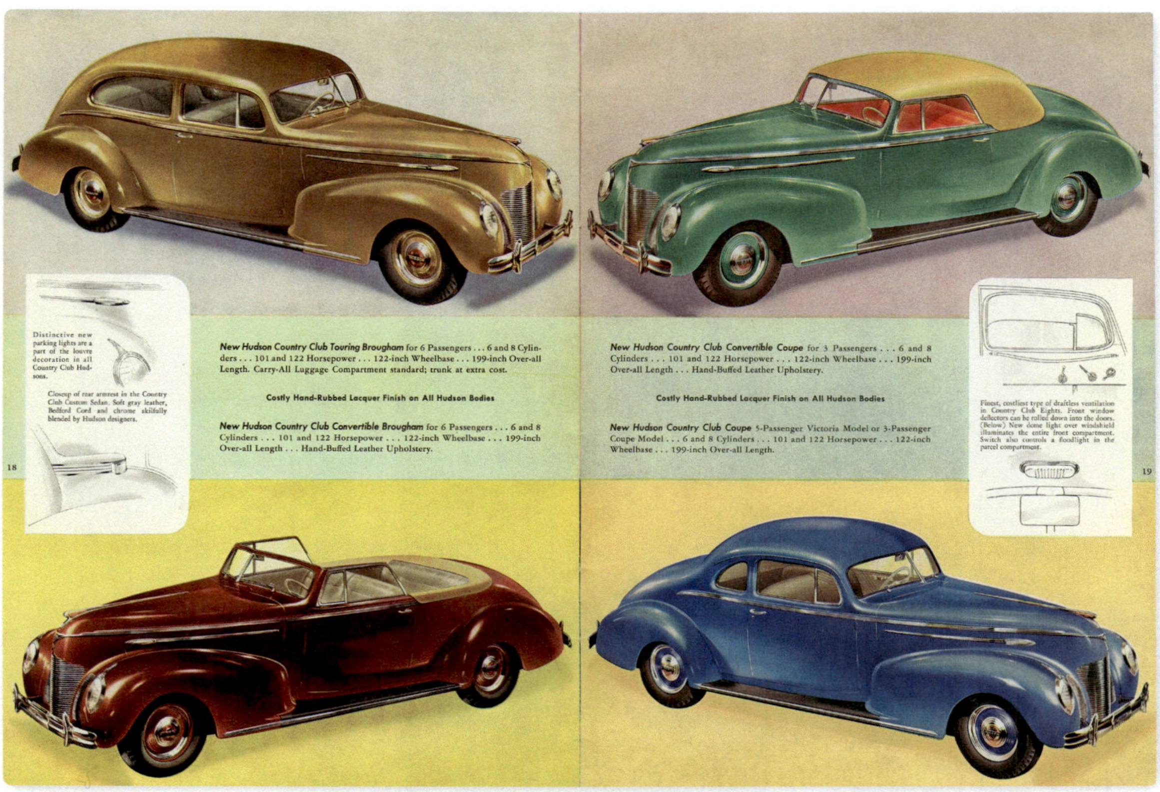

Hupmobile

Hupmobile 1934 *(right)*, **1935** *(below)*

Two of the industry's acclaimed designers, Raymond Loewy and Amos Northup, had a hand in creating the Aerodynamic Hupmobiles, whose rollout is saluted in this elegant artwork. Unfortunately, a destructive battle for control of the company was going on at the same time.

Zwei der angesehensten Konstrukteure der Branche – Raymond Loewy und Amos Northup – waren am Design der Aerodynamic Hupmobiles beteiligt, deren Markteinführung mit diesen eleganten Zeichnungen angekündigt wurde. Leider fand zur gleichen Zeit auch ein zerstörerischer Kampf um die Vorherrschaft im Unternehmen statt.

Deux des plus importants designers du monde de l'automobile, Raymond Loewy et Amos Northup, ont participé à la création des Hupmobile Aerodynamic, dont le lancement est salué par cette élégante illustration. Hélas, une bataille destructrice pour le contrôle de l'entreprise faisait rage au même moment.

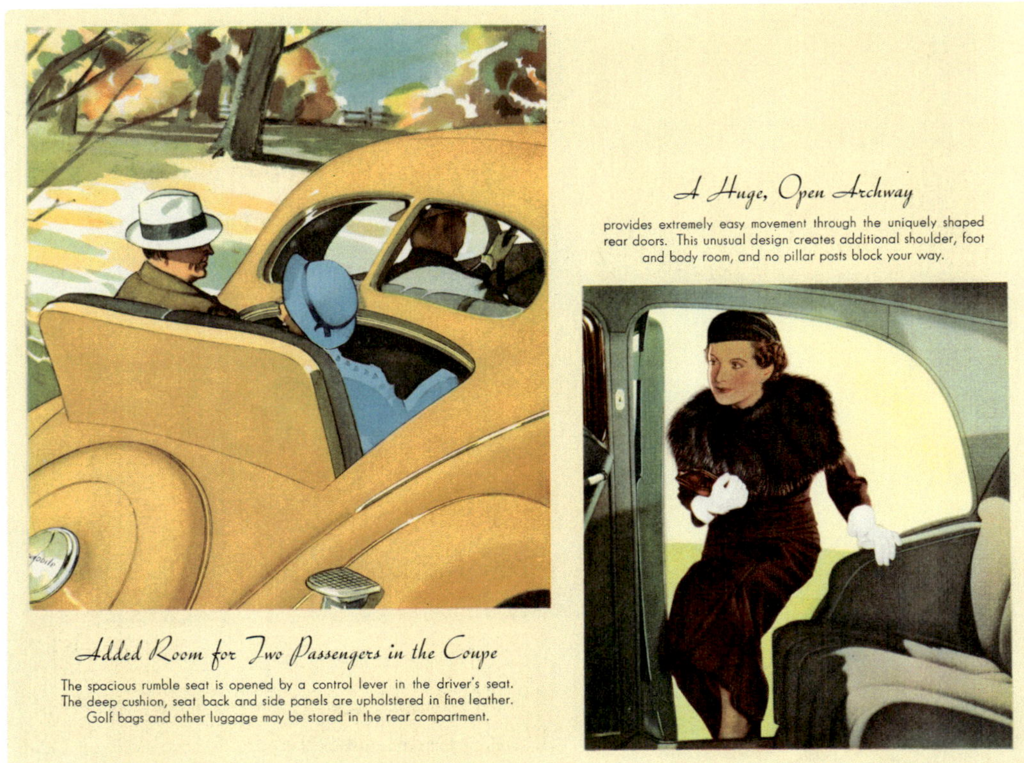

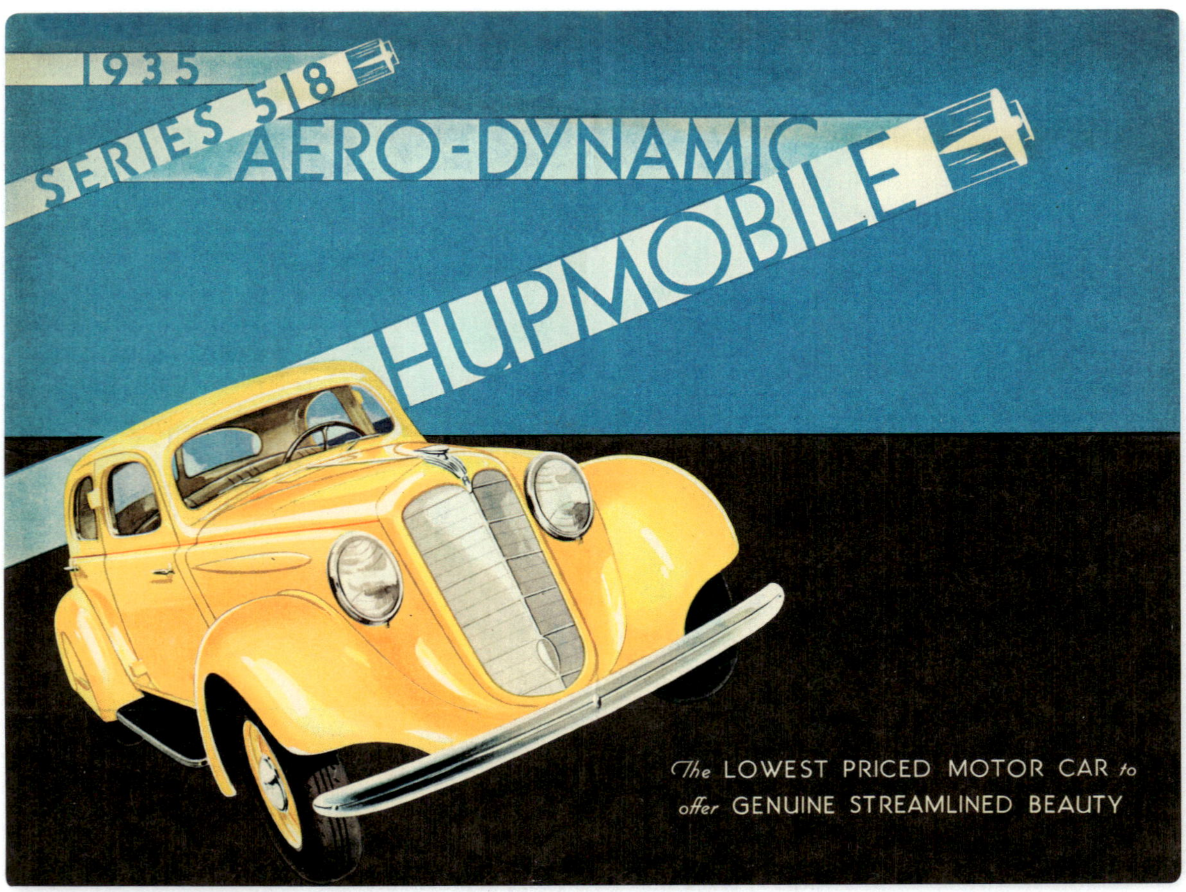

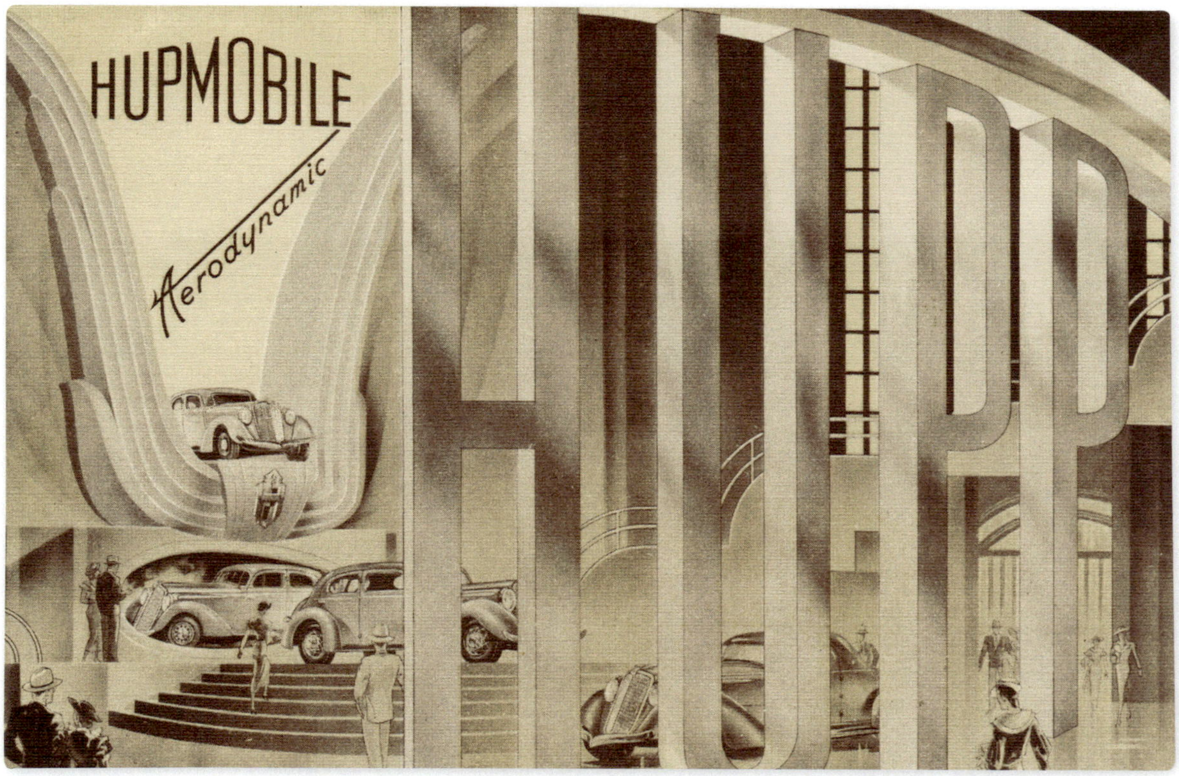

Hupmobile 1934 *(above)*, **1935** *(right)*

Hupmobile suspended production in 1935, out of funds. Desperate, the firm acquired the 810/812 body dies from the recently departed Cord to build the 1940 Skylark, its final car. Graham used the same dies — and also closed in 1940.

Mangels Kapitals musste Hupmobile die Produktion 1935 vorübergehend einstellen. In ihrer Verzweiflung erwarb die Firma von dem kürzlich geschlossenen Unternehmen Cord die Formen für die Karosserie 810/812, um den 1940er-Skylark zu bauen, ihr letztes Modell. Graham verwendete die gleichen Pressformen – und stellte ebenfalls 1940 die Produktion ein.

À court de trésorerie, Hupmobile suspendit sa production en 1935. Aux abois, la firme fit l'acquisition des moules des Cord 810/812 après la disparition récente de la marque afin de construire la Skylark de 1940, qui sera sa dernière voiture. Graham utilisa les mêmes moules et mit également la clé sous la porte en 1940.

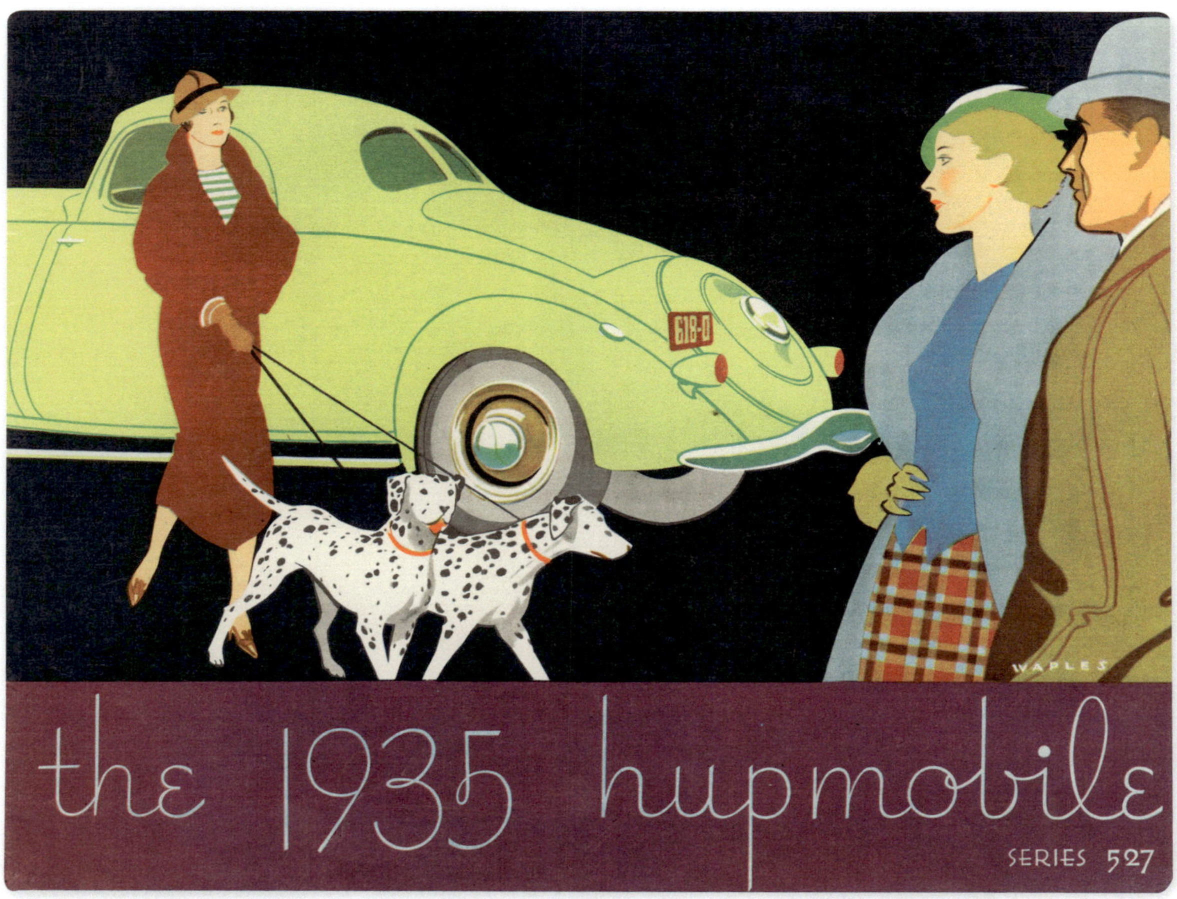

LaSalle

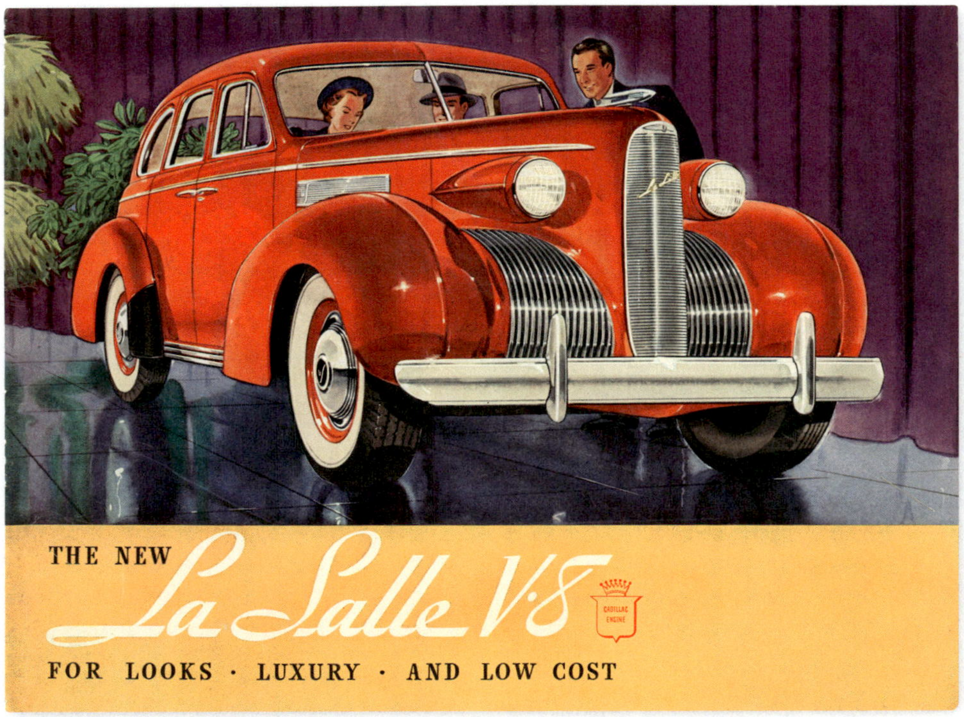

LaSalle 1939 *(this spread)*

Despite a short life, from 1927 through 1940, the LaSalle is an important car in General Motors' history. A brighter, less expensive Cadillac, it was the first "styled" car created by Harley Earl's new staff. The 1939 models offered a sliding steel sunroof.

Trotz seiner kurzen Lebensspanne – von 1927 bis 1940 – war der LaSalle ein bedeutendes Automobil in der Geschichte von General Motors. Der grellere und preisgünstigere Cadillac war das erste Fahrzeug, dessen „Styling" von der neuen Mannschaft um Harley Earl stammte. Die Modelle des Jahres 1939 waren mit stählernen Schiebedächern ausgestattet.

Malgré son existence éphèmère, de 1927 à 1940, cette LaSalle est restée comme une voiture importante dans l'histoire de la General Motors. Envisagée comme une Cadillac plus clinquante et moins chère, elle fut la première automobile « stylée » créée par la nouvelle équipe d'Harley Earl. Les modèles de 1939 proposaient un toit ouvrant coulissant en acier.

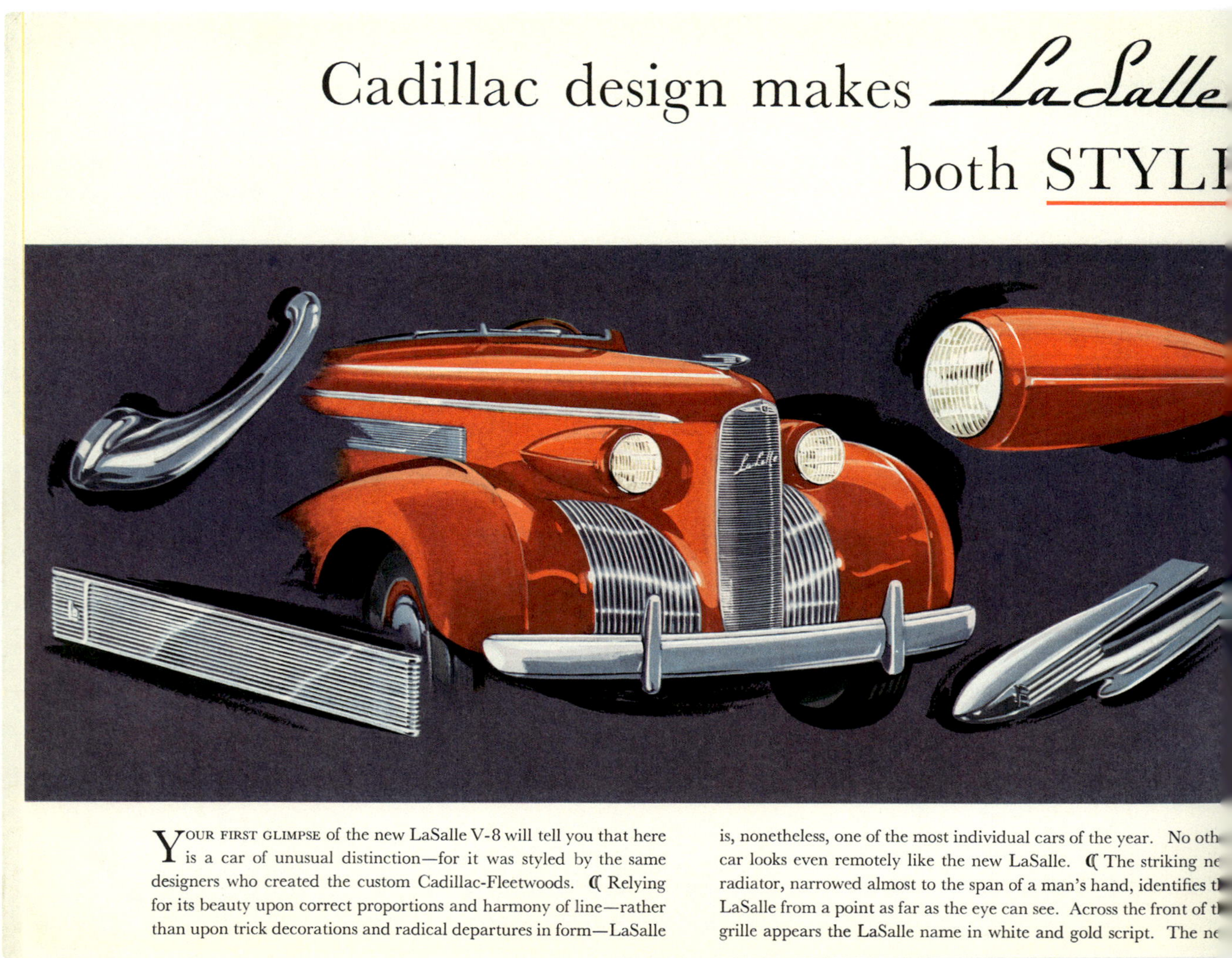

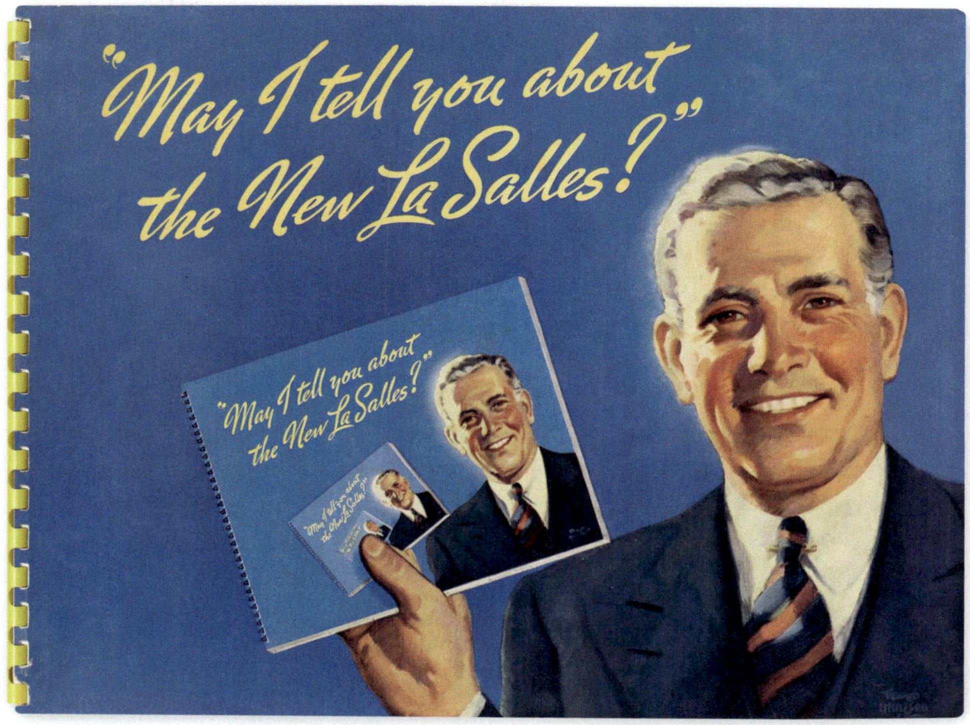

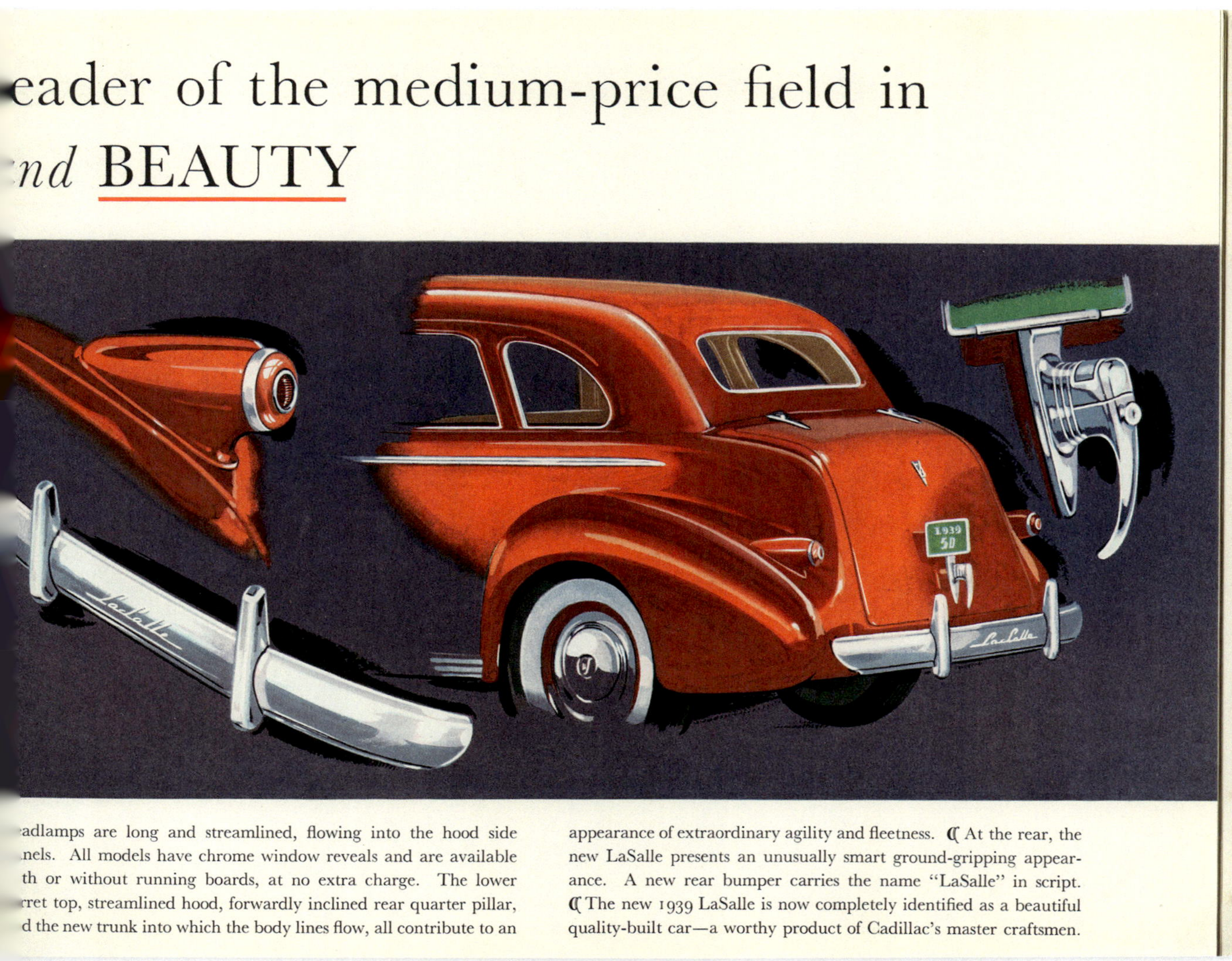
eader of the medium-price field in
nd BEAUTY

adlamps are long and streamlined, flowing into the hood side nels. All models have chrome window reveals and are available th or without running boards, at no extra charge. The lower rret top, streamlined hood, forwardly inclined rear quarter pillar, d the new trunk into which the body lines flow, all contribute to an appearance of extraordinary agility and fleetness. ℭ At the rear, the new LaSalle presents an unusually smart ground-gripping appearance. A new rear bumper carries the name "LaSalle" in script. ℭ The new 1939 LaSalle is now completely identified as a beautiful quality-built car—a worthy product of Cadillac's master craftsmen.

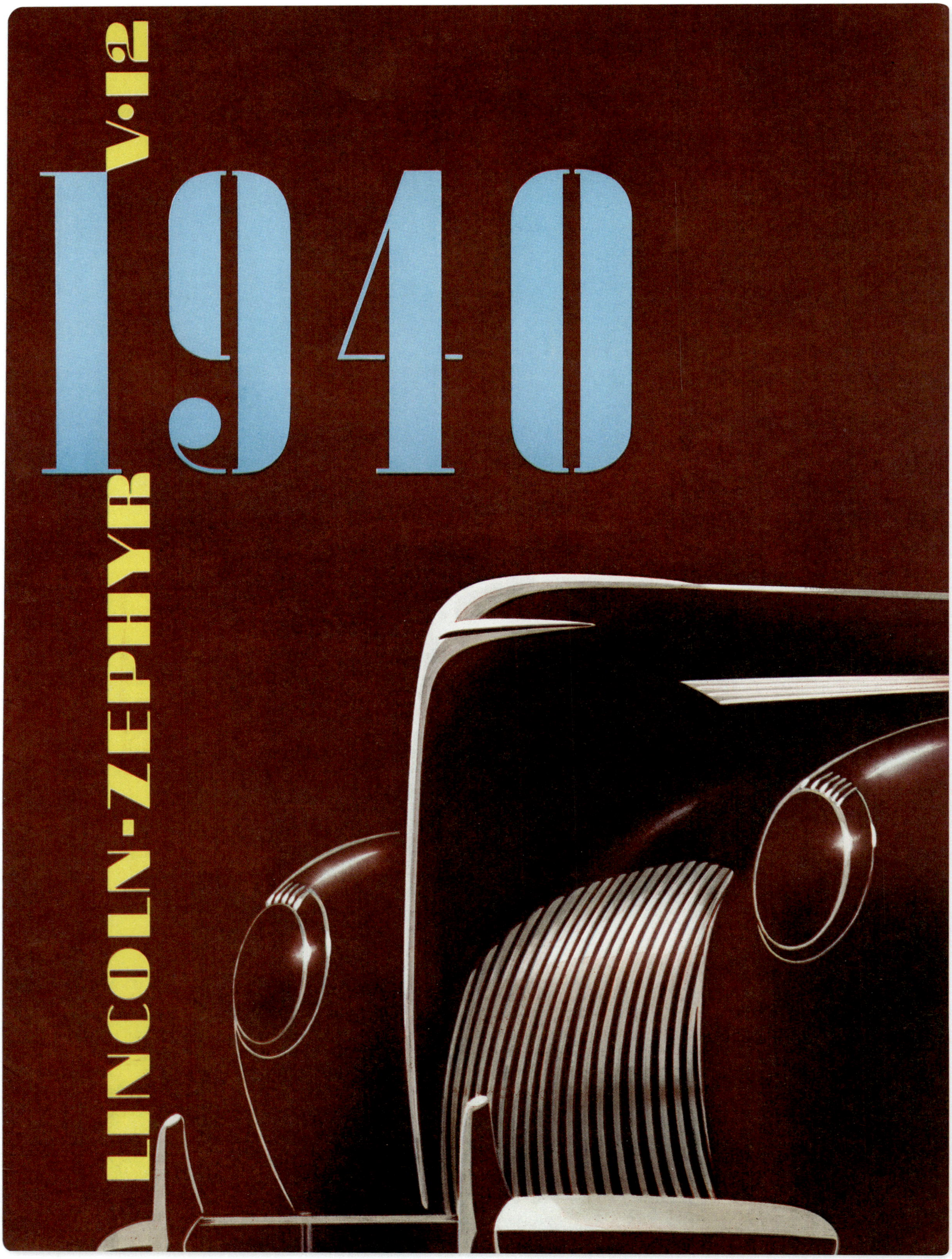
1940
V-12
LINCOLN-ZEPHYR

Lincoln

Lincoln Zephyr 1940 *(opposite)*, **1936** *(right)*, **1933** *(below)*

This revolutionary car was a benchmark 1930s design, with a radical teardrop shape, front and rear, along with a horizontal-bar grille that resembled a liner's prow cutting through the waves. Lincoln was originally an independent brand, founded in 1917 by the machining genius Henry M. Leland, but was acquired by the Ford Motor Company in 1922.

Mit seiner radikalen Tropfenform – vorn wie hinten – und einem Kühlergrill mit waagerechten Lamellen, der an einen Ozeanriesen erinnerte, wie er durch die Wellen pflügte, setzte das Design dieses revolutionären Fahrzeugs in den 1930er-Jahren Maßstäbe. Lincoln war ursprünglich eine eigenständige Marke, die 1917 von dem genialen Mechaniker und Maschinenbauer Henry M. Leland gegründet worden war, 1922 aber an die Ford Motor Company veräußert wurde.

Cette voiture révolutionnaire fait figure de référence dans le design des années 1930, avec sa forme radicale en goutte d'eau – face avant et arrière – et sa paire de grilles de radiateur qui évoque la proue d'un paquebot fendant les flots. À l'origine Lincoln était une marque indépendante, fondée en 1917 par le génie de la mécanique Henry M. Leland, avant d'être rachetée par Ford en 1922.

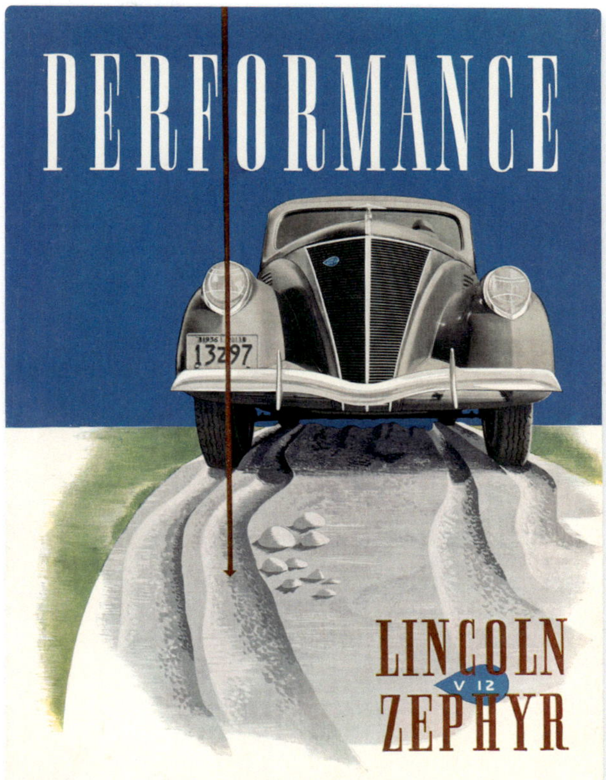

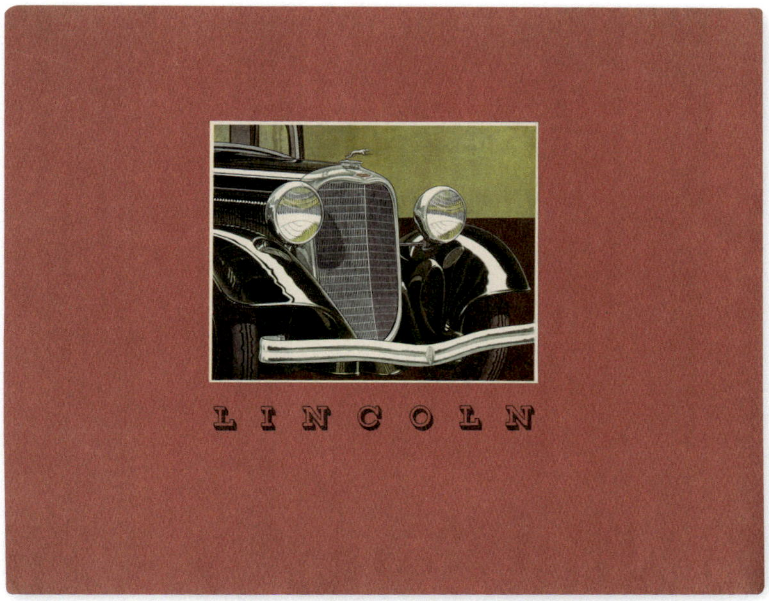

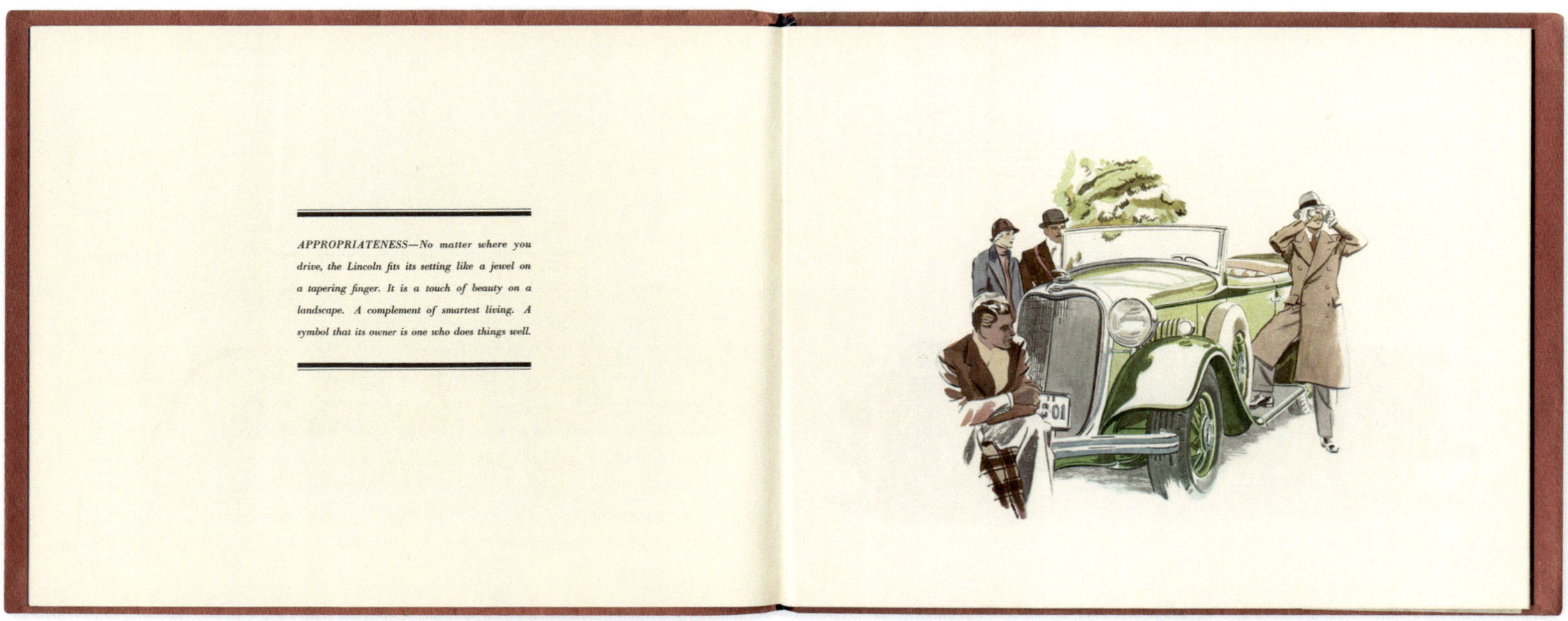

Marmon

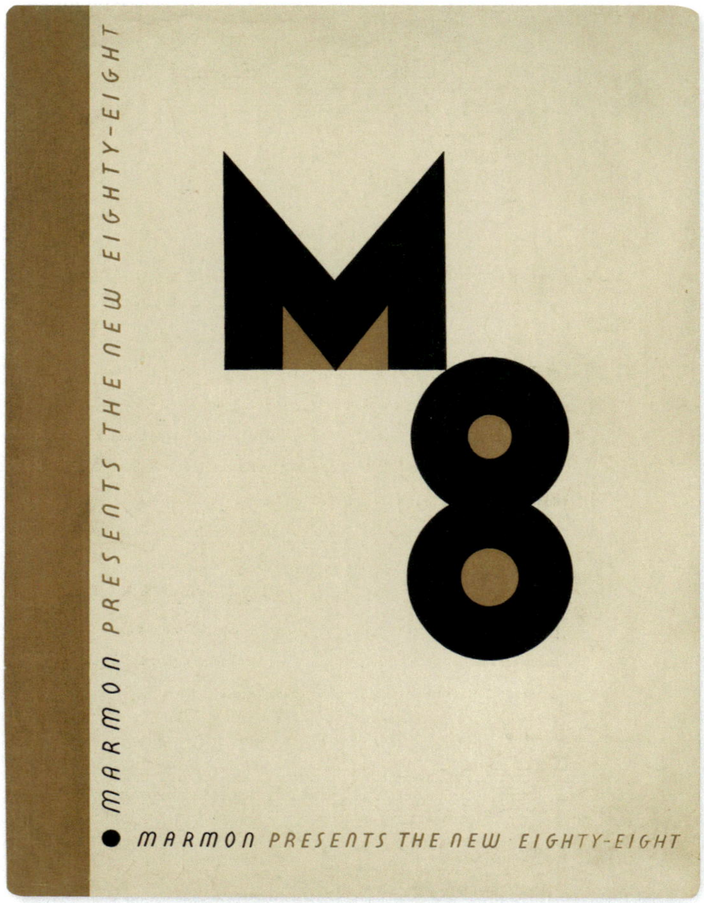

Marmon 1931, 1930 *(opposite)*

Proudly from Indiana, where it won the first Indianapolis 500, Marmon was an august car. The Model 70 wasn't small, given its straight-eight power, but the Model 88 had a longer wheelbase. The simple-but-dramatic handouts don't hint that in 1931 Marmon sales would plummet about 70 percent, leading to its demise in 1933, despite the introduction of a massive 16-cylinder engine.

Marmon, der stolz aus Indiana kam, wo er das erste 500-Meilen-Rennen von Indianapolis gewonnen hatte, war ein illustres Fahrzeug. Das Model 70 war schon nicht klein, denn er brauchte Platz für seinen Achtzylinderreihenmotor, aber das Model 88 besaß einen noch längeren Radstand. Die schlichten, aber eindrucksvollen Informationsblätter ließen noch nicht ahnen, dass der Absatz 1931 um 70 Prozent zurückgehen würde, was 1933 zum Untergang führte – trotz der Einführung eines klotzigen Sechzehnzylindermotors.

Fière de sa naissance dans l'Indiana, où elle remporta les premiers 500 miles d'Indianapolis, la Marmon était une voiture des plus prestigieuses. Le modèle 70 prenait certes un peu de place, compte tenu du 8 cylindres en ligne qui le motorisait, mais la version 88 affichait un empattement plus long. Cette documentation qui réussit à conjuguer sobriété et sensations ne dit pas qu'en 1931 les ventes de la marque plongeront de 70%, ce qui conduira à son déclin en 1933 malgré l'introduction d'un moteur à seize cylindres particulièrement massif.

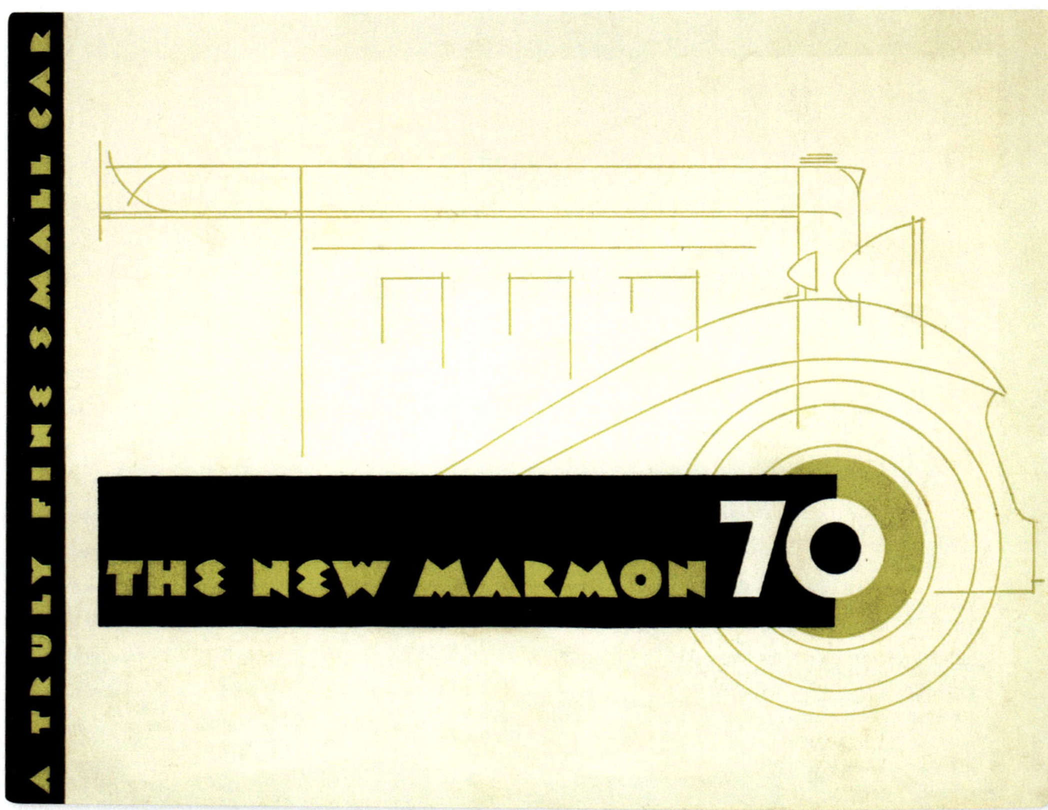

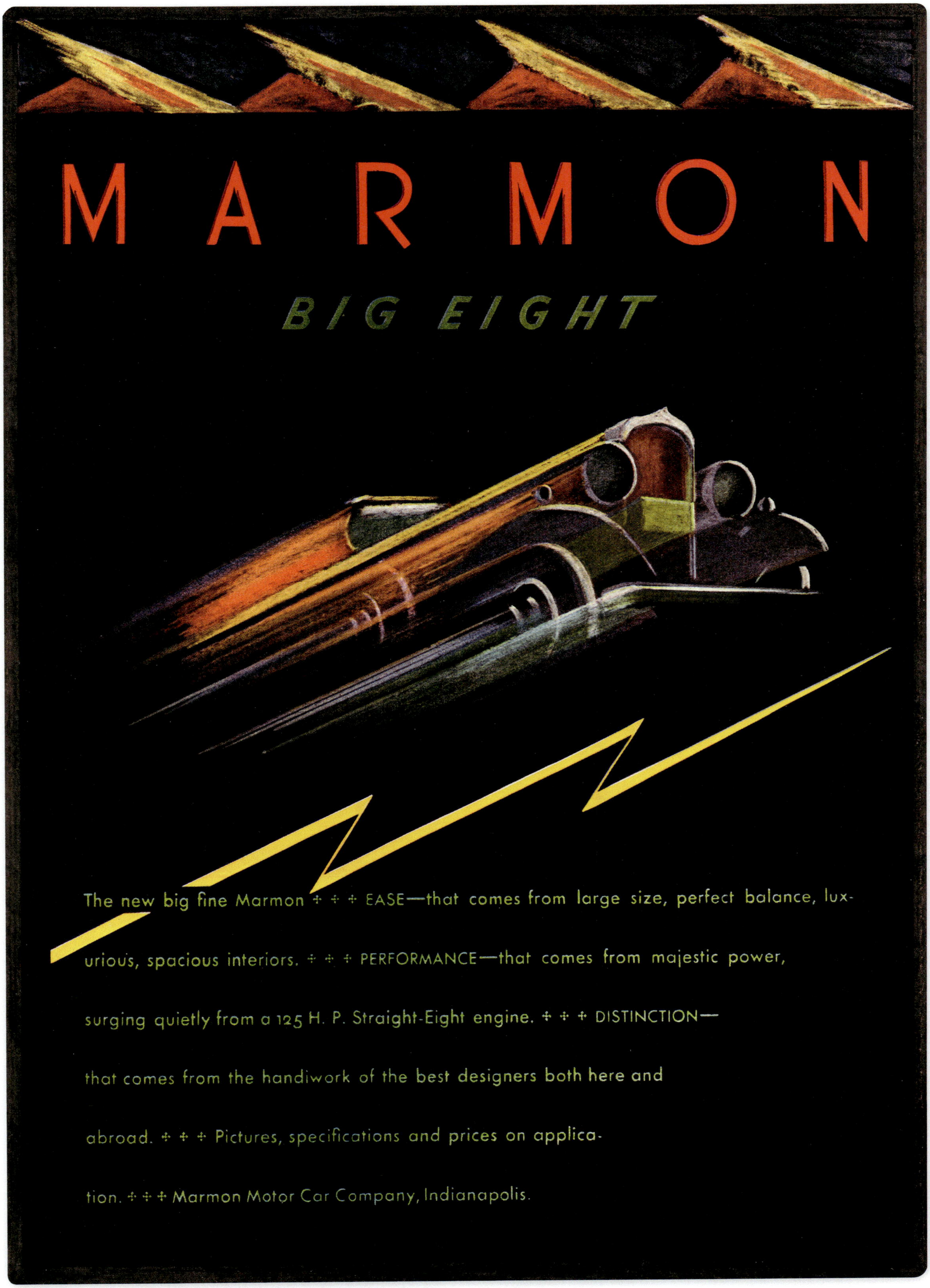
MARMON
BIG EIGHT
The new big fine Marmon ∻ ∻ ∻ EASE—that comes from large size, perfect balance, luxurious, spacious interiors. ∻ ∻ ∻ PERFORMANCE—that comes from majestic power, surging quietly from a 125 H. P. Straight-Eight engine. ∻ ∻ ∻ DISTINCTION—that comes from the handiwork of the best designers both here and abroad. ∻ ∻ ∻ Pictures, specifications and prices on application. ∻ ∻ ∻ Marmon Motor Car Company, Indianapolis.

Mercury

Mercury 1940 *(right)*, **1939** *(below)*, **1940** *(opposite)*

Here is a simple idea that worked: a slightly upscale Ford with more horsepower that could compete against other moderately priced entries like Oldsmobile or Dodge. The Mercury was an immediate success. The artwork shows the tight rear-wheel openings that identified its early models.

Hier sieht man eine einfache Idee, die funktionierte: ein leicht aufgepeppter, etwas leistungsstärkerer Ford, der sich mit Konkurrenten wie Oldsmobile oder Dodge in den niedrigen Preislagen messen konnte. Der Mercury war von Anfang an ein Erfolg. Die Zeichnung zeigt die knapp bemessene hintere Radöffnung, an der die frühen Modelle gut zu erkennen waren.

Voilà une idée toute simple et qui a formidablement bien marché : une Ford légèrement haut de gamme et gonflée en chevaux vapeur, capable de rivaliser avec d'autres concurrents pas trop onéreux comme Oldsmobile ou Dodge. La Mercury a été un succès immédiat.
Le dessin montre les passages de roue arrière resserrés, véritable signature des premiers modèles.

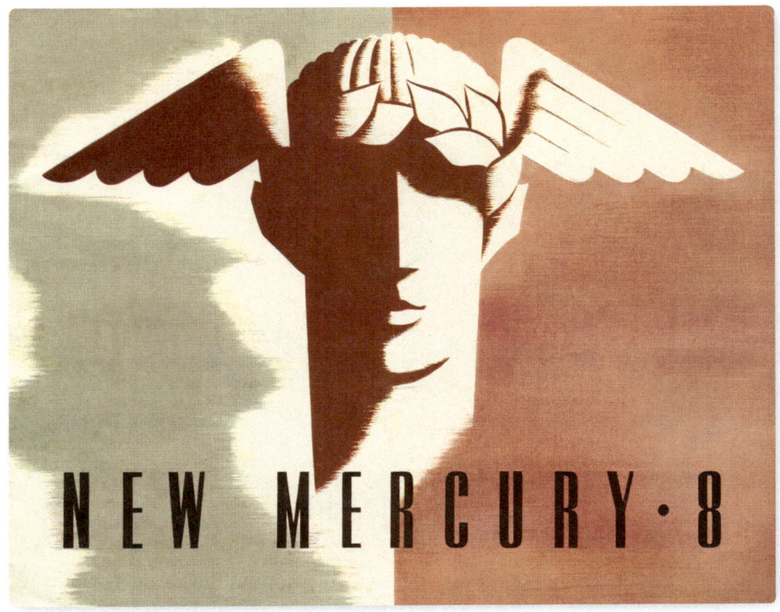

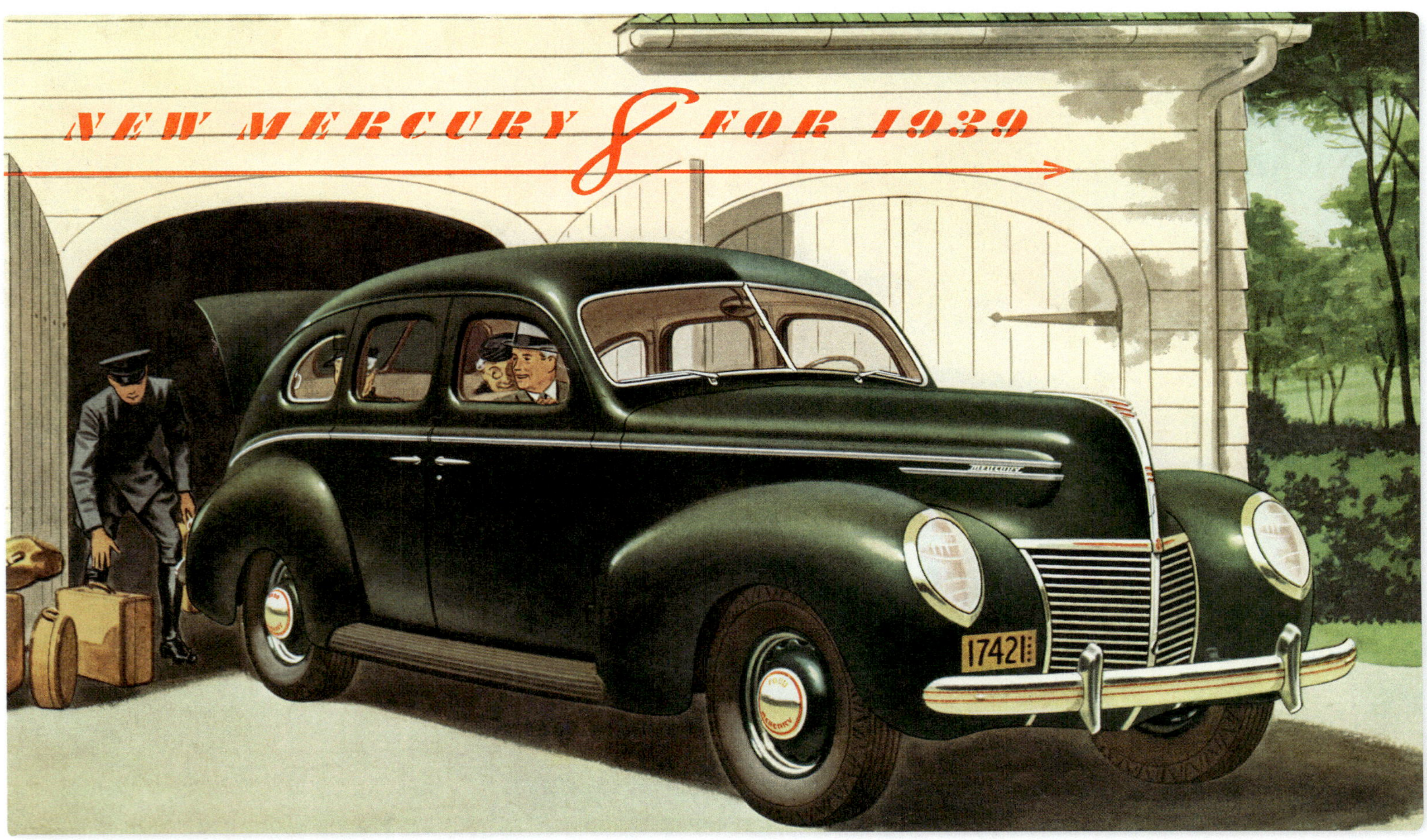

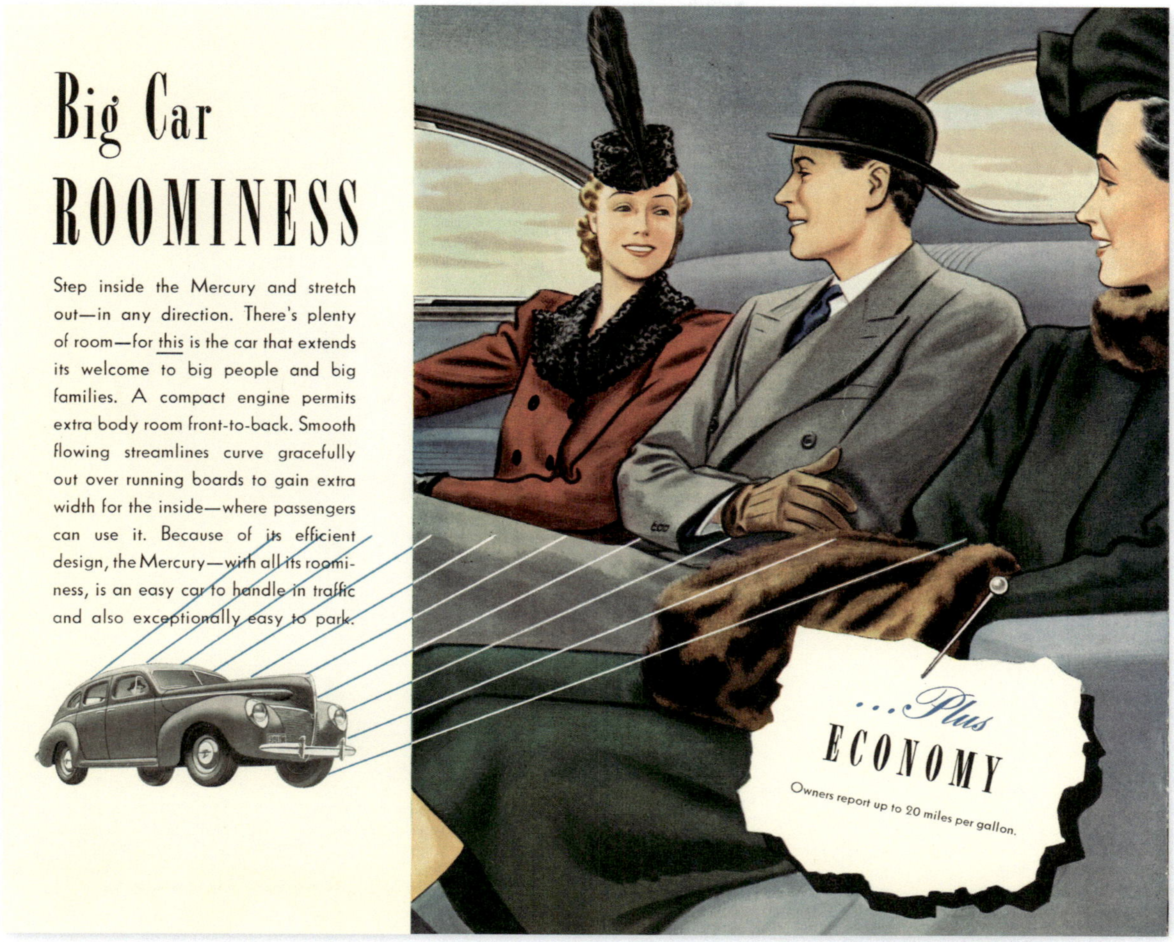
Big Car
ROOMINESS
Step inside the Mercury and stretch out—in any direction. There's plenty of room—for this is the car that extends its welcome to big people and big families. A compact engine permits extra body room front-to-back. Smooth flowing streamlines curve gracefully out over running boards to gain extra width for the inside—where passengers can use it. Because of its efficient design, the Mercury—with all its roominess, is an easy car to handle in traffic and also exceptionally easy to park.
...Plus
ECONOMY
Owners report up to 20 miles per gallon.

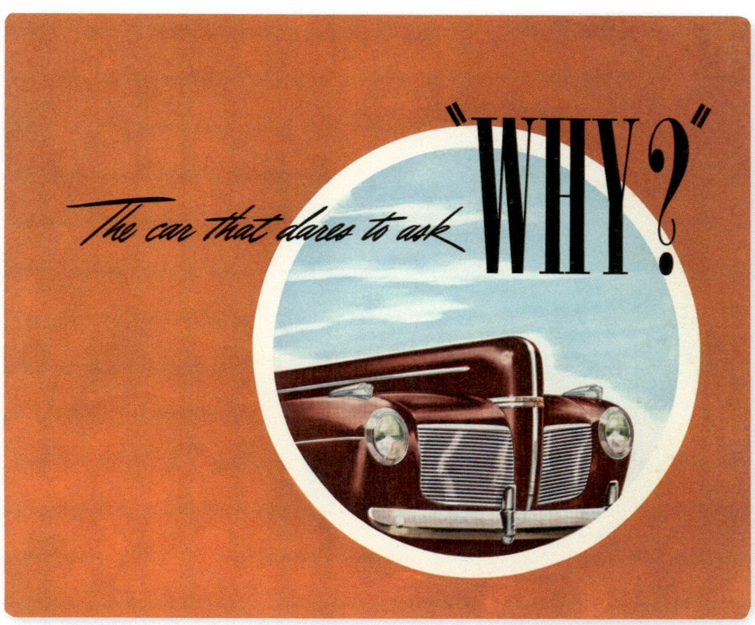
The car that dares to ask "WHY?"

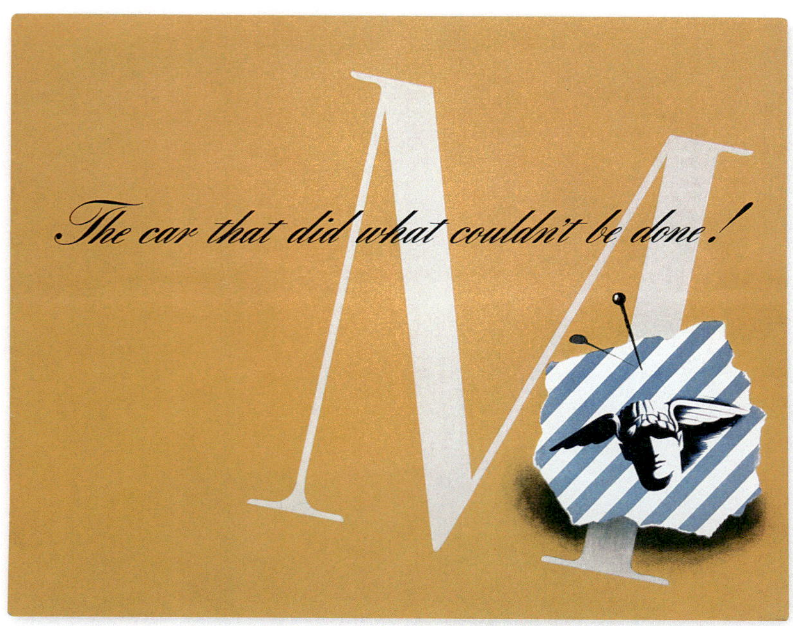
The car that did what couldn't be done!

Nash

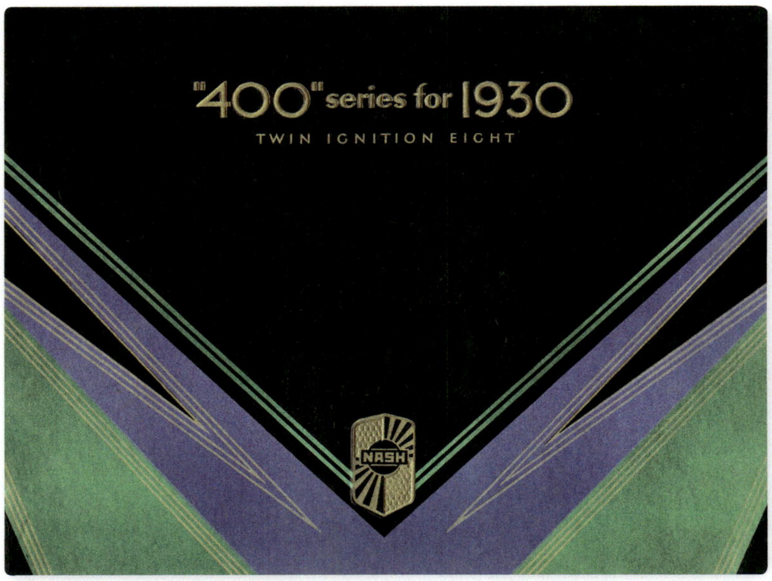

"400" series for 1930
TWIN IGNITION EIGHT
NASH

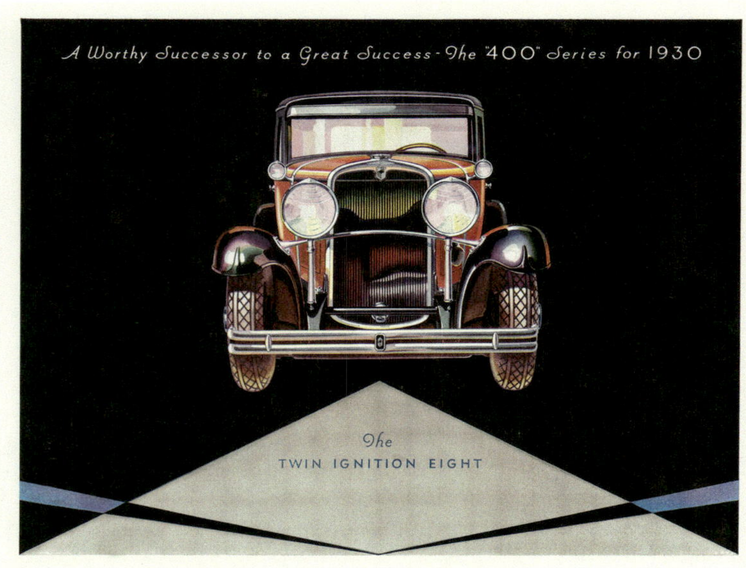

A Worthy Successor to a Great Success - The "400" Series for 1930
The
TWIN IGNITION EIGHT

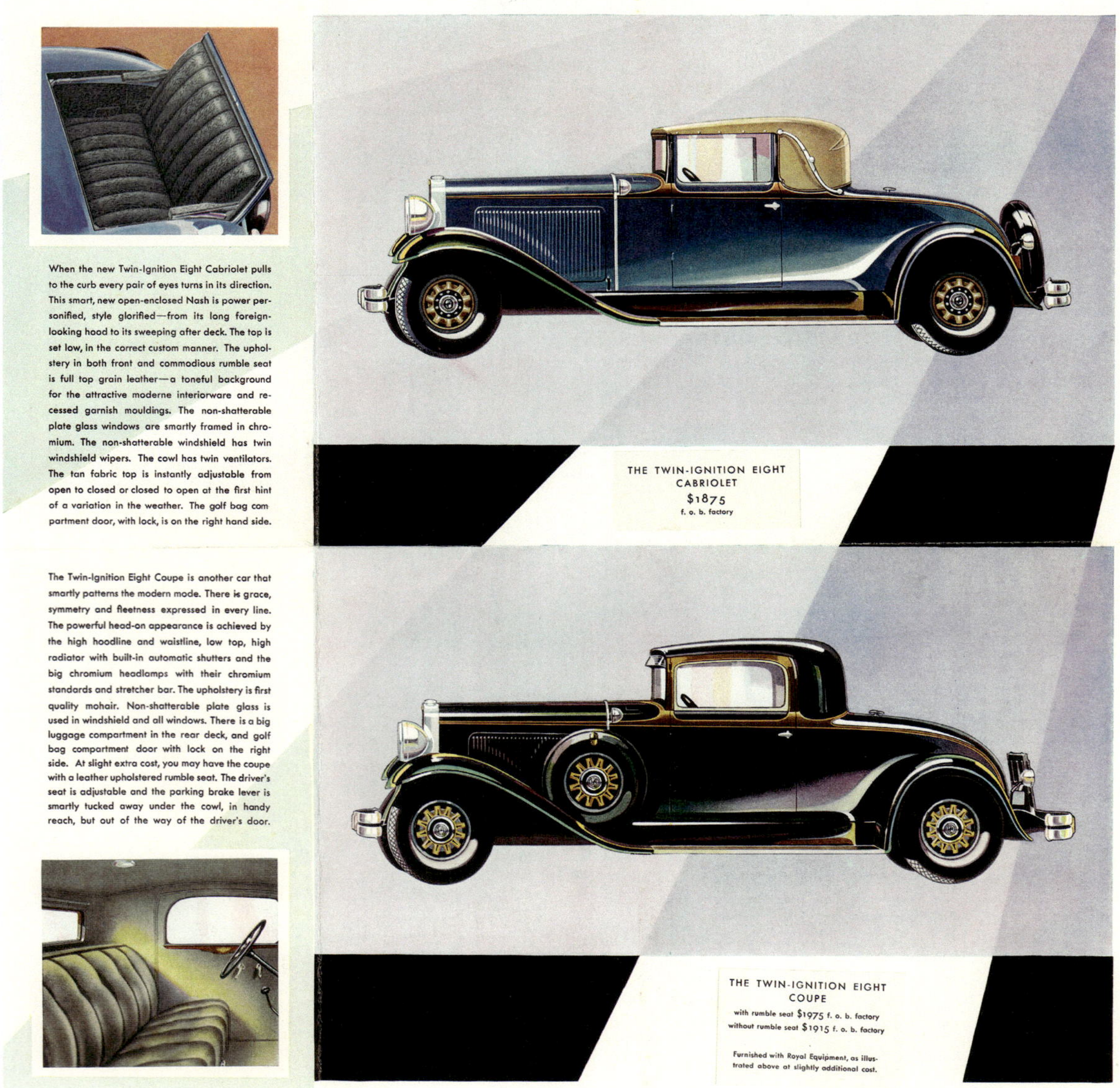

When the new Twin-Ignition Eight Cabriolet pulls to the curb every pair of eyes turns in its direction. This smart, new open-enclosed Nash is power personified, style glorified—from its long foreign-looking hood to its sweeping after deck. The top is set low, in the correct custom manner. The upholstery in both front and commodious rumble seat is full top grain leather—a toneful background for the attractive moderne interiorware and recessed garnish mouldings. The non-shatterable plate glass windows are smartly framed in chromium. The non-shatterable windshield has twin windshield wipers. The cowl has twin ventilators. The tan fabric top is instantly adjustable from open to closed or closed to open at the first hint of a variation in the weather. The golf bag compartment door, with lock, is on the right hand side.
THE TWIN-IGNITION EIGHT
CABRIOLET
$1875
f. o. b. factory
The Twin-Ignition Eight Coupe is another car that smartly patterns the modern mode. There is grace, symmetry and fleetness expressed in every line. The powerful head-on appearance is achieved by the high hoodline and waistline, low top, high radiator with built-in automatic shutters and the big chromium headlamps with their chromium standards and stretcher bar. The upholstery is first quality mohair. Non-shatterable plate glass is used in windshield and all windows. There is a big luggage compartment in the rear deck, and golf bag compartment door with lock on the right side. At slight extra cost, you may have the coupe with a leather upholstered rumble seat. The driver's seat is adjustable and the parking brake lever is smartly tucked away under the cowl, in handy reach, but out of the way of the driver's door.
THE TWIN-IGNITION EIGHT
COUPE
with rumble seat $1975 f. o. b. factory
without rumble seat $1915 f. o. b. factory
Furnished with Royal Equipment, as illustrated above at slightly additional cost.

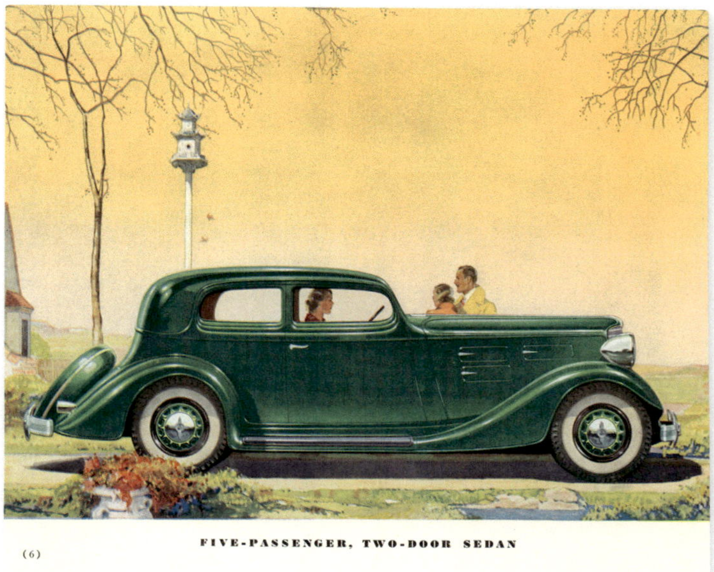

Nash Twin Ignition Eight 1930 *(opposite)*, **Nash 1936** *(above)*, **1934** *(right)*

Charles Nash bought out the remaining inventory and plant assets of Jeffery and renamed the company after himself. In their heyday, Nashes were spectacular cars. One of the best was the Twin-Ignition Eight, a car that could easily approach 100 mph. They were large, lavish, and respected for their power.

Charles Nash kaufte von Jeffery das verbleibende Inventar und die Fertigungsanlagen auf und gab der neuen Firma seinen eigenen Namen. In ihrer Glanzzeit waren die Fahrzeuge echte Hingucker. Einer der besten war der Twin-Ignition Eight, ein Auto, das problemlos 160 km/h erreichte. Diese Autos waren groß, nobel und wurden aufgrund ihrer Leistung respektiert.

Charles Nash a racheté ce qui restait des stocks et des actifs constitués de la marque Jeffery et lui a donné son nom. À leurs heures les plus fastes, les Nash furent des voitures spectaculaires. L'une des plus performantes était sans aucun doute le modèle Eight à double allumage, qui flirtait sans se forcer avec les 160 km/h. Elles étaient spacieuses, luxueuses et respectées pour leur puissance.

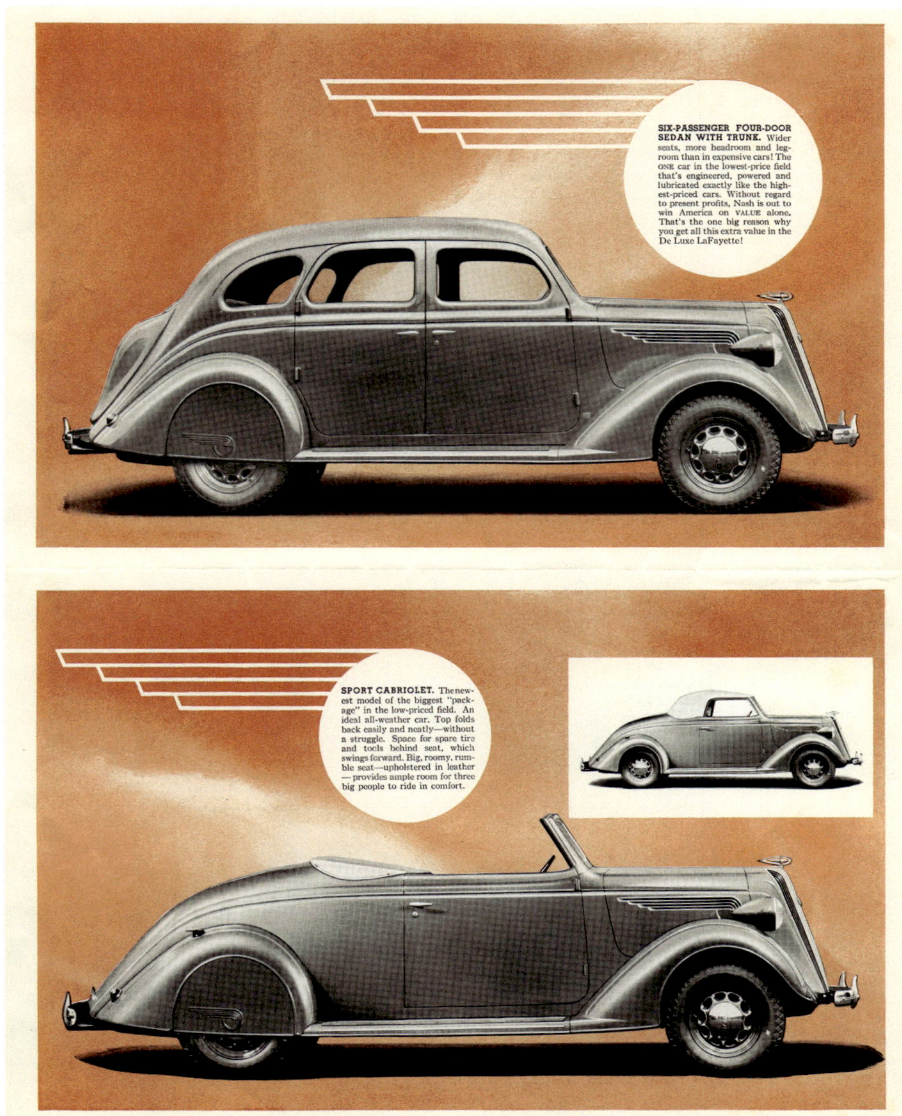

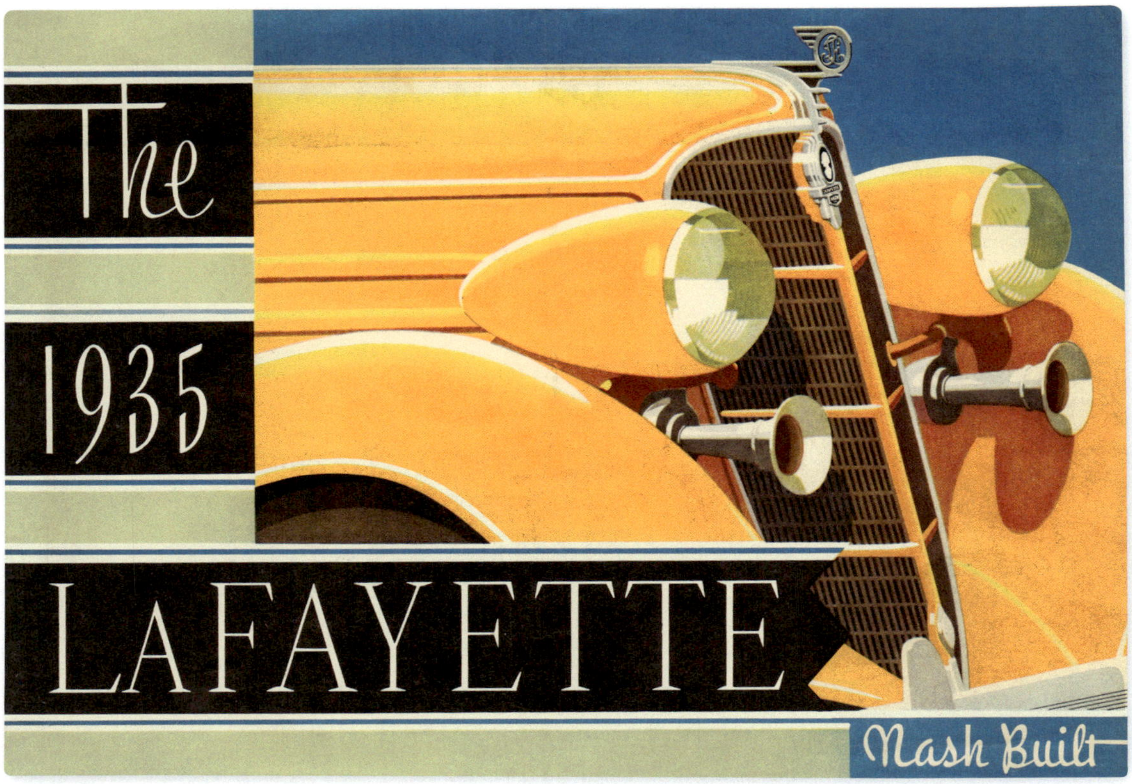

LaFayette 1935, 1940
(opposite), **Nash 1932** *(pages 176–177)*

Besides building excellent cars, Nash practiced fiscal discipline and preached the same to buyers in the Depression. The LaFayette, a very expensive Nash from the 1920s, was revived in 1934 for a new, low-priced Nash. They were good, handsome cars, but tooling costs forced Nash into the red for the first time. That was disappointing, given that the car so effectively mimicked the streamline look that was all the rage among aircraft and passenger railroad equipment during the 1930s.

Nash baute nicht nur hervorragende Autos, sondern konnte auch diszipliniert haushalten und predigte seinen Kunden während der Weltwirtschaftskrise das gleiche Verhalten. LaFayette, ein sehr teurer Nash aus den 1920er-Jahren, wurde 1934 als Modellbezeichnung für einen neuen, preisgünstigen Nash wiederbelebt. Es waren gute und hübsche Autos, doch die Anlagekosten brachten die Firma zum ersten Mal in die roten Zahlen – eine große Enttäuschung, zumal das Fahrzeug so wirkungsvoll die Stromlinienform nachahmte, die in den 1930er-Jahren bei Luftfahrzeugen und Personenzügen als der letzte Schrei galt.

Nash ne se contentait pas de fabriquer d'excellentes automobiles : l'entreprise pratiquait la discipline budgétaire et encourageait ses acheteurs à en faire de même. Ainsi le modèle LaFayette, une Nash très coûteuse des années 1920, ressuscita-t-elle en 1934 sous les traits d'une nouvelle Nash particulièrement bon marché. C'était une belle et bonne voiture mais les frais d'outillage firent plonger le constructeur dans le rouge pour la première fois de son histoire. Une déception, d'autant qu'elle revisitait brillamment le look aérodynamique qui faisait fureur dans l'aéronautique et les chemins de fer au cours des années 1930.

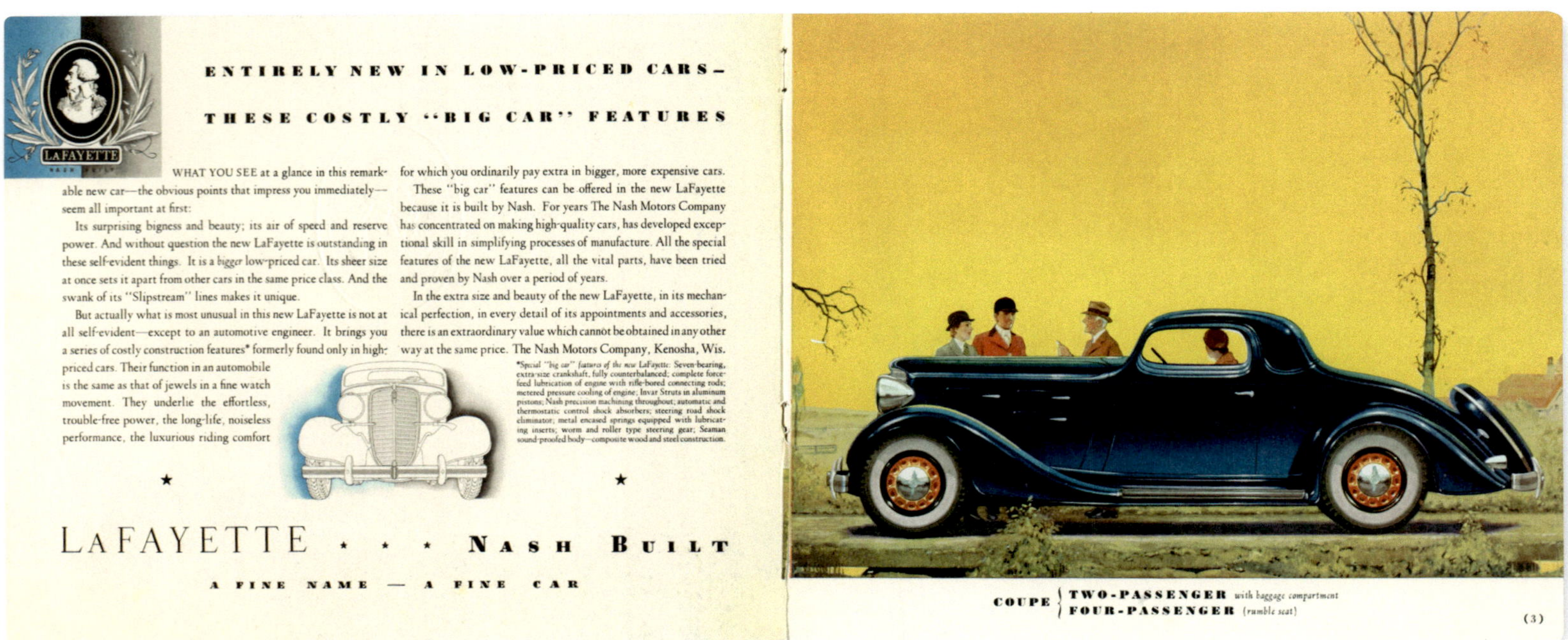

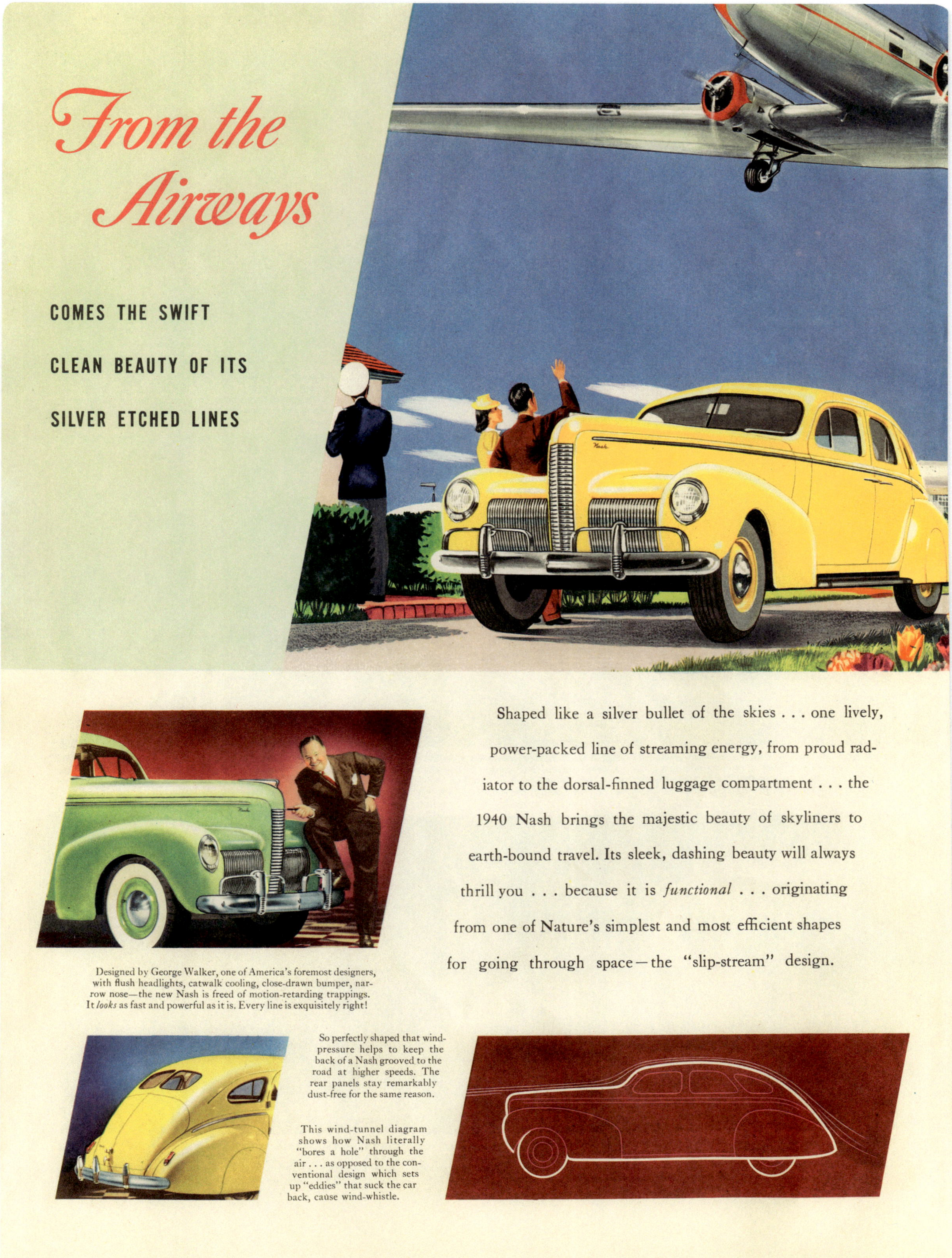
From the Airways
COMES THE SWIFT CLEAN BEAUTY OF ITS SILVER ETCHED LINES
Shaped like a silver bullet of the skies . . . one lively, power-packed line of streaming energy, from proud radiator to the dorsal-finned luggage compartment . . . the 1940 Nash brings the majestic beauty of skyliners to earth-bound travel. Its sleek, dashing beauty will always thrill you . . . because it is *functional* . . . originating from one of Nature's simplest and most efficient shapes for going through space—the "slip-stream" design.
Designed by George Walker, one of America's foremost designers, with flush headlights, catwalk cooling, close-drawn bumper, narrow nose—the new Nash is freed of motion-retarding trappings. It *looks* as fast and powerful as it is. Every line is exquisitely right!
So perfectly shaped that wind-pressure helps to keep the back of a Nash grooved to the road at higher speeds. The rear panels stay remarkably dust-free for the same reason.
This wind-tunnel diagram shows how Nash literally "bores a hole" through the air . . . as opposed to the conventional design which sets up "eddies" that suck the car back, cause wind-whistle.

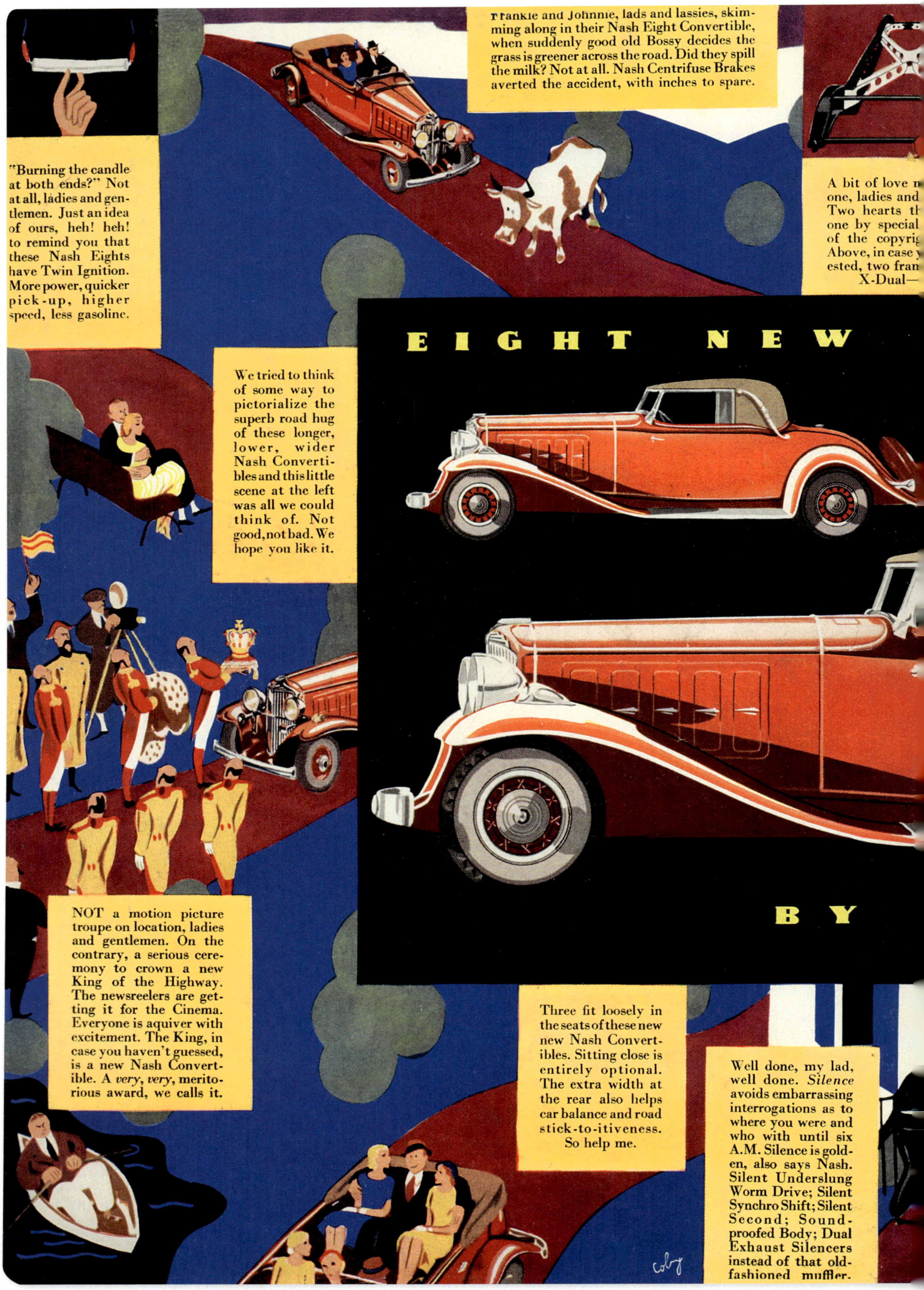

Frankie and Johnnie, lads and lassies, skimming along in their Nash Eight Convertible, when suddenly good old Bossy decides the grass is greener across the road. Did they spill the milk? Not at all. Nash Centrifuse Brakes averted the accident, with inches to spare.
"Burning the candle at both ends?" Not at all, ladies and gentlemen. Just an idea of ours, heh! heh! to remind you that these Nash Eights have Twin Ignition. More power, quicker pick-up, higher speed, less gasoline.
A bit of love n
one, ladies and
Two hearts th
one by special
of the copyrig
Above, in case y
ested, two fran
X-Dual—
EIGHT NEW
We tried to think of some way to pictorialize the superb road hug of these longer, lower, wider Nash Convertibles and this little scene at the left was all we could think of. Not good, not bad. We hope you like it.
BY
NOT a motion picture troupe on location, ladies and gentlemen. On the contrary, a serious ceremony to crown a new King of the Highway. The newsreelers are getting it for the Cinema. Everyone is aquiver with excitement. The King, in case you haven't guessed, is a new Nash Convertible. A *very, very*, meritorious award, we calls it.
Three fit loosely in the seats of these new new Nash Convertibles. Sitting close is entirely optional. The extra width at the rear also helps car balance and road stick-to-itiveness. So help me.
Well done, my lad, well done. *Silence* avoids embarrassing interrogations as to where you were and who with until six A.M. Silence is golden, also says Nash. Silent Underslung Worm Drive; Silent Synchro Shift; Silent Second; Soundproofed Body; Dual Exhaust Silencers instead of that old-fashioned muffler.

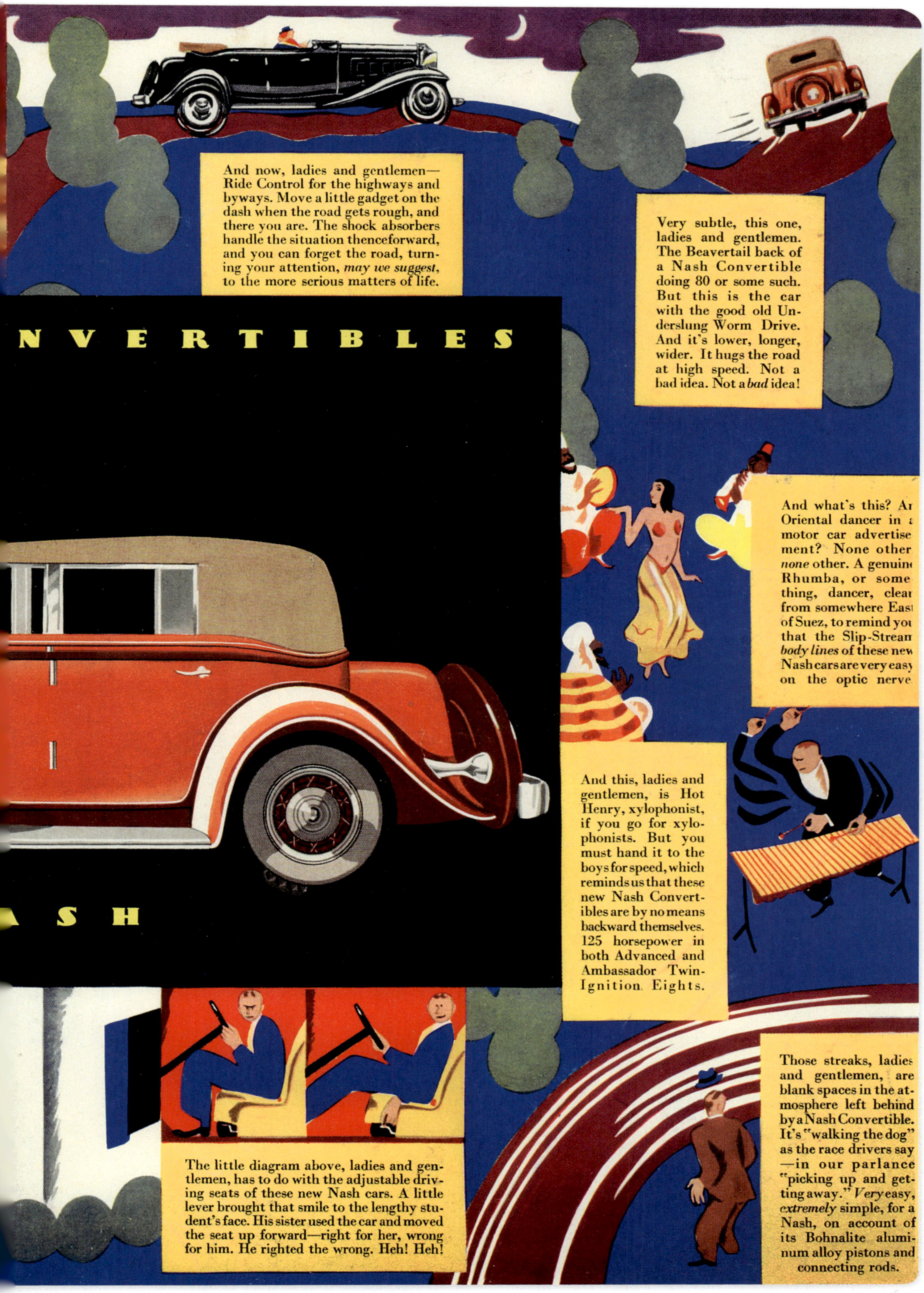
And now, ladies and gentlemen—Ride Control for the highways and byways. Move a little gadget on the dash when the road gets rough, and there you are. The shock absorbers handle the situation thenceforward, and you can forget the road, turning your attention, *may we suggest*, to the more serious matters of life.
Very subtle, this one, ladies and gentlemen. The Beavertail back of a Nash Convertible doing 80 or some such. But this is the car with the good old Underslung Worm Drive. And it's lower, longer, wider. It hugs the road at high speed. Not a bad idea. Not a *bad* idea!
NVERTIBLES
ASH
And what's this? Ar Oriental dancer in a motor car advertisement? None other, *none* other. A genuine Rhumba, or something, dancer, clear from somewhere East of Suez, to remind you that the Slip-Stream *body lines* of these new Nash cars are very easy on the optic nerve.
And this, ladies and gentlemen, is Hot Henry, xylophonist, if you go for xylophonists. But you must hand it to the boys for speed, which reminds us that these new Nash Convertibles are by no means backward themselves. 125 horsepower in both Advanced and Ambassador Twin-Ignition Eights.
The little diagram above, ladies and gentlemen, has to do with the adjustable driving seats of these new Nash cars. A little lever brought that smile to the lengthy student's face. His sister used the car and moved the seat up forward—right for her, wrong for him. He righted the wrong. Heh! Heh!
Those streaks, ladies and gentlemen, are blank spaces in the atmosphere left behind by a Nash Convertible. It's "walking the dog" as the race drivers say—in our parlance "picking up and getting away." *Very* easy, *extremely* simple, for a Nash, on account of its Bohnalite aluminum alloy pistons and connecting rods.

Oldsmobile

Oldsmobile 1935, 1936 *(opposite)*

Oldsmobile's transition to the latter half of the 1930s was built around its new, rounded grille, but buyers still enjoyed its other virtues such as a deep-channel steel frame, expanding hydraulic brakes, and interior ventilation.

Den Übergang in die späten 1930er-Jahre markierte Oldsmobile mit einem neuen, gerundeten Grill, aber die Käufer erfreuten sich noch immer an den traditionellen Tugenden des Fahrzeugs, wie etwa einem Profilstahlrahmen, Hydraulikbremsen und Innenraumbelüftung.

La transition d'Oldsmobile vers la seconde moitié des années 1930 reposait sur sa nouvelle calandre arrondie, ce qui n'empêchait pas ses clients d'apprécier d'autres de ses vertus, comme ses freins à segments internes et à commande hydraulique, son châssis en acier allongé et sa ventilation de l'habitacle.

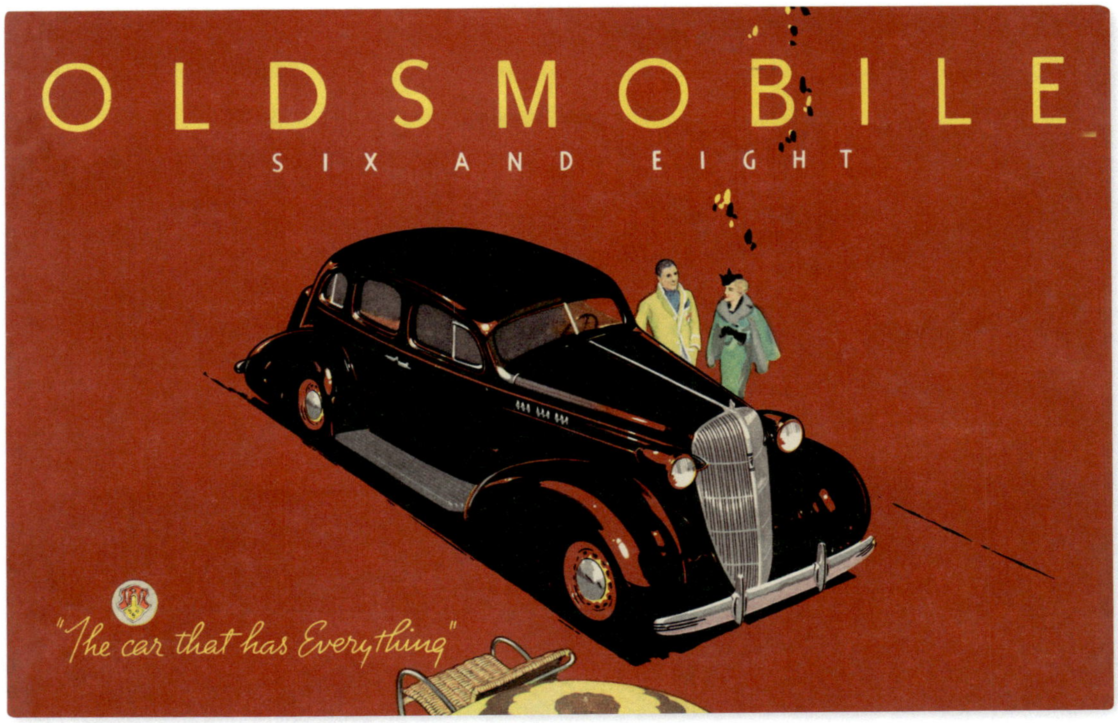

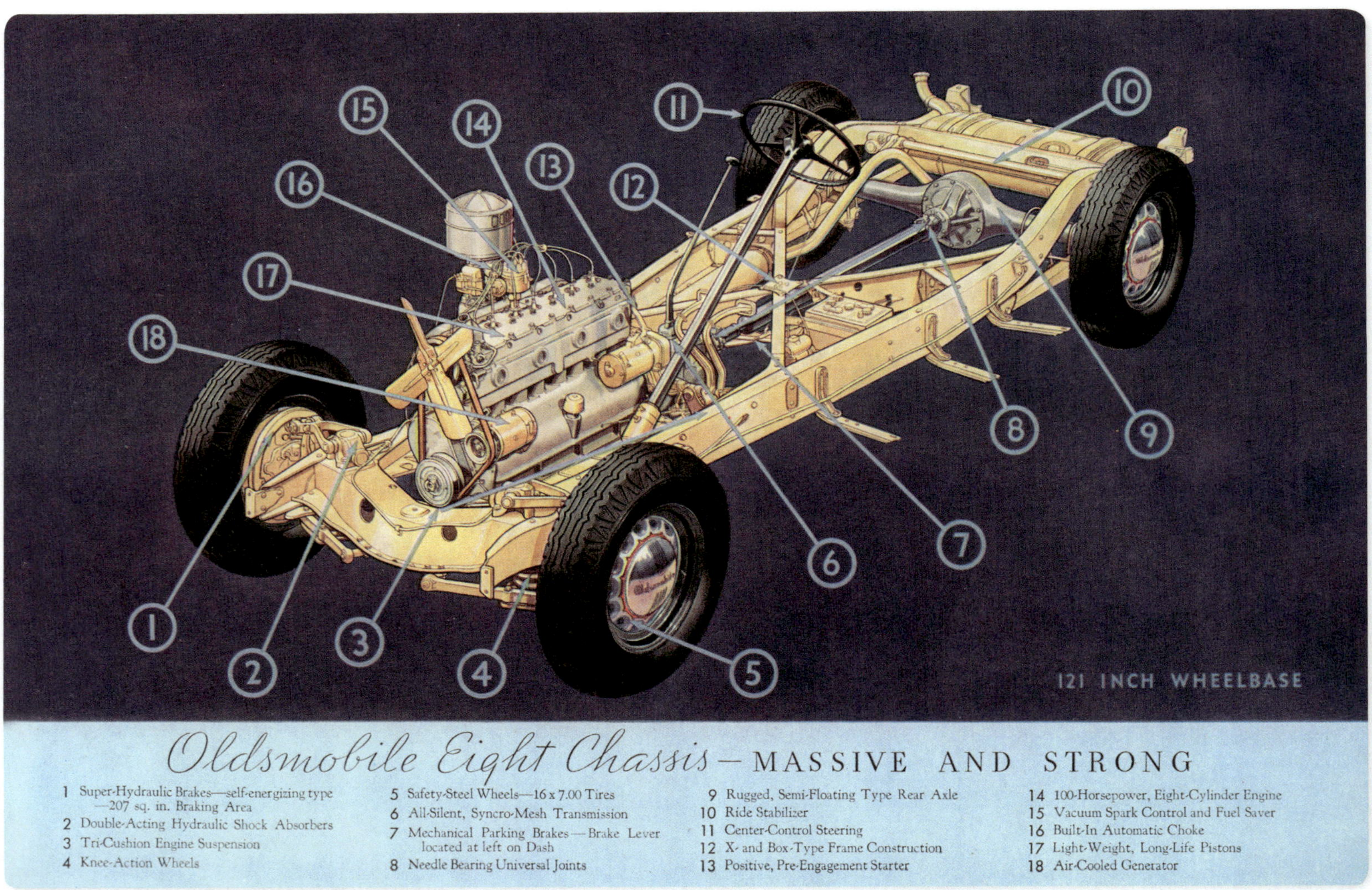

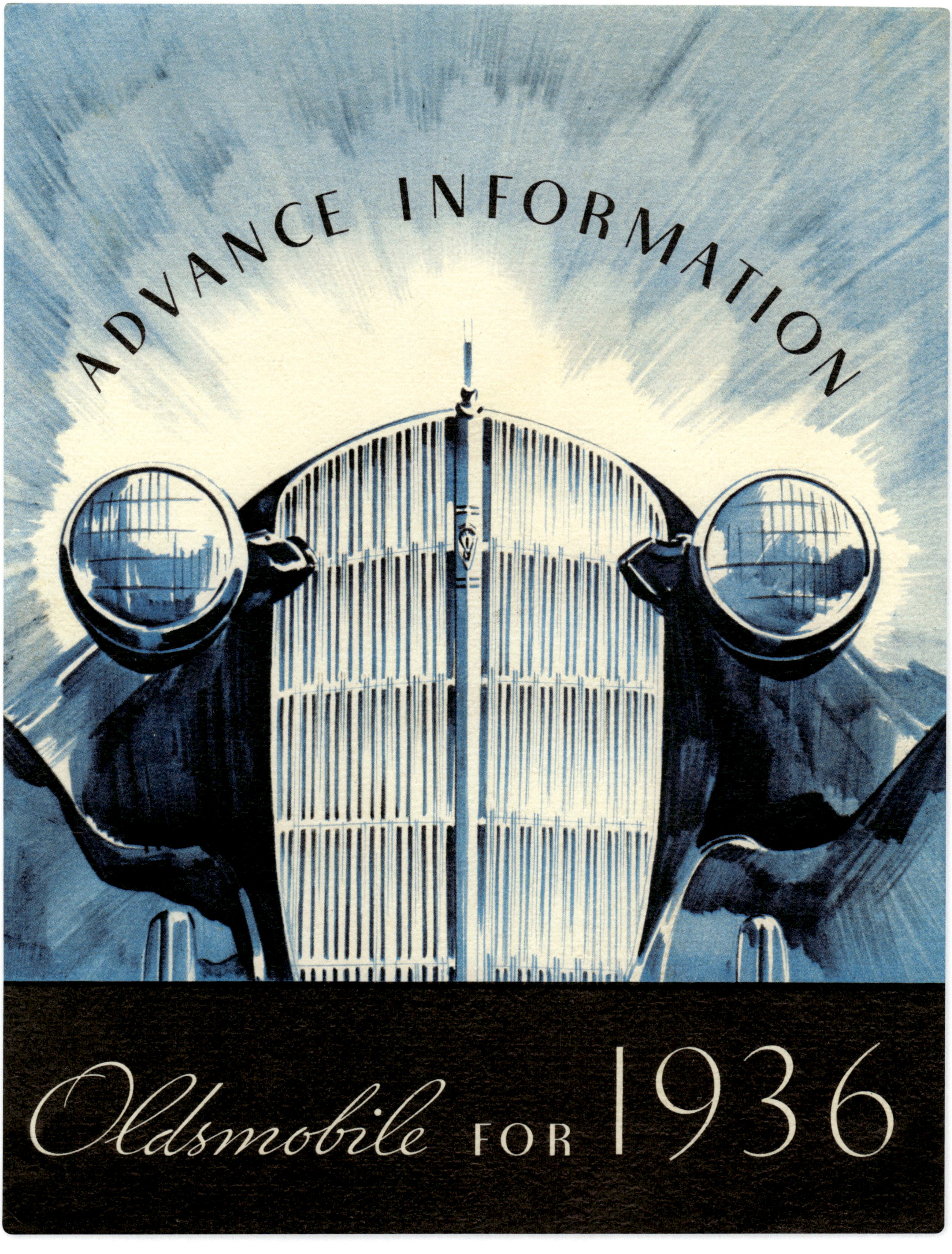
ADVANCE INFORMATION
Oldsmobile FOR 1936

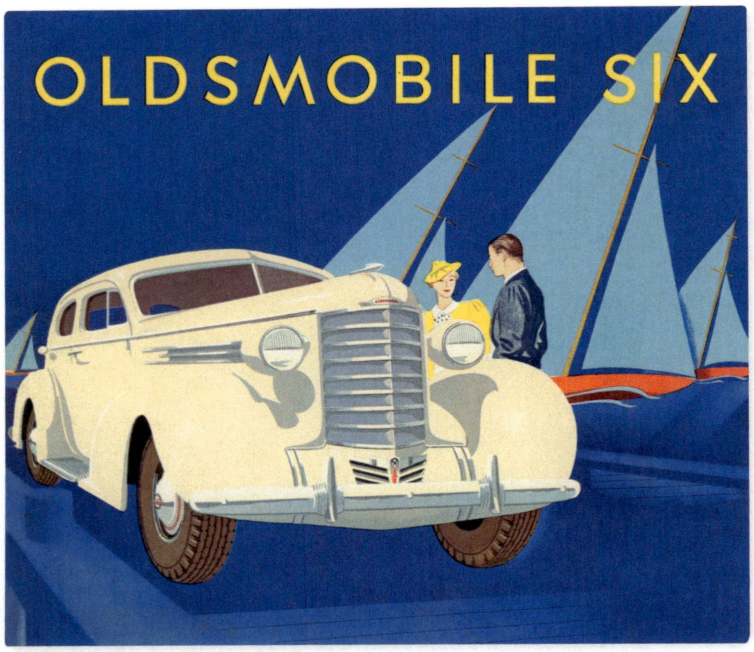

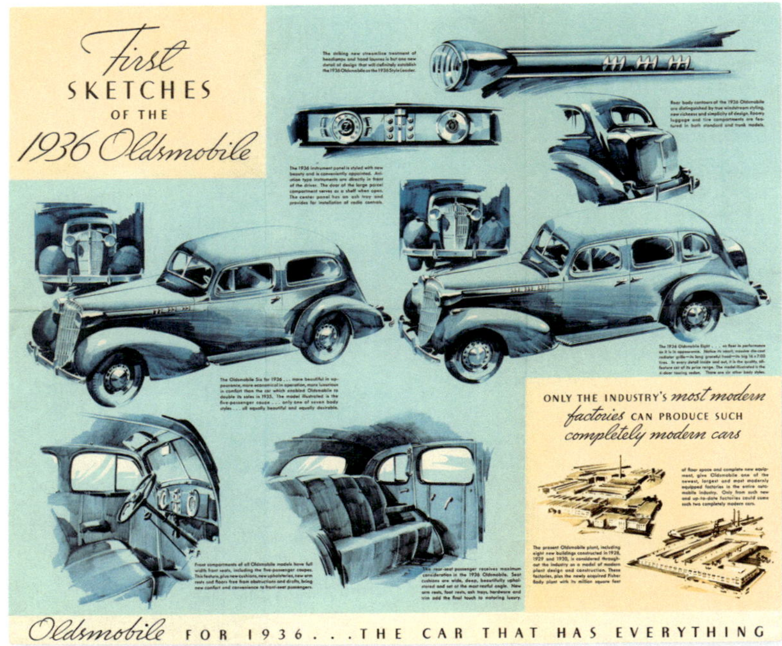

Oldsmobile 1936

In the 1930s, coaxing 95 hp from a valve-in-head straight-six engine was a notable accomplishment. The Oldsmobile Six's full-length water jackets and high-capacity pump prevent the cylinder head or engine block from cracking under hot running. Full-pressure lubrication, which did a lot to lengthen engine life, was another selling point.

In den 1930er-Jahren war es eine bemerkenswerte Leistung, einem Sechszylinderreihenmotor mit OHV-Ventilsteuerung 129 kW Leistung zu entlocken. Der Kühlwassermantel, der sich im Oldsmobile Six über die gesamte Länge des Motors zog, sowie die großvolumige Pumpe verhinderten, dass Zylinderköpfe oder Motorblock heiß liefen und rissen. Ein weiteres Verkaufsargument war die Druckschmierung, die die Lebensdauer des Motors erheblich verlängerte.

Dans les années 1930, réussir à tirer 95 chevaux vapeur d'un moteur six cylindres en ligne à soupapes en tête était un véritable exploit. Le système de refroidissement de l'Oldsmobile Six, grâce à la longueur de ses chemises et à sa pompe à eau très performante, empêchait la culasse ou le bloc moteur de se fissurer à haute température. La lubrification sous pression, qui fit beaucoup pour prolonger la vie des moteurs, constituait un autre argument de vente.

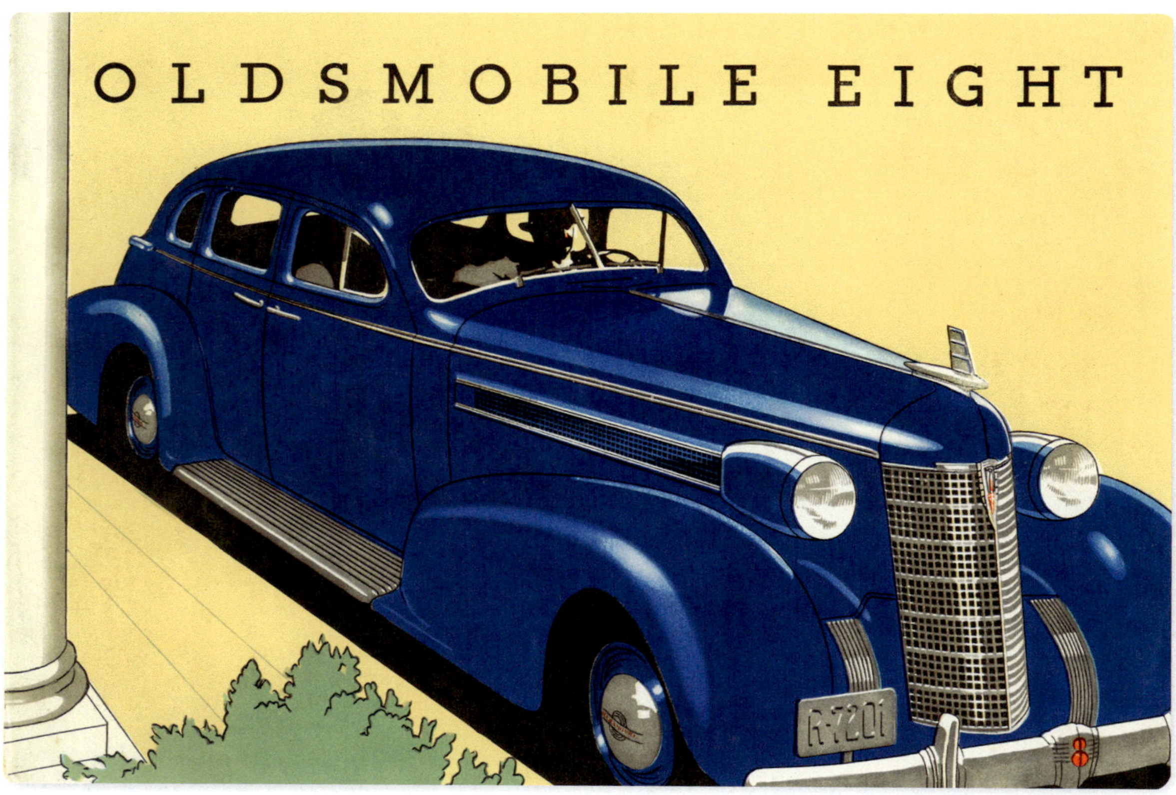

Oldsmobile 1940

The great age of General Motors styling from Harley Earl's Art & Colour section was approaching its zenith. Closely examine the grille and headlamp treatment, and you will understand why some enthusiasts consider the 1940 Oldsmobile one of the most attractive of all prewar cars. Earl labored to give all GM cars, regardless of price, their own visual identities that were updated on an annual basis, creating the yearly model change that transformed the industry.

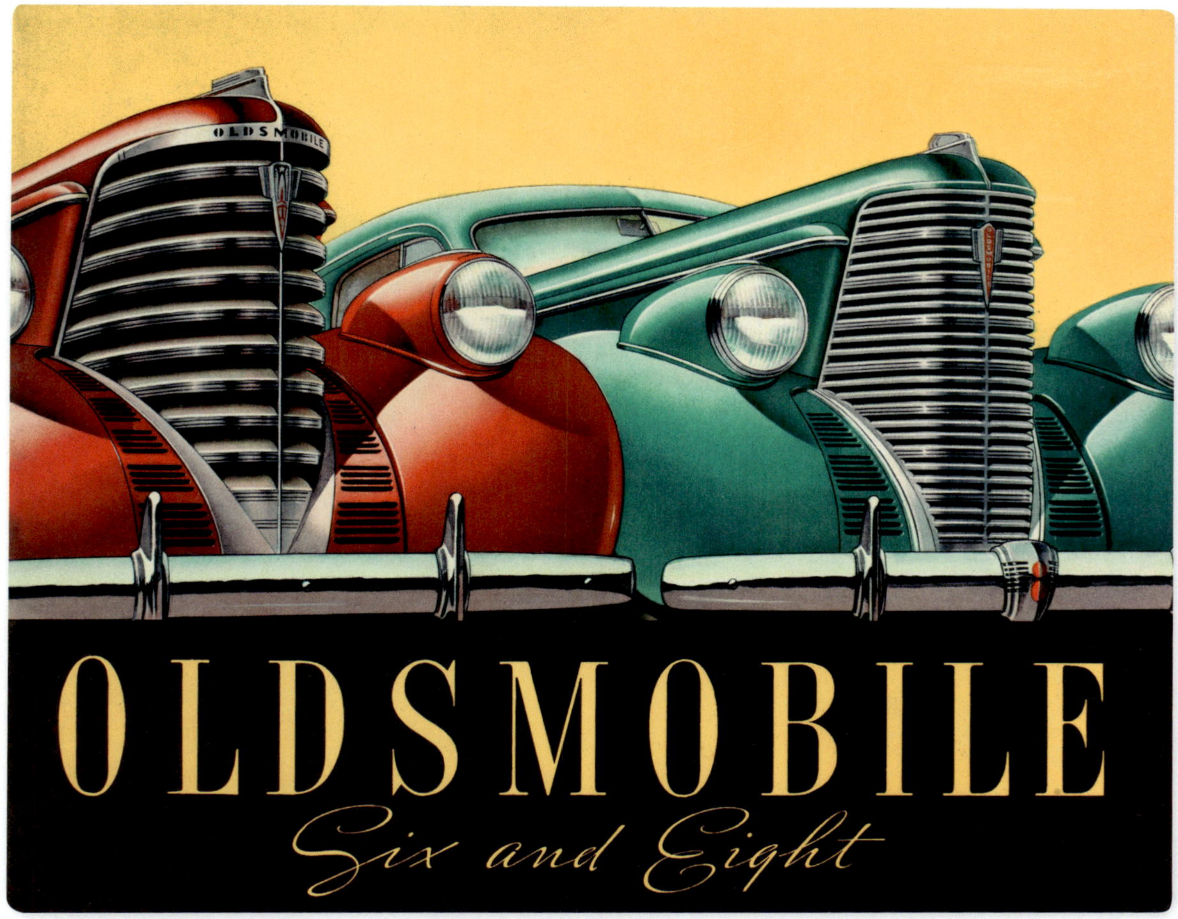

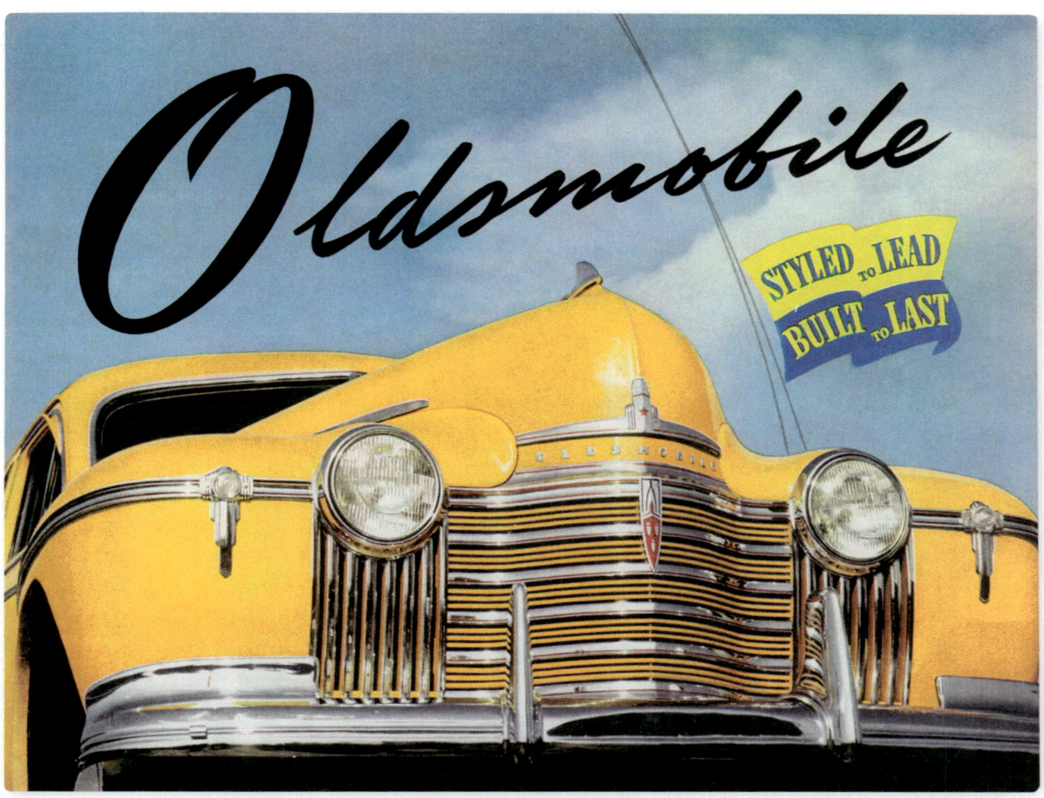

Das große Zeitalter des GM-Designs aus Harley Earls Abteilung Art & Colour näherte sich seinem Zenit. Wenn man den Kühlergrill und die Anordnung der Scheinwerfer unter die Lupe nimmt, versteht man, weshalb einige Enthusiasten den 1940er-Oldsmobile für eines der attraktivsten Automobile aus der Zeit zwischen den beiden Weltkriegen halten. Earl bemühte sich, allen Fahrzeugen von General Motors, unabhängig vom Preis, eine eigenständige Optik zu verleihen, die Jahr für Jahr aktualisiert wurde und jenen jährlichen Modellwechsel einläutete, der die Industrie grundlegend veränderte.

L'influence du département Art & Colour d'Harley Earl sur le style de la General Motors approchait de son apogée. Observez de près le traitement de la calandre et des phares, et vous comprendrez pourquoi certains passionnés considèrent l'Oldsmobile de 1940 comme l'une des voitures les plus séduisantes de l'avant-guerre. Earl s'évertua à conférer à toutes les voitures de la gamme GM, quel que fût leur prix, leur propre identité visuelle, et à l'actualiser chaque année. Cette politique de renouvellement transforma le visage de l'industrie automobile.

Packard

Packard 1930, 1935 *(opposite)*

The notion of a Packard chassis in a luxuriant showroom really isn't far off base. A Standard Eight in the 7-33 series produced 90 hp. Packard had prestige to spare but attracted intelligent buyers who appreciated good engineering.

Die Vorstellung von einem Packard-Fahrgestell in einem prunkvollen Verkaufsraum war gar nicht so weit hergeholt. Der Reihenachter in der Baureihe 7–33 lieferte 122 kW Leistung. Dem Packard mochte es zwar an Prestige fehlen, doch er lockte intelligente Kunden an, die solide Ingenieurskunst zu schätzen wussten.

Cette image de châssis Packard exposé dans un showroom exhalant l'opulence n'est pas si éloignée de la réalité. Une Standard Eight de la série 7–33 développait 90 chevaux. Packard avait du chic à revendre mais n'en séduisait pas moins des clients intelligents, amateurs d'habile ingénierie.

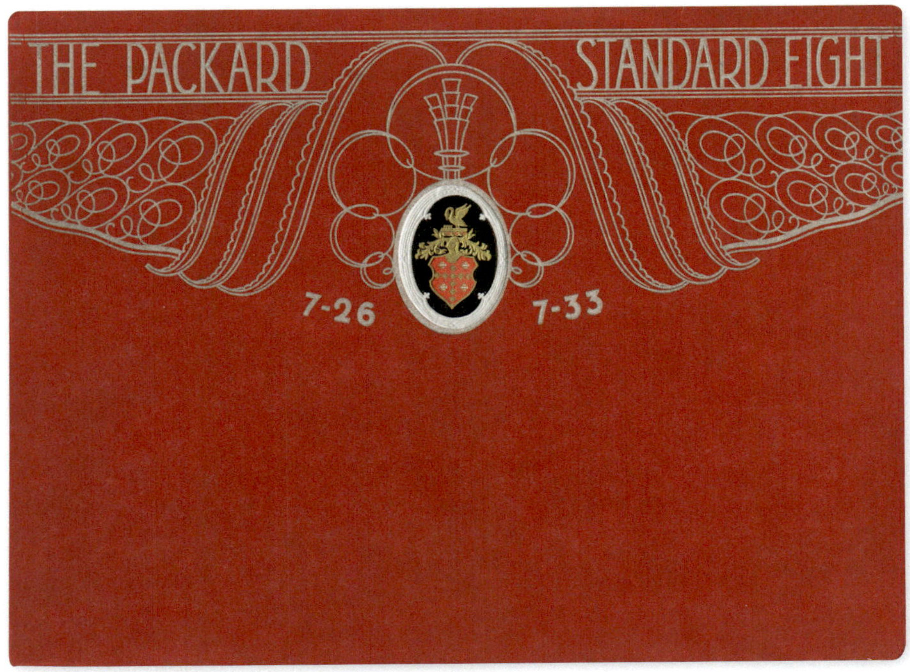

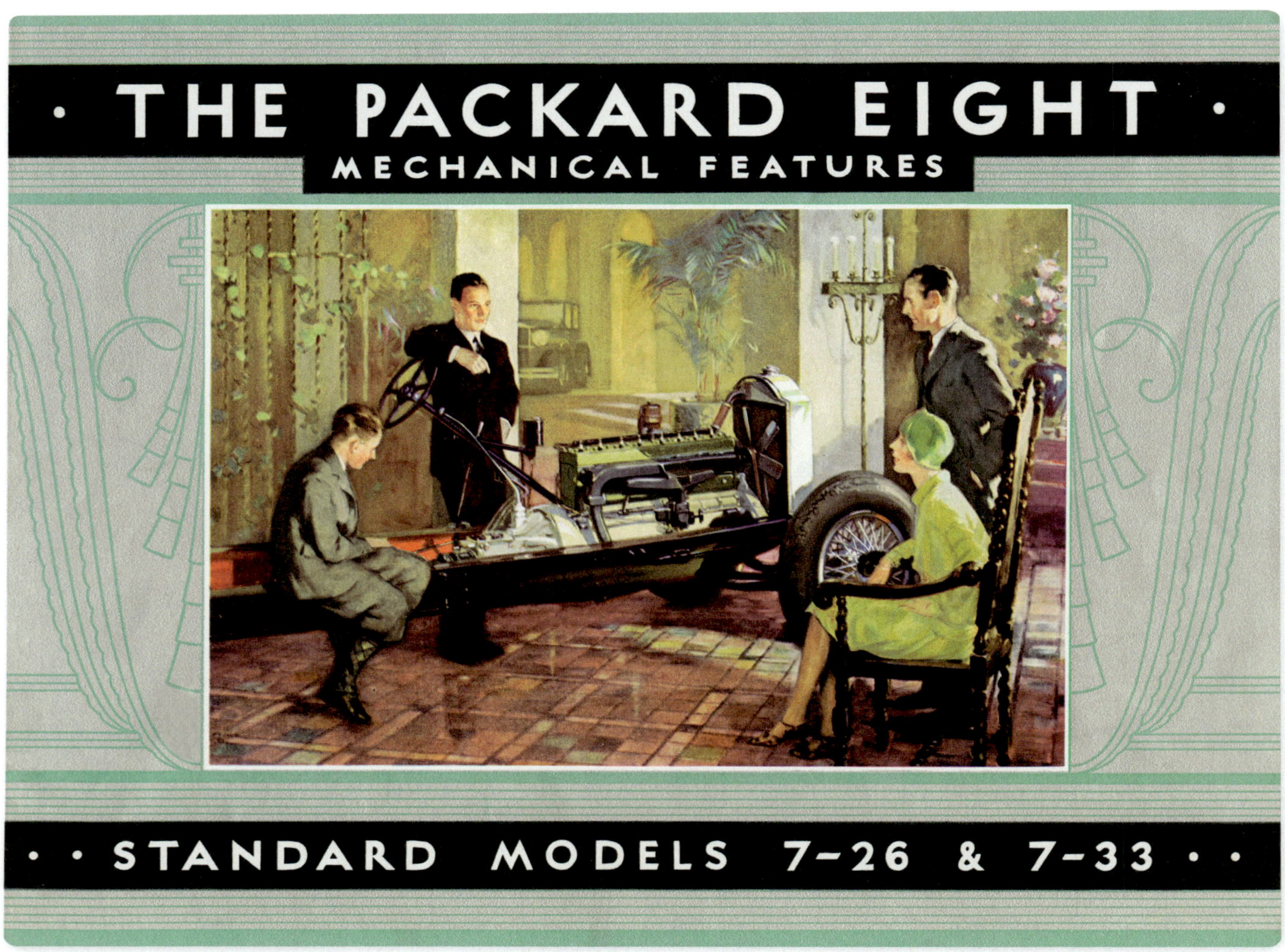

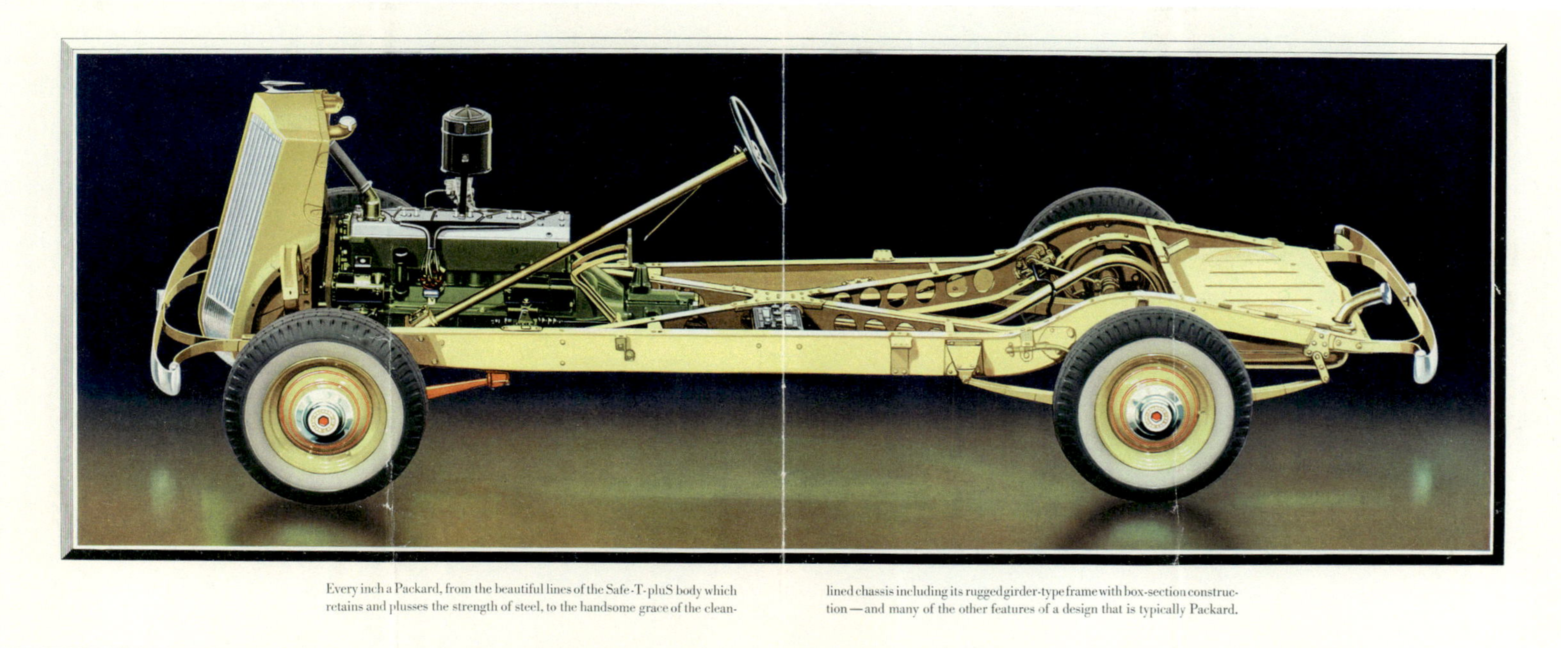
Every inch a Packard, from the beautiful lines of the Safe-T-pluS body which retains and plusses the strength of steel, to the handsome grace of the clean-lined chassis including its rugged girder-type frame with box-section construction — and many of the other features of a design that is typically Packard.

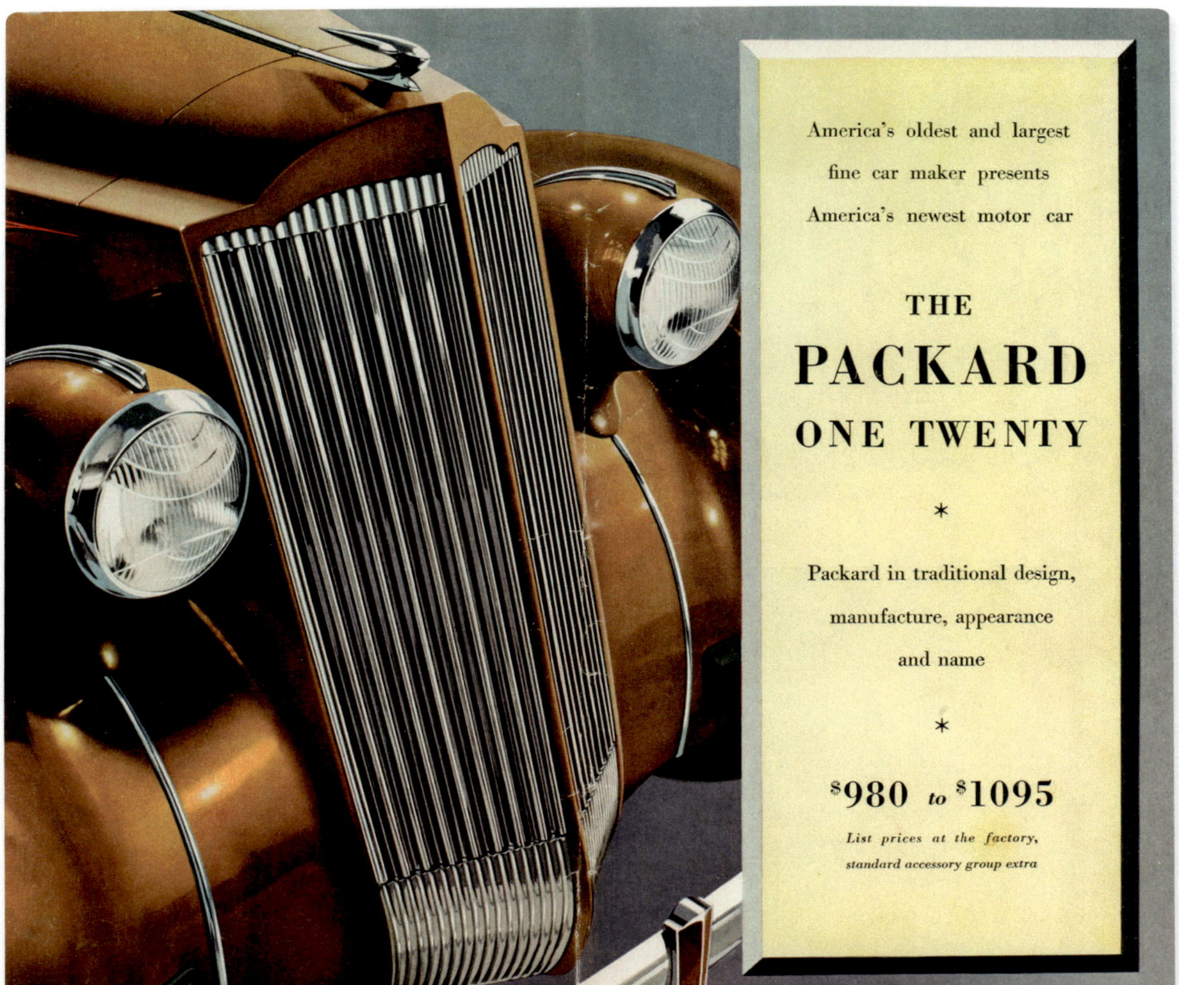
America's oldest and largest fine car maker presents America's newest motor car
THE
PACKARD
ONE TWENTY
*
Packard in traditional design, manufacture, appearance and name
*
$980 to $1095
List prices at the factory, standard accessory group extra

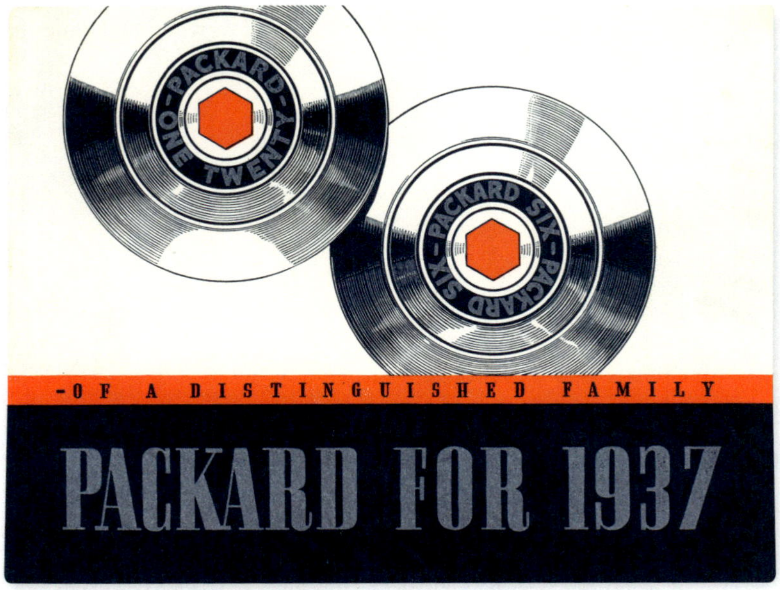

Packard 1937 *(left)*, **1938** *(below and opposite)*

Packard offered straight-six, straight-eight, and V-12 engines in 1938. Massive steel beams gave the 12-cylinder chassis strength and rigidity to rival the steel framework of a skyscraper.

Packard bot 1938 Reihensechs- und Reihenachtzylinder- sowie V12-Motoren an. Massive Stahlverstrebungen verliehen dem Fahrgestell des Zwölfzylinders eine ähnliche Steifigkeit und Festigkeit wie das Stahlskelett einem Wolkenkratzer.

En 1938, Packard proposait des moteurs six cylindres en ligne, huit cylindres en ligne et V12. Ses longerons massifs en tôle d'acier emboutie conféraient au châssis des 12 cylindres une solidité et une rigidité manifestement dignes de la charpente d'un gratte-ciel.

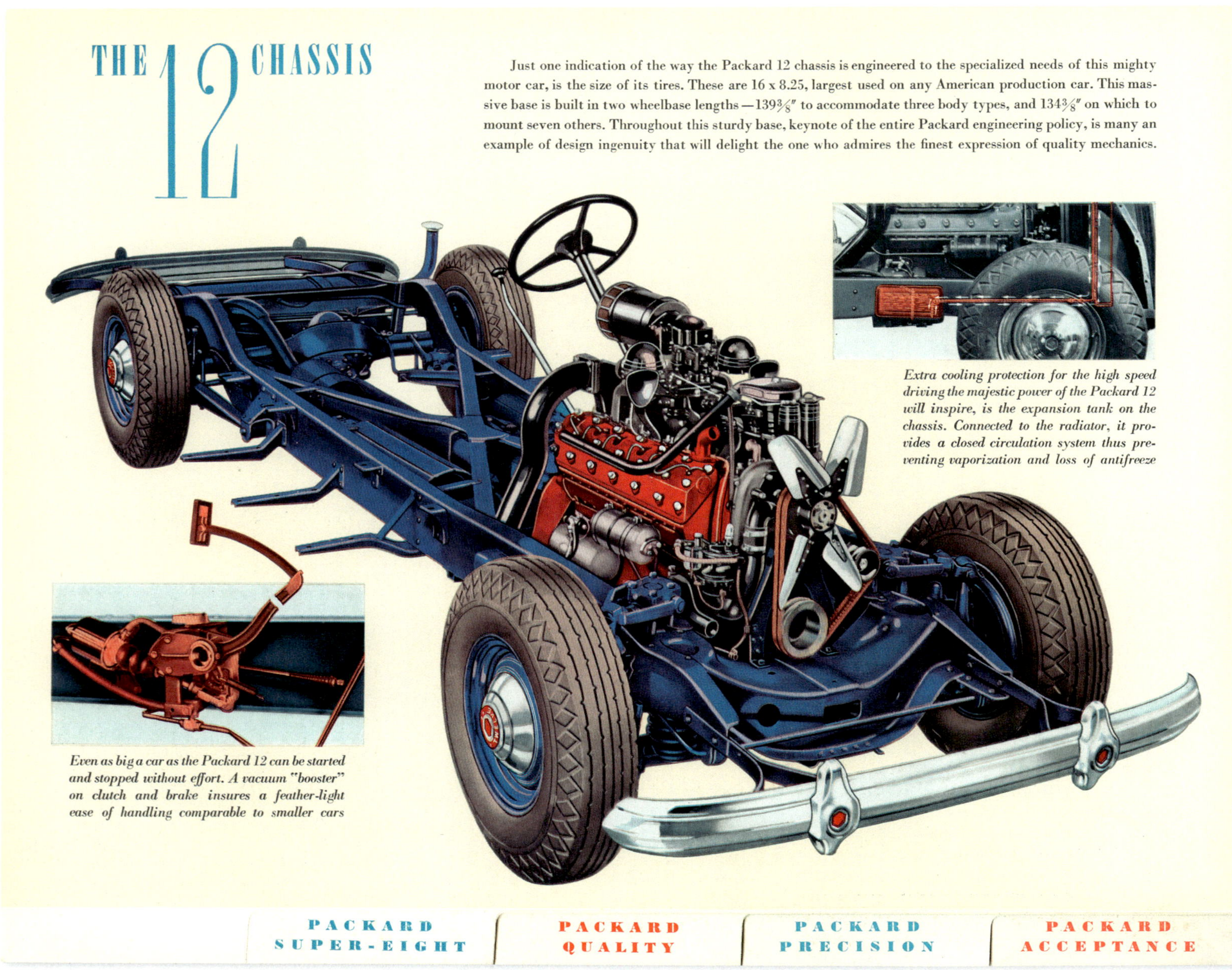

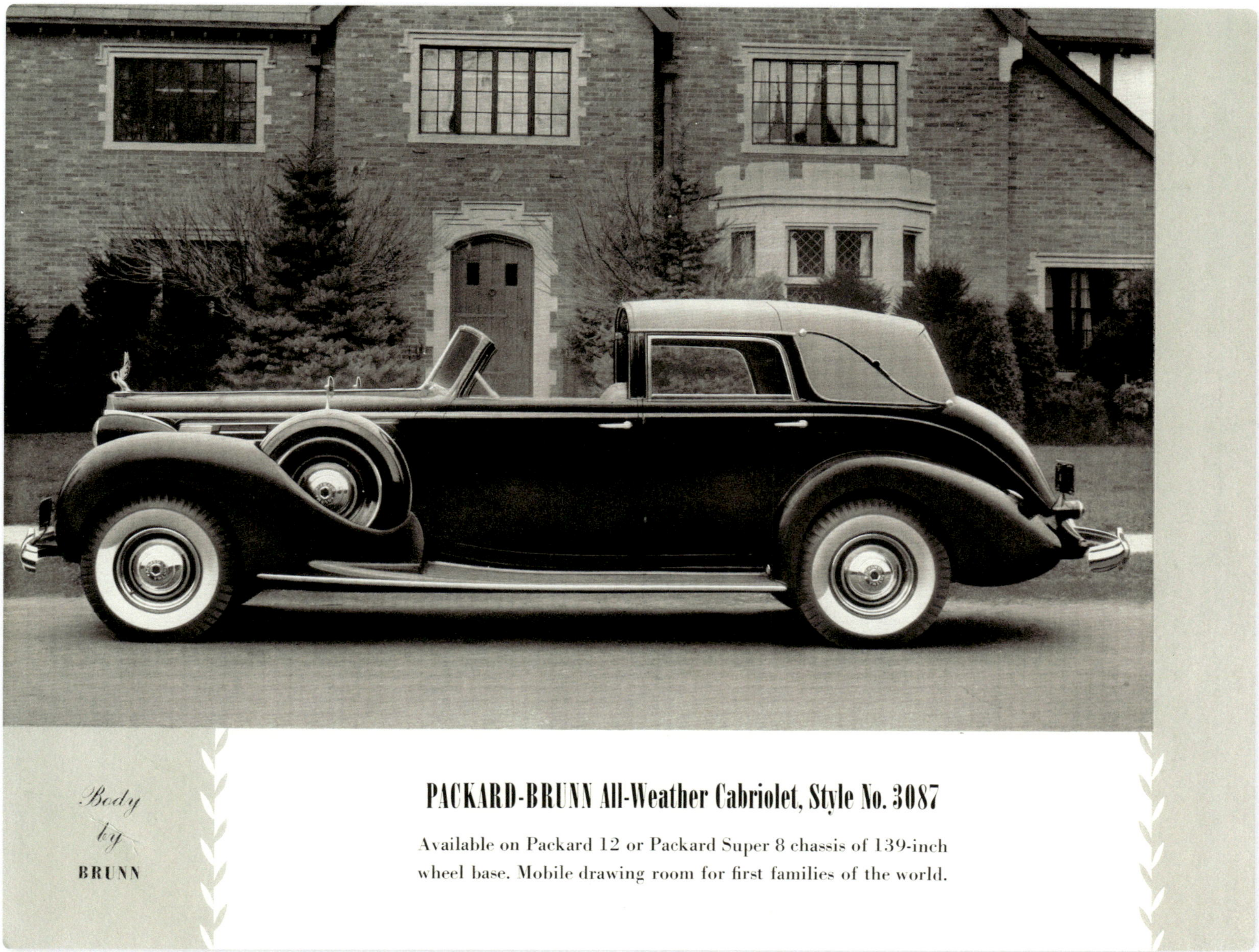
Body by BRUNN
PACKARD-BRUNN All-Weather Cabriolet, Style No. 3087
Available on Packard 12 or Packard Super 8 chassis of 139-inch wheel base. Mobile drawing room for first families of the world.

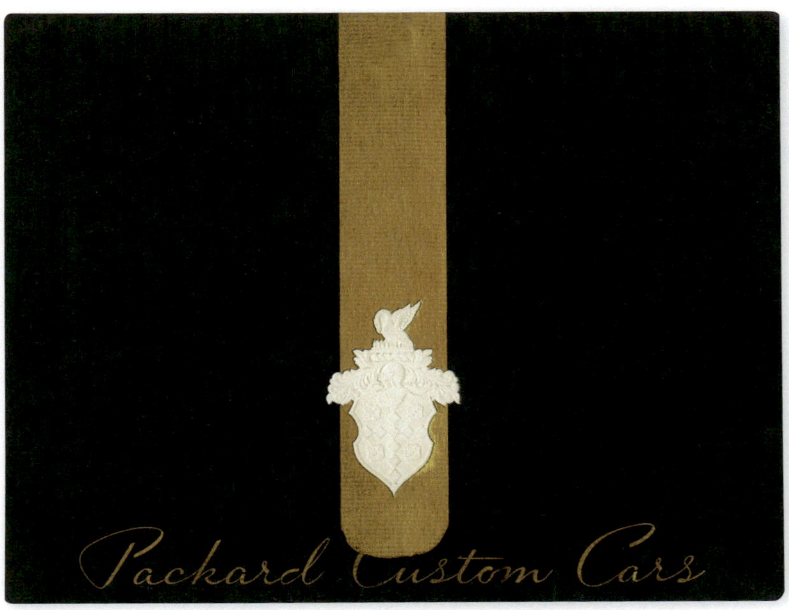
Packard Custom Cars

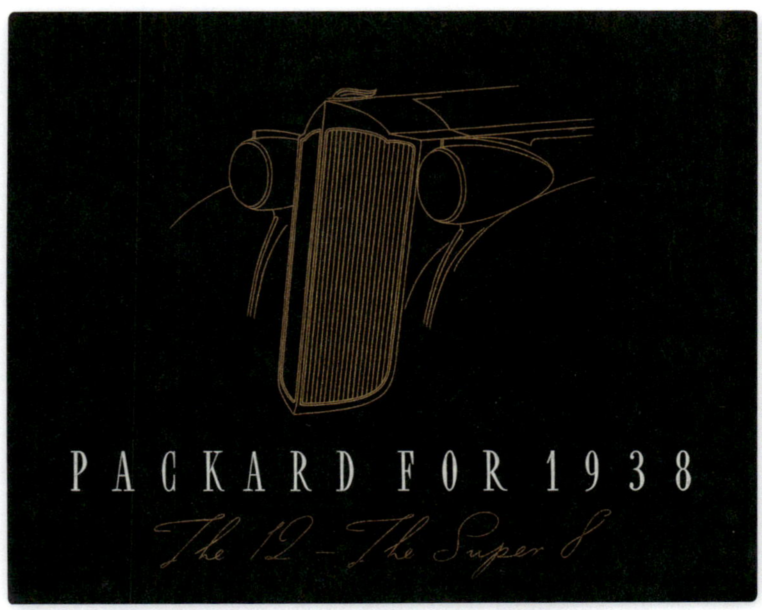
PACKARD FOR 1938
The 12 – The Super 8

Phantom

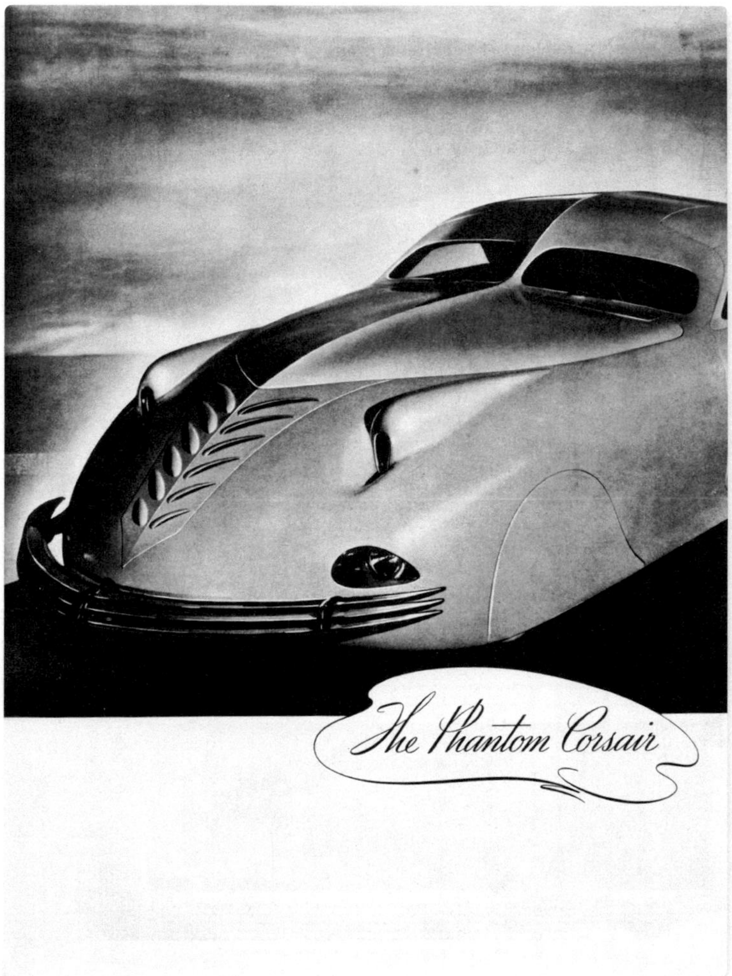

Phantom Corsair 1938

Commissioned by Rust Heinz, heir to the Heinz fortune, Maurice Schwartz of Bohman & Schwartz constructed the Corsair on a front-drive Cord chassis with Lycoming V-8 power, electric doors, and a fully independent suspension. Due to Heinz's death in a car accident, only the prototype was built.

Im Auftrag von Rust Heinz, dem Erben des Heinz-Vermögens, konstruierte Maurice Schwartz von Bohman & Schwartz den Corsair auf dem Chassis eines Cord mit Vorderradantrieb, ausgestattet mit einem V8-Motor von Lycoming, elektrischen Türen und Einzelradaufhängung. Da Heinz bei einem Autounfall ums Leben kam, blieb es bei diesem Prototypen.

Répondant à une commande de Rust Heinz, héritier de la fortune du même nom, Maurice Schwartz, pour Bohman & Schwartz, a construit la Corsair sur un châssis Cord à traction avant avec une motorisation Lycoming V8, des portes électriques et une suspension indépendante aux quatre roues. Heinz trouva la mort dans un accident de voiture, aussi seul un prototype de la Corsair a-t-il vu le jour.

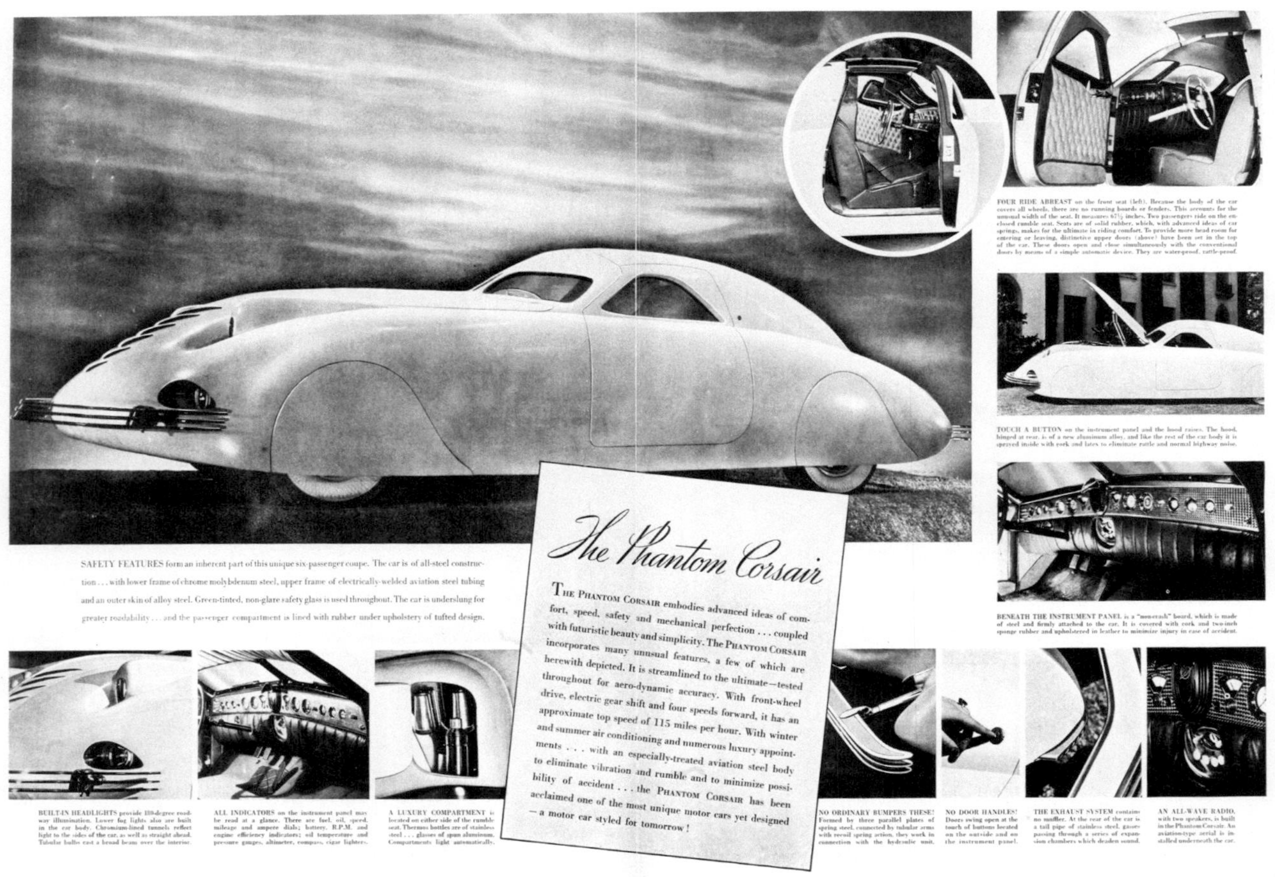

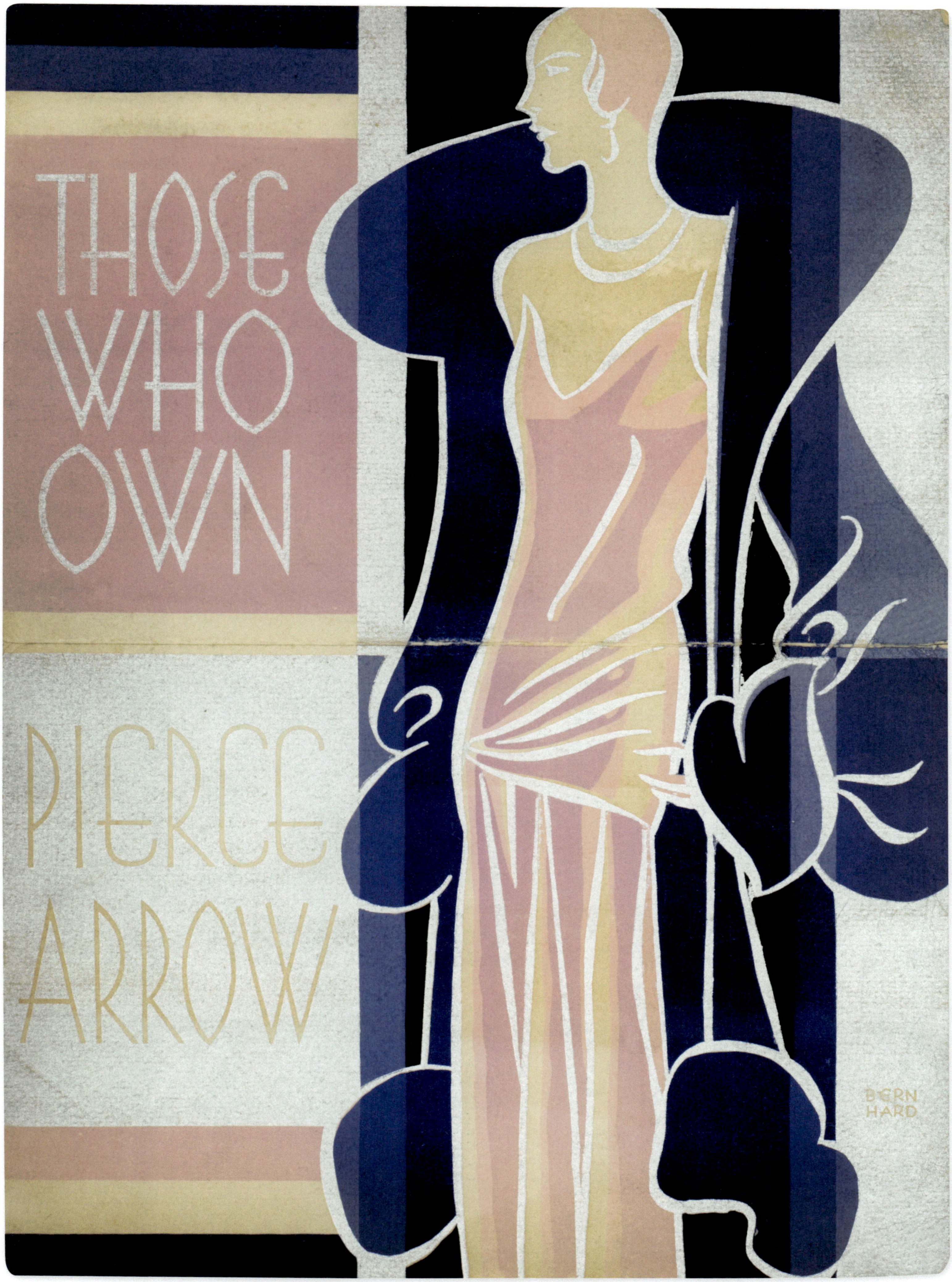
THOSE
WHO
OWN
PIERCE
ARROW
BERN
HARD

Pierce-Arrow

Pierce-Arrow 1930 *(previous page and this page)*, **1929** *(opposite)*

As if the splendidly attired and coiffed lady on the cover wasn't enough, the Pierce-Arrow copywriters made sure everybody understood that only "patricians," in their words, made suitable owners. Pierce-Arrow was a beautifully engineered and appointed car, with an equally strong image.

Als ob die vornehm gekleidete und frisierte Dame auf der Titelseite noch nicht ausreichte, ließen auch die Werbetexter von Pierce-Arrow keinen Zweifel daran, dass nur „Patrizier" – wie sie es ausdrückten – geeignete Besitzer eines solchen Automobils sein könnten. Der Pierce-Arrow war ein wunderschön gebautes und ausgestattetes Fahrzeug mit einem entsprechend starken Image.

Comme si les créatures magnifiquement apprêtées et coiffées de la couverture du catalogue ne suffisaient pas, les rédacteurs de Pierce-Arrow se sont évertués à faire clairement passer leur message : seuls les « patriciens » (sic) faisaient des acheteurs convenables. La Pierce-Arrow était une voiture superbement conçue et équipée, qui jouissait d'une image tout aussi puissante.

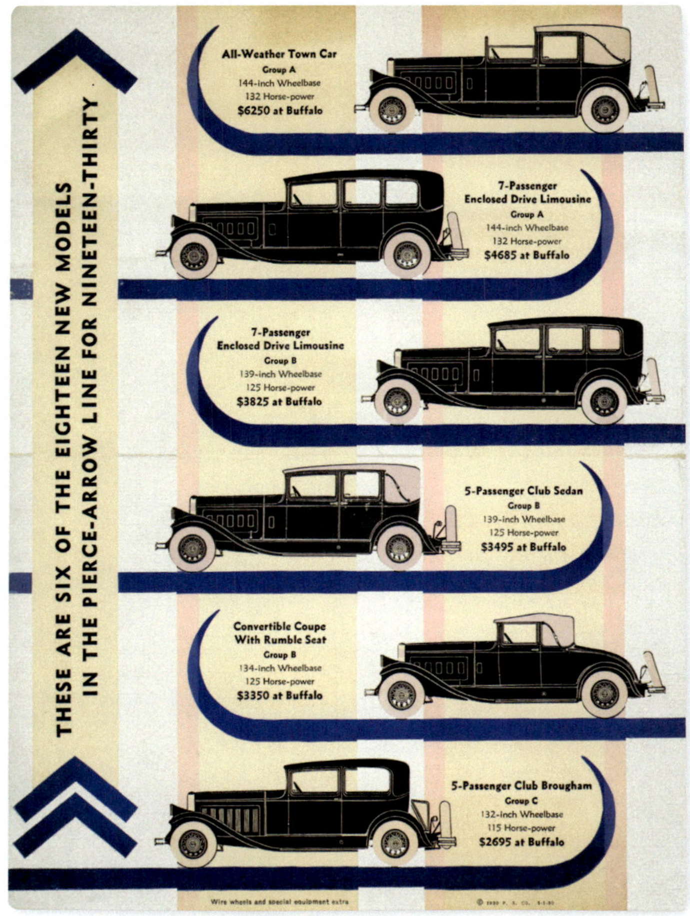

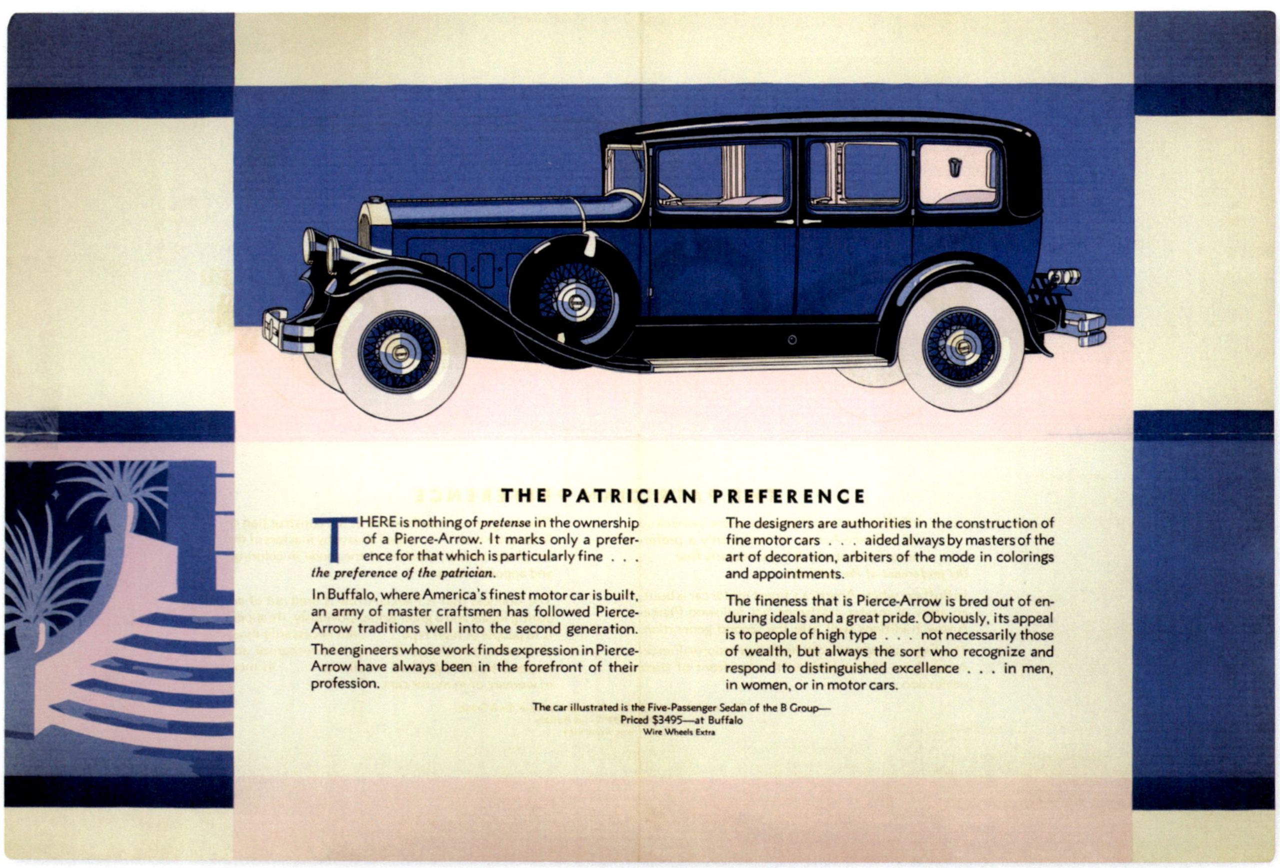

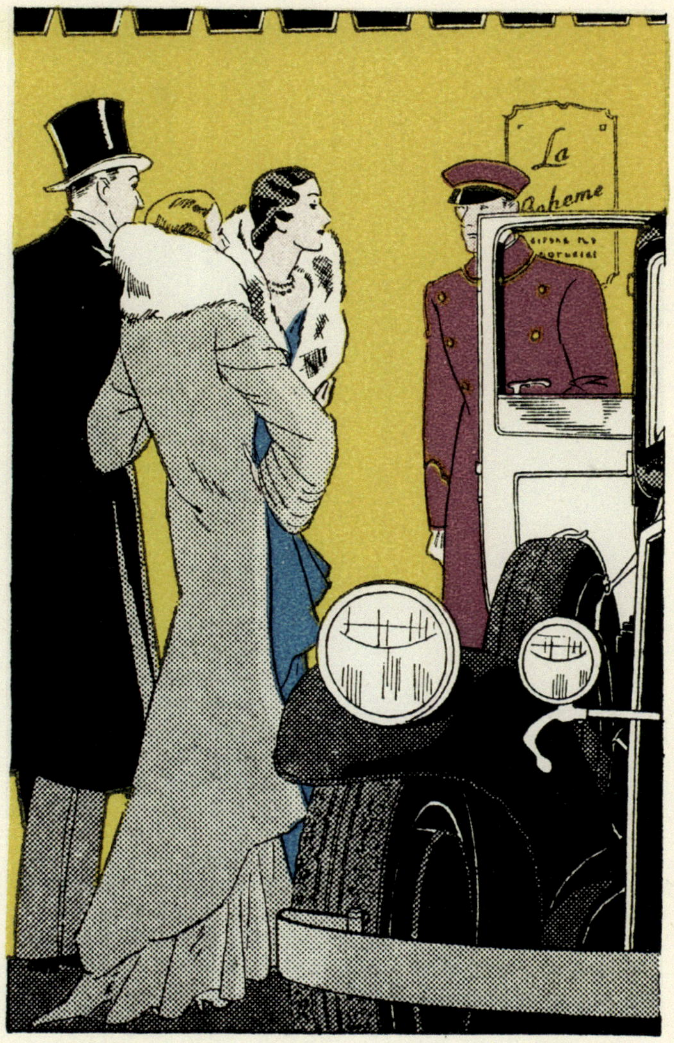

The Pierce-Arrow Sort of People

The Pierce-Arrow sort of people are to be recognized by the excellence of taste which marks always the well-bred . . . never the commonplace.

It was natural that such people should find affinity with those traditions which inspire fineness . . . with the exquisite hand-wrought detail of coachcraft and the meticulous hand-tooled precisions of engine-building . . . for which Pierce-Arrow has been famous from the beginning.

Plymouth

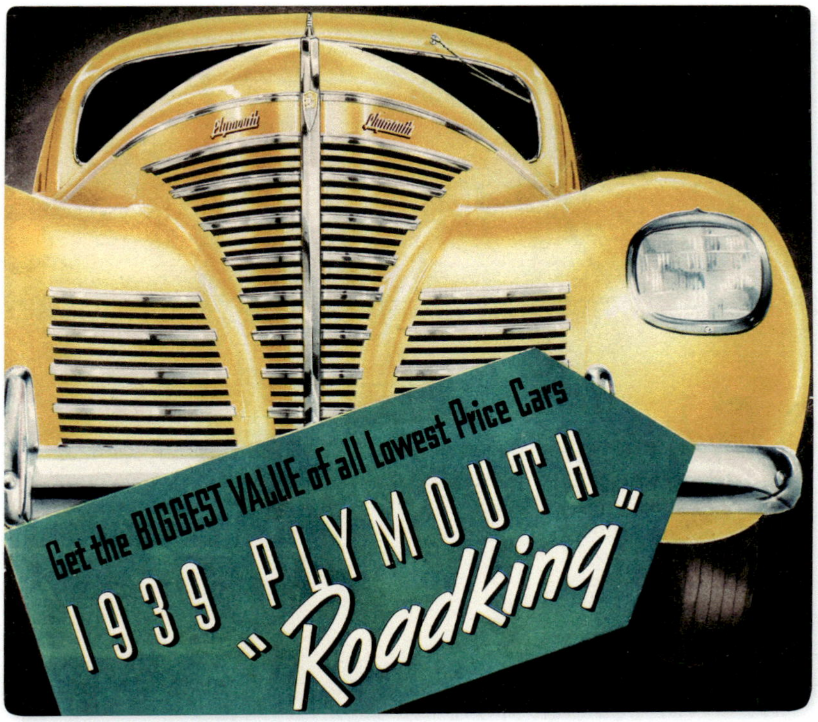

Plymouth Deluxe 1939 *(above)*,
1936 *(right)*, **1942** *(opposite)*

These classic prewar car brochures did their job selling Plymouth's core values as an economical, durable car. Liberal bearing surfaces, hardened valve seats, double-acting brakes, and roomy all-steel bodies made shoppers take notice.

Diese klassischen Broschüren aus der Zeit zwischen den Weltkriegen erfüllten ihren Zweck, die Kerntugenden des Plymouth als sparsames und haltbares Auto zu vermitteln. Großzügige Auflagerflächen, gehärtete Ventilsitze, doppelt wirkende Bremsen und geräumige Ganzstahlkarosserien weckten die Aufmerksamkeit der Kundschaft.

Ces brochures caractéristiques des années d'avant-guerre ont rempli leur office, à savoir transmettre les valeurs d'économie et de longévité chères à Plymouth. Généreuses surfaces de portée, sièges de soupape renforcés, freins à double action et carrosseries spacieuses tout en acier : l'acheteur potentiel avait de quoi tendre l'oreille.

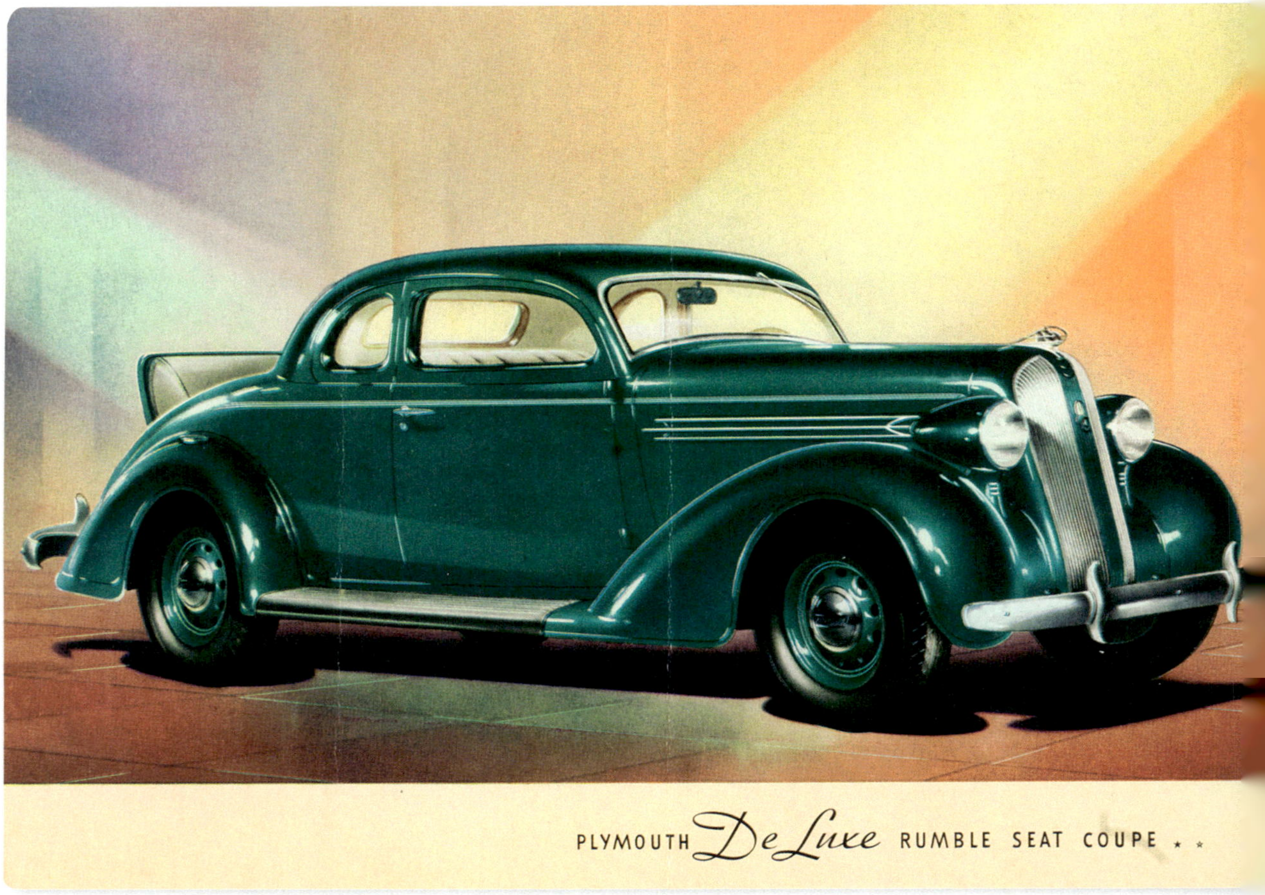

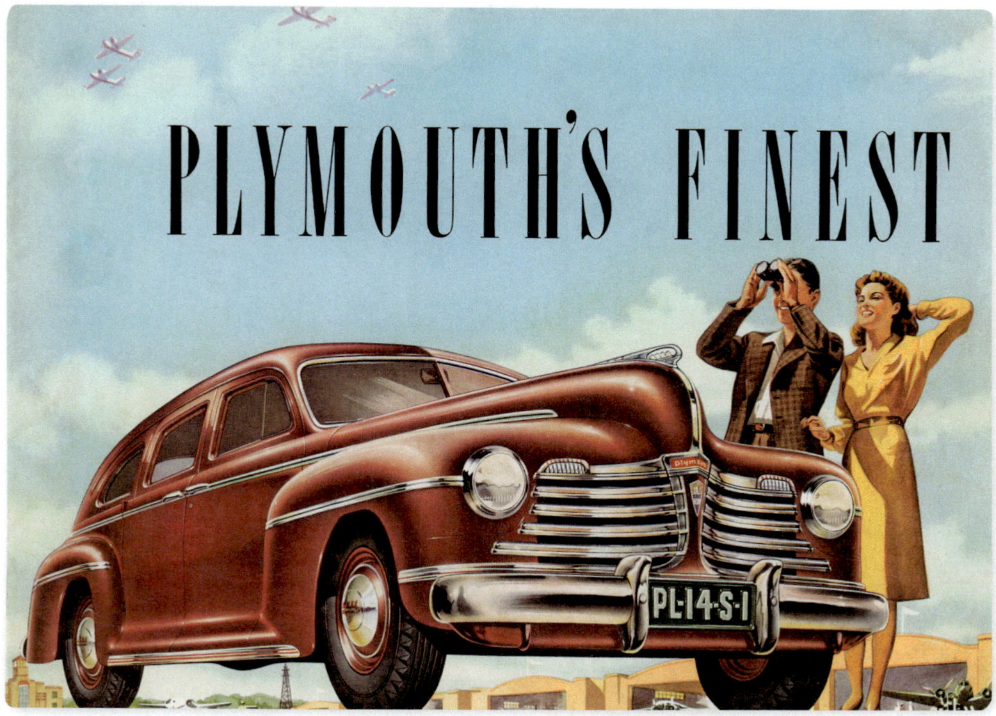
PLYMOUTH'S FINEST
PL-14-S-1

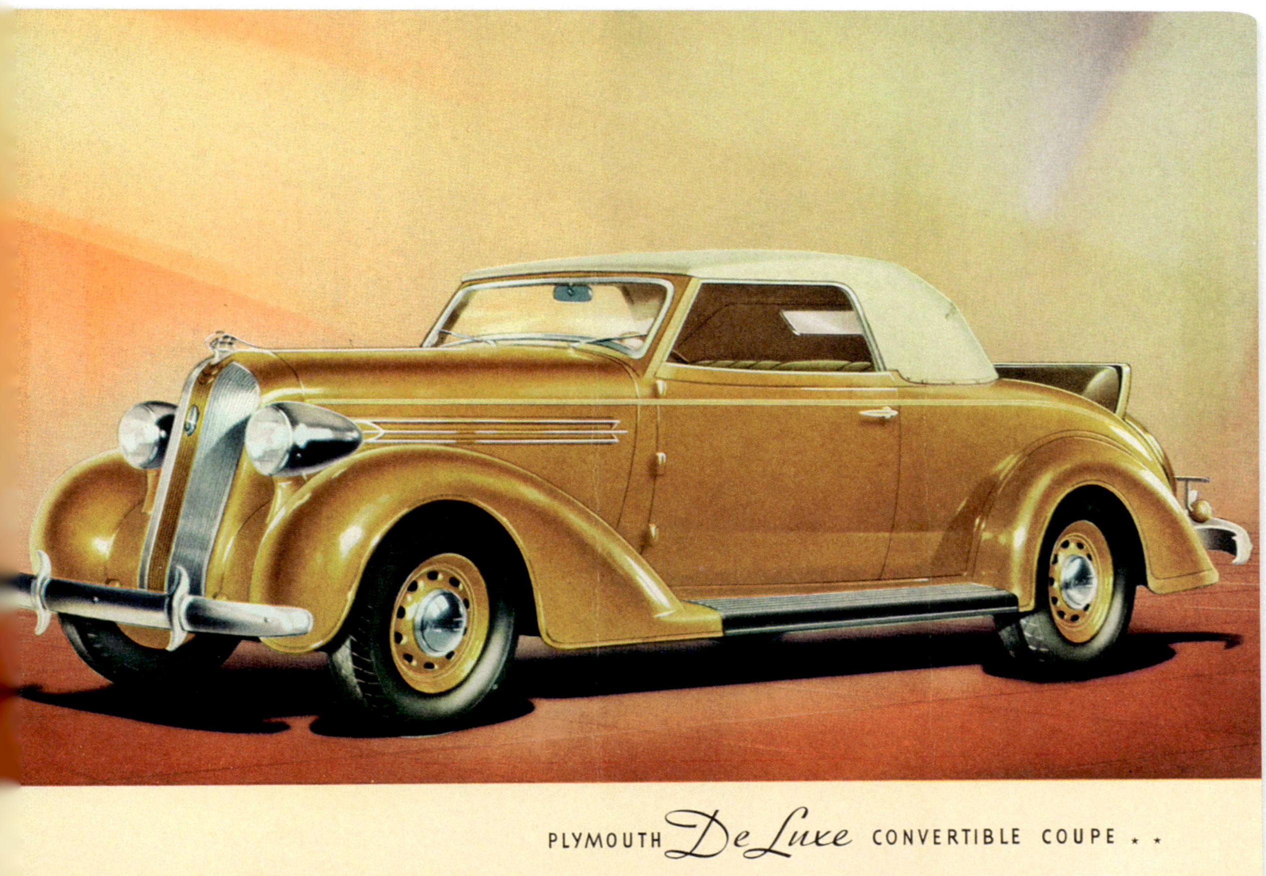
PLYMOUTH De Luxe CONVERTIBLE COUPE . .

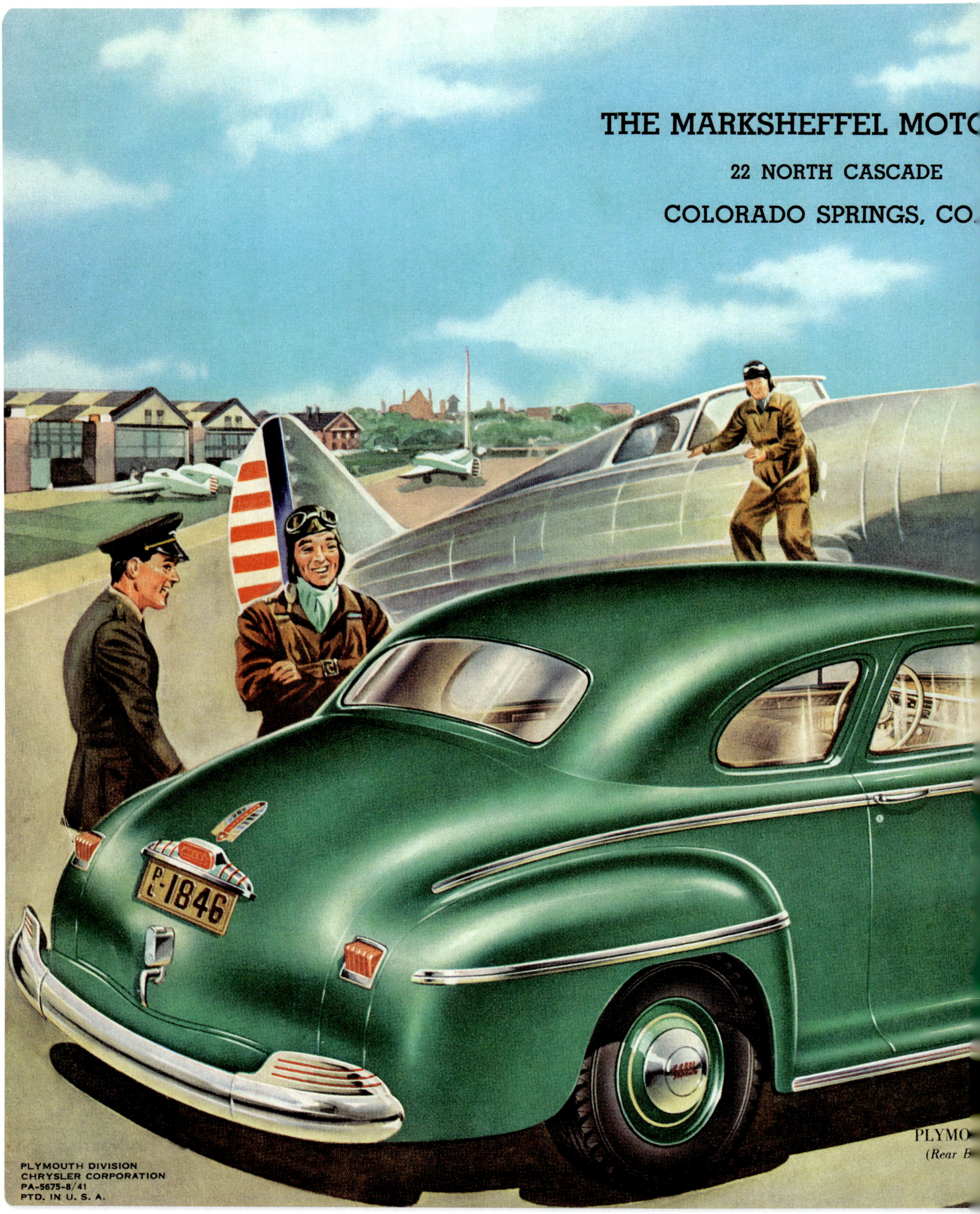
THE MARKSHEFFEL MOTO
22 NORTH CASCADE
COLORADO SPRINGS, CO.
P L 1846
PLYMO
(Rear B
PLYMOUTH DIVISION
CHRYSLER CORPORATION
PA-5675-8/41
PTD. IN U. S. A.

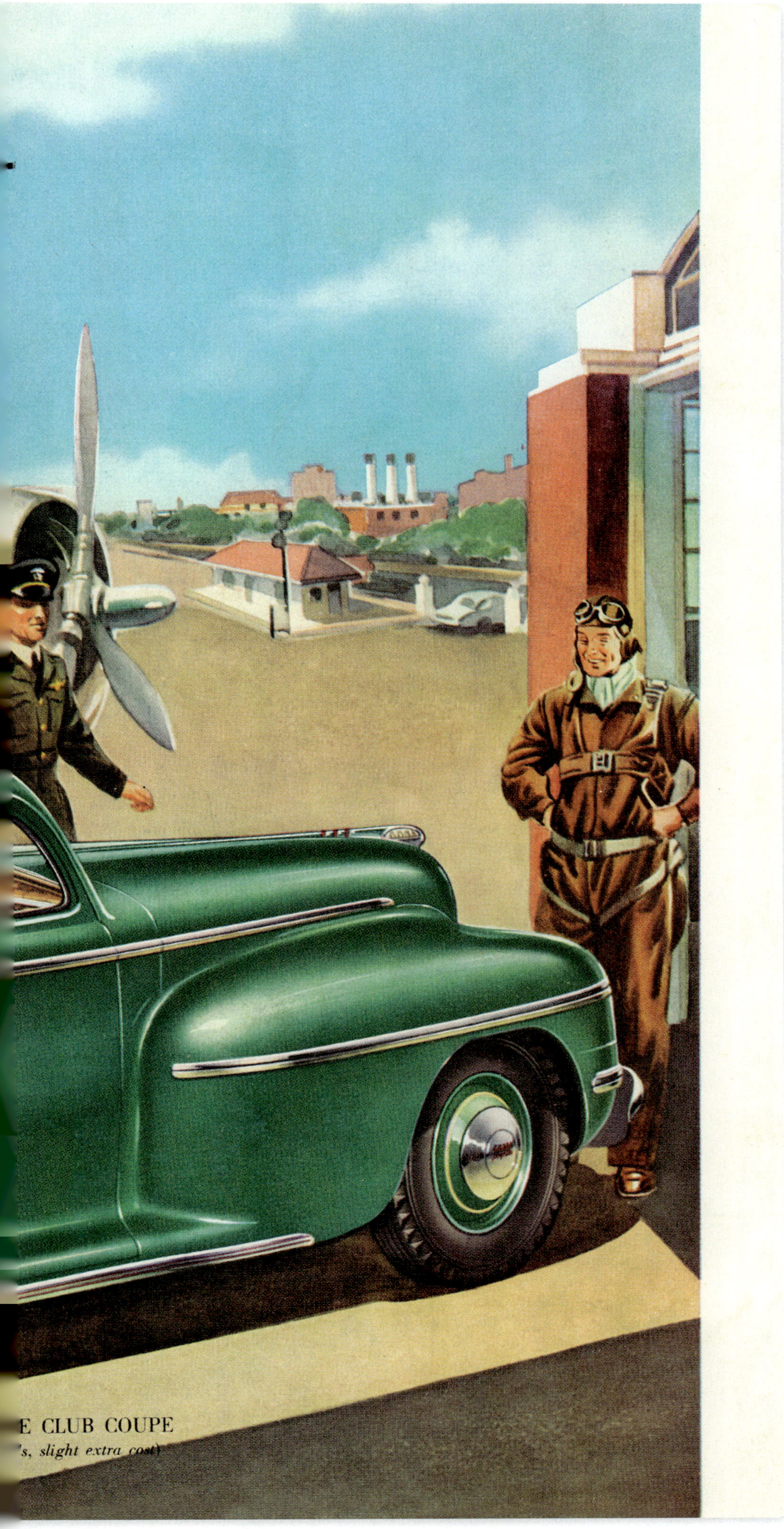

Plymouth 1942

The smiling men in uniform would have been extremely lucky to find one of these cars. Yet the new styling, combined with a new box-perimeter frame, meant crisp sales for Plymouth until World War II interrupted production. Some had painted trim due to chrome shortages.

Die lächelnden Männer in Uniform müssten schon sehr viel Glück gehabt haben, um eines dieser Automobile zu finden. Das neue Design, zusammen mit einem neuen Kastenrahmen, kurbelte zwar den Umsatz bei Plymouth an, doch der Zweite Weltkrieg legte die Produktion still. Weil Chrom Mangelware wurde, waren die Zierleisten bei einigen Fahrzeugen aufgemalt.

Nos souriants militaires auraient effectivement eu beaucoup de chance s'ils avaient pu tomber sur l'une de ces voitures. Le nouveau style, combiné à un nouveau châssis-caisson, généra d'excellents résultats commerciaux pour Plymouth avant que la Seconde Guerre mondiale n'impose une interruption de la production. Certaines garnitures étaient directement peintes, faute de chrome en ces temps difficiles.

Pontiac

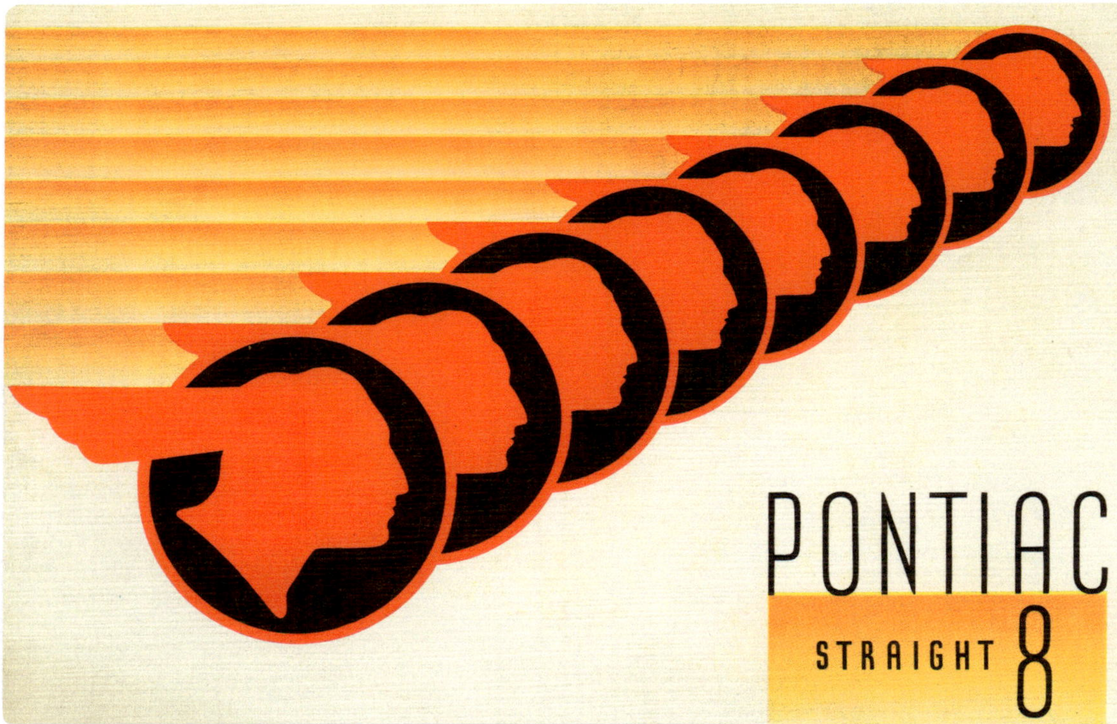

Pontiac 1933, 1936 *(opposite)*

Pontiac's iconic styling theme, the Silver Streak, appeared in 1935 and immediately became synonymous with the company. The streak trim moldings flowed down the hood and onto the grille. Silver Streak styling helped identify Pontiac well into the 1950s.

Das optische Erkennungszeichen von Pontiac, der „Silver Streak" („Silberstreif"), tauchte 1935 auf und wurde umgehend zum Inbegriff der Marke. Die Zierstreifen flossen über die Motorhaube und auf den Kühlergrill. Am Silver Streak waren die Pontiacs bis weit in die 1950er-Jahre hinein leicht zu identifizieren.

Le style Pontiac emblématique, dit « Silver Streak », apparu en 1935 et immédiatement devenu synonyme de la marque, était reconnaissable aux bandes chromées partant du capot moteur pour descendre vers la calandre. Le style Silver Streak servira de signature à Pontiac pratiquement jusqu'à la fin des années 1950.

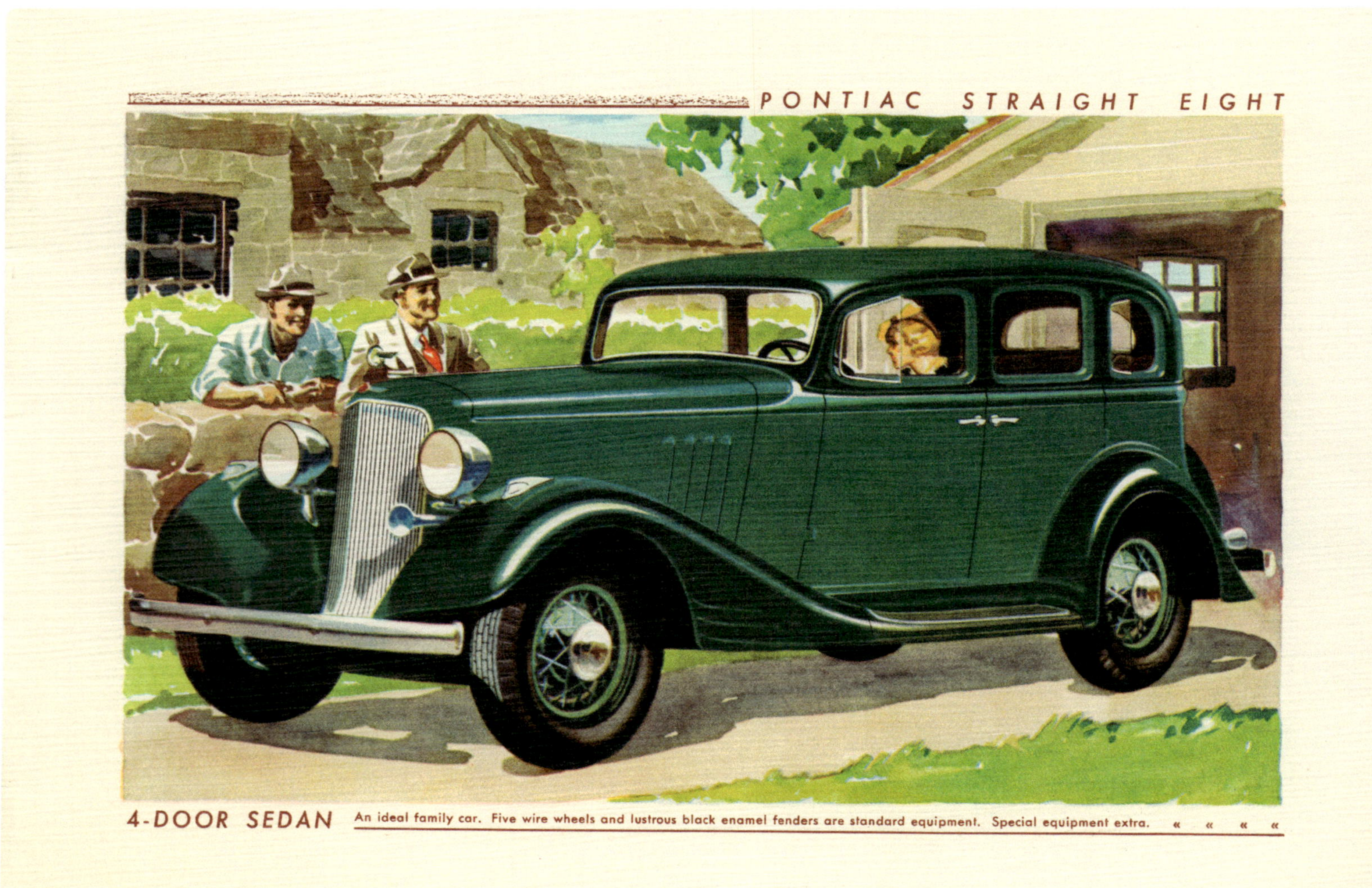

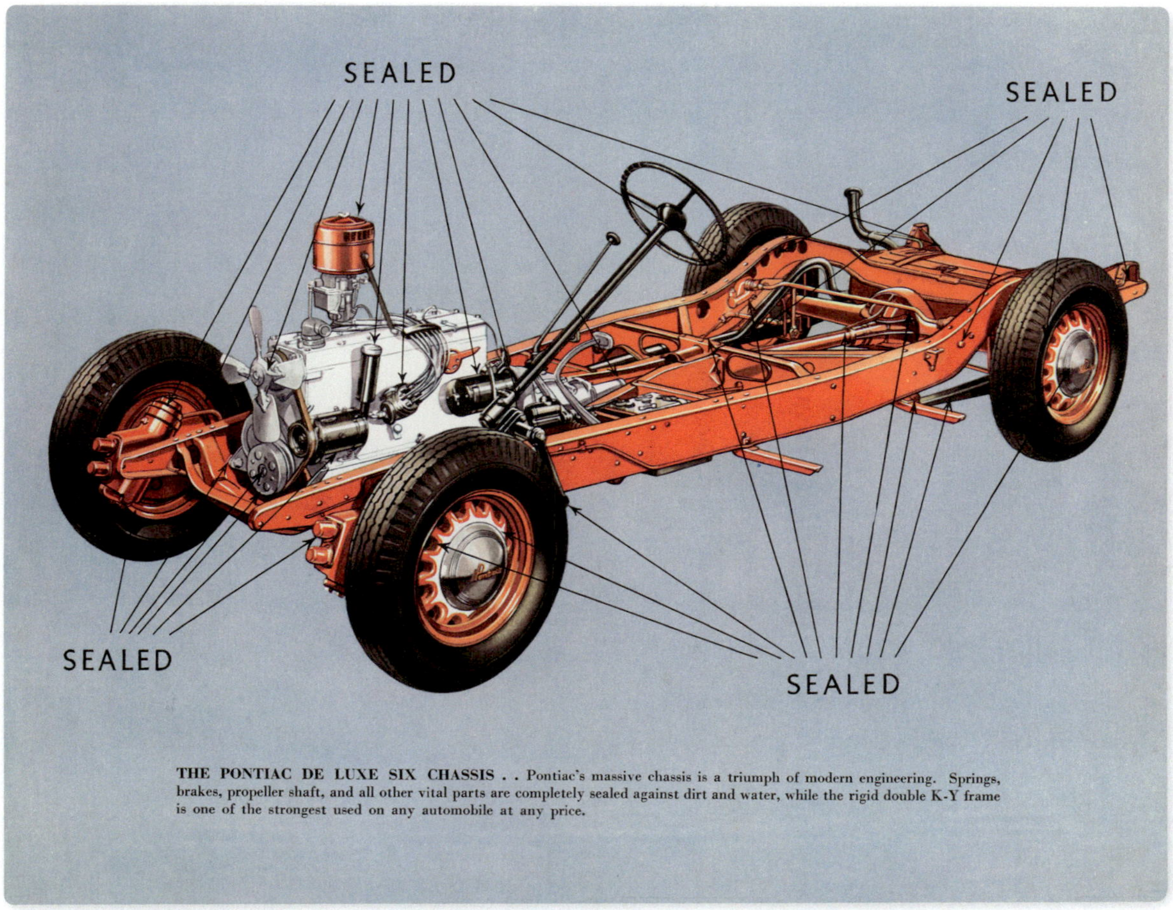
SEALED
SEALED
SEALED
SEALED
THE PONTIAC DE LUXE SIX CHASSIS . . Pontiac's massive chassis is a triumph of modern engineering. Springs, brakes, propeller shaft, and all other vital parts are completely sealed against dirt and water, while the rigid double K-Y frame is one of the strongest used on any automobile at any price.

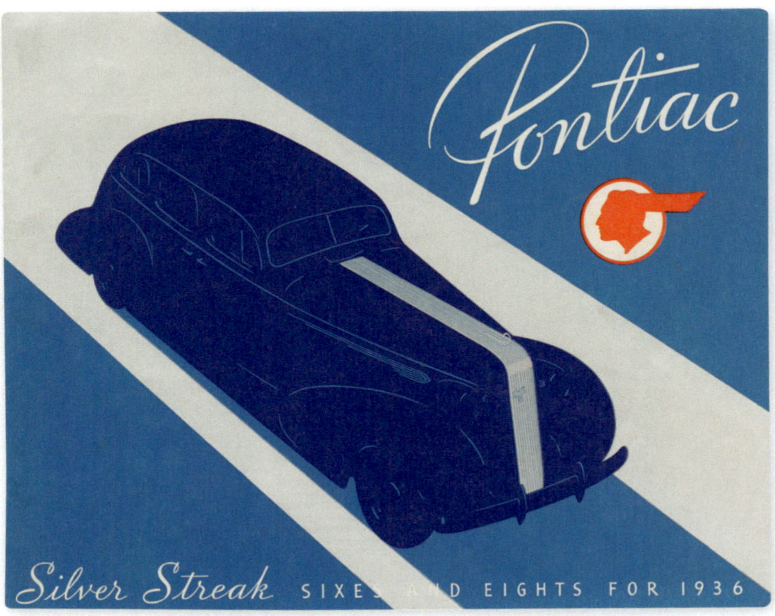
Pontiac
Silver Streak SIXES AND EIGHTS FOR 1936

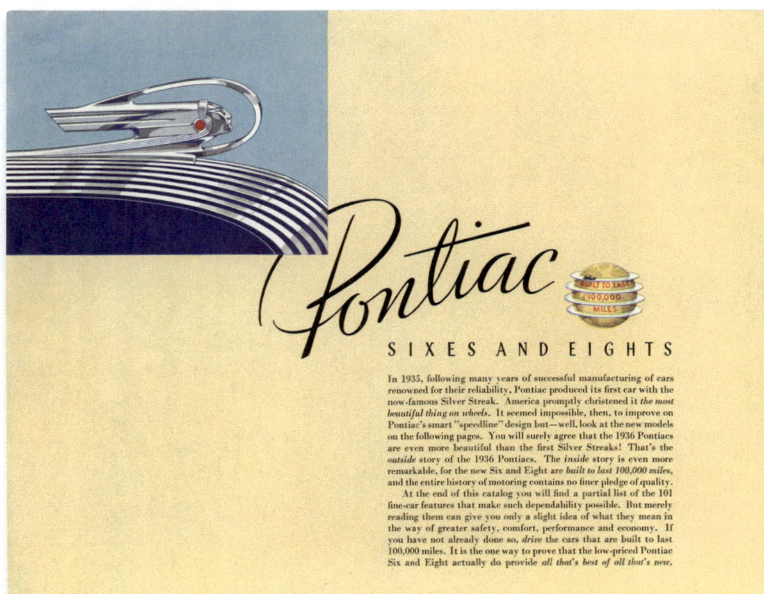
Pontiac
BUILT TO LAST 100,000 MILES
SIXES AND EIGHTS
In 1935, following many years of successful manufacturing of cars renowned for their reliability, Pontiac produced its first car with the now-famous Silver Streak. America promptly christened it *the most beautiful thing on wheels.* It seemed impossible, then, to improve on Pontiac's smart "speedline" design but—well, look at the new models on the following pages. You will surely agree that the 1936 Pontiacs are even more beautiful than the first Silver Streaks! That's the *outside* story of the 1936 Pontiacs. The *inside* story is even more remarkable, for the new Six and Eight are *built to last 100,000 miles,* and the entire history of motoring contains no finer pledge of quality.
At the end of this catalog you will find a partial list of the 101 fine-car features that make such dependability possible. But merely reading them can give you only a slight idea of what they mean in the way of greater safety, comfort, performance and economy. If you have not already done so, *drive* the cars that are built to last 100,000 miles. It is the one way to prove that the low-priced Pontiac Six and Eight actually do provide *all that's best of all that's new.*

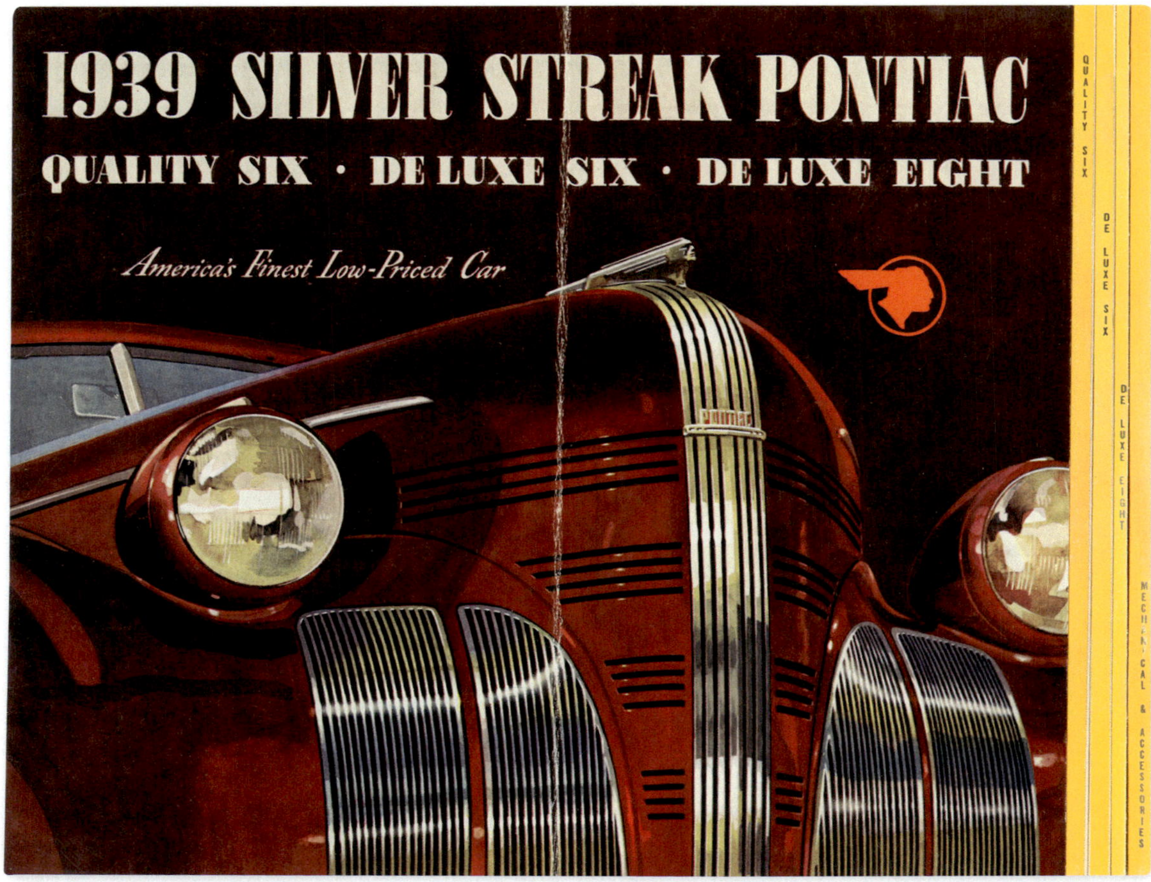

Pontiac 1939 *(above)*,
1940 *(below and opposite)*

Pontiac advertising in the prewar years exhibited variety in its themes, running from undiluted artwork to outtakes on the car's technical advances and style. Naturally, the iconic Silver Streak loomed over all else. Pontiac first appeared in 1926 as the eventual successor to Oakland in the mid-priced category. The name and logo take their identity from Pontiac, chief of the Ottawa tribe—namesake of Pontiac, Michigan, where the car was built.

In den Jahren zwischen den Kriegen thematisierte die Werbung von Pontiac sehr unterschiedliche Aspekte und reichte von reiner Kunst bis zu Details im Design und in den technischen Neuerungen des Fahrzeugs. Natürlich stand über allem der unverkennbare Silver Streak. Pontiac tauchte erstmals 1926 auf und löste schließlich den Oakland im mittleren Preissegment ab. Der Name und das Markenemblem gehen auf Obwandiyag (anglisiert Pontiac) zurück, einen Häuptling der Ottawa-Indianer, nach dem die Stadt Pontiac in Michigan benannt wurde, in der die Fahrzeuge gebaut wurden.

Pendant l'avant-guerre, la publicité Pontiac fit preuve d'une grande variété d'inspiration, passant de l'illustration brute aux saynètes chargées de mettre en valeur le style et les avancées techniques de la voiture. Naturellement, l'emblématique Silver Streak planait sur l'ensemble de cette communication. C'est en 1926 que Pontiac s'est imposé comme l'évident successeur d'Oakland sur le marché du milieu de gamme à petit prix. La marque empruntait son nom et le motif de son logo à Pontiac, chef de la tribu Ottawa et homonyme de la ville de Pontiac, dans le Michigan, où la voiture était fabriquée.

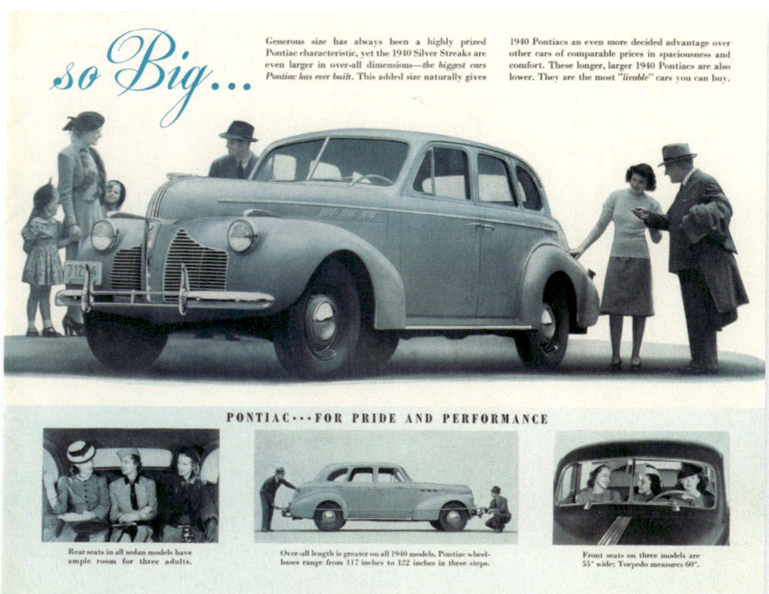

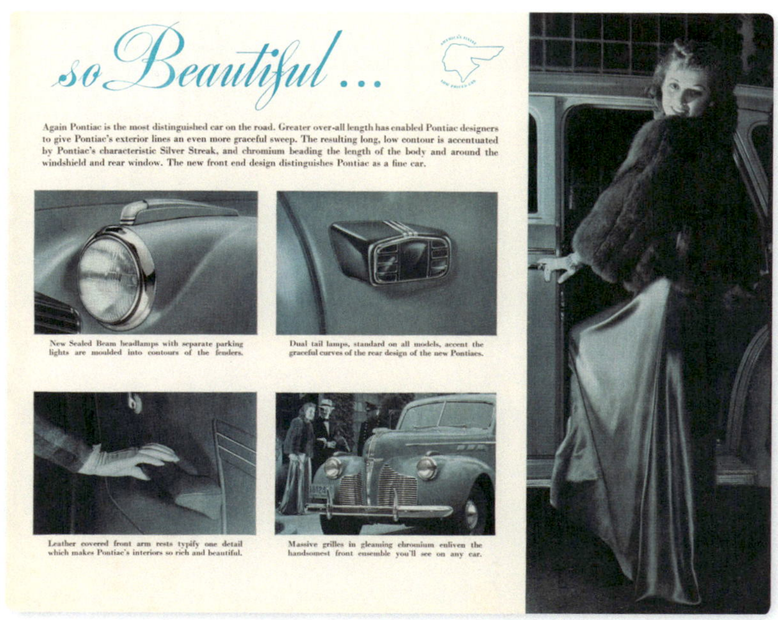

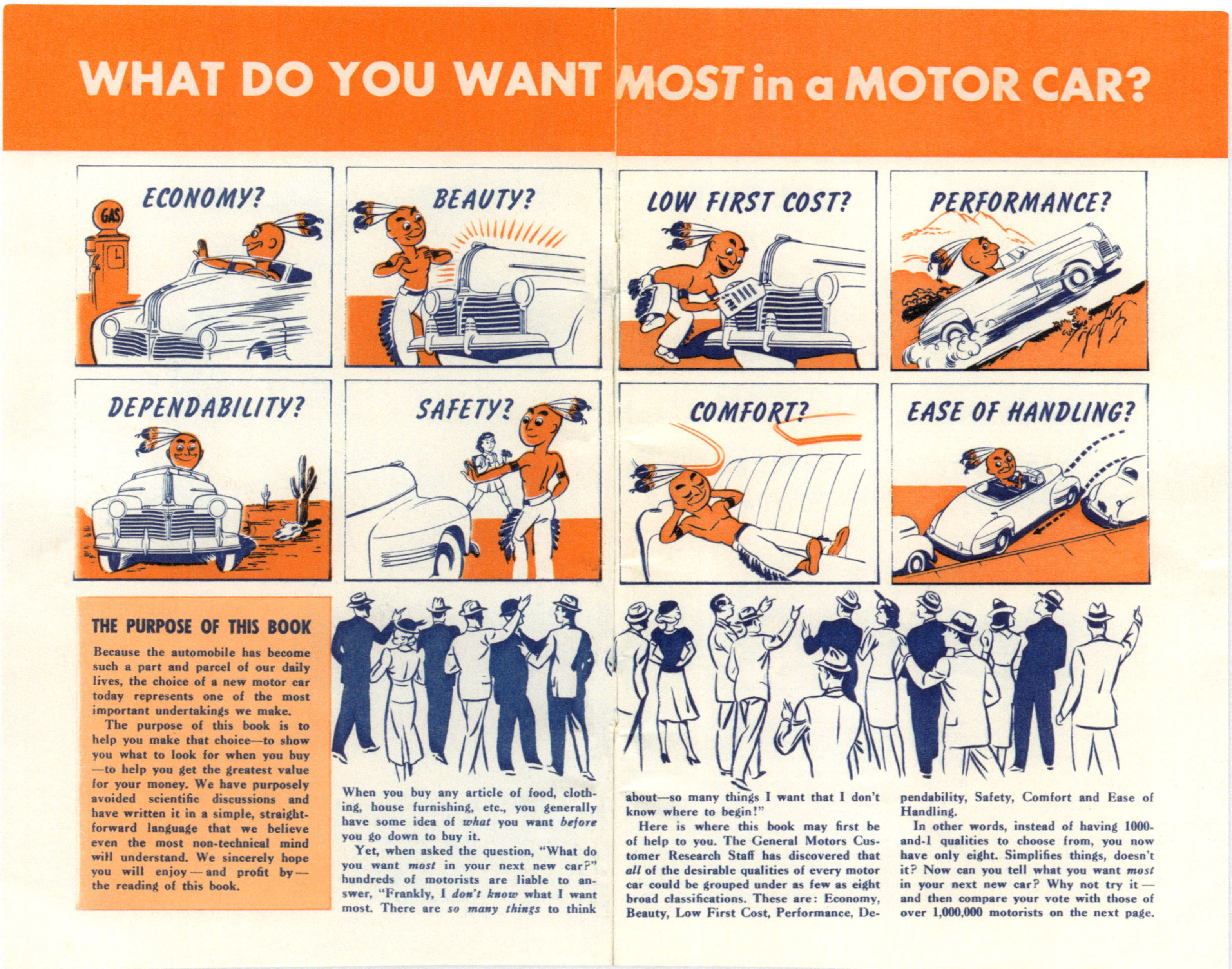

THE PURPOSE OF THIS BOOK

Because the automobile has become such a part and parcel of our daily lives, the choice of a new motor car today represents one of the most important undertakings we make.

The purpose of this book is to help you make that choice—to show you what to look for when you buy—to help you get the greatest value for your money. We have purposely avoided scientific discussions and have written it in a simple, straightforward language that we believe even the most non-technical mind will understand. We sincerely hope you will enjoy—and profit by—the reading of this book.

When you buy any article of food, clothing, house furnishing, etc., you generally have some idea of *what* you want *before* you go down to buy it.

Yet, when asked the question, "What do you want *most* in your next new car?" hundreds of motorists are liable to answer, "Frankly, I *don't know* what I want most. There are *so many things* to think about—so many things I want that I don't know where to begin!"

Here is where this book may first be of help to you. The General Motors Customer Research Staff has discovered that *all* of the desirable qualities of every motor car could be grouped under as few as eight broad classifications. These are: Economy, Beauty, Low First Cost, Performance, Dependability, Safety, Comfort and Ease of Handling.

In other words, instead of having 1000-and-1 qualities to choose from, you now have only eight. Simplifies things, doesn't it? Now can you tell what you want *most* in your next new car? Why not try it—and then compare your vote with those of over 1,000,000 motorists on the next page.

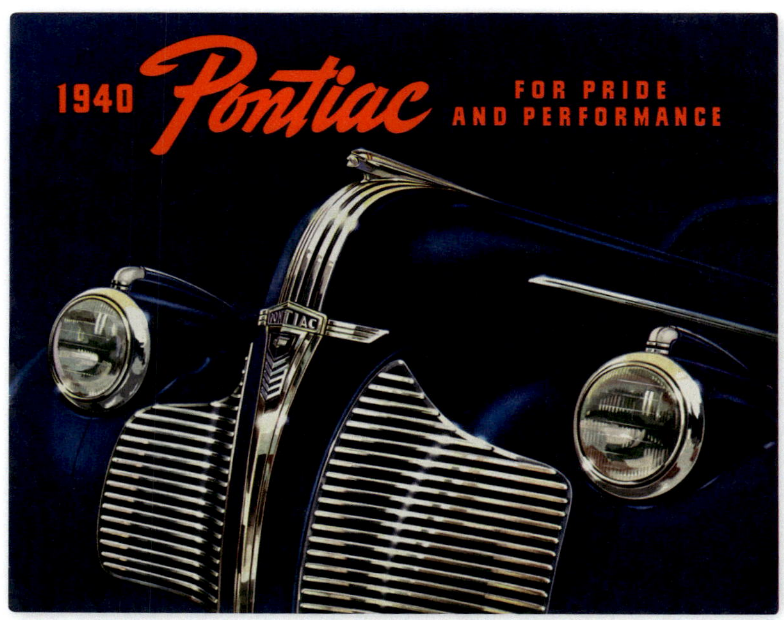

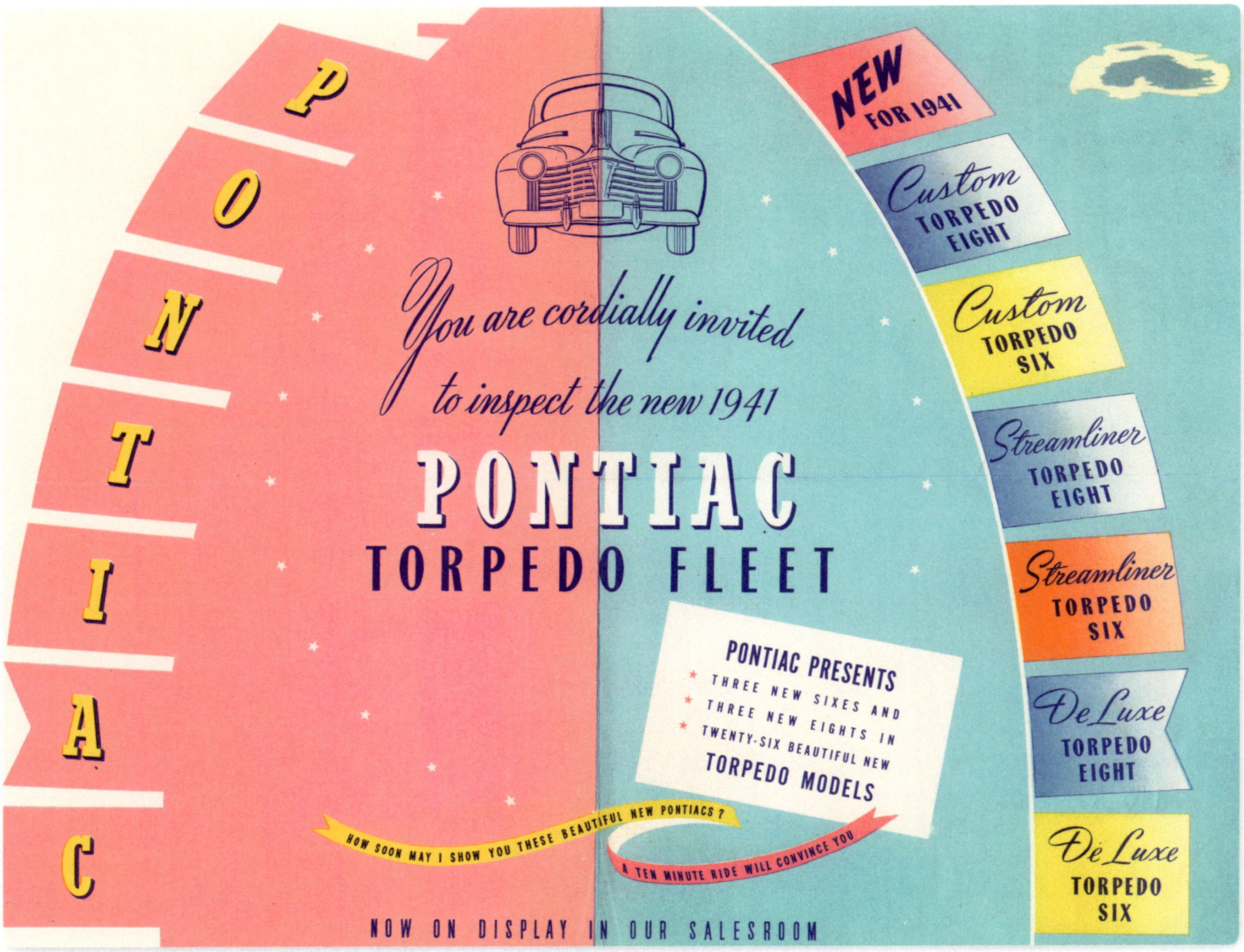

Pontiac 1941

With a creative approach, the nautical signal flags, salty-dressed lady, and the Torpedo nameplate suggest America's impending military future, and also gently references a nautical theme.

Mit diesem kreativen Ansatz, den Signalflaggen aus der Seefahrt, der jungen Dame im Matrosenoutfit und dem Wort „Torpedo" im Modellnamen stellte Pontiac seine Fahrzeuge nicht nur ganz allgemein in einen nautischen Zusammenhang, sondern deutete auch in die nahe Zukunft der amerikanischen Streitkräfte.

Dans une ambiance tout en créativité, les pavillons inspirés par le code maritime, la tenue iodée de la matelote et le nom « Torpedo » (torpille) suggèrent autant l'imminence de l'intervention militaire des États-Unis qu'une thématique nautique plus apaisée.

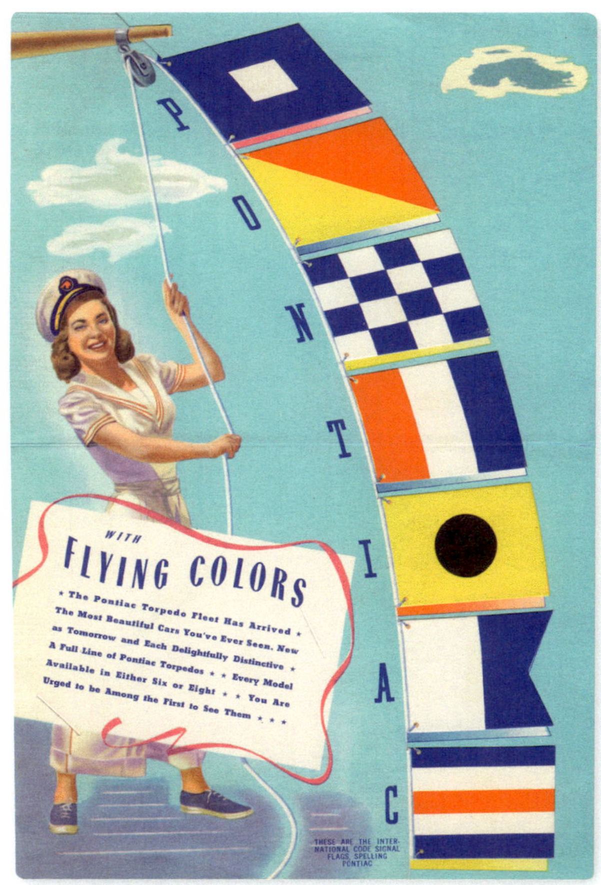

REO

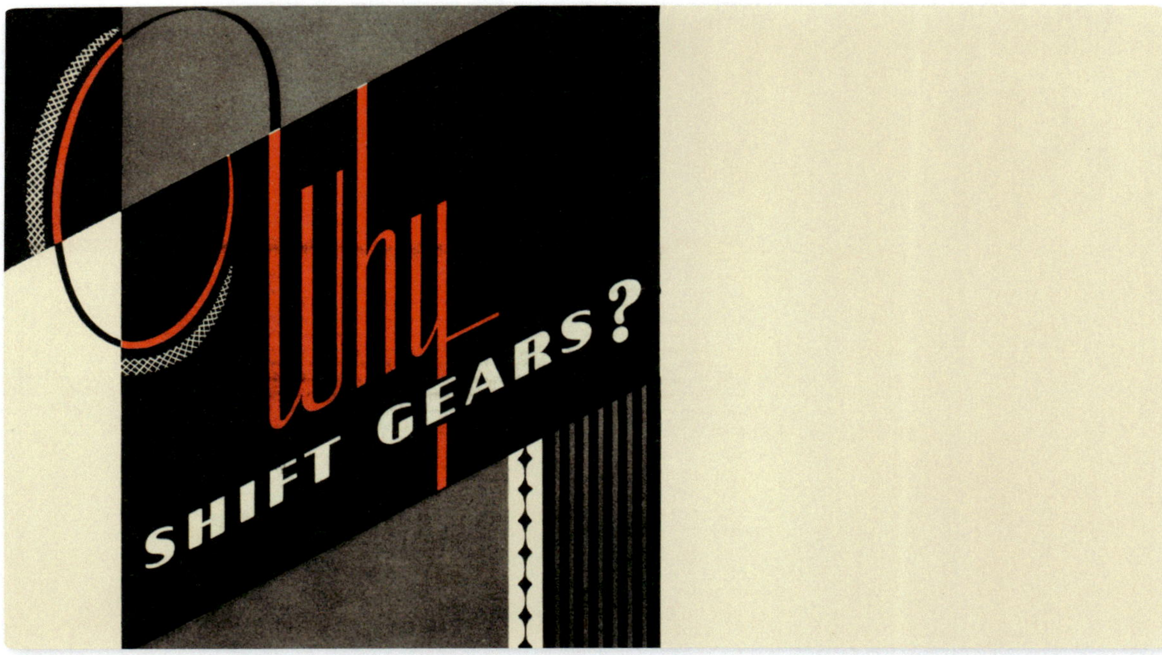

REO Royale 1934

You could say that Ransom Eli Olds saved his best car for last. The Royale was magnificent, using a 125 hp straight-eight-cylinder engine with nine main bearings attached to the patented Self-Shifter. Promoted in this brochure, it was one of the first workable designs for an automatic transmission.

Man könnte sagen, dass sich Ransom Eli Olds sein bestes Automobil für den Schluss aufgehoben hatte. Der Royale war ein prächtiges Fahrzeug mit einem 170-Kilowatt-Reihenachtzylindermotor mit neun Hauptlagern, die mit dem patentierten „Self-Shifter“ verbunden waren. Die Schaltung, die in dieser Broschüre beworben wird, war eines der ersten brauchbaren Automatikgetriebe.

On pourra dire que Ransom Eli Olds avait gardé sa meilleure voiture pour la fin. La Royale était magnifique, avec son moteur à huit cylindres en ligne qui développait fièrement 125 chevaux, son vilebrequin à neuf paliers et la transmission semi-automatique brevetée maison dite « Self-Shifter », l'une des premières du genre : cette brochure la célèbre à sa juste valeur.

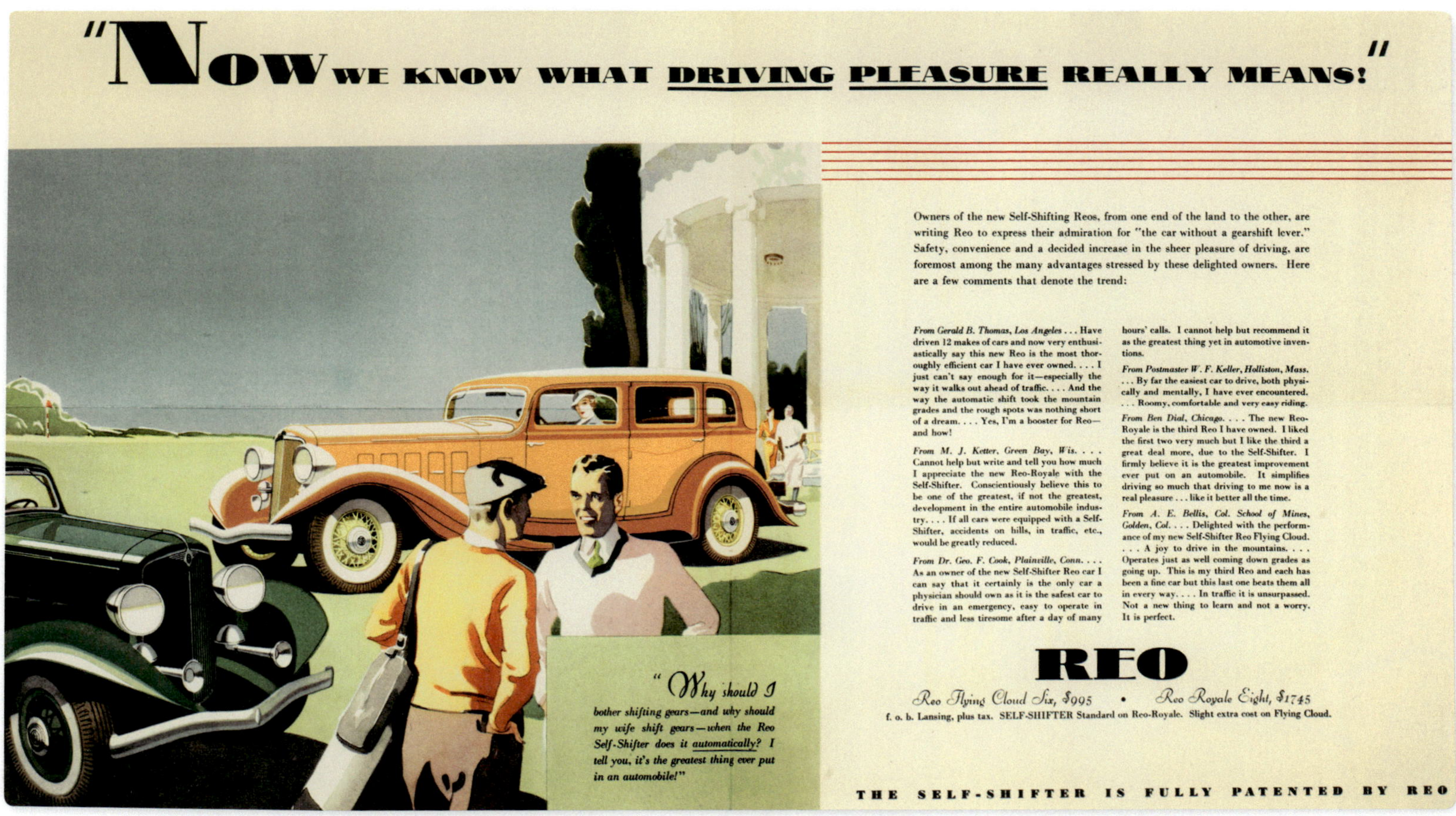

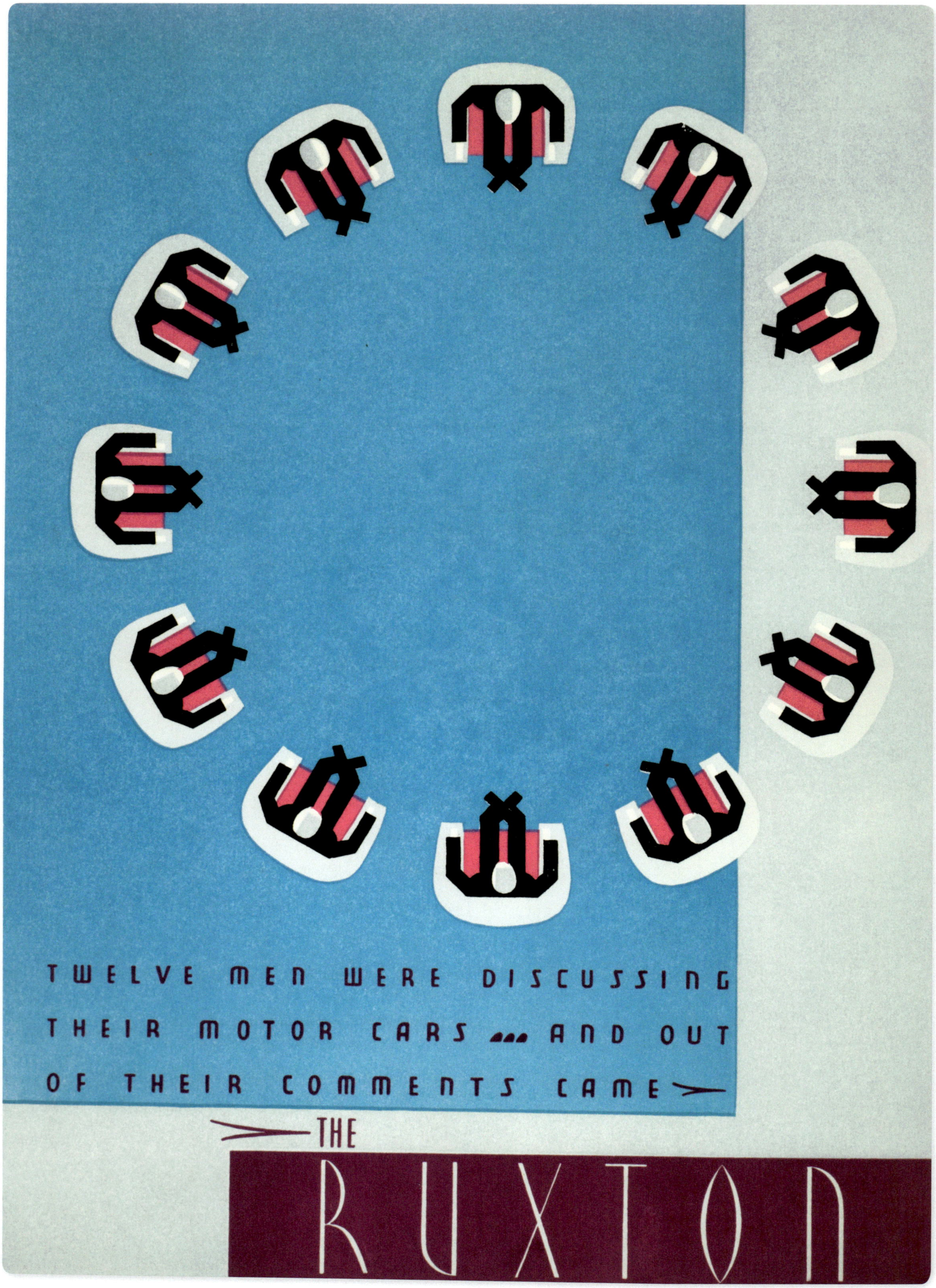
TWELVE MEN WERE DISCUSSING
THEIR MOTOR CARS ... AND OUT
OF THEIR COMMENTS CAME
THE
RUXTON

Ruxton

Ruxton 1929 *(this spread)*

Due to numerous legal and production hurdles, barely 500 Ruxton automobiles were built. The brochure concentrates instead on the luxurious fabrics and eye-popping, multihued paints offered on the radically low, Studebaker-powered Ruxton. Today the Ruxton is considered a highly advanced car, buried by bad management.

Aufgrund zahlreicher rechtlicher und produktionstechnischer Hindernisse wurden nur knapp 500 Ruxton-Automobile gebaut. Die Broschüre konzentriert sich aber auf die luxuriösen Stoffe und die augenfälligen mehrfarbigen Lackierungen, die für den radikal flachen Ruxton mit Studebaker-Motor angeboten wurden. Heutzutage gilt der Ruxton als äußerst fortschrittliches Fahrzeug, das einer schlechten Geschäftsführung zum Opfer fiel.

Par la faute de multiples obstacles juridiques et industriels, la production des automobiles Ruxton excèdera à peine 500 modèles. Le catalogue préfère se concentrer sur les luxueux revêtements intérieurs et la palette de couleurs éblouissante proposés pour la très basse Ruxton à moteur Studebaker. On la regarde aujourd'hui comme une voiture en avance sur son temps, mais sabotée par une mauvaise gestion.

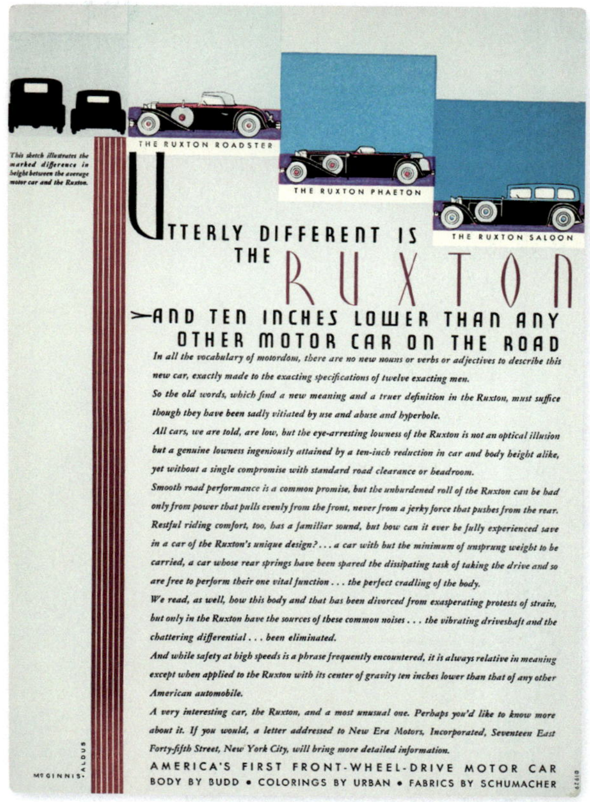

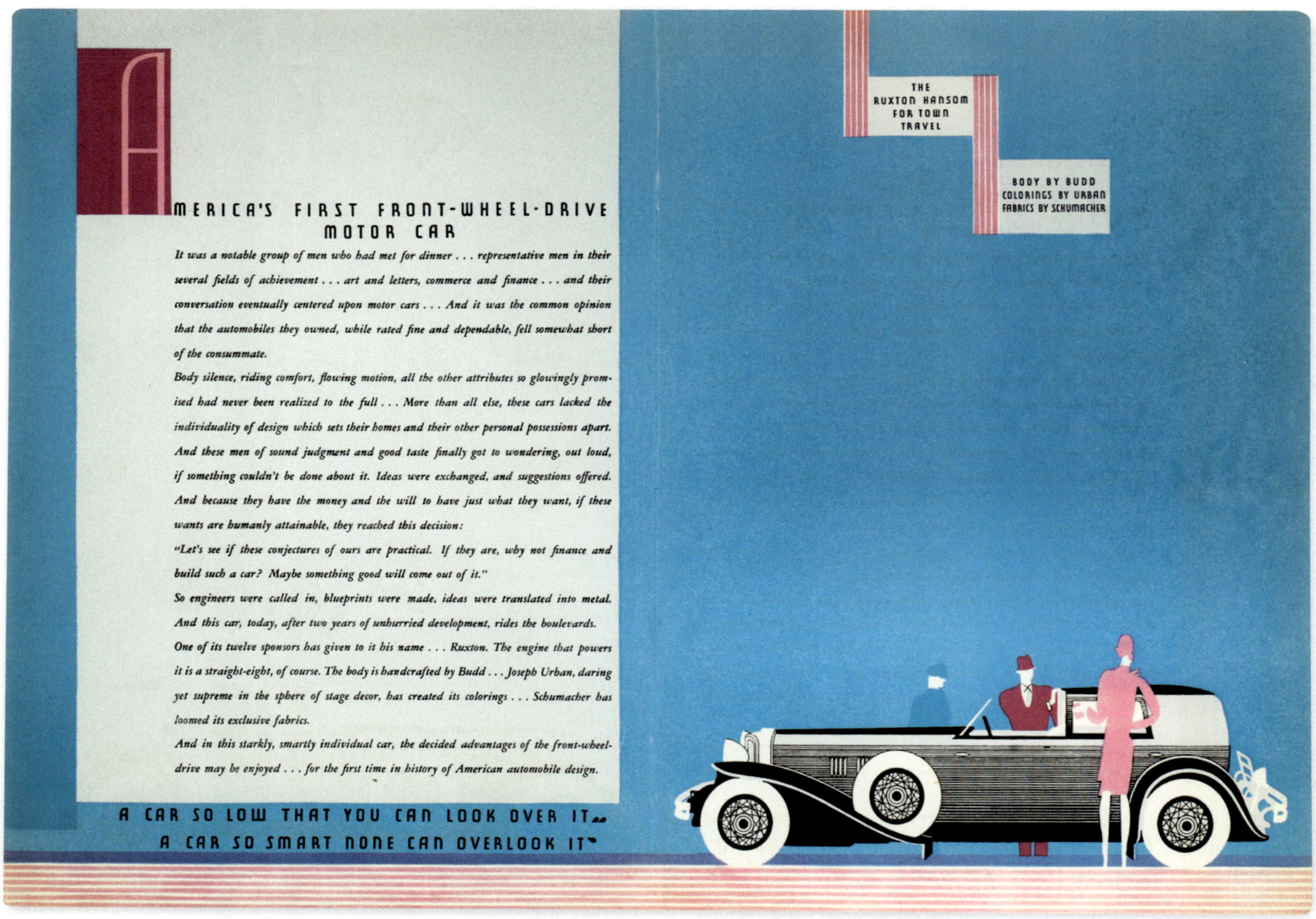

Studebaker

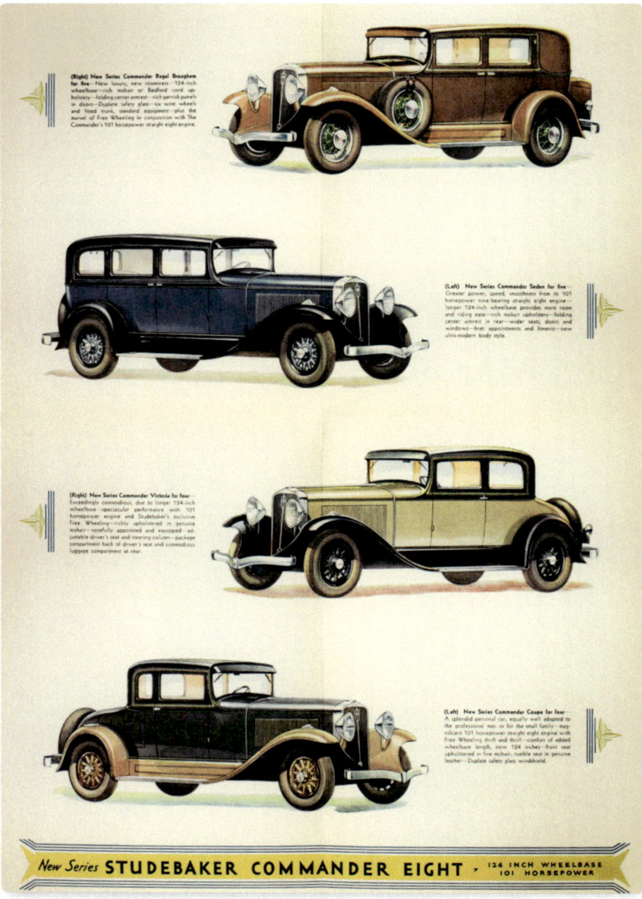

Studebaker 1931 *(left)*, **1936** *(below)*, **1931** *(opposite)*

By 1934 Studebaker had set speed and endurance records on speedways from Indianapolis to Atlantic City. Despite the somewhat florid, airy prose of its brochures, Studebaker took its reputation for performance very seriously.

Im Jahre 1934 hatte der Studebaker schon Geschwindigkeits- und Ausdauerrekorde auf Rennstrecken von Indianapolis bis Atlantic City aufgestellt. Trotz der etwas blumigen und luftigen Prosa seiner Broschüren nahm Studebaker seinen Ruf als Hersteller leistungsstarker Autos sehr ernst.

En 1934, Studebaker avait déjà battu des records de vitesse et d'endurance sur des anneaux de vitesse comme Indianapolis ou Atlantic City. En dépit de la prose quelque peu ampoulée et éthérée de ses catalogues, Studebaker prenait très au sérieux sa réputation en matière de performance.

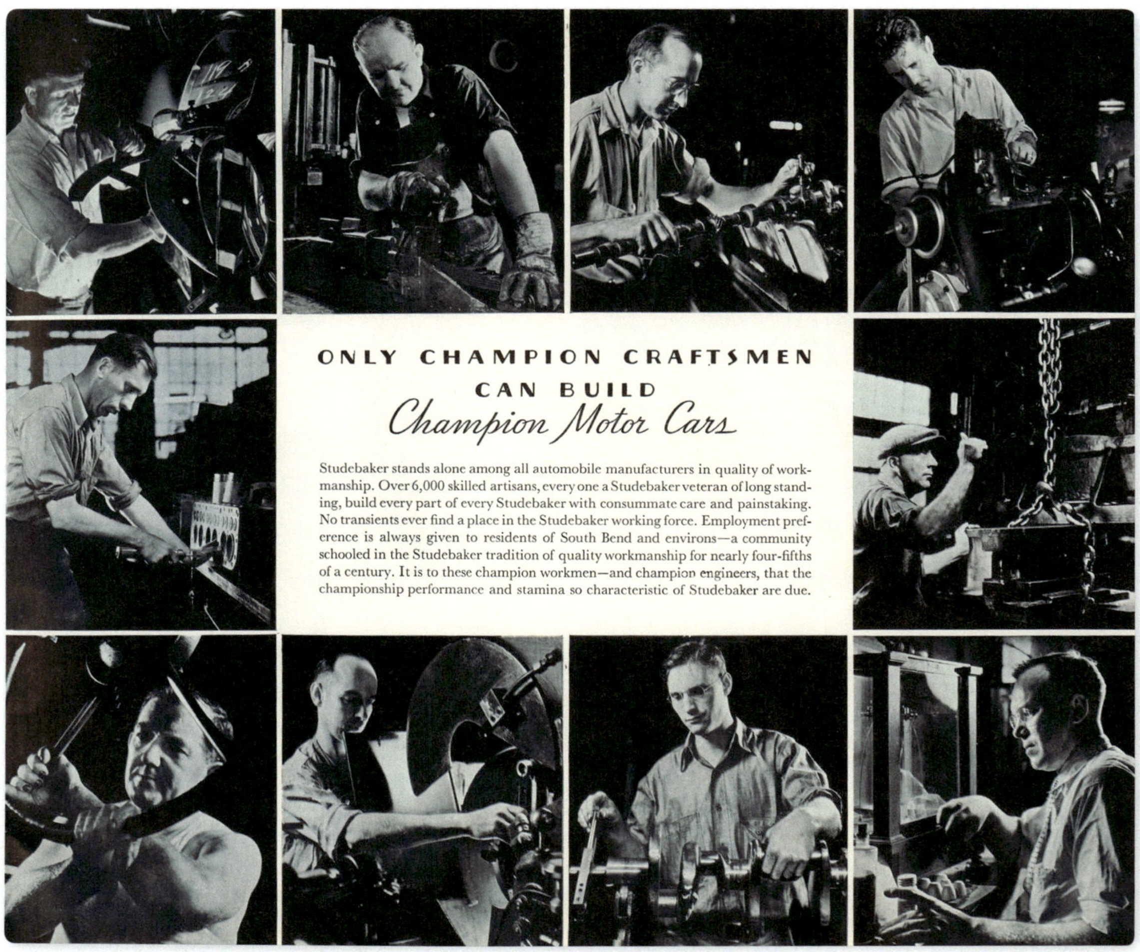

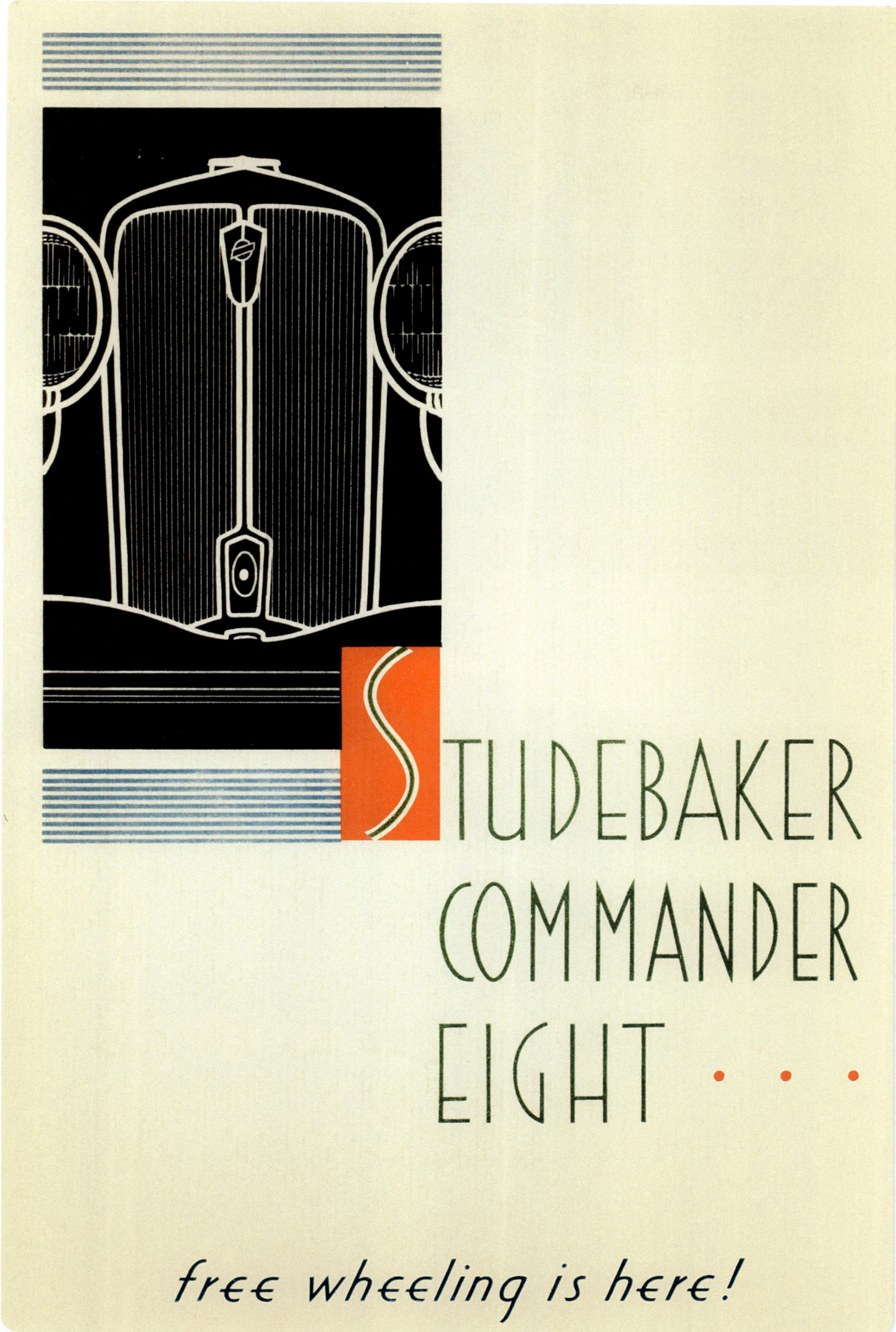
STUDEBAKER
COMMANDER
EIGHT . . .
free wheeling is here!

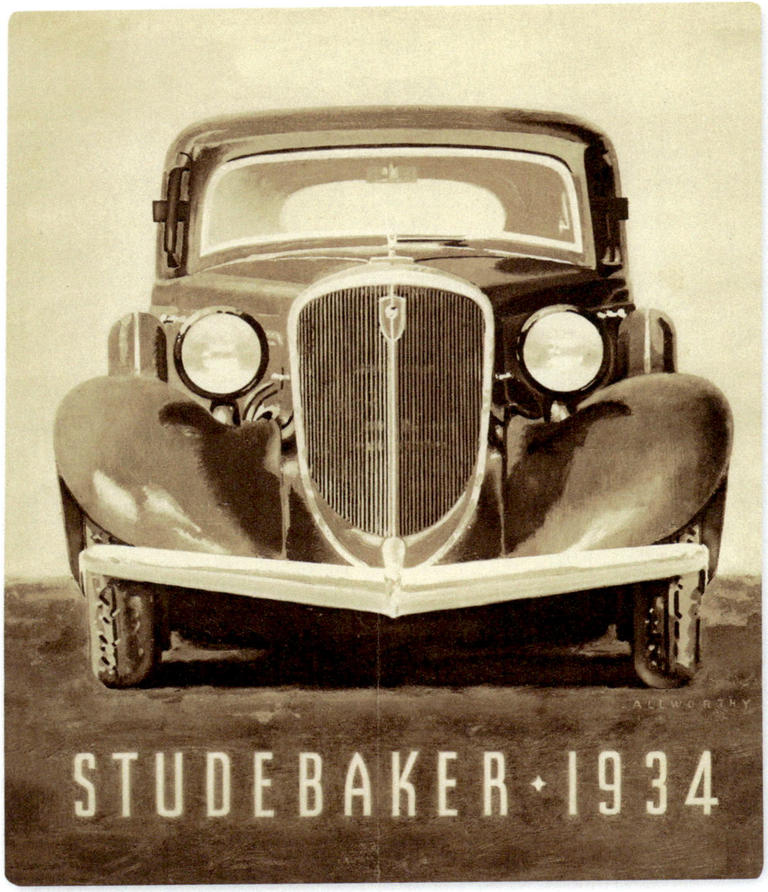

Studebaker 1934 *(left)*, **1936** *(below)*, **1941** *(opposite top)*, **1936** *(opposite bottom)*

Not long before gasoline and tires for pleasure travel became unobtainable, Studebaker encouraged people to buy a car and travel coast to coast, and perhaps even encounter a Navajo merchant.

Kurz bevor es keinen Treibstoff und keine Reifen für Vergnügungsfahrten mehr gab, ermunterte Studebaker seine Kunden noch einmal, ein Auto zu kaufen und damit von einer Küste der USA zur anderen zu fahren – und dabei vielleicht sogar einer Navajo-Indianerin zu begegnen, die ihre Waren feilbietet.

Peu de temps avant que le carburant et les pneumatiques dédiés au voyage d'agrément ne deviennent inaccessibles en raison du conflit mondial, Studebaker encourageait les Américains à acheter une voiture et à visiter leur pays d'est en ouest, « coast to coast ». Et peut-être même à rencontrer en route une colporteuse navajo.

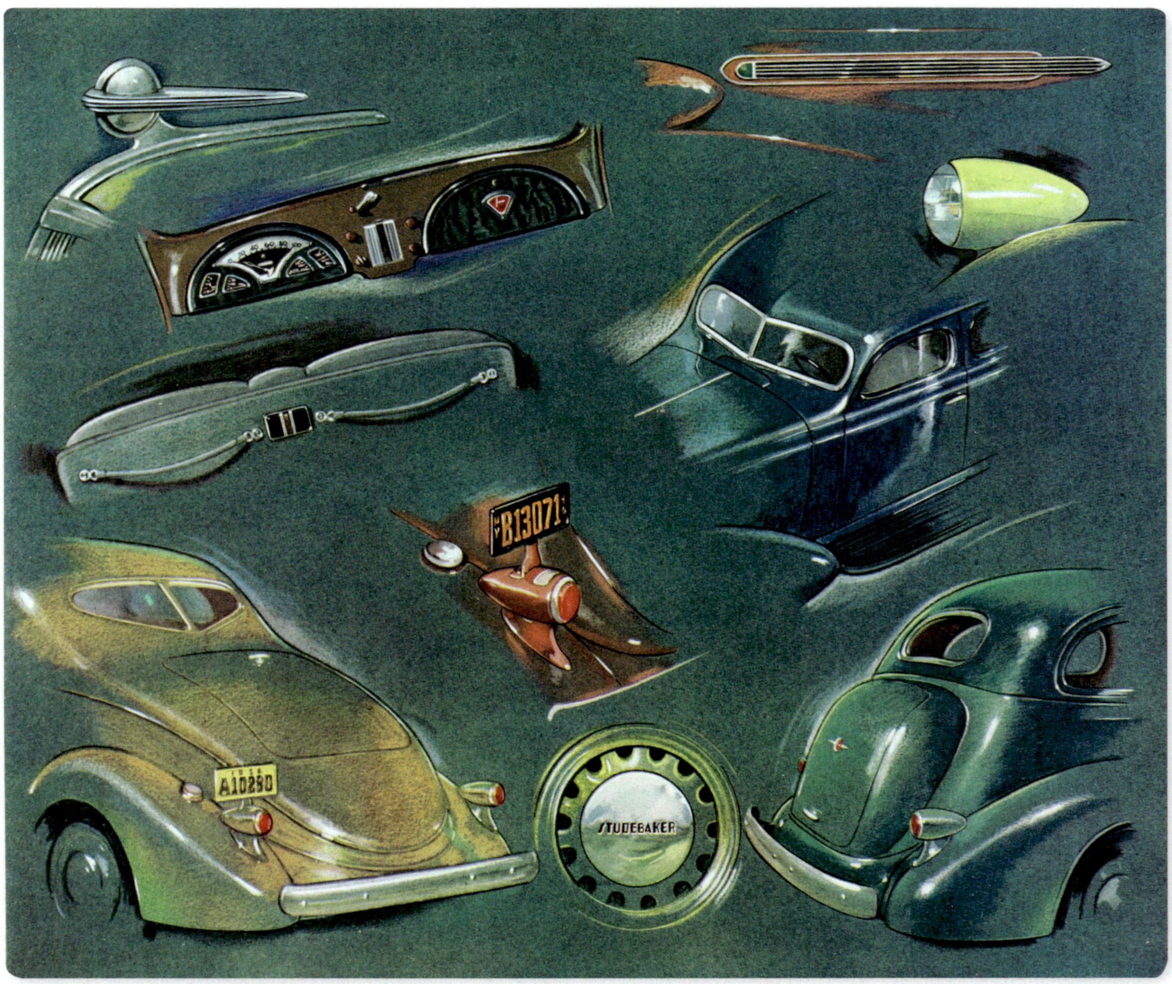

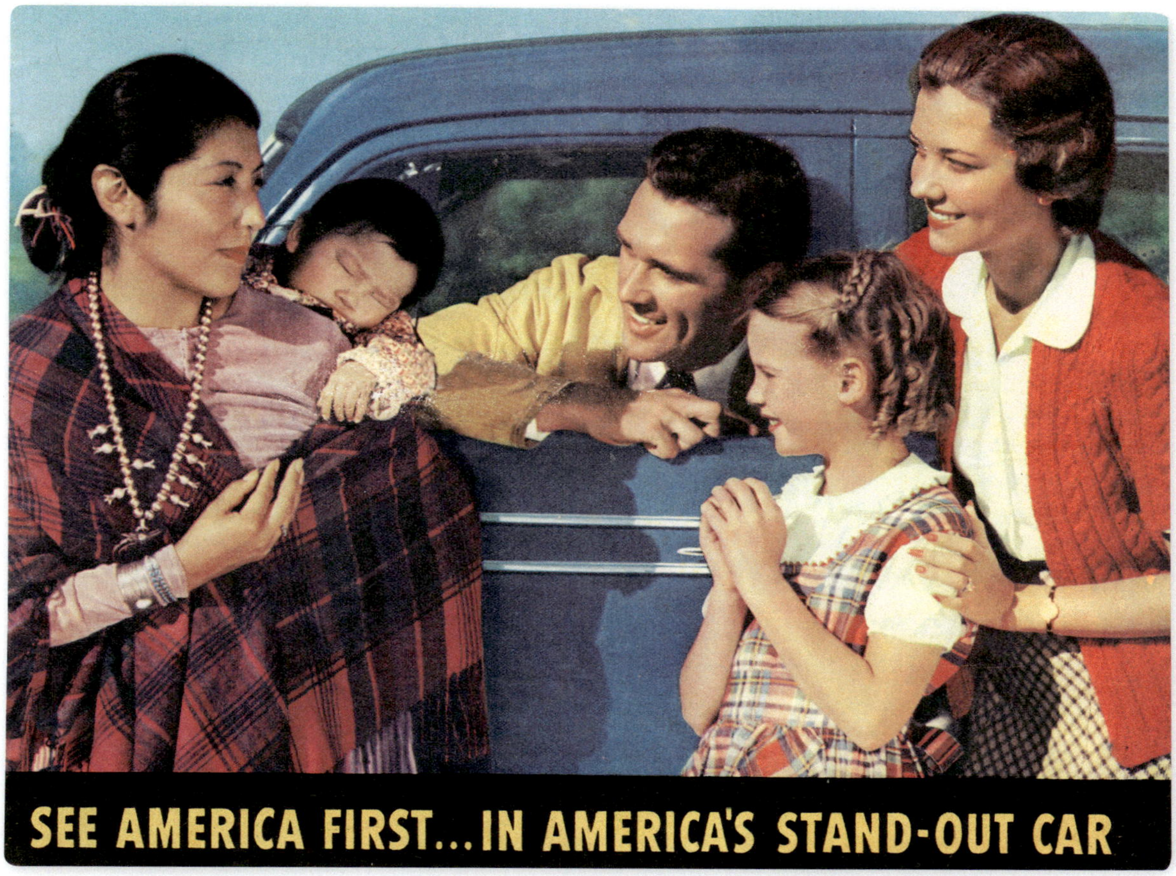
SEE AMERICA FIRST...IN AMERICA'S STAND-OUT CAR

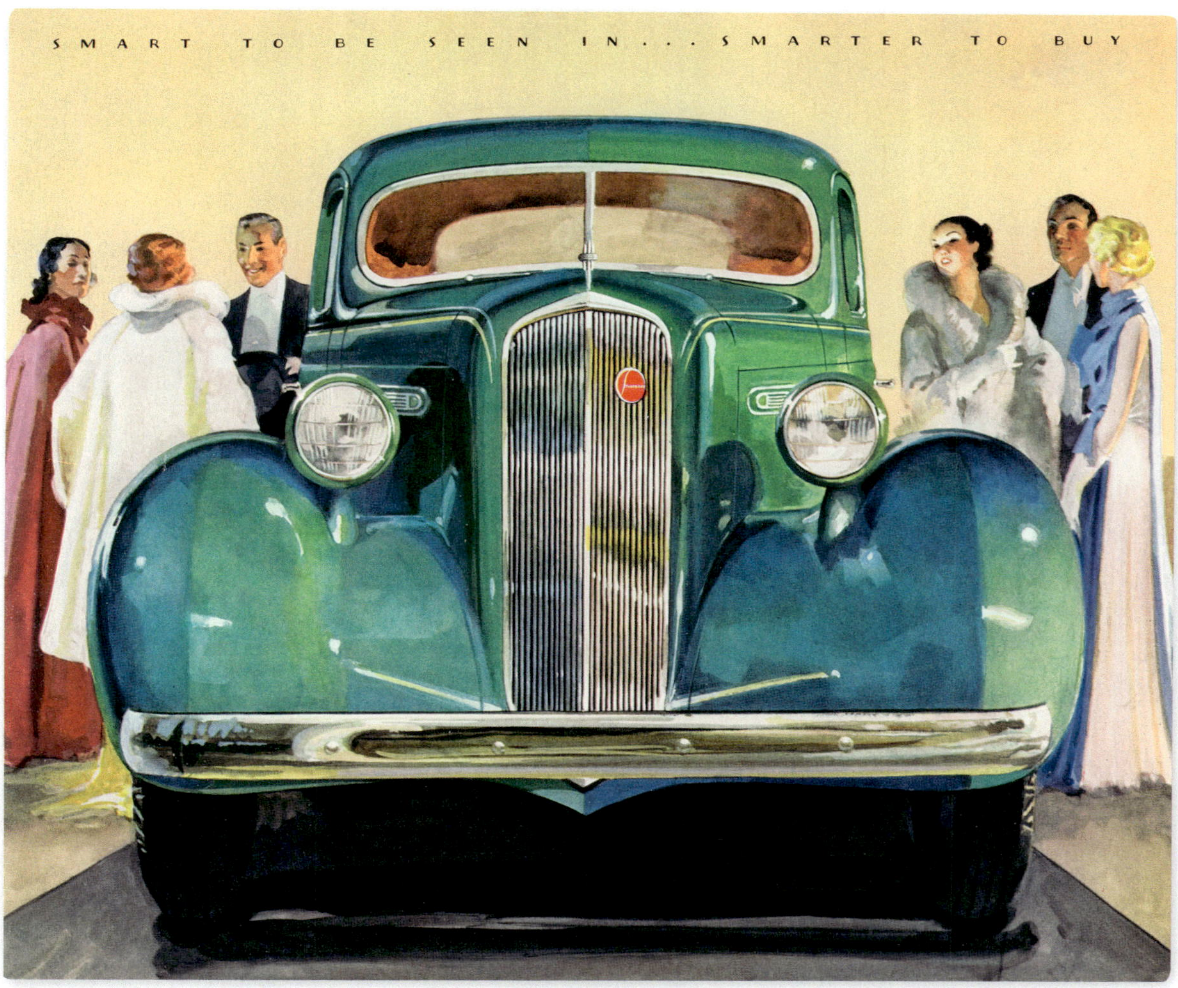
SMART TO BE SEEN IN... SMARTER TO BUY

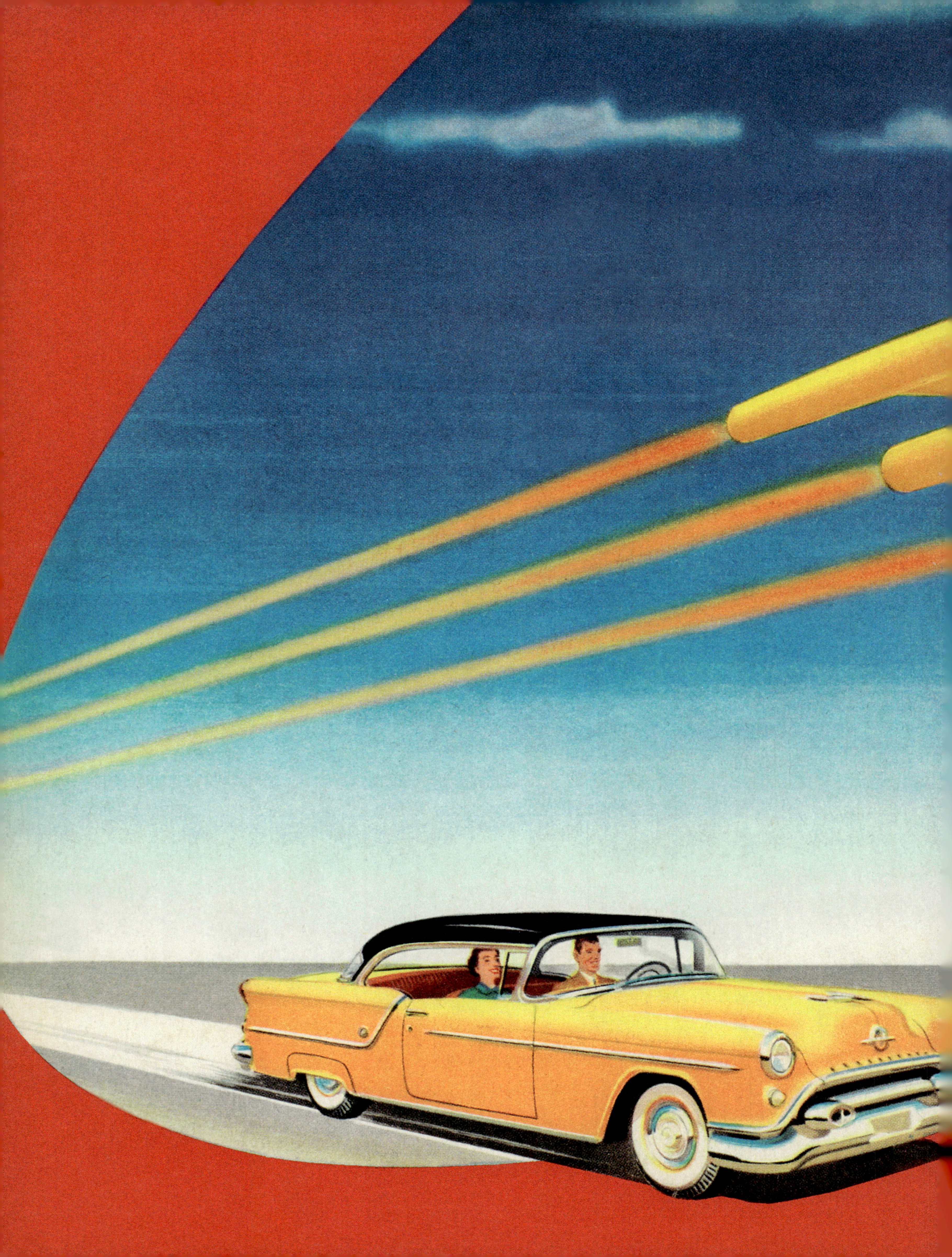

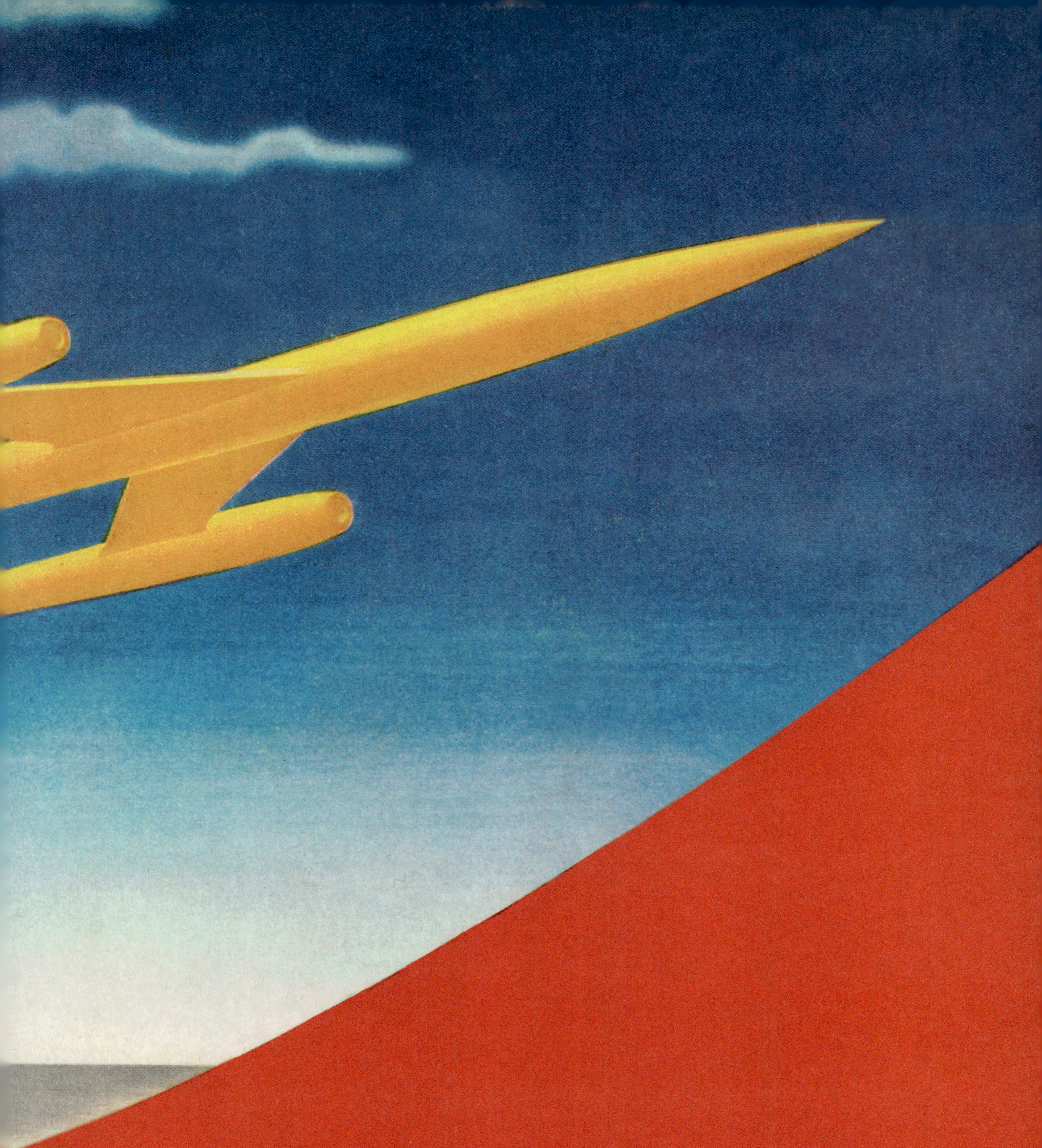

Stalwarts of Marketing
1947 – 1961

American Motors

American Motors Rambler 1955, 1960
(opposite)

If you purchased a 1956 Rambler, you were proudly different. These cars were not only sturdy, but designed not to be mistaken for any other car. Highlights included their wraparound rear windows and up to three loud colors. While the copy spoke to utility, the images said something very different.

Wenn Sie einen 1956er-Rambler kauften, dann waren Sie garantiert anders – und stolz darauf. Diese Fahrzeuge waren nicht nur robust, sondern so gestaltet, dass man sie mit keinem anderen Auto verwechseln konnte. Zu den Glanzlichtern gehörten gebogene Heckfenster und bis zu drei grelle Farben. Während der Text hauptsächlich auf den Nutzen des Fahrzeugs abhob, sprachen die Bilder eine ganz andere Sprache.

Si vous aviez fait l'acquisition d'une Rambler 1956, alors vous étiez différent et vous pouviez en être fier. Ces automobiles n'étaient pas seulement robustes mais également dessinées pour qu'on ne puisse pas les confondre avec d'autres. On remarquait notamment leur lunette arrière panoramique et jusqu'à trois couleurs bien voyantes. Si le texte insistait sur la fonctionnalité de la voiture, les images disaient tout autre chose.

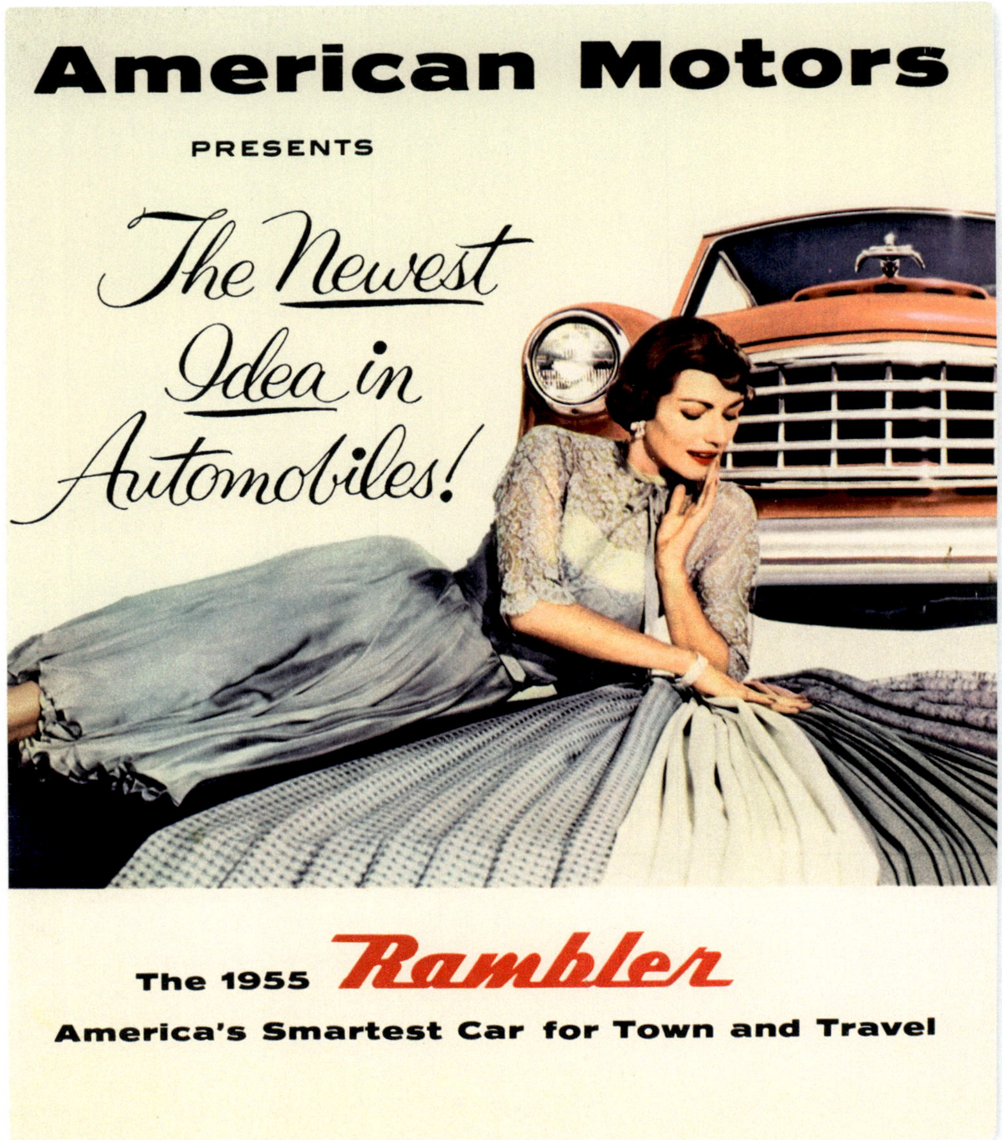

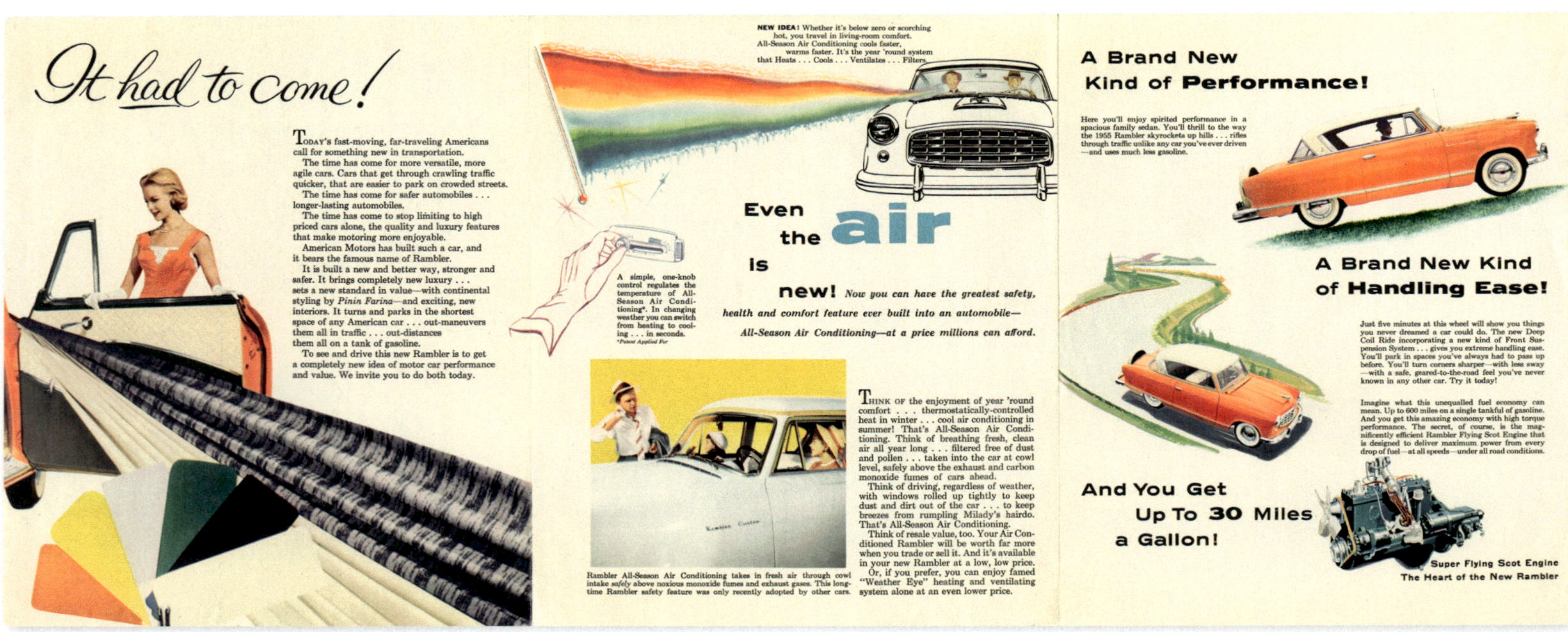

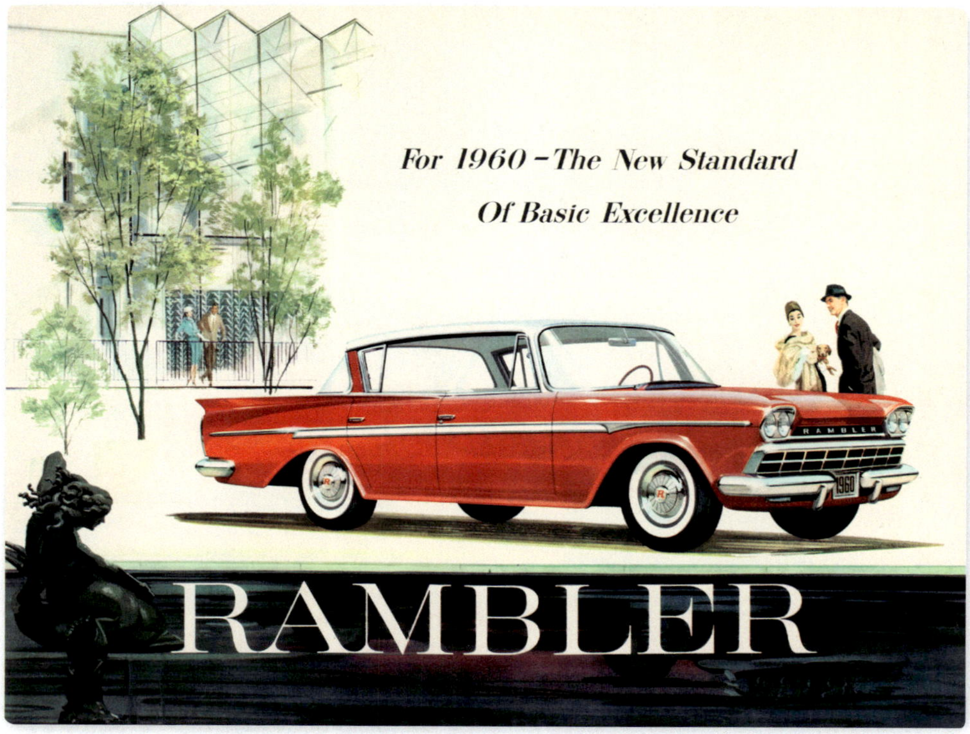
For 1960 – The New Standard
Of Basic Excellence
RAMBLER

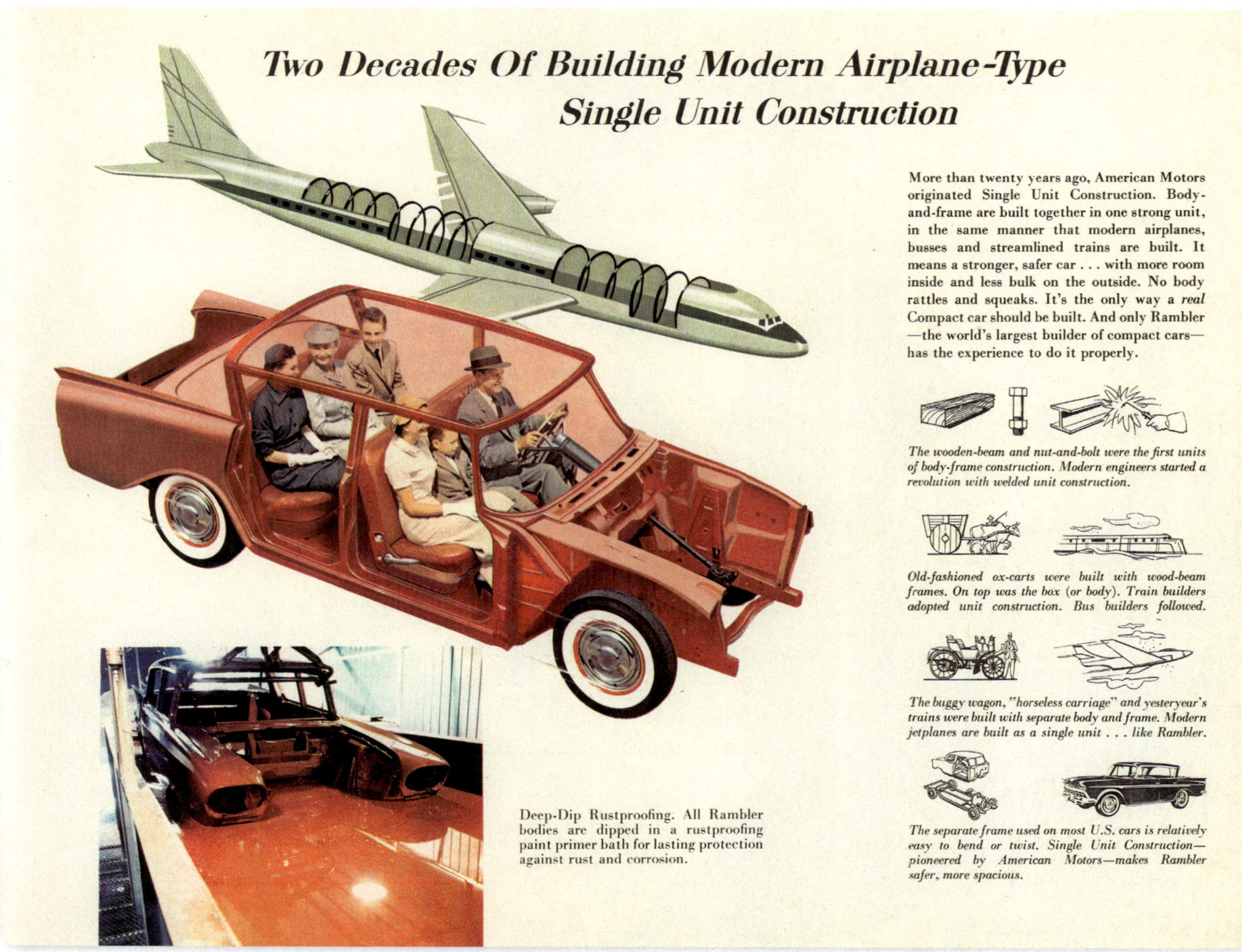
Two Decades Of Building Modern Airplane-Type Single Unit Construction
More than twenty years ago, American Motors originated Single Unit Construction. Body-and-frame are built together in one strong unit, in the same manner that modern airplanes, busses and streamlined trains are built. It means a stronger, safer car . . . with more room inside and less bulk on the outside. No body rattles and squeaks. It's the only way a real Compact car should be built. And only Rambler —the world's largest builder of compact cars— has the experience to do it properly.
The wooden-beam and nut-and-bolt were the first units of body-frame construction. Modern engineers started a revolution with welded unit construction.
Old-fashioned ox-carts were built with wood-beam frames. On top was the box (or body). Train builders adopted unit construction. Bus builders followed.
The buggy wagon, "horseless carriage" and yesteryear's trains were built with separate body and frame. Modern jetplanes are built as a single unit . . . like Rambler.
The separate frame used on most U.S. cars is relatively easy to bend or twist. Single Unit Construction— pioneered by American Motors—makes Rambler safer, more spacious.
Deep-Dip Rustproofing. All Rambler bodies are dipped in a rustproofing paint primer bath for lasting protection against rust and corrosion.

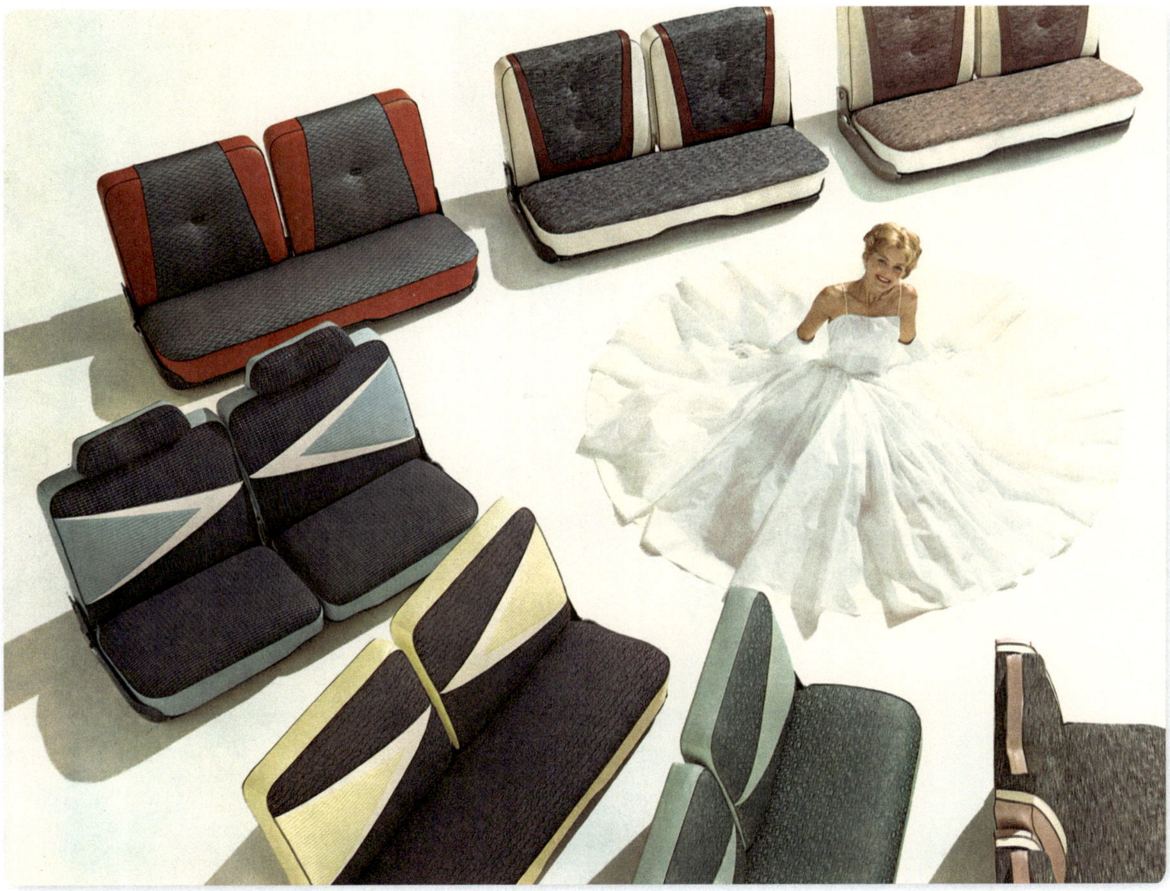

American Motors Rambler 1960

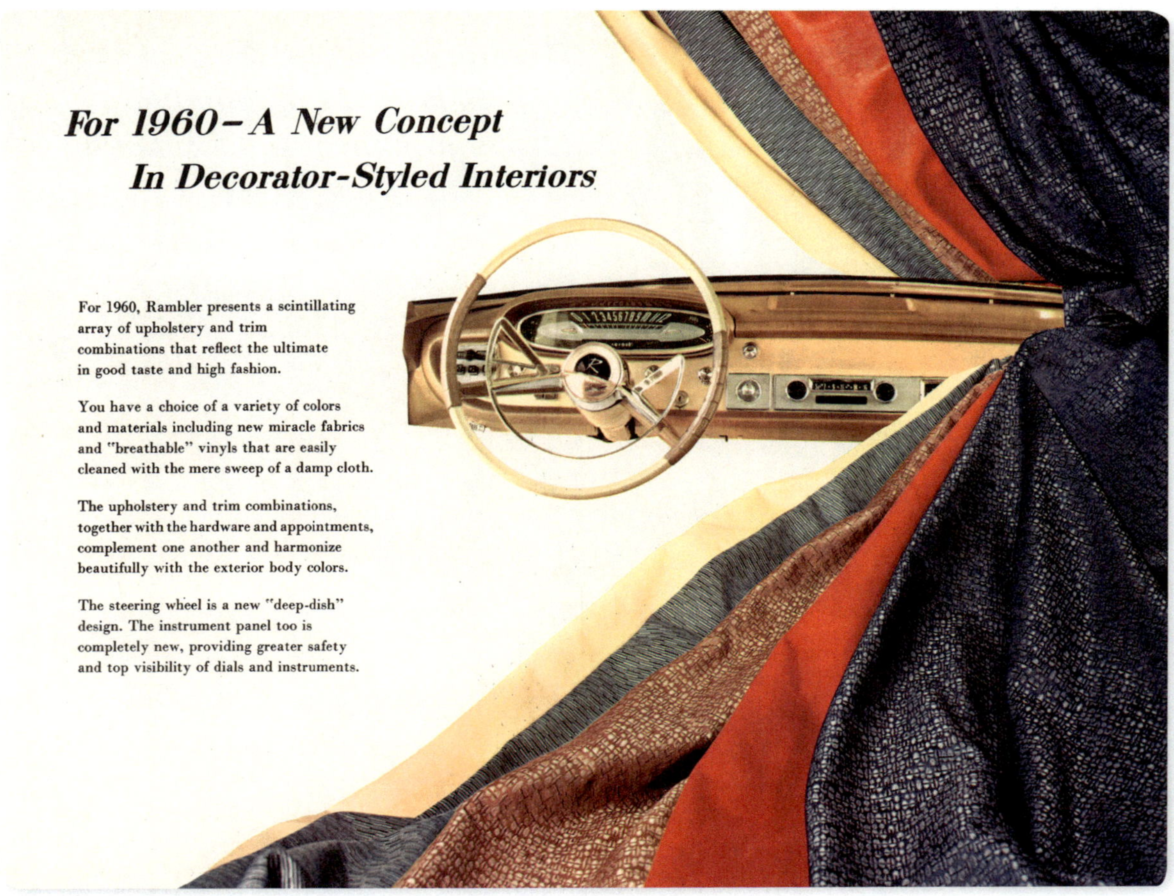

If some people considered the Rambler behind the Big Three when it came to pizzazz, American Motors was having none of it. It touted structural strength and cabin accommodations that were both worthy of the new Jet Age.

Manche Menschen mochten der Meinung gewesen sein, dass der Rambler den „Big Three“ der amerikanischen Automobilindustrie in Sachen Eleganz hinterherhinkte, aber davon wollte American Motors nichts wissen. Stattdessen hob der Hersteller eine Strukturfestigkeit und einen Innenraum hervor, die dem neuen Düsenzeitalter durchaus würdig waren.

Si certaines personnes plaçaient la Rambler, pour le panache, derrière les « big three », American Motors ne l'entendait pas de cette oreille. La marque faisait en effet valoir une solidité structurelle et une habitabilité dignes l'une comme l'autre de l'ère de l'avion à réaction.

Borgward

Borgward Isabella 1960

In 1960 most Americans could only name two German cars, Mercedes-Benz and Volkswagen, but the attributes of this handsomely proportioned coupe included chairlike seats, something no domestic car could boast, and strong braking power.

Im Jahre 1960 konnten die meisten Amerikaner wohl nur zwei deutsche Automarken nennen, nämlich Mercedes-Benz und Volkswagen, doch zu den Besonderheiten dieses hübsch proportionierten Coupés zählten sesselähnliche Sitze, wie kein amerikanisches Auto sie besaß und starke Bremsen.

En 1960, la plupart des Américains auraient été bien incapables de citer plus de deux marques allemandes, à savoir Mercedes-Benz et Volkswagen. Or ce coupé joliment proportionné offrait par exemple des sièges-couchettes aux rembourrages généreux – le genre d'option qu'aucune voiture *made in USA* ne pouvait proposer – et une grande puissance de freinage.

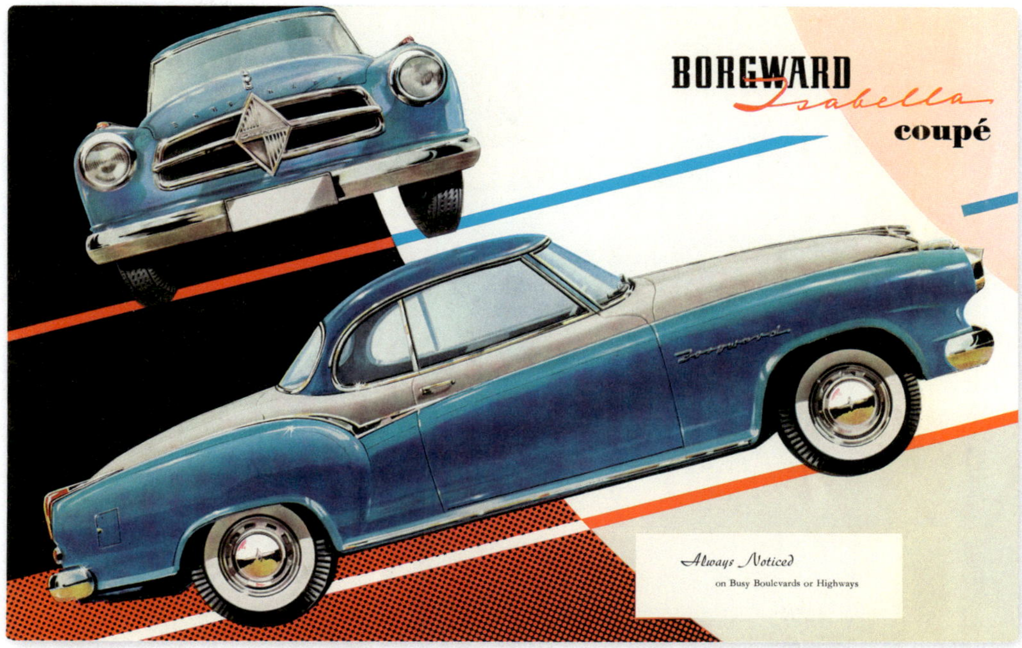

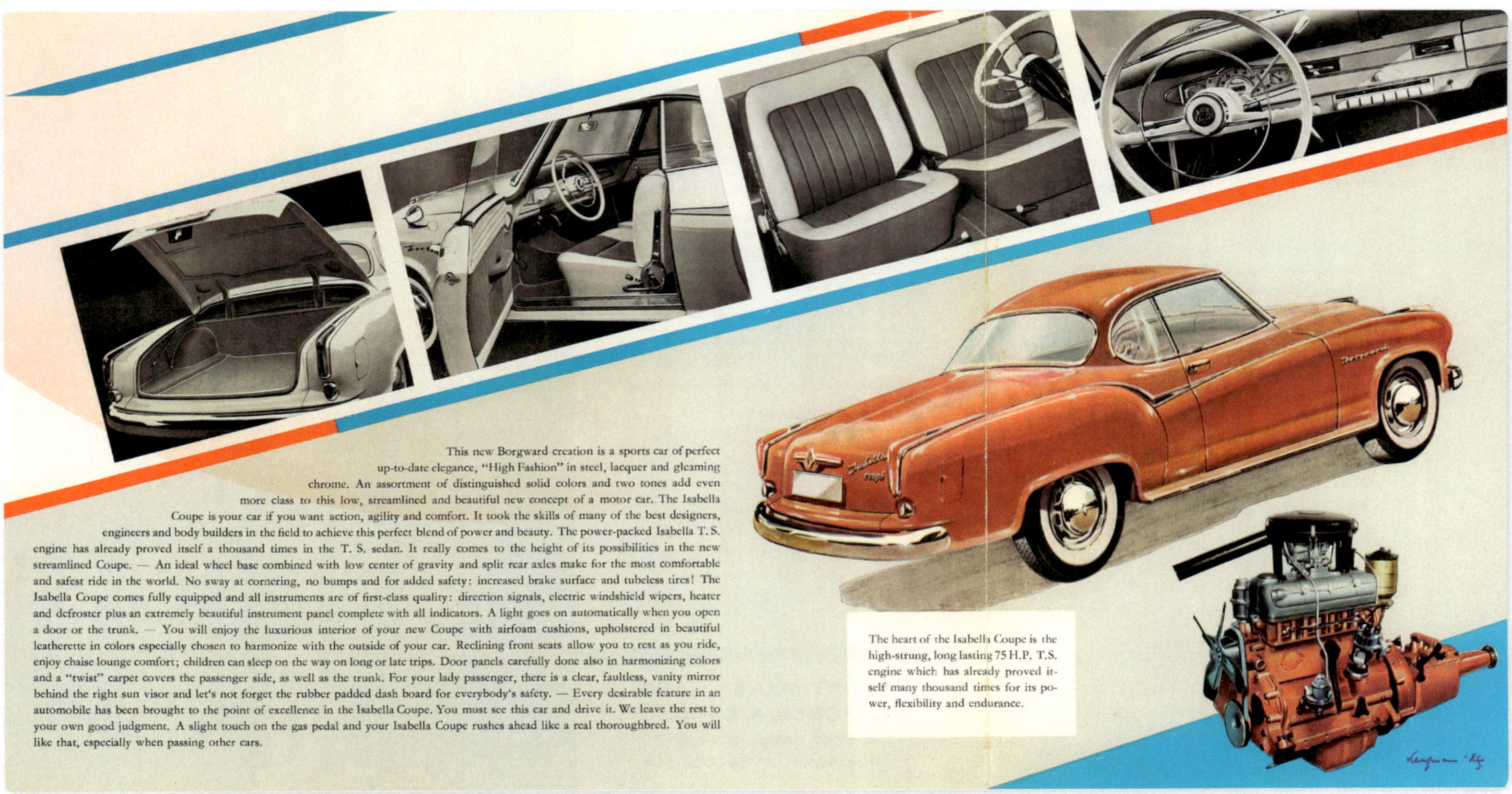

Buick

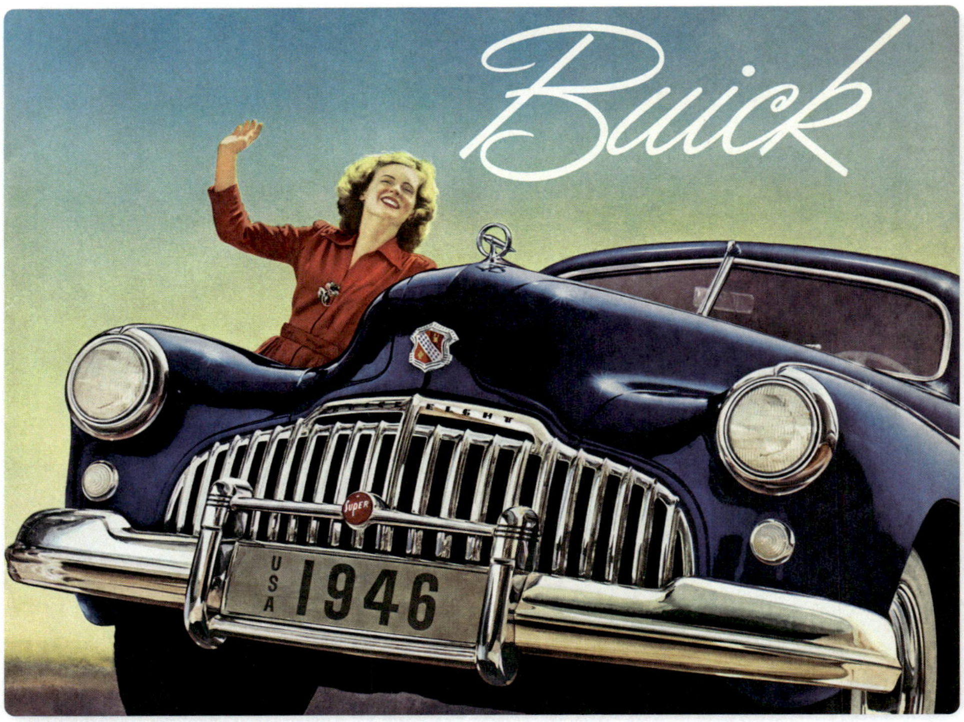

Buick 1946, 1954 *(opposite)*

Optimism was rampant in the postwar years. Cars were being built again, and they were there for anyone who could get onto a dealer's waiting list to purchase one. Buick was about to enjoy some of its best years from a sales standpoint.

In den Nachkriegsjahren blühte der Optimismus allerorten. Es wurden wieder Autos gebaut, und jeder konnte sich beim Händler auf eine Warteliste setzen lassen, um eines zu erwerben. Aus Verkaufssicht hatte Buick zu dieser Zeit einige seiner besten Jahre.

Dans les années d'après-guerre, l'optimisme avait le vent en poupe. On construisait de nouveau des voitures, destinées à quiconque avait les moyens de s'inscrire sur la liste d'attente d'un concessionnaire. Sur le plan commercial, Buick s'apprêtait à connaître quelques-unes de ses plus belles années.

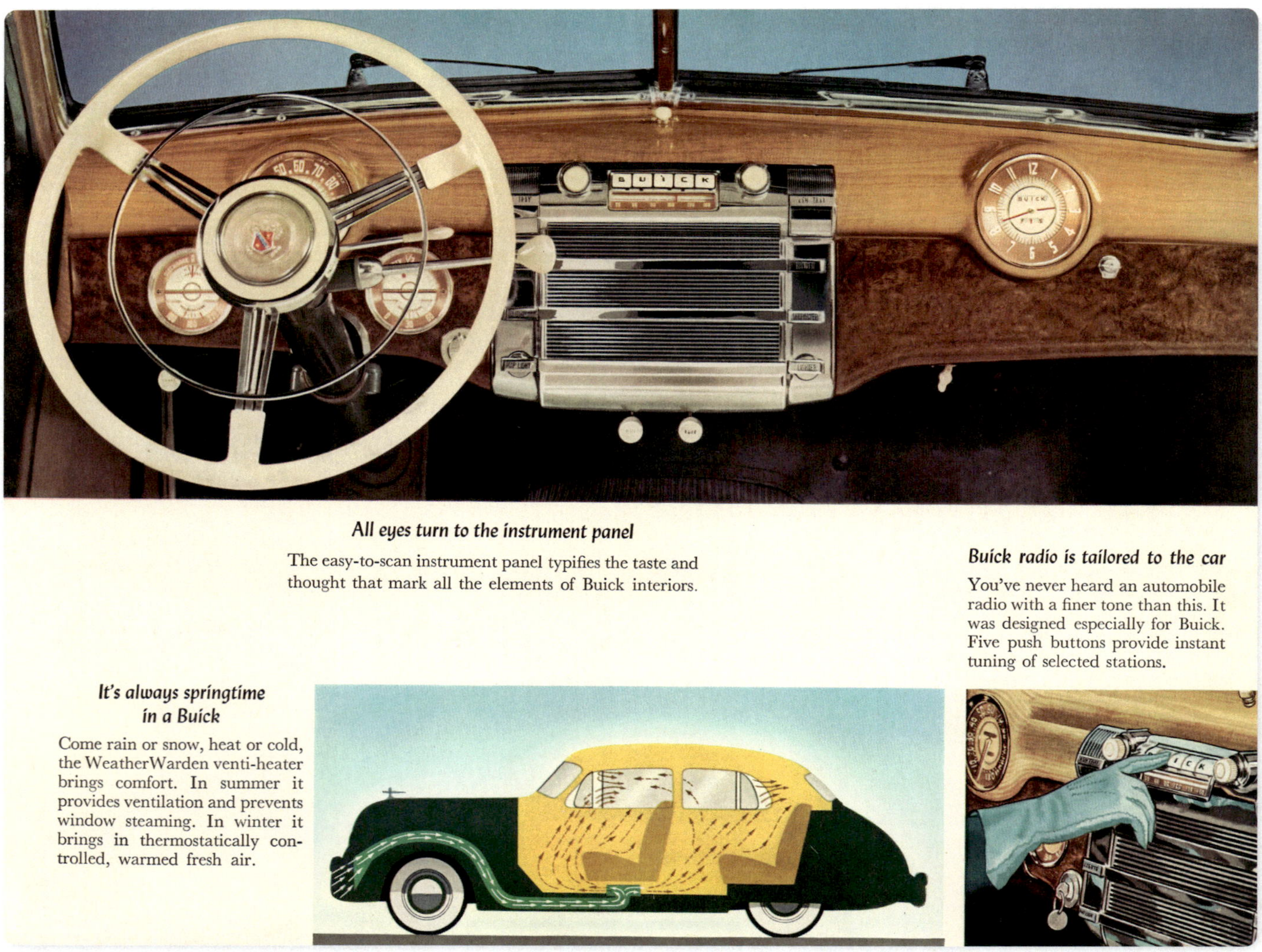

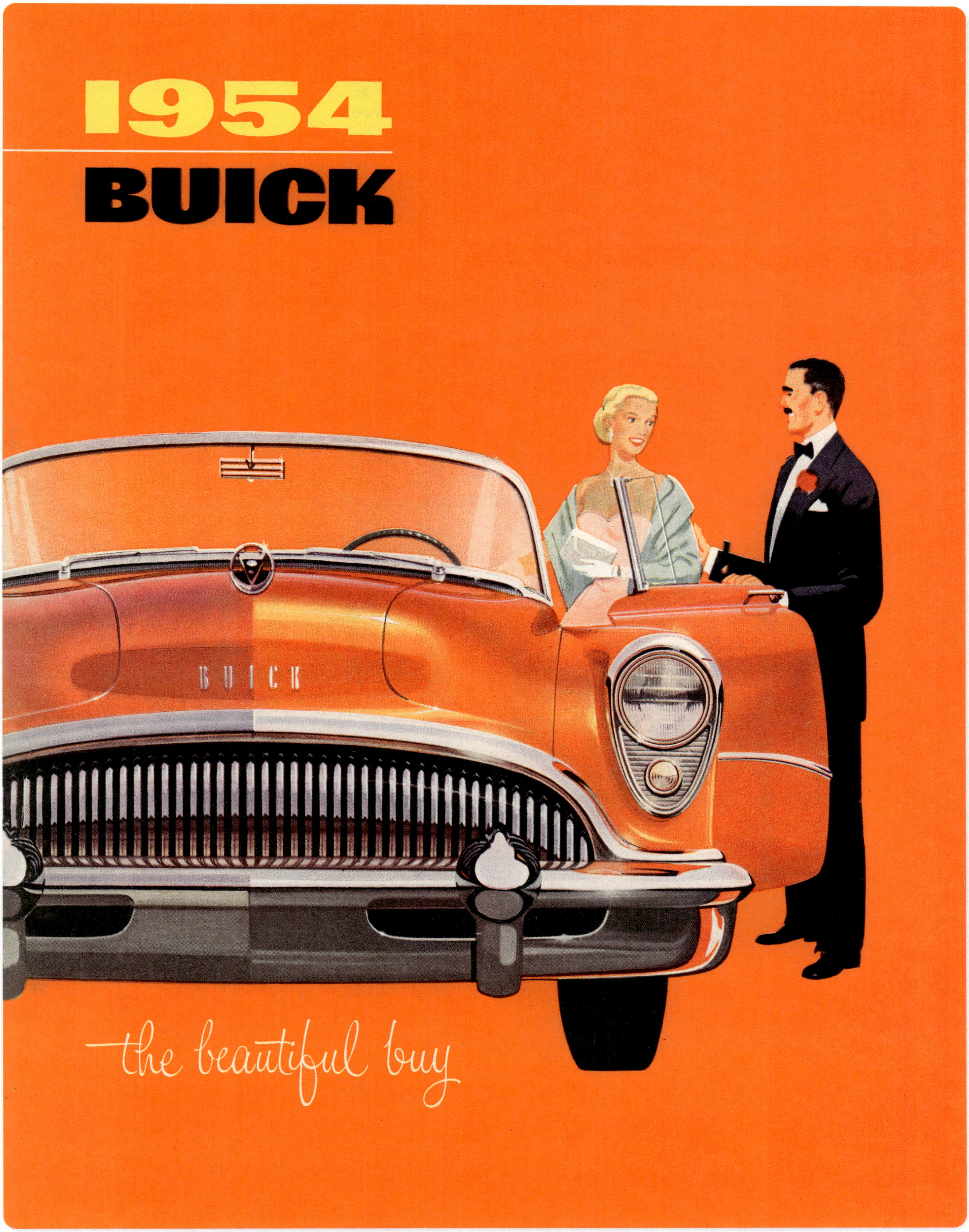
1954
BUICK
BUICK
the beautiful buy

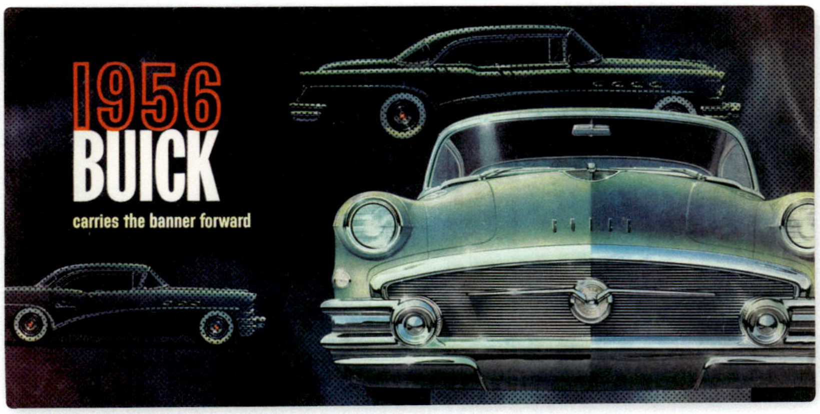

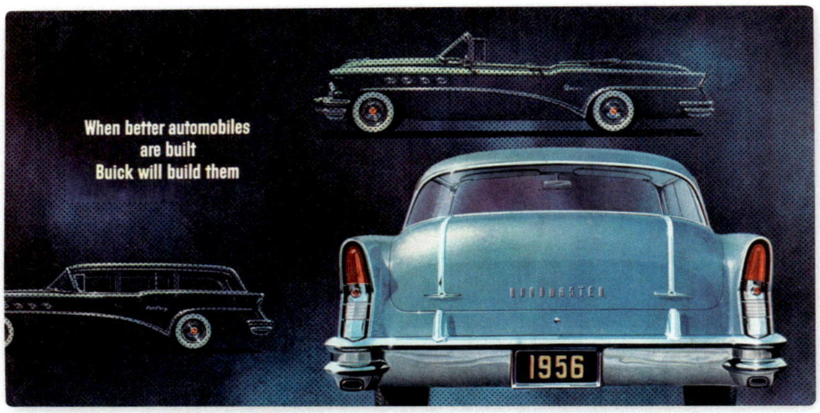

Buick 1956 *(left and opposite)*,
1958 *(below)*

This Dali-esque, surrealistic landscape was a provocative approach for the handsomely conservative 1956 Buick.

Diese surrealistische Landschaft im Stile von Dalí war ein provokanter Ansatz für den hübsch konservativen Buick von 1956.

Ce paysage surréaliste à la Salvador Dalí faisait office d'approche plutôt provocante pour la Buick de 1956, à l'élégance conservatrice.

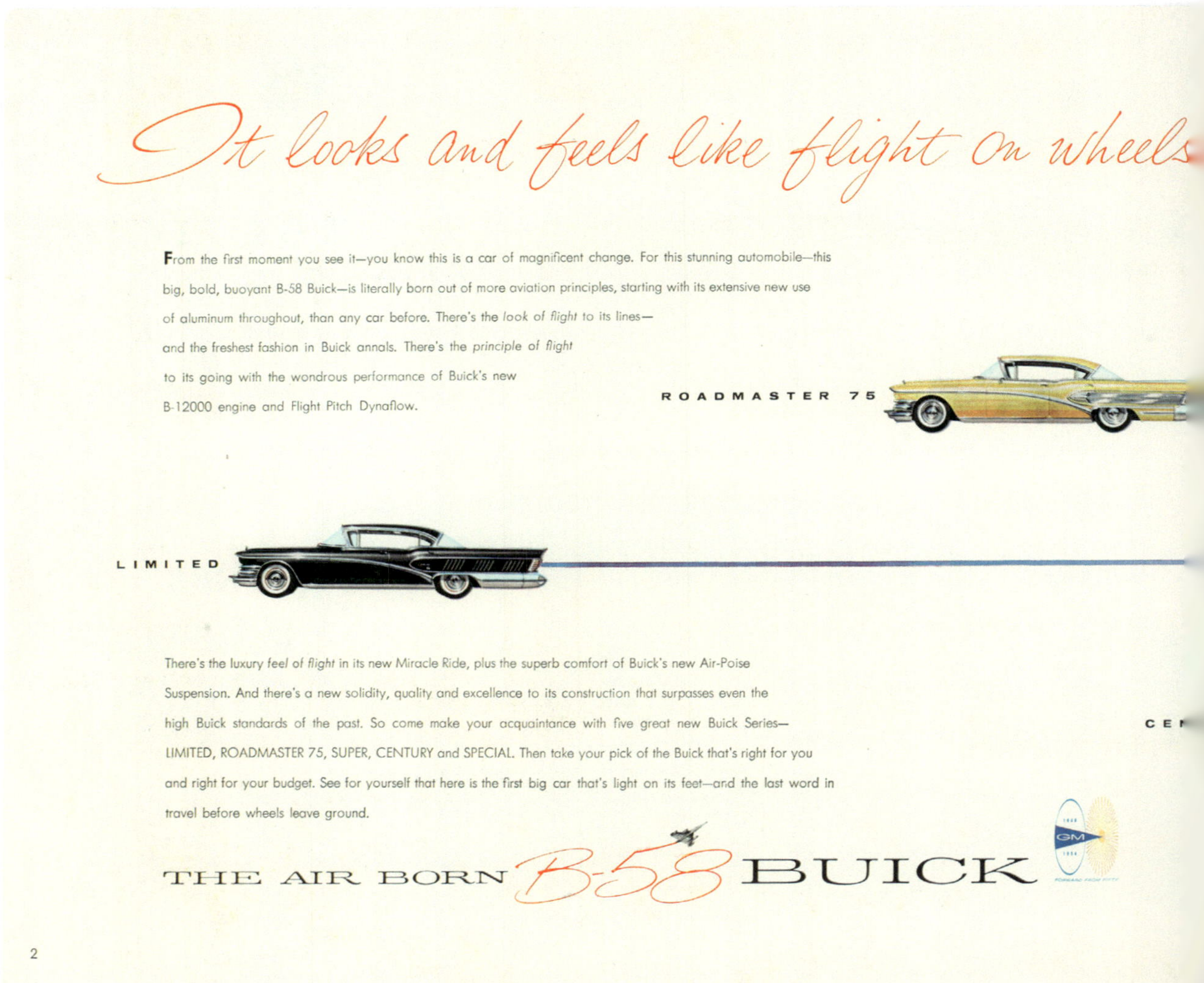

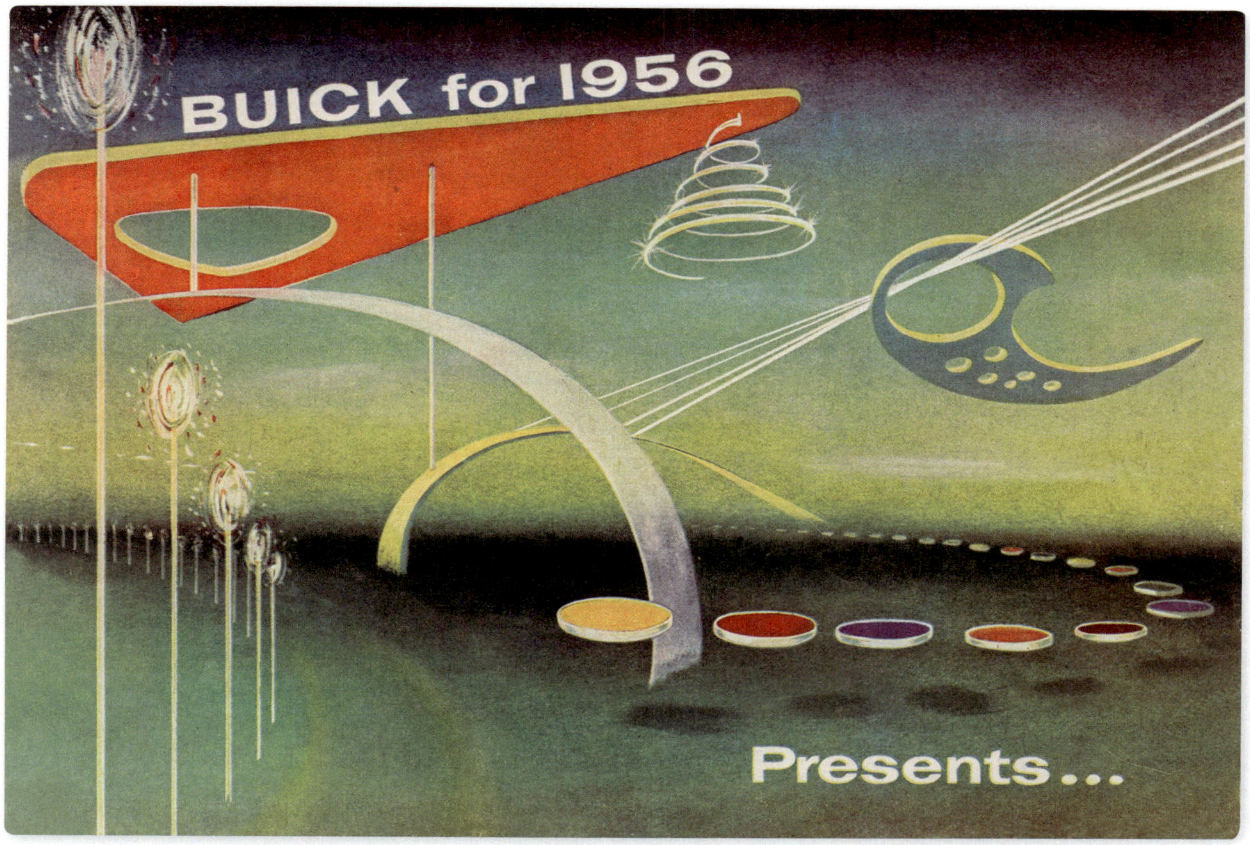
BUICK for 1956
Presents...

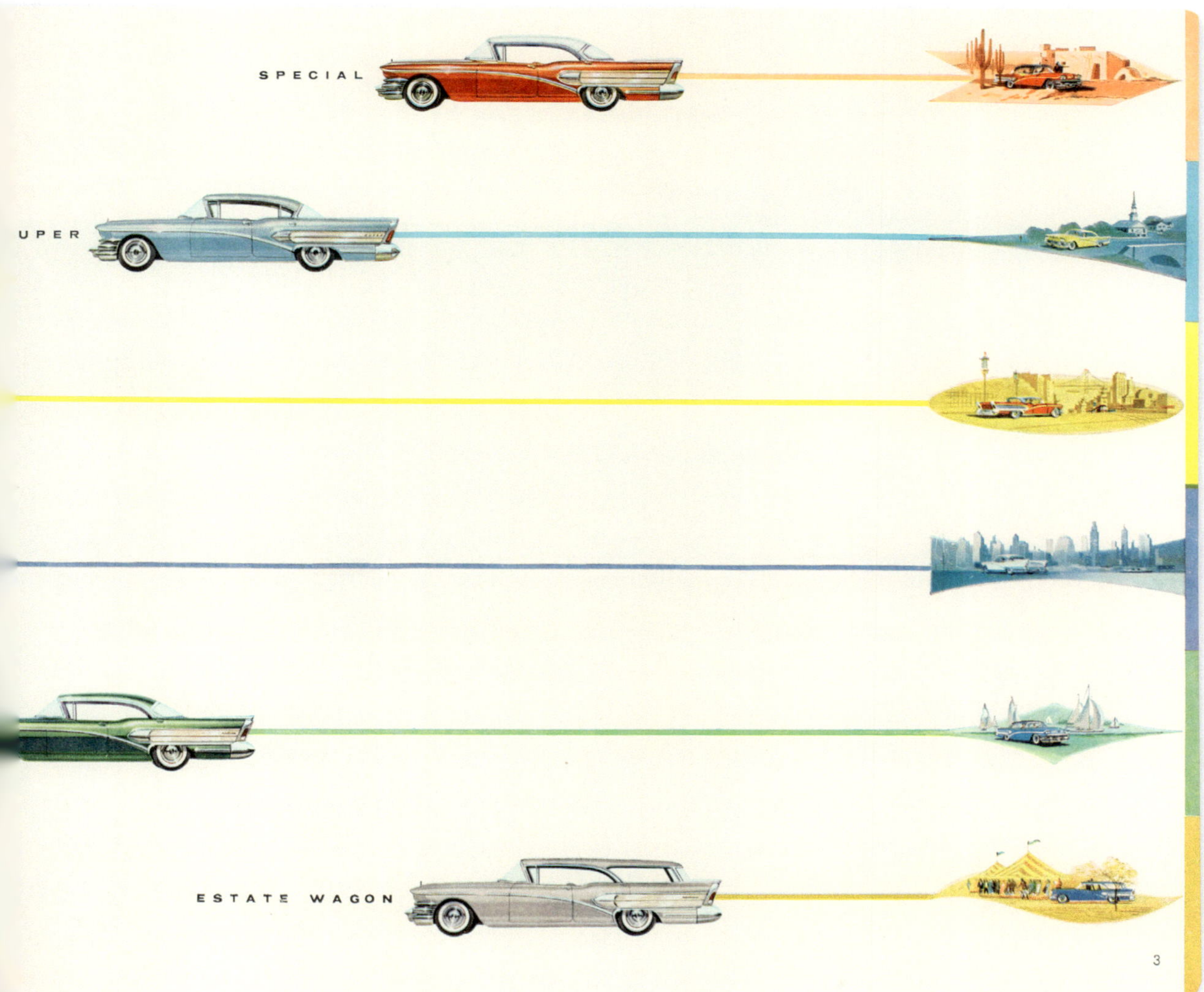
SPECIAL
UPER
ESTATE WAGON
3

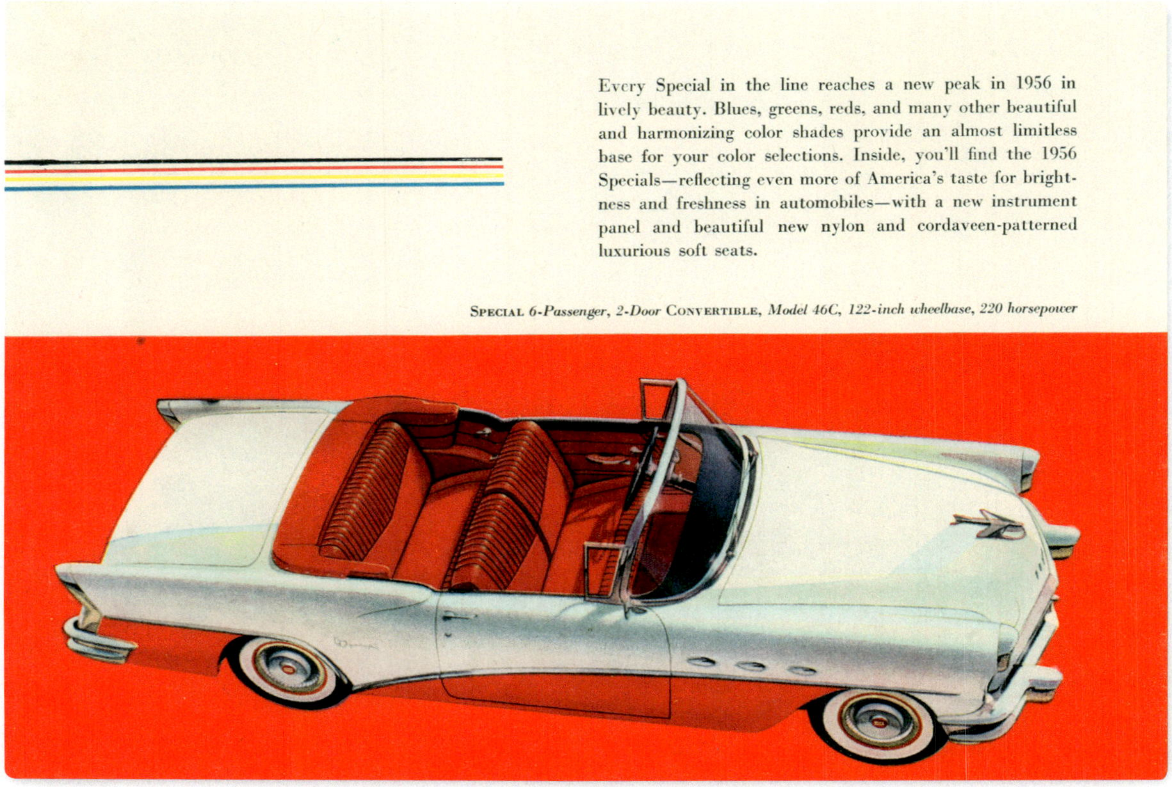
Every Special in the line reaches a new peak in 1956 in lively beauty. Blues, greens, reds, and many other beautiful and harmonizing color shades provide an almost limitless base for your color selections. Inside, you'll find the 1956 Specials—reflecting even more of America's taste for brightness and freshness in automobiles—with a new instrument panel and beautiful new nylon and cordaveen-patterned luxurious soft seats.
SPECIAL 6-Passenger, 2-Door CONVERTIBLE, Model 46C, 122-inch wheelbase, 220 horsepower

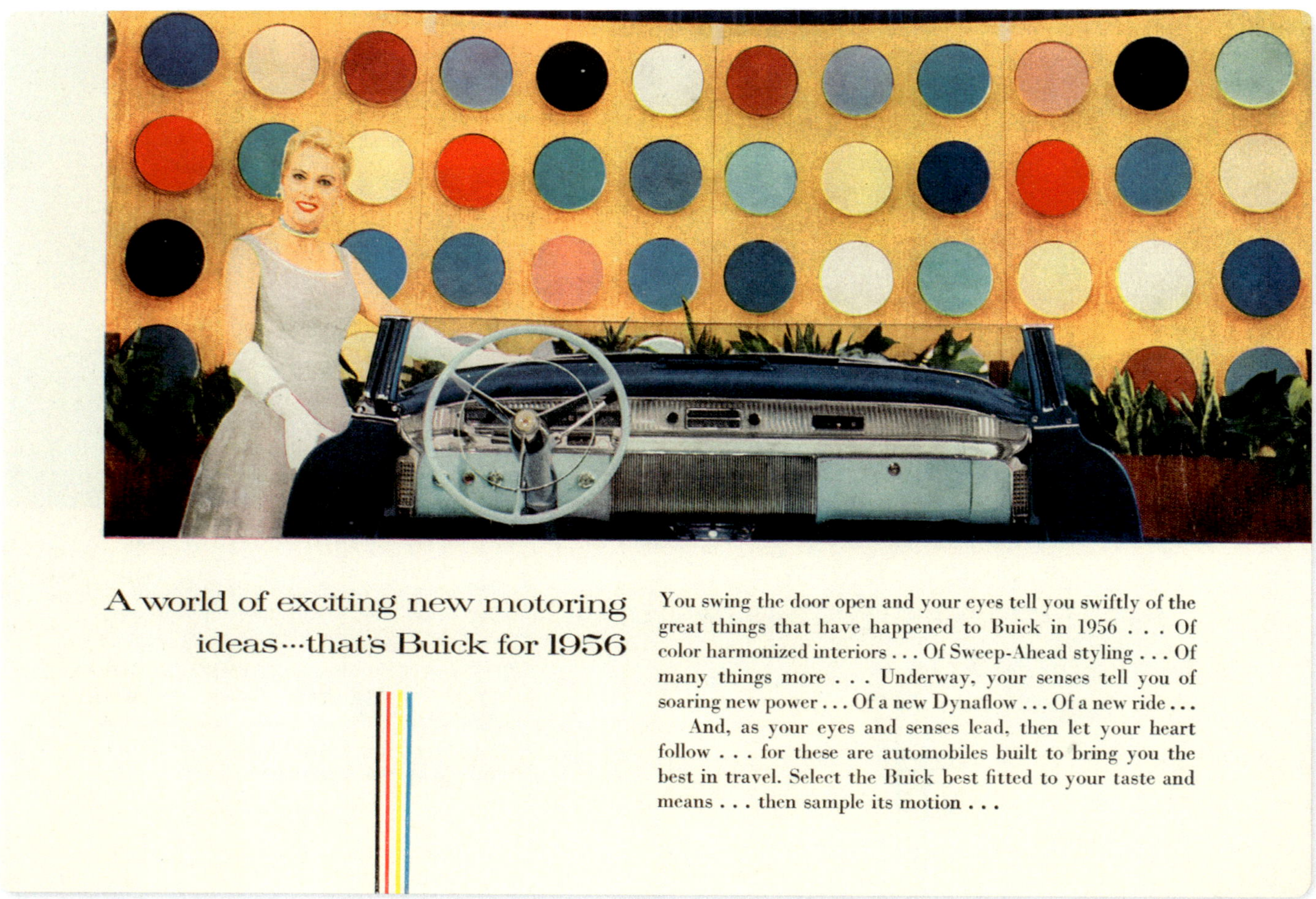
A world of exciting new motoring ideas···that's Buick for 1956
You swing the door open and your eyes tell you swiftly of the great things that have happened to Buick in 1956 . . . Of color harmonized interiors . . . Of Sweep-Ahead styling . . . Of many things more . . . Underway, your senses tell you of soaring new power . . . Of a new Dynaflow . . . Of a new ride . . .
And, as your eyes and senses lead, then let your heart follow . . . for these are automobiles built to bring you the best in travel. Select the Buick best fitted to your taste and means . . . then sample its motion . . .

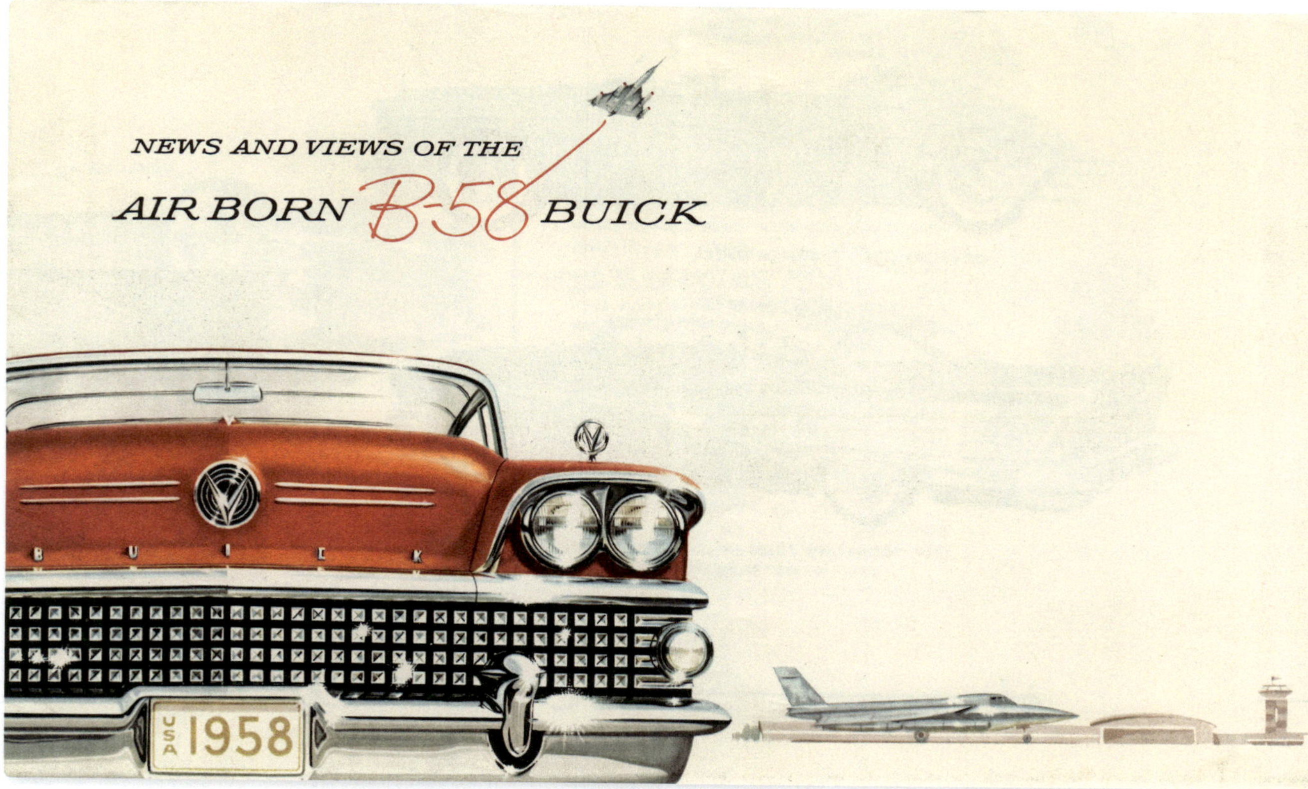

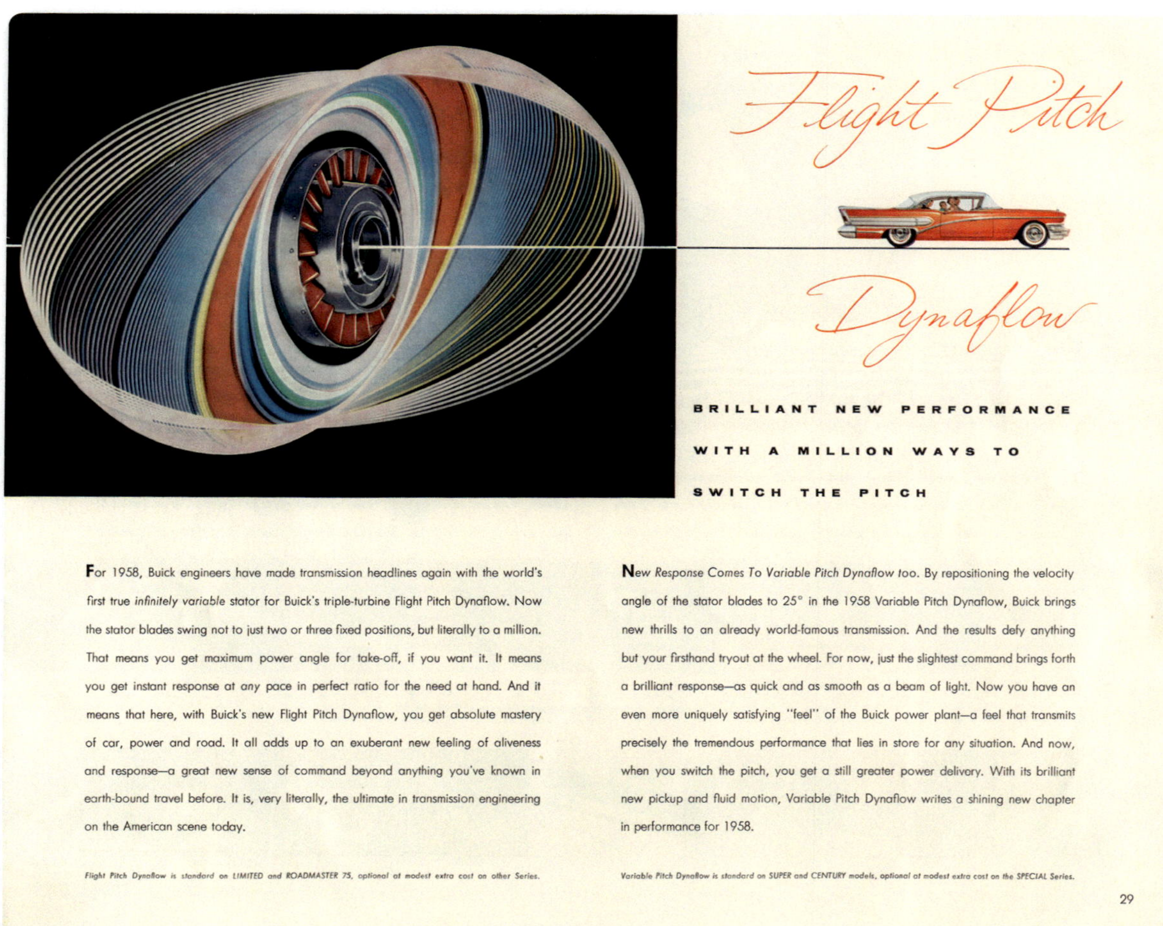

Buick 1956 *(opposite)*, **1958**

Buick had made a quantum leap from 1957 in terms of fins and chrome. It invoked the supersonic Convair B-58 bomber when showing its technical improvements, including variable-pitch Dynaflow automatic gear changing and adjustable ride height.

In Sachen Heckflossen und Chrom vollzog Buick 1957 einen gewaltigen Sprung nach vorn. Man zog bewusst Parallelen zum Überschallbomber Convair B-58 in der Darstellung der technischen Ver-besserungen, zu denen unter anderem die Automatikschaltung „Dynaflow" (mit einem Drehmomentwandler mit verstellbaren Turbinenradschaufeln) und ein höhenverstellbares Fahrgestell gehörten.

Dans le registre des ailerons et des chromes, la Buick effectua un saut qualitatif à partir de 1957. C'est d'ailleurs sous l'autorité du bombardier supersonique Convair B-58 qu'elle se range au moment d'exhiber ses innovations techniques, parmi lesquelles sa transmission automatique Dynaflow à convertisseur de couple et sa hauteur de caisse réglable.

Buick 1959, 1961 *(opposite)*

The radically revised 1959 Buicks shared their full-size bodies with Chevrolet, along with their Vista (sedan) and "bubble top" (hardtop) rooflines. The canted headlamps were used this year only. Besides Cordaveen and Balfour interior facings, a new 401-cubic-inch V-8 engine, commonly called the Nailhead, punched the Buick's output up to 325 hp.

Die radikal überarbeiteten Buicks von 1959 verwendeten die gleichen Karosserieformen wie Chevrolet, inklusive der Varianten „Vista" (Limousine) und „bubble top" (Coupé). Die versetzten Scheinwerfer kamen nur in diesem einen Jahr zum Einsatz. Neben Innenverkleidungen in „Cordaveen" und „Balfour" jagte ein neuer 6½-Liter-V8-Motor (6 571 cm³) – allgemein bekannt unter dem Namen „Nailhead" – die Leistung des Buick auf 441 Kilowatt hoch.

Radicalement remaniées, les Buick de 1959 partageaient avec Chevrolet – et avec ses déclinaisons berline (Vista) et coupé (Bubble Top) – leurs imposantes carrosseries « full size » . Les phares obliques ne seront utilisés que cette année-là. Outre les intérieurs en vinyle Cordaveen et tissu Balfour, un nouveau moteur V8 de 6 litres de cylindrée, communément appelé « Nailhead » (tête de clou), hissait la puissance de la Buick à 325 chevaux.

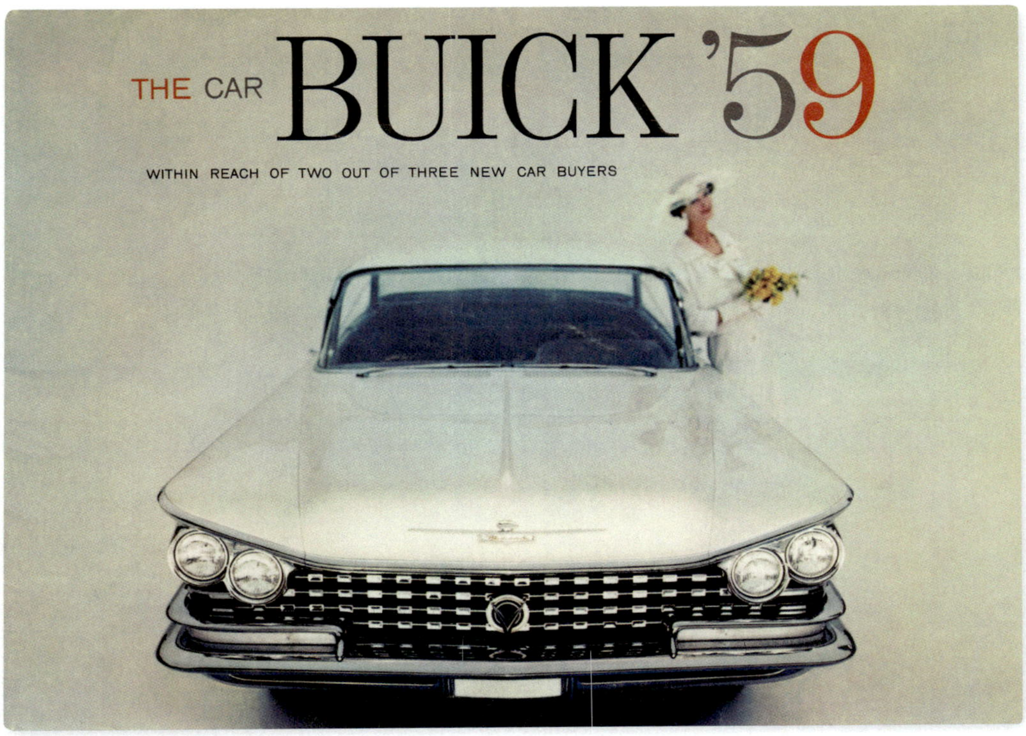

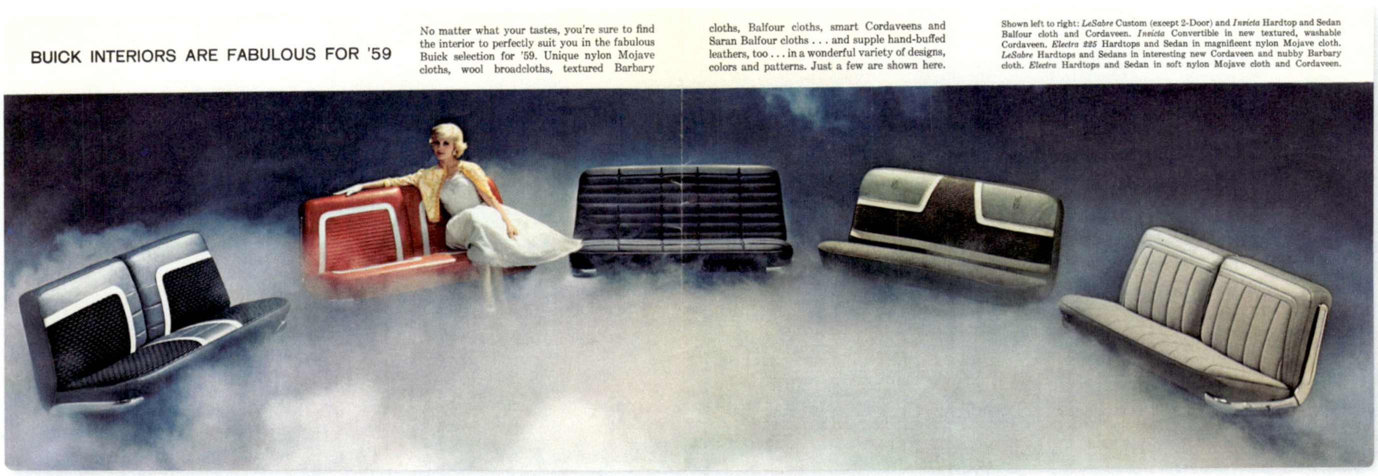

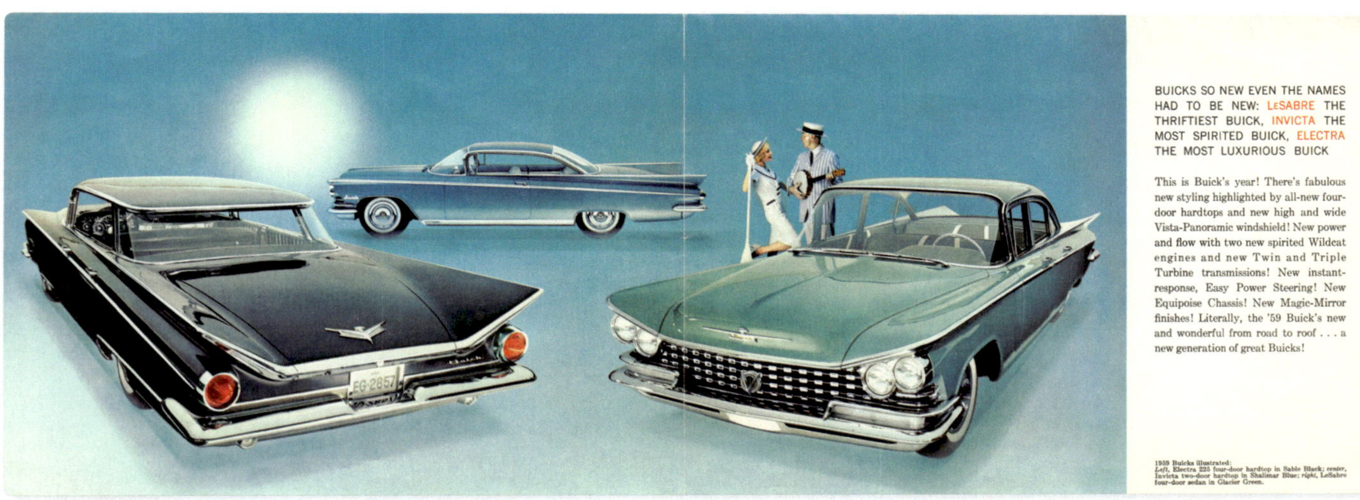

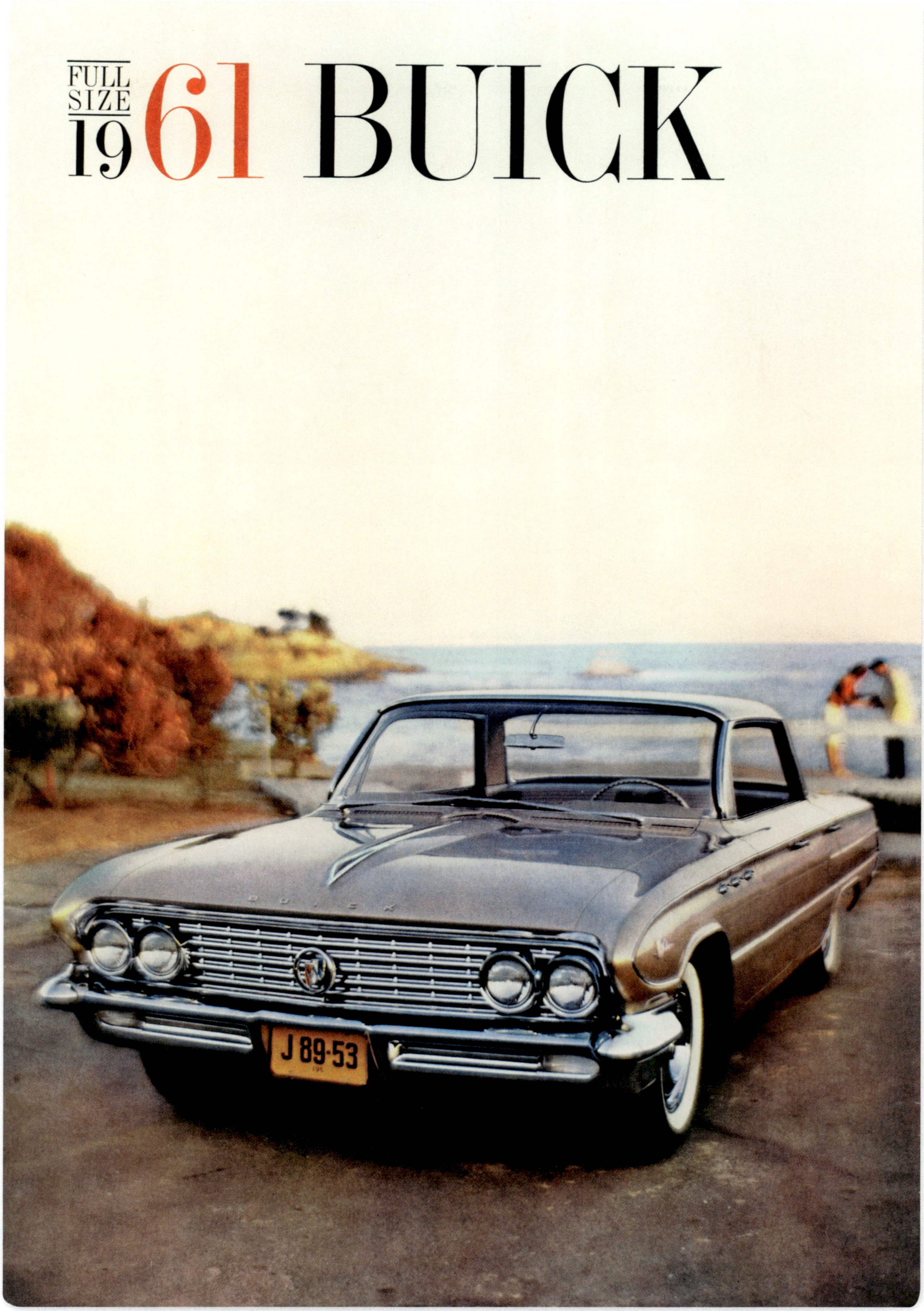
FULL SIZE
1961 BUICK
J 89-53

Cadillac

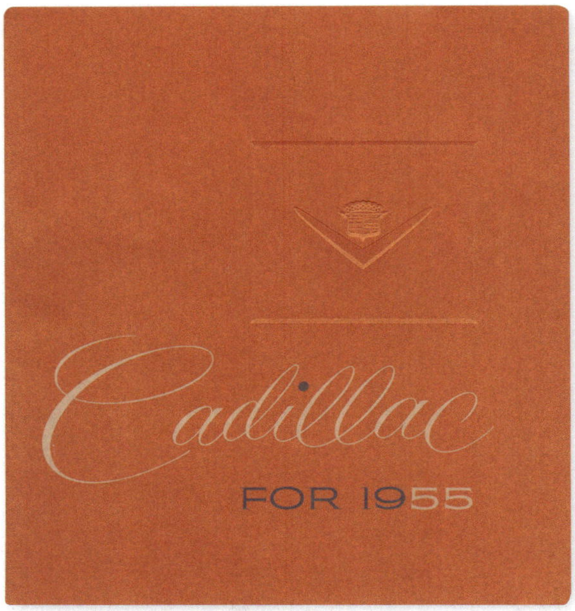

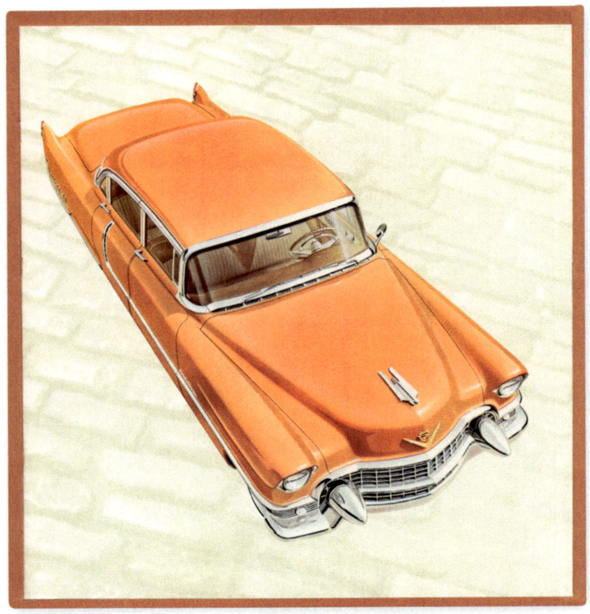

Cadillac 1955 *(above)*, **1954** *(below)*

Exquisitely subtle, the embossed cover draws the reader in with a gentle shade of salmon pink, an unexpectedly popular automotive color in these years.

Der edel anmutende Umschlag mit Prägedruck weckt die Aufmerksamkeit des Lesers mit einem zarten Lachsrosa, einer zu dieser Zeit überraschend beliebten Lackfarbe für Autos.

D'une exquise subtilité, cette couverture gaufrée attire le lecteur de toute la délicatesse de son rose saumon, une couleur d'automobile étonnamment populaire dans ces années cinquante.

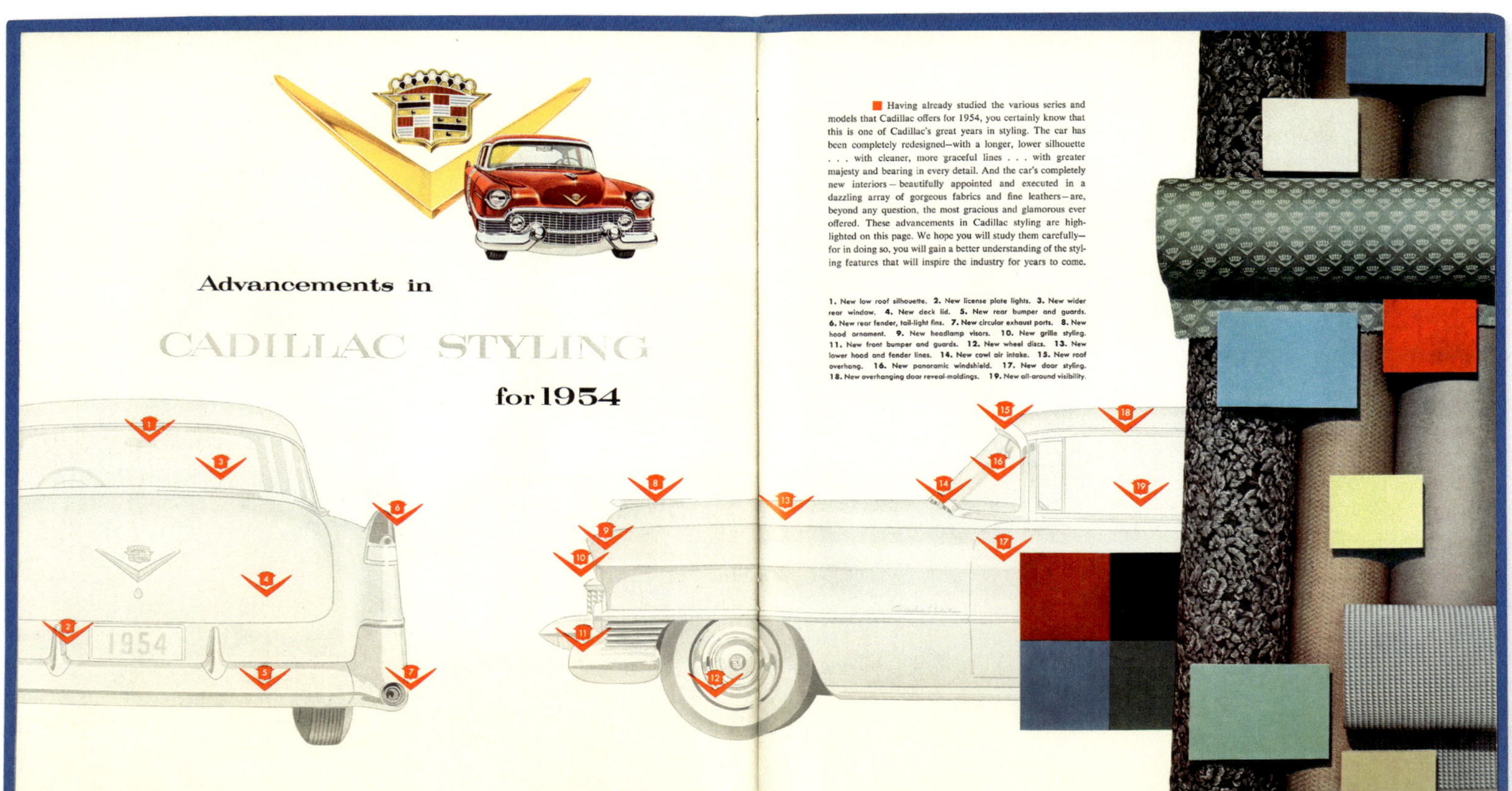

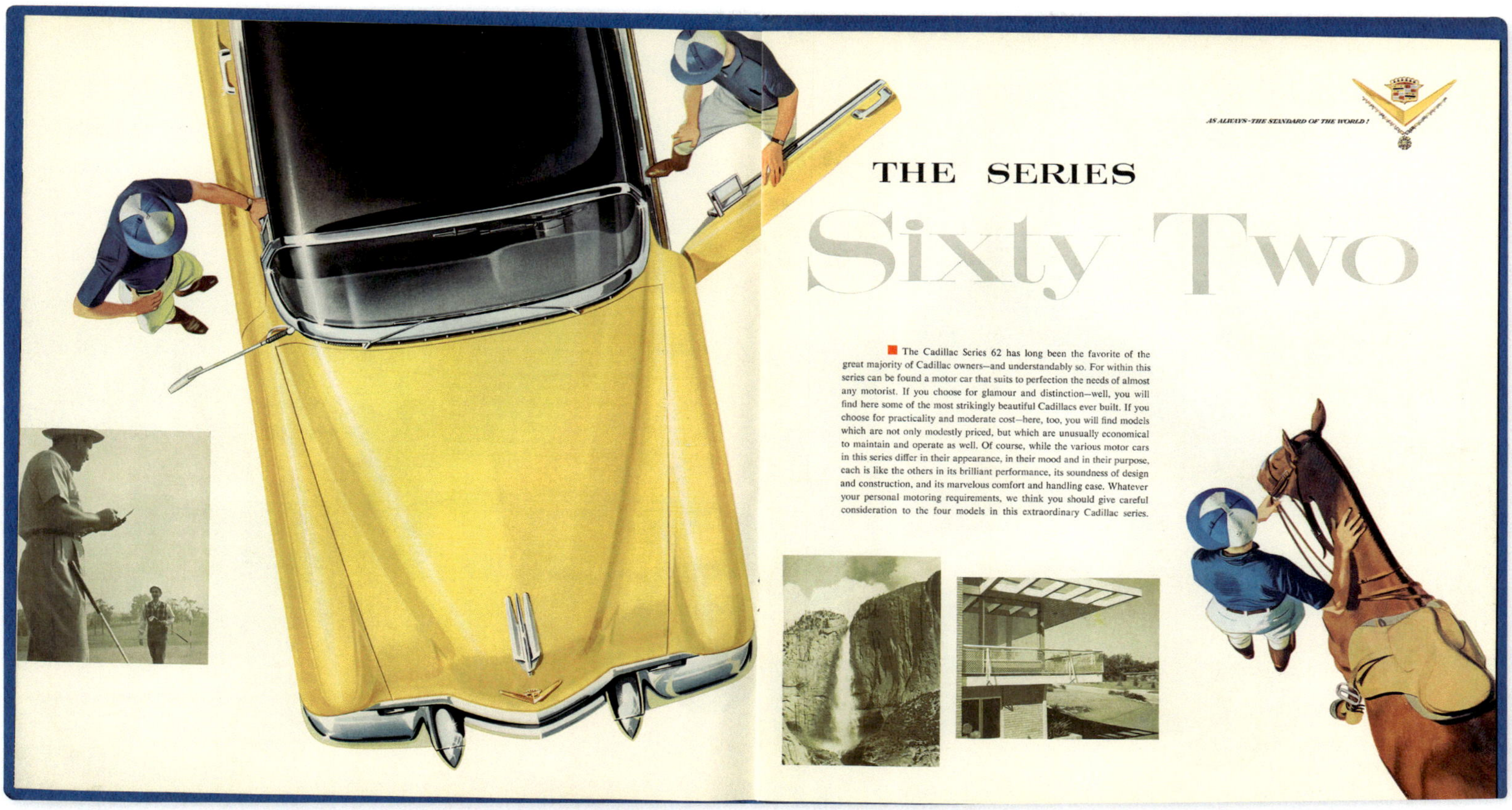

Cadillac 1954

Some consider this the first true post-war Cadillac, fully lavished with 1950s styling cues including the huge wraparound windshield borrowed from the high-luxury Eldorado line, much taller and sharper tailfins, and deeply hooded headlamps.

Für manche ist dies der erste wirkliche Nachkriegs-Cadillac, mit typischen Merkmalen des Designs der 1950er-Jahre, wie etwa den gewaltigen umlaufenden Windschutzscheiben, die von der Luxusreihe Eldorado entliehen waren, deutlich höheren und schärferen Heckflossen und tief eingebetteten Scheinwerfern.

Certains voient en elle la première vraie Cadillac d'après-guerre, toute ornée d'éléments de style propres aux années 1950 comme l'énorme parebrise panoramique emprunté à la gamme de grand luxe Eldorado, les ailerons agrandis et plus effilés et les feux avant profondément encastrés.

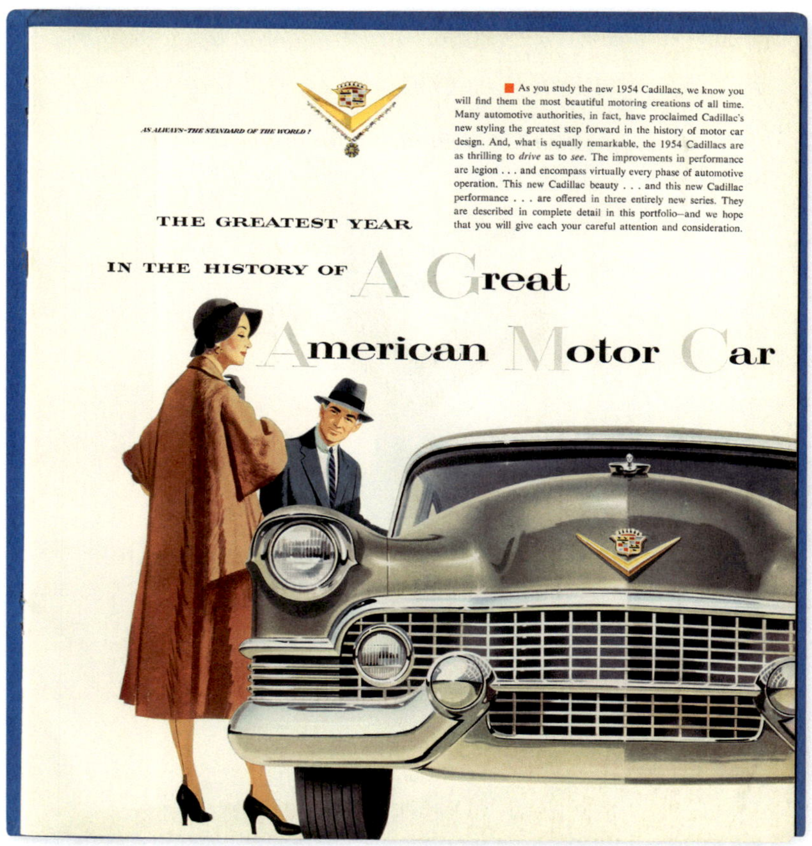

Cadillac 1956 *(right)*, **1960** *(far right)*, **1961** *(below)*

When you were a man or woman of serious means, you capped a hard day of capitalism with a ride, sunk deeply into the Bedford cloth of a new Fleetwood 75 limousine, en route to an evening that was sure to include dinner and cocktails.

Wer es sich leisten konnte, ließ sich am Abend eines harten „kapitalistischen" Arbeitstages in die bequemen Bedford-Bezüge der neuen Limousine Fleetwood 75 sinken und fuhr damit zu Dinner und Cocktails.

Homme ou femme, pour peu que vous en ayez les moyens, vous avez droit à une dure « journée de capitalisme » sur quatre roues, profondément enfoncé dans les sièges en laine Bedford de la nouvelle limousine Fleetwood 75 – en route vers une soirée qui ne manquera pas d'inclure cocktail et dîner fin.

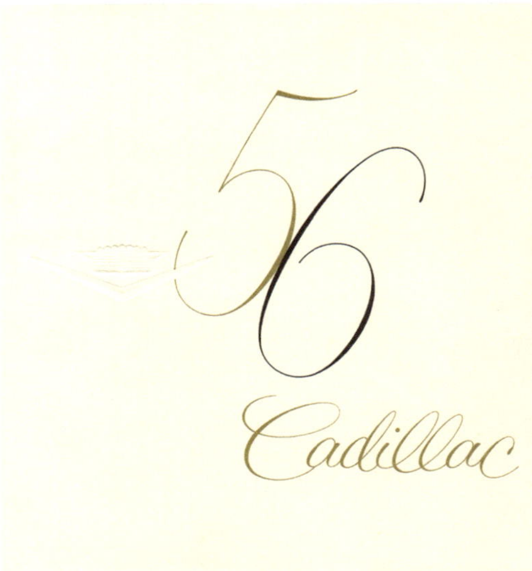

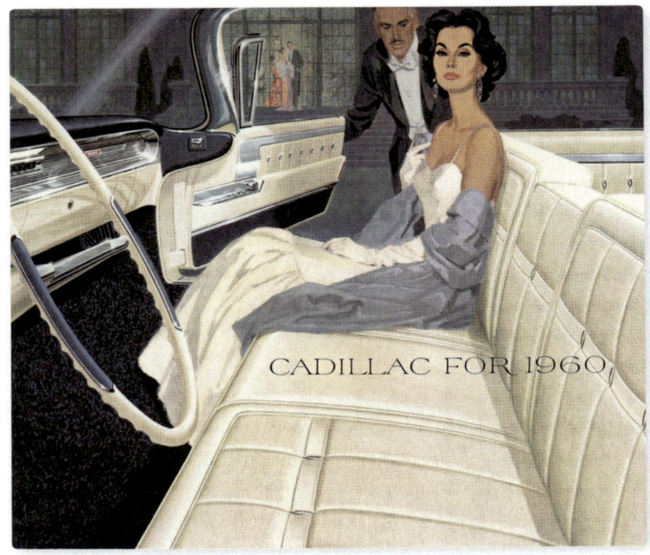

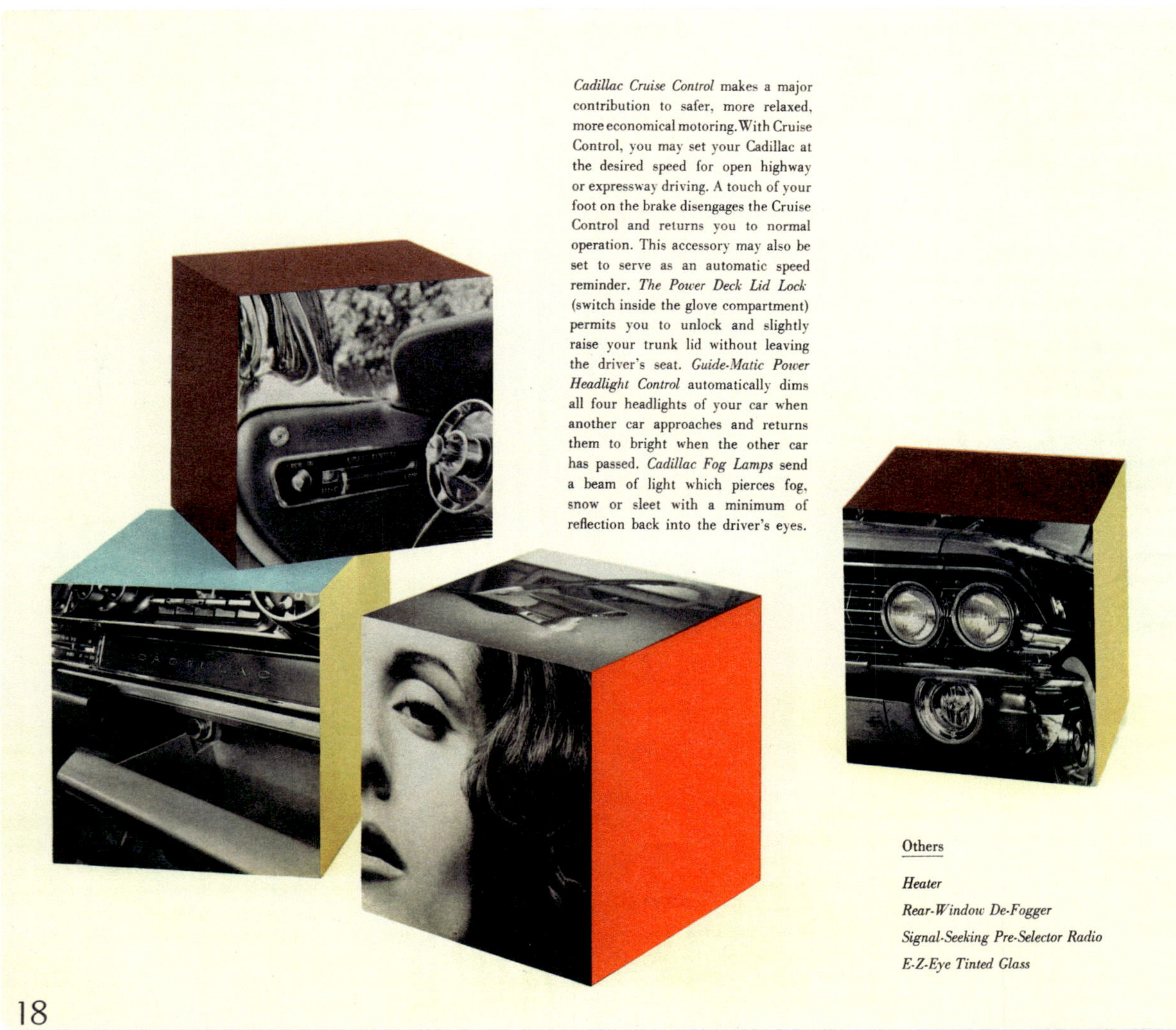

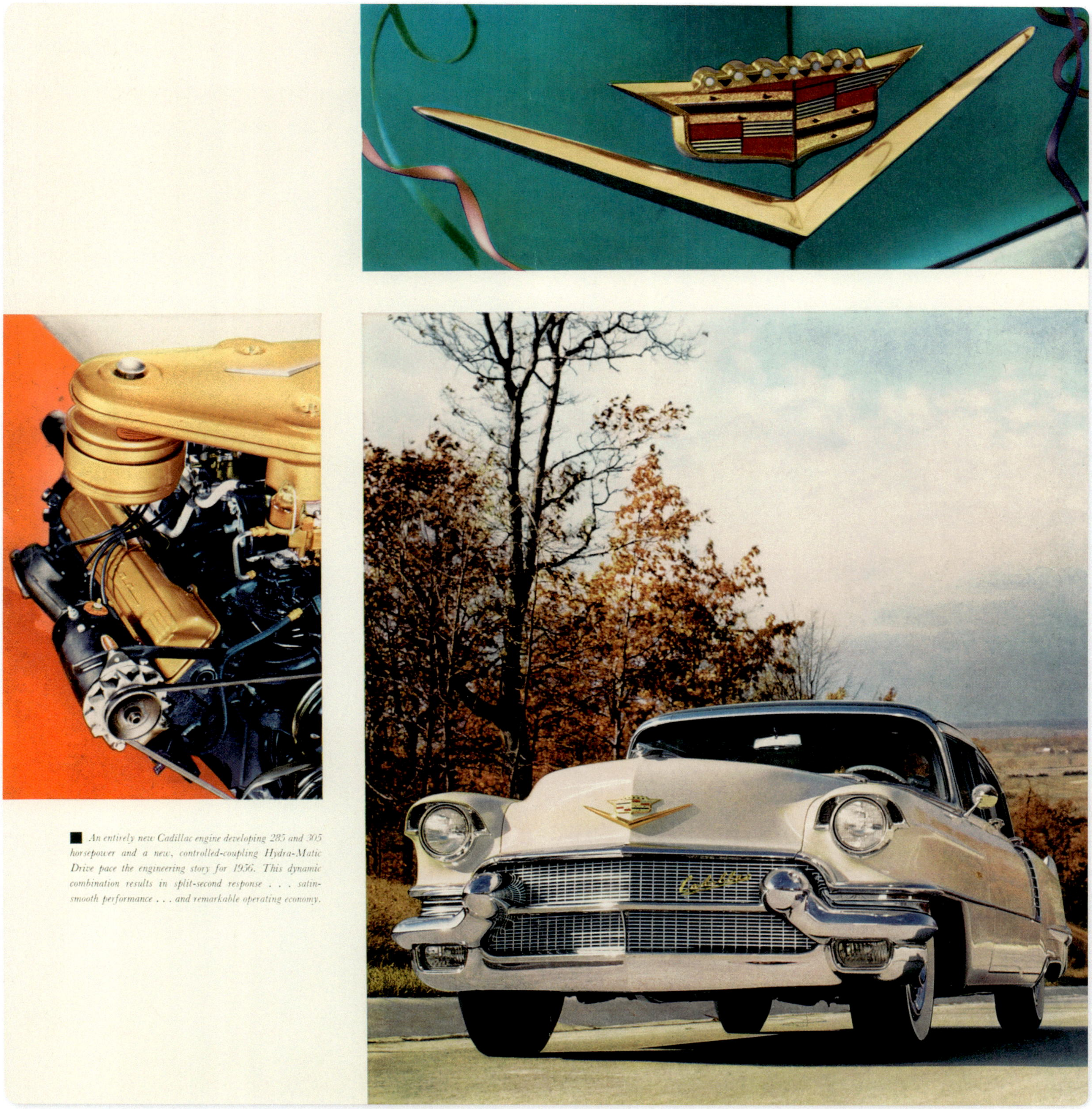

Cadillac 1956

These were massive, gorgeous luxury cars. Most of the Cadillac line still used the vestigial fins that dated back to the start of the decade, but the exclusive Eldorado substituted elongated bullet-style taillights, predicting a future Cadillac styling theme.

Dies waren massive, umwerfend schöne Luxusautos. Die meisten Cadillacs besaßen noch Stummel jener Heckflossen, mit denen das Jahrzehnt begonnen hatte, doch der exklusive Eldorado zeigte schon die lang gezogenen Heckleuchten in Geschossform, die bereits ein künftiges Designthema von Cadillac vorwegnahmen.

Ces voitures de luxe étaient massives et magnifiques. La plupart des modèles de la gamme Cadillac utilisaient toujours les ailerons de requin rescapés du début de la décennie, mais la très exclusive Eldorado voyait ses feux arrière s'effiler comme des obus, préfigurant une approche stylistique que la marque ne tardera pas à décliner.

Chevrolet

Chevrolet 1947 *(right)*,
1952 *(below and opposite)*

With an ancient six-cylinder engine and other basic mechanical systems, the Chevrolet was still preaching a sermon of economy even as the postwar boom was under way.

Während schon der Wirtschaftsaufschwung der Nachkriegszeit in vollem Gange war, mahnte der Chevrolet mit seinem alten Sechszylindermotor und anderen einfachen mechanischen Systemen noch immer zu Sparsamkeit.

Agrémenté d'un vieux moteur six cylindres et de quelque autres aménagements mécaniques pour le moins basiques, la Chevrolet prêchait l'économie en toute circonstance, même pendant les années du boom de l'après-guerre.

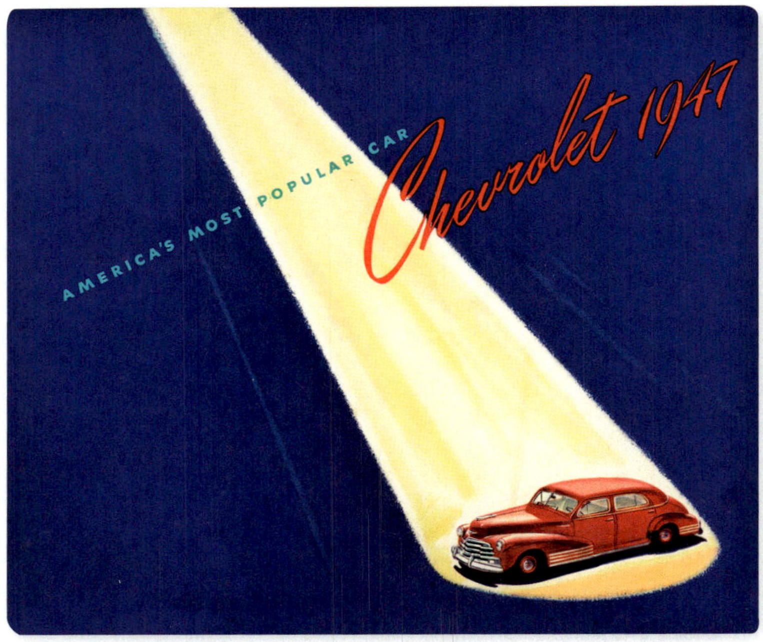

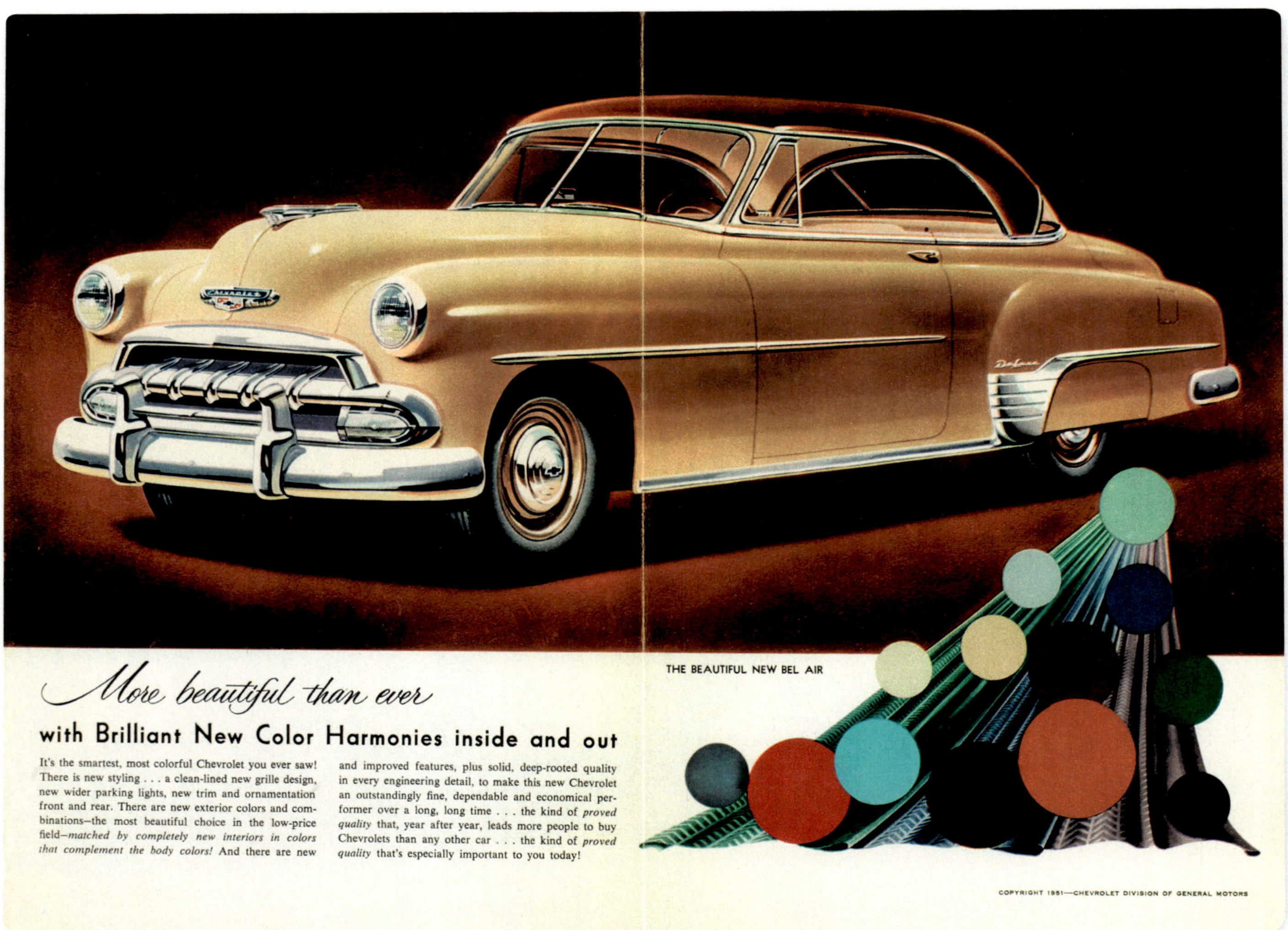

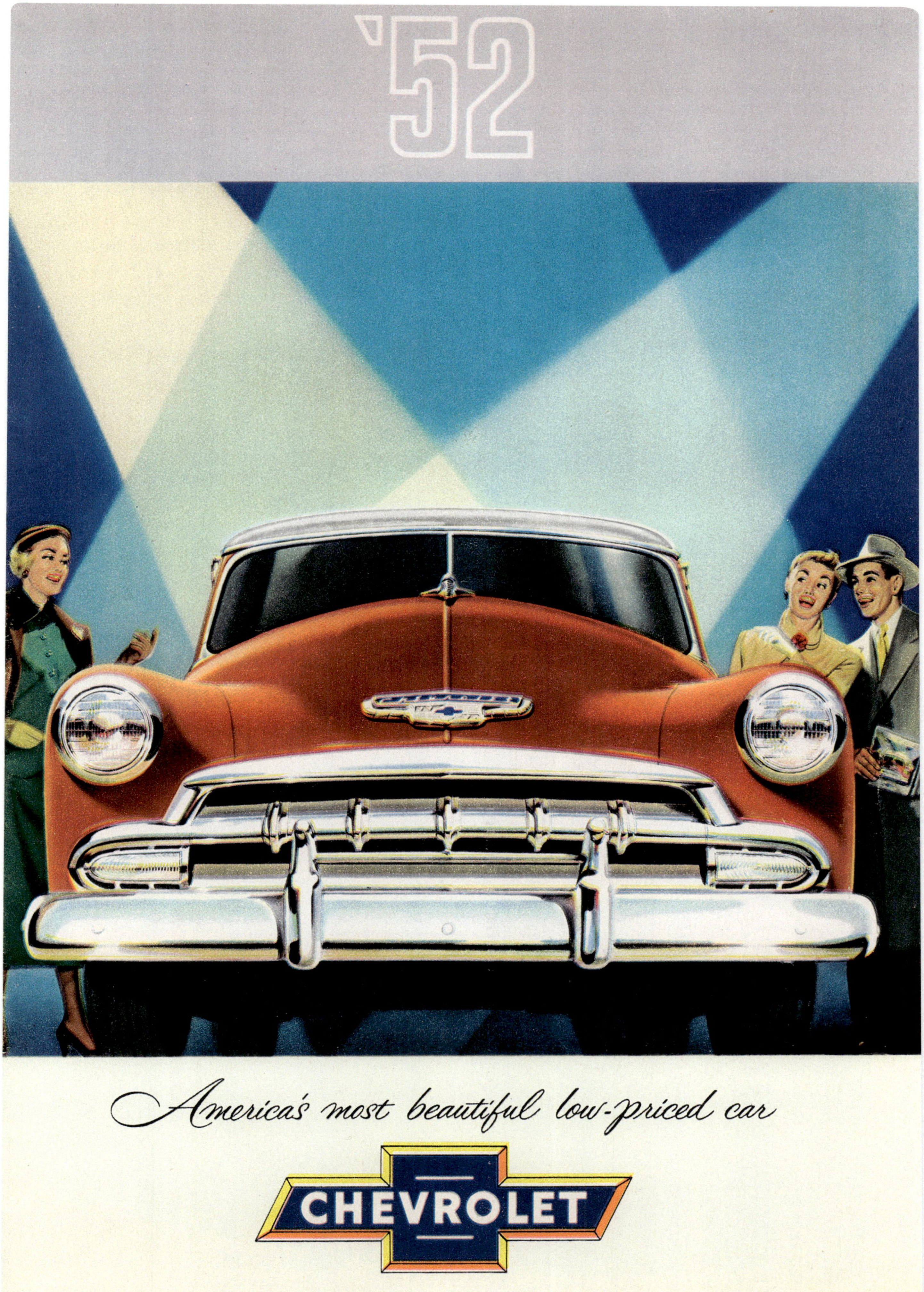
'52
America's most beautiful low-priced car
CHEVROLET

Chevrolet 1955 *(right and below)*, **1954** *(opposite)*

This two-door sport station wagon, the Nomad, evolved from a Corvette show car called the Corvair (Corvette + Bel Air), a name later used for a very different car. Most Nomads used the new, lightweight, small-block V-8 engine, displacing 265 cubic inches in its initial configuration. The Nomad was never a high-production car, and lasted only three years, but examples of the two-door wagon are prized today by Chevrolet enthusiasts and collectors of 1950s cars.

Dieser dreitürige Sportkombi, der Nomad, war aus einem Corvette-Konzeptauto namens Corvair (aus Corvette und Bel Air) hervorgegangen, dessen Name später für ein ganz anderes Fahrzeugmodell verwendet wurde. Die meisten Nomads besaßen einen neuen, kompakten und leichten V8-Motor, der in seiner ursprünglichen Konfiguration einen Hubraum von 4,3 Litern (4342 cm³) hatte. Der Nomad wurde nie in großen Stückzahlen und nur drei Jahre lang gebaut, doch bei Chevrolet-Fans und Sammlern von Automobilen der 1950er-Jahre sind diese Kombis heute sehr begehrt.

Ce break sportif deux portes baptisé Nomad était une évolution d'un modèle d'exposition Corvette appelé Corvair (Corvette et Bel Air), dénomination qui sera ultérieurement utilisée pour une tout autre voiture. La plupart des Nomad adoptèrent le nouveau moteur léger V8 small-block (4,3 l tout de même dans sa configuration originelle). La Nomad n'a jamais été produite en nombre, et son existence n'excéda pas trois ans, mais elle reste très prisée des amoureux de Chevrolet et des collectionneurs de voitures des années 1950.

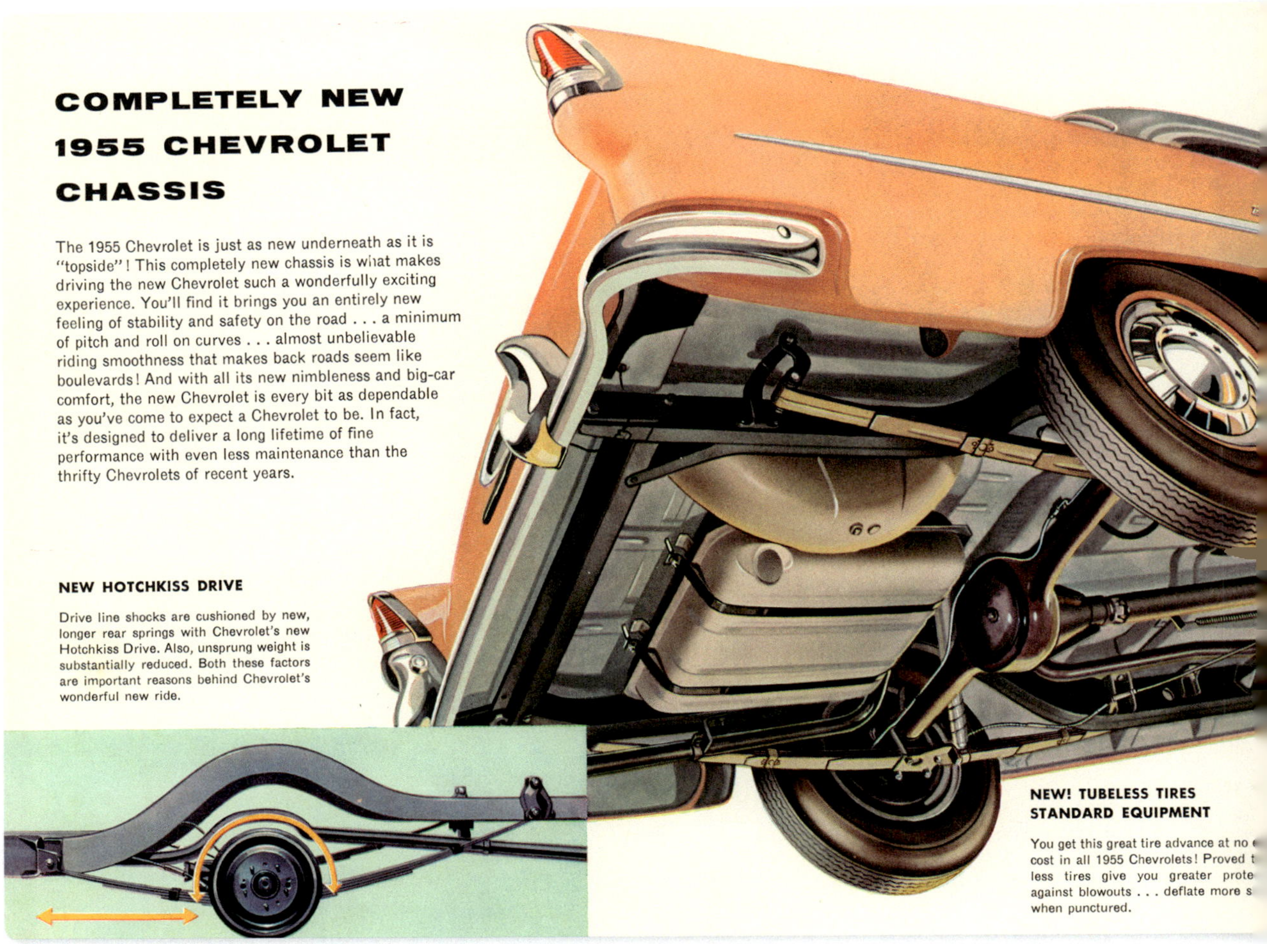

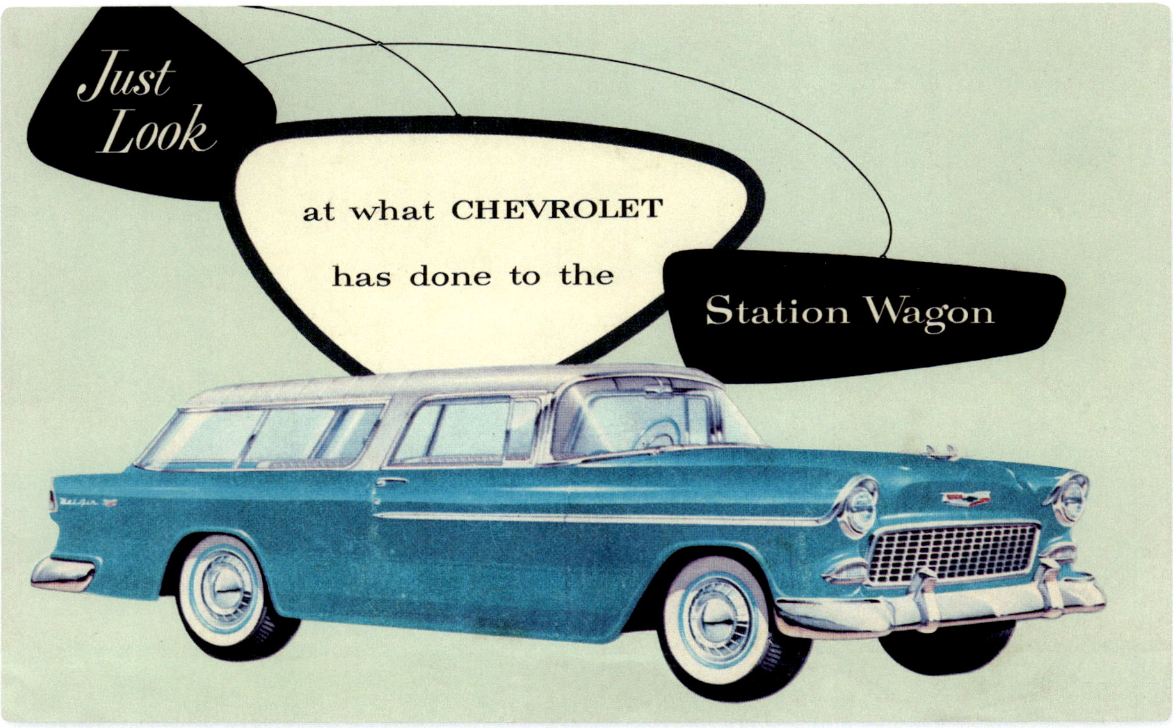

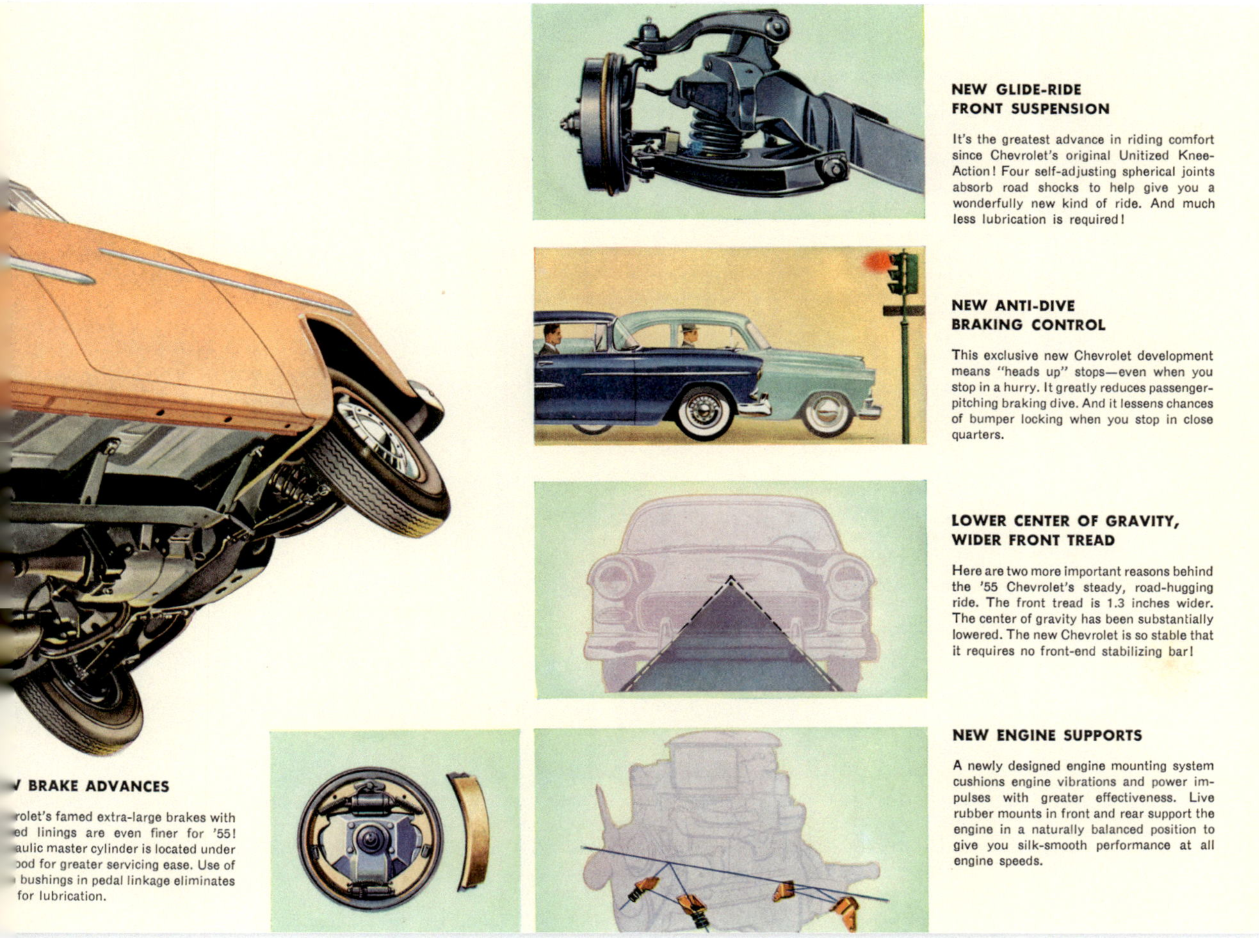
NEW GLIDE-RIDE FRONT SUSPENSION
It's the greatest advance in riding comfort since Chevrolet's original Unitized Knee-Action! Four self-adjusting spherical joints absorb road shocks to help give you a wonderfully new kind of ride. And much less lubrication is required!
NEW ANTI-DIVE BRAKING CONTROL
This exclusive new Chevrolet development means "heads up" stops—even when you stop in a hurry. It greatly reduces passenger-pitching braking dive. And it lessens chances of bumper locking when you stop in close quarters.
LOWER CENTER OF GRAVITY, WIDER FRONT TREAD
Here are two more important reasons behind the '55 Chevrolet's steady, road-hugging ride. The front tread is 1.3 inches wider. The center of gravity has been substantially lowered. The new Chevrolet is so stable that it requires no front-end stabilizing bar!
NEW ENGINE SUPPORTS
A newly designed engine mounting system cushions engine vibrations and power impulses with greater effectiveness. Live rubber mounts in front and rear support the engine in a naturally balanced position to give you silk-smooth performance at all engine speeds.
V BRAKE ADVANCES
rolet's famed extra-large brakes with
ed linings are even finer for '55!
aulic master cylinder is located under
ood for greater servicing ease. Use of
bushings in pedal linkage eliminates
for lubrication.

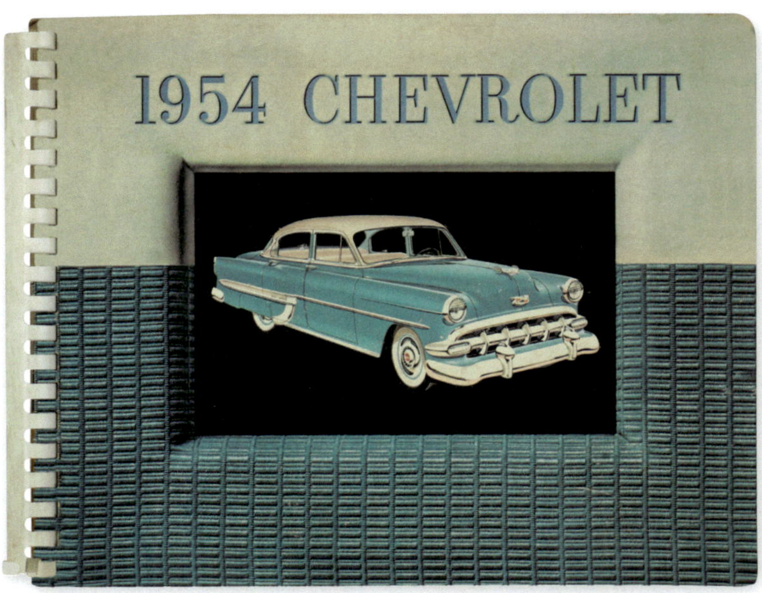
1954 CHEVROLET

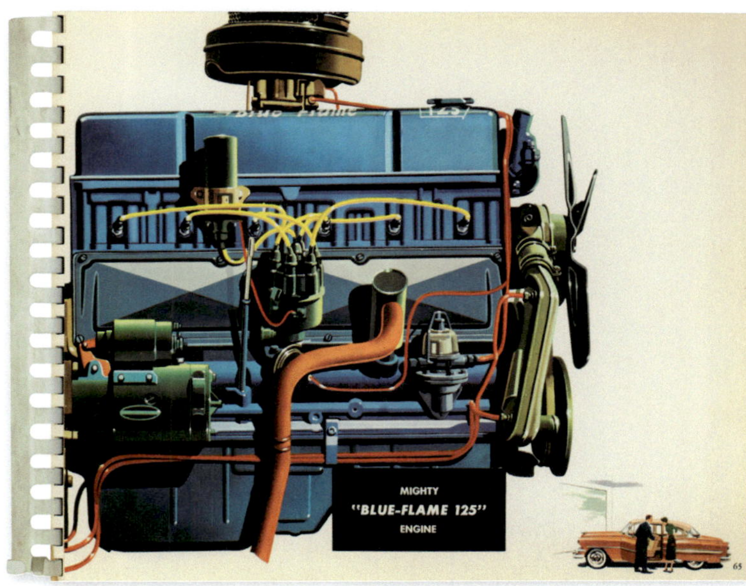
MIGHTY
"BLUE-FLAME 125"
ENGINE

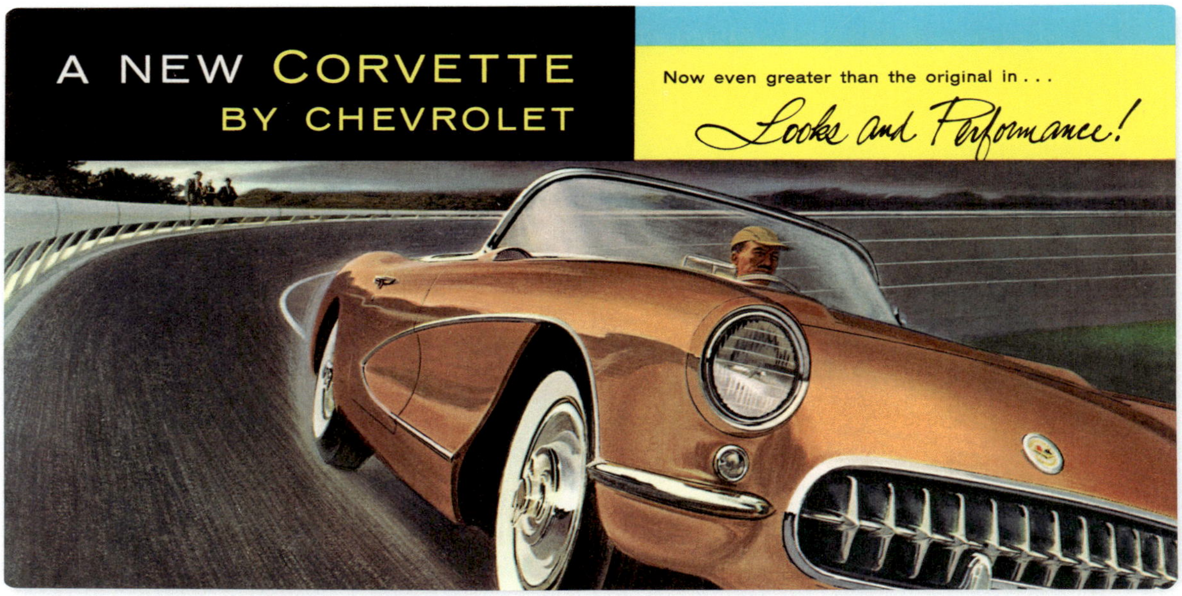

Chevrolet 1955 *(above)*, **1958** *(below)*

For all its later legend, the Corvette began as a somewhat dowdy six-cylinder roadster. All that changed in 1956 with new front and rear treatments, plus a dual-carburetor small-block V-8 engine with up to 240 hp. GM deserved credit for adding a true sports car to its lineup. A corporate plan to discontinue the car after its first several years was single-handedly blocked by chief Corvette engineer Zora Arkus-Duntov. More than half a century later, the Corvette is Chevrolet's greatest icon.

Trotz ihrer legendären Karriere begann die Corvette als recht unspektakulärer Roadster mit Sechszylindermotor. All dies änderte sich 1956 mit neuer Front und neuem Heck sowie kompaktem V8-Motor mit Doppelvergaser und bis zu 326 kW Leistung. Damit hatte GM nun einen echten Sportwagen in der Modellpalette. Als das Unternehmen plante, die Corvette nach den ersten Jahren wieder einzustellen, wehrte sich ihr Konstrukteur Zora Arkus-Duntov mit Händen und Füßen – und Erfolg: Über ein halbes Jahrhundert später ist die Corvette die größte Ikone aus dem Hause Chevrolet.

Avant de se parer durablement d'une auréole de légende, la Corvette fut d'abord un roadster six cylindres vaguement ringard. Tout changea en 1956 avec la rectification avant et arrière dont elle bénéficia, sans oublier l'adjonction d'un V8 small-block à carburateur double corps qui développait jusqu'à 240 chevaux. Saluons la General Motors pour avoir ajouté à sa gamme une vraie voiture de sport, même s'il était initialement prévu d'en interrompre la production au bout de quelques années – projet bloqué par l'ingénieur en chef du programme, Zora Arkus-Duntov. Plus d'un demi-siècle s'est écoulé depuis et la Corvette demeure l'icône emblématique de Chevrolet.

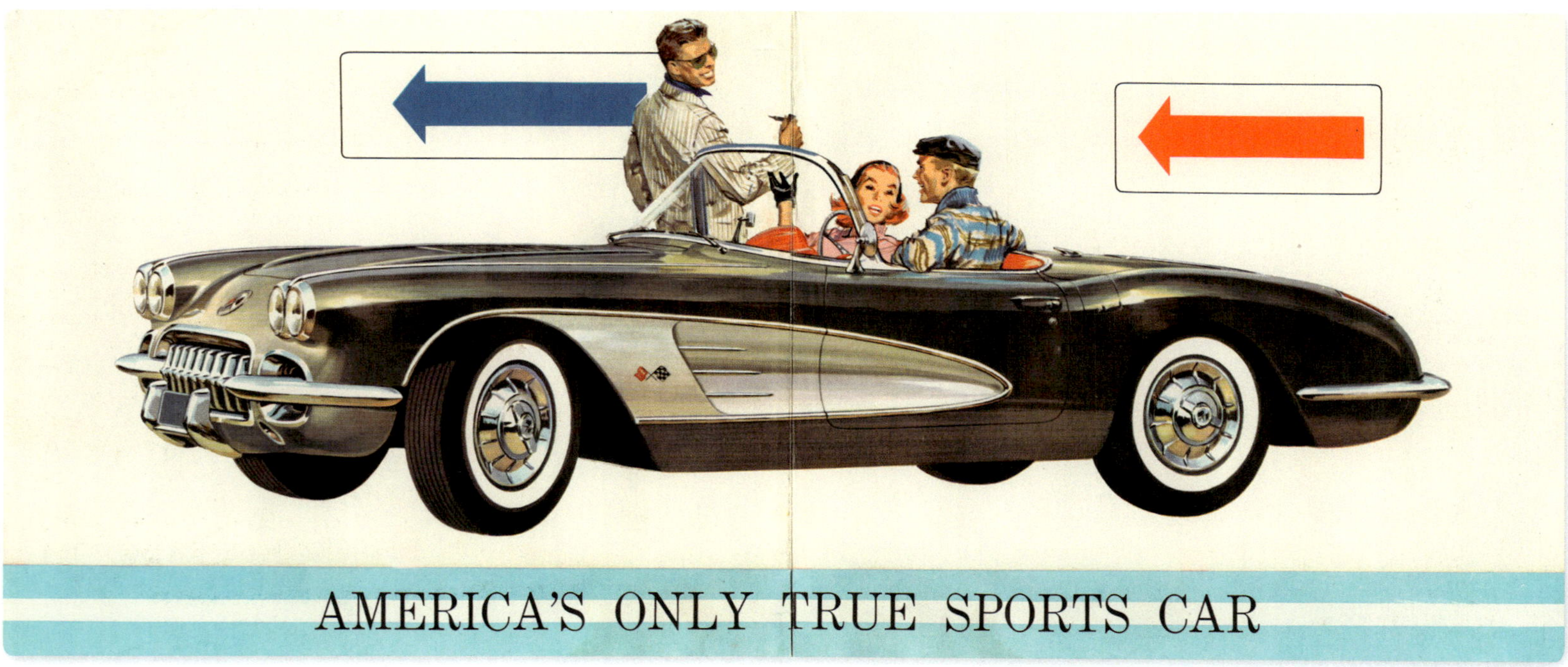

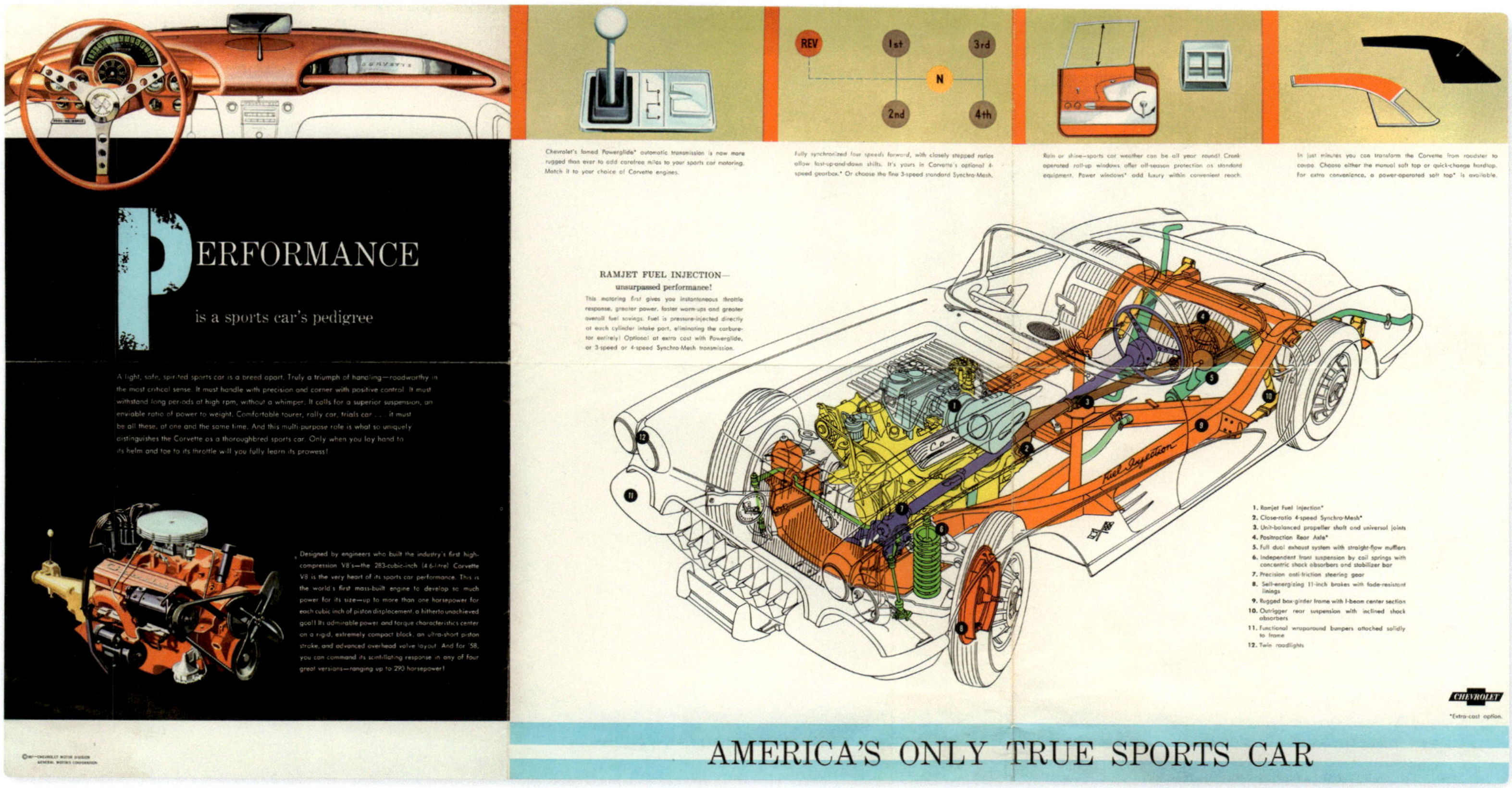

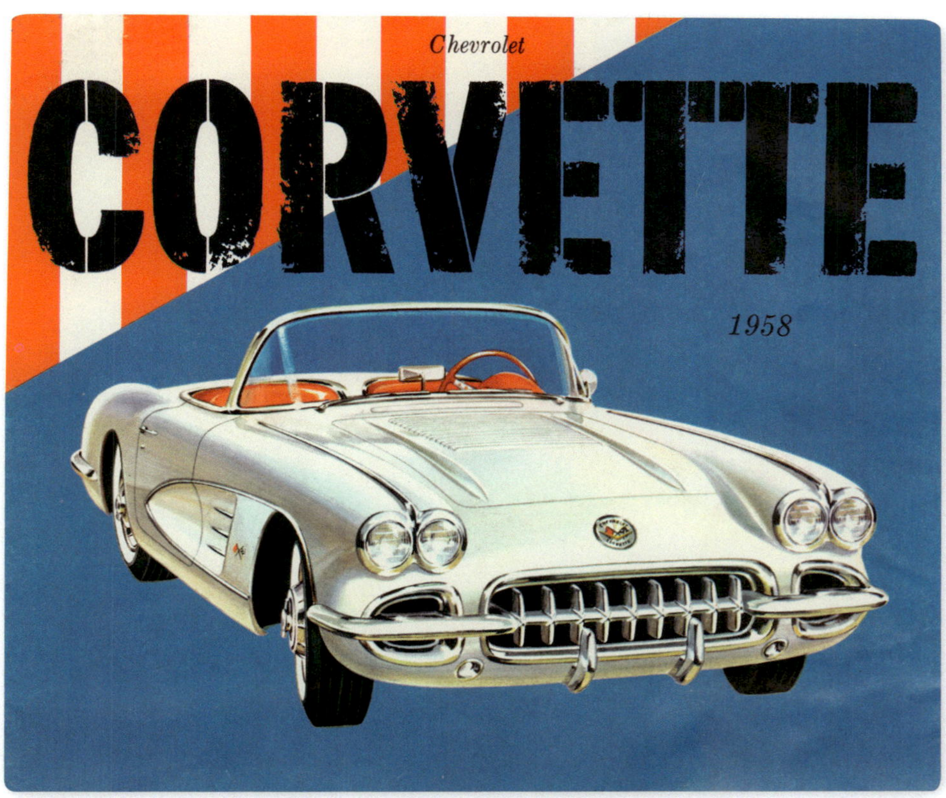

Chevrolet 1958

Despite somewhat awkward quad headlamps, which came into widespread use beginning in 1958, you could indeed blast out of town in one of these. Corvette power from the dual-quad V-8 was now up to 270 hp; a Rochester fuel-injected version introduced in 1957 had promised 250 hp.

Trotz der etwas plump wirkenden Vierfachscheinwerfer, die von 1958 an weit verbreitet waren, ging diese Corvette tatsächlich ab wie eine Rakete – dank ihrem V8-Motor mit doppeltem Vierfachvergaser, der 367 Kilowatt Leistung auf die Straße brachte. Eine Variante mit Rochester-Einspritzer, die 1957 auf den Markt kam, versprach 339 kW.

Malgré ses quatre phares pas forcément très heureux, adoptés un peu partout à partir de 1958, cette Corvette avait fière allure. La puissance de son V8 Dual Quad à quatre carburateurs double corps et injection atteignait désormais 270 chevaux. Une version à injection Rochester, introduite en 1957, avait déjà atteint la barre des 250 chevaux.

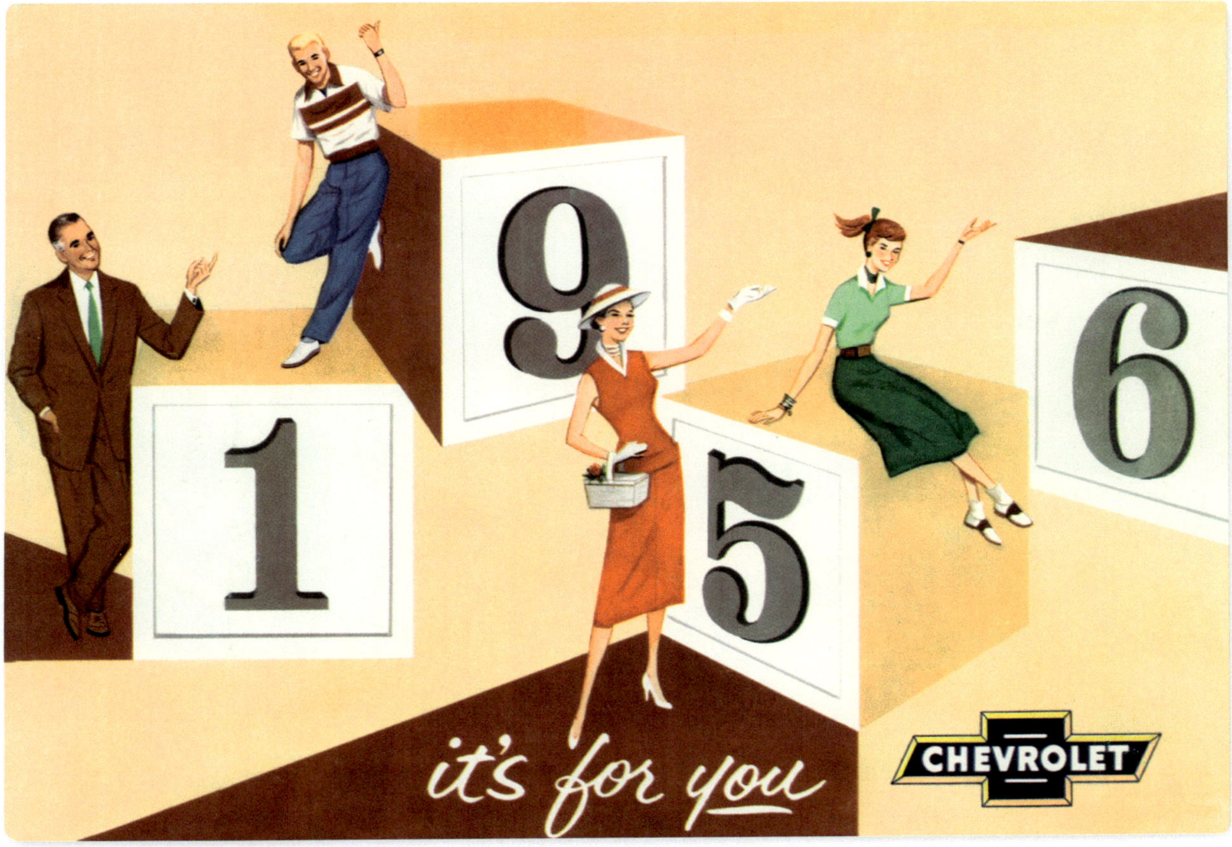

Chevrolet 1956 *(above)*, **1959** *(right)*, **1957** *(opposite)*

In 1957 Ford eked out a rare, small win over Chevrolet in annual sales. But with its soaring fins, the 1957 Chevrolet is now widely considered to be the most iconic 1950s car of all, from any manufacturer.

Im Jahre 1957 konnte Ford Chevrolet im Jahresabsatz geringfügig ausstechen, doch mit seinen hohen, geschwungenen Heckflossen gilt der 1957er-Chevrolet heutzutage unter den Fahrzeugen sämtlicher Hersteller als typischstes Auto der 1950er-Jahre.

Au classement des ventes de l'année 1957, Ford parvint – une fois n'est pas coutume – à distancer Chevrolet, d'un souffle. Mais avec ses fulgurants ailerons, la Chevy de cette année-là restera comme l'automobile la plus emblématique des Fifties, toutes marques confondues.

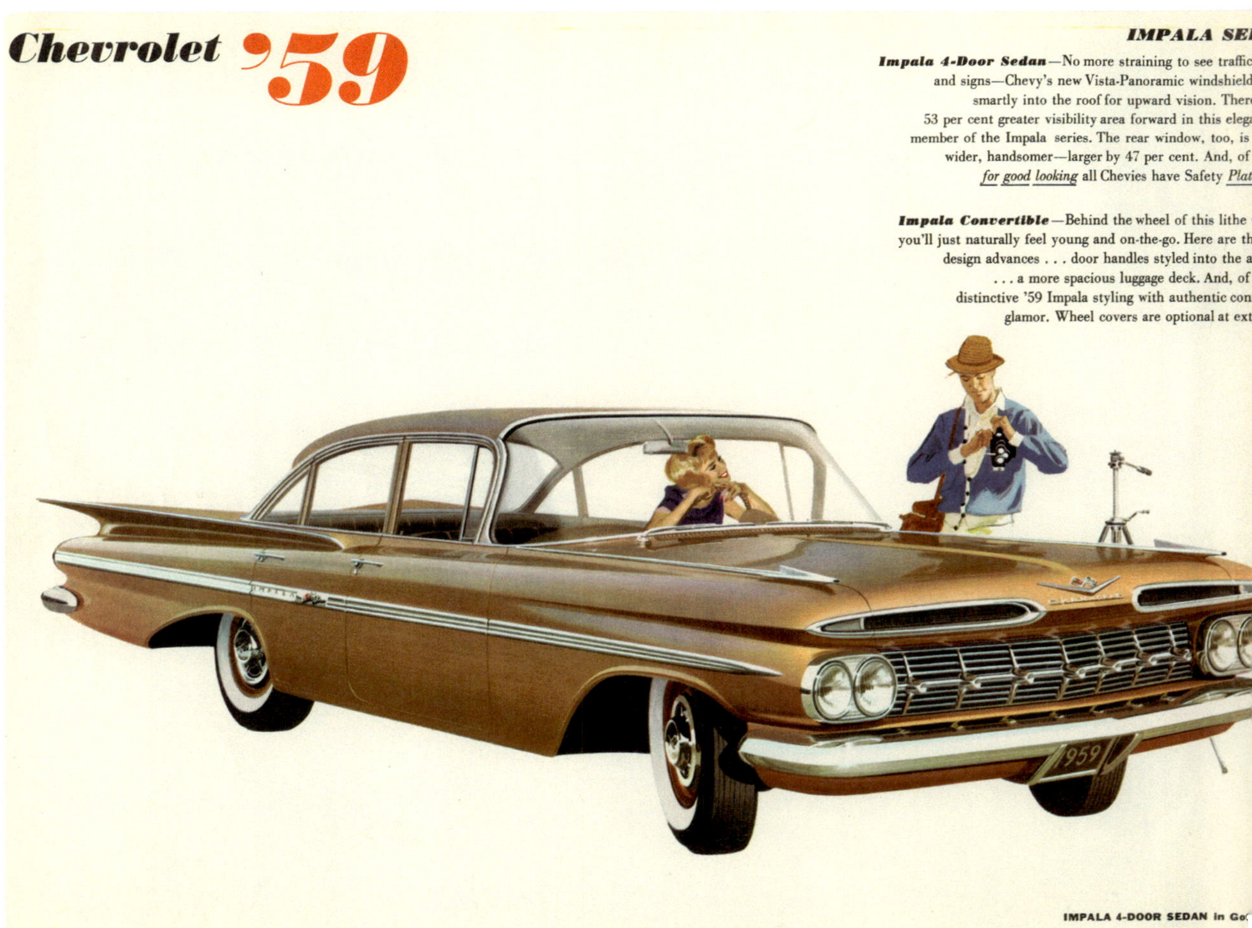

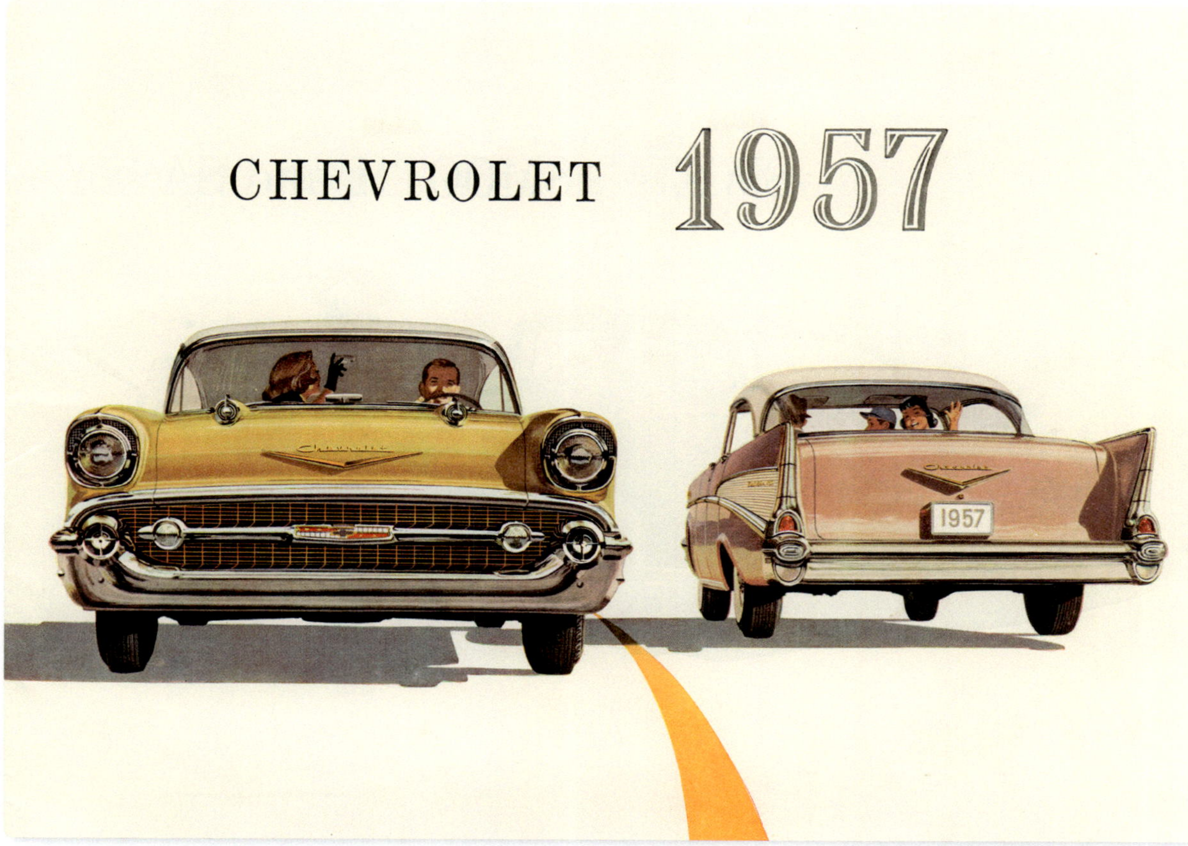
CHEVROLET 1957
1957

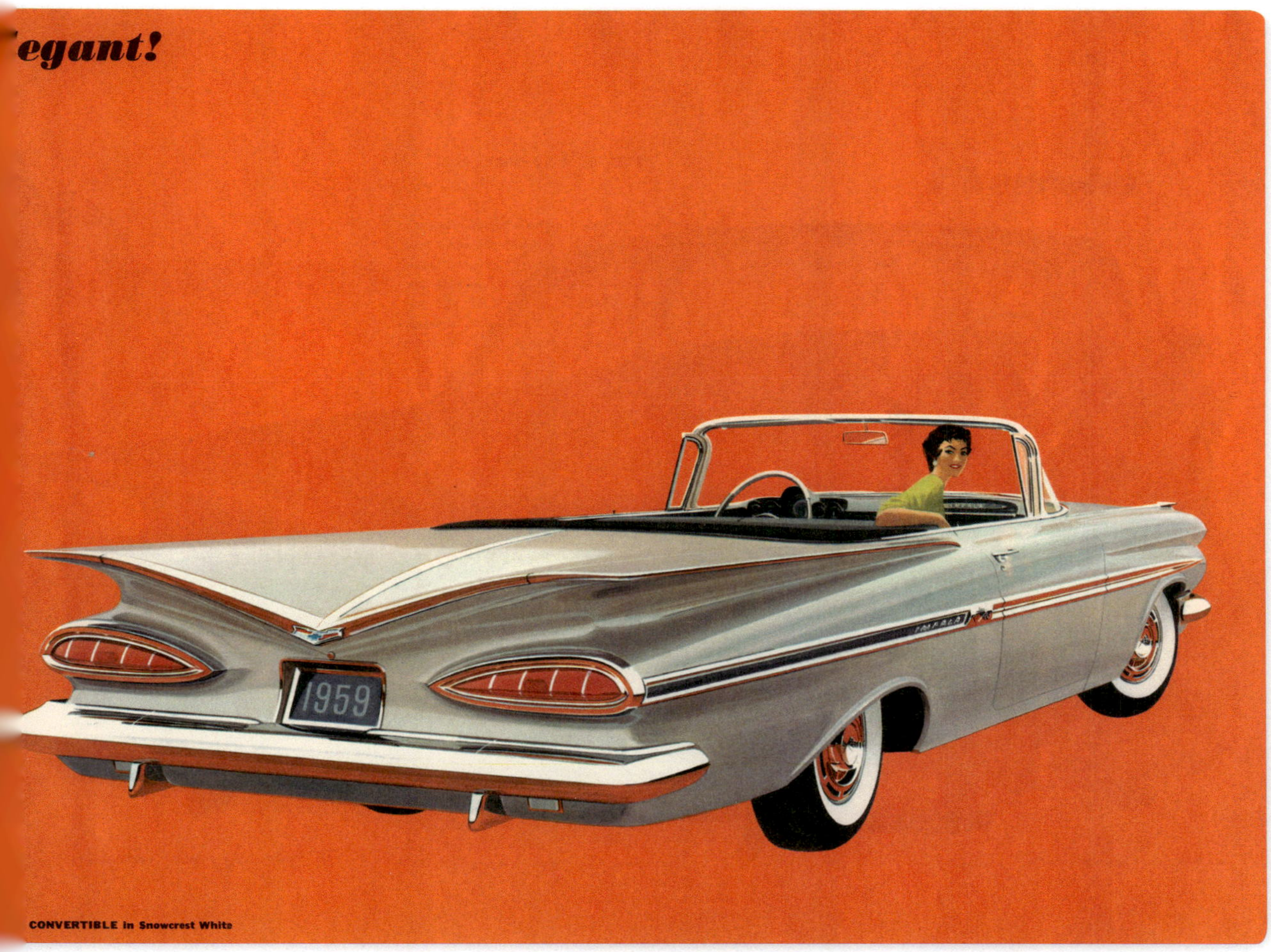
'egant!
1959
CONVERTIBLE in Snowcrest White

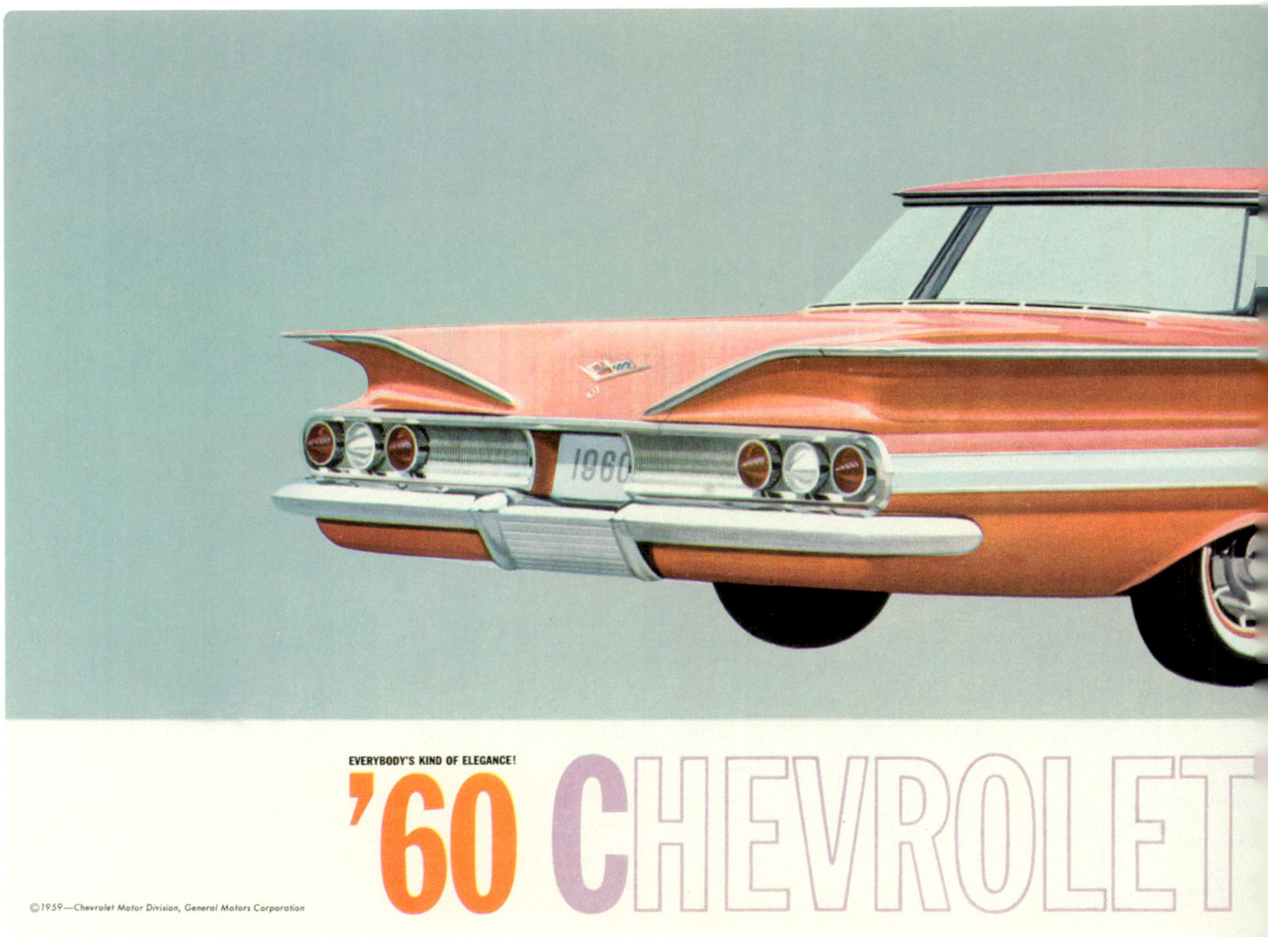
1960
EVERYBODY'S KIND OF ELEGANCE!
'60 CHEVROLET
©1959—Chevrolet Motor Division, General Motors Corporation

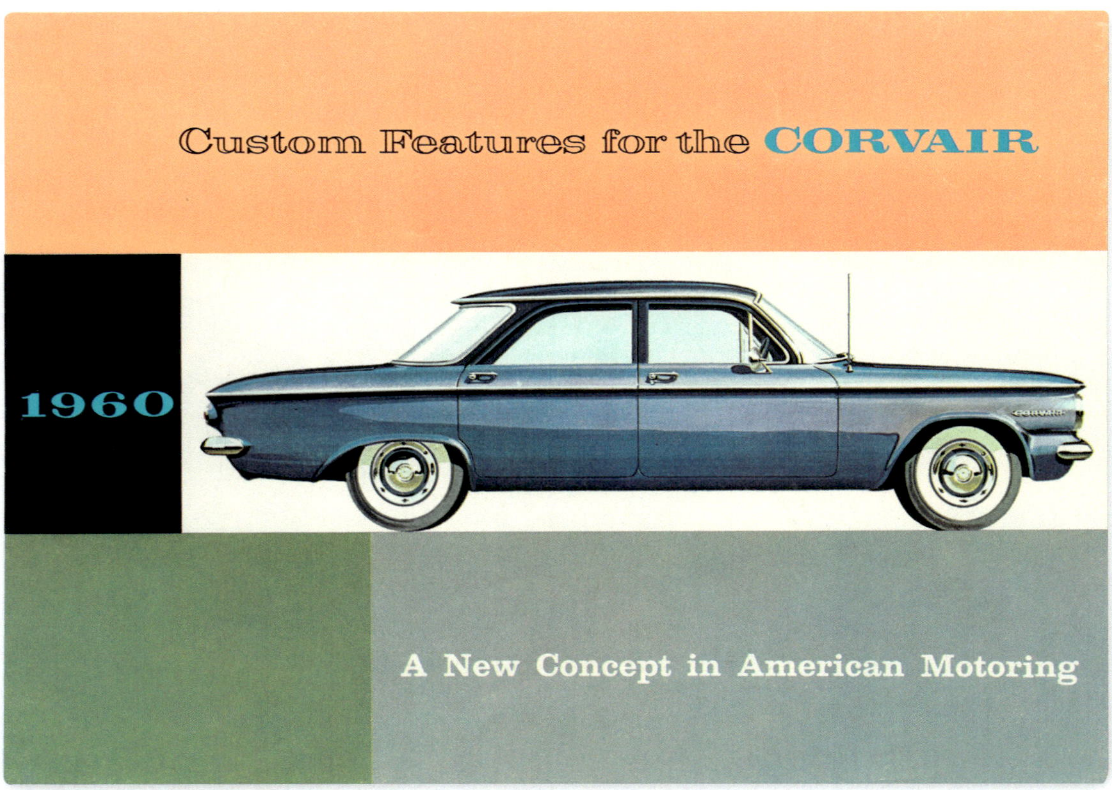
Custom Features for the CORVAIR
1960
A New Concept in American Motoring

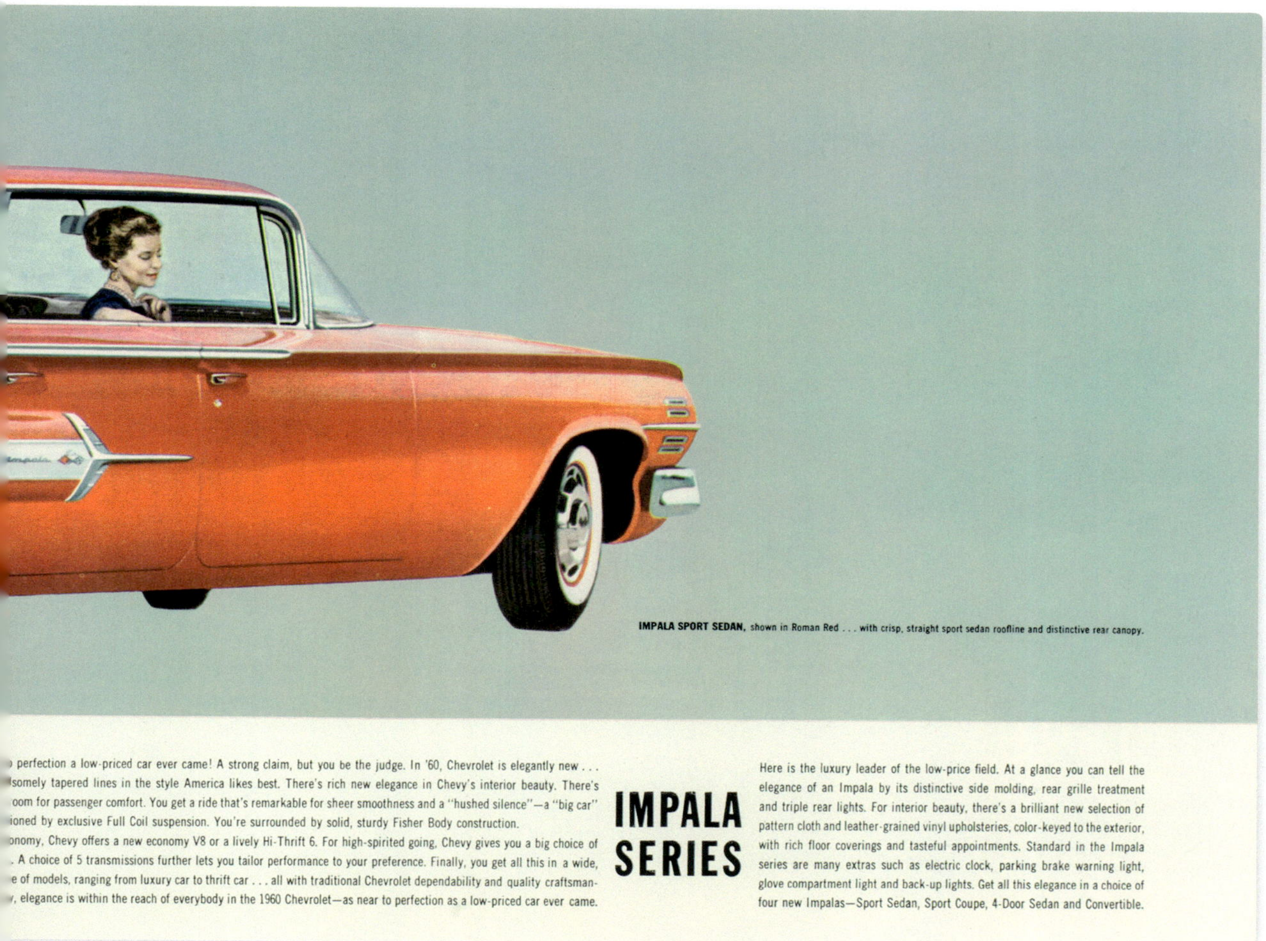

Chevrolet 1960 *(this spread)*

Edward Cole, the former chief engineer of Chevrolet, ran the division in 1960 and personally approved the Corvair, a forward-thinking compact. It had an air-cooled flat-six engine and transaxle in the rear plus independent suspension. The Corvair fell victim to bad publicity and was dropped from the line in 1969, but it is revered today for smart design.

Edward Cole, ehemals Chefingenieur bei Chevrolet, leitete die Abteilung 1960 und gab dem Corvair, einem zukunftsträchtigen Kompaktwagen, persönlich grünes Licht. Der Corvair besaß einen luftgekühlten Sechszylinderboxermotor, Transaxle (Motor vorn, Getriebe hinten) und Einzelradaufhängung. Der Corvair machte seinerzeit negative Schlagzeilen und wurde daraufhin 1969 eingestellt, doch heute wird er wegen seiner schnittigen Form wieder bewundert.

Edward Cole, jusqu'alors ingénieur en chef de Chevrolet, prit la tête de la division en 1960 et dirigea de très près la création de la Corvair, un modèle compact et avant-gardiste. Elle disposait d'un moteur six cylindres à plat refroidi par air et placé à l'arrière avec la transmission, ainsi que d'une suspension à quatre roues indépendantes. La Corvair souffrit d'une mauvaise publicité et fut retirée du catalogue en 1969. Elle reste toutefois encore adulée aujourd'hui pour son élégance.

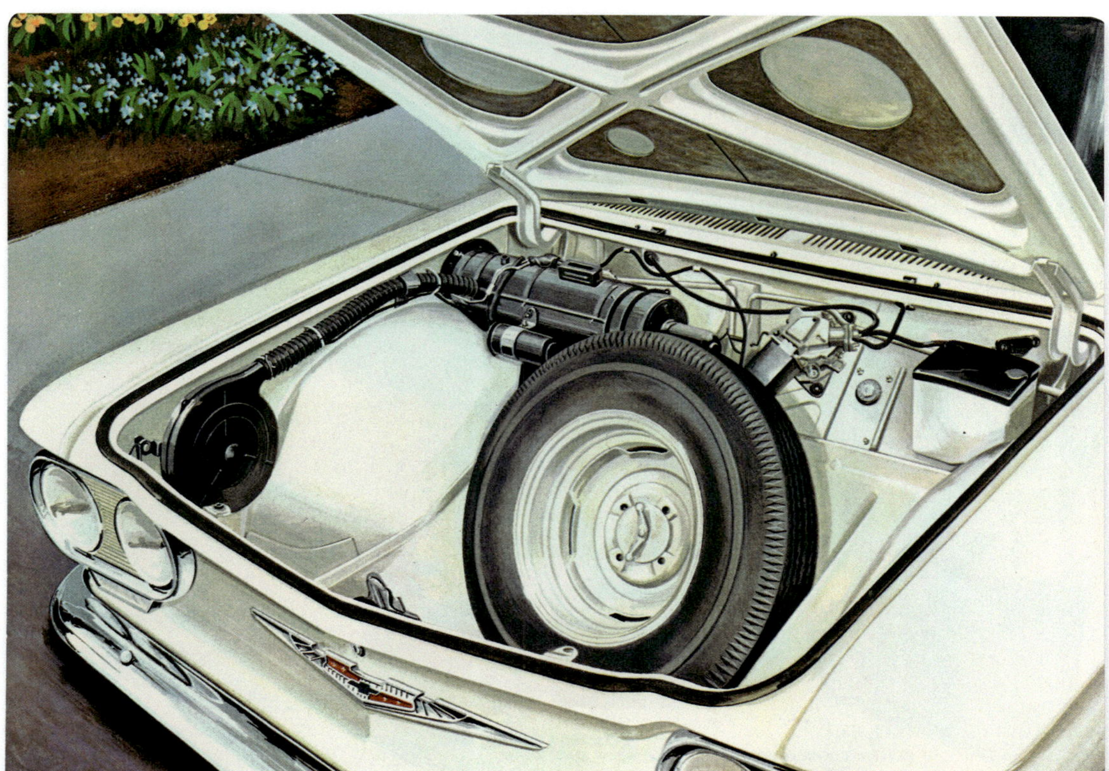

Chrysler

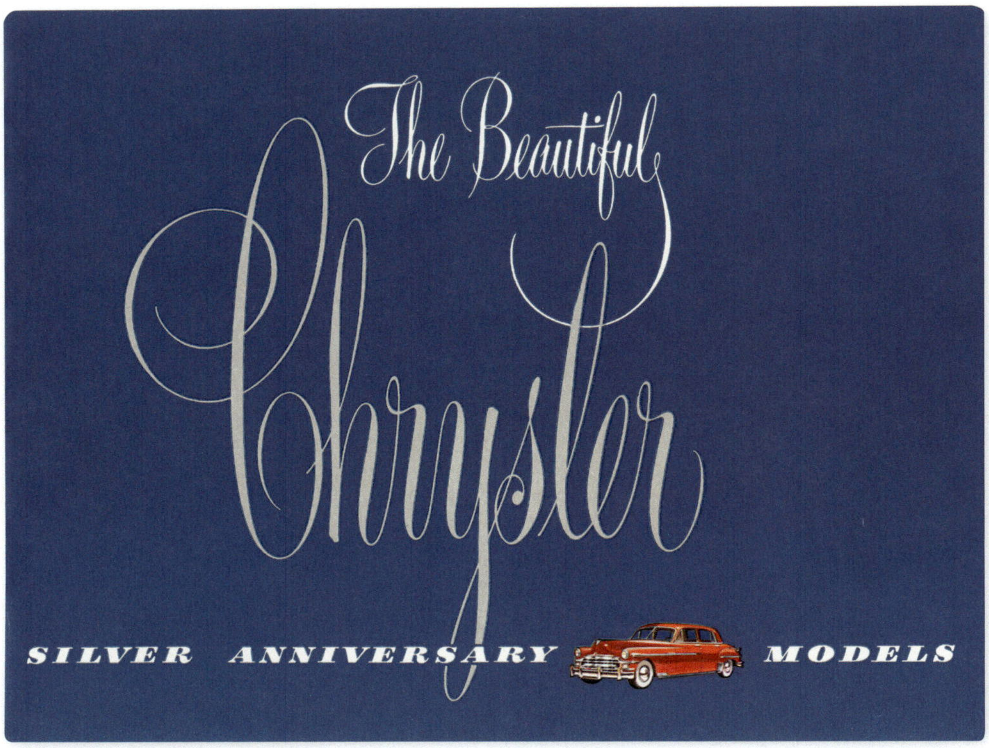

Chrysler 1949 *(above)*, **1950** *(right)*

For its 25th anniversary, Chrysler got all-new bodies, but the revised looks were eclipsed in 1951 by Chrysler's first V-8. It became known as the Hemi for its combustion chamber shape and produced 180 hp. Hemi engines have been synonymous with Chrysler ever since.

Zum 25-jährigen Firmenjubiläum gönnte sich Chrysler komplett neue Karosserien, doch noch aufsehenerregender als das neue Äußere war 1951 Chryslers erster V8-Motor, der aufgrund der halbkugelförmigen Form seines Brennraums als „Hemi" (für *hemispherical combustion chamber*, nämlich „hemisphärischer Brennraum") bekannt wurde und 244 kW Leistung lieferte. Seither sind Hemi-Motoren untrennbar mit Chrysler verknüpft.

Pour son 25ᵉ anniversaire Chrysler hérita de carrosseries dernier cri, mais les nouveaux looks se virent éclipser en 1951 par le premier V8 maison, qui développait 180 chevaux. La forme en hémisphère de ses chambres de combustion lui a valu le nom de « Hemi », qui subsiste aujourd'hui et demeure indéfectiblement associée à la marque.

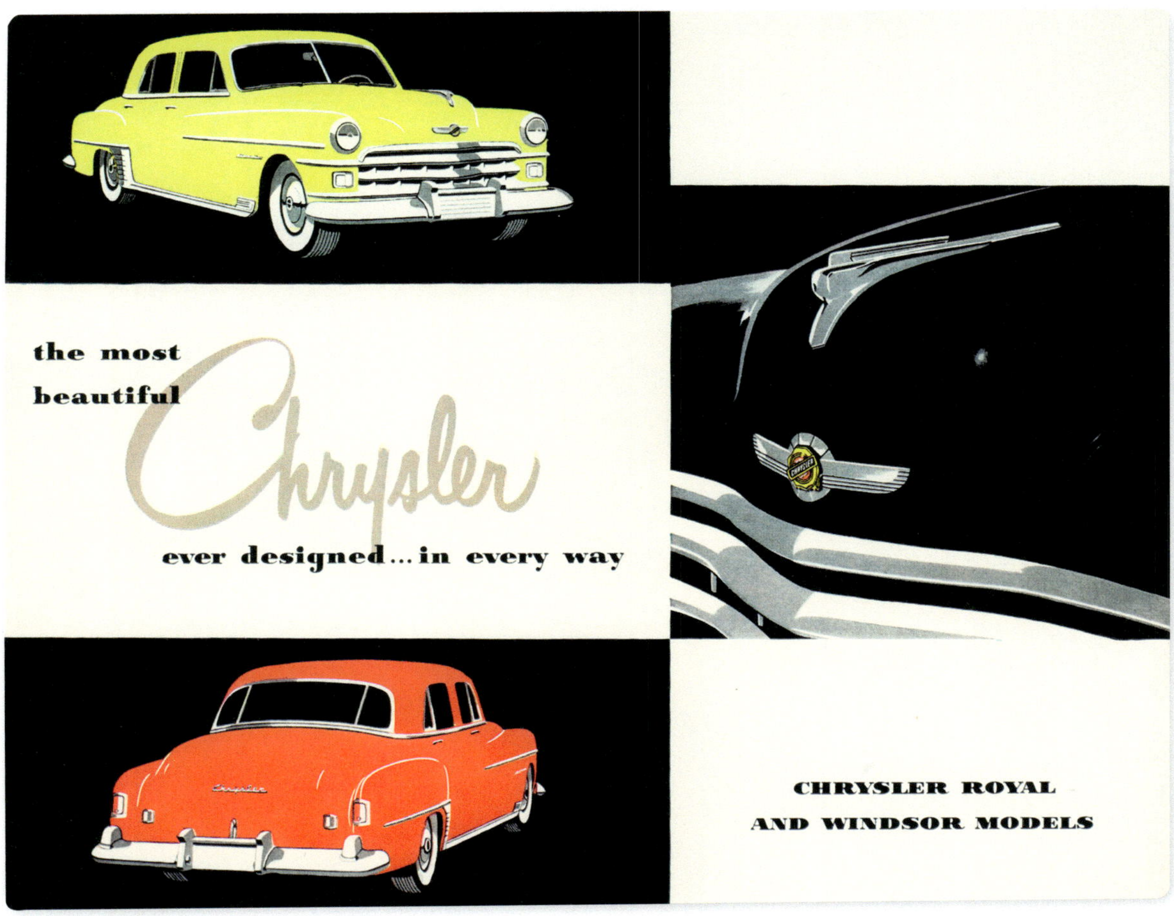

Chrysler 1950 *(right)*, **1948** *(below)*

Chrysler entered the 1950s as a reliable, but conservative, premium automobile. Straight-six and straight-eight engines were the norm until Chrysler transformed the engineering landscape with its Hemi V-8 for 1951.

Chrysler begann die 1950er-Jahre als Hersteller zuverlässiger, aber konservativer Premium-Automobile. Sechs- und Achtzylinderreihenmotoren waren die Norm, bis Chrysler die Motorenlandschaft 1951 mit seinem Hemi V8 grundlegend veränderte.

Chrysler aborda les années 1950 avec la réputation d'une marque certes fiable, mais conservatrice et haut de gamme. Les moteurs 6 et 8 cylindres en ligne étaient la norme jusqu'à ce que Chrysler bouleverse le paysage de la technologie des moteurs avec la sortie de son Hemi V8 en 1951.

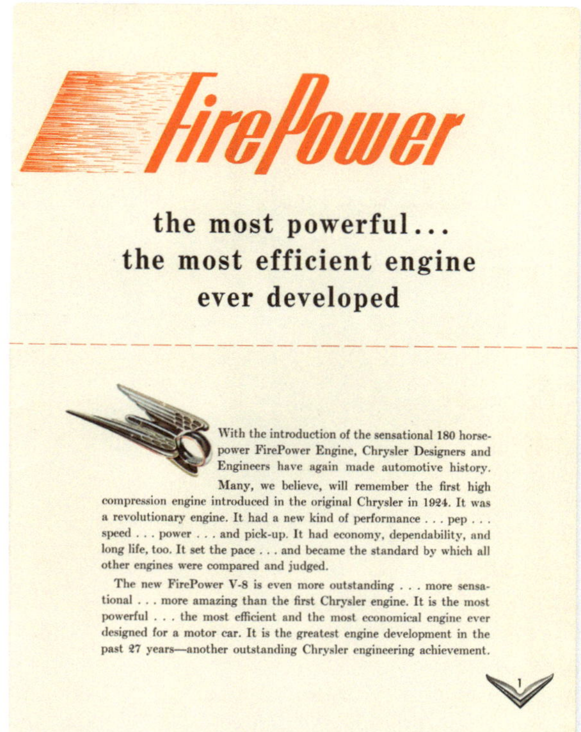

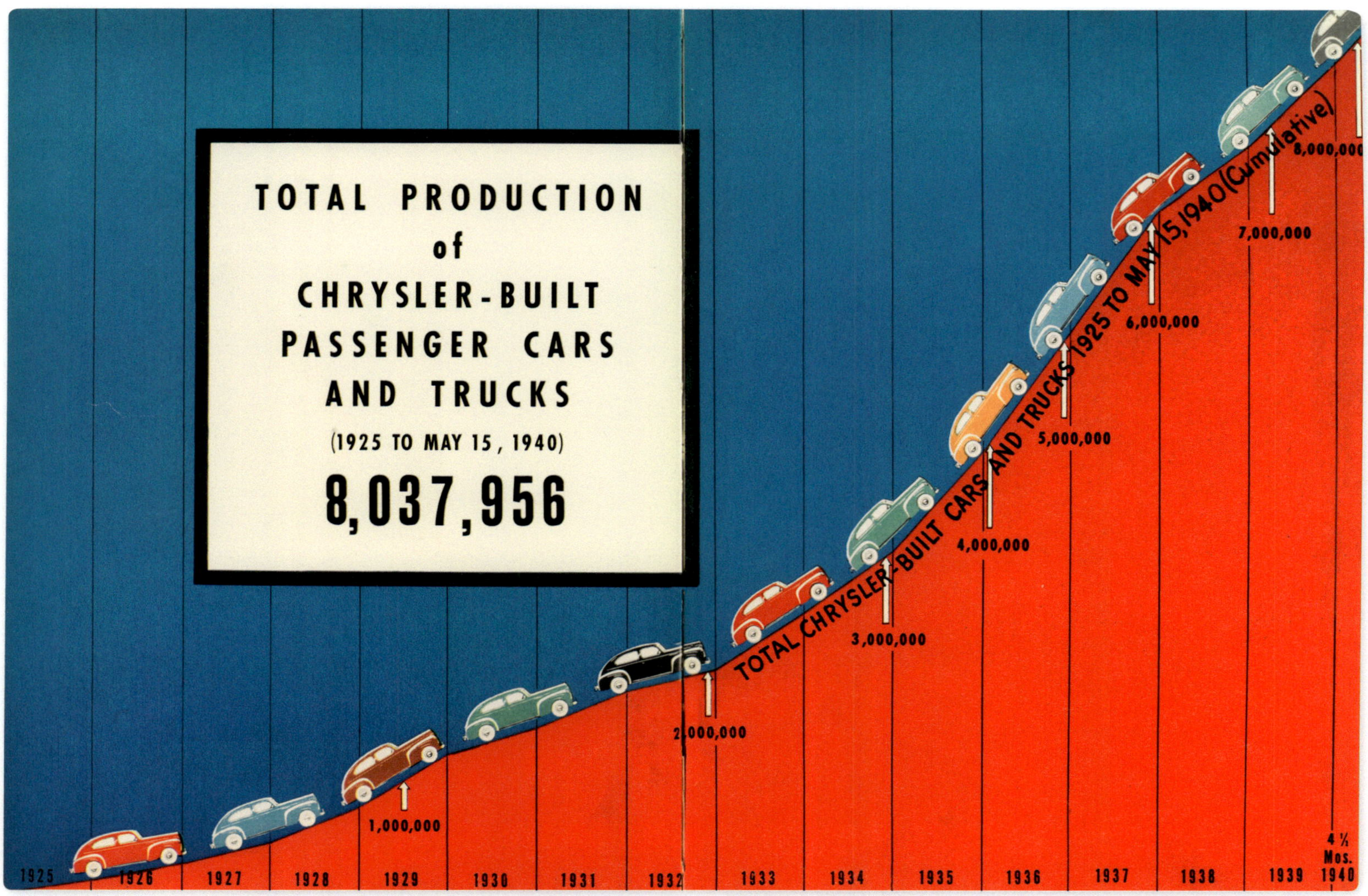

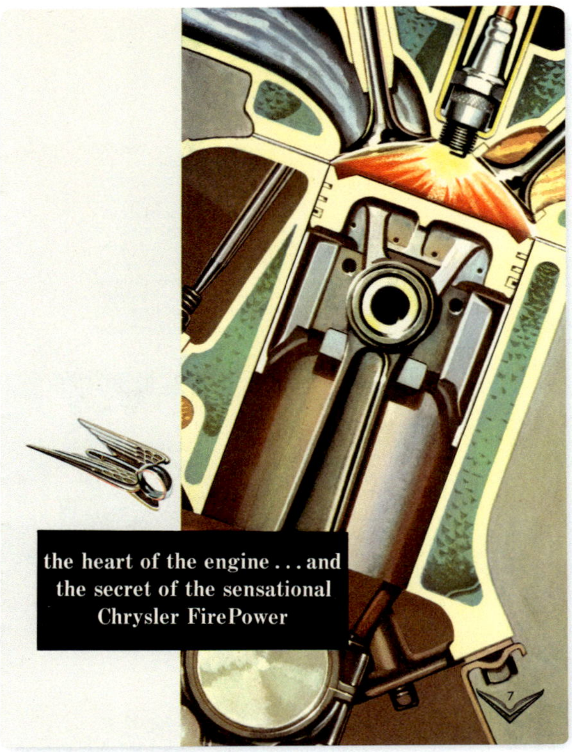

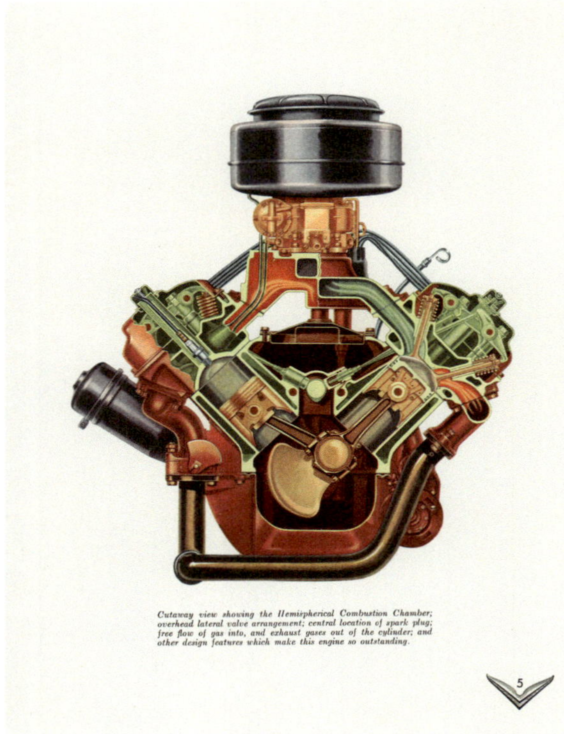

Chrysler 1950 *(above)*, **1953** *(right)*, **1950** *(opposite)*

As the 1950s commenced, Chrysler was a solid, sensible luxury car, particularly with more than two doors, the kind of car a bank president might find appropriate. That stolidity would change soon. Chrysler reestablished its fine engineering credentials with the hemispherical-combustion V-8, and soon went on a styling binge led by Virgil Exner. The conservative Chrysler was about to be no more.

Zu Beginn der 1950er-Jahre war ein Chrysler ein solides, vernünftiges Luxusauto, vor allem mit mehr als zwei Türen – die Art von Limousine, die der Vorstandsvorsitzende einer Bank als angemessen empfand. Aber diese Behäbigkeit war nicht von Dauer. Chrysler unterstrich seinen Ruf in der Ingenieurskunst mit dem V8-Hemi-Motor und leistete sich anschließend unter Leitung von Virgil Exner etliche extravagante Exzesse im Design – und begrub damit endgültig seinen konservativen Ruf.

À l'orée des années 1950, la Chrysler était une voiture solide au luxe étudié, surtout quand elle comptait plus de deux portes ; tout ce que recherche un banquier, en somme. Ce flegme anglo-saxon n'allait pas tarder à passer de mode. Avec son fameux V8 à combustion hémisphérique, la marque fit de nouveau honneur à sa réputation en matière d'ingénierie mécanique et se jeta dans une débauche d'efforts stylistiques, sous la direction de Virgil Exner. Le conservatisme de Chrysler ne serait bientôt plus qu'un souvenir.

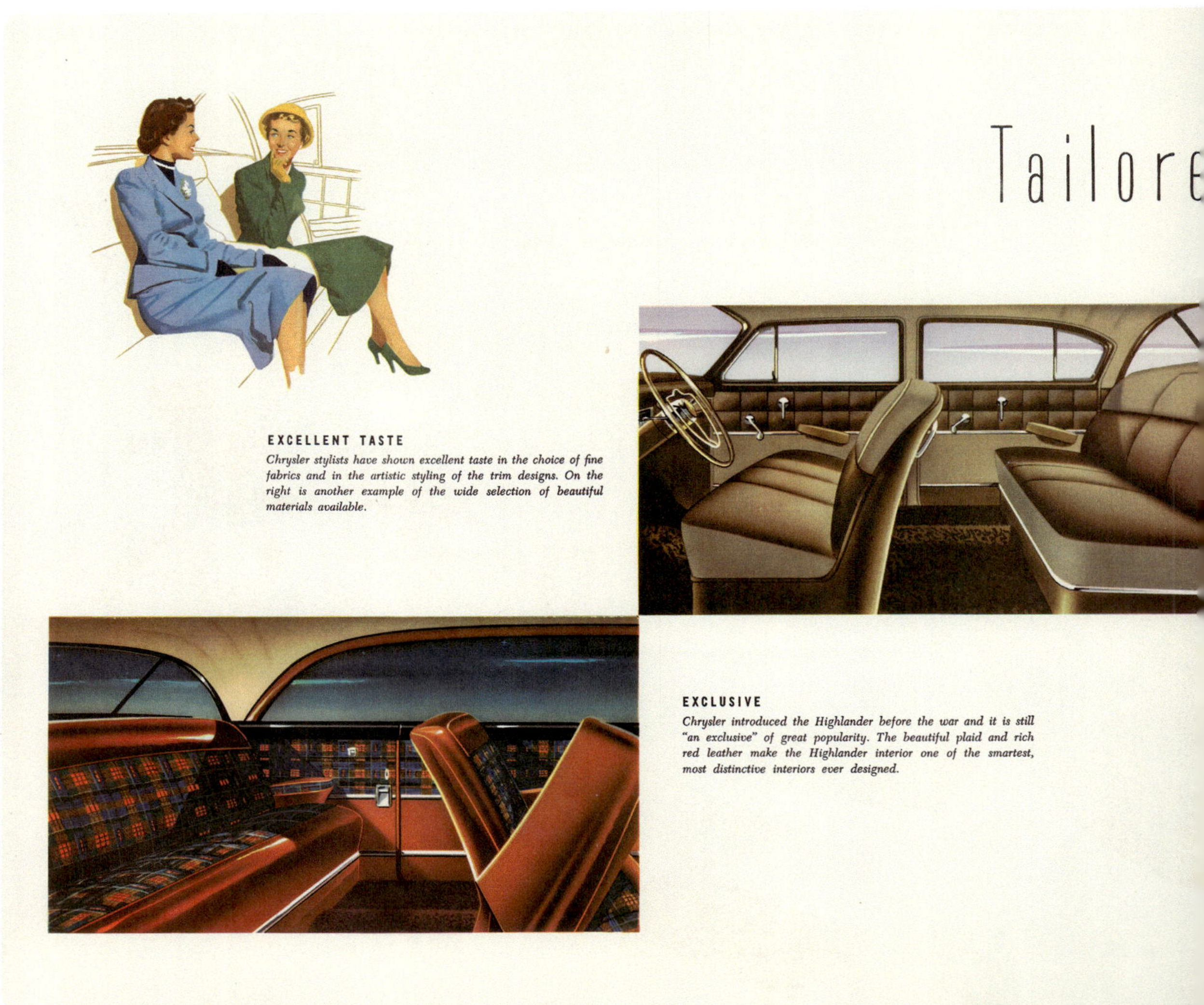

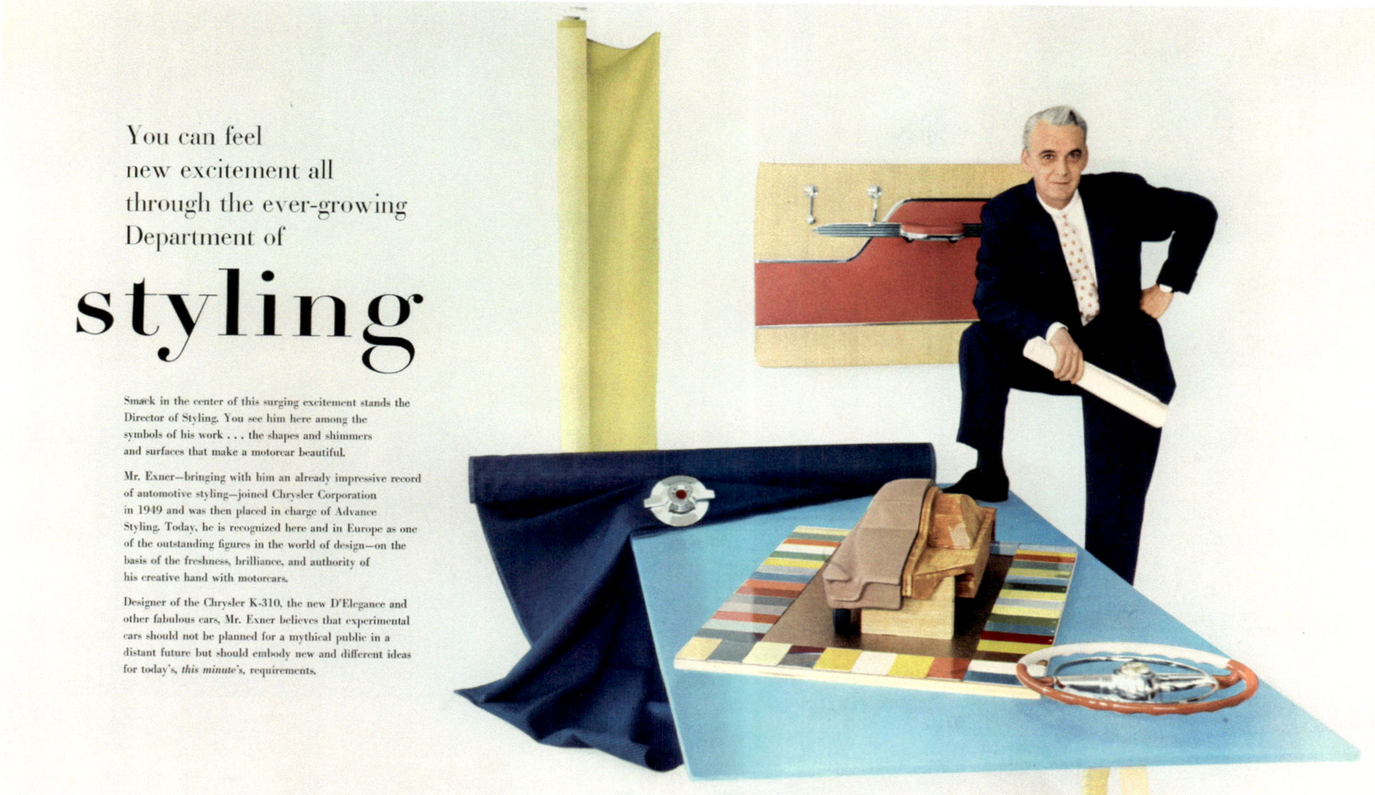

You can feel new excitement all through the ever-growing Department of
styling
Smack in the center of this surging excitement stands the Director of Styling. You see him here among the symbols of his work . . . the shapes and shimmers and surfaces that make a motorcar beautiful.
Mr. Exner—bringing with him an already impressive record of automotive styling—joined Chrysler Corporation in 1949 and was then placed in charge of Advance Styling. Today, he is recognized here and in Europe as one of the outstanding figures in the world of design—on the basis of the freshness, brilliance, and authority of his creative hand with motorcars.
Designer of the Chrysler K-310, the new D'Elegance and other fabulous cars, Mr. Exner believes that experimental cars should not be planned for a mythical public in a distant future but should embody new and different ideas for today's, *this minute's*, requirements.

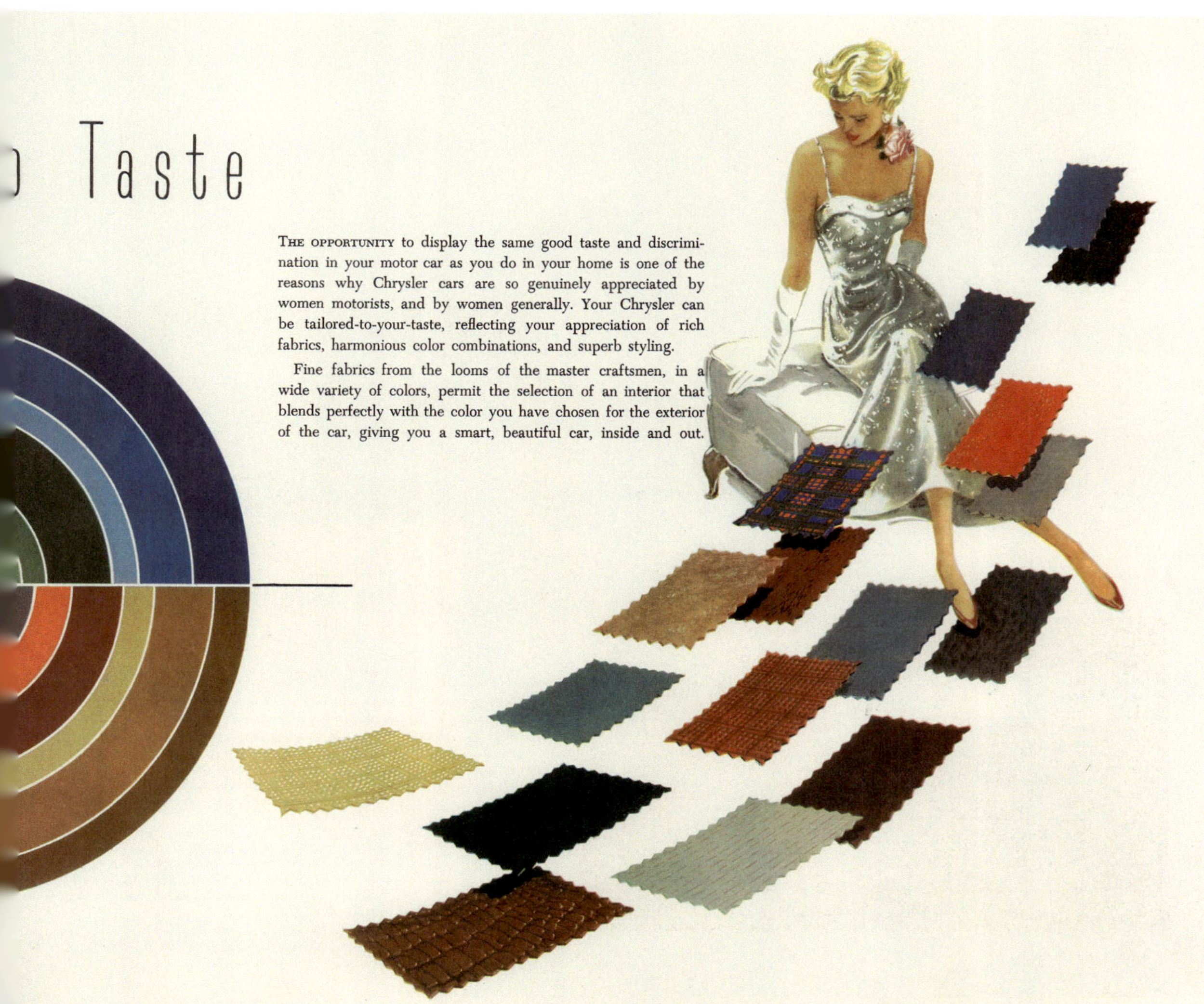

Taste
THE OPPORTUNITY to display the same good taste and discrimination in your motor car as you do in your home is one of the reasons why Chrysler cars are so genuinely appreciated by women motorists, and by women generally. Your Chrysler can be tailored-to-your-taste, reflecting your appreciation of rich fabrics, harmonious color combinations, and superb styling.
Fine fabrics from the looms of the master craftsmen, in a wide variety of colors, permit the selection of an interior that blends perfectly with the color you have chosen for the exterior of the car, giving you a smart, beautiful car, inside and out.

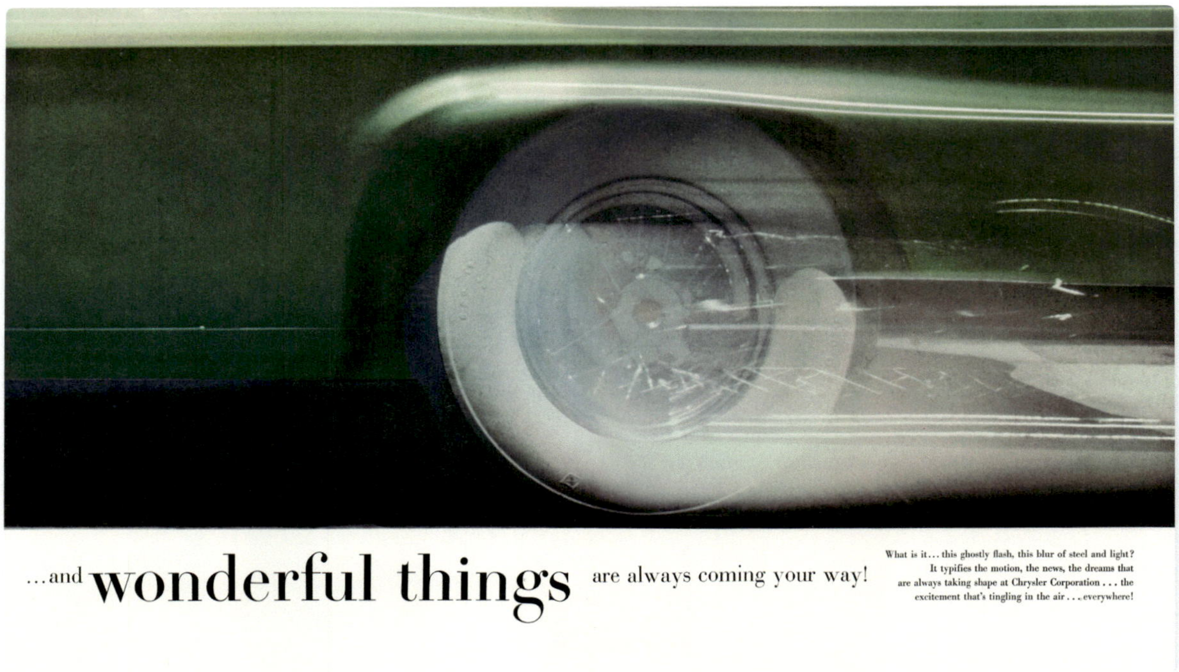

Chrysler 1953, 1954 *(opposite)*

In an atypical move, this brochure pays homage to the design teams who labor in secret, turning artist's clay into prototypes of new cars for executives to assess. By this time, cars like the Town & Country station wagon had all-steel bodies.

Ganz untypisch war an dieser Broschüre, dass sie den Konstrukteuren Tribut zollte, die üblicherweise im Verborgenen arbeiteten und die Tonmodelle der Künstler in Prototypen für neue Autos umsetzten, die dann von der Geschäftsleitung begutachtet werden konnten. Zu dieser Zeit besaßen Autos wie der Kombi Town & Country noch Ganzstahlkarosserien.

De façon assez inaccoutumée, ce catalogue rend hommage aux équipes de design qui œuvrèrent dans le plus grand secret pour donner naissance à ces prototypes soumis à l'appréciation des cadres de la maison. À cette époque, des voitures comme le break Town & Country bénéficiaient de carrosseries tout acier.

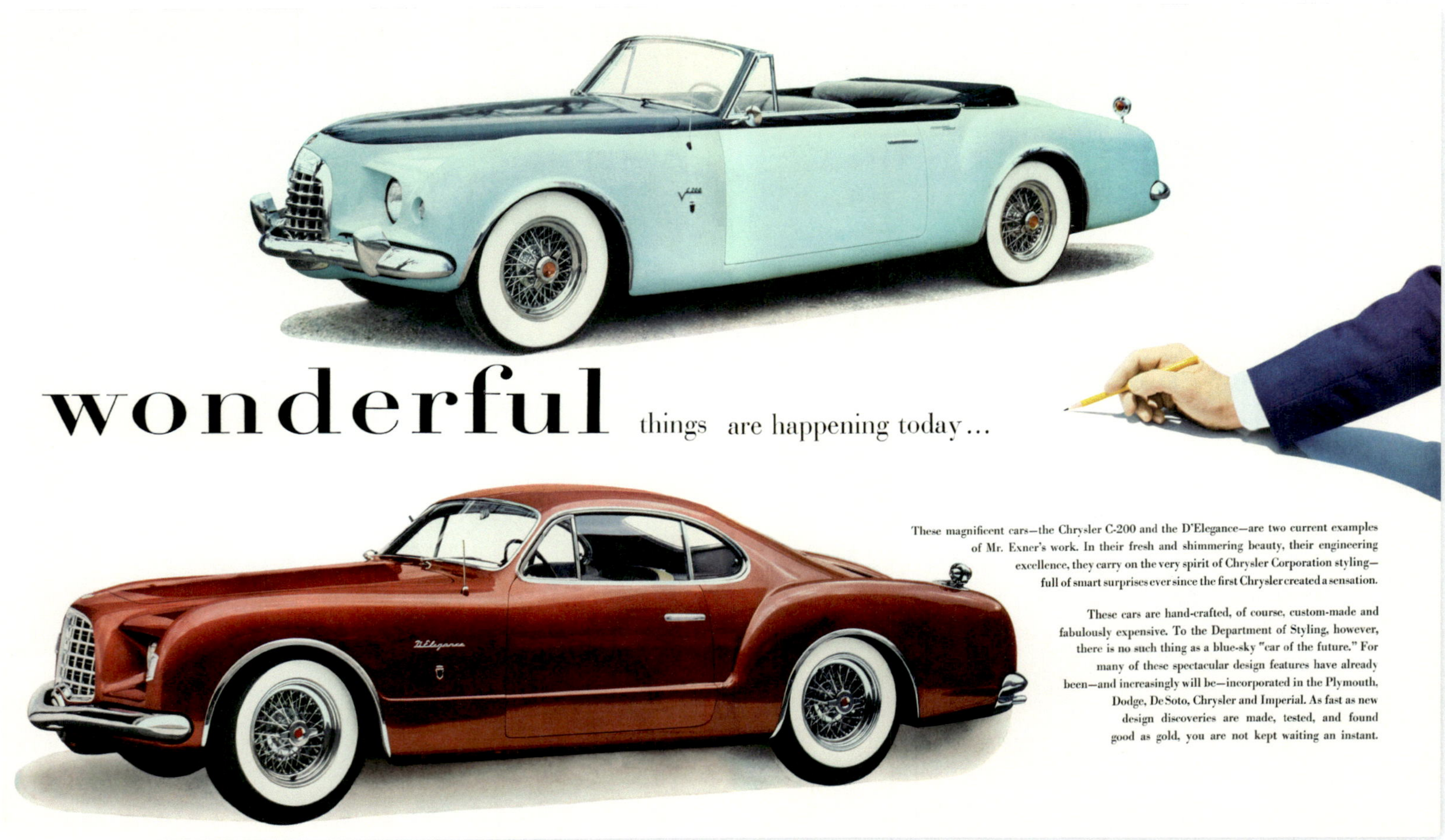

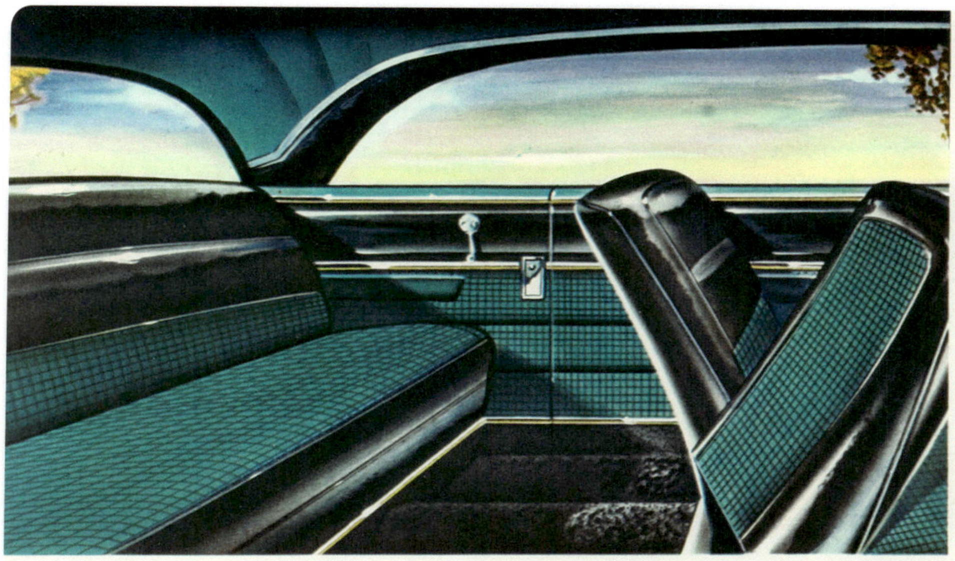

On the left *is one of the gorgeous tailored-to-taste upholstery combinations offered on the New Yorker Deluxe Newport. It is Turquoise Nyplaid with Black Leather. Six Two-Tone Leather combinations are also available on special order.*

Interiors of Superb Styling

The luxurious interior *of the New Yorker Deluxe Six Passenger Sedan is illustrated below. This harmonious combination is Two-Tone Green Broadcloth, one of four beautiful options for your selection. The center arm rest is standard in both front and rear.*

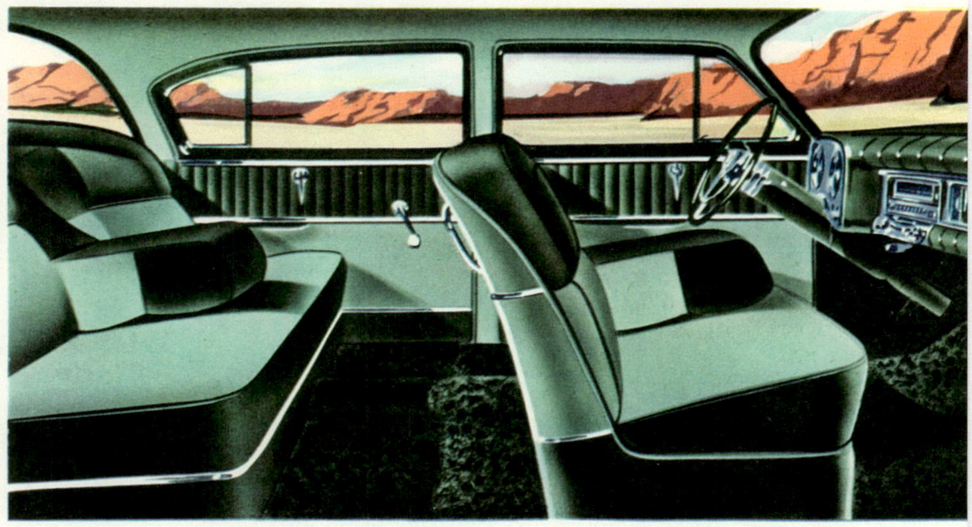

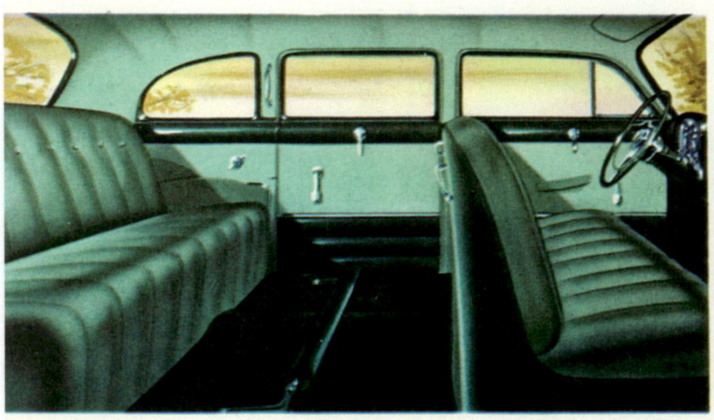

The beauty and charm *of Chrysler's tailored-to-taste interiors are very much in evidence in this illustration of the New Yorker Eight Passenger Sedan. This combination is Light Green Pillow Puff Nylon.*

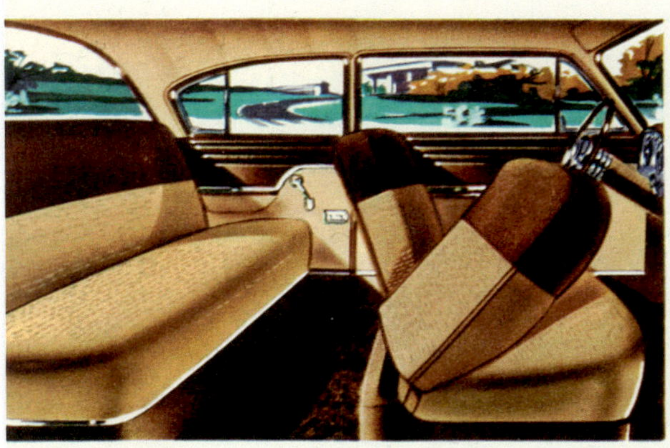

There is beauty, *roominess, and solid comfort in the spacious interior of the New Yorker Club Coupe. Illustrated is the Tan Pillow Puff Nylon and Dark Brown Gabardine upholstery combination, one of four beautiful selections on the New Yorker Club Coupe.*

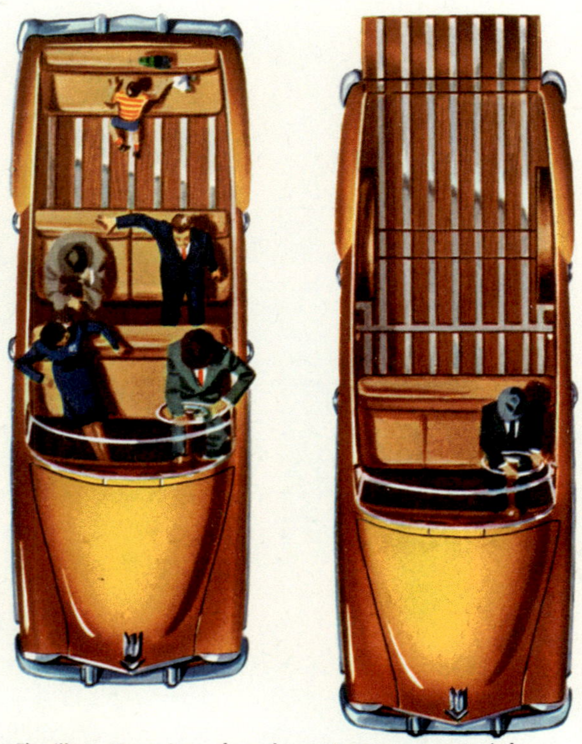

The illustrations above *show the seating arrangement of the Town & Country Wagon. Either half, or the entire rear seat can be folded forward, making storage space 47 inches wide, 35 inches high, and 75 inches long (116 inches long with the tail-gate open). The removable seat is available on special order.*

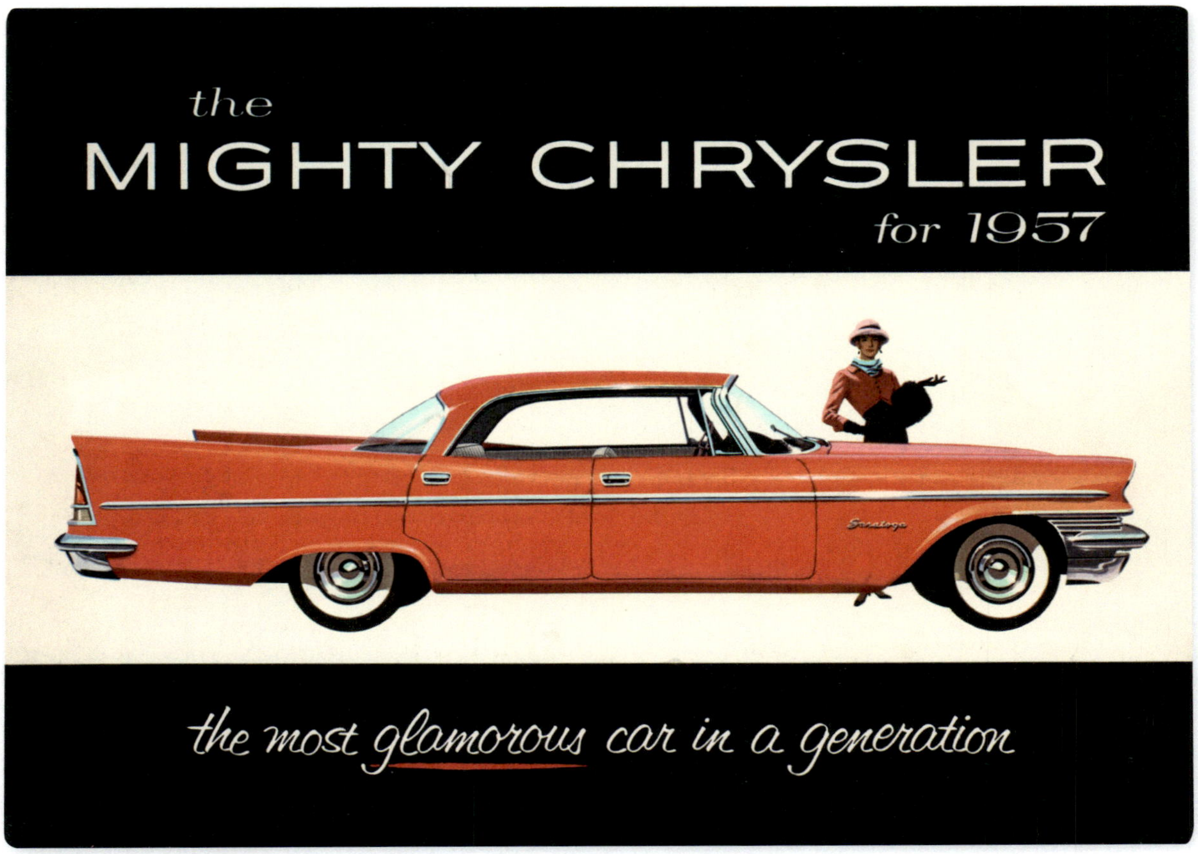

Chrysler 1957 *(above)*, **1961** *(right)*

Fins literally measured the magnificence of a car when Chrysler introduced its 1957 models. Then buyer tastes changed abruptly. The 1961 Chrysler boasted bold, canted quad headlamps, but the tailfins are notably lower in altitude. In another year, they would be gone entirely.

Als Chrysler 1957 die Heckflossen einführte, wurden sie buchstäblich zum Maßstab für die Herrlichkeit eines Autos, doch der Geschmack der Käufer änderte sich rasch wieder. Der 1961er-Chrysler besaß noch kühne, schräg angeordnete Vierfachscheinwerfer, aber die Heckflossen waren schon deutlich niedriger geworden und waren ein Jahr später komplett von der Bildfläche verschwunden.

À l'époque où Chrysler lança ses modèles 1957, le prestige d'une voiture se mesurait à ses ailerons arrière. Puis les goûts changèrent brutalement. La Chrysler de 1961 affichait quatre gros phares inclinés, mais les ailes de requin étaient notablement moins hautes. Un an plus tard elles auront complètement disparu..

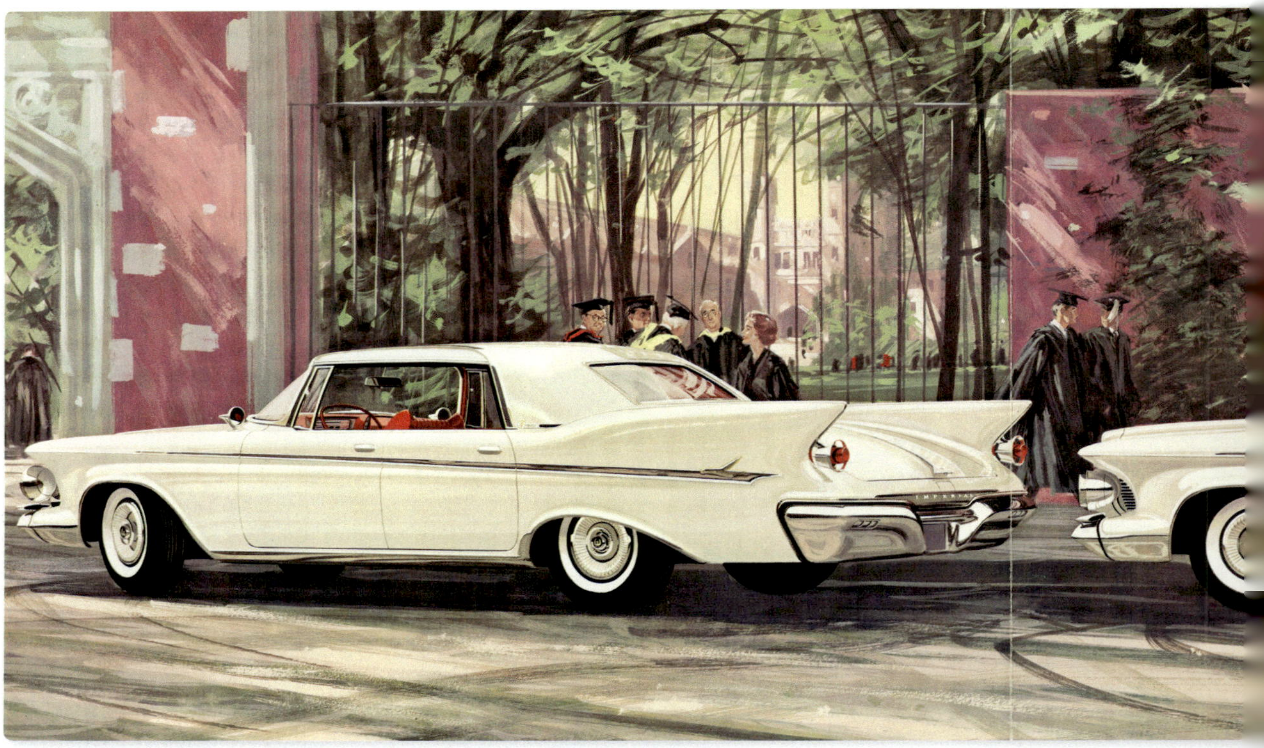

Chrysler 1959

The deep eyebrows atop the fenders that hooded the headlamps helped to define the styling of Chryslers in the final year of the 1950s. The ensuing years would prove to be more unconventional.

Die markanten „Augenbrauen" über den Scheinwerferschutzblechen gehörten zu den optischen Erkennungszeichnen eines Chrysler im letzten Jahr der Fünfziger. In den Folgejahren ging es dann weniger konventionell zu.

Dans cette ultime année de la décennie, les « sourcilleuses » visières de phares aidaient à définir le style des Chrysler. Les années suivantes seront plus conventionnelles.

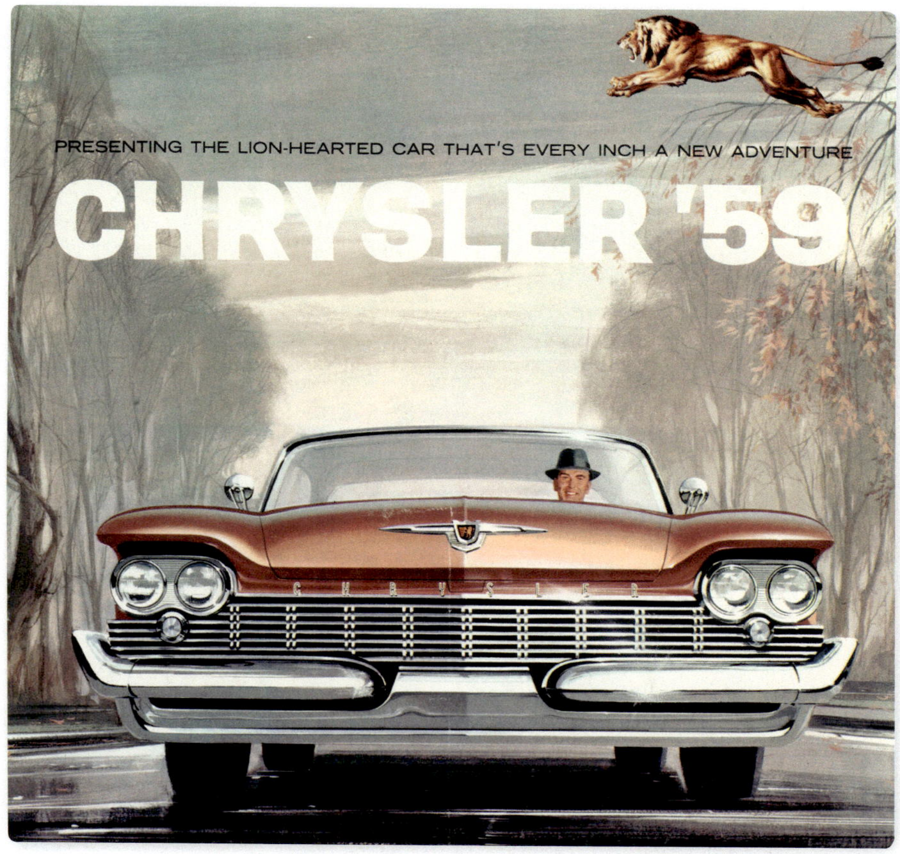

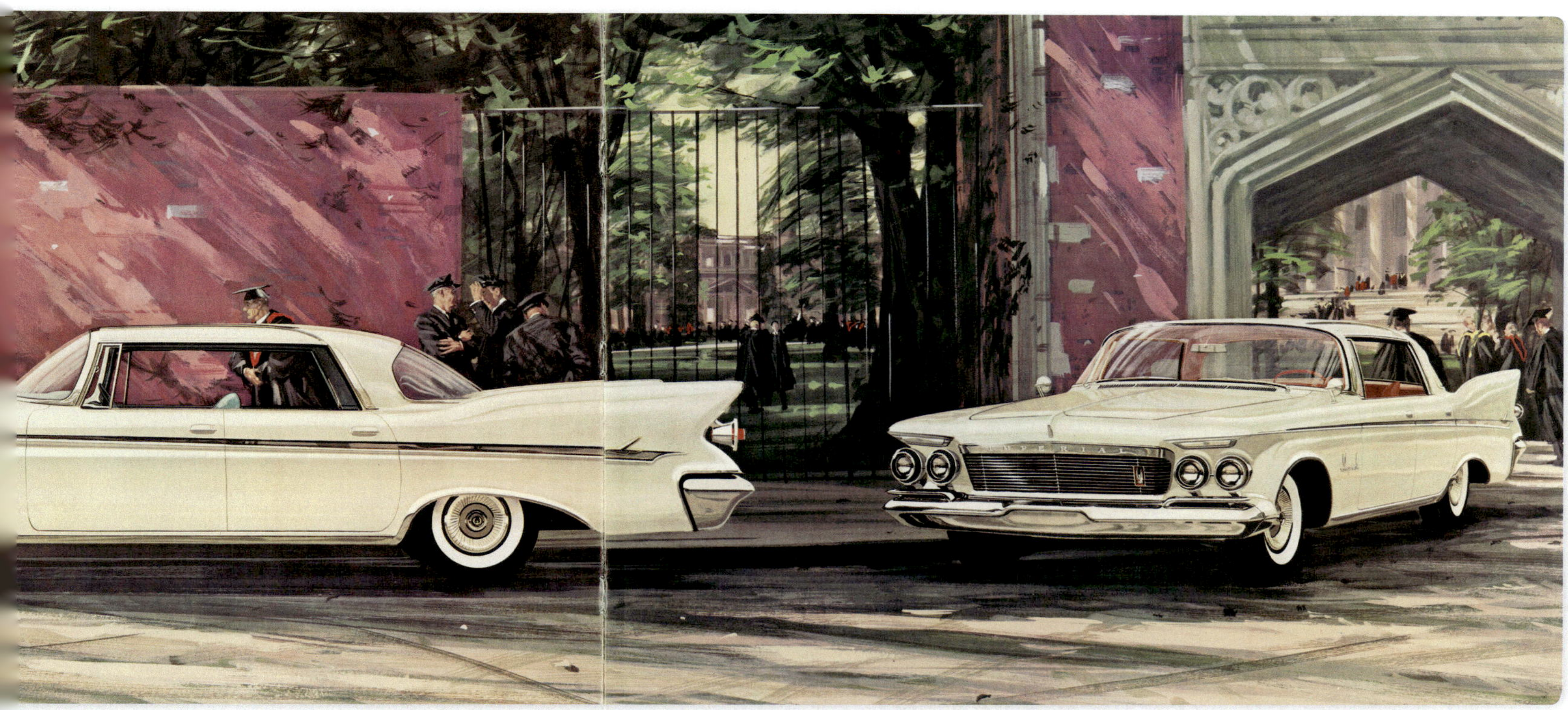

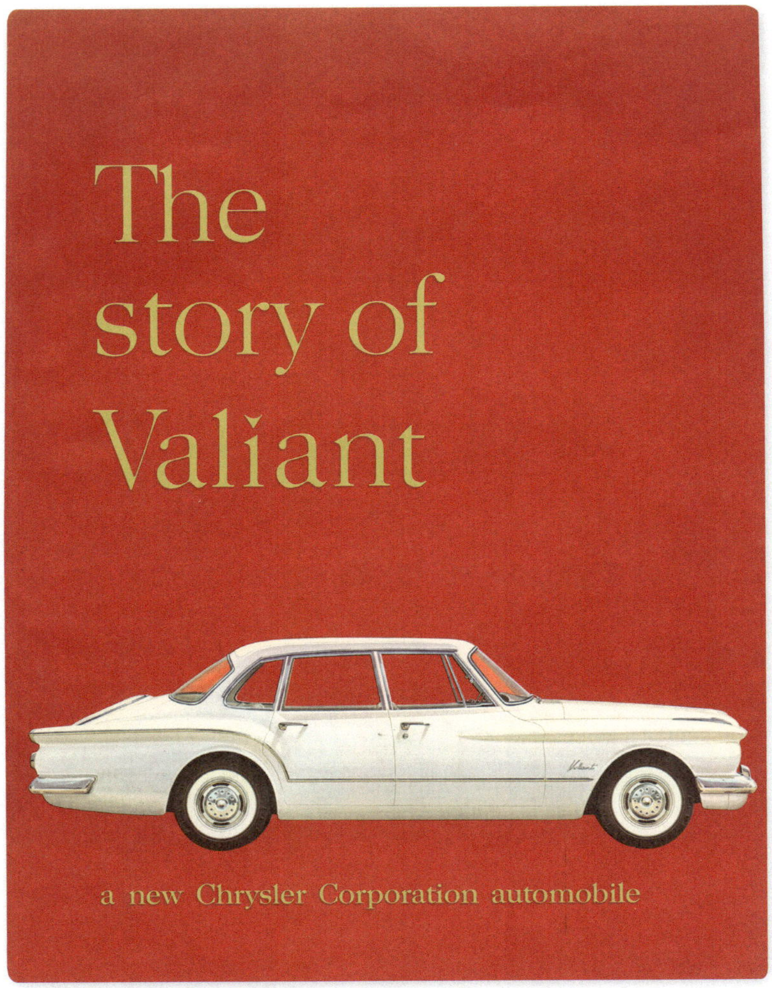

Chrysler 1957 *(left)*, **1956** *(below)*, **1961** *(bottom)*, **1960** *(opposite)*

The end of the '50s marked a profound change for Chrysler as it transitioned from heavy full-frame to lighter unitized-body construction architecture. The bigger news was that in 1960, Chrysler introduced its wildly sculpted but mechanically conventional compact car, the Valiant, to be sold by Plymouth dealers. It remained in Chrysler's lineup for decades.

Das Ende der 1950er-Jahre brachte einen tief greifenden Wandel für Chrysler mit sich, als man vom schweren Vollrahmen zur leichteren selbsttragenden Karosserie überging. Die größere Neuigkeit war aber, dass Chrysler 1960 seinen verwegen aussehenden, aber technisch recht konventionellen Kompaktwagen Valiant vorstellte, der über Plymouth-Händler vertrieben wurde. Dieser Wagen blieb über Jahrzehnte Bestandteil der Modellpalette von Chrysler.

La fin des années 1950 marqua un profond changement pour Chrysler, qui passa du lourd châssis séparé à une architecture monocoque plus légère. Plus important encore, la marque introduisit en 1960 la Valiant, une citadine compacte au style aussi fougueux que sa mécanique était classique, et qui sera distribuée par le réseau Plymouth. Elle figurera plusieurs décennies d'affilée dans le catalogue Chrysler.

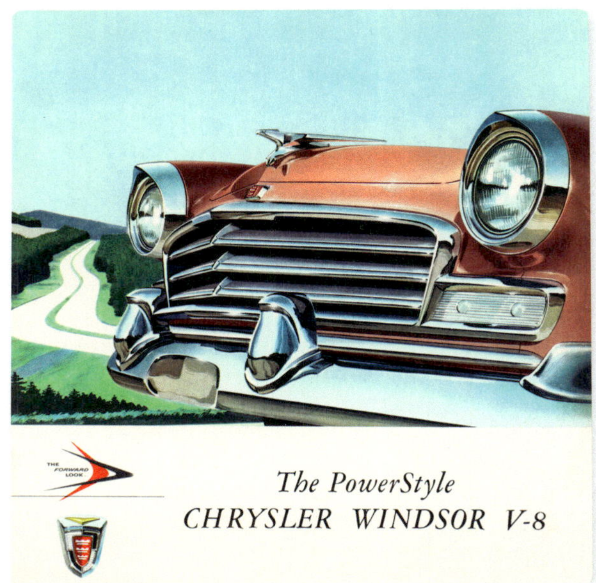

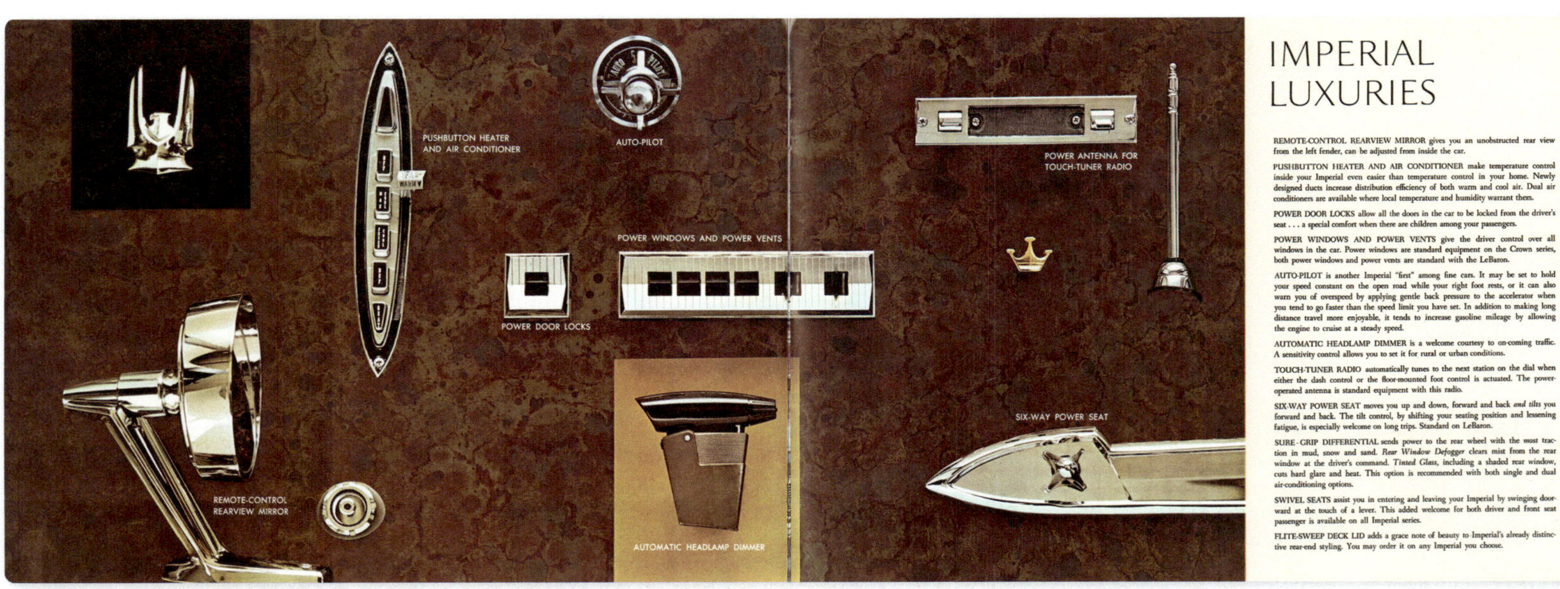

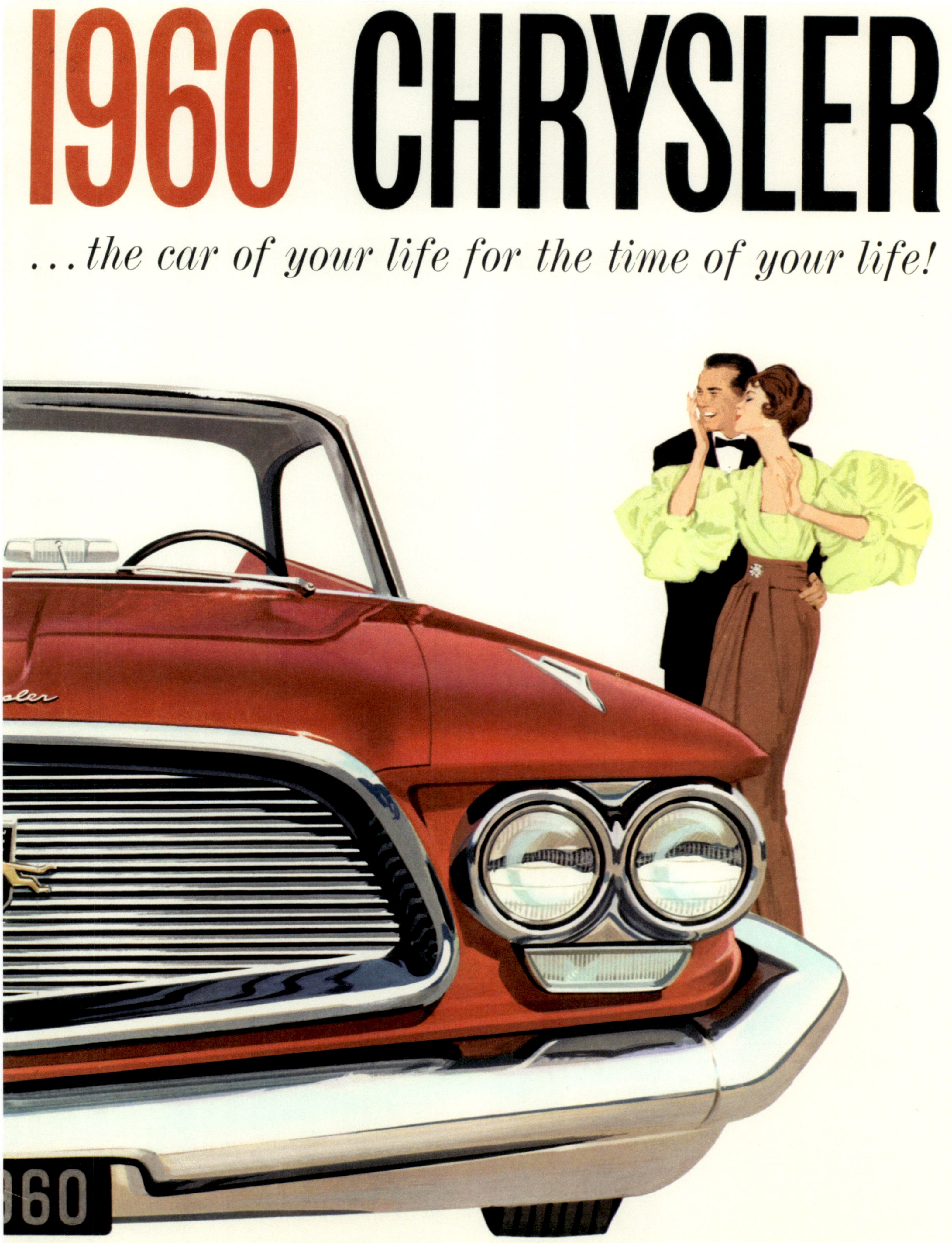
1960 CHRYSLER
...the car of your life for the time of your life!
960

Crosley

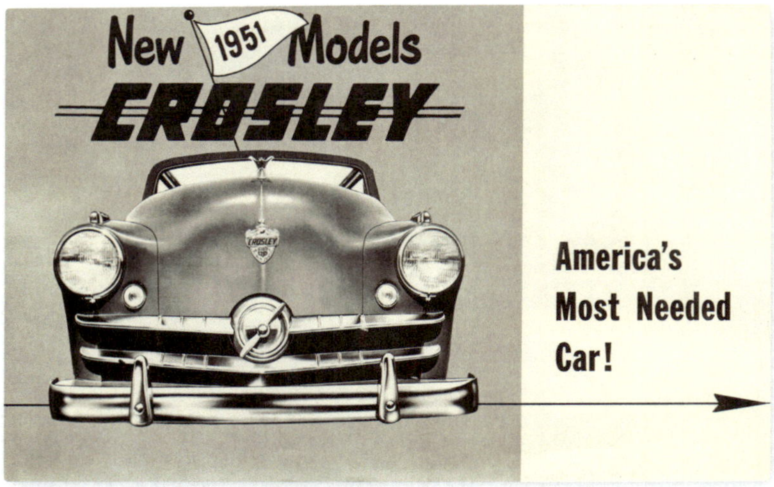

Crosley 1951

The tiny cars built by Ohio industrialist Powel Crosley were anything but toys. The Hotshot Roadster with a sheet-metal engine led the very first Sebring race in 1950. Another raced at Le Mans. Around 7,000 were sold before the firm ended automobile production in 1952. Today the Crosley automobile stands as a forerunner of the subcompacts that started appearing in the 1970s, such as the original Honda Civic, a lineage that stretches all the way up to today's Chevrolet Sonic.

Die winzigen Autos, die der Industrielle Powel Crosley aus Ohio baute, waren alles andere als Spielzeuge. Der Hotshot Roadster mit Blechmotor führte 1950 das allererste Rennen in Sebring an, ein anderer ging in Le Mans an den Start. Rund 7 000 Fahrzeuge wurden verkauft, bevor die Firma 1952 die Automobilproduktion bereits wieder einstellte. Heute gilt der Crosley als Vorläufer der Kleinwagen, die in den 1970er-Jahren aufkamen, wie etwa des ursprünglichen Honda Civic. Ein modernes Beispiel für diese Klasse auf dem US-Markt wäre der Chevrolet Sonic, der hierzulande als Aveo vermarktet wird.

Les petites voitures fabriquées par l'industriel de l'Ohio Powel Crosley étaient tout sauf des jouets. En 1950 le roadster Hotshot à bloc moteur en fonte remporta la toute première course de Sebring. Une autre participa aux 24 Heures du Mans 1951. Il se vendit près de 7 000 Crosley avant que l'entreprise ne cesse toute production automobile en 1952. Aujourd'hui la marque fait figure de pionnière sur le marché des microvoitures ou petites citadines qui pointeront le bout de leur nez dans les années 1970, comme l'originale Honda Civic. La Chevrolet Sonic d'aujourd'hui est l'héritière de cette lignée.

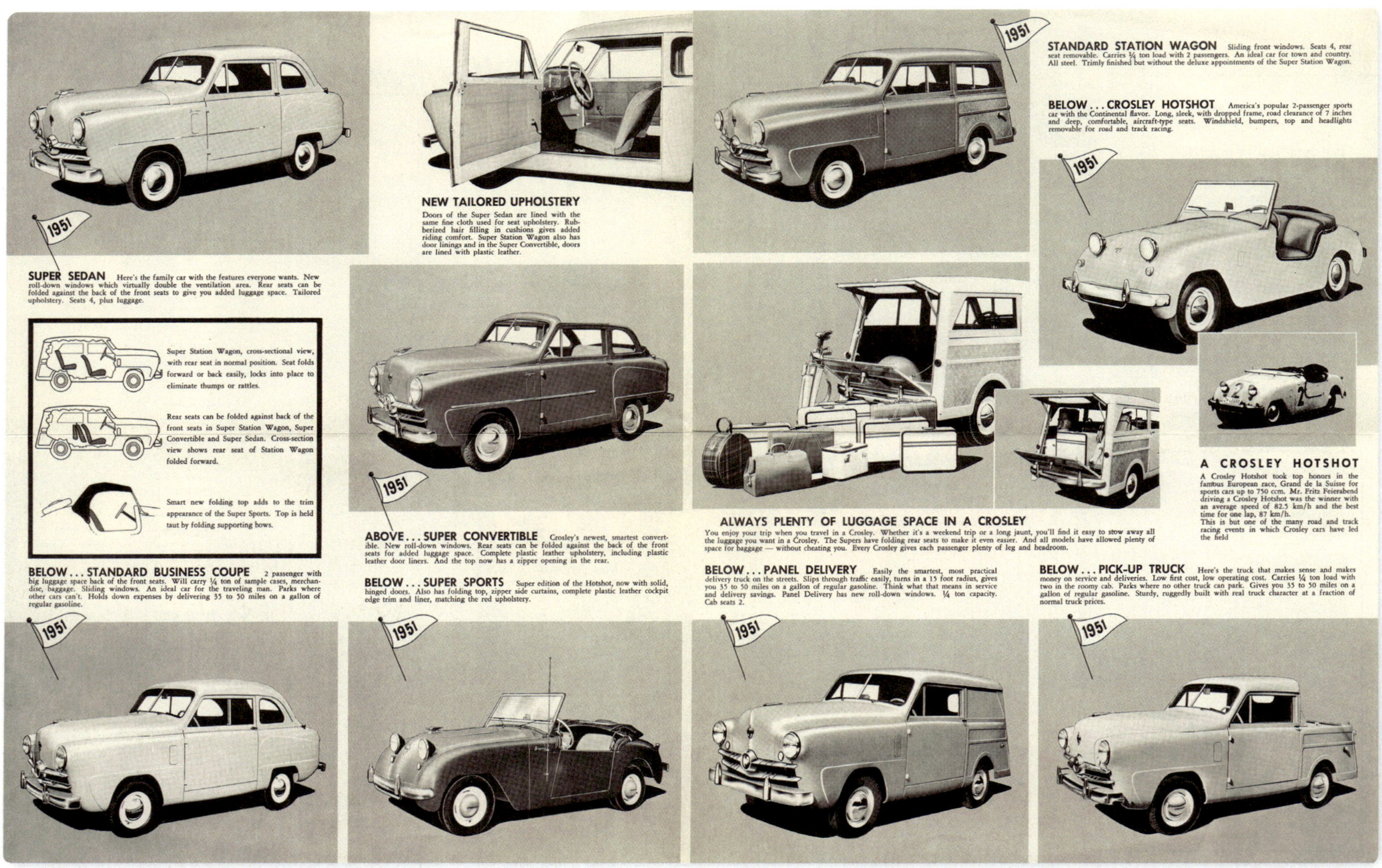

Datsun

Datsun Bluebird 1961

Nissan imported very few cars into the United States before 1958, including the Bluebird 310 sedan with a 1.2-liter, 60 hp engine. Even the brochure copy was misspelled. The Bluebird 510, arriving in 1967, would prove more successful.

Nissan führte vor 1958 nur sehr wenige Fahrzeuge in die Vereinigten Staaten ein. Eines der Modelle war die Limousine Bluebird 310 mit einem 1,2-Liter-Motor, der 81 Kilowatt Leistung auf die Straße brachte. Selbst die Broschüre enthielt noch Rechtschreibfehler. Der Bluebird 510, der 1967 auf den US-Markt kam, war dort erfolgreicher.

Avant 1958, Nissan n'avait exporté aux États-Unis que quelques rares modèles, dont sa berline Bluebird 310 à moteur 1,2 litre délivrant 60 chevaux (le catalogue comportait encore des coquilles typographiques !). La Bluebird 510, débarquée en 1967, rencontrera davantage de succès.

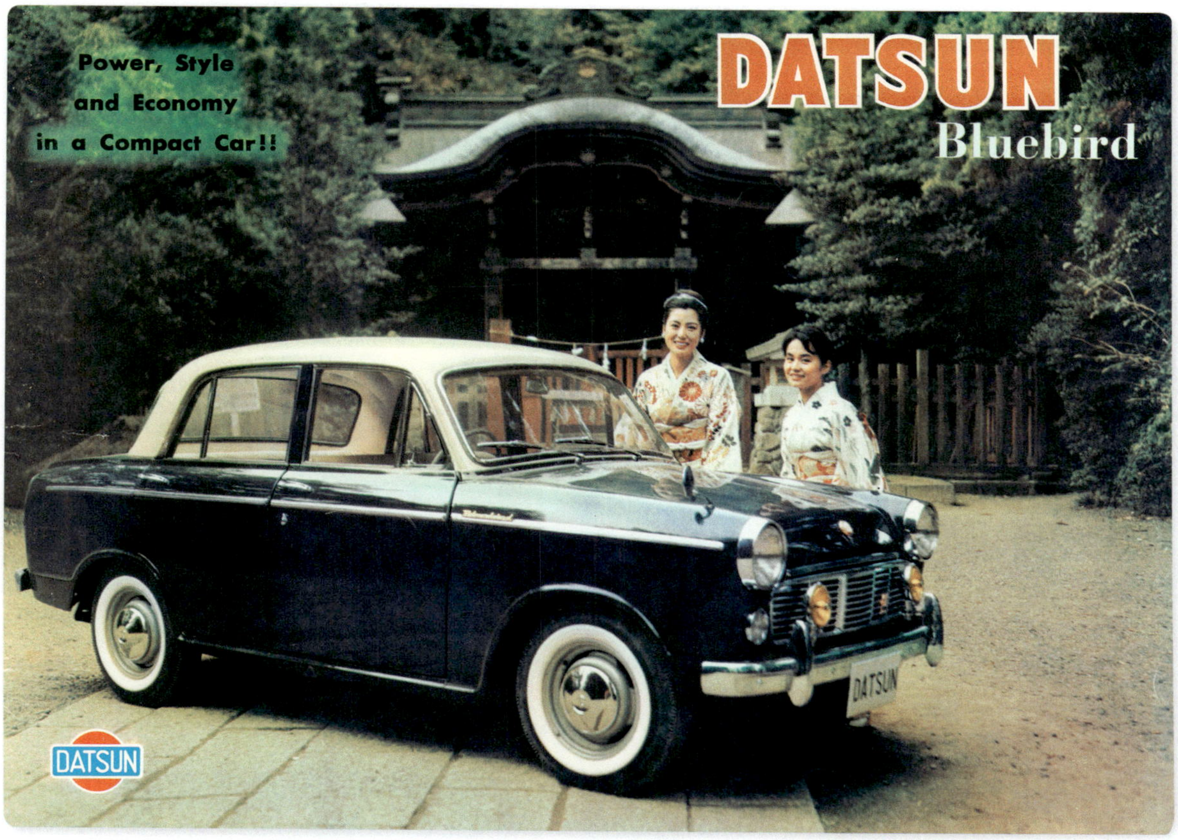

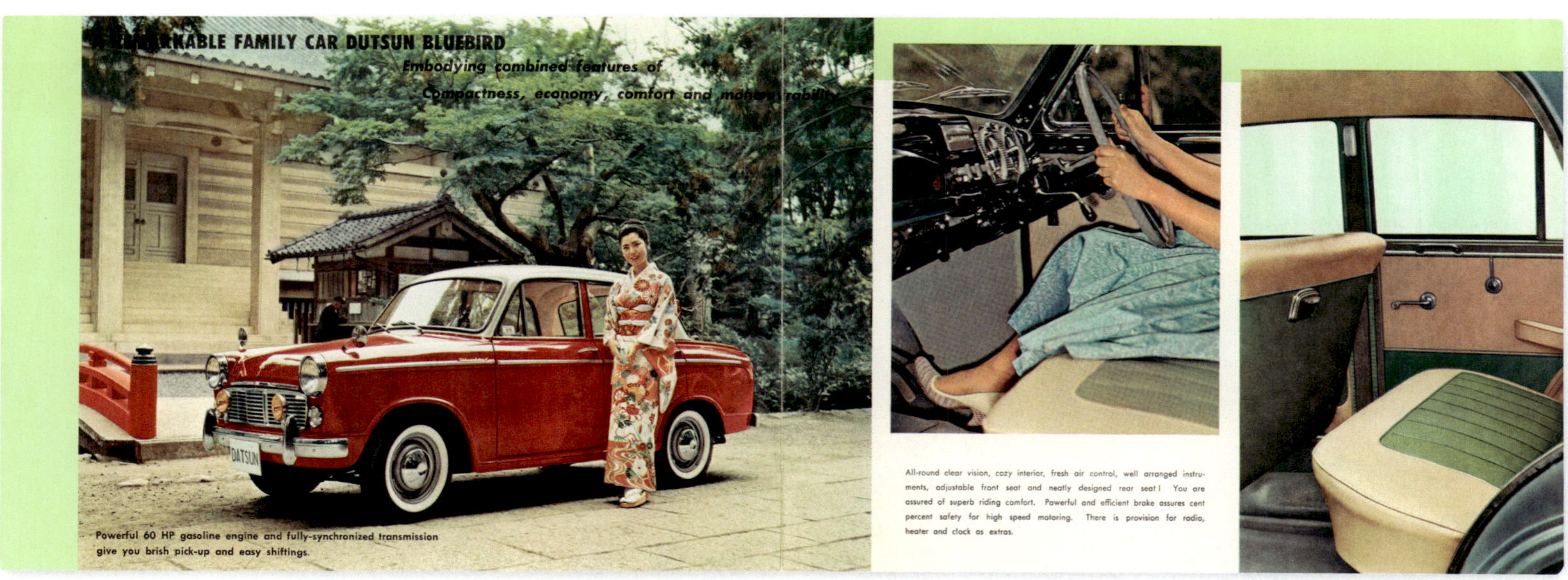

De Soto

De Soto 1952 *(right)*, **1957** *(below)*, **1958** *(below top)*, **1956** *(below bottom)*

Despite sales numbers that trailed other Chrysler brands, the toothy vertical grilles that characterized De Soto for many years made them some of the corporation's most visually appealing cars.

Wenngleich die Absatzzahlen anderen Chrysler-Marken hinterherhinkten, war der De Soto dank seinem „bissigen" senkrechten Kühlergrill viele Jahre lang eines der auffälligsten und optisch gefälligsten Autos dieses Herstellers.

Malgré des chiffres de ventes inférieurs à ceux des autres marques de la famille Chrysler, les calandres verticales aux grilles dentées qui ont caractérisé la De Soto pendant de nombreuses années en ont fait l'une des voitures à l'esthétique la plus séduisante du groupe.

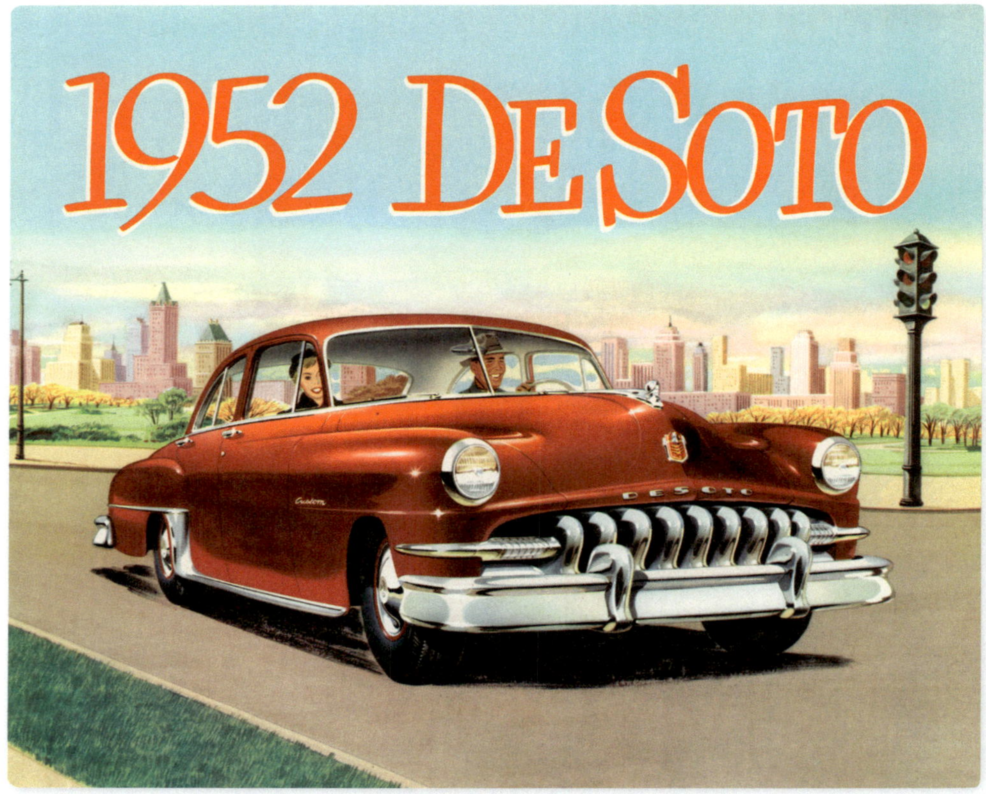

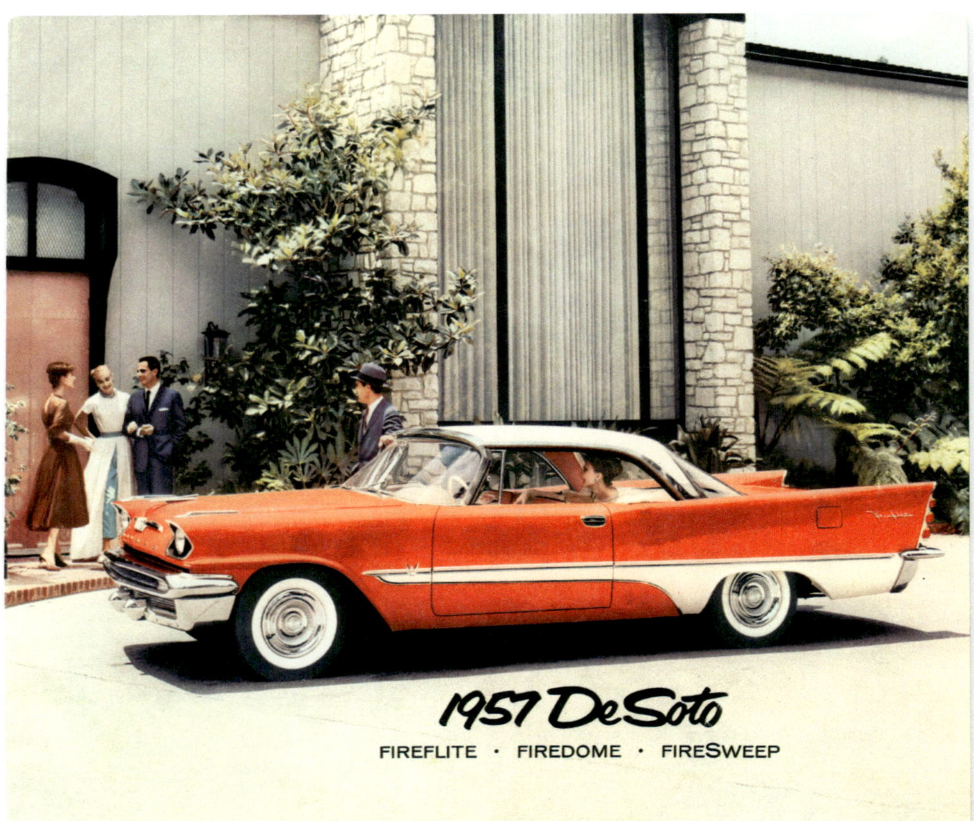

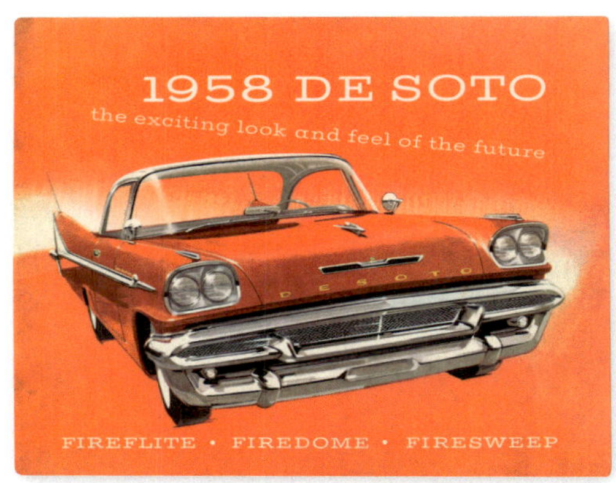

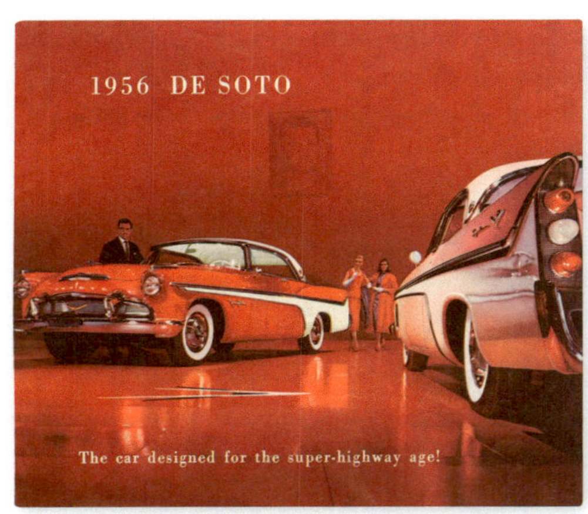

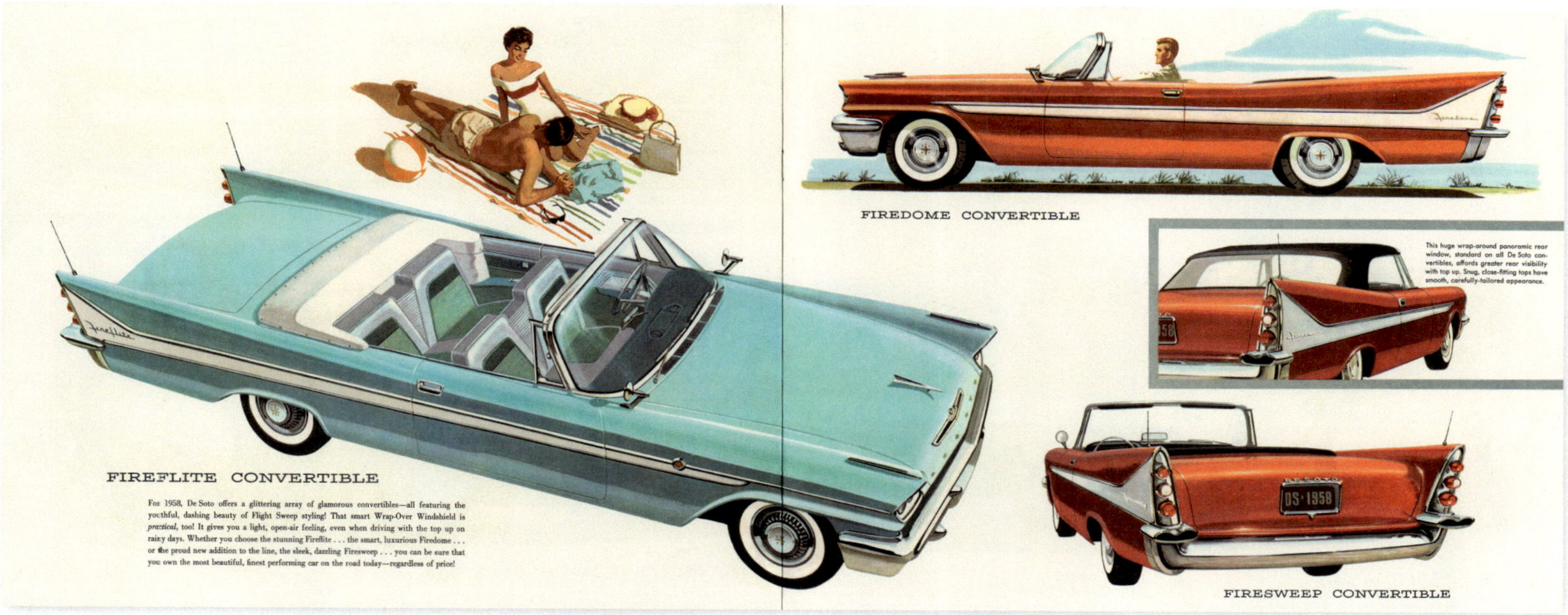

De Soto 1958 *(above)*, **1960** *(below)*

Late in its existence, De Soto had a fling with the Space Age. A short-lived 1958 option was Bendix electronic fuel injection on the Adventurer model. It worked so poorly that the cars were recalled and the new wedge-head V-8 engine was refitted with carburetors. De Soto died as an automotive nameplate in 1961.

Gegen Ende seines Daseins flirtete De Soto noch einmal mit dem Raumfahrtzeitalter. Eine kurze Zeit lang war 1958 eine elektronische Benzineinspritzung von Bendix als Extra für das Modell Adventurer erhältlich. Sie funktionierte so schlecht, dass die Fahrzeuge zurückgerufen werden mussten und der neue V8-Motor mit keilförmigem Brennraum („wedge-head") nachträglich mit Vergasern ausgestattet wurde. Als Marke segnete De Soto 1961 das Zeitliche.

Vers la fin de son existence, De Soto s'est offert une petite escapade dans l'ère spatiale… À titre d'éphémère option proposée en 1958 sur le modèle Adventurer figurait ainsi une injection électronique Bendix. Elle fonctionnait si mal que les voitures furent rappelées et le nouveau moteur V8 rééquipé de carburateurs. De Soto disparut en tant que marque automobile en 1961.

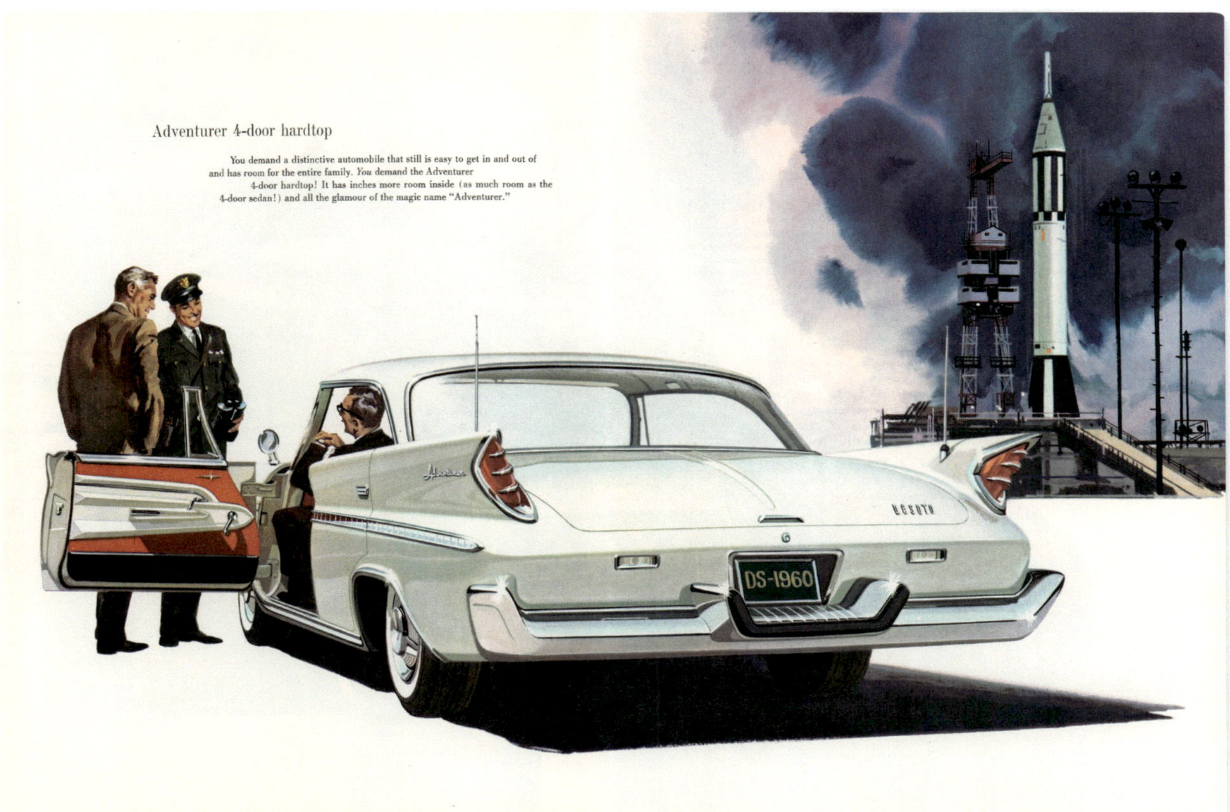

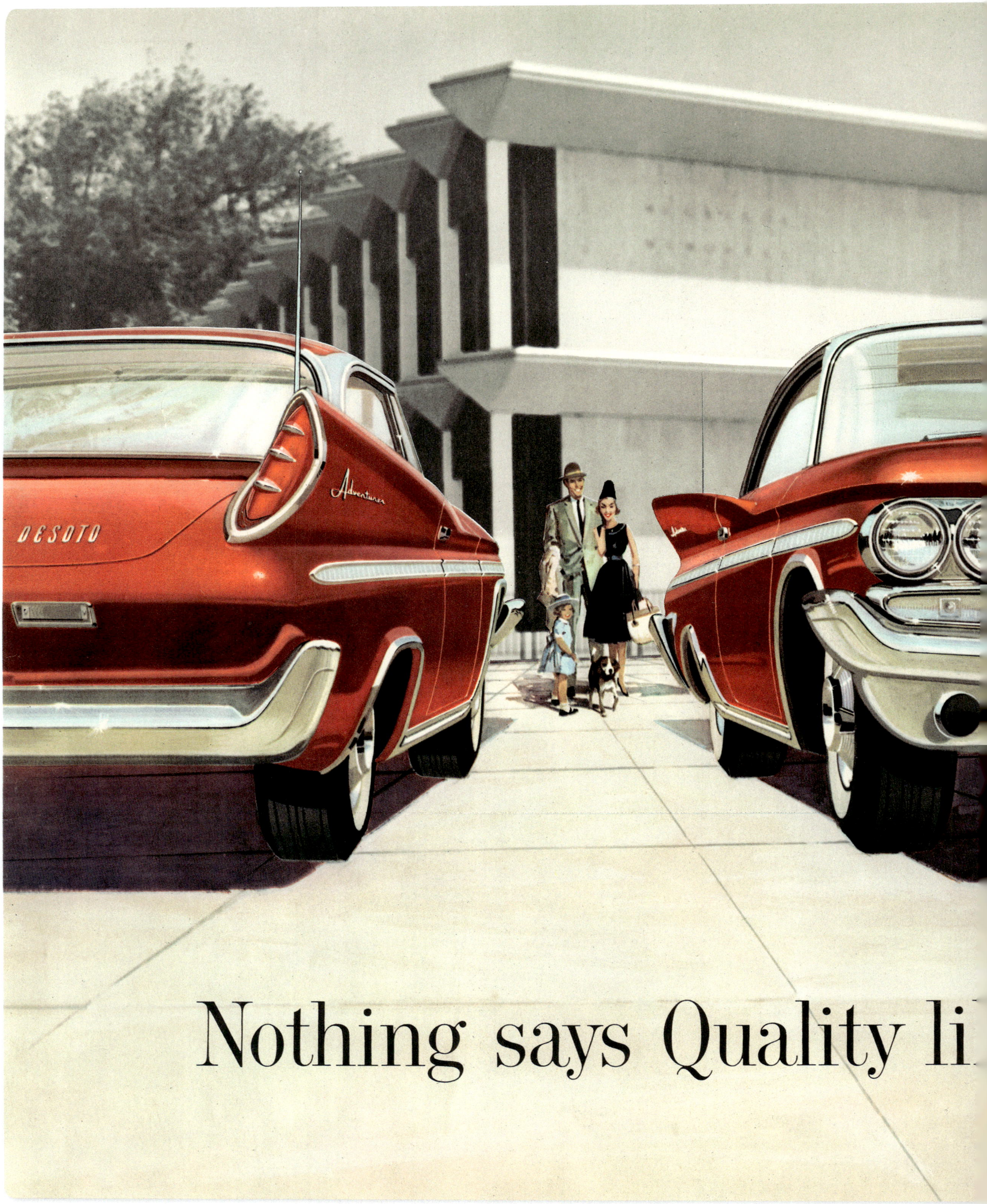
DESOTO
Adventurer
Nothing says Quality li

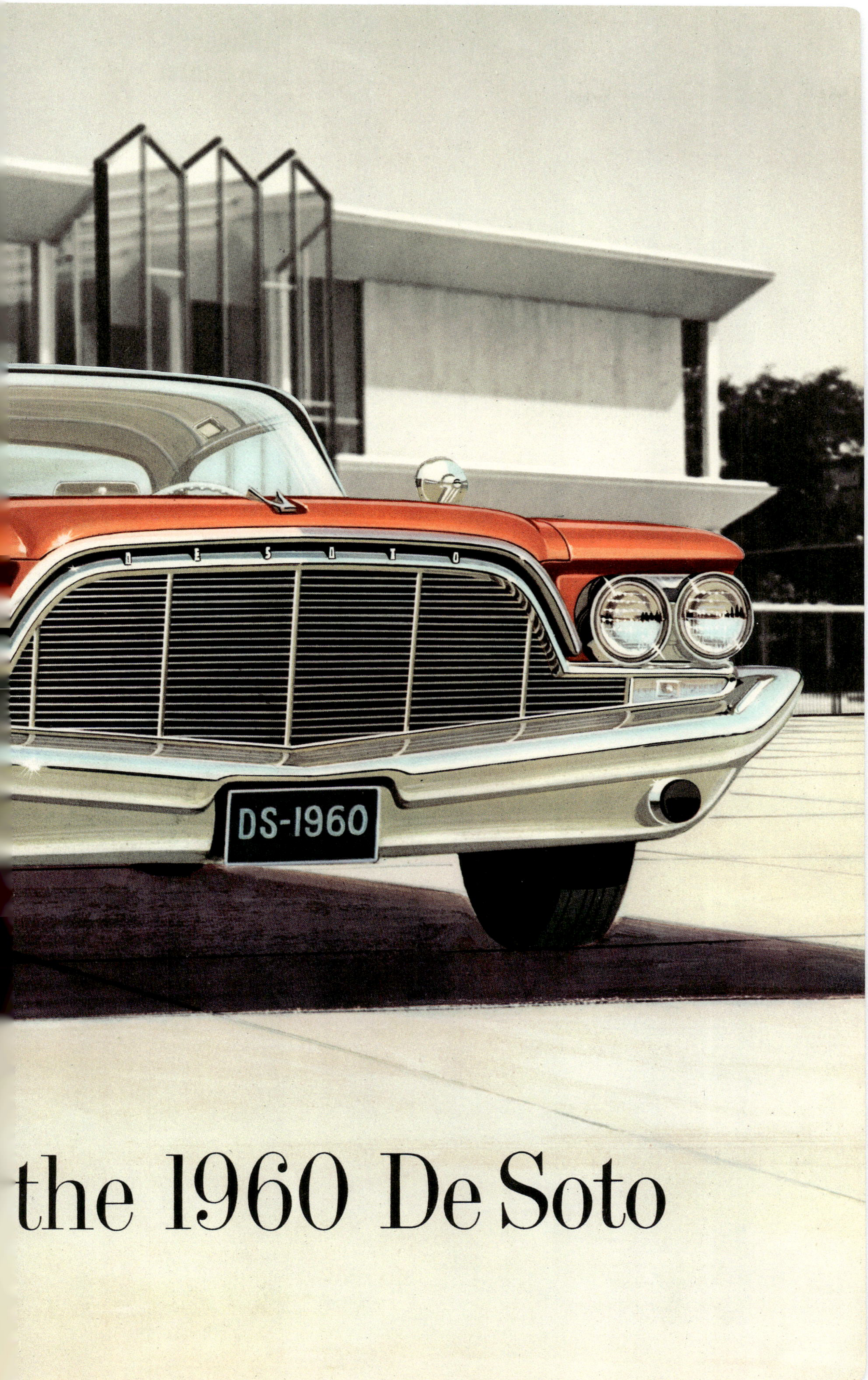

De Soto 1960

If smart styling alone sold cars, De Soto would have likely survived. Instead, 1960 marked the next-to-last year of its existence in the Chrysler orbit. Ultimately, De Soto was done in by its price positioning just below Chrysler; most buyers opted to step up for a few hundred dollars. The failure of Edsel during the same years could only have hastened De Soto's departure.

Wenn man Autos allein über schnittiges Aussehen hätte verkaufen können, dann hätte der De Soto vermutlich überlebt, doch 1960 sollte schon das vorletzte Jahr seines Daseins im Chrysler-Universum sein. Letzten Endes ging der De Soto durch seine Preispositionierung knapp unterhalb von Chrysler ein: Die meisten Käufer entscheiden sich, ein paar hundert Dollar draufzulegen. Der Flop des Edsel bei Ford im gleichen Zeitraum beschleunigte den Untergang des De Soto nur noch.

Si l'élégance du style suffisait à faire vendre les voitures, De Soto aurait probablement survécu. Faute de quoi 1960 marqua l'avant-dernière année de son existence dans l'orbite de Chrysler. Finalement De Soto périclitera par la faute de ses prix, positionnés juste au-dessous de Chrysler ; pour quelques (centaines de) dollars de plus, la plupart des acheteurs choisissaient la montée en gamme. L'échec d'Edsel au cours de la même période ne fit que précipiter la disparition de De Soto.

Dodge

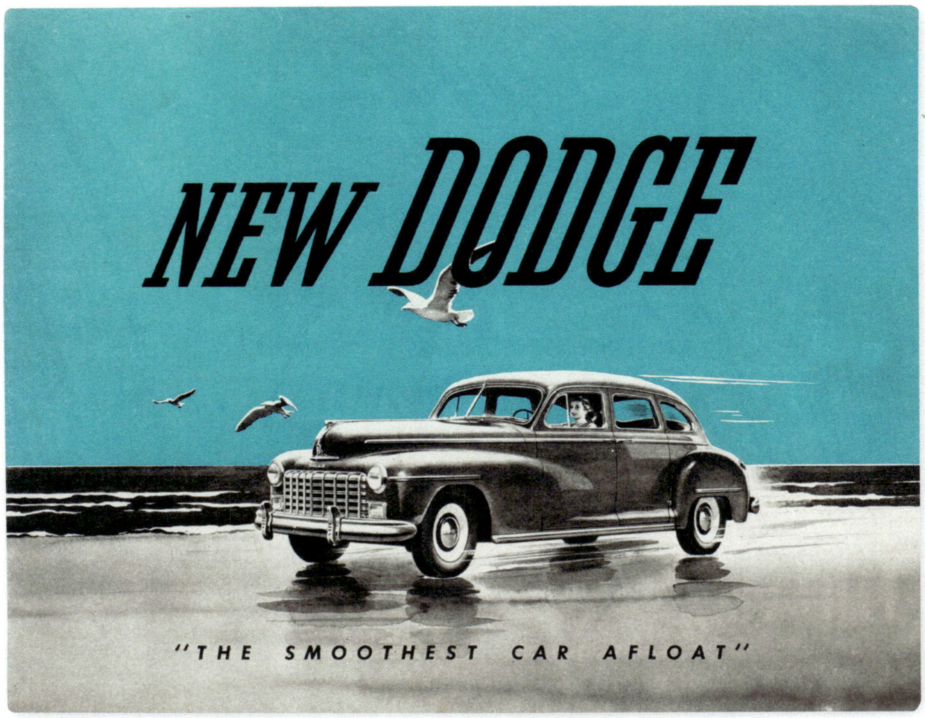

Dodge 1946 *(left)*, **1951** *(below)*, **1955** *(opposite top)*, **1960** *(opposite bottom)*

The "new" engine was really a continuation of the prewar L-head six with 102 hp, so the young at heart doubtlessly enjoyed open-air breezes more than power. A few seasons' worth of appearance revisions didn't change much. Still, look at the dealer manual from 1955. It's made of plastic, a material that would become standard in auto manufacturing within just a few years.

Der „neue" Motor war lediglich eine Weiterentwicklung des L-Kopf-Sechszylinders mit 138 kW aus der Zeit vor dem Zweiten Weltkrieg, sodass die Junggebliebenen wohl eher die frische Luft als einen leistungsstarken Motor genießen konnten. Daran änderten auch ein paar kosmetische Verfeinerungen in den Folgejahren nichts. Bemerkenswert ist allerdings die Händlerbroschüre aus dem Jahre 1955. Sie bestand aus Kunststoff, einem Material, das ein paar Jahre später in der Automobilherstellung zur Norm wurde.

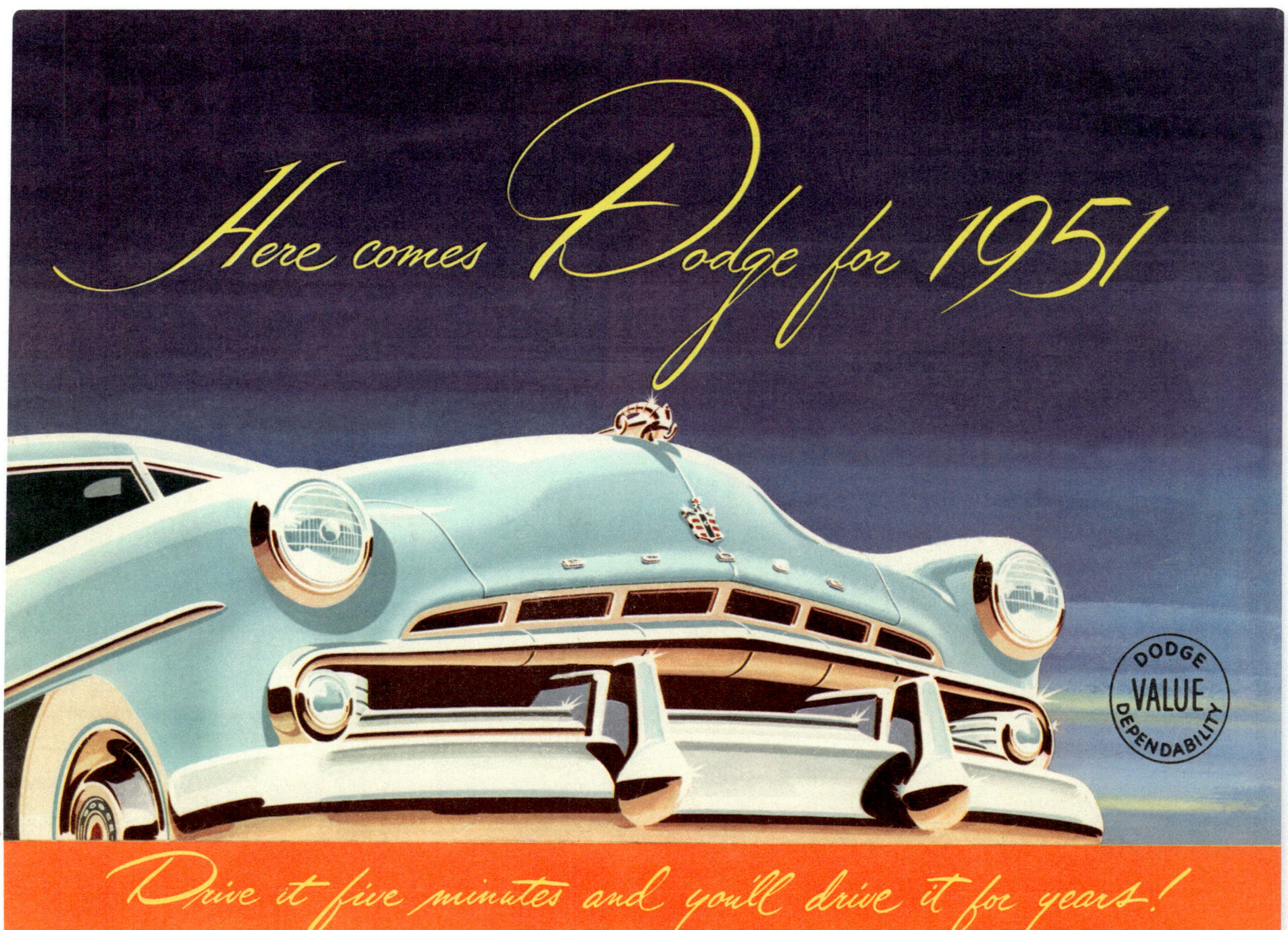

Le « nouveau » moteur ne faisait en réalité que prolonger le six cylindres à soupapes latérales d'avant-guerre et ses 102 chevaux, aussi les cœurs ardents goûtaient-ils le grand air plus sûrement que la puissance brute. Et ce ne sont pas quelques retouches extérieures qui changeront radicalement la donne. Observons néanmoins le manuel d'utilisation de 1955 : il est en plastique, une matière qui en à peine quelques années va se généraliser dans l'industrie automobile.

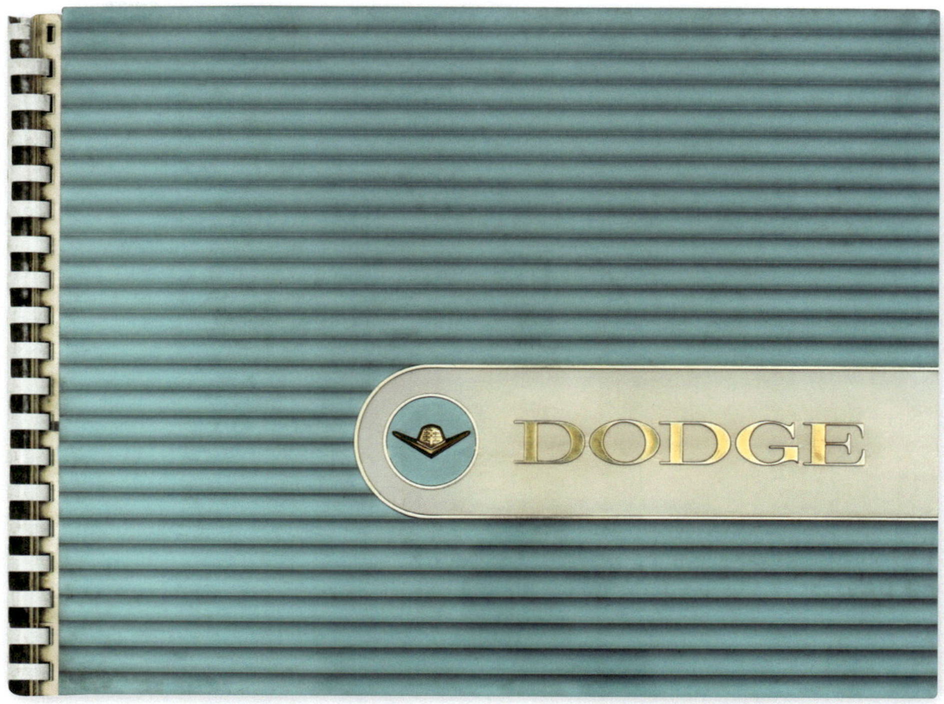

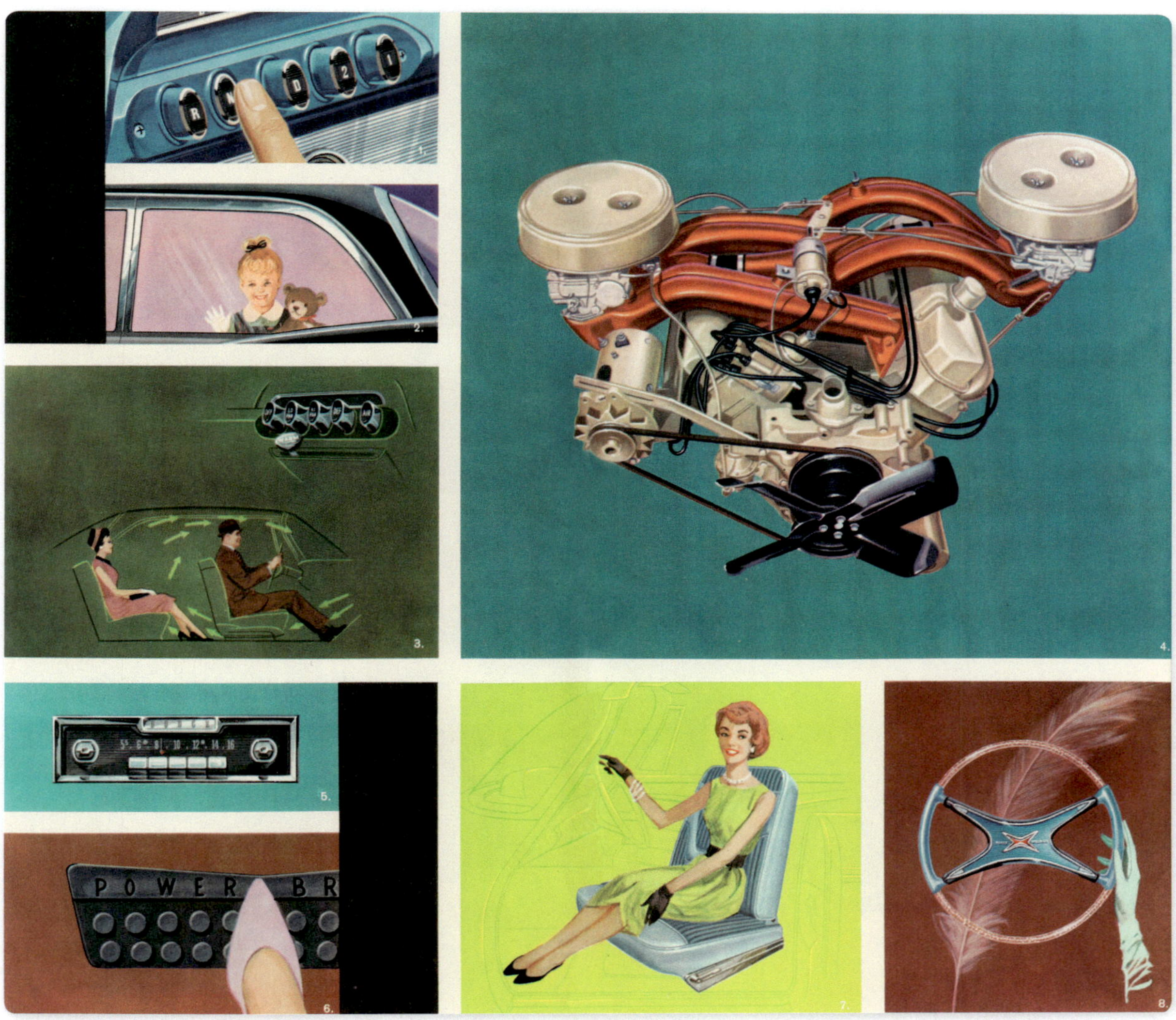

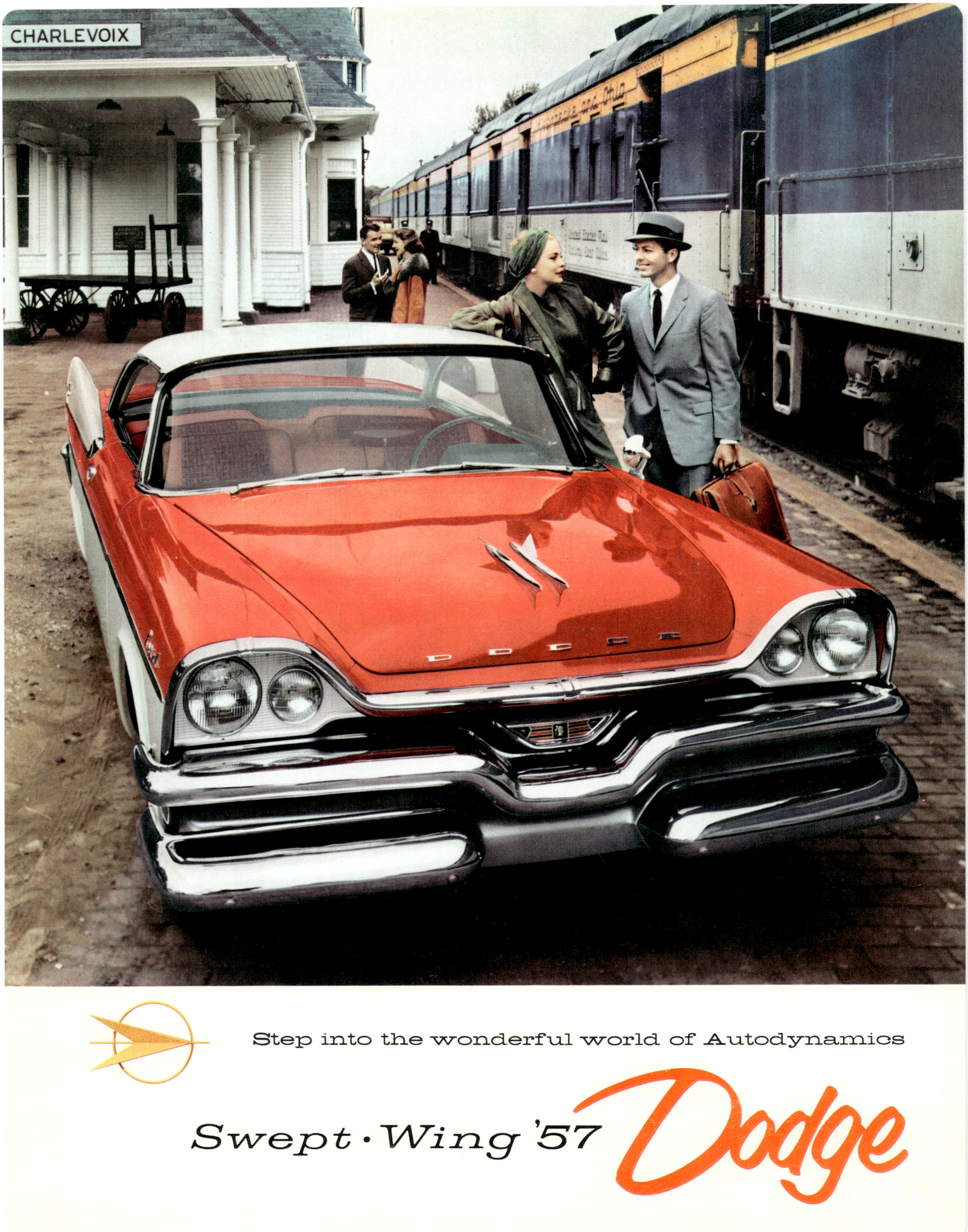

CHARLEVOIX
DODGE
Step into the wonderful world of Autodynamics
Swept · Wing '57 Dodge

Dodge 1957 *(opposite)*, **1959** *(right)*, **1956** *(below)*

The Custom Royal convertible added swivel seats with an asymmetrical upholstery pattern, but performance was a more significant matter. The 1957 Dodges gained wheelbase and torsion bar suspension. The savage D-501 cross-ram Hemi engine produced 340 hp.

In dem Kabriolett Custom Royal gab es drehbare Sitze mit asymmetrischem Bezugsmuster, aber wichtiger war die Leistung. Die 1957er-Dodges erhielten einen erweiterten Radstand und eine Torsionsstabfederung. Der brutale D-501-Hemi-Motor mit „cross-ram"-Ansaugkrümmer lieferte 462 Kilowatt Leistung ab.

La Custom Royal décapotable a beau ajouter des sièges pivotants à une sellerie friande d'asymétrie, ce sont ses performances qui se montrent les plus éloquentes. Les Dodge de 1957 croissent en empattement et gagnent une suspension à barres de torsion. Le sauvage Hemi D-501 à culasse cross-ram développe 340 chevaux.

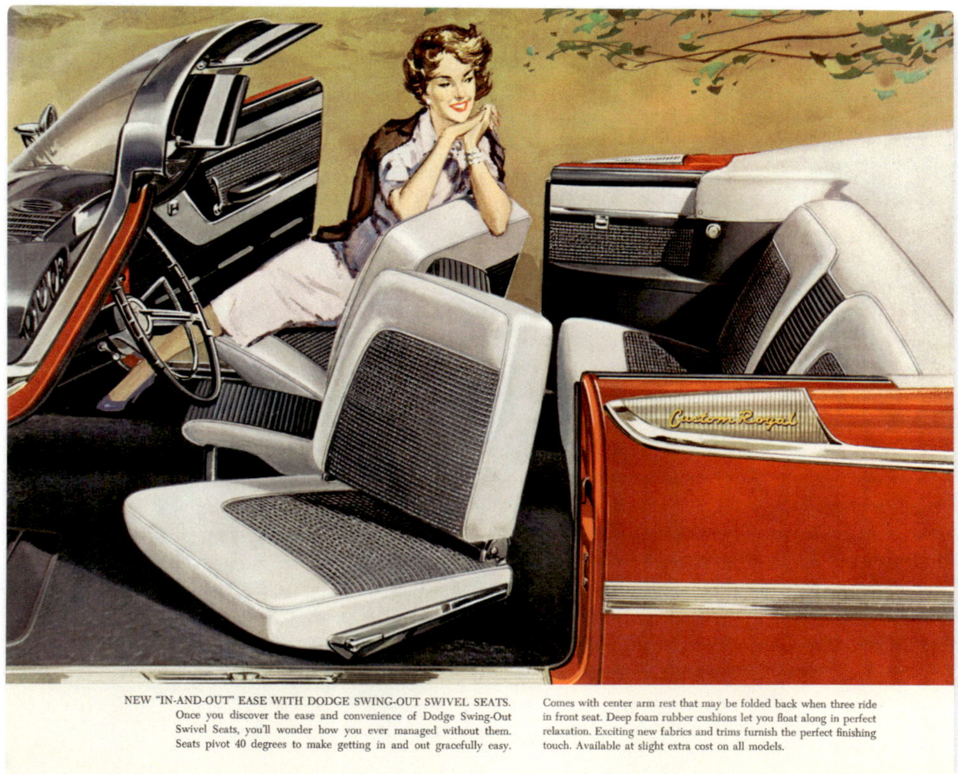

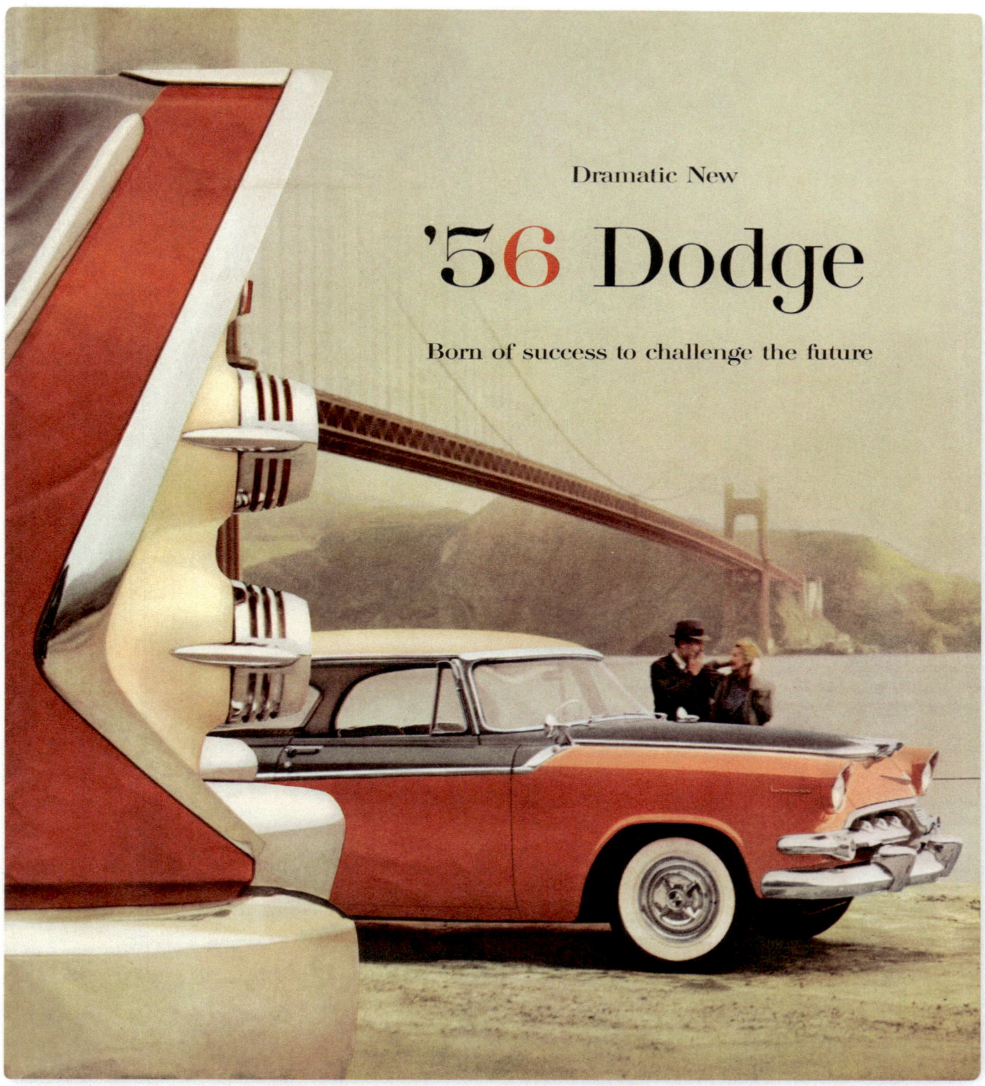

Edsel

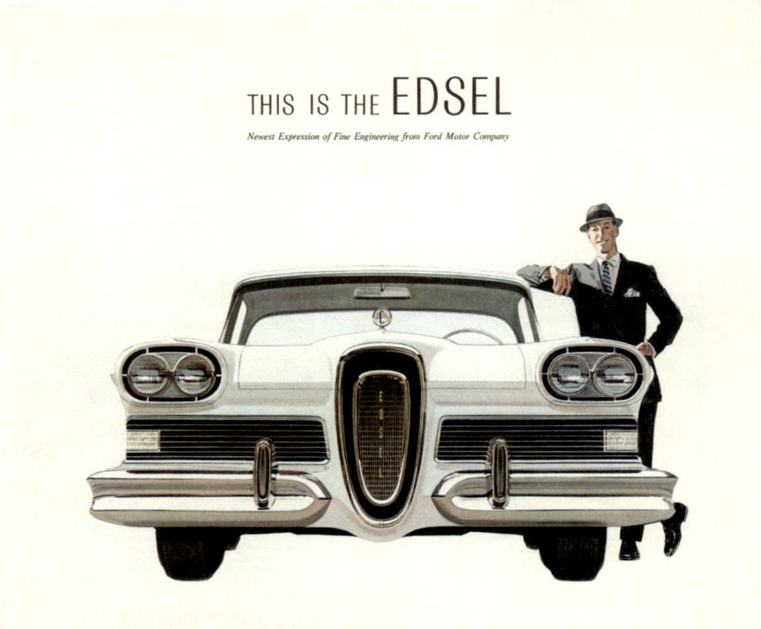

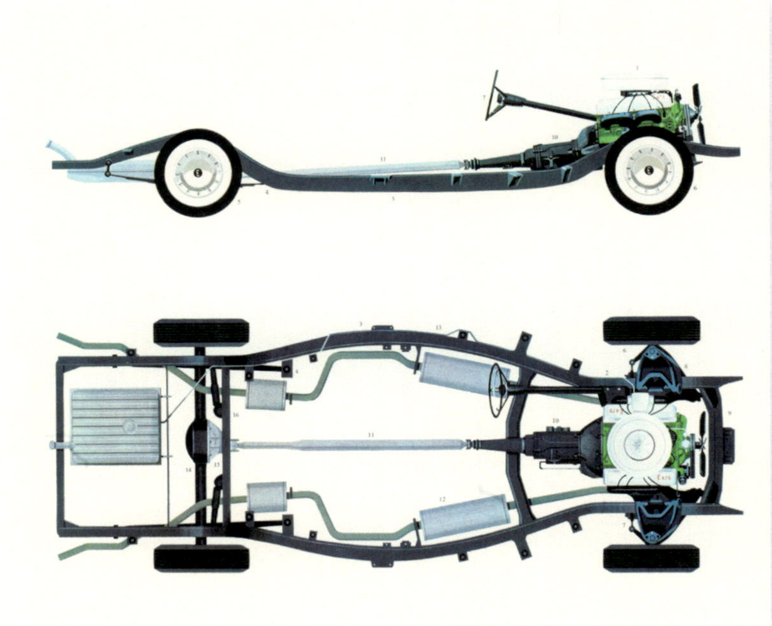

Edsel 1958 *(above)*, **1960** *(right)*

Transmission buttons in the steering hub were the least of the Edsel's oddities. The design was misplaced and a solid dud with the public. This weird-looking car was positioned between Ford and Mercury, a sales niche that didn't exist.

Schaltknöpfe im Lenkradtopf waren noch die geringste der zahlreichen Merkwürdigkeiten des Edsel. Das Design war ein kompletter Schuss in den Ofen und fiel bei den potenziellen Käufern völlig durch. Zudem war dieses abartig aussehende Auto zwischen Ford und Mercury angesiedelt – also in einer Nische, die es gar nicht gab.

La sélection des vitesses au centre du volant constituait l'une des curiosités de l'Edsel, et pas la plus étonnante. Son design était raté et lui valut un fiasco auprès du public. Cette voiture d'apparence étrange entendait se positionner entre Ford et Mercury, une niche commerciale qui présentait l'inconvénient de ne pas exister.

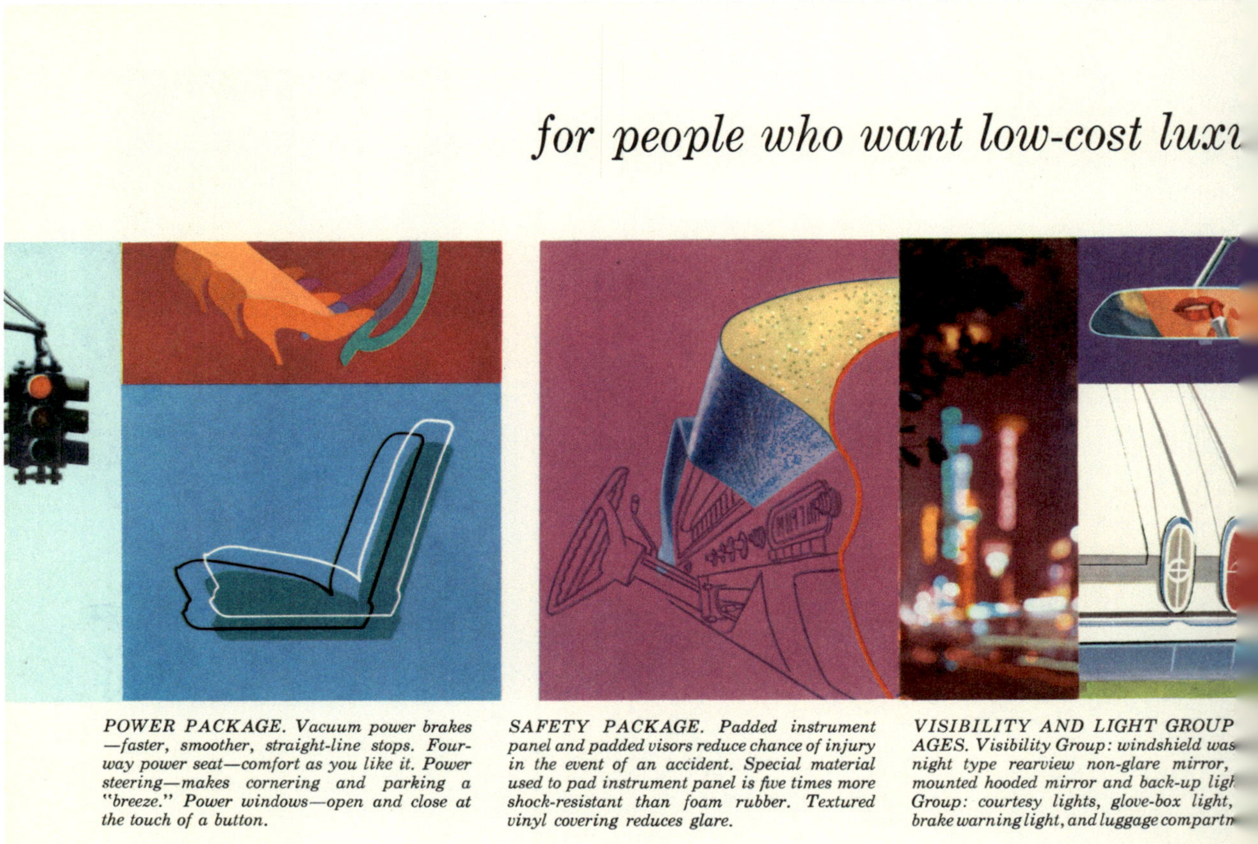

Edsel 1960

A botched launch and reliability problems bedeviled the company. Ford eliminated it in 1960, but its memory affected product decisions for years.

Eine misslungene Markteinführung und Probleme mit der Zuverlässigkeit machten dem Unternehmen das Leben schwer. Ford entledigte sich 1960 des ungeliebten Modells, doch das Trauma wirkte sich noch über viele Jahre auf Produktentscheidungen aus.

Un lancement bâclé et des problèmes de fiabilité coûtèrent cher à l'entreprise. Ford s'en débarrassa en 1960 mais son souvenir affectera longtemps les décisions industrielles du groupe.

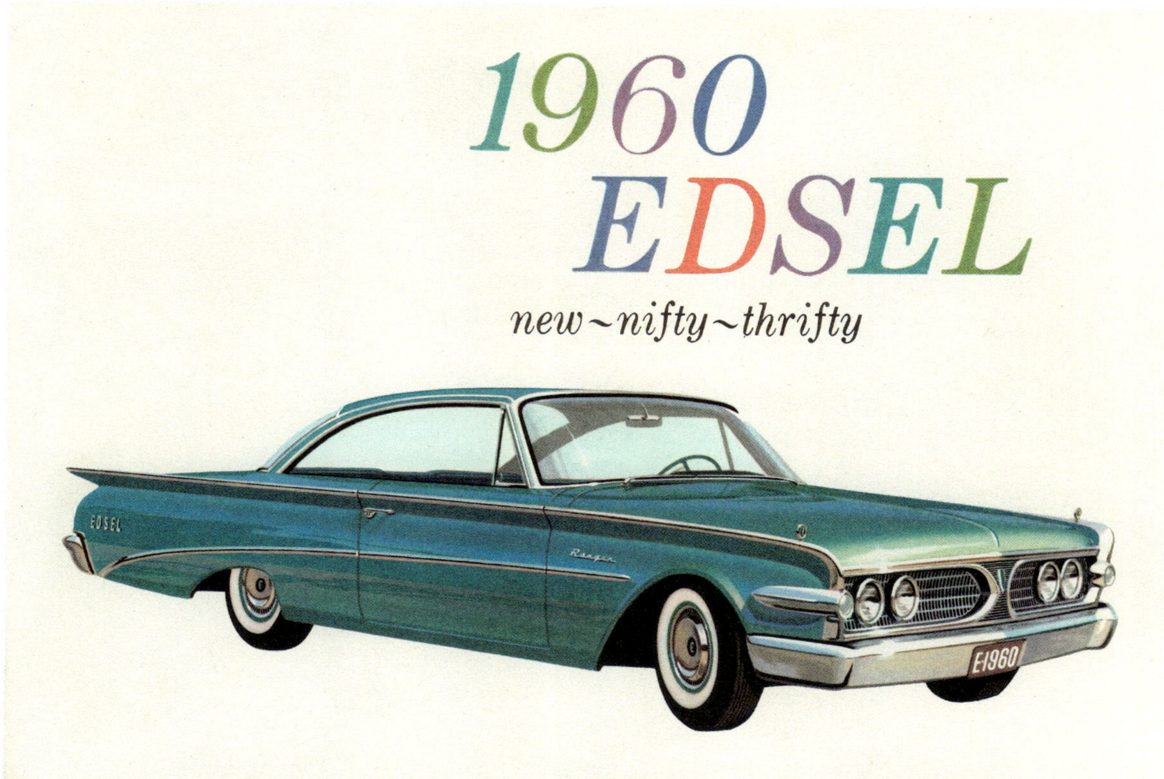

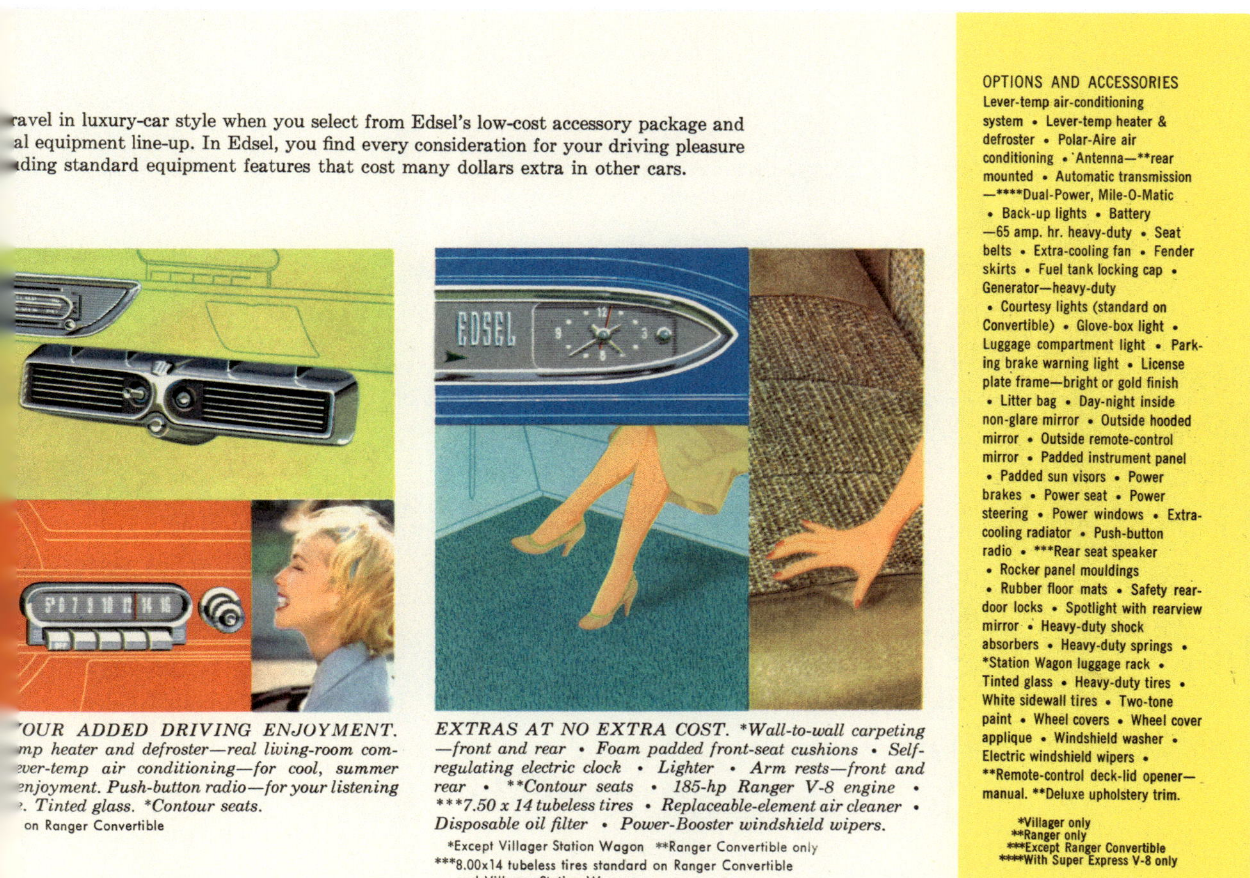

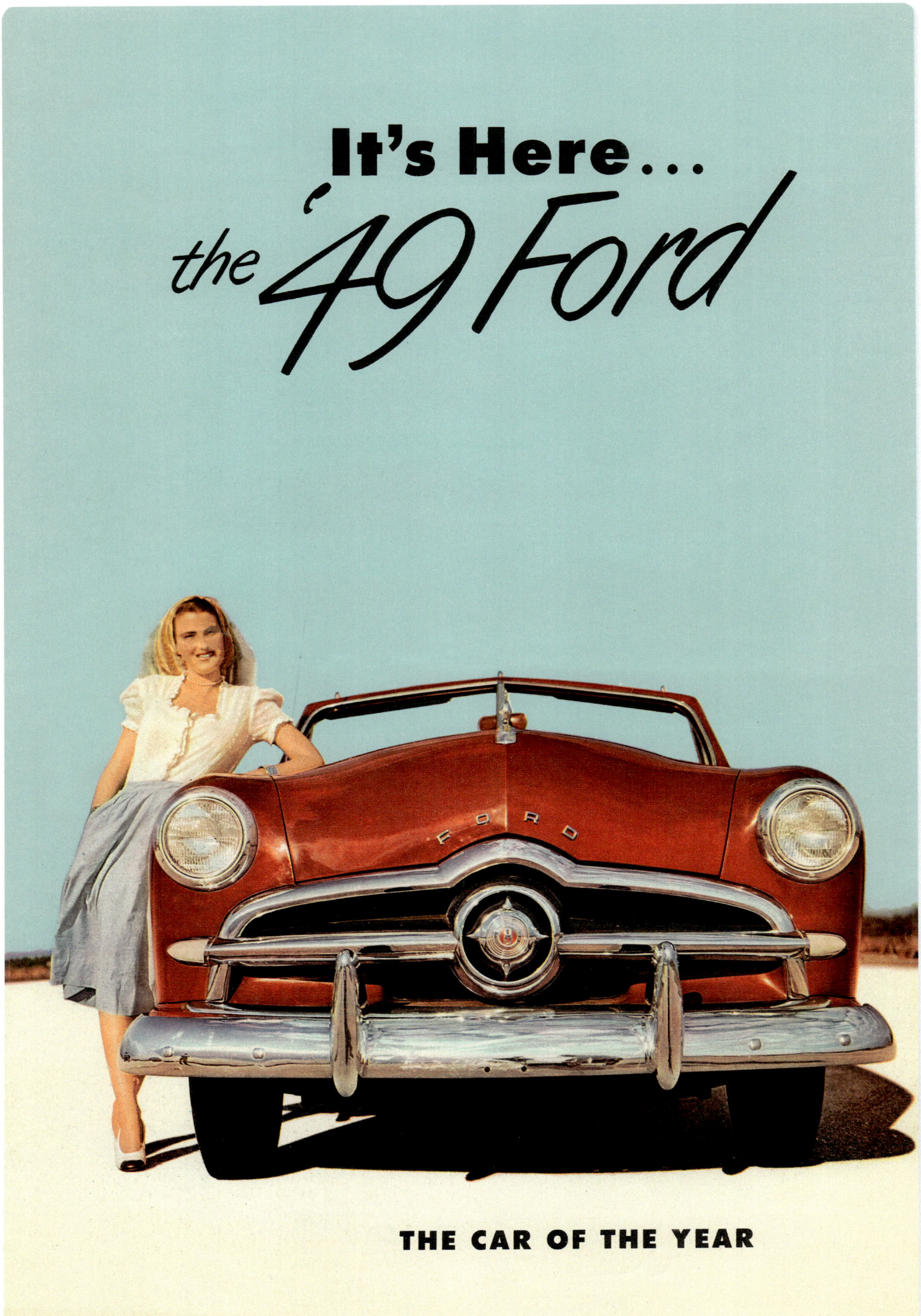
It's Here...
the '49 Ford
FORD
THE CAR OF THE YEAR

Ford

Ford 1949 *(opposite and below)*,
1951 *(right)*

The Ford Motor Company might have ceased to exist if these cars had been failures. The George Walker design completely redefined Ford, with integrated steel bodies, independent front suspension, and longitudinal leaf springs at the rear. More than a million were sold in their opening year.

Wenn diese Autos gefloppt hätten, hätte dies leicht das Aus für die Ford Motor Company bedeuten können. Das Design von George Walker gab Ford buchstäblich ein ganz neues Gesicht: mit integrierter Ganzstahlkarosserie, Einzelradaufhängung vorn und Längsblattfedern hinten. Im Jahr der Markteinführung wurden mehr als eine Million dieser Fahrzeuge verkauft.

La Ford Motor Company aurait pu fermer boutique si ces modèles avaient été des échecs. Le design de George Walker redéfinissait complètement l'identité de la marque, avec ses carrosseries tout acier, ses suspensions avant à roues indépendantes et des ressorts longitudinaux à lame à l'arrière. Il s'en vendit plus d'un million la première année.

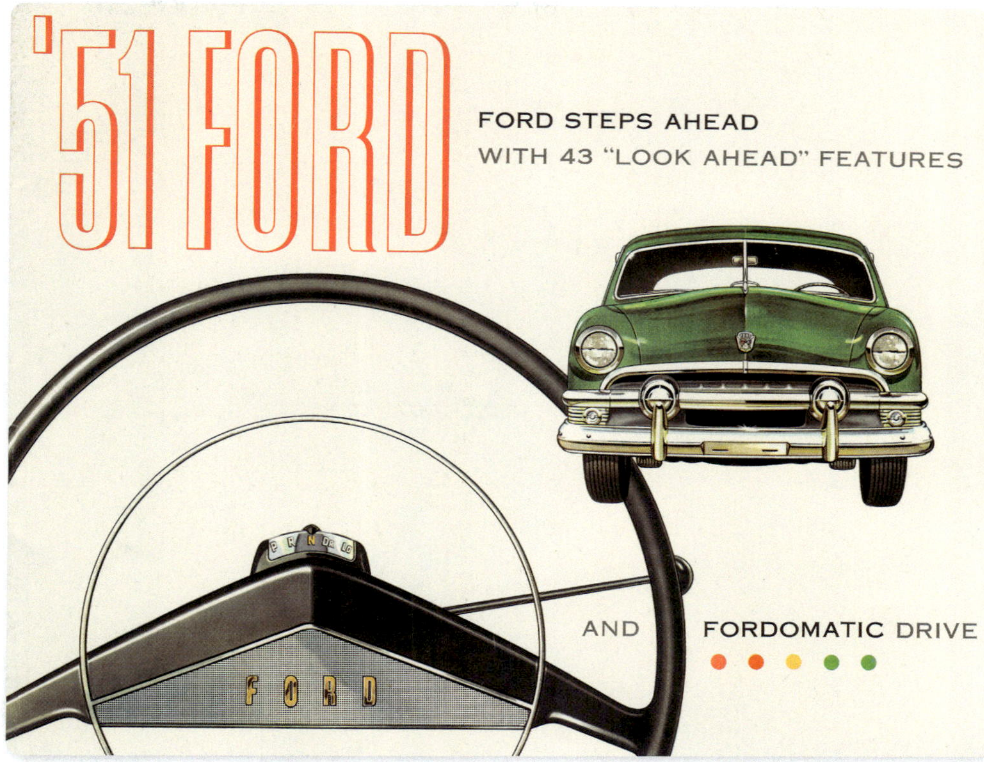

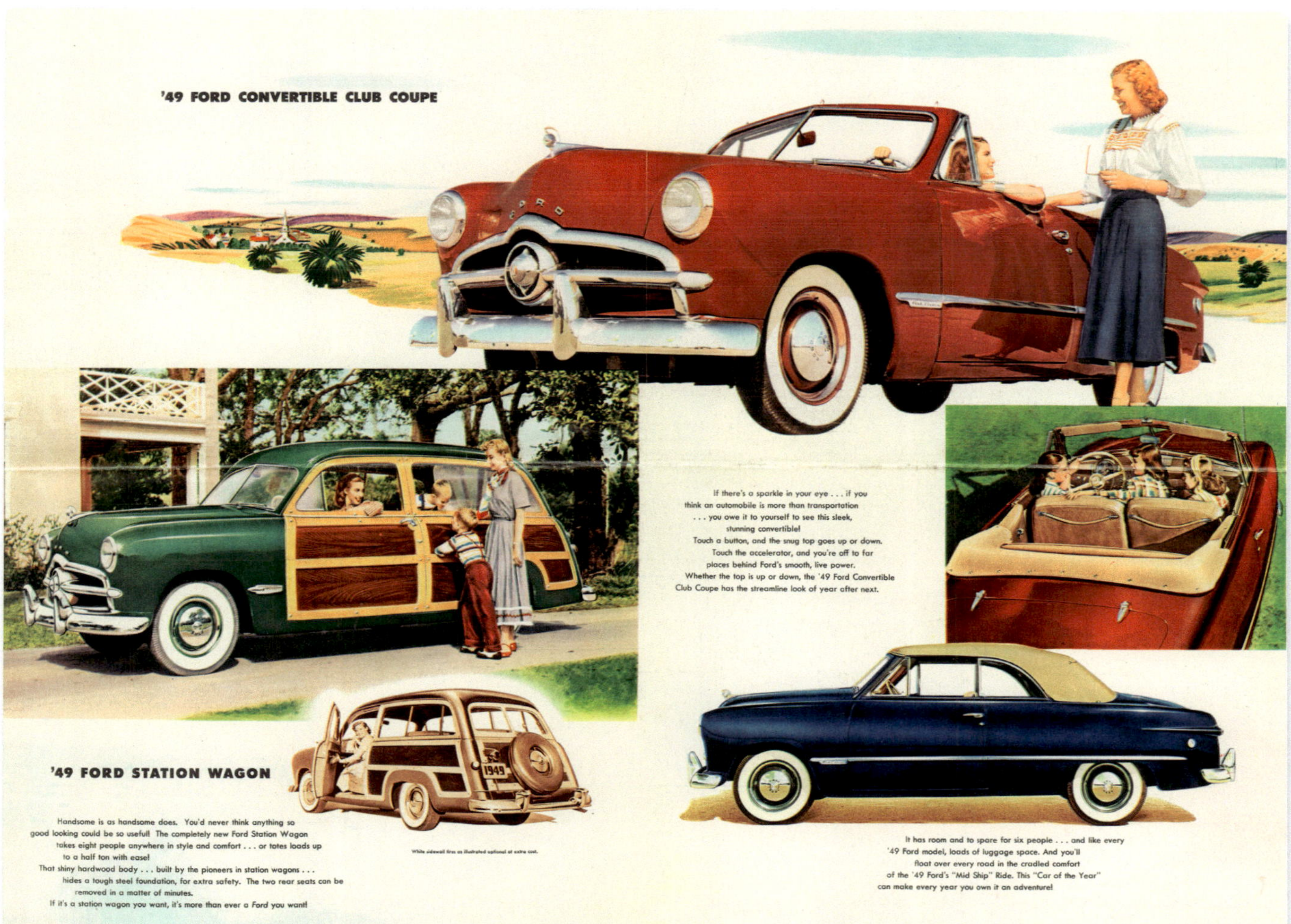

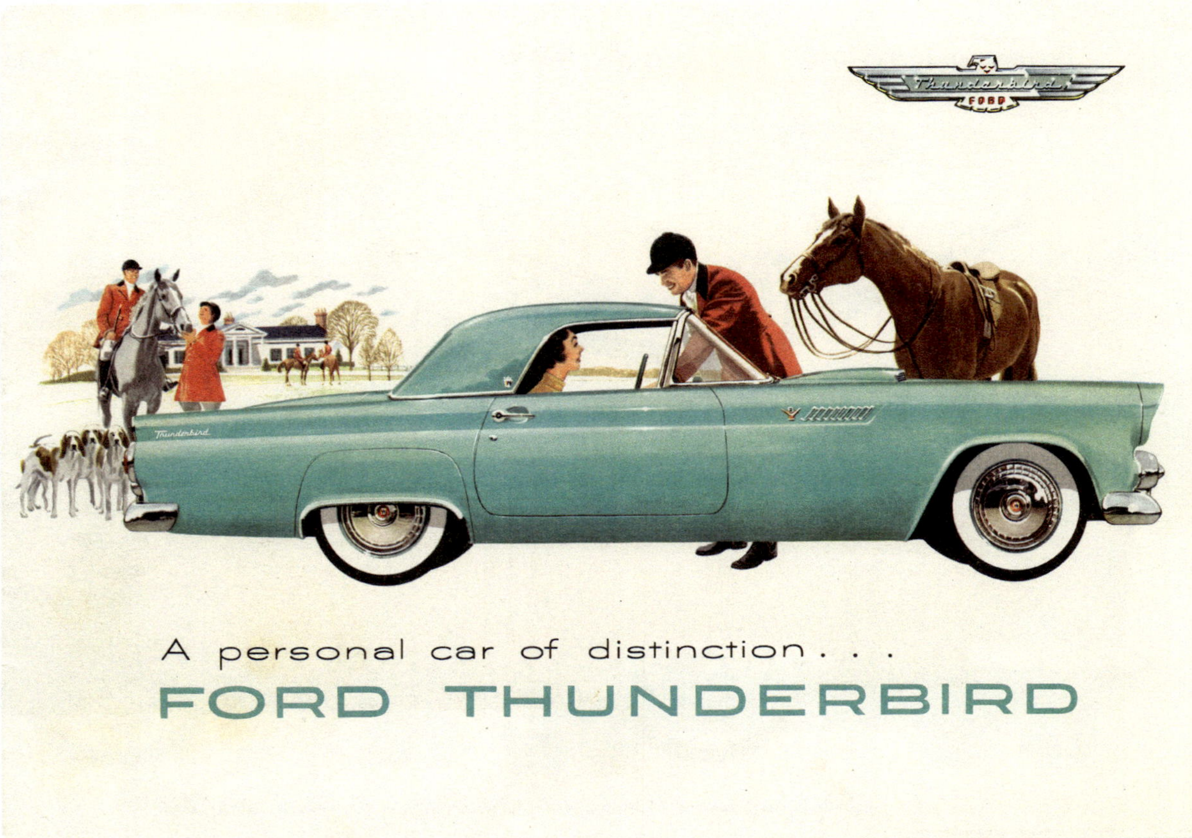
Thunderbird
FORD
A personal car of distinction . . .
FORD THUNDERBIRD

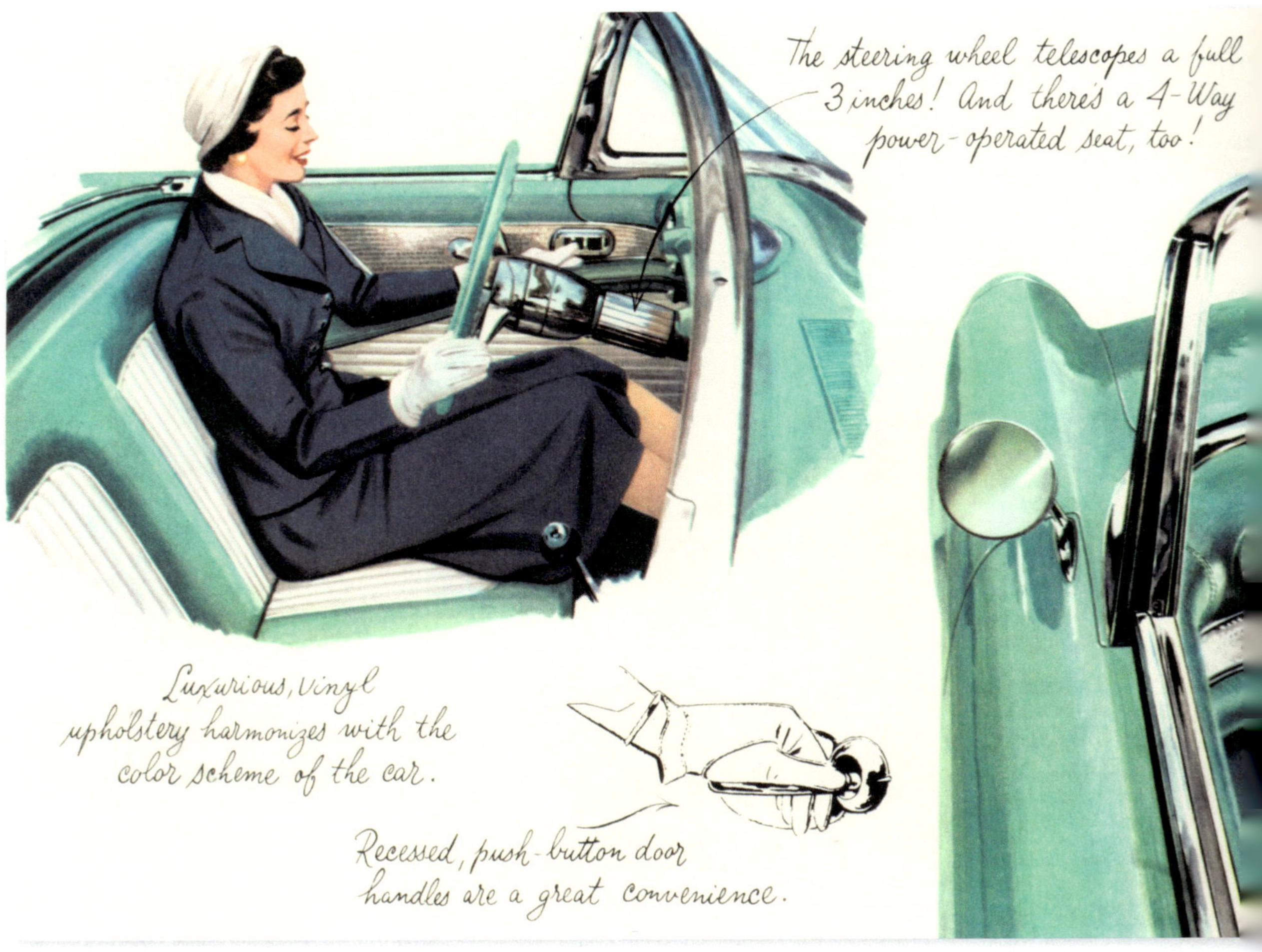
The steering wheel telescopes a full 3 inches! And there's a 4-Way power-operated seat, too!
Luxurious, vinyl upholstery harmonizes with the color scheme of the car.
Recessed, push-button door handles are a great convenience.

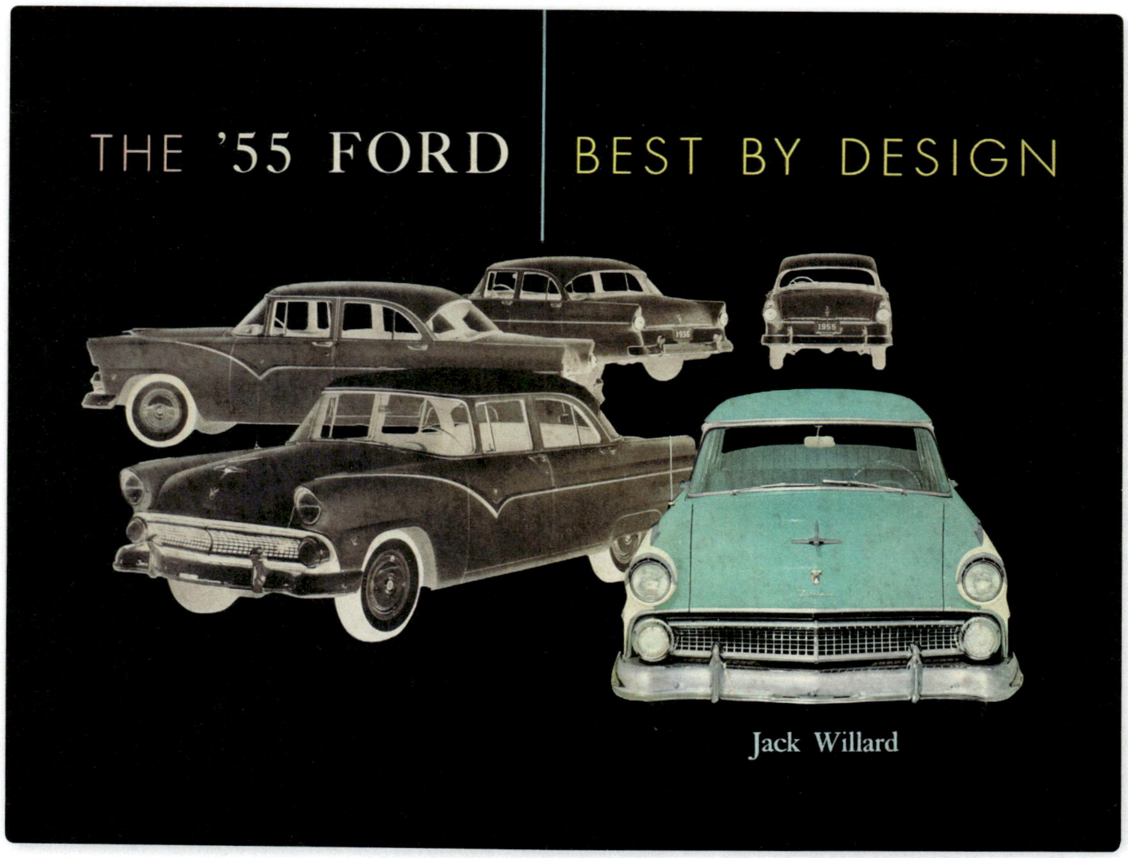

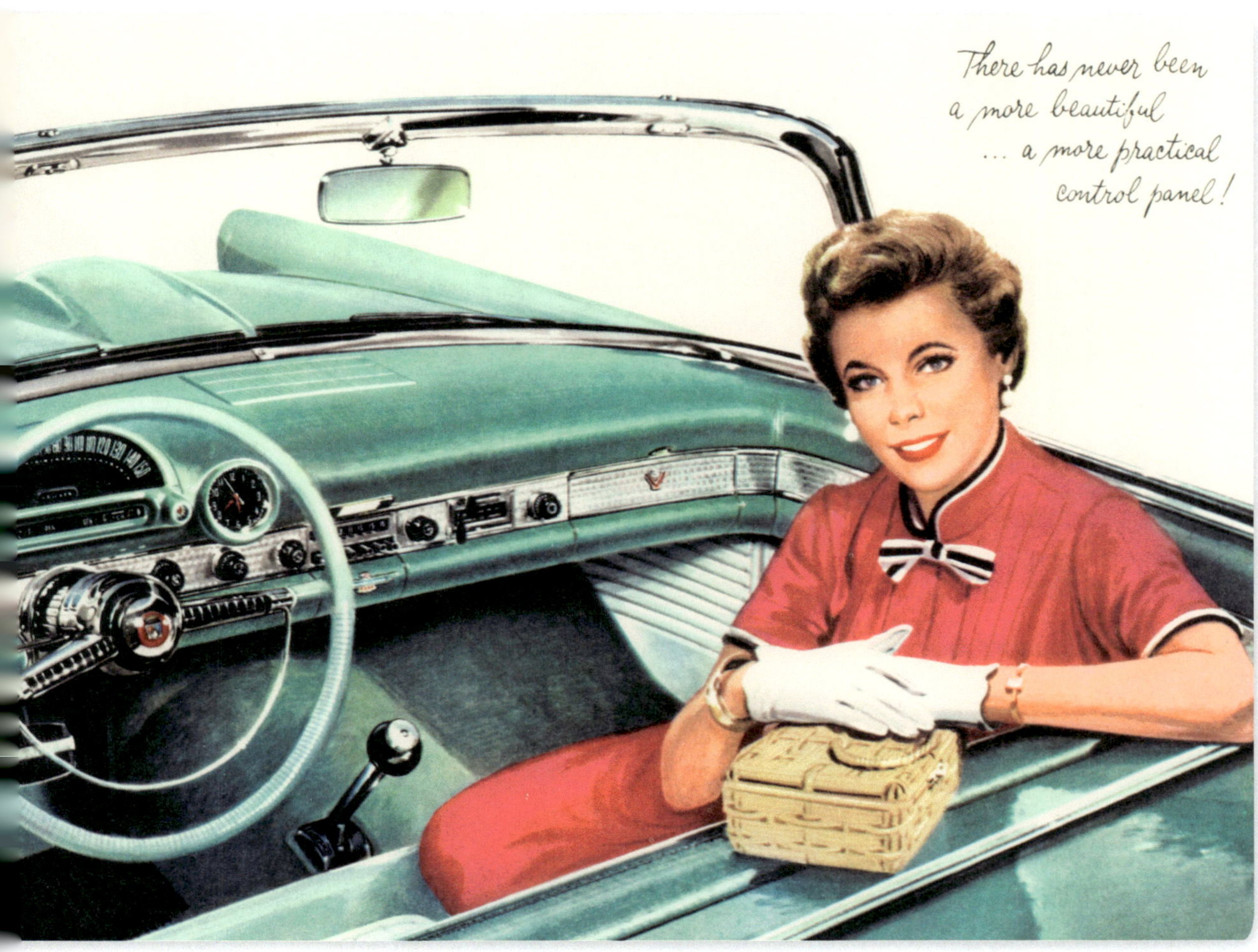

Ford 1954 *(opposite and left)*,
1955 *(above)*

Like Chevrolet, Ford restyled its entire lineup for 1955 and was rewarded with its second-best sales year ever, surpassing all since the mighty Model T era. A Sunliner convertible with an overdrive for cruising had great appeal. The biggest news was Ford introducing the two-seat Thunderbird for 1955 with V-8 power to match that of the Chevrolet Corvette.

Wie Chevrolet gestaltete auch Ford für das Modelljahr 1955 seine gesamte Palette um und wurde dafür mit seinem zweitbesten Verkaufsjahr aller Zeiten belohnt – dem besten seit der Zeit des unschlagbaren Model T. Besonders gut kam das Kabriolett Sunliner mit einem Schongang für entspanntes Fahren an. Den größten Nachrichtenwert besaß jedoch die Einführung des zweisitzigen Thunderbird für 1955 mit V8-Motor als Konkurrenz zur Chevrolet Corvette.

À l'instar de Chevrolet, Ford relooka l'intégralité de sa gamme en 1955. Récompense : la deuxième meilleure année de la marque, supérieure même aux saisons fastes de la Ford T. La Sunliner décapotable, avec son overdrive pour soulager le moteur en vitesse de croisière, avait fort belle allure. Mais la grande nouvelle était la présentation de la Thunderbird deux places à moteur V8, lancée à la poursuite de la Chevrolet Corvette.

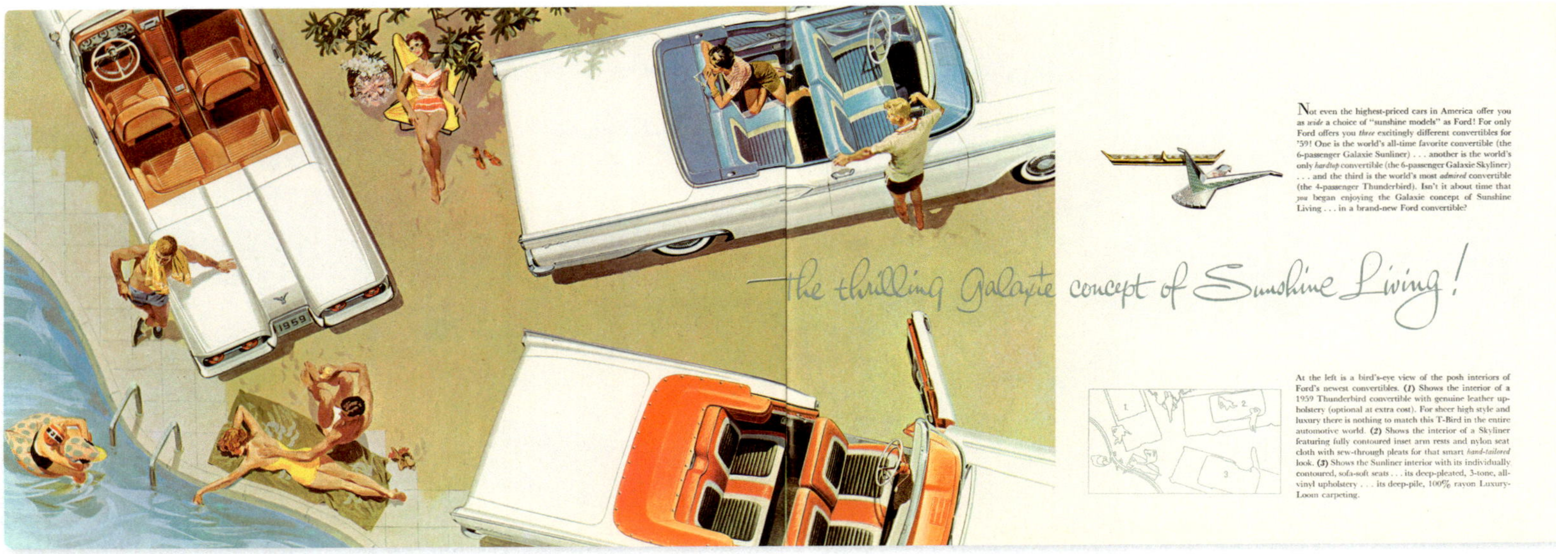

Ford 1959 *(above)*, **1957** *(below)*, **1956** *(opposite top)*, **1959** *(opposite bottom)*

The year 1959 was one of transition. The new Galaxie, first a trim level of the Fairlane 500, shared some visual cues with the four-seat Thunderbird, which had replaced the original two-seater in 1958.

Das Jahr 1959 war ein Jahr des Übergangs. Der neue Galaxie, der zunächst nur eine Ausstattungsvariante des Fairlane 500 gewesen war, teilte sich ein paar äußerliche Merkmale mit dem viersitzigen Thunderbird, der den ursprünglichen Zweisitzer 1958 ersetzt hatte.

1959 fut une année de transition. La nouvelle Galaxie, à l'origine une variante de la Fairlane 500, partageait certains éléments visuels avec la Thunderbird à quatre places, qui avait remplacé la deux places originelle en 1958.

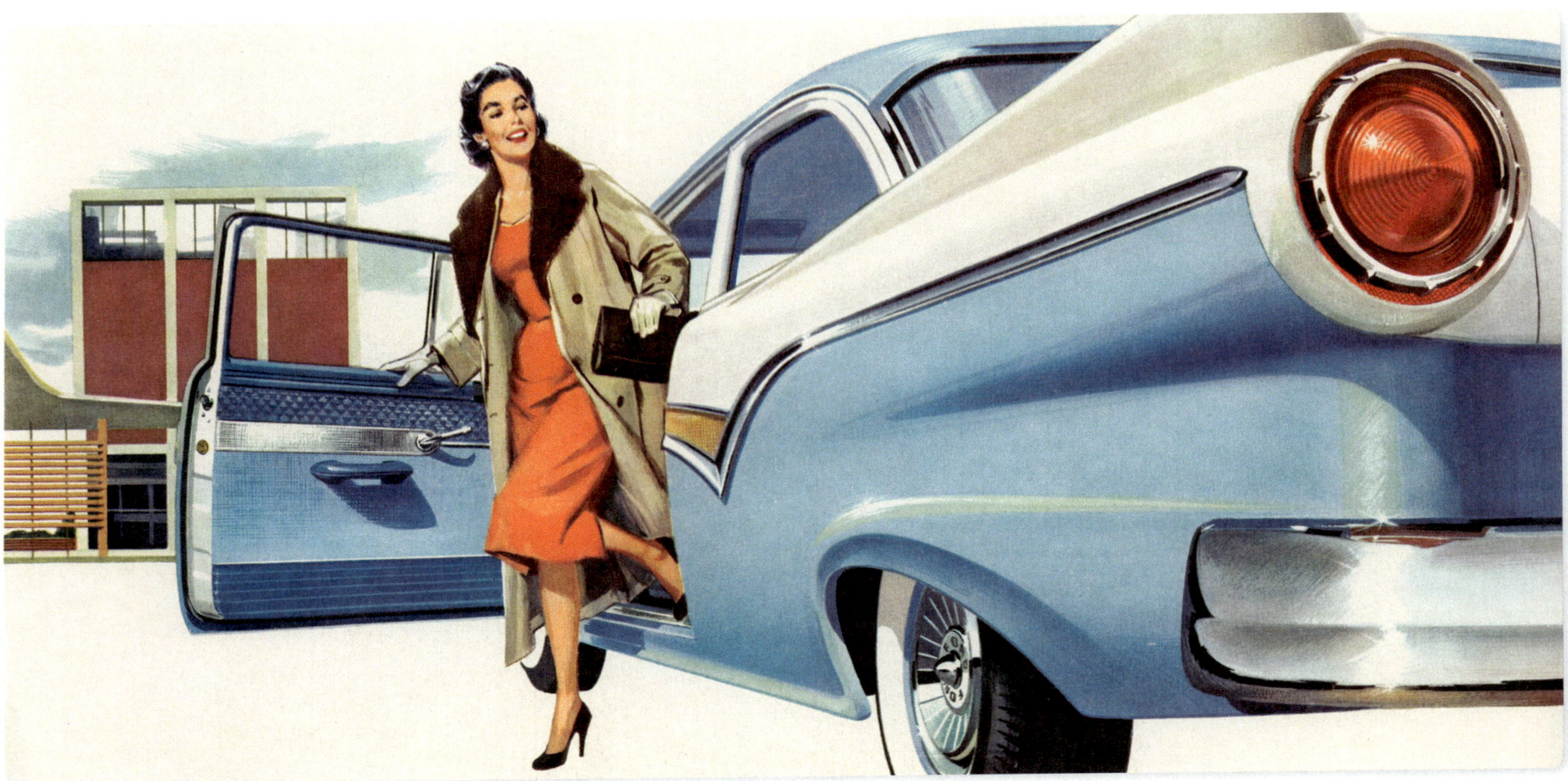

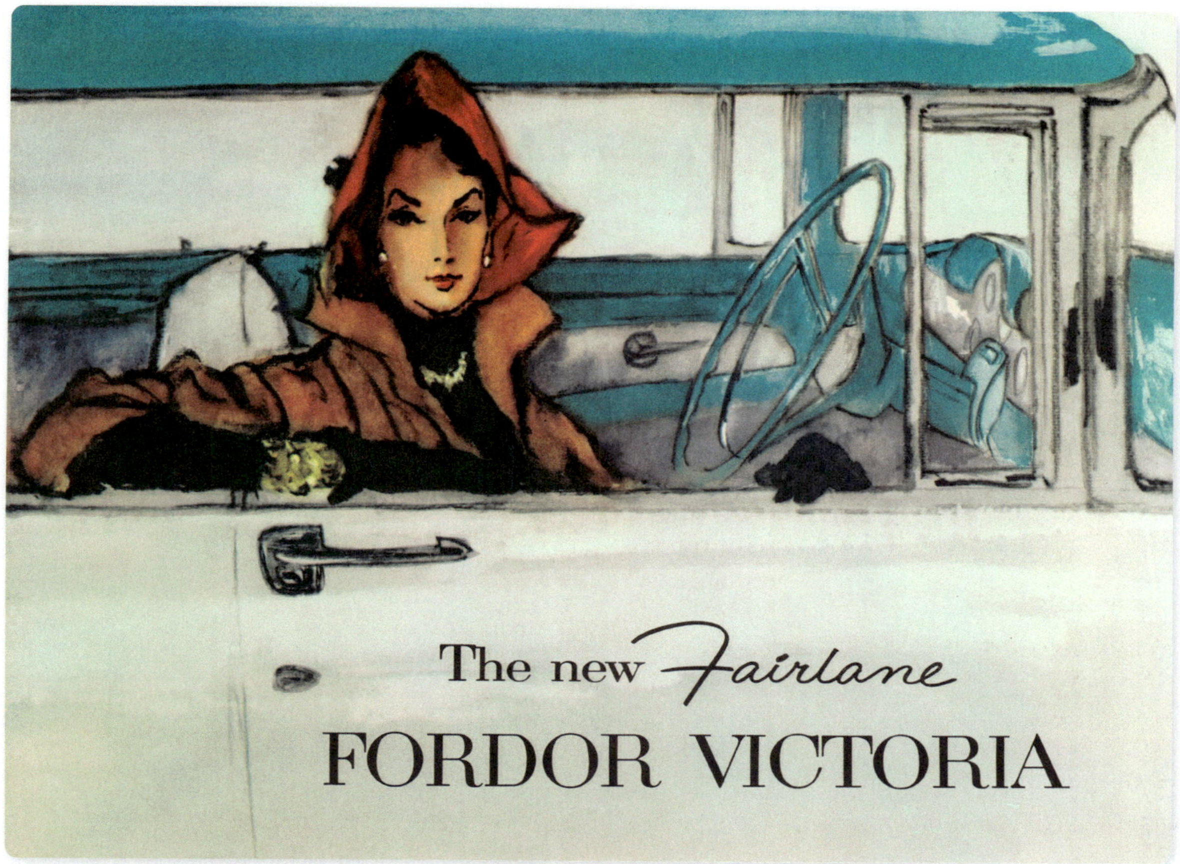
The new Fairlane
FORDOR VICTORIA

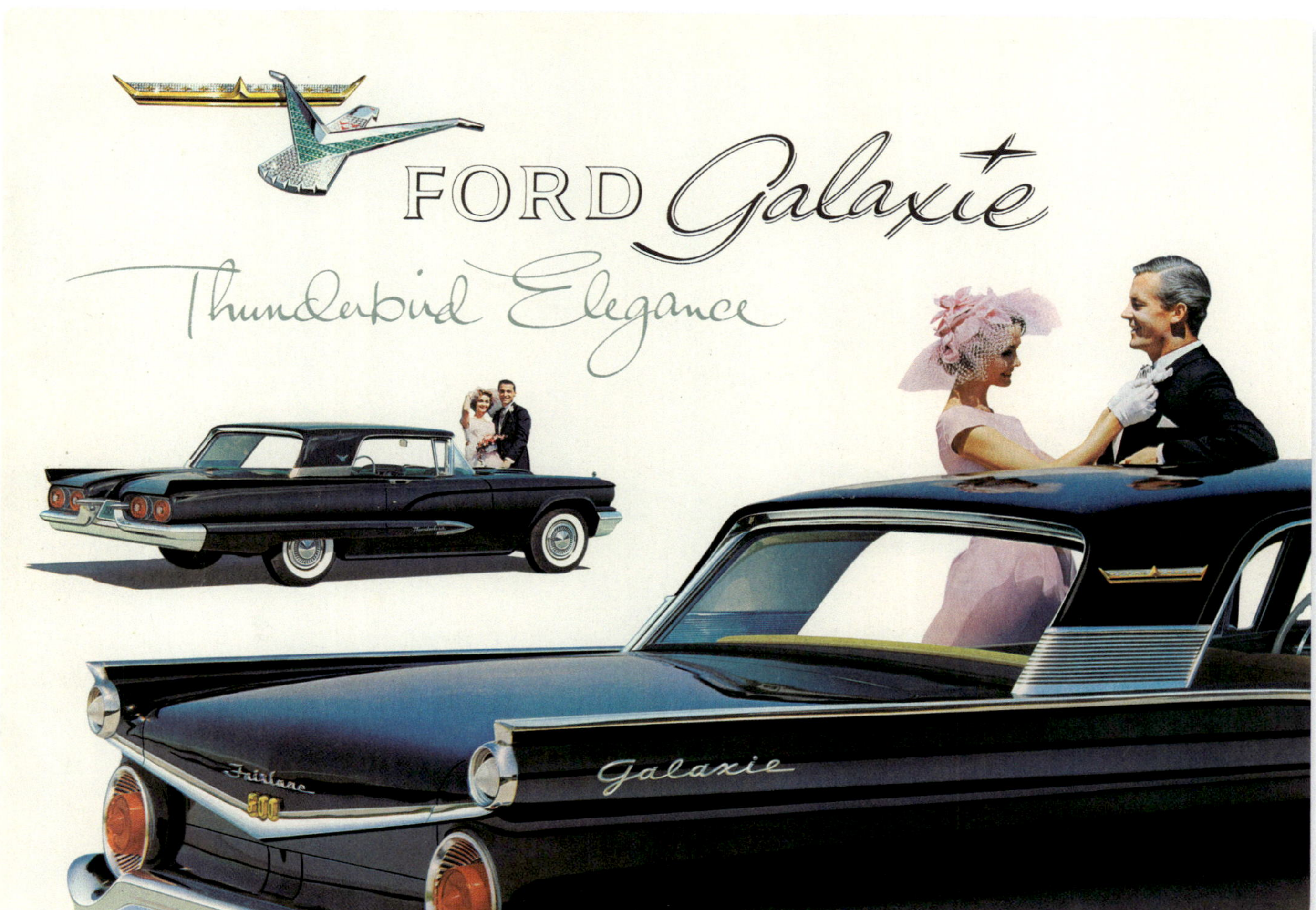
FORD Galaxie
Thunderbird Elegance
Fairlane
500
Galaxie

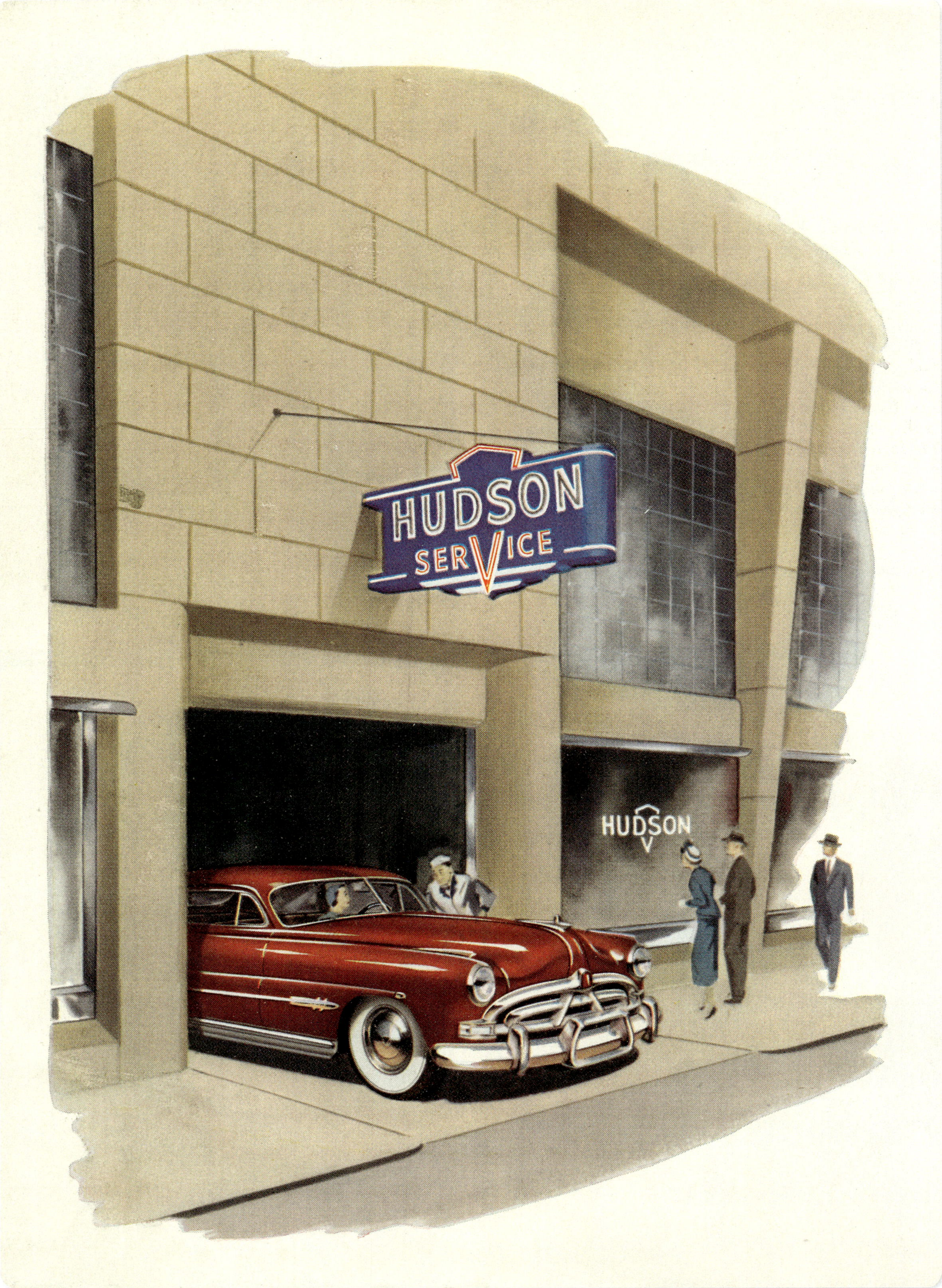
HUDSON
SERVICE
HUDSON

Hudson

Hudson 1951 *(opposite)*, **1949** *(right)*, **1952** *(below)*

These cars shocked the world. A team of engineers under Frank Spring created the lowered chassis that gave the Hudson its "step-down" layout. Combined with a dual-carburetor Twin-H engine, the Hornets crushed their competition in early NASCAR races, and were a favorite among U.S. highway patrols.

Diese Autos schockierten die Welt. Ein Team von Ingenieuren unter Frank Spring schuf das tiefer gelegte Fahrgestell, das dem Hudson sein typisches „step-down"-Layout verlieh. Dank ihrem „Twin-H"-Motor – mit doppeltem Einfachvergaser – schlugen die Hornets ihre Konkurrenz in frühen NASCAR-Rennen haushoch und gehörten auch zu den Lieblingsfahrzeugen der US-amerikanischen Highwaystreifen.

Ces voitures ont ébranlé le monde. Une équipe d'ingénieurs dirigée par Frank Spring est à l'origine de ce châssis surbaissé qui donnait à la Hudson son architecture particulière dite « step-down » (le plancher plus bas que le châssis). Avec leur moteur six cylindres Twin-H à deux carburateurs indépendants (à partir de 1952), les Hornets (« frelons ») écraseront la concurrence dans les premières courses du championnat NASCAR et infiltreront massivement la flotte de la police autoroutière.

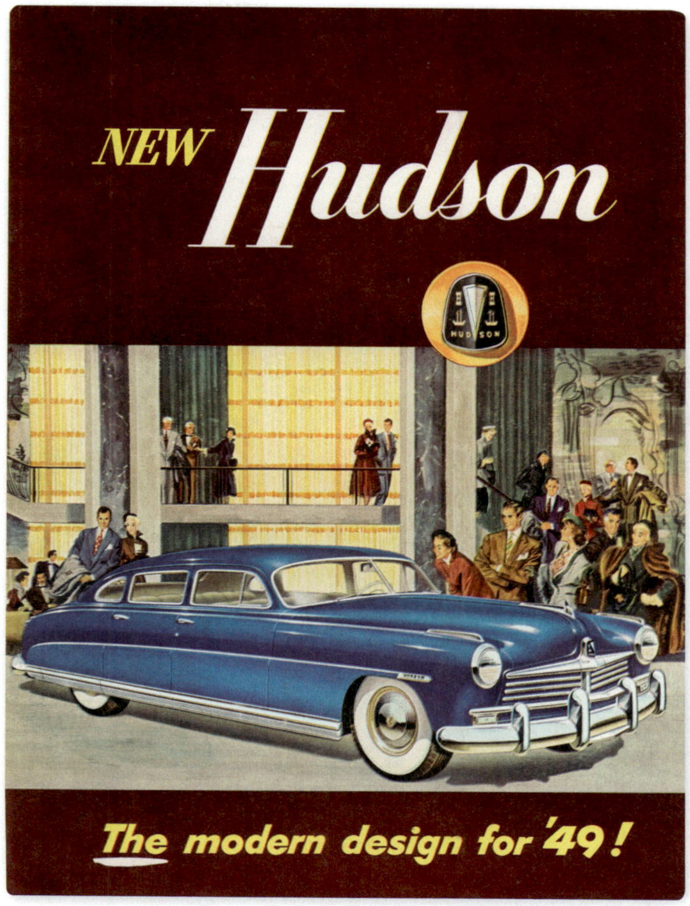

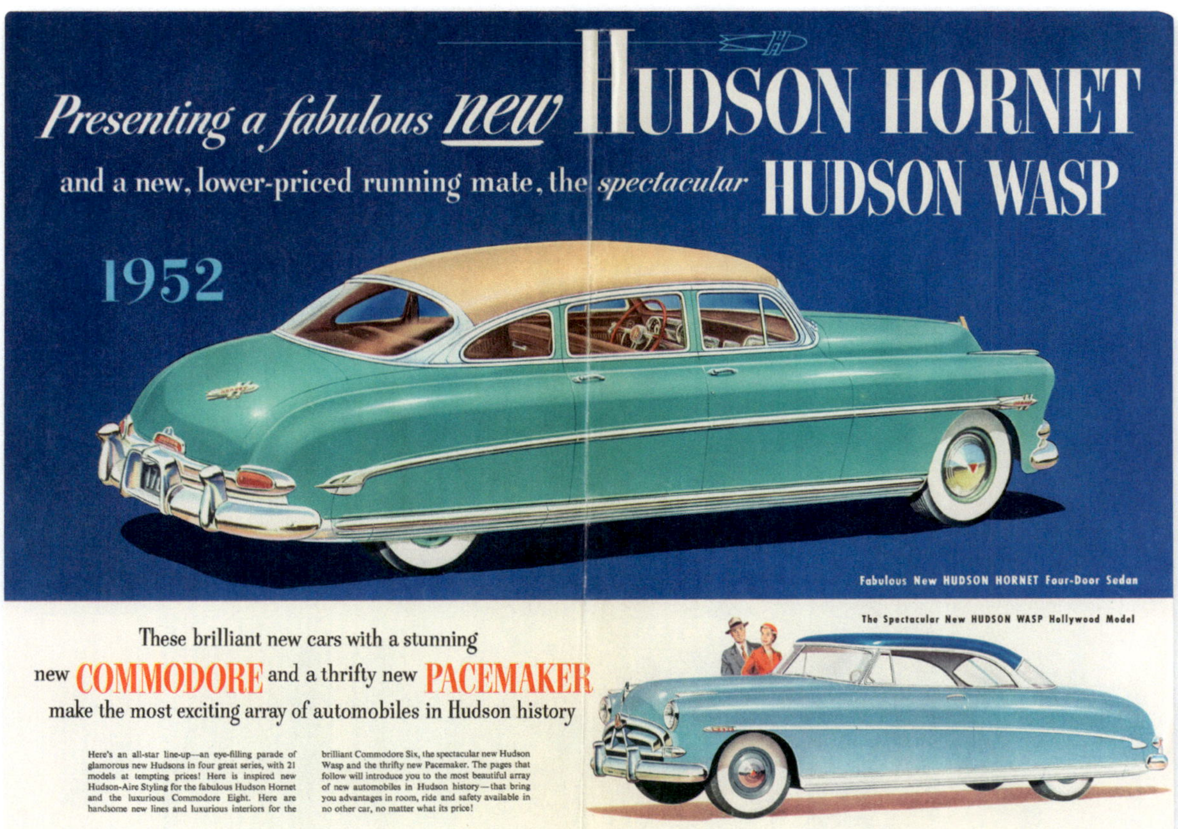

Kaiser

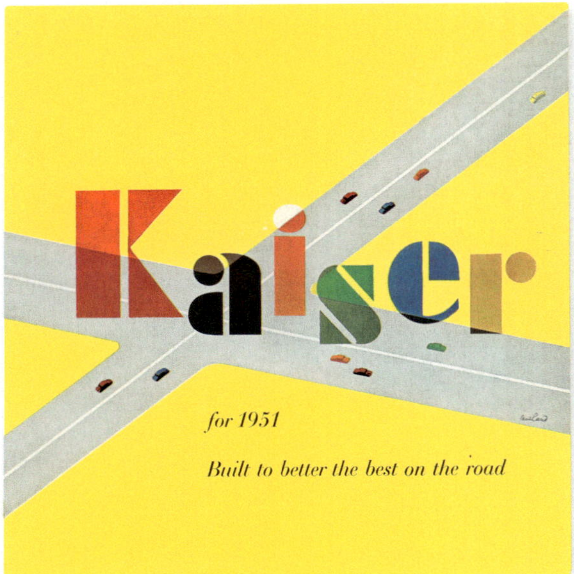

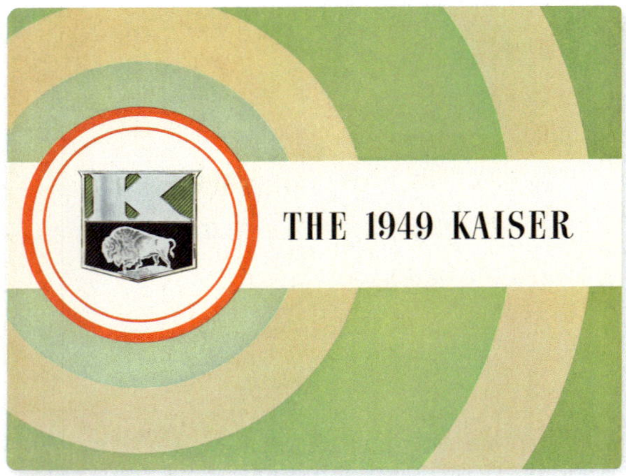

Kaiser 1951 *(far left)*, **1949** *(left)*, **1952** *(below and opposite)*

Henry J. Kaiser built the Hoover Dam and World War II ships before tackling cars. The Kaiser was a strong, well-appointed car, which used a straight-six engine supplied by Continental, a builder of engines for many independent makes. (Not coincidentally, Continental was also owned by Kaiser.) A dramatic 1951 restyling and the lovely 1954 Kaiser-Darrin sports car are memorable. Ultimately, Kaiser bought Willys-Overland and built Jeeps.

Bevor er sich an Automobile wagte, hatte Henry J. Kaiser schon den Hoover-Damm und Schiffe für den Einsatz im Zweiten Weltkrieg gebaut. Der Kaiser war ein starkes, gut ausgestattetes Auto, das einen zugelieferten Sechszylinderreihenmotor von Continental verwendete, einem Motorenbauer, der viele unabhängige Marken versorgte. (Es war kein Zufall, das Continental auch Kaiser gehörte.) Eine drastische Umgestaltung von 1951 und der anmutige Sportwagen Kaiser-Darrin von 1954 sind unvergessen. Letztendlich kaufte Kaiser Willys-Overland auf und baute Jeeps.

Avant de se lancer dans les voitures, Henry J. Kaiser avait construit le barrage Hoover sur le fleuve Colorado, ainsi que des navires pour la Seconde Guerre mondiale. La Kaiser était une automobile robuste et bien équipée qui s'appuyait sur un moteur six cylindres en ligne de Continental, motoriste fournisseur de nombreuses marques indépendantes (et également propriété de Kaiser, ce qui n'était pas fortuit). Un spectaculaire relooking en 1951 et la jolie sportive Kaiser-Darrin en 1954 sont restés dans les mémoires. Kaiser finira par racheter Willys-Overland et produira dès lors des Jeep.

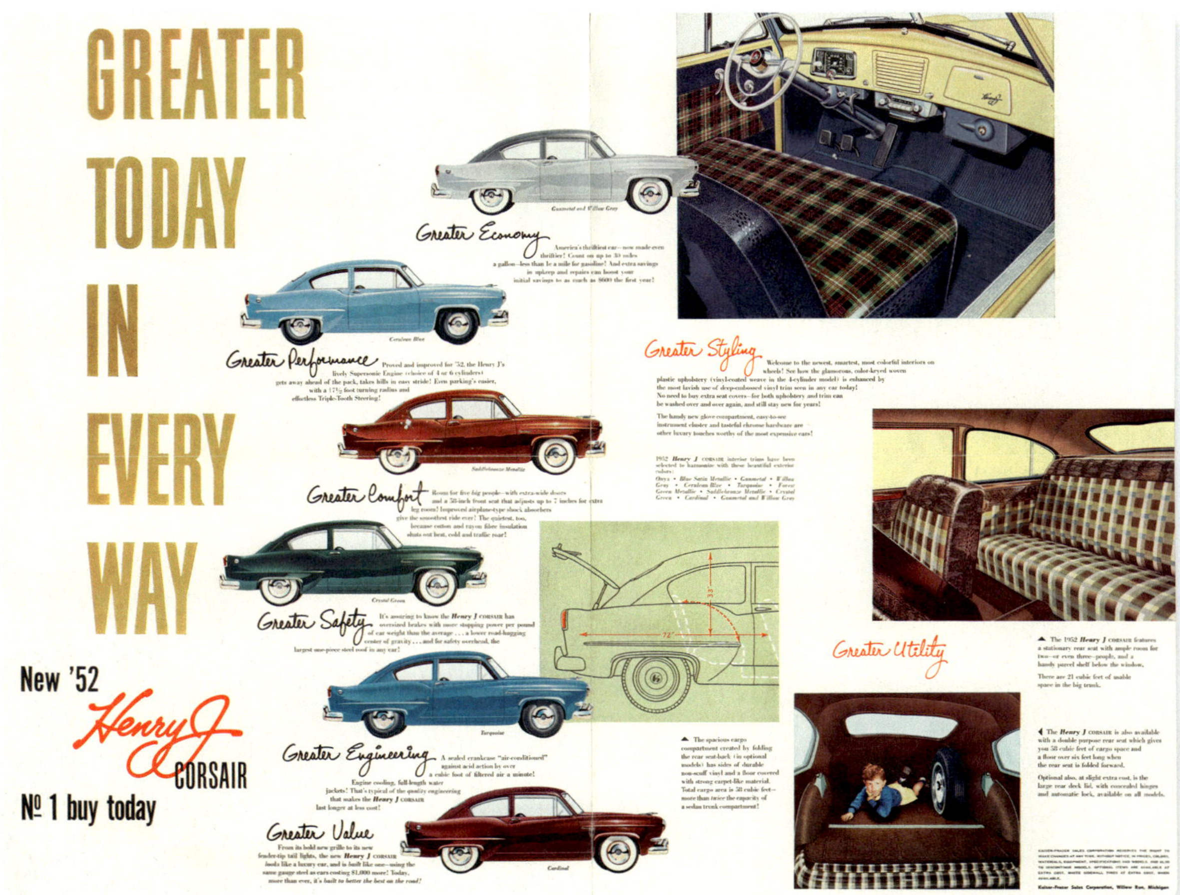

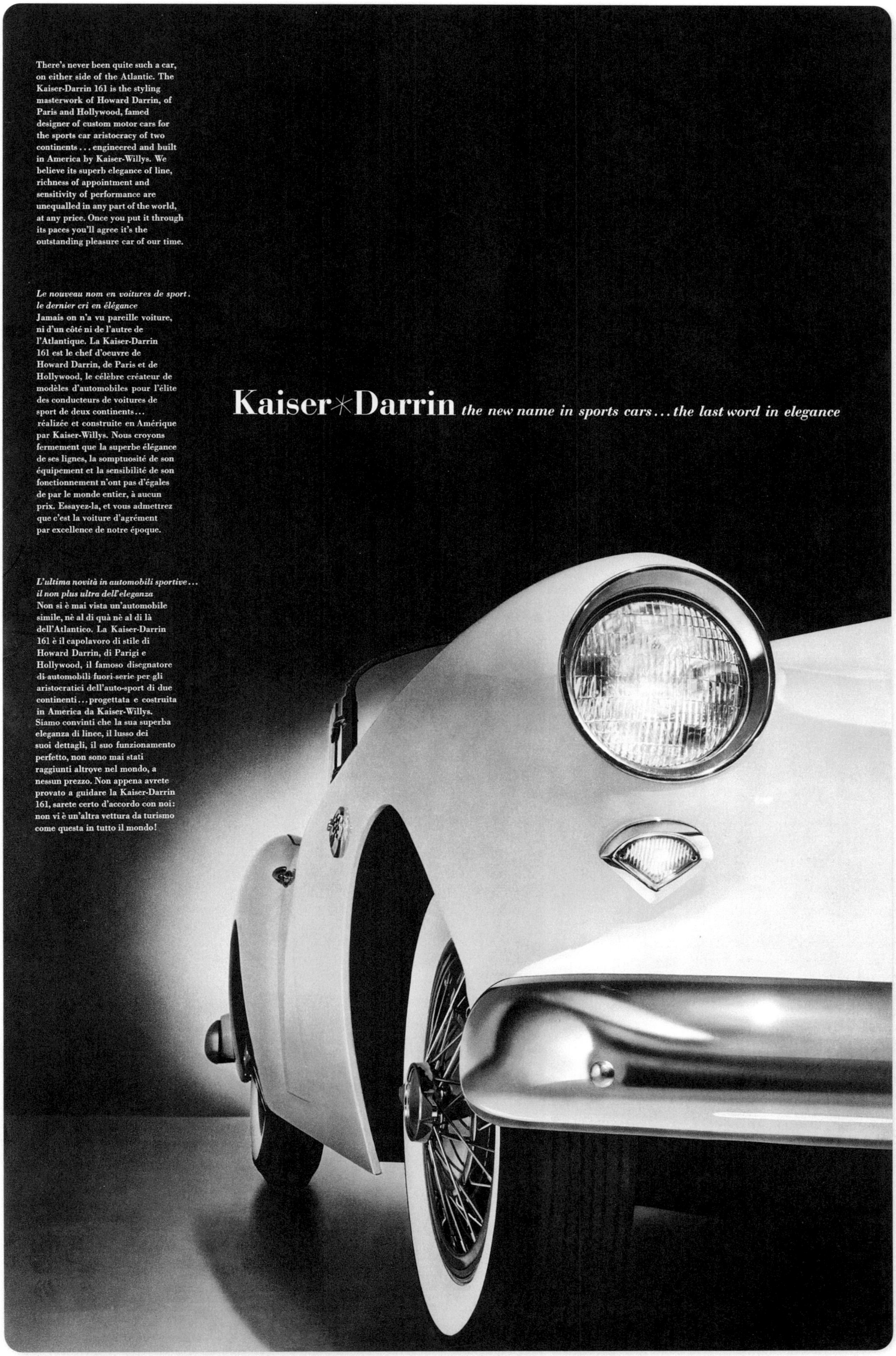
There's never been quite such a car, on either side of the Atlantic. The Kaiser-Darrin 161 is the styling masterwork of Howard Darrin, of Paris and Hollywood, famed designer of custom motor cars for the sports car aristocracy of two continents . . . engineered and built in America by Kaiser-Willys. We believe its superb elegance of line, richness of appointment and sensitivity of performance are unequalled in any part of the world, at any price. Once you put it through its paces you'll agree it's the outstanding pleasure car of our time.
Le nouveau nom en voitures de sport. le dernier cri en élégance
Jamais on n'a vu pareille voiture, ni d'un côté ni de l'autre de l'Atlantique. La Kaiser-Darrin 161 est le chef d'oeuvre de Howard Darrin, de Paris et de Hollywood, le célèbre créateur de modèles d'automobiles pour l'élite des conducteurs de voitures de sport de deux continents... réalizée et construite en Amérique par Kaiser-Willys. Nous croyons fermement que la superbe élégance de ses lignes, la somptuosité de son équipement et la sensibilité de son fonctionnement n'ont pas d'égales de par le monde entier, à aucun prix. Essayez-la, et vous admettrez que c'est la voiture d'agrément par excellence de notre époque.
L'ultima novità in automobili sportive... il non plus ultra dell'eleganza
Non si è mai vista un'automobile simile, nè al di quà nè al di là dell'Atlantico. La Kaiser-Darrin 161 è il capolavoro di stile di Howard Darrin, di Parigi e Hollywood, il famoso disegnatore di automobili fuori serie per gli aristocratici dell'auto-sport di due continenti... progettata e costruita in America da Kaiser-Willys. Siamo convinti che la sua superba eleganza di linee, il lusso dei suoi dettagli, il suo funzionamento perfetto, non sono mai stati raggiunti altrove nel mondo, a nessun prezzo. Non appena avrete provato a guidare la Kaiser-Darrin 161, sarete certo d'accordo con noi: non vi è un'altra vettura da turismo come questa in tutto il mondo!
Kaiser*Darrin *the new name in sports cars... the last word in elegance*

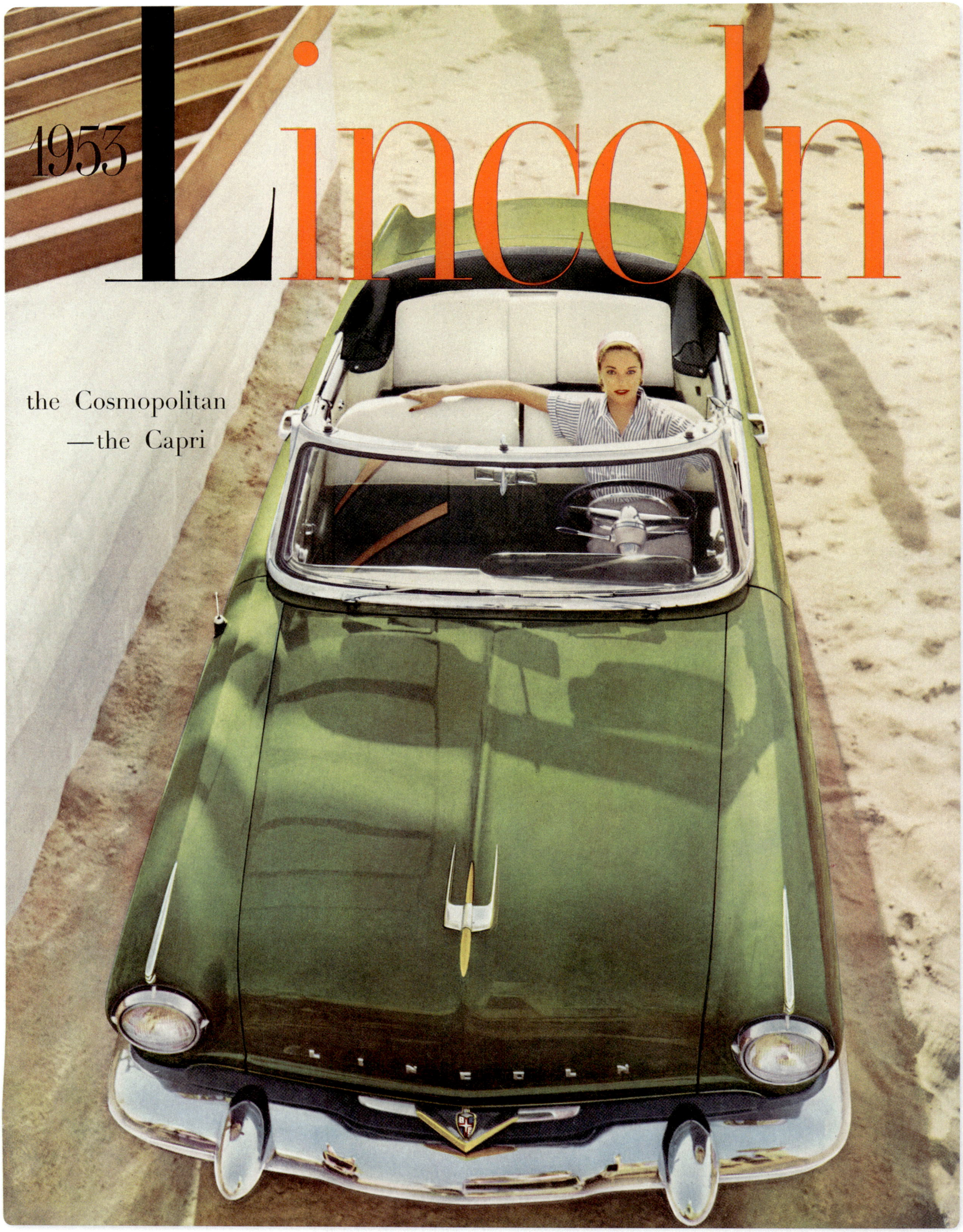

1953
Lincoln
the Cosmopolitan
—the Capri
LINCOLN

Lincoln

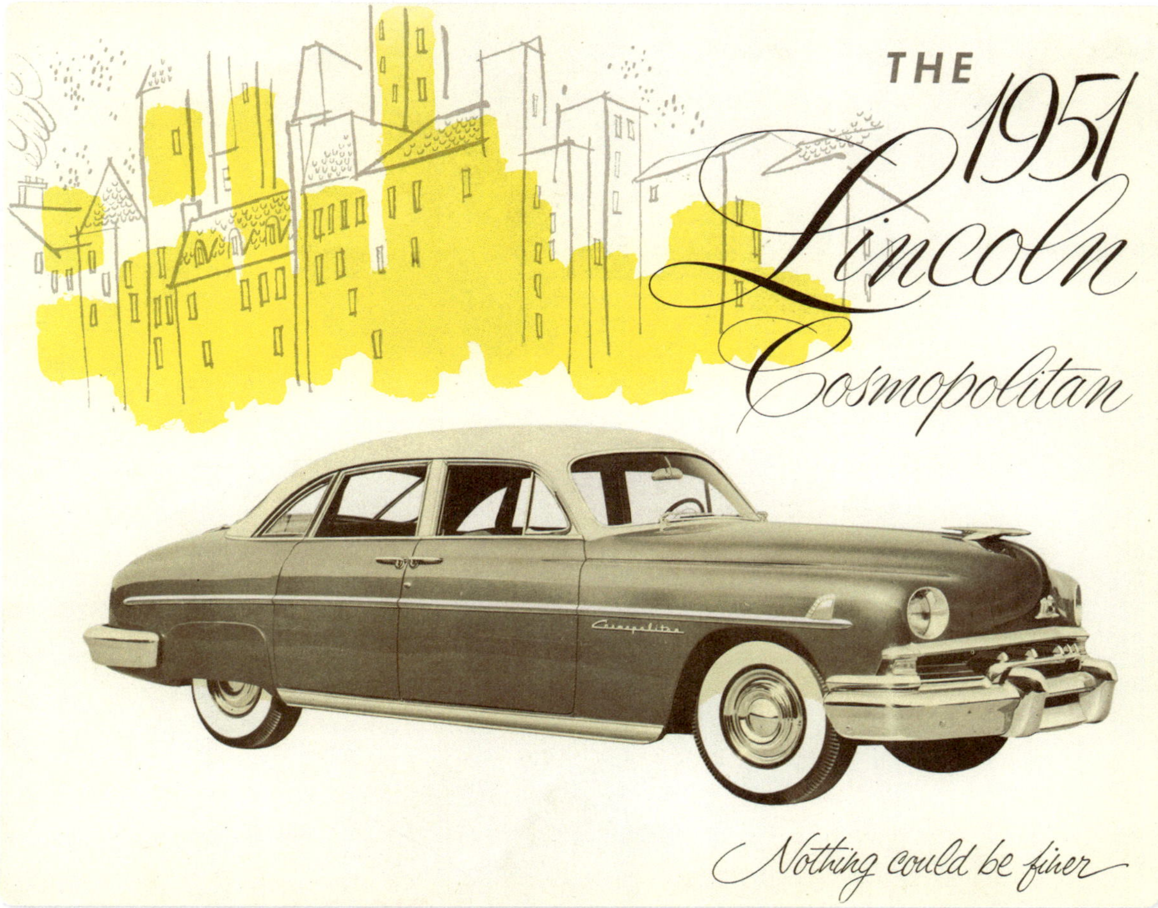

Lincoln 1953 *(opposite)*, **1951** *(left)*, **1952** *(below)*

There was a long-running quip in the car business that Lincoln always forecast the appearance of the following year's Mercury, and there was some truth to it. The postwar redesign included a 336-cubic-inch version of Ford's venerable flathead V-8 engine.

Jahrelang scherzte man in der Automobilbranche, der Lincoln nehme das Aussehen des Mercury im folgenden Jahr vorweg, und daran war durchaus etwas Wahres. Zur Neugestaltung nach dem Zweiten Weltkrieg gehörte auch eine 5½-Liter-Version des altehrwürdigen V8-Seitenventilmotors von Ford.

Dans le milieu automobile, il s'est longtemps raconté avec malice que Lincoln annonçait systématiquement le look de la Mercury à venir l'année suivante… Ce n'était pas tout à fait faux. La refonte générale opérée dans l'après-guerre comprenait notamment une version 5,5 litres du vénérable moteur Ford V8 à soupapes latérales.

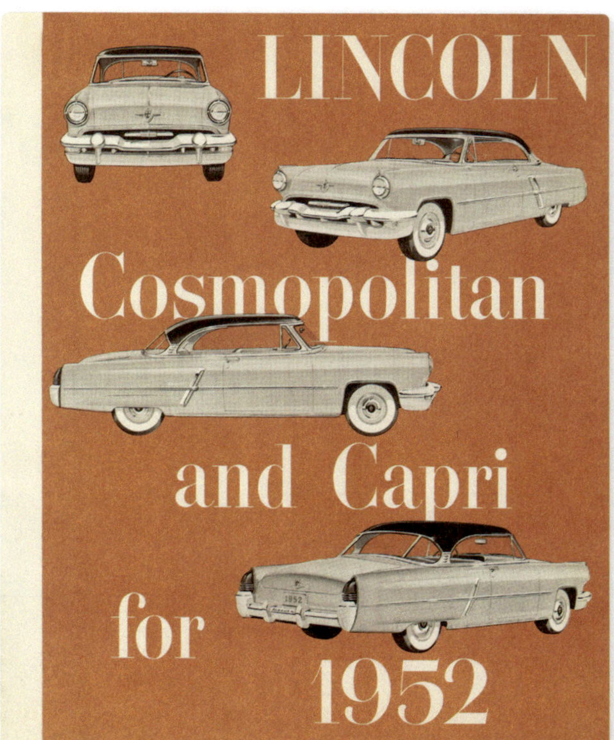

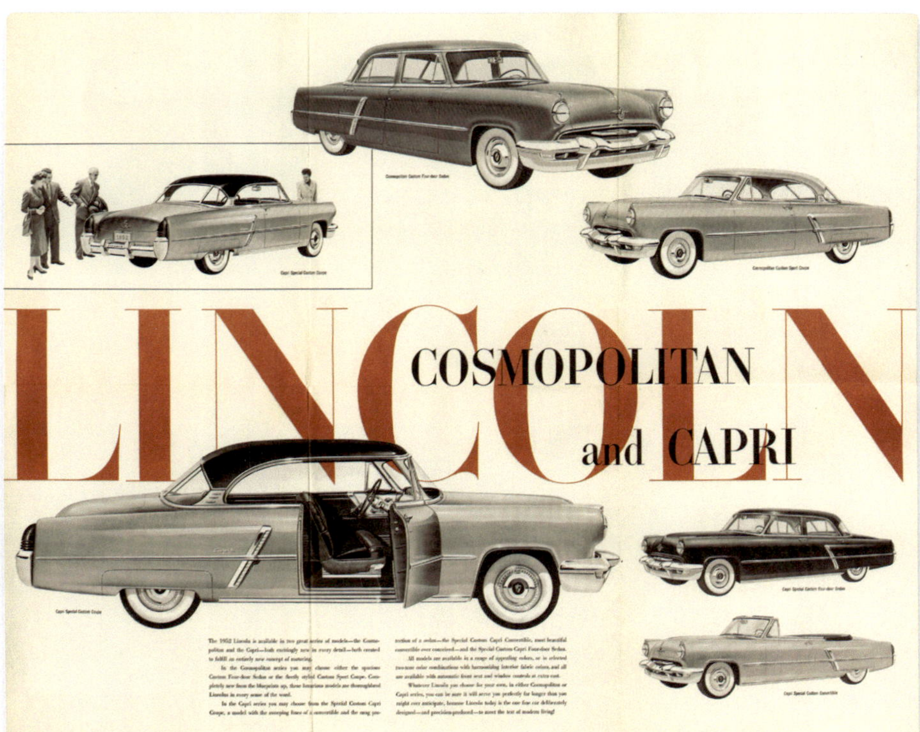

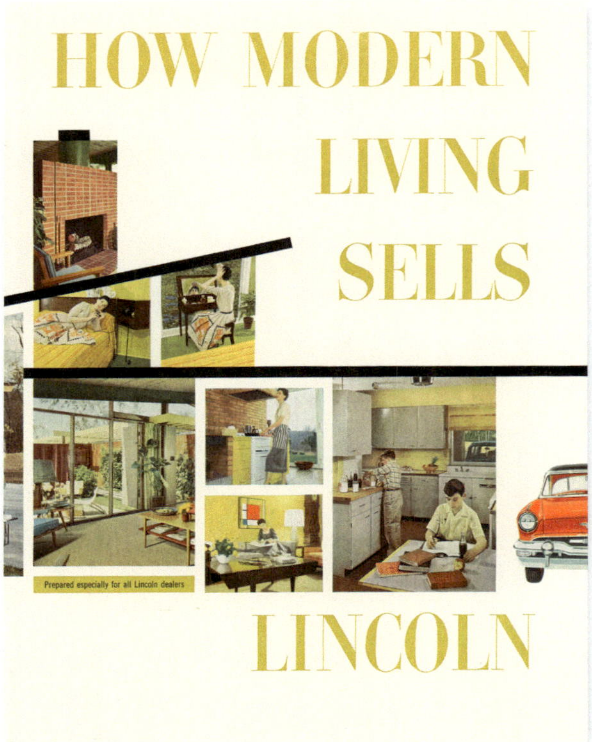

Lincoln 1951

This smart, forward-thinking brochure concentrated much more strongly on the buyer's lifestyle than on the Capri. It predicted a 1950s America that was focused on the consumerist good life: home conveniences, antiquing, boating, and other mid-century pleasures. A Lincoln's exclusivity fit just right.

Diese clevere, zukunftsweisende Broschüre konzentrierte sich viel stärker auf die Lebensweise der Käufer als auf den Capri. Sie sagte ein Amerika der 1950er-Jahre ganz im Zeichen von Konsumdenken und Wohlstand voraus: moderne Annehmlichkeiten, Antiquitätensammeln, Bootfahren und andere Vergnügungen der Jahrhundertmitte. Dazu passte die Exklusivität eines Lincoln wie die Faust aufs Auge.

Cette brochure élégante et avant-gardiste s'intéressait beaucoup plus au style de vie du conducteur qu'à la Capri elle-même. Elle préfigurait une Amérique des Fifties focalisée sur les agréments offerts par la société de consommation : commodités domestiques, course aux antiquités, navigation de plaisance, entre autres plaisirs très « milieu de siècle ». Le parti-pris d'exclusivité de la Lincoln tombait à point nommé.

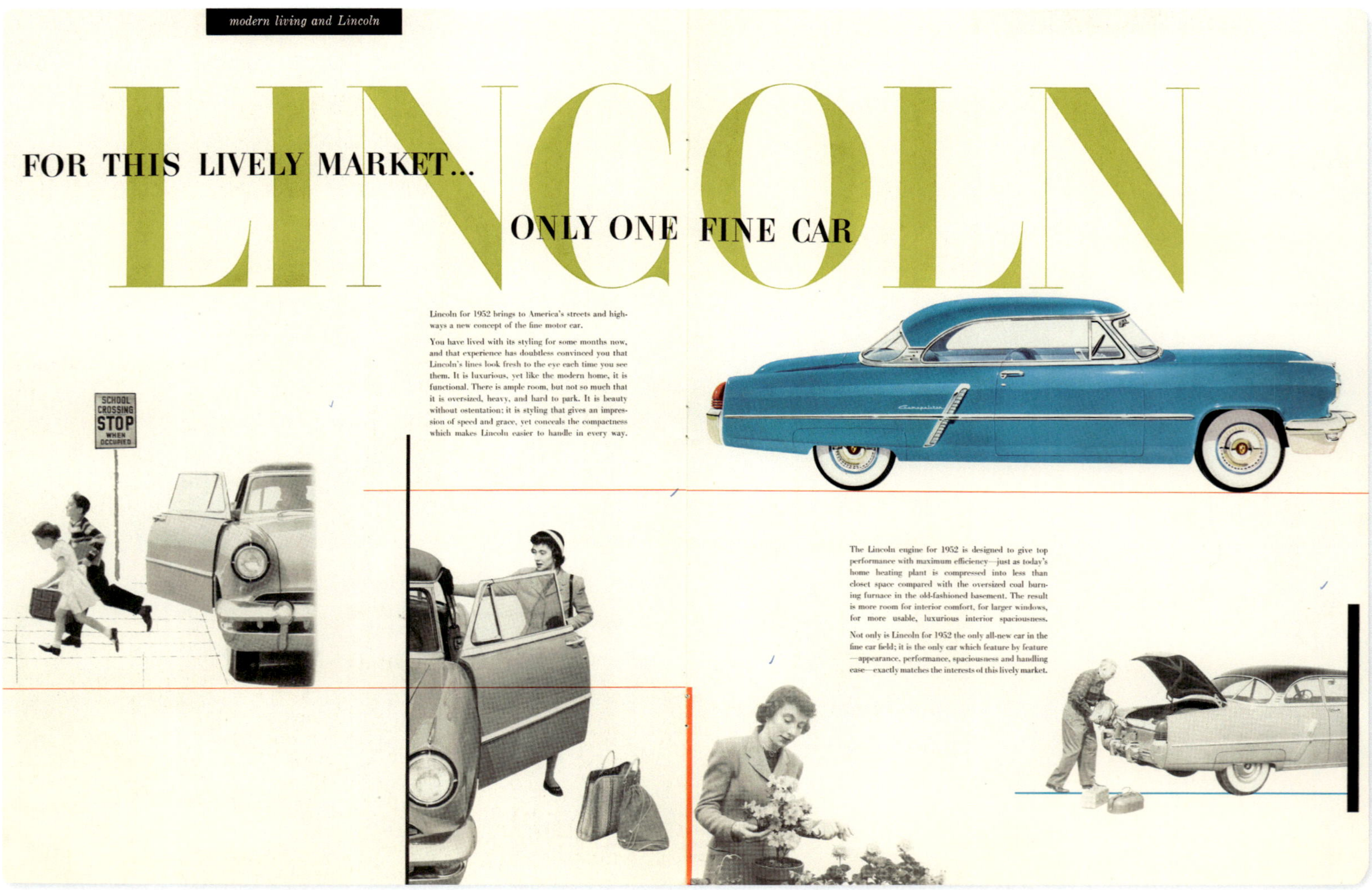

Lincoln 1951

The sophisticated and up-to-the-minute concept of "Modern Living" suggested the Capri's colors and fabrics could satisfy the most discerning customer. Accessorizing the car, another personal statement, would come to be known as a "profit center" among dealers.

Das kultivierte und hochaktuelle Konzept von „Modern Living" suggerierte, dass die Lackierungen und Bezüge des Capri auch den anspruchsvollsten Kunden zufriedenstellen würden. Die Sonderausstattung eines Autos – ein weiterer Ausdruck des persönlichen Geschmacks – wurde für Händler bald zu einer einträglichen Einnahmequelle.

À en croire le concept hautement sophistiqué et tout récent du « modern living » (la vie moderne), les couleurs et tissus associés à la Capri combleront le plus exigeant des clients. Les concessionnaires comprendront vite que l'accessoirisation de la voiture, qui elle aussi permet aux personnalités de s'affirmer, représente un vrai « centre de profit ».

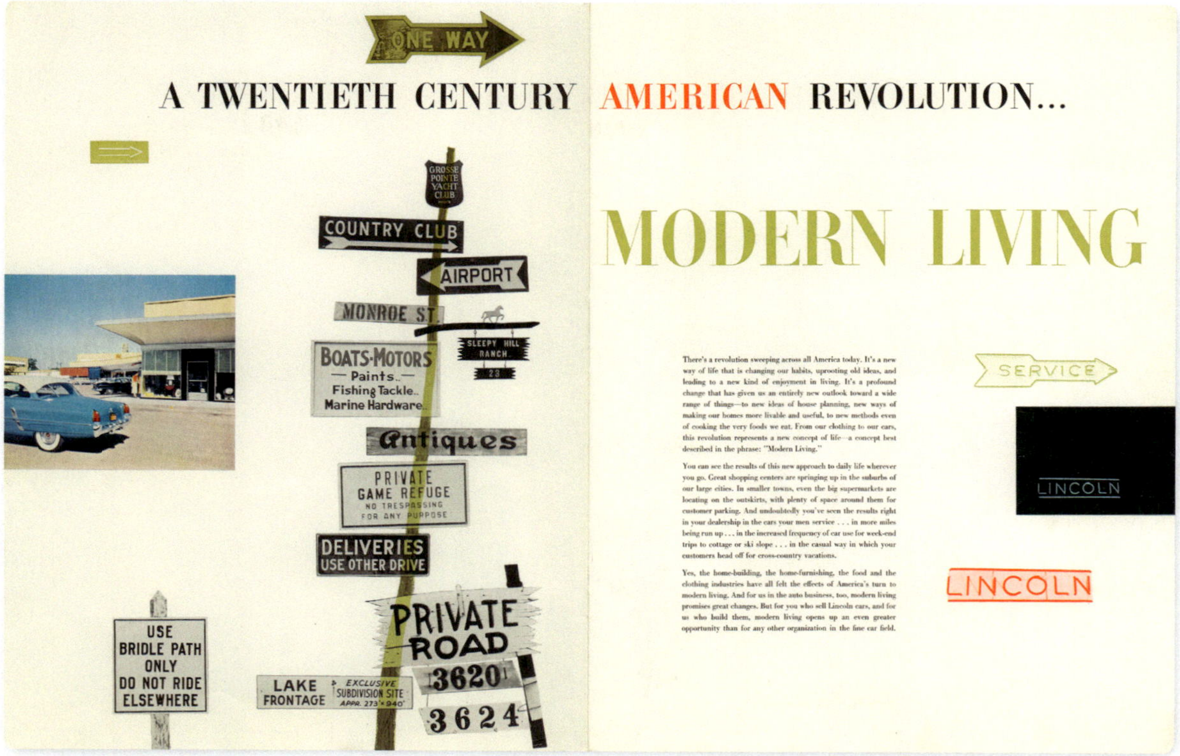

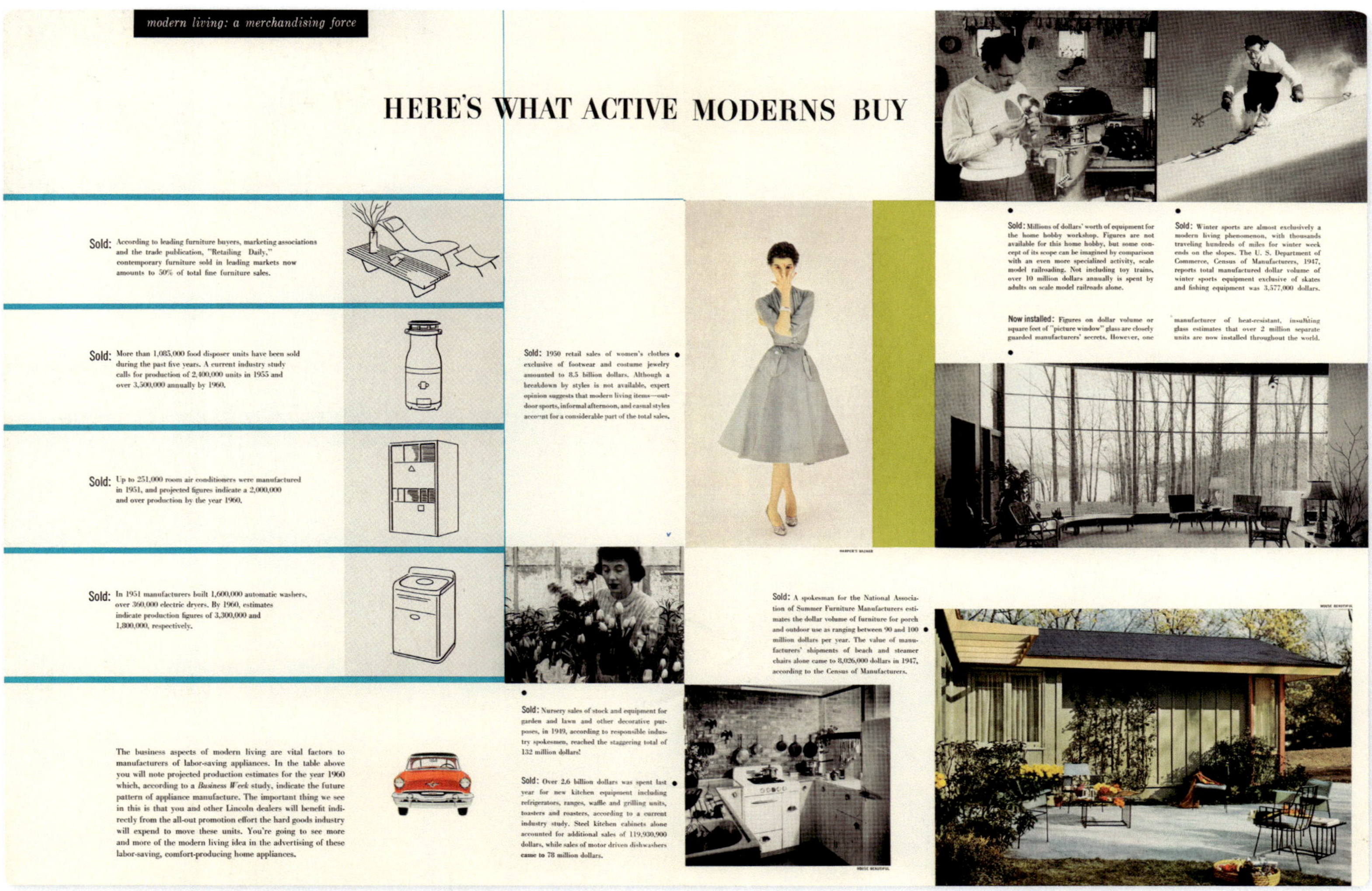

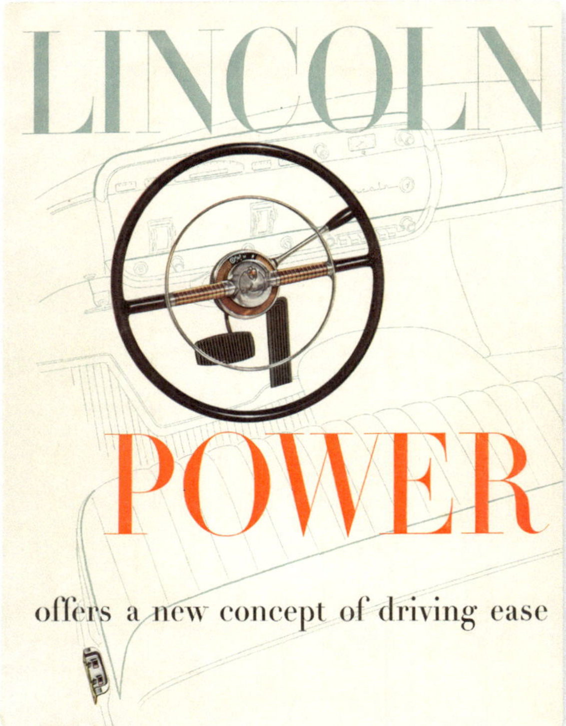

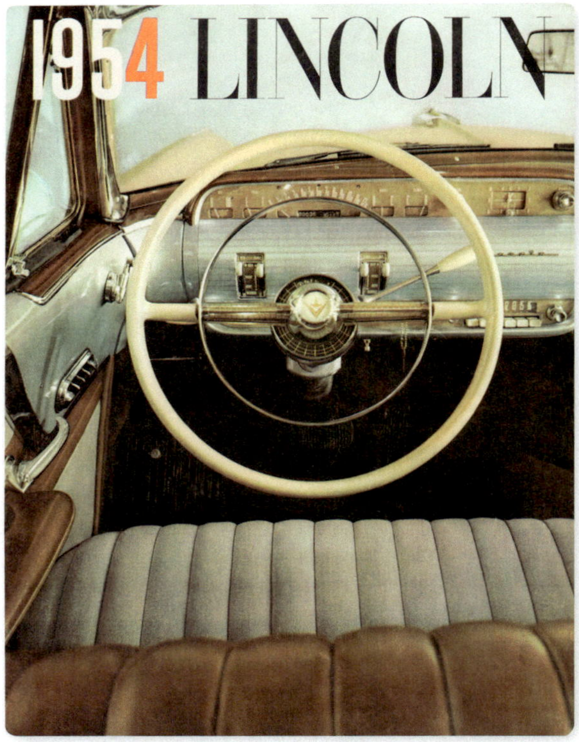

Lincoln 1953 *(far left and opposite)*, **1954** *(left and below)*

The cars are one thing, but the advertising and marketing materials for Ford's 1950s Lincoln lineup project a clearly superior flair for graphics and design. These were huge, powerful cars bred by racing in international events, such as the Carrera Panamericana of Mexico a brutal long-distance road race. Performance was a selling point for luxury cars until later in the 1950s.

Die Fahrzeuge selbst sind eine Sache, aber die Materialien für die Werbung und Vermarktung der 1950er-Lincolns zeigen, dass Ford auch ein überragendes Gespür für Grafik und Design besaß. Die beworbenen Fahrzeuge waren leistungsstarke Autos, die bei internationalen Motorsportveranstaltungen herangezüchtet worden waren, wie etwa bei der Carrera Panamericana in Mexiko, einem brutalen Langstreckenstraßenrennen. Bis in die späten 1950er-Jahre wurden Luxusautos über ihre Leistungsstärke verkauft.

La publicité et le matériel marketing utilisé par Ford pour la gamme Lincoln des années 1950 révèlent des dispositions hors-pair pour le graphisme et le design. Ces voitures étaient aussi massives que puissantes, endurcies par leur participation à des compétitions internationales comme la Carrera Panamericana mexicaine, particulièrement éprouvante. Jusqu'à la fin des années 1950, la performance constituera un argument de vente sur le segment du luxe.

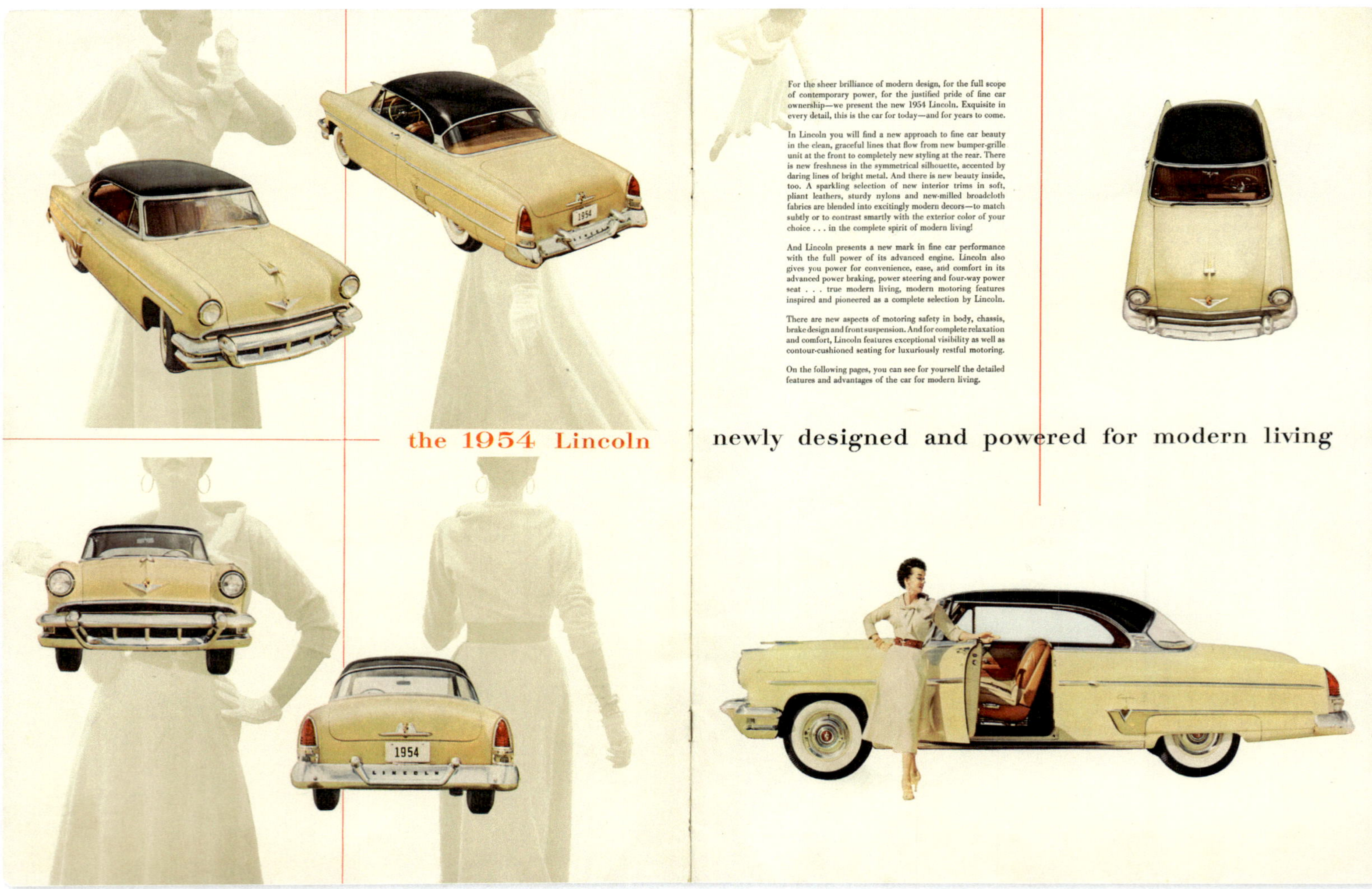

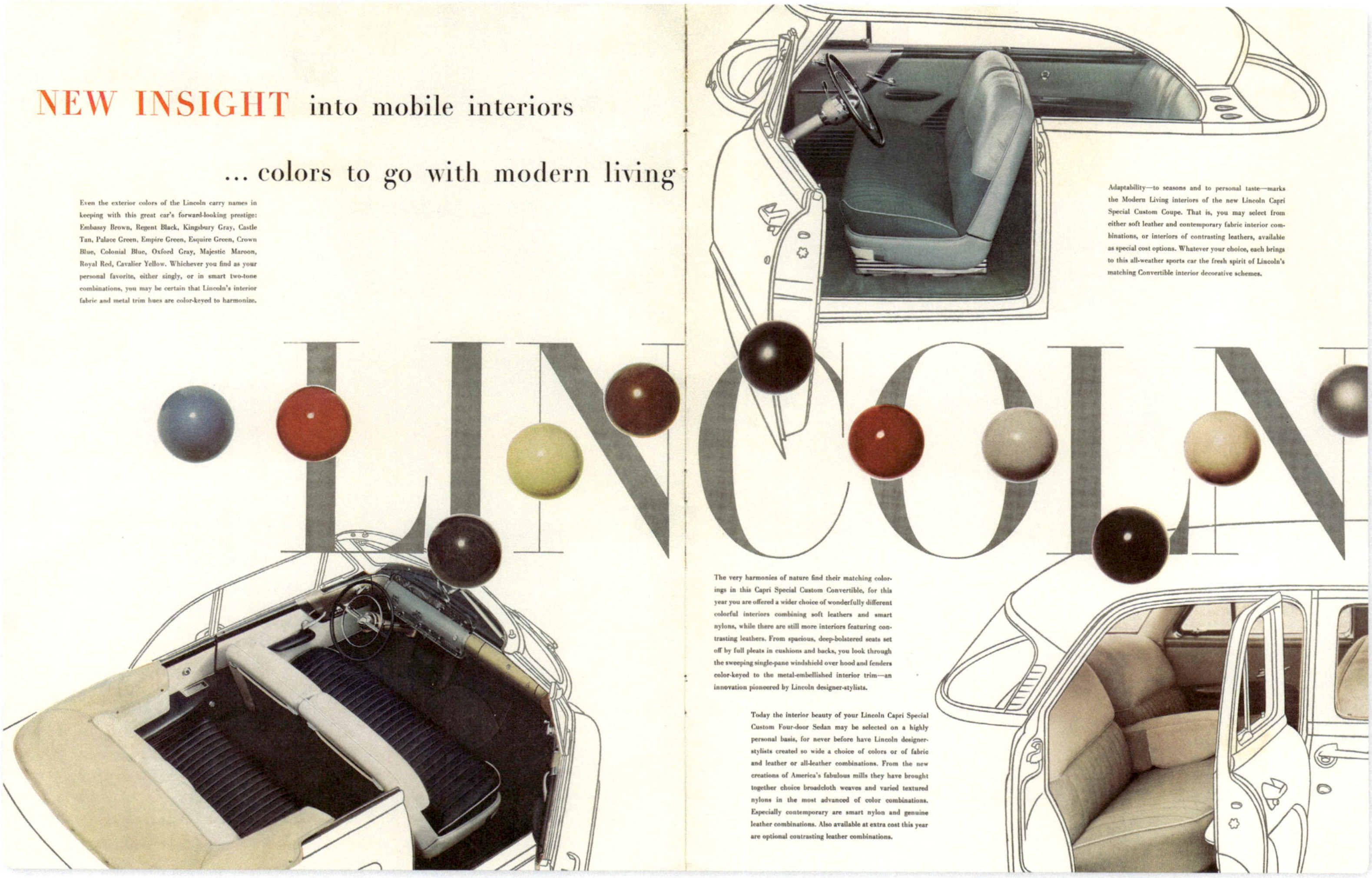
NEW INSIGHT into mobile interiors
... colors to go with modern living
Even the exterior colors of the Lincoln carry names in keeping with this great car's forward-looking prestige: Embassy Brown, Regent Black, Kingsbury Gray, Castle Tan, Palace Green, Empire Green, Esquire Green, Crown Blue, Colonial Blue, Oxford Gray, Majestic Maroon, Royal Red, Cavalier Yellow. Whichever you find as your personal favorite, either singly, or in smart two-tone combinations, you may be certain that Lincoln's interior fabric and metal trim hues are color-keyed to harmonize.
Adaptability—to seasons and to personal taste—marks the Modern Living interiors of the new Lincoln Capri Special Custom Coupe. That is, you may select from either soft leather and contemporary fabric interior combinations, or interiors of contrasting leathers, available as special cost options. Whatever your choice, each brings to this all-weather sports car the fresh spirit of Lincoln's matching Convertible interior decorative schemes.
LINCOLN
The very harmonies of nature find their matching colorings in this Capri Special Custom Convertible, for this year you are offered a wider choice of wonderfully different colorful interiors combining soft leathers and smart nylons, while there are still more interiors featuring contrasting leathers. From spacious, deep-bolstered seats set off by full pleats in cushions and backs, you look through the sweeping single-pane windshield over hood and fenders color-keyed to the metal-embellished interior trim—an innovation pioneered by Lincoln designer-stylists.
Today the interior beauty of your Lincoln Capri Special Custom Four-door Sedan may be selected on a highly personal basis, for never before have Lincoln designer-stylists created so wide a choice of colors or of fabric and leather or all-leather combinations. From the new creations of America's fabulous mills they have brought together choice broadcloth weaves and varied textured nylons in the most advanced of color combinations. Especially contemporary are smart nylon and genuine leather combinations. Also available at extra cost this year are optional contrasting leather combinations.

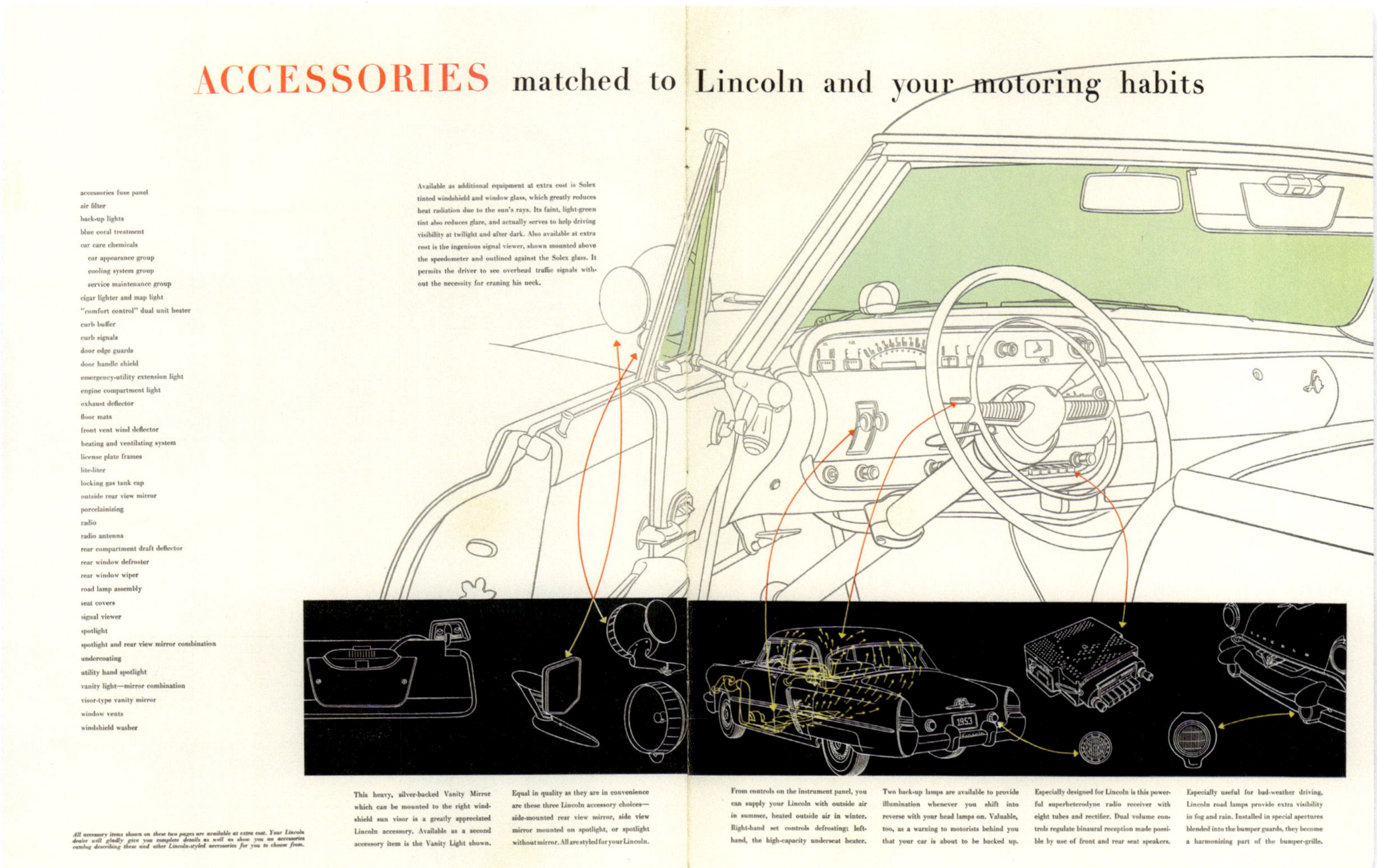
ACCESSORIES matched to Lincoln and your motoring habits
accessories fuse panel
air filter
back-up lights
blue coral treatment
car care chemicals
car appearance group
cooling system group
service maintenance group
cigar lighter and map light
"comfort control" dual unit heater
curb buffer
curb signals
door edge guards
door handle shield
emergency-utility extension light
engine compartment light
exhaust deflector
floor mats
front vent wind deflector
heating and ventilating system
license plate frames
lite-liter
locking gas tank cap
outside rear view mirror
porcelainizing
radio
radio antenna
rear compartment draft deflector
rear window defroster
rear window wiper
road lamp assembly
seat covers
signal viewer
spotlight
spotlight and rear view mirror combination
undercoating
utility hand spotlight
vanity light—mirror combination
visor-type vanity mirror
window vents
windshield washer
Available as additional equipment at extra cost is Solex tinted windshield and window glass, which greatly reduces heat radiation due to the sun's rays. Its faint, light-green tint also reduces glare, and actually serves to help driving visibility at twilight and after dark. Also available at extra cost is the ingenious signal viewer, shown mounted above the speedometer and outlined against the Solex glass. It permits the driver to see overhead traffic signals without the necessity for craning his neck.
1953
All accessory items shown on these two pages are available at extra cost. Your Lincoln dealer will gladly give you complete details as well as show you an accessories catalog describing these and other Lincoln-styled accessories for you to choose from.
This heavy, silver-backed Vanity Mirror which can be mounted to the right windshield sun visor is a greatly appreciated Lincoln accessory. Available as a second accessory item is the Vanity Light shown.
Equal in quality as they are in convenience are these three Lincoln accessory choices—side-mounted rear view mirror, side view mirror mounted on spotlight, or spotlight without mirror. All are styled for your Lincoln.
From controls on the instrument panel, you can supply your Lincoln with outside air in summer, heated outside air in winter. Right-hand set controls defrosting; left-hand, the high-capacity underseat heater.
Two back-up lamps are available to provide illumination whenever you shift into reverse with your head lamps on. Valuable, too, as a warning to motorists behind you that your car is about to be backed up.
Especially designed for Lincoln is this powerful superheterodyne radio receiver with eight tubes and rectifier. Dual volume controls regulate binaural reception made possible by use of front and rear seat speakers.
Especially useful for bad-weather driving, Lincoln road lamps provide extra visibility in fog and rain. Installed in special apertures blended into the bumper guards, they become a harmonizing part of the bumper-grille.

Lincoln 1958 *(left)*, **1960** *(below)*

Lincoln saved its biggest, most ornate cars for the end of the 1950s. The Mark IV featured a rearward-sloping back window that could be power-lowered. The canted quad headlamps remain unforgettable. These enormous cars easily weighed more than 5,000 pounds.

Lincoln hob sich seine größten und verschnörkeltsten Autos für die späten 1950er-Jahre auf. Der Continental Mark IV besaß ein leicht nach oben hin schräg gestelltes Heckfenster, das sich elektrisch absenken ließ. Die versetzten Vierfachscheinwerfer sind bis heute unvergesslich geblieben. Diese Kolosse brachten locker 2,3 Tonnen und mehr auf die Waage.

Lincoln avait gardé ses modèles les plus imposants et les plus ornementés pour la fin des années 1950. La Mark IV comportait ainsi une lunette arrière à pente inversée, réglable électriquement, et quatre phares de guingois qui demeurent inoubliables. Ces colossales automobiles pesaient facilement plus de 2,2 tonnes.

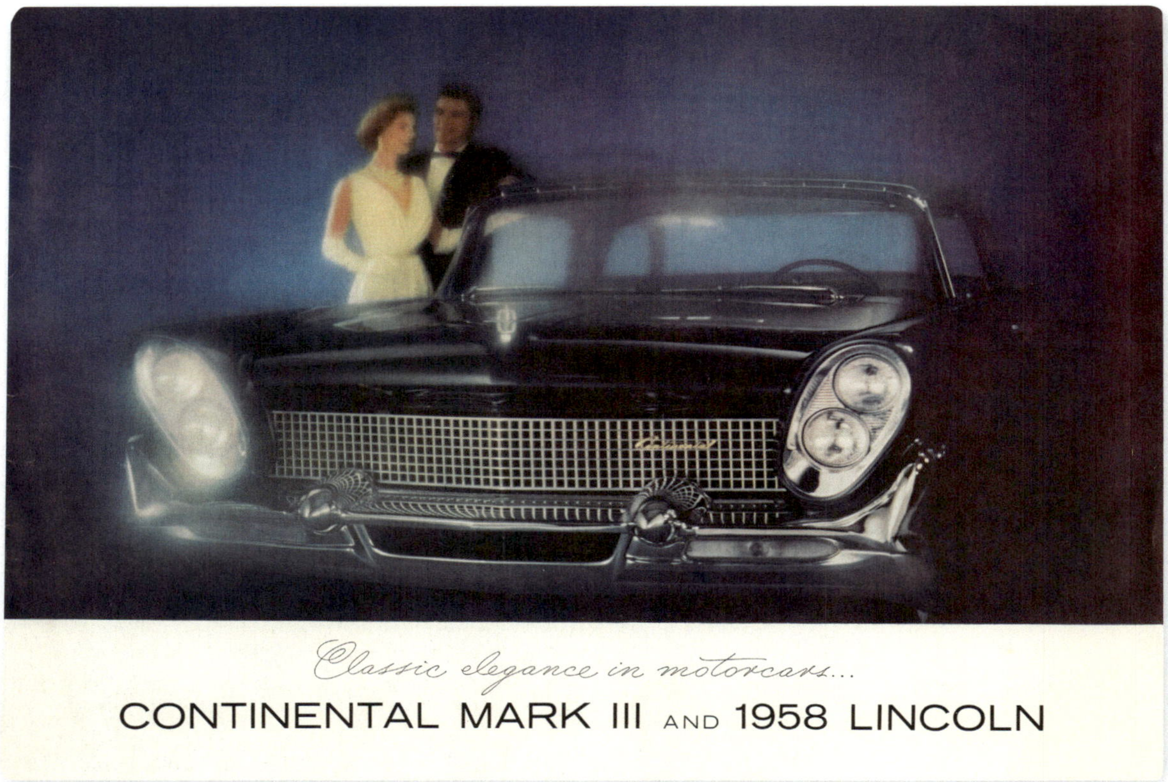

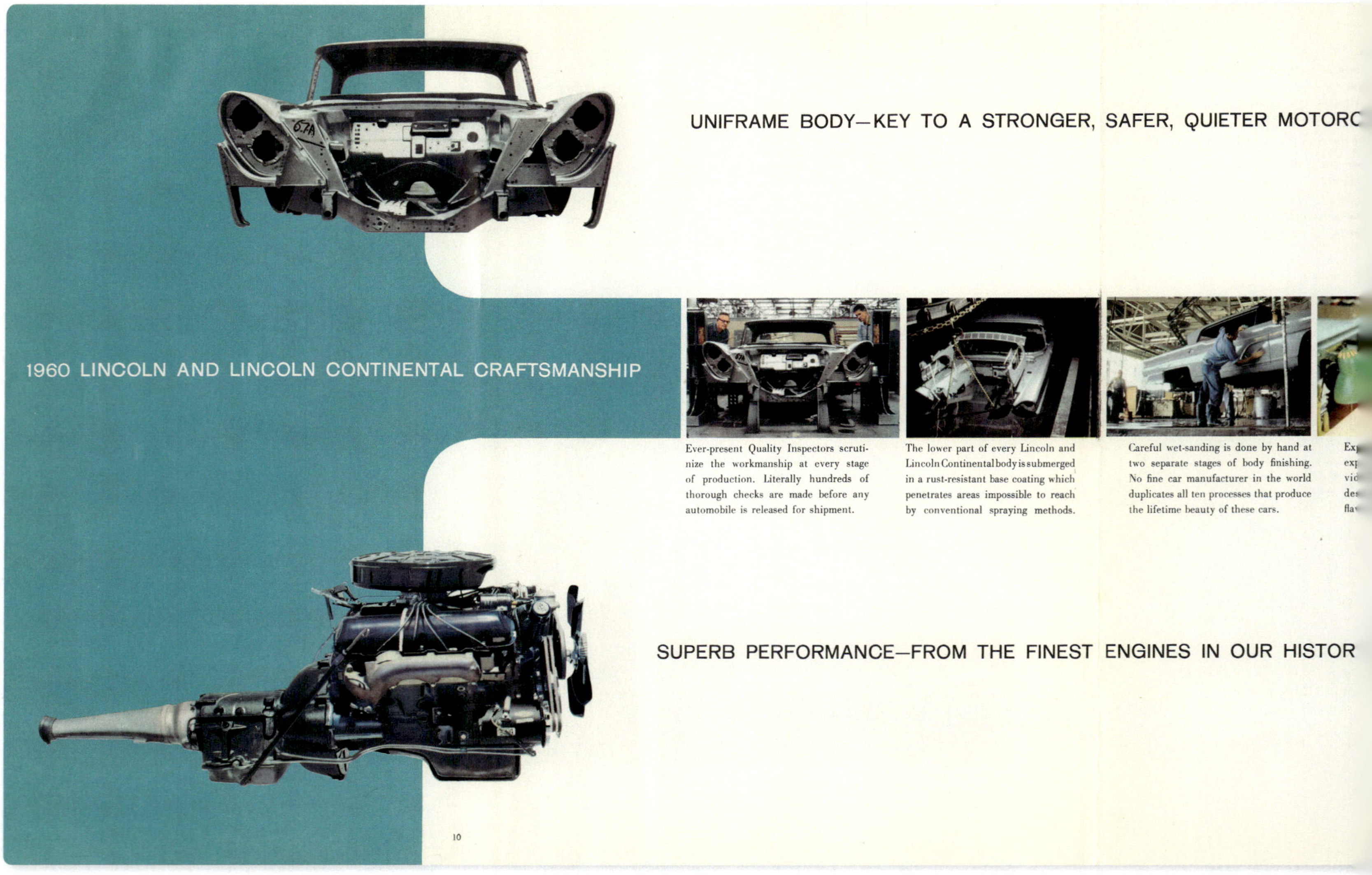

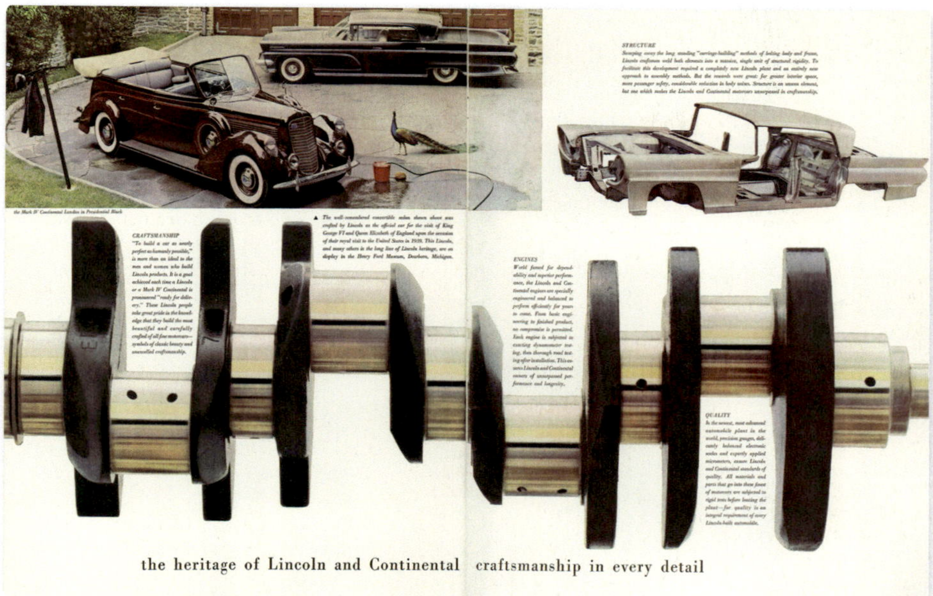

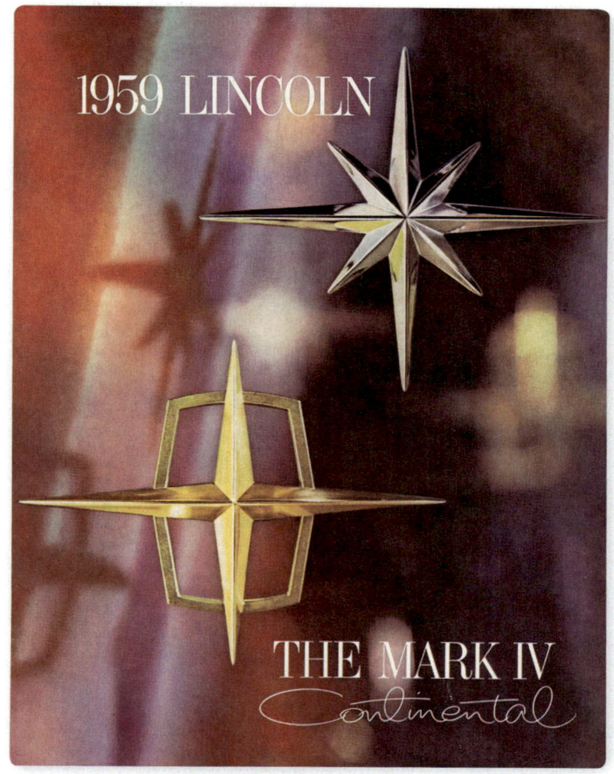

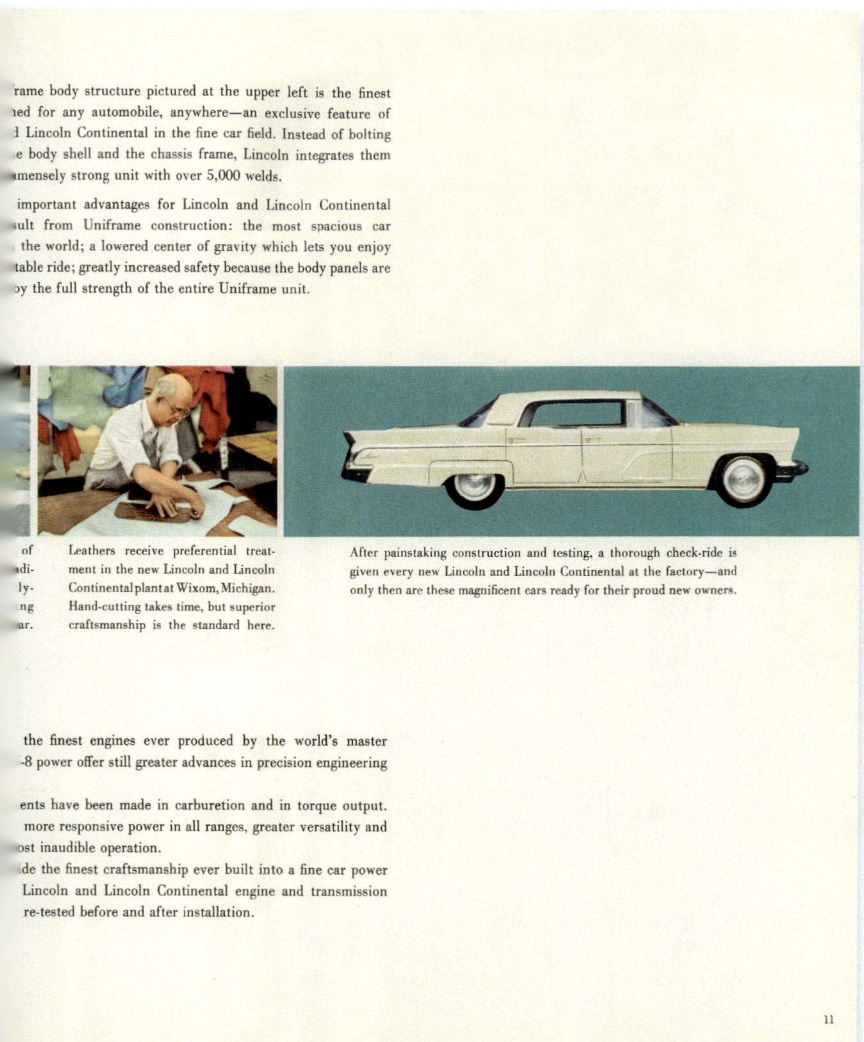
rame body structure pictured at the upper left is the finest
ıed for any automobile, anywhere—an exclusive feature of
l Lincoln Continental in the fine car field. Instead of bolting
e body shell and the chassis frame, Lincoln integrates them
ımensely strong unit with over 5,000 welds.

important advantages for Lincoln and Lincoln Continental
ult from Uniframe construction: the most spacious car
the world; a lowered center of gravity which lets you enjoy
table ride; greatly increased safety because the body panels are
by the full strength of the entire Uniframe unit.

of
adi-
ly-
ng
ar.

Leathers receive preferential treatment in the new Lincoln and Lincoln Continental plant at Wixom, Michigan. Hand-cutting takes time, but superior craftsmanship is the standard here.

After painstaking construction and testing, a thorough check-ride is given every new Lincoln and Lincoln Continental at the factory—and only then are these magnificent cars ready for their proud new owners.

the finest engines ever produced by the world's master
-8 power offer still greater advances in precision engineering

ents have been made in carburetion and in torque output.
more responsive power in all ranges, greater versatility and
ost inaudible operation.
de the finest craftsmanship ever built into a fine car power
Lincoln and Lincoln Continental engine and transmission
re-tested before and after installation.

11

Lincoln 1959 *(above)*

Lincoln proudly proclaimed the links not just to this Model K phaeton of the late 1930s, but to the attention paid to paint, trim, and interior work during assembly. Given the Lincoln's weight, it needed a huge engine, a 430-cubic-inch workhorse producing 350 hp.

Lincoln verwies nicht nur voller Stolz auf die Verbindungen zu diesem Phaeton mit der Bezeichnung Model K aus den 1930er-Jahren, sondern auch auf die Sorgfalt, die man Lackierung, Verkleidung und Innenausstattung bei der Montage angedeihen ließ. Angesichts seines hohen Gewichts brauchte der Lincoln freilich auch einen kräftigen Motor – und bekam daher ein 7-Liter-Arbeitspferd verpasst, das 475 kW auf die Straße brachte.

Avec fierté, Lincoln revendiquait non seulement la filiation avec cette Model K Phaeton de la fin des années 1930, mais aussi le soin extrême accordé à la peinture, aux finitions et à l'aménagement intérieur. Compte tenu du poids de la Lincoln, il lui fallait un énorme moteur, en l'occurrence une bête de somme de 7 litres capable de délivrer 350 chevaux.

Mercedes-Benz

Mercedes-Benz 1959 *(this spread)*

Introduced in 1953, these rounded-bodied cars were called "pontons" for their pontoon shape. They were sophisticated, with unibody construction and a swing-axle rear suspension. The tone of the company's brochures would soon get more aggressive as Mercedes-Benz would claim to be the world's best-engineered car.

Die abgerundete Karosserie, die Mercedes-Benz 1953 bei seinen Modellen einführte, nannte sich aufgrund ihrer Form „Pontonkarosserie". Es waren elegante Fahrzeuge mit selbsttragender Karosserie und Pendelachse hinten. Der Tonfall der Firmenbroschüren wurde bald danach schärfer, als Mercedes-Benz behauptete, die technisch ausgereiftesten Automobile der Welt zu bauen.

Présentées en 1953, ces berlines tout en rondeurs étaient baptisées « pontons », en référence à la forme des pontons flottants. Elles étaient particulièrement sophistiquées, avec leur structure autoporteuse – châssis-carrosserie solidaires – et une suspension arrière à un essieu oscillant. La tonalité des brochures de Mercedes-Benz se fera bientôt plus agressive lorsque la marque se targuera de produire les voitures les mieux conçues au monde.

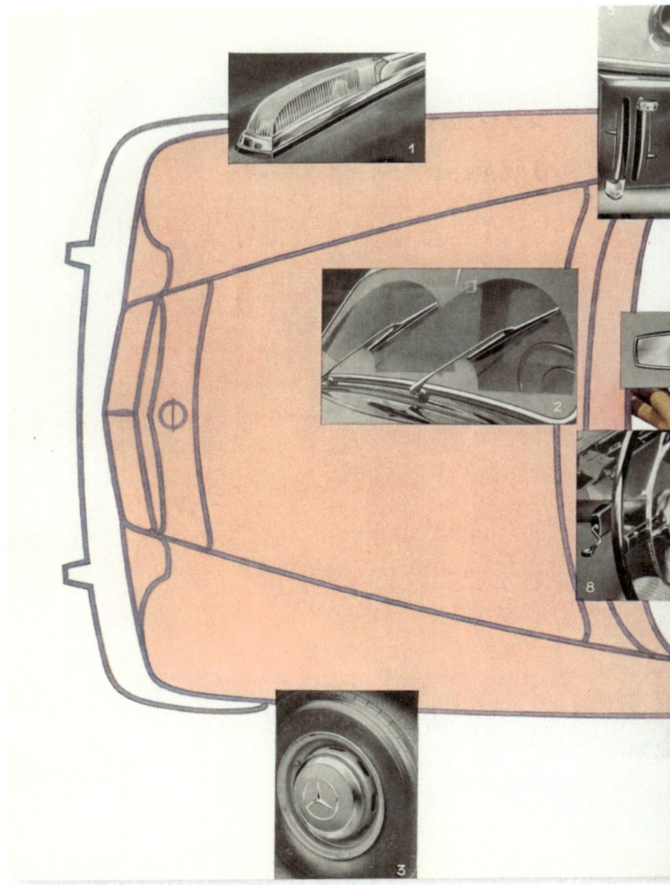

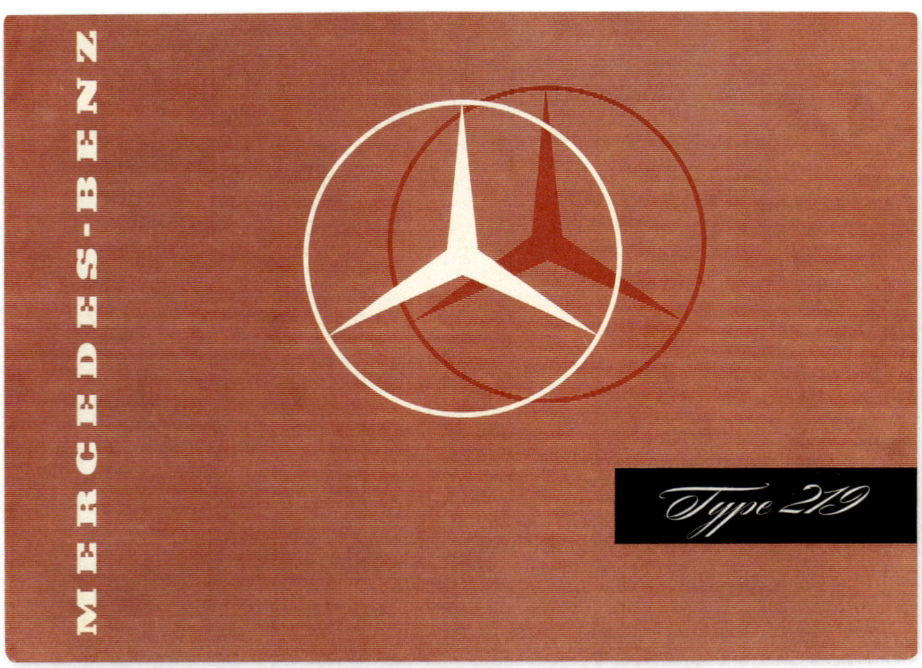

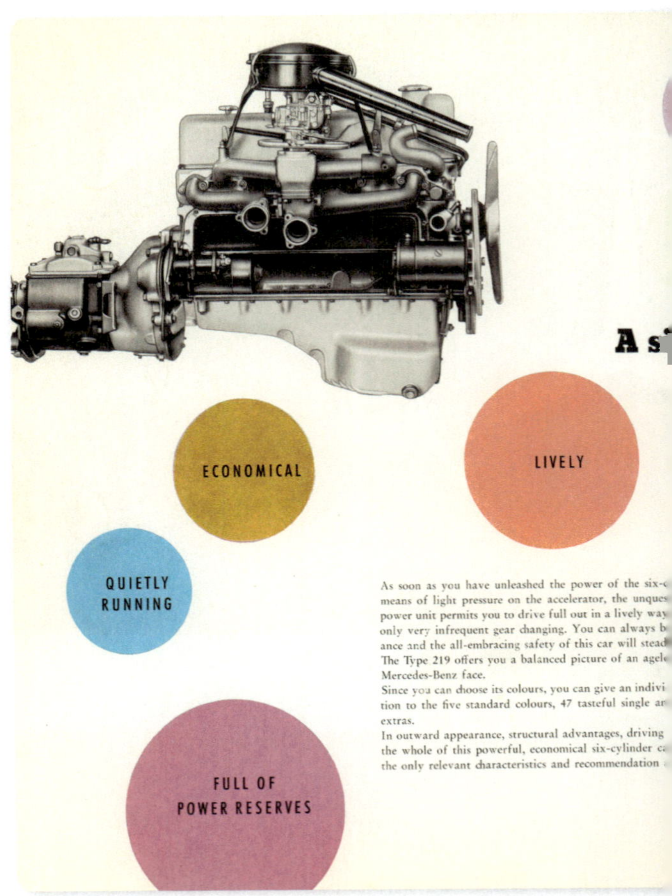

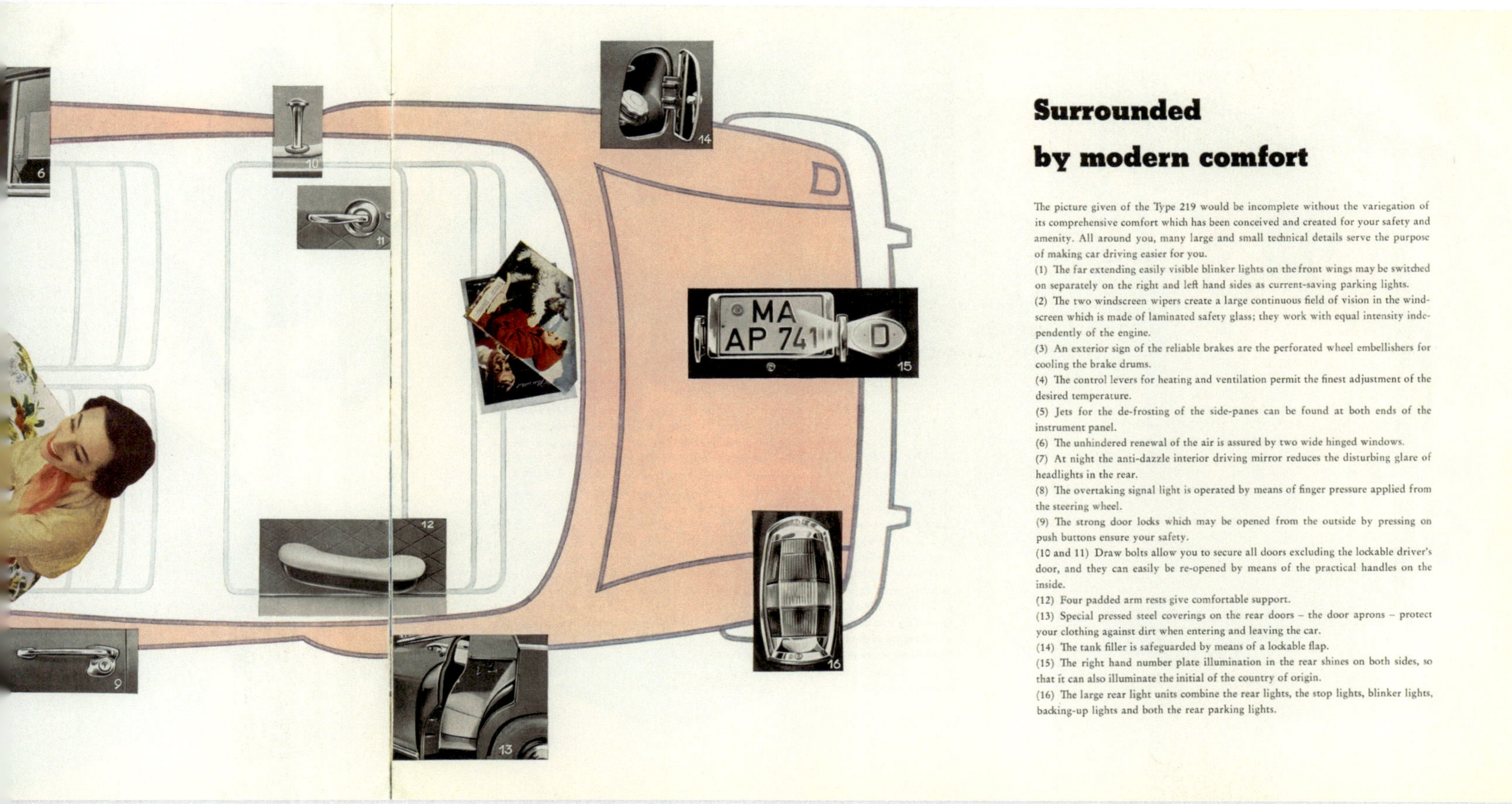

Surrounded by modern comfort

The picture given of the Type 219 would be incomplete without the variegation of its comprehensive comfort which has been conceived and created for your safety and amenity. All around you, many large and small technical details serve the purpose of making car driving easier for you.

(1) The far extending easily visible blinker lights on the front wings may be switched on separately on the right and left hand sides as current-saving parking lights.

(2) The two windscreen wipers create a large continuous field of vision in the windscreen which is made of laminated safety glass; they work with equal intensity independently of the engine.

(3) An exterior sign of the reliable brakes are the perforated wheel embellishers for cooling the brake drums.

(4) The control levers for heating and ventilation permit the finest adjustment of the desired temperature.

(5) Jets for the de-frosting of the side-panes can be found at both ends of the instrument panel.

(6) The unhindered renewal of the air is assured by two wide hinged windows.

(7) At night the anti-dazzle interior driving mirror reduces the disturbing glare of headlights in the rear.

(8) The overtaking signal light is operated by means of finger pressure applied from the steering wheel.

(9) The strong door locks which may be opened from the outside by pressing on push buttons ensure your safety.

(10 and 11) Draw bolts allow you to secure all doors excluding the lockable driver's door, and they can easily be re-opened by means of the practical handles on the inside.

(12) Four padded arm rests give comfortable support.

(13) Special pressed steel coverings on the rear doors – the door aprons – protect your clothing against dirt when entering and leaving the car.

(14) The tank filler is safeguarded by means of a lockable flap.

(15) The right hand number plate illumination in the rear shines on both sides, so that it can also illuminate the initial of the country of origin.

(16) The large rear light units combine the rear lights, the stop lights, blinker lights, backing-up lights and both the rear parking lights.

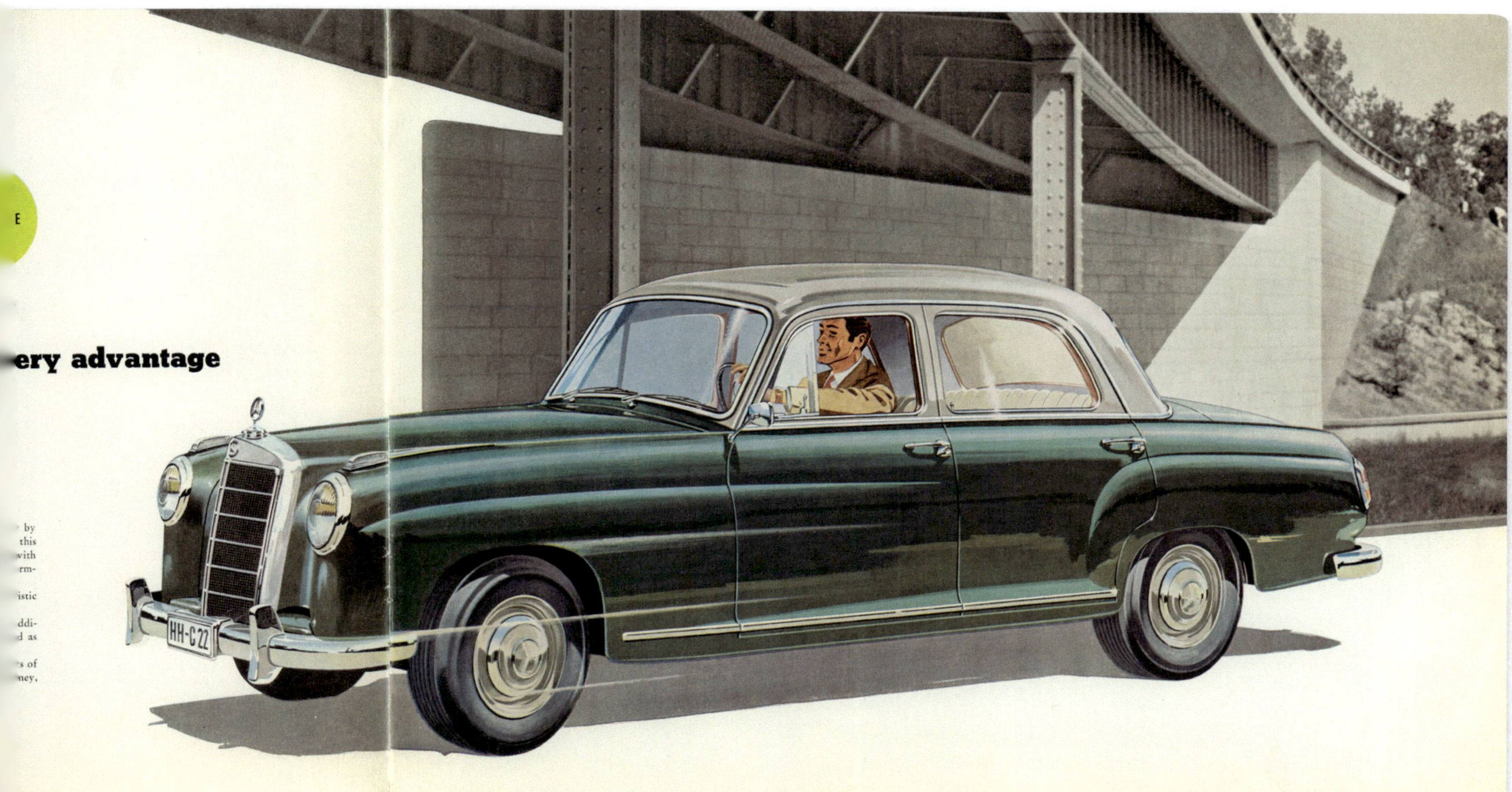

Mercury

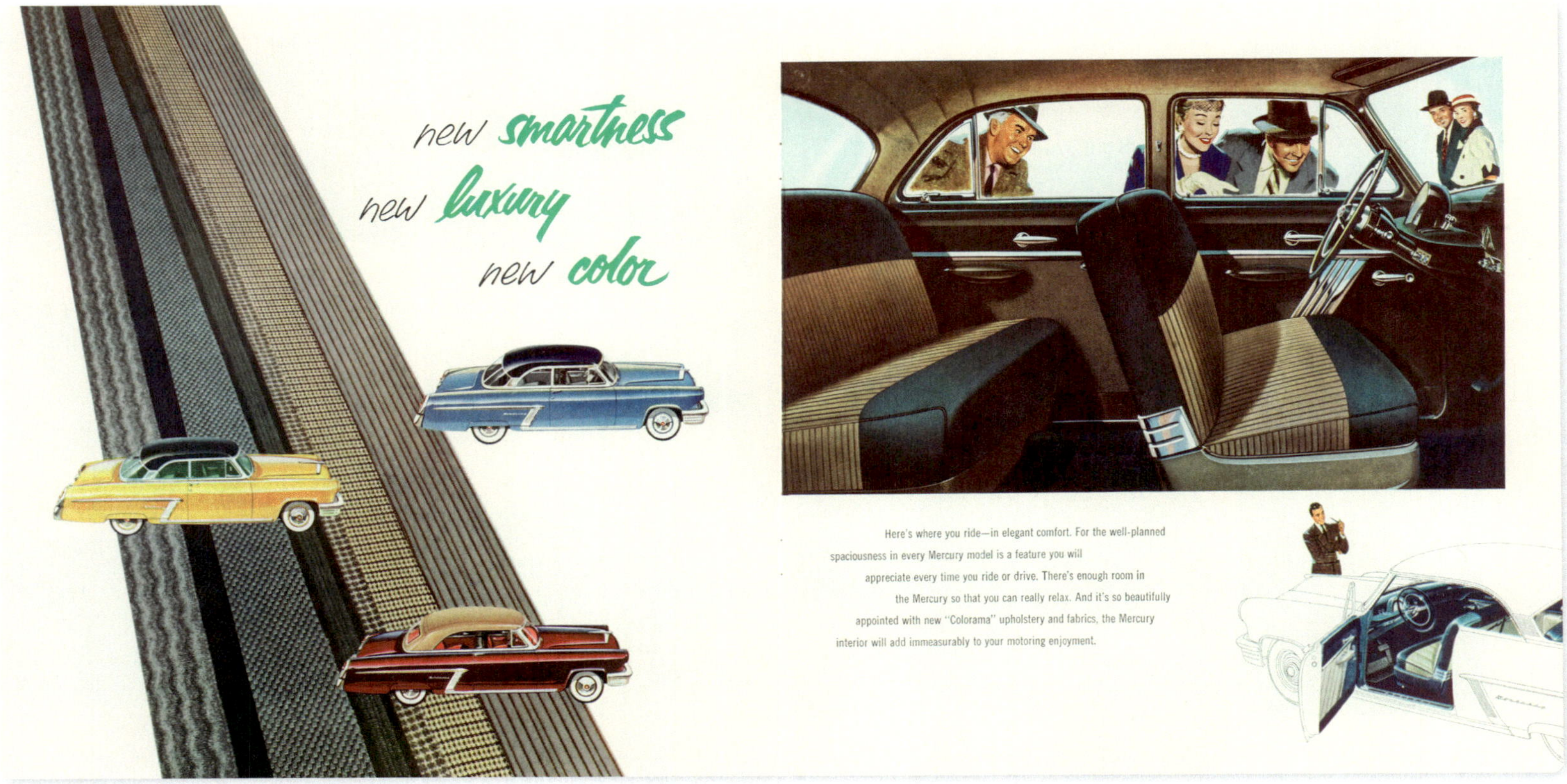

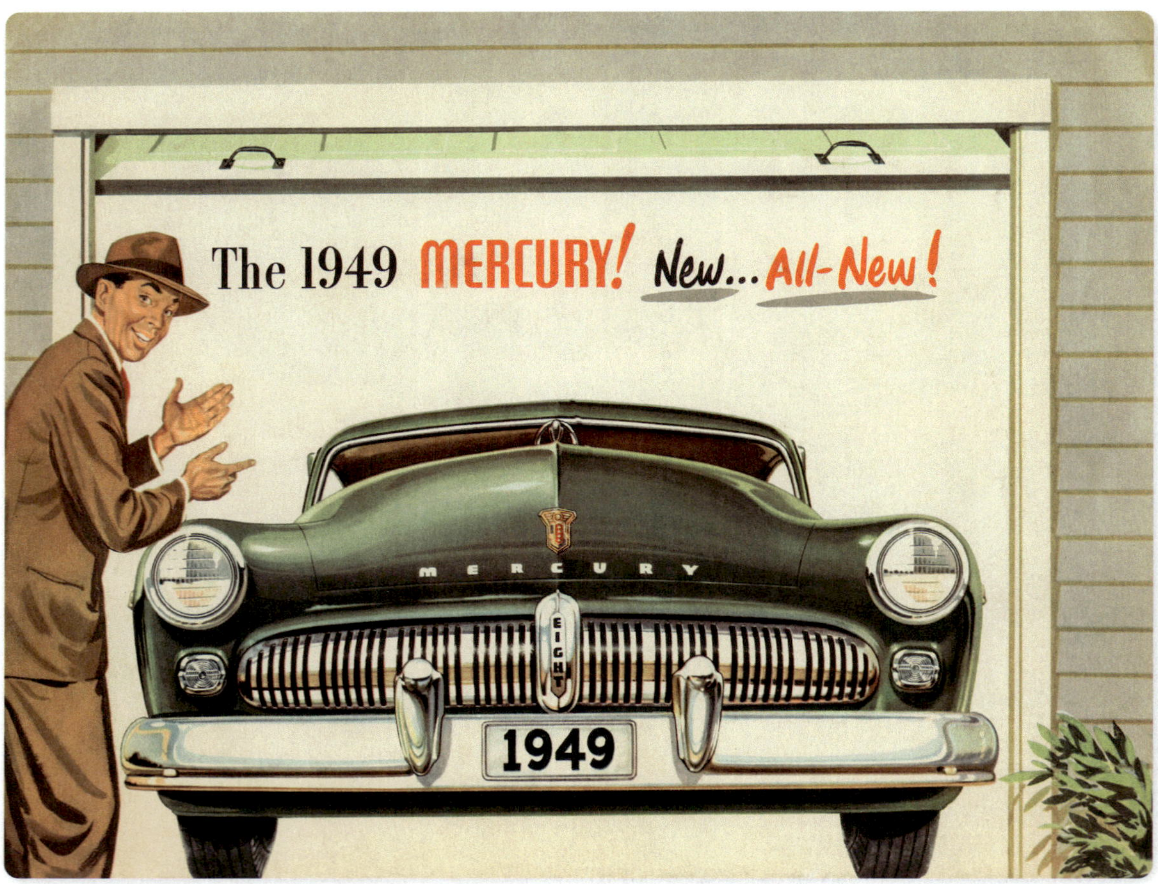

Mercury 1952 *(above)*, **1949** *(left)*, **1946** *(opposite top)*, **1957** *(opposite bottom)*

Mercury's stunning 1949 models no longer looked like bechromed Fords, but instead shared the new Lincoln's body shell. Despite a nearly perfect shape, they have been customizing favorites since they first appeared.

Die atemberaubenden Mercury-Modelle von 1949 sahen nicht mehr wie chrombeschlagene Fords aus, sondern waren mit Karosserien von Lincoln ausgestattet. Trotz ihrer nahezu perfekten Form wurden sie seit ihrem ersten Erscheinen von den Käufern gerne individuell modifiziert.

Les saisissantes Mercury de 1949 ne ressemblaient plus à des Ford auxquelles on aurait prescrit une cure de chrome mais partageaient leur caisse avec la nouvelle Lincoln. Malgré une allure proche de la perfection, elles sont devenues dès leur apparition les proies favorites des customiseurs.

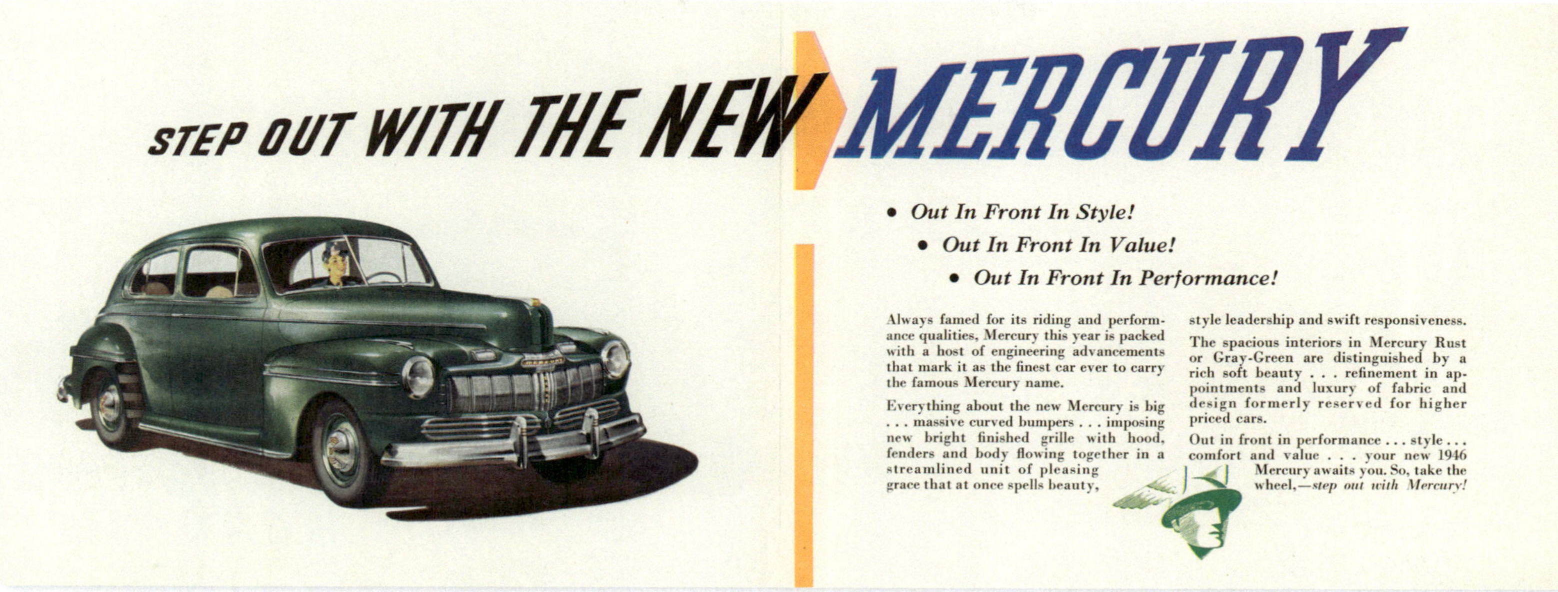

STEP OUT WITH THE NEW MERCURY
• Out In Front In Style!
• Out In Front In Value!
• Out In Front In Performance!
Always famed for its riding and performance qualities, Mercury this year is packed with a host of engineering advancements that mark it as the finest car ever to carry the famous Mercury name.
Everything about the new Mercury is big . . . massive curved bumpers . . . imposing new bright finished grille with hood, fenders and body flowing together in a streamlined unit of pleasing grace that at once spells beauty, style leadership and swift responsiveness.
The spacious interiors in Mercury Rust or Gray-Green are distinguished by a rich soft beauty . . . refinement in appointments and luxury of fabric and design formerly reserved for higher priced cars.
Out in front in performance . . . style . . . comfort and value . . . your new 1946 Mercury awaits you. So, take the wheel,—step out with Mercury!

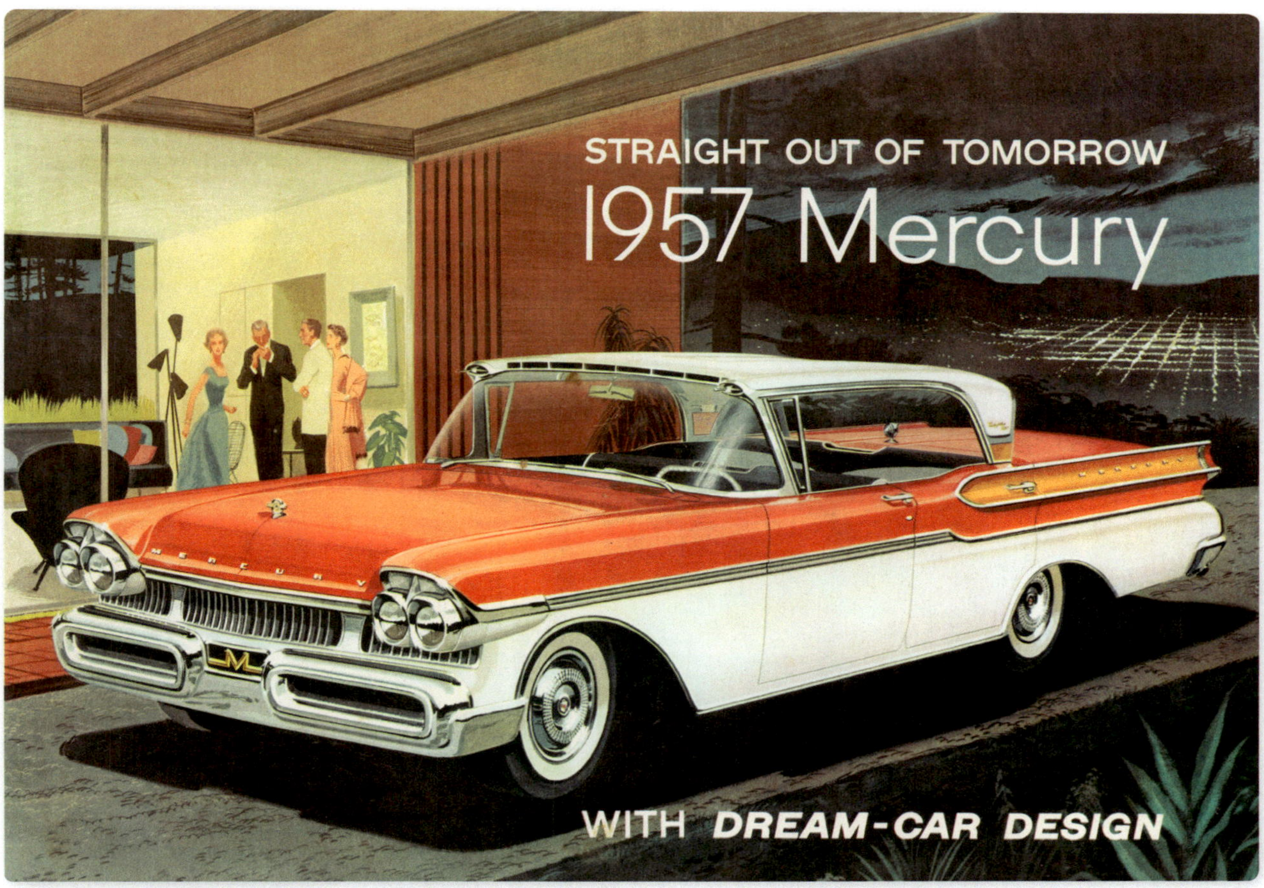

STRAIGHT OUT OF TOMORROW
1957 Mercury
WITH DREAM-CAR DESIGN

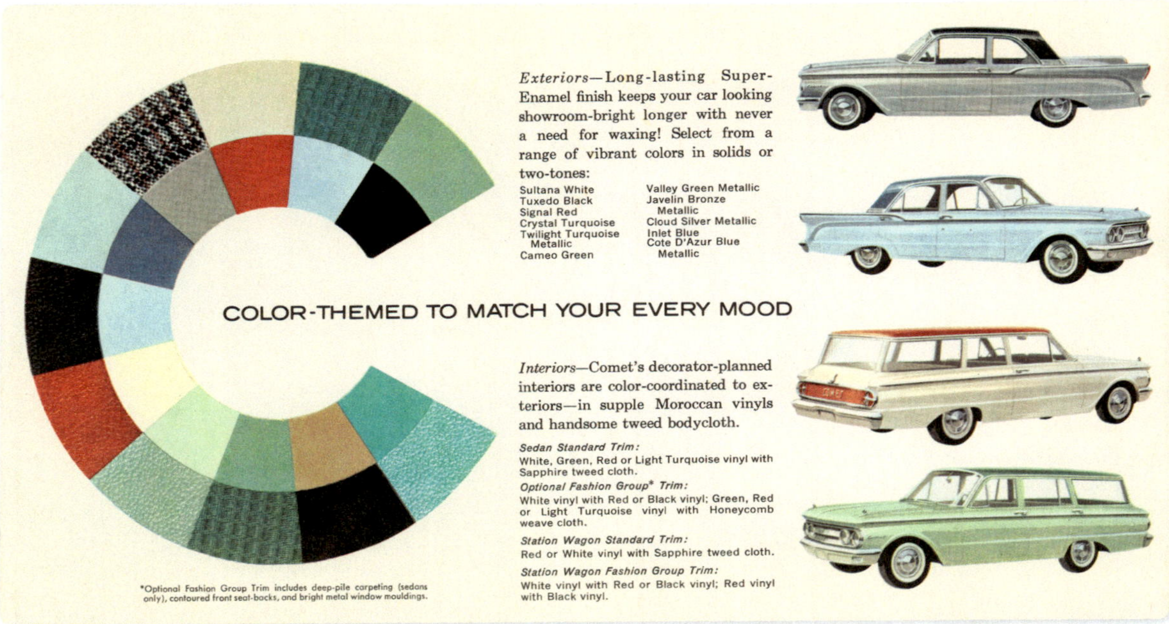

Mercury 1960

Mercury occupied a challenging slot between Ford and Lincoln. As 1960 neared, its cars grew ever larger and more complex. Mercury introduced the Comet in 1960, its first compact, with generous color and fabric choices.

Mercury besetzte eine reizvolle Nische zwischen Ford und Lincoln. Als die 1960er-Jahre näher rückten, wurden die Fahrzeuge noch größer und komplizierter. Im Jahr 1960 stellte Mercury seinen ersten Kompaktwagen, den Comet, vor, für den es eine breite Palette an Lackierungen und Bezügen gab.

Mercury occupait un créneau délicat entre Ford et Lincoln. Plus on se rapprochera de 1960, plus ses voitures gagneront en volume et en sophistication. Cette année-là Mercury présentera la Comet, sa première voiture compacte, déclinée dans de très riches gammes de couleurs et de tissus..

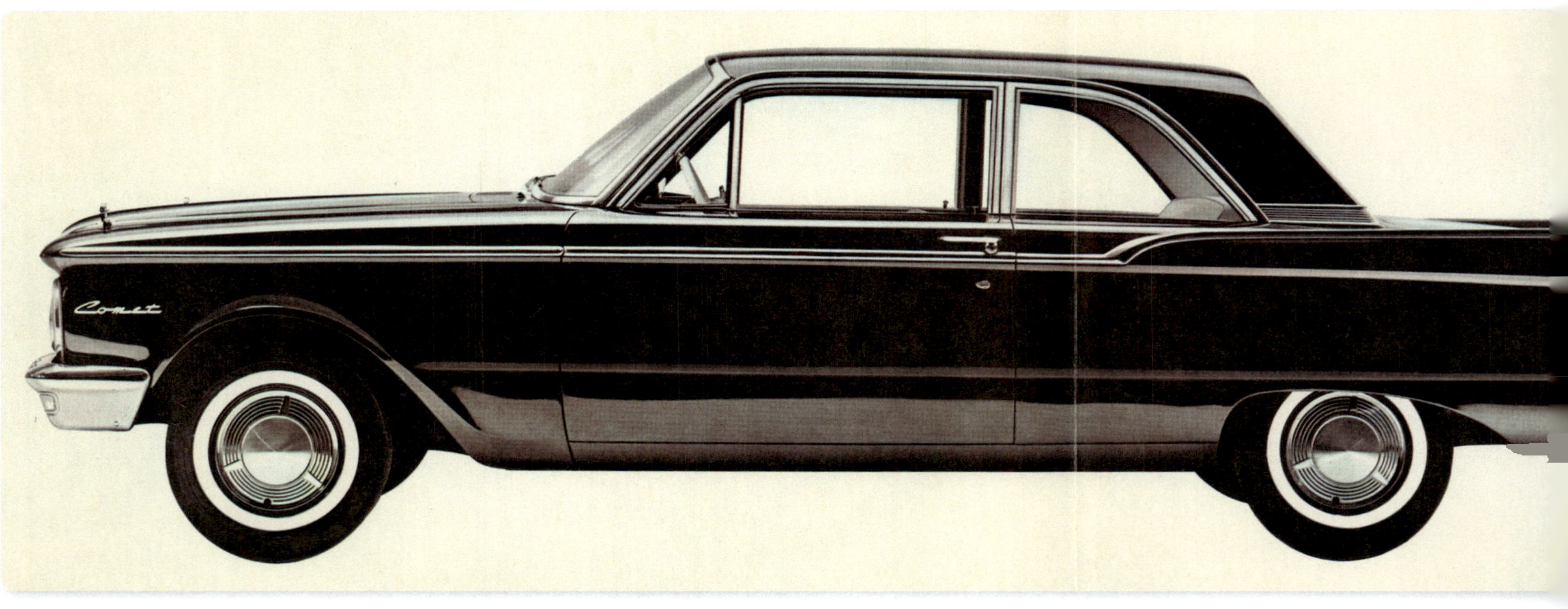

Mercury 1961 *(right)*, **1956** *(right bottom)*

Mercury built some highly imposing cars as Ford's intermediate brand, and in a variety of sizes. The traditional high-end hardtop existed in 1956 in the form of the Montclair, a durable model range. But by 1961, Mercury wanted in on the expanding compact market, and got there by spinning off a more heavily styled version of the existing Ford Falcon, which was dubbed the Mercury Comet.

Als Zwischenmarke von Ford baute Mercury ein paar äußerst eindrucksvolle Autos unterschiedlicher Größe. Das traditionell hochpreisige Coupé kam 1956 in Form des Montclair daher, einer langlebigen Modellreihe. Doch 1961 wollte Mercury unbedingt auch in die wachsende Kompaktklasse einsteigen und tat dies mit einer aufgemotzten Version des bereits existierenden Ford Falcon, die als Mercury Comet auf den Markt kam.

Marque intermédiaire du groupe Ford, Mercury a construit certaines voitures qui savaient en imposer, dans les tailles les plus diverses. On retrouvait par exemple en 1956, sur la Montclair, le traditionnel hard-top haut de gamme. Mais en 1961, pour pénétrer le marché émergent des modèles plus compacts, Mercury déclinera une version stylistiquement alourdie de la Ford Falcon existante, qui sera baptisée Mercury Comet.

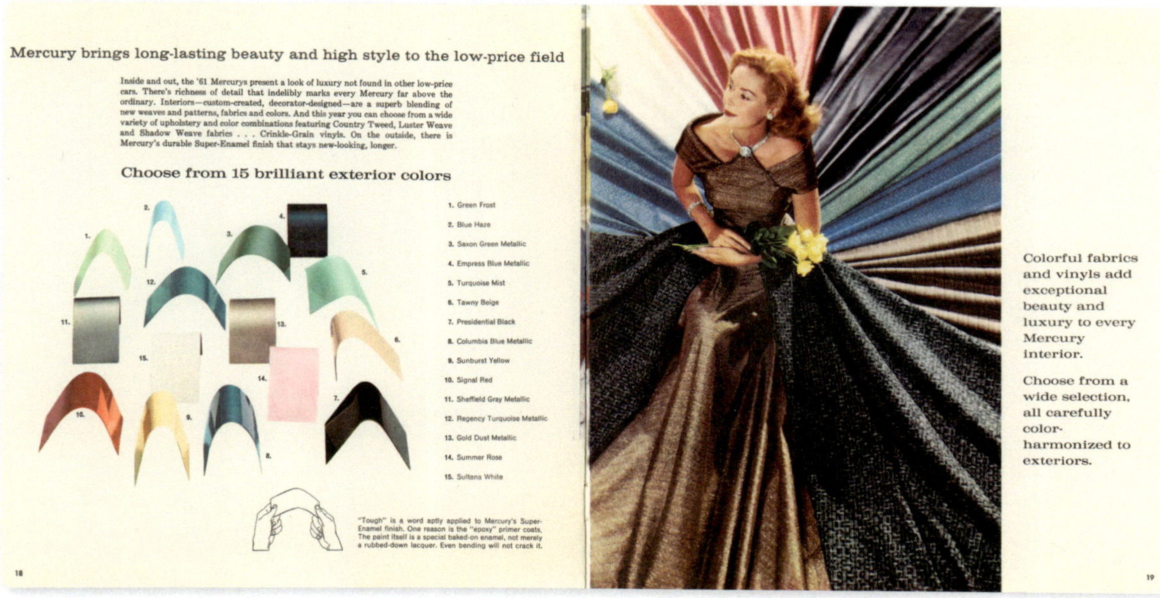

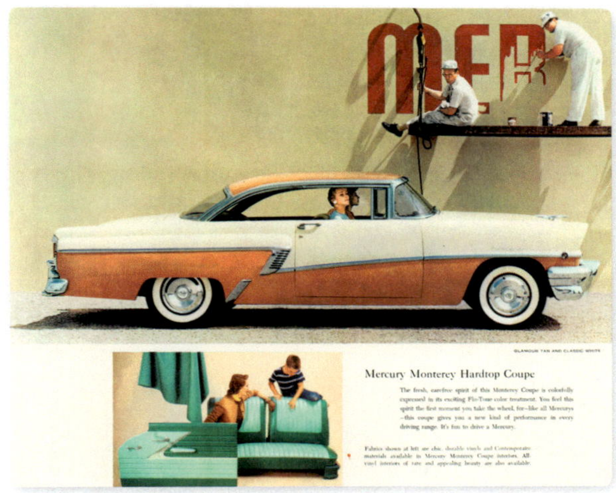

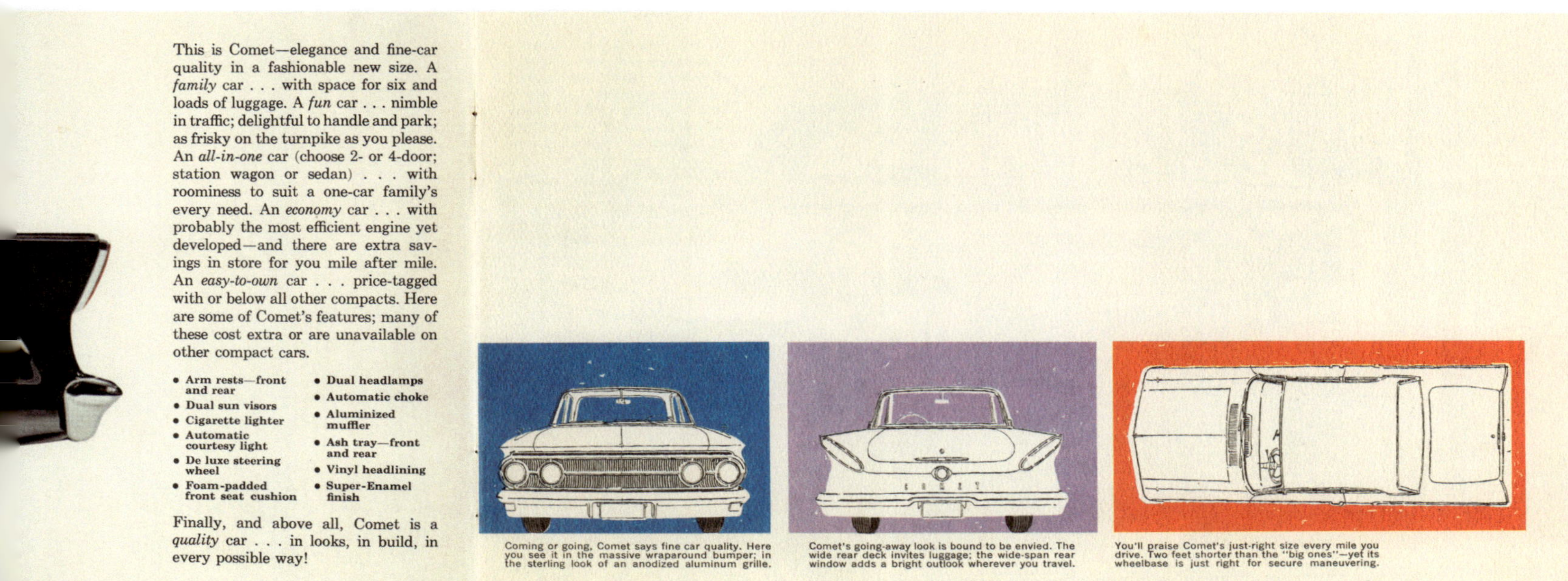

Nash

Nash 1948 *(right)*, **1951** *(below)*

Nash referred to its technology as Airflyte, and was the first U.S. brand to embrace fully unitized construction. The Airflyte-series cars also had one-piece windshields and instruments grouped high in front of the driver, aircraft style.

Nash bezeichnete die eigene Technologie als „Airflyte" und war die erste Marke in den USA, die ganz auf die selbsttragende Karosserie setzte. Die Fahrzeuge der Airflyte-Reihe besaßen zudem Windschutzscheiben aus einem Stück und Instrumente, die wie in einem Flugzeug zentral vor dem Fahrer angeordnet waren.

Nash, qui avait baptisé sa technologie « Airflyte », fut la première marque américaine à adopter la construction monocoque. Les voitures de la série Airflyte disposaient aussi d'un pare-brise fait d'une seule pièce et incurvé ainsi que d'une planche de bord à compteur surélevé devant le conducteur, à l'inspiration très aéronautique.

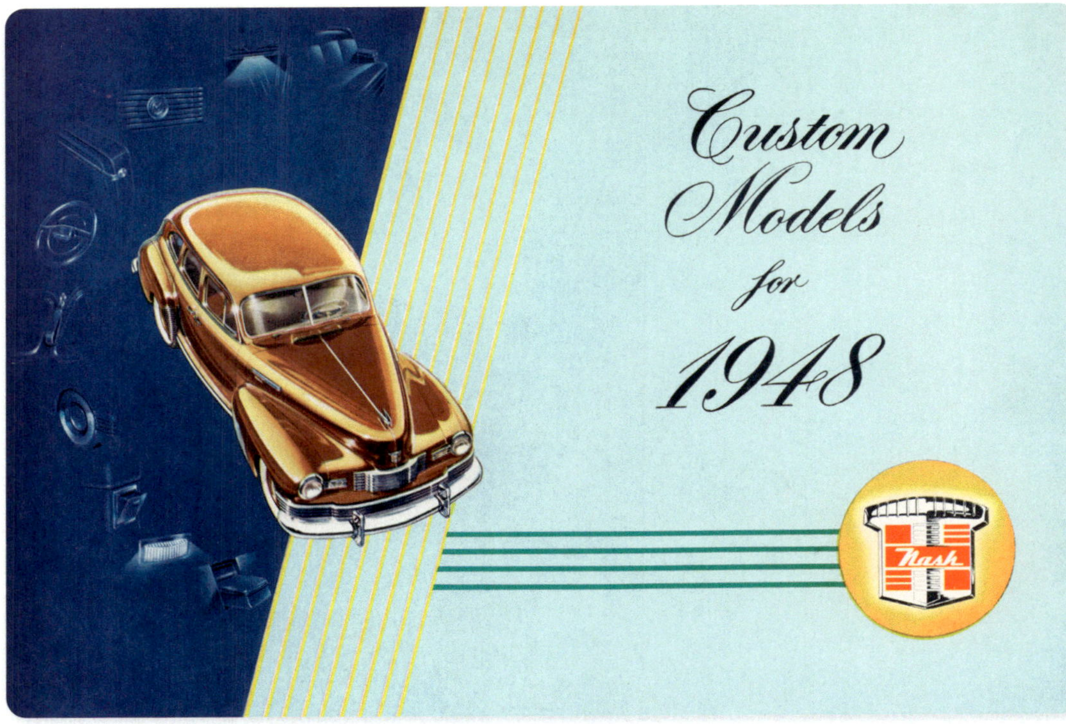

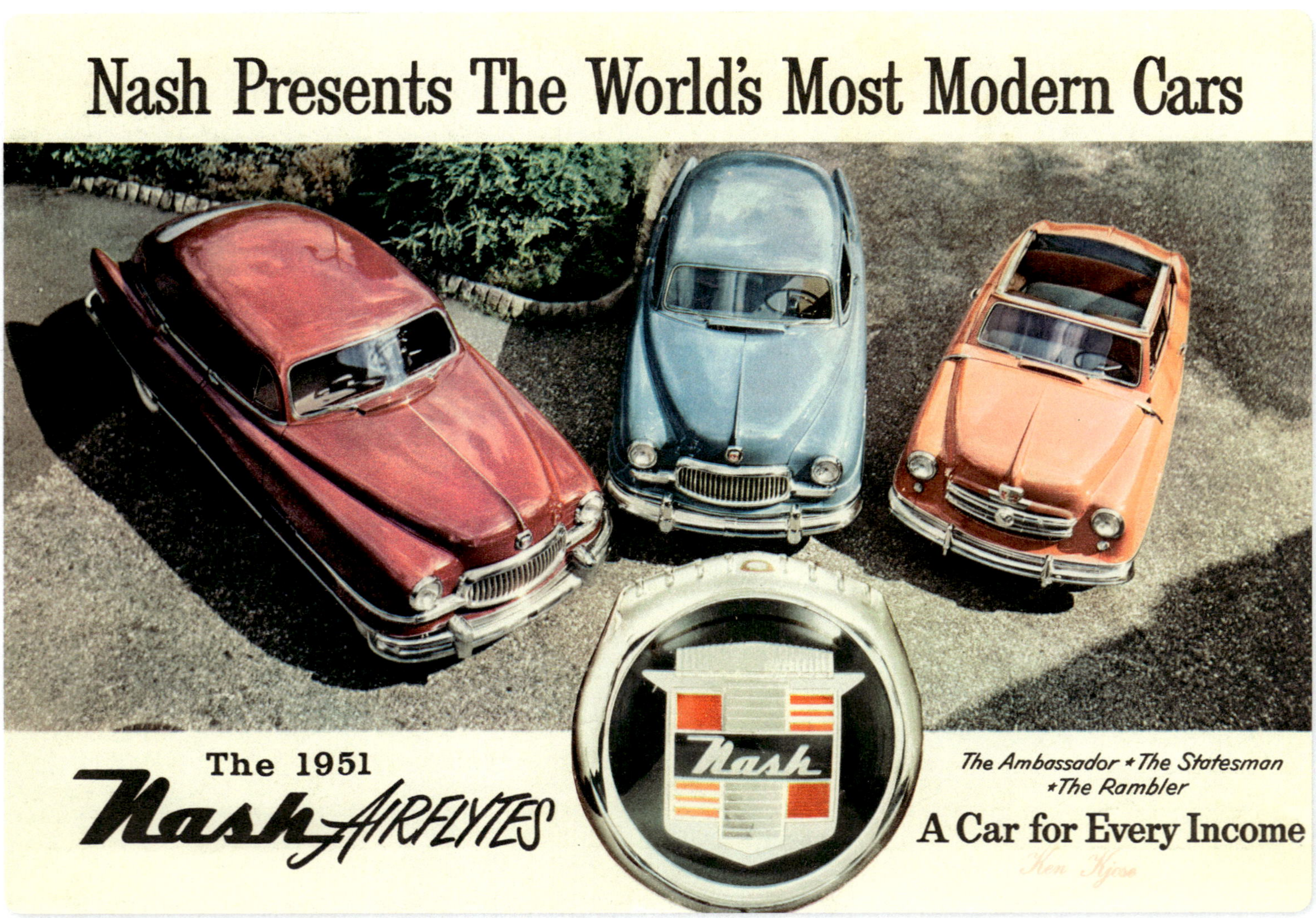

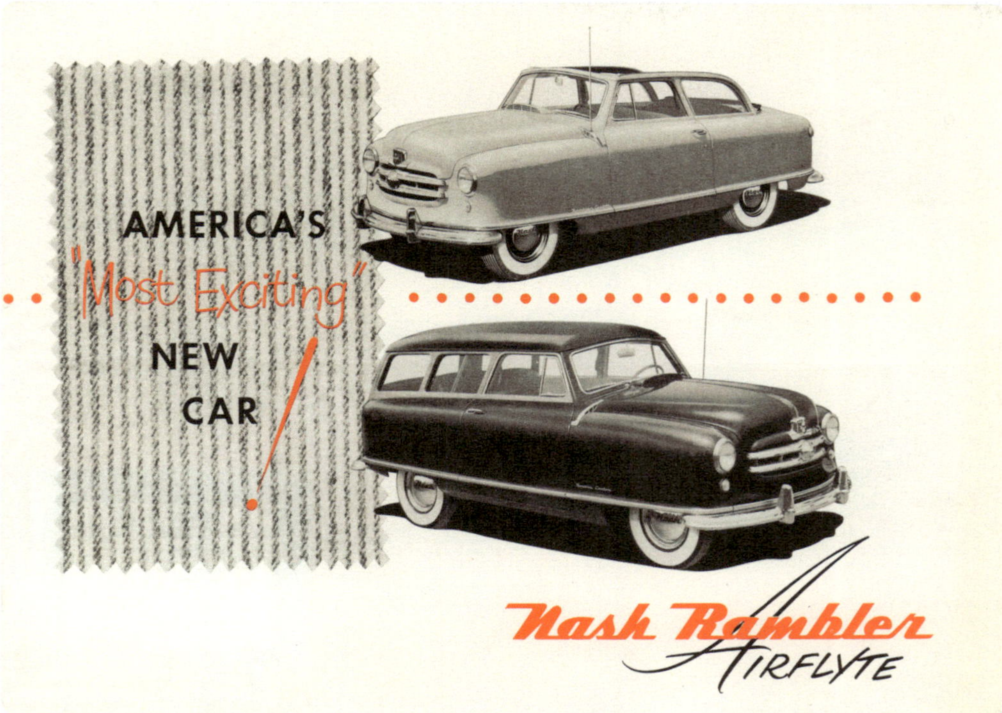

Nash 1951 *(left)*, **1955** *(below)*

Nash began to plan an ultracompact at a time of diminishing resources. It needed a car that was cheap to build and to buy. The Metropolitan was the first car designed in the United States and built overseas, by Austin in Britain, which also furnished its A40 engine. About 104,000 were built.

In einer Zeit schwindender Rohstoffe begann Nash mit der Planung eines Kleinstwagens. Man brauchte ein Auto, das preiswert in der Herstellung und billig für die Käufer war. Der Metropolitan war das erste in den Vereinigten Staaten entworfene Automobil, das in Übersee gefertigt wurde, nämlich in Großbritannien bei Austin, die auch den A40-Motor beisteuerten. Es wurden rund 104 000 Fahrzeuge gebaut.

En ces temps de ceintures serrées, Nash décida de lancer une petite auto ultra-compacte, aussi économique à fabriquer qu'à acquérir. La Metropolitan fut ainsi la première voiture dessinée aux États-Unis et construite outre-mer, en l'occurrence en Grande-Bretagne par Austin qui fournissait également son moteur A40. Il s'en produira près de 104 000 exemplaires.

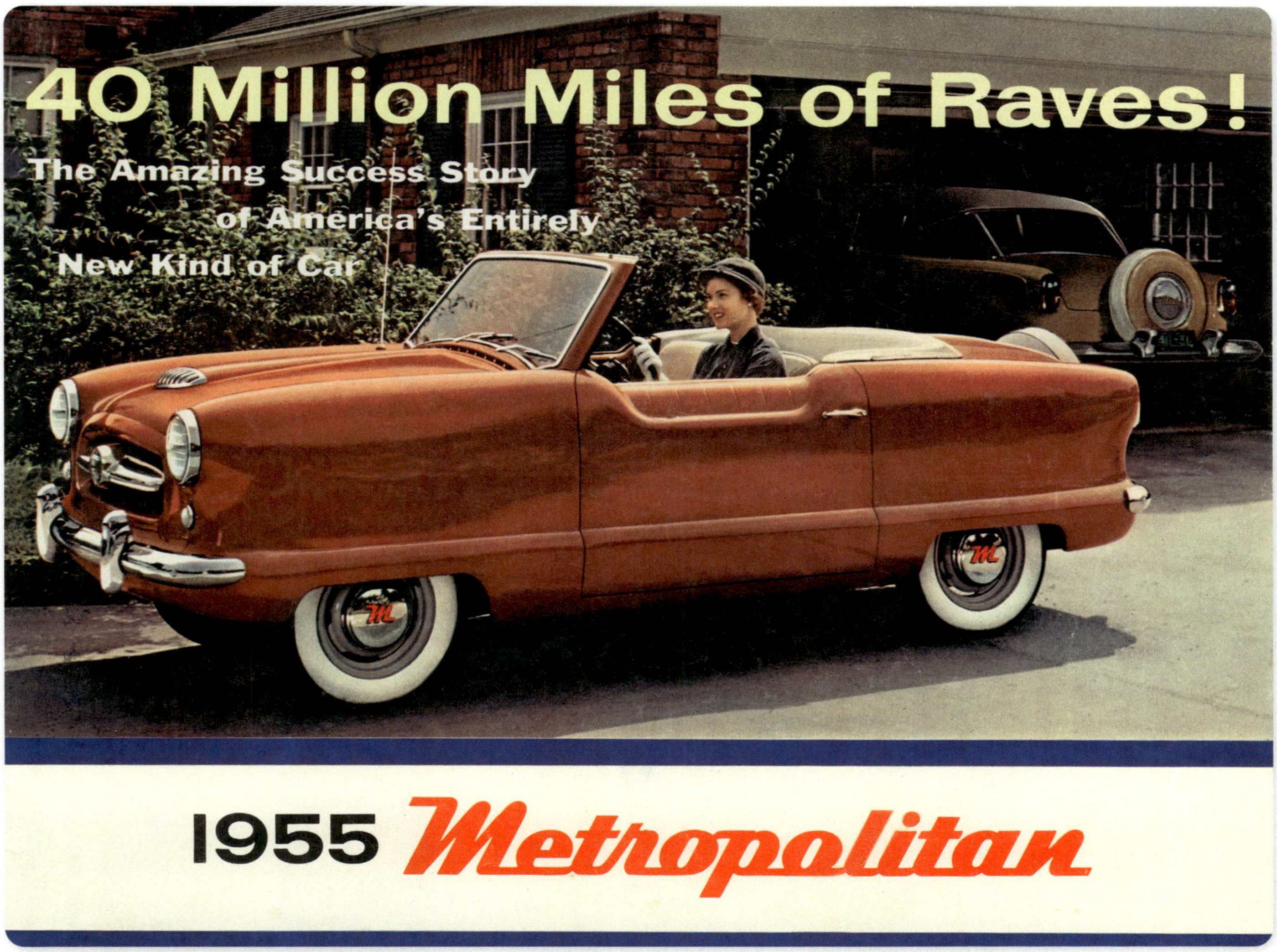

Oldsmobile

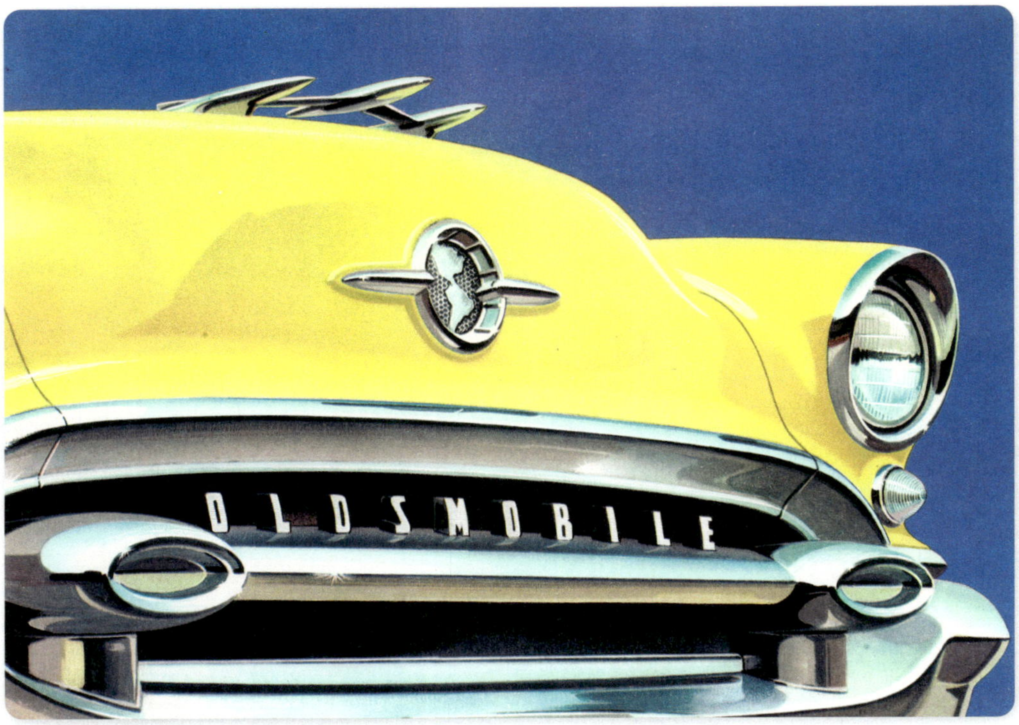

Oldsmobile 1955 *(left)*, **1953** *(below)*, **1954** *(opposite)*

Oldsmobile's imagery in the early 1950s firmly reflected styling and power. Compare the deeply hooded headlamp on this Oldsmobile with the dramatic matching instrument pod in its polished steel dashboard. Lastly, the Oldsmobile Rocket V-8, with overhead valves, is considered by many to be the first true postwar American performance engine.

Das Erscheinungsbild eines typischen Oldsmobile der frühen 1950er-Jahre war ganz eindeutig auf Kraft und Design ausgerichtet. Man vergleiche einmal die tief eingebetteten Scheinwerfer dieses Oldsmobile mit dem dazu passenden Kombiinstrument in seiner Armaturentafel aus poliertem Stahl. Und schließlich gilt der Oldsmobile Rocket V8 mit OHV-Ventilsteuerung für viele als erster wirklich leistungsfähiger amerikanischer Motor der Nachkriegszeit.

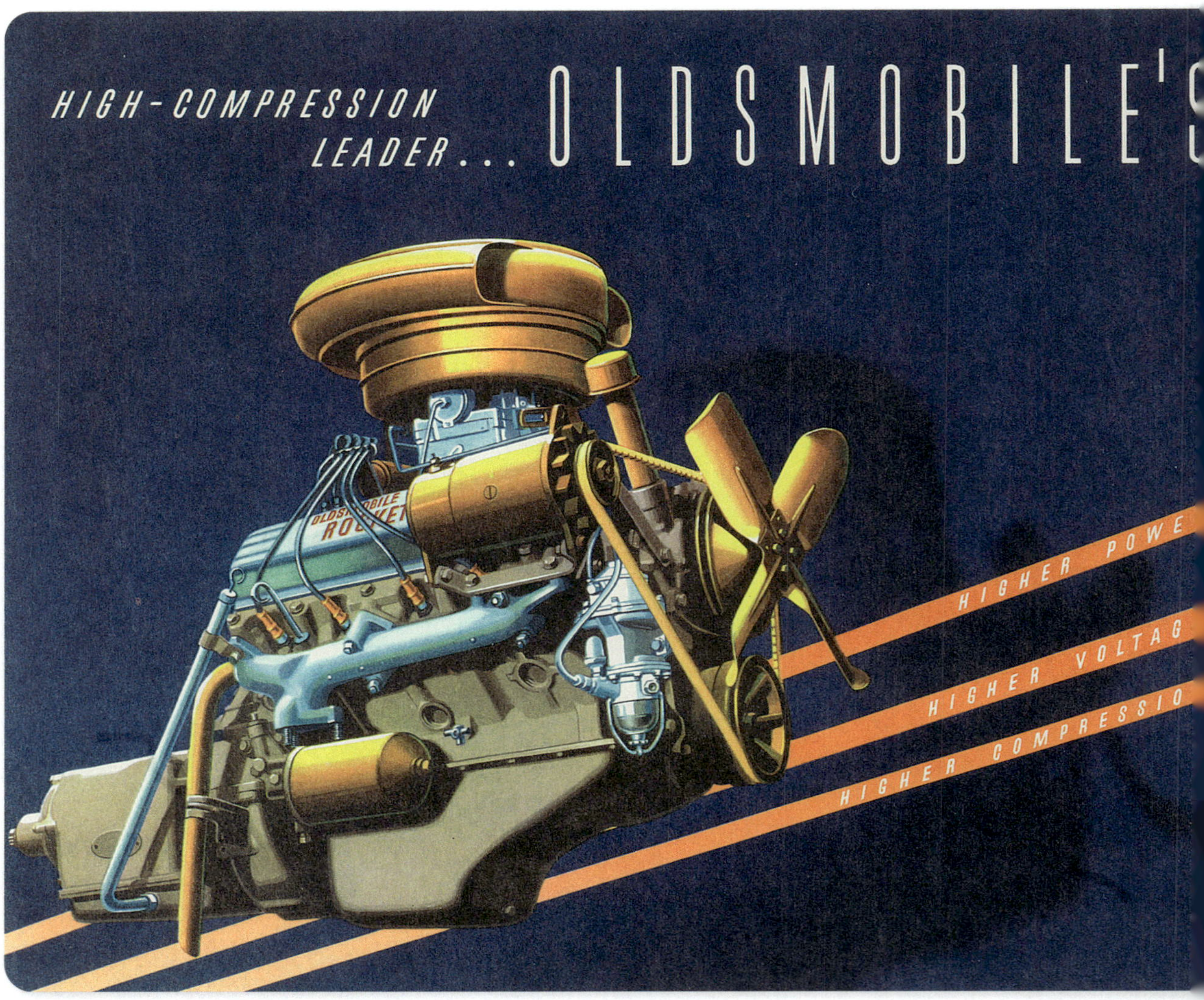

L'imagerie Oldsmobile, au début des années 1950, entendait clairement refléter le style et la puissance des productions maison. On remarquera ainsi la cohérence étudiée entre le phare profondément encastré de cette Oldsmobile et le compteur de sa planche de bord en acier poli. Enfin le V8 Rocket à soupapes en tête est considéré par beaucoup comme le premier moteur américain de l'après-guerre réellement performant.

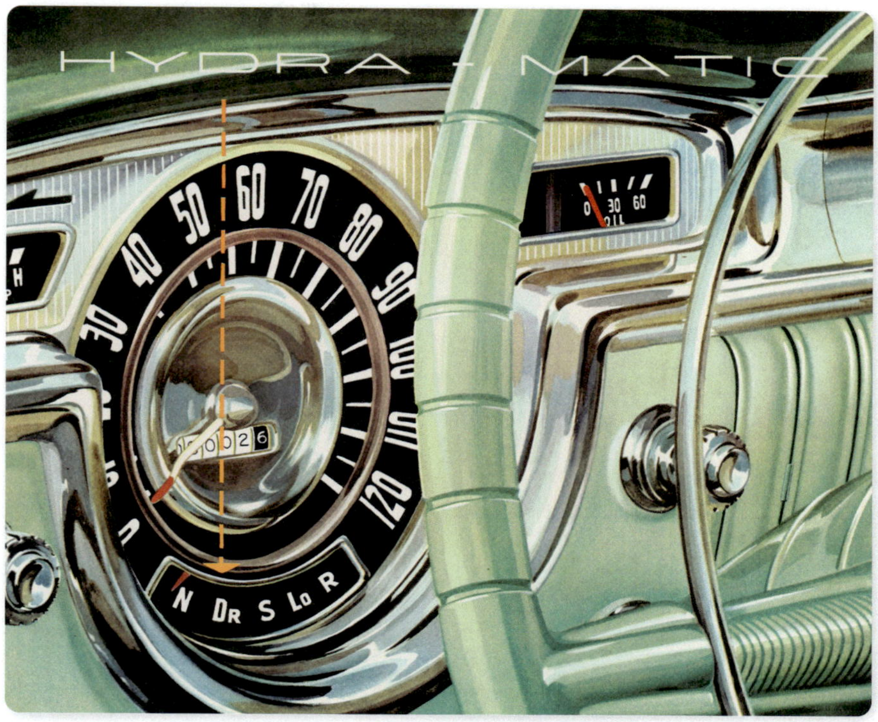

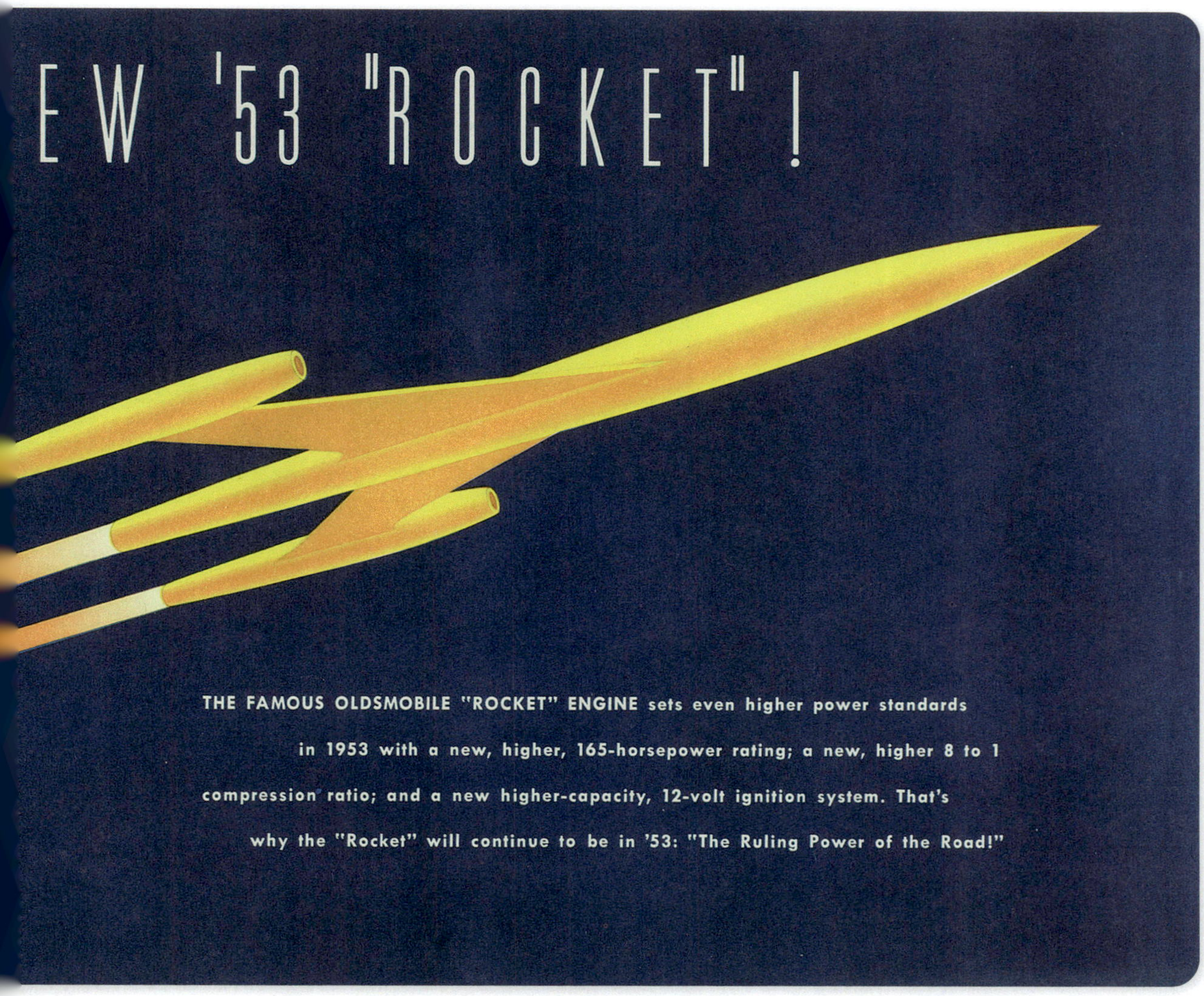

Oldsmobile 1956, 1957 *(opposite)*

Fun in the sun, tropical pastel colors, and summery clothing implied some of the pleasures you could find once a new Oldsmobile got you to your destination behind its 240 hp Rocket V-8 engine.

Spaß in der Sonne, Tropenflair in Pastellfarben und sommerliche Kleidung deuteten einen Teil der Vergnügungen an, die Käufer eines neuen Oldsmobile genießen konnten, wenn sie ihr Auto mit dem 326 kW starken Rocket-V8-Motor komfortabel ans Ziel ihrer Reise gebracht hatten.

Tenues d'été, joies estivales et couleurs tropicales pastel, voilà, entre autres plaisirs, ce qui attendait l'heureux propriétaire d'une nouvelle Oldsmobile une fois arrivé à destination. Car les 240 chevaux de son moteur V8 Rocket ne demandaient qu'à l'emmener loin.

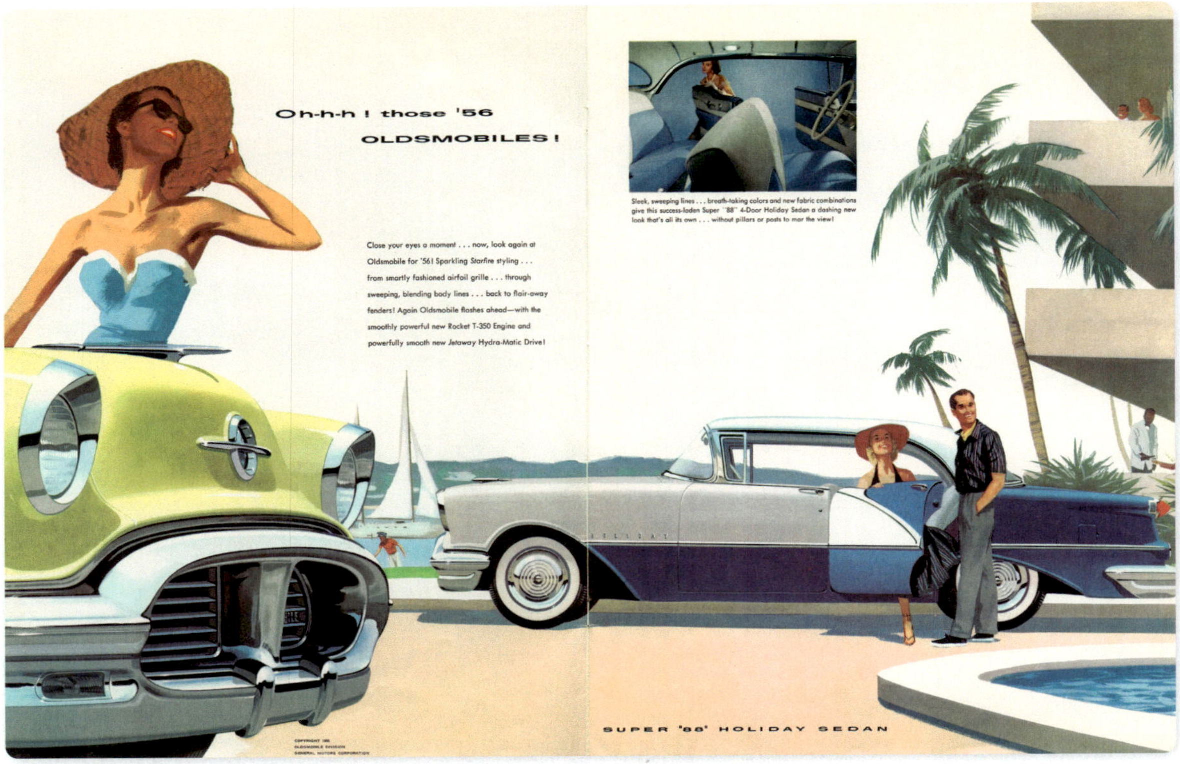

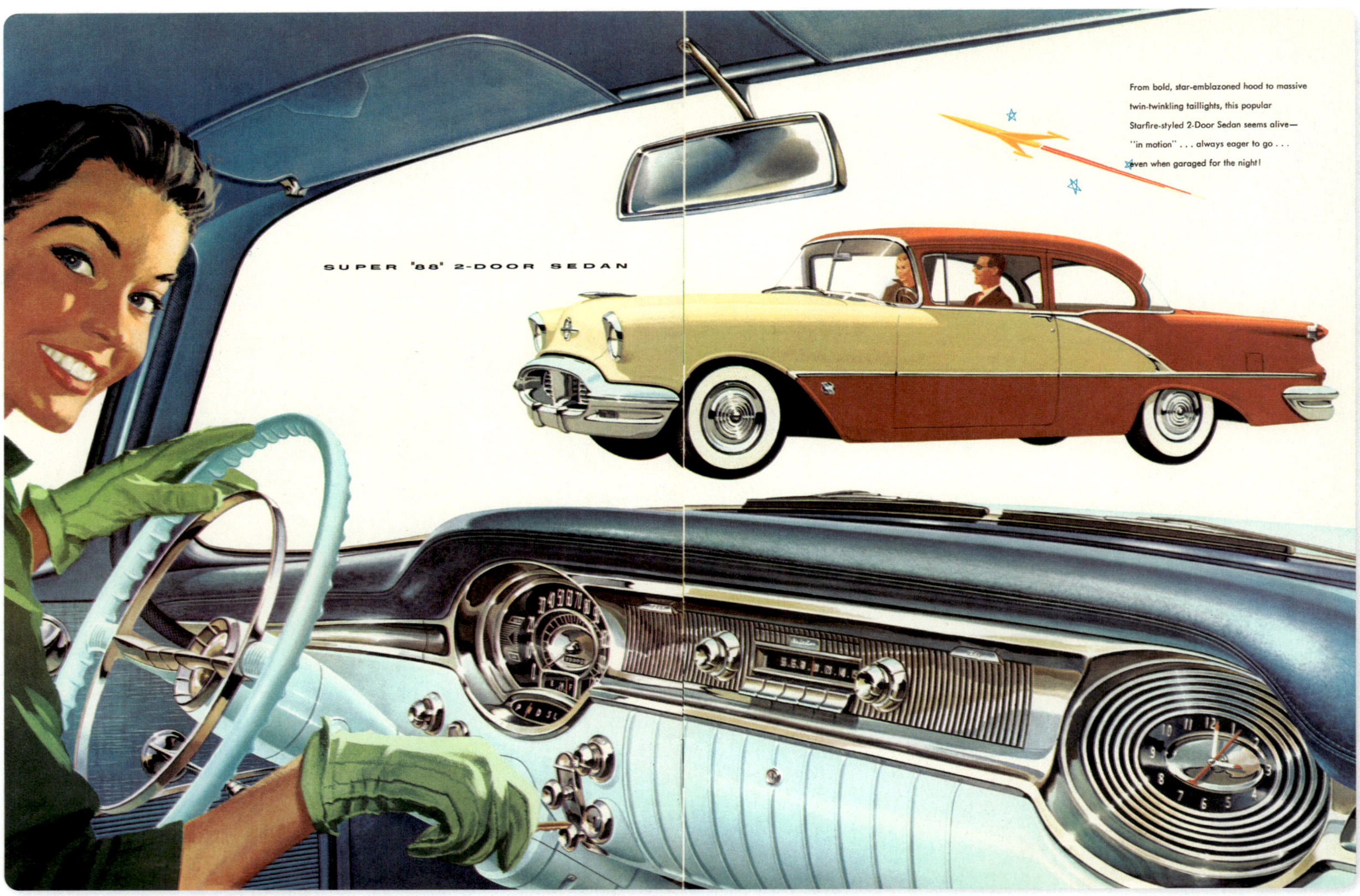

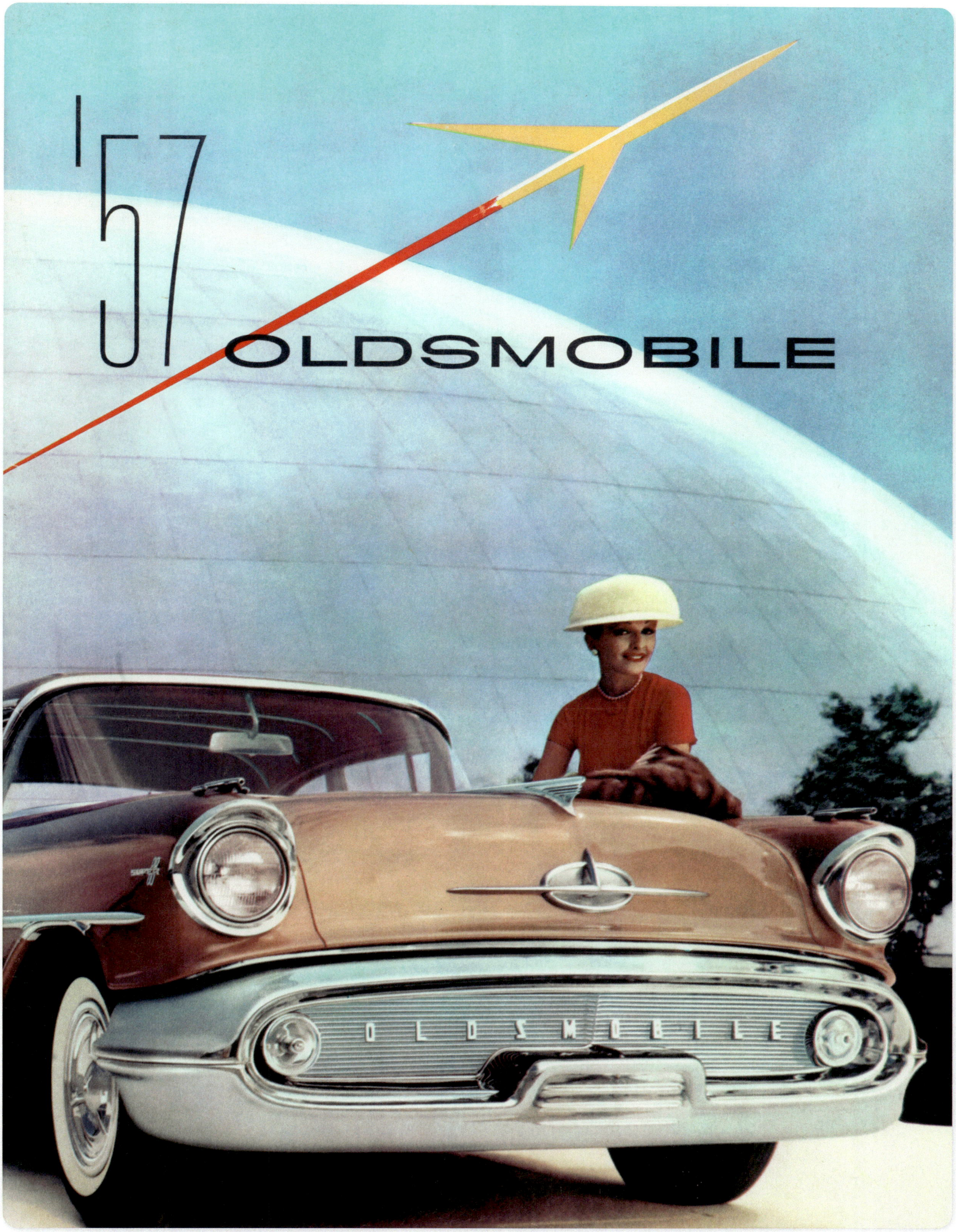
'57 OLDSMOBILE
OLDSMOBILE

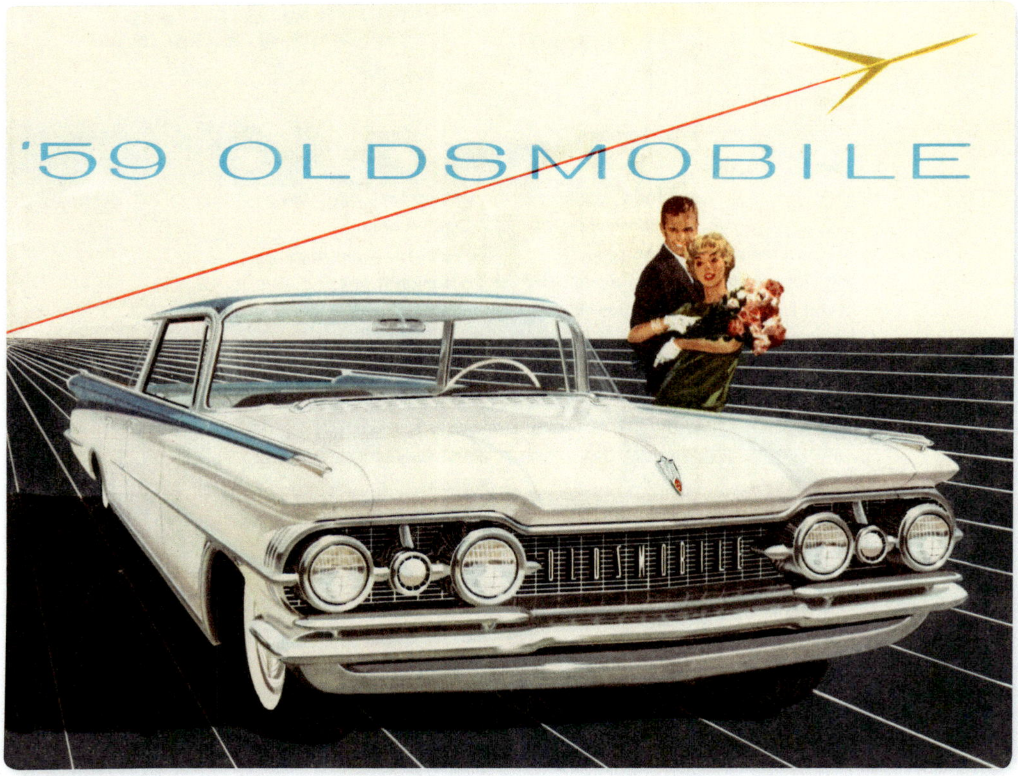

Oldsmobile 1959 *(left)*, **1960** *(below)*, **1954** *(opposite top)*, **1961** *(opposite bottom)*

The image-makers of Detroit and beyond expanded the lexicon of car buying during the 1950s. Certainly, Oldsmobile engineered additional strength into its deep-channel frame and suspension, but calling it the "Power-Ride" added marked creativity.

In den 1950er-Jahren erweiterten die Imagezauberer in Detroit und anderswo ihr Vokabular mit sehr viel Fantasie. Freilich hatte Oldsmobile Rahmen und Aufhängung tatsächlich verstärkt, doch mit dem kreativen Begriff „Power-Ride" klang es nach viel mehr.

Au cours des années 1950, les têtes pensantes du marketing de Detroit et d'ailleurs ont enrichi le lexique de l'achat automobile. Certes Oldsmobile a su intelligemment renforcer son large châssis-caisson et sa suspension, mais lui donner le nom de « Power-Ride » témoignait d'une créativité volontiers débordante.

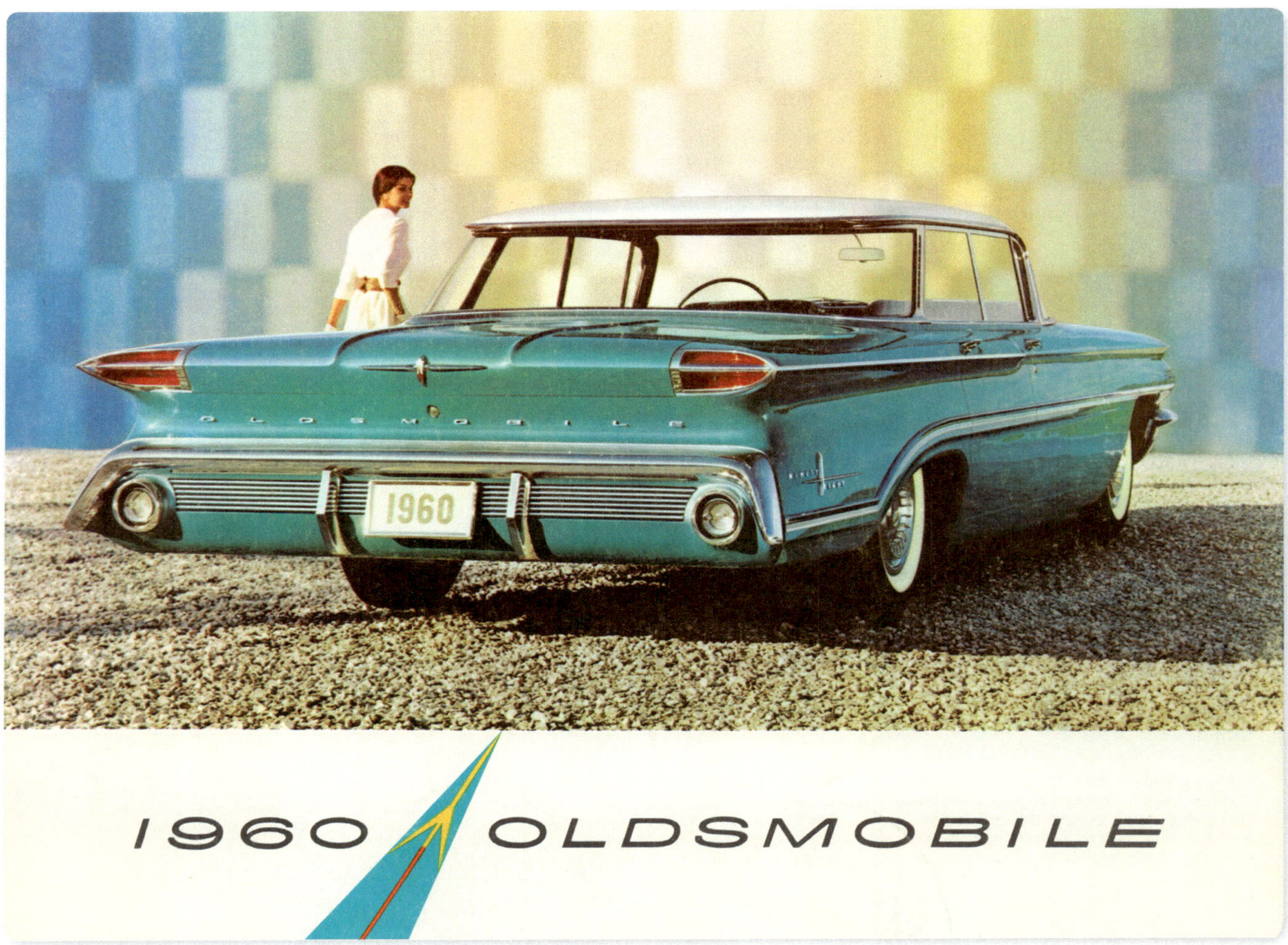

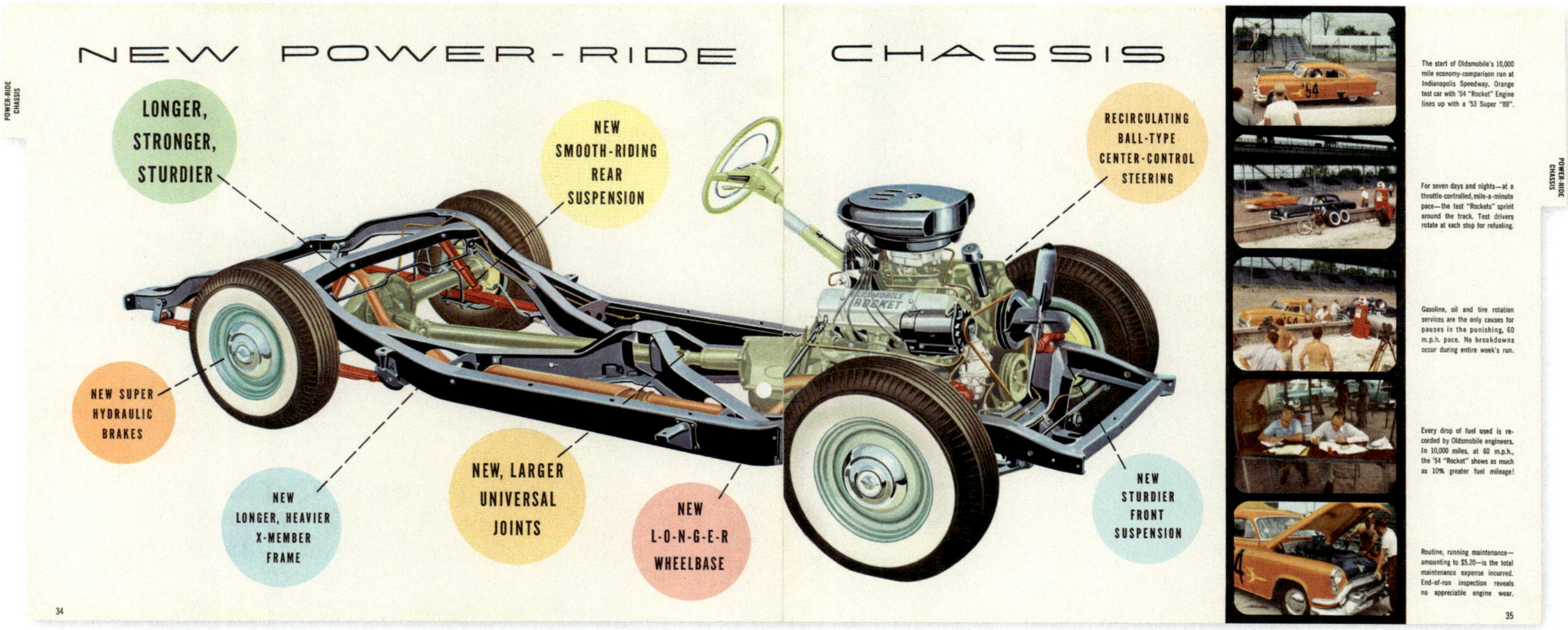
NEW POWER-RIDE CHASSIS
LONGER, STRONGER, STURDIER
NEW SMOOTH-RIDING REAR SUSPENSION
RECIRCULATING BALL-TYPE CENTER-CONTROL STEERING
NEW SUPER HYDRAULIC BRAKES
NEW LONGER, HEAVIER X-MEMBER FRAME
NEW, LARGER UNIVERSAL JOINTS
NEW L-O-N-G-E-R WHEELBASE
NEW STURDIER FRONT SUSPENSION
The start of Oldsmobile's 10,000 mile economy-comparison run at Indianapolis Speedway. Orange test car with '54 "Rocket" Engine lines up with a '53 Super "88".
For seven days and nights—at a throttle-controlled, mile-a-minute pace—the test "Rockets" sprint around the track. Test drivers rotate at each stop for refueling.
Gasoline, oil and tire rotation services are the only causes for pauses in the punishing, 60 m.p.h. pace. No breakdowns occur during entire week's run.
Every drop of fuel used is recorded by Oldsmobile engineers. In 10,000 miles, at 60 m.p.h., the '54 "Rocket" shows as much as 10% greater fuel mileage!
Routine, running maintenance—amounting to $5.20—is the total maintenance expense incurred. End-of-run inspection reveals no appreciable engine wear.
34
35

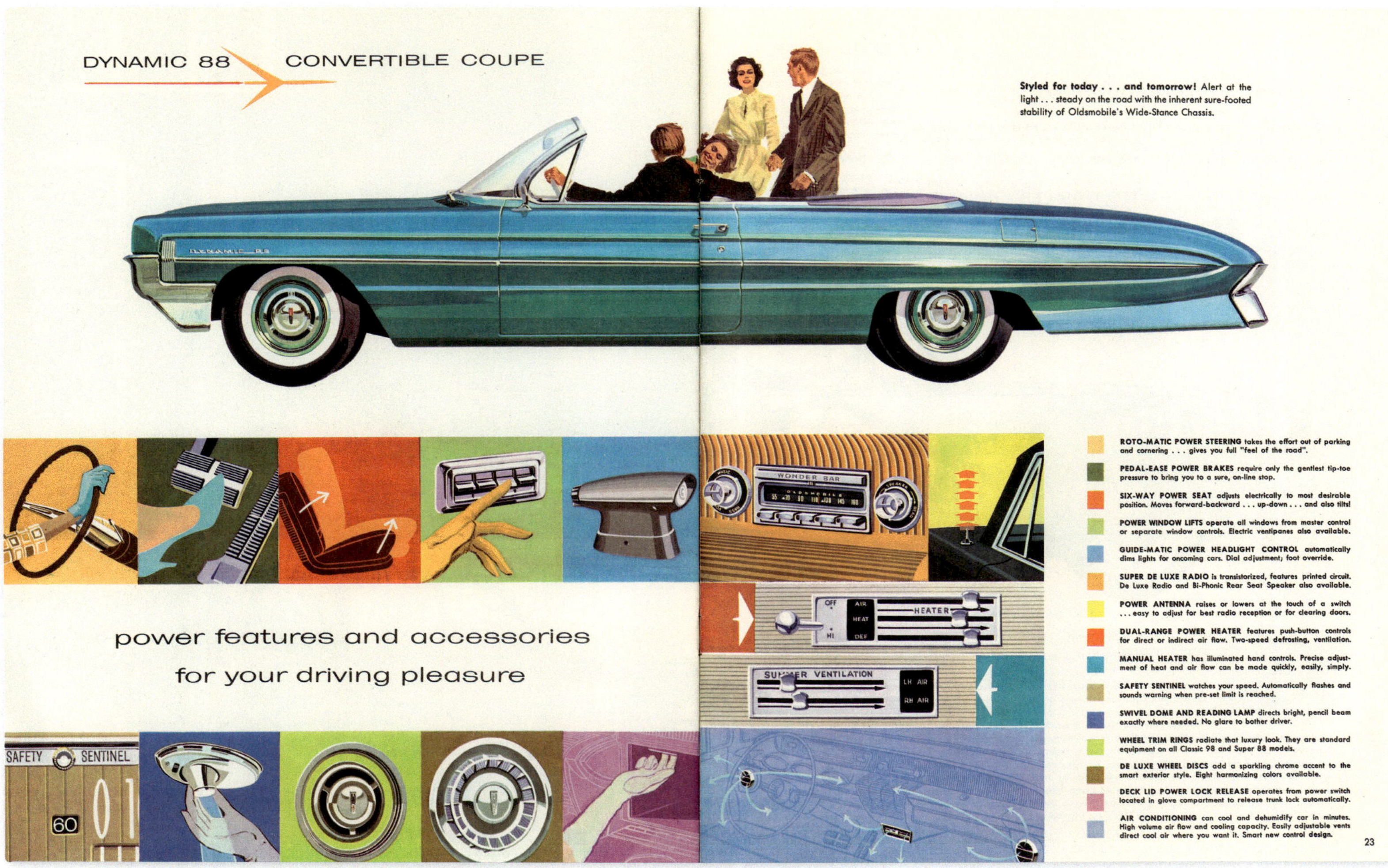
DYNAMIC 88 CONVERTIBLE COUPE
Styled for today . . . and tomorrow! Alert at the light . . . steady on the road with the inherent sure-footed stability of Oldsmobile's Wide-Stance Chassis.
power features and accessories for your driving pleasure
ROTO-MATIC POWER STEERING takes the effort out of parking and cornering . . . gives you full "feel of the road".
PEDAL-EASE POWER BRAKES require only the gentlest tip-toe pressure to bring you to a sure, on-line stop.
SIX-WAY POWER SEAT adjusts electrically to most desirable position. Moves forward-backward . . . up-down . . . and also tilts!
POWER WINDOW LIFTS operate all windows from master control or separate window controls. Electric ventipanes also available.
GUIDE-MATIC POWER HEADLIGHT CONTROL automatically dims lights for oncoming cars. Dial adjustment; foot override.
SUPER DE LUXE RADIO is transistorized, features printed circuit. De Luxe Radio and Bi-Phonic Rear Seat Speaker also available.
POWER ANTENNA raises or lowers at the touch of a switch . . . easy to adjust for best radio reception or for clearing doors.
DUAL-RANGE POWER HEATER features push-button controls for direct or indirect air flow. Two-speed defrosting, ventilation.
MANUAL HEATER has illuminated hand controls. Precise adjustment of heat and air flow can be made quickly, easily, simply.
SAFETY SENTINEL watches your speed. Automatically flashes and sounds warning when pre-set limit is reached.
SWIVEL DOME AND READING LAMP directs bright, pencil beam exactly where needed. No glare to bother driver.
WHEEL TRIM RINGS radiate that luxury look. They are standard equipment on all Classic 98 and Super 88 models.
DE LUXE WHEEL DISCS add a sparkling chrome accent to the smart exterior style. Eight harmonizing colors available.
DECK LID POWER LOCK RELEASE operates from power switch located in glove compartment to release trunk lock automatically.
AIR CONDITIONING can cool and dehumidify car in minutes. High volume air flow and cooling capacity. Easily adjustable vents direct cool air where you want it. Smart new control design.
23

Packard

Packard 1948 *(right)*, **1956** *(below and opposite top)*, **1957** *(opposite bottom)*

The largest problem this august car manufacturer faced was eroding funds, especially for engine development. Packard did not gain a V-8 engine until 1955. Rapidly changing model names may have confused buyers. While Packards were lavish, it was obvious the company, now merged with Studebaker, was in trouble.

Das Hauptproblem für diesen Hersteller herrlicher Automobile war, dass ihm das Geld ausging, vor allem durch die Motorenentwicklung. Der Packard erhielt erst 1955 einen V8-Motor, und ein rascher Wechsel der Modellbezeichnungen mag die Käufer verwirrt haben. So nobel die Packards auch waren – es war offensichtlich, dass das Unternehmen, das inzwischen mit Studebaker fusioniert hatte, in Schwierigkeiten steckte.

Le principal problème rencontré par cet auguste constructeur, surtout pour développer ses moteurs, c'était l'argent. Packhard devra ainsi attendre 1955 avant de disposer d'un V8. La fréquence des changements de dénomination appliqués aux modèles de la gamme pouvait par ailleurs désorienter les acheteurs potentiels. Autant les voitures étaient superbes, autant l'entreprise, alliée à Studebaker depuis la fusion de 1954, se portait manifestement mal.

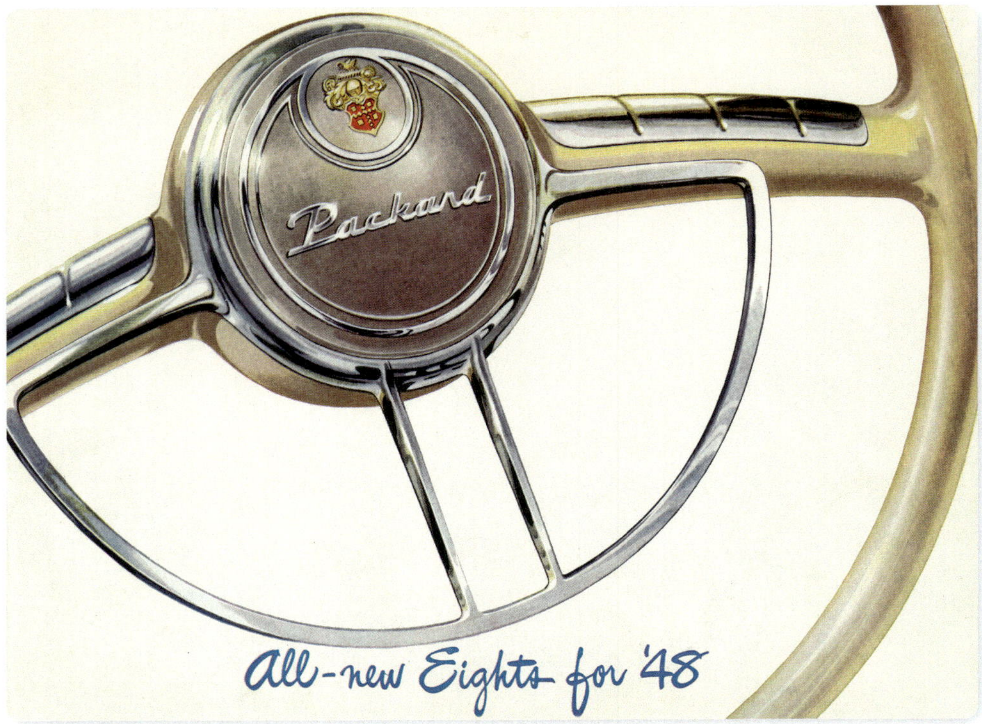

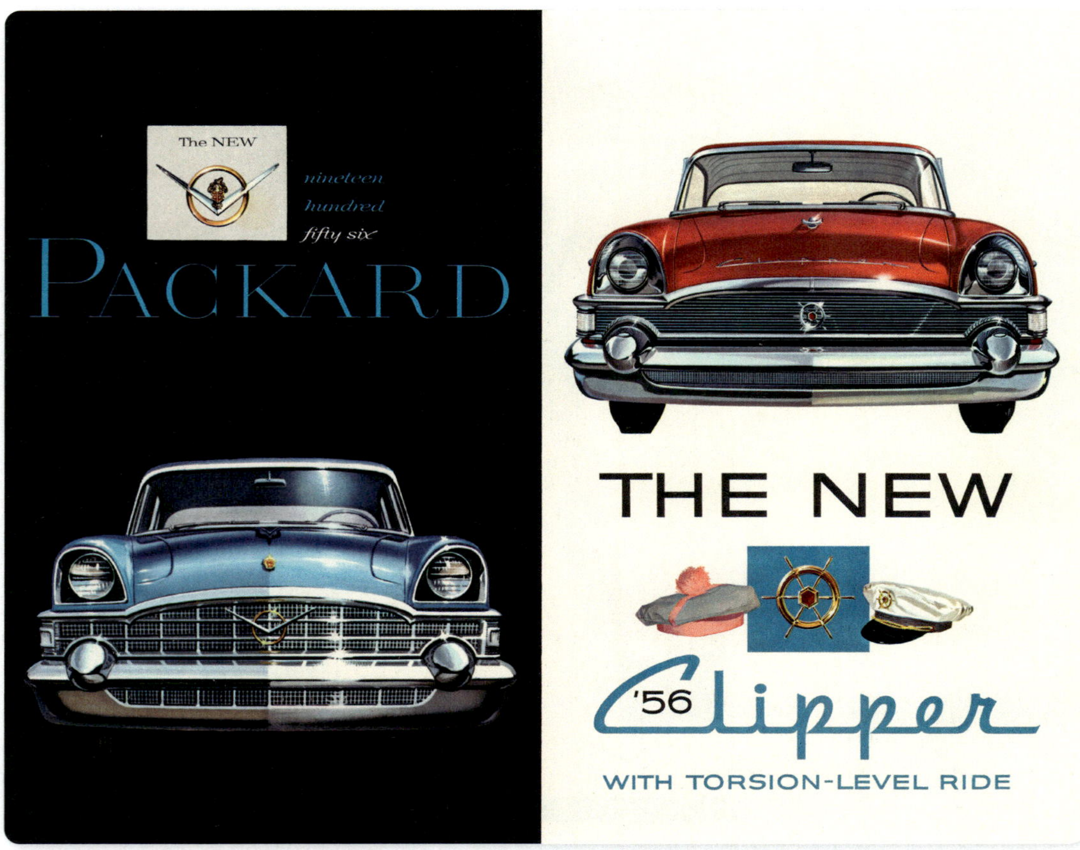

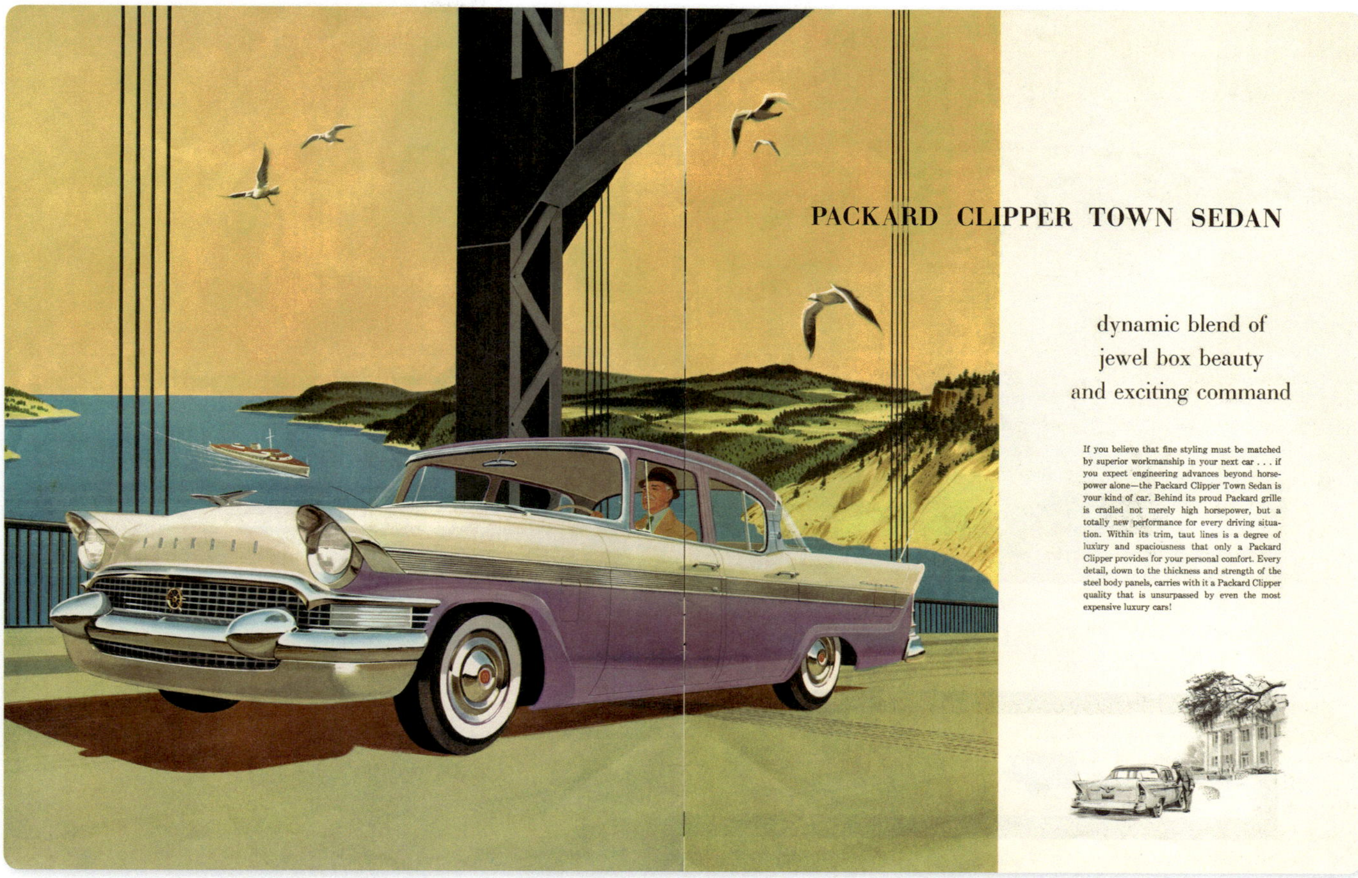
PACKARD CLIPPER TOWN SEDAN
dynamic blend of jewel box beauty and exciting command
If you believe that fine styling must be matched by superior workmanship in your next car . . . if you expect engineering advances beyond horsepower alone—the Packard Clipper Town Sedan is your kind of car. Behind its proud Packard grille is cradled not merely high horsepower, but a totally new performance for every driving situation. Within its trim, taut lines is a degree of luxury and spaciousness that only a Packard Clipper provides for your personal comfort. Every detail, down to the thickness and strength of the steel body panels, carries with it a Packard Clipper quality that is unsurpassed by even the most expensive luxury cars!

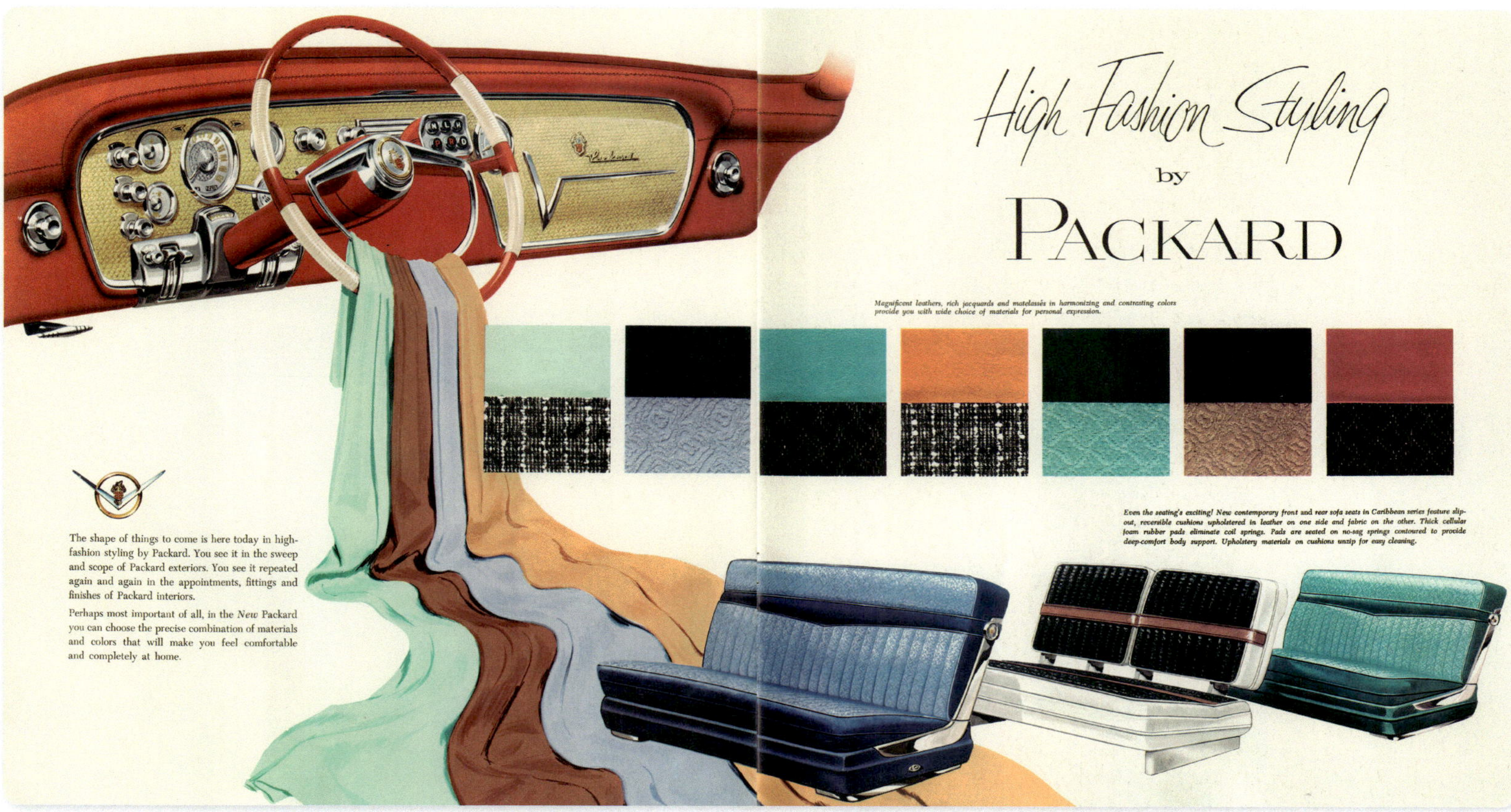
High Fashion Styling by PACKARD
Magnificent leathers, rich jacquards and matelassés in harmonizing and contrasting colors provide you with wide choice of materials for personal expression.
Even the seating's exciting! New contemporary front and rear sofa seats in Caribbean series feature slip-out, reversible cushions upholstered in leather on one side and fabric on the other. Thick cellular foam rubber pads eliminate coil springs. Pads are seated on no-sag springs contoured to provide deep-comfort body support. Upholstery materials on cushions unzip for easy cleaning.
The shape of things to come is here today in high-fashion styling by Packard. You see it in the sweep and scope of Packard exteriors. You see it repeated again and again in the appointments, fittings and finishes of Packard interiors.
Perhaps most important of all, in the New Packard you can choose the precise combination of materials and colors that will make you feel comfortable and completely at home.

Plymouth

Plymouth 1956, 1953 *(opposite)*

The brochure covers dramatically demonstrate the scope of change at Plymouth in just three years. The 1953 line was virtuous but somewhat boring with aged six-cylinder engines. By 1956 Plymouth had grown taillamps, jetlike trim, and a new V-8.

Die Titelseiten der Broschüren zeigen eindrucksvoll, wie drastisch sich der Plymouth in nur drei Jahren gewandelt hatte. Das Aufgebot von 1953 war mit seinen in die Jahre gekommenen Sechszylindermotoren zwar brav, aber auch langweilig. Doch 1956 hatte der Plymouth auffällige Heckleuchten und -flossen erhalten, Zierelemente, die an ein Düsenflugzeug erinnerten, und einen neuen V8-Motor.

Les couvertures de catalogue illustrent à merveille l'étendue des changements intervenus chez Plymouth en seulement trois ans. La gamme 1953 paraissait aussi irréprochable que légèrement soporifique, avec ses moteurs six cylindres qui n'étaient plus de première jeunesse. En 1956, Plymouth arborera de nouveaux feux arrière, une finition très « jet » et un nouveau V8.

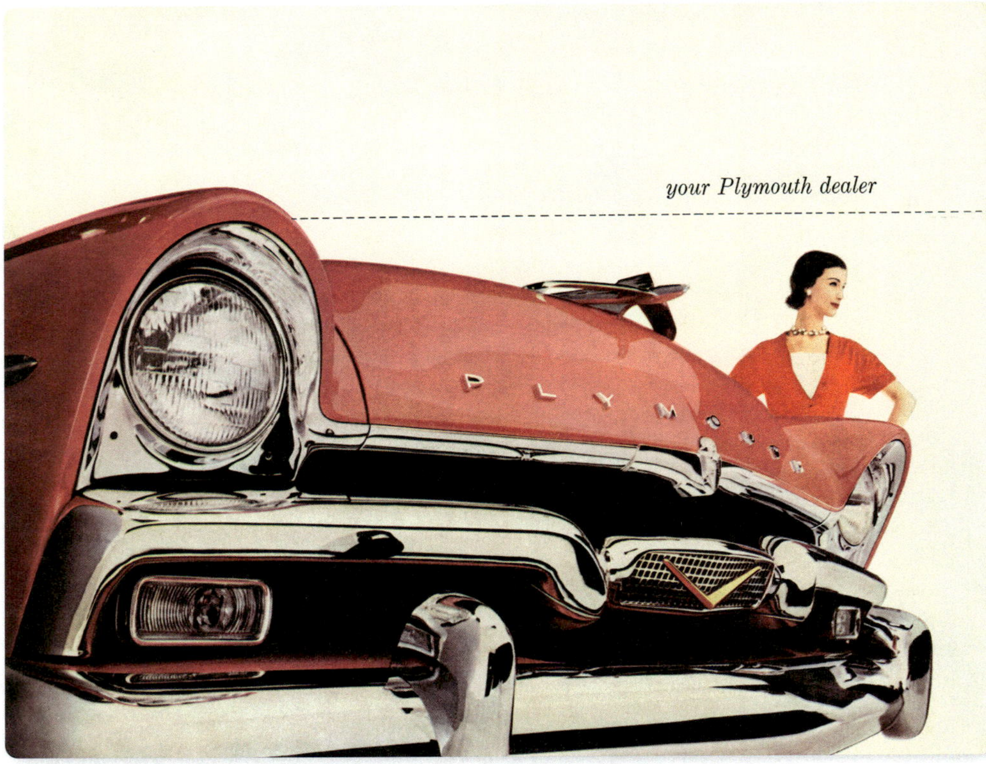

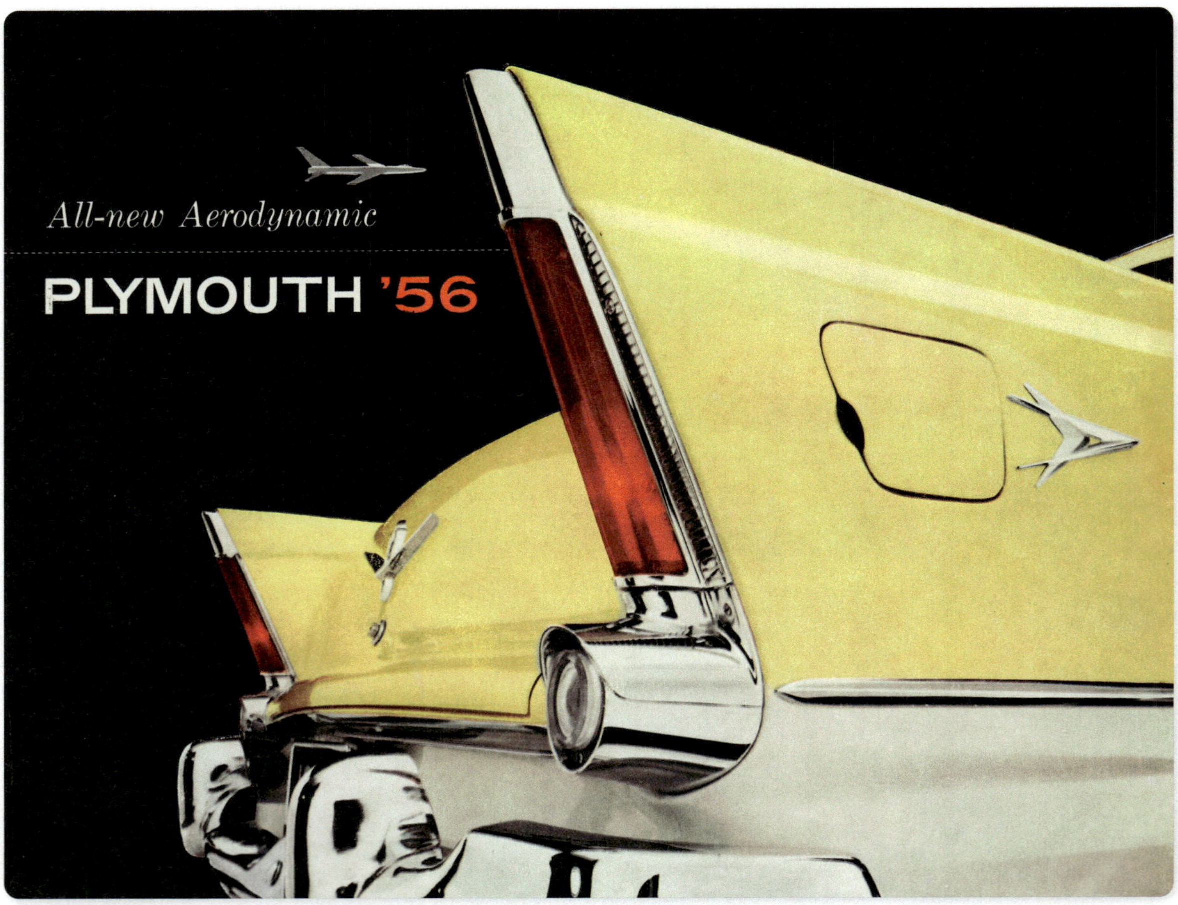

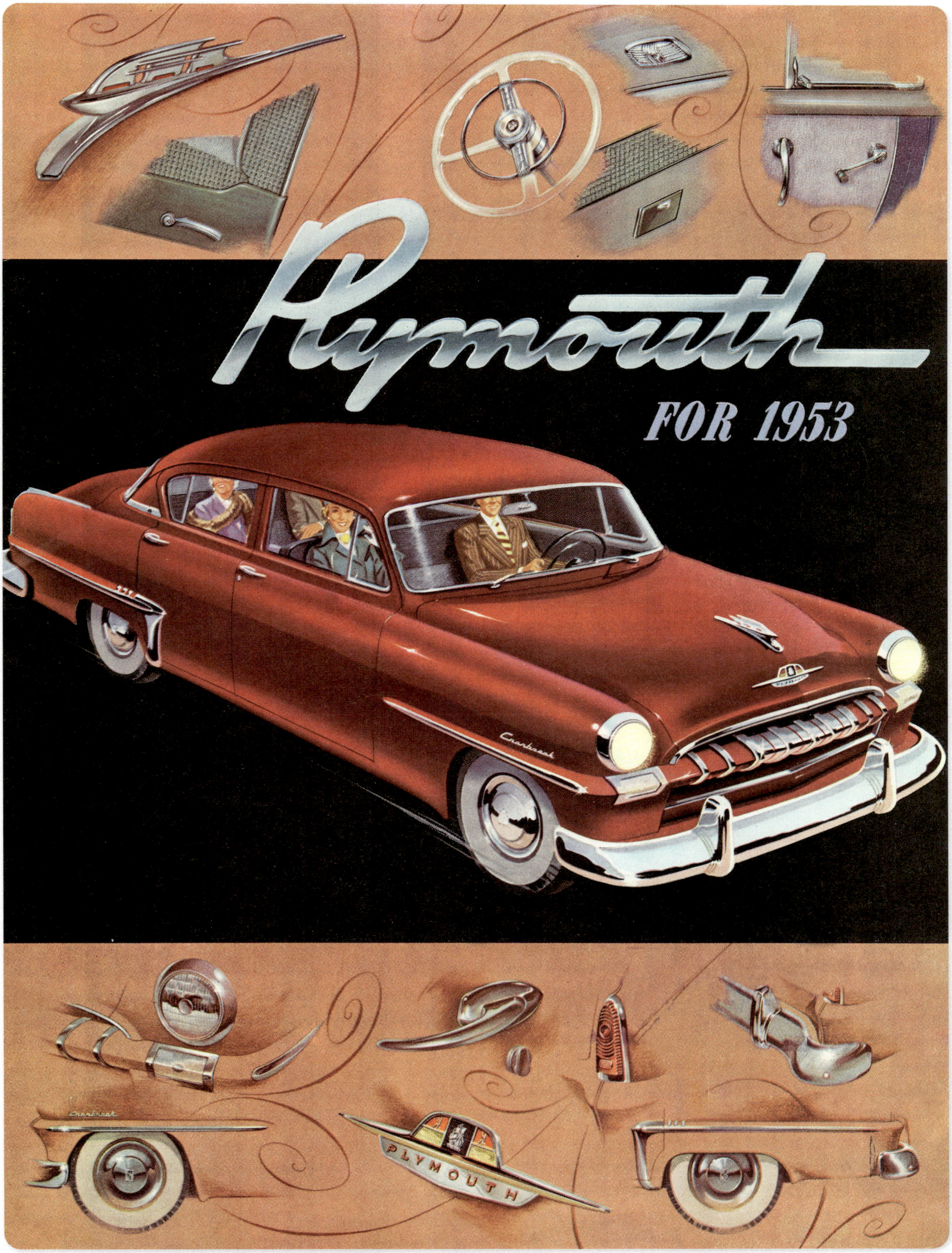
Plymouth
FOR 1953
PLYMOUTH

Plymouth 1957, 1958 *(opposite)*

A powertrain advance saw Plymouth introduce its version of the new Chrysler A series V-8, a more modern and efficient engine that lasted well into the 1960s. They earned a reputation for being nearly unbreakable. A 301-cubic-inch A series was a perfect fit for a 1957 Plymouth Suburban station wagon, whose three rows of seats could accommodate a big family.

Nach einer Weiterentwicklung des Antriebsstrangs legte Plymouth seine Version des neuen Chrysler-V8-Motors der Serie A vor, eines moderneren und wirtschaftlicheren Motors, der bis weit in die 1960er-Jahre hinein gebaut wurde. Sie erwarben sich den Ruf, unzerstörbar zu sein. Ein 5-Liter-Motor der A-Serie passte perfekt zum 1957er-Plymouth Suburban, einem Kombi mit drei Sitzreihen, in dem auch eine große Familie bequem Platz fand.

Côté moteurs, Plymouth fit un pas en avant lorsque la marque présenta sa version du nouveau V8 Chrysler « série A » : un bloc plus moderne et efficace dont la carrière se prolongera tout au long des années 1960. Ce moteur avait la réputation d'être quasiment incassable. Le 4,9 litres Série A convenait parfaitement au break Plymouth Suburban de 1957, dont la triple rangée de sièges pouvait accueillir une famille nombreuse.

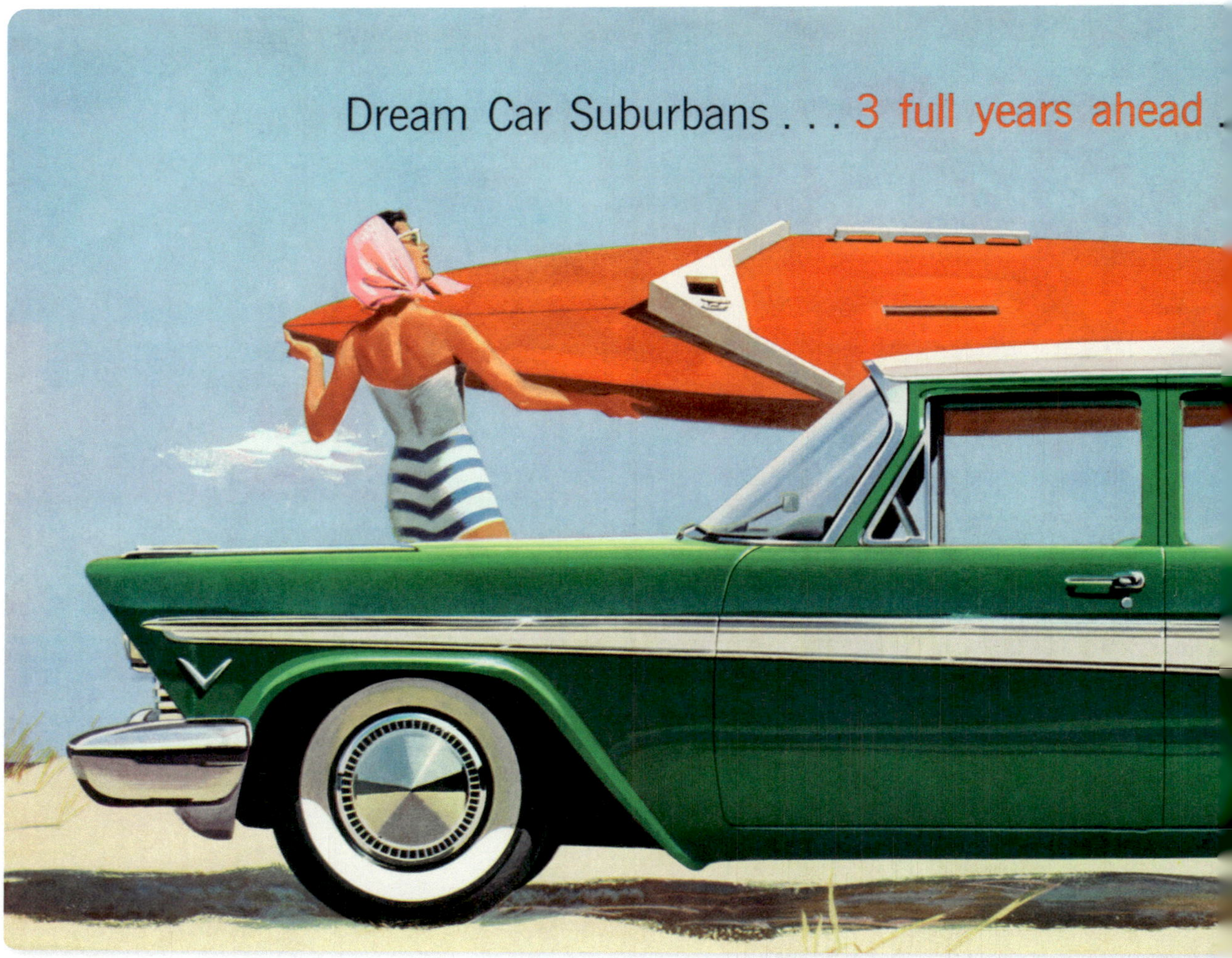

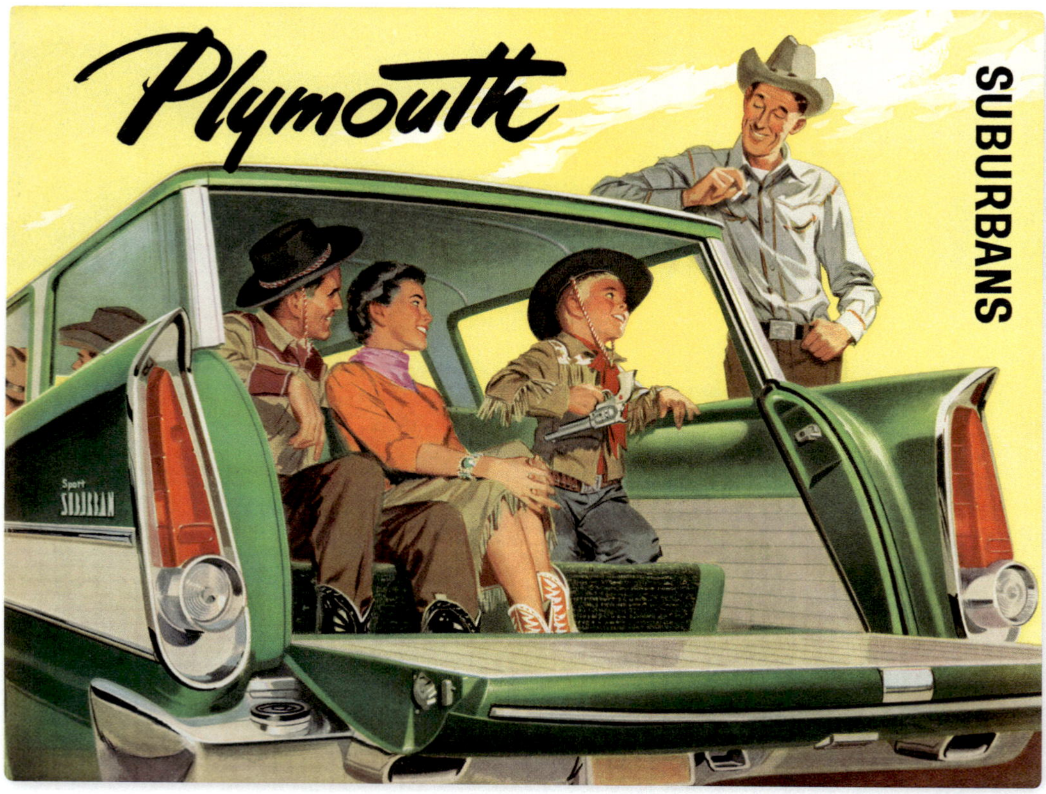

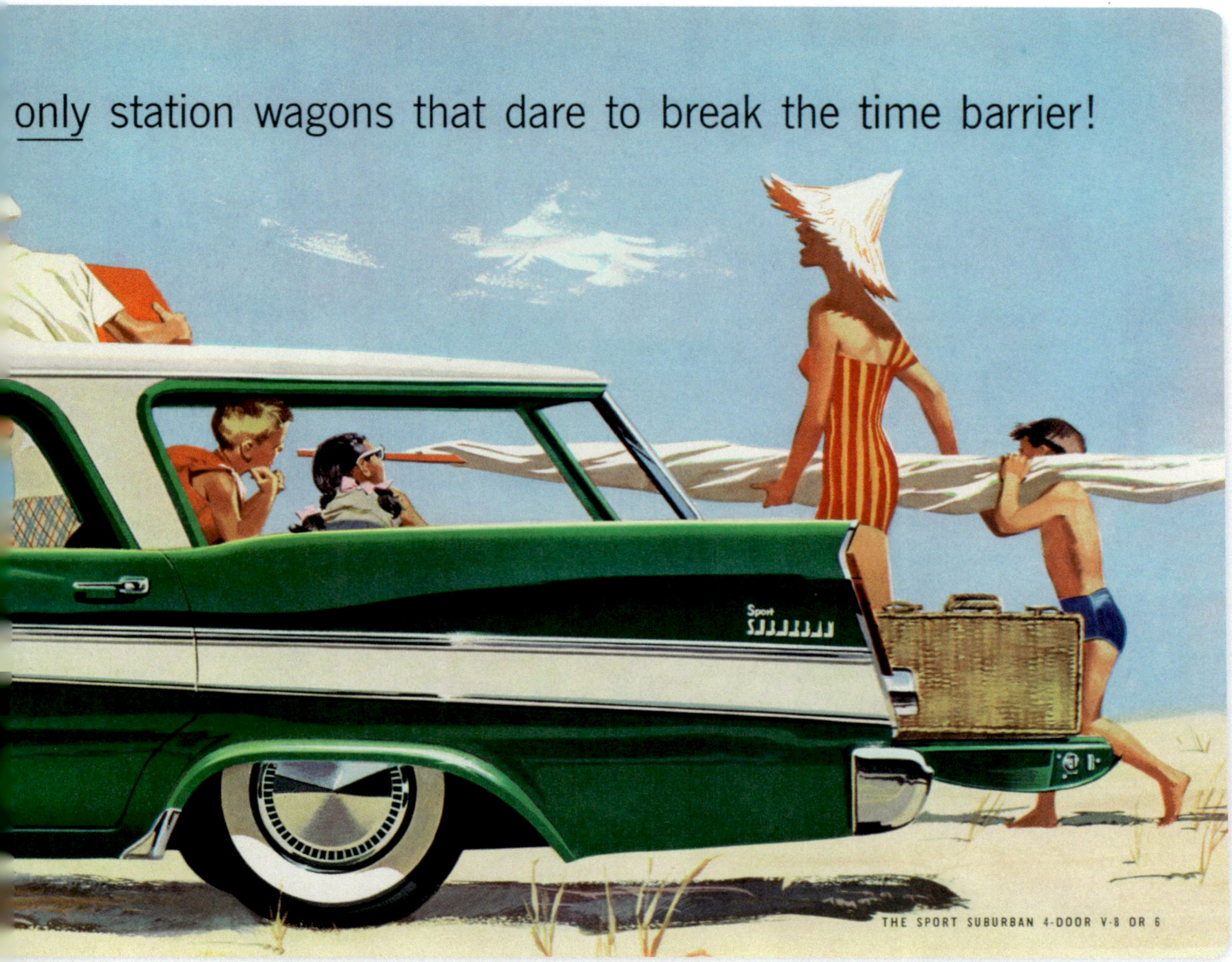

only station wagons that dare to break the time barrier!
THE SPORT SUBURBAN 4-DOOR V-8 OR 6

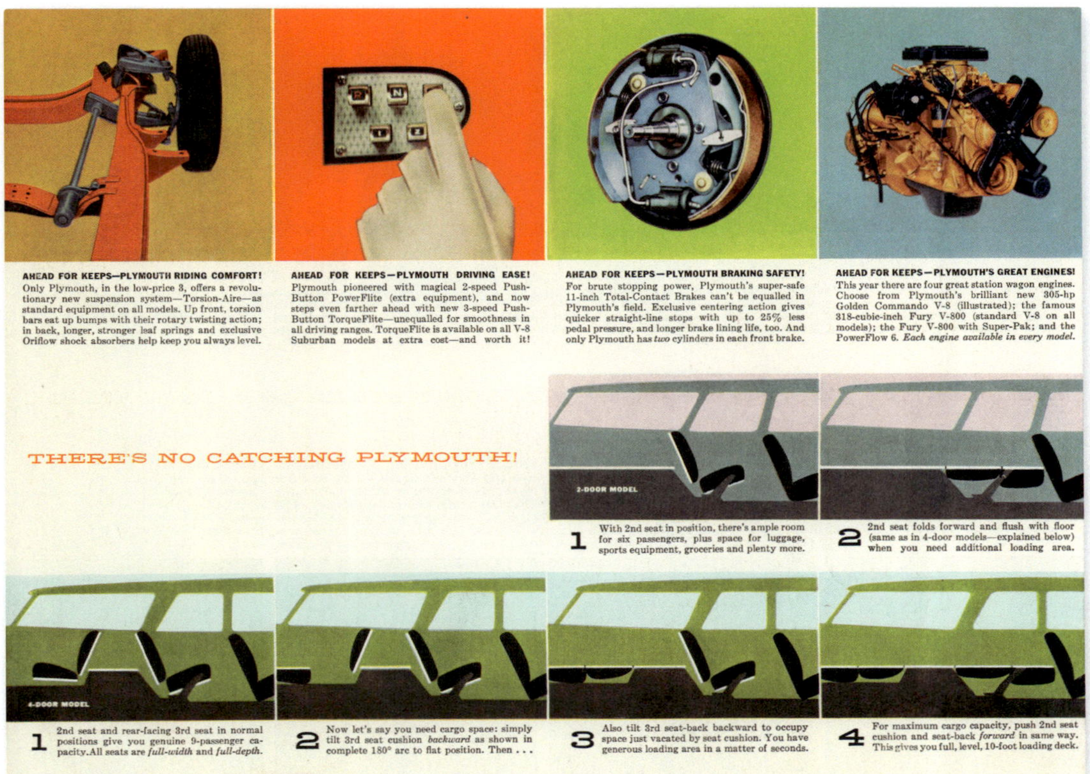

AHEAD FOR KEEPS—PLYMOUTH RIDING COMFORT!
Only Plymouth, in the low-price 3, offers a revolutionary new suspension system—Torsion-Aire—as standard equipment on all models. Up front, torsion bars eat up bumps with their rotary twisting action; in back, longer, stronger leaf springs and exclusive Oriflow shock absorbers help keep you always level.
AHEAD FOR KEEPS—PLYMOUTH DRIVING EASE!
Plymouth pioneered with magical 2-speed Push-Button PowerFlite (extra equipment), and now steps even farther ahead with new 3-speed Push-Button TorqueFlite—unequalled for smoothness in all driving ranges. TorqueFlite is available on all V-8 Suburban models at extra cost—and worth it!
AHEAD FOR KEEPS—PLYMOUTH BRAKING SAFETY!
For brute stopping power, Plymouth's super-safe 11-inch Total-Contact Brakes can't be equalled in Plymouth's field. Exclusive centering action gives quicker straight-line stops with up to 25% less pedal pressure, and longer brake lining life, too. And only Plymouth has *two* cylinders in each front brake.
AHEAD FOR KEEPS—PLYMOUTH'S GREAT ENGINES!
This year there are four great station wagon engines. Choose from Plymouth's brilliant new 305-hp Golden Commando V-8 (illustrated); the famous 318-cubic-inch Fury V-800 (standard V-8 on all models); the Fury V-800 with Super-Pak; and the PowerFlow 6. *Each engine available in every model.*
THERE'S NO CATCHING PLYMOUTH!
2-DOOR MODEL
1 With 2nd seat in position, there's ample room for six passengers, plus space for luggage, sports equipment, groceries and plenty more.
2 2nd seat folds forward and flush with floor (same as in 4-door models—explained below) when you need additional loading area.
4-DOOR MODEL
1 2nd seat and rear-facing 3rd seat in normal positions give you genuine 9-passenger capacity. All seats are *full-width* and *full-depth*.
2 Now let's say you need cargo space: simply tilt 3rd seat cushion *backward* as shown in complete 180° arc to flat position. Then . . .
3 Also tilt 3rd seat-back backward to occupy space just vacated by seat cushion. You have generous loading area in a matter of seconds.
4 For maximum cargo capacity, push 2nd seat cushion and seat-back *forward* in same way. This gives you full, level, 10-foot loading deck.

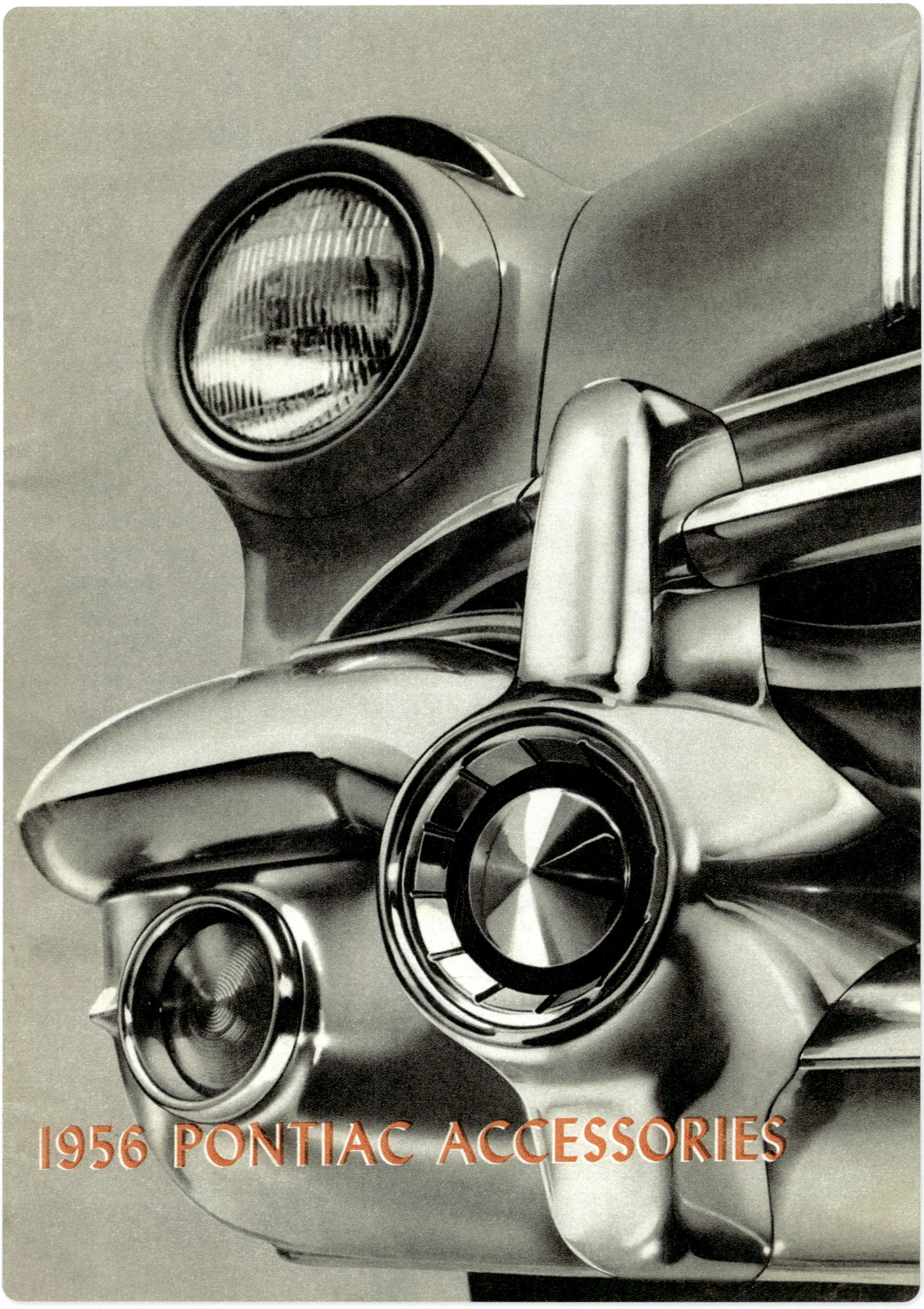
1956 PONTIAC ACCESSORIES

Pontiac

Pontiac 1956 *(opposite)*, **1950** *(right)*, **1955** *(below)*

Pontiac hadn't offered a V-8 engine since 1932. The new one that arrived in 1955 had a short-stroke configuration with overhead valves and produced 180 hp. Pontiac rightly celebrated by showing off its Strato Star and Club de Mer dream cars from the GM Motorama displays.

Pontiac hatte seit 1932 keinen V8-Motor mehr angeboten. Der neue, der 1955 auf den Markt kam, war ein Kurzhuber mit OHV-Ventilsteuerung, der 244 kW Leistung brachte. Pontiac feierte den Motor zu Recht, indem man die Konzeptfahrzeuge Strato Star und Club de Mer aus der Motorama-Ausstellung von GM zeigte.

Pontiac n'avait pas proposé de moteur V8 depuis 1932. Celui qui arriva en 1955, à soupapes en tête et course courte, développait 180 chevaux. Pontiac le célébra dignement en exhibant ses dream-cars Strato Star et Club de Mer dans les salons Motorama de la GM.

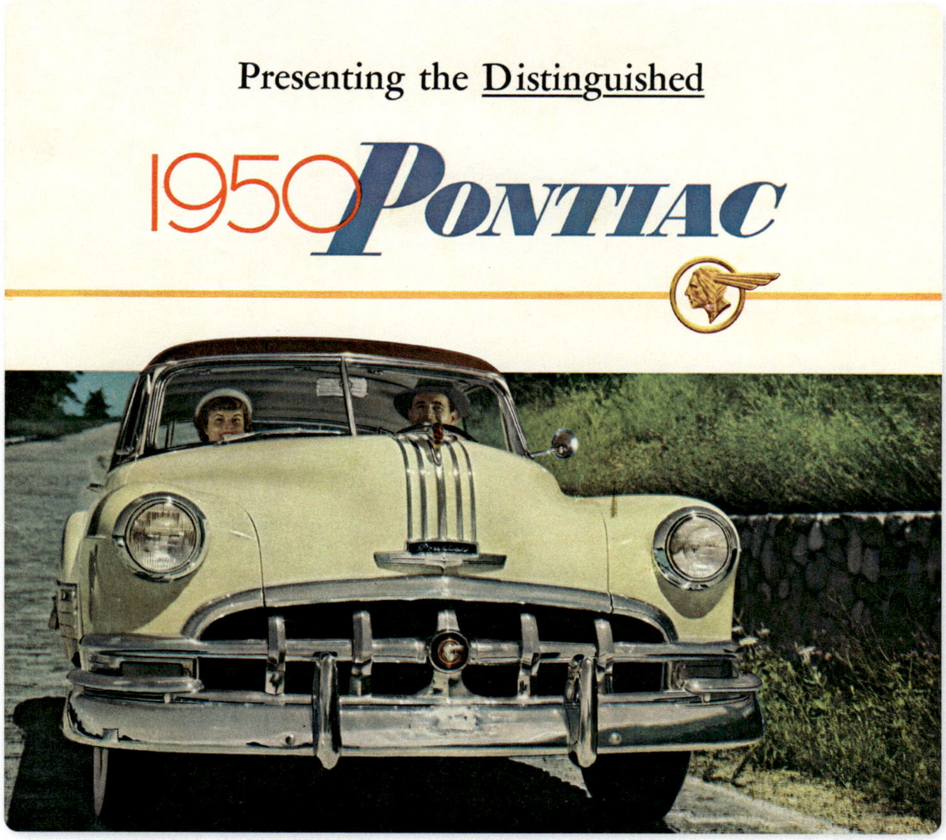

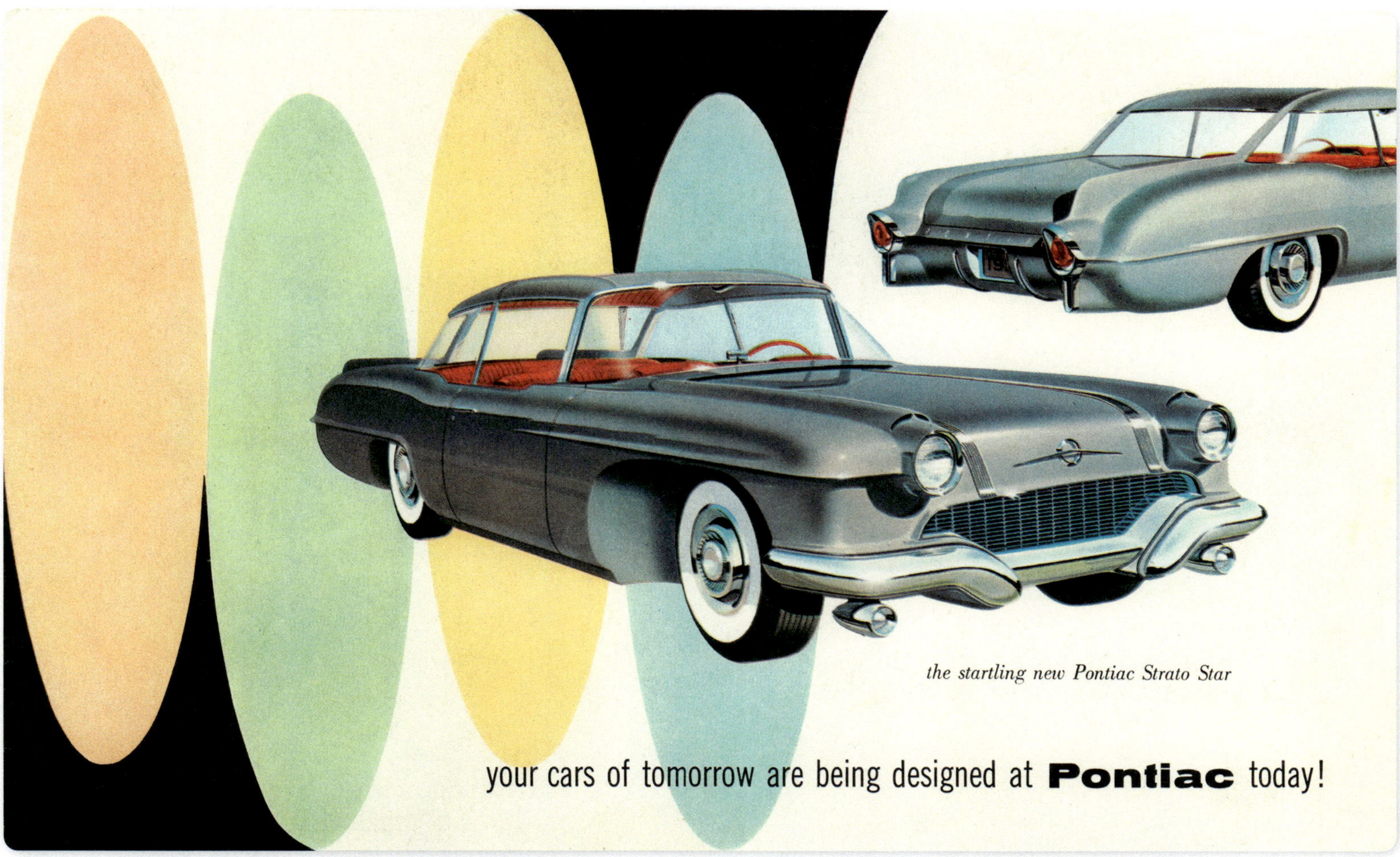

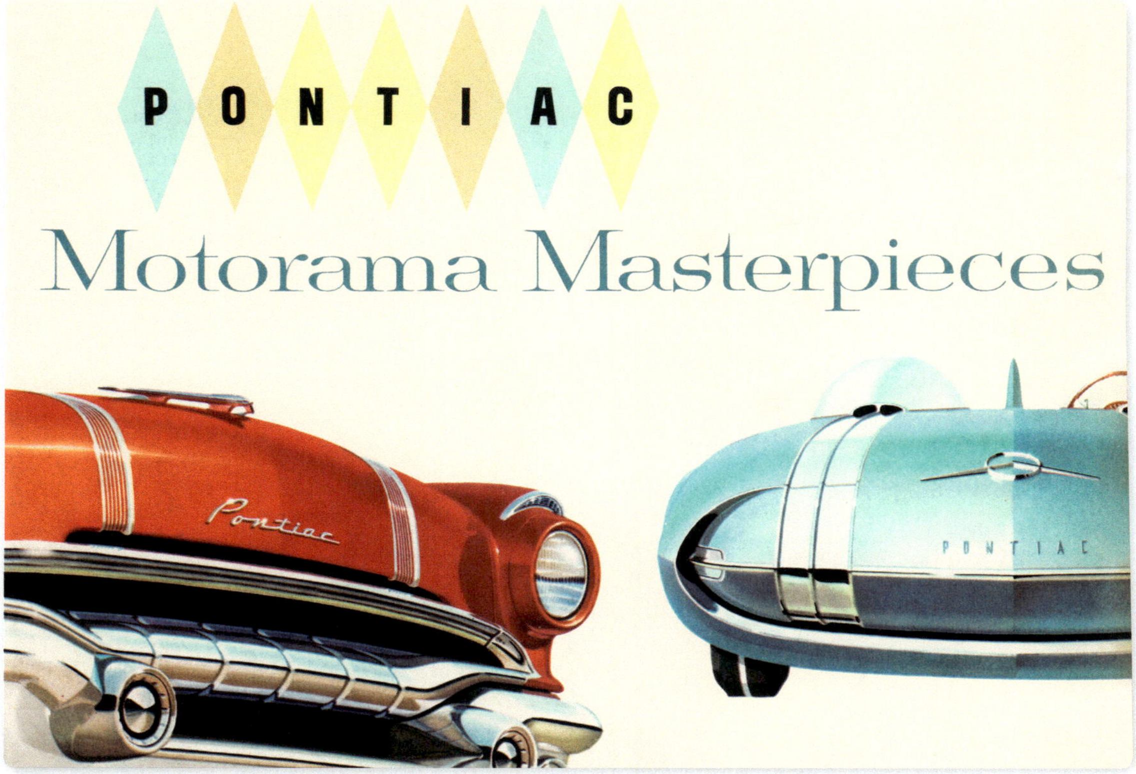

Pontiac 1956 *(this spread)*

One of the great model names and image creators in Pontiac history appeared in 1950. The Catalina was a handsomely proportioned two-door hardtop, with no center roof pillars, that was added to the Chieftain line. It was offered only in combinations of these colors, Sierra Rust and San Pedro Ivory.

Einer der gelungensten Modellnamen und eines der für das Image der Marke einflussreichsten Fahrzeuge erschien 1950 auf der Bildfläche. Der Catalina war ein gefällig proportioniertes zweitüriges Coupé ohne B-Säule, das in die Chieftain-Reihe eingeordnet wurde. Es war nur in einer Kombination dieser beiden Farben erhältlich, „Sierra Rust“ (rostfarben) und „San Pedro Ivory“ (elfenbeinfarben).

C'est en 1950 qu'apparut ce grand nom de l'histoire de Pontiac, qui contribua fortement à son image. La Catalina était un coupé deux portes hard-top joliment proportionné, sans montants centraux, ajouté à la série Chieftain. Seules couleurs proposées (et combinables) : « Sierra Rust » et « San Pedro Ivory » (rouille sierra et ivoire de San Pedro).

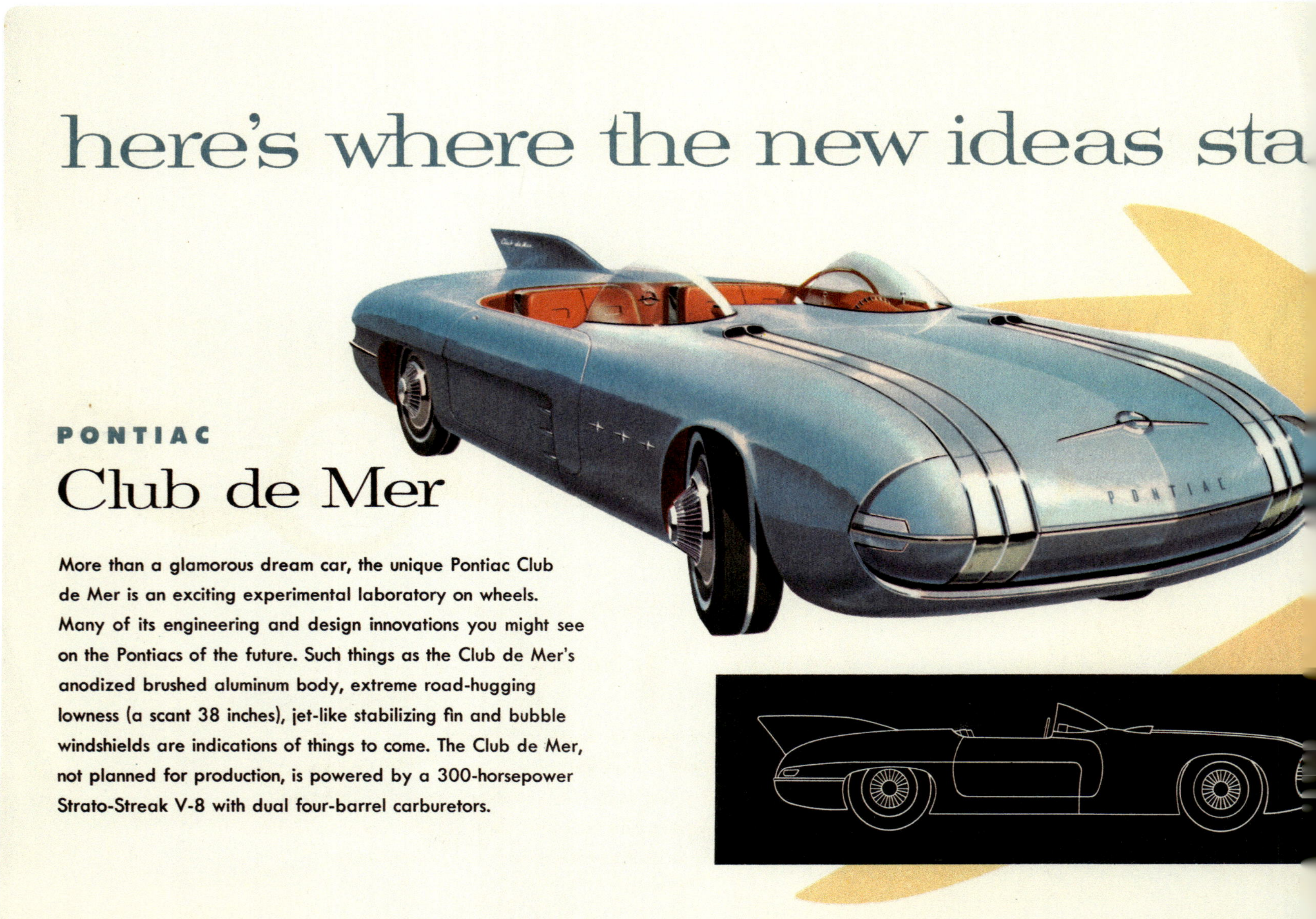

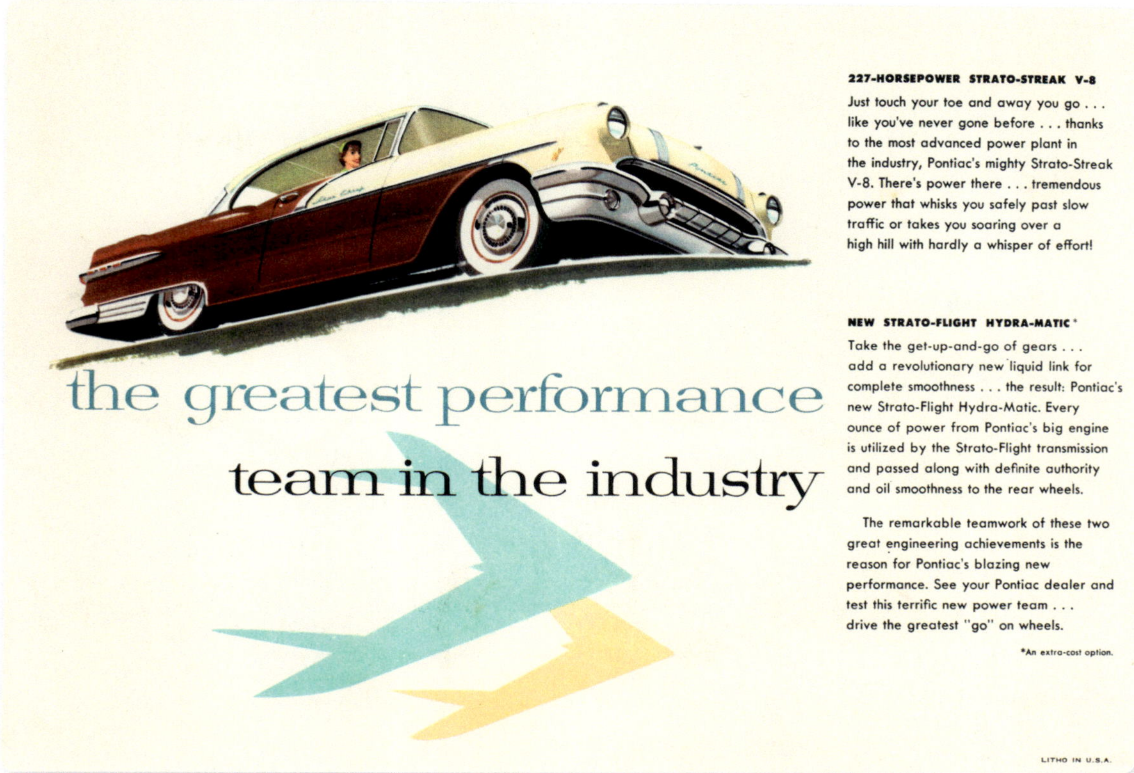
the greatest performance
team in the industry
227-HORSEPOWER STRATO-STREAK V-8
Just touch your toe and away you go . . . like you've never gone before . . . thanks to the most advanced power plant in the industry, Pontiac's mighty Strato-Streak V-8. There's power there . . . tremendous power that whisks you safely past slow traffic or takes you soaring over a high hill with hardly a whisper of effort!
NEW STRATO-FLIGHT HYDRA-MATIC*
Take the get-up-and-go of gears . . . add a revolutionary new liquid link for complete smoothness . . . the result: Pontiac's new Strato-Flight Hydra-Matic. Every ounce of power from Pontiac's big engine is utilized by the Strato-Flight transmission and passed along with definite authority and oil smoothness to the rear wheels.
The remarkable teamwork of these two great engineering achievements is the reason for Pontiac's blazing new performance. See your Pontiac dealer and test this terrific new power team . . . drive the greatest "go" on wheels.
*An extra-cost option.
LITHO IN U.S.A.

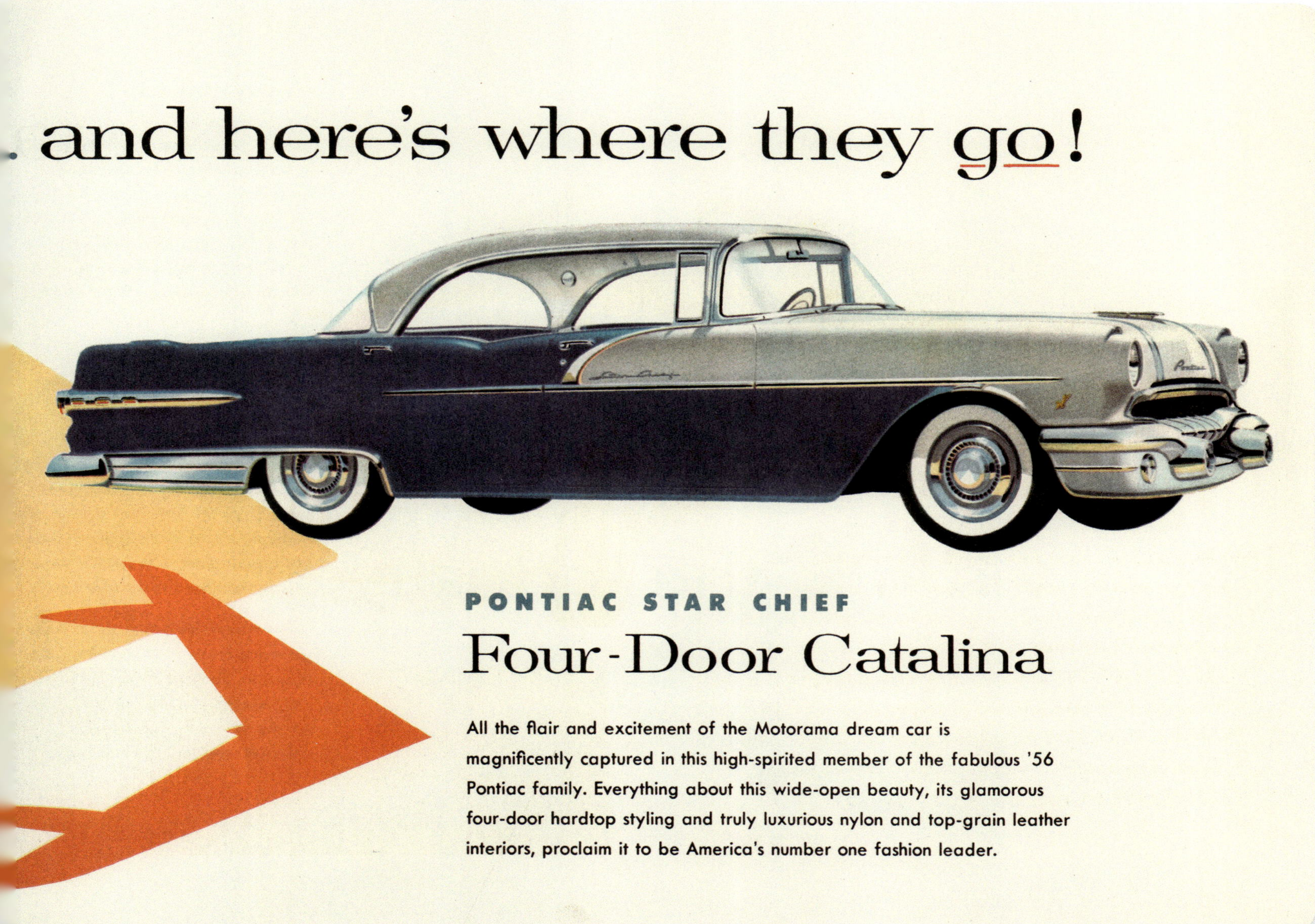
. and here's where they go!
PONTIAC STAR CHIEF
Four-Door Catalina
All the flair and excitement of the Motorama dream car is magnificently captured in this high-spirited member of the fabulous '56 Pontiac family. Everything about this wide-open beauty, its glamorous four-door hardtop styling and truly luxurious nylon and top-grain leather interiors, proclaim it to be America's number one fashion leader.

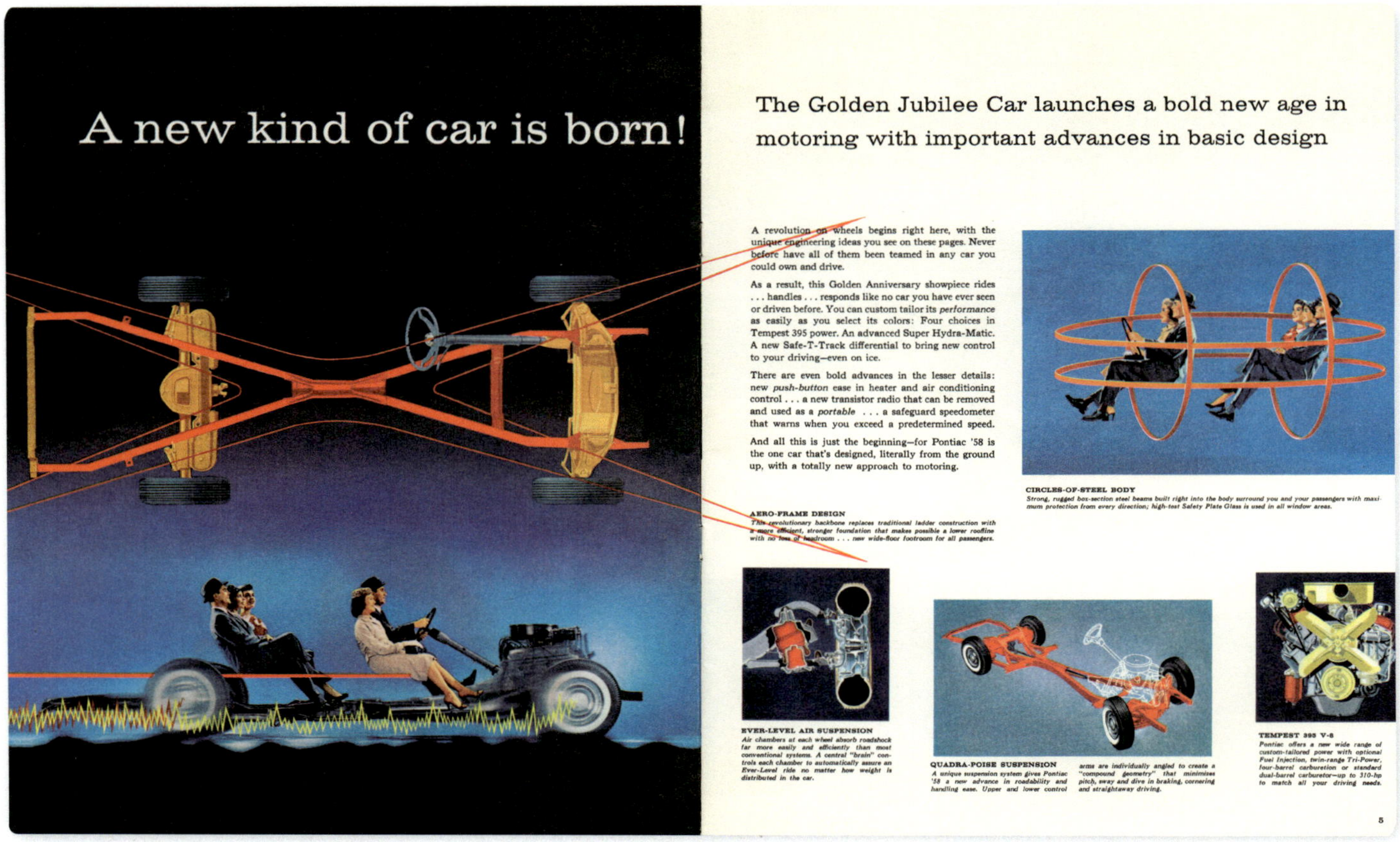

Pontiac 1958 *(above and opposite)*, **1960** *(right)*

The signature innovation that defined Pontiac during the 1960s was its Wide Track stance. Pontiacs were being restyled, and division president Bunkie Knudsen thought the new bodies looked too wide atop their chassis. He ordered the distance between the wheels increased five inches.

Die markanteste Neuerung, die das Erscheinungsbild des Pontiac in den 1960er-Jahren prägte, war die große Spurweite. Als die Pontiacs nämlich ein neues Design erhielten, war der für die Marke zuständige Semon „Bunkie" Knudsen der Meinung, die neuen Karosserien wirkten auf den bisherigen Fahrgestellen zu breit, und so ordnete er eine Erweiterung des Abstands zwischen den Rädern um 13 Zentimeter an.

L'innovation emblématique qui définira le mieux Pontiac pendant les Sixties restera l'adoption de voies élargies, baptisée « Wide Track ». Les Pontiac étaient en plein relooking et le président de la division Bunkie Knudsen jugeait que les nouvelles carrosseries semblaient flotter un peu trop au-dessus du châssis : il exigea un allongement de douze centimètres de la distance séparant les roues.

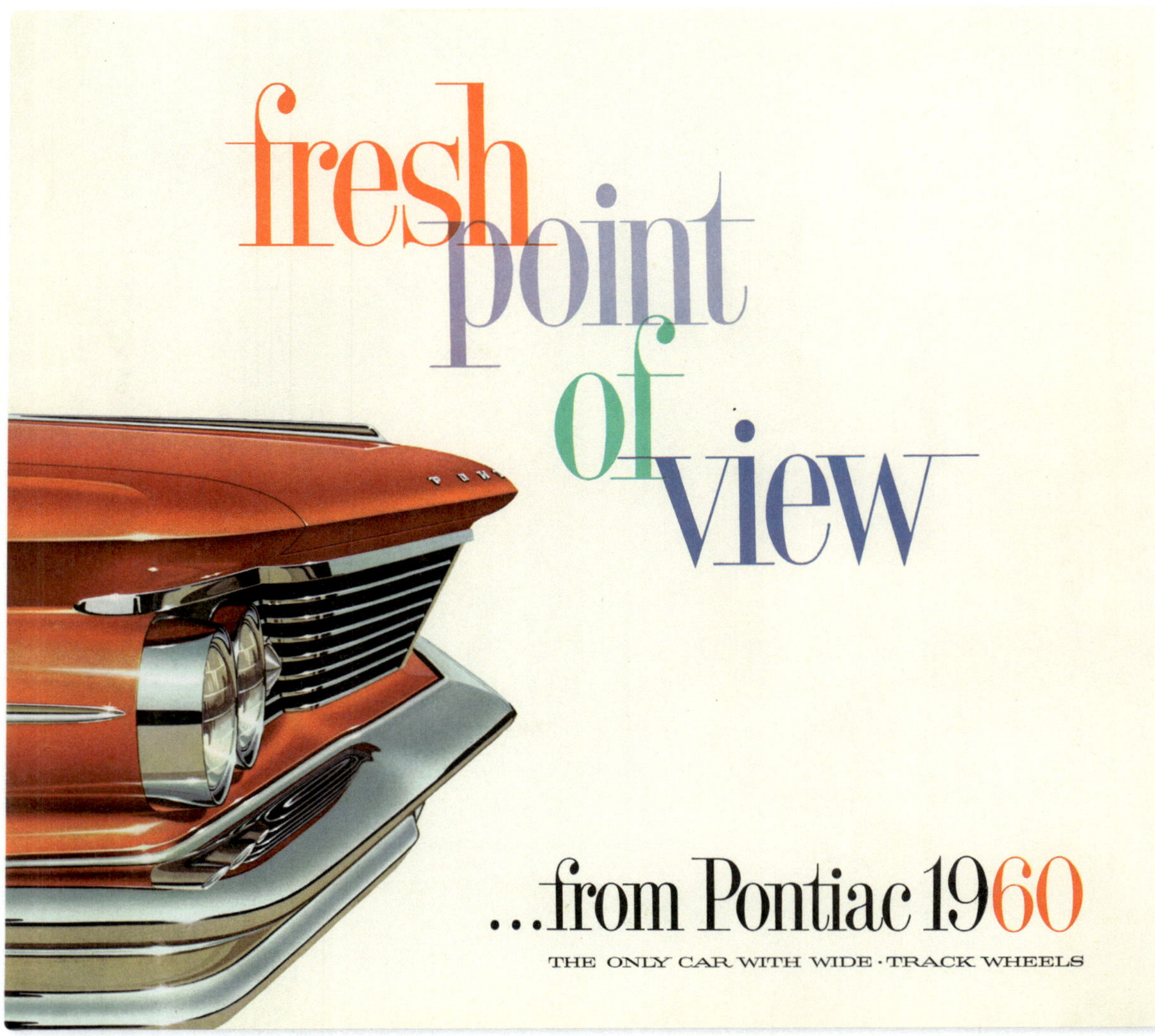

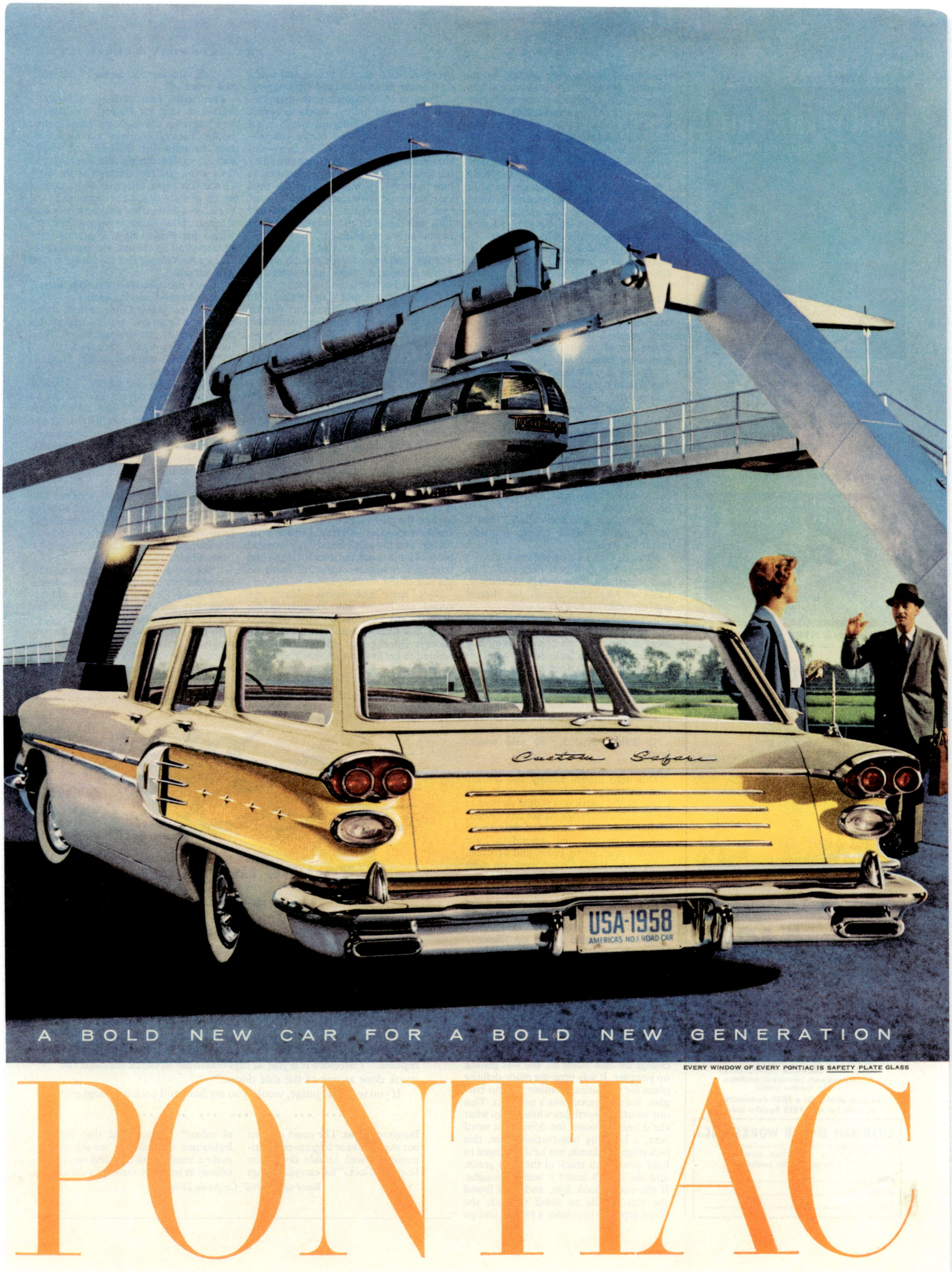
Custom Safari
USA-1958
AMERICAS NO.1 ROAD CAR
A BOLD NEW CAR FOR A BOLD NEW GENERATION
EVERY WINDOW OF EVERY PONTIAC IS SAFETY PLATE GLASS
PONTIAC

Renault

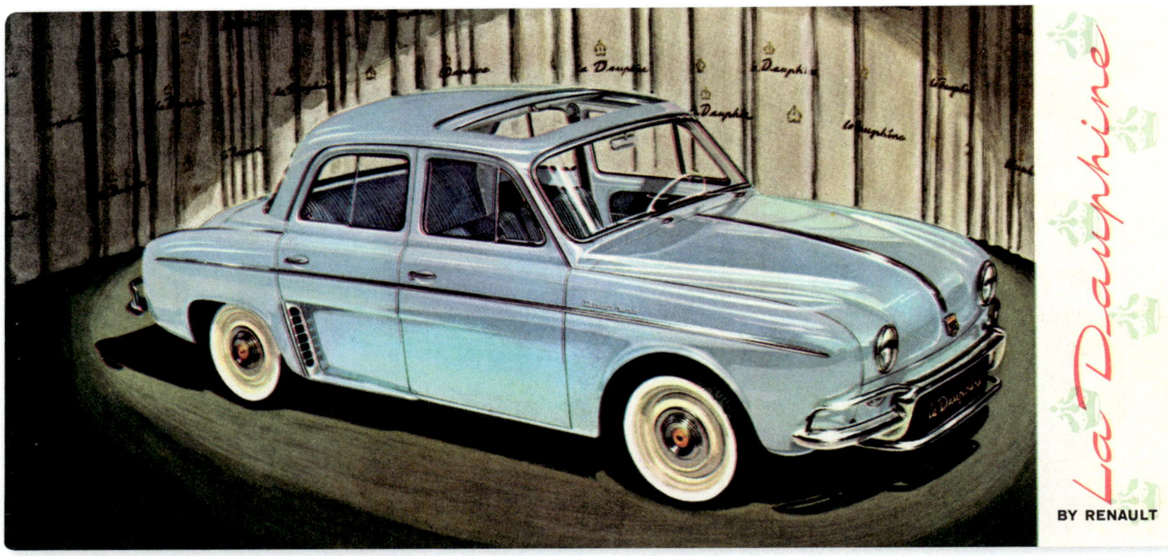

Renault 1960

It took a lot of gall to pitch this tiny four-door to power-crazed American buyers — regardless of how colorful it was — even as a second car. The Dauphine used a 845 cc engine, and its swing rear axles could make for scary handling.

Es bedurfte schon einiger Unverfrorenheit, hubraumgeilen amerikanischen Autokäufern diesen winzigen Viertürer schmackhaft machen zu wollen, selbst als Zweitwagen – ganz gleich, wie bunt er daherkam. Die Dauphine besaß einen 845-cm³-Motor, und die hintere Pendelachse sorgte zeitweise für ein abenteuerliches Fahrverhalten.

Même comme deuxième voiture, il fallait beaucoup de culot pour proposer cette petite quatre portes à un public américain affamé de puissance plus que de pittoresque. La Dauphine était propulsée par un moteur de 845 cm³, et sa suspension arrière à demi-essieux oscillants pouvait rendre sa tenue de route légèrement effrayante.

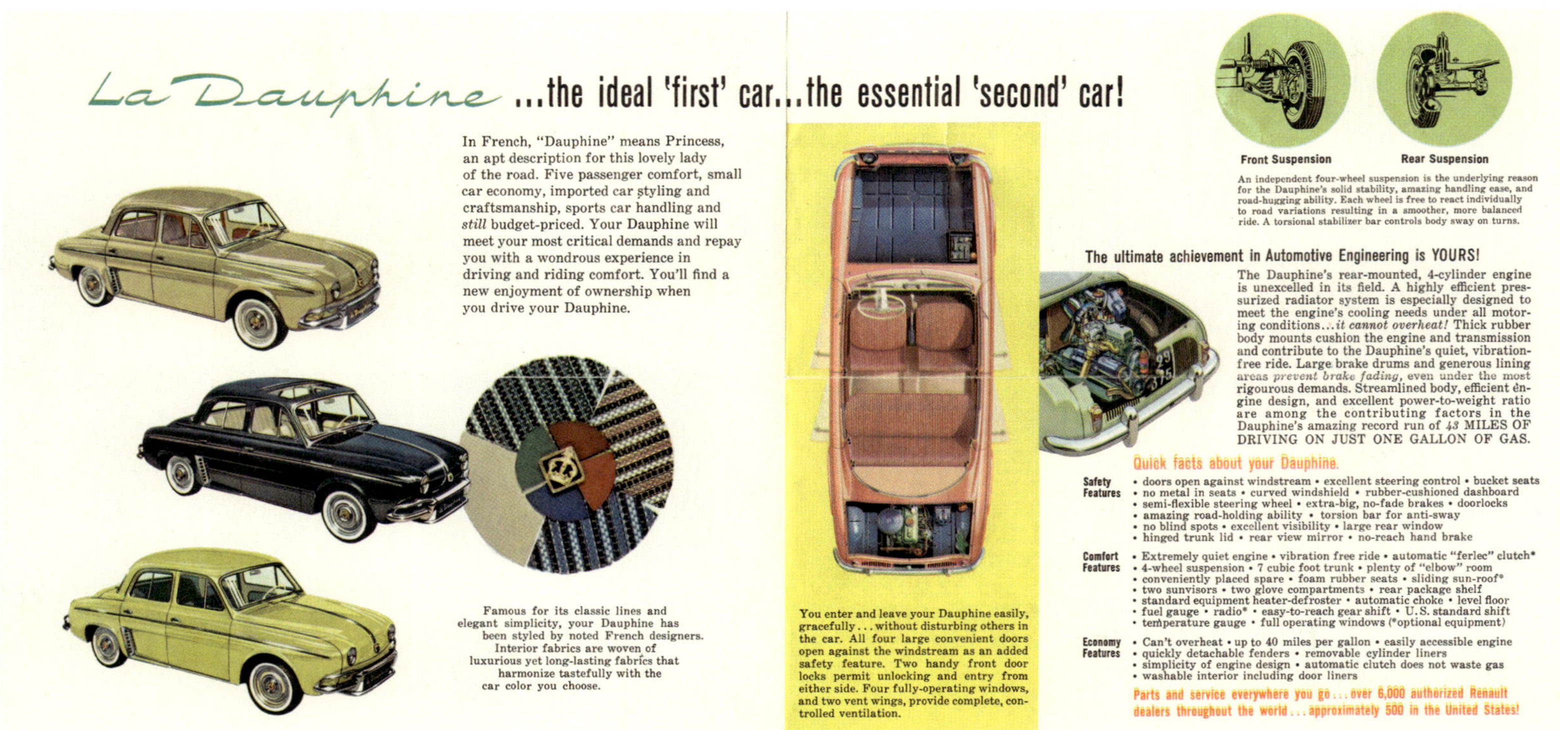

Rolls-Royce

Rolls-Royce 1960

Beginning in 1959, the signature Rolls-Royce had a V-8 engine, replacing the venerable straight-six of British yore. These were some of the world's most costly and desired automobiles. Rolls-Royce was also skilled in building highly exacting aircraft engines.

Ab 1959 erhielt der Rolls-Royce einen V8-Motor, der den altehrwürdigen britischen Reihensechszylinder ablöste. Diese Automobile gehörten zu den teuersten und begehrtesten der Welt. Rolls-Royce baute außerdem äußerst anspruchsvolle Flugzeugtriebwerke.

À partir de 1959, l'emblématique Rolls-Royce disposera d'un moteur V8 qui remplaçait l'honorable et typiquement « british » six cylindres d'antan. Ces automobiles figuraient parmi les plus onéreuses et les plus convoitées du monde. Rolls-Royce savait aussi fabriquer brillamment de très élaborés moteurs d'avion.

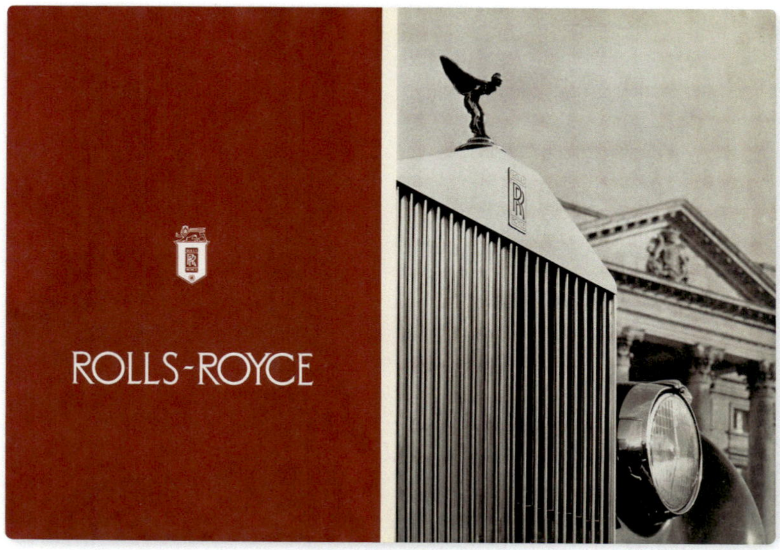

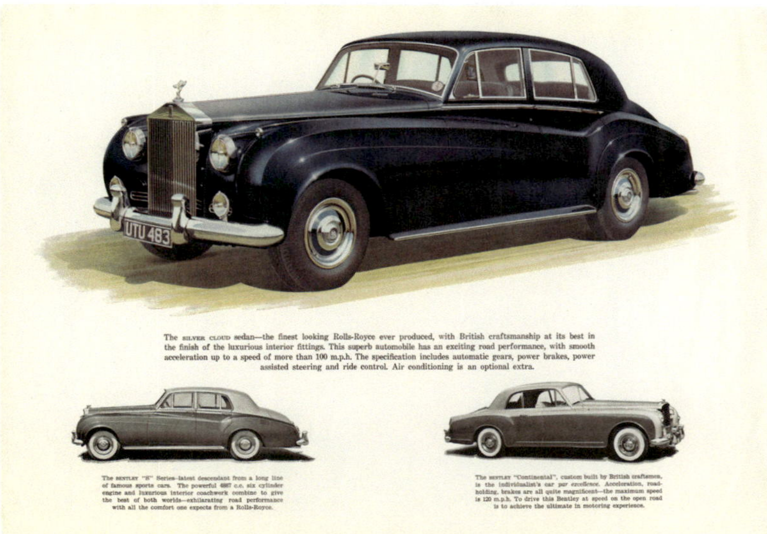

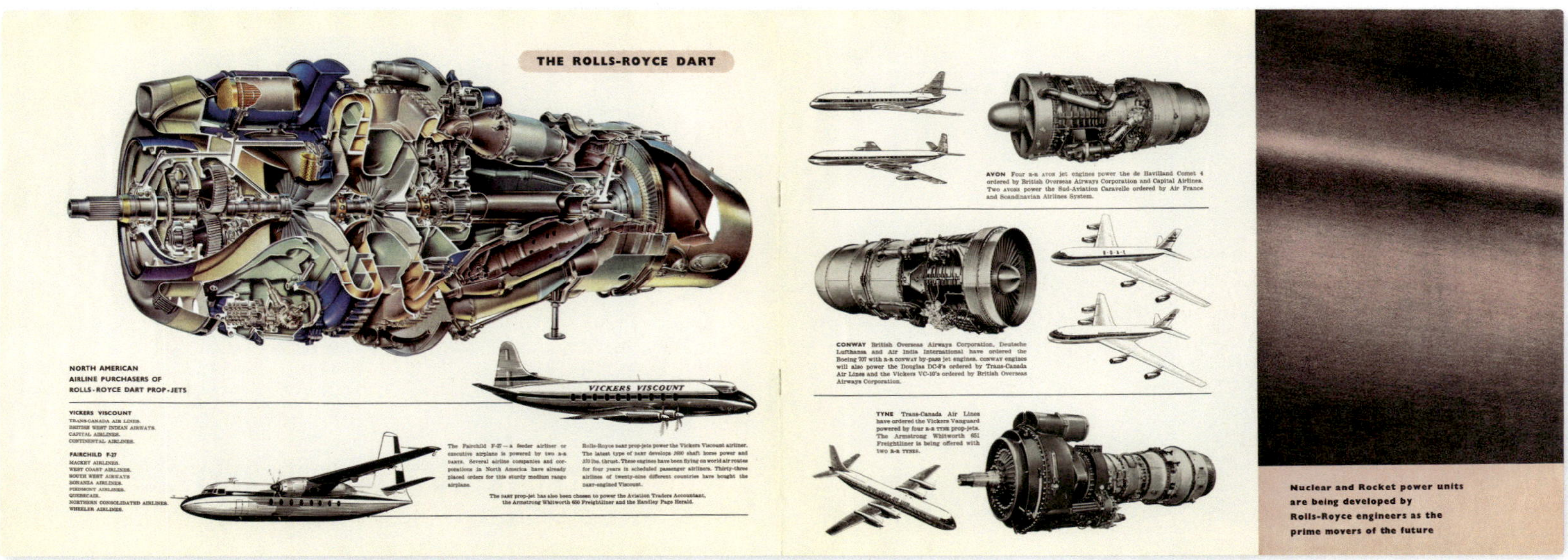

Rover

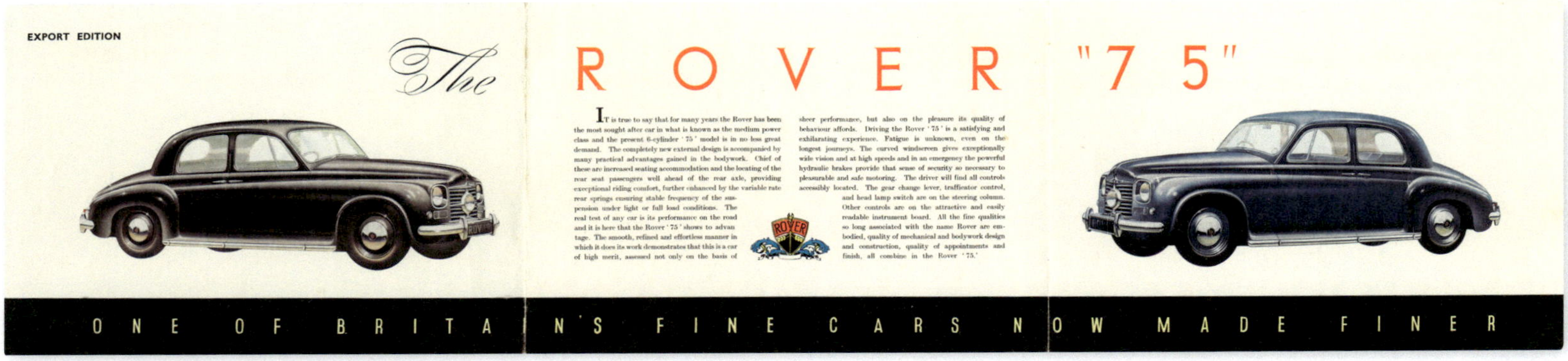

Rover P4 75 1949

These gentleman's four-door sedans were around in various forms from 1949 through 1962. At first they used a Rover IOE or F-head six, meaning the intake valves were located in the cylinder head but the exhaust valves were in the engine block, like in early Jeeps.

Diese Herrenlimousinen mit vier Türen kamen zwischen 1949 und 1962 in unterschiedlichen Formen auf den Markt. Zunächst verwendeten sie einen IOE-Sechszylindermotor von Rover, in den USA auch als „F-head" („F-Kopf") bekannt, was bedeutet, dass die Einlassventile im Zylinderkopf lagen, die Auslassventile hingegen darunter im Motorblock (IOE = „inlet over exhaust"), wie bei frühen Jeeps.

De 1949 à 1962, ces très distinguées berlines quatre portes se sont présentées sous diverses formes. Au début elles utilisaient un moteur Rover IOE six cylindres à soupapes en tête et latérales, c'est-à-dire que les soupapes d'admission étaient situées dans la culasse et les soupapes d'échappement dans le bloc moteur, comme dans les premières Jeep.

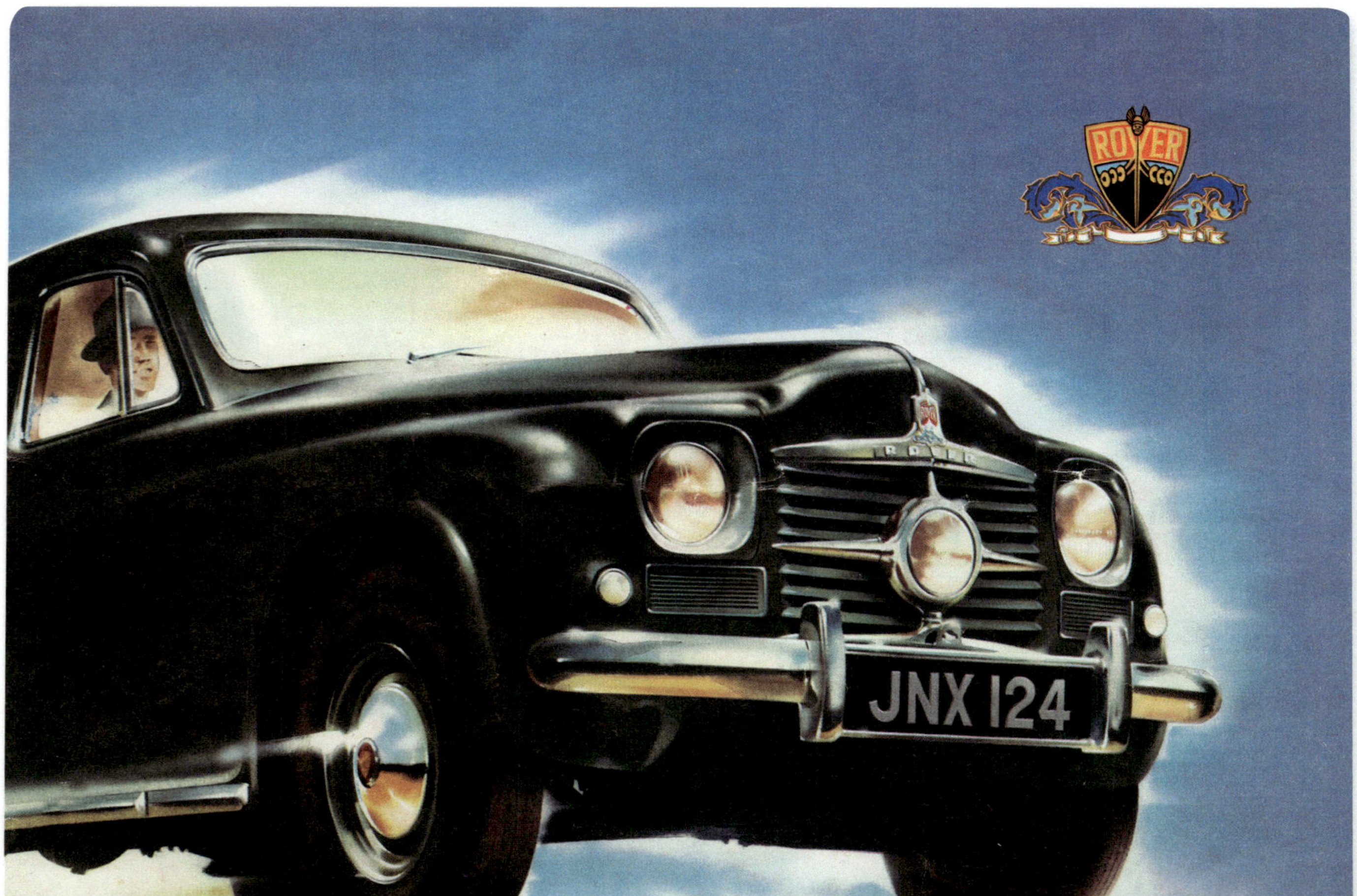

Studebaker

Studebaker 1949 *(right)*, **1950** *(below)*

As well-built and famous as these cars were, almost nobody noticed in 1950. Studebaker had added a very noticeable wraparound rear window the previous year, but then followed it up with a bullet-nose frontal treatment. It didn't become the next look in cars, but the bullet-front Studebakers were unforgettable.

So solide diese Autos auch gebaut waren und so berühmt sie später wurden, nahm 1950 kaum jemand Notiz von ihnen. Studebaker hatte seinen Fahrzeugen im Jahr zuvor ein sehr auffälliges umlaufendes Heckfenster verpasst und zog nun mit einer geschossförmigen Nase nach. Der Look setzte sich zwar im Automobilbau nicht durch, machte die solchermaßen gestalteten Studebakers aber unvergesslich.

Elles avaient beau être célèbres et fort bien conçues, en 1950 personne ou presque ne remarqua ces voitures. Studebaker les avait pourtant pourvues l'année précédente d'une ostensible lunette arrière panoramique avant de leur ajouter ce que certains qualifieront de « nez d'avion ». Un look qui ne fera pas beaucoup d'adeptes dans le monde automobile mais rendra ces Studebaker inoubliables.

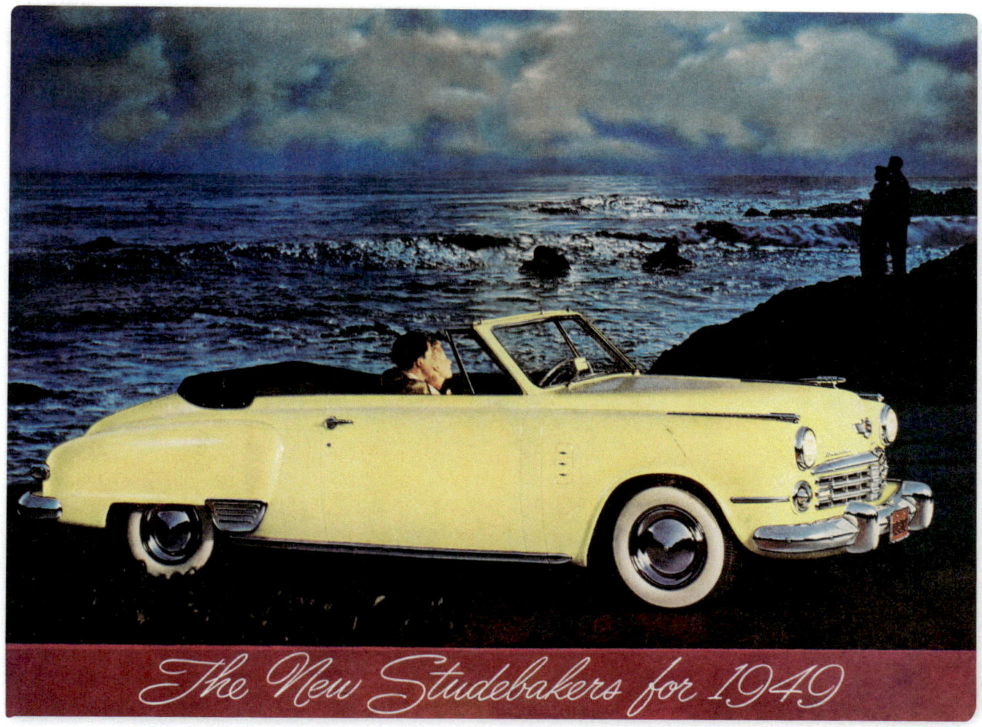

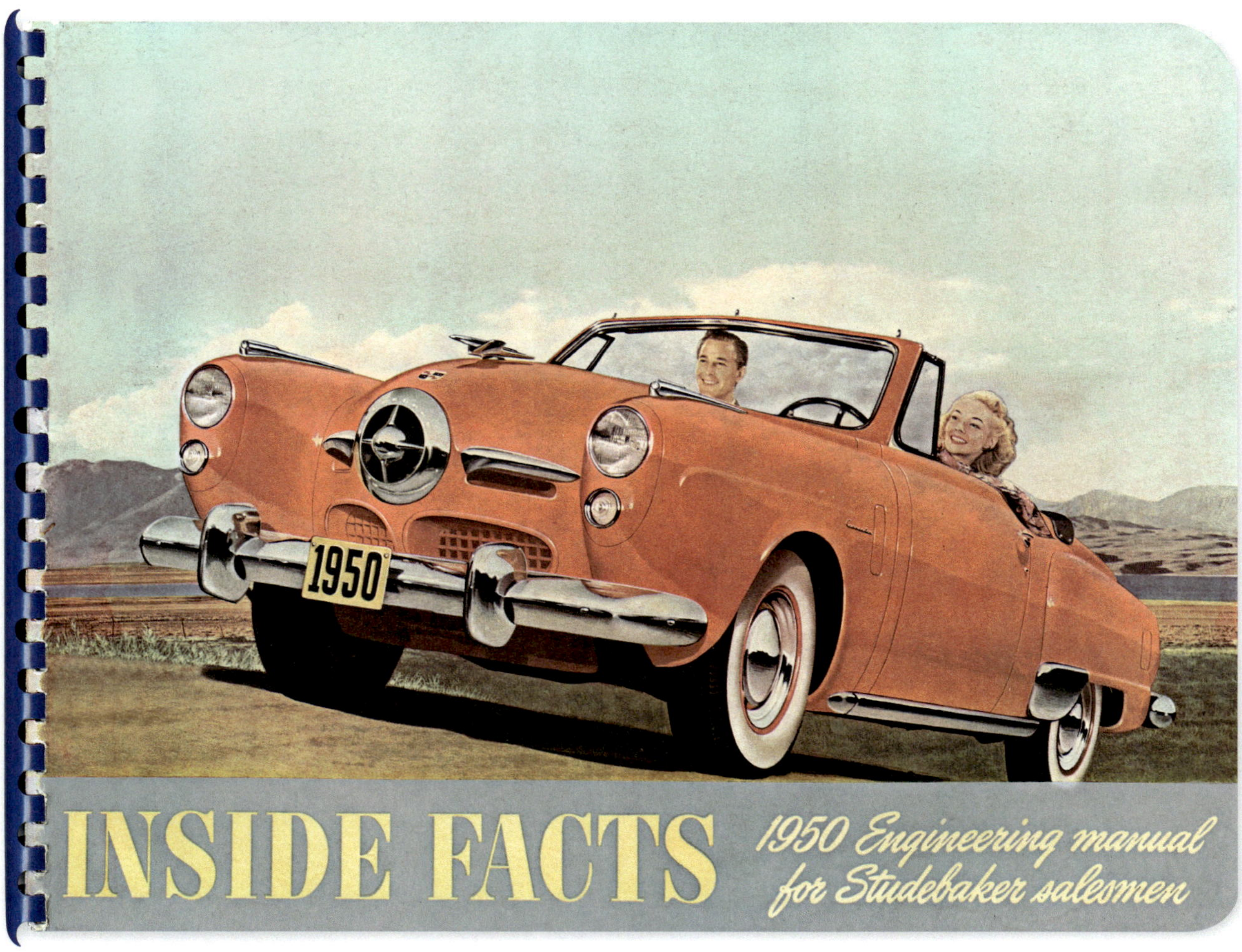

Studebaker 1956 *(right)*, **1957**
(below and opposite)

Two Hawk models were offered in 1957. The interior view shows the Silver Hawk, which was a pillared coupe only. Both bodies dated back to 1953, but were freshened by industrial designer Brooks Stevens. Studebaker-Packard was the U.S. distributor for Mercedes-Benz, so the grille design was no coincidence.

Im Jahre 1957 wurden zwei Modelle des Hawk angeboten. Die Innenansicht zeigt den Silver Hawk, ein Coupé mit B-Säule. Beide Karosserieformen gehen bis ins Jahr 1953 zurück, wurden aber von Industriedesigner Brooks Stevens aufgepeppt. Studebaker-Packard war auch Vertriebspartner von Mercedes-Benz in den USA, sodass die Form des Kühlergrills kein Zufall ist.

Deux modèles Hawk ont été présentés en 1957. On voit ci-dessous le coupé Silver Hawk, avec montant central. Les deux carrosseries remontaient en fait à 1953 mais avaient été rafraîchies par le designer industriel Brooks Stevens. Studebaker-Packard était aussi le distributeur de Mercedes-Benz pour les États-Unis, d'où le dessin modérément fortuit de la calandre.

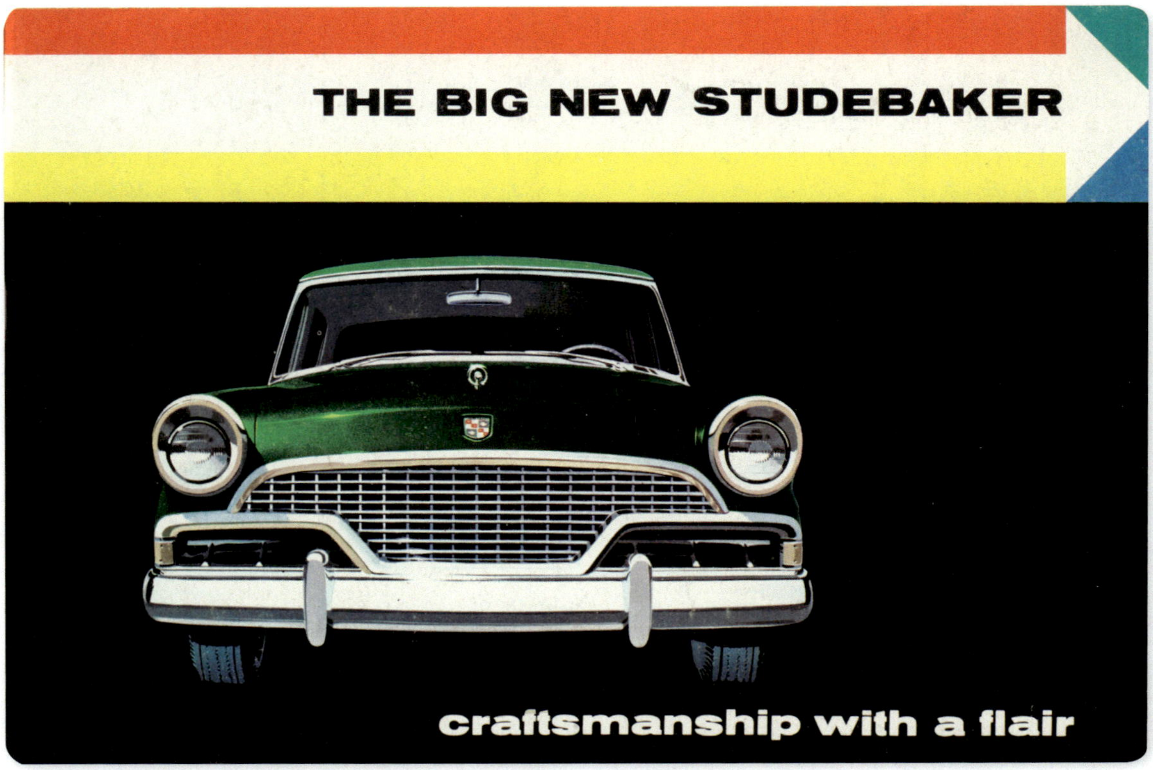

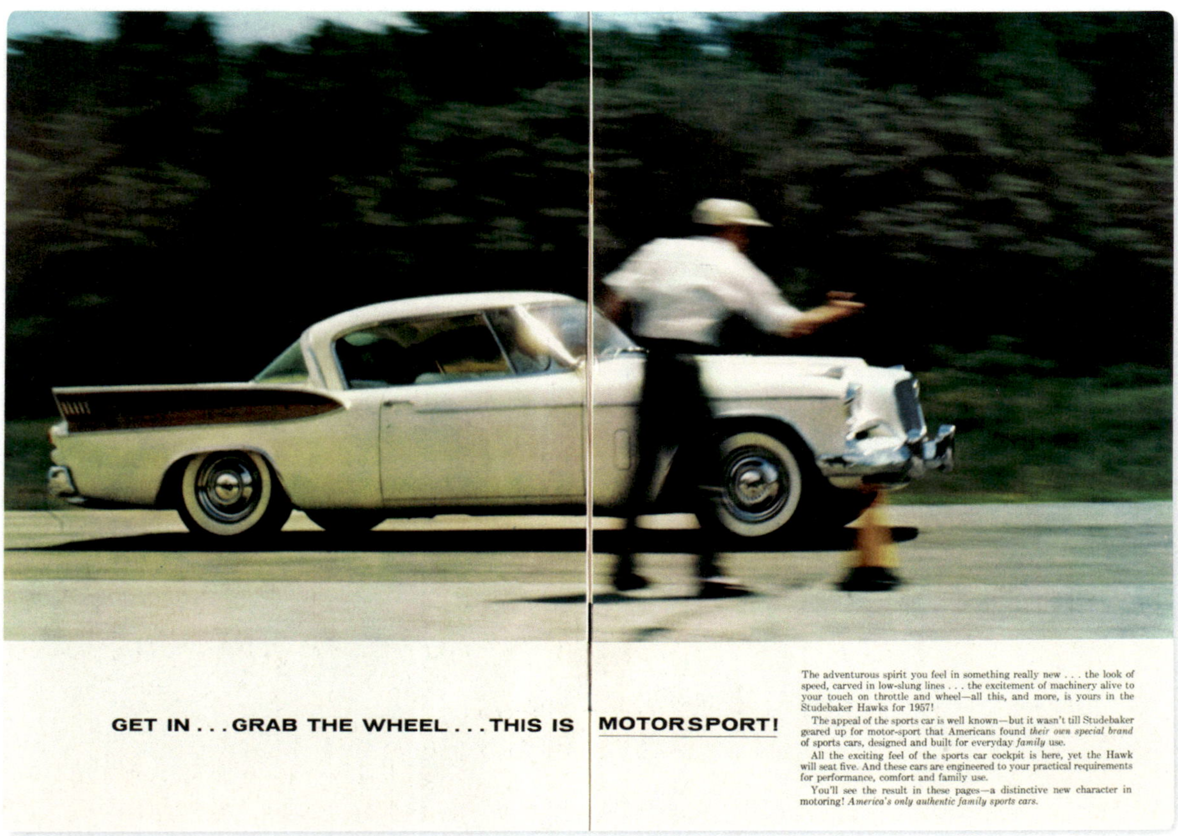

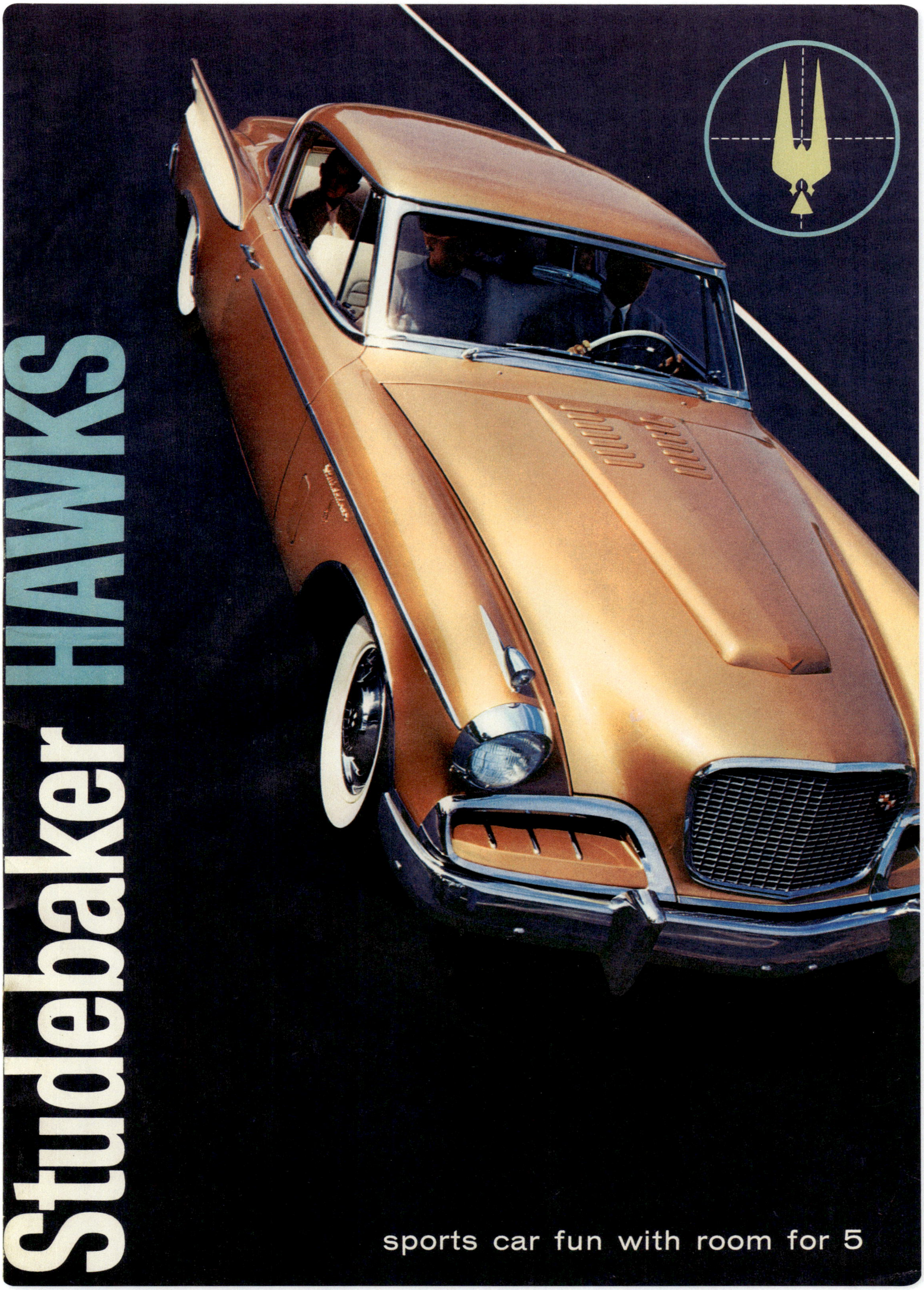
Studebaker HAWKS
sports car fun with room for 5

Tucker

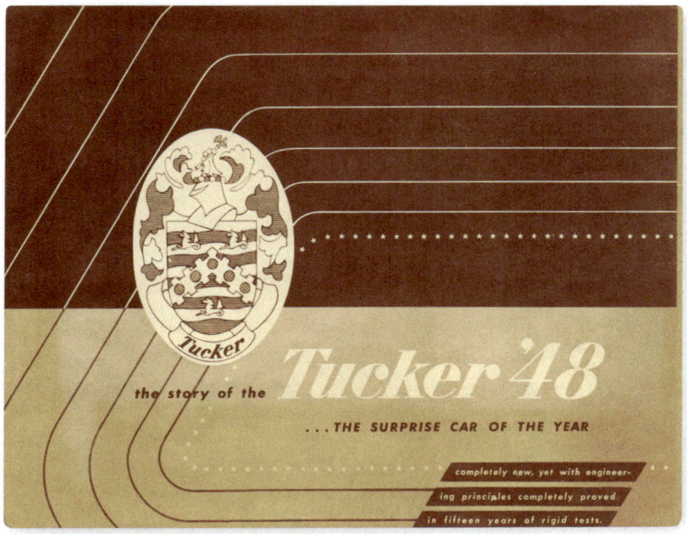

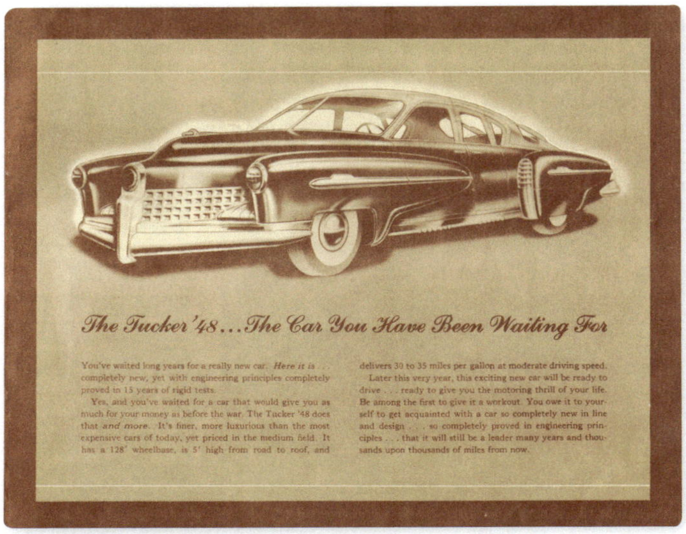

Tucker 1948 *(this spread)*

Preston Tucker was a promoter and entrepreneur. He engaged Alex Tremulis to design this wildly different sedan with a rear-mounted Continental flat six engine. Only 51 were built before a stock fraud trial sank the firm, although Tucker was cleared.

Preston Tucker war ein Unternehmer und Organisator. Er engagierte Alex Tremulis, der ihm diese völlig andersartige Limousine mit einem Sechszylinderboxerheckmotor von Continental konstruierte. Es wurden nur 51 Fahrzeuge gebaut, bevor das Unternehmen im Zuge eines Verfahrens wegen angeblichen Aktienbetrugs unterging – auch wenn Tucker von allen Vorwürfen freigesprochen wurde.

Entrepreneur passionné d'automobile depuis son plus jeune âge, Preston Tucker engagea Alex Tremulis pour lui confier la conception de cette berline assez follement anticonformiste, propulsée par un moteur six cylindres à plat Continental. 51 exemplaires seulement seront fabriqués avant qu'un procès pour malversations boursières (entre autres choses) n'entraîne la chute de la firme, même si Tucker sera lavé de tout soupçon.

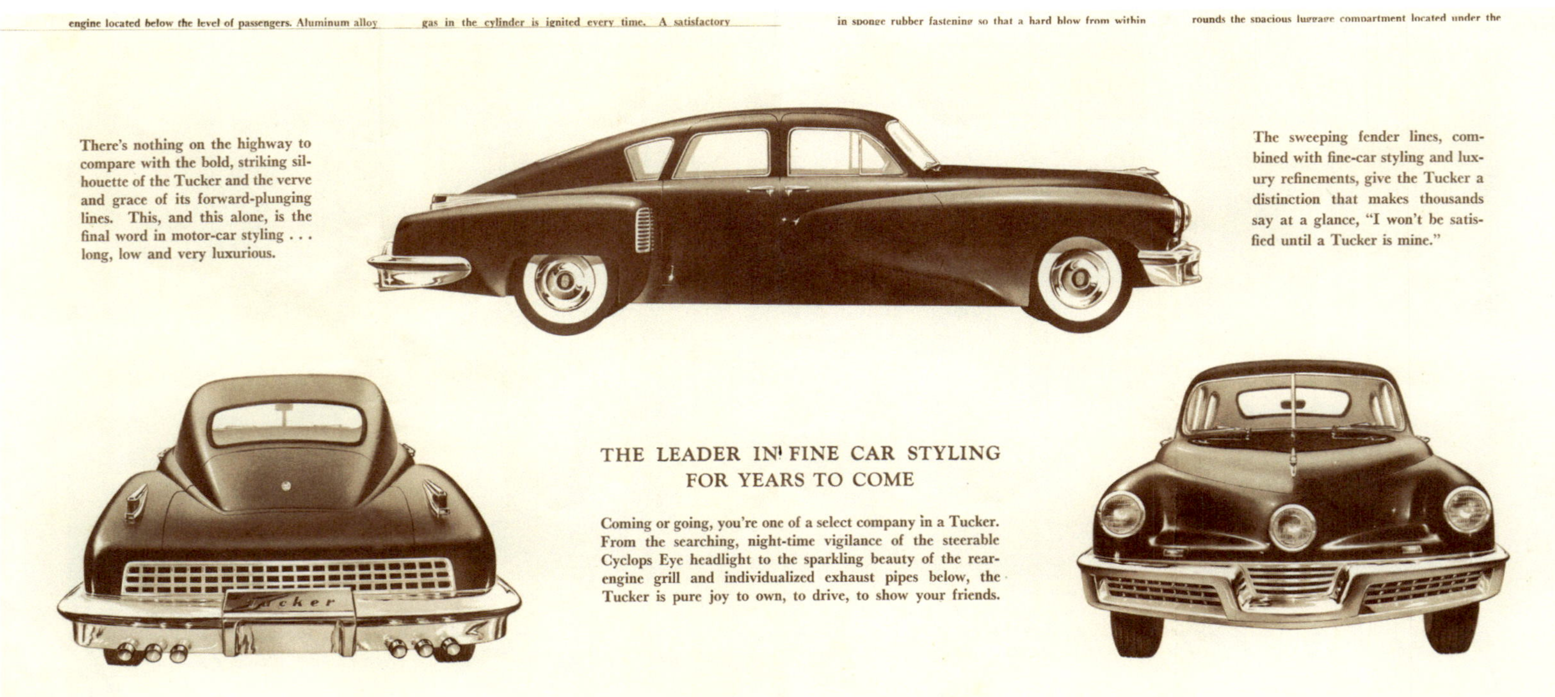

Step into a New Automotive Age in the Rear Engine Tucker

A FEW OF MANY NEW Engineering Features

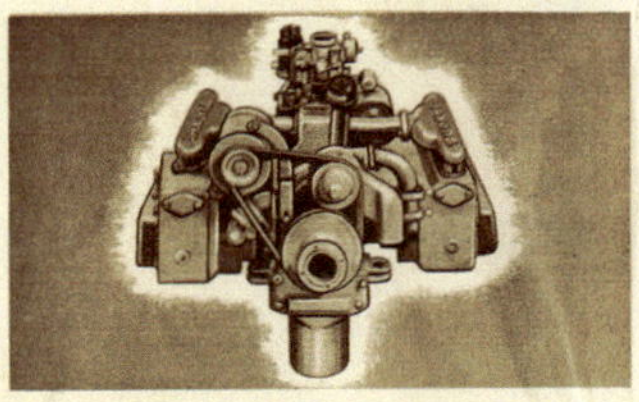

Rear Engine—166 horsepower, flat opposed 6-cylinder engine located below the level of passengers. Aluminum alloy construction. More power for the weight of the car than any volume production automotive engine ever built. Eliminates fumes, heat, noise in passenger compartment.

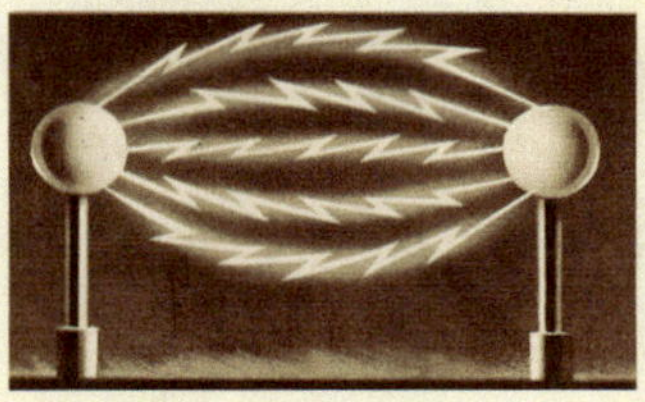

Tucker Ignition—A hot, lasting, ignition-spark. All the gas in the cylinder is ignited every time. A satisfactory answer to engine "pings" and power knocks. Real assurance of all-weather push-button speed in starting.

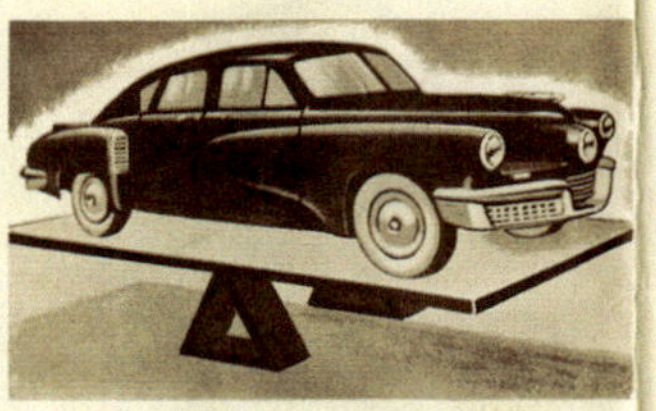

Precision Balance—The unique Tucker design distributes weight to give maximum safety, maximum power transmission, feather-light steering and driving control, and—for the first time—insures complete four-wheel traction in braking. Only a rear engine car can achieve this precision balance—for years the goal of all automotive engineers.

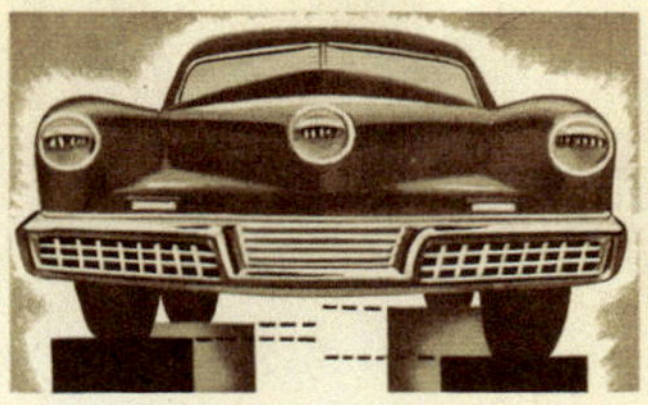

Individual Wheel Suspension—The new Tucker individual rubber torsional wheel-suspension cushions each wheel by its own resilient action arm, actually erasing shock instead of simply softening it. Also eliminates all gyroscopic forces which frequently cause conventional cars to veer with wind, and weave or pitch at touring speeds.

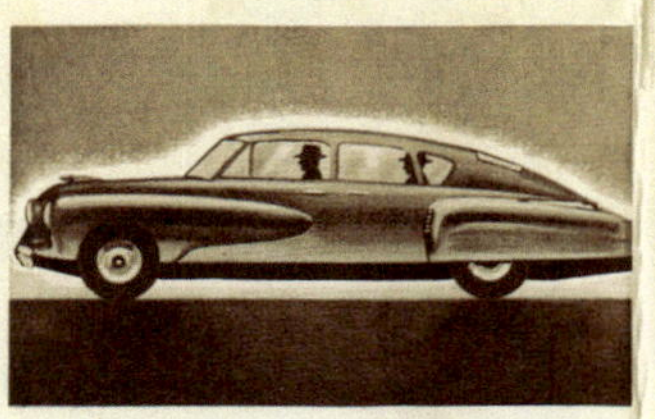

Frame Lower Than Center Line of Wheels—An exclusive feature made possible by locating engine in rear and eliminating conventional drive shaft. This, combined with Tucker suspension system, greatly reduces chances of skidding or overturning.

A FEW OF MANY NEW Safety Features

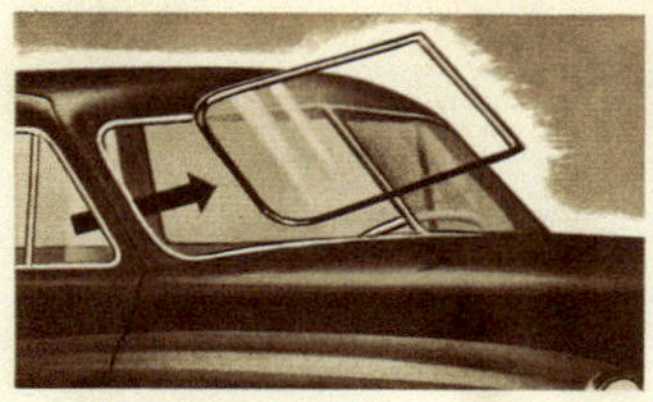

Safety Windshield—Laminated safety glass is mounted in sponge rubber fastening so that a hard blow from within will eject it in one piece. Thus, greatest collision hazard—lacerations or fractured skull from striking windshield—is entirely eliminated. Windows are armor-plate glass which disintegrates without cutting edges or slivers.

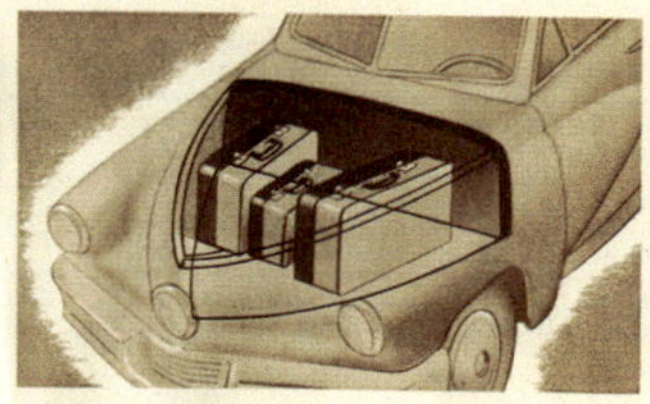

Safety Steel Bulkheads—A steel safety bulkhead surrounds the spacious luggage compartment located under the hood, thus shielding the passenger compartment from head-on collisions far more effectively than the conventional front engine construction. A second steel safety bulkhead walls off the rear engine.

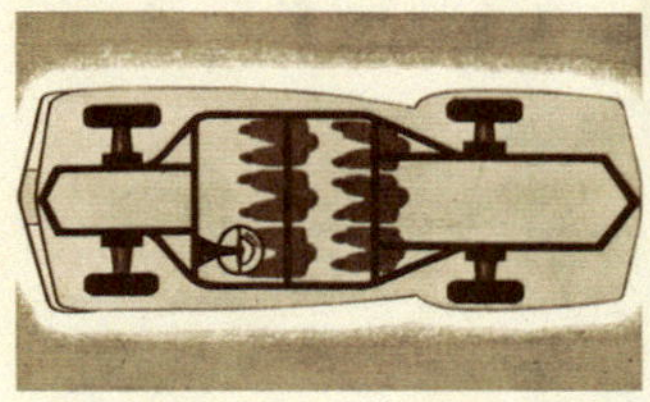

Safety Frame Surrounding Passenger Compartment—Vital protection against injury in case of collision. And protection for car, too, because frame is tapered front and rear like the prow of a ship. Thus a slanting blow—as in 90% of all collisions—is deflected sidewards with minimum damage.

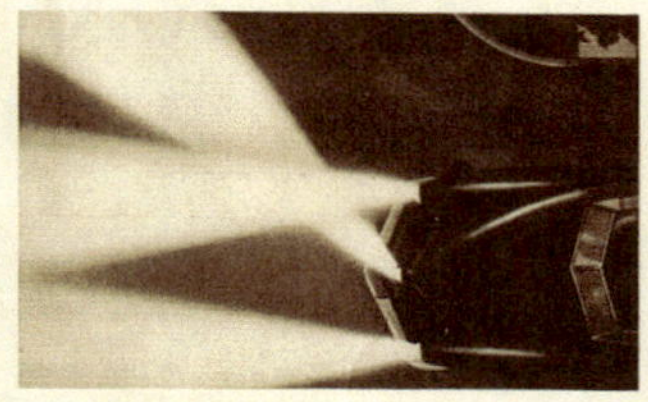

Cyclops Eye—In addition to regular fixed headlights, the Tucker has a center cyclops eye which turns with front wheels. Result: Your Cyclops beam is around the corner before you are, lighting the way ahead, giving you precious seconds to avoid accidents.

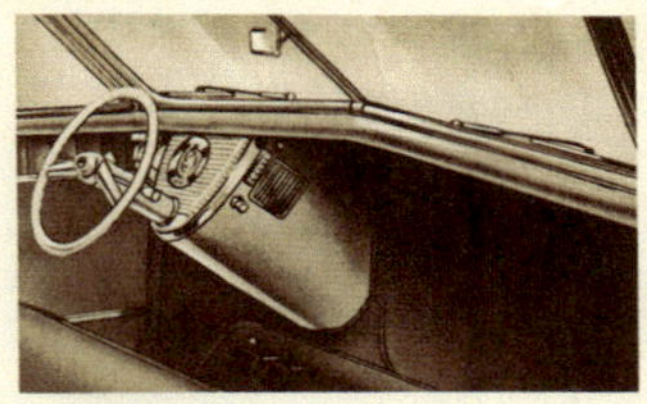

Crash Board Cowl and Safety Chamber—Conventional instrument panel is replaced by attractive sponge rubber crash board cowl. Instruments in steering column. Under cowl is spacious safety chamber, protected by steel bulkheads, which driver and front seat occupants can drop into, in a split second, in case of impending collision.

A Word To Women Who Drive Or Ride

The Tucker is built with women's own particular needs in mind. When you drive, you frequently have children in the car. Tucker safety features give you the EXTRA protection that means peace of mind in traffic and on the highway. For they help you avoid accidents as well as give you and yours added security in case of unavoidable mishaps.

You'll glory in the effortless ease of driving the new Tucker. It has true fingertip steering control. Ordinary traction jolts and jars are either eliminated or unbelievably softened. There are no fumes or heat to bother you because the engine is in the rear.

And what luxury inside! Seats you literally sink into and relax. A "shoppers' shelf" behind the rear seat, designed specially for your parcels. "No-stoop" doors that open up into the roof for graceful entrances and exits . . . even when wearing your frilly new hat.

Yes, when you see the new Tucker it will be a case of love for life. For the Tucker combines pulse-stirring beauty of line with the very things you've always wanted in a car.

NOTE: This folder highlights only a few of the new and exclusive features of the Tucker. There are many others now being refined, improved and adopted for mass production. Consequently the Tucker Corporation must reserve the right to make mechanical changes.

Luxury beyond comparison in the Medium Priced Field

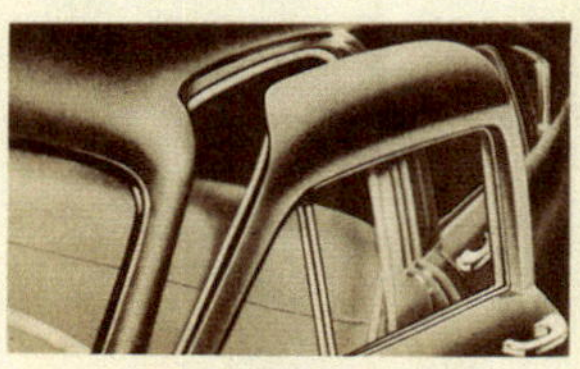

Roof Top doors are another exclusive Tucker comfort feature. The doors are curved at the top so that they become part of the roof itself. As a result, getting in and out of the Tucker is a new experience in ease and comfort. Even from a high curb, exit or entrance is made without the usual jack-knife bend.

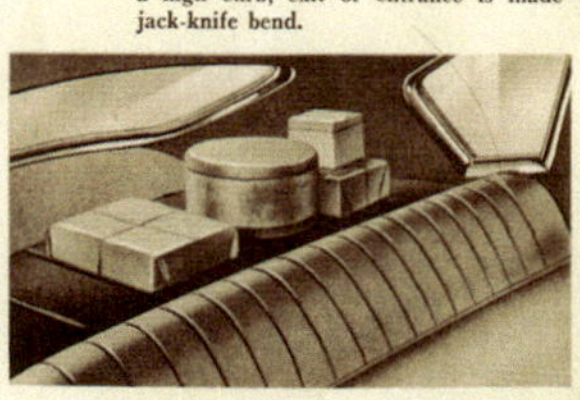

At last a car that does something about women's parcels. The Tucker shoppers' shelf, right behind the rear seat, is 52 inches wide, large enough to hold the weekend purchases for the whole family . . . out of the way of passengers, but within easy reach.

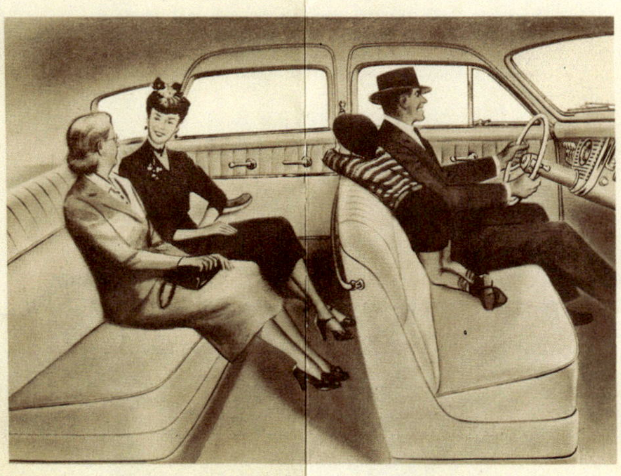

In fifty years you've never seen a car interior like this! A floor as flat as your living room's, with no unsightly and awkward hump in the center where the drive shaft runs back in conventional cars. And space, plenty of space to spread out!

A sixty-four-inch rear seat is more than ample for four people in complete comfort. And the front seat is just as wide as the rear . . . a full sixty-four inches. This means front seat cushion and rear seat cushion are interchangeable.

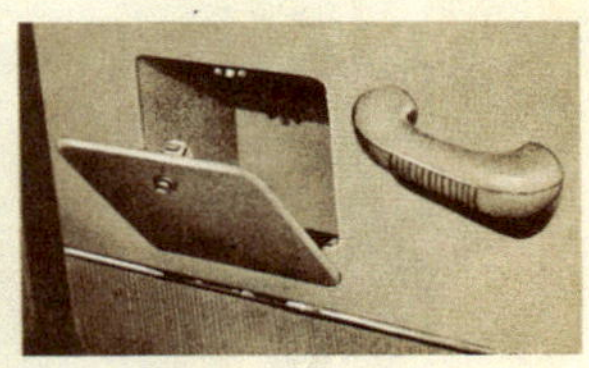

You'll say, "Why hasn't someone thought of that before," when you see the Tucker glove compartment. It's recessed in the front door on the passenger's side, upholstery-lined. At the touch of the button clasp it opens and there are your gloves, your papers, your maps.

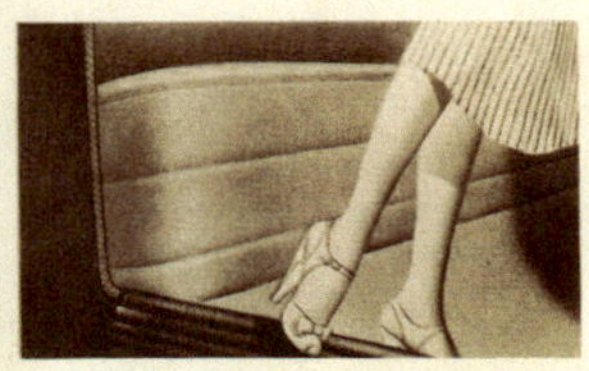

You step down when you step in a Tucker . . . down to a floor just NINE INCHES above the pavement . . . into a car with greater road clearance than many comparable cars and greater headroom than any other car on the road. Yet Tucker is only five feet high from road to roof!

LITHO IN U.S.A.

Volkswagen

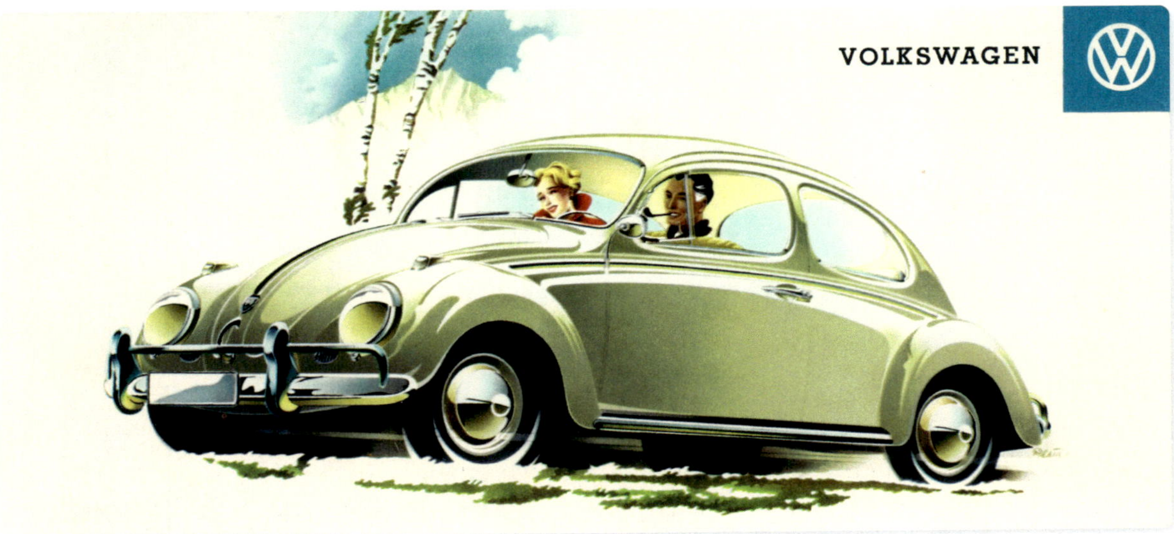

Volkswagen 1960 *(above)*, **1961** *(below and opposite)*

The handsome artwork of Bernd Reuters helped tell Volkswagen's story to a skeptical American audience. The tiny rear-engine car's message of simplicity, ease of parts replacement, and functionality began to find listeners after several years.

Das hübsche Aquarell von Bernd Reuters half, einem skeptischen amerikanischen Publikum die Geschichte des Volkswagens zu erzählen. Die Botschaft des Kleinwagens mit Heckmotor – einfache Bauweise, Funktionalität und leicht austauschbare Ersatzteile – fand nach einigen Jahren Anlauf tatsächlich offene Ohren.

Le magnifique travail visuel de Bernd Reuters aidait à raconter l'histoire de Volkswagen à un public américain sceptique. Le message de simplicité de la petite voiture à moteur arrière, la facilité des remplacements de pièces et son caractère fonctionnel commenceront à trouver des oreilles attentives au bout de quelques années.

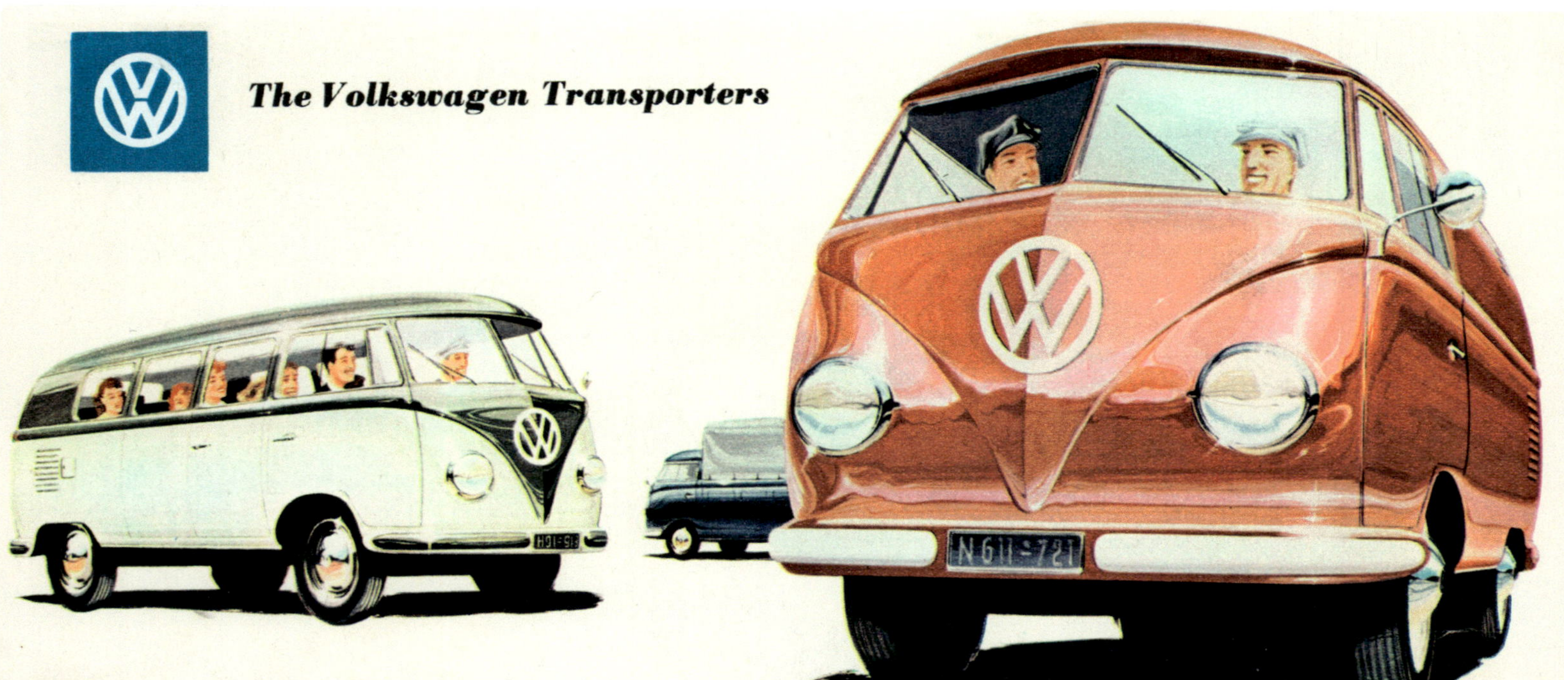

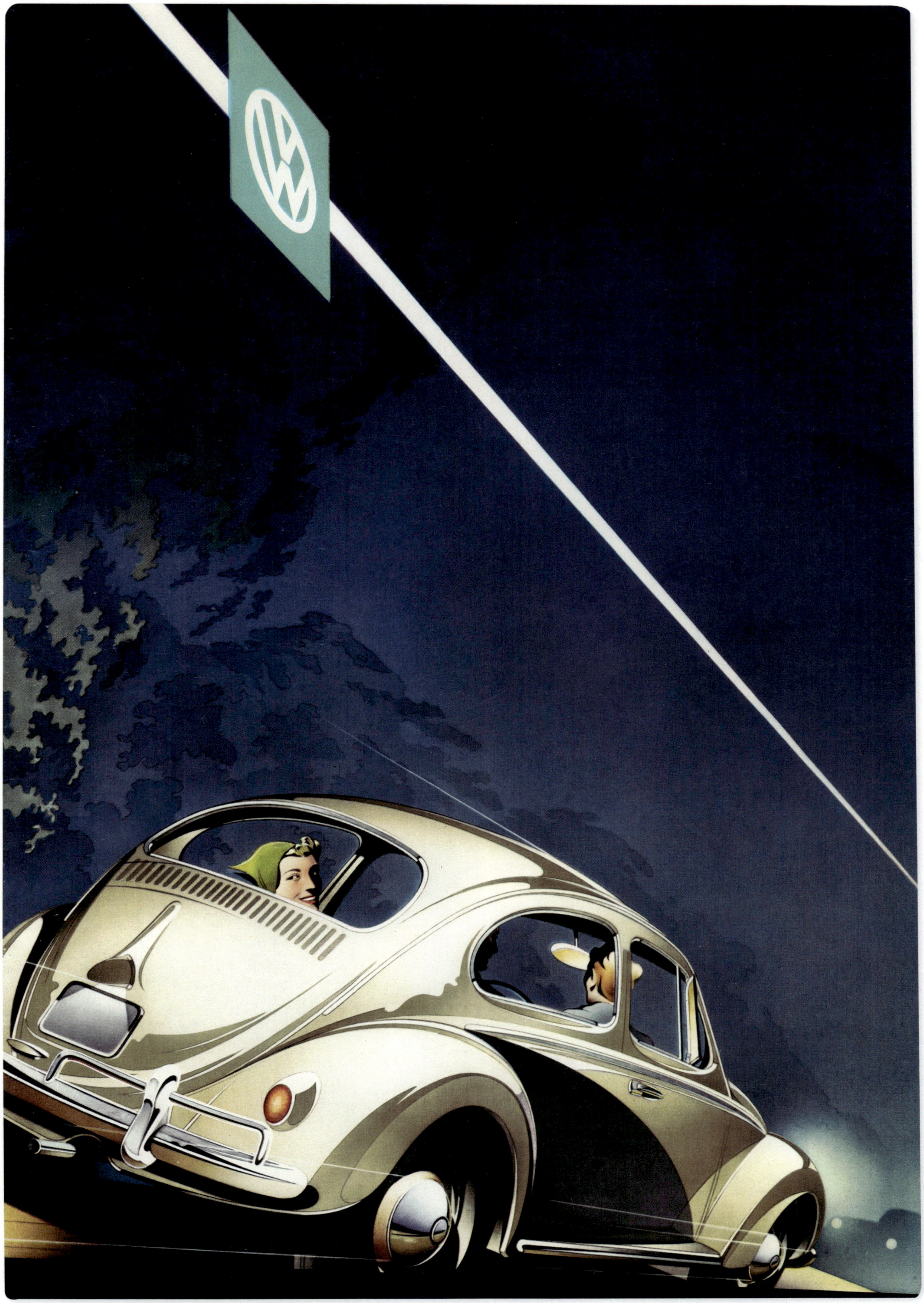

Suspension

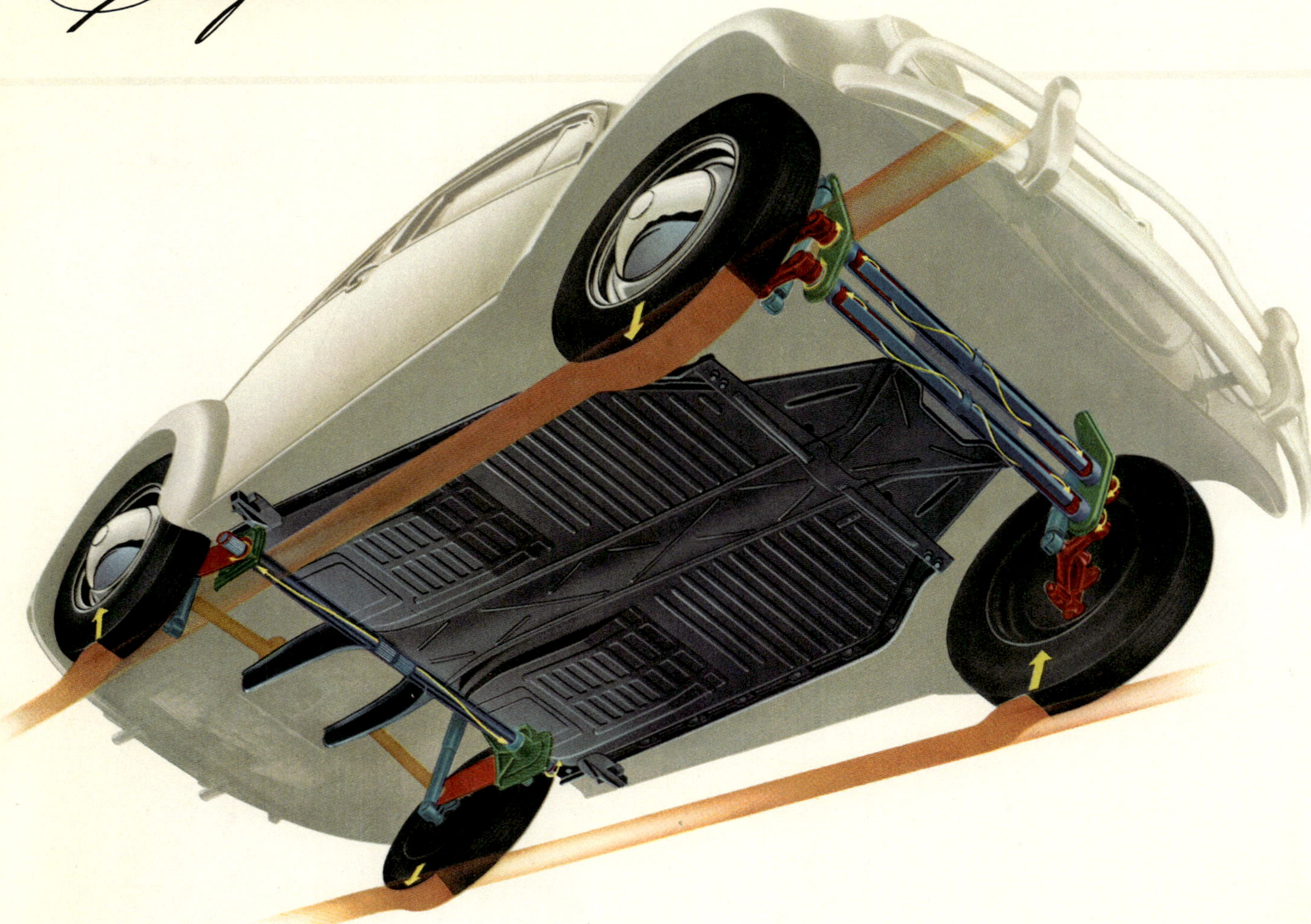

SUSPENSION All four wheels of the Volkswagen are individually suspended, and each has its indepedent torsion bar. In front, two parallel connected transverse tubes each enclose a full-length square torsion bar consisting of eight spring leaves; each is attached to the appropriate trailing arm. These trailing arms form parallelograms which give ideal spring geometry and progressively increasing spring power, no matter what the driving conditions. Bumps actuating the springs do not alter the wheel track.

REAR SUSPENSION The rear axle consists of two swing axle shafts, whose springing is likewise mutually independent. A full-length transverse tube encloses two torsion bars supported at the center of the frame; these, connected by spring braces with the axle shafts, give each wheel individual springing. This torsion-bar springing, carefully balanced in itself and progressively snubbed by long-stroke hydraulic, telescopic shock absorbers, keeps the car from nodding or bouncing.

The smooth underbody of the Volkswagen offers the least possible wind resistance. It is stamped of high-grade sheet steel, and down its longitudinal axis is an electrically welded center beam that forks at the rear to receive the engine and transmission assembly. Body and underbody, joined and rubber sealed, form a single torsion-resistant unit.

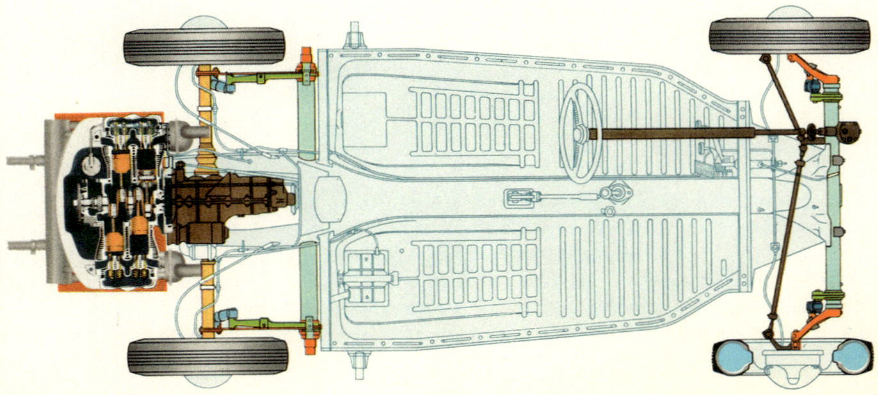

Volkswagen 1954 *(opposite)*, **1955** *(right)*, **1961** *(right bottom)*, **1956** *(below)*

A large part of Volkswagen's incredible success was due to the fact that nearly anyone could understand how it worked. It also gave economy-minded buyers an escape route from Detroit's ever-intensifying excess, bulk, and thirst. For those who wanted more, for many years Volkswagen also offered the fetching Karmann-Ghia coupe and sports roadster, which was something of a budget Porsche.

Ein Großteil des unglaublichen Erfolgs von Volkswagen war darauf zurückzuführen, dass jeder dessen Funktionsweise verstehen konnte. Außerdem fanden preisbewusste Käufer hierin einen Ausweg aus der endlosen Spirale von Gigantomanie, Benzindurst und anderen Auswüchsen aus den Detroiter Autofabriken. Für die, die etwas mehr wünschten, bot Volkswagen viele Jahre lang auch das schicke Karmann-Ghia-Sportcoupé und -Cabriolet an, eine Art Porsche des kleinen Mannes.

L'incroyable succès de la Volkswagen est largement dû au fait que que personne, ou presque, ne réussissait à comprendre son fonctionnement. Elle offrait aussi aux automobilistes soucieux d'économies une alternative aux excès, à la corpulence et à la soif inextinguible des productions de Detroit. Pour ceux qui en voulaient davantage, pendant de nombreuses années Volkswagen proposera également le ravissant coupé et roadster sportif Karmann-Ghia, qu'on pourrait qualifier de « Porsche bon marché ».

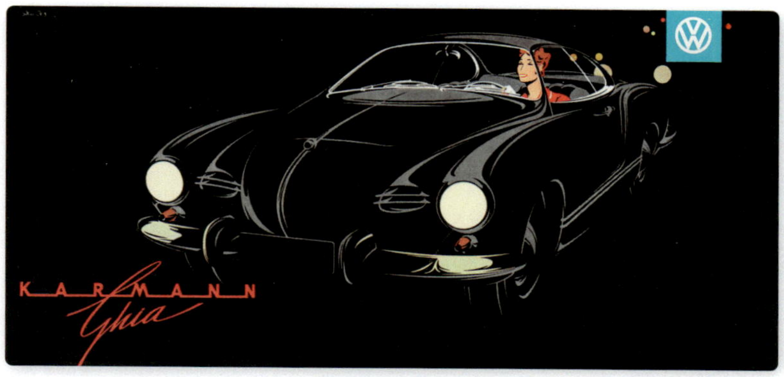

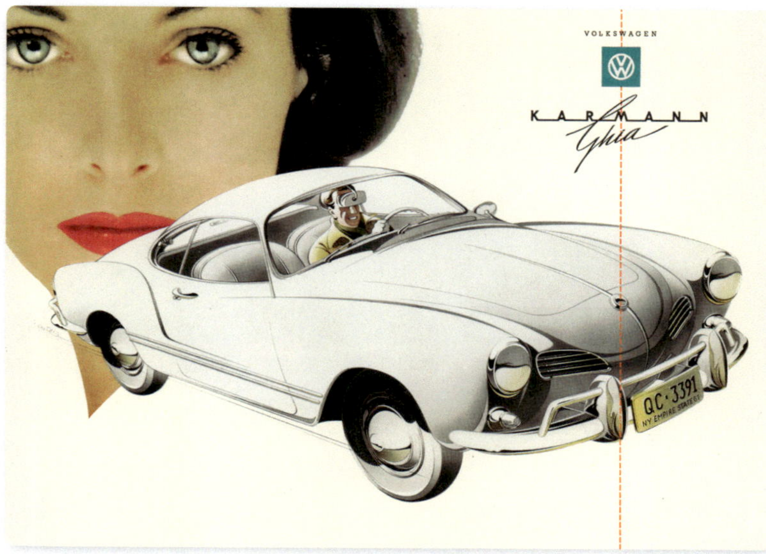

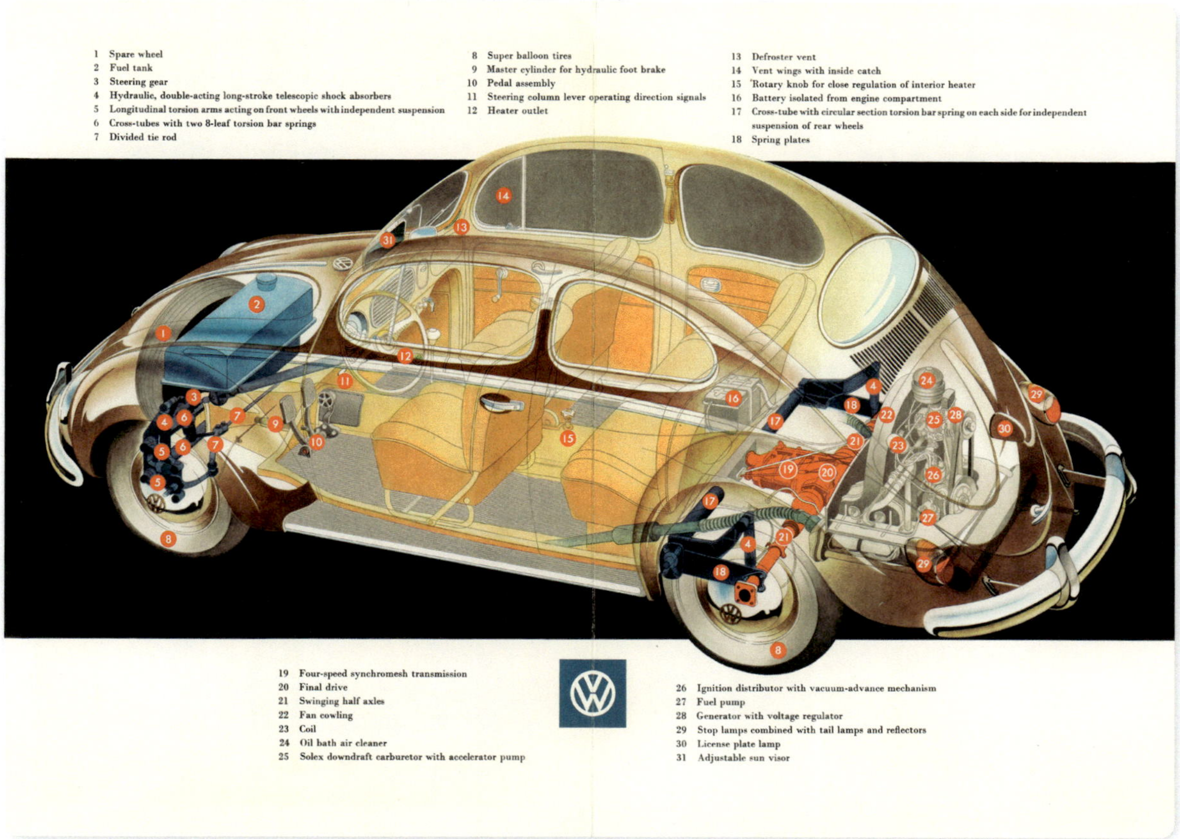

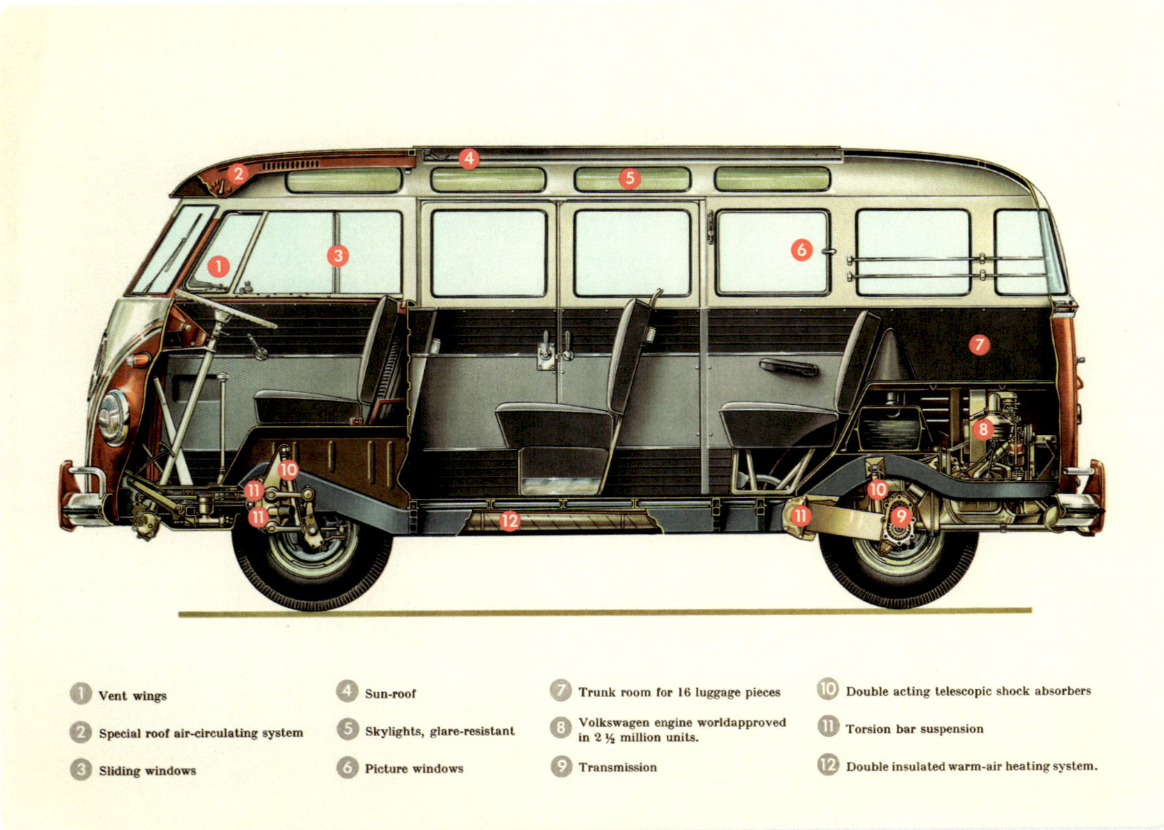

Volkswagen 1958 *(left)*, **1960** *(below)*, **1961** *(opposite)*

Sometimes it's called the Bus, or the Kombi, but the early Volkswagen van was officially the Type 2, and unofficially the Transporter. They buzzed their way into folklore as pickups with side-drop gates (very useful at curbside) and, stereotypically, for hauling hippies to rock festivals during the 1960s.

Manchmal wird er „Bus" oder „Bulli" genannt, aber offiziell trug der frühe Kleintransporter von Volkswagen die Bezeichnung „Typ 2", inoffiziell „Transporter". Als Pritschenwagen waren sie wegen ihrer seitlichen Ladeklappe beliebt (sehr nützlich beim Beladen am Bürgersteig) und als Bus, wenn sie – völlig klischeehaft – Hippies in den 1960er-Jahren zu Rockfestivals karrten.

Certains l'appelaient le bus, ou le Combi, mais le premier van Volkswagen portait officiellement le nom de Type 2 et officieusement celui de Transporter. Reconnaissable à ses portes latérales battantes (très pratiques en bordure de trottoir), il a fait son petit bonhomme de chemin dans le folklore des Sixties, évidemment (!) rempli de hippies qu'il convoyait joyeusement vers les festivals pop et rock.

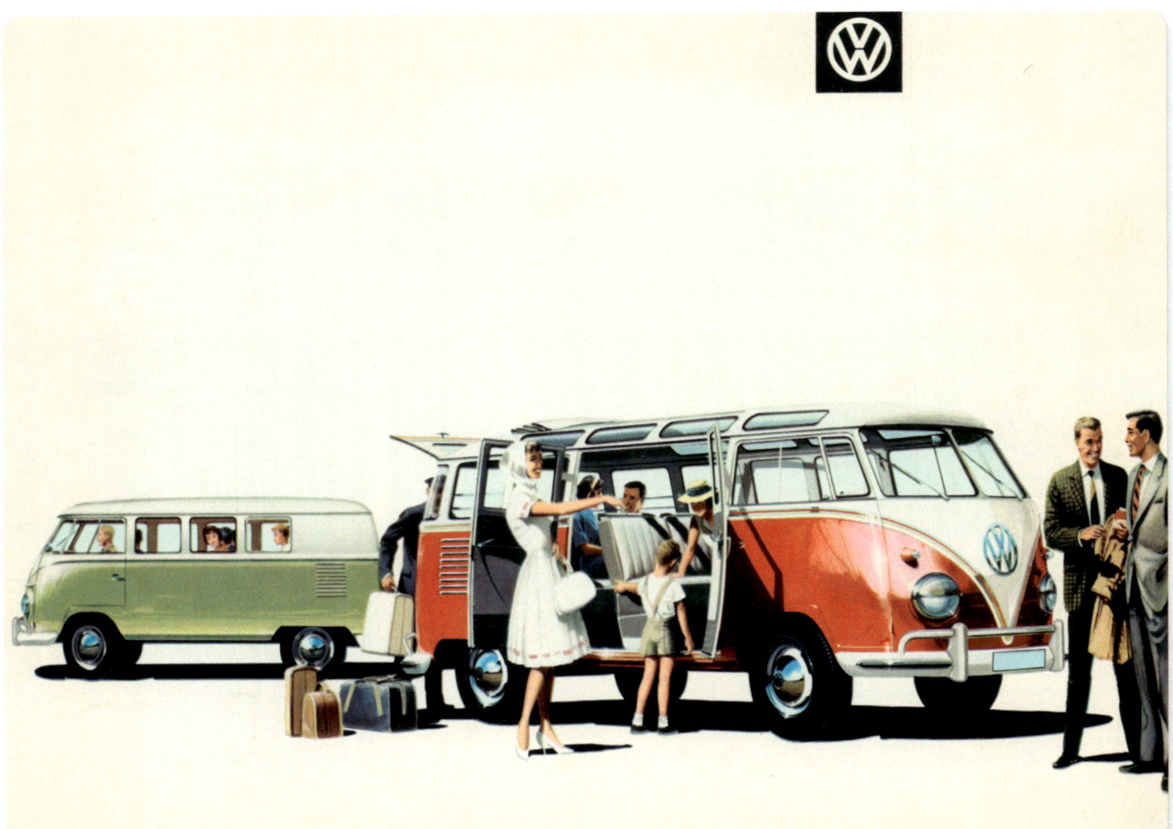

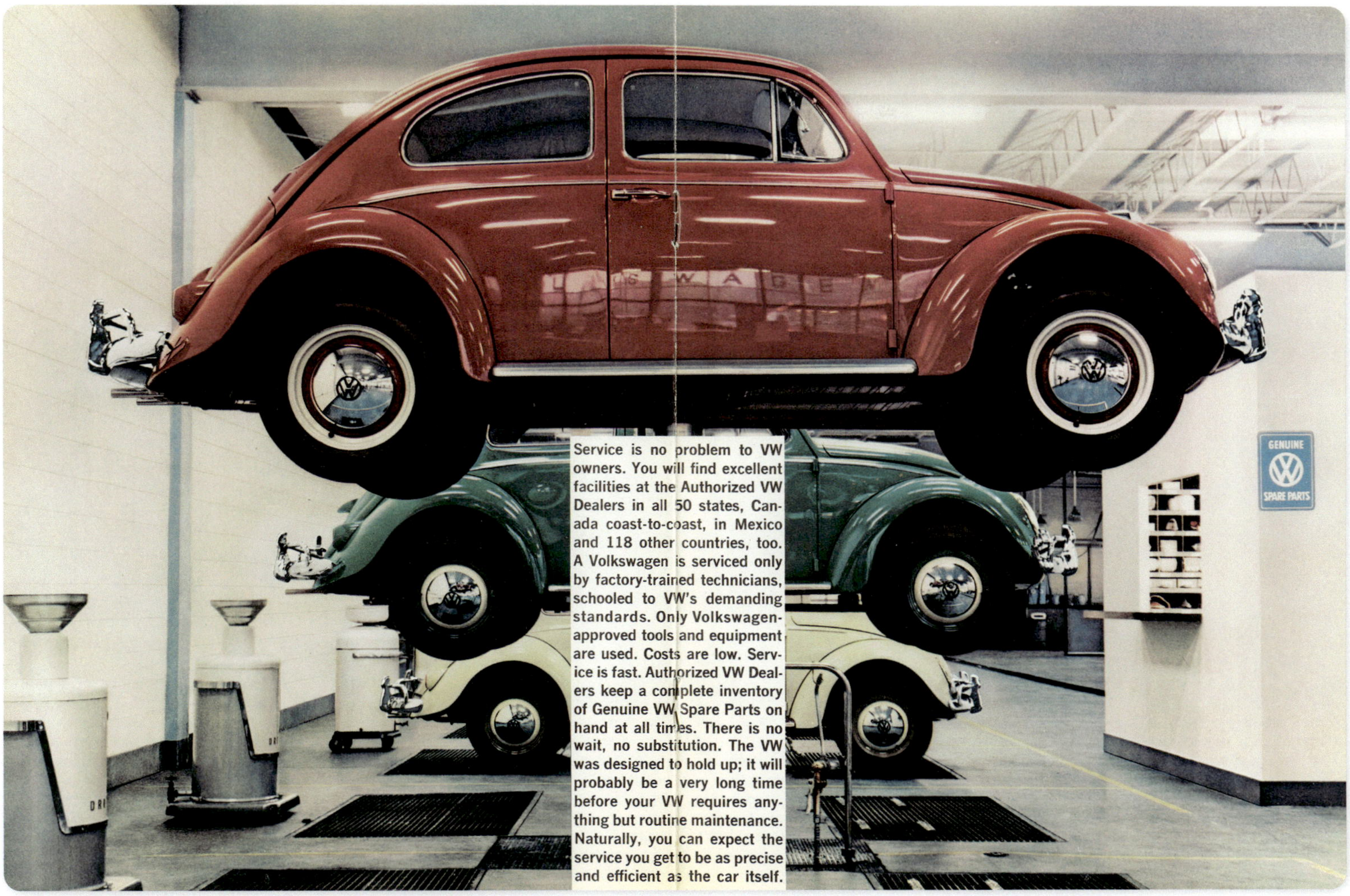

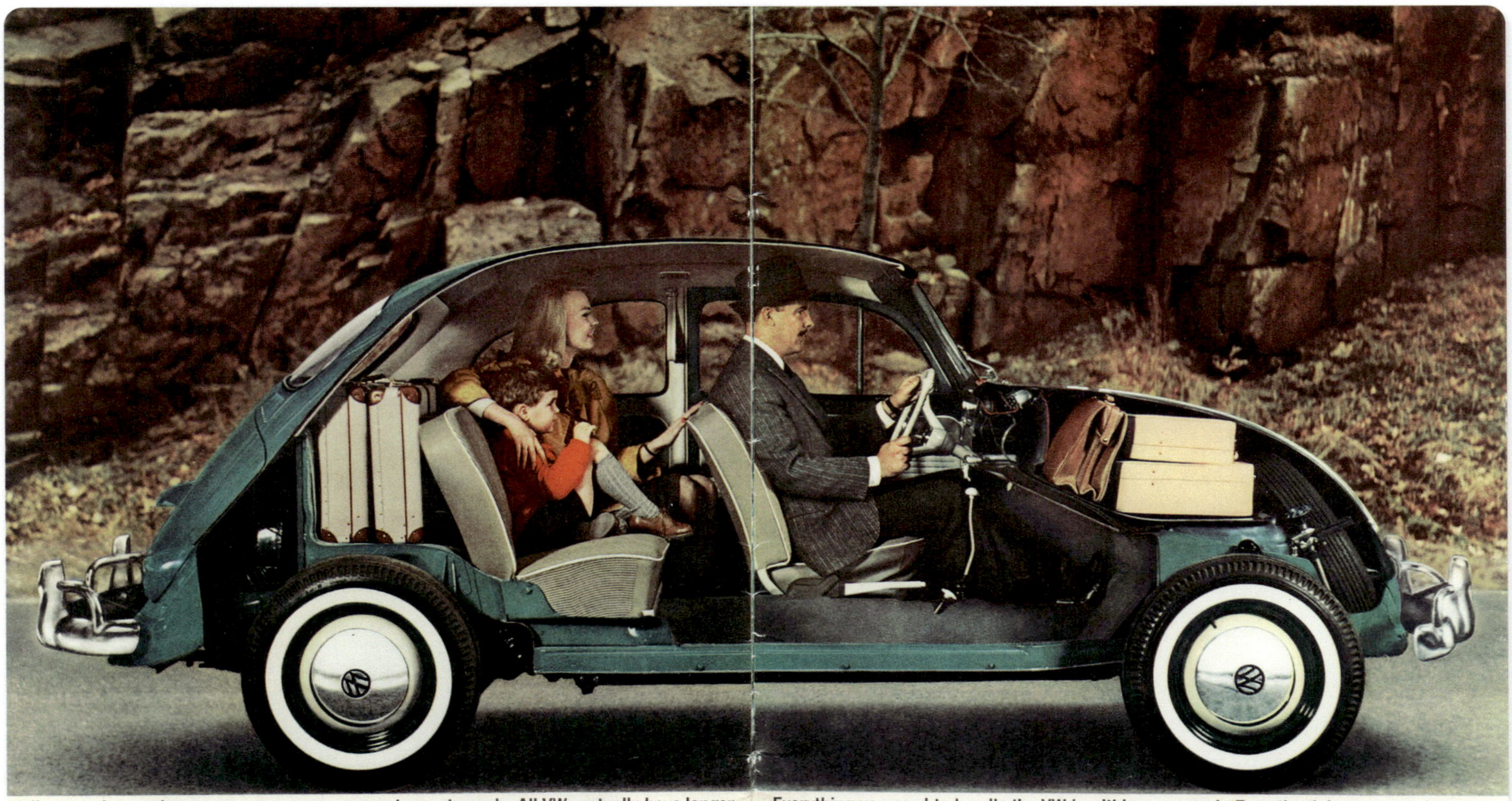

Volkswagen's roominess comes as an eye-opener to most people. All VWs actually have longer leg room in the front seat than many big cars — more head room, too. There is also space for a surprising amount of luggage. Two good-sized suitcases go under the hood. (So does the spare, and you don't have to move anything to get at it.) Other luggage gets stowed in the carpeted, 3-foot compartment behind the rear seat. The best-riding section of the car (the cradle between the 4 wheels) is reserved for the VW's passengers who travel in soft, form-fitting seats. The bucket seats in front adjust back and forth easily and independently, even while the car is moving. And the backs of the VW's seats recline at 3 different angles for a change of pace on long trips or a chance for a passenger to snooze.

Everything you need to handle the VW is within easy reach. Even the tiniest woman driver you know can reach the glove compartment without taking her eyes off the road. All the VW's instruments are in a convenient cluster directly in front of the driver. The gear shift lever, the parking brake and the heater control fall to hand as naturally as your pocket. When you drive a VW, all you stretch are your legs. And you get all the niceties you expect a fine car to have: automatic windshield washer, sun visors, coat hooks, assist straps, ashtrays—even a hand grip for the right front seat and a huge in-the-door pocket for the driver. Open VW's extra-wide door and a bright overhead light goes on; you get an impression of trimness and serviceability. Drive it and you confirm the world-wide opinion: the Volkswagen is like no other car on the road.

Volvo

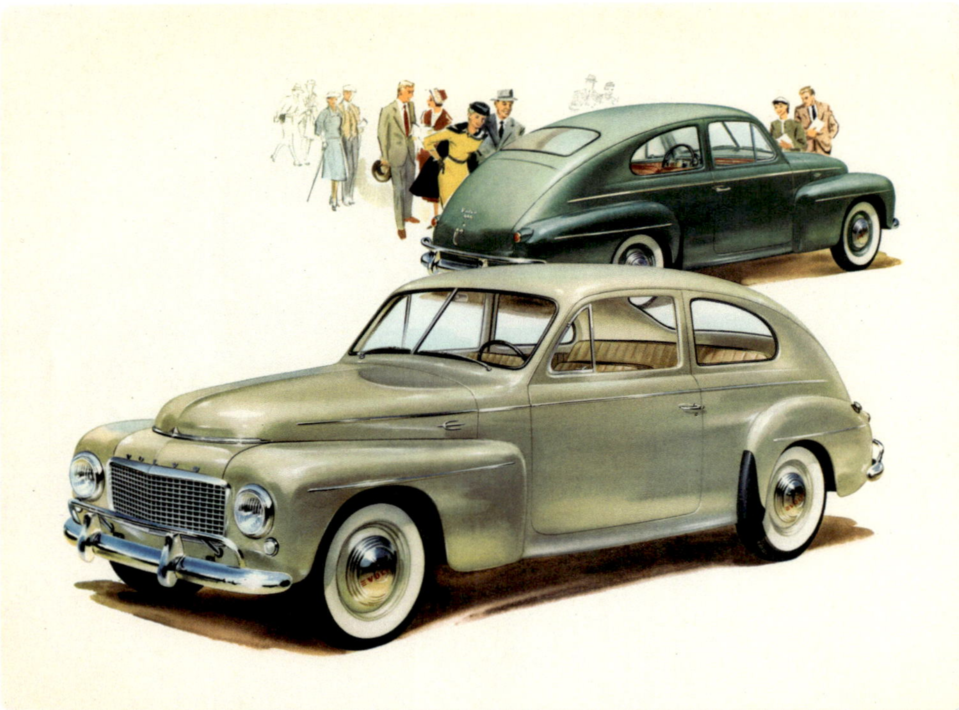

Volvo 1956 *(above)*, **1962** *(right)*

Its somewhat antiquated appearance resembled a Ford Tudor from the 1940s, but the PV 544 won acceptance almost immediately, including in the United States. It was a rock-solid sedan with excellent safety features and a very sturdy B18 engine. These Volvos won rallies everywhere from Africa to the Arctic.

Das leicht antiquiert wirkende Äußere glich einem Ford Tudor aus den 1940er-Jahren, aber der PV 544 kam sofort bei den Kunden an, auch in den Vereinigten Staaten. Es handelte sich um eine grundsolide Limousine mit überragender Sicherheit und einem sehr robusten B18-Motor. Diese Volvos gewannen Rallyes von Afrika bis in die Arktis.

Son apparence un peu désuète rappelait la Ford Tudor des années 1940, mais la PV 544 sut conquérir presque instantanément l'approbation générale, y compris auprès des consommateurs des États-Unis. Cette berline à toute épreuve pouvait se prévaloir d'excellents éléments de sécurité et d'un moteur B18 très robuste. Ces Volvo ont gagné des rallyes un peu partout, des pistes africaines jusqu'au cercle Arctique.

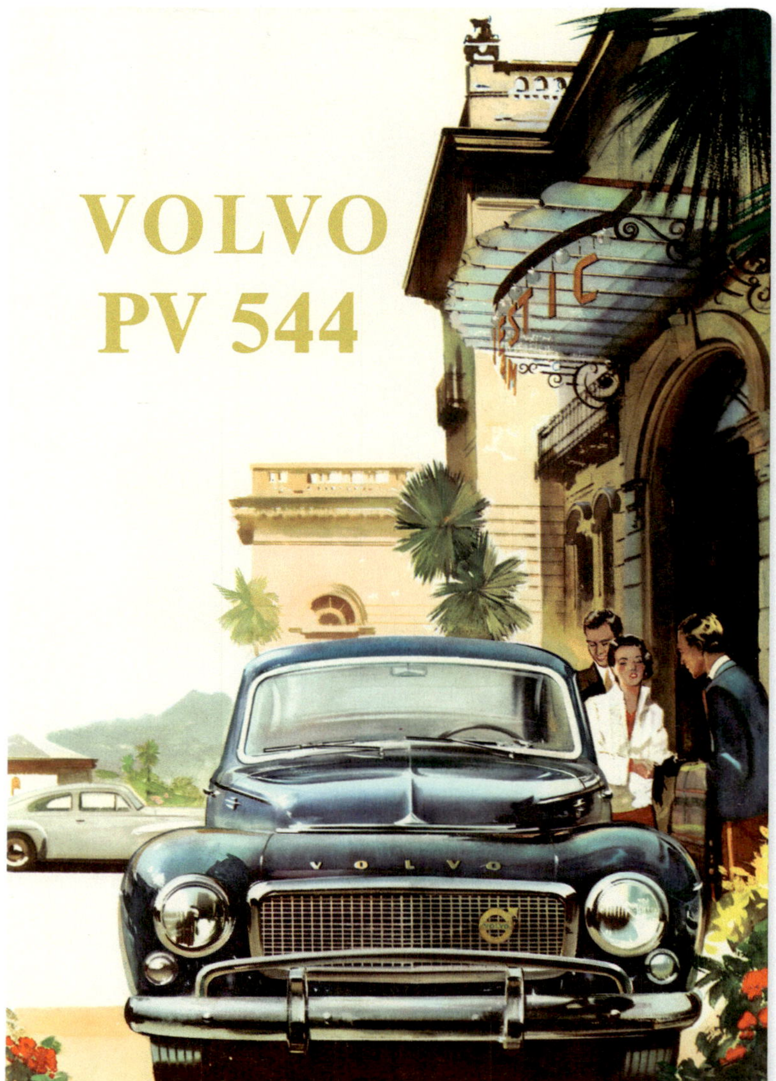

Willys

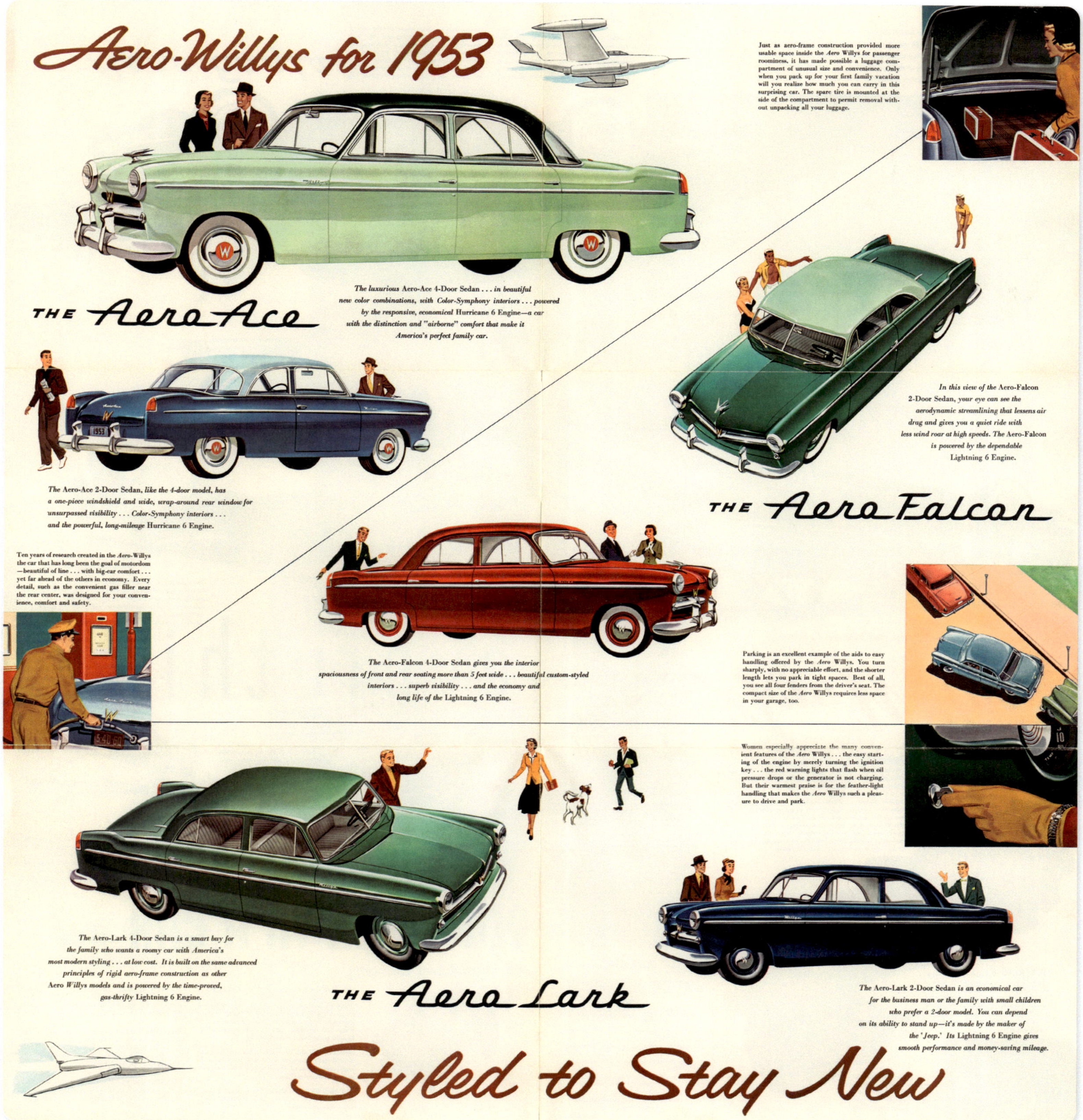

Willys 1953

These are largely forgotten cars, unfortunate because they represented a complete line of compact cars with unitized construction and uncluttered looks. The F-head, Jeep-derived engine was standard, but a straight-six was optional. Aeros came in four different lines, and a four-door sedan was new.

Diese Autos sind fast vergessen, was schade ist, denn sie bildeten eine Reihe Kompaktwagen mit selbsttragender Karosserie und aufgeräumtem Erscheinungsbild. Der vom Jeep abgeleitete IOE-Motor oder „F-head" gehörte zur Grundausstattung, aber ein Sechszylinderreihenmotor war als Alternative erhältlich. Die Aeros gab es in vier Modellreihen, neu war die viertürige Limousine.

Ces voitures aujourd'hui méconnues s'inscrivaient dans une gamme de modèles compacts et monocoques aux lignes épurées. Un moteur six cylindres en ligne était proposé en option, alternativement au F-head à soupapes en tête et latérales livré en série et dérivé du Jeep. Les Aero étaient déclinées en quatre versions, accompagnées par une toute nouvelle berline quatre portes.

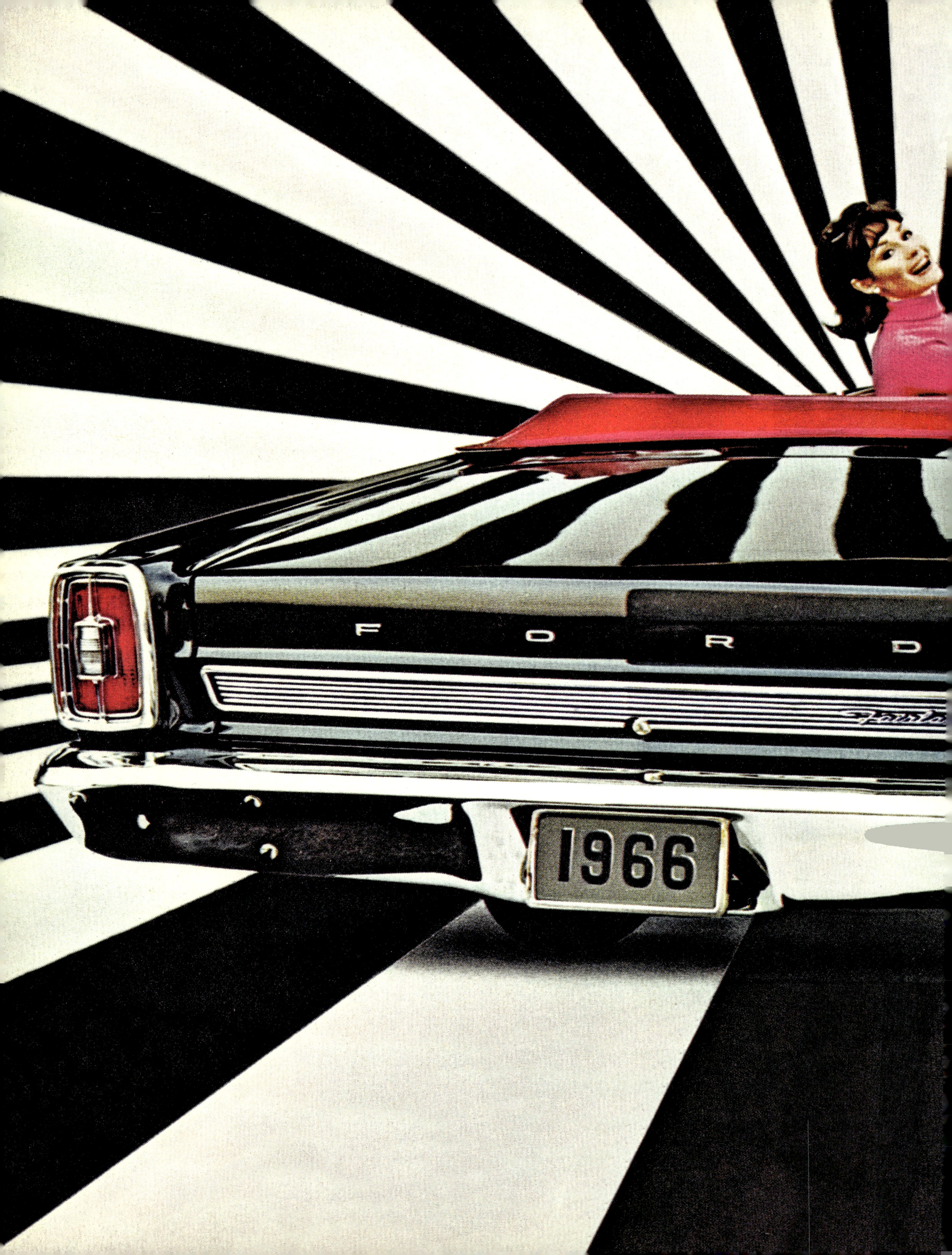

FORD
1966

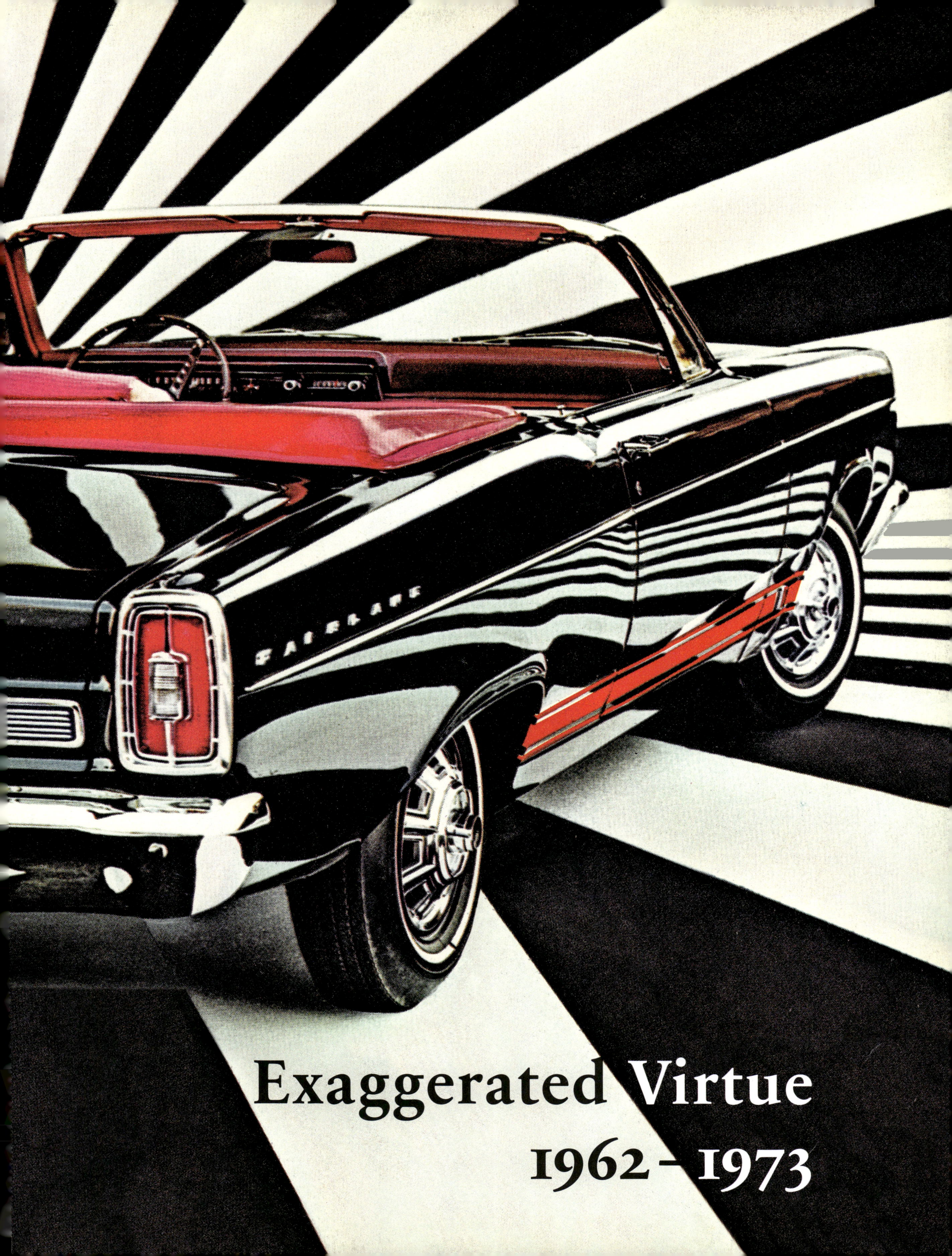

Exaggerated Virtue
1962–1973

American Motors

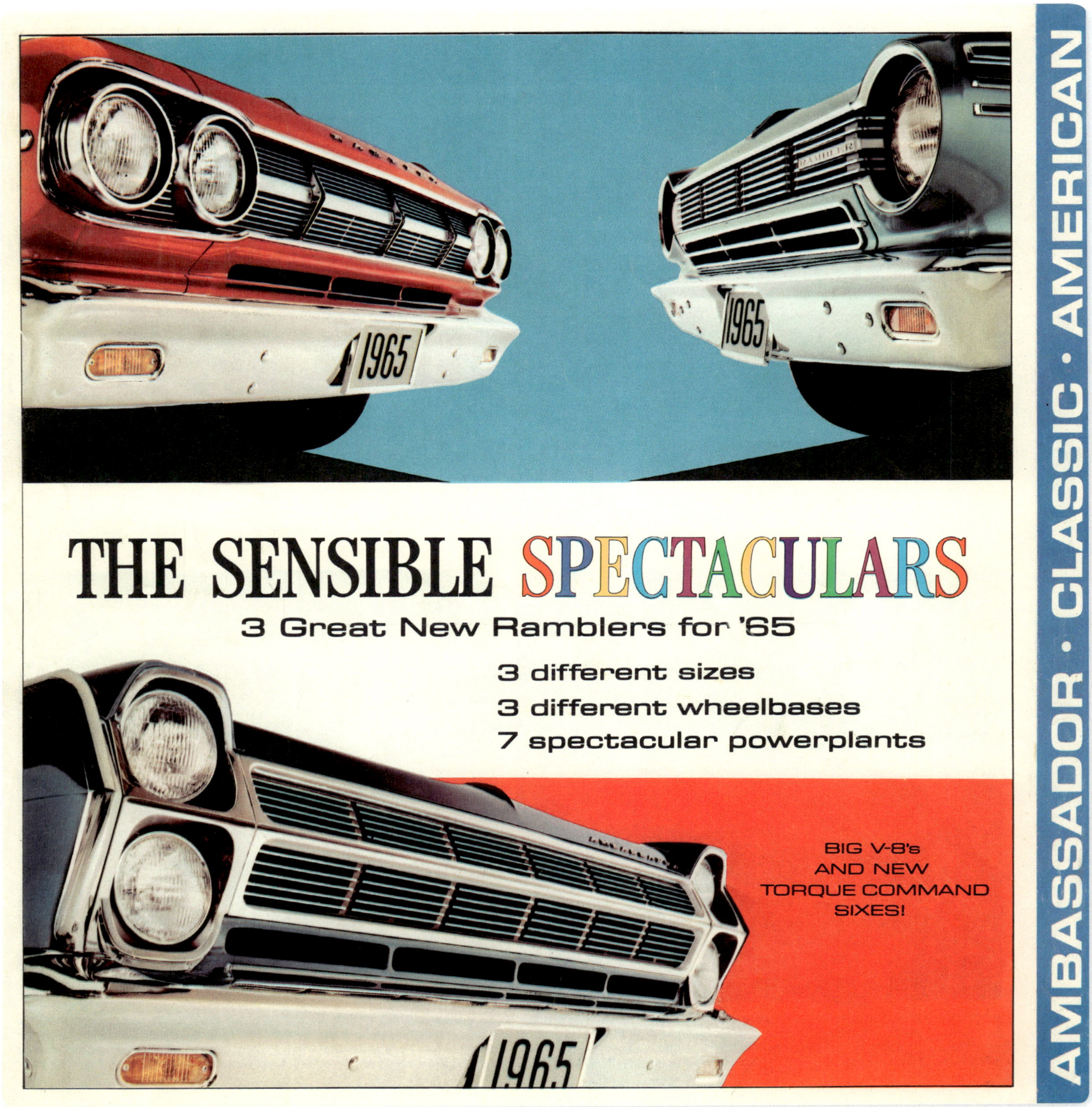

American Motors 1965

With very simple imagery, American Motors forcefully showed that even though it was the smallest U.S. automaker, it had product lines in three sizes with distinctly different personalities.

In sehr einfachen Bildern verdeutlichte American Motors, dass man auch als kleinster der großen US-Automobilhersteller Produktreihen in drei Größen mit jeweils einer ausgeprägten eigenen Persönlichkeit zu bieten hatte.

À travers une imagerie d'une grande simplicité, American Motors montrait avec conviction que même le plus modeste des constructeurs automobiles américains pouvait proposer trois tailles différentes de modèles, chacun doté d'une forte personnalité.

Buick

Buick 1964

Buick had the novel idea of making its 1964 dealer brochure into something of an art volume. Further, it underscored the growing role of women in car shopping by suggesting an appropriate feminine profile for each of its models.

Buick hatte den neuartigen Einfall, aus seiner Händlerbroschüre für 1964 eine Art Kunstband zu machen. Zudem wurde die wachsende Bedeutung von Frauen beim Autokauf dadurch unterstrichen, dass man für jedes seiner Modelle ein dazu passendes weibliches Profil vorlegte.

Buick a eu l'idée novatrice de concevoir son catalogue 1964 à la manière d'un livre d'art. Le rôle croissant des femmes dans l'achat automobile y était souligné par la suggestion, pour chaque modèle, d'un profil féminin approprié.

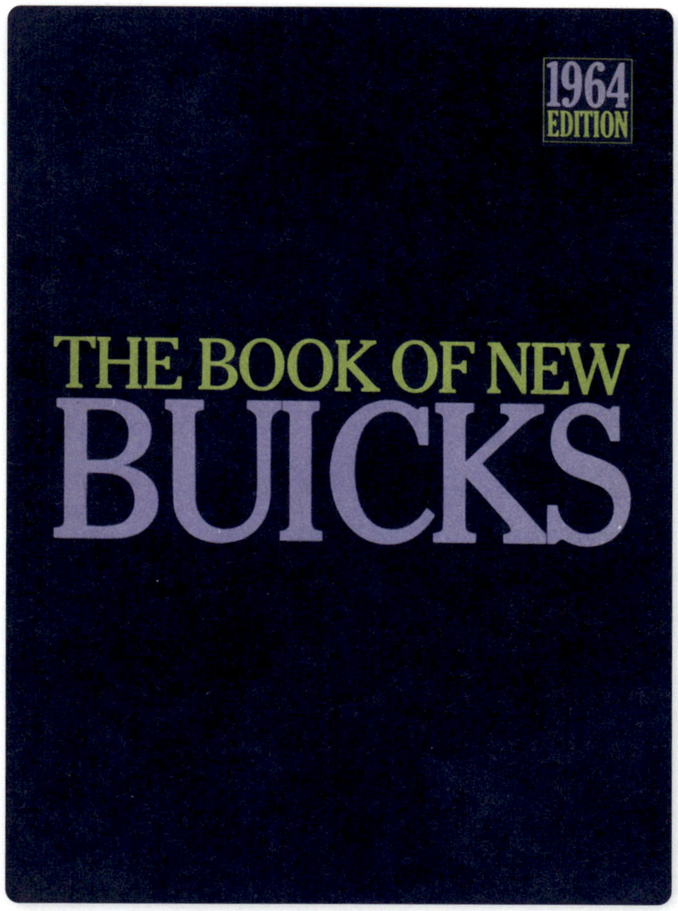

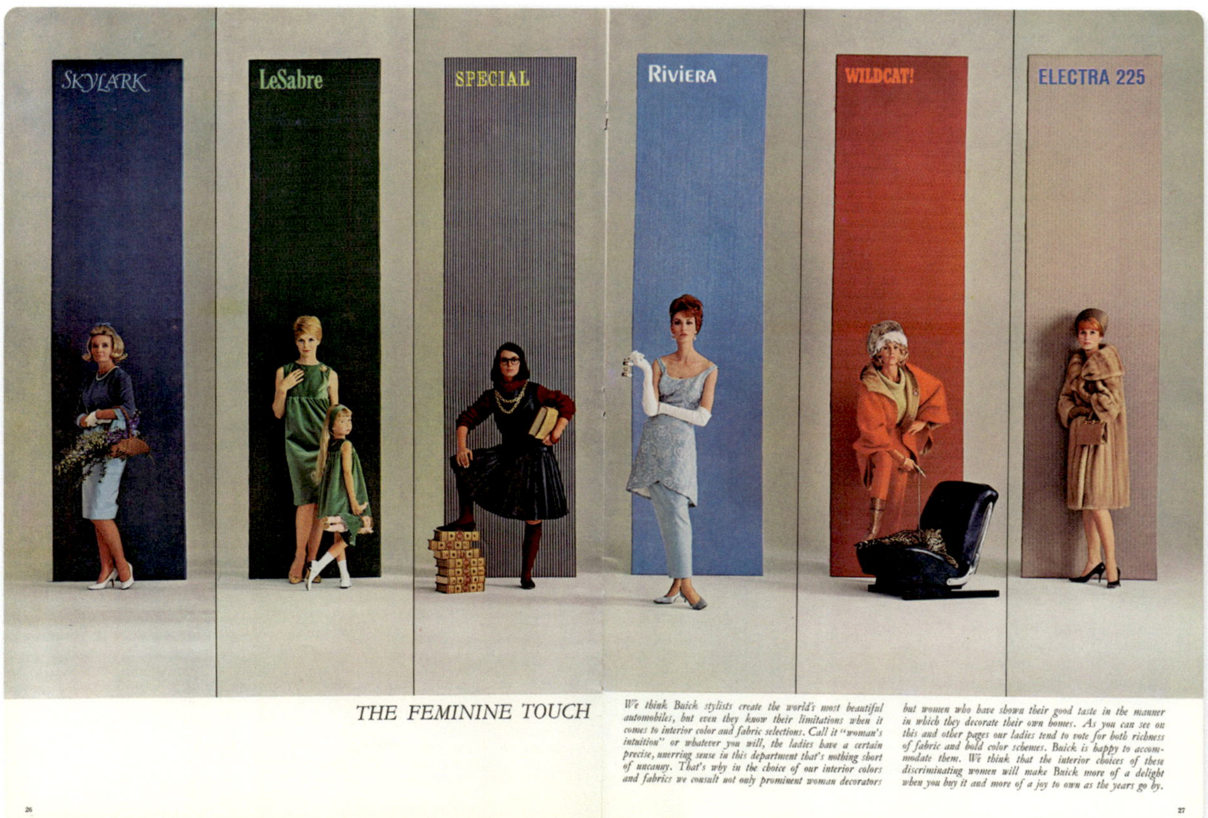

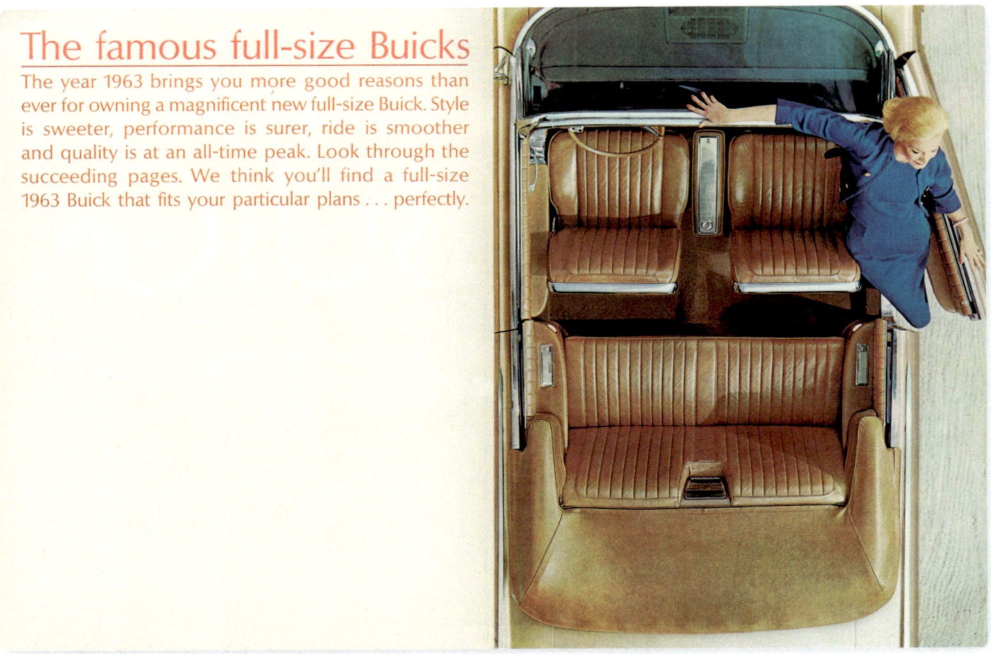

Buick 1963 *(left)*, **1964,** *(below and opposite bottom)*, **1965** *(opposite top)*

Buick offered a welter of engine and transmission choices, from a novel V-6 to monstrous V-8s, plus a quartet of transmissions. Eventually, General Motors figured out that so many options increased build costs and cut profits. Most buyers picked big V-8s and automatics anyway.

Buick bot eine Unzahl von Motoren und Getrieben zur Wahl an, von einem neuartigen V6 zu einem monströsen V8, und dazu ein ganzes Getriebequartett. Schließlich erkannte aber auch General Motors, dass so viele Wahlmöglichkeiten die Baukosten erhöhten und die Gewinne schmälerten – und die meisten Käufer entschieden sich ohnehin für den V8 mit Automatikgetriebe.

Buick proposait une multitude de choix de groupes propulseurs : la gamme allait d'un novateur V6 aux monstrueux V8, sans oublier un quatuor de transmissions. General Motors considéra finalement que tant d'options renchérissaient les coûts de production et limitaient les bénéfices. La plupart des acheteurs optaient de toute façon pour les gros V8 et les boîtes automatiques.

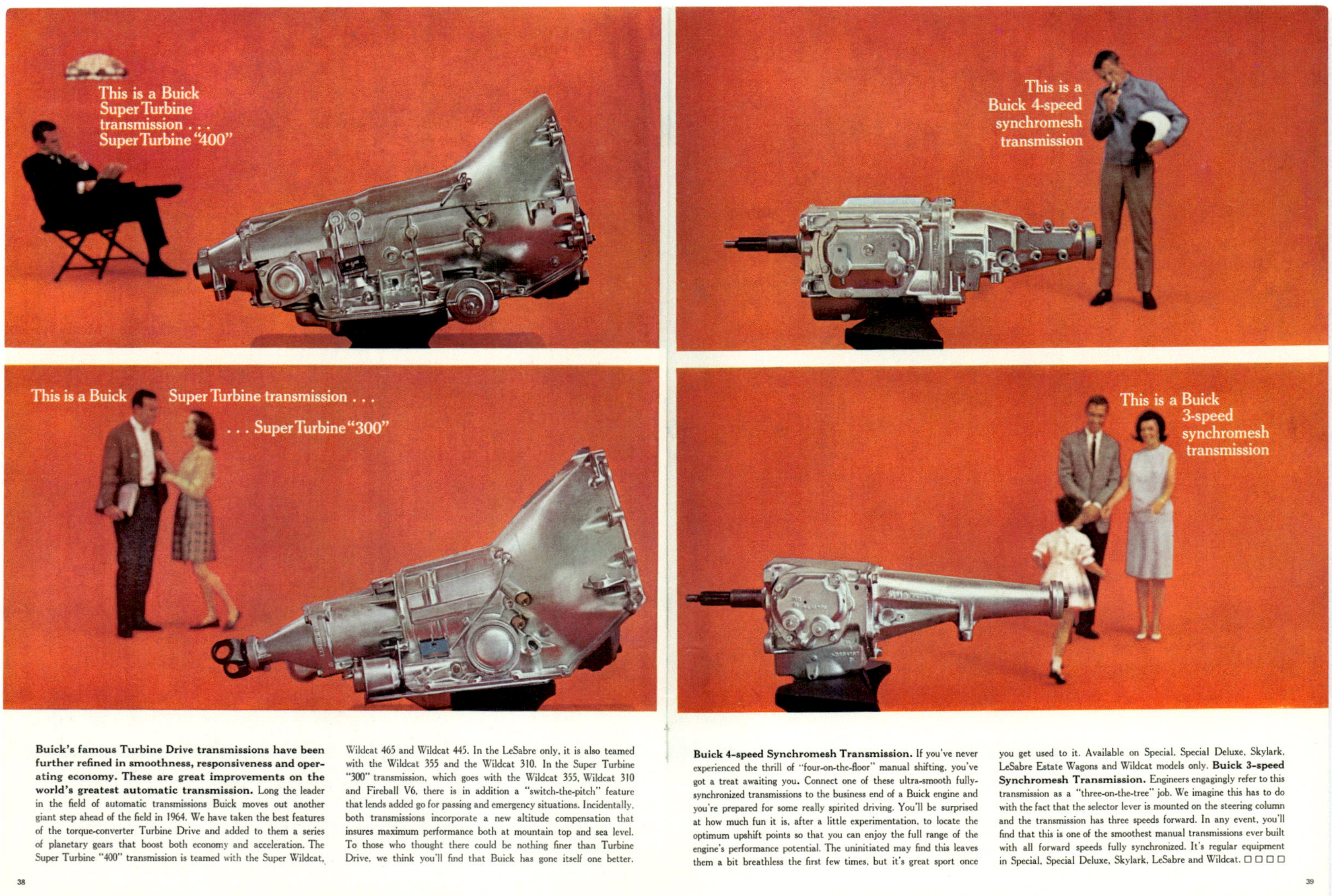

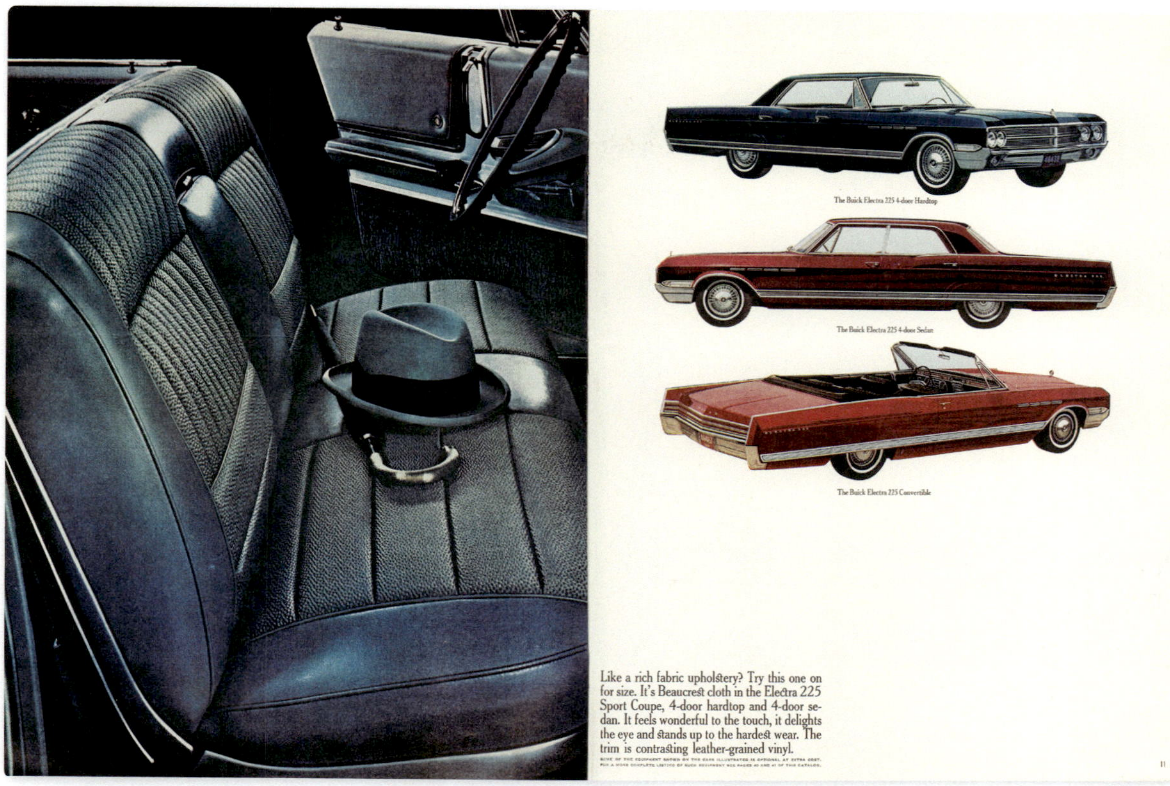
The Buick Electra 225 4-door Hardtop
The Buick Electra 225 4-door Sedan
The Buick Electra 225 Convertible
Like a rich fabric upholstery? Try this one on for size. It's Beaucrest cloth in the Electra 225 Sport Coupe, 4-door hardtop and 4-door sedan. It feels wonderful to the touch, it delights the eye and stands up to the hardest wear. The trim is contrasting leather-grained vinyl.
11

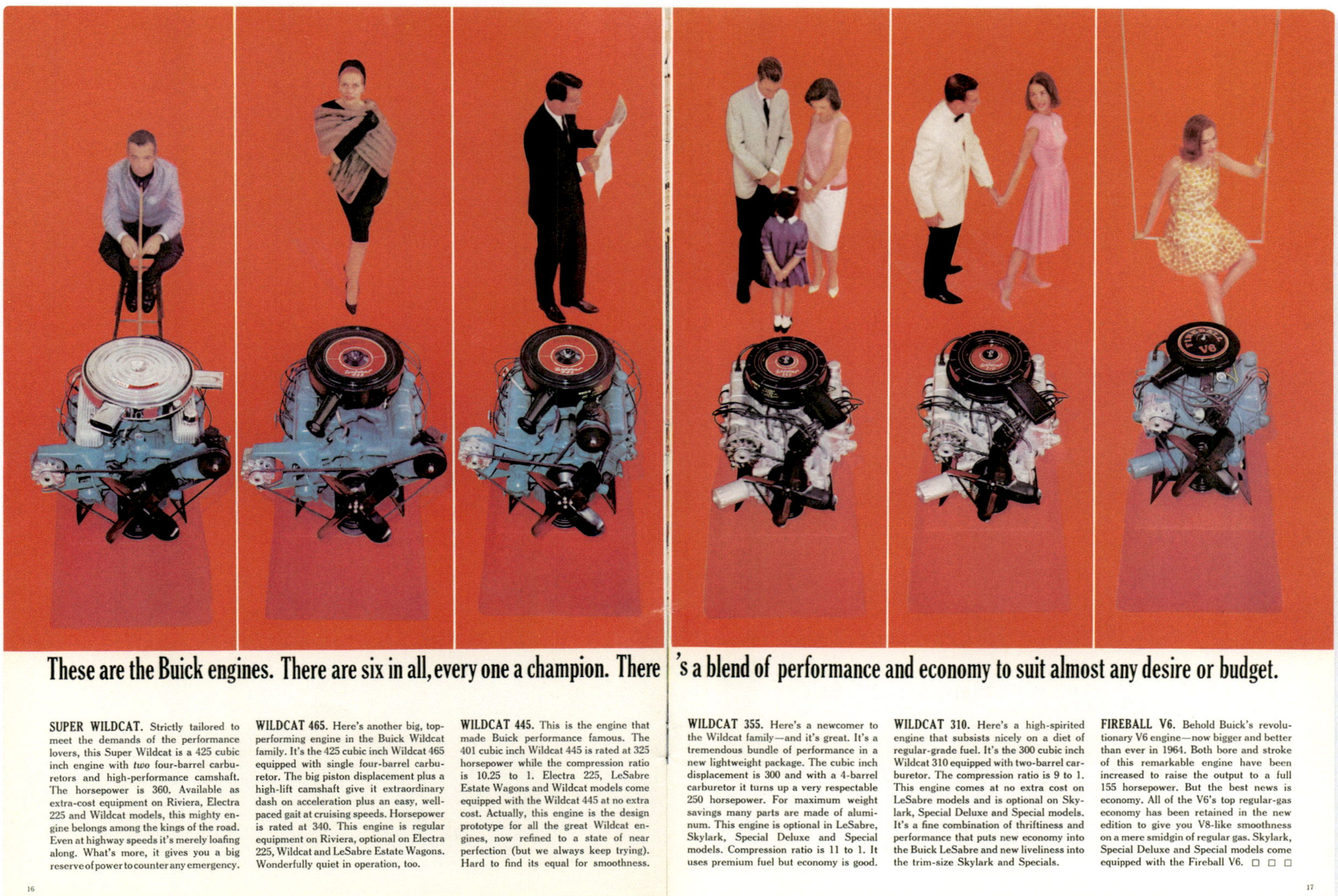
These are the Buick engines. There are six in all, every one a champion. There's a blend of performance and economy to suit almost any desire or budget.
SUPER WILDCAT. Strictly tailored to meet the demands of the performance lovers, this Super Wildcat is a 425 cubic inch engine with two four-barrel carburetors and high-performance camshaft. The horsepower is 360. Available as extra-cost equipment on Riviera, Electra 225 and Wildcat models, this mighty engine belongs among the kings of the road. Even at highway speeds it's merely loafing along. What's more, it gives you a big reserve of power to counter any emergency.
WILDCAT 465. Here's another big, top-performing engine in the Buick Wildcat family. It's the 425 cubic inch Wildcat 465 equipped with single four-barrel carburetor. The big piston displacement plus a high-lift camshaft give it extraordinary dash on acceleration plus an easy, well-paced gait at cruising speeds. Horsepower is rated at 340. This engine is regular equipment on Riviera, optional on Electra 225, Wildcat and LeSabre Estate Wagons. Wonderfully quiet in operation, too.
WILDCAT 445. This is the engine that made Buick performance famous. The 401 cubic inch Wildcat 445 is rated at 325 horsepower while the compression ratio is 10.25 to 1. Electra 225, LeSabre Estate Wagons and Wildcat models come equipped with the Wildcat 445 at no extra cost. Actually, this engine is the design prototype for all the great Wildcat engines, now refined to a state of near perfection (but we always keep trying). Hard to find its equal for smoothness.
WILDCAT 355. Here's a newcomer to the Wildcat family—and it's great. It's a tremendous bundle of performance in a new lightweight package. The cubic inch displacement is 300 and with a 4-barrel carburetor it turns up a very respectable 250 horsepower. For maximum weight savings many parts are made of aluminum. This engine is optional in LeSabre, Skylark, Special Deluxe and Special models. Compression ratio is 11 to 1. It uses premium fuel but economy is good.
WILDCAT 310. Here's a high-spirited engine that subsists nicely on a diet of regular-grade fuel. It's the 300 cubic inch Wildcat 310 equipped with two-barrel carburetor. The compression ratio is 9 to 1. This engine comes at no extra cost on LeSabre models and is optional on Skylark, Special Deluxe and Special models. It's a fine combination of thriftiness and performance that puts new economy into the Buick LeSabre and new liveliness into the trim-size Skylark and Specials.
FIREBALL V6. Behold Buick's revolutionary V6 engine—now bigger and better than ever in 1964. Both bore and stroke of this remarkable engine have been increased to raise the output to a full 155 horsepower. But the best news is economy. All of the V6's top regular-gas economy has been retained in the new edition to give you V8-like smoothness on a mere smidgin of regular gas. Skylark, Special Deluxe and Special models come equipped with the Fireball V6. □ □ □
16
17

Cadillac

Cadillac 1963, 1964 *(opposite)*,
1962 *(opposite bottom)*

The fashions, the clothing, the models, and the Cadillacs were intended to reflect each others' glory. The gentleman in Madison Avenue attire is an ideal fit for the car, as Cadillacs grew increasingly square-cornered and conservative as the 1960s unwound.

Die Mode, die Kleidung, die Fotomodelle und die Cadillacs sollten im gegenseitigen Glanz erstrahlen und aufeinander wirken. Der Herr im Anzug der Werbebranche aus der Madison Avenue passte ideal zu diesem Auto, denn die Cadillacs wurden im Laufe der 1960er-Jahre immer kantiger und konservativer.

La mode, la beauté, les mannequins – et les Cadillac pour en refléter tout l'éclat. L'élégance du gentleman de la très chic Madison Avenue allait d'autant mieux à ces Cadillac qu'elles ne cesseront de s'embourgeoiser au fil des années 1960.

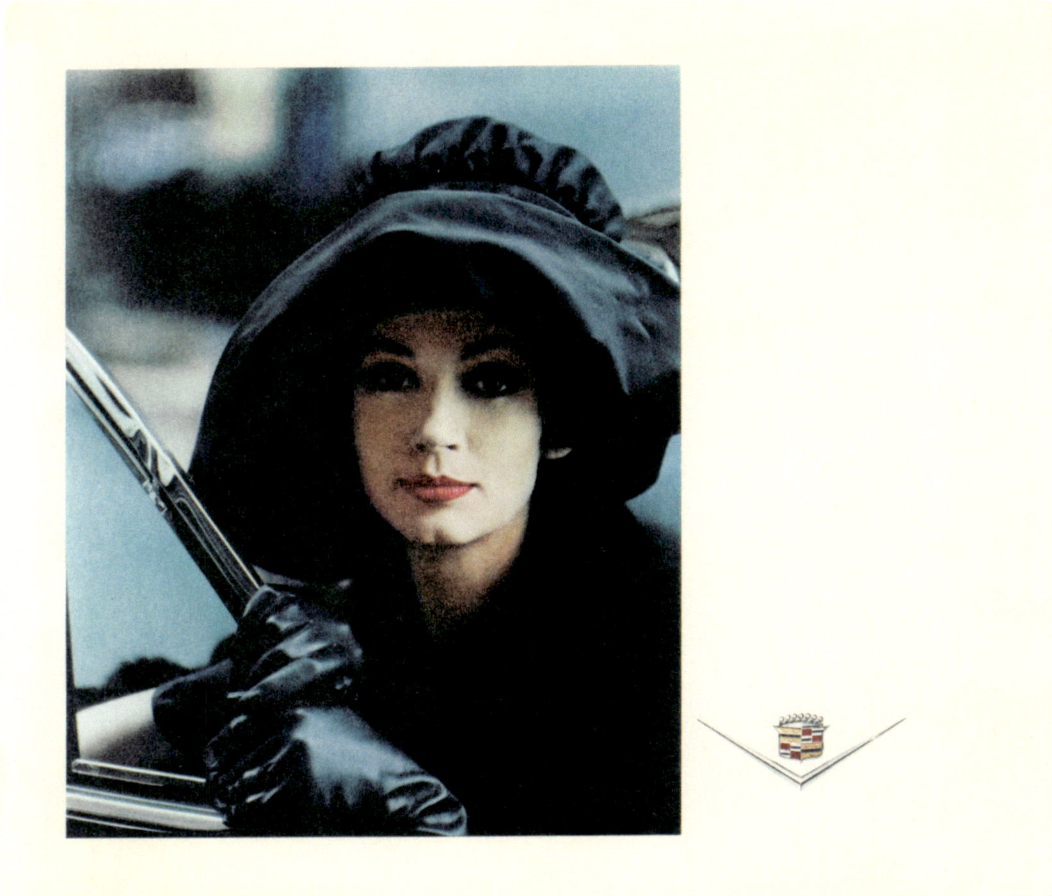

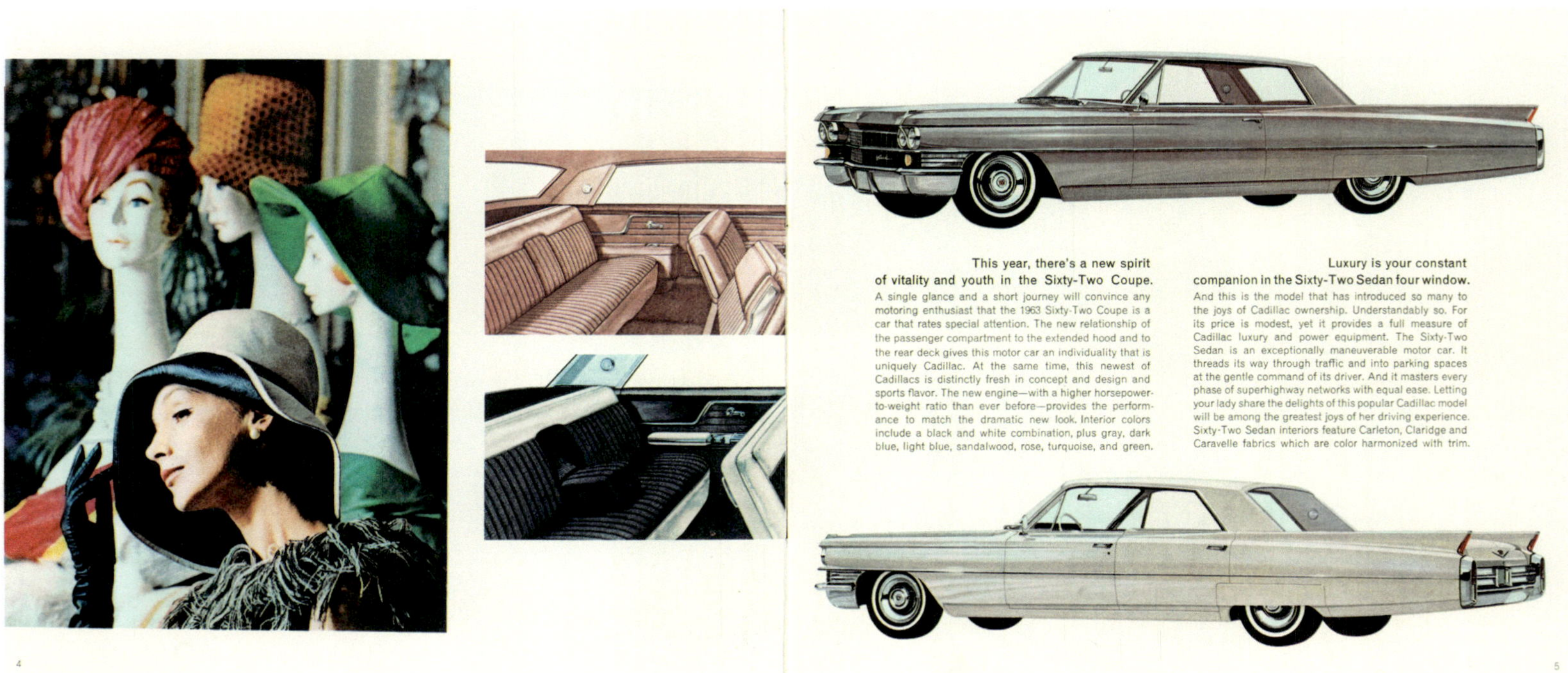

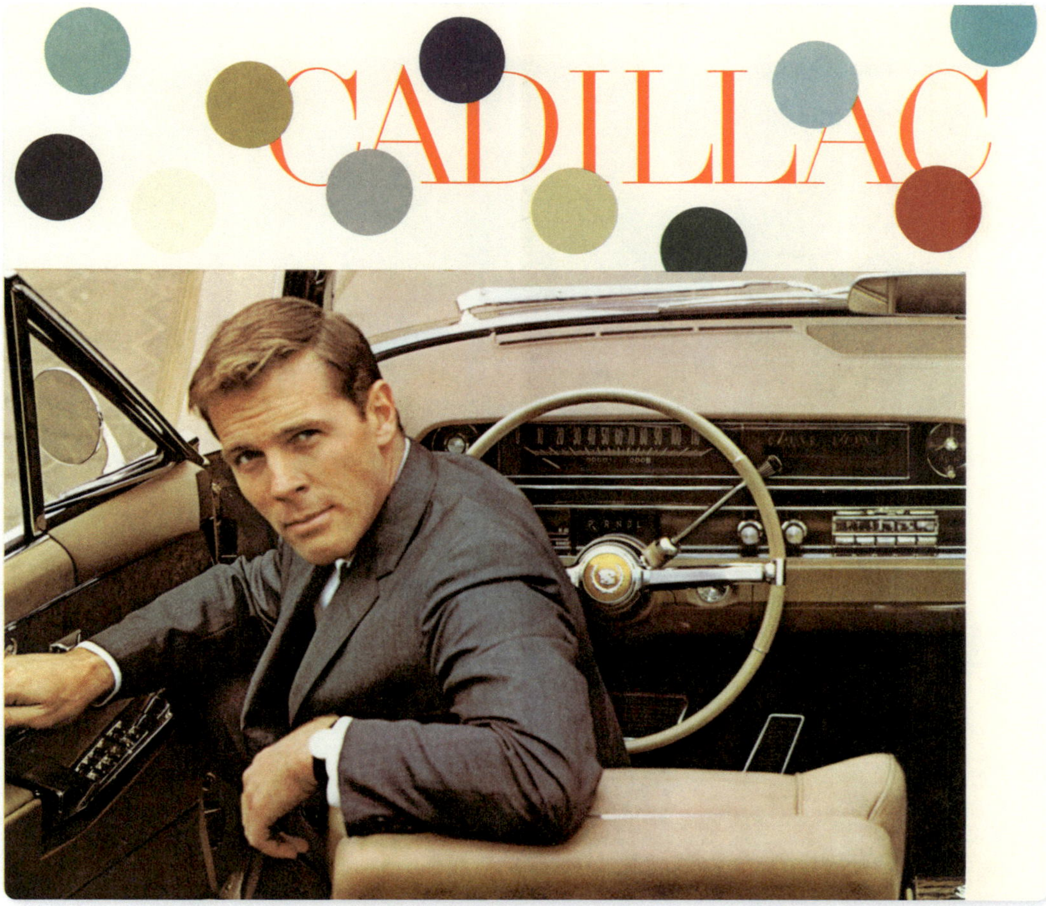
CADILLAC

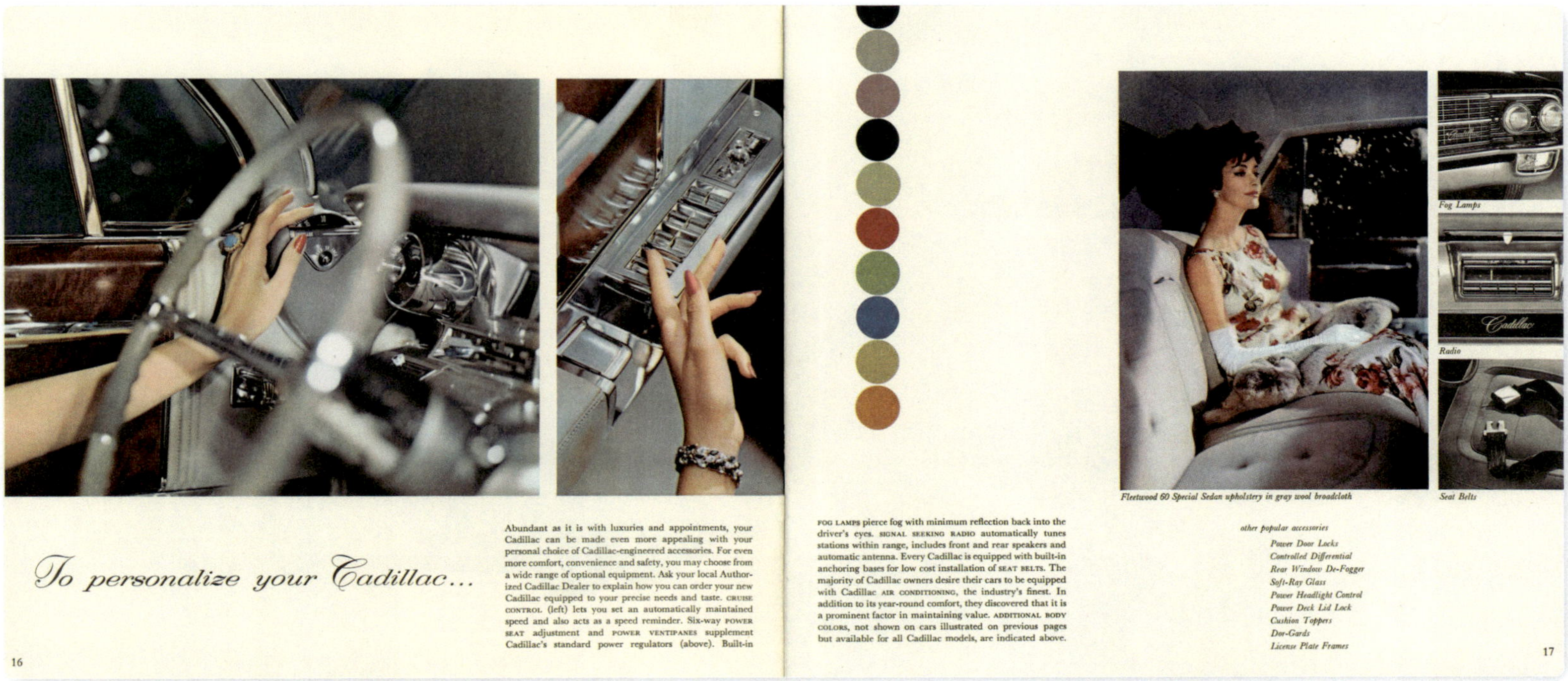
To personalize your Cadillac...
Abundant as it is with luxuries and appointments, your Cadillac can be made even more appealing with your personal choice of Cadillac-engineered accessories. For even more comfort, convenience and safety, you may choose from a wide range of optional equipment. Ask your local Authorized Cadillac Dealer to explain how you can order your new Cadillac equipped to your precise needs and taste. CRUISE CONTROL (left) lets you set an automatically maintained speed and also acts as a speed reminder. Six-way POWER SEAT adjustment and POWER VENTIPANES supplement Cadillac's standard power regulators (above). Built-in
16
FOG LAMPS pierce fog with minimum reflection back into the driver's eyes. SIGNAL SEEKING RADIO automatically tunes stations within range, includes front and rear speakers and automatic antenna. Every Cadillac is equipped with built-in anchoring bases for low cost installation of SEAT BELTS. The majority of Cadillac owners desire their cars to be equipped with Cadillac AIR CONDITIONING, the industry's finest. In addition to its year-round comfort, they discovered that it is a prominent factor in maintaining value. ADDITIONAL BODY COLORS, not shown on cars illustrated on previous pages but available for all Cadillac models, are indicated above.
Fleetwood 60 Special Sedan upholstery in gray wool broadcloth
Fog Lamps
Radio
Seat Belts
other popular accessories
Power Door Locks
Controlled Differential
Rear Window De-Fogger
Soft-Ray Glass
Power Headlight Control
Power Deck Lid Lock
Cushion Toppers
Dor-Gards
License Plate Frames
17

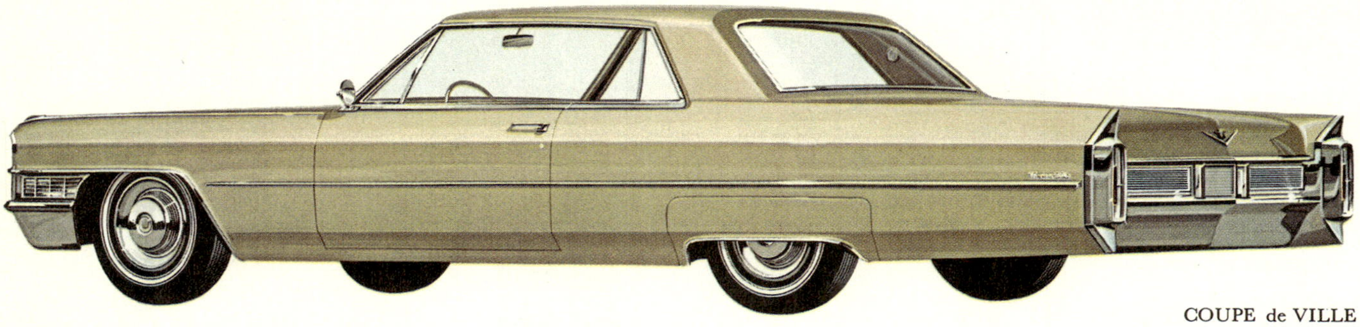

COUPE de VILLE

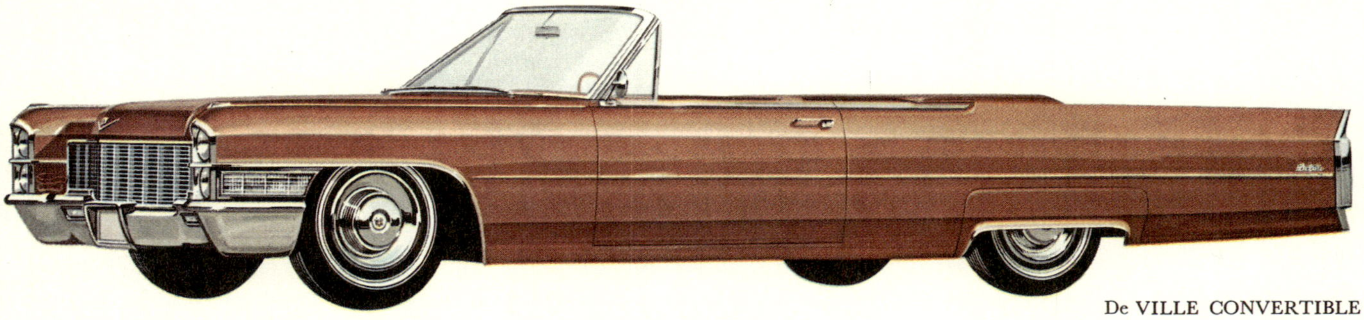

De VILLE CONVERTIBLE

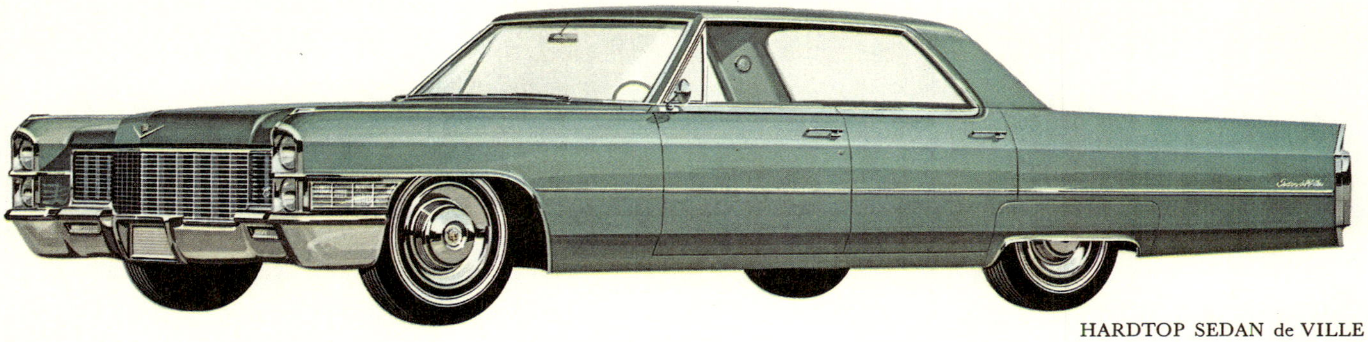

HARDTOP SEDAN de VILLE

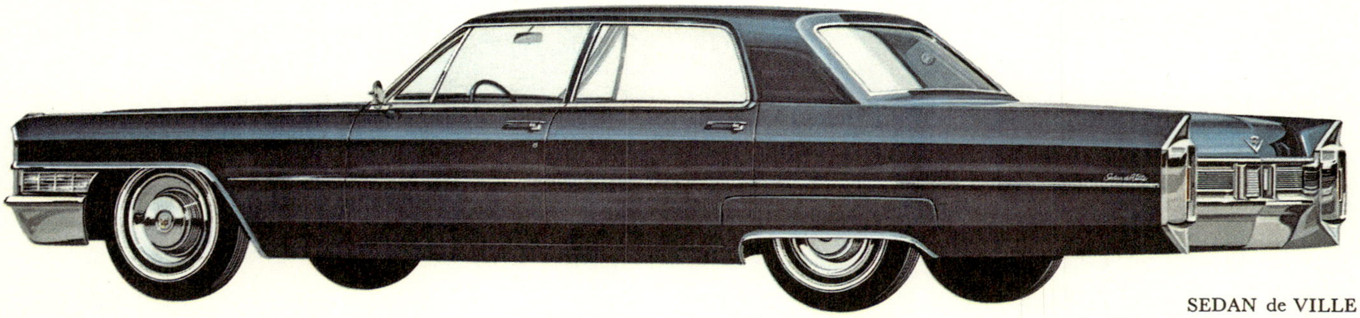

SEDAN de VILLE

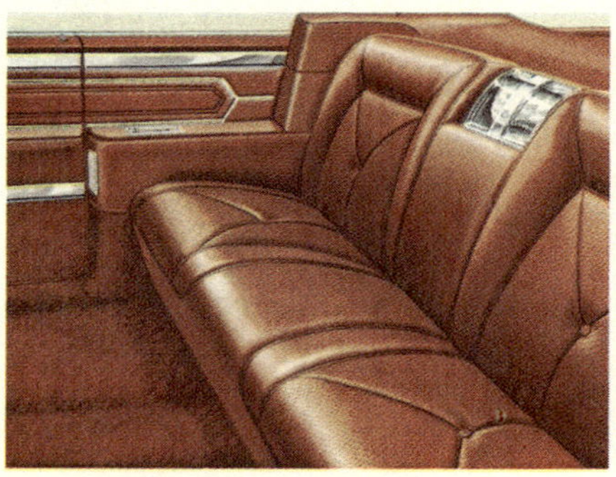

This De Ville Convertible interior is bronze leather, one of eleven interior color combinations. Front bucket seats with padded console are available at extra cost. Convertible tops are available in white, black, blue, green and sandalwood.

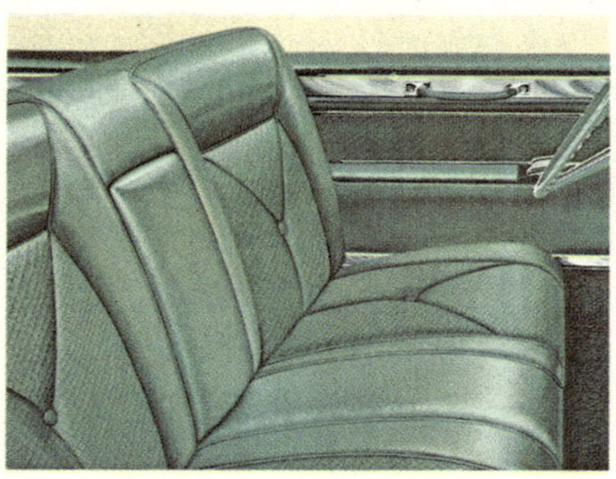

Hardtop Sedan de Ville fabric and leather interiors are offered in eight color selections. Seven perforated leather combinations, available in bench or bucket seats, are extra-cost options. Shown is honeycomb weave Drummond cloth.

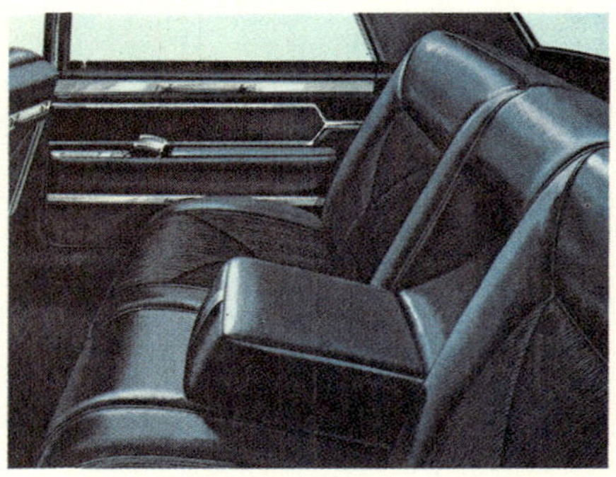

The Sedan de Ville offers the same interior fabric and leather selections as the Hardtop Sedan de Ville shown at left. The fabric illustrated here is random weave Delta cloth in midnight blue, a combination of extraordinary richness.

Cadillac 1965 *(opposite and right)*, **1967** *(below)*

A timeless shape and elegant colors that form a pleasing contrast against any background defined the Cadillacs of the middle 1960s. Options to entice buyers included automatically dimming headlamps and leveling control.

Eine zeitlose Form und elegante Farben, die sich von jedem Hintergrund angenehm abhoben, prägten die Cadillacs Mitte der 1960er-Jahre. Zu den Extras, mit denen man Kunden zu ködern versuchte, gehörten automatisch abblendende Scheinwerfer und eine Niveauregulierung.

Une ligne intemporelle et des couleurs élégantes qui se détachent agréablement dans tous les environnements, voilà à quoi l'on reconnaissait les Cadillac du mitan des années 1960. Parmi les options destinées à séduire les acheteurs, on remarquait l'éclairage automatique des feux et une suspension avec correction d'assiette automatique.

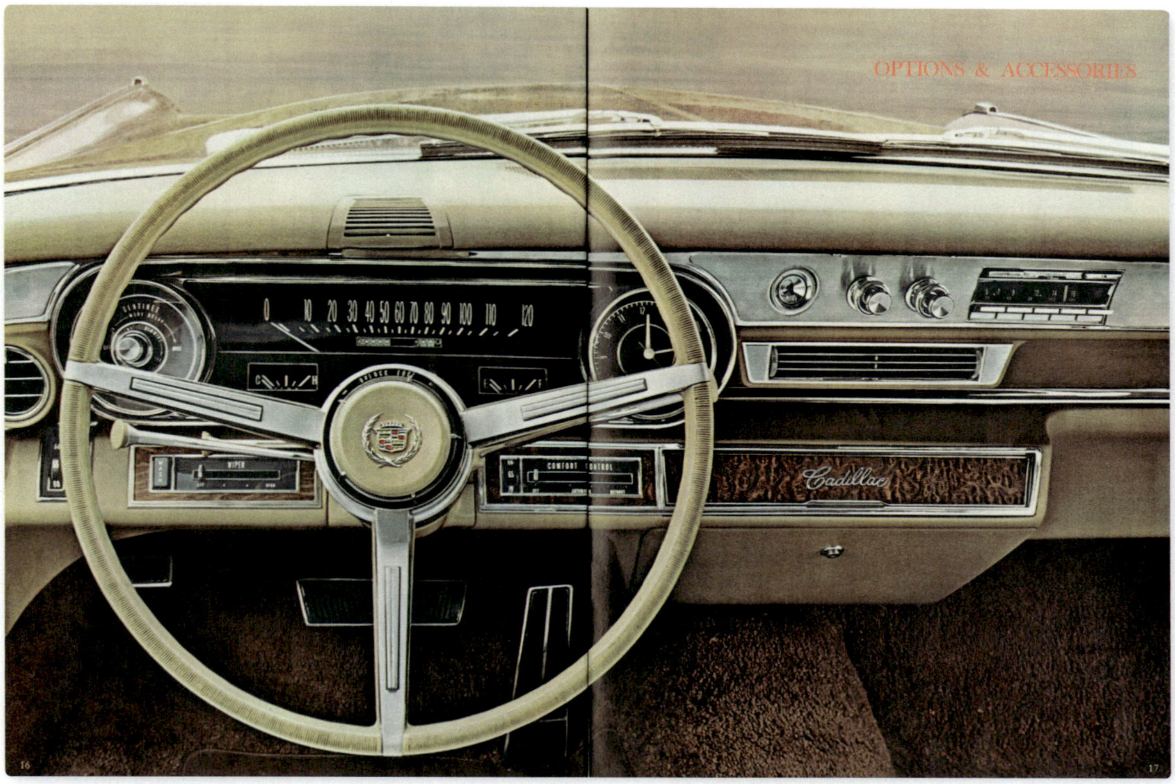

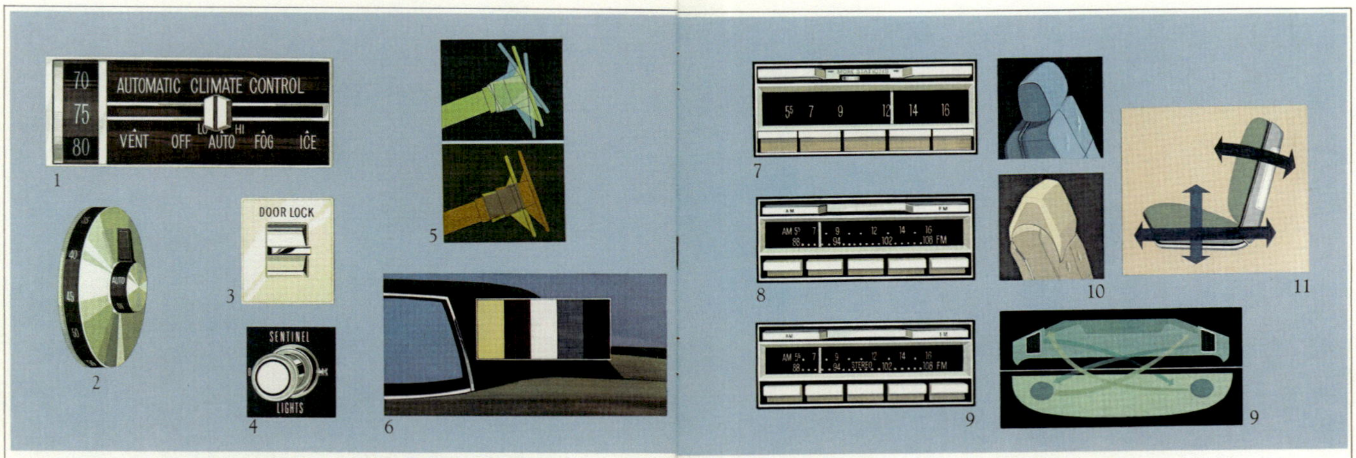

Personalize your Cadillac for greater motoring pleasure

No luxury car built offers a wider or more attractive selection of options and accessories than Cadillac. In addition to all its standard equipment, Cadillac caters still further to your individual needs and desires with an array of special conveniences to make each Cadillac the most personalized car possible.

1 Automatic climate control maintains a preselected temperature as automatically and easily as you set your home thermostat. The system can be bypassed by a new vent position for filtered outside air.

2 Cruise Control permits setting and maintaining a desired speed without using the accelerator pedal. It also acts as a speed reminder by a gentle back pressure to the accelerator when selected speed is reached.

3 All doors can be locked or unlocked from either front door, a feature especially appreciated by Cadillac families.

4 Twilight Sentinel turns lights on and off automatically according to light conditions and can be set as you leave the car to keep lights on up to ninety seconds for a path of light at night.

5 Tilt and Telescope steering wheel adjusts in four directions for personal comfort and swings up for easier entry and exit.

6 A padded vinyl roof enhances the beauty of Cadillac styling. It is standard in five colors on the Brougham and available on De Ville closed models, the Seventy-Five series and the Eldorado.

7 The Cadillac AM Radio has superb tone with wide bass or treble adjustment. Sound may be directed through front or rear speaker or both. Tuning is manual by pushbutton or automatic with signal bar selector.

8 AM-FM Radio adds the enjoyment of fine FM and features two speakers, automatic frequency control and an automatic antenna which is standard with all Cadillac radios.

9 The Cadillac AM-FM Stereo Radio brings you the full richness of stereo FM. Two speakers in front and two in the rear are cross-balanced—left front with right rear and right front with left rear for true stereo sound for driver and passengers.

10 Adjustable front seat headrests are available for both conventional bench type seat, above, and bucket seats, below.

11 Six-way power seat control gives horizontal, vertical and angle adjustment for maximum comfort.

Chevrolet

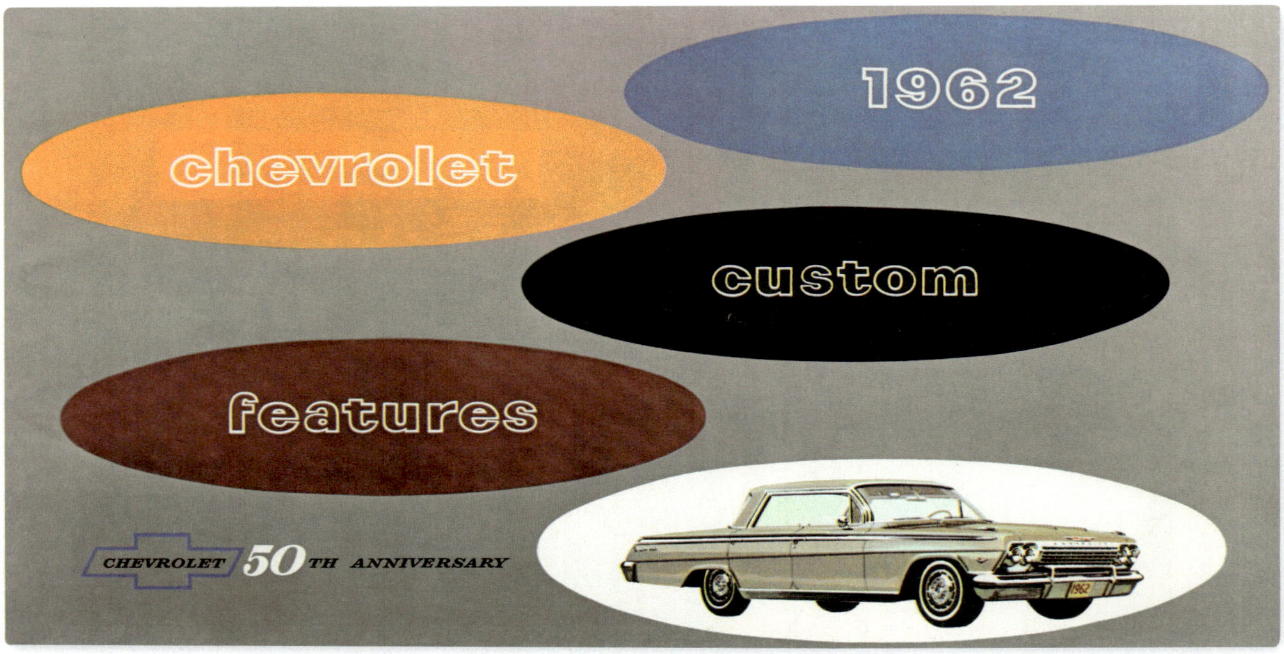

Chevrolet 1962, 1963 *(opposite)*

Manufacturers like Chevrolet made money with major options such as larger engines or air conditioning, but dealers also had opportunities selling and installing decorative extras such as bumper guards, window shades, and spotlights.

Hersteller wie Chevrolet machten mit umfangreichen Extras wie größeren Motoren und Klimaanlagen Geld, aber Händler hatten auch die Möglichkeit, dazu noch schmückendes Beiwerk wie Rammschutzleisten, Fensterrollos und Suchscheinwerfer zu verkaufen und einzubauen.

Des constructeurs tels que Chevrolet gagnaient de l'argent avec d'importantes options comme les moteurs musclés ou la climatisation, mais les concessionnaires avaient également la possibilité de vendre et d'installer des petits plus décoratifs : butoirs de pare-chocs, vitres teintées, projecteurs supplémentaires…

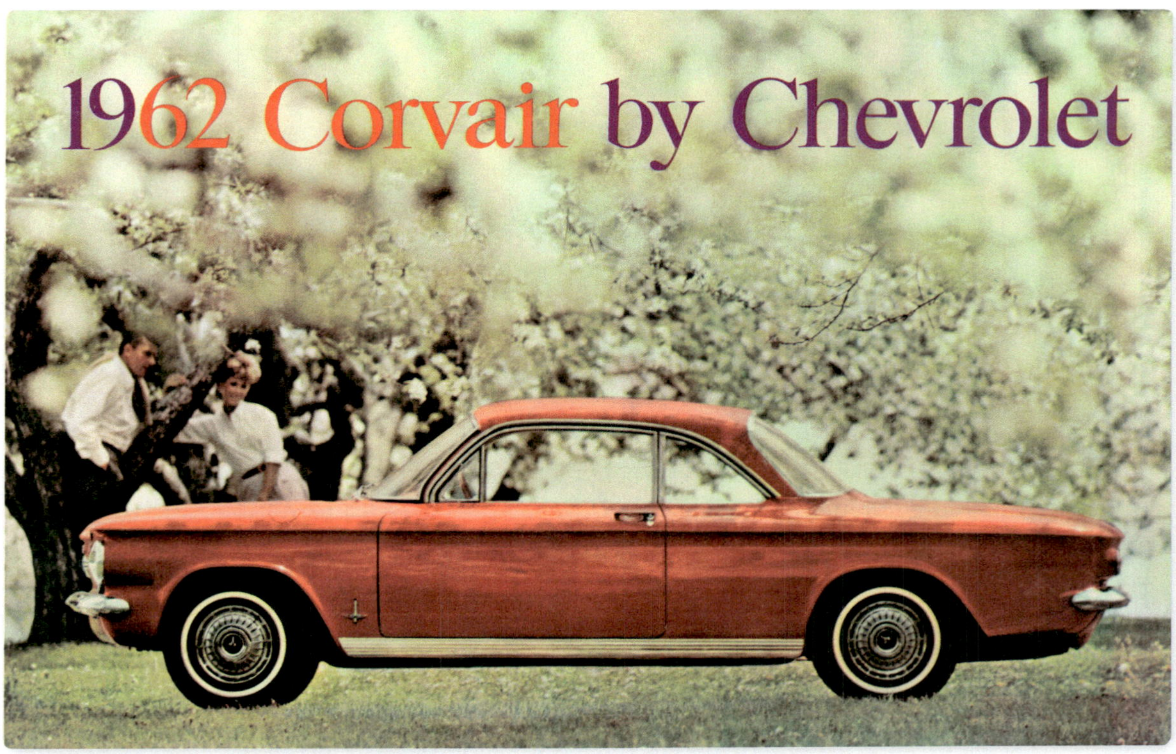

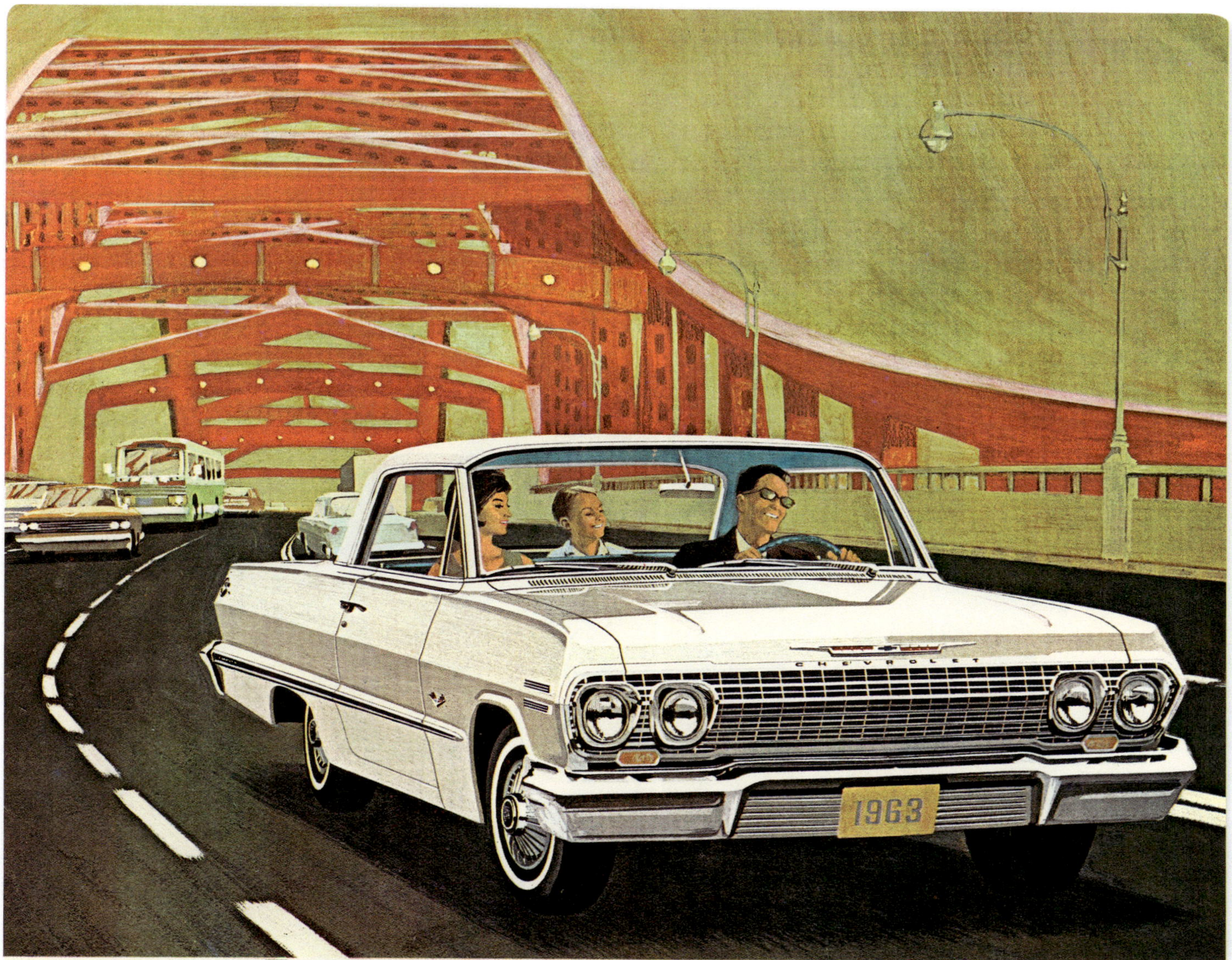

Impala Sport Coupe in Ermine White.

'63 CHEVROLET

CLEAN-CUT AS A JEWEL, SMOOTH-RIDING AS A JET—Here's the car that holds title to first-class travel any way you look at it. A single glance shows you Chevrolet's new beauty. Crisp, fresh, elegant. A step inside puts you in a spacious, luxuriously appointed interior. Comfort is highlighted by Chevrolet's famous Jet-smooth ride. New money-saving features include self-adjusting Safety-Master brakes and new battery-saving Delcotron generator. There's a new standard 6 and a refined standard V8 engine for economy and pleasurable performance. And you'll find Chevrolet's three popular series—Impala, Bel Air and Biscayne—are about as comfortably priced as full-size cars can be.

IMPALA INTERIORS lead the Chevrolet line for luxury and extra touches. Rich, roomy, fashionably finished interiors in a variety of beautifully blended fabrics compliment every '63 Chevrolet.

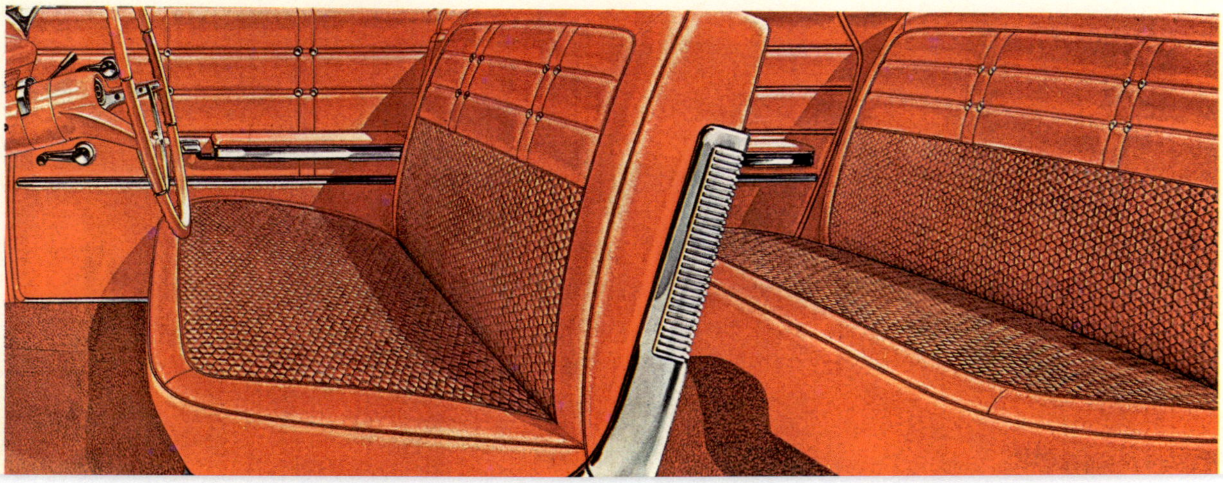

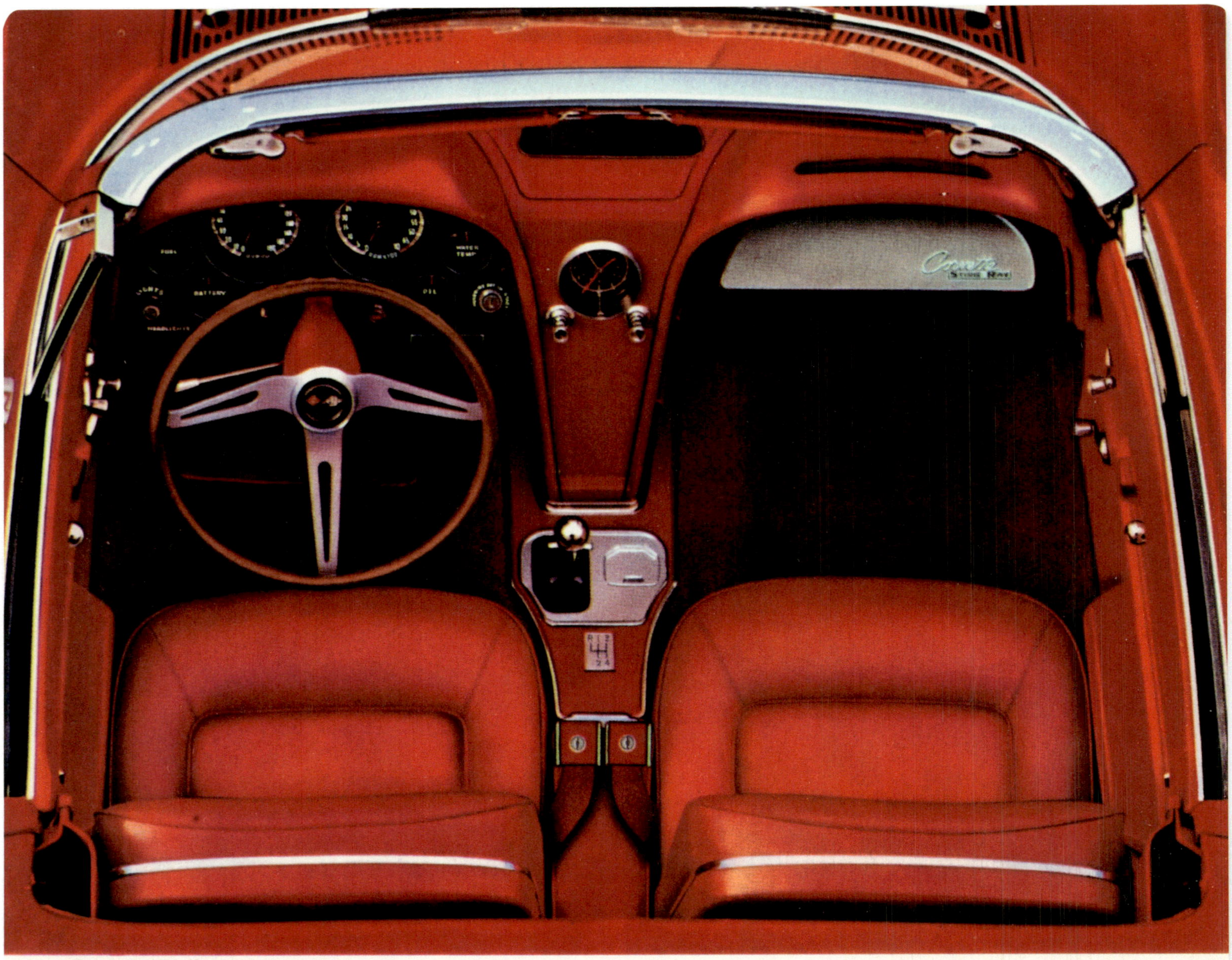

Corvette Sting Ray Convertible with red interior

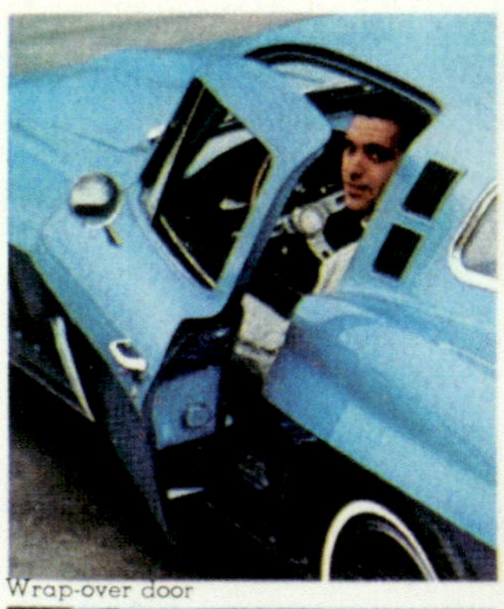

Wrap-over door

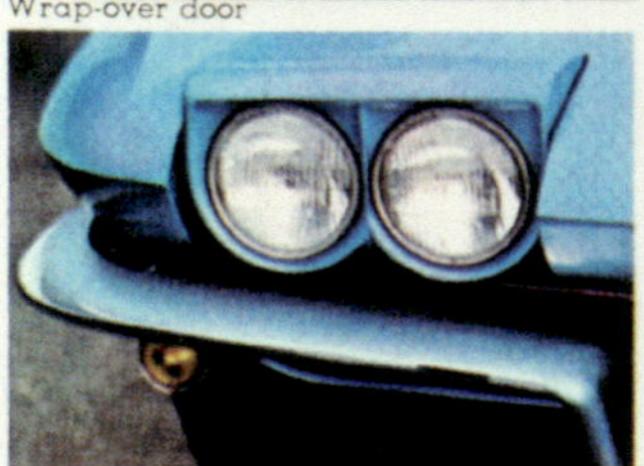

Retractable headlights

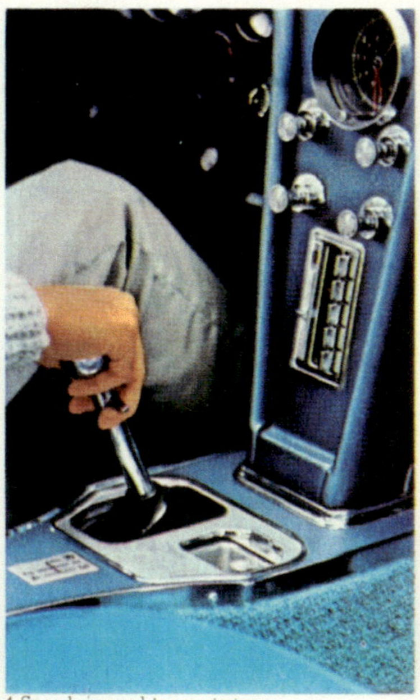

4-Speed manual transmission

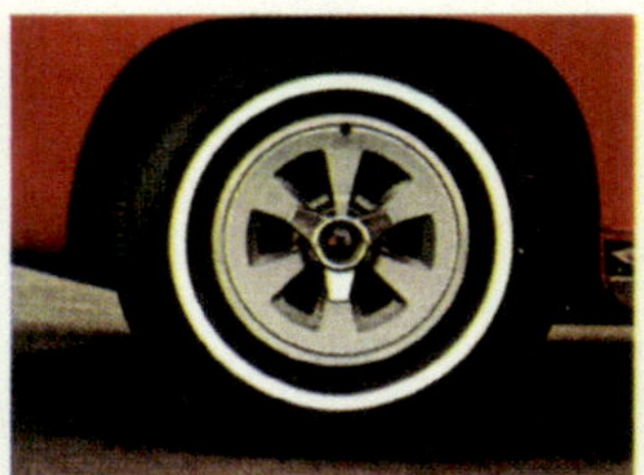

Standard wheel cover

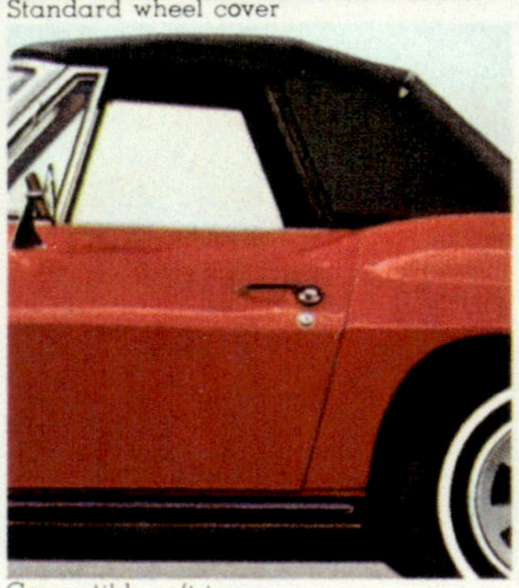

Convertible soft top

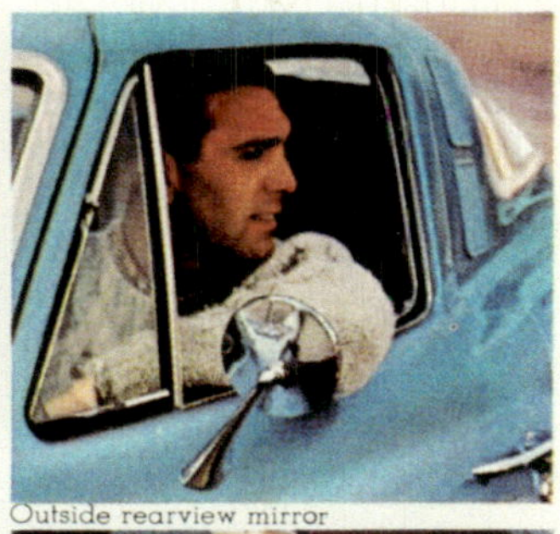

Outside rearview mirror

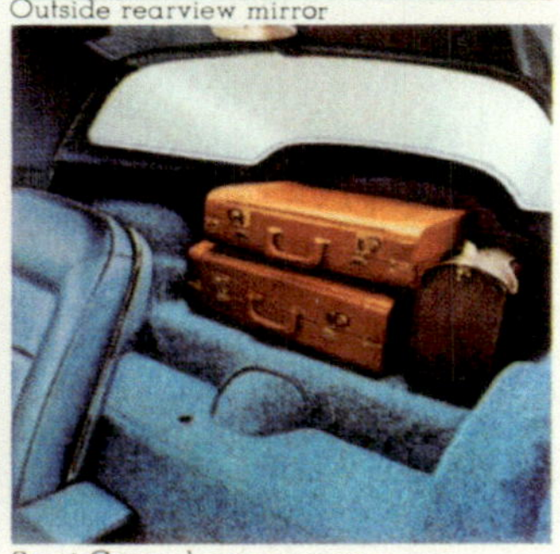

Sport Coupe luggage space

10

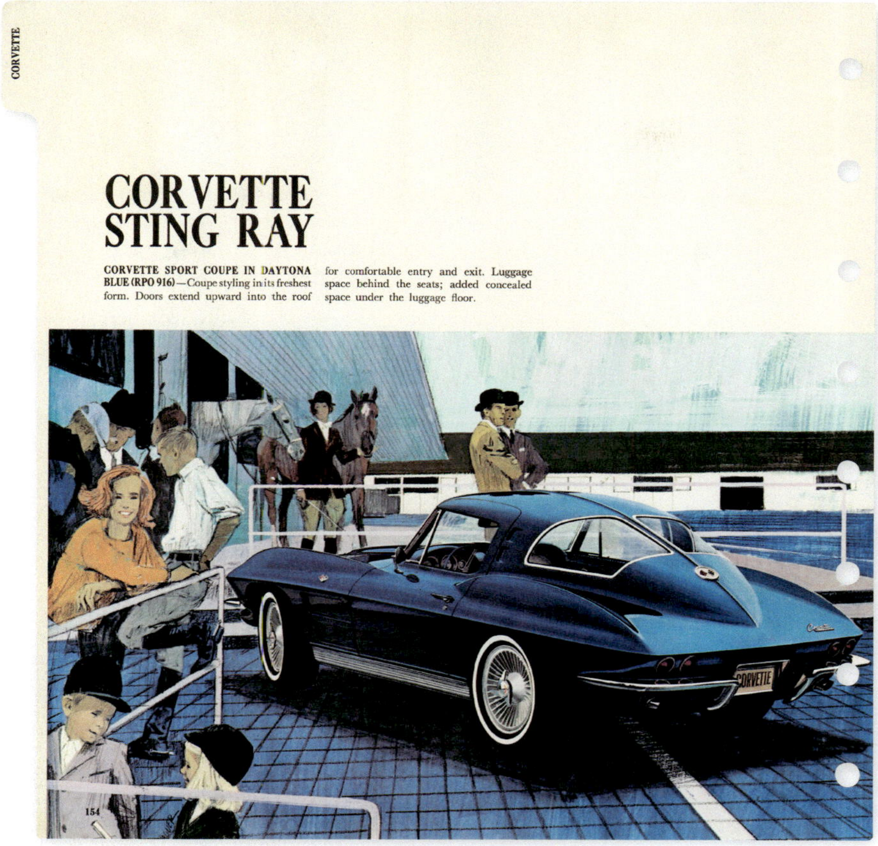

Chevrolet 1965 *(opposite and below)*, **1963** *(left)*

A car, and a look, for the ages. The Sting Ray's lines were based loosely on a Corvette race car. Designer Larry Shinoda created its body for the street under supervision of GM styling chief Bill Mitchell. These cars, also called "mid-year" Corvettes, gained independent rear suspension. Sales jumped 25 percent.

Ein Auto und ein Design für die Ewigkeit: Die Linienführung des Sting Ray basierte grob auf der Rennausführung der Corvette. Konstrukteur Larry Shinoda entwarf die Karosserie für die Straße unter Aufsicht von GM-Styling-Chef Bill Mitchell. Diese Fahrzeuge, die auch „Mittjahres"-Corvettes genannt wurden, erhielten an der Hinterachse eine Einzelradaufhängung. Der Absatz schoss um 25 Prozent nach oben.

Une voiture et un look légendaires. Les lignes de la Sting Ray s'inspiraient assez librement de la Corvette de compétition. Le designer Larry Shinoda lui créa une robe civile sous la supervision du patron du style à la GM, Bill Mitchell. Ces voitures, également appelées Corvette « mid-year », recevaient des suspensions arrière à roues indépendantes. Les ventes bondirent de 25 %.

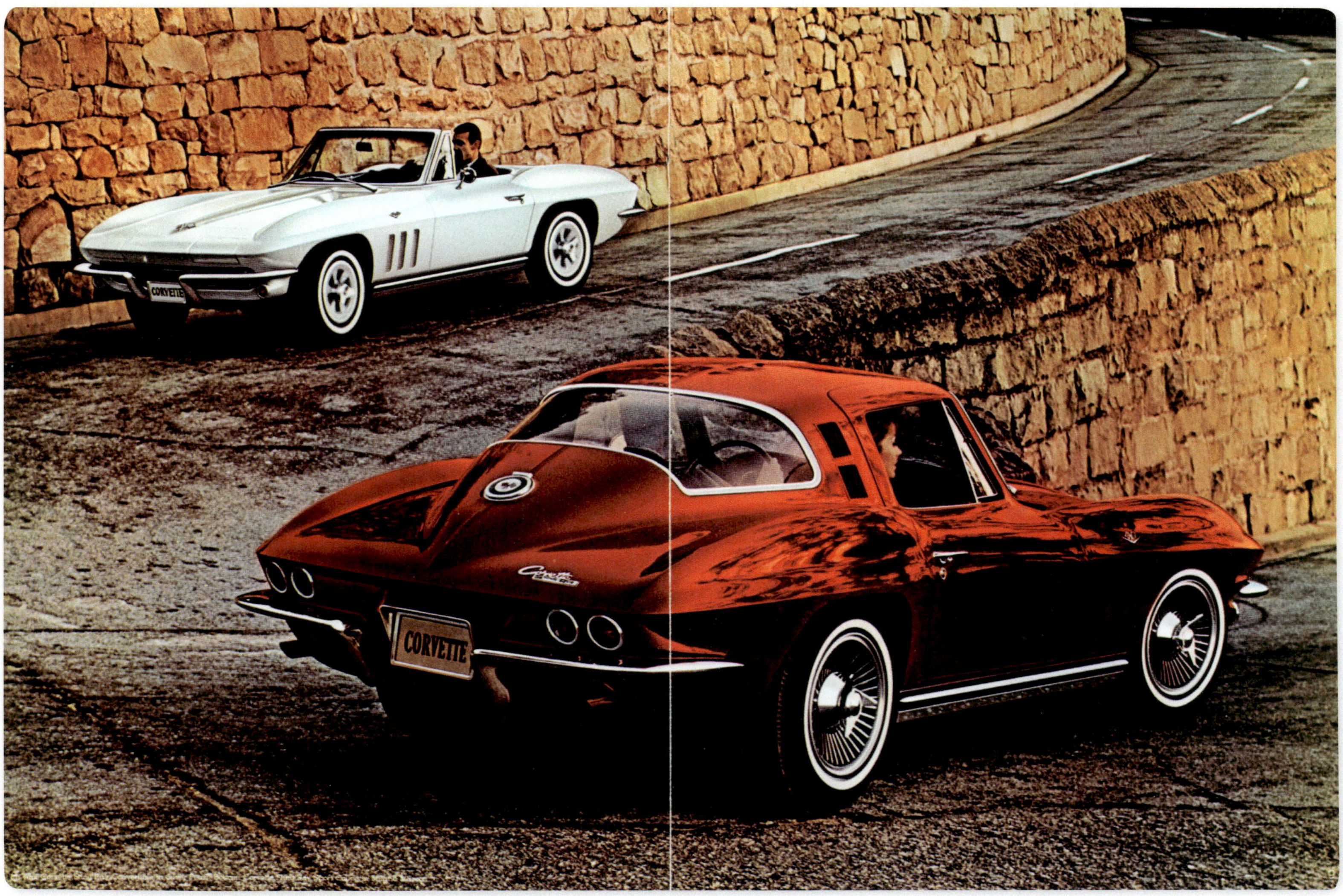

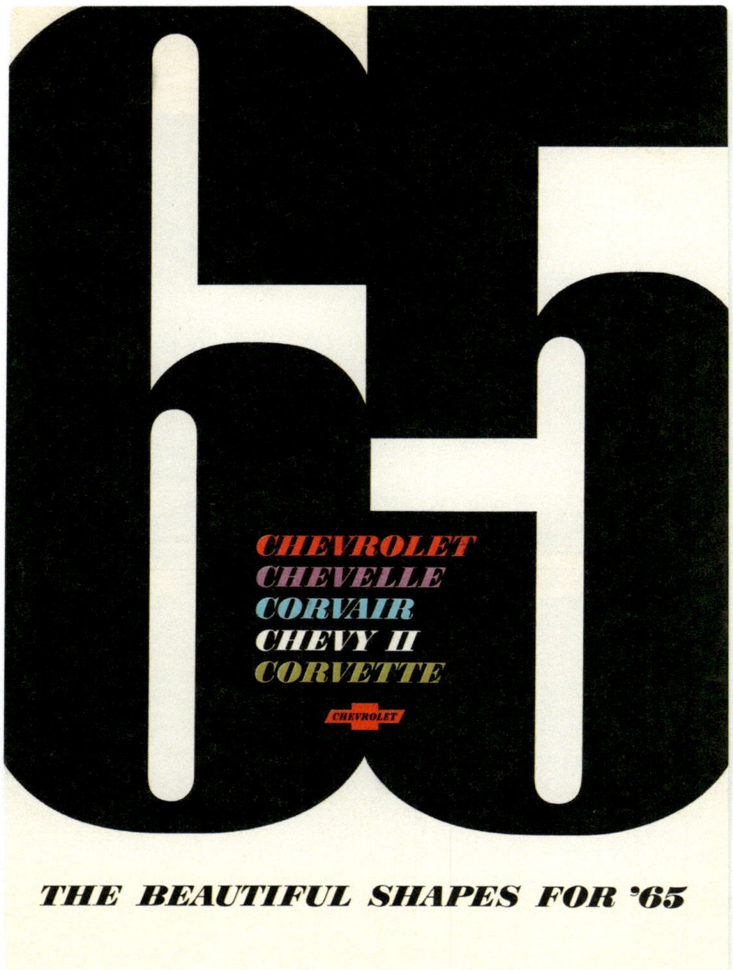

Chevrolet 1965 *(left)*, **1966** *(below)*, **1969** *(opposite top)*, **1967** *(opposite bottom)*

During the 1960s, you could add accessories to a new Chevrolet in practically every imaginable way. Four-speed manual transmissions could turn any full-size Chevy into a muscle car. Dealer-installed extras for the 1967 Nova line included a tissue dispenser and a compass.

Während der 1960er-Jahre erhielt man für einen neuen Chevrolet auf Wunsch praktisch alle erdenklichen Extras. Ein Viergangschaltgetriebe konnte einen normal großen Chevy in ein „muscle car" verwandeln. Zu den Extras, die die Händler noch in den 1967er-Nova einbauen konnten, gehörten auch ein Papiertuchspender und ein Kompass.

Au cours des années 1960, il était loisible d'ajouter toutes sortes d'accessoires à une nouvelle Chevrolet. Les boîtes de vitesses manuelles à quatre rapports pouvaient transformer n'importe quelle grosse Chevy en « muscle car ». Et parmi les extras installés par le concessionnaire pour la gamme Nova 1967, on trouvait un distributeur de mouchoirs et une boussole.

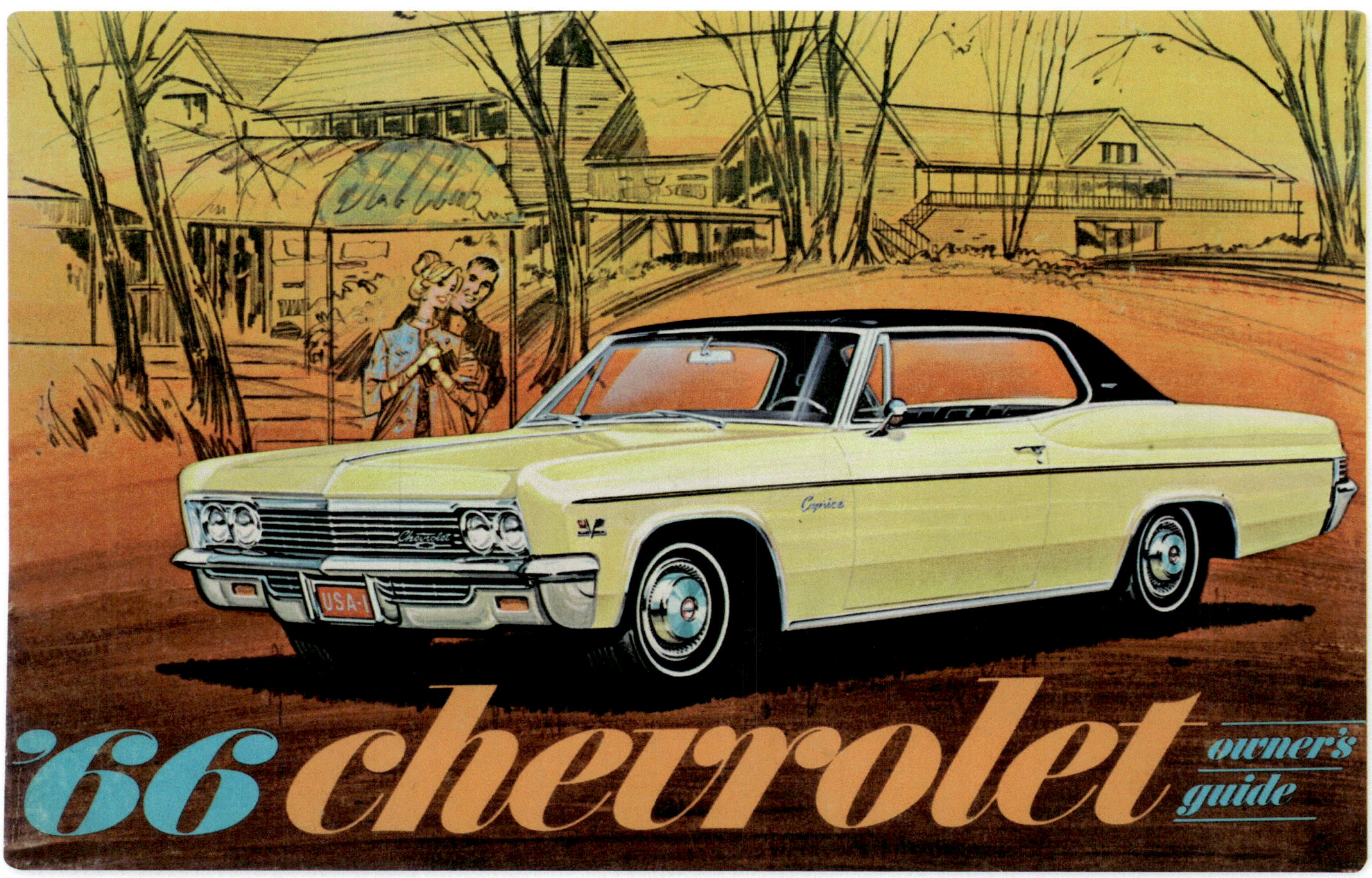

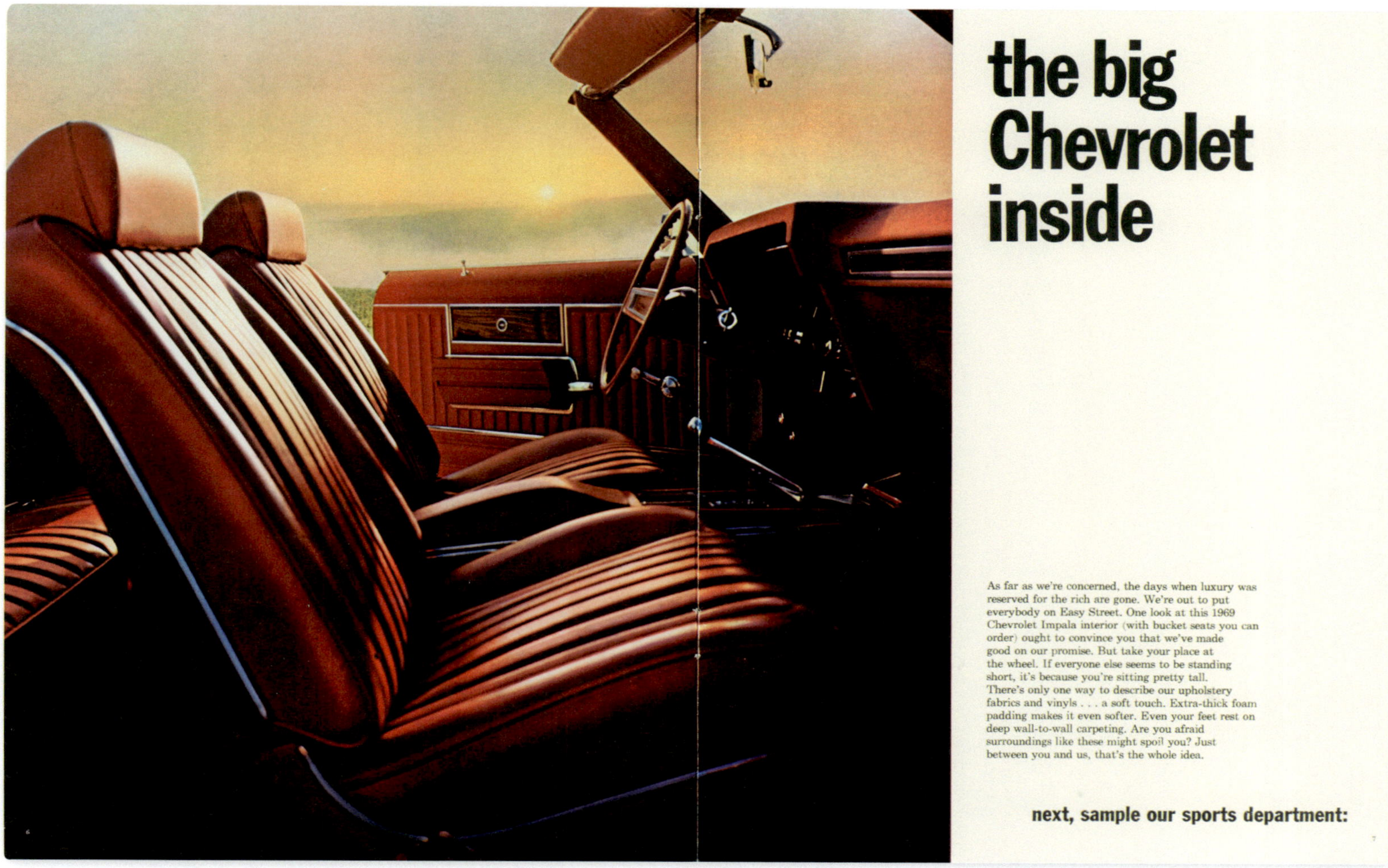

the big Chevrolet inside
As far as we're concerned, the days when luxury was reserved for the rich are gone. We're out to put everybody on Easy Street. One look at this 1969 Chevrolet Impala interior (with bucket seats you can order) ought to convince you that we've made good on our promise. But take your place at the wheel. If everyone else seems to be standing short, it's because you're sitting pretty tall. There's only one way to describe our upholstery fabrics and vinyls . . . a soft touch. Extra-thick foam padding makes it even softer. Even your feet rest on deep wall-to-wall carpeting. Are you afraid surroundings like these might spoil you? Just between you and us, that's the whole idea.
next, sample our sports department:

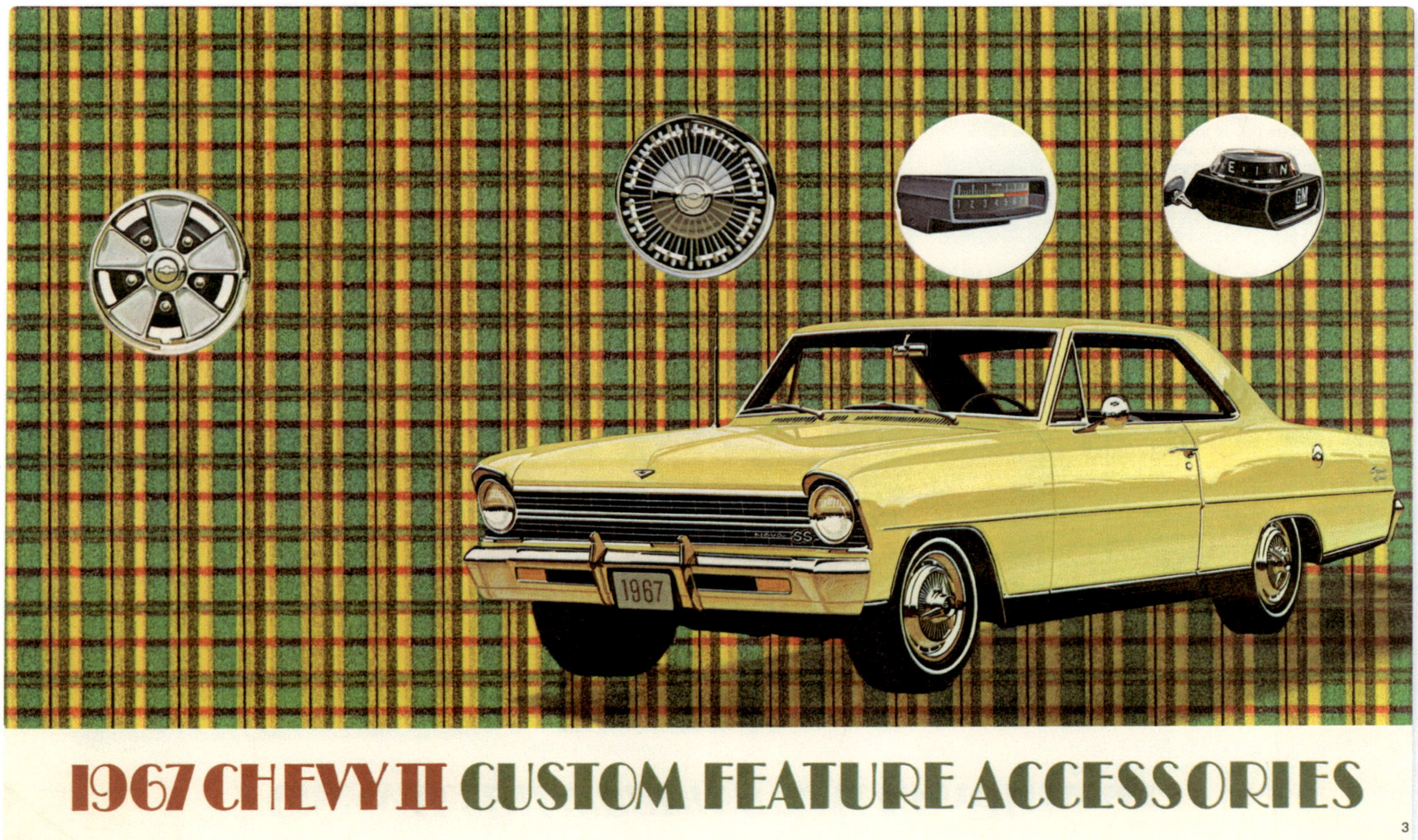

1967
1967 CHEVY II CUSTOM FEATURE ACCESSORIES
3

Chrysler

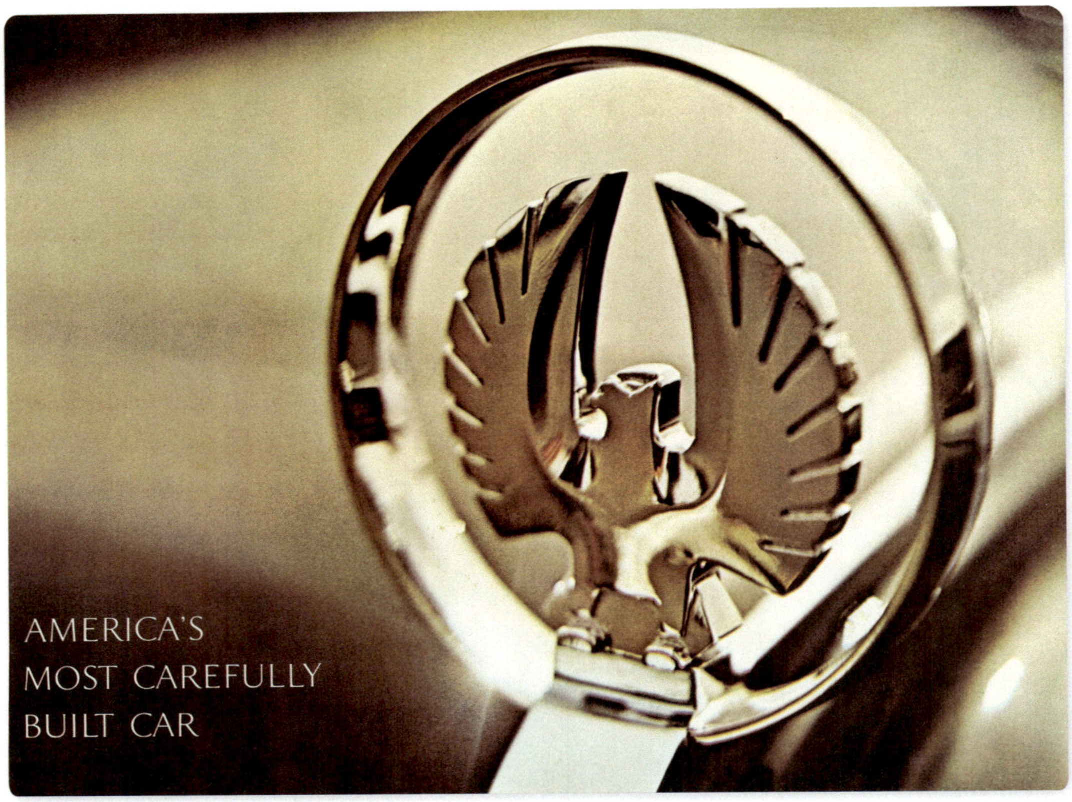

Chrysler 1962 *(left)*, **1966** *(below)*, **1968** *(opposite)*

A not-so-subtle reference to how good the car's luxury seating felt was communicated by the brochure, which suggested that the truly fortunate would put a front or rear seat in the living room next to the tiger-hide rug.

Damit auch der Einfältigste verstand, wie bequem die Luxussitze dieses Autos angeblich waren, zeigte die Broschüre, dass man sich die Vorder- oder Rücksitze auch gleich ins Wohnzimmer aufs Tigerfell stellen konnte.

La brochure risque ici une allusion modestement subtile aux sensations procurées par les voitures de luxe. Elle encourage même leurs heureux propriétaires à installer la banquette arrière au salon, devant la carpette en peau de tigre…

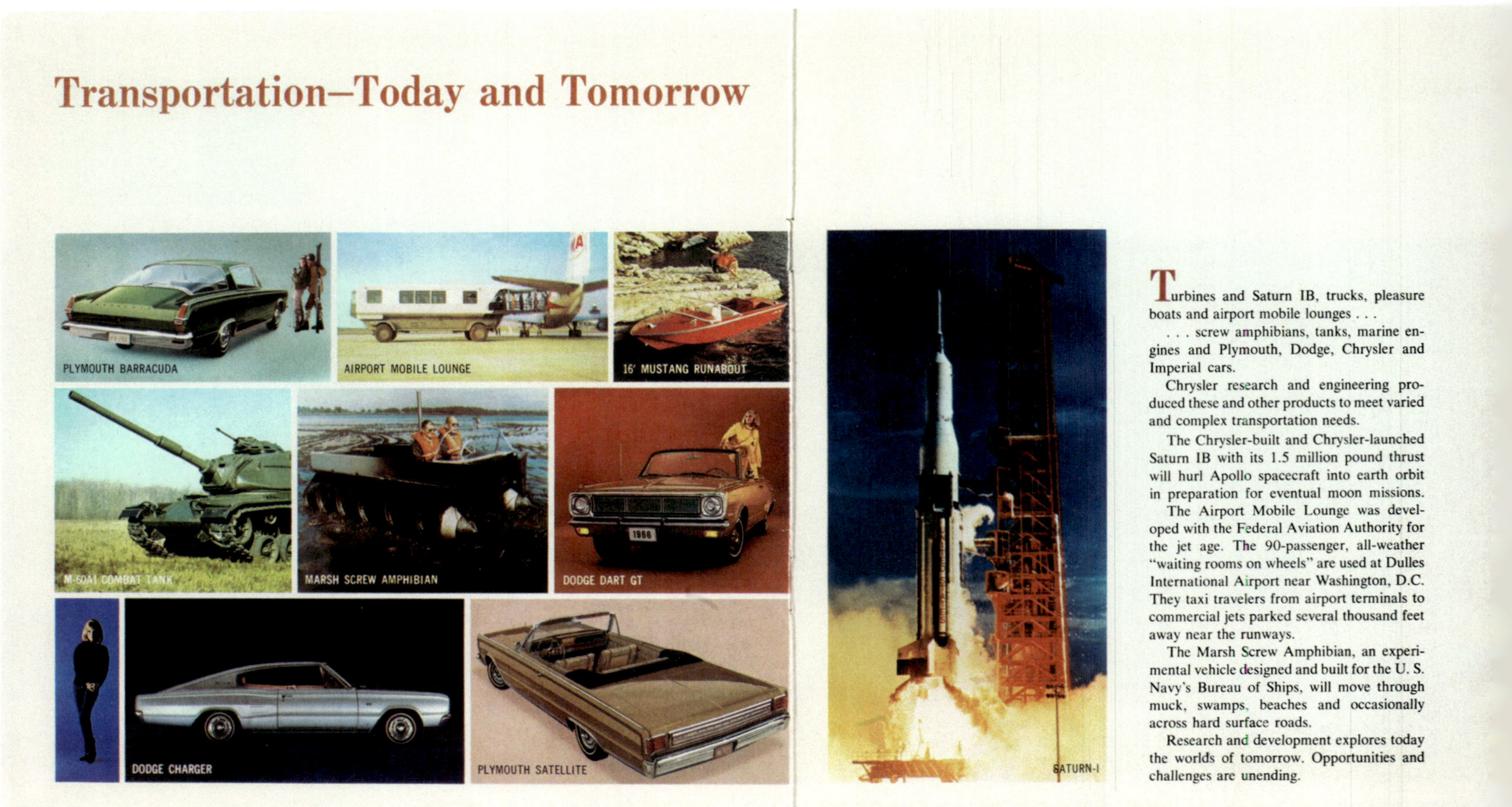

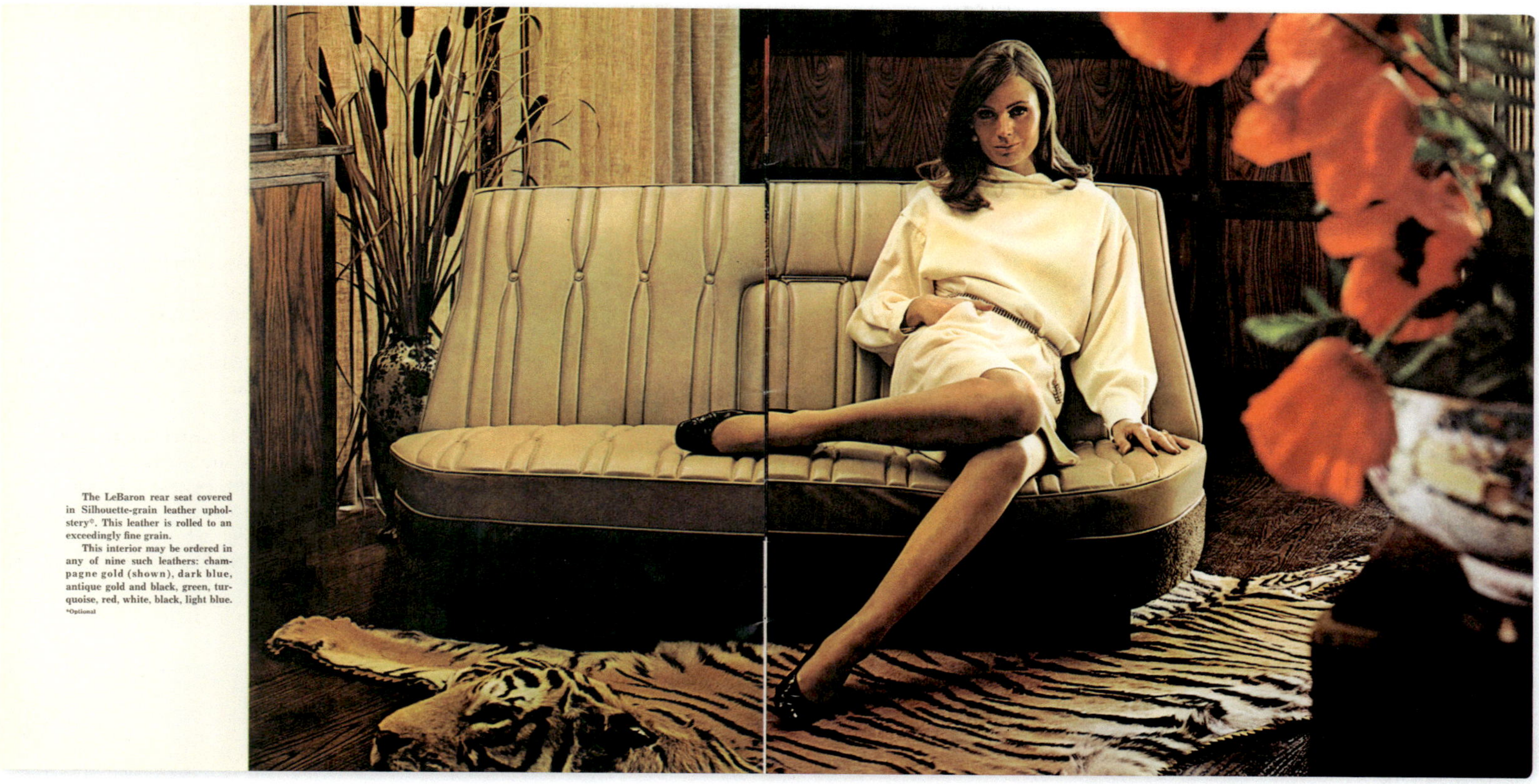

The LeBaron rear seat covered in Silhouette-grain leather upholstery*. This leather is rolled to an exceedingly fine grain.
This interior may be ordered in any of nine such leathers: champagne gold (shown), dark blue, antique gold and black, green, turquoise, red, white, black, light blue.
*Optional

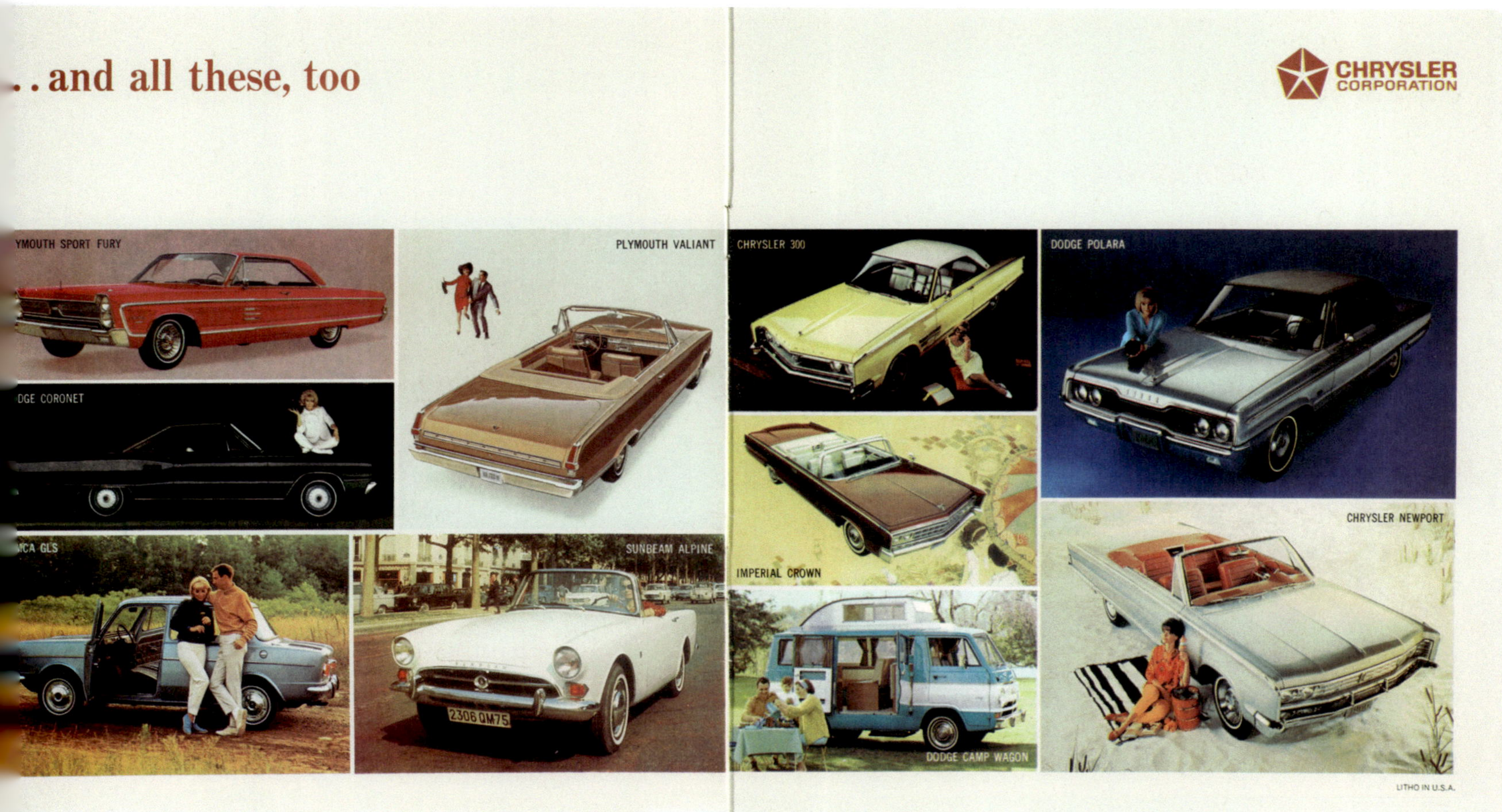

..and all these, too
CHRYSLER CORPORATION
YMOUTH SPORT FURY
PLYMOUTH VALIANT
CHRYSLER 300
DODGE POLARA
DGE CORONET
MCA GLS
SUNBEAM ALPINE
IMPERIAL CROWN
CHRYSLER NEWPORT
2306 QM75
DODGE CAMP WAGON
LITHO IN U.S.A.

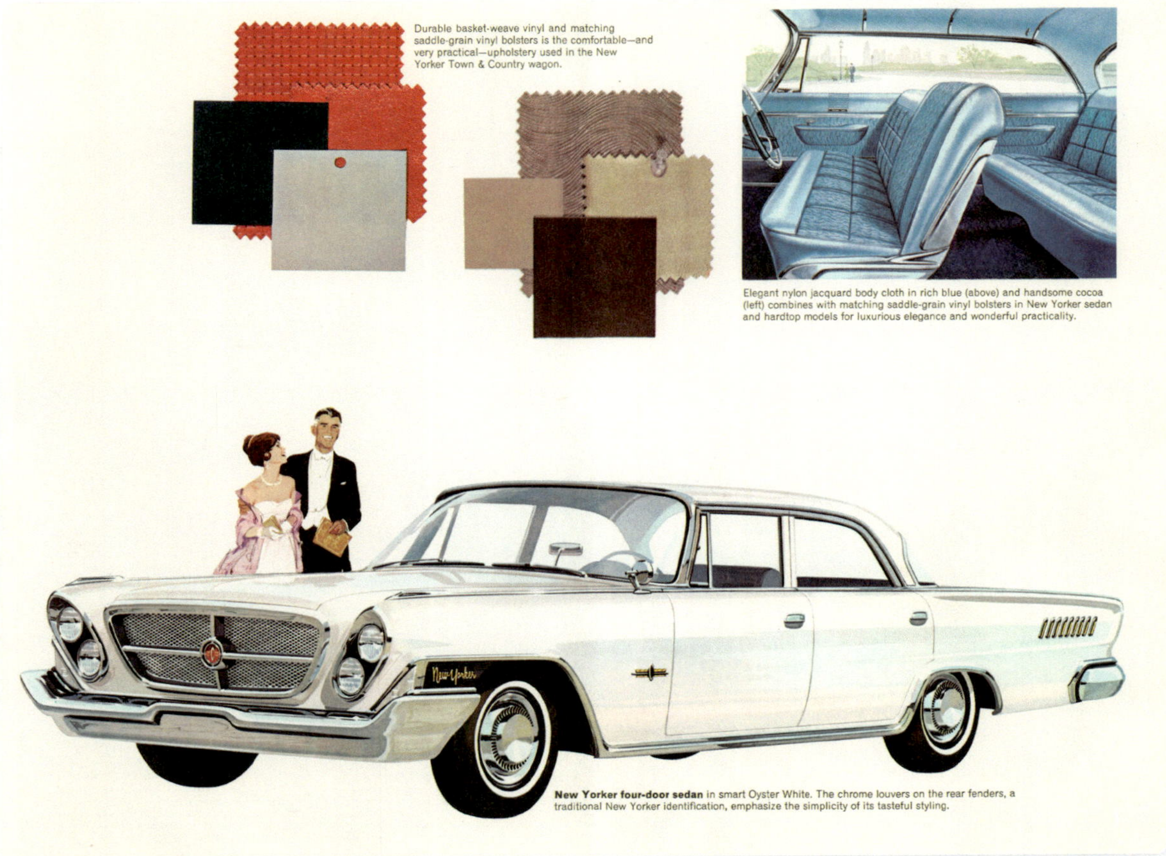

Chrysler 1962

The brochure perfectly illustrates why in 1962 the Chrysler New Yorker was considered one of the most immediately recognizable cars on the road. The styling team made the incongruously canted headlamps work effectively as a design feature.

Die Broschüre verdeutlicht, weshalb der 1962er-Chrysler New Yorker als eines der markantesten Autos im Straßenverkehr galt: Die Designer hatten die merkwürdig versetzten Scheinwerfer effektvoll in das Gesamtbild integriert.

La brochure illustre parfaitement pourquoi la Chrysler New Yorker, en 1962, était considérée comme l'une des voitures les plus immédiatement reconnaissables. Ainsi ces phares incongrûment inclinés, dont l'équipe de style a réussi à faire un élément probant du design de la voiture.

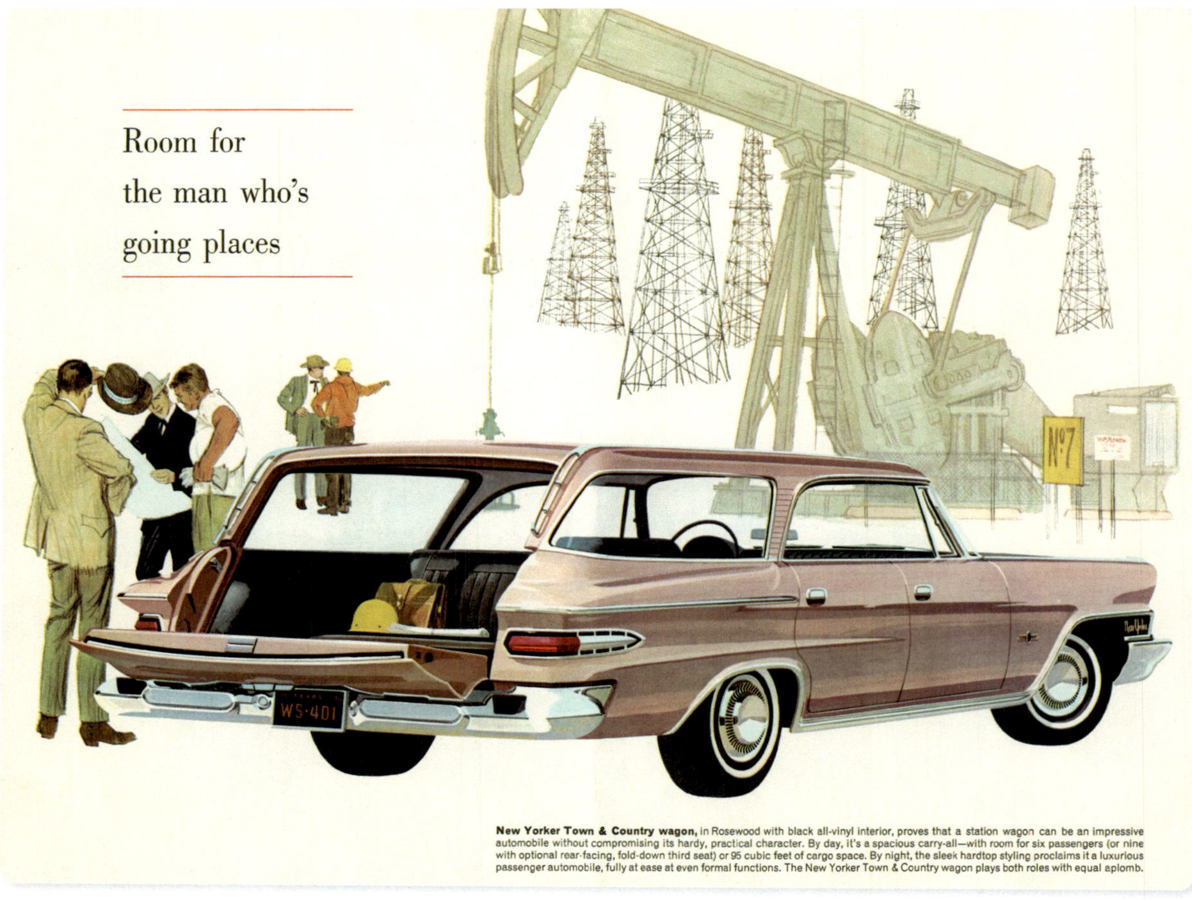

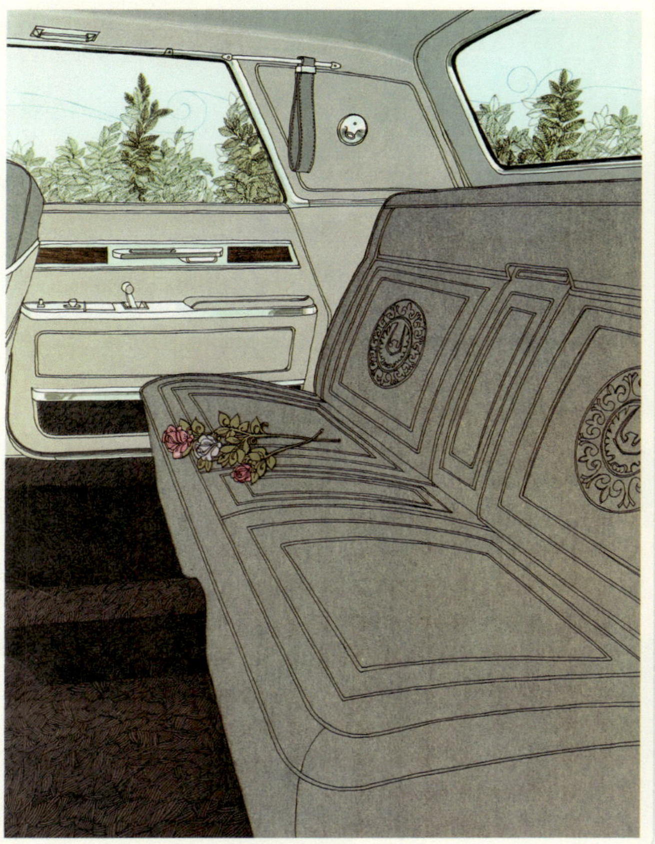

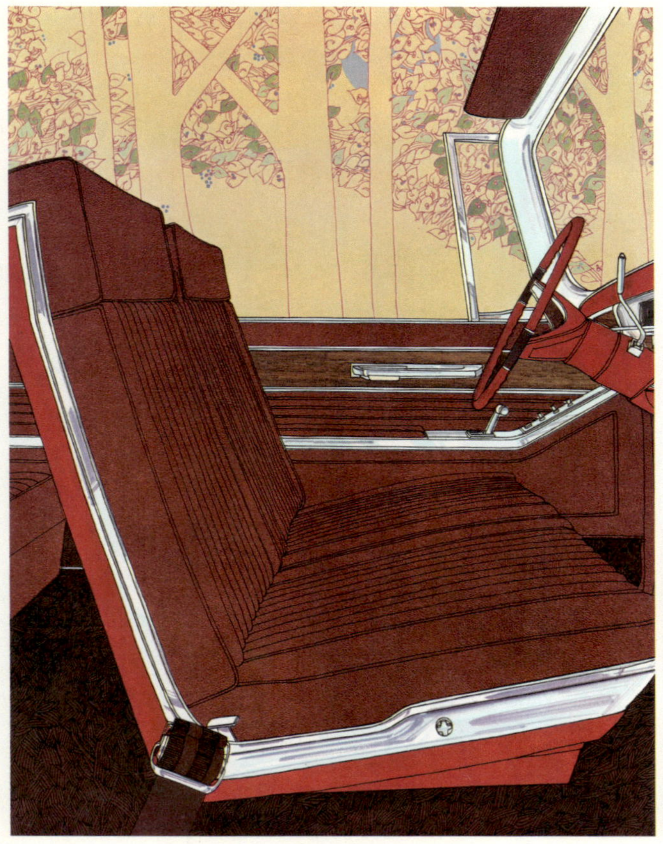

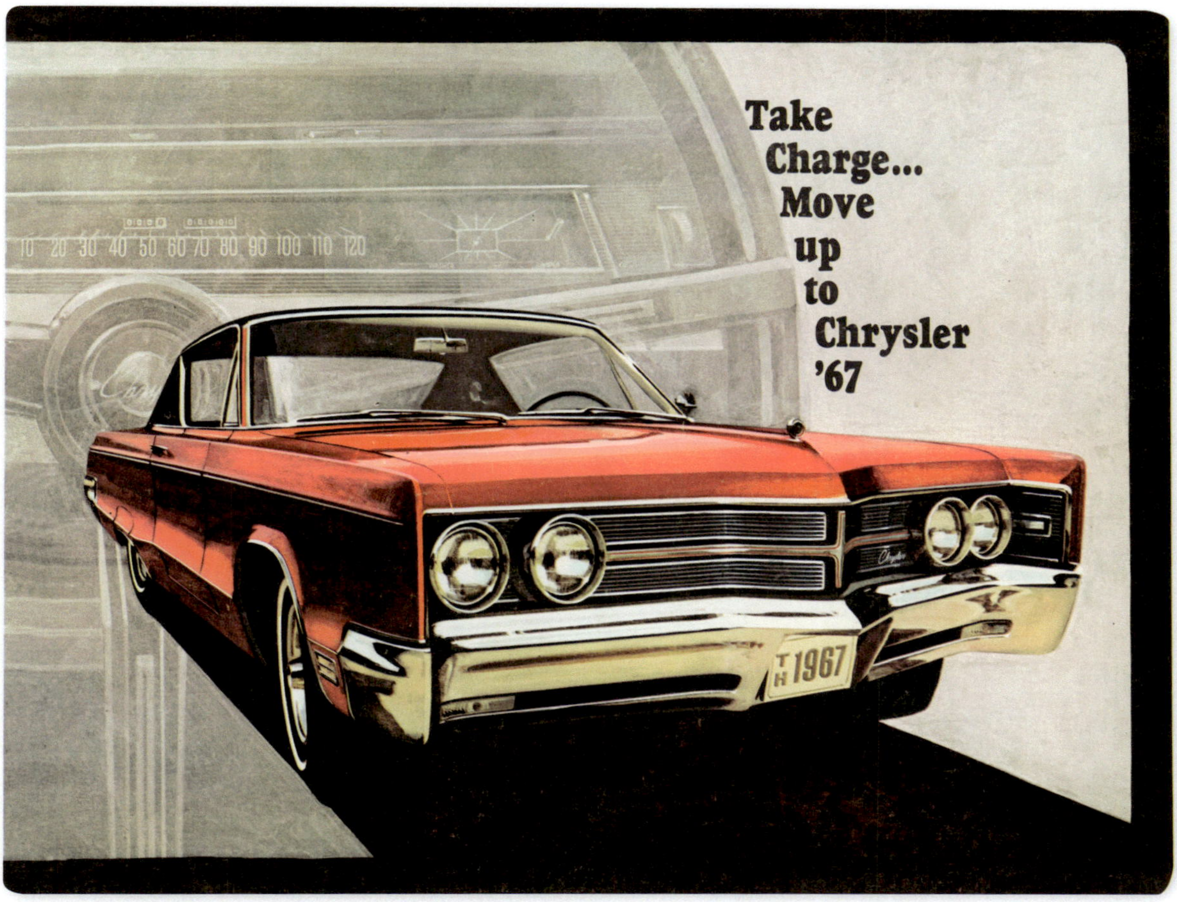

Chrysler 1966 *(above)*, **1967** *(left)*

Brocaded interior fabrics with deep patterning benchmarked Chrysler's offerings as true luxury automobiles. The hardtop-style station wagons are prized rarities among Chrysler enthusiasts today.

In Brokat gefasste Bezüge mit plastischen Mustern wiesen die Fahrzeuge von Chrysler als echte Luxusautos aus. Die Kombis im Coupé-Look sind unter Chrysler-Enthusiasten heute kostbare Raritäten.

Ses tissus de sièges aux motifs brochés classaient les modèles de la gamme Chrysler parmi les authentiques automobiles de luxe. Les breaks d'inspiration hard-top sont aujourd'hui des raretés très recherchées par les amoureux de la marque.

Dodge

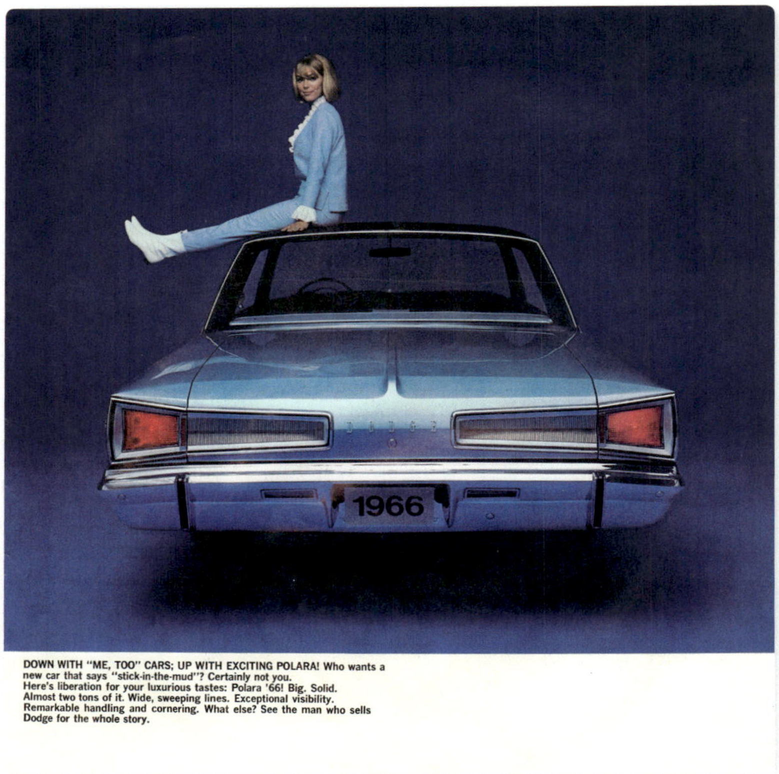

Dodge 1966 *(this spread)*

Recognizable by its nearly full-width taillights, the new Dodge Monaco for 1966 took the make squarely into Chrysler territory in terms of pure luxury. The model was a big success.

Der neue Dodge Monaco für das Jahr 1966 ist an seinen breiten Rückleuchten, die fast über die gesamte Wagenbreite laufen, gut zu erkennen und katapultierte die Marke in Sachen Luxus auf einen Schlag in Chrysler-Gefilde. Das Modell war ein Riesenerfolg.

Reconnaissable à ses feux arrière déployés sur presque toute sa largeur, la nouvelle Dodge Monaco de 1966, en matière de luxe pur, conduisait directement la marque sur le terrain de Chrysler. Le modèle connut un grand succès.

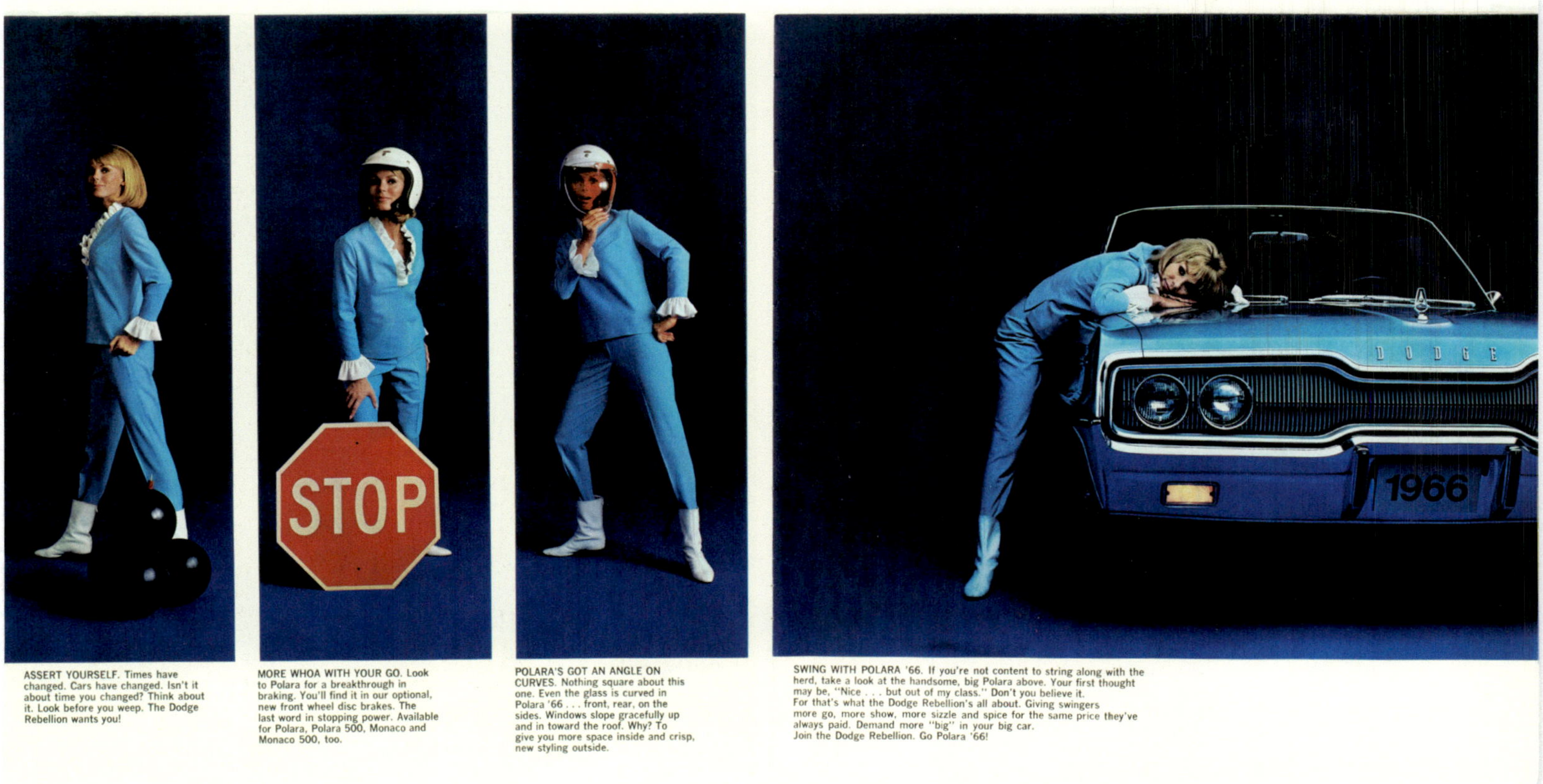

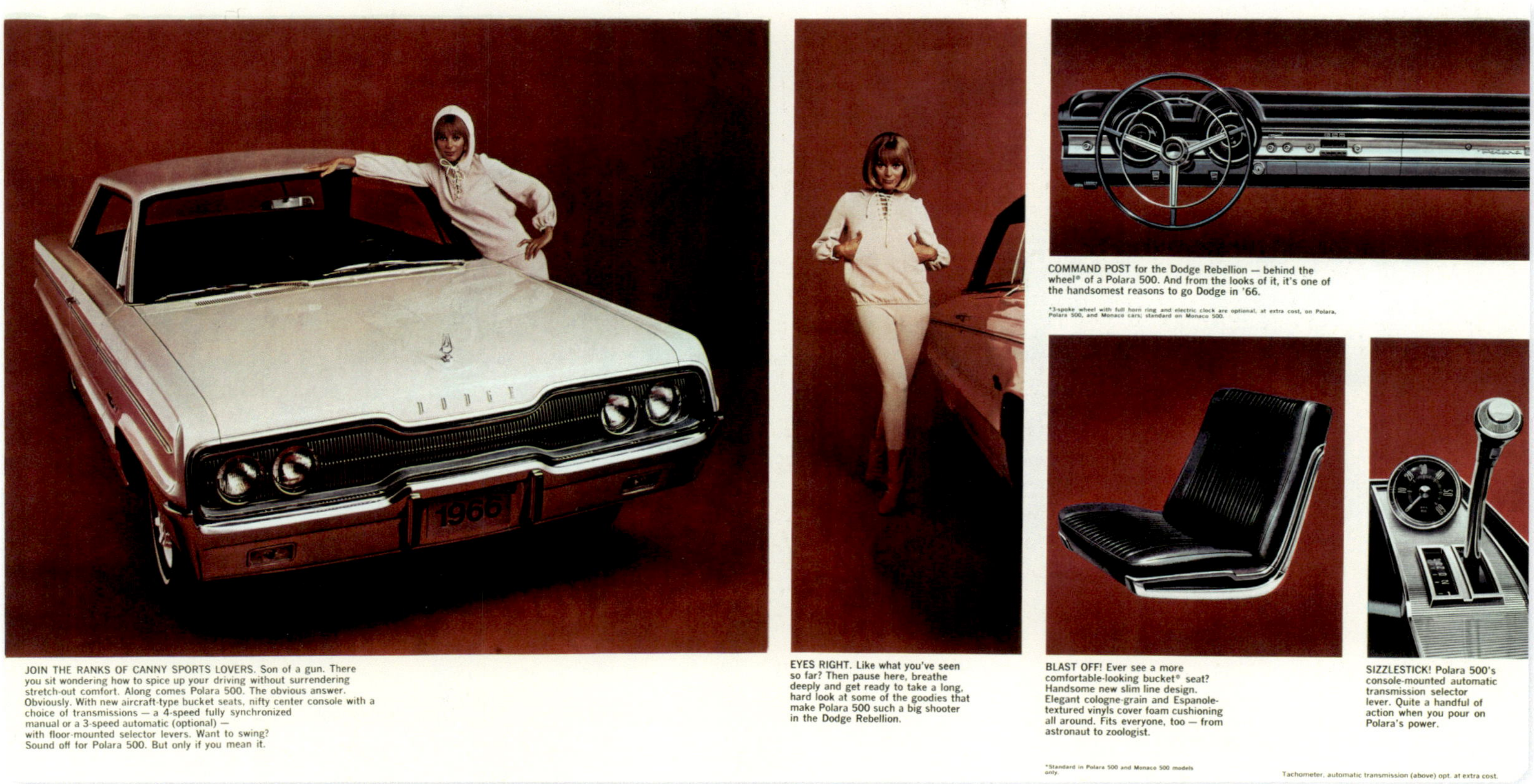
JOIN THE RANKS OF CANNY SPORTS LOVERS. Son of a gun. There you sit wondering how to spice up your driving without surrendering stretch-out comfort. Along comes Polara 500. The obvious answer. Obviously. With new aircraft-type bucket seats, nifty center console with a choice of transmissions — a 4-speed fully synchronized manual or a 3-speed automatic (optional) — with floor-mounted selector levers. Want to swing? Sound off for Polara 500. But only if you mean it.
EYES RIGHT. Like what you've seen so far? Then pause here, breathe deeply and get ready to take a long, hard look at some of the goodies that make Polara 500 such a big shooter in the Dodge Rebellion.
COMMAND POST for the Dodge Rebellion — behind the wheel* of a Polara 500. And from the looks of it, it's one of the handsomest reasons to go Dodge in '66.
*3-spoke wheel with full horn ring and electric clock are optional, at extra cost, on Polara, Polara 500, and Monaco cars; standard on Monaco 500.
BLAST OFF! Ever see a more comfortable-looking bucket* seat? Handsome new slim line design. Elegant cologne-grain and Espanole-textured vinyls cover foam cushioning all around. Fits everyone, too — from astronaut to zoologist.
*Standard in Polara 500 and Monaco 500 models only.
SIZZLESTICK! Polara 500's console-mounted automatic transmission selector lever. Quite a handful of action when you pour on Polara's power.
Tachometer, automatic transmission (above) opt. at extra cost.

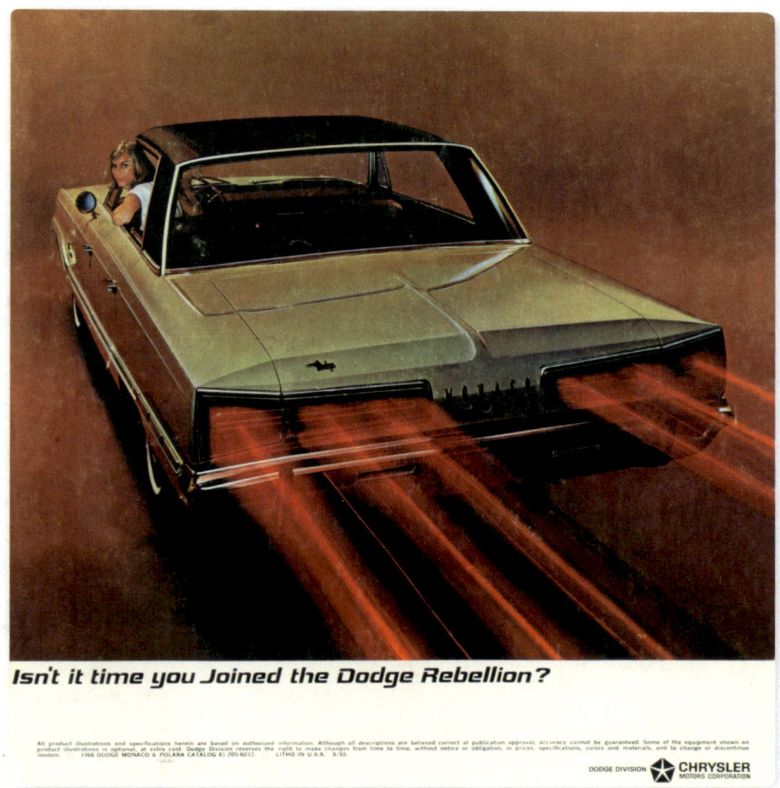
Isn't it time you Joined the Dodge Rebellion?
DODGE DIVISION
CHRYSLER
MOTORS CORPORATION

Duesenberg

Duesenberg 1966 *(this spread)*

A descendant of the original Duesenberg brothers commissioned ex-Chrysler stylist Virgil Exner to style a revival of the great 1930s classic. The fender sweeps and grille shell make broad hints as to the car's origins. Only one prototype was built, powered by a Chrysler V-8 engine.

Ein Nachfahre der Brüder Friedrich und August Düsenberg bestellte bei Virgil Exner, einem ehemaligen Stylisten bei Chrysler, eine Neuauflage des Klassikers der 1930er-Jahre. Die geschwungenen Kotflügel und der markante Rahmen um den Kühlergrill erinnern an die Ursprünge des Fahrzeugs, doch es wurde nur ein Prototyp gebaut, der von einem Chrysler-V8-Motor angetrieben wurde.

Un descendant des frères Duesenberg mandata l'ex-styliste de Chrysler Virgil Exner pour faire revivre les grandes classiques des années 1930. Les ailes saillantes et la calandre verticale font ici clairement référence aux origines de la voiture. Seul un prototype, équipé d'un moteur Chrysler V8, sortit des ateliers de montage.

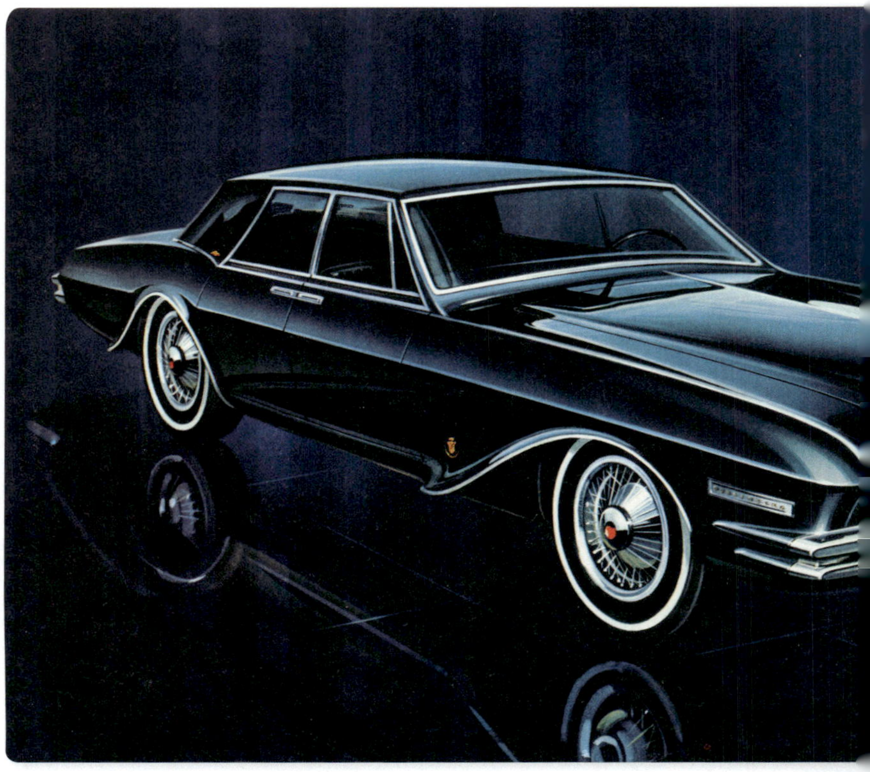

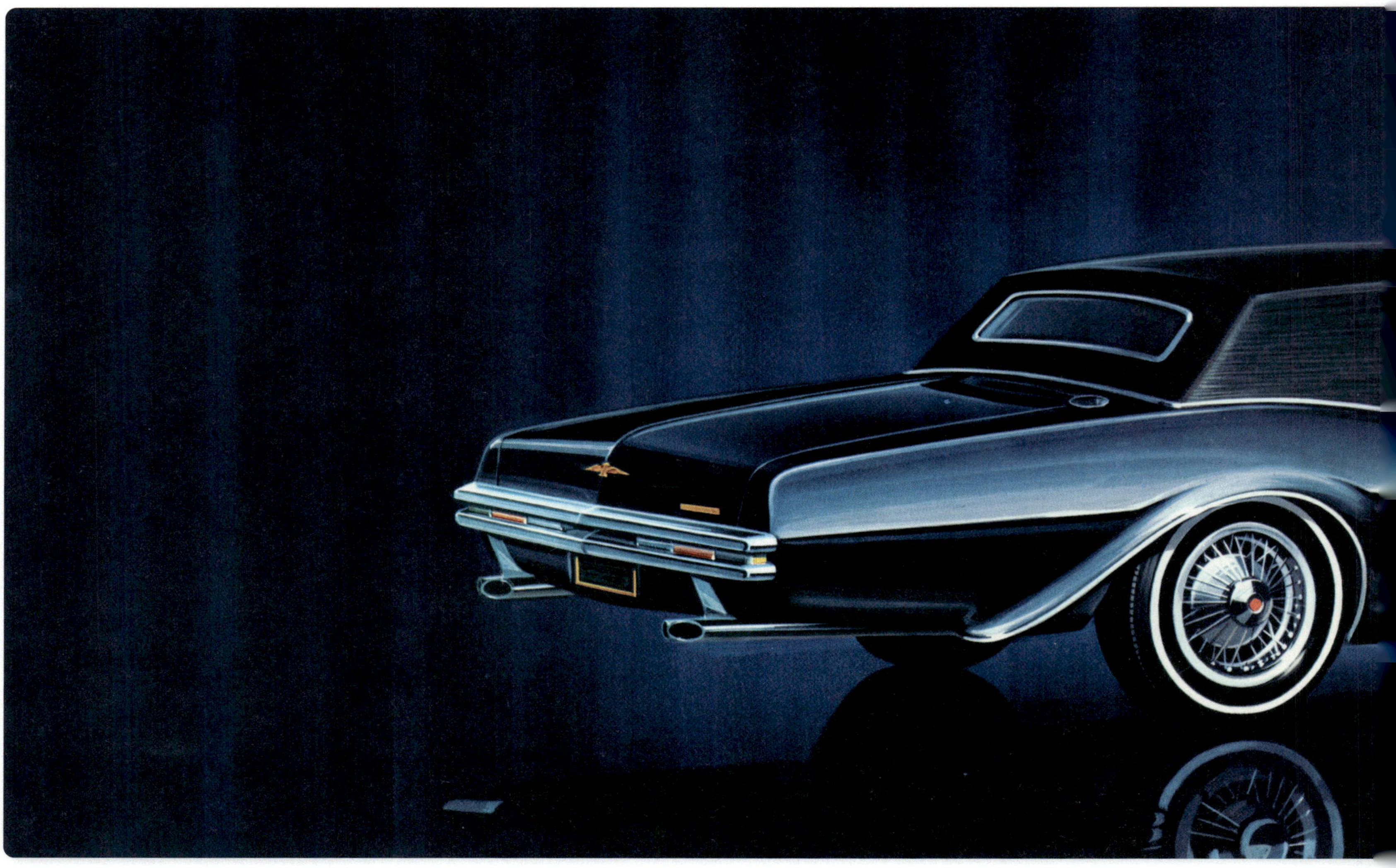

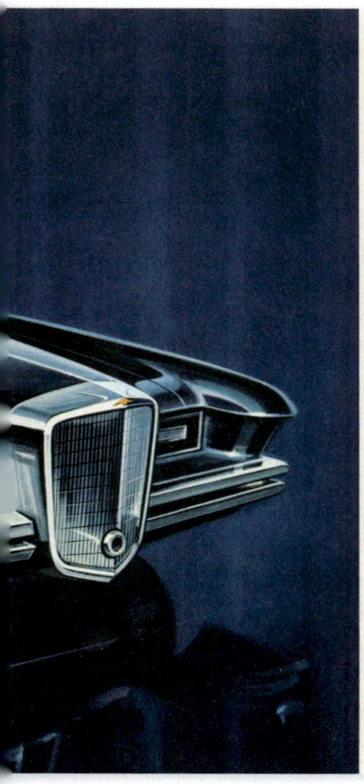

The Duesenberg is the result of over five years of careful research into today's automotive design, engineering, manufacture and craftsmanship.

The body and coachwork are hand-crafted in Italy by Carrozzeria Ghia, SpA. It is entirely of the most suitable steels available. Ghia was chosen to execute designer Virgil M. Exner's plans because of its worldwide reputation for superb quality coachwork.

The Duesenberg is powered by a 440 cubic inch Chrysler built V-8 engine. This power plant develops 425 horsepower and is the most powerful, most responsive engine manufactured today for personal automobiles.

The automobile is assembled in the modern Duesenberg Facility in Indianapolis, Indiana, the birth state of the original Duesenberg. Craftsmen of the highest skills assemble each Duesenberg. Many of these craftsmen worked with the original cars. Each Duesenberg is driven and tuned so that the automobile is delivered to its owner in perfect condition ready for optimum performance.

The first Duesenbergs are four-door brougham sedans. But since each Duesenberg is virtually a custom automobile, special body designs can be executed to meet an individual's preferences. All equipment, instrumentation and accessories usually listed as optional on luxury cars are standard with the Duesenberg. In addition, there are many optional equipment choices to be made such as built-in television and refreshment bars.

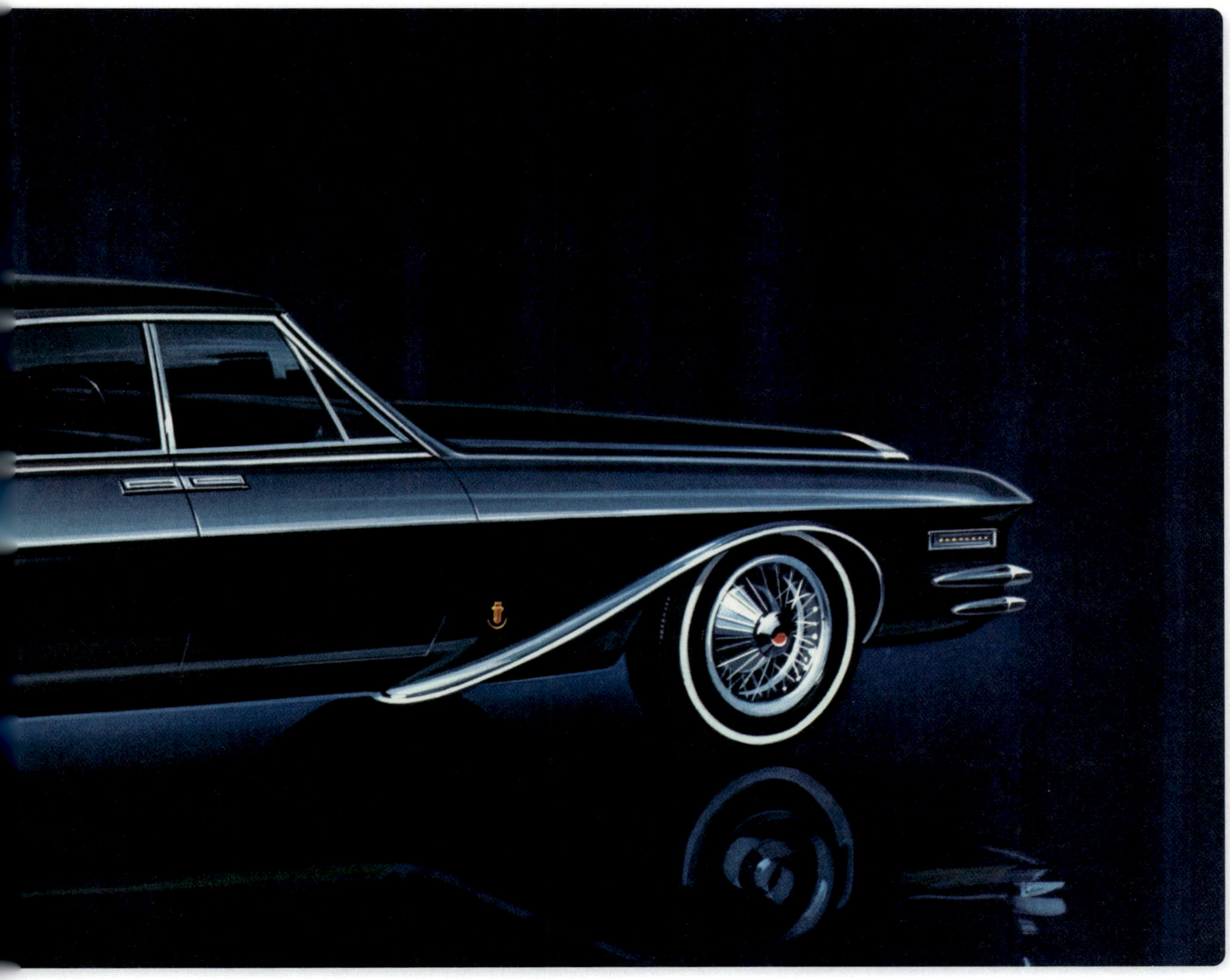

Ford

Ford 1962 *(right)*, **1965** *(below)*, **1964** *(opposite)*

From humble beginnings, legends are sometimes born. Ford dug deeply into the Falcon parts box to create and sell the Mustang, the offspring of a gifted design team, which was promoted by its master salesman, Lee Iacocca. They weren't sophisticated cars, but they triggered dreams by the millions.

Manchmal entstehen aus bescheidenen Anfängen echte Legenden. Ford griff tief in die Ersatzteilkiste des Falcon, um den Mustang zu konstruieren und zu verkaufen. Er war das Kind eines talentierten Designteams und wurde von einem wahren Verkaufsgenie vermarktet: Lee Iacocca. Die Fahrzeuge waren nicht sonderlich kultiviert, aber sie ließen Millionen träumen.

Les légendes naissent parfois de petits riens. Ford fouilla ainsi consciencieusement dans la caisse de pièces de la Falcon pour créer et commercialiser cette Mustang, progéniture d'une équipe de design particulièrement douée et lancée par Lee Iacocca, génie des ventes. Ces voitures n'avaient rien de sophistiqué mais elles feront rêver des millions de gens.

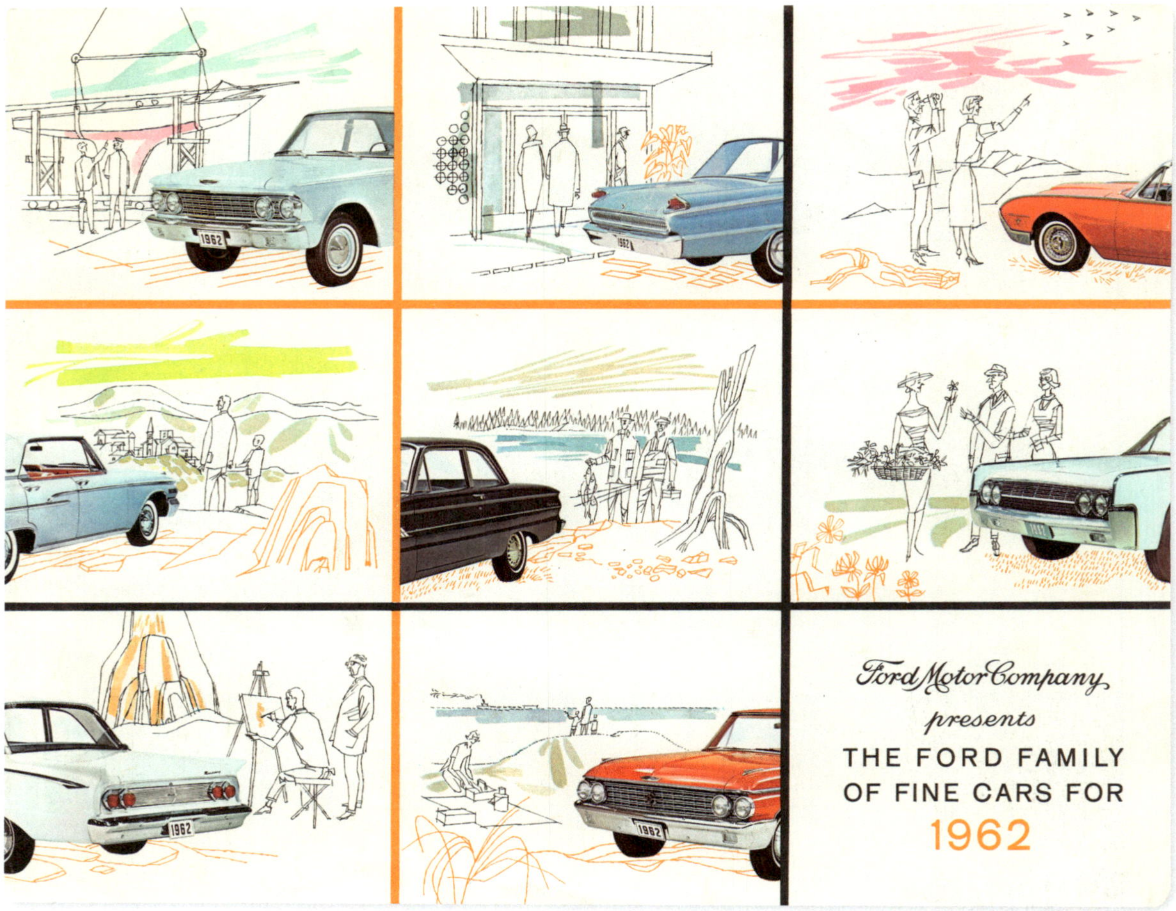

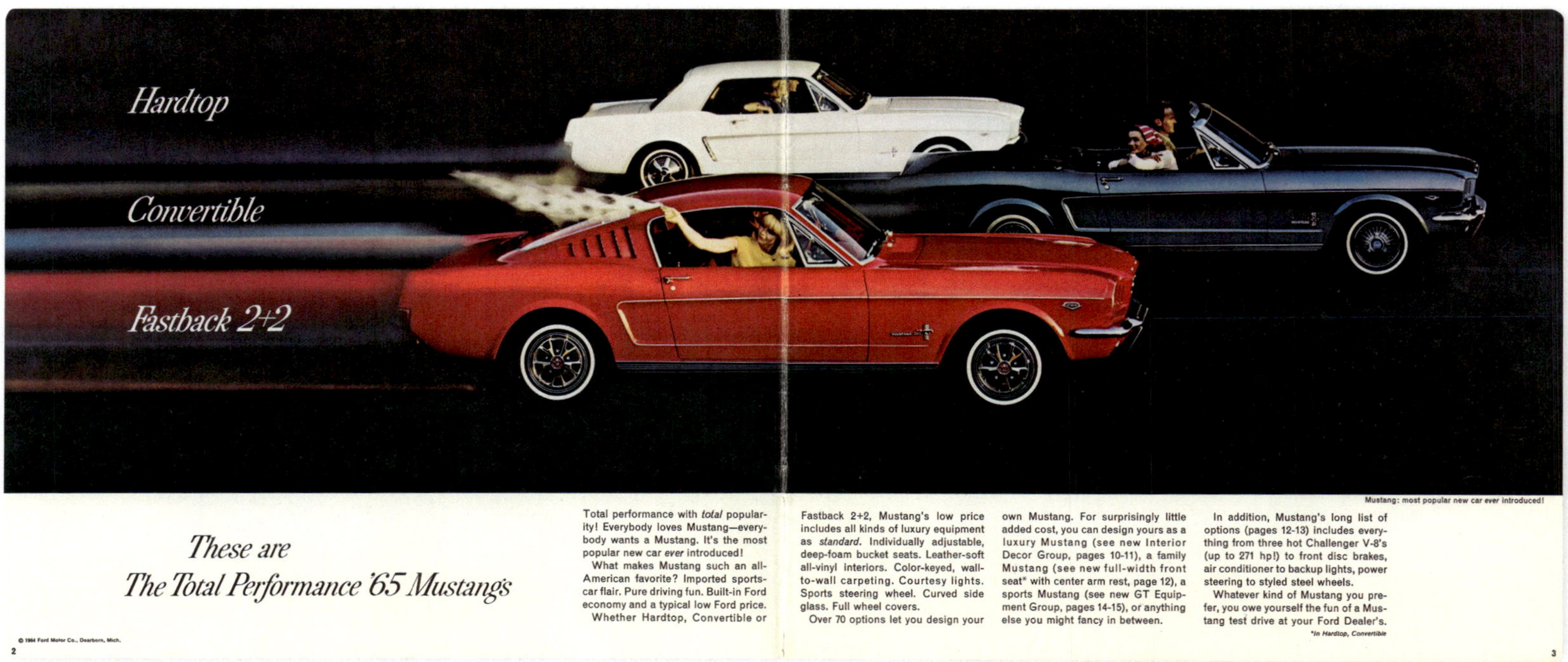

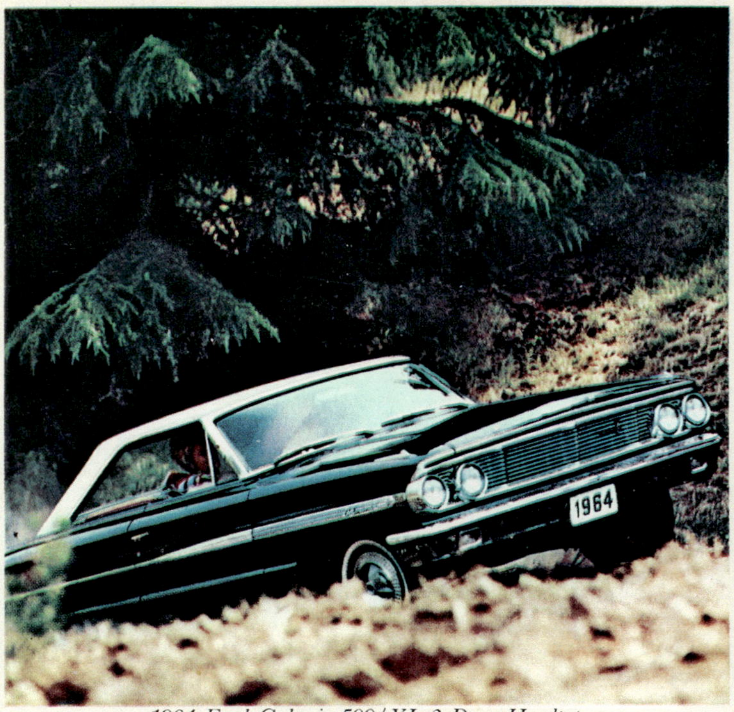

1964 Ford Galaxie 500/XL 2-Door Hardtop

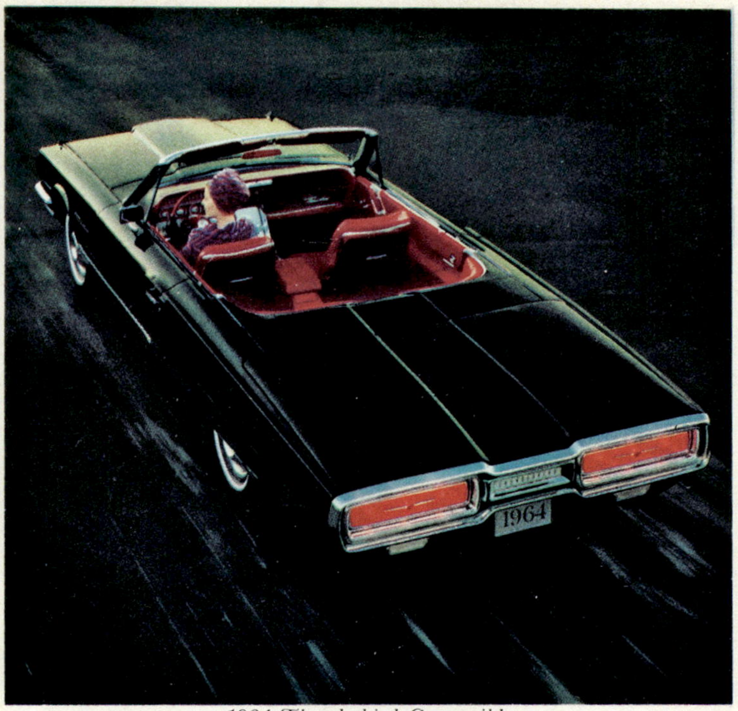

1964 Thunderbird Convertible

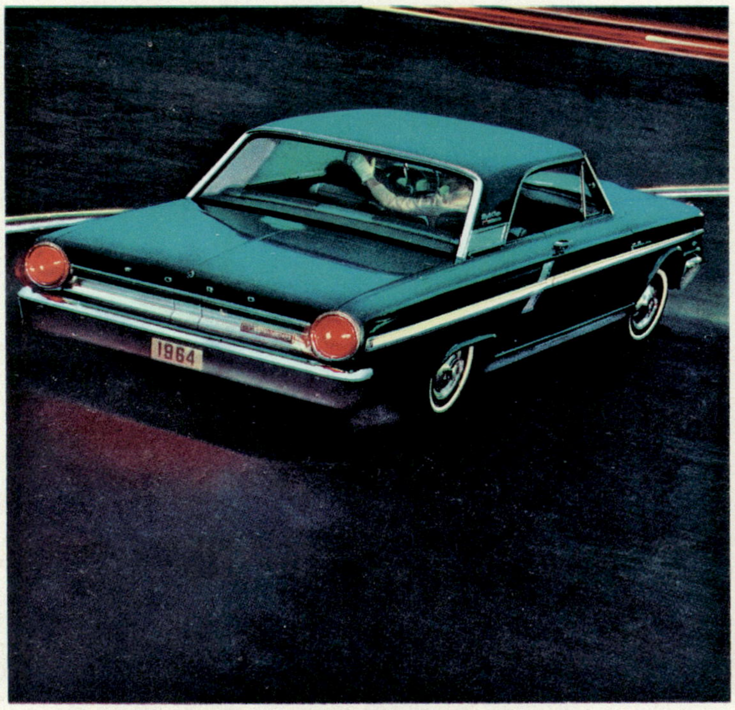

1964 Fairlane 500 Sports Coupe

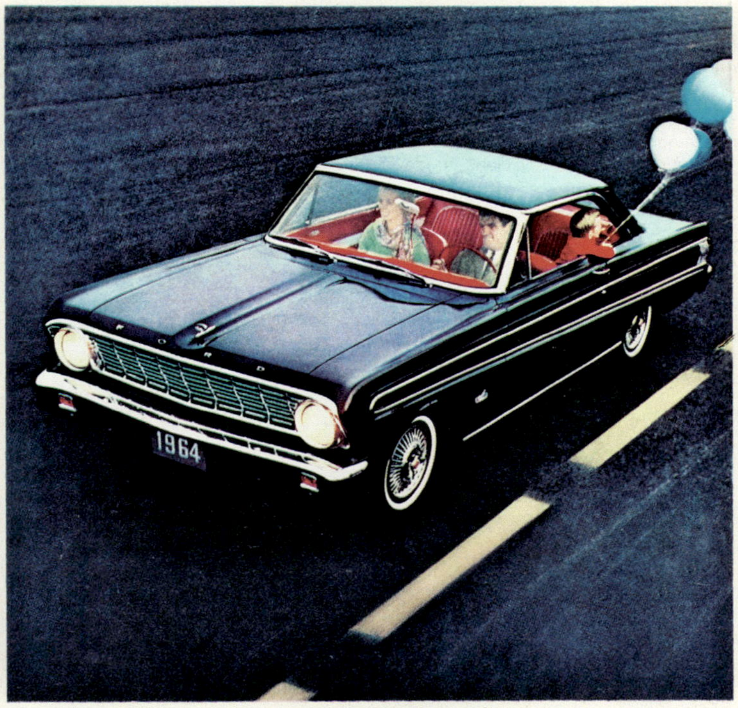

1964 Falcon Futura Sports Coupe

The TOTAL PERFORMANCE Cars for 1964

FORD

FALCON · FAIRLANE · FORD · THUNDERBIRD

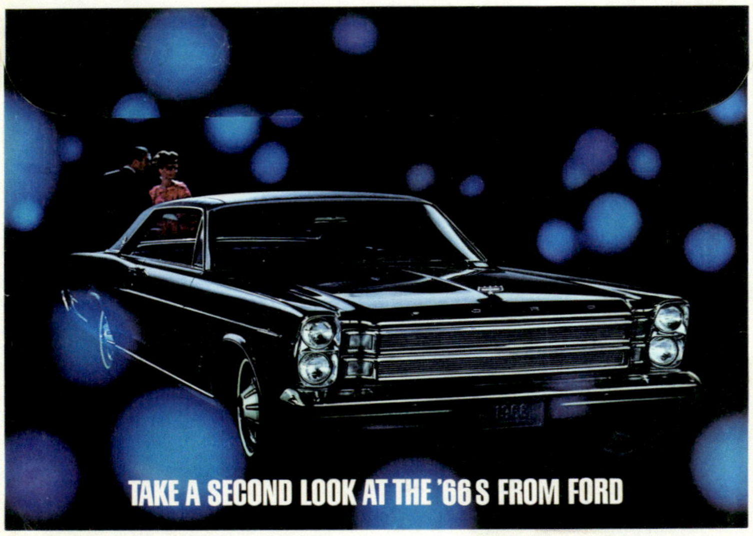

Ford 1966, 1967 *(opposite)*

The advertisements for Ford's upscale-inching compact have a strong flower-child motif and feature women; male buyers having stampeded off in a herd to chase the Mustang. About 65,000 Falcons were produced, compared with 472,000 Mustangs. The luxury Thunderbird, meanwhile, was re-invented once more. Having begun as a two-seat sports car and then grown into a four-place performance vehicle, it assumed its longest-lived identity as a personal coupe and, from 1967 to 1970, a four-door sedan.

In der Werbung für den langsam aufstrebenden Kompaktwagen von Ford wimmelte es von Blumenkindermotiven und Frauen, nachdem die Männer in Scharen dem Mustang hinterherliefen. Von den Falcons wurden rund 65 000 Stück gebaut – im Vergleich zu 472 000 Mustangs. Der luxuriöse Thunderbird hingegen wurde wieder einmal neu erfunden. Was als zweisitziger Sportwagen begonnen hatte und dann zu einem viersitzigen Hochgeschwindigkeitsfahrzeug heranwuchs, lebte am längsten als Luxus-Coupé und dann von 1967 bis 1970 als viertürige Limousine.

La publicité pour les compactes toujours plus élégantes du groupe Ford fait assaut de petites fleurs, de petites filles et de jolies femmes : la clientèle masculine, en effet, s'était carapatée comme un seul homme sur les talons de la Mustang. 65 000 Falcon ont été produites, à comparer aux 472 000 Mustang. Entre-temps la luxueuse Thunderbird était une fois de plus réinventée. Après avoir démarré sous la forme d'un petit roadster deux places, puis évolué vers la berline sportive, c'est en tant que coupé au style très affirmé qu'elle se forgera l'identité la plus durable – avant de redevenir une berline quatre portes de 1967 à 1970.

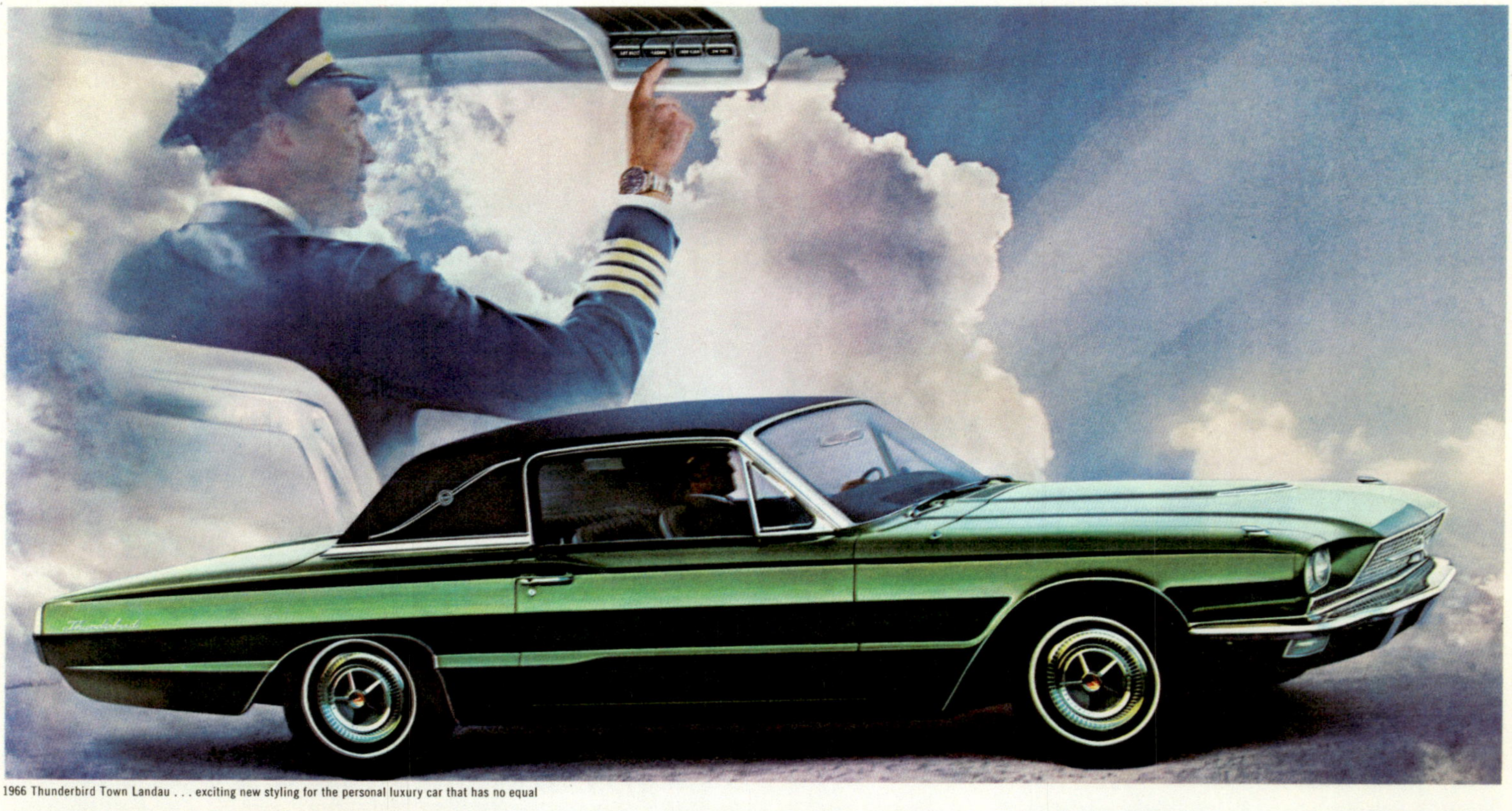

1966 Thunderbird Town Landau . . . exciting new styling for the personal luxury car that has no equal

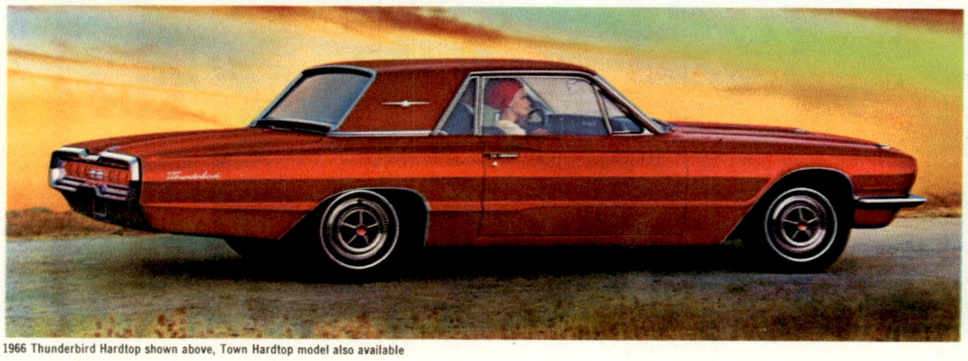

1966 Thunderbird Hardtop shown above, Town Hardtop model also available

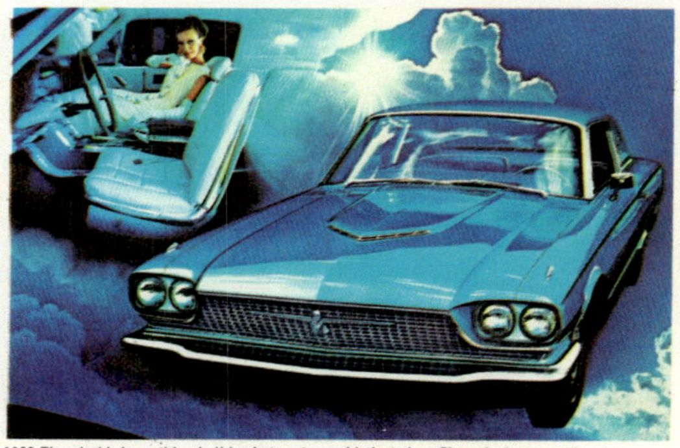

1966 Thunderbird . . . thin-shell bucket seats, sophisticated roofline, sleek new grille

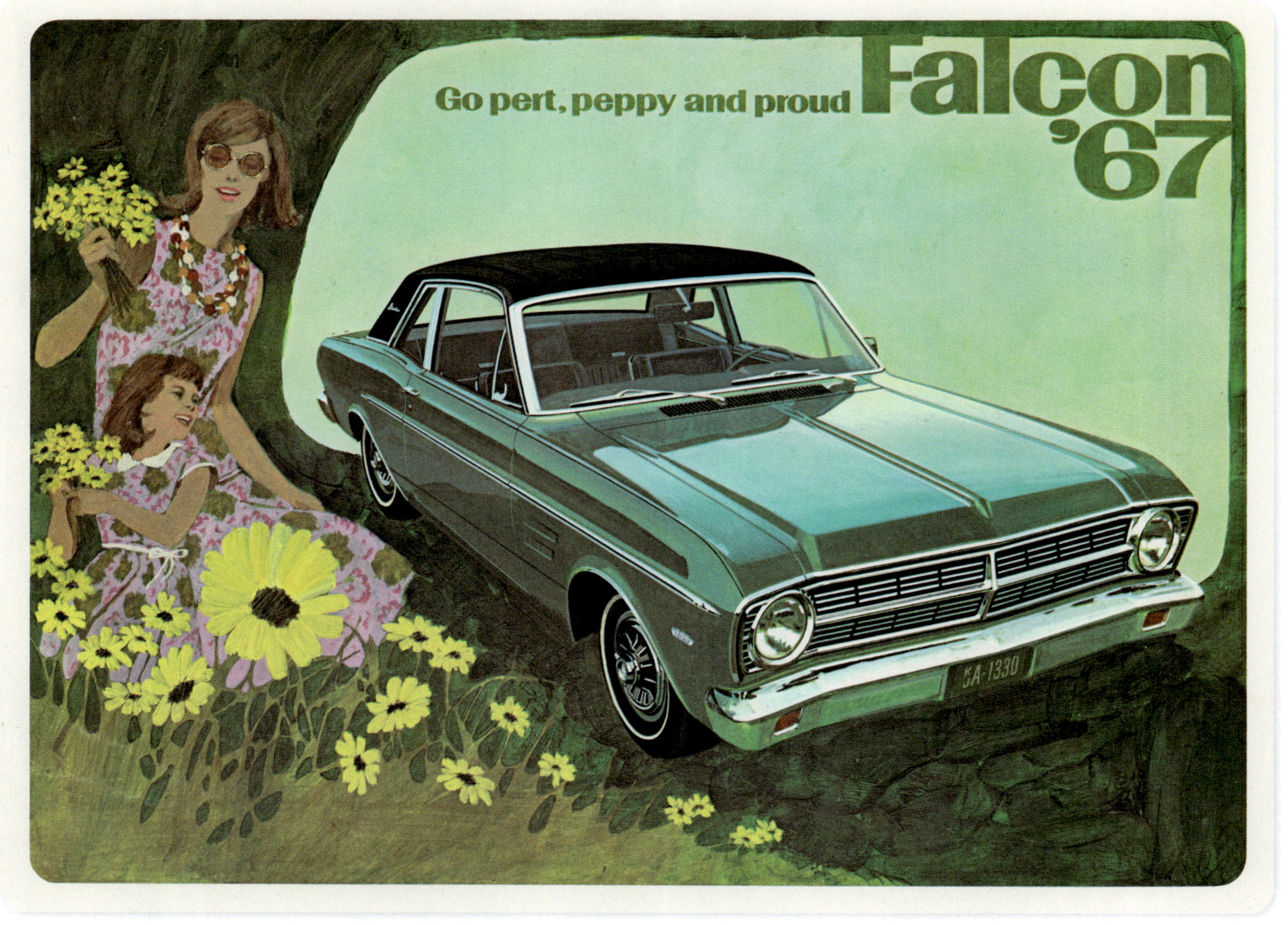
Go pert, peppy and proud
Falcon
'67

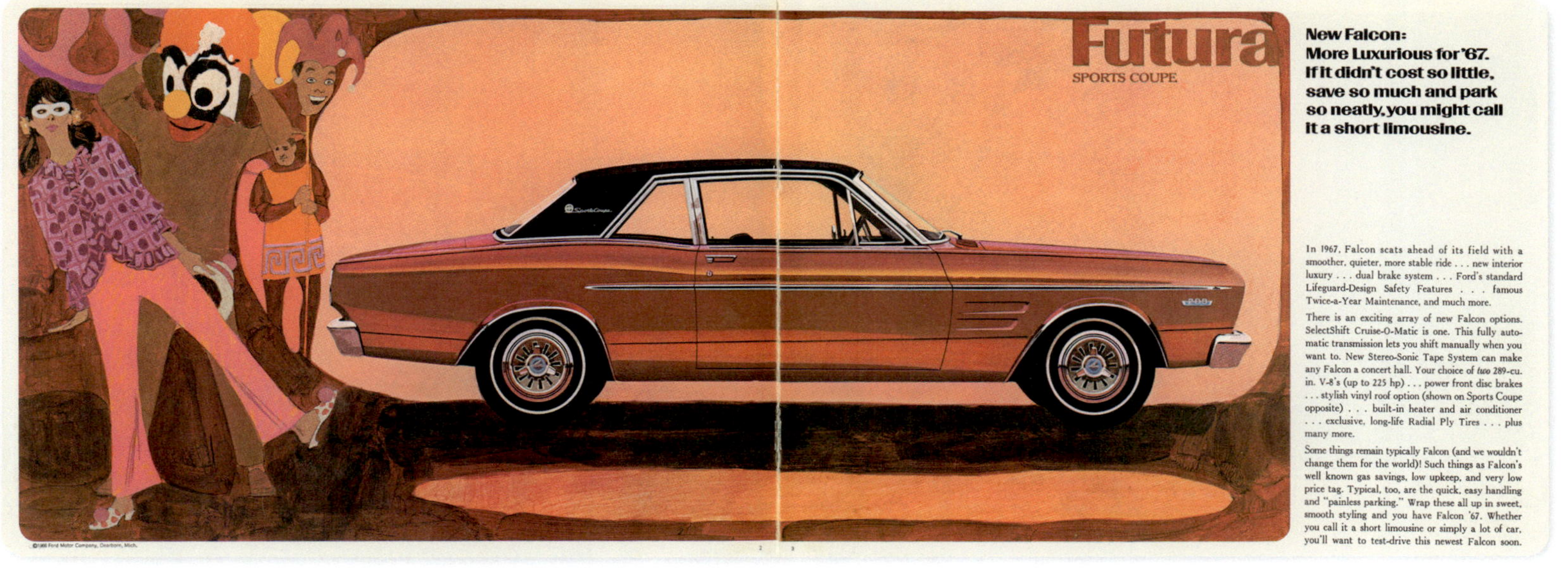
Futura
SPORTS COUPE
New Falcon:
More Luxurious for '67.
If it didn't cost so little, save so much and park so neatly, you might call it a short limousine.
In 1967, Falcon scats ahead of its field with a smoother, quieter, more stable ride . . . new interior luxury . . . dual brake system . . . Ford's standard Lifeguard-Design Safety Features . . . famous Twice-a-Year Maintenance, and much more.
There is an exciting array of new Falcon options. SelectShift Cruise-O-Matic is one. This fully automatic transmission lets you shift manually when you want to. New Stereo-Sonic Tape System can make any Falcon a concert hall. Your choice of *two* 289-cu. in. V-8's (up to 225 hp) . . . power front disc brakes . . . stylish vinyl roof option (shown on Sports Coupe opposite) . . . built-in heater and air conditioner . . . exclusive, long-life Radial Ply Tires . . . plus many more.
Some things remain typically Falcon (and we wouldn't change them for the world)! Such things as Falcon's well known gas savings, low upkeep, and very low price tag. Typical, too, are the quick, easy handling and "painless parking." Wrap these all up in sweet, smooth styling and you have Falcon '67. Whether you call it a short limousine or simply a lot of car, you'll want to test-drive this newest Falcon soon.

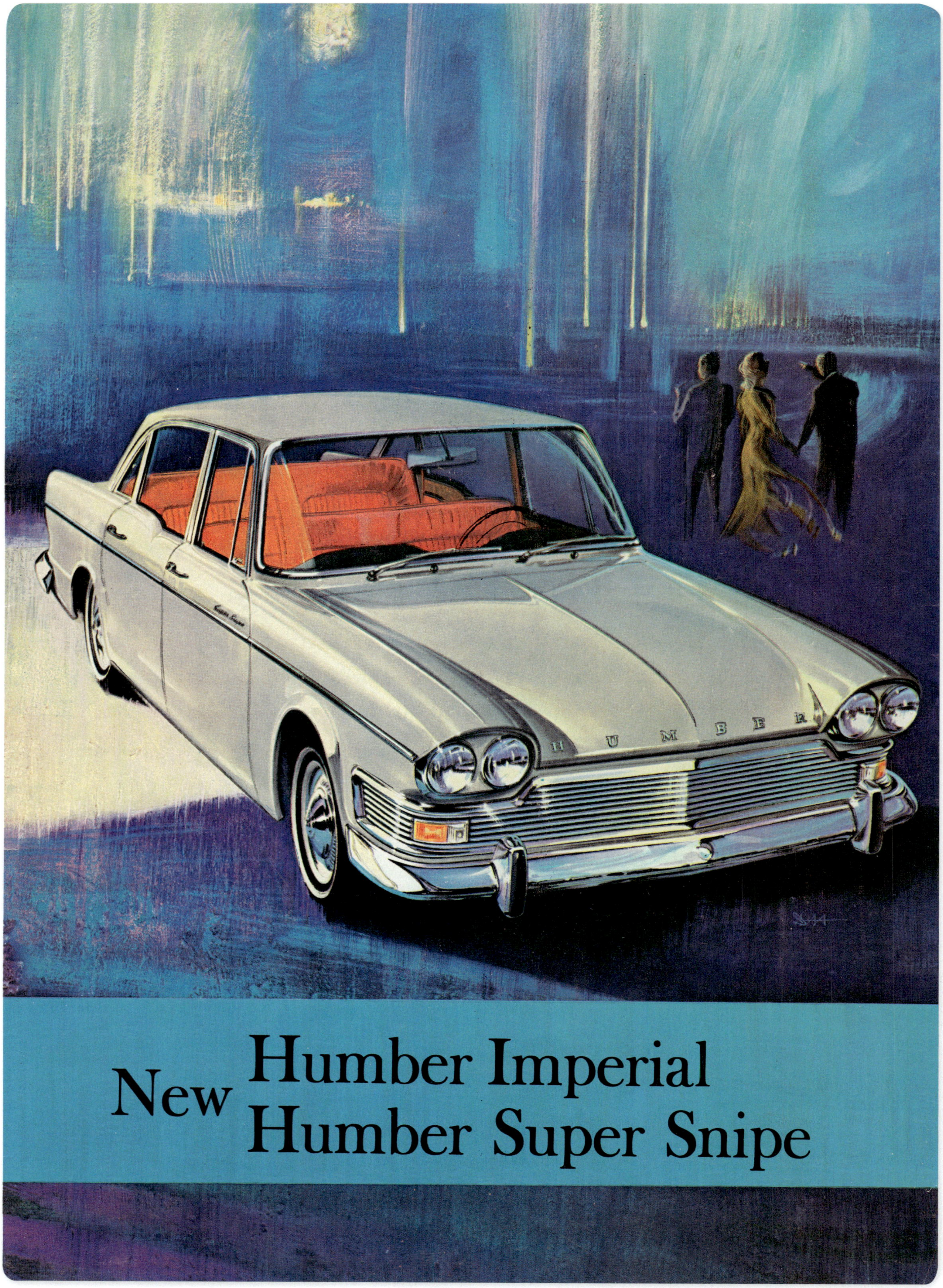
HUMBER
New
Humber Imperial
Humber Super Snipe

Humber

Humber 1967 *(this spread)*

The left-hand drive indicates this brochure is intended to introduce the Super Snipe to U.S. buyers. Sadly, very few of them existed. Humber was a proud British name in the Rootes Group, and the Super Snipe was an upscale saloon with an OHV six-cylinder engine and a walnut and leather interior, affording excellent comfort.

Die Linkslenkung zeigt, dass diese Broschüre den Super Snipe US-amerikanischen Käufern nahebringen sollte. Leider gab es nur sehr wenige davon. Humber war ein stolzer britischer Name in der Rootes Group, und der Super Snipe war eine Limousine der gehobenen Preisklasse mit OHV-Sechszylindermotor und einer Innenausstattung in Nussbaum und Leder, die außerordentlich komfortabel war.

La conduite à gauche indique que la brochure entend proposer la Super Snipe au public américain. Il n'en existera hélas qu'un tout petit nombre. Humber était un orgueilleux blason britannique intégré au groupe Rootes, et la Super Snipe une berline de prestige équipée d'un moteur six cylindres à soupapes en tête, avec intérieur cuir et placages en noyer, garants d'un confort accompli.

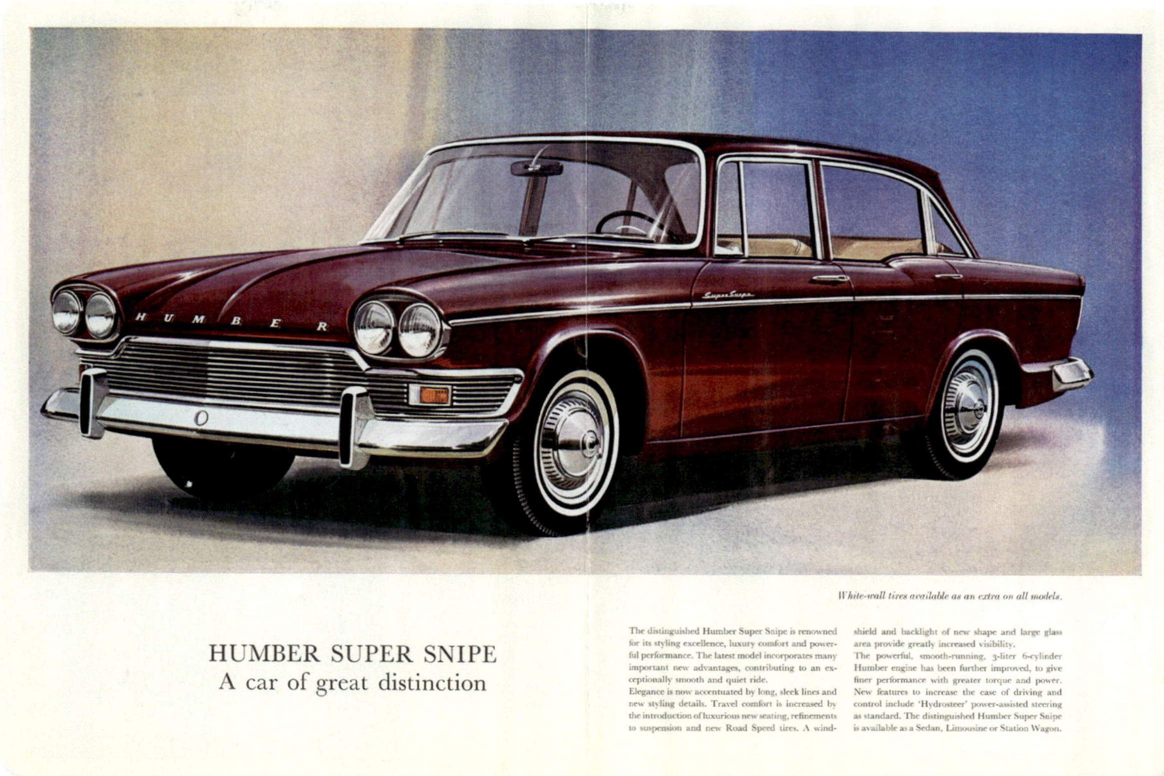

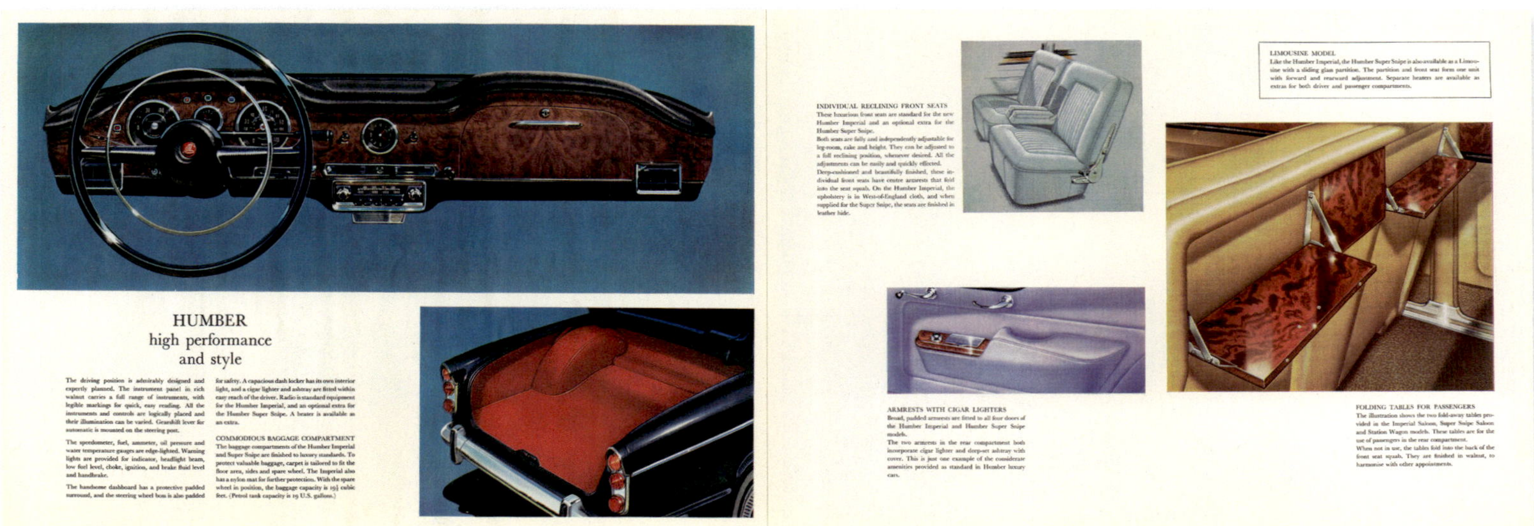

Lincoln

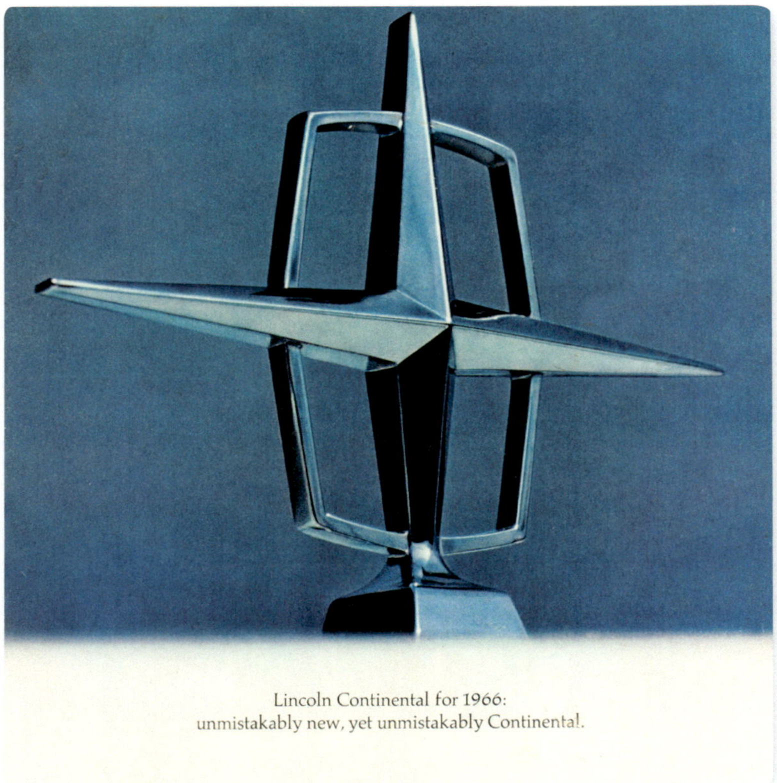

Lincoln 1966

In the 1960s, the four-pointed star that denoted Lincoln was one of the auto industry's most immediately recognizable trademarks. The Continental sedan continued with its trend of forward-opening rear doors, which dated back to 1961.

In den 1960er-Jahren war der vierzackige Stern von Lincoln eines der Markenzeichen in der amerikanischen Automobilindustrie mit dem höchsten Wiedererkennungswert. Die Continental-Limousine setzte den Trend der Hintertüren mit Anschlag an der C-Säule fort, der 1961 begonnen hatte.

Dans les années 1960, l'étoile à quatre branches identifiant Lincoln était l'une des signatures automobiles les plus immédiatement reconnaissables. La berline Continental persistait dans son goût des portes arrière à ouverture antagoniste, apparues en 1961.

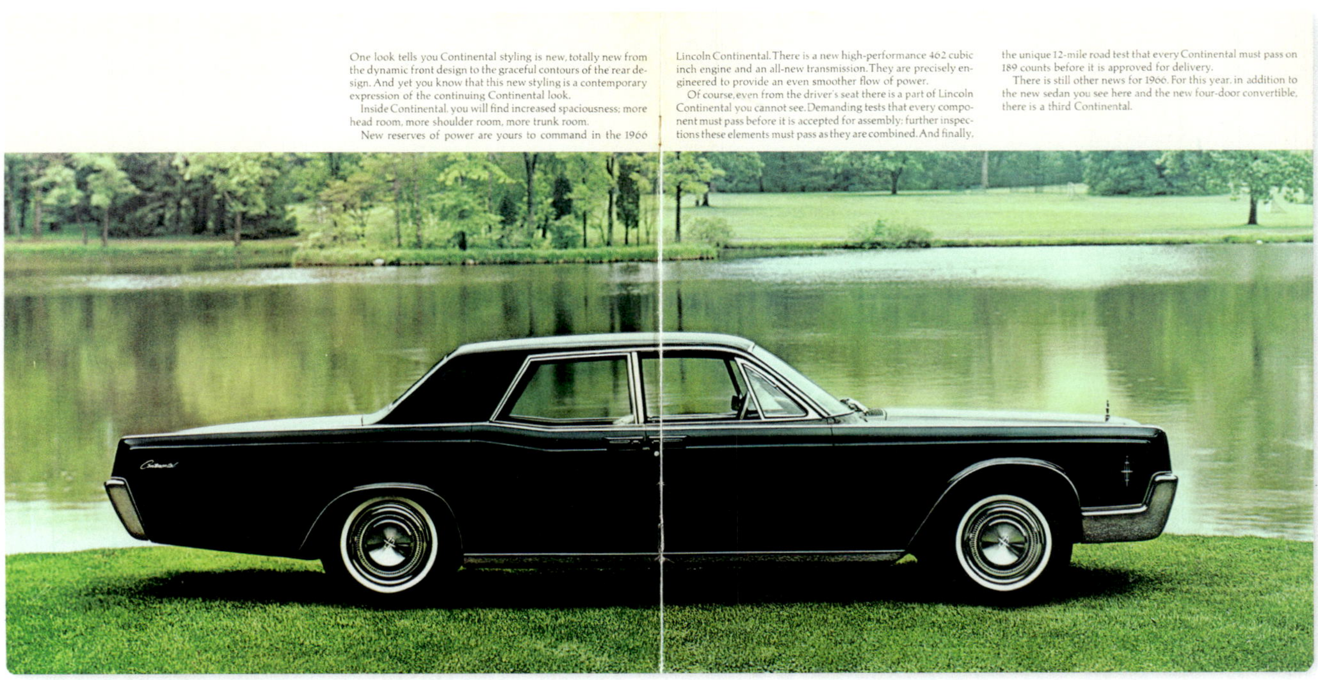

Mercury

Mercury 1964 *(right)*, **1962** *(below)*, **1966** *(bottom)*

The Comet was essentially a restyled Ford Falcon. The Meteor is the more interesting car. At first a full-size Mercury with mid-level trim, it was re-created for 1962 as a Ford Fairlane–based midsize sedan line with rocket-shaped taillights. A model reshuffle in 1964 took the Comet upscale, and the Meteor fell.

Der Comet war im Grunde ein umgemodelter Ford Falcon. Das interessantere Auto ist der Meteor. Er war zunächst ein großer Mercury in Mittelklasseaufmachung und wurde dann für den Jahrgang 1962 neu erfunden als Limousinenreihe mittlerer Größe auf der Grundlage des Ford Fairlane mit raketenförmigen Rückleuchten. Eine Neuordnung der Modellreihen drückte den Comet 1964 nach oben, während der Meteor abstürzte – passend zu den Namen.

La Comet était, pour l'essentiel, une Ford Falcon relookée. La Meteor est ici la voiture la plus intéressante. Née Mercury « full-size » avec des finitions milieu de gamme, elle fut recréée en 1962 pour devenir une berline intermédiaire (« midsize ») aux feux arrière en obus, sur une base de Ford Fairlane. En 1964 une refonte complète du modèle fit monter la Comet en gamme, et la Meteor périclita.

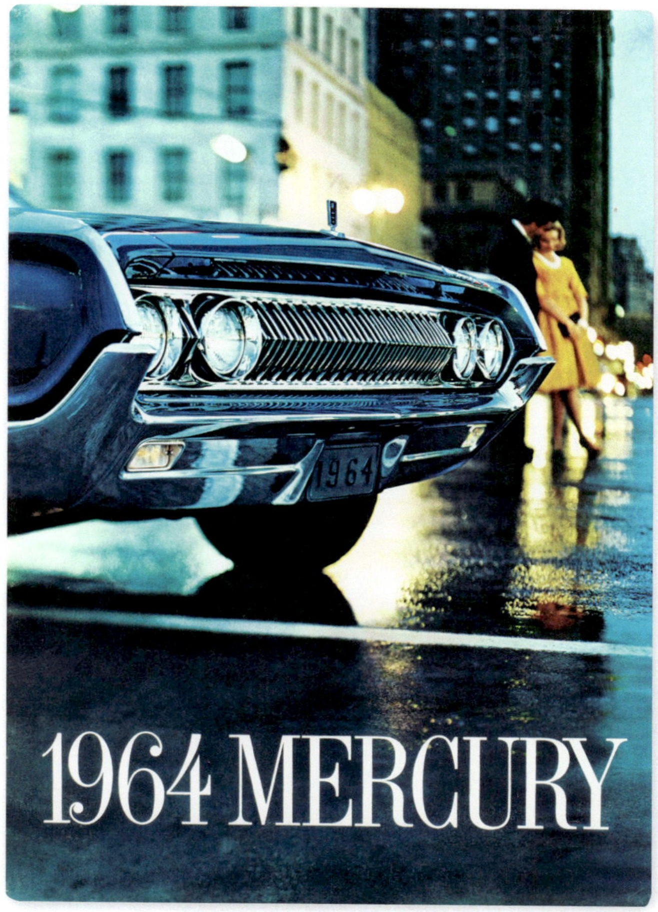

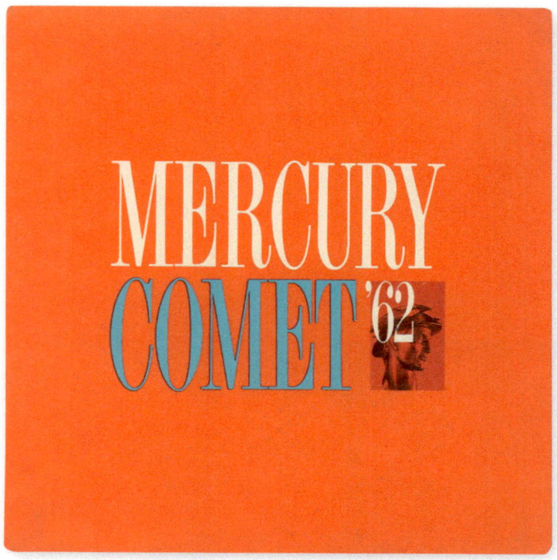

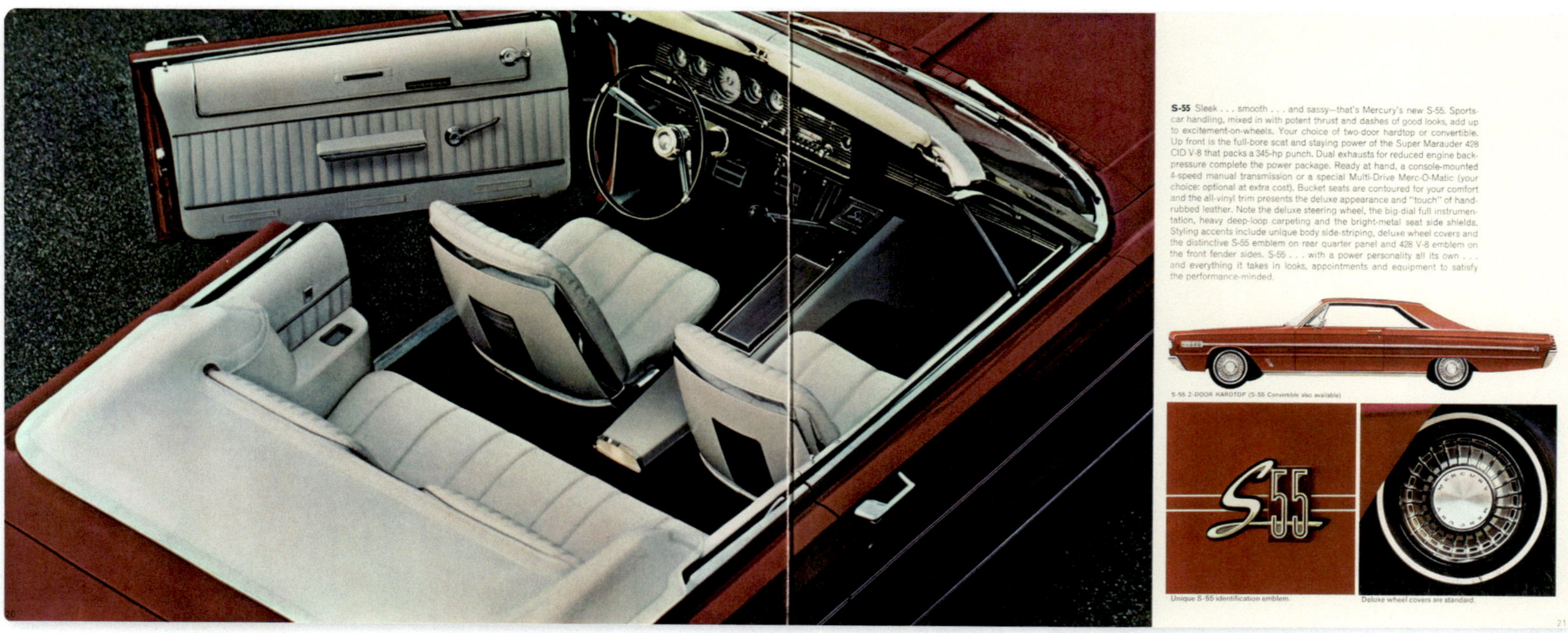

Oldsmobile

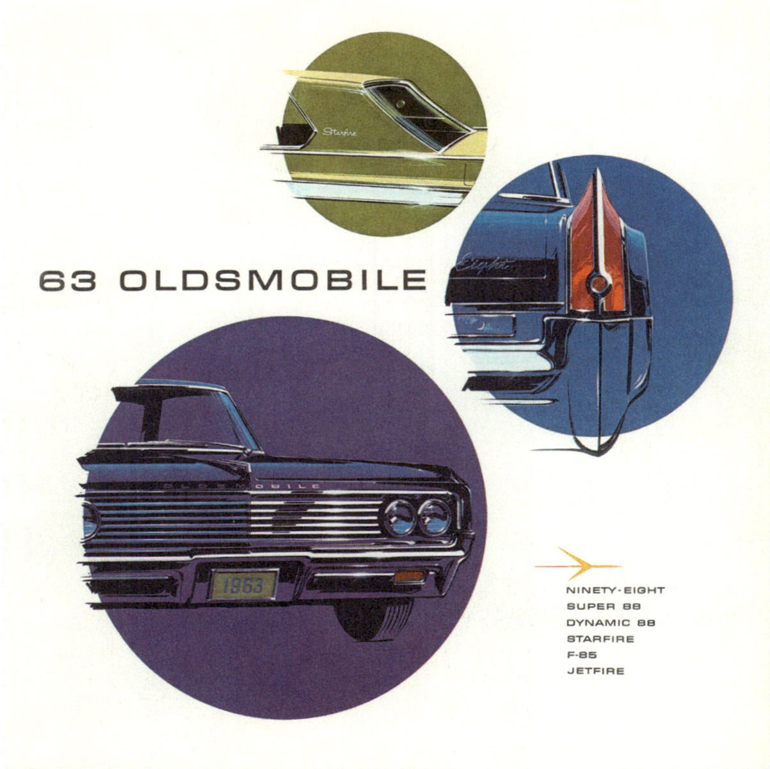

Oldsmobile 1963 *(left and opposite)*, **1965** *(below)*

Oldsmobiles were deeply respected during these years for their quality, good looks, and prodigious horsepower. For 1963 full-size cars led by the Dynamic 88 received new bodywork below the window glass. Oldsmobile rightly stressed its engines, up to 345 hp, and also made self-adjusting brakes standard.

Die Oldsmobiles standen seinerzeit aufgrund ihrer Qualität, ihres guten Aussehens und ihrer ungeheuren Leistungsstärke hoch im Kurs. Für das Modelljahr 1963 erhielten die großen Modelle, allen voran der Dynamic 88, unterhalb der Verglasung neue Karosserien. Oldsmobile stellte zu Recht vor allem die Motorisierung heraus, die bis zu 468 Kilowatt erreichte, und führte die selbstnachstellenden Bremsen als Standardausstattung ein.

Ces années-là, les Oldsmobile jouissaient d'un profond respect en raison de leur qualité générale, de leur beauté et d'une puissance moteur prodigieuse. Les plus grosses d'entre elles, emmenées par la Dynamic 88, reçurent en 1963 une nouvelle carrosserie. Oldsmobile soulignait à bon droit la vigueur de ses moteurs (jusqu'à 345 chevaux) et proposait également des freins auto-réglables de série.

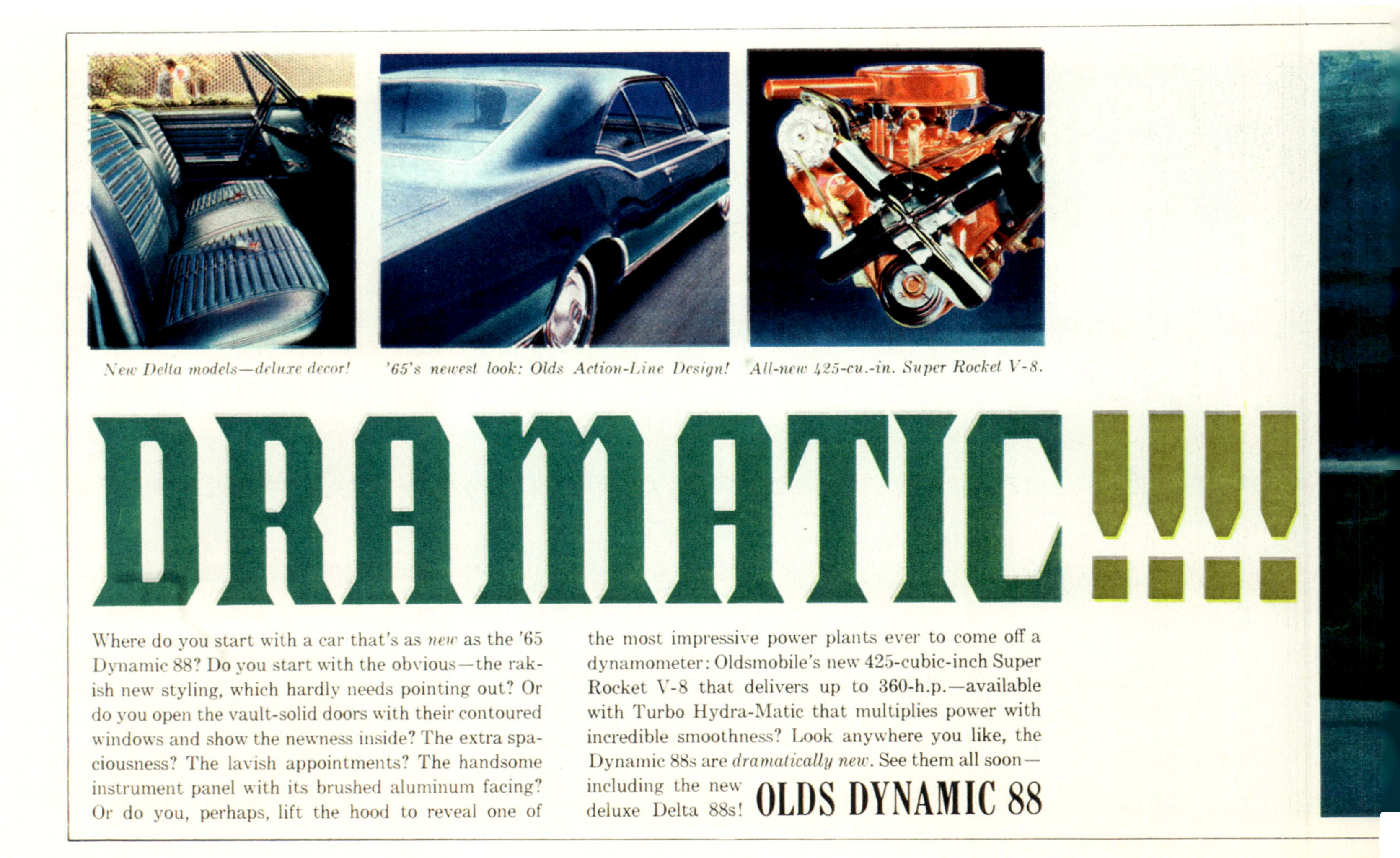

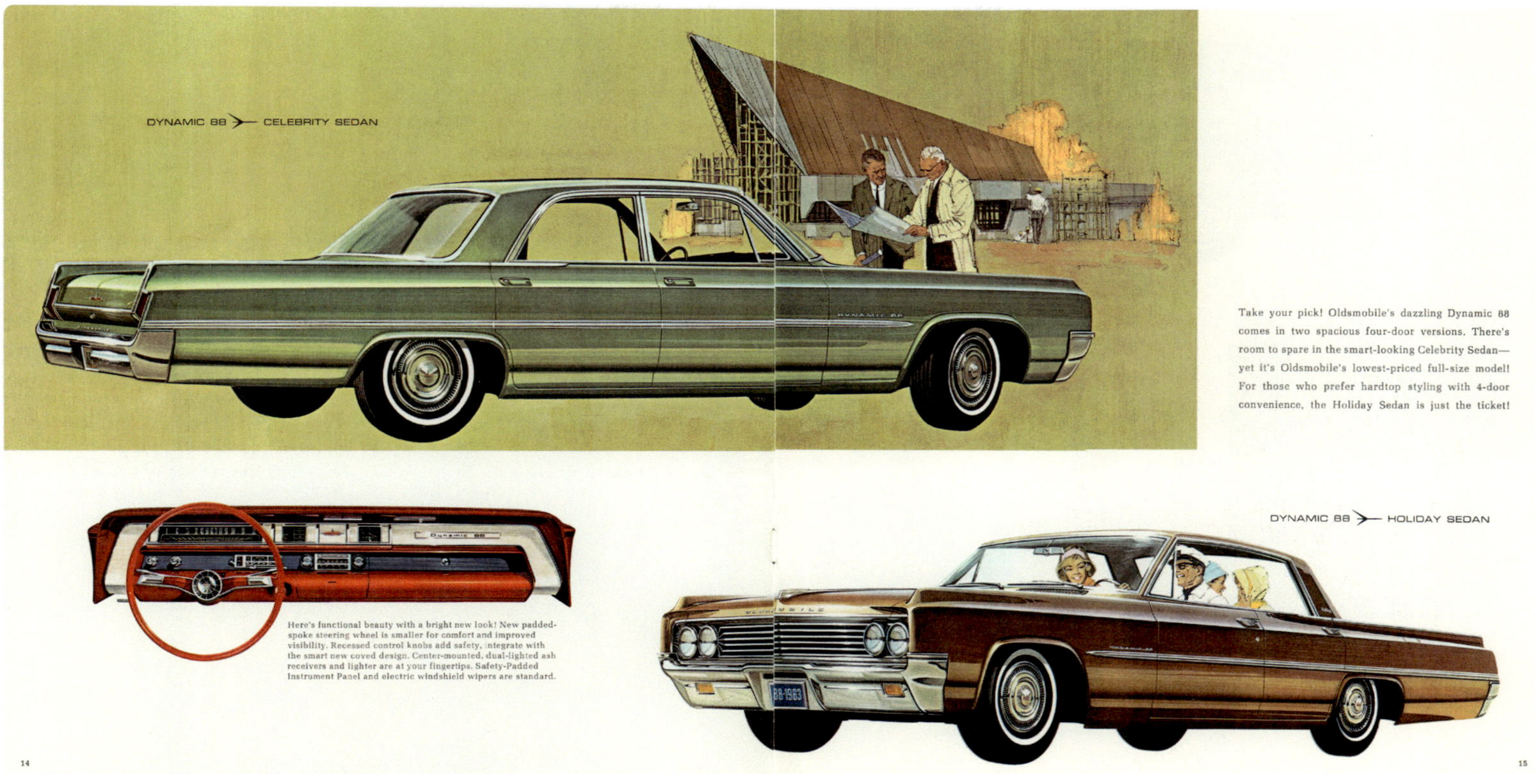
DYNAMIC 88 CELEBRITY SEDAN
Take your pick! Oldsmobile's dazzling Dynamic 88 comes in two spacious four-door versions. There's room to spare in the smart-looking Celebrity Sedan—yet it's Oldsmobile's lowest-priced full-size model! For those who prefer hardtop styling with 4-door convenience, the Holiday Sedan is just the ticket!
DYNAMIC 88 HOLIDAY SEDAN
Here's functional beauty with a bright new look! New padded-spoke steering wheel is smaller for comfort and improved visibility. Recessed control knobs add safety, integrate with the smart new coved design. Center-mounted, dual-lighted ash receivers and lighter are at your fingertips. Safety-Padded Instrument Panel and electric windshield wipers are standard.
14
15

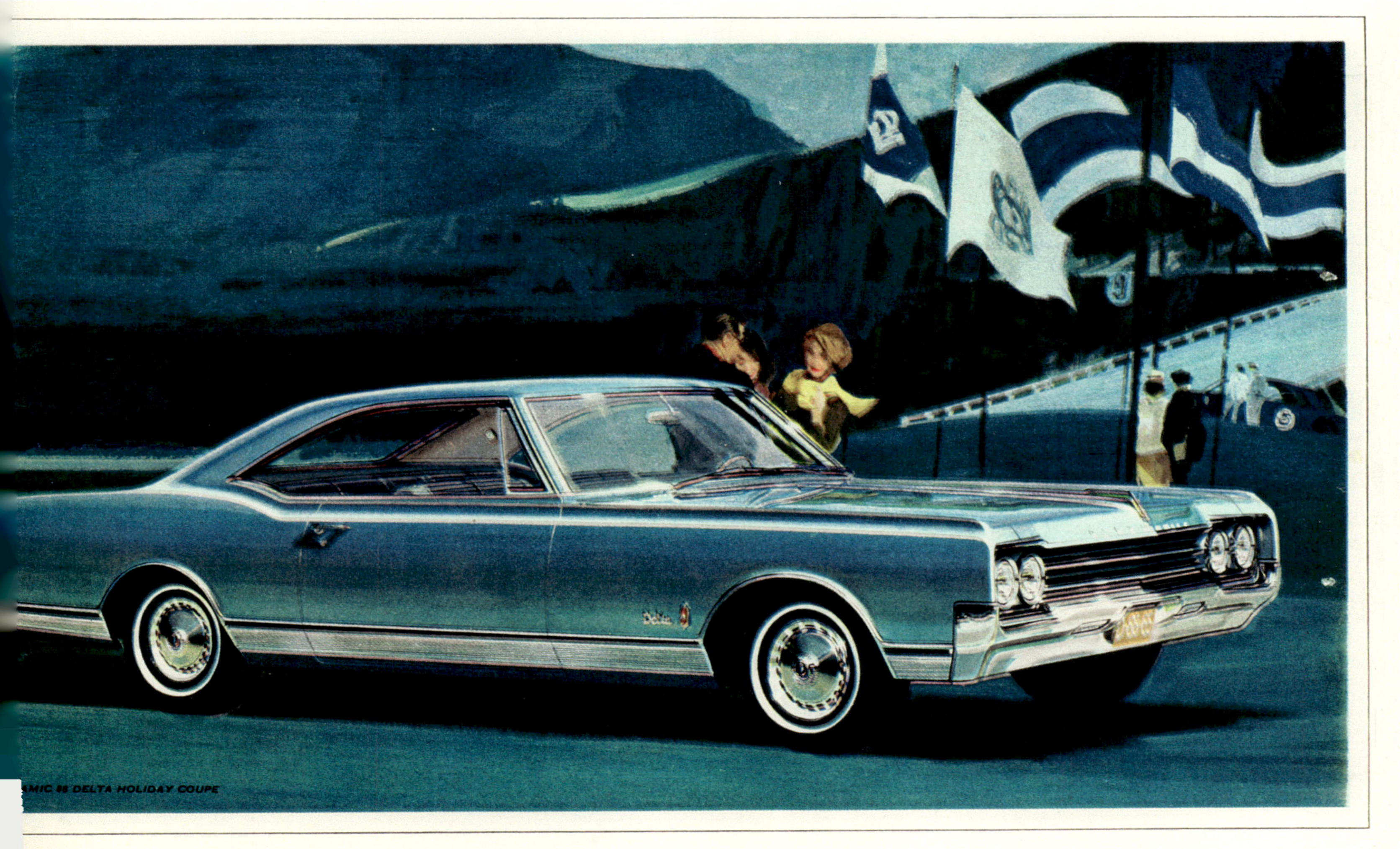
MIC 88 DELTA HOLIDAY COUPE

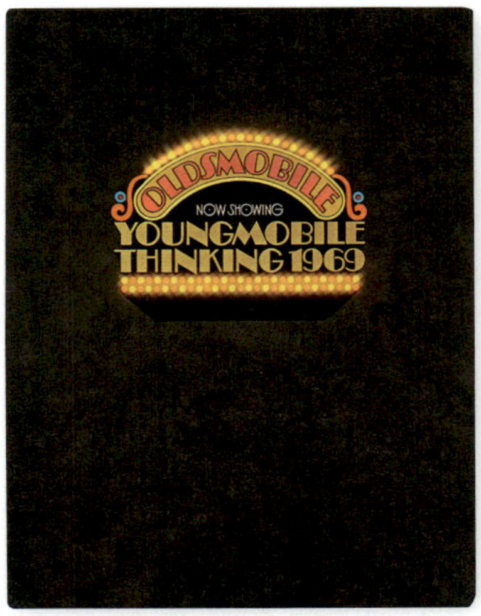

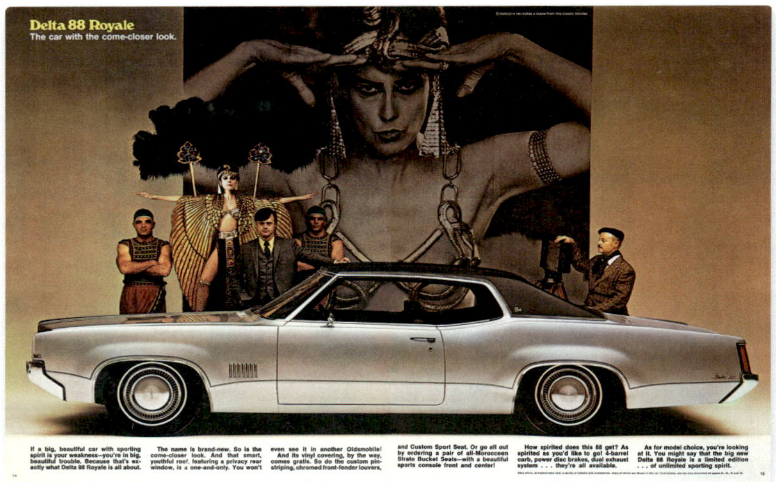

Oldsmobile 1969, 1973 *(opposite)*

In a year of massive rock festivals and the shocking Broadway musical *Hair*, Oldsmobile courted youth culture with a whimsical wordplay on its name. The Cutlass S two-door was both handsome and "young."

In der Zeit riesiger Rockfestivals und des schockierenden Broadway-Musicals *Hair* umwarb Oldsmobile die Jugend mit einem Wortspiel: Aus „Oldsmobile" wurde „Youngmobile". Der Zweitürer Cutlass S sah sowohl schick als auch „jung" aus.

Dans une année marquée par de gigantesques festivals rock et par la comédie musicale à scandale *Hair*, Oldsmobile devenait « Youngmobile » dans l'espoir d'infiltrer la culture jeune son nom en clin d'œil à la jeune génération. De fait la Cutlass S deux portes était tout aussi séduisante que « juvénile ».

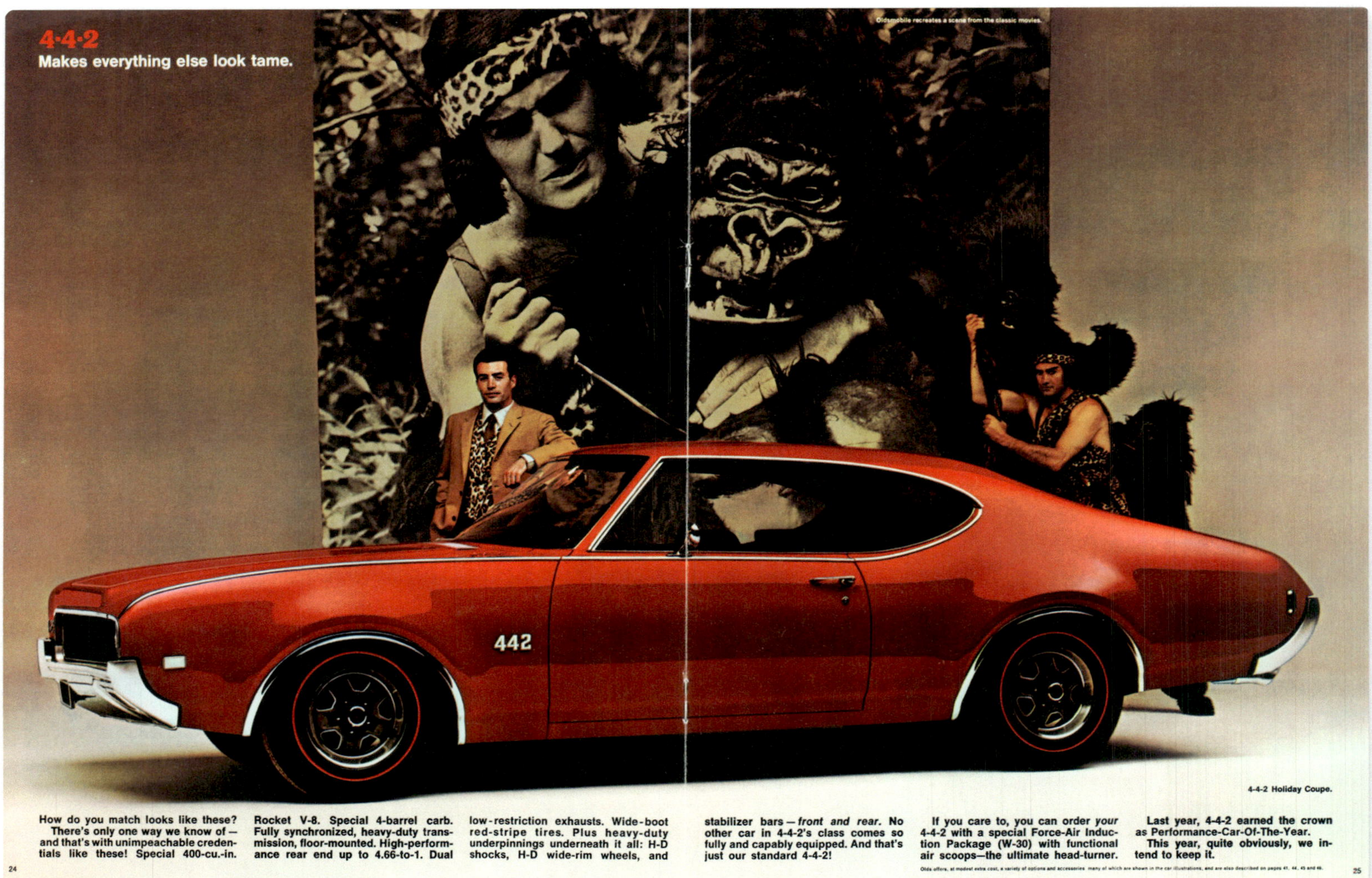

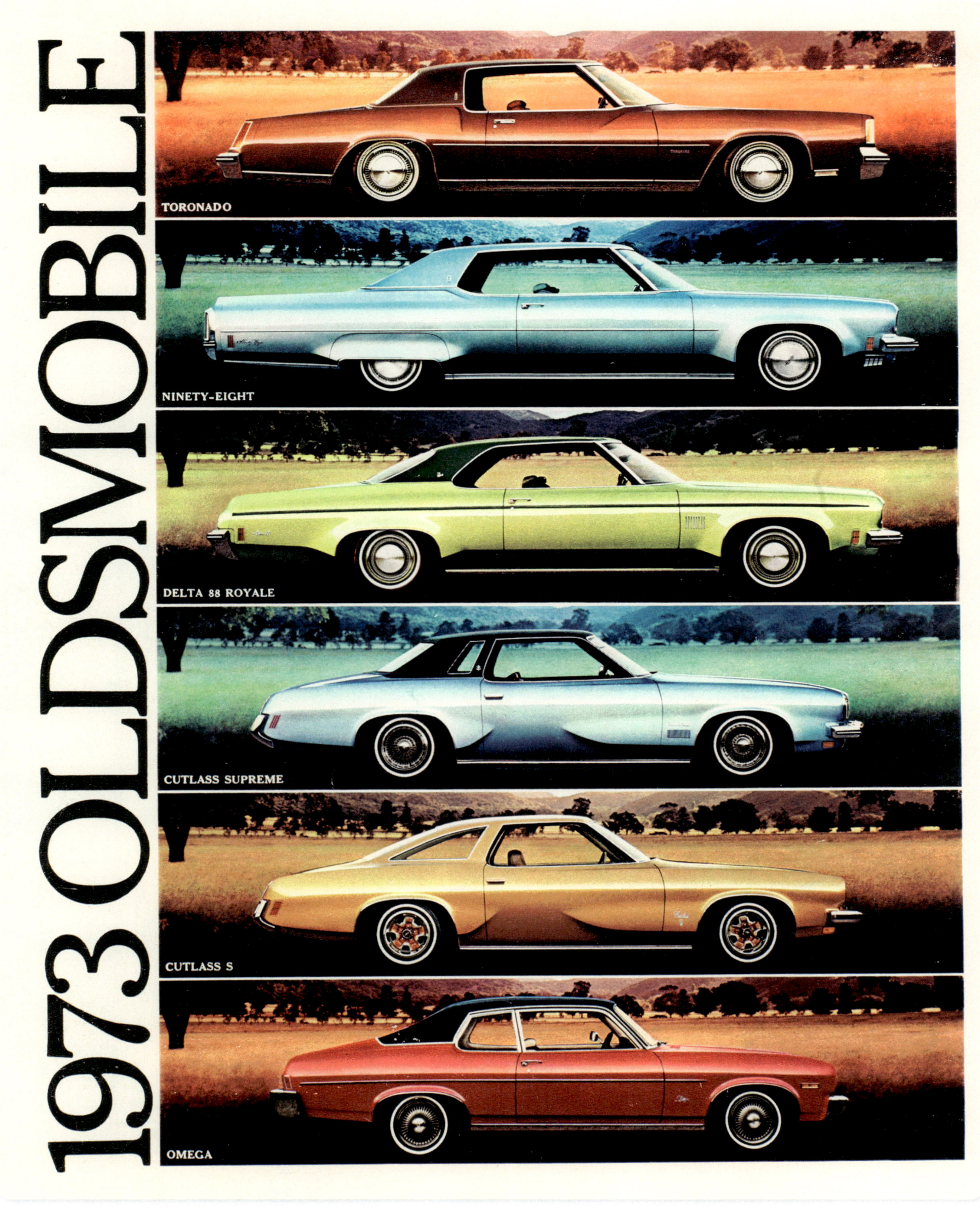
1973 OLDSMOBILE
TORONADO
NINETY-EIGHT
DELTA 88 ROYALE
CUTLASS SUPREME
CUTLASS S
OMEGA

Pontiac

Pontiac 1963 *(right and opposite top)*, **1962** *(below)*, **1965** *(opposite bottom)*

Pontiac was in its glory years during the 1960s. It was also one of the American auto industry's final manufacturers to employ illustrative artwork of the car and its features in its advertising materials. By now brochure art was exclusively becoming the realm of photographers, not artists.

Pontiac war in den 1960er-Jahren in seiner Glanzzeit. Er war aber auch einer der letzten Hersteller in der amerikanischen Automobilindustrie, der in seinem Werbematerial gezeichnete und gemalte Illustrationen verwendete. Von nun an überließ man die Bilder der Broschüren mehr und mehr den Fotografen.

Les Sixties marquèrent l'âge d'or de Pontiac. Dans ses supports promotionnels, la marque était aussi l'un des derniers constructeurs américains à confier à des dessinateurs le soin d'illustrer modèles et équipements. À cette époque, en effet, les photographes régnaient sans partage sur les brochures et catalogues.

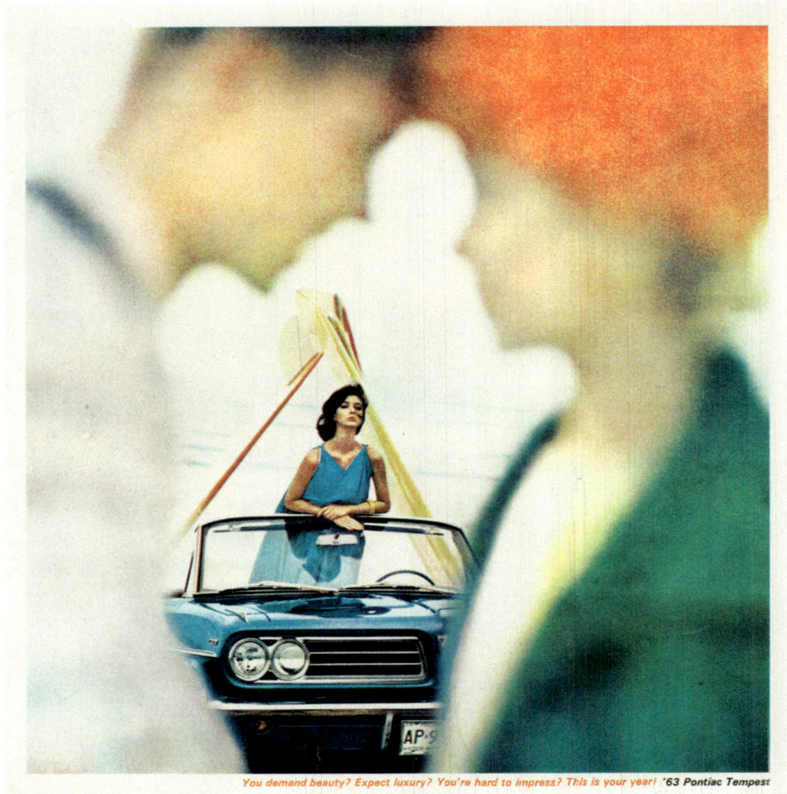

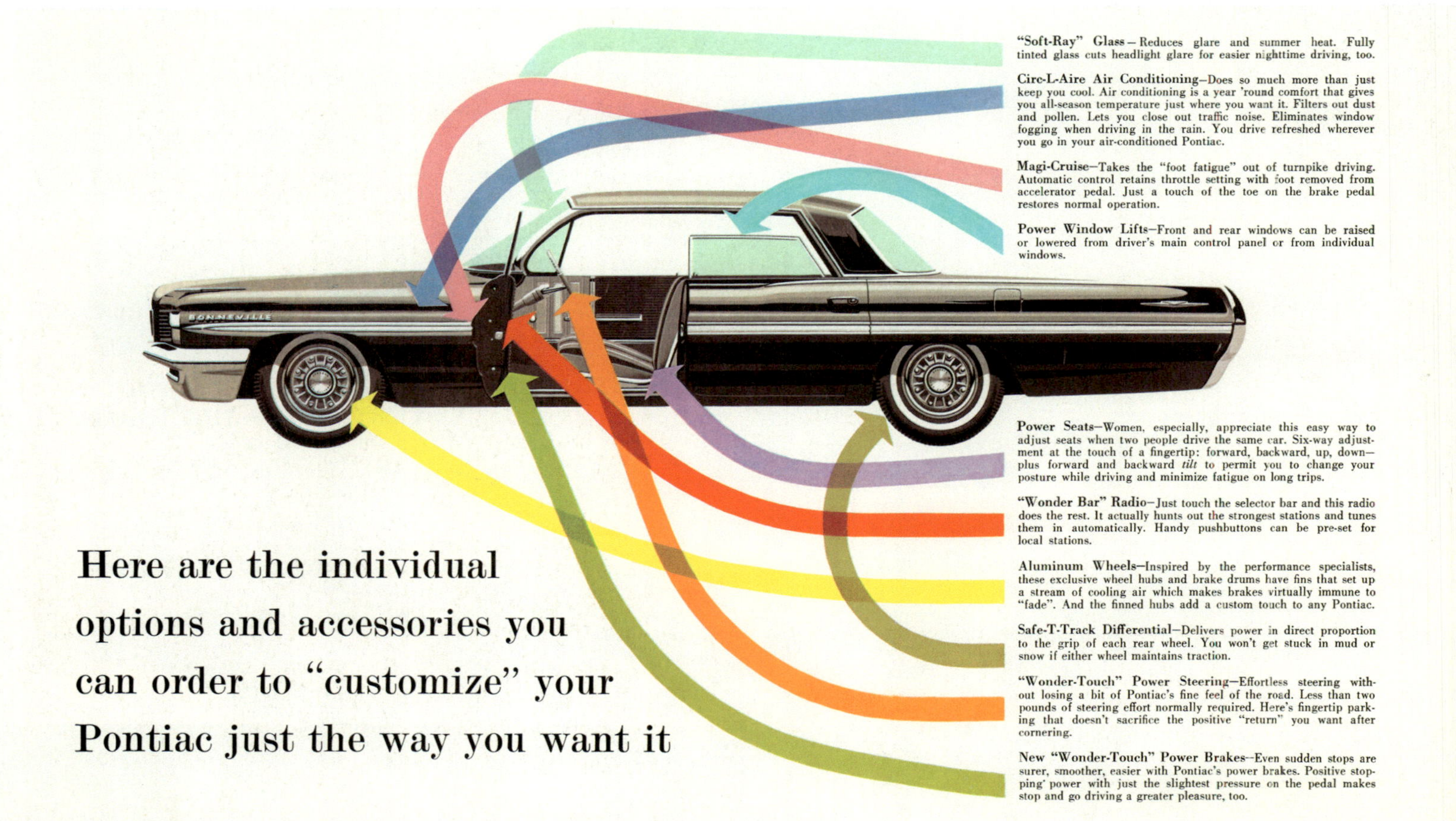

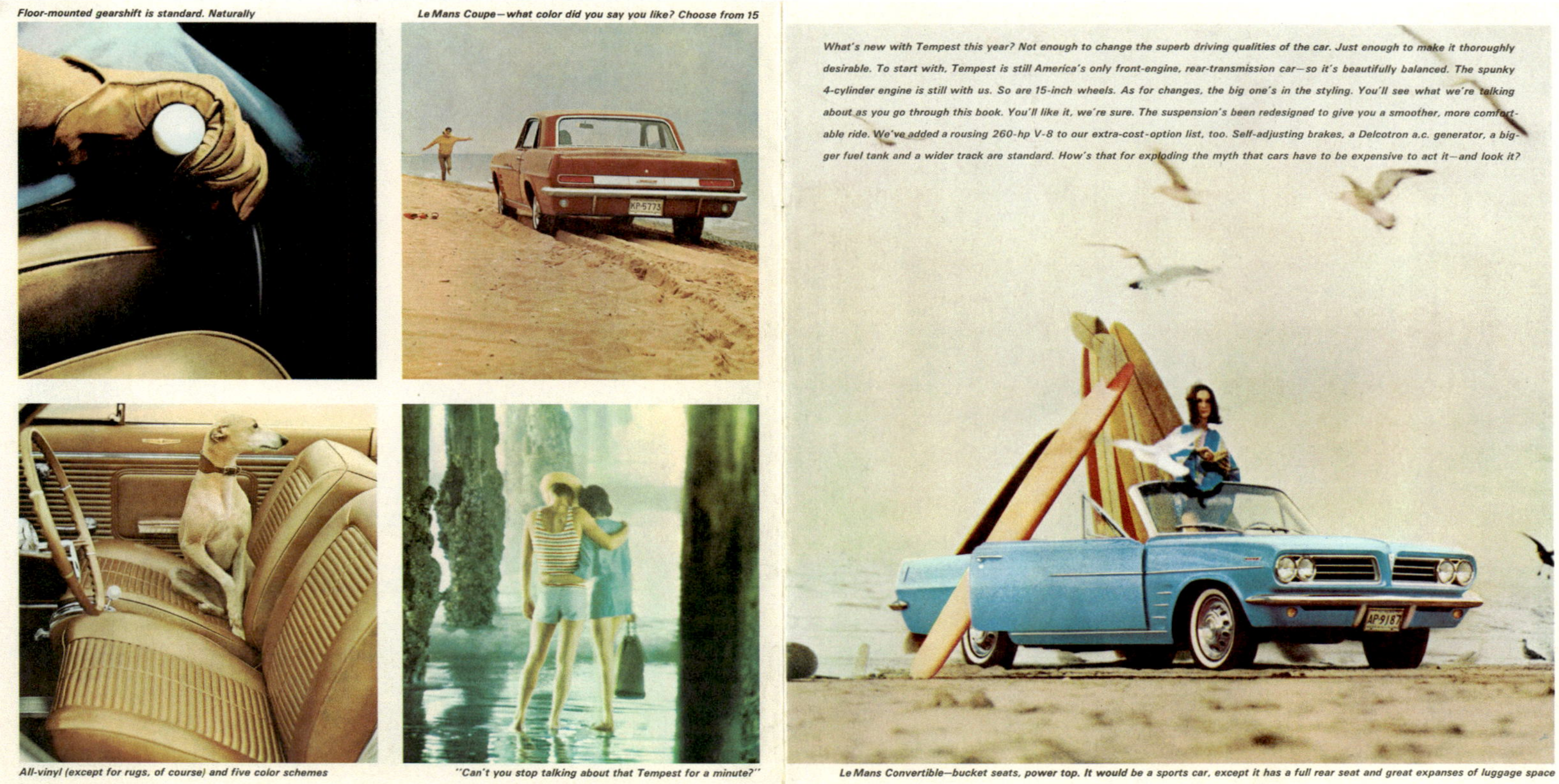
Floor-mounted gearshift is standard. Naturally
Le Mans Coupe—what color did you say you like? Choose from 15
What's new with Tempest this year? Not enough to change the superb driving qualities of the car. Just enough to make it thoroughly desirable. To start with, Tempest is still America's only front-engine, rear-transmission car—so it's beautifully balanced. The spunky 4-cylinder engine is still with us. So are 15-inch wheels. As for changes, the big one's in the styling. You'll see what we're talking about as you go through this book. You'll like it, we're sure. The suspension's been redesigned to give you a smoother, more comfortable ride. We've added a rousing 260-hp V-8 to our extra-cost-option list, too. Self-adjusting brakes, a Delcotron a.c. generator, a bigger fuel tank and a wider track are standard. How's that for exploding the myth that cars have to be expensive to act it—and look it?
All-vinyl (except for rugs, of course) and five color schemes
"Can't you stop talking about that Tempest for a minute?"
Le Mans Convertible—bucket seats, power top. It would be a sports car, except it has a full rear seat and great expanses of luggage space

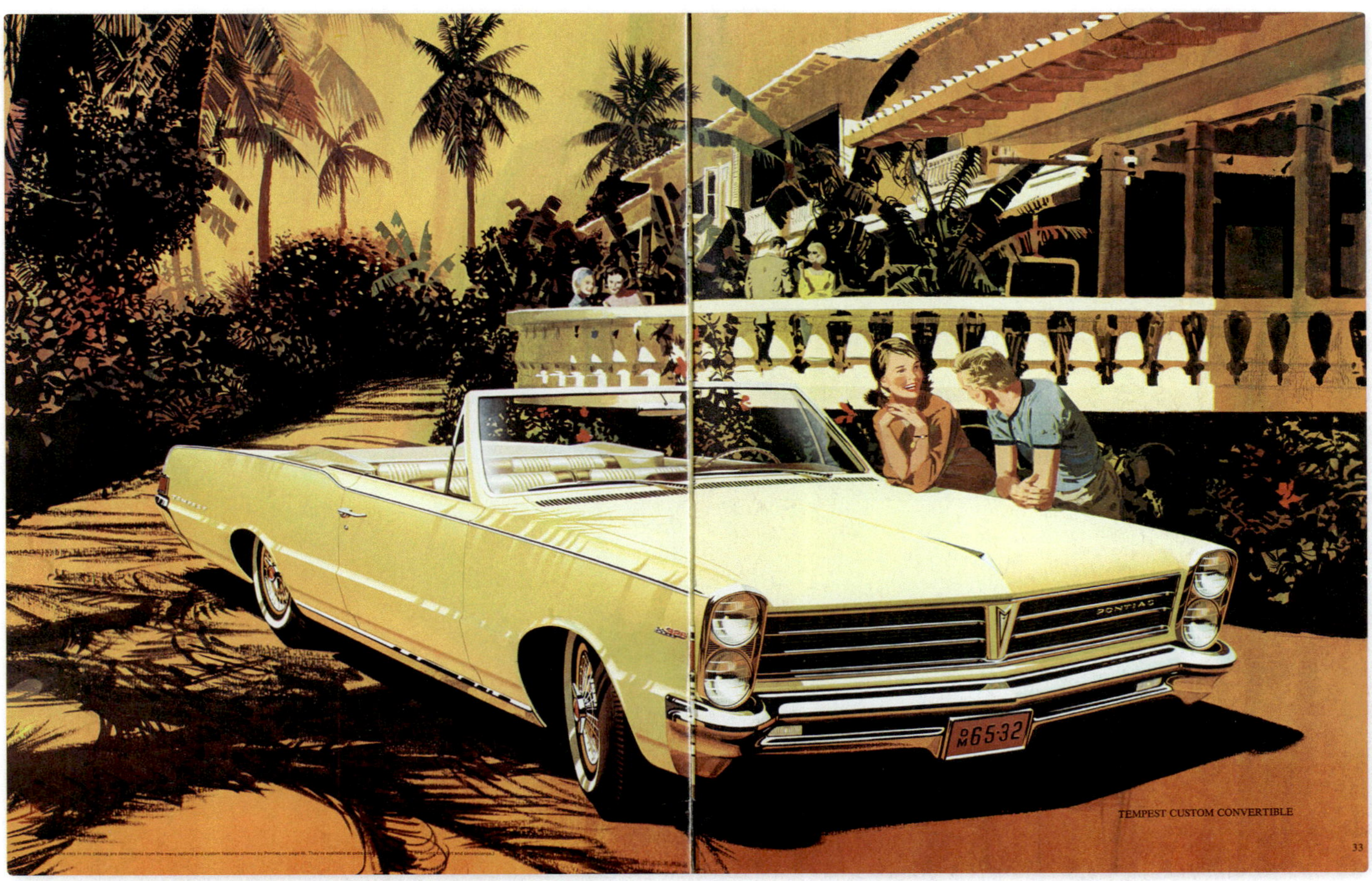
TEMPEST CUSTOM CONVERTIBLE
33

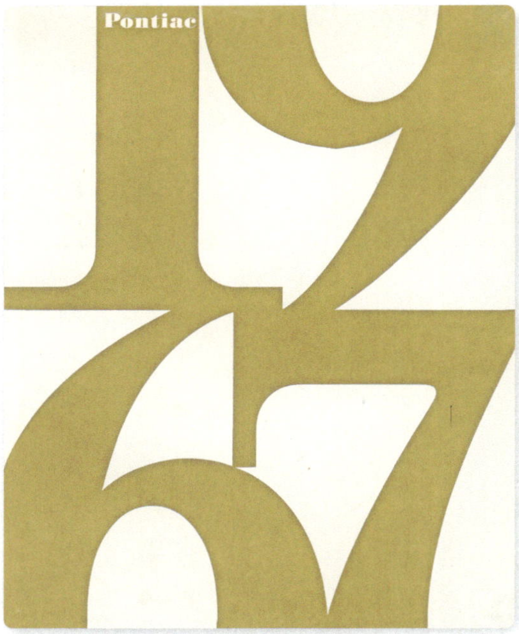

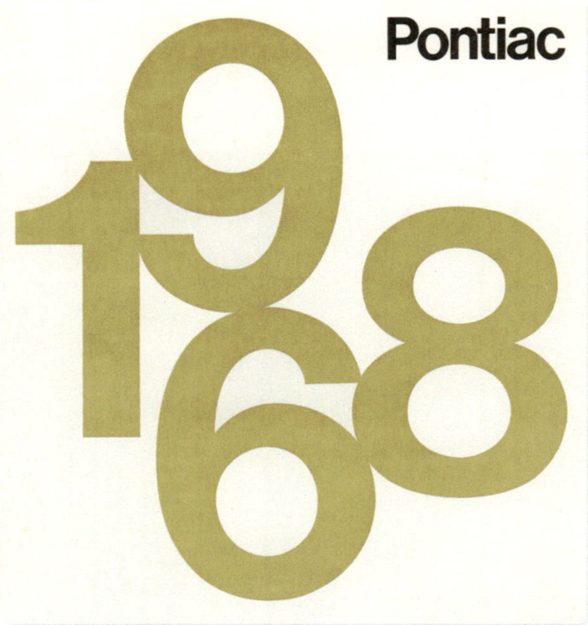

Pontiac 1967 *(above)*, **1968** *(above right and opposite)*, **1970** *(right)*

John Z. DeLorean's influence on his beloved Pontiac was vast. He approved the GTO muscle car, mating a very large V-8 engine to a midsize body. The frontal "beak" treatment shared by all these cars became a longstanding Pontiac visual signature.

Der Einfluss von John Z. DeLorean auf seinen geliebten Pontiac war gewaltig. Er gab grünes Licht für das „muscle car" GTO, bei dem ein sehr großer V8-Motor in eine Mittelklassekarosserie gezwängt wurde. Der „Schnabel" an der Vorderseite, den all diese Fahrzeuge gemeinsam hatten, blieb lange Zeit ein äußerliches Erkennungsmerkmal des Pontiac.

L'influence de John Z. DeLorean sur cette marque Pontiac qu'il aimait tant s'est manifestée de mille façons. Il œuvra ainsi au développement de la GTO, une « muscle car » qu'équipait un très gros moteur V8 caché sous une carrosserie de taille intermédiaire (celle de la Tempest). Le « bec » de la face avant, partagé par toutes ces voitures, deviendra le signe distinctif durable de Pontiac.

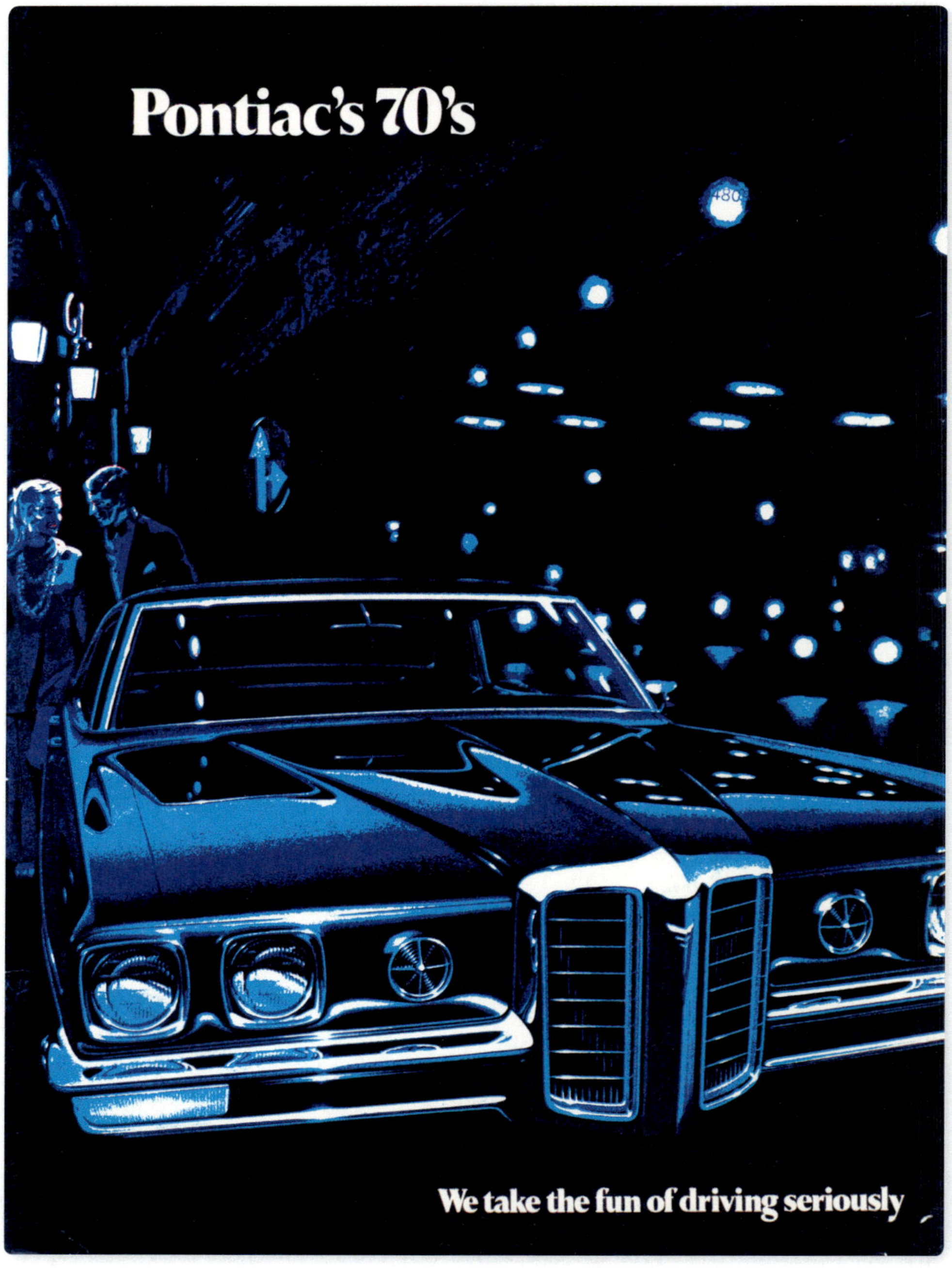

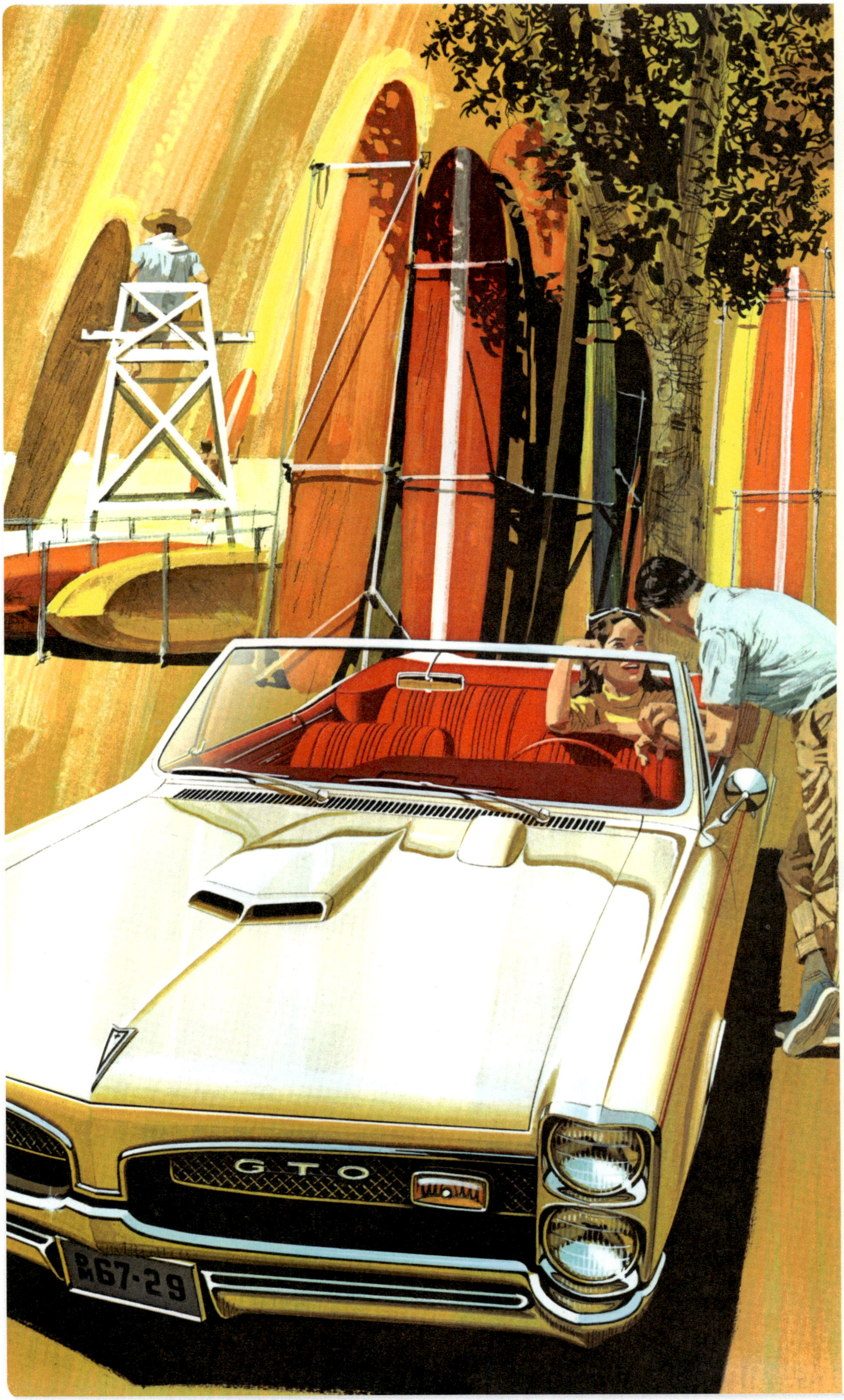

GTO

GTO CONVERTIBLE

Porsche

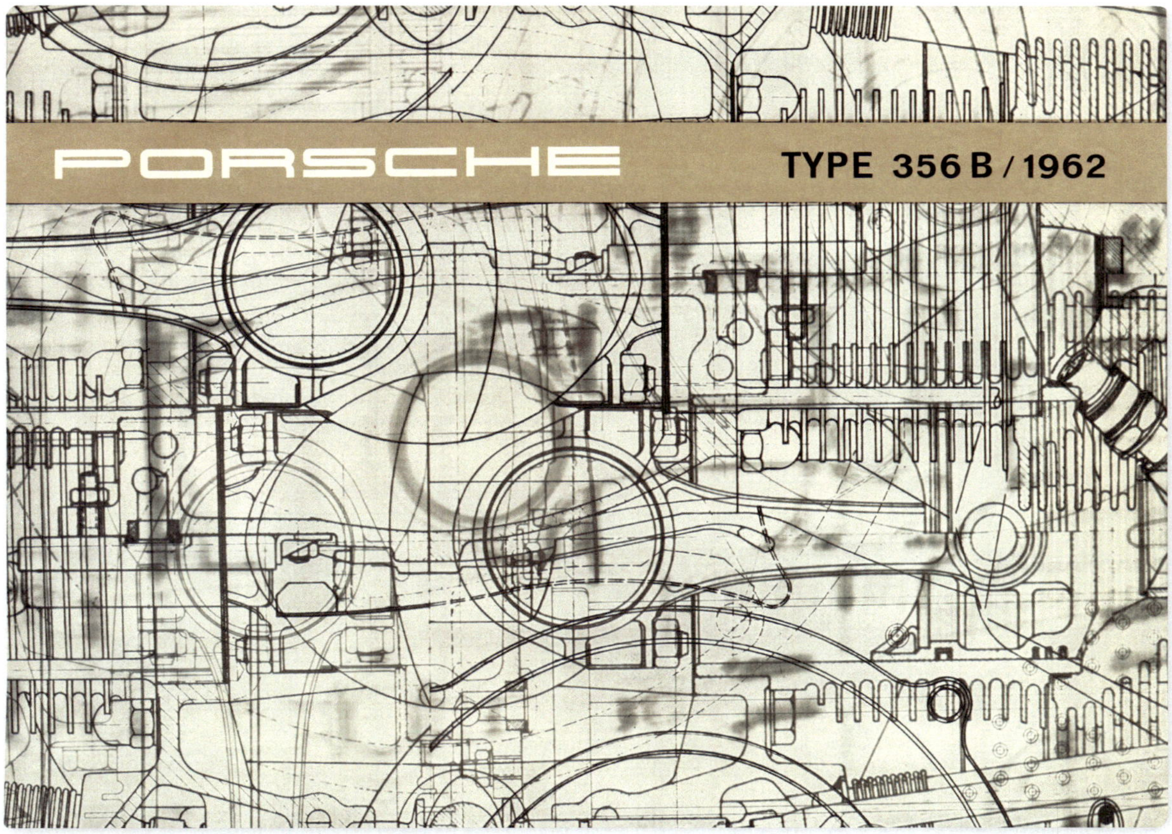

Porsche 1962 *(this spread)*

Though it still had a flat floor pan and air-cooled engine in the rear, nobody was calling Porsche a modified Volkswagen anymore. The twin rear grilles mark this Cabriolet as one with the lightweight T-6 body and the 1,600 cc pushrod engine.

Obwohl er noch immer ein flaches Bodenblech und einen luftgekühlten Motor im Heck besaß, nannte den Porsche niemand mehr einen umgebauten Volkswagen. An dem doppelten Lüftungsgrill auf der hinten liegenden Motorklappe erkennt man, dass es sich bei diesem Cabriolet um die leichtgewichtige T6-Karosserie mit 1,6-Liter-OHV-Motor handelt.

Bien qu'elle soit restée fidèle à son plancher plat et au quatre cylindres en porte-à-faux arrière refroidi par air, plus personne ne qualifiait la Porsche de Volkswagen modifiée. La double grille de refroidissement sur le capot moteur révèle que cette 356 Cabriolet est l'une des Porsche que propulsait le 1 600 cm³ à soupapes culbutées, sous une carrosserie légère de type T6.

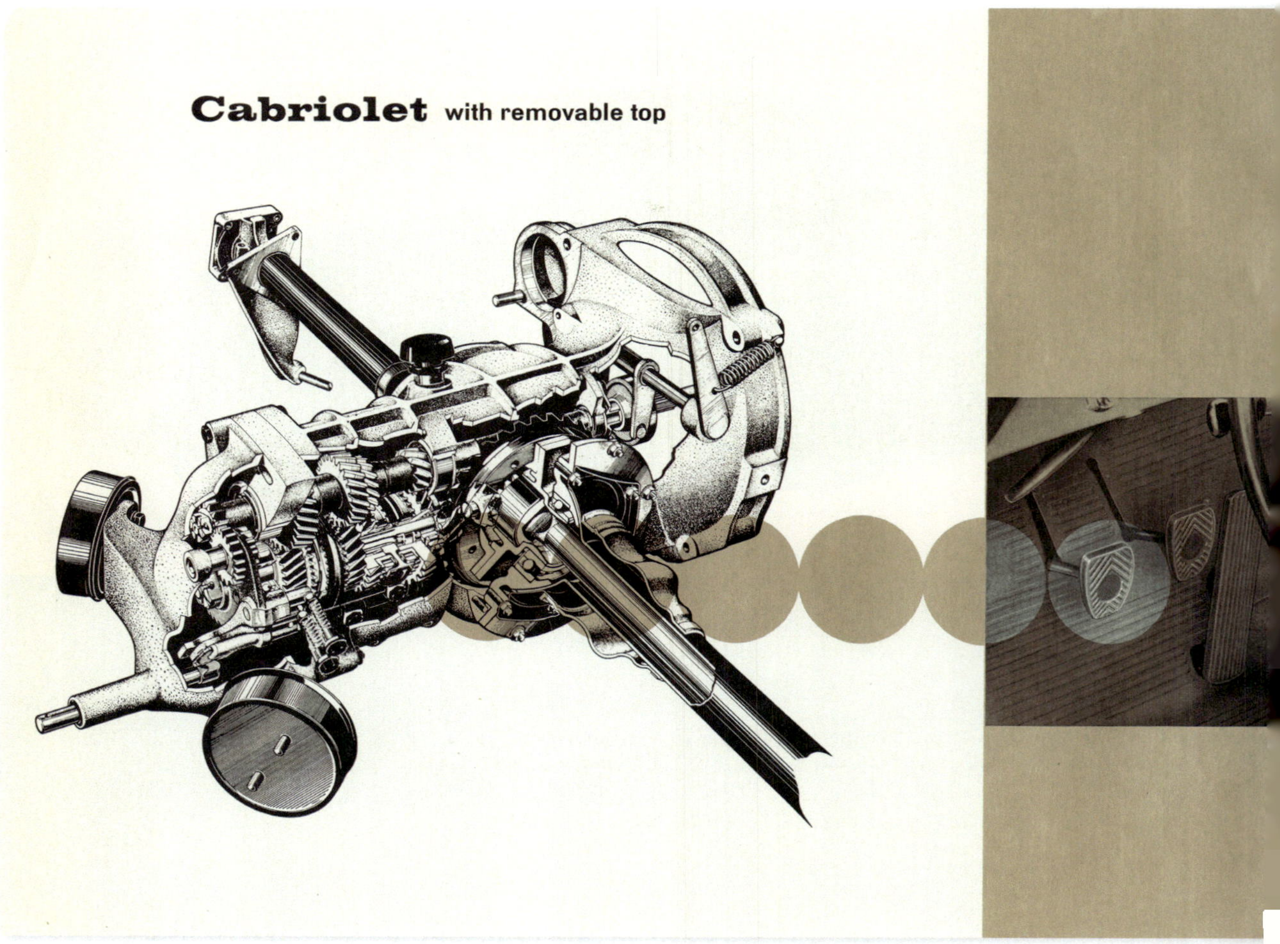

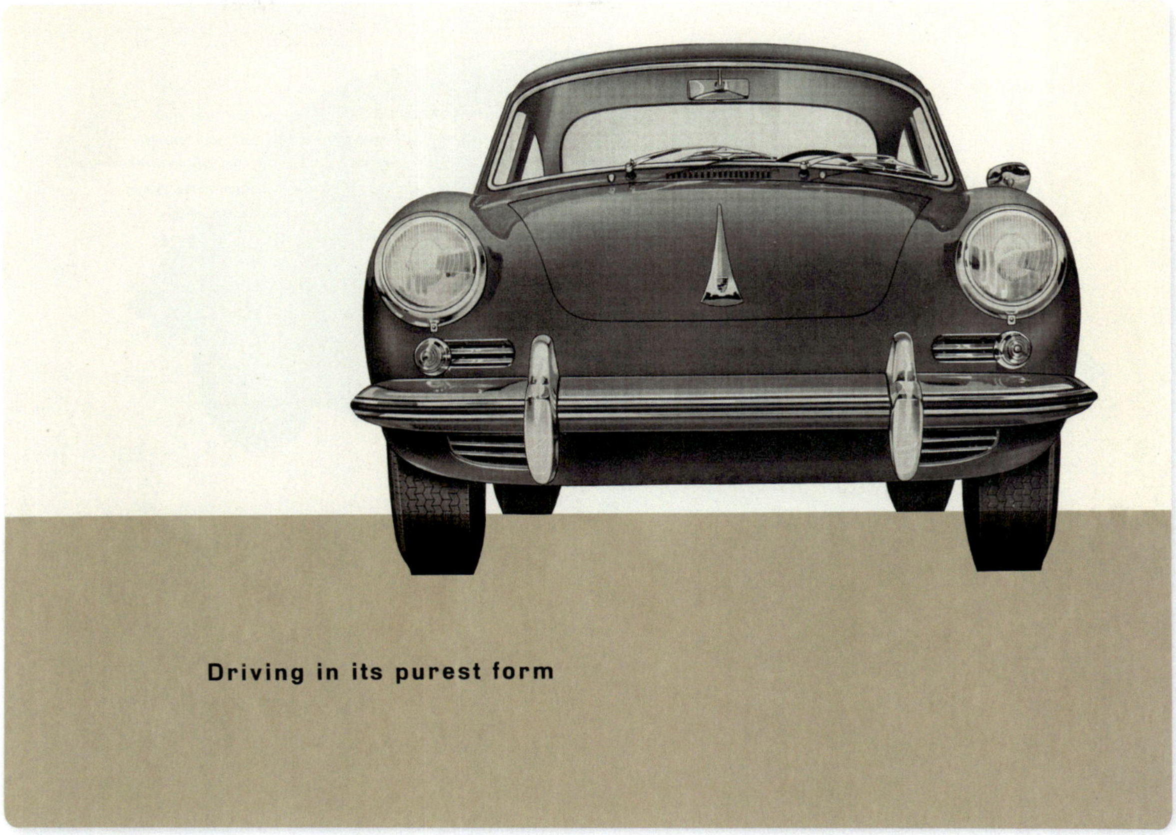
Driving in its purest form

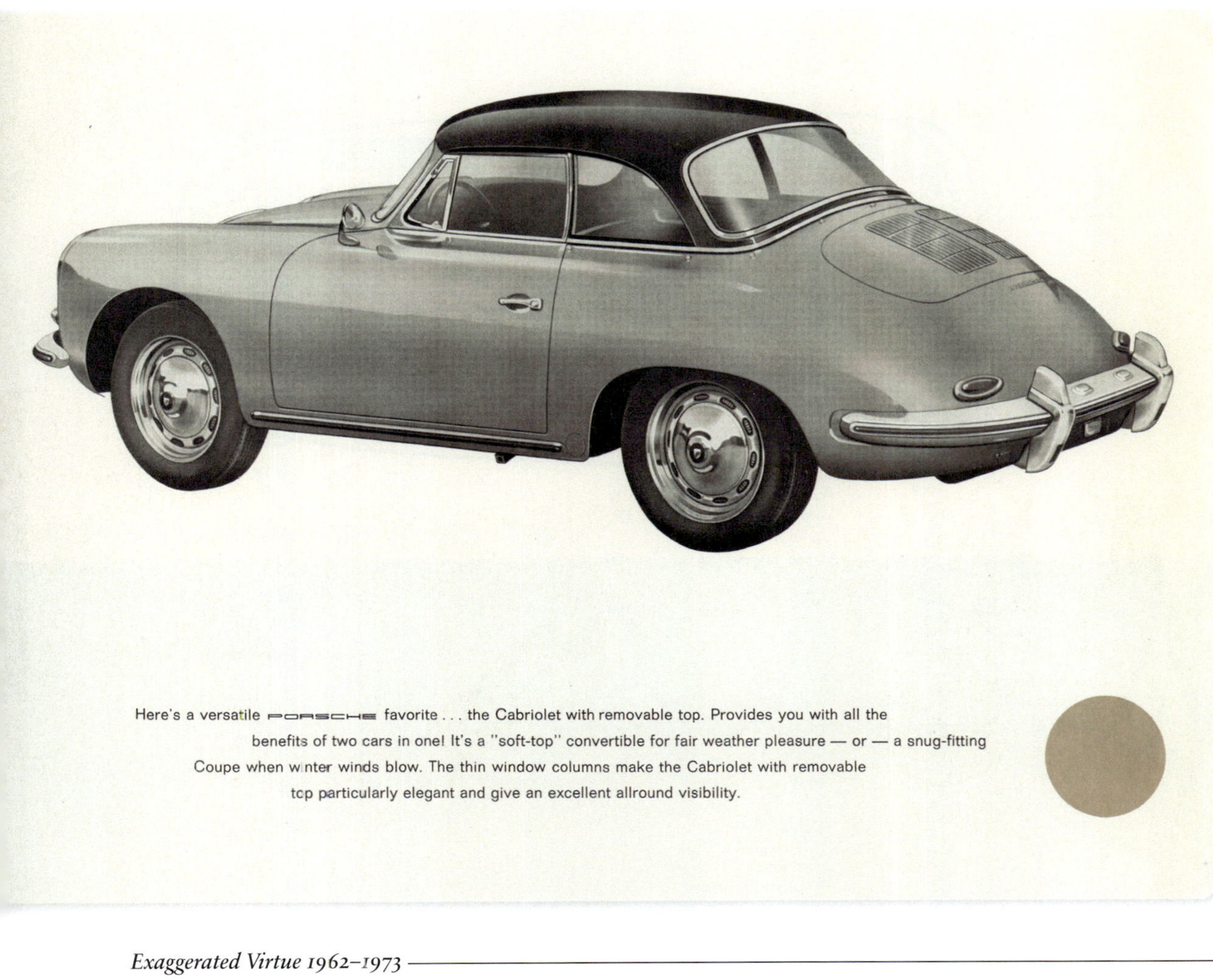
Here's a versatile PORSCHE favorite . . . the Cabriolet with removable top. Provides you with all the
benefits of two cars in one! It's a "soft-top" convertible for fair weather pleasure — or — a snug-fitting
Coupe when winter winds blow. The thin window columns make the Cabriolet with removable
top particularly elegant and give an excellent allround visibility.

Porsche 1968 *(this spread)*

To many who love sport motoring, this is the greatest automobile ever conceived. After four years in the marketplace, the 911 was reaching its stride with a four-cylinder 912 version, a base flat-six engine that had just been called the 911L, and the removable-roof Targa. All could be had with Fuchs wheels.

Für viele, die sportliches Fahren lieben, ist dies das großartigste Auto, das je gebaut wurde. Nach vier Jahren am Markt kam der 911 buchstäblich in Fahrt mit der abgeschwächten Vierzylinder version 912, dem Sechszylinderboxer-motor 911L und dem abnehmbaren Targadach. Alle Typen waren mit Fuchs-felgen erhältlich.

Pour bien des amoureux de l'automobile sportive, voici la meilleure voiture jamais conçue. Après quatre années d'existence, la 911 atteignait son plein régime avec une version à quatre cylindres baptisée 912, un modèle de base à moteur six cylindres à plat tout juste nommé 911L et la fameuse Targa au toit pliant amovible. Toutes étaient disponibles avec des jantes Fuchs.

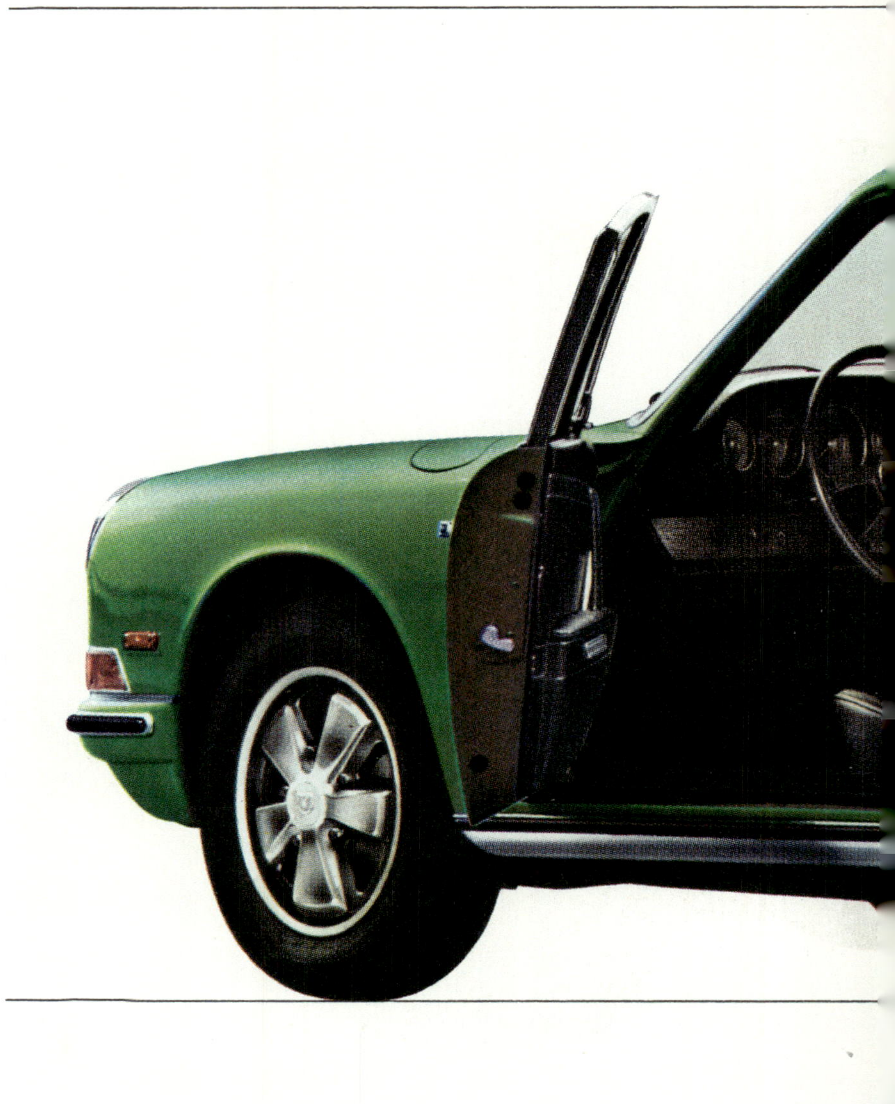

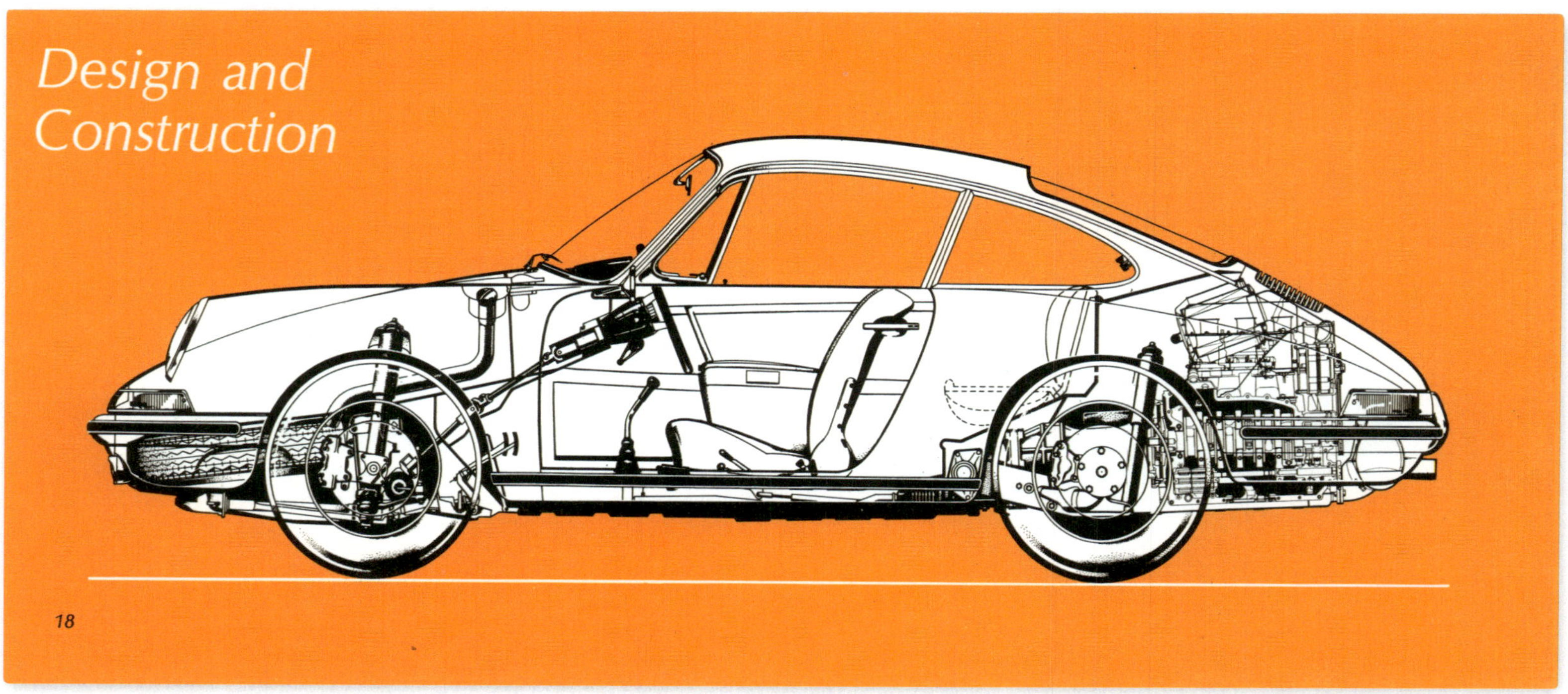

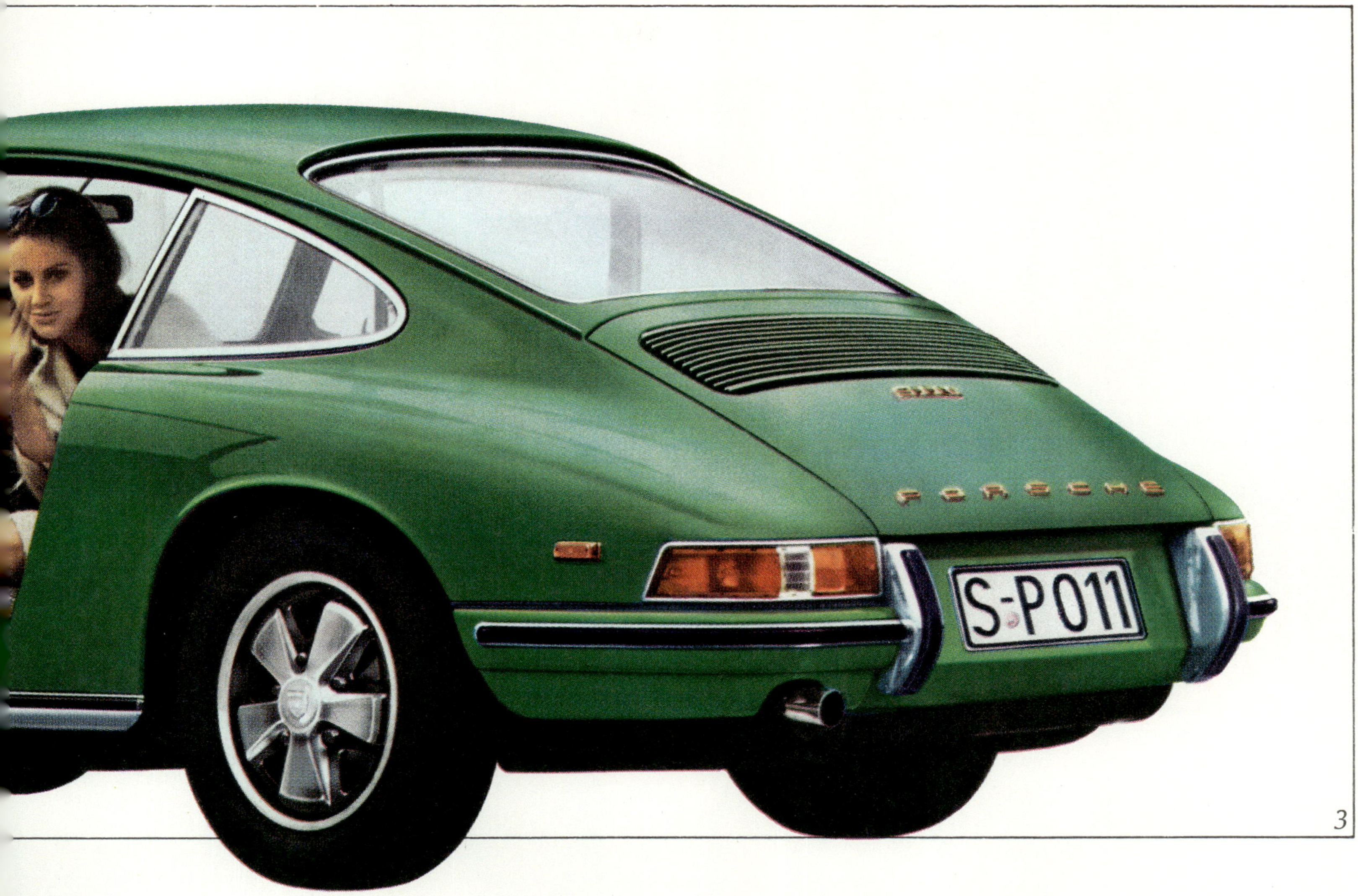

3

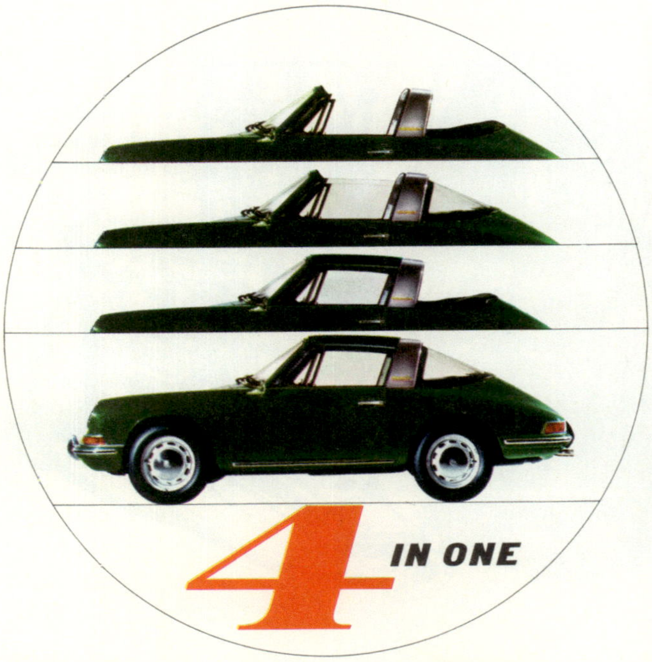

For years, roll bars have been standard equipment on all racing cars. In the Targa—appropriately named for the grueling Sicilian mountain road race in which so many of Porsche's advanced-engineering ideas have been proven—this remarkably logical racing-car ingredient provides vital protection for driver and passengers. It gives the Targa's unitized body-chassis combination the strength and rigidity of the Porsche coupe. It also makes possible the unique variations in openness provided by the Targa: fully open, top down, rear window down, fully closed.

The new Porsche Targa, available in the 912, 911 and 911L models, has the same confidence-inspiring stability and maneuverability inherent in every Porsche. Like all Porsches, it is a rugged, competition-developed vehicle, designed to be driven hard and fast—with maximum safety.

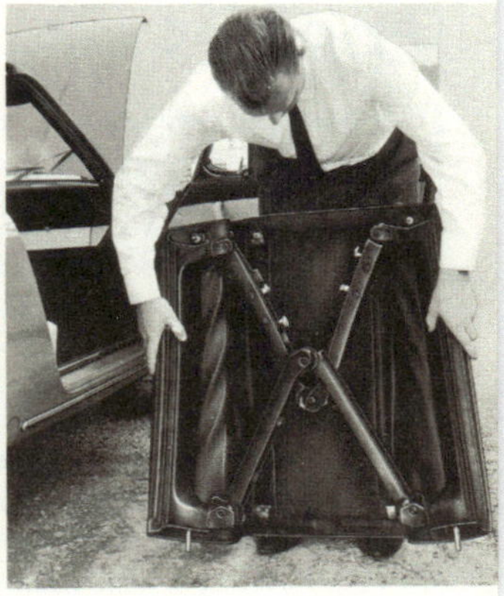

7

Studebaker

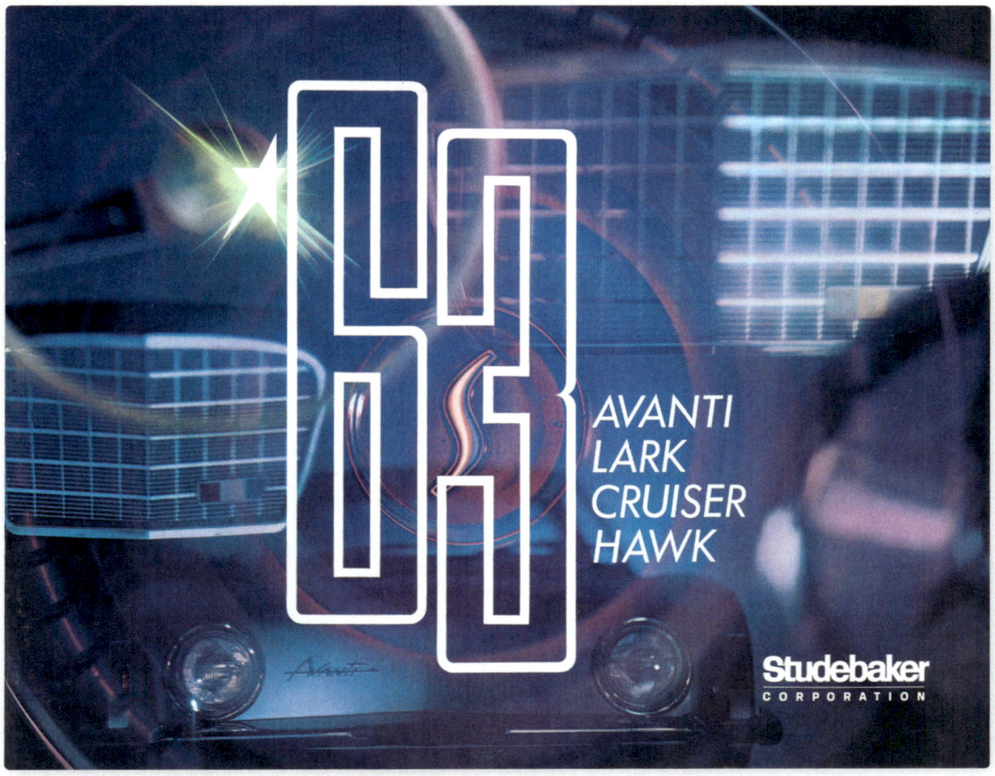

Studebaker 1963 *(this spread)*

Studebaker president Sherwood Egbert gambled on a stunning sports coupe to save his company. A team led by Raymond Loewy designed the Avanti in secret. With power up to an R-2 supercharged V-8 engine with 289 hp, it set multiple Bonneville records. Various owners have built Avantis since Studebaker collapsed.

Studebaker-Chef Sherwood Egbert setzte auf ein atemberaubendes Sportcoupé, um sein Unternehmen zu retten. Ein Team unter der Leitung von Raymond Loewy konstruierte den Avanti still und heimlich. Mit einer Motorisierung, die bis hinauf zum V8-Turbomotor R2 mit 392 kW reichte, stellte er mehrere Geschwindigkeitsrekorde auf den Salzebenen von Bonneville (Utah) auf. Diverse Besitzer haben Avantis noch nach dem Untergang von Studebaker weitergebaut.

Pour sauver son entreprise, le président de Studebaker, Sherwood Egbert, fit le pari d'un coupé sportif décoiffant. Ainsi naquit l'Avanti sous la houlette de Raymond Loewy. Avec un moteur V8 R3 suralimenté développant 335 chevaux, l'Avanti battra de nombreux records de vitesse sur le lac salé de Bonneville. Après la fin de Studebaker, le modèle continuera d'être construit par les repreneurs de la marque.

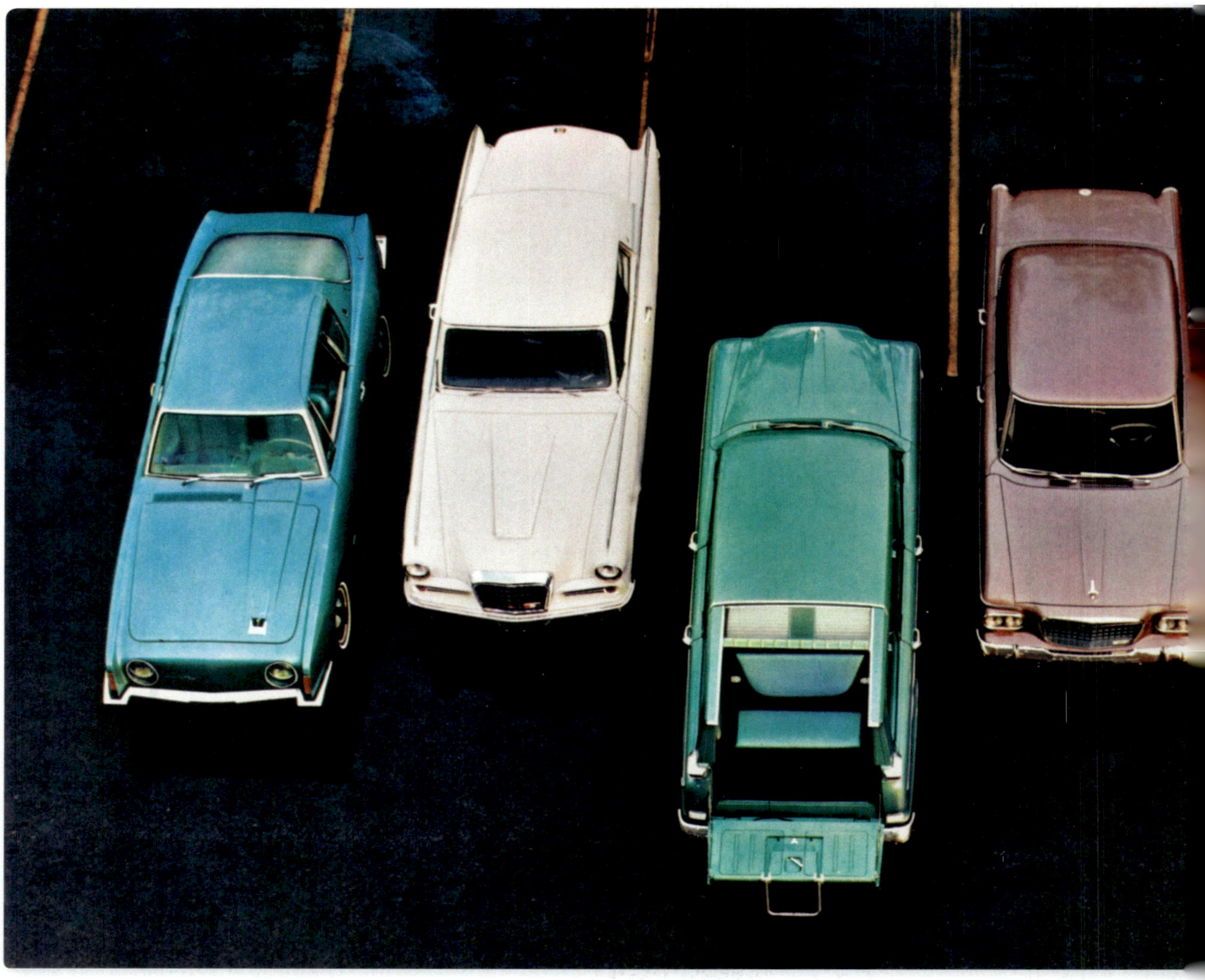

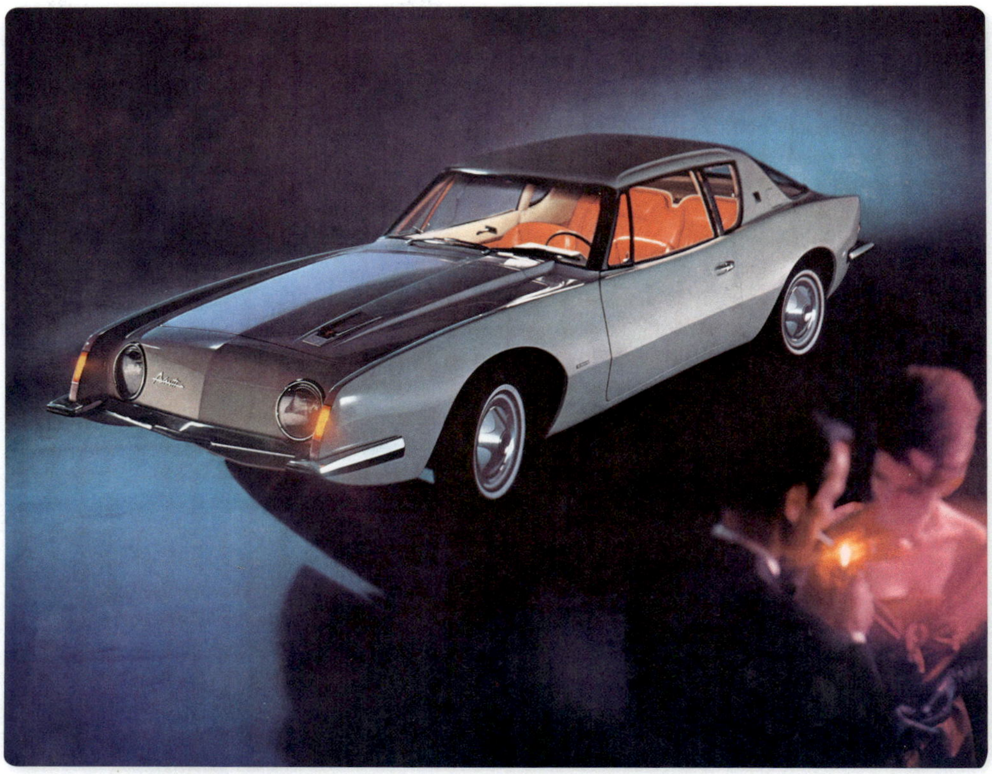

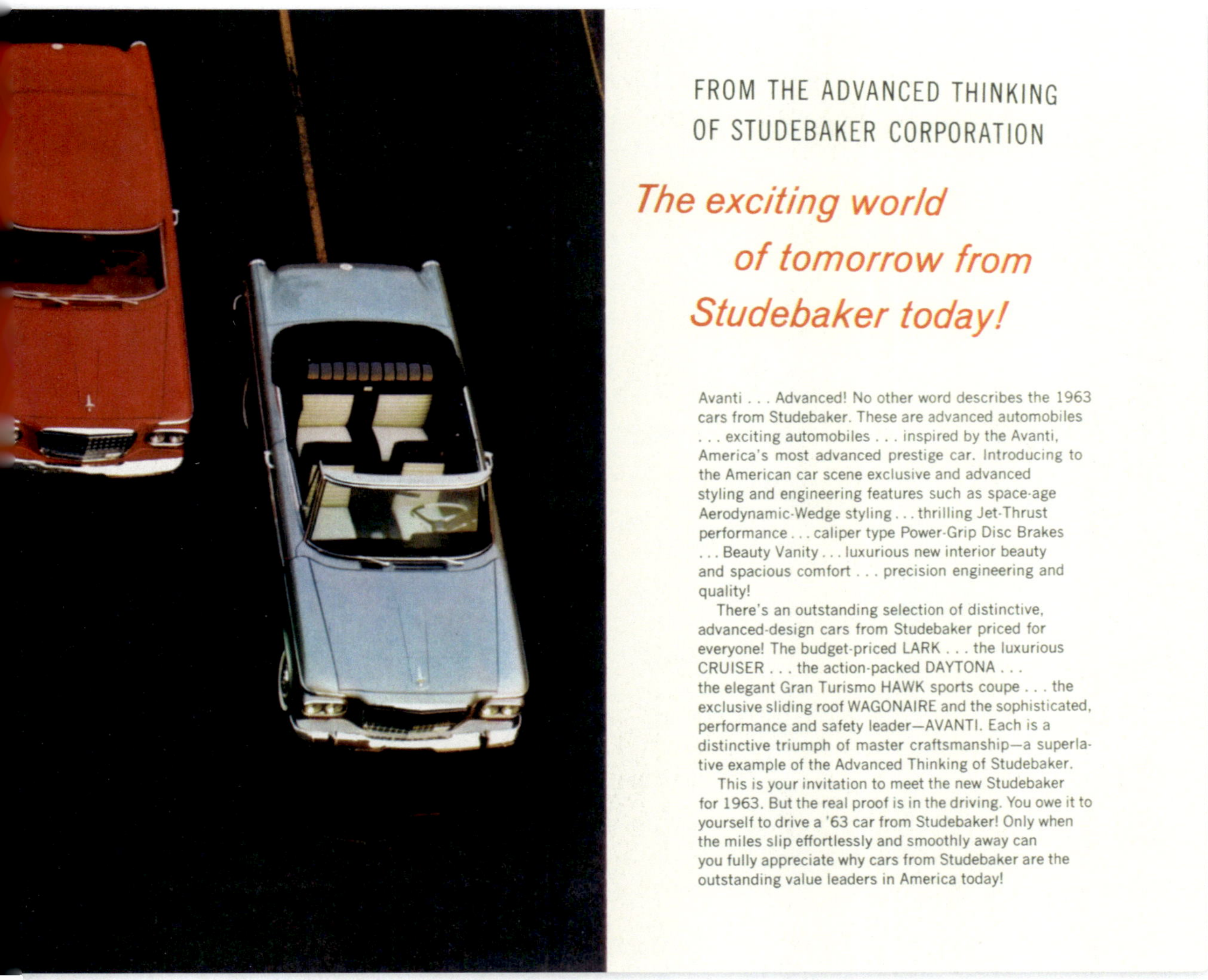

FROM THE ADVANCED THINKING
OF STUDEBAKER CORPORATION
The exciting world of tomorrow from Studebaker today!
Avanti . . . Advanced! No other word describes the 1963 cars from Studebaker. These are advanced automobiles . . . exciting automobiles . . . inspired by the Avanti, America's most advanced prestige car. Introducing to the American car scene exclusive and advanced styling and engineering features such as space-age Aerodynamic-Wedge styling . . . thrilling Jet-Thrust performance . . . caliper type Power-Grip Disc Brakes . . . Beauty Vanity . . . luxurious new interior beauty and spacious comfort . . . precision engineering and quality!
There's an outstanding selection of distinctive, advanced-design cars from Studebaker priced for everyone! The budget-priced LARK . . . the luxurious CRUISER . . . the action-packed DAYTONA . . . the elegant Gran Turismo HAWK sports coupe . . . the exclusive sliding roof WAGONAIRE and the sophisticated, performance and safety leader—AVANTI. Each is a distinctive triumph of master craftsmanship—a superlative example of the Advanced Thinking of Studebaker.
This is your invitation to meet the new Studebaker for 1963. But the real proof is in the driving. You owe it to yourself to drive a '63 car from Studebaker! Only when the miles slip effortlessly and smoothly away can you fully appreciate why cars from Studebaker are the outstanding value leaders in America today!

Volkswagen

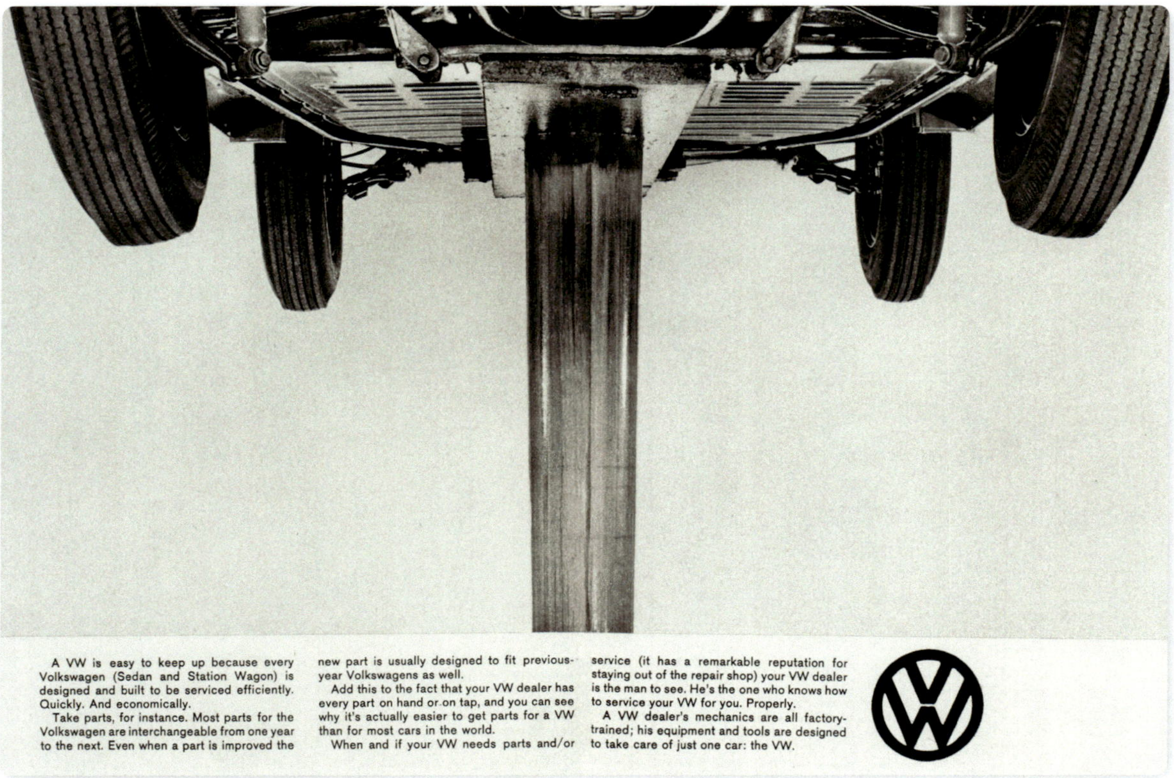

Volkswagen 1965 *(left and opposite)*, **1964** *(below)*

By now Volkswagen was referring to its Transporter as the Truck, since it was being marketed to any number of small businesses and their owners. The wonderfully self-effacing and direct advertising that Volkswagen used through most of the 1960s is considered utter genius today.

Inzwischen nannte Volkswagen seinen Transporter in den USA „Truck“, weil er an kleine Betriebe und ihre Inhaber vermarktet wurde. Die wunderbar zurückhaltende und direkte Werbung, die Volkswagen während des größten Teils der 1960er-Jahre verwendete, gilt heute als geniale Werbestrategie.

Volkswagen parlait désormais de son Transporter comme du « Truck » (le fourgon, le van), car il s'adressait aux petites entreprises et à leurs propriétaires. Pour les observateurs d'aujourd'hui, la publicité merveilleusement modeste et directe déployée par Volkswagen pendant la quasi-totalité des années 1960 confinait au génie.

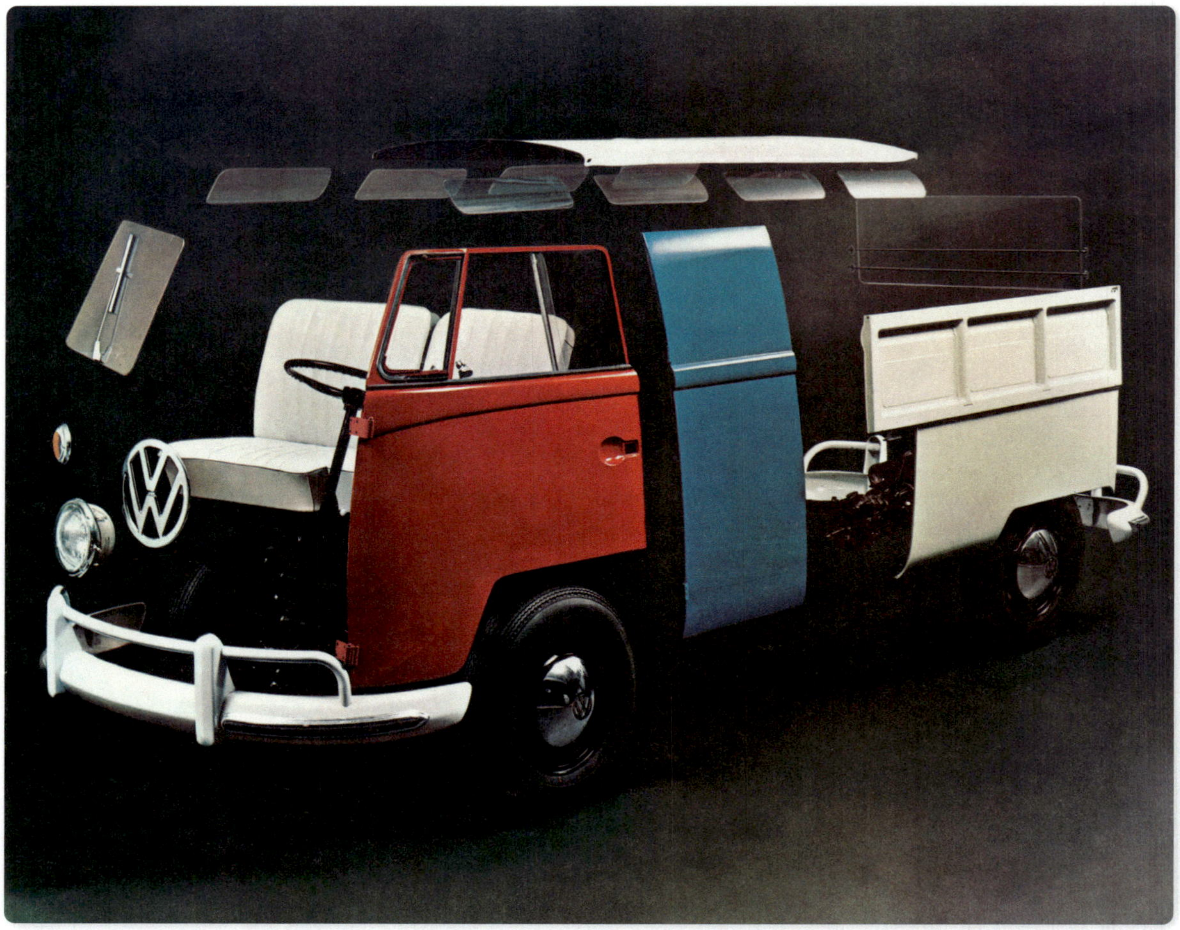

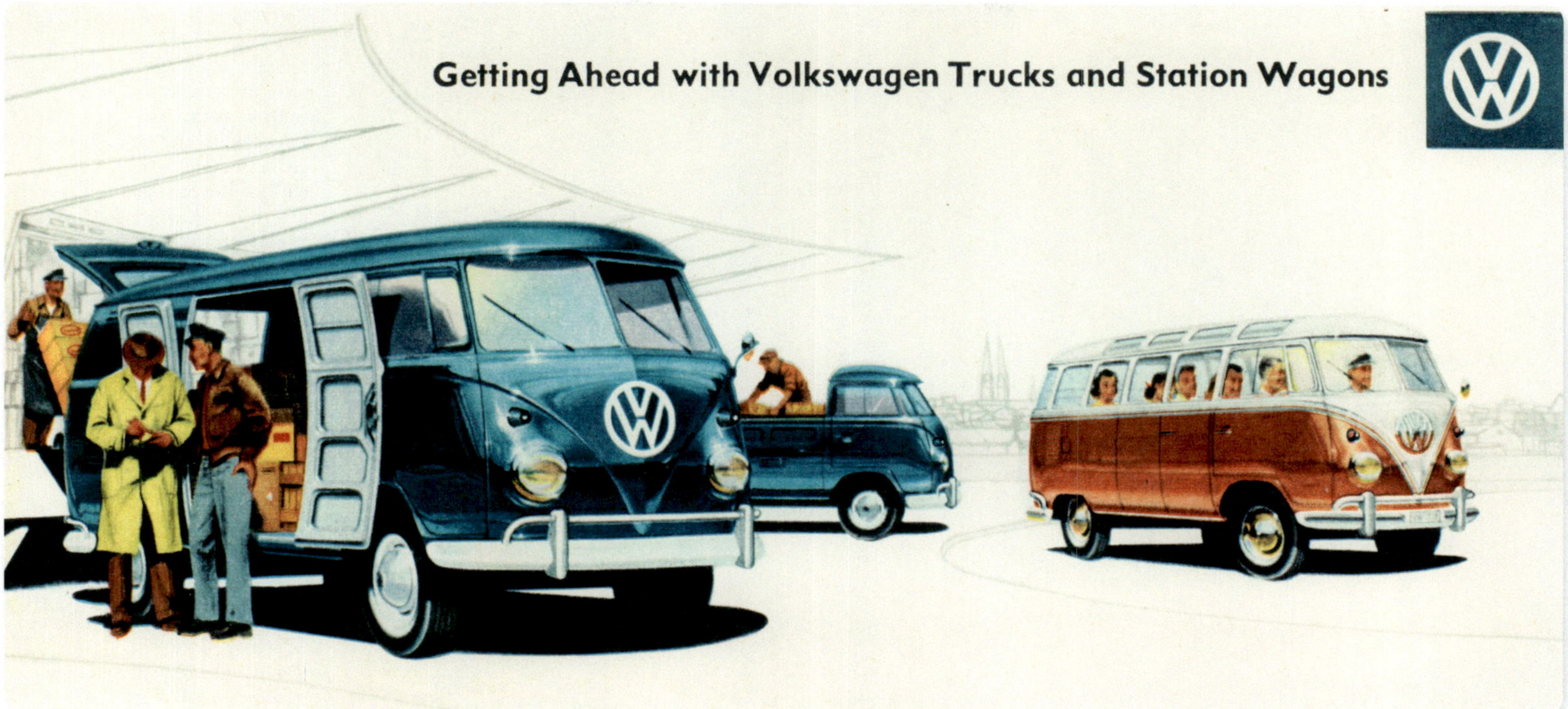
Getting Ahead with Volkswagen Trucks and Station Wagons

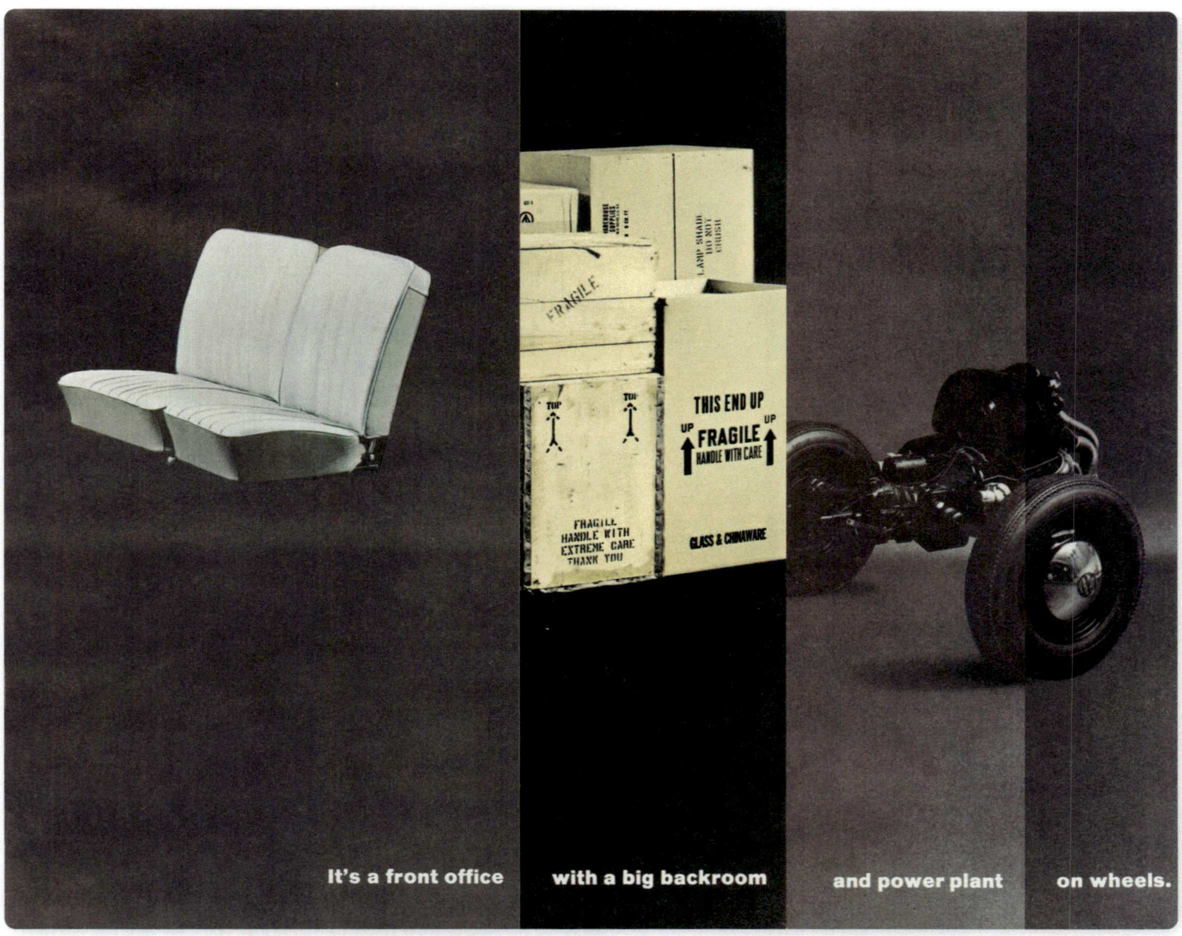
THIS END UP
FRAGILE
HANDLE WITH CARE
GLASS & CHINAWARE
FRAGILE
HANDLE WITH
EXTREME CARE
THANK YOU
It's a front office
with a big backroom
and power plant
on wheels.

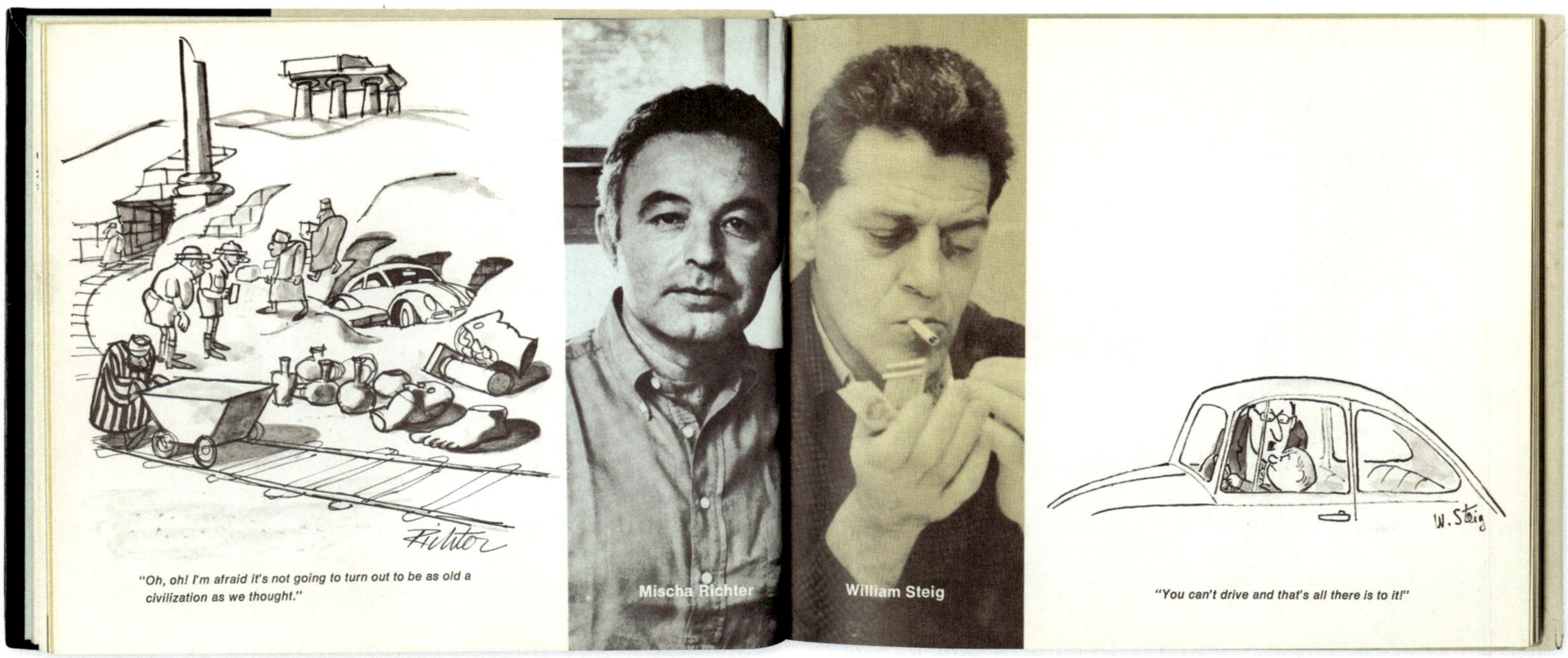

Volkswagen 1967, 1973 *(opposite and pages 366–367)*

Officially, it was the Volkswagen Type 181, but nobody called it that. The Thing, as it became known, was the modernized reincarnation of a German military scout car from World War II, riding on a modified Transporter pan. Regulators went crazy, claiming the simple vehicle was inherently unsafe, and Volkswagen dropped it after 1974.

Offiziell hieß er VW Typ 181 „Kurierwagen“, aber niemand nannte ihn so. Was ursprünglich für die Bundeswehr entwickelt und in den USA salopp als „Das Ding“ („The Thing“) bezeichnet wurde, war eine modernisierte Ausgabe des VW-Kübelwagens der Wehrmacht aus dem Zweiten Weltkrieg mit dem leicht modifizierten Getriebe eines Bulli. Die US-Aufsichtsbehörden sprangen im Dreieck, weil das Fahrzeug angeblich nicht sicher war, und so wurde der Export nach Amerika 1974 eingestellt.

« The Thing » (la chose), officiellement, Volkswagen Type 181, était la réincarnation moderne d'un engin de reconnaissance militaire allemand datant de la Seconde Guerre mondiale et monté sur un châssis de Transporter modifié. Les autorités des États-Unis s'en émurent vigoureusement, jugeant le véhicule intrinsèquement dangereux. Les importations cessèrent après 1974.

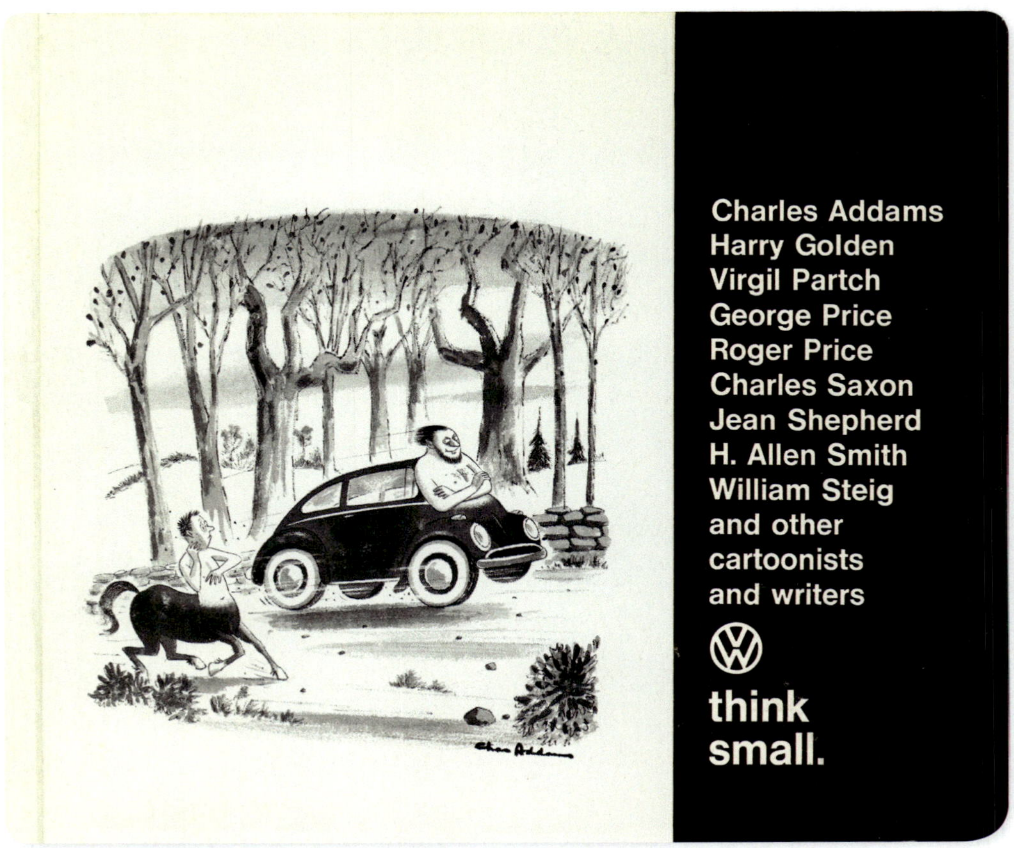

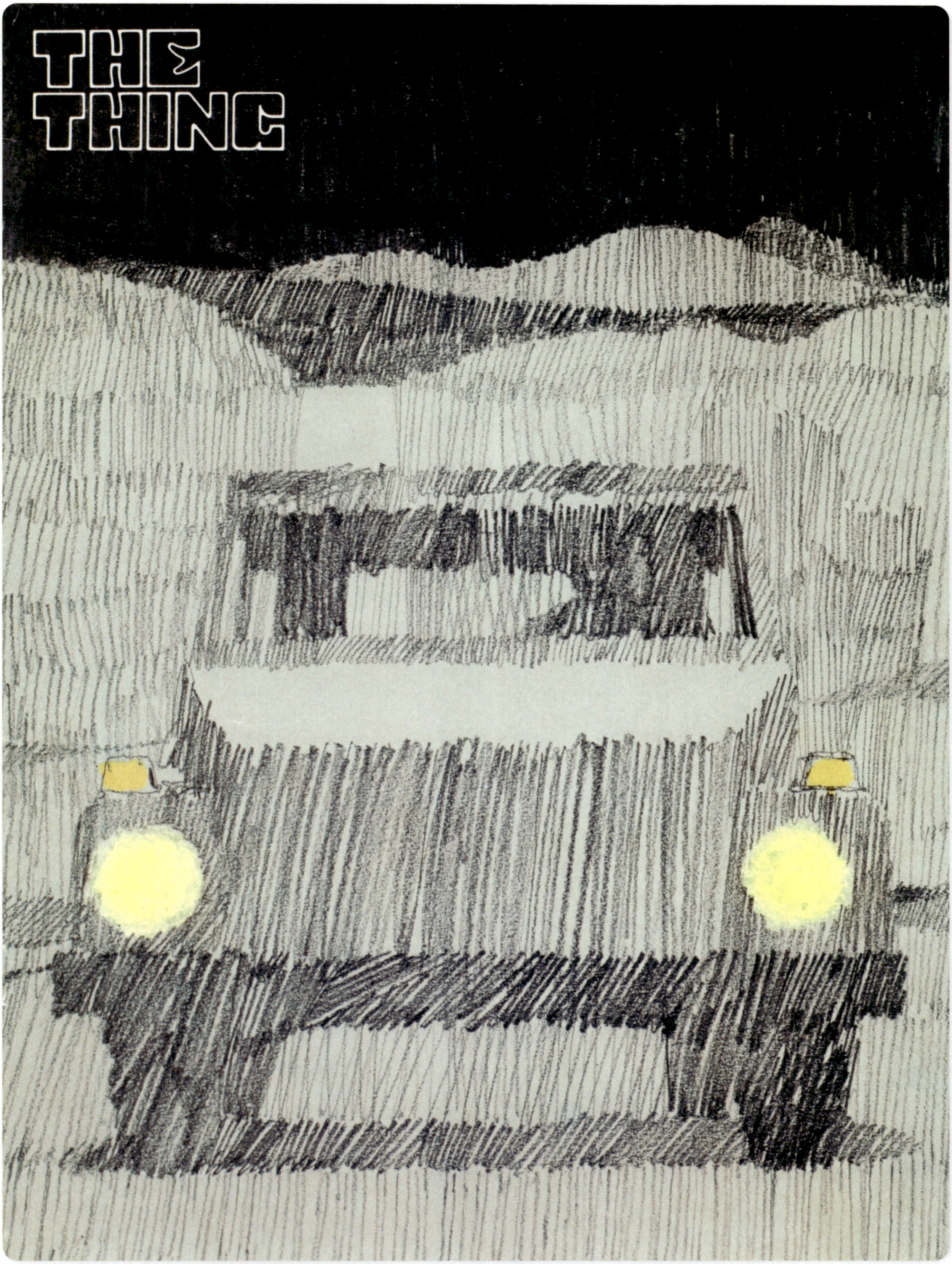
THE
THING

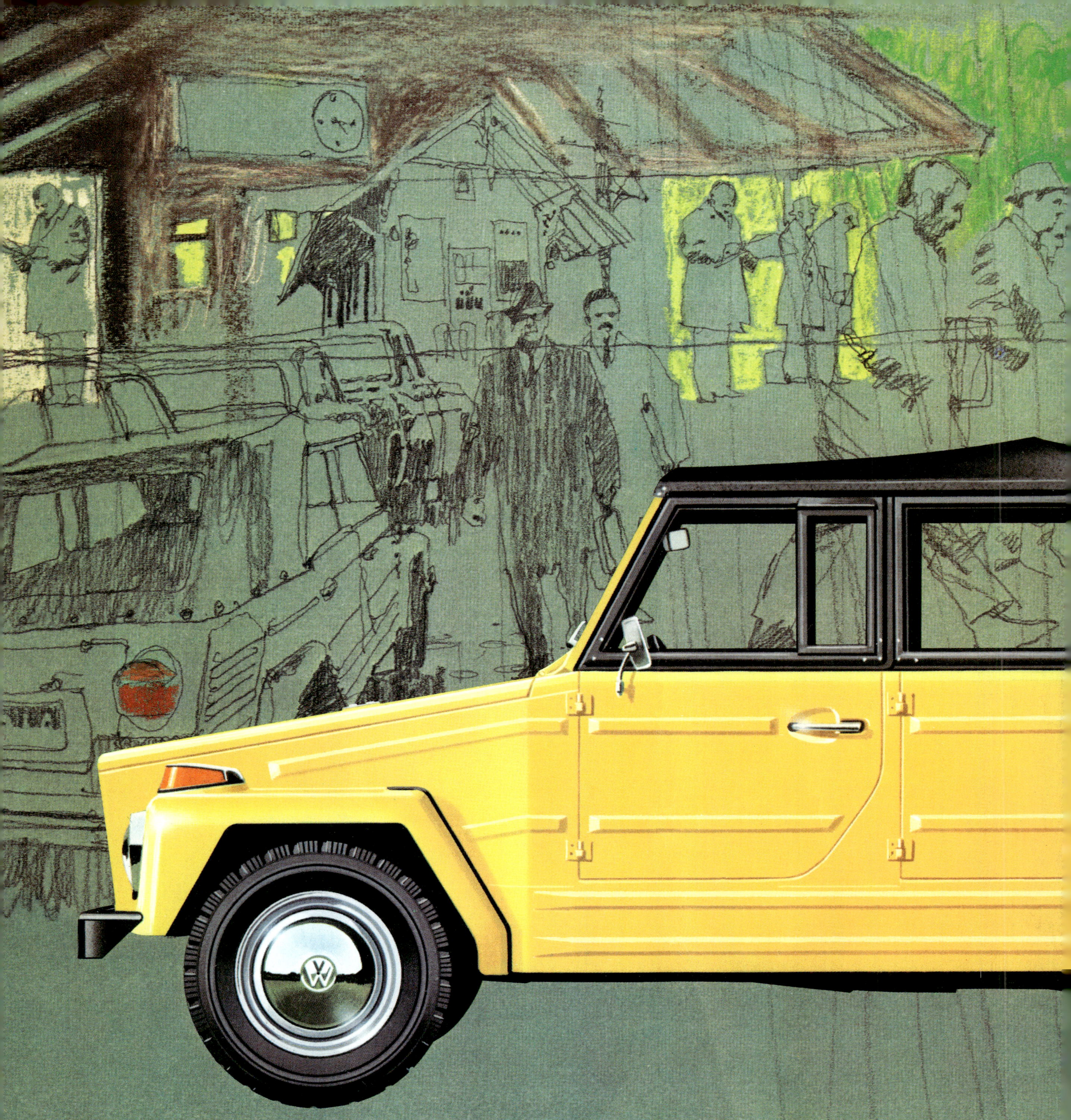

Good for fun and good for work. Good for the city and good for the country. Good off-the-road and good on-the-road. For the shopping, the chauffeuring, the hauling—all the things you need a good passenger car for. "The Thing" parks like the Beetle. (It has the same suspension and quick responsive steering.) Averages 21 miles per gallon.* And it's covered by the Volkswagen Owner's Security Blanket like all Volkswagens.† It's "The Thing"—the car that can be anything.

*Based on DIN(70030)

†See specifications on back cove

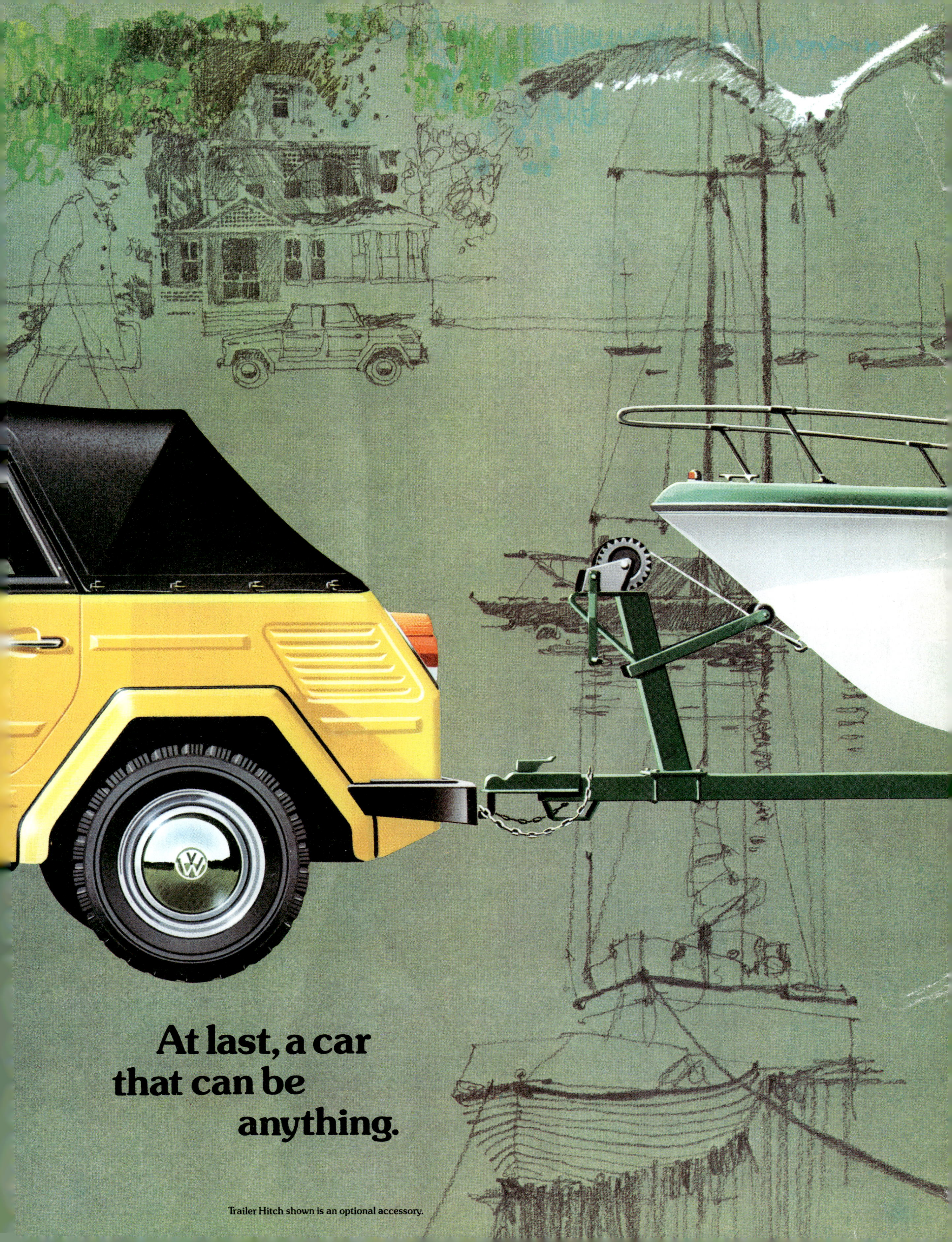
At last, a car
that can be
anything.
Trailer Hitch shown is an optional accessory.

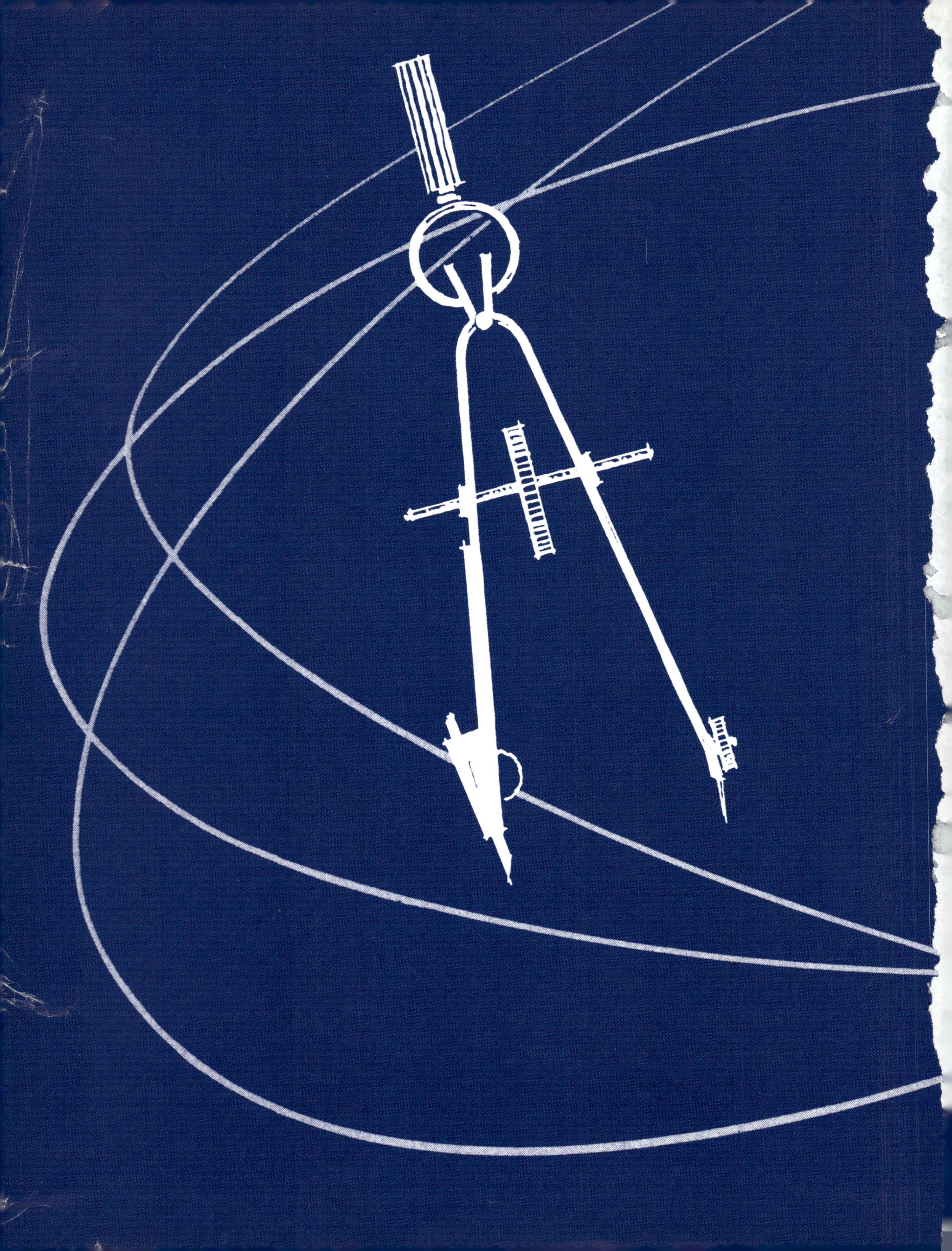

As always, a special thanks to Benedikt Taschen—who wholeheartedly continues to support subjects both mainstream and arcane, adding to the depth of knowledge of the visual world. This volume represents another example of his prescience in the book world.

Special thanks to the institutions and collectors who made available material and research for this book, including Jeff Carr and Jim Cherry for supplying outstanding selections of auto brochures. Additional thanks for assistance to Jacob P. Cooley, Leslie Kendall at the Petersen Automotive Museum, Coco Shinomiya, and Tony Thacker.

Writers Steven Heller and Jim Donnelly both provided the exact tone and knowledge to validate this subject, which has slipped through the annals of both design and automotive history.

Hohenzollernring 53, D–50672 Köln
www.taschen.com

Art direction: Josh Baker, Oakland
Design: Jess Sappenfield, Los Angeles
Editorial coordination: Amanda Horn, New York; and Kathrin Murr, Cologne
Production: Stefan Klatte, Cologne
German translation: Thomas J. Kinne, Nauheim
French translation: François Dirdans, Paris

ISBN 978-3-8228-5371-9
Printed in Slovenia

Also thanks to the TASCHEN team: Doug Adrianson, Josh Baker, Lindsey Dole, Camellia Ford, Nemuel DePaula, Isabelle Gioffredi, Joey Heller, Amanda Horn, Stefan Klatte, Ryan Mungia, Kathrin Murr, Jennifer Patrick, Jess Sappenfield, Eric Schwartau, Anna Skinner, and Nina Wiener.

—Jim Heimann, Executive Editor, TASCHEN America

Collector's note: While most of the brochures in this book retain their original character and integrity, some of the brochures have been altered for production and design considerations.

All images are from the collection of Jim Heimann unless otherwise noted. Any omissions for copy or credit will be given in future editions if such copyright holders contact the publisher.

Endpapers: Chevrolet 1953
Cover and pages 2–3: Buick 1939
Back cover: Volkswagen 1961
Pages 4–5: Pontiac 1962
Pages 54–55: Hupmobile 1929
Pages 118–119: Marmon 1930
Pages 206–207: Oldsmobile 1954
Pages 316–317: Ford 1966

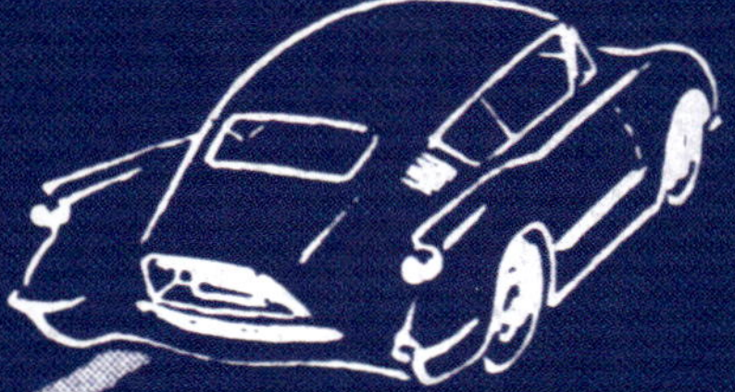

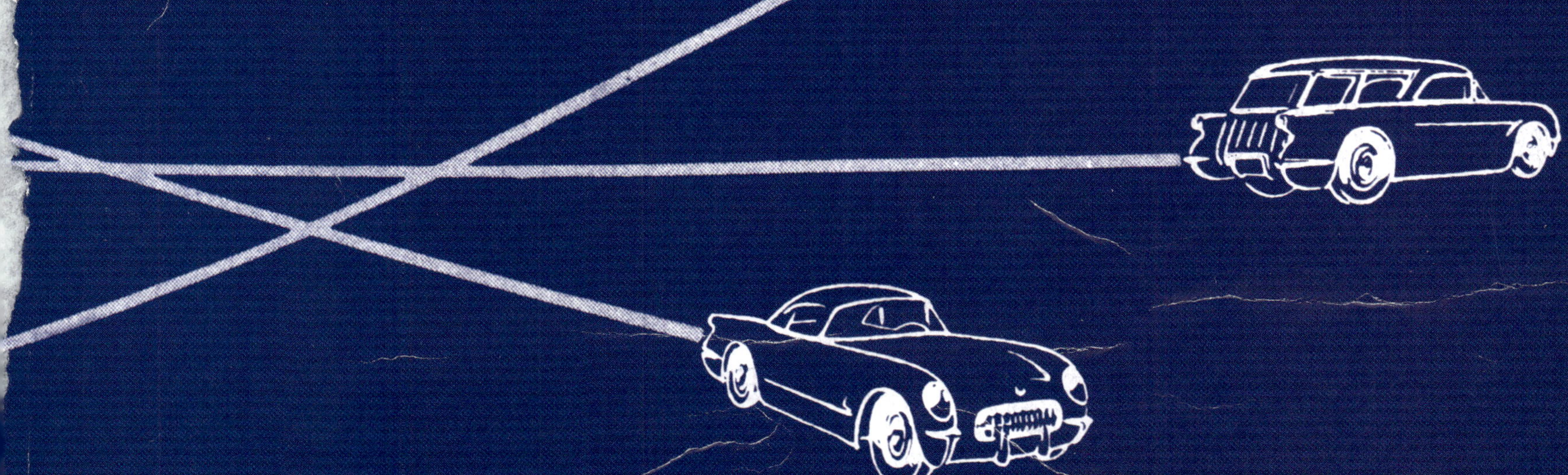